Population Genomics

Editor-in-Chief

Om P. Rajora, Faculty of Forestry and Environmental Management,
University of New Brunswick, Fredericton, NB, Canada

This pioneering Population Genomics Series deals with the concepts and approaches of population genomics and their applications in addressing fundamental and applied topics in a wide variety of organisms. Population genomics is a fast emerging discipline, which has created a paradigm shift in many fields of life and medical sciences, including population biology, ecology, evolution, conservation, agriculture, horticulture, forestry, fisheries, human health and medicine.

Population genomics has revolutionized various disciplines of biology including population, evolutionary, ecological and conservation genetics, plant and animal breeding, human health, genetic medicine, and pharmacology by allowing to address novel and long-standing intractable questions with unprecedented power and accuracy. It employs large-scale or genome-wide genetic information across individuals and populations and bioinformatics, and provides a comprehensive genome-wide perspective and new insights that were not possible before.

Population genomics has provided novel conceptual approaches, and is tremendously advancing our understanding the roles of evolutionary processes, such as mutation, genetic drift, gene flow, and natural selection, in shaping up genetic variation at individual loci and across the genome and populations, disentangling the locus-specific effects from the genome-wide effects, detecting and localizing the functional genomic elements, improving the assessment of population genetic parameters or processes such as adaptive evolution, effective population size, gene flow, admixture, inbreeding and outbreeding depression, demography, and biogeography, and resolving evolutionary histories and phylogenetic relationships of extant and extinct species. Population genomics research is also providing key insights into the genomic basis of fitness, local adaptation, ecological and climate acclimation and adaptation, speciation, complex ecologically and economically important traits, and disease and insect resistance in plants, animals and/or humans. In fact, population genomics research has enabled the identification of genes and genetic variants associated with many disease conditions in humans, and it is facilitating genetic medicine and pharmacology. Furthermore, application of population genomics concepts and approaches facilitates plant and animal breeding, forensics, delineation of conservation genetic units, understanding evolutionary and genetic impacts of resource management practices and climate and environmental change, and conservation and sustainable management of plant and animal genetic resources.

The volume editors in this Series have been carefully selected and topics written by leading scholars from around the world.

Om P. Rajora
Editor

Population Genomics: Crop Plants

 Springer

Editor
Om P. Rajora
Faculty of Forestry and Environmental Management
University of New Brunswick
Fredericton, NB, Canada

ISSN 2364-6764 ISSN 2364-6772 (electronic)
Population Genomics
ISBN 978-3-031-63001-9 ISBN 978-3-031-63002-6 (eBook)
https://doi.org/10.1007/978-3-031-63002-6

© Springer Nature Switzerland AG 2024
Chapters "Population Genomics of Yams: Evolution and Domestication of *Dioscorea* Species" and "Population Genomics Along With Quantitative Genetics Provides a More Efficient Valorization of Crop Plant Genetic Diversity in Breeding and Pre-breeding Programs" are licensed under the terms of the Creative Commons Attribution 4.0 International License (http://creativecommons.org/licenses/by/4.0/). For further details, see license information in the chapters.

This work is subject to copyright. All rights are solely and exclusively licensed by the Publisher, whether the whole or part of the material is concerned, specifically the rights of translation, reprinting, reuse of illustrations, recitation, broadcasting, reproduction on microfilms or in any other physical way, and transmission or information storage and retrieval, electronic adaptation, computer software, or by similar or dissimilar methodology now known or hereafter developed.

The use of general descriptive names, registered names, trademarks, service marks, etc. in this publication does not imply, even in the absence of a specific statement, that such names are exempt from the relevant protective laws and regulations and therefore free for general use.

The publisher, the authors and the editors are safe to assume that the advice and information in this book are believed to be true and accurate at the date of publication. Neither the publisher nor the authors or the editors give a warranty, expressed or implied, with respect to the material contained herein or for any errors or omissions that may have been made. The publisher remains neutral with regard to jurisdictional claims in published maps and institutional affiliations.

This Springer imprint is published by the registered company Springer Nature Switzerland AG
The registered company address is: Gewerbestrasse 11, 6330 Cham, Switzerland

If disposing of this product, please recycle the paper.

Population Genomics Book Series
This Population Genomics book series is dedicated to my (late) parents, and my wife Malti and children Apoorva, Anu, and Maneesha.

Om P. Rajora

Population Genomics: Crop Plants
This book is respectfully dedicated to my (late) father, Shri Than Singh, who was a prominent farmer, a prominent person, and a leader, well known for his honesty, integrity and generosity.

Om P. Rajora

Preface

Population genomics has revolutionized various disciplines of plant biology, especially population, evolutionary, ecological and conservation genetics, plant breeding, and conservation and sustainable management of plant genetic resources by allowing to address novel and long-standing intractable questions with unprecedented power and accuracy. Population genomics provides comprehensive genome-wide information and new insights that were not possible earlier.

Population genomics concepts and approaches along with bioinformatics tools and models have created a paradigm shift in several disciplines of crop plants biology by making significant, unprecedented advances in both basic and applied research. Crop plants have been domesticated from their wild progenitors over several centuries and have undergone severe genetic bottlenecks and selection sweeps through selection and breeding. This has resulted in a narrow genetic base of crop plants. Sustainability and enhancement of crop production is essential for food security to meet increasing demands of increasing world population. Climate change is also posing challenges to sustainability of crop plants and crop production. Sustainable crop production can be facilitated by conservation and enhancement of plant genetic resources, greater and accelerated genetic improvement of crop plants, genetic diversity enrichment in their pre-breeding and breeding programs, and improved ecological, environmental, and climate adaptation of crop plants. Genetic diversity provides the raw material for evolution and adaptation of organisms, especially under changing environmental and disease conditions. Therefore, a deeper and precise understanding of genetic/genomic diversity and structure of wild and domesticated populations and species, origin, evolution, demographic history, center of diversity, domestication history, genetic/genomic basis of domestication syndrome, genomic footprints of domestication, selection and breeding, interspecific and intraspecific phylogenetic relationships, introgression from wild species, genes and genomic regions underlying traits of interest (including productivity, biotic and abiotic stress tolerance), and ecological and climate adaptation of crop plants is required. Population genomics concepts and approaches are unraveling key, novel, and deeper insights into these above aspects of crop plants with unprecedented power and accuracy. In addition, population genomics approaches are revealing key insights into speciation, systematics, and de-domestication and allowing to construct pangenomes. Population genomics has also facilitated identifying genotype-phenotype associations and building prediction models for estimating genetic/breeding value of individuals, thus facilitating genomics-assisted early selection and accelerated breeding of crop plants. Moreover, population epigenomics and population transcriptomics studies have begun in crop plants which can contribute to enhancing our understanding of acclimation, adaptation, and disease and insect resistance of populations.

As a part of the pioneering Population Genomics book series, this book discusses the progress and perspectives of population genomics in addressing various fundamental and applied crucial aspects in crop plants. The book provides insights into a range of emerging topics in crop plants including pangenomes, genomic diversity and population structure, demography, evolution, domestication, de-domestication, speciation, taxonomy, phylogenomics, molecular breeding, population epigenomics, population transcriptomics, biotic and abiotic stress tolerance, ecological and climate adaptation, gene banks genomics, and conservation of plant genetic resources. The chapters are written by leading and emerging research scholars in crop plants population genomics.

The book has 22 chapters which are organized into three parts. The first part has six chapters which discuss population genomics aspects of pangenomes, organellar genomes, demographic history, evolution, speciation, adaptation, gene banks, and genetic resource conservation in crop plants. The second part includes four chapters which discuss the application of population genomics concepts and approaches in enhancing crop plants breeding and valorization of genetic diversity in breeding and pre-breeding programs. The third part consists of 12 chapters covering the progress and prospects of population genomics research and application in major crop plants. Each chapter focuses on a crop plant and first provides a review of genomic, transcriptomic, epigenomic, and plant resources available for population genomics studies and then discusses the progress made in population genomics aspects including pangenome, genetic diversity and population structure, origin, evolution, domestication, speciation and admixture, phylogenomics, genome-wide association studies, genomic selection, population epigenomics, population transcriptomics, and conservation and sustainable management of plant genetic resources. Finally, future perspectives are discussed.

The book is envisioned for a wide readership, including undergraduate and graduate students, research scholars, and professionals and experts in the field. It fills a vacuum in the field and is expected to become a primary reference in population genomics of crop plants worldwide.

I thank all the authors who have contributed to this volume.

Fredericton, NB, Canada Om P. Rajora

Contents

Part I Pangenomes, Demography, Evolution, Speciation, and Conservation

Pangenomics in Crop Plants 3
Cécile Monat and François Sabot

Population Genomics of Organelle Genomes in Crop Plants 37
Nora Scarcelli

Coalescent Models of Demographic History: Application to Plant Domestication 65
Olivier François, Philippe Cubry, Concetta Burgarella, and Yves Vigouroux

Population Genomics of Weedy Crop Relatives: Insights from Weedy Rice 87
Lin-Feng Li and Kenneth M. Olsen

Population Genomics of Speciation and Adaptation in Sunflowers 113
Dan G. Bock, Michael B. Kantar, and Loren H. Rieseberg

Genomics of Plant Gene Banks: Prospects for Managing and Delivering Diversity in the Digital Age 143
Chris Richards

Part II Population Genomics for Crop Plant Breeding

Enhancing Crop Breeding Using Population Genomics Approaches 179
Ryan J. Andres, Jeffrey C. Dunne, Luis Fernando Samayoa, and James B. Holland

Population Genomics Along With Quantitative Genetics Provides a More Efficient Valorization of Crop Plant Genetic Diversity in Breeding and Pre-breeding Programs 225
Peter Civan, Renaud Rincent, Alice Danguy-Des-Deserts, Jean-Michel Elsen, and Sophie Bouchet

Population Genomics and Molecular Breeding of Sorghum 289
Arthur Bernardeli, Cynthia Maria Borges Damasceno, Jurandir Vieira de Magalhães, Vander Fillipe de Souza, Janaína de Oliveira Melo, Amanda Avelar de Oliveira, Maria Lúcia Ferreira Simeone, Aluízio Borém, Robert Eugene Schaffert, Rafael Augusto da Costa Parrella, and Maria Marta Pastina

Population Genomics and Genomics-Assisted Trait Improvement in Tea (*Camellia sinensis* (L.) O. Kuntze) 341
Tony Maritim, Romit Seth, Ashlesha Holkar, and Ram Kumar Sharma

Part III Population Genomics of Major Crop Plants

Population Genomics of Maize 377
Marcela Pedroso Mendes Resende, Ailton José Crispim Filho, Adriana Maria Antunes, Bruna Mendes de Oliveira, and Renato Gonçalves de Oliveira

Population Genomics of Pearl Millet 457
Ndjido Ardo Kane and Cécile Berthouly-Salazar

Potato Population Genomics 477
Xiaoxi Meng, Heather Tuttle, and Laura M. Shannon

Population Genomics of Tomato 533
Christopher Sauvage, Stéphanie Arnoux, and Mathilde Causse

Population Genomics of Soybean 573
Milind B. Ratnaparkhe, Rishiraj Raghuvanshi, Vennampally Nataraj, Shivakumar Maranna, Subhash Chandra, Giriraj Kumawat, Rucha Kavishwar, Prashant Suravajhala, Shri Hari Prasad, Dalia Vishnudasan, Subulakshmi Subramanian, Pranita Bhatele, Supriya M. Ratnaparkhe, Ajay K. Singh, Gyanesh K. Satpute, Sanjay Gupta, Kunwar Harendra Singh, and Om P. Rajora

Population Genomics of *Phaseolus* spp.: A Domestication Hotspot ... 607
Travis A. Parker and Paul Gepts

Population Genomics of Cotton 691
Lavanya Mendu, Kaushik Ghose, and Venugopal Mendu

Population Genomics of *Brassica* Species 741
Yonghai Fan, Yue Niu, Xiaodong Li, Shengting Li,
Cunmin Qu, Jiana Li, and Kun Lu

Population Genomics of Peanut 793
Ramesh S. Bhat, Kenta Shirasawa, Vinay Sharma, Sachiko N. Isobe,
Hideki Hirakawa, Chikara Kuwata, Manish K. Pandey,
Rajeev K. Varshney, and M. V. Channabyre Gowda

**Population Genomics of Yams: Evolution and Domestication
of *Dioscorea* Species** 837
Yu Sugihara, Aoi Kudoh, Muluneh Tamiru Oli, Hiroki Takagi,
Satoshi Natsume, Motoki Shimizu, Akira Abe, Robert Asiedu,
Asrat Asfaw, Patrick Adebola, and Ryohei Terauchi

Population Genomics of Sweet Watermelon 865
Padma Nimmakayala, Purushothaman Natarajan, Carlos Lopez-Ortiz,
Sudip K. Dutta, Amnon Levi, and Umesh K. Reddy

Population Genomics of Perennial Temperate Forage Legumes 903
Muhammet Şakiroğlu

Index 943

Contributors

Akira Abe Iwate Biotechnology Research Center, Kitakami, Iwate, Japan

Patrick Adebola International Institute of Tropical Agriculture (IITA), Ibadan, Nigeria

Ryan J. Andres USDA-ARS Plant Science Research Unit, Raleigh, NC, USA
Department of Crop and Soil Sciences, North Carolina State University, Raleigh, NC, USA

Adriana Maria Antunes Escola de Agronomia, Universidade Federal de Goiás, Goiânia, GO, Brazil

Stéphanie Arnoux INRAE, GAFL, Avignon, France

Asrat Asfaw International Institute of Tropical Agriculture (IITA), Ibadan, Nigeria

Robert Asiedu International Institute of Tropical Agriculture (IITA), Ibadan, Nigeria

Arthur Bernardeli Universidade Federal de Viçosa (UFV), Viçosa, MG, Brazil
University of Nebraska-Lincoln (UNL), Lincoln, NE, USA

Cécile Berthouly-Salazar Laboratoire mixte international Adaptation des Plantes et microorganismes associés aux Stress Environnementaux (LAPSE), Dakar, Senegal
DIADE, Université Montpellier, Institut de Recherche pour le Développement, Montpellier, France

Ramesh S. Bhat Department of Biotechnology, University of Agricultural Sciences, Dharwad, India

Pranita Bhatele M. H. College of Home Science & Science for Women, Autonomous, Jabalpur, Madhya Pradesh, India

Dan G. Bock Department of Biology, Washington University in St. Louis, St. Louis, MO, USA

Aluízio Borém Universidade Federal de Viçosa (UFV), Viçosa, MG, Brazil

Sophie Bouchet INRAE, Université Clermont Auvergne, Clermont-Ferrand, France

Concetta Burgarella Université de Montpellier, Institut pour la Recherche et le Développement, Unité Diversité, Adaptation, Développement des plantes, Montpellier, France
Human Evolution, Department of Organismal Biology, Uppsala University, Uppsala, Sweden

Mathilde Causse INRAE, GAFL, Avignon, France

Subhash Chandra ICAR-Indian Institute of Soybean Research, Indore, Madhya Pradesh, India

Peter Civan GDEC, INRAE, Université Clermont Auvergne, Clermont-Ferrand, France

Rafael Augusto da Costa Parrella Embrapa Milho e Sorgo, Sete Lagoas, MG, Brazil

Philippe Cubry Université de Montpellier, Institut pour la Recherche et le Développement, Unité Diversité, Adaptation, Développement des plantes, Montpellier, France

Cynthia Maria Borges Damasceno Embrapa Milho e Sorgo, Sete Lagoas, MG, Brazil

Alice Danguy-Des-Deserts GDEC, Université Clermont Auvergne, Clermont-Ferrand, France
GenPhySE (Génétique Physiologie et Systèmes d'Elevage), Université de Toulouse, INRAE, ENVT, Castanet-Tolosan, France

Jeffrey C. Dunne Department of Crop and Soil Sciences, North Carolina State University, Raleigh, NC, USA

Sudip K. Dutta Gus R. Douglass Institute, Department of Biology, West Virginia State University, Institute, WV, USA

Jean-Michel Elsen GenPhySE (Génétique Physiologie et Systèmes d'Elevage), Université de Toulouse, INRAE, ENVT, Castanet-Tolosan, France

Yonghai Fan College of Agronomy and Biotechnology, Southwest University, Chongqing, China

Ailton José Crispim Filho Escola de Agronomia, Universidade Federal de Goiás, Goiânia, GO, Brazil

Olivier François Université Grenoble-Alpes, Centre National de la Recherche Scientifique, Grenoble, France

Paul Gepts Department of Plant Sciences/MS1, Section of Crop and Ecosystem Sciences, University of California, Davis, CA, USA

Kaushik Ghose Department of Plant and Soil Science, Texas Tech University, Lubbock, TX, USA

M. V. Channabyre Gowda Department of Genetics and Plant Breeding, University of Agricultural Sciences, Dharwad, India

Sanjay Gupta ICAR-Indian Institute of Soybean Research, Indore, Madhya Pradesh, India

Hideki Hirakawa Facility for Genome Informatics, Kazusa DNA Research Institute, Chiba, Japan

Ashlesha Holkar Biotechnology Department, CSIR-Institute of Himalayan Bioresource Technology (CSIR-IHBT), Palampur, Himachal Pradesh, India
Academy of Scientific and Innovative Research (AcSIR), Ghaziabad, India

James B. Holland USDA-ARS Plant Science Research Unit, Raleigh, NC, USA
Department of Crop and Soil Sciences, North Carolina State University, Raleigh, NC, USA

Sachiko N. Isobe Department of Frontier Research and Development, Kazusa DNA Research Institute, Chiba, Japan

Ndjido Ardo Kane Institut Sénégalais de Recherches Agricoles (ISRA), Centre d'Étude Régional pour l'amélioration de l'Adaptation à la Sécheresse, Thiès, Senegal
Laboratoire mixte international Adaptation des Plantes et microorganismes associés aux Stress Environnementaux (LAPSE), Dakar, Senegal

Michael B. Kantar Department of Tropical Plant and Soil Sciences, University of Hawaii, Honolulu, HI, USA

Rucha Kavishwar ICAR-Indian Institute of Soybean Research, Indore, Madhya Pradesh, India

Aoi Kudoh Laboratory of Crop Evolution, Graduate School of Agriculture, Kyoto University, Kyoto, Japan

Giriraj Kumawat ICAR-Indian Institute of Soybean Research, Indore, Madhya Pradesh, India

Chikara Kuwata Chiba Prefectural Agriculture and Forestry Research Center, Chiba, Japan

Amnon Levi USDA, ARS, U.S. Vegetable Laboratory, Charleston, SC, USA

Jiana Li College of Agronomy and Biotechnology, Southwest University, Chongqing, China
Academy of Agricultural Sciences, Southwest University, Chongqing, China
Engineering Research Center of South Upland Agriculture, Ministry of Education, Chongqing, China

Lin-Feng Li Ministry of Education Key Laboratory for Biodiversity Science and Ecological Engineering, and Coastal Ecosystems Research Station of the Yangtze River Estuary, School of Life Sciences, Fudan University, Shanghai, China

Shengting Li College of Agronomy and Biotechnology, Southwest University, Chongqing, China

Xiaodong Li College of Agronomy and Biotechnology, Southwest University, Chongqing, China

Carlos Lopez-Ortiz Gus R. Douglass Institute, Department of Biology, West Virginia State University, Institute, WV, USA

Kun Lu College of Agronomy and Biotechnology, Southwest University, Chongqing, China
Academy of Agricultural Sciences, Southwest University, Chongqing, China
Engineering Research Center of South Upland Agriculture, Ministry of Education, Chongqing, China

Jurandir Vieira de Magalhães Embrapa Milho e Sorgo, Sete Lagoas, MG, Brazil

Shivakumar Maranna ICAR-Indian Institute of Soybean Research, Indore, Madhya Pradesh, India

Tony Maritim Biotechnology Department, CSIR-Institute of Himalayan Bioresource Technology (CSIR-IHBT), Palampur, Himachal Pradesh, India
Tea Breeding and Genetic Improvement Division, KALRO-Tea Research Institute, Kericho, Kenya
Academy of Scientific and Innovative Research (AcSIR), Ghaziabad, India

Lavanya Mendu Department of Plant Sciences and Plant Pathology, Montana State University, Bozeman, MT, USA

Venugopal Mendu Department of Plant Sciences and Plant Pathology, Montana State University, Bozeman, MT, USA

Xiaoxi Meng Department of Horticultural Science, University of Minnesota, St. Paul, MN, USA

Cécile Monat DIADE, University of Montpellier, IRD, Montpellier, France
South Green Bioinformatics Platform, IRD, Bioversity, CIRAD, INRAE, Montpellier, France
Université Clermont Auvergne, INRAE, GDEC, Clermont-Ferrand, France

Vennampally Nataraj ICAR-Indian Institute of Soybean Research, Indore, Madhya Pradesh, India

Purushothaman Natarajan Gus R. Douglass Institute, Department of Biology, West Virginia State University, Institute, WV, USA

Satoshi Natsume Iwate Biotechnology Research Center, Kitakami, Iwate, Japan

Padma Nimmakayala Gus R. Douglass Institute, Department of Biology, West Virginia State University, Institute, WV, USA

Yue Niu College of Agronomy and Biotechnology, Southwest University, Chongqing, China

Muluneh Tamiru Oli Iwate Biotechnology Research Center, Kitakami, Iwate, Japan
Department of Animal, Plant, and Soil Sciences, School of Life Sciences, La Trobe University, Melbourne, VIC, Australia

Amanda Avelar de Oliveira Escola Superior de Agricultura Luiz de Queiroz/Universidade de São Paulo (ESALQ/USP), Piracicaba, Brazil

Bruna Mendes de Oliveira Escola de Agronomia, Universidade Federal de Goiás, Goiânia, GO, Brazil

Renato Gonçalves de Oliveira Escola de Agronomia, Universidade Federal de Goiás, Goiânia, GO, Brazil

Janaína de Oliveira Melo Universidade Federal dos Vales do Jequitinhonha e Mucuri (UFVJM), Diamantina, MG, Brazil

Kenneth M. Olsen Department of Biology, Washington University in St. Louis, St. Louis, MO, USA

Manish K. Pandey Center of Excellence in Genomics and Systems Biology, International Crops Research Institute for the Semi-Arid Tropics, Hyderabad, India

Travis A. Parker Department of Plant Sciences/MS1, Section of Crop and Ecosystem Sciences, University of California, Davis, CA, USA

Maria Marta Pastina Embrapa Milho e Sorgo, Sete Lagoas, MG, Brazil

Shri Hari Prasad Centre for Plant Biotechnology and Molecular Biology, Kerala Agricultural University, Thrissur, Kerala, India

Cunmin Qu College of Agronomy and Biotechnology, Southwest University, Chongqing, China
Academy of Agricultural Sciences, Southwest University, Chongqing, China
Engineering Research Center of South Upland Agriculture, Ministry of Education, Chongqing, China

Rishiraj Raghuvanshi ICAR-Indian Institute of Soybean Research, Indore, Madhya Pradesh, India

Om P. Rajora Faculty of Forestry and Environmental Management, University of New Brunswick, Fredericton, NB, Canada

Milind B. Ratnaparkhe ICAR-Indian Institute of Soybean Research, Indore, Madhya Pradesh, India

Supriya M. Ratnaparkhe Indore Biotech Inputs and Research (P) Ltd., Indore, Madhya Pradesh, India

Umesh K. Reddy Gus R. Douglass Institute, Department of Biology, West Virginia State University, Institute, WV, USA

Marcela Pedroso Mendes Resende Escola de Agronomia, Universidade Federal de Goiás, Goiânia, GO, Brazil

Chris Richards USDA National Laboratory for Genetic Resources Preservation, Fort Collins, CO, USA

Loren H. Rieseberg Department of Botany and Biodiversity Research Centre, University of British Columbia, Vancouver, BC, Canada

Renaud Rincent GQE-Le Moulon, INRAE, Univ. Paris-Sud, CNRS, AgroParisTech, Université Paris-Saclay, Clermont-Ferrand, France

François Sabot DIADE, University of Montpellier, IRD, Montpellier, France
South Green Bioinformatics Platform, IRD, Bioversity, CIRAD, INRAE, Montpellier, France

Muhammet Şakiroğlu Department of Bioengineering, Adana Alparslan Türkeş Science and Technology University, Adana, Turkey

Luis Fernando Samayoa Department of Crop and Soil Sciences, North Carolina State University, Raleigh, NC, USA

Gyanesh K. Satpute ICAR-Indian Institute of Soybean Research, Indore, Madhya Pradesh, India

Christopher Sauvage INRAE, GAFL, Avignon, France

Nora Scarcelli DIADE, Univ Montpellier, IRD, Montpellier, France

Robert Eugene Schaffert Embrapa Milho e Sorgo, Sete Lagoas, MG, Brazil

Romit Seth Biotechnology Department, CSIR-Institute of Himalayan Bioresource Technology (CSIR-IHBT), Palampur, Himachal Pradesh, India
Plants for Human Health Institute – North Carolina State University, Kannapolis, NC, USA

Laura M. Shannon Department of Horticultural Science, University of Minnesota, St. Paul, MN, USA

Contributors

Ram Kumar Sharma Biotechnology Department, CSIR-Institute of Himalayan Bioresource Technology (CSIR-IHBT), Palampur, Himachal Pradesh, India
Academy of Scientific and Innovative Research (AcSIR), Ghaziabad, India

Vinay Sharma Center of Excellence in Genomics and Systems Biology, International Crops Research Institute for the Semi-Arid Tropics, Hyderabad, India

Motoki Shimizu Iwate Biotechnology Research Center, Kitakami, Iwate, Japan

Kenta Shirasawa Department of Frontier Research and Development, Kazusa DNA Research Institute, Chiba, Japan

Maria Lúcia Ferreira Simeone Embrapa Milho e Sorgo, Sete Lagoas, MG, Brazil

Ajay K. Singh ICAR-National Institute of Abiotic Stress Management, Khurd, Baramati, Maharashtra, India

Kunwar Harendra Singh ICAR-Indian Institute of Soybean Research, Indore, Madhya Pradesh, India

Vander Fillipe de Souza Universidade Federal de Lavras (UFLA), Lavras, MG, Brazil

Subulakshmi Subramanian Amrita School of Biotechnology, Amrita University, Kollam, Kerala, India

Yu Sugihara Laboratory of Crop Evolution, Graduate School of Agriculture, Kyoto University, Kyoto, Japan

Prashant Suravajhala Amrita School of Biotechnology, Amrita University, Kollam, Kerala, India

Hiroki Takagi Iwate Biotechnology Research Center, Kitakami, Iwate, Japan
Ishikawa Prefectural University, Nonoichi, Ishikawa, Japan

Ryohei Terauchi Laboratory of Crop Evolution, Graduate School of Agriculture, Kyoto University, Kyoto, Japan
Iwate Biotechnology Research Center, Kitakami, Iwate, Japan

Heather Tuttle Department of Horticultural Science, University of Minnesota, St. Paul, MN, USA

Rajeev K. Varshney Center of Excellence in Genomics and Systems Biology, International Crops Research Institute for the Semi-Arid Tropics, Hyderabad, India

Yves Vigouroux Université de Montpellier, Institut pour la Recherche et le Développement, Unité Diversité, Adaptation, Développement des plantes, Montpellier, France

Dalia Vishnudasan Amrita School of Biotechnology, Amrita University, Kollam, Kerala, India

Part I
Pangenomes, Demography, Evolution, Speciation, and Conservation

Part I

Pangenomics in Crop Plants

Cécile Monat and François Sabot

Abstract Identifying diversity at many scales is the keystone to understand how genomes evolved and plants are so adaptable, even if they are under extreme constraint. Understanding the processes that create diversity and how the genome manages this variability is of high importance to be able to face today's challenges for breeding and genetic resource conservation of crop plants. For about past 15 years, we ushered into the era of pangenomics and started to learn what the true intra-species genomic diversity is and how it shapes the structure and the expression of the genome, especially in crop plants. The pangenome is the complete repertoire of sequences for a given population (often a species). It can be divided in compartments, the first one is the core genome which contains the sequences shared by the whole population. The second compartment is the dispensable genome which regroups the sequences shared by some individuals of the population but not all of them. As a sub-part of the dispensable genome, an individual-specific-genome regroups the sequences that are uniquely found in one individual of the population. By comparing individual gene and sequence contents within a population, pangenomics is a great tool to investigate how sequences evolve within a species and how genetic diversity is shaped within it. More and more studies are including relative wild species to have a larger overview of how genetic and phenotypic diversity was shaped through domestication and selection processes. In this chapter, we will present this concept of pangenomics and the associated methods and progress made, particularly in crop plants and future perspective.

C. Monat
DIADE, University of Montpellier, IRD, Montpellier, France

South Green Bioinformatics Platform, IRD, Bioversity, CIRAD, INRAE, Montpellier, France

Université Clermont Auvergne, INRAE, GDEC, Clermont-Ferrand, France

F. Sabot (✉)
DIADE, University of Montpellier, IRD, Montpellier, France

South Green Bioinformatics Platform, IRD, Bioversity, CIRAD, INRAE, Montpellier, France
e-mail: Francois.sabot@ird.fr

Keywords Copy number variations · Crops · Genomic diversity · Pangenome · Pangenomics · Plants · Presence/absence variations · Structural variations

1 Introduction

Understanding genetic diversity within a species is required for various population, evolutionary and conservation genetic studies, and this knowledge is essential to understand species' phenotypic variation, as well as mechanisms allowing environmental adaptation (Hirsch et al. 2014; Carlos Guimaraes et al. 2015). However, estimation of genetic diversity is based on the data, methods, and technologies available at the time of analysis. Thus, while Carl Von Linné or Georges-Louis de Buffon (Lawrence Farber 2000) classified animals and plants based on visual phenotypic observations, nowadays biologists deal with large molecular dataset, and diversity estimations are based on genetics and genomics approaches and dataset.

Recent genetic diversity studies are mostly performed using SNP (single nucleotide polymorphism) markers (Springer et al. 2009; Hirsch et al. 2014) detected through sequencing approaches, and the allelic diversity in coding and non-coding sequences is generally thought to be the main cause of phenotypic variations. Up to recently it was widely assumed that individuals from the same species have a similar genetic constitution (Swanson-Wagner et al. 2010) and differ only by these (few) SNPs. However, since the rise of NGS (next generation sequencing) and of the subsequent massive sequencing efforts, it appeared that structural variation (SV), in the form of not only insertion/deletions (InDel) ranging from few bp and up to tens of kb or even megabase-sized, but also chromosomal rearrangements (translocations, inversions, duplications), is much more important than expected (Beyter et al. 2019) (Fig. 1). These variations might be due to CNV (copy number variation) of the same sequence, or even missing sequences in some individuals (PAV, presence/absence variations) (Da Silva et al. 2013; Montenegro et al. 2017). The different levels of diversity (SNP, SV, CNV, and PAV) add a challenge in the accurate representation of the genome. It is now clear that we need more assembled individual genomes (and more variation information) to capture the whole- genetic/genomic diversity of a given species. Thus, we need to have a wider overview of the genomic diversity by sequencing more than one genome per species; for that, approaches

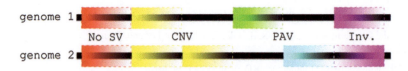

Fig. 1 Schematic representation of no structural variation (no SV), copy number variation (CNV), presence/absence variation (PAV), and example of Inversion (Inv.) as a structural variation between two sequences. Each colored bloc represents a genomic segment

using pangenomes have been deployed (Gan et al. 2011; Li et al. 2014; Hirsch et al. 2014; Schatz et al. 2014a).

The purpose of this chapter is to present an overview of what is pangenomics as applied to crop plants and why it is important for the future. We will first present the pangenome concept, how it was initially developed in the microorganisms world and then deployed in other biological domains. We will focus afterward on pangenomics in plants and more precisely on crop plants. Finally, we will briefly present the methodologies to create and analyze pangenomes.

2 Pangenome Concept

The access to DNA sequences, *i.e.* to the whole genetic and genomic information, has not evolved at the same pace for different phyla, mainly because of the disparity of genome size and complexity between them. For instance, basal eukaryotes (ex: *Saccharomyces cerevisiae*, 13 Mb) and bacteria (ex: *Escherichia coli*, 4.6 Mb) mostly have small simple genomes, easy to sequence and assemble. On the other hand, plants have large (Gb sized or more) and complex genomes; thus, only the true first pangenomic studies were performed on small bacterial genomes (Tettelin et al. 2005, 2008) before plants.

The concept was introduced first by Tettelin et al. (2005), as follows: the pangenome consists of a "core genome containing genes present in all strains and a dispensable genome composed of genes absent from one or more strains and genes that are unique to each strain" (Fig. 2) (Tettelin et al. 2005). This definition has been since then re-used many times (Tettelin et al. 2008; Liang et al. 2012; Lukjancenko et al. 2012; Mann et al. 2013; Lukjancenko 2013; Kahlke 2013; Soares et al. 2013) and extended to transcripts, proteins, pseudogenes as well as to TEs (Transposable Elements) (Lukjancenko 2013; Nguyen 2014; Hirsch et al. 2014; Tranchant-Dubreuil et al. 2019). A pangenomic analysis allows to characterize the folder of genes/sequences available within a given group, as well as to estimate the genomic missing part (*i.e.*, the number of new individuals to be sequenced to reach this almost complete folder) (Tettelin et al. 2008).

Pangenomic studies are performed on a set of organisms/individuals supposed to be representative of a family, a genus, or more generally a species. These analyses provide a large array of phylogenomic information at different levels, from general implications for the group to individual-specific genetic information (Kahlke 2013; Soares et al. 2013). As a corollary, two individuals may not be similar to each other because of the genes they share, but also for the ones they missed (Snipen and Ussery 2010).

Pangenomics is also a way to characterize how conserved are the genomic sequences between individuals within a population. We may, by such means, better understand the genetic variation underlying ecological adaptations and have a more complete view on how genetic variation is important for adaptation and survival of a given group of organisms (Hansen et al. 2012). Indeed, as shown by Caputo et al.

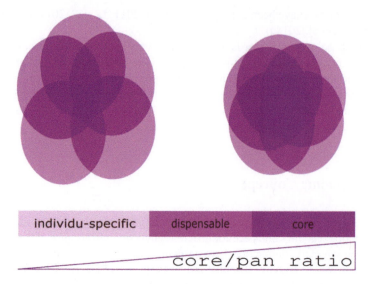

Fig. 2 Pangenome with high or low core/pan ratio. Each individual is represented by a different circle. Intersection between all circles represents the core genome, parts without intersection represent the individual-specific genome, and intersections with not all circle the dispensable genome

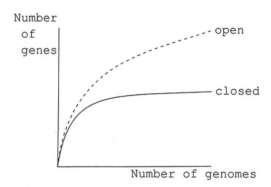

Fig. 3 Open or closed pangenome

(2015), a core/pan ratio (*i.e.,* size of core *vs* size of pangenome, in bp) greater than 90% indicates a high rate of genome conservation between sampled individuals. On the other hand, if this ratio is lower than 85%, then the species may present a certain potential of adaptability because of a higher genomic diversity. Indeed, in this case, if we consider the population, individuals are different enough from each other to be able to respond to rapid and/or abrupt changes and increase the chances of survival for the species as a whole (Fig. 3).

Some studies found that analyzing many genomes of the same species allows to discover more intra-specific genomic diversity than expected (Tettelin et al. 2008). Working on many individuals from the same species leads to identify genomic diversity which might not be discovered if we look only at the inter-species variability (Lukjancenko et al. 2012).

2.1 Core Genome

The core genome is made of genes/sequences present in all individuals of the studied group (Lukjancenko et al. 2012; Lukjancenko 2013; Soares et al. 2013). In bacteria, those genes are conserved and seem to be important genes for basic biological processes (Tettelin et al. 2005; Soares et al. 2013), and thus have been qualified as essential genes (Kahlke 2013). However, as more individuals are added to the study, the less core genes are found (Tettelin et al. 2005; Lukjancenko et al. 2012), and the core genome is generally over-estimated. In the same way, Collins and Higgs (2012) have shown that genes considered as essential (because their function seems to be essential for the survival of the individual) might be absent from the core genome. So, the level of definition of essential is critical in the way of interpretation of a core genome. In this regard, the core genome would compose not only of putative essential (functionally essential) genes but also of genes evolving slowly enough to be present in each genome ("pangenomically" defined as essential), not required for their host survival. Finally, complementation of function can occur through different genes in different individuals, and thus, if considering the sequence, the core genome may be really low, while considering function, the same core genome may be larger.

2.2 Dispensable Genome

Dispensable genes are found in at least two individuals among those studied, but not in all (Tettelin et al. 2005; Liang et al. 2012). These genes mostly belong to secondary biological processes, and are usually associated with adaptation, resistance to biotic and abiotic stresses or, for microorganisms, new host colonization (Tettelin et al. 2008; Mann et al. 2013; Hirsch et al. 2014). For instance, in plants, and more particularly in rice *Oryza sativa*, Schatz et al (2014a) showed that some of them are related to disease resistance genes.

The dispensable genes are supposed to be the main contributors of the variability among individuals within species (Kahlke 2013). Their origin is supposed to be distinct between different phyla: in bacteria, they probably originate from horizontal transfers (Medini et al. 2005), while in eukaryotes they may be related to genes lost, segmental or whole-genome duplications, neogenesis, or TEs activity (Hirsch et al. 2014).

2.3 Individual-Specific Genome

Genes belonging to a single individual are grouped under the term individual-specific genome which is a sub-part of the dispensable genome. (Tettelin et al. 2005; Lukjancenko 2013; Soares et al. 2013). In microorganisms, including bacteria, they are mostly related to phages, horizontal transfers or TEs (Kahlke 2013). For plants, as in rice (*Oryza sativa*), Schatz et al. (2014a) have shown that most of these sequences are repetitive sequences, some being related to genes. They also found that the distribution of these unique regions is not precisely located in a specific region, but they are well distributed across the genome. Different studies have shown that new genes, or genes recently evolved, have generally smaller CDS (coding DNA sequence) compared to older/strictly conserved genes (Cai and Petrov 2010; Capra et al. 2010; Lipman et al. 2002), such as those of the individual-specific-genome, which suggests they might be new or recently evolved (Schatz et al. 2014a). However, the classification as a genome-specific sequence is quite debatable (see (Tranchant-Dubreuil et al. 2019)), as its attribution may be related only to sampling effect or sequencing limits.

2.4 Pangenomics Goes from Bacteria to Plants

Since the beginning of pangenomics (Tettelin et al. 2005), many projects have been performed on different organisms (bacteria, *archaea*, animals, plants, virus, and recently fungi) with a high variation in the range of studied genomes (from 2 individuals to 14,129; Lu et al. 2015) and the number of species involved (between 1 and 3; Darling et al. 2010; Gordienko et al. 2013; Boussaha et al. 2015; Ghatak et al. 2016). Many different methods have been used to be adapted for each data set (see Vernikos et al. 2015 and later in this chapter for more details).

2.5 Pangenomics in Plants

In plants, there have been many pangenomics studies listed in Table 1, 2, and 3. Plants pangenomes seem to be very dynamic and variable depending on species, but their large genome size poses the difficulties for representative sequencing as well as scalable analyses. To override these problems, early studies were limited to chromosomal regions and not the whole genome (Wang and Dooner 2006). Morgante et al. (2007) have first translated the pangenome definition from bacteria to plants, in order to better understand genomic variation. According to this study, performed with two maize (*Zea mays*) lines on four genomic regions, they have shown that only 50% of the sequences are shared, which might constitute the core genome. Similar observations were obtained in barley (*Hordeum vulgare*) and rice (Morgante et al.

Table 1 Pangenomics studies in major crop plants

Species	Number of genotypes	Strategy	References
Zea mays	2 inbred lines	Comparison of BAC assemblies for a few loci	Brunner et al. (2005)
Zea mays	2 inbred lines	Comparison of BAC assemblies on whole genome	Morgante et al. (2005)
Zea mays	27 inbred lines	Reference-based mapping	Gore et al. (2009)
Zea mays	Inbred lines	Comparative genomic hybridization	Springer et al. (2009)
Zea mays	6 inbred lines	Reference-based mapping	Lai et al. (2010)
Zea mays	Inbred lines	Comparative genomic hybridization	Belo et al. (2010)
Zea mays and *Zea mays* ssp. *parviglumis*	Inbred lines	Comparative genomic hybridization	Swanson-Wagner et al. (2010)
Zea mays	103 inbred lines	Reference-based mapping	Chia et al. (2012)
Zea mays	21 inbred lines	Reference-based mapping on RNA samples	Hansey et al. (2012)
Zea mays	503 inbred lines	*De novo* transcriptome assembly	Hirsch et al. (2014)
Zea mays	14,129 inbred lines	Genetic mapping followed by machine learning approaches	Lu et al. (2015)
Zea mays	1 inbred line	Reference-quality *de novo* genome assembly	Hirsch et al. (2016)
Triticum aestivum	2 accessions (chr3B)	Reference-based mapping and draft assembly of unmapped reads	Liu et al. (2016)
Triticum aestivum	19 accessions	Reference-based mapping and draft assembly of unmapped reads	Montenegro et al. (2017)
Triticum aestivum	2 accessions (700 Mb of chr2D)	Reference-based mapping	Thind et al. (2018)
Oryza sativa	2 accessions	Comparative genomic hybridization	Yu et al. (2011)
Oryza sativa	50 accessions	Reference-based mapping and draft assembly of unmapped reads	Xu et al. (2011)
Oryza sativa	3 inbred lines	High-quality *de novo* genome assemblies	Schatz et al. (2014b)
Oryza sativa	1,483 accessions	Metagenome-like *de novo* assembly	Yao et al. (2015)

(continued)

Table 1 (continued)

Species	Number of genotypes	Strategy	References
Oryza sativa	2 accessions	Comparison of BAC assemblies on whole genome	Zhang et al. (2016)
Oryza glaberrima	3 accessions	*De novo* draft genome assemblies	Monat et al. (2016)
Oryza sativa	3,010 accessions	*De novo* draft genome assemblies and reference-based mapping	Sun et al. (2017)
Oryza sativa	66 accessions	*De novo* draft genome assemblies	Zhao et al. (2018)
Oryza sativa	3,010 accessions	Reference-guided *de novo* assembly	Wang et al. (2018)
Oryza sativa	12 accessions	High-quality *de novo* genome assemblies	Zhou et al. (2020)
Brassica napus and *B. rapa*	BAC segments from 17 accessions	*De novo* draft assembly	Cheung et al. (2009)
Brassica rapa	1 doubled haploid line, 1 inbred line and the reference genome	*De novo* draft genome assembly	Lin et al. (2014)
Brassica oleracea and *B. macrocarpa*	10 lines	Iterative mapping and assembly	Golicz et al. (2016a, b)
Brassica napus	2 accessions	*De novo* draft genome assembly then reference-based guided	Bayer et al. (2017)
		Pseudomolecule-level assembly	
Brassica napus	53 accessions	Iterative mapping and assembly	Hurgobin et al. (2018)
Cicer arietinum	35 accessions	Draft reference-based mapping	Thudi et al. (2016a, b)
Cicer arietinum	129 accessions	Draft reference-based mapping	Thudi et al. (2016b)
Glycine soja and *Glycine max*	31 accessions	Reference-based mapping	Lam et al. (2010)
Gycine max	4 accessions	Comparative genomic hybridization and reference-based mapping on exome data	Haun et al. (2011)
Glycine max	4 accessions	Comparative genomic hybridization and reference-based mapping on exome data	McHale et al. (2012)
Glycine soja	7 accessions	*De novo* draft genome assembly	Li et al. (2014)

(continued)

Table 1 (continued)

Species	Number of genotypes	Strategy	References
Glycine max	41 accessions	Comparative genomic hybridization and reference-based mapping	Anderson et al. (2014)
Sorghum bicolor	3 inbred lines	Reference-based mapping	Zheng et al. (2011)
Hordeum vulgare ssp. vulgare and *H. vulgare ssp. spontaneum*	14 accessions	Comparative genomic hybridization	Muñoz-Amatriaín et al. (2013)
Ipomoea trifida	2 accessions	*De novo* draft genome assemblies and cross mapping	Hirakawa et al. (2015)
Solanum tuberosum	12 simple and double monoploid clones	Reference-based mapping	Hardigan et al. (2016)
Capsicum annuum, *C. baccatum*, *C. chinense* and *C. frutescens*	383 accessions	*De novo* draft genome assemblies and reference-based mapping	Lijun et al. (2018)
Solanum lycopersicum, *S. pimpinellifolium*, *S. cheesmaniae* and *S. galapagense*	725 accessions	Map-to-pan	Gao et al. (2019)
Sesamum indicum	5 accessions	Reference-guided and *de novo* assembly	Yu et al. (2019)
Cajanus cajan	90 accessions	Reference-based mapping and draft assembly of unmapped reads	Zhao et al. (2020)

2007). In another study, this time with 8 lines of maize on the *bz1* locus, Wang and Dooner (2006) estimated that core genome is not as large as it was expected earlier from the analysis of two individuals (from 25 to 84%), and that the dispensable genome, for its part, increases when more genomes are included, as seen in bacteria.

Plants genomes have some specificity making their analysis more complex than for bacteria or microorganisms, explaining why pangenomics studies came later to plants. Among those, we can cite the high level of polyploidy, *i.e.* partial or even whole-genome duplication, particularly frequent in angiosperm species (Cheung et al. 2009). Some of these recent duplications may impair the pangenome identification: thus, in rapeseed (*Brassica napus*), various subgenome loci are distinguishable only through a few SNP and InDels (Cheung et al. 2009). However, Lin et al. (2014) performed a pangenomics study with three genomes of *Brassica rapa*, the diploid progenitor of *Brassica napus* and shown that on average 1,200 genes are individual-specific (and thus subgenome-specific), which provided a supplementary evidence of increasing interest in pangenomics studies.

Table 2 Key results from pangenomics studies in crop plants

Species	Reference	Number of pan-genes	% of core genes	Pangenome size	% core size	Example of SVs
Oryza sativa	Schatz et al. (2014b)	40,362	92.16%	341 Mb	88.82%	PAV of Phosphorus uptake1 (*Pup1*), major rice QTL associated with tolerance to phosphorus deficiency in soils. CNV and InDel of S5, major locus for hybrid sterility in rice that affects embryo sac fertility. CNV of Submergence 1 (*Sub1*), major QTL on chromosome 9 determining submergence tolerance in rice
Brassica ssp.	Golicz et al. (2016a, b)	61,379	81.3%	587 Mb	NA	PAV of auxin-related genes (*AUX, IAA, GH3, PIN, SAUR, TIR, TPL, and YUCCA* PAV of flowering related genes (*MAF5, SEP2, ARP4, GID1B, FPF1-like, FHY1, GA2, GA3, and CO* CNV of Flowering locus C (*FLC*), an important regulator of vernalization and flowering time. PAV of glucosinolate-related genes *CYP79A2, SUR1, and SOT18*
Capsicum ssp.	Lijun et al. (2018)	51,757	55.7%	4.31 Gb	NA	SV of capsanthin/capsorubin synthase (*Ccs*), locus involved in carotenoid contents. SV of phytoene synthase gene (*Psy*), locus involved in carotenoid contents. Sv of Pungent gene 1 (*Pun1*), locus involved in capsaicinoid biosynthetic pathway
Sesamum indicum	Yu et al. (2019)	26,472	58.21%	554 Mb	46.70%	PAV between modern cultivars and landraces for genes with functions related to energy metabolism, growth and development, as well as biomass accumulation PAV between landraces and modern cultivars for genes with functions related to environmental adaptation, signal transduction, protein folding, sorting, degradation, transport, and catabolism

Table 3 Pangenomics studies in Arabidopsis and other plants

Species	Number of genotypes	Strategy	References
Arabidopsis thaliana	3 accessions	Reference-based mapping	Ossowski et al. (2008)
Arabidopsis thaliana	4 accessions	Reference-based mapping	Santuari et al. (2010)
Arabidopsis thaliana	80 accessions	Reference-based mapping and draft assembly of unmapped reads	Cao et al. (2011)
Arabidopsis thaliana	18 accessions	Iterative reads mapping combined with *de novo* assembly and reference-based mapping	Gan et al. (2011)
Arabidopsis thaliana	2 accessions	Reference-based mapping	Lu et al. (2012)
Arabidopsis thaliana	3 accessions	High-quality *de novo* genome assembly	Pucker et al. (2016)
Arabidopsis thaliana	2 accessions	High-quality *de novo* genome assembly	Pucker et al. (2019)
Populus nigra, Populus deltoides and Populus trichocarpa	22 accessions	Reference-based mapping on both DNA and RNA samples	Pinosio et al. (2016)
Populus alba, P. davidiana, P. euphratica, P. lasiocarpa, P. nigra, P. deltoides, P. cathayana, P. simonii, P. ussuriensis and P. maximowiczii	10 accessions	Reference-based mapping	Zhang et al. (2019)
Utricularia gibba	13 specimens	Reference- and other species-based mapping and draft assembly of unmapped reads	Alcaraz et al. (2016)
Medicago truncatula	16 accessions	*De novo* assemblies	Zhou et al. (2017)
Brachypodium distachyon	54 inbred lines	*De novo* draft genome assembly	Gordon et al. (2017)

3 Pangenomics in Crop Plants

Crop plants are of key importance for food sustainability of the whole human population. Depending on the regions of the world, population are relying on different crop species to get their nutrient uptakes every day, such as: wheat, maize, rice, barley. This is why it is very important to better characterize the genomic diversity of these species, so better schemes could be developed for future breeding and genetic resource conservation programs.

Indeed, crop breeding programs benefit more and more from marker-assisted or genomic selection. To make this last one more efficient, it is necessary to have a better and more precise view of the species genetic/genomic diversity, which makes crop plants pangenomics studies a high priority.

3.1 Maize

Maize (*Zea mays ssp. mays* L.) is both an important model organism for fundamental research and an important crop species of high genetic diversity. The maize genome went through many rounds of genome duplication, including a polyploidization about 70 million years ago and a whole-genome duplication 5–12 million years ago making it very different from its closest wild relative, the teosinte (Paterson et al. 2004; Blanc and Wolfe 2004; Swigon et al. 2004; Paterson et al. 2009). As a result of all these changes and also due to a strong proliferation of LTR-retrotransposon, the genome size has expanded to 2.3 Gb over the last 3 million years (Sanmiguel et al. 1996). In addition to all these rearrangements, maize was domesticated over the past 10,000 years in Central America and since then is under cultivation and selection by breeders.

The first insights into plant pangenomics were on maize by Brunner et al. (2005) and Morgante et al. (2007) through a comparison of four genomic regions between two inbred line (B73 and Mo17). They revealed that only 50% of the sequences were shared by them. They also showed that non-shared sequences are mostly composed of TEs, which have potential phenotypic consequences other than those caused by genes.

The first whole-genome scale analysis between these two previous inbreed line was done by Springer et al. (2009). They used comparative genome hybridization to investigate PAV and CNV inside maize genome and have provided several evidences that a 2.6 Mb region on chromosome 6 present in B73 but absent in Mo17. These results were confirmed a year later by Belo et al. (2010). With an expanded panel of 34 maize and teosinte lines, Swanson-Wagner et al. (2010) used the same strategy to observe nearly 4,000 CNVs/PAVs.

The first-generation HapMap (haplotype map) developed for 27 maize lines estimated that the B73 genome represents only 70% of the total low-copy sequence in the maize pangenome (Gore et al. 2009). A few years later, the second-generation

HapMap identified that SVs are enriched at important traits associated loci (Chia et al. 2012). Meanwhile, the resequencing of six elite inbred lines important for commercial hybrid production was done by Lai et al. (2010) and identified many complete dispensable genes and, in some cases, PAV with heterotic group specificity.

The high proportion of TEs in maize genome makes it difficult to obtain a sufficient genome representation through short-reads data. To overcome this, Hansey et al. (2012) used a transcriptome profiling analysis of 21 inbred lines as a proxy for the pangenome and identified 1,321 non-reference loci, about 11% of which being heterotic group specific, thus confirming the results from Lai et al. (2010).

Another transcriptome profiling analysis was then performed by Hirsch et al. (2014) from 503 inbred lines to define the pan-transcriptome. They identified 8,681 non-reference transcripts. To enlarge this, they latter assembled the PH207 inbreed line genome, and a comparison to the B73 reference genome revealed PAV for more than 100 gene families (Hirsch et al. 2016).

To facilitate the maize pangenome construction, Lu et al. (2015) developed a new methodology consisting of genetic mapping followed by machine learning approaches. As a proof of concept, they implemented it on more than 14,000 inbred lines and selected 4.4 million of anchored tags of which a quarter are presence/absence variations.

3.2 Wheat

Bread wheat (*Triticum aestivum*, 2n = 6x = 42, AABBDD genome) is one of the most important food crops in the world (Abberton et al. 2016). Wheat has a complex genome due to different factors: (1) it is huge (15.4 – 15.8 Gb); (2) composed of more than 85% of highly repetitive sequences (Paux et al. 2006; Wicker et al. 2010, 2018); and (3) wheat is a young allopolyploid species that arose through two recent, natural polyploidization events that involved three diploid grass species (Chantret et al. 2005). The first hybridization event occurred probably 0.58–0.82 million years ago between the A genome donor (*Triticum urartu*) and a yet unidentified B genome donor close to the actual *Aegilops speltoides*, creating the wild tetraploid emmer wheat (*Triticum turgidum ssp. dicoccoides*) (Jordan et al. 2015). The second event occurred around 10,000 years ago between the domesticated emmer wheat and the D genome donor (wild goatgrass, *Aegilops tauschii*), creating the hexaploid bread wheat (Avni et al. 2017). Due to this complex genome, obtaining a high-quality reference genome for wheat was a difficult task. The International Wheat Genome Sequencing Consortium (IWGSC) decided to start with physical mapping and individual chromosomes sequencing of the genotype Chinese Spring (CS), and achieving the high-quality complete reference genome sequence took a long time (Appels et al. 2018).

However, some sequences were released before the release of the complete sequence (Choulet et al. 2014), and the partial genome sequences were compared with other datasets. As an example, Liu et al. (2016) isolated and sequenced the 3Bchromosome from a *Fusarium* crown rot (FCR) resistant line CRNIL1A and compared it to CS. They used a reference-based mapping and draft assembly of unmapped reads strategy to have the first insight into wheat pangenomics. By doing so, they identified about 8.3 Mb of CRNIL1A missing sequences in CS, validated through PCR. The validated primers were then used to assess the distribution of these sequences among genotypes distributed worldwide, and for most of them could be classified as either CRNIL1A-type or CS-type.

Another study focused on a specific part of the wheat genome (Thind et al. 2018), where they made a comparative analysis of the ~700-Mb chromosome 2D between CS and another hexaploid accession, CH Campala Lr22a, hereunder Campala. They also used a reference-based mapping strategy, with some differences: they used previously generated scaffolds (Putnam et al. 2016; Thind et al. 2017) (and not reads) and mapped them against CS. In order to identify large ($\geq$100 Kb that were not identifiable with previous analyses) InDels, they compared 10 Mb windows between the pseudomolecules of CS and that of Campala. They found 26 putative large InDels and validated four of them by checking the upstream and downstream regions of the breakpoints: three of them showed interesting CNV for nucleotide binding site-leucine-rich repeat (NLR; known to be involved in resistance) genes.

The first whole-genome pangenomics study for wheat was performed by Montenegro et al. (2017). In this study, they compared 19 accessions of bread wheat, including a re-assembly of CS. Beside SNP calling and analyses, they used a reference-based mapping and draft assembly of unmapped reads strategy to build the pangenome. The core genome seems to carry 64.3% of genes for these 19 accessions. Based on these numbers, the authors extrapolate that modern wheat cultivars contain 140,500 genes on average with around 50 unique gene per genotype. As already shown in Liu et al. (2016), in this study, they have also identified that CS wheat is a genomic outlier compared to the actual modern wheat varieties.

Last but not least, in all these three studies (Liu et al. 2016; Montenegro et al. 2017; Thind et al. 2018), the GO or similar analyses performed gave all the same conclusion: dispensable genes are enriched in environmental stress and defense response genes, mostly.

3.3 Rice

Rice is the most important food crop resource for human, with 80% of the population relying on it for 20% of its daily nutrient intake (Ammiraju et al. 2008). Because of its small genome (about 390 Mb), it is an excellent model for functional genomic research of monocots. Asian rice (*Oryza sativa*) is cultivated worldwide and is divided in two main subspecies (*O. sativa ssp. indica* and *O. sativa ssp. japonica*). Another cultivated rice is endemic and restricted to West Africa, the African rice (*Oryza glaberrima*).

The first comparison between the reference genome (Nipponbare accession) and another genotype was done by Yu et al. (2011), through a comparative genomic hybridization strategy, where they identified 641 CNVs. Most of them were small ($\leq$10 Kb) but a few proportion (about 2.5%) were quite large ($\geq$50 Kb). From them, 85 were randomly chosen to be PCR analyzed: 90% of them have actually shown PAV between the two genotypes.

In their study, Zhang et al. (2016) made a comparison of two other accessions using BAC assemblies of the whole genome and compared them to the reference sequence of Nipponbare. They identified about 21.5 Mb present only in one genotype and 23.3 Mb specific to the other one. These sequences are mostly composed of TEs, but around 4,000 gene loci were identified in these regions and a Pfam analysis showed a high proportion of the genes belonging to the NB-ARC domain, protein kinase or LRR known to be disease and stress resistance related.

In their analysis, Xu et al. (2011) compared 50 accessions, each being a representative of the variability of *Oryza sativa* species. They used a reference-based mapping and draft assembly of unmapped reads strategy. From the unmapped reads they predicted 1,415 new genes with homology in other plants, almost half of those (48%) being found only in one accession. They also identified 1,676 CNVs with more copies in the 50 accessions than in the reference genome, and almost a quarter of them occurring in more than five accessions. This gave the first insight into Asian rice pangenome.

Subsequently, Schatz et al. (2014a) made a higher-quality data comparison but focusing only on three accessions, representing the *indica, aus,* and temperate *japonica* subpopulations of Asian rice. Around 302 Mb per genome was shared between the three accessions, but a sizeable fraction was found specific to each of them (from 4.8 to 8.2 Mb). Among those, the genes related to disease resistance were the most prevalent. They focused on four agronomically relevant regions (S5 hybrid sterility locus, *Sub1*, LRK gene cluster and *Pup1*) known to have subpopulations diversity.

Similar to Schatz et al. (2014a, b), Monat et al. (2016) developed the first insight into the African rice pangenome by focusing on three accessions using a *de novo* assembly strategy. Even if the quality was not as good as for reference genome, they were able to obtain better sequences for some important loci, such as RYMV resistance genes or SWEET14 duplications (Hutin et al. 2015).

A metagenome-like strategy was applied to rice by Yao et al. (2015), by comparing 1,483 accessions. As expected, the mapping rates against the Nipponbare (*japonica*) genome were higher for *japonica* accessions compared to *indica* ones. The dispensable genomes for each subspecies were assembled and then aligned against the reference genome. Only 12.1 and 10.9% of the *indica* and *japonica* dispensable genome, respectively, were aligned. The redundancy between these two dispensable genomes was also evaluated by cross-alignment; from the *indica* dispensable genome, 34% had hits on *japonica* dispensable genome. On the other hand,

54% of *japonica* sequences had hits on *indica* dispensable genome showing that the dispensable genomes are mostly subspecies-specific and probably subject to fast evolution.

In their study Sun et al. (2017) focused on visualization part of the analysis of the 3,000 Rice Genome Project (3K RGP; The 3K Rice Genome Project 2014). They provided a brief overview of the pangenome results: 23,914 core genes and 22,094 ± 4,986 dispensable genes were identified. From the dispensable genes, 853 genes were found to be subspecies specific. A more detailed analysis on the 3K RGP data was performed by Wang et al. (2018). They showed how high genetic diversity present in domesticated rice population gave a strong base for the improvement of cultivars.

Re-using the same dataset, Zhou et al. (2020) showed that Asian rice is probably divided into 15 subpopulations. They generated 12 new high-quality genomes to have a representation as close as possible to these newly identified subpopulations.

With 66 accessions included in their panel, Zhao et al. (2018) identified 10,872 genes that were at least partially absent in the Nipponbare reference genome. After excluding redundancy, 42,580 genes were annotated in at least one accession, including Nipponbare. Interestingly, this value corresponds to the beginning of a plateau in the stepwise addition curve, suggesting a close pangenome for Asian rice. As expected, the dispensable genes were enriched for disease and stress response.

3.4 Brassica

Brassica is a genus which displays extreme morphological diversity and has economically important crop species (including mustards, broccoli, Brussels sprouts, cabbage, cauliflower, turnips, kohlrabi, and kale). Almost all plant parts are used for consumption: roots, inflorescences, axillary buds, leaves and tuberized stems/hypocotyls. Even seeds and seedpods are used for consumer oil types or biofuel. Besides their economic value, *Brassica* species have a complex genome structure due to multiple rounds of duplication making them a group of particular interest to study genome evolution. Three diploid species with three genome types (*B. rapa*, genome AA, *B. nigra*, genome BB and *B. oleracea*, genome CC) evolved from a common ancestor, which underwent at least one whole-genome triplication (Lysak et al. 2005). Then, three amphidiploid species (*B. juncea*, genome AABB, *B. carinata*, genome BBCC, and *B. napus*, genome AACC) combined two subgenomes into one species. The *Brassica* species are also the closest relative of *Arabidopsis* cultivated as crops.To shape the structure of polyploid genomes over relatively short timescales Cheung et al. (2009) analyzed at sequence-level the homoeologous segments of *B. napus* and the representatives of its ancestral diploid donor (*B. rapa* and *B. oleracea*). They found a higher gene density in the A genome and a higher transposon density in the C genome. They made a comparison of the genome segments with *Arabidopsis* on the basis of gene annotations, showing 21 cases where the *Brassica* sequences are in collinear positions with *Arabidopsis* gene models, but where the annotation process did not predict any gene models at this position on *Brassica* sequences: this could represent partial gene loss in *Brassica* compared to *Arabidopsis*.

To investigate genetic variation underlying high diversity in morphology in the *Brassica* crops species, Lin et al. (2014) used a *de novo* draft genome assembly strategy on two *B. rapa* subspecies. This analysis revealed around 1,224 unique genes in each of the three genomes (turnip, rapid cycling, and Chinese cabbage).

Similarly, Bayer et al. (2017) compared two *B. napus* accessions using a *de novo* draft genome assembly then reference-based guided pseudomolecules level assembly strategy. However, in this study, between the two final assemblies, only a few real and putative genes were identified as lost in one of the two accessions.

By using ten accessions, Golicz et al. (2016a, b) build the first insight into the *Brassica* pangenome. They used an iterative mapping and assembly strategy and identified 61,379 genes, from which 18.7% exhibited PAV among the ten individuals. This study also demonstrated the greater diversity in synthetic *B. napus* than non-synthetic accessions.

In a wider study, Hurgobin et al. (2018) increased the pangenome of *B. napus* and investigated the role of homoeologous exchanges (HE) in its genomic diversity. They used the same strategy as Golicz et al. (2016a, b) and identified a core genome composing 62% of the pangenome; they also showed that dispensable genes are shorter with fewer exons per genes. They investigated the difference between the HE-related and non-HE PAVs and found non-HE PAVs appearing in all accessions, while only 30 of the 53 accessions also had HE-related PAVs. These PAVs seem to be more frequently occurring at the chromosomal ends and more often in synthetics (recent) accessions.

3.5 Chickpea

Chickpea (*Cicer arietinum*) is the second most important pulse crop, cultivated mostly on residual soil moisture in the arid and semi-arid regions of the world and, by doing so, improves soil health through symbiotic nitrogen fixation. Nevertheless, many biotic (*Fusarium* wilt, *Ascochyta* blight, *Botrytis* gray mold and *Helicoverpa* pod borer) and abiotic (drought, heat and salinity) stresses affect its productivity and improving it is a hard task due to the low genetic diversity in cultivated gene pool.

Two recent studies by Thudi et al. (2016a, b) were performed on chickpea to better characterize this cultivated gene pool diversity, in order to improve future breeding programs. Both of them have used a draft genome reference-based mapping strategy.

The first one (Thudi et al. 2016b) used 35 genotypes of high agronomical importance (parental lines of 16 mapping populations) and was aimed to find small InDels ($\leq$60 bp), CNVs, and PAVs in case of genes longer than 1 Kb. Some genes were found duplicated, as the gene Ca27299 presenting function for cell membrane, which might be related to defense response, that was found duplicated in 23 genotypes. Another striking example is the salt tolerant ICC 1431 genotype, for which the authors identified 27 defense-related duplicated genes. A final example could be that of Ca13947, a gene coding for a PPR protein which seems to be involved in transcriptional and epigenetics regulation, absent in seven genotypes. They also found a large number of specific variants that segregate for major biotic stresses (blight, wilt, and gray mold), and for salinity; these variants could be used as future markers in breeding programs. Furthermore, the annotation of these line-specific variants revealed a lot of transcription factors, disease resistance protein, heat shock proteins, and DNA-damage repair proteins, correlating with dispensable genes mostly involved in stresses and defense responses.

The second study (Thudi et al. 2016a) reports the analysis of 129 genotypes grouped according to their release time: group one for the varieties released before 1993; group two for the varieties released between 1993 and 2002; and group three for the varieties released after 2002. The study revealed 3,822 CNVs between these varieties, but no significant differences were observed for the two earlier groups (2,315 for group one and 2,318 CNVs for group two) compared to the third group of the recently released varieties (3,511 CNVs). Similar to their first study, they found that GO for the CNV and PAV impacted genes are related to response to stimulus. As an illustration, the Ca14192 gene, encoding an activator of flowering locus, is absent in 92 genotypes. In the same way, a Spanish breeding line susceptible to *Ascochyta* blight can be linked to the presence of the Ca1095 gene in this genotype.

3.6 Soybean

Soybean (*Glycine max*) is a self-pollinating species exhibiting stringent cleistogamy (closed flower pollination), with a comparatively narrow genetic base due to strong bottlenecks during domestication and modern genetic improvement (Hyten et al. 2006; Li et al. 2013). Its closest related species is the annual wild soybean (*Glycine soja*), distributed across broad geographical and ecological ranges, providing a potential source of genetic diversity for soybean improvement.

Genome-wide genetic variation in soybean was first reported by Lam et al. (2010). Through resequencing of 17 wild and of 14 cultivated soybean, and using a reference-based mapping strategy, they identified specific features of the soybean genome: indeed, the linkage disequilibrium (LD) is exceptionally high compared to other plants (cultivated soybean: 150 Kb; wild soybean: 75 Kb; maize: <1 Kb; wild and cultivated rice: <1 Kb; *Arabidopsis thaliana*: 3–4 Kb) which as expected is due to the stringent cleistogamy, for instance. But this study revealed a different LD pattern in wild and in cultivated soybeans: the frequency of short ($\leq$20 Kb) LD blocks was higher in wild soybean and the number of small LD blocks ($\leq$1–2 Kb) was doubled in wild soybean. When investigating larger blocks ($\geq$150 Kb), the number found in wild soybean was this time half that of cultivated soybean. The second specific feature for soybean genome is that the non-synonymous versus synonymous substitution ratio (dN/dS) is on average higher than in other plants (wild total: 1.36; wild-specific SNPs: 1.36; cultivated total: 1.38; cultivated-specific SNPs: 1.61; rice: 1.2; *A. thaliana*: 0.83).

In another study Haun et al. (2011) used a comparative genomic hybridization strategy and confirmed the high rate of SVs in soybean genome. In addition, they used a reference-based mapping approach on exome data and identified sub-lineage-specific regions of SV including regions with PAV.

In their study McHale et al. (2012) used a similar strategy, but, instead of comparing the reference genome (Williams 82) with its parental lineages (as in (Haun et al. 2011)), they compared the reference genome with three other accessions. They identified a high-confidence (but probably underestimated) number of PAVs between these four genotypes. The GO associated with these genes is mostly related to "plant-type hypersensitive response," "programmed cell death," "apoptosis," and "defense response." For this last category, NB and RL protein were more particularly enriched and are involved in disease resistance. The literature showed that these type of genes might be transposed or deleted with the help of TEs; however, no significant enrichment for TE was observed in this study in the CNVs regions nearby these genes (probably due to technical limitations inherent to use of short-reads).

A third study by Anderson et al. (2014) used the same strategy again but with a higher number of accessions included. They confirmed the major trends previously observed on smaller analysis, including enrichment of SV genes in tandem duplicated blocks and SV genes contributing to biotic stress responses. The larger dataset nonetheless allowed for more details, such as genes with retained paralogs from the most recent whole-genome duplication event are underrepresented for association with SVs.

Li et al. (2014) used a *de novo* draft genome assembly strategy for seven accessions to obtain a better insight into the soybean pangenome. The dispensable genes were found to evolve more rapidly and being more variable than the core genes. The GO analysis confirmed the previous results (Haun et al. 2011) for the GO enrichment. Some features have been highly selected in the cultivated soybean, for example from the 1,332 genes related to acyl lipid sub-pathways, about 16% contained SVs between cultivated and wild soybean accessions.

3.7 *Other Crops Studies and Main Results*

Whatever is the used strategy to study the genome diversity of a given species, the different crops pangenomics studies, at different scales, since almost 15 years, highlighted some common trends, on the major crops (see above) or minor ones (see below).

The first major one is that GO associated with dispensable genes are enriched for disease and stress resistance. Indeed, genes impacted by PAVs and SVs are frequently encoding for disease resistance proteins, as shown earlier and also in poplar

(*Populus spp;* Pinosio et al. 2016; Zhang et al. 2019), pigeon pea (*Cajanus cajan*); Zhao et al. 2020), barley (*Hordeum vulgare;* Muñoz-Amatriaín et al. 2013), or sorghum (*Sorghum bicolor;* Zheng et al. 2011). Those gene families are involved in pathogen recognition and signal initiation, and this enrichment is important to understand how plants genomes evolve to respond to biotic and abiotic changes, and how plants genomes can create and use their diversity to respond to abiotic and biotic stresses. As an example, in *Medicago spp* (Zhou et al. 2017), a tandem duplication of an NBS-LRR gene was found: from the two copies of the gene, one is altered and, interestingly, RNA expression of this copy is absent. This result means that SVs might be a possible way to pseudogenization. Zhou et al. (2017) hypothesized that it might be a way to maintain a reservoir of genetic diversity preventing a future pathogen attack. Detailed explanation of the impact of SVs on these environmentally sensitive gene families is reviewed in detail by Young et al. (2016).

Such genes families also have been identified with higher ratios of non-synonymous to synonymous SNPs (dN/dS), for example in sorghum (Zheng et al. 2011) and poplar (Pinosio et al. 2016). This is an indication of the relaxed selective pressure due to biotic and abiotic stresses acting on the diversification of these plant disease resistance genes families or correspond to pseudogenes. In contrast, the genes encoding for essential biological functions, such as ubiquitin (regulatory protein targeting proteins to the proteasome for degradation or recycling), elongation factors (set of proteins facilitating translational elongation during protein synthesis), or binding domain proteins (protein domain which binds to a specific atom or molecule, essential to help proteins to splice, assemble, and/or translate) had the lowest dN/dS ratio.

The importance of wild species is another trend emerged from studies which have added such groups for a better diversity comprehension. For instance, in barley (Muñoz-Amatriaín et al. 2013), the inclusion of wild accessions gave insight into domestication and selection processes and their effects on genome structure. They have shown that regions exhibiting SVs are fewer in the cultivated pool of accessions than in the wild one. Nevertheless, they were able to identify wild-specific as well as cultivated-specific variants, but to a lesser extent for the cultivated barley. Similarly, for *Brachypodium* (Gordon et al. 2017), which has not experienced any domestication bottleneck, dispensable genes are of high importance for understanding the segmentation of subpopulations. For tomato (*Solanum spp;* Gao et al. 2019), even if only one wild accession was included in the study, there is a general trend for gene loss during domestication. In any cases, including wild accession will always provide more information to understand structure of genomic diversity and its evolution. For future pangenomics studies, we think it is of major importance to include more wild relatives, or landraces (when available) and accessions, to have a better overview on how the genome is shaped by anthropogenic effects.

An interesting study was done on the carnivorous plant *Utricularia gibba* by Alcaraz et al. (2016), in which they crossed pangenomic and metagenomic analyses. They have shown a high level of complementation between the plant and its environment, with some reads belonging to microorganisms able to map on the plant pangenome reference, for example. This study is a move forward to better understand plant–environment interactions.

At the genome scale, complementation might also be something to investigate more in detail. It is particularly important to understand complementation especially within the framework of polyploidy that occurs quite often in plant genomes (Cheung et al. 2009). Causes or consequences, genes for which complementation is found elsewhere in the genome are part of the dispensable compartment. Polyploidy is estimated to range between 30 and 70% of flowering plants genome. An often-given reason for these species staying at this level of ploidy is that the increased diversity and the gain it can bring to the species to survive (Soltis et al. 2015). SVs are speculated to allow complementation of missing genes, and by doing so be involved in heterosis and hybrid vigor (Hardigan et al. 2016). In many cases, dispensable genes are enriched with putative adaptive functions (see above), suggesting that they will be more easily retained when they acquire new beneficial functions (Gordon et al. 2017); otherwise duplicated genes will either complement each other or one copy will be probably pseudogenized. Better understanding of complementation between homologous and homeologous genes is an important part for future pangenomics studies in plants.Finally, and in direct link with the agronomic interest, the crop plant community has many examples that have shown the relationship between SVs and phenotype. The characterization of these variations is thus of high importance for trait mapping and crop improvement (Thudi et al. 2016b) and encourages to continue our efforts in pangenomics (non-exhaustive list gathered by Anderson et al. (2014) and we complement): glyphosate resistance, palmer amaranth (Gaines et al. 2010, 2011); boron toxicity tolerance and winter hardiness, barley (Sutton 2013; Knox et al. 2010); seed coat pigmentation, seed weight, flowers color and cyst nematode resistance, soybean (Todd and Vodkin 1996; Zabala and Vodkin 2005, 2007; Cook et al. 2012); female gamete fitness and growth and cell division, potato (*Solanum tuberosum;* Iovene et al. 2013; Hardigan et al. 2016); flavor quality, strawberry (*Fragaria spp;* Chambers et al. 2014); dwarfism and flowering time, wheat (Pearce et al. 2011; Dıaz et al. 2012; Li et al. 2012); submergence tolerance, Asian rice (Xu et al. 2006); aluminum tolerance and glume formation, maize (Han et al. 2012; Wingen et al. 2012; Maron et al. 2013); seed weight regulation, pigeon pea (*Cajanus cajan;* Zhao et al. 2020); delayed flowering, *Brachypodium* (Gordon et al. 2017); fruit flavor, tomato (Gao et al. 2019). All of these traits are of significant importance for plant breeders, having an impact on yield, but also on consumers (fruit flavor or quality). It is thus important to include more and more accessions in pangenomics studies to better characterize these variations that are often not identifiable through basic SNPs analyses.

All of the pangenomics studies also help to unravel the whole mechanisms by which SVs can be created. The first mechanism is TE-related as dispensable is mostly composed of TEs, and the SVs repartition coincides mostly with the TE one (Zhou et al. 2017; Gordon et al. 2017). LTR-retrotransposons are the major type of TEs represented in large plants genome, and since SVs are correlated with the presence of TEs, it has been hypothesized that TEs might be involved through the retrotransposition in the events by which dispensable genes are lost or gained. Indeed, as Pinosio et al. (2016) pointed out, non-homologous recombination rearrangements can be created with the help of TEs. Other mechanisms can create structural variability (either CNVs, PAVs, translocations, inversions, and InDels) (Muñoz-Amatriaín et al. 2013), in addition to the mechanisms related to the TEs, including (non-exhaustive): non-allelic homologous recombination (NAHR) occurring between sequence with similarity but are not alleles; non-homologous end-joining (NHEJ) or the similar but more error prone pathway micro-homology-mediated end-joining (MMEJ), occurring for repairing double-strand breaks in DNA between sequences with little or no similarity; micro-homology-mediated break-induced replication (MMBIR), replication-error mechanisms occurring between sequences with few or no similarity. CNVs could also be due to non-allelic homologs (SNH) segregation among F_2 siblings or recombinant inbred lines (RILs) (Liu et al. 2012). For more details about CNVs creation, we recommend the reading of the paper by Hastings et al. (2010) and the review by Saxena et al. (2014) focusing more particularly on plants genome.

3.8 Limitations/Constraints

Pangenomics results are mainly influenced by six main factors (Vernikos et al. 2015):

1. The alignment algorithm and the parameters used to define similarity, as different tools use different *a priori* and, thus, provide different results for the same dataset.
2. Phylogenetic resolution for the studied population, as a larger population may provide different patterns than a smaller one.
3. Sampled individuals selected to represent the studied population, because of the well-known sampling error effect (Anderson and Gerbing 1984).
4. Which model is used to estimate the number of new genes according to the number of genomes, as its efficiency depends greatly upon the dataset itself.
5. Type and quality of the annotation, *i.e.* the level of completion and curation.
6. The level of comparison between the studied individuals (based, for instance, on the sequence similarity or on PAV profile of each gene independent of the sequence similarity).

Pangenomic analysis allows not only to determine the genomic diversity of a dataset, but also to predict how many extra genomes have to be added to characterize the whole pangenome. For instance, extrapolation would be robust only if a high number of genomes are taken into account (Vernikos et al. 2015).

4 Pangenomics Methodologies

Two types of methods dedicated to pangenomics studies can be distinguished. The first method is for building the pangenomes from the dataset, and the second method is for analyzing the pangenome already built from the dataset, and to identify the core, dispensable, and individual-specific genome.

4.1 Methodologies to Build Pangenome

Once the individuals to be included in the study have been selected based on thorough and representative criteria, it is time to choose the strategy to build the pangenome. So far, four strategies have been deployed, depending on the dataset and the data type:

1. Comparative *de novo* approach involves high-depth sequencing of all the study individuals, the generation of all the corresponding assemblies, and then comparing the assemblies to each other. This approach is the least biased but requires a significant budget to obtain quality and quantity of sequences, as well as a mix of sequencing technologies. A "cheaper" version exists, the map-to-pan approach, which consists of draft assemblies of all individuals then mapping contigs against the reference genome, but the results are not as qualitative.
2. Iterative assembly approach (also known as map-then-assemble) starts with the mapping of the samples reads (generally short-reads) to a reference genome. The unmapped reads are then recovered to be *de novo* assembled by individuals and integrated into a chimerical pan-reference. This approach is less expensive, because the sequencing depth has no need to be as great as in the previous approach, but it is partly biased due to the initial use of a reference genome.
3. Metagenomics-like approach starts with *de novo* assembly of sequences of all individuals into a pan-reference then performs a reverse mapping of each individual assembly against the pan-reference. This method is efficient on low-coverage sequencing but is chimera-prone.
4. Graph and k-mers based approaches use k-mers algorithm to build a graph representation of the pangenome. This approach is robust and all variants could be included. So far, this approach has only been used in prokaryotic studies since it needs huge computational capacities.

Many reviews have been written on this subject among which we recommend the following: Golicz et al. (2016a), Tao et al. (2019), Tranchant-Dubreuil et al. (2019), Khan et al. (2020).

4.2 Methodologies for Pangenome Analyses

To identify shared and non-shared sequences, we need to compare two or more genomes from the same group. There might be variability between studies of the same set of genomes, because of the differences in bioinformatics treatments and the sequence comparison methods.

Vernikos et al. (2015) inventoried 8 different methods dedicated to pangenomic analyses (in bacteria mainly):

1. ORFsim: ORF alignment similarity;
2. OG: method based on cluster of orthologs;

3. Comb: combinatorial approach with successive addition of genomes;
4. Gene freq: frequency of gene presence/absence;
5. Ref: generation of a reference pangenome using a subset of individual genomes;
6. FSM: finite supragenome model;
7. BMM: binomial mixture model;
8. IMGM: infinitely many genes model.

ORFsim and OG methods are both based on an extrapolation model (Baumdicker et al. 2012), *i.e.* they use low quantity data to compute a larger set of information. They are genome-oriented approaches (Lapierre and Gogarten 2009), purely descriptive, and very well adapted to a dataset with a few individuals. Their use is highly computationally intensive when more genomes are included in the analysis, as they search how many genes are unique among the studied genomes through successions of BLAST (Basic Local Alignment Search Tool) (Lapierre and Gogarten 2009). The Comb method, consisting of adding each genome one by one, is always associated with one of these two methods.

Gene freq is a gene-oriented approach, more computationally adapted with numerous genomes (Lapierre and Gogarten 2009); the principle is to look for similarity of one gene through each genome. Those three approaches (ORFsim, OG, and Gene freq) require assembling all the studied genomes and to annotate them, which becomes a limiting factor when working on complex genomes. Moreover, they do not allow to differentiate orthologous and paralogous genes (Lapierre and Gogarten 2009).

Ref method gathers information from many reference genomes to a "reference pangenome," which becomes the base for the pangenomics study with the sampled individuals, and covers more genes than a simple reference genome. The result is presented as the presence/absence matrix for each gene compared to the "reference pangenome" (Méric et al. 2014).

FSM and BMM methods are both based on the mixed model (binomial or not), well adapted when many genomes are studied (Snipen and Ussery 2010). These approaches are predictive as well as descriptive (Boissy et al. 2011).

The IMGM method is based on the idea that one gene does not have a single origin inside the population, and that core genes are absolutely mandatory for the survival of the individual (Baumdicker et al. 2012). Collins and Higgs (2012) extended the model, assuming that dispensable genes could be classified into two categories, each one having a different gain and loss rate. FSM, BMM, and IMGM methods provide tools for simulating PAV gene data in order to validate/invalidate different models.

We can group together the ORFsim and OG under the term "extrapolation model" and the FSM and BMM as "supragenome model." The differences between extrapolation, supragenome, and IMGM models are more striking when comparing the pangenome size predictions, but the results are highly dependent on the size of the sample studied (Baumdicker et al. 2012). For instance, the supragenome model is not suitable to predict very low frequency genes (<1%), while the IMGM is less efficient than supragenome for approximations of medium gene frequencies (Baumdicker et al. 2012). Finally, the state of the pangenome (open of close) depends greatly on the population size and on the model used (Baumdicker et al. 2012).

Whatever, results obtained with a particular method might be conclusive or not, and each approach is more or less adapted to a given dataset (Baumdicker et al. 2012; Vernikos et al. 2015). That means that even if a method might be extremely close to the reality for one specific dataset, it might also be really far away from it for another one. Each species has some genetic and demographic characteristics (e.g., effective population size, allelic diversity, birth/death rate, and so on) that influence all methods in one way or another (Vernikos et al. 2015).

5 Conclusions

In plant pangenomics, the current research aims on the detection of variations with or without gene annotation (e.g. Hübner et al. 2019; reviewed in Tranchant-Dubreuil et al. 2019). In addition, more and more projects focus on the use of long reads technologies, such as Oxford Nanopore Technology or Pacific Biosciences, as well as long range approaches such as Hi-C or BioNano to identify PAV and SV, for instance, for the 10+ Wheat Genome Project (http://www.10wheatgenomes.com/). This would help us to better understand the way pangenome evolution will impact the evolution of species (e.g., the effect of domestication reviewed in Lye and Purugganan 2019) and be a great resource to support biology, genetics, and breeding of major crop plants (reviewed in Tao et al. 2019; Gabur et al. 2019; Khan et al. 2020).

6 Future Perspectives

Pangenomics, the study of genomes of many individuals from the same species, rises population genomics to a higher level, the species one. It is analogous to metagenomics, which includes studying many genomes from the same environment, at the ecosystem level.

The future of population genomics might be the combination of these two aspects by studying many genomes of many species occupying the same environment in order to better understand how the species interact. Population genomics would then be at the global level, allowing us to better understand the evolution of species.

References

3,000 Rice Genomes Project. The 3,000 rice genomes project. Gigascience. 2014;3(1):7. https://doi.org/10.1186/2047-217X-3-7.

Abberton M, Batley J, Bentley A, Bryant J, Cai H, Cockram J, et al. Global agricultural intensification during climate change: a role for genomics. Plant Biotechnol J. 2016;14(4):1095–8. https://doi.org/10.1111/pbi.12467.

Alcaraz LD, Martínez-Sánchez S, Torres I, Ibarra-Laclette E, Herrera-Estrella L. The metagenome of Utricularia gibba's traps: into the microbial input to a carnivorous plant. PLoS One. 2016;11(2):e0148979. http://dx.doi.org/10.1371%2Fjournal.pone.0148979.

Ammiraju JSS, Lu F, Sanyal A, Yu Y, Song X, Jiang N, et al. Dynamic evolution of Oryza genomes is revealed by comparative genomic analysis of a genus-wide vertical data set. Plant Cell. 2008;20(12):3191–209. https://doi.org/10.1105/tpc.108.063727.

Anderson JC, Gerbing DW. The effect of sampling error on convergence, improper solutions, and goodness-of-fit indices for maximum likelihood confirmatory factor analysis. Psychometrika. 1984;49(2):155–73.

Anderson JE, Kantar MB, Kono TY, Fu F, Stec AO, Song Q, et al. A roadmap for functional structural variants in the soybean genome. G3 (Bethesda). 2014;4(7):1307–18. https://doi.org/10.1534/g3.114.011551.

Appels R, Eversole K, Feuillet C, Keller B, Rogers J, Stein N, et al. Shifting the limits in wheat research and breeding using a fully annotated reference genome. Science. 2018;361(6403):eaar7191. https://doi.org/10.1126/science.aar7191.

Avni R, Nave M, Barad O, Baruch K, Twardziok SO, Gundlach H, et al. Wild emmer genome architecture and diversity elucidate wheat evolution and domestication. Science. 2017;357:93–7.

Baumdicker F, Hess WR, Pfaffelhuber P. The infinitely many genes model for the distributed genome of bacteria. Genome Biol Evol. 2012;4(4):443–56. https://doi.org/10.1093/gbe/evs016.

Bayer PE, Hurgobin B, Golicz AA, Chan CKK, Yuan Y, Lee HT, et al. Assembly and comparison of two closely related Brassica napus genomes. Plant Biotechnol J. 2017;15(12):1602–10. https://doi.org/10.1111/pbi.12742.

Belo A, Beatty MK, Hondred D, Fengler KA, Li B, Rafalski A. Allelic genome structural variations in maize detected by array comparative genome hybridization. Theor Appl Genet. 2010:355–67. https://doi.org/10.1007/s00122-009-1128-9.

Beyter D, Ingimundardottir H, Eggertsson HP, Bjornsson E, Kristmundsdottir S, Mehringer S, et al. Long read sequencing of 1,817 Icelanders provides insight into the role of structural variants in human disease. bioRxiv. 2019; https://doi.org/10.1101/848366.

Blanc G, Wolfe KH. Widespread paleopolyploidy in model plant species inferred from age distributions of duplicate genes. Plant Cell. 2004;16:1667–78. https://doi.org/10.1105/tpc.021345.formed.

Boissy R, Ahmed A, Janto B, Earl J, Hall BG, Hogg JS, et al. Comparative supragenomic analyses among the pathogens Staphylococcus aureus, Streptococcus pneumoniae, and Haemophilus influenzae using a modification of the finite supragenome model. BMC Genomics. 2011;12:187. https://doi.org/10.1186/1471-2164-12-187.

Boussaha M, Esquerré D, Barbieri J, Djari A, Pinton A, Letaief R, et al. Genome-wide study of structural variants in bovine Holstein, Montbéliarde and Normande dairy breeds. Plos One. 2015;10(8):e0135931. https://doi.org/10.1371/journal.pone.0135931.

Brunner S, Fengler K, Morgante M, Tingey S, Rafalski A. Evolution of DNA sequence nonhomologies among maize inbreds. Plant Cell. 2005;17:343–60. https://doi.org/10.1105/tpc.104.025627.1.

Cai JJ, Petrov D a. Relaxed purifying selection and possibly high rate of adaptation in primate lineage-specific genes. Genome Biol Evol. 2010;2(1):393–409. https://doi.org/10.1093/gbe/evq019.

Cao J, Schneeberger K, Ossowski S, Günther T, Bender S, Fitz J, et al. Whole-genome sequencing of multiple Arabidopsis thaliana populations. Nat Genet. 2011;43(10):956–63. https://doi.org/10.1038/ng.911.

Capra JA, Pollard KS, Singh M. Novel genes exhibit distinct patterns of function acquisition and network integration. Genome Biol. 2010;11(12):R127. https://doi.org/10.1186/gb-2010-11-12-r127.

Caputo A, Merhej V, Georgiades K, Fournier P-E, Croce O, Robert C, et al. Pan-genomic analysis to redefine species and subspecies based on quantum discontinuous variation: the Klebsiella paradigm. Biol Direct. 2015;10(1):55. https://doi.org/10.1186/s13062-015-0085-2.

Carlos Guimaraes L, Benevides de Jesus L, Vinicius Canario Viana M, Silva A, Thiago Juca Ramos R, de Castro Soares S, et al. Inside the pan-genome – methods and software overview. Curr Genomics. 2015;16(4):245–52. https://doi.org/10.2174/1389202916666150423002311.

Chambers AH, Pillet J, Plotto A, Bai J, Whitaker VM, Folta KM. Identification of a strawberry flavor gene candidate using an integrated genetic-genomic-analytical chemistry approach. BMC Genomics. 2014;15(1):1–15. https://doi.org/10.1186/1471-2164-15-217.

Chantret N, Salse J, Sabot F, Rahman S, Bellec A, Laubin B, et al. Molecular basis of evolutionary events that shaped the hardness locus in diploid and polyploid wheat species (Triticum and Aegilops). Plant Cell. 2005;17(4):1033–45. https://doi.org/10.1105/tpc.104.029181.

Cheung F, Trick M, Drou N, Lim YP, Park J-Y, Kwon S-J, et al. Comparative analysis between homoeologous genome segments of Brassica napus and its progenitor species reveals extensive sequence-level divergence. Plant Cell. 2009;21(7):1912–28. https://doi.org/10.1105/tpc.108.060376.

Chia J-M, Song C, Bradbury PJ, Costich D, De Leon N, Doebley J, et al. Maize HapMap2 identifies extant variation from a genome in flux. Nat Genet. 2012;44(7):803–7. https://doi.org/10.1038/ng.2313.

Choulet F, Alberti A, Theil S, Glover N, Barbe V, Daron J, et al. Structural and functional partitioning of bread wheat chromosome 3B. Science. 2014;345(6194):1250092. https://doi.org/10.1126/science.1251788.

Collins RE, Higgs PG. Testing the infinitely many genes model for the evolution of the bacterial core genome and pangenome. Mol Biol Evol. 2012;29(11):3413–25. https://doi.org/10.1093/molbev/mss163.

Cook DE, Lee TG, Guo X, Melito S, Wang K, Bayless A, et al. Copy number variation of multiple genes at Rhg1 mediates nematode resistance in soybean. Science. 2012;338(6111):1206–9. https://doi.org/10.1126/science.1228746.

Da Silva C, Zamperin G, Ferrarini A, Minio A, Dal Molin A, Venturini L, et al. The high polyphenol content of grapevine cultivar Tannat berries is conferred primarily by genes that are not shared with the reference genome. Plant Cell. 2013;25(12):4777–88. https://doi.org/10.1105/tpc.113.118810.

Darling AE, Mau B, Perna NT. progressiveMauve: multiple genome alignment with gene gain, loss and rearrangement. PLoS One. 2010;5(6):e11147. https://doi.org/10.1371/journal.pone.0011147.

Diaz A, Zikhali M, Turner AS, Isaac P, Laurie DA. Copy number variation affecting the Photoperiod-B1 and Vernalization-A1 genes is associated with altered flowering time in wheat (Triticum aestivum). PLoS One. 2012;7(3):e33234. https://doi.org/10.1371/journal.pone.0033234.

Gabur I, Chawla HS, Snowdon RJ, Parkin IAP. Connecting genome structural variation with complex traits in crop plants. Theor Appl Genet. 2019;132(3):733–50. https://doi.org/10.1007/s00122-018-3233-0.

Gaines TA, Zhang W, Wang D, Bukun B, Chisholm ST, Shaner DL, et al. Gene amplification confers glyphosate resistance in Amaranthus palmeri. Proc Natl Acad Sci U S A. 2010;107(3) https://doi.org/10.1073/pnas.0906649107.

Gaines TA, Shaner DL, Ward SM, Leach JE, Preston C, Westra P. Mechanism of resistance of evolved glyphosate-resistant Palmer amaranth (Amaranthus palmeri). J Agric Food Chem. 2011;59:5886–9.

Gan X, Stegle O, Behr J, Steffen JG, Drewe P, Hildebrand KL, et al. Multiple reference genomes and transcriptomes for Arabidopsis thaliana. Nature. 2011;477(7365):419–23. https://doi.org/10.1038/nature10414.

Gao L, Gonda I, Sun H, Bao K, Tieman DM, Fish TL, et al. The tomato pan-genome uncovers new genes and a rare allele regulating fruit flavor. Nat Genet. 2019; https://doi.org/10.1038/s41588-019-0410-2.

Ghatak S, Blom J, Das S, Sanjukta R, Puro K, Mawlong M, et al. Pan-genome analysis of Aeromonas hydrophila, Aeromonas veronii and Aeromonas caviae indicates phylogenomic diversity and greater pathogenic potential for Aeromonas hydrophila. Antonie Van Leeuwenhoek. 2016;109(7):945–56. https://doi.org/10.1007/s10482-016-0693-6.

Golicz AA, Batley J, Edwards D. Towards plant pangenomics. Plant Biotechnol J. 2016a;14(4):1099–105. https://doi.org/10.1111/pbi.12499.

Golicz AA, Bayer PE, Barker GC, Edger PP, Kim HR, Martinez PA, et al. The pangenome of an agronomically important crop plant Brassica oleracea. Nat Commun. 2016b;7:13390. https://doi.org/10.1038/ncomms13390.

Gordienko EN, Kazanov MD, Gelfand MS, Gelfand S. Evolution of pan-genomes of Escherichia coli, Shigella spp., and Salmonella enterica. J Bacteriol. 2013;195(12):2786–92. https://doi.org/10.1128/JB.02285-12.

Gordon SP, Contreras-Moreira B, Woods DP, Marais DLD, Burgess D, Shu S, et al. Extensive gene content variation in the Brachypodium distachyon pan-genome correlates with population structure. Nat Commun. 2017;8(1):2184. https://doi.org/10.1038/s41467-017-02292-8.

Gore M a, Chia J-M, Elshire RJ, Qi S, Ersoz ES, Hurwitz BL, et al. A first-generation haplotype map of maize. Science. 2009;326(5956):1115–7. https://doi.org/10.1126/science.1177837.

Han J-j, Jackson D, Martienssen R. Pod corn is caused by rearrangement at the Tunicate1 locus. Plant Cell. 2012;24:2733–44. https://doi.org/10.1105/tpc.112.100537.

Hansen E, Amend J, Hansen EE. Comparative and functional genomic analysis of the Methanobrevibacter smithii pan genome. PhD thesis. 2012.

Hansey CN, Vaillancourt B, Sekhon RS, De Leon N, Shawn M, Robin Buell C. Maize (Zea mays L.) genome diversity as revealed by RNA-sequencing. PLoS One. 2012;7(3):1–10. https://doi.org/10.1371/journal.pone.0033071.

Hardigan MA, Crisovan E, Hamiltion JP, Kim J, Laimbeer P, Leisner CP, et al. Genome reduction uncovers a large dispensable genome and adaptive role for copy number variation in asexually propagated Solanum tuberosum. Plant Cell. 2016;28:388–405. https://doi.org/10.1105/tpc.15.00538.

Hastings PJ, Lupski JR, Rosenberg SM, Ira G. Mechanisms of change in gene copy number. Nat Rev Genet. 2010;10(8):551–64. https://doi.org/10.1038/nrg2593.Mechanisms.

Haun WJ, Hyten DL, Xu WW, Gerhardt DJ, Albert TJ, Richmond T, et al. The composition and origins of genomic variation among individuals of the soybean reference. Plant Physiol. 2011;155:645–55. https://doi.org/10.1104/pp.110.166736.

Hirakawa H, Okada Y, Tabuchi H, Shirasawa K, Watanabe A, Tsuruoka H, et al. Survey of genome sequences in a wild sweet potato, Ipomoea trifida (H. B. K.) G. Don. DNA Res. 2015;22:171–9. https://doi.org/10.1093/dnares/dsv002.

Hirsch CN, Foerster JM, Johnson JM, Sekhon RS, Muttoni G, Vaillancourt B, et al. Insights into the maize pan-genome and pan-transcriptome. Plant Cell. 2014;26(1):121–35. https://doi.org/10.1105/tpc.113.119982.

Hirsch CN, Hirsch CD, Brohammer AB, Bowman MJ, Soifer I, Barad O, et al. Draft assembly of elite inbred line PH207 provides insights into genomic and transcriptome diversity in maize. Plant Cell. 2016;28(11):2700–14. https://doi.org/10.1105/tpc.16.00353.

Hübner S, Bercovich N, Todesco M, Mandel JR, Odenheimer J, Ziegler E, et al. Sunflower pan-genome analysis shows that hybridization altered gene content and disease resistance. Nat Plants. 2019;5(1):54.

Hurgobin B, Golicz AA, Bayer PE, Chan CKK, Tirnaz S, Dolatabadian A, et al. Homoeologous exchange is a major cause of gene presence/absence variation in the amphidiploid Brassica napus. Plant Biotechnol J. 2018;16(7):1265–74. https://doi.org/10.1111/pbi.12867.

Hutin M, Sabot F, Ghesquière A, Koebnik R, Szurek B. A knowledge-based molecular screen uncovers a broad-spectrum OsSWEET14 resistance allele to bacterial blight from wild rice. Plant J. 2015;84(4):694–703. https://doi.org/10.1111/tpj.13042.

Hyten DL, Song Q, Zhu Y, Choi I-y, Nelson RL, Costa JM, et al. Impacts of genetic bottlenecks on soybean genome diversity. Proc Natl Acad Sci U S A. 2006;103(45):16666–71.

Iovene M, Zhang T, Lou Q, Buell CR, Jiang J. Copy number variation in potato – an asexually propagated autotetraploid species. Plant J. 2013;75:80–9. https://doi.org/10.1111/tpj.12200.

Jordan KW, Wang S, Lun Y, Gardiner L-j, Maclachlan R, Hucl P, et al. A haplotype map of allohexaploid wheat reveals distinct patterns of selection on homoeologous genomes. Genome Biol. 2015;16:1–18. https://doi.org/10.1186/s13059-015-0606-4.

Kahlke T. Analysis of the vibrionaceae pan-genome. PhD thesis. 2013.

Khan AW, Garg V, Roorkiwal M, Golicz AA, Edwards D, Varshney RK. Super-pangenome by integrating the wild side of a species for accelerated crop improvement. Trends Plant Sci. 2020;25(2):148–58. https://doi.org/10.1016/j.tplants.2019.10.012.

Knox AK, Dhillon T, Cheng H, Tondelli A, Pecchioni N, Stockinger EJ. CBF gene copy number variation at Frost Resistance-2 is associated with levels of freezing tolerance in temperate-climate cereals. Theor Appl Genet. 2010;121(1):21–35. https://doi.org/10.1007/s00122-010-1288-7.

Lai J, Li R, Xu X, Jin W, Xu M, Zhao H, et al. Genome-wide patterns of genetic variation among elite maize inbred lines. Nat Genet. 2010;42(11):1027–30. https://doi.org/10.1038/ng.684.

Lam H-M, Xu X, Liu X, Chen W, Yang G, Wong F-L, et al. Resequencing of 31 wild and cultivated soybean genomes identifies patterns of genetic diversity and selection. Nat Genet. 2010;42 (12):1053–9. https://doi.org/10.1038/ng.715.

Lapierre P, Gogarten JP. Estimating the size of the bacterial pan-genome. Trends Genet. 2009;25 (3):107–10. https://doi.org/10.1016/j.tig.2008.12.004.

Lawrence Farber P. Finding order in nature: the naturalist tradition from Linnaeus to E. O. Wilson. London: Johns Hopkins University Press; 2000.

Li Y-h, Zhao S-c, Ma J-x, Li D, Yan L, Li J, et al. Molecular footprints of domestication and improvement in soybean revealed by whole genome re-sequencing. BMC Genomics. 2013;14:579.

Li Y-h, Zhou G, Ma J, Jiang W, Jin L-g, Zhang Z, et al. De novo assembly of soybean wild relatives for pan-genome analysis of diversity and agronomic traits. Nat Biotechnol. 2014;32 (10):1045–52. https://doi.org/10.1038/nbt.2979.

Li Y, Xiao J, Jiajie W, Duan J, Liu Y, Ye X, et al. A tandem segmental duplication (TSD) in green revolution gene Rht-D1b region underlies plant height variation. New Phytol. 2012;196 (1):282–91. https://doi.org/10.1111/j.1469-8137.2012.04243.x.

Liang W, Zhao Y, Chen C, Cui X, Yu J, Xiao J, et al. Pan-genomic analysis provides insights into the genomic variation and evolution of Salmonella Paratyphi A. PLoS One. 2012;7(9):e45346. https://doi.org/10.1371/journal.pone.0045346.

Lijun O, Li D, Lv J, Wenchao C, Zhuqing Z, Li X, et al. Pan-genome of cultivated pepper (Capsicum) and its use in gene presence – absence variation analyses. New Phytol. 2018;220 (2):360–3. https://doi.org/10.1111/nph.15413.

Lin K, Zhang N, Severing EI, Nijveen H, Cheng F, Visser RGF, et al. Beyond genomic variation – comparison and functional annotation of three Brassica rapa genomes: a turnip, a rapid cycling and a Chinese cabbage. BMC Genomics. 2014;15(1):250. https://doi.org/10.1186/1471-2164-15-250.

Lipman DJ, Souvorov A, Koonin EV, Panchenko AR, Tatusova TA. The relationship of protein conservation and sequence length. BMC Evol Biol. 2002;2:20. https://doi.org/10.1186/1471-2148-2-20.

Liu M, Stiller J, Holušová K, Vrána J, Liu D, Doležel J, et al. Chromosome-specific sequencing reveals an extensive dispensable genome component in wheat. Sci Rep. 2016;6:1–9. https://doi.org/10.1038/srep36398.

Liu S, Ying K, Yeh C-t, Yang J, Swanson-wagner R, Wei W, et al. Changes in genome content generated via segregation of non-allelic homologs. Plant J. 2012;72:390–9. https://doi.org/10.1111/j.1365-313X.2012.05087.x.

Lu F, Romay MC, Glaubitz JC, Bradbury PJ, Elshire RJ, Wang T, et al. High-resolution genetic mapping of maize pan-genome sequence anchors. Nat Commun. 2015;6 https://doi.org/10.1038/ncomms7914.

Lu P, Han X, Ji Q, Yang J, Wijeratne AJ, Li T, et al. Analysis of Arabidopsis genome-wide variations before and after meiosis and meiotic recombination by resequencing Landsberg erecta and all four products of a single meiosis. Genome Res. 2012;22:508–18. https://doi.org/10.1101/gr.127522.111.Freely.

Lukjancenko O. Analysis of pan-genome content and its application in microbial identification. PhD thesis. 2013.

Lukjancenko O, Ussery DW, Wassenaar TM. Comparative genomics of Bifidobacterium, Lactobacillus and related probiotic genera. Microb Ecol. 2012;63(3):651–73. https://doi.org/10.1007/s00248-011-9948-y.

Lye ZN, Purugganan MD. Copy number variation in domestication. Trends Plant Sci. 2019; https://doi.org/10.1016/J.TPLANTS.2019.01.003.

Lysak MA, Koch MA, Pecinka A, Schubert I. Chromosome triplication found across the tribe Brassiceae. Genome Res. 2005;15:516–25. https://doi.org/10.1101/gr.3531105

Mann RA, Smits THM, Bühlmann A, Blom J, Goesmann A, Frey JE, et al. Comparative genomics of 12 strains of Erwinia amylovora identifies a pan-genome with a large conserved core. PLoS One. 2013;8(2):e55644. https://doi.org/10.1371/journal.pone.0055644.

Maron LG, Guimarães CT, Kirst M, Albert PS, Birchler JA, Bradbury PJ. Aluminum tolerance in maize is associated with higher MATE1 gene copy number. Proc Natl Acad Sci U S A. 2013;110(13):5241–6. https://doi.org/10.1073/pnas.1220766110.

McHale LK, Haun WJ, Wayne WX, Bhaskar PB, Anderson JE, Hyten DL, et al. Structural variants in the soybean genome localize to clusters of biotic stress-response genes. Plant Physiol. 2012;159(4):1295–308. https://doi.org/10.1104/pp.112.194605.

Medini D, Donati C, Tettelin H, Masignani V, Rappuoli R. The microbial pan-genome. Curr Opin Genet Dev. 2005;15(6):589–94. https://doi.org/10.1016/j.gde.2005.09.006.

Méric G, Yahara K, Mageiros L, Pascoe B, Maiden MCJ, Jolley KA, et al. A reference pan-genome approach to comparative bacterial genomics: identification of novel epidemiological markers in pathogenic Campylobacter. PLoS One. 2014;9(3) https://doi.org/10.1371/journal.pone.0092798.

Monat C, Pera B, Ndjiondjop M-N, Sow M, Tranchant-Dubreuil C, Bastianelli L, et al. De novo assemblies of three Oryza glaberrima accessions provide first insights about pan-genome of African rices. Genome Biol Evol. 2016;1(1):evw253. https://doi.org/10.1093/gbe/evw253.

Montenegro JD, Golicz AA, Bayer PE, Hurgobin B, Lee HT, Chan C-KK, et al. The pangenome of hexaploid bread wheat. Int J Lab Hematol. 2017;38(1):42–9. https://doi.org/10.1111/ijlh.12426.

Morgante M, Brunner S, Pea G, Fengler K, Zuccolo A, Rafalski A. Gene duplication and exon shuffling by helitron-like transposons generate intraspecies diversity in maize. Nat Genet. 2005;37(9):997–1002. https://doi.org/10.1038/ng1615.

Morgante M, De Paoli E, Radovic S. Transposable elements and the plant pan-genomes. Curr Opin Plant Biol. 2007;10(2):149–55. https://doi.org/10.1016/j.pbi.2007.02.001.

Muñoz-Amatriaín M, Eichten SR, Wicker T, Richmond TA, Mascher M, Steuernagel B, et al. Distribution, functional impact, and origin mechanisms of copy number variation in the barley genome. Genome Biol. 2013;14(6):R58. https://doi.org/10.1186/gb-2013-14-6-r58.

Nguyen NK. Addressing the omics data explosion: a comprehensive reference genome representation and the democratization of comparative genomics and immunogenomics. PhD thesis. 2014.

Ossowski S, Schneeberger K, Clark RM, Lanz C, Warthmann N, Weigel D. Sequencing of natural strains of Arabidopsis thaliana with short reads. Genome Res. 2008;18(12):2024–33. https://doi.org/10.1101/gr.080200.108.

Paterson AH, Bowers JE, Chapman BA. Ancient polyploidization predating divergence of the cereals, and its consequences for comparative genomics. Proc Natl Acad Sci U S A. 2004;101(26):9903–8.

Paterson AH, Bowers JE, Bruggmann R, Dubchak I, Grimwood J, Gundlach H, et al. The Sorghum bicolor genome and the diversification of grasses. Nature. 2009;457(7229):551–6. https://doi.org/10.1038/nature07723.

Paux E, Roger D, Badaeva E, Gay G, Bernard M, Sourdille P, et al. Characterizing the composition and evolution of homoeologous genomes in hexaploid wheat through BAC-end sequencing on chromosome 3B. Plant J. 2006;48(3):463–74. https://doi.org/10.1111/j.1365-313X.2006.02891.x.

Pearce S, Saville R, Vaughan SP, Chandler PM, Wilhelm EP, Sparks CA, et al. Molecular characterization of Rht-1 dwarfing genes in hexaploid wheat. Plant Physiol. 2011;157:1820–31. https://doi.org/10.1104/pp.111.183657.

Pinosio S, Giacomello S, Faivre-rampant P, Taylor G, Jorge V, Christine M, et al. Characterization of the poplar pan-genome by genome-wide identification of structural variation. Mol Biol Evol. 2016;33(10):2706–19.

Pucker B, Holtgrawe D, Rosleff Sorensen T, Stracke R, Viehover P, Weisshaar B. A de novo genome sequence assembly of the Arabidopsis thaliana accession niederzenz-1 displays presence/absence variation and strong synteny. PLoS One. 2016;11(10):1–23. https://doi.org/10.1371/journal.pone.0164321.

Pucker B, Holtgräwe D, Stadermann KB, Frey K, Huettel B, Reinhardt R, et al. A chromosome-level sequence assembly reveals the structure of the Arabidopsis thaliana Nd-1 genome and its gene set. Plos One. 2019;14(5):1–23. https://doi.org/10.1371/journal.pone.0216233.

Putnam NH, Connell BO, Stites JC, Rice BJ, Hartley PD, Sugnet CW, et al. Chromosome-scale shotgun assembly using an in vitro method for long-range linkage arXiv: 1502. 05331v1 [q-bio.GN] 18 Feb 2015. Genome Res. 2016;26:342–50. https://doi.org/10.1101/gr.193474.115. Freely.

Sanmiguel P, Tikhonov A, Jin Y-k, Motchoulskaia N, Zakharov D, Melake-berhan A, et al. Nested retrotransposons in the intergenic regions of the maize genome. Science. 1996;274(5288):765–8.

Santuari L, Pradervand S, Thomas J, Dorcey E, Harshman K, Xenarios I, et al. Substantial deletion overlap among divergent Arabidopsis genomes revealed by intersection of short reads and tiling arrays. Genome Biol. 2010;11:R4.

Saxena RK, Edwards D, Varshney RK. Structural variations in plant genomes. Brief Funct Genomics. 2014;13(4):296–307. https://doi.org/10.1093/bfgp/elu016.

Schatz MC, Maron LG, Stein JC, Hernandez Wences A, Gurtowski J, Biggers E, et al. Whole genome de novo assemblies of three divergent strains of rice, Oryza sativa, document novel gene space of aus and indica. Genome Biol. 2014a;15(11):506. https://doi.org/10.1101/003764.

Schatz MC, Maron LG, Stein JC, Wences AH, Gurtowski J, Biggers E, et al. Whole genome de novo assemblies of three divergent strains of rice, Oryza sativa, document novel gene space of aus and indica. Genome Biol. 2014b;15(11):506. https://doi.org/10.1186/s13059-014-0506-z.

Snipen L, Ussery DW. Standard operating procedure for computing pangenome trees. Stand Genomic Sci. 2010;2(1):135–41. https://doi.org/10.4056/sigs.38923.

Soares SC, Silva A, Trost E, Blom J, Ramos R, Carneiro A, et al. The pan-genome of the animal pathogen Corynebacterium pseudotuberculosis reveals differences in genome plasticity between the biovar ovis and equi strains. PLoS One. 2013;8(1) https://doi.org/10.1371/journal.pone.0053818.

Soltis PS, Blaine Marchant D, Van de Peer Y, Soltis DE. Polyploidy and genome evolution in plants. Curr Opin Genet Dev. 2015;35:119–25. https://doi.org/10.1016/j.gde.2015.11.003.

Springer NM, Ying K, Fu Y, Ji T, Yeh C-T, Jia Y, et al. Maize inbreds exhibit high levels of copy number variation (CNV) and presence/absence variation (PAV) in genome content. PLoS Genet. 2009;5(11):e1000734. https://doi.org/10.1371/journal.pgen.1000734.

Sun C, Hu Z, Zheng T, Lu K, Zhao Y, Wang W, et al. RPAN: rice pan-genome browser for ∼3000 rice genomes. Nucleic Acids Res. 2017;45(2):597–605. https://doi.org/10.1093/nar/gkw958.

Sutton T. Boron-toxicity tolerance in barley arising from efflux transporter amplification. Science. 2013:1446. https://doi.org/10.1126/science.1146853.

Swanson-Wagner R a, Eichten SR, Kumari S, Tiffin P, Stein JC, Ware D, et al. Pervasive gene content variation and copy number variation in maize and its undomesticated progenitor. Genome Res. 2010;20(12):1689–99. https://doi.org/10.1101/gr.109165.110.

Swigon Z, Lai J, Ma J, Ramakrishna W, Llaca V, Bennetzen JL, et al. Close split of sorghum and maize genome progenitors. Genome Res. 2004;14:1916–23. https://doi.org/10.1101/gr.2332504.maize.

Tao Y, Zhao X, Mace E, Henry R, Jordan D. Exploring and exploiting pan-genomics for crop improvement. Mol Plant. 2019;12(2):156–69. https://doi.org/10.1016/j.molp.2018.12.016.

Tettelin H, Masignani V, Cieslewicz MJ, Donati C, Medini D, Ward NL, et al. Genome analysis of multiple pathogenic isolates of Streptococcus agalactiae: implications for the microbial "pan-genome". Proc Natl Acad Sci U S A. 2005;102(39):13950–5. https://doi.org/10.1073/pnas.0506758102.

Tettelin H, Riley D, Cattuto C, Medini D. Comparative genomics: the bacterial pan-genome. Curr Opin Microbiol. 2008;11(5):472–7. https://doi.org/10.1016/j.mib.2008.09.006.

Thind AK, Wicker T, Mueller T, Ackermann PM, Steuernagel B, Wulff BBH, et al. Chromosome-scale comparative sequence analysis unravels molecular mechanisms of genome dynamics between two wheat cultivars. Genome Biol. 2018;19(1):104. https://doi.org/10.1101/260406.

Thind AK, Wicker T, Šimková H, Fossati D, Moullet O, Brabant C, et al. Rapid cloning of genes in hexaploid wheat using cultivar-specific long-range chromosome assembly. Nat Biotechnol. 2017;35(8):793–6. https://doi.org/10.1038/nbt.3877.

Thudi M, Chitikineni A, Liu X, He W, Roorkiwal M. Recent breeding programs enhanced genetic diversity in both desi and kabuli varieties of chickpea (Cicer arietinum L.). Sci Rep. 2016a;6:38636. https://doi.org/10.1038/srep38636.

Thudi M, Khan AW, Kumar V, Gaur PM, Katta K, Garg V, et al. Whole genome re-sequencing reveals genome-wide variations among parental lines of 16 mapping populations in chickpea (Cicer arietinum L.). BMC Plant Biol. 2016b;16(Suppl 1):10. https://doi.org/10.1186/s12870-015-0690-3.

Todd JJ, Vodkin LO. Duplications that suppress and deletions that restore expression from chalcone synthase multigene family. Plant Cell. 1996;8:687–99.

Tranchant-Dubreuil C, Rouard M, Sabot F. Plant pangenome: impacts on phenotypes and evolution. Annu Plant Rev. 2019; https://hal.archives-ouvertes.fr/hal-02053647

Vernikos G, Medini D, Riley DR, Tettelin H. Ten years of pan-genome analyses. Curr Opin Microbiol. 2015;23:148–54. https://doi.org/10.1016/j.mib.2014.11.016.

Wang Q, Dooner HK. Remarkable variation in maize genome structure inferred from haplotype diversity at the bz locus. Proc Natl Acad Sci U S A. 2006;103(47):17644–9. https://doi.org/10.1073/pnas.0603080103.

Wang W, Mauleon R, Hu Z, Chebotarov D, Tai S, Wu Z, et al. Genomic variation in 3,010 diverse accessions of Asian cultivated rice. Nature. 2018;557(7703):43–9. https://doi.org/10.1038/s41586-018-0063-9.

Wicker T, Gundlach H, Spannagl M, Uauy C, Borrill P, Ramírez-González RH, et al. Impact of transposable elements on genome structure and evolution in bread wheat. Genome Biol. 2018;19(1):1–18. https://doi.org/10.1186/s13059-018-1479-0.

Wicker T, Rustenholz C, Paux E, Leroy P, Budak H, Breen J, et al. Megabase level sequencing reveals contrasted organization and evolution patterns of the wheat gene and transposable element spaces. Plant Cell. 2010;22:1686–701. https://doi.org/10.1105/tpc.110.074187.

Wingen LU, Münster T, Faigl W, Deleu W, Sommer H, Saedler H, et al. Molecular genetic basis of pod corn (Tunicate maize). Proc Natl Acad Sci U S A. 2012;109(18):7115–20. https://doi.org/10.1073/pnas.1111670109.

Xu K, Xia X, Fukao T, Canlas P, Maghirang-rodriguez R, Heuer S, et al. Sub1A is an ethylene-response-factor-like gene that confers submergence tolerance to rice. Nature. 2006;442:705–8. https://doi.org/10.1038/nature04920.

Xu X, Liu X, Ge S, Jensen JD, Hu F, Li X, et al. Resequencing 50 accessions of cultivated and wild rice yields markers for identifying agronomically important genes. Nat Biotechnol. 2011;30:105–11. https://doi.org/10.1038/nbt.2050.

Yao W, Li G, Zhao H, Wang G, Lian X, Xie W. Exploring the rice dispensable genome using a metagenome-like assembly strategy. Genome Biol. 2015;16(1):1–20. https://doi.org/10.1186/s13059-015-0757-3.

Young ND, Zhou P, Silverstein KA. Exploring structural variants in environmentally sensitive gene families. Curr Opin Plant Biol. 2016;30:19–24. https://doi.org/10.1016/j.pbi.2015.12.012.

Yu J, Golicz AA, Kun L, Dossa K, Zhang Y, Chen J, et al. Insight into the evolution and functional characteristics of the pan-genome assembly from sesame landraces and modern cultivars. Plant Biotechnol J. 2019;17(5):881–92. https://doi.org/10.1111/pbi.13022.

Yu P, Wang C, Xu Q, Feng Y, Yuan X, Yu H, et al. Detection of copy number variations in rice using array-based comparative genomic hybridization. BMC Genomics. 2011;12(1):372. https://doi.org/10.1186/1471-2164-12-372.

Zabala G, Vodkin LO. The wp mutation of glycine max carries a gene-fragment-rich transposon of the CACTA superfamily. 2005;17:2619–32. https://doi.org/10.1105/tpc.105.033506.1.

Zabala G, Vodkin L. A rearrangement resulting in small tandem repeats in the F3′5′H gene of white flower genotypes is associated with the soybean W1 locus. Crop Sci. 2007;2:113–124. https://doi.org/10.2135/cropsci2006.12.0838tpg.

Zhang B, Zhu W, Diao S, Wu X, Lu J, Ding CJ, et al. The poplar pangenome provides insights into the evolutionary history of the genus. Commun Biol. 2019;2(1) https://doi.org/10.1038/s42003-019-0474-7.

Zhang J, Chen LL, Xing F, Kudrna DA, Yao W, Copetti D, et al. Extensive sequence divergence between the reference genomes of two elite indica rice varieties Zhenshan 97 and Minghui 63. Proc Natl Acad Sci U S A. 2016;113(35):E5163–71. https://doi.org/10.1073/pnas.1611012113.

Zhao J, Bayer PE, Ruperao P, Saxena RK, Khan AW, Golicz AA, et al. Trait associations in the pangenome of pigeon pea (Cajanus cajan). Plant Biotechnol J. 2020:1–9. https://doi.org/10.1111/pbi.13354.

Zhao Q, Feng Q, Lu H, Li Y, Wang A, Tian Q, et al. Pan-genome analysis highlights the extent of genomic variation in cultivated and wild rice. Nat Genet. 2018;50(2):278–84. https://doi.org/10.1038/s41588-018-0041-z.

Zheng L-Y, Guo X-S, He B, Sun L-J, Peng Y, Dong S-S, et al. Genome-wide patterns of genetic variation in sweet and grain sorghum (Sorghum bicolor). Genome Biol. 2011;12(11):R114. https://doi.org/10.1186/gb-2011-12-11-r114.

Zhou P, Silverstein KAT, Ramaraj T, Guhlin J, Denny R, Liu J, et al. Exploring structural variation and gene family architecture with De Novo assemblies of 15 Medicago genomes. BMC Genomics. 2017;18:261. https://doi.org/10.1186/s12864-017-3654-1.

Zhou Y, Chebotarov D, Kudrna D, Llaca V, Lee S, Rajasekar S, et al. A platinum standard pan-genome resource that represents the population structure of Asian rice. Sci Data. 2020;7:113. https://doi.org/10.1038/s41597-020-0438-2.

Population Genomics of Organelle Genomes in Crop Plants

Nora Scarcelli

Abstract Chloroplast and mitochondria are specialized structure located in the cytoplasm of plant cells which possess their own genomes and are usually transmitted by one parent only. Organelle DNA markers have been widely used for phylogeny and population genetics studies because their low evolutionary rates allow to look further back in the past than nuclear data. This low evolutionary rate, however, was also a drawback because the diversity revealed by traditional methods (Sanger sequencing, microsatellites, RFLP, etc.) was low and was not necessarily discriminatory enough to study genetic structure and genetic relationships. Nowadays, the fast development of next-generation sequencing methods has changed the way organelle genomes sequences are obtained. It is now easy to get whole organelle genome sequences, even for an orphan species. Because the quantity of data and the amount of diversity generated by these new methods have strongly increased, the way organelle genetic data can be used has also changed.

This chapter provides a review on how to retrieve organelle genomic polymorphism from next-generation sequencing technology, with an emphasis on the specificity of organelles compared to nuclear data: how to efficiently sequence this highly repeated DNA resulting from multiple copies of the organelle genome within a cell, how to deal with the intra-individual diversity generated, and how to reconstruct a whole organelle genome sequence. This chapter will then review how next-generation sequencing technology has changed the main fields of organelle population genomics, i.e. population genetics, phylogeny, phylogeography and DNA barcoding. These new possibilities will be analysed in the light of the new drawbacks going along with the "big data".

Keywords Bioinformatics · Chloroplast · Gene flow · Mitochondria · NGS · Phylogeny · Population genetics

---

N. Scarcelli (✉)
DIADE, Univ Montpellier, IRD, Montpellier, France
e-mail: nora.scarcelli@ird.fr

Abbreviation

cpDNA	Chloroplast DNA
mtDNA	Mitochondrial DNA
NGS	Next generation sequencing
nucDNA	Nuclear DNA
orgDNA	Organelle DNA
PCR	Polymerase chain reaction
SNP	Single nucleotide polymorphism
SSR	Simple sequence repeats (= microsatellites)

1 Introduction

Organelles are specialized structures located in eukaryote cells. From a broad point of view, organelles include several structures like the Golgi apparatus, but a strict definition reduces them to those structures that have their own genome and are capable of multiplication. This strict definition includes only the mitochondrion and the chloroplast. Mitochondria are shared by all eukaryotes and are the seat of cellular respiration. Chloroplasts are specialized in photosynthesis and are shared by all plants and algae.

The endosymbiotic theory (Mereschkowsky 1905) suggests that organelles evolved from free-living prokaryotic cells (Fig. 1), presumably as an aerobic proteobacterium for the mitochondrion and as a photosynthetic cyanobacterium for the chloroplast (Dyall 2004). During the assimilation process, a large part of the genome of the original prokaryote cells was transferred to the nuclear genome (Martin and Herrmann 1998; Race et al. 1999). As a result, organelles depend entirely on the eukaryote cells to survive, as well as the eukaryote cells depend entirely on organelles for cell respiration and photosynthesis.

Compared to nuclear genomes, organelle genomes exhibit special features that were applied in both fundamental and applied research. For example, organelle genomes evolve slower than nuclear genomes and can therefore be used to look further back in the past. The shorter coalescent time also allows a better separation between species (Birky 1995). From an agronomical point of view, cytoplasmic male sterility has been used for decades by breeders to create commercial varieties (Chen et al. 2017; Duvick 1965). The recent and fast development of new sequencing technologies brought about major changes in the organelle genomics. Now, we have easy access to whole organelle genomes, even for orphan crops.

In this chapter, I will discuss the contributions of next-generation sequencing technologies to organelle genomics. After a brief overview of organelle genomes, I will discuss how genomics technology and bioinformatics need to be adapted to the specificities of organelle genomes in order to (1) sequence organelle genomes, (2) reveal organelle genomes' diversity and (3) assemble whole organelle genomes.

Population Genomics of Organelle Genomes in Crop Plants 39

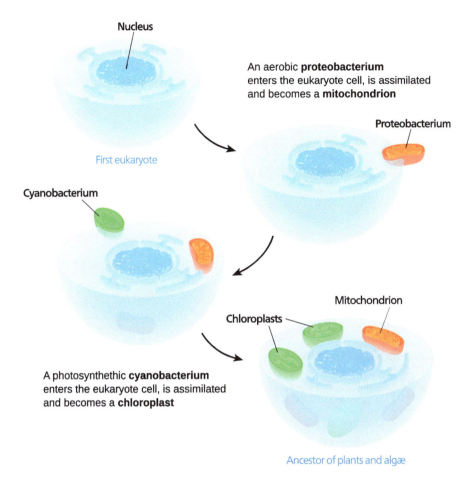

Fig. 1 Illustration of the endosymbiotic theory. Figure adapted from Kelvinsong – Own work, CC BY-SA 3.0

Then I will provide a review of how the whole organelle genome sequence data have provided new insights and contributed to important research fields, such as population genetics, phylogeny and phylogeography, and DNA barcoding.

2 The Chloroplasts and the Chloroplast Genome in Crop Plants

Chloroplasts are located in the cytoplasm of plant cells. They are constituted of a double membrane (Fig. 2a) holding the thylakoid system, where the photosynthesis takes place. The chloroplast DNA (cpDNA or plastome) is located in the stroma.

(A) Chloroplast structure

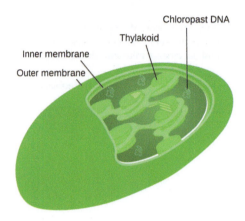

(B) Chloroplast genome

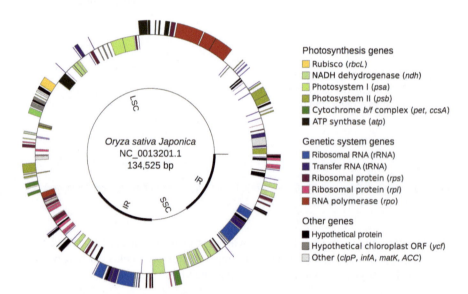

Fig. 2 Structure of the chloroplast (**a**) and example of the chloroplast genome (**b**). The rice chloroplast genome has been drawn using GenomeVx (Conant and Wolfe 2008). Genes located on the positive strand are drawn on the outside and genes located on the negative strand on the inside. Figure adapted from Kelvinsong – Own work, CC BY-SA 3.0

The number of chloroplasts per cell is highly variable and depends on several factors, including tissue function, developmental stage and environmental conditions (Bendich 1987; Kumar et al. 2014). Each chloroplast contains several cpDNA molecules; therefore, one single cell can contain a large number of cpDNA copies (more than 50,000 in wheat leaves, *Triticum aestivum*, Boffey and Leech 1982).

During reproduction, chloroplasts are passed down through the cytoplasm. They are usually inherited from one parent only, most often the female in angiosperms (Jansen and Ruhlman 2012). Several pre- and postfertilization systems exist to prevent biparental transmission (Birky 1995), but cases of biparental inheritance have been reported in nearly 20% of angiosperms species, though not necessarily at a high rate (Corriveau and Coleman 1988). Plastomes generally do not recombine, even in case of biparental transmission; therefore, they are asexually inherited.

The first complete plastome to be published was that of tobacco, *Nicotiana tabacum*, in 1986 (Shinozaki et al. 1986). Since then, thousands of complete plastome sequences were published as a result of the development of next-generation sequencing technologies (NGS), which has allowed a better understanding of the plastome organization. Basically, cpDNA is a double-stranded and haploid molecule, which can exhibit as a circular or a linear form (Oldenburg and Bendich 2016). A single molecule is approximately 150 kb long in crop plants and contains approximately 4 rRNA, 30 tRNA and 100 protein-coding genes. Compared to nuclear genes, number and order of the chloroplast genes are well conserved among the plants, but some gene losses and rearrangements can be found (Jansen and Ruhlman 2012). A common specificity of the cpDNA is the presence of duplicated and inverted regions. These two regions contain the same genes, notably the rRNA, and are named IRa and IRb for inverted repeat. They delimit a large and two small sections, named LSC and SSC for long single copy and short single copy (Fig. 2b). The size of the inverted regions varies according to the species. They can be reduced to few hundred base pairs or even absent in conifers and some other species (e.g. Asaf et al. 2018).

Chloroplast genes are involved in photosynthesis (Rubisco, NADH dehydrogenase, photosystem, cytochrome and ATP synthase genes) and in the genetic system (rRNA, tRNA, ribosomal protein and RNA polymerase genes). However, normal functioning of the chloroplast requires importing proteins expressed by ancient chloroplast genes transferred to the nuclear genome.

Apart from the genes completely relocated into the nuclear DNA, cpDNA sequences are frequently copied and transferred to the nuclear DNA (Leister 2005; Wolfe and Randle 2004). These copies are named NUPT (nuclear plastid DNA). NUPT does not necessarily involve coding cpDNA. They vary in size and in proportion, from 200 to 20,000 bp and from 0.01% to 0.27% of the nuclear genome (Michalovova et al. 2013; Richly and Leister 2004; Yoshida et al. 2014). cpDNA sequences can also be copied to the mitochondrial genome and are named mtptDNA, for mitochondrial plastid DNA (Bock and Timmis 2008; Leister 2005; Wolfe and Randle 2004). They represent up to 10% of the mitochondrial genome (Wang et al. 2012; Zhang et al. 2012).

3 The Mitochondria and the Mitochondrion Genome in Crop Plants

Like chloroplasts, mitochondria are located in the cytoplasm of plant cells and possess their own DNA/genome (mtDNA or chondriome, Fig. 3a). The number of mitochondria per cell as well as the mtDNA copies per mitochondrion is highly variable but is lower than for the chloroplast (up to 450 mtDNA copies in *Arabidopsis* leaves (Preuten et al. 2010)).

Like for the chloroplast, most of the ancient proteobacterium genes were transferred to the nuclear genome. Genes remaining in the chondriome are involved in cell respiration (NADH dehydrogenase, cytochrome and ATP synthase genes) and in the genetic system (rRNA, tRNA, ribosomal protein and RNA polymerase genes). Mitochondrion DNA sequences can also be transferred to the nuclear genomes (Hazkani-Covo et al. 2010; Leister 2005; Michalovova et al. 2013). These NUMT (nuclear mitochondrial DNA) are very common in plants. They are usually highly duplicated and vary in size and in proportion. They can represent up to 0.25% of the nuclear genome.

Like chloroplasts, mitochondria are generally maternally inherited, but paternal inheritance and biparental inheritance are also frequently reported (Petit and Vendramin 2007).

However, the mitochondrion genome structure is much more complex and diverse than the plastome structure. First of all, the chondriome is frequently represented as a double-stranded DNA single ring (the "master chromosome", Fig. 3b) but rarely exists in this state in the plant cells. The total mtDNA is split in several structures like linear forms, open circles of variable size, supercoiled molecules and other complex structures (Burger et al. 2003; Morley and Nielse 2017). Those subgenomic DNA molecules partly originate from replication and recombination (Preuten et al. 2010). Secondly, the chondriome greatly varies in size, even within the same species, generally from 200 to 750 kb in crop plants. However, gene number is very constant with 50 to 60 genes (Gualberto et al. 2014; Kubo and Newton 2008). Most of the mtDNA is actually non-coding DNA and explains the observed differences in size. Non-coding DNA includes repeated regions, large introns, but also chloroplast DNA (mtptDNA), nuclear DNA and other species mtDNA incorporated via horizontal gene transfers (Kubo and Newton 2008). As suggested by Preuten et al. (2010), each mitochondrion only contains a part of the total mtDNA. Indeed, the number of mitochondria present in one cell is higher than the number of mtDNA present in the cell. It also suggests that there are specific mechanisms to ensure normal functioning of mitochondria with incomplete gene set, as well as to insure the transmission of a complete gene set during the meiosis (Gualberto and Newton 2017).

(A) Mitochondrion structure

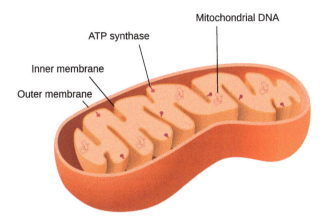

(B) Mitochondrion genome

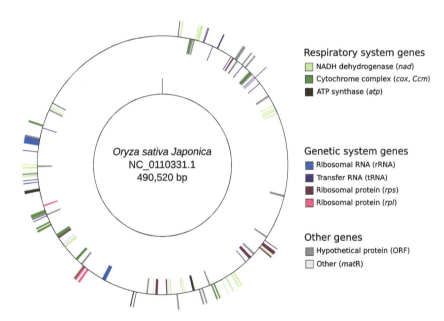

Fig. 3 Structure of the mitochondrion (**a**) and example of the mitochondrial genome (**b**). The rice mitochondrial genome has been drawn using GenomeVx (Conant and Wolfe 2008). Genes located on the positive strand are drawn on the outside and genes located on the negative strand on the inside. Figure adapted from Kelvinsong – Own work, CC BY-SA 3.0

4 Revealing Organelles Genomes Diversity

4.1 Sequencing the Organelles Genomes

For a long time, the sequencing of organelle genomes was performed by Sanger sequencing of PCR-amplified genes or intergenic spacers (IGS) using universal primers. A large set of primers was published for the chloroplast genome parts, which allows the amplification of virtually the whole plastome (e.g. Prince 2015; Scarcelli et al. 2011; Shaw et al. 2005, 2007; Taberlet et al. 1991). However, relatively few primers are available for the mitochondrial genome (e.g. Duminil et al. 2002), probably reflecting the high structural variability of the mitochondrial genome and, therefore, the difficulty to design universal primers.

NGS methods provide a good alternative to Sanger sequencing and a good opportunity to analyse quickly the whole organelle genome rather than a limited number of genes or IGS. NGS short-read sequencing technologies basically consist of cutting DNA into small fragments, either using restriction enzyme or mechanical shearing. A library of DNA fragments is then created by adding specific adapters and selecting the optimal fragment sizes for sequencing. Fragments are individually sequenced in parallel and later assembled to reconstruct a longer DNA fragment (van Dijk et al. 2014). Alternatively, long-read technologies based on reading the sequence of single molecules could directly produce longer DNA fragment sequences (Gupta 2008).

NGS methods, however, are not necessarily efficient when working on the organelle genomes only. Indeed, whole genome sequencing based on the total cellular DNA will generate little orgDNA reads compared to the nucDNA reads. As organelle genomes are highly repeated, shallow whole genome sequencing (named genome skimming, Straub et al. 2012) will result in a good coverage of organelle genomes (e.g. Mariac et al. 2014). However, this strategy might be very costly if organelle sequences of several individuals are needed. To get around this problem, several methods of orgDNA enrichment are available. They can be classified in four categories (Jansen et al. 2005; Twyford and Ness 2017):

1. *Specific extraction of orgDNA.* The most common methods use gradients to first separate organelles from the nucleus and then to separate orgDNA from remaining nucDNA. Other methods specifically destroy non-circular DNA (Jansen et al. 2005; Lang and Burger 2007; Mackenzie 1999). These methods are usually very laborious and rarely produce pure organelle DNA, even when optimized (e.g. Diekmann et al. 2008; Takamatsu et al. 2018).

2. *Cloning orgDNA into a vector.* In these methods, whole DNA is inserted into a bacterial vector, usually a BAC (bacterial artificial chromosome) or a fosmid (Jansen et al. 2005). Clones are then hybridized with organelle-specific probes to retain only orgDNA. The advantage is the possibility to clone very long fragments (up to 40 kb for fosmid and 150 kb for BAC). However, these methods are tedious, and a large number of clones may be needed, depending on the size of the organelle genome and the number of samples analysed.

3. *PCR amplification of long orgDNA fragments.* Using specific set of primers and a *Taq* polymerase combined to a proofreading enzyme, the long-range PCR method produces DNA fragments up to 20 kb (Davies and Gray 2002). Alternatively, the rolling circle amplification (RCA) takes advantage of the bacteriophage phi29 DNA polymerase to amplify >70 kb using one (or few) primer pair (Johne et al. 2009). Both methods necessitate upstream work to test universal primers or to design specific primers. Because they are based on PCR amplification, they can generate bias in the sequences produced (PCR error) as well as in the coverage (PCR failure) (Cronn et al. 2012).

4. *Capture of orgDNA.* Here, probes are designed to specifically hybridize with orgDNA and to isolate orgDNA from nucDNA (Cronn et al. 2012). The technique works with both RNA probes (Stull et al. 2013) and DNA probes (Mariac et al. 2014). A drawback of the capture method is the design of the probes, which requires a minimal knowledge of the species or closely related species. However, by adjusting hybridization conditions, it is possible to relax stringency to facilitate hybridization with more distant species (Cronn et al. 2012; Li et al. 2013).

   One method consists of using the difference in methylation of nucDNA vs. orgDNA, but this method is highly species-specific (Yigit et al. 2014). Other methods consist of capturing long fragments using short probes, in order to take advantage of the development of long-read sequencing technology (Bethune et al. 2019; Dapprich et al. 2016; Gasc and Peyret 2017). However, we still lack perspectives on these new methods, especially on the representation bias induced by the technique itself and on the effect of the high error rate of long-read sequencing technology (Goodwin et al. 2016; Jung et al. 2019).

All these methods have been applied alone or in combination to sequence whole plastomes (e.g. Atherton et al. 2010; Bethune et al. 2019; Diekmann et al. 2008; Mariac et al. 2014; Uribe-Convers et al. 2014; Yang et al. 2014). However, the whole chondriome sequencing is more difficult due to its high genetic and structural variability. As a result, few whole chondriomes are available compared to whole plastomes (318 chondriomes vs 4,845 plastomes, Genbank https://www.ncbi.nlm.nih.gov/genome/organelle/, on 04/2020). mtDNA is usually retrieved from specific mtDNA extraction (e.g. Clifton 2004; Hisano et al. 2016; Notsu et al. 2002) or directly using whole genome sequencing (e.g. Kim et al. 2016).

4.2 Genetic Diversity of the Organelle Genomes

NGS sequencing produces a large amount of reads that need to be analysed using bioinformatics tools to reveal the genetic diversity. Basically, raw reads are first analysed to remove adapter sequences (and tags or indexes in case of multiplexing) and then filtered for base quality. They are aligned to a reference (a step named "mapping"), and then variants to the reference are retrieved (the "calling" step). There is a huge literature available on different software and pipelines specifically developed for analysing NGS reads and on how to calibrate them. The point here is not to review this literature but to tackle the specificities of organelles NGS analysis, i.e. the variant calling step. Indeed, pre-mapping analysis steps are the same as when dealing with nuclear sequences. Mapping needs to be adjusted according to the length of the reads and the divergence between the target species and the reference, exactly like for nuclear sequences. Variant calling, however, is a crucial step because a single organism holds a large population of organelles sequences with potential intra-organism diversity/heterogeneity (e.g. Al-Attas et al. 2020). The intra-organism organelle diversity revealed during the variant calling step can have three different origins:

1. *Errors.* NGS technology generates sequencing errors, rate of which varies according to the platform used (Goodwin et al. 2016; Jung et al. 2019): from less than 0.1% for Illumina HiSeq and SOLiD, short-read technologies, to 15% for PacBio and MinION, long-read technologies. Sequencing errors therefore artificially produce intra-organism diversity that must not be considered in analyses.
2. *Duplication.* Because they are duplicated sequences, NUPT, NUMT and mtptDNA can generate diversity by mutation, especially when they are duplicated in the nuclear genome (NUPT and NUMT), where the mutation rate is higher. This diversity observed is the result of parallel evolution of different sequences and therefore is not representative of the evolution of the organelle genomes. Therefore, it should be ignored for analyses.
3. *Heteroplasmy.* Heteroplasmy is a particular state when genetically different organelles co-exist within the same organism (Wolfe and Randle 2004). Heteroplasmy arises when mutations appear in the organelle population, or because of biparental transmission of genetically different organelles. Unless it provides a selective advantage, heteroplasmy is expected to be a transitional state only, because diversity will be progressively lost by genetic drift (Birky 2001; Greiner et al. 2015). One can choose to specifically pay attention to heteroplasmy, but from a population genetics point of view, it may be easier to consider only one lineage and therefore to ignore the diversity created by heteroplasmy.

It has been shown that the intra-organism chloroplast diversity mainly originated in NUPTs (Scarcelli et al. 2016) and should therefore be ignored. A simplistic method would be to delete all heterozygous SNP after calling. A good alternative is to adjust calling parameters to call only high-frequency variants and to force SNP to be homozygous. However, the calling software and its parameters need to be carefully chosen, because not all software are able to call homozygous SNP without increasing false-positive and false-negative SNP (Scarcelli et al. 2016).

Alternatively, long-read sequencing is expected to reduce the quantity of NUPTs and NUMTs sequenced, because inclusions in the nuclear genome are mainly shorter than the long-read size. Still, some very large NUPTs have been observed (Michalovova et al. 2013), and the long-read technology also suffers from high error rate.

4.3 Assembling the Whole Organelles Genomes

To analyse the genetic diversity, a simple SNP calling is usually enough. However, when one wants to generate a new organelle genome reference or to use the whole organelle sequence, he/she will have to assemble the complete sequence from raw reads. Several commercial and non-commercial algorithms are available to perform genome assembly. The most appropriate method will mainly depend on whether the reference genomes exist or not. There are two classes of genome assembly:

1. *Genome resequencing* (Fig. 4a). The idea is to sequence a whole genome then to map reads against a reference genome in order to retrieve the complete sequence of the newly sequenced genome. This technique is particularly adapted to short-read sequencing technology. If the reference genome is very close to the target genome (usually from the same species), a mapping and variant calling as described above is usually enough to reveal the few genetic differences and to construct the new reference genome (Dierckxsens et al. 2016). However, if the reference genome is more distant, one needs to use algorithms able to deal with more important mismatches (e.g. MITObim, Hahn et al. 2013).
2. De novo assembly (Fig. 4b). Here, no reference genome is used. The method consists of finding overlap between reads, assembling the overlapping reads into

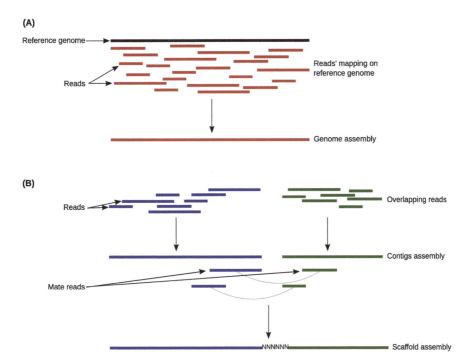

Fig. 4 Illustration of two genome assembly procedures: (**a**) genome resequencing assembly and (**b**) de novo genome assembly

contigs and then filling the gap to produce scaffolds. If the reads were obtained without enrichment, it may be necessary to filter organelle reads using an organelle sequence database in order to decrease computational time. Even if it is possible to use de novo assemblers adapted to large or small genomes (e.g. SOAPdenovo2, Luo et al. 2012; Velvet, Zerbino and Birney 2008), some assemblers were specifically developed to produce mitochondrial and chloroplast genomes from whole genome sequencing (e.g. NOVOplasty, Dierckxsens et al. 2016; IOGA, Bakker et al. 2016).

The development of long-read sequencing technology was expected to facilitate de novo genome assembly. However, genome assembly still remains challenging because the long-read technology is associated with a high error rate (de Lannoy et al. 2017; Goodwin et al. 2016). New methods are currently developed to specifically deal with this problem (e.g. Soorni et al. 2017; Vaser et al. 2017). Alternatively, new algorithms combining both long- and short-read technology have emerged and are able to take advantage of both methods to obtain high-quality genome assembly (Jung et al. 2019; Wang et al. 2018).

With the advancement of NGS technology and the development of many assemblers, it becomes easy to obtain enough data to assemble a new organelle genome, even for non-model plants. It is more challenging, however, to obtain a genome assembly of high quality. The sequencing depth, for example, is a key point to genome assembly. Assemblers need a minimal depth to reconstruct a genome (Desai et al. 2013) but increasing the depth to a certain point will decrease the quality of the assembly (Lonardi et al. 2015). To overcome this problem, Bakker et al. (2016) recommend to rely on genome skimming (Straub et al. 2012) to cover the highly repeated organelle genomes without producing too many reads. This observation, however, is not necessarily true when working with long-read technologies. For example, Ferrarini et al. (2013) has shown that 320× depth produced the same results as 35× depth when reconstructing the *Potentilla micrantha* plastome from PacBio sequencing.

A second problem may arise when filtering whole genome sequencing to reduce the complexity of the assembly, either by differential coverage or with an organelle database. Indeed, some regions could be excluded just because it is technically more difficult to obtain them at a high coverage (e.g. due to the presence of GC-rich or GC-poor regions) (Benjamini and Speed 2012) or because they are too divergent from the database used as reference (Dierckxsens et al. 2016).

A third problem, arising specifically with plastome, is the assembly of both inverted repeat regions (IR). Because they are identical, most assemblers struggle to reconstruct them separately, and it is still necessary to use Sanger sequencing to obtain the sequences of the IR boundaries (Twyford and Ness 2017). The development of long-read sequencing only partially solved this problem because it is still challenging to obtain and sequence reads that are longer than the IR regions (Bethune et al. 2019).

Usually, once a new genome has been assembled, it needs to be annotated for protein coding genes, tRNA and rRNA. For the chloroplast genome, this process is facilitated by the highly conserved structure of the plastome. It is more difficult, however, for the chondriome. Several tools have been developed to automatically retrieve annotations from comparison with the public database (e.g. Liu et al. 2012; Wyman et al. 2004). The results still need to be manually checked, especially for start and stop codon and for intron/exon boundaries.

5 Applications of Organelles Genomes in Population Genomics

5.1 Population Genetics

Population genetics is the branch studying the evolution over space and time of allele frequency at the population level. Models are based on the analysis of population genetic diversity and structure. They take into account the five major evolutionary forces – mutation, drift, selection, recombination and migration – in order to estimate, for example, gene flow, detect selection or evaluate demographic scenarios (bottleneck, expansion, size of the ancestral population, migration, etc.). The fast development of NGS technology has strongly impacted population genetics because thousands or millions of SNP are now routinely used, instead of the tens to hundreds of markers used 10 years ago (e.g. 30 SSR in rice (*Oryza sativa*), Thomson et al. 2007; vs. ~30,000,000 SNPs in rice, Li et al. 2017a). Analyses at the genome level allow to test more complex scenarios and to refine parameter estimations (Pool et al. 2010). However, by using increasingly larger dataset, our modelling abilities tend to be overwhelmed (Sousa and Hey 2013). Some problems may also appear when using data generated from low-coverage sequencing because genotype retrieval of diploid genomes is not error-free (Nielsen et al. 2011). Fortunately, chloroplast NGS data are expected to be less affected than nuclear NGS data because they are limited in size and represent haploid genomes.

Chloroplast DNA is commonly used to study gene flow and introgression (McCauley 1995). By contrasting biparental gene flow (nuclear markers) with maternal gene flow (chloroplast markers), it was possible to trace pollen vs. seed gene flow (e.g. Grassi et al. 2003; McCauley 1997; Oddou-Muratorio et al. 2001) or to trace crops migration (e.g. Balfourier et al. 2000; Besnard et al. 2002; Boccacci and Botta 2009). A common criticism, however, was that chloroplast diversity was very low and could not reveal efficiently intra-specific diversity (McCauley 1995). Nowadays, whole chloroplast genome sequences, therefore whole chloroplast polymorphisms, are available, and they do reveal intra-specific diversity. For example, whole chloroplast sequence reveals 3,016 SNPs and 385 SNPs for 19 African wild rice, *Oryza barthii*, and 19 African cultivated rice, *O. glaberrima* samples, respectively (Tong et al. 2016). A comparison of these polymorphisms with Asian rice chloroplast diversity, *O. sativa*, *O. rufipogon* and *O. nivara* (Tong et al. 2016), strongly supports the hypothesis of an independent domestication of African rice (Li et al. 2011; Wang et al. 2014).

In the past, low or lack of genetic diversity observed in chloroplast DNA genes or regions prevented to establish detailed scenario of domestication in crop plants. In 2006, for example, Londo et al. (2006) analysed the genetic diversity of Asian rice (211 wild *O. rufipogon* and 677 cultivated *O. sativa*) using a single chloroplast intergenic spacer (*atp*B-*rbc*L). They observed a total of 44 haplotypes, among which 36 were specific to one sample, i.e. only eight haplotypes were found among the remaining 852 samples. In fact, they observed only two main haplotypes, roughly corresponding to *O. sativa indica* and *O. sativa japonica* subspecies. Wild *O. rufipogon* fell in both main haplotypes. If the authors could then support the old hypothesis that Asian cultivated rice originated from *O. rufipogon*, they could not decide, based on chloroplast data alone, if rice was domesticated once or twice independently. They needed to add nuclear data to conclude about independent origins. In 2017, Wang et al. also analysed a large set of Asian rice (435 wild and 203 cultivated samples) but using NGS technology. They analysed the whole chloroplast sequences and retrieved 480 polymorphic sites. After removing missing data and low-frequency SNPs, they analysed 74 high-quality SNPs and obtained 56 haplotypes, among them 28 haplotypes were frequent, i.e. shared by more than six individuals. A detailed comparison between the nuclear and the fine structure obtained at the chloroplast genome level allowed the authors to conclude about multiple domestications followed by extensive crop-to-wild gene flows.

Similar results were found in African yam, *Dioscorea rotundata* (Scarcelli et al. 2017), where a comparison of nuclear microsatellites and whole plastome diversity revealed that wild yam relatives, *D. abyssinica* and *D. praehensilis*, are contaminated by gene flow from cultivated yam. Both the yam and rice cases involve a feralization of cultivated seeds followed by recurrent backcrosses with the wild relatives. The demonstration of these patterns of gene flow was possible only because whole plastome sequences were used. Indeed, other traditional markers used in previous studies revealed little to no diversity (e.g. cpSSR on rice, Provan et al. 1996; cpSSR on yam, Chaïr et al. 2005).

5.2 Phylogenetics and Phylogeography

Phylogenetics is the study of the relationship between taxa through the reconstruction of phylogenies, i.e. tree representing the evolutionary histories of taxa. Phylogenetics is used to classify life (e.g. The Angiosperm Phylogeny Group 2016) and also for various analyses, such as the study of species diversification (e.g. Eiserhardt et al. 2017; Morlon 2014), ecosystem characteristics (e.g. Cadotte et al. 2012) or the prediction of gene functions (e.g. Mi et al. 2013).

Phylogeography aims at understanding the geographical distribution of species, by analysing phylogenies in the light of the species geographical distribution (Emerson and Hewitt 2005). It can be used to understand species migration over time and was particularly used to infer post-glacial colonization routes (e.g. Faye et al. 2016; Petit et al. 2002). As it is based on phylogeny and population genetics, phylogeography will benefit from the use of the whole organelle genome sequences as phylogenetics and population genetics (Edwards et al. 2015) but may also suffer from the same drawbacks, such as a lack of discriminating power due to a lack of diversity.

Due to its high structural variability, the mitochondrion is not usually used when trying to understand the relationship between plant species. Historically, plant phylogenies were based on chloroplast DNA markers (e.g. Chase et al. 1993; Gielly and Taberlet 1994) because their slow evolution facilitates the use of universal primers. The flip side of this slow evolution is that chloroplast DNA markers are not necessarily polymorphic enough to get a good resolution when markers for only a few chloroplast genes are used. Thanks to the development of NGS, it is now possible to sequence the whole plastome, which can significantly improve the resolution of phylogenies (Harrison and Kidner 2011), especially at low taxonomic levels (e.g. Magwé-Tindo et al. 2018; Parks et al. 2009).

Despite the enthusiasm created by the huge amount of data generated by NGS, many authors started to worry that too much information may decrease the quality of phylogenetic inferences. Indeed, increasing data also increases the amount of non-informative or heterogeneous sites and the difficulty to accurately model DNA substitution rates across the whole region (Lemmon and Lemmon 2013). Moreover, increasing the amount of information also increases the number of errors produced by evolutionary models (Olmstead and Bedoya 2019). However, these problems are not expected to strongly affect phylogenies based on the whole plastomes. Indeed, chloroplast genome size and polymorphisms are limited, the plastome is haploid, and there are minimal structural rearrangements like gene duplications.

With the development of NGS, it is tempting to concatenate hundreds of SNPs to construct phylogenies (e.g. Carbonell-Caballero et al. 2015). This, however, will produce biased phylogenies (Bertels et al. 2014). Indeed, estimation of branch lengths under maximum likelihood are affected by the exclusion of monomorphic (non-variable) positions (Lewis 2001). Moreover, the mapping ability decreases as the divergence between reads and the reference increases (Straub et al. 2012). Therefore, by mapping different species to a single reference, SNP calling will produce non-random missing data, which is known to have a negative impact on phylogeny estimation (Xi et al. 2016). A good strategy to produce chloroplast phylogenies from the NGS data is to work on the whole plastome assembly for each species or group of species. An important drawback, however, is that the majority of programmes developed for phylogenetic inferences are not adapted to chloroplast genome size datasets and analyses (Sanderson and Driskell 2003). Recently, the maximum likelihood software RAxML has been adapted to work on genome-size datasets (Stamatakis et al. 2012), and a new and faster software, ExaML (Stamatakis and Aberer 2013), has been released. They can generate phylogenies with millions of base pairs on thousands of taxa. However, they also require huge informatics facilities (~1,500 cores) that are not available in most labs.

Whole chloroplast, or very large plastome portions, phylogenies have already been used to clarify the relationships between crops and their wild relatives (e.g. Cheng et al. 2019; Sherman-Broyles et al. 2014; Wambugu et al. 2015), to

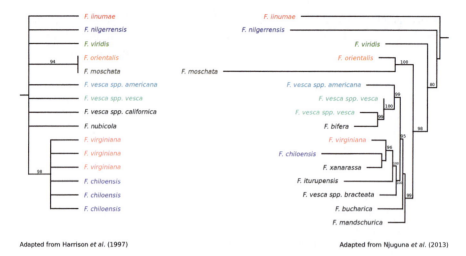

Fig. 5 Comparison of the phylogenies of the genus *Fragaria* obtained by Harrison et al. (1997), using chloroplast RFLP, and Njuguna et al. (2013), using whole chloroplast genome. Trees have been edited to remove distant species and to highlight the correspondence between both phylogenies

date the divergence time of crops and their wild relatives (e.g. Middleton et al. 2014; Nikiforova et al. 2013) and to elucidate crop domestication (e.g. Kistler et al. 2014). As expected, these phylogenies provide higher resolution than phylogenies constructed from limited chloroplast DNA loci.

As an example, two phylogenies of the strawberry genus (*Fragaria* spp.) were produced 16 years apart by Harrison et al. (1997) and Njuguna et al. (2013). In 1997, Harrison et al. used chloroplast RFLP to analyse the relationships between 26 samples belonging to 16 (sub-)species of the genus *Fragaria*. Using a combination of 16 restriction enzymes and 5 probes derived from the tobacco chloroplast, they observed a total of 66 mutations (polymorphisms). In the tree built from these data, most of the species relationships were unresolved and unsupported (Fig. 5). Sixteen years later, Njuguna et al. (2013) analysed 25 samples from 21 (sub-)species of *Fragaria* using whole chloroplast sequencing. They identified a total of 4,188 polymorphic sites. The phylogenetic tree built was fully resolved and most of the (sub-)species were strongly supported (Fig. 5).

In another study using whole chloroplast sequencing, Wambugu et al. (2015) obtained 484 polymorphic sites when comparing 10 closely related *Oryza* species (wild and cultivated rice species from Asia and Africa). This can be compared to the 166 polymorphic sites found by Ge et al. (1999) on one single chloroplast gene (*mat*K), but for 27 very different species, including outgroups from a different genus. Ge et al. (1999) could not distinguish African from Asian rice species, while Wambugu et al. (2015) clearly demonstrated the relationships between African wild and cultivated species and between Asian wild and cultivated species, suggesting that African rice was domesticated independently from Asian rice.

Despite the interest of whole chloroplast sequencing, chloroplast-based phylogenies will continue to reflect one parent lineage only (generally the mother in angiosperms) and will still need to use biparental (nuclear) markers. This may be a reason why the use of chloroplast-based phylogenies only has strongly decreased in phylogeography studies (Garrick et al. 2015).

5.3 Barcoding

DNA barcoding aims at identifying unknown individuals using diagnostic sequences (barcodes), by comparing these sequences to a barcode database. DNA barcoding can be used to identify a single individual or to identify many species from a bulk of different individuals (metabarcoding, Taberlet et al. 2012). The latter method is particularly adapted to retrieve species composition of environmental samples, like soil or ice samples, faeces, stomach contents, etc.

DNA barcoding success relies on to three major points:

1. *The choice of NGS technique*. Even if there has been a continuous improvement in the NGS technology, the choice between NGS methods is still dictated by cost. Genome skimming and obviously whole genome sequencing are very efficient to retrieve barcode sequences but can be out of the question depending on the number of samples to analyse (Coissac et al. 2016). A cost-effective alternative is the use of barcode enrichment, but the portability of the probes used may not allow to identify distant species (Hollingsworth et al. 2016).
2. *The database quality*. A critical point in the DNA barcoding method is the development of an accurate and comprehensive database (Collins and Cruickshank 2013). Indeed, only species included in the database would be identified in the DNA barcode studies, and any taxonomic misidentification in the database will produce a wrong genetic identification.
3. *The barcode choice*. The efficiency of DNA barcoding studies also strongly relies on the quality of the barcode chosen. A good barcode needs to show high interspecific diversity but little to no intra-specific diversity. It needs to be portable to other taxonomic groups and to be short enough to be retrieved from degraded DNA (Valentini et al. 2009). The CBOL (Consortium for the Barcode of Life) considers the mitochondrial region *CO*1 (cytochrome *c* oxidase 1) to be the default barcode for eukaryote. But if *CO*1 is a very powerful barcode for animal, it is not polymorphic enough for use in plants. The CBOL has therefore approved the use of chloroplast genes *rbc*L and *mat*K for plants (CBOL Plant Working Group et al. 2009). Unfortunately, *rbc*L and *mat*K are not as discriminating in plant as *CO*1 is in animals, and other regions need to be used to complement them (Dong et al. 2014; Hollingsworth et al. 2016). Moreover, the frequent duplication of organelle portions to the nuclear genome may also screw up the genetic identification, as it was shown for arthropods with *CO*1 barcode (Song et al. 2008). Similar observations were reported because of the insertion of chloroplast

genome into the mitochondrion (Park et al. 2020). Some authors suggest to use the whole chloroplast sequence as an ultra-barcode to overcomes all these problems (Brozynska et al. 2014; Kane et al. 2012; Li et al. 2015; Nock et al. 2011). The main problem, however, is that the chloroplast (partly or in totality) may not be the best candidate for DNA barcoding because it is not always capable of defining species, especially because of horizontal transfers, inter-specific hybridization, polyploidy and limited seed dispersal (Hollingsworth et al. 2011; Stegemann et al. 2012).

Presently, chloroplast or mitochondrial barcodes are commonly used for plant identification, for example, to monitor invasive species (e.g. Hoveka et al. 2016), to characterize unknown flora (e.g. Le Gall and Saunders 2010), for food traceability (e.g. Prosser and Hebert 2017) or even to estimate herbivore diet (e.g. Erickson et al. 2017). DNA barcoding in general (i.e. using nuclear, mitochondrial or ribosomal DNA and with or without the use of NGS) was recently applied to archaeological DNA in order to reconstruct past vegetation (e.g. Gismondi et al. 2013; Jørgensen et al. 2012; Pedersen et al. 2013) and to identify crops in archaeological remains (e.g. Fernández et al. 2013; Gismondi et al. 2016; Pollmann et al. 2005).

In their study of cocoa (*Theobroma cacao*) varieties, Kane et al. (2012) showed that whole chloroplast genome sequences can be used as ultra-barcode to distinguish cocoa varieties. In their study, they analysed three cocoa varieties using whole chloroplast genome sequencing, and they obtained 78 variant sites that allowed them to distinguish each individual, but more importantly to distinguish between the two varieties Criollo and Forastero. On the other hand, when looking at the sequences of the barcodes approved by the CBOL for plants (genes *rbc*L and *matK*), they found only one variant site that did not allow to distinguish different varieties. Such a result seems promising in terms of varietal identification for breeding and commercial traceability. However, in their study, Kane et al. (2012) were unable to separate a third variety, Trinitario. Even if that result was expected due to the hybrid origin of the variety, this pinpoints one of the limits of chloroplast ultra-barcodes, i.e. that chloroplast generally represents only one parental lineage.

6 Future Perspectives and Conclusions

In the NGS era, the development of whole organelle genome analyses opens new possibilities to study crop diversity and evolution, that can be used for conservation and breeding. However, because of the development of whole nuclear genomes, organelle genomes are frequently considered as a by-product that is not worthy of attention. Yet, the use of whole chloroplast genomes could be very helpful to understand crop evolutionary histories. It can also provide details about crop wild relatives (CWR), that can be used later for crop improvement (Brozynska et al. 2016), especially for climate adaptation and disease resistance. Basic knowledge is usually lacking for orphan or understudied crops, which restrain efficient breeding and conservation. It is therefore particularly interesting for orphan crops to study whole chloroplast genome because the construction of the chloroplast genome does not involve as much scientific, financial and time efforts as the construction of the nuclear genome does. The chloroplast genome also provides a good gateway to start analysing complex, polyploidy genomes. Some studies already made good use of chloroplast genome sequencing to reveal wild progenitor, geographical structure or origin of domestication (e.g. P. Li et al. 2017b; Muñoz-Rodríguez et al. 2018; Nock et al. 2019), and it is hoped that researchers will continue to use chloroplast genome data for a better understanding of orphan crops rather than for producing mass data only.

The mitochondrial genome, although less studied than the chloroplast genome, can be of great interest for crop breeding because it is the seat of cytoplasmic male sterility (CMS), i.e. the maternal inherited failure to produce functional pollen. It has been extensively used by breeders to produce F1 hybrid seeds for cereals, cotton (*Gossypium* spp.) and vegetables (Bohra et al. 2016). CMS is a complex character, involving mitochondrial chimeric recombination that prevent the normal development of the male gametophyte and that can be regulated by complex interactions with nuclear genes, named restorer-of-fertility genes (Chen et al. 2017). With the development of NGS, there is a great hope that detailed analyses of the mitochondrial genomes will help understanding and using the CMS phenomenon even for orphan crops.

With the fast development of NGS technologies, it becomes easy to generate organelle genome sequences. Although the rate publication of whole mitochondrion genomes remains low (6 in 2010 *vs*. 50 in 2019 published on NCBI), publications of whole chloroplast genomes have shot up (23 in 2010 *vs*. 1,275 in 2019 published on NCBI). However, NGS technologies also come with their own drawbacks and only a careful bioinformatics treatment of NGS data will ensure a high-quality dataset, an essential part to any accurate analysis. Unfortunately, the current mass production mostly relies on automated processes and does not take into account the specificities of the crops or the impact of inappropriate settings on the final results. More disappointing is the fact that most of organelle genomes are produced and published without scientific questionings. Let's hope a more careful use of organelle genomes will return their use and respectability back.

References

Al-Attas BE, Ali HM, Khan T, Edris S, Gadalla NO, Ramadan AM, et al. Mitochondrial genome SNPs analysis of eight barley genotypes displaying hot spot regions, phylogenetic relationships and heteroplasmy. Plant Biosyst. 2020:1–11. https://doi.org/10.1080/11263504.2020.1727977.

Asaf S, Khan AL, Khan MA, Shahzad R, Lubna, Kang SM, et al. Complete chloroplast genome sequence and comparative analysis of loblolly pine (*Pinus taeda* L.) with related species. PLoS One. 2018;13:e0192966. https://doi.org/10.1371/journal.pone.0192966.

Atherton RA, McComish BJ, Shepherd LD, Berry LA, Albert NW, Lockhart PJ. Whole genome sequencing of enriched chloroplast DNA using the Illumina GAII platform. Plant Methods. 2010;6:22. https://doi.org/10.1186/1746-4811-6-22.

Bakker FT, Lei D, Yu J, Mohammadin S, Wei Z, van de Kerke S, et al. Herbarium genomics: plastome sequence assembly from a range of herbarium specimens using an Iterative Organelle Genome Assembly pipeline. Biol J Linn Soc. 2016;117:33–43. https://doi.org/10.1111/bij.12642.

Balfourier F, Imbert C, Charmet G. Evidence for phylogeographic structure in *Lolium* species related to the spread of agriculture in Europe. A cpDNA study. Theor Appl Genet. 2000;101:131–8. https://doi.org/10.1007/s001220051461.

Bendich AJ. Why do chloroplasts and mitochondria contain so many copies of their genome? Bioessays. 1987;6:279–82. https://doi.org/10.1002/bies.950060608.

Benjamini Y, Speed TP. Summarizing and correcting the GC content bias in high-throughput sequencing. Nucleic Acids Res. 2012;40:e72. https://doi.org/10.1093/nar/gks001.

Bertels F, Silander OK, Pachkov M, Rainey PB, van Nimwegen E. Automated reconstruction of whole-genome phylogenies from short-sequence reads. Mol Biol Evol. 2014;31:1077–88. https://doi.org/10.1093/molbev/msu088.

Besnard G, Khadari B, Baradat P, Bervillé A. *Olea europaea* (Oleaceae) phylogeography based on chloroplast DNA polymorphism. Theor Appl Genet. 2002;104:1353–61. https://doi.org/10.1007/s00122-001-0832-x.

Bethune K, Mariac C, Couderc M, Scarcelli N, Ardisson M, Martin JF, et al. Long-fragment targeted capture for long read sequencing of plastomes. Appl Plant Sci. 2019;7(5):e1243.

Birky CW. Uniparental inheritance of mitochondrial and chloroplast genes: mechanisms and evolution. Proc Natl Acad Sci U S A. 1995;92:11331–8. https://doi.org/10.1073/pnas.92.25.11331.

Birky CW. The inheritance of genes in mitochondria and chloroplasts: laws, mechanisms, and models. Annu Rev Genet. 2001;35:125–48. https://doi.org/10.1146/annurev.genet.35.102401.090231.

Boccacci P, Botta R. Investigating the origin of hazelnut (*Corylus avellana* L.) cultivars using chloroplast microsatellites. Genet Resour Crop Evol. 2009;56:851–9. https://doi.org/10.1007/s10722-009-9406-6.

Bock R, Timmis JN. Reconstructing evolution: gene transfer from plastids to the nucleus. Bioessays. 2008;30:556–66. https://doi.org/10.1002/bies.20761.

Boffey SA, Leech RM. Chloroplast DNA levels and the control of chloroplast division in light-grown wheat leaves. Plant Physiol. 1982;69:1387–91. https://doi.org/10.1104/pp.69.6.1387.

Bohra A, Jha UC, Adhimoolam P, Bisht D, Singh NP. Cytoplasmic male sterility (CMS) in hybrid breeding in field crops. Plant Cell Rep. 2016;35:967–93. https://doi.org/10.1007/s00299-016-1949-3.

Brozynska M, Furtado A, Henry RJ. Direct chloroplast sequencing: comparison of sequencing platforms and analysis yools for whole chloroplast barcoding. PLoS One. 2014;9:e110387. https://doi.org/10.1371/journal.pone.0110387.

Brozynska M, Furtado A, Henry RJ. Genomics of crop wild relatives: expanding the gene pool for crop improvement. Plant Biotechnol J. 2016;14:1070–85. https://doi.org/10.1111/pbi.12454.

Burger G, Gray MW, Franz Lang B. Mitochondrial genomes: anything goes. Trends Genet. 2003;19:709–16. https://doi.org/10.1016/j.tig.2003.10.012.

Cadotte MW, Dinnage R, Tilman D. Phylogenetic diversity promotes ecosystem stability. Ecology. 2012;93:S223–33. https://doi.org/10.1890/11-0426.1.

Carbonell-Caballero J, Alonso R, Ibañez V, Terol J, Talon M, Dopazo J. A phylogenetic analysis of 34 chloroplast genomes elucidates the relationships between wild and domestic species within the genus *Citrus*. Mol Biol Evol. 2015;32:2015–35. https://doi.org/10.1093/molbev/msv082.

CBOL Plant Working Group, Hollingsworth PM, Forrest LL, Spouge JL, Hajibabaei M, Ratnasingham S, et al. A DNA barcode for land plants. Proc Natl Acad Sci U S A. 2009;106:12794–7. https://doi.org/10.1073/pnas.0905845106.

Chaïr H, Perrier X, Agbangla C, Marchand JL, Dainou O, Noyer JL. Use of cpSSRs for the characterisation of yam phylogeny in Benin. Genome. 2005;48:674–84. https://doi.org/10.1139/g05-018.

Chase MW, Soltis DE, Olmstead RG, Morgan D, Les DH, Mishler BD, et al. Phylogenetics of seed plants: an analysis of nucleotide sequences from the plastid gene *rbc*L. Ann Missouri Bot Gard. 1993;80:528. https://doi.org/10.2307/2399846.

Chen Z, Zhao N, Li S, Grover CE, Nie H, Wendel JF, et al. Plant mitochondrial genome evolution and cytoplasmic male sterility. Crit Rev Plant Sci. 2017;36:55–69. https://doi.org/10.1080/07352689.2017.1327762.

Cheng L, Nam J, Chu S-H, Rungnapa P, Min M, Cao Y, et al. Signatures of differential selection in chloroplast genome between *japonica* and *indica*. Rice. 2019;12:65. https://doi.org/10.1186/s12284-019-0322-x.

Clifton SW. Sequence and comparative analysis of the maize NB mitochondrial genome. Plant Physiol. 2004;136:3486–503. https://doi.org/10.1104/pp.104.044602.

Coissac E, Hollingsworth PM, Lavergne S, Taberlet P. From barcodes to genomes: extending the concept of DNA barcoding. Mol Ecol. 2016;25:1423–8. https://doi.org/10.1111/mec.13549.

Collins RA, Cruickshank RH. The seven deadly sins of DNA barcoding. Mol Ecol Resour. 2013;13:969–75. https://doi.org/10.1111/1755-0998.12046.

Conant GC, Wolfe KH. GenomeVx: simple web-based creation of editable circular chromosome maps. Bioinformatics. 2008;24:861–2. https://doi.org/10.1093/bioinformatics/btm598.

Corriveau JL, Coleman AW. Rapid screening method to detect potential biparental inheritance of plastid DNA and results for over 200 angiosperm species. Am J Bot. 1988;75:1443–58. https://doi.org/10.1002/j.1537-2197.1988.tb11219.x.

Cronn R, Knaus BJ, Liston A, Maughan PJ, Parks M, Syring JV, et al. Targeted enrichment strategies for next-generation plant biology. Am J Bot. 2012;99:291–311. https://doi.org/10.3732/ajb.1100356.

Dapprich J, Ferriola D, Mackiewicz K, Clark PM, Rappaport E, D'Arcy M, et al. The next generation of target capture technologies – large DNA fragment enrichment and sequencing determines regional genomic variation of high complexity. BMC Genomics. 2016;17:486. https://doi.org/10.1186/s12864-016-2836-6.

Davies PA, Gray G. Long-range PCR. In: Theophilus BDM, Rapley R, editors. PCR mutation detection protocols. Totowa: Humana Press; 2002. p. 51–5.

de Lannoy C, de Ridder D, Risse J. The long reads ahead: *de novo* genome assembly using the MinION. F1000Res. 2017;6:1083. https://doi.org/10.12688/f1000research.12012.2.

Desai A, Marwah VS, Yadav A, Jha V, Dhaygude K, Bangar U, et al. Identification of optimum sequencing depth especially for *de novo* genome assembly of small genomes using next generation sequencing data. PLoS One. 2013;8:e60204. https://doi.org/10.1371/journal.pone.0060204.

Diekmann K, Hodkinson TR, Fricke E, Barth S. An optimized chloroplast DNA extraction protocol for grasses (Poaceae) proves suitable for whole plastid genome sequencing and SNP detection. PLoS One. 2008;3:e2813. https://doi.org/10.1371/journal.pone.0002813.

Dierckxsens N, Mardulyn P, Smits G. NOVOPlasty: *de novo* assembly of organelle genomes from whole genome data. Nucleic Acids Res. 2016;45:gkw955. https://doi.org/10.1093/nar/gkw955.

Dong W, Cheng T, Li C, Xu C, Long P, Chen C, et al. Discriminating plants using the DNA barcode *rbc*L b: an appraisal based on a large data set. Mol Ecol Resour. 2014;14:336–43. https://doi.org/10.1111/1755-0998.12185.

Duminil J, Pemonge M-H, Petit RJ. A set of 35 consensus primer pairs amplifying genes and introns of plant mitochondrial DNA. Mol Ecol Notes. 2002;2:428–30. https://doi.org/10.1046/j.1471-8286.2002.00263.x.

Duvick DN. Cytoplasmic pollen sterility in corn. Adv Genet. 1965;13:1–56. https://doi.org/10.1016/S0065-2660(08)60046-2.

Dyall SD. Ancient invasions: from endosymbionts to organelles. Science. 2004;304:253–7. https://doi.org/10.1126/science.1094884.

Edwards SV, Shultz AJ, Campbell-Staton SC. Next-generation sequencing and the expanding domain of phylogeography. Folia Zool. 2015;64:187–206. https://doi.org/10.25225/fozo.v64.i3.a2.2015.

Eiserhardt WL, Couvreur TLP, Baker WJ. Plant phylogeny as a window on the evolution of hyperdiversity in the tropical rainforest biome. New Phytol. 2017;214:1408–22. https://doi.org/10.1111/nph.14516.

Emerson BC, Hewitt GM. Phylogeography. Curr Biol. 2005;15:R367–71. https://doi.org/10.1016/j.cub.2005.05.016.

Erickson DL, Reed E, Ramachandran P, Bourg NA, McShea WJ, Ottesen A. Reconstructing a herbivore's diet using a novel *rbc*L DNA mini-barcode for plants. AoBP. 2017;9:plx015. https://doi.org/10.1093/aobpla/plx015.

Faye A, Deblauwe V, Mariac C, Richard D, Sonké B, Vigouroux Y, et al. Phylogeography of the genus *Podococcus* (Palmae/Arecaceae) in Central African rain forests: climate stability predicts unique genetic diversity. Mol Phylogenet Evol. 2016;105:126–38. https://doi.org/10.1016/j.ympev.2016.08.005.

Fernández E, Thaw S, Brown TA, Arroyo-Pardo E, Buxó R, Serret MD, et al. DNA analysis in charred grains of naked wheat from several archaeological sites in Spain. J Archaeol Sci. 2013;40:659–70. https://doi.org/10.1016/j.jas.2012.07.014.

Ferrarini M, Moretto M, Ward JA, Šurbanovski N, Stevanović V, Giongo L, et al. An evaluation of the PacBio RS platform for sequencing and *de novo* assembly of a chloroplast genome. BMC Genomics. 2013;14:670. https://doi.org/10.1186/1471-2164-14-670.

Garrick RC, Bonatelli IAS, Hyseni C, Morales A, Pelletier TA, Perez MF, et al. The evolution of phylogeographic data sets. Mol Ecol. 2015;24:1164–71. https://doi.org/10.1111/mec.13108.

Gasc C, Peyret P. Revealing large metagenomic regions through long DNA fragment hybridization capture. Microbiome. 2017;5:33. https://doi.org/10.1186/s40168-017-0251-0.

Ge S, Sang T, Lu B-R, Hong D-Y. Phylogeny of rice genomes with emphasis on origins of allotetraploid species. Proc Natl Acad Sci U S A. 1999;96:14400–5. https://doi.org/10.1073/pnas.96.25.14400.

Gielly L, Taberlet P. The use of chloroplast DNA to resolve plant phylogenies: noncoding versus *rbc*L sequences. Mol Biol Evol. 1994;11:769–77. https://doi.org/10.1093/oxfordjournals.molbev.a040157.

Gismondi A, Leonardi D, Enei F, Canini A. Identification of plant remains in underwater archaeological areas by morphological analysis and DNA barcoding. Adv Anthropol. 2013;03:240–8. https://doi.org/10.4236/aa.2013.34034.

Gismondi A, Di Marco G, Martini F, Sarti L, Crespan M, Martínez-Labarga C, et al. Grapevine carpological remains revealed the existence of a Neolithic domesticated *Vitis vinifera* L. specimen containing ancient DNA partially preserved in modern ecotypes. J Archaeol Sci. 2016;69:75–84. https://doi.org/10.1016/j.jas.2016.04.014.

Goodwin S, McPherson JD, McCombie WR. Coming of age: ten years of next-generation sequencing technologies. Nat Rev Genet. 2016;17:333–51. https://doi.org/10.1038/nrg.2016.49.

Grassi F, Imazio S, Failla O, Scienza A, Ocete Rubio R, Lopez MA, et al. Genetic isolation and diffusion of wild grapevine Italian and Spanish populations as estimated by nuclear and chloroplast SSR analysis. Plant Biol. 2003;5:608–14. https://doi.org/10.1055/s-2003-44689.

Greiner S, Sobanski J, Bock R. Why are most organelle genomes transmitted maternally? Bioessays. 2015;37:80–94. https://doi.org/10.1002/bies.201400110.

Gualberto JM, Newton KJ. Plant mitochondrial genomes: dynamics and mechanisms of mutation. Annu Rev Plant Biol. 2017;68:225–52. https://doi.org/10.1146/annurev-arplant-043015-112232.

Gualberto JM, Mileshina D, Wallet C, Niazi AK, Weber-Lotfi F, Dietrich A. The plant mitochondrial genome: dynamics and maintenance. Biochimie. 2014;100:107–20. https://doi.org/10.1016/j.biochi.2013.09.016.

Gupta PK. Single-molecule DNA sequencing technologies for future genomics research. Trends Biotechnol. 2008;26:602–11. https://doi.org/10.1016/j.tibtech.2008.07.003.

Hahn C, Bachmann L, Chevreux B. Reconstructing mitochondrial genomes directly from genomic next-generation sequencing reads – a baiting and iterative mapping approach. Nucleic Acids Res. 2013;41:e129. https://doi.org/10.1093/nar/gkt371.

Harrison N, Kidner CA. Next-generation sequencing and systematics: what can a billion base pairs of DNA sequence data do for you? Taxon. 2011;60:1552–66. https://doi.org/10.1002/tax.606002.

Harrison RE, Luby JJ, Furnier GR. Chloroplast DNA restriction fragment variation among strawberry (*Fragaria* spp.) taxa. J Am Soc Hortic Sci. 1997;122:63–8. https://doi.org/10.21273/JASHS.122.1.63.

Hazkani-Covo E, Zeller RM, Martin W. Molecular poltergeists: mitochondrial DNA copies (numts) in sequenced nuclear genomes. PLoS Genet. 2010;6:e1000834. https://doi.org/10.1371/journal.pgen.1000834.

Hisano H, Tsujimura M, Yoshida H, Terachi T, Sato K. Mitochondrial genome sequences from wild and cultivated barley (*Hordeum vulgare*). BMC Genomics. 2016;17:824. https://doi.org/10.1186/s12864-016-3159-3.

Hollingsworth PM, Graham SW, Little DP. Choosing and using a plant DNA barcode. PLoS One. 2011;6:e19254. https://doi.org/10.1371/journal.pone.0019254.

Hollingsworth PM, Li D-Z, van der Bank, M, Twyford AD. Telling plant species apart with DNA: from barcodes to genomes. Philos Trans R Soc B. 2016;371:20150338. https://doi.org/10.1098/rstb.2015.0338.

Hoveka LN, van der Bank, M, Boatwright JS, Bezeng BS, Yessoufou K. The noncoding *trn*H-*psb*A spacer, as an effective DNA barcode for aquatic freshwater plants, reveals prohibited invasive species in aquarium trade in South Africa. S Afr J Bot. 2016;102:208–16. https://doi.org/10.1016/j.sajb.2015.06.014.

Jansen RK, Ruhlman TA. Plastid genomes of seed plants. In: Bock R, Knoop V, editors. Genomics of chloroplasts and mitochondria. Dordrecht: Springer International Publishing AG, part of Springer Nature; 2012. p. 103–26.

Jansen RK, Raubeson LA, Boore JL, dePamphilis CW, Chumley TW, Haberle RC, et al. Methods for obtaining and analyzing whole chloroplast genome sequences. Meth Enzymol. 2005;395:348–84. https://doi.org/10.1016/S0076-6879(05)95020-9.

Johne R, Müller H, Rector A, van Ranst M, Stevens H. Rolling-circle amplification of viral DNA genomes using *phi29* polymerase. Trends Microbiol. 2009;17:205–11. https://doi.org/10.1016/j.tim.2009.02.004.

Jørgensen T, Haile J, Möller P, Andreev A, Boessenkool S, Rasmussen M, et al. A comparative study of ancient sedimentary DNA, pollen and macrofossils from permafrost sediments of northern Siberia reveals long-term vegetational stability. Mol Ecol. 2012;21:1989–2003. https://doi.org/10.1111/j.1365-294X.2011.05287.x.

Jung H, Winefield C, Bombarely A, Prentis P, Waterhouse P. Tools and strategies for long-read sequencing and *de novo* assembly of plant genomes. Trends Plant Sci. 2019;24:700–24. https://doi.org/10.1016/j.tplants.2019.05.003.

Kane N, Sveinsson S, Dempewolf H, Yang JY, Zhang D, Engels JMM, et al. Ultra-barcoding in cacao (*Theobroma* spp.; Malvaceae) using whole chloroplast genomes and nuclear ribosomal DNA. Am J Bot. 2012;99:320–9. https://doi.org/10.3732/ajb.1100570.

Kim B, Kim K, Yang T-J, Kim S. Completion of the mitochondrial genome sequence of onion (*Allium cepa* L.) containing the CMS-S male-sterile cytoplasm and identification of an independent event of the $ccmF_N$ gene split. Curr Genet. 2016;62:873–85. https://doi.org/10.1007/s00294-016-0595-1.

Kistler L, Montenegro A, Smith BD, Gifford JA, Green RE, Newsom LA, et al. Transoceanic drift and the domestication of African bottle gourds in the Americas. Proc Natl Acad Sci U S A. 2014;111:2937–41. https://doi.org/10.1073/pnas.1318678111.

Kubo T, Newton KJ. Angiosperm mitochondrial genomes and mutations. Mitochondrion. 2008;8:5–14. https://doi.org/10.1016/j.mito.2007.10.006.

Kumar RA, Oldenburg DJ, Bendich AJ. Changes in DNA damage, molecular integrity, and copy number for plastid DNA and mitochondrial DNA during maize development. J Exp Bot. 2014;65:6425–39. https://doi.org/10.1093/jxb/eru359.

Lang BF, Burger G. Purification of mitochondrial and plastid DNA. Nat Protoc. 2007;2:652–60. https://doi.org/10.1038/nprot.2007.58.

Le Gall L, Saunders GW. DNA barcoding is a powerful tool to uncover algal diversity: a case study of the Phyllophoraceae (Gigartinales, Rhodophyta) in the Canadian flora. J Phycol. 2010;46:374–89. https://doi.org/10.1111/j.1529-8817.2010.00807.x.

Leister D. Origin, evolution and genetic effects of nuclear insertions of organelle DNA. Trends Genet. 2005;21:655–63. https://doi.org/10.1016/j.tig.2005.09.004.

Lemmon EM, Lemmon AR. High-throughput genomic data in systematics and phylogenetics. Annu Rev Ecol Evol Syst. 2013;44:99–121. https://doi.org/10.1146/annurev-ecolsys-110512-135822.

Lewis PO. A likelihood approach to estimating phylogeny from discrete morphological character data. Syst Biol. 2001;50:913–25. https://doi.org/10.1080/106351501753462876.

Li Z-M, Zheng X-M, Ge S. Genetic diversity and domestication history of African rice (*Oryza glaberrima*) as inferred from multiple gene sequences. Theor Appl Genet. 2011;123:21–31. https://doi.org/10.1007/s00122-011-1563-2.

Li C, Hofreiter M, Straube N, Corrigan S, Naylor GJP. Capturing protein-coding genes across highly divergent species. Biotechniques. 2013;54:321–6. https://doi.org/10.2144/000114039.

Li X, Yang Y, Henry RJ, Rossetto M, Wang Y, Chen S. Plant DNA barcoding: from gene to genome. Biol Rev. 2015;90:157–66. https://doi.org/10.1111/brv.12104.

Li L-F, Li Y-L, Jia Y, Caicedo AL, Olsen KM. Signatures of adaptation in the weedy rice genome. Nat Genet. 2017a;49:811–4. https://doi.org/10.1038/ng.3825.

Li P, Zhang S, Li F, Zhang S, Zhang H, Wang X, et al. A phylogenetic analysis of chloroplast genomes elucidates the relationships of the six economically important *Brassica* species comprising the triangle of U. Front Plant Sci. 2017b;8:111. https://doi.org/10.3389/fpls.2017.00111.

Liu C, Shi L, Zhu Y, Chen H, Zhang J, Lin X, et al. CpGAVAS, an integrated web server for the annotation, visualization, analysis, and GenBank submission of completely sequenced chloroplast genome sequences. BMC Genomics. 2012;13:715. https://doi.org/10.1186/1471-2164-13-715.

Lonardi S, Mirebrahim H, Wanamaker S, Alpert M, Ciardo G, Duma D, et al. When less is more: 'slicing' sequencing data improves read decoding accuracy and *de novo* assembly quality. Bioinformatics. 2015;31:2972–80. https://doi.org/10.1093/bioinformatics/btv311.

Londo JP, Chiang Y-C, Hung K-H, Chiang T-Y, Schaal BA. Phylogeography of Asian wild rice, *Oryza rufipogon*, reveals multiple independent domestications of cultivated rice, *Oryza sativa*. Proc Natl Acad Sci U S A. 2006;103:9578–83. https://doi.org/10.1073/pnas.0603152103.

Luo R, Liu B, Xie Y, Li Z, Huang W, Yuan J, et al. SOAPdenovo2: an empirically improved memory-efficient short-read *de novo* assembler. GigaScience. 2012;1:18. https://doi.org/10.1186/2047-217X-1-18.

Mackenzie S. Higher plant mitochondria. Plant Cell. 1999;11:571–86. https://doi.org/10.1105/tpc.11.4.571.

Magwé-Tindo J, Wieringa JJ, Sonké B, Zapfack L, Vigouroux Y, Couvreur TLP, et al. Guinea yam (*Dioscorea* spp., Dioscoreaceae) wild relatives identified using whole plastome phylogenetic analyses. Taxon. 2018;67:905–15. https://doi.org/10.12705/675.4.

Mariac C, Scarcelli N, Pouzadou J, Barnaud A, Billot C, Faye A, et al. Cost-effective enrichment hybridization capture of chloroplast genomes at deep multiplexing levels for population genetics and phylogeography studies. Mol Ecol Resour. 2014;14:1103–13. https://doi.org/10.1111/1755-0998.12258.

Martin W, Herrmann RG. Gene transfer from organelles to the nucleus: how much, what happens, and why? Plant Physiol. 1998;118:9–17. https://doi.org/10.1104/pp.118.1.9.

McCauley DE. The use of chloroplast DNA polymorphism in studies of gene flow in plants. Trends Ecol Evol. 1995;10:198–202. https://doi.org/10.1016/S0169-5347(00)89052-7.

McCauley DE. The relative contributions of seed and pollen movement to the local genetic structure of *Silene alba*. J Hered. 1997;88:257–63. https://doi.org/10.1093/oxfordjournals.jhered.a023103.

Mereschkowsky C. Über Natur und Ursprung der Chromatophoren im Pflanzenreiche. Biol Zent. 1905;25:593–604.

Mi H, Muruganujan A, Thomas PD. PANTHER in 2013: modeling the evolution of gene function, and other gene attributes, in the context of phylogenetic trees. Nucleic Acids Res. 2013;41: D377–86. https://doi.org/10.1093/nar/gks1118.

Michalovova M, Vyskot B, Kejnovsky E. Analysis of plastid and mitochondrial DNA insertions in the nucleus (NUPTs and NUMTs) of six plant species: size, relative age and chromosomal localization. Heredity. 2013;111:314–20. https://doi.org/10.1038/hdy.2013.51.

Middleton CP, Senerchia N, Stein N, Akhunov ED, Keller B, Wicker T, et al. Sequencing of chloroplast genomes from wheat, barley, rye and their relatives provides a detailed insight into the evolution of the Triticeae tribe. PLoS One. 2014;9:e85761. https://doi.org/10.1371/journal.pone.0085761.

Morley SA, Nielse BL. Plant mitochondrial DNA. Front Biosci. 2017;22:1023–32. https://doi.org/10.2741/4531.

Morlon H. Phylogenetic approaches for studying diversification. Ecol Lett. 2014;17:508–25. https://doi.org/10.1111/ele.12251.

Muñoz-Rodríguez P, Carruthers T, Wood JRI, Williams BRM, Weitemier K, Kronmiller B, et al. Reconciling conflicting phylogenies in the origin of sweet potato and dispersal to Polynesia. Curr Biol. 2018;28:1246–1256.e12. https://doi.org/10.1016/j.cub.2018.03.020.

Nielsen R, Paul JS, Albrechtsen A, Song YS. Genotype and SNP calling from next-generation sequencing data. Nat Rev Genet. 2011;12:443–51. https://doi.org/10.1038/nrg2986.

Nikiforova SV, Cavalieri D, Velasco R, Goremykin V. Phylogenetic analysis of 47 chloroplast genomes clarifies the contribution of wild species to the domesticated apple maternal line. Mol Biol Evol. 2013;30:1751–60. https://doi.org/10.1093/molbev/mst092.

Njuguna W, Liston A, Cronn R, Ashman T-L, Bassil N. Insights into phylogeny, sex function and age of *Fragaria* based on whole chloroplast genome sequencing. Mol Phylogenet Evol. 2013;66:17–29. https://doi.org/10.1016/j.ympev.2012.08.026.

Nock CJ, Waters DLE, Edwards MA, Bowen SG, Rice N, Cordeiro GM, et al. Chloroplast genome sequences from total DNA for plant identification. Plant Biotechnol J. 2011;9:328–33. https://doi.org/10.1111/j.1467-7652.2010.00558.x.

Nock CJ, Hardner CM, Montenegro JD, Ahmad Termizi AA, Hayashi S, Playford J, et al. Wild origins of macadamia domestication identified through intraspecific chloroplast genome sequencing. Front Plant Sci. 2019;10:334. https://doi.org/10.3389/fpls.2019.00334.

Notsu Y, Masood S, Nishikawa T, Kubo N, Akiduki G, Nakazono M, et al. The complete sequence of the rice (*Oryza sativa* L.) mitochondrial genome: frequent DNA sequence acquisition and loss during the evolution of flowering plants. Mol Genet Genomics. 2002;268:434–45. https://doi.org/10.1007/s00438-002-0767-1.

Oddou-Muratorio S, Petit RJ, Le Guerroue B, Guesnet D, Demesure B. Pollen- versus seed-mediated gene flow in a scattered forest tree species. Evolution. 2001;55:1123. https://doi.org/10.1554/0014-3820(2001)055[1123:PVSMGF]2.0.CO;2.

Oldenburg DJ, Bendich AJ. The linear plastid chromosomes of maize: terminal sequences, structures, and implications for DNA replication. Curr Genet. 2016;62:431–42. https://doi.org/10.1007/s00294-015-0548-0.

Olmstead RG, Bedoya AM. Whole genomes: the holy grail. A commentary on: molecular phylogenomics of the tribe Shoreeae (Dipterocarpaceae) using whole plastidgenomes. Ann Bot. 2019;123:iv–v. https://doi.org/10.1093/aob/mcz055.

Park H-S, Jayakodi M, Lee SH, Jeon J-H, Lee H-O, Park JY, et al. Mitochondrial plastid DNA can cause DNA barcoding paradox in plants. Sci Rep. 2020;10:6112. https://doi.org/10.1038/s41598-020-63233-y.

Parks M, Cronn R, Liston A. Increasing phylogenetic resolution at low taxonomic levels using massively parallel sequencing of chloroplast genomes. BMC Biol. 2009;7:84. https://doi.org/10.1186/1741-7007-7-84.

Pedersen MW, Ginolhac A, Orlando L, Olsen J, Andersen K, Holm J, et al. A comparative study of ancient environmental DNA to pollen and macrofossils from lake sediments reveals taxonomic overlap and additional plant taxa. Quaternary Sci Rev. 2013;75:161–8. https://doi.org/10.1016/j.quascirev.2013.06.006.

Petit RJ, Vendramin GG. Plant phylogeography based on organelle genes: an introduction. In: Weiss S, Ferrand N, editors. Phylogeography of Southern European Refugia. Dordrecht: Springer International Publishing AG, part of Springer Nature; 2007. p. 23–97.

Petit RJ, Brewer S, Bordács S, Burg K, Cheddadi R, Coart E, et al. Identification of refugia and post-glacial colonisation routes of European white oaks based on chloroplast DNA and fossil pollen evidence. For Ecol Manage. 2002;156:49–74. https://doi.org/10.1016/S0378-1127(01)00634-X.

Pollmann B, Jacomet S, Schlumbaum A. Morphological and genetic studies of waterlogged *Prunus* species from the Roman *vicus Tasgetium* (Eschenz, Switzerland). J Archaeol Sci. 2005;32:1471–80. https://doi.org/10.1016/j.jas.2005.04.002.

Pool JE, Hellmann I, Jensen JD, Nielsen R. Population genetic inference from genomic sequence variation. Genome Res. 2010;20:291–300. https://doi.org/10.1101/gr.079509.108.

Preuten T, Cincu E, Fuchs J, Zoschke R, Liere K, Börner T. Fewer genes than organelles: extremely low and variable gene copy numbers in mitochondria of somatic plant cells: gene copy numbers in mitochondria. Plant J. 2010;64:948–59. https://doi.org/10.1111/j.1365-313X.2010.04389.x.

Prince LM. Plastid primers for angiosperm phylogenetics and phylogeography. Appl Plant Sci. 2015;3:1400085. https://doi.org/10.3732/apps.1400085.

Prosser SWJ, Hebert PDN. Rapid identification of the botanical and entomological sources of honey using DNA metabarcoding. Food Chem. 2017;214:183–91. https://doi.org/10.1016/j.foodchem.2016.07.077.

Provan J, Corbett G, Waugh R, McNicol JW, Morgante M, Powell W. DNA fingerprints of rice (*Oryza sativa*) obtained from hypervariable chloroplast simple sequence repeats. Proc R Soc Lond B Biol Sci. 1996;263:1275–81. https://doi.org/10.1098/rspb.1996.0187.

Race HL, Herrmann RG, Martin W. Why have organelles retained genomes? Trends Genet. 1999;15:364–70. https://doi.org/10.1016/S0168-9525(99)01766-7.

Richly E, Leister D. NUPTs in sequenced eukaryotes and their genomic organization in relation to NUMTs. Mol Biol Evol. 2004;21:1972–80. https://doi.org/10.1093/molbev/msh210.

Sanderson MJ, Driskell AC. The challenge of constructing large phylogenetic trees. Trends Plant Sci. 2003;8:374–9. https://doi.org/10.1016/S1360-1385(03)00165-1.

Scarcelli N, Barnaud A, Eiserhardt W, Treier UA, Seveno M, d'Anfray A, et al. A set of 100 chloroplast DNA primer pairs to study population genetics and phylogeny in monocotyledons. PLoS One. 2011;6:e19954. https://doi.org/10.1371/journal.pone.0019954.

Scarcelli N, Mariac C, Couvreur TLP, Faye A, Richard D, Sabot F, et al. Intra-individual polymorphism in chloroplasts from NGS data: where does it come from and how to handle it? Mol Ecol Resour. 2016;16:434–45. https://doi.org/10.1111/1755-0998.12462.

Scarcelli N, Chaïr H, Causse S, Vesta R, Couvreur TLP, Vigouroux Y. Crop wild relative conservation: wild yams are not that wild. Biol Conserv. 2017;210:325–33. https://doi.org/10.1016/j.biocon.2017.05.001.

Shaw J, Lickey EB, Beck JT, Farmer SB, Liu W, Miller J, et al. The tortoise and the hare II: relative utility of 21 noncoding chloroplast DNA sequences for phylogenetic analysis. Am J Bot. 2005;92:142–66. https://doi.org/10.3732/ajb.92.1.142.

Shaw J, Lickey EB, Schilling EE, Small RL. Comparison of whole chloroplast genome sequences to choose noncoding regions for phylogenetic studies in angiosperms: the tortoise and the hare III. Am J Bot. 2007;94:275–88. https://doi.org/10.3732/ajb.94.3.275.

Sherman-Broyles S, Bombarely A, Grimwood J, Schmutz J, Doyle J. Complete plastome sequences from *Glycine syndetika* and six additional perennial wild relatives of soybean. G3. 2014;4:2023–33. https://doi.org/10.1534/g3.114.012690.

Shinozaki K, Ohme M, Tanaka M, Wakasugi T, Hayshida N, Matsubayasha T, et al. The complete nucleotide sequence of the tobacco chloroplast genome. Plant Mol Biol Rep. 1986;4:111–48. https://doi.org/10.1007/BF02669253.

Song H, Buhay JE, Whiting MF, Crandall KA. Many species in one: DNA barcoding overestimates the number of species when nuclear mitochondrial pseudogenes are coamplified. Proc Natl Acad Sci U S A. 2008;105:13486–91. https://doi.org/10.1073/pnas.0803076105.

Soorni A, Haak D, Zaitlin D, Bombarely A. Organelle_PBA, a pipeline for assembling chloroplast and mitochondrial genomes from PacBio DNA sequencing data. BMC Genomics. 2017;18:49. https://doi.org/10.1186/s12864-016-3412-9.

Sousa V, Hey J. Understanding the origin of species with genome-scale data: modelling gene flow. Nat Rev Genet. 2013;14:404–14. https://doi.org/10.1038/nrg3446.

Stamatakis A, Aberer AJ. Novel parallelization schemes for large-scale likelihood-based phylogenetic inference. In: Presented at the 2013 IEEE 27th international symposium on parallel and distributed processing, IEEE, Cambridge. 2013. p. 1195–204. https://doi.org/10.1109/IPDPS.2013.70

Stamatakis A, Aberer AJ, Goll C, Smith SA, Berger SA, Izquierdo-Carrasco F. RAxML-light: a tool for computing terabyte phylogenies. Bioinformatics. 2012;28:2064–6. https://doi.org/10.1093/bioinformatics/bts309.

Stegemann S, Keuthe M, Greiner S, Bock R. Horizontal transfer of chloroplast genomes between plant species. Proc Natl Acad Sci U S A. 2012;109:2434–8. https://doi.org/10.1073/pnas.1114076109.

Straub SCK, Parks M, Weitemier K, Fishbein M, Cronn RC, Liston A. Navigating the tip of the genomic iceberg: next-generation sequencing for plant systematics. Am J Bot. 2012;99:349–64. https://doi.org/10.3732/ajb.1100335.

Stull GW, Moore MJ, Mandala VS, Douglas NA, Kates H-R, Qi X, et al. A targeted enrichment strategy for massively parallel sequencing of Angiosperm plastid genomes. Appl Plant Sci. 2013;1:1200497. https://doi.org/10.3732/apps.1200497.

Taberlet P, Gielly L, Pautou G, Bouvet J. Universal primers for amplification of three non-coding regions of chloroplast DNA. Plant Mol Biol. 1991;17:1105–9. https://doi.org/10.1007/BF00037152.

Taberlet P, Coissac E, Pompanon F, Brochmann C, Willerslev E. Towards next-generation biodiversity assessment using DNA metabarcoding. Mol Ecol. 2012;21:2045–50. https://doi.org/10.1111/j.1365-294X.2012.05470.x.

Takamatsu T, Baslam M, Inomata T, Oikawa K, Itoh K, Ohnishi T, et al. Optimized method of extracting rice chloroplast DNA for high-quality plastome resequencing and *de novo* assembly. Front Plant Sci. 2018;9:266. https://doi.org/10.3389/fpls.2018.00266.

The Angiosperm Phylogeny Group. An update of the Angiosperm Phylogeny Group classification for the orders and families of flowering plants: APG IV. Bot J Linn Soc. 2016;181:1–20. https://doi.org/10.1111/boj.12385.

Thomson MJ, Septiningsih EM, Suwardjo F, Santoso TJ, Silitonga TS, McCouch SR. Genetic diversity analysis of traditional and improved Indonesian rice (*Oryza sativa* L.) germplasm using microsatellite markers. Theor Appl Genet. 2007;114:559–68. https://doi.org/10.1007/s00122-006-0457-1.

Tong W, Kim T-S, Park Y-J. Rice chloroplast genome variation architecture and phylogenetic dissection in diverse *Oryza* species assessed by whole-genome resequencing. Rice. 2016;9:57. https://doi.org/10.1186/s12284-016-0129-y.

Twyford AD, Ness RW. Strategies for complete plastid genome sequencing. Mol Ecol Resour. 2017;17:858–68. https://doi.org/10.1111/1755-0998.12626.

Uribe-Convers S, Duke JR, Moore MJ, Tank DC. A long PCR-based approach for DNA enrichment prior to next-generation sequencing for systematic studies. Appl Plant Sci. 2014;2:1300063. https://doi.org/10.3732/apps.1300063.

Valentini A, Pompanon F, Taberlet P. DNA barcoding for ecologists. Trends Ecol Evol. 2009;24:110–7. https://doi.org/10.1016/j.tree.2008.09.011.

van Dijk EL, Auger H, Jaszczyszyn Y, Thermes C. Ten years of next-generation sequencing technology. Trends Genet. 2014;30:418–26. https://doi.org/10.1016/j.tig.2014.07.001.

Vaser R, Sović I, Nagarajan N, Šikić M. Fast and accurate *de novo* genome assembly from long uncorrected reads. Genome Res. 2017;27:737–46. https://doi.org/10.1101/gr.214270.116.

Wambugu PW, Brozynska M, Furtado A, Waters DL, Henry RJ. Relationships of wild and domesticated rices (*Oryza* AA genome species) based upon whole chloroplast genome sequences. Sci Rep. 2015;5:13957. https://doi.org/10.1038/srep13957.

Wang D, Rousseau-Gueutin M, Timmis JN. Plastid sequences contribute to some plant mitochondrial genes. Mol Biol Evol. 2012;29:1707–11. https://doi.org/10.1093/molbev/mss016.

Wang M, Yu Y, Haberer G, Marri PR, Fan C, Goicoechea JL, et al. The genome sequence of African rice (*Oryza glaberrima*) and evidence for independent domestication. Nat Genet. 2014;46:982–8. https://doi.org/10.1038/ng.3044.

Wang H, Vieira FG, Crawford JE, Chu C, Nielsen R. Asian wild rice is a hybrid swarm with extensive gene flow and feralization from domesticated rice. Genome Res. 2017;27:1029–38. https://doi.org/10.1101/gr.204800.116.

Wang W, Schalamun M, Morales-Suarez A, Kainer D, Schwessinger B, Lanfear R. Assembly of chloroplast genomes with long- and short-read data: a comparison of approaches using *Eucalyptus pauciflora* as a test case. BMC Genomics. 2018;19:977. https://doi.org/10.1186/s12864-018-5348-8.

Wolfe AD, Randle CP. Recombination, heteroplasmy, haplotype polymorphism, and paralogy in plastid genes: implications for plant molecular systematics. Syst Bot. 2004;29:1011–20. https://doi.org/10.1600/036364404042451008.

Wyman SK, Jansen RK, Boore JL. Automatic annotation of organellar genomes with DOGMA. Bioinformatics. 2004;20:3252–5. https://doi.org/10.1093/bioinformatics/bth352.

Xi Z, Liu L, Davis CC. The impact of missing data on species tree estimation. Mol Biol Evol. 2016;33:838–60. https://doi.org/10.1093/molbev/msv266.

Yang J-B, Li D-Z, Li H-T. Highly effective sequencing whole chloroplast genomes of angiosperms by nine novel universal primer pairs. Mol Ecol Resour. 2014;14:1024–31. https://doi.org/10.1111/1755-0998.12251.

Yigit E, Hernandez DI, Trujillo JT, Dimalanta E, Bailey CD. Genome and metagenome sequencing: using the Human methyl-binding domain to partition genomic DNA derived from plant tissues. Appl Plant Sci. 2014;2:1400064. https://doi.org/10.3732/apps.1400064.

Yoshida T, Furihata HY, Kawabe A. Patterns of genomic integration of nuclear chloroplast DNA fragments in plant species. DNA Res. 2014;21:127–40. https://doi.org/10.1093/dnares/dst045.

Zerbino DR, Birney E. Velvet: algorithms for *de novo* short read assembly using de Bruijn graphs. Genome Res. 2008;18:821–9. https://doi.org/10.1101/gr.074492.107.

Zhang T, Fang Y, Wang X, Deng X, Zhang X, Hu S, et al. The complete chloroplast and mitochondrial genome sequences of *Boea hygrometrica*: insights into the evolution of plant organellar genomes. PLoS One. 2012;7:e30531. https://doi.org/10.1371/journal.pone.0030531.

Coalescent Models of Demographic History: Application to Plant Domestication

Olivier François, Philippe Cubry, Concetta Burgarella, and Yves Vigouroux

Abstract A detailed understanding of the origins of domesticated species is important for many disciplines. Recent advances in this field have been made with the use of genome-wide polymorphisms and improved statistical methods. In this chapter, we review the most important developments of coalescent models for the inference of demographic history from genome-wide data. These methods include sequential Markovian models, range expansion models, and approximate Bayesian computation. We summarize the applications of the methods to some major cultivated cereals, including rice, maize, and millet. Then we discuss extensions of these methods that would better incorporate the interaction of demographic processes with selection and inclusion of gene flow with wild related species.

Keywords African and Asian rice · Approximate Bayesian computation · Bottleneck · Coalescent models · Demographic history · Domestication · Maize · Pearl millet · Sequential Markovian coalescent

---

O. François (✉)
Université Grenoble-Alpes, Centre National de la Recherche Scientifique, Grenoble, France
e-mail: Olivier.Francois@univ-grenoble-alpes.fr

P. Cubry · Y. Vigouroux
Université de Montpellier, Institut pour la Recherche et le Développement, Unité Diversité, Adaptation, Développement des plantes, Montpellier, France

C. Burgarella
Université de Montpellier, Institut pour la Recherche et le Développement, Unité Diversité, Adaptation, Développement des plantes, Montpellier, France

Human Evolution, Department of Organismal Biology, Uppsala University, Uppsala, Sweden

1 Introduction

Domestication of plants and animals has triggered a major transition in human history and allowed the development of human civilizations in different parts of the globe. Domestication is characterized as a complex process in which the human use of plant and animal species leads to morphological divergence of the domesticated taxa from their wild progenitor (Doebley et al. 2006; Meyer and Purugganan 2013). From an evolutionary perspective, domesticated species are considered ideal model systems for studying the impact of selection and explaining the rise of species differences (Darwin 1882). In anthropology, studying the origins of domesticated species provides insights on the historical and cultural contexts by which human societies become dependent on particular cultivated species. Where, when, and how many times domestication took place in human history have been questions of primary interest within a wide range of disciplines (Zeder 2015).

Reconstructing the history and geographic origins of cultivated species is now possible with the advent of genome-wide polymorphisms and accurate demographic inference methods (Meyer et al. 2012; Larson et al. 2014; Meyer and Purugganan 2013). A detailed understanding of domestication history indeed requires a large number of loci in conjunction with accurate estimation of population demography. Early work on demographic history used mean patterns of genetic diversity to fit a bottleneck model, a sharp but temporary reduction in the size of a population during domestication (Eyre-Walker et al. 1998). That modeling approach was later extended to include an explicit likelihood framework. Recently, investigators have used methods that incorporate more detailed information, such as the site frequency spectrum, recombination, and geographic information, to distinguish among different evolutionary models (Beichman et al. 2018). With genome-wide polymorphisms, the origins of domestication have been recently re-evaluated in the light of models of population genomics for several important crop species, including Asian and African rice, maize, pearl millet and yam (Huang et al. 2012; Meyer et al. 2016; Beissinger et al. 2016; Wang et al. 2017a; Burgarella et al. 2018; Scarcelli et al. 2019).

In this chapter, we survey the recent developments of coalescent models for the inference of demographic history from genome-wide data and their application to major cultivated cereals. The second section of this chapter provides a short background on coalescent theory and presents modern approaches to the inference of past demographic events. The third section reviews the results of coalescent models regarding demographic history and origins of four examples of plant species: Asian and African rice, maize, and pearl millet. The interpretation of the results is discussed in relation to the cost of domestication, wild species gene flow, and adaptation to new environments during agricultural range expansion. Future directions of research are then highlighted.

2 Coalescent Models and Demographic History

2.1 Background on the Coalescent

Introduced by Kingman (1982), the coalescent is a stochastic model for the genealogy of a sample of DNA sequences at a particular locus in the genome of a species. All the existing copies of this locus are related to each other and to a most recent common ancestor at random in the genealogy. Polymorphism at the locus is due to mutations that occur along the branches of the genealogical tree, and the frequency of each genetic variant is determined by the length of the branches that inherits the mutation (Rosenberg and Nordborg 2002). In the construction of the stochastic process, the rate at which two lineages track back to a common ancestor is inversely proportional to the effective population size at the time considered in the past. The coalescent model is an extension of classical population genetic models and can accommodate a wide variety of demographic scenarios, including population growth and bottlenecks, structured populations, and split or mixture events (Fig. 1).

Recombination can also be incorporated into the coalescent framework (Hein et al. 2004). The main effect of recombination is that it allows linked loci to have different genealogical trees. In the coalescent model, recombination may be viewed from a spatial (along the chromosome) rather than a temporal perspective. More precisely, the genealogy of a sample of sequences with recombination can be considered as a random process along the chromosomes, with each recombination event affecting a subset of the tree branches (Hein et al. 2004). Because the genealogies of unlinked loci are conditionally independent given the demographic history of the sample, inference methods based on genealogical processes with recombination can uncover past population history in detail and are essential to the study of recent domesticated crop species.

Coalescent models, which describe the genealogy of DNA sequences for one species, have been extended to multispecies coalescent models (Yang 2014). In multispecies coalescent models, gene trees are estimated simultaneously with a species tree as a means of estimating phylogenetic relationships (Edwards et al. 2016). A difficulty with domesticated species is that their recent origin, often earlier than 10,000 years, increases the likelihood that ancestral polymorphisms persist in domesticated taxa through incomplete lineage sorting. Incomplete lineage sorting leads to conflicting gene trees when multiple loci are analyzed independently, and data concatenation can generate misleading phylogenetic relationships (Degnan and Rosenberg 2009). Conflicting gene trees because of incomplete lineage sorting have been a problem for species delimitation, but phylogenetic analyses using the multispecies coalescent models are able to detect signals of species differentiation even before their gene trees are reciprocally monophyletic (Heled and Drummond 2009). Table 1 provides a summary of the main coalescent-based methods described in this chapter.

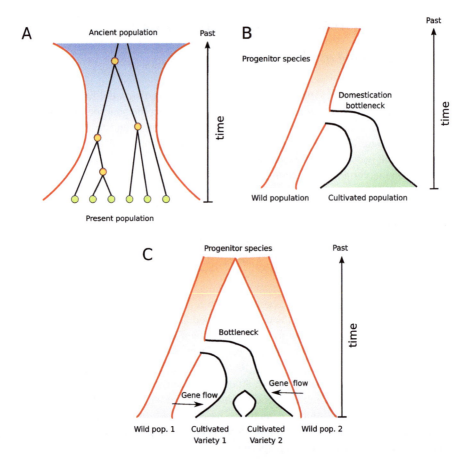

Fig. 1 Demographic models. (**a**) A simple bottleneck model. Most sampled lineages find their common ancestor at a date more recent than the bottleneck initiation, but a few lineages may escape the bottleneck and coalesce at more ancient dates. (**b**) Domestication from a wild progenitor species is associated with a genetic bottleneck and followed by population expansion of the domesticated species (African rice). (**c**) A more complex scenario in which domestication is accompanied by gene flow from wild relatives and leads to two domesticated subspecies (Asian rice)

2.2 Sequential Markovian Coalescent Models

Studying the coalescent with recombination as a process along the genome, Wiuf and Hein (1999) defined the *sequential coalescent* in which the genealogy at a given locus depends on the genealogies at adjacent loci. The sequential coalescent is a complex stochastic process, not easily amenable to statistical analysis (Wiuf and Hein 1999). Simpler stochastic processes – called sequentially Markovian coalescent (SMC) – were proposed as approximations of the original model (McVean and Cardin 2005; Spence et al. 2018). Under the SMC, the genome data could be viewed as coming from a hidden Markov model for which the genealogy of the sampled

Table 1 List of software used in demographic analyses

Program	Description	Reference
SMC		
SFS methods		
msmc	Estimates of ancestral effective population size for more than one diploid genome	Schiffels and Durbin (2014)
smc++	Inference of population size histories and split times in diverged populations	Terhorst et al. (2017)
∂a∂i	Computing the likelihood of a model based on the SFS and a diffusion equation	Gutenkunst et al. (2009)
Population splits		
Admixture		
TreeMix	Inference of population splits and mixtures in multiple populations	Pickrell and Pritchard (2012)
Admixture	Likelihood-based estimates of ancestry based on the *structure* model	Alexander et al. (2009)
snmf	Estimates of ancestry coefficients allowing departures from model assumptions	Frichot et al. (2014)
Monte Carlo		
Simulation		
ms	Coalescent sampler under a Wright-Fisher model of genetic variation	Hudson (2002)
msprime	A fast and efficient coalescent sampler for large sample sizes	Kelleher et al. (2016)
fastsimcoal	Coalescent framework to infer demographic parameters from SFS	Excoffier et al. (2013)
abc	Bayesian inference based on summary statistics and machine learning models	Blum and François (2010)

individuals at a given locus is the hidden variable. Because demography impacts the shape of genealogies and linkage disequilibrium, SMC methods are extremely powerful at filtering the underlying demographic process (Li and Durbin 2011; Schiffels and Durbin 2014).

Among the first SMC models, the pairwise sequentially Markovian coalescent (PSMC) used a hidden Markov model that identifies recombination events across a single diploid genome (Li and Durbin 2011). PSMC is based on coalescent theory according to which the rate of the inferred pairwise coalescent events at a given time is inversely proportional to effective population size. PSMC infers the time to the most recent common ancestor for each genomic segment. Then, based on the rate of coalescence events, it infers ancestral effective population size at a given time in the past. Under the PSMC, long runs of low heterozygosity correspond to recent coalescent events, and short runs of high heterozygosity correspond to more ancient coalescent events. Multiple SMC (MSMC) was proposed as an extension of PSMC to more than one diploid genome (Schiffels and Durbin 2014).

The PSMC and MSMC methods can accurately estimate changes in effective population size, such as population growth, population decline, and bottlenecks in a random mating population. Recent improvements of the SMC methods include models than can jointly infer population size histories and split times in diverged populations (SMC^{++}, Terhorst et al. (2017)), allowing analysis of larger sets of genomes than with MSMC. While SMC models are powerful approaches to the inference of demographic history, their results should nevertheless be interpreted cautiously. The outputs of SMC methods could be biased by the presence of migrant individuals in the target population (Mazet et al. 2016), and bottleneck inference might be confounded by gene flow from cryptic populations (Nielsen and Beaumont 2009).

2.3 Frequency-Based Methods

Frequency-based approaches estimate the demographic history of multiple populations from multidimensional single nucleotide polymorphism (SNP) frequency data, by using the data contained in the multi-population joint site frequency spectrum (SFS). The joint SFS is the joint distribution of allele frequencies of genetic variants from a genomic region sequenced in multiple individuals from each population. The method proposed by Gutenkunst et al. (2009) is based on a diffusion approximation of allele frequencies, implemented in the program $\partial a \partial i$. The program $\partial a \partial i$ calculates the likelihood of a demographic model given an observed SFS. It computes the expected SFS under various demographic scenarios by solving a Wright-Fisher diffusion equation describing the effects of mutation, drift, and migration. The likelihood of each model is computed based on the product of the likelihoods for each entry of the SFS. Following a similar approach, Excoffier et al. (2013) introduced a coalescent framework to infer demographic parameters from SFS computed on large genomic datasets. Their composite-likelihood approach extends $\partial a \partial i$ and allows studying evolutionary models of greater complexity than likelihood-based methods (Excoffier et al. 2013). Frequency-based approaches were also recently extended to incorporate linkage disequilibrium statistics (Boitard et al. 2016).

2.4 Admixture and Tree-Based Methods

Most studies of population genetic variation start with an ascertainment of population structure and admixture. A popular approach to studying population genetic structure is implemented in the computer program STRUCTURE (Pritchard et al. 2000). In STRUCTURE, the models suppose that the data originate from the admixture of K ancestral populations that may be unavailable to the study. In these models, the parameters of interest are the ancestry coefficients, also termed admixture proportions, computed for each individual in the sample. These coefficients are stored in a matrix, Q, in which each element, (q_{ik}), represents the proportion of individual i's genome that originates from the kth ancestral population. The program STRUCTURE implements a variant of the admixture model with correlated allele frequencies (Pritchard et al. 2000). In this model extension, allele frequencies in the K populations have drifted away from frequencies in an ancestral population according to a starlike tree. With large dimensional SNP data, fast and efficient versions of structure have been proposed, including the program ADMIXTURE, which is based on the same equilibrium assumptions as STRUCTURE (Alexander et al. 2009), and nonparametric approaches that are appropriate to deal with large levels of inbreeding or geographic proximity in plants (Frichot et al. 2014; Caye et al. 2018). Like principal component analysis (Patterson et al. 2006), the nonparametric approaches make no prior assumption on the evolutionary processes that have generated the data. Genetic introgression and admixture with related species during the domestication process can also be tested statistically with f-statistics or with the ABBA-ABBA statistic (Patterson et al. 2012; Durand et al. 2011). The most interesting f-statistics are f_3 and f_4, which is similar to ABBA-ABBA (Peter 2016). Under a population tree model for the derived (domesticated) and progenitor populations plus a related species, the f_3 statistic corresponds to the length of the external branch connecting the derived species lineage to the tree. Negative values of f_3 indicate departure from treelike evolution caused by an admixture event. With an additional out-group population, the f_4 statistic corresponds to the length of branches shared between the derived and progenitor populations on one side and the related and out-group species on the other side. Without admixture, the expected value of f_4 is zero, representing a basis for null hypothesis testing with the f_4 test (Peter 2016).

Using similar ideas, Pickrell and Pritchard (2012) developed *TreeMix*, a statistical model for inferring the patterns of population splits and mixtures in multiple populations. In the TreeMix model, the sampled populations in a species are related to their common ancestor through a graph of ancestral populations. Using genome-wide allele frequency data and a Brownian approximation to genetic drift, a covariance matrix for the observed allele frequencies is estimated and used to infer the structure of the admixture graph. The covariance matrix is formulated with statistics that are closely connected to the f-statistics, and both approaches are based on the same underlying assumption of having a constant population size and genetic drift along all evolutionary paths.

2.5 Approximate Bayesian Computation

One of the most widespread uses of coalescent models is as simulation tools (Hudson 2002; Kelleher et al. 2016). Using coalescent trees, it is possible to simulate samples from almost any demographic models. Compared with population genetic simulations which run forwards in time, coalescent simulations that run backwards in time are much faster and more efficient. The extensive use of simulations has allowed researchers to perform inference in coalescent models without resorting to likelihood calculations. This strategy has given rise to approximate Bayesian computation (ABC) approaches that bypass exact likelihood calculations by using summary statistics (Beaumont 2010; Csilléry et al. 2010). Summary statistics are values calculated from the data to represent the maximum amount of information on model parameters. The principle of ABC relies on the simulation of large numbers of datasets using parameters drawn from a prior distribution. Summary statistics are then calculated for each simulated sample and compared with the values for the observed sample. Parameters that have generated summary statistics close enough to the observed data are retained to form an approximate sample from the posterior distribution (Blum and François 2010).

In an early study of maize domestication, Eyre-Walker et al. (1998) investigated a bottleneck model by using coalescent simulations and a method similar to ABC (Fig. 1). ABC is particularly useful for inference on the complex demographic processes of plant domestication and extends the collection of models that can be tested by the researchers with exact likelihood calculations. An important aspect of ABC methods is their ability to incorporate more realistic characteristics of domestication processes, such as gene flow with wild species, isolation-by-distance patterns, or recurrent founder effects that occur during range expansion (Edmonds et al. 2004; Excoffier et al. 2009). Classical measures in population genetics, such as estimates of nucleotide diversity and fixation indices, are most often used as summary statistics in ABC analyses. With genomic data, these measures are replaced by direct estimates of SFS which can be more informative about the demographic processes that led to the observed sequences (François et al. 2008; Cubry et al. 2018). ABC can also infer the geographical origin of a domesticated species by incorporating spatially explicit models and summary statistics reflecting genetic diversity in the geographic sample (Box 1, Fig. 2).

Box 1 Spatial ABC

Range expansion model. To evaluate the geographic origin of African rice, Cubry et al. (2018) applied an ABC approach based on simulations of spatially explicit coalescent models. In their study, the demography of the cultivated species was simulated in a nonequilibrium stepping-stone model defined on a lattice of regularly spaced populations covering the African continent. Geographic information was encoded into a resistance value for each local population, which sent migrants to their neighboring demes at rate m inversely proportional to the resistance value. Once a site was colonized, its population increased according to a logistic growth model with rate r and carrying capacity C (Currat et al. 2004). The model resulted in a wave-of-advance, with speed depending on the model parameters, r, C, and m. Backward in time, the simulations were used to generate gene genealogies for samples taken at observed sample sites.

Summary statistics. To provide statistical estimates for the geographic origin of expansion, uniform prior distributions were assumed. Histograms of

(continued)

Box 1 (continued)
posterior distribution were obtained for the latitude and longitude of the origin by using ABC and machine learning models. One key point of ABC is the choice of informative summary statistics. The relative values of parameters related to migration and expansion (m and r) were evaluated on the basis of the SFS. Cubry et al. (2017) proposed to estimate the geographic origin by using the spatial density of singletons, defined as genetic variants represented only once in the whole sample. Singletons carry information on the external branch lengths of coalescent genealogies and provide local estimates of genetic diversity. The distribution of singletons in space can be summarized as a two-dimensional histogram. The use of this histogram in ABC approaches allows accurate estimation of the origin of expansion (Fig. 2).

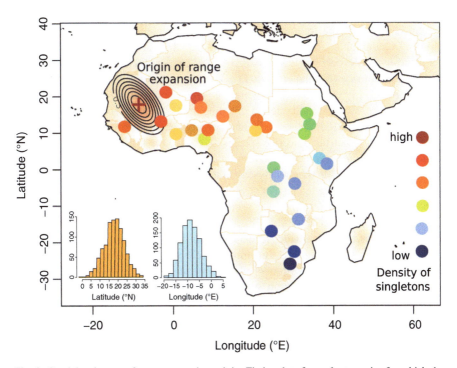

Fig. 2 Spatial estimates of range expansion origin. Fictive data for a plant species for which the density of singletons decreases from a West African origin. A Bayesian analysis using ABC provides estimates for the onset of range expansion (contour lines of posterior density and histograms). The red cross corresponds to the origin of expansion estimated from a spatial coalescent model

3 Inferences on Origins of Cultivated Plants

In this section, we summarize results obtained from modern coalescent methods describing the demographic history of four important cultivated plants: Asian and African rice, maize, and pearl millet.

3.1 Asian Rice

Asian rice, *Oryza sativa*, is one of world's oldest and most important crop species. Asian rice is believed to have been domesticated 9,000 years ago, but the details of its domestication process are still debated. A single-origin model suggests that two main subspecies of Asian rice, *indica* and *japonica*, were domesticated from the wild rice *O. rufipogon*. In contrast, a multiple independent domestication model proposes that these two major rice types were domesticated separately and in different parts of the geographic range of the wild species (Sang and Ge 2007).

Molina et al. (2011) examined the evolutionary history of domesticated rice by resequencing 630 gene fragments from a set of wild and domesticated rice accessions. Demographic modeling based on the diffusion approach $\partial a \partial i$ and multispecies coalescent models provided strong support for a single domestication origin of Asian rice (Molina et al. 2011). The origin of domestication was dated at around 8,200–13,500 years BP, which is consistent with archaeological data suggesting rice was first cultivated at this time in the Yangtze valley of China.

Huang et al. (2012) generated genome sequences from 446 geographically diverse accessions of the wild rice species *O. rufipogon* and from 1,083 cultivated *indica* and *japonica* varieties to construct a comprehensive map of rice genome variation (Huang et al. 2012). Analyses of the genome-wide patterns conducted with phylogenetic methods revealed that *O. sativa japonica* was first domesticated from a specific population of *O. rufipogon* around the middle area of the Pearl River in southern China and that *O. sativa indica* was subsequently developed from crosses between *japonica* and local wild rice when the initial cultivars spread into Southeast and South Asia. By examining signatures of selection in the genomes of different cultivated rice, an alternative model of domestication proposed three independent domestication events in different parts of Asia (Civáň et al. 2015). This study identified wild populations in southern China and the Yangtze valley as the source of the *japonica* gene pool and populations in Indochina and the Brahmaputra valley as the source of the *indica* gene pool. A third variety *O. sativa* ssp. *aus* originated in India. According to Civáň et al. (2015), the tropical and temperate versions of *japonica* are later adaptations of one crop. This three-event hypothesis was re-evaluated by Choi et al. (2017), using a multispecies coalescent model. The multispecies coalescent model estimated that the first domesticated rice population to diverge from the ancestor *O. rufipogon* was *japonica* around 13,000 to 24,000 years BP. This date was two or three times older than the earliest archaeological date of domestication, but consistent with the recovery of ancestral populations after the Last Glacial Maximum and evidence for early wild rice management in China. The results supported a demographic model in which each variety had a separate origin, but in which de novo domestication occurred only once, in *japonica*. According to the results of ABBA-BABA tests, introgressive hybridization from early *japonica* to proto-varieties of *indica* and *aus* led to *indica* and *aus* (Fig. 1c).

3.2 African Rice

The cultivation of rice in Africa dates back more than 3,000 years. The African rice, *Oryza glaberrima* Steud., is an African species of rice that was independently domesticated from the wild progenitor *Oryza barthii* A. Chev about 6,000 years after the domestication of Asian rice. The geographical origins of crop domestication in Africa have been seen through centric and noncentric perspectives. Portères (1962) was the first to propose that *O. glaberrima* was domesticated in the inland delta of the upper Niger River and diffused to two secondary centers of diversification, one along the Senegambian coast and the other in the interior Guinea Highlands. In contrast, Harlan (1971) proposed a noncentric model of domestication for most African plants. Wang et al. (2014) resequenced 20 *O. glaberrima* and 94 *O. barthii* accessions. Population structure estimated by the program ADMIXTURE partitioned the ancestral population samples into four subgroups and an admixed group. The phylogenetic relationships of *O. glaberrima* accessions with its ancestor supported the hypothesis that *O. glaberrima* was domesticated in a single region along the Niger River (Wang et al. 2014).

Meyer et al. (2016) used paired-end Illumina sequencing to resequence the genomes of 93 traditional *O. glaberrima* landraces from West and Central sub-Saharan Africa. The authors used TreeMix to examine the topology of relationships and migration history among populations of *O. glaberrima*. Using the wild progenitor *O. barthii* as an out-group, they observed an older split between coastal and inland populations and a more recent separation of northern and southern populations. To examine whether there was a domestication bottleneck in *O. glaberrima*, Meyer et al. (2016) applied a PSMC model on two haplotypes. The PSMC profiles indicated a reduction in effective population size starting around 13,000–15,000 years ago ($N_e \approx 60,000$) and ending approximately 3,500 years ago ($N_e \approx 3,000$). In contrast, no severe bottleneck was evident in wild *O. barthii* (Meyer et al. 2016). The PSMC results indicated that the minimum plateau for the African rice bottleneck occurred close to the dates corresponding to the earliest archaeological evidence.

Cubry et al. (2018) sequenced 163 cultivated *O. glaberrima* and 83 *O. barthii* individual genomes collected in the Sahel area and in East Africa. This study performed a spatially explicit analysis using ABC, which inferred the geographical origin of domestication in the Inner Niger Delta (Box 1). This result, alongside with the origins of pearl millet and yam, supported the hypothesis that the Inner Niger Delta region was a major cradle of African agriculture (Burgarella et al. 2018; Scarcelli et al. 2019). The demographic history of African rice was also reconstructed by using several SMC methods. The results indicated that domestication was preceded by a sharp decline of wild populations during the drying of the Sahara around 10,000 years ago. The effective population size of *O. glaberrima* populations increased 2,000 years ago, and this increase was followed by a sharp decline that coincided with the introduction of Asian rice in Africa. Overall, the genomic data revealed a complex history of African rice domestication influenced by important climatic changes in the Sahara, by the expansion of African agricultural society, and by recent replacement of African rice by Asian rice.

3.3 Maize

Maize, *Zea mays* ssp. *mays*, was domesticated in a narrow region of southwest Mexico from the wild plant teosinte (*Zea mays* ssp. *parviglumis*). Archaeological evidence suggests that after initial domestication, maize spread across the Americas, reaching the southwestern USA by approximately 4,500 years BP and coastal South America as early as 6,700 years BP. Early analyses based on multiallelic markers proposed that maize arose from a single domestication event in southern Mexico about 9,000 years ago and experienced a bottleneck that removed a substantial proportion of the diversity found in its progenitor (Matsuoka et al. 2002). Wright et al. (2005) studied SNP diversity in 774 gene fragments in a sample of 14 maize inbred lines and 16 inbred teosintes. Using ABC and coalescent simulations, the authors estimated the *bottleneck severity* – the ratio of the size of the bottlenecked population to the duration of the bottleneck. The SNP data were consistent with a domestication bottleneck of moderate size, dating the domestication event 2,800 years ago. Under this time scale, fewer than 3,500 individuals, or 10% of the teosinte population, contributed to the genetic diversity captured in the maize sample (Wright et al. 2005).

To investigate demography in maize and teosinte, Beissinger et al. (2016) analyzed data from 23 maize and 13 teosinte genomes. Their analysis used $\partial a \partial i$ and MSMC to estimate the parameters of the domestication bottleneck from a joint maize-teosinte SFS. Their coalescent approach modeled a teosinte population of constant effective size that, in the past, gave rise to a smaller maize population which grew exponentially to its present size. The model included gene flow between teosinte and maize populations. The maximum likelihood estimates for the ancestral effective population size were 123,000 teosinte individuals, and maize split from teosinte was estimated to occur around 15,000 generations in the past, with an initial size of only 5% of the ancestral effective population size (Beissinger et al. 2016).

Using high-depth resequencing data from 31 maize landraces spanning the pre-Columbian distribution of maize and 4 wild teosinte individuals, Wang et al. (2017a) estimated historical changes in effective population size of maize and

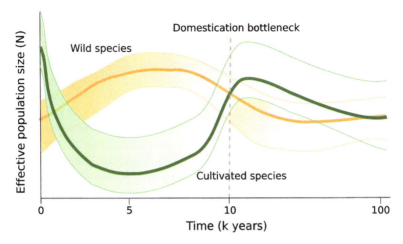

Fig. 3 Sequential Markovian coalescent analysis of past demography. Inspired by the maize data, SMC plots are shown for a domesticated species (green color) and its wild progenitor (orange color). Wild and domesticated populations diverged from one another approximately 10ky BP. The cultivated populations showed a decline in diversity followed by extremely rapid population growth

teosinte using the multiple sequentially Markovian coalescent (MSMC). Consistent with archaeological evidence, they found that the demographic histories of the maize populations diverged from one another approximately 10,000 years BP and that divergence from teosinte occurred much earlier (75,000 years BP). All maize populations showed a gradual decline in diversity concomitant with divergence from teosinte, but the slope became more pronounced around the time of domestication. The period of declining diversity continued until the recent past (1,100–2,400 years BP), and it was followed by extremely rapid population growth (Fig. 3). The result suggested that recovery from the domestication bottleneck postdated the expansion of maize across the Americas.

3.4 Pearl Millet

Pearl millet (*Cenchrus americanus (L.) Morrone* syn. *Pennisetum glaucum (L.) R. Br.*) is the oldest African cereal in the archaeological records and is associated with the dawn of agriculture in West Africa. It is today the most nutritious low-input staple cereal cultivated in the arid and low-fertility soils of Africa and Asia. The wild progenitor of pearl millet, *P. glaucum* ssp. *monodii (L.) R. Br.*, has a natural range that spans the Sahel region. Clotault et al. (2011) used a random set of 20 genes to model pearl millet domestication using approximate Bayesian computation. Their analysis showed that a model with exponential growth and gene flow from the wild species was well supported by their data. Under this model, the domestication date of pearl millet was estimated at around 4,800 years ago (Clotault et al. 2011).

Burgarella et al. (2018) analyzed whole-genome sequences of 221 accessions of wild forms and traditional varieties of pearl millet to infer the domestication origin for this plant. Using coalescent models of domestication bottleneck, population growth, and gene flow, these authors found that cultivated pearl millet originated in a wild population from central Sahel. To investigate hybridization with other groups of wild relatives during expansion and rule out incomplete lineage sorting, they employed *f*-statistics and detected signatures of admixture with sympatric wild populations in both western and eastern Sahel. Admixture events were confirmed with TreeMix. Estimates of the origin of domestication and the timing of the expansion of pearl millet cultivation were obtained from a spatially explicit coalescent model (Fig. 2). Calibrating ABC with dates from archaeological remains, the onset of diffusion of pearl millet agriculture was estimated around 4,650 years BP. The spatial model identified an origin corresponding to the Taoudeni Basin in Western Sahara at latitude higher than the current range of wild populations in central Sahel, an origin that matches the wetter climate which prevailed in the Sahara 6,000 years ago (Burgarella et al. 2018).

4 Interaction of Demography with Other Processes

The details of domestication scenarios are complex, and the interpretation of inference results may be obscured by the existence of processes not included in the demographic models. The main difficulties arise as domestication is an evolutionarily recent phenomenon, and most of the genealogical history at any locus is shared between a domesticate and its wild progenitor. In several studies, comparisons of alleles within and between domesticated and wild taxa revealed divergence times that greatly predate the origin of the cultivated form. Those estimates might reflect the time to most recent common ancestor of the species rather than the time of divergence of the domesticate.

4.1 Some Difficulties

Analyses from recent genome-wide resequencing have failed to reach a consensus on the number of domestication events of Asian rice. Whether rice was domesticated once and subsequent varieties were formed by introgression with different wild progenitors, or whether each variety was domesticated independently in different parts of Asia, is hotly debated. The debate mainly arose from two studies analyzing the same data but arriving at two different domestication scenarios: Huang et al. (2012) supported a single domestication event with introgression model, while Civáň et al. (2015) supported a multiple domestication model. Genetic differentiation modeling supported separate domestications followed by introgression at agronomically important loci, whereas the site frequency spectrum and phylogenetic analysis provided evidence for a single origin. The debate is still ongoing. Extensive gene flow from wild relatives may be one explanation for the continued controversy regarding the origins of the domesticated *indica* and *japonica* subspecies of Asian rice.

4.2 Domestication Bottlenecks

Domestication is typically accompanied by a population bottleneck, and the amplitude and duration before recovery of the bottleneck have been intensively studied for cultivated plants. For maize, Eyre-Walker et al. (1998) estimated that maize contains 75% of the variation found in its progenitor and reported that sequence diversity in maize can be explained by a bottleneck of short duration and very small size. With whole-genome estimates of the demographic history of maize, Beissinger et al. (2016) showed that maize was reduced to approximately 5% the population size of teosinte before it experienced rapid expansion post-domestication to population sizes much larger than its progenitor. The results produced an estimate for the timing of the bottleneck at 15,000 years BP, conflicting with archaeological estimates. The genetic split between populations likely preceded anatomical changes that can be identified in the archaeological record, but another explanation is inflation due to population structure, as the teosinte samples included populations diverged from those that gave rise to maize.

Sequentially Markovian coalescent methods that infer the effective population size suggest the possibility of long protracted extended human impacts on domesticated lineages prior to their purposeful cultivation. Genome-wide SMC demographic analyses conducted by Wang et al. (2017a) revealed that maize experienced more pronounced declines in effective population size and a prolonged population size reduction. This prolonged population size reduction in maize following the onset of domestication would be explained by repeated colonization bottlenecks during the spread of maize across the Americas. Genome-wide levels of heterozygosity in maize samples decreased with distance from the center of maize domestication in the Balsas River Basin and supported this idea.

In African rice, an SMC analysis of genome-wide variation inferred an early onset of the bottleneck and a prolonged decrease in effective population size (Cubry et al. 2018). The bottleneck could have resulted from regional climate change, or from a protracted period of low-intensity cultivation and management before full domestication, just after the growth of western African human population that occurred 4,000–5,000 years ago. This estimated slow decrease in effective population size contrasts with findings based on nonspatial ABC and on RNA-seq technology that estimated a severe bottleneck during domestication of African rice (Nabholz et al. 2014). Surveying ten gene sequences, a severe bottleneck was also inferred from population data during the domestication of Asian rice (Zhu et al. 2007). Whether domestication occurred quickly or slowly is an important question for most cultivated plants (Gaut et al. 2018). Yet evidence from various population genetic data has provided contradictory hypotheses, and evidence on temporal duration of the domestication event remains speculative for most species.

4.3 Cost of Domestication

Genetic bottlenecks during domestication are coupled with selection and limited recombination, which explain the reduction in effective population size. In turn, those evolutionary processes contribute to the empirical observation that domesticated crops have more deleterious alleles than their wild relatives. This observation is often called the *cost of domestication*, a hypothesis assuming that the process of domestication increases the number of deleterious variants in the genome (Lu et al. 2006; Moyers et al. 2017). The accumulation of deleterious mutations was observed in Asian and African rice (Lu et al. 2006; Liu et al. 2017; Nabholz et al. 2014), maize Wang et al. (2017a), and several other domesticated species (Moyers et al. 2017), resulting from the interplay of demography and selection (Wang et al. 2017a). Population demographic history exhibiting slow effective population size decline could exacerbate the accumulation of deleterious mutations (Fig. 4). But this accumulation also depends on the strength and duration of the domestication bottleneck (Kono et al. 2016; Gaut et al. 2018).

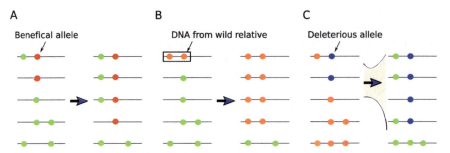

Fig. 4 Interactions of selection and demography during domestication. (**a**) Selective sweep: A beneficial trait was selected by early farmers during domestication, leading a genetic variant to near fixation in the population. (**b**) Adaptive introgression: A chunk of DNA received from a closely related taxon (orange color) was beneficial and increased in frequency during range expansion. (**c**) Domestication cost: A deleterious allele present in the progenitor population (orange) increased in frequency due to genetic drift during the domestication bottleneck

4.4 Genetic Introgression and Adaptation to New Environments

One way for domesticated species to alleviate their deleterious burden is through genetic introgression with wild relatives (Ellstrand et al. 1999). Introgression by hybridization between crops and wild relatives has been indeed recognized as a major source of gene flow and likely happens during dispersion from domestication centers to new geographic ranges. Range expansion requires adaptation to new environments, which can be assisted by farmer selection or by introgression of alleles from native populations (Fig. 4). For example, introgression between cultivated maize (*Zea mays*) and its wild progenitor teosinte (*Zea mays* ssp. *mexicana*) was detected in both directions of gene flow (Hufford et al. 2013; Wang et al. 2017a). Gene flow was also detected between *Oryza sativa* and its wild progenitor *Oryza rufipogon* based on admixture analyses (Wang et al. 2017b). In pearl millet, wild gene flow has shaped genetic diversity and favored local adaptation of western African crop populations (Burgarella et al. 2018). Those studies suggest that introgression could have influenced the cost of domestication by counterbalancing the accumulation of deleterious mutations in domesticated plants. In support of this hypothesis, Wang et al. (2017a) have detected a genomic region of domesticated maize with fewer deleterious variants introgressed into a crop from wild populations. This important result suggested that adaptive introgression may be more important at reducing genetic burden than at contributing better adapted alleles to the domesticated gene pool. To know whether introgression has played a role in lessening the domestication burden is an important open question for future genomic studies (Gaut et al. 2018).

4.5 Future Directions and Conclusions

Deciphering the origin and spread of domesticated crops has triggered sustained interest from plant biologists, population geneticists, and archaeologists for several decades. The demographic history of crops has been investigated intensively, and the recent period has seen an emergence of new powerful statistical methods that increase our understanding of domestication processes. Coalescent models and sophisticated approximate Bayesian computation approaches have proposed new answers to some crucial questions: Where did domestication occur? Was it rapid or slow? When did agricultural expansion start and which paths did it follow? The answers to these questions help us to better understand the history of humans and their cultivated plants. However, many questions remain open, and coalescent methods may not be able to address them with their current implementation.

Coupling powerful inference methods, such as SMC or ABC, with models that integrate interactions between geographic expansion, demography, and selection would be key direction for future improvements of coalescent models of crop domestication. Natural selection deeply affects the gene genealogies and the observed patterns of genetic polymorphisms. For example, the effect of background selection induced by the cost of domestication seems pervasive across domesticated crops. It is expected that background selection could bias inference from demographic models that assume selective neutrality. The effect of background selection resembles a reduction in effective population size similar to a bottleneck, and it probably accounts for the observation that genetic diversity is reduced in low recombination regions (Charlesworth 2009). Coalescent models of plant domestication should better incorporate selective and biogeographic processes on genomes and account for regions with low recombination. These model improvements would allow a better understanding of the tempo and mode of domestication for many species. In addition, integration of data from archaeology might help better calibrate the demographic models. ABC methods are particularly suitable to these objectives. The improved models would allow us to better evaluate the relative contributions of selection by farmers, drift induced by range expansions, and genetic introgression from wild relatives in the pattern of polymorphisms of cultivated plants.

Acknowledgments This work has been supported by a grant from French National Research Agency ANR-13-BSV7-0017, AFRICROP. It was developed in the framework of the Grenoble Alpes Data Institute, supported by the French National Research Agency ANR-15-IDEX-02, "Investissements d'avenir."

References

Alexander DH, Novembre J, Lange K. Fast model-based estimation of ancestry in unrelated individuals. Genome Res. 2009;19(9):1655–64.
Beaumont MA. Approximate Bayesian computation in evolution and ecology. Annu Rev Ecol Evol Syst. 2010;41:379–406.
Beichman AC, Huerta-Sanchez E, Lohmueller KE. Using genomic data to infer historic population dynamics of nonmodel organisms. Annu Rev Ecol Evol Syst. 2018;49:433–56.
Beissinger TM, Wang L, Crosby K, Durvasula A, Hufford MB, Ross-Ibarra J. Recent demography drives changes in linked selection across the maize genome. Nat Plants. 2016;2(7):16084.
Blum MG, François O. Non-linear regression models for approximate Bayesian computation. Stat Comput. 2010;20(1):63–73.
Boitard S, Rodríguez W, Jay F, Mona S, Austerlitz F. Inferring population size history from large samples of genome-wide molecular data-an approximate Bayesian computation approach. PLoS Genet. 2016;12(3):e1005877.
Burgarella C, Cubry P, Kane NA, Varshney RK, Mariac C, et al. A Western Sahara origin of African agriculture inferred from pearl millet genomes. Nat Ecol Evol. 2018;2:1377–80.
Caye K, Jay F, Michel O, François O. Fast inference of individual admixture coefficients using geographic data. Ann Appl Stat. 2018;12(1):586–608.
Charlesworth B. Effective population size and patterns of molecular evolution and variation. Nat Rev Genet. 2009;10(3):195–205.
Choi JY, Platts AE, Fuller DQ, Wing RA, Purugganan MD. The rice paradox: multiple origins but single domestication in Asian rice. Mol Biol Evol. 2017;34(4):969–79.
Civáň P, Craig H, Cox CJ, Brown TA. Three geographically separate domestications of Asian rice. Nat Plants. 2015;1:15164.

Clotault J, Thuillet AC, Buiron M, De Mita S, Couderc M, et al. Evolutionary history of pearl millet (*Pennisetum glaucum [L.] R. Br.*) and selection on flowering genes since its domestication. Mol Biol Evol. 2011;29(4):1199–212.

Csilléry K, Blum MG, Gaggiotti OE, François O. Approximate Bayesian computation (ABC) in practice. Trends Ecol Evol. 2010;25(7):410–8.

Cubry P, Vigouroux Y, François O. The empirical distribution of singletons for geographic samples of DNA sequences. Front Genet. 2017;8:139.

Cubry P, et al. The rise and fall of African rice cultivation revealed by analysis of 246 new genomes. Curr Biol. 2018;28(14):2274–82.

Currat M, Ray N, Excoffier L. SPLATCHE: a program to simulate genetic diversity taking into account environmental heterogeneity. Mol Ecol Notes. 2004;4(1):139–42.

Darwin C. The variation of animals and plants under domestication. London: John Murray; 1882.

Degnan JH, Rosenberg NA. Gene tree discordance, phylogenetic inference and the multispecies coalescent. Trends Ecol Evol. 2009;24:332–40.

Doebley JF, Gaut BS, Smith BD. The molecular genetics of crop domestication. Cell. 2006;127:1309–21.

Durand EY, Patterson N, Reich D, Slatkin M. Testing for ancient admixture between closely related populations. Mol Biol Evol. 2011;28(8):2239–52.

Edmonds CA, Lillie AS, Cavalli-Sforza LL. Mutations arising in the wave front of an expanding population. Proc Natl Acad Sci. 2004;101(4):975–9.

Edwards SV, Xi Z, Janke A, Faircloth BC, McCormack JE, Glenn TC, et al. Implementing and testing the multispecies coalescent model: a valuable paradigm for phylogenomics. Mol Phylogenet Evol. 2016;94:447–62.

Ellstrand NC, Prentice HC, Hancock JF. Gene flow and introgression from domesticated plants into their wild relatives. Annu Rev Ecol Syst. 1999;30(1):539–63.

Excoffier L, Foll M, Petit RJ. Genetic consequences of range expansions. Annu Rev Ecol Evol Syst. 2009;40:481–501.

Excoffier L, Dupanloup I, Huerta-Sánchez E, Sousa VC, Foll M. Robust demographic inference from genomic and SNP data. PLoS Genet. 2013;9(10):e1003905.

Eyre-Walker A, Gaut RL, Hilton H, Feldman DL, Gaut BS. Investigation of the bottleneck leading to the domestication of maize. Proc Natl Acad Sci. 1998;95(8):4441–6.

François O, Blum MGB, Jakobsson M, Rosenberg NA. Demographic history of European populations of *Arabidopsis thaliana*. PLoS Genet. 2008;4(5):e1000075.

Frichot E, Mathieu F, Trouillon T, Bouchard G, François O. Fast and efficient estimation of individual ancestry coefficients. Genetics. 2014;196(4):973–83.

Gaut BS, Seymour DK, Liu Q, Zhou Y. Demography and its effects on genomic variation in crop domestication. Nat Plants. 2018;4:512–20.

Gutenkunst RN, Hernandez RD, Williamson SH, Bustamante CD. Inferring the joint demographic history of multiple populations from multidimensional SNP frequency data. PLoS Genet. 2009;5(10):e1000695.

Harlan JR. Agricultural origins: centers and noncenters. Science. 1971;174:468–74.

Hein J, Schierup M, Wiuf C. Gene genealogies, variation and evolution: a primer in coalescent theory. Oxford: Oxford University Press; 2004.

Heled J, Drummond AJ. Bayesian inference of species trees from multilocus data. Mol Biol Evol. 2009;27(3):570–80.

Huang X, Kurata N, Wei X, Wang Z-X, Wang A, Zhao Q, Zhao Y, Liu K, Lu H, Li W, et al. A map of rice genome variation reveals the origin of cultivated rice. Nature. 2012;490:497–501.

Hudson RR. Generating samples under a Wright-Fisher neutral model of genetic variation. Bioinformatics. 2002;18:337–8.

Hufford MB, Lubinksy P, Pyhäjärvi T, Devengenzo MT, Ellstrand NC, Ross-Ibarra J. The genomic signature of crop-wild introgression in maize. PLoS Genet. 2013;9(5):e1003477.

Kelleher J, Etheridge AM, McVean G. Efficient coalescent simulation and genealogical analysis for large sample sizes. PLoS Comput Biol. 2016;12(5):e1004842.

Kingman JFC. The coalescent. Stoch Process Appl. 1982;13(3):235–48.

Kono TJ, Fu F, Mohammadi M, Hoffman PJ, Liu C, Stupar RM, Smith KP, Tiffin P, Fay JC, Morrell PL. The role of deleterious substitutions in crop genomes. Mol Biol Evol. 2016;33(9):2307–17.

Larson G, Piperno DR, Allaby RG, Purugganan MD, Andersson L, Arroyo-Kalin M, et al. Current perspectives and the future of domestication studies. Proc Natl Acad Sci. 2014;111(17):6139–46.

Li H, Durbin R. Inference of human population history from individual whole-genome sequences. Nature. 2011;475(7357):493–6.

Liu Q, Zhou Y, Morrell PL, Gaut BS. Deleterious variants in Asian rice and the potential cost of domestication. Mol Biol Evol. 2017;34(4):908–24.

Lu J, Tang T, Tang H, Huang J, Shi S, Wu CI. The accumulation of deleterious mutations in rice genomes: a hypothesis on the cost of domestication. Trends Genet. 2006;22(3):126–31.

Matsuoka Y, Vigouroux Y, Goodman MM, Sanchez J, Buckler E, Doebley J. A single domestication for maize shown by multilocus microsatellite genotyping. Proc Natl Acad Sci. 2002;99(9):6080–4.

Mazet O, Rodríguez W, Grusea S, Boitard S, Chikhi L. On the importance of being structured: instantaneous coalescence rates and human evolution – lessons for ancestral population size inference? Heredity. 2016;116(4):362–71.

McVean GA, Cardin NJ. Approximating the coalescent with recombination. Philos Trans R Soc B Biol Sci. 2005;360(1459):1387–93.

Meyer RS, Purugganan MD. Evolution of crop species: genetics of domestication and diversification. Nat Rev Genet. 2013;14:840–52.

Meyer RS, DuVal AE, Jensen HR. Patterns and processes in crop domestication: an historical review and quantitative analysis of 203 global food crops. New Phytol. 2012;196(1):29–48.

Meyer RS, Choi JY, Sanches M, Plessis A, Flowers JM, Amas J, Dorph K, Barretto A, Gross B, Fuller DQ, et al. Domestication history and geographical adaptation inferred from a SNP map of African rice. Nat Genet. 2016;48:1083–8.

Molina J, Sikora M, Garud N, Flowers JM, Rubinstein S, Reynolds A, et al. Molecular evidence for a single evolutionary origin of domesticated rice. Proc Natl Acad Sci. 2011;108(20):8351–6.

Moyers BT, Morrell PL, McKay JK. Genetic costs of domestication and improvement. J Hered. 2017;109(2):103–16.

Nabholz B, Sarah G, Sabot F, Ruiz M, Adam H, Nidelet S, et al. Transcriptome population genomics reveals severe bottleneck and domestication cost in the African rice (*Oryza glaberrima*). Mol Ecol. 2014;23(9):2210–27.

Nielsen R, Beaumont MA. Statistical inferences in phylogeography. Mol Ecol. 2009;18(6):1034–47.

Patterson N, Price AL, Reich D. Population structure and eigenanalysis. PLoS Genet. 2006;2(12):e190.

Patterson N, Moorjani P, Luo Y, Mallick S, Rohland N, Zhan Y, et al. Ancient admixture in human history. Genetics. 2012;192(3):1065–93.

Peter BM. Admixture, population structure, and F-statistics. Genetics. 2016;202(4):1485–501.

Pickrell JK, Pritchard JK. Inference of population splits and mixtures from genome-wide allele frequency data. PLoS Genet. 2012;8(11):e1002967.

Portères R. Berceaux agricoles primaires sur le continent africain. J Afr Hist. 1962;3(2):195–210.

Pritchard JK, Stephens M, Donnelly P. Inference of population structure using multilocus genotype data. Genetics. 2000;155(2):945–59.

Rosenberg NA, Nordborg M. Genealogical trees, coalescent theory and the analysis of genetic polymorphisms. Nat Rev Genet. 2002;3(5):380–90.

Sang T, Ge S. The puzzle of rice domestication. J Integr Plant Biol. 2007;49:760–8.

Scarcelli N, Cubry P, Akakpo R, Thuillet AC, Obidiegwu J, Baco MN, et al. Yam genomics supports West Africa as a major cradle of crop domestication. Sci Adv. 2019;5(5):eaaw1947.

Schiffels S, Durbin R. Inferring human population size and separation history from multiple genome sequences. Nat Genet. 2014;46(8):919–25.

Spence JP, Steinrücken M, Terhorst J, Song YS. Inference of population history using coalescent HMMs: review and outlook. Curr Opin Genet Dev. 2018;53:70–6.

Terhorst J, Kamm JA, Song YS. Robust and scalable inference of population history from hundreds of unphased whole genomes. Nat Genet. 2017;49(2):303–9.

Wang M, Yu Y, Haberer G, Marri PR, Fan C, Goicoechea JL, Zuccolo A, Song X, Kudrna D, Ammiraju JSS, et al. The genome sequence of African rice (*Oryza glaberrima*) and evidence for independent domestication. Nat Genet. 2014;46:982–8.

Wang L, Beissinger TM, Lorant A, Ross-Ibarra C, Ross-Ibarra J, Hufford MB. The interplay of demography and selection during maize domestication and expansion. Genome Biol. 2017a;18(1):215.

Wang H, Vieira FG, Crawford JE, Chu C, Nielsen R. Asian wild rice is a hybrid swarm with extensive gene flow and feralization from domesticated rice. Genome Res. 2017b;27:1029–38.

Wiuf C, Hein J. Recombination as a point process along sequences. Theor Popul Biol. 1999;55(3):248–59.

Wright SI, Bi IV, Schroeder SG, Yamasaki M, Doebley JF, McMullen MD, Gaut BS. The effects of artificial selection on the maize genome. Science. 2005;308(5726):1310–4.

Yang Z. Molecular evolution: a statistical approach. Oxford: Oxford University Press; 2014.

Zeder MA. Core questions in domestication research. Proc Natl Acad Sci. 2015;112(11):3191–8.

Zhu Q, Zheng X, Luo J, Gaut BS, Ge S. Multilocus analysis of nucleotide variation of *Oryza sativa* and its wild relatives: severe bottleneck during domestication of rice. Mol Biol Evol. 2007;24:875–88.

Population Genomics of Weedy Crop Relatives: Insights from Weedy Rice

Lin-Feng Li and Kenneth M. Olsen

Abstract Weedy crop relatives can evolve as a byproduct of the domestication process, and most modern crop species have conspecific or congeneric weedy relatives. These weedy relatives invade crop fields and aggressively outcompete desirable cultivars through a suite of weediness traits. Weedy rice (*Oryza sativa* f. *spontanea*) is a highly morphologically diverse group of undesirable rice strains that infest rice production areas and causes heavy yield loss worldwide. In this chapter, we focus on recent phenotypic and genomic characterizations of weedy rice, with special emphasis on its evolutionary origins and adaptive mechanisms. While weedy rice strains can be broadly divided into "crop mimic" and "wild-like" forms, genetic surveys around the world have revealed that most strains have evolved through de-domestication from cultivated rice and that this process has occurred multiple times from different genetically distinct rice varieties. Weedy strains have further evolved and adapted with varying degrees of input from domesticated and wild *Oryza* populations. The weediness traits can evolve independently in weedy rice through a combination of de novo mutation, standing genetic variation, and adaptive introgression. Recent advances in high-throughput sequencing platforms have made it possible to perform genome-enabled QTL mapping and comparative population genomics approaches to identify weediness-related genes in independently evolved weedy strains. We hope that perspectives raised in this review can provide a point of comparison for future studies of other weedy species.

Keywords Adaptive introgression · Agricultural weeds · Comparative population genomics · De novo mutation · De-domestication · Standing genetic variation · Weediness trait · Weedy rice

---

L.-F. Li
Ministry of Education Key Laboratory for Biodiversity Science and Ecological Engineering, and Coastal Ecosystems Research Station of the Yangtze River Estuary, School of Life Sciences, Fudan University, Shanghai, China

K. M. Olsen (✉)
Department of Biology, Washington University in St. Louis, St. Louis, MO, USA
e-mail: kolsen@wustl.edu

1 Introduction

1.1 Weedy Crop Relatives: Plants at the Margins of the Domestication Process

The domestication of wild plants is one of the most pivotal achievements in human history, having led to the development of agricultural settlements and ultimately the emergence of urban centers and civilization (Hilbert et al. 2017). Modern humans still rely almost entirely on a handful of staple food crop species that were domesticated approximately 10,000 years ago in several geographical centers of domestication around the world (Larson et al. 2014). Archaeobotanical remains and other data suggest that domestication was likely a protracted process (Allaby et al. 2008; Larson et al. 2014), with its origins in the harvesting of wild cereals and other plants (Hillman et al. 2001; Weiss et al. 2006; Willcox et al. 2008). Over time, wild plant gathering would have shifted from reliance on wild populations, whose growth was tolerated or encouraged (e.g., through burning or clearing of competing vegetation), to active cultivation of wild species, to farming of protodomesticates bearing some traits that distinguished them from their wild relatives (e.g., reduced seed shattering and dormancy), to farming of fully domesticated landraces that were dependent on human cultivation for their perpetuation. Subsequent to these initial steps in the domestication process, later selective breeding gave rise to crop varieties with varietal-specific improvement and diversification traits (e.g., fruit color, starch and cooking quality, local climatic adaptation) (Meyer and Purugganan 2013; Li et al. 2017).

Humans can take credit not only for crop domestication but also for the emergence of agricultural weeds that have plagued farmers' fields ever since. Agricultural weeds arose as byproducts of the shift to subsistence farming and have evolved to exploit the agricultural habitat (De Wet and Harlan 1975; Warwick and Stewart 2005; Stewart 2017). Among agricultural weed species, those that are closely related to crop species can pose especially difficult challenges for weed detection and control. It is very likely that these weedy crop relatives have formed a part of the agricultural weed assemblage since the earliest days of agriculture (see, e.g., Fig. 1). Weedy crop relatives infest crop fields and aggressively outcompete desirable cultivars through a combination of weediness traits, such as rapid growth, efficient seed dispersal, and persistent seed dormancy (Baker 1965; Vigueira et al. 2013). Most contemporary crop species have conspecific or congeneric weedy relatives (Ellstrand et al. 1999); some of the best known examples include weedy beet (*Beta vulgaris*), weedy sorghum (shattercane) (*Sorghum bicolor*), and weedy rice (*Oryza sativa* f. *spontanea*) (Arnaud et al. 2003; Barnaud et al. 2009; Stewart 2017). While weedy crop relatives often possess some traits that are more characteristic of wild species than domesticates (e.g., seed shattering and dormancy), they are specifically adapted to agroecosystems and typically are not found outside of crop fields (Stewart 2017).

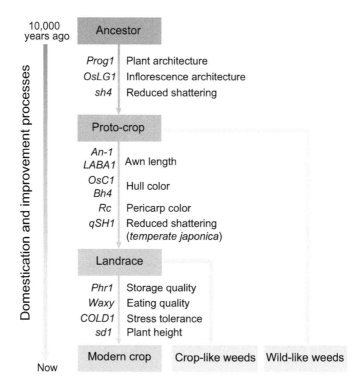

Fig. 1 Origin models of different weedy rice types

Among weedy crop relatives, weedy rice is the species that is by far the best characterized through genomic analyses. For this reason, we focus specifically on weedy rice in this chapter, with the hope that the research reviewed here can provide a point of reference for future genomic characterizations of other weedy crop relatives. Weedy rice infests rice production areas and causes crop yield losses worldwide (Delouche et al. 2007; Qiu et al. 2014; Song et al. 2014). As described below, it includes various forms of undesirable rice strains that occur in cultivated rice fields and possess characteristics intermediate between wild and domesticated rice (De Wet and Harlan 1975; Oka 1988; Delouche et al. 2007). Key features of weedy rice include highly shattering seed, deep seed dormancy, dark-pigmented pericarps (a trait associated with seed dormancy), rapid vegetative growth, and the ability to aggressively outcompete the crop for nutrients and light (De Wet and Harlan 1975; Basu et al. 2004; Estorninos et al. 2005; Burgos et al. 2006; Xia et al. 2011).

Weedy forms of rice have likely been present at the margins of rice fields for thousands of years. Written documentation of weedy rice extends back at least as far as the ancient Chinese book "Huainanzi"《淮南子》, which dates to 2,200 years ago. However, it has only gained worldwide attention as a major agricultural weed in recent decades (Gressel and Valverde 2009). This change is primarily due to the worldwide agricultural shifts away from hand-transplanting of paddy-grown rice seedlings (which, while time-consuming and laborious, has the advantage of providing ample opportunities for hand weeding of rice fields), to mechanized, direct-seeded farming (Hill et al. 1994; Chauhan 2013; Ziska et al. 2015). In China, for example, weedy rice infestations now affect up to three million ha of rice production areas, with annual yield losses exceeding 3.4 billion kg (Liang and Qiang 2011). Similarly, weedy rice infestations have been estimated to cause grain yield losses of more than $20 and $45 million in the United States and Peninsular Malaysia, respectively (Estorninos et al. 2005; Gealy et al. 2012).

Weedy rice strains are highly morphologically variable, but strains have been proposed to fall into two broad adaptive strategies (Ling-Hwa and Morishima 1997; Suh et al. 1997; Federici et al. 2001): a "crop mimic" form (characterized by short stature, low degree of shattering, and overall similarity to the crop), which persists via crypsis in rice fields, and a more wild-like form, which aggressively outgrows its domesticated relative and then lodges, scattering its seed into the soil where they can remain viable for 20 years or more. Morphological and molecular-based surveys around the world suggest that different weedy rice types have evolved independently from genetically distinct domesticated and wild relatives and that diverse genetic mechanisms have contributed to the formation of weediness traits (Song et al. 2014; Qi et al. 2015; Huang et al. 2017; Li et al. 2017; Qiu et al. 2017). This chapter focuses on recent phenotypic and population genomic insights into the evolutionary origins of weedy rice and introductions out of the center of rice domestication in Asia to rice production areas worldwide.

2 Evolutionary Origins of Domesticated and Weedy Rice

2.1 Origins of Domesticated Rice

As the evolution of weedy rice is fundamentally intertwined with the domestication history of its cultivated relative, we first provide a brief overview of the current state of knowledge on rice domestication origins. Asian cultivated rice (*Oryza sativa* L.) is one of the world's oldest and most important crop species, serving as the primary staple crop for more than one-third of the global population. As the first crop species to have a reference genome published (Goff et al. 2002; Yu et al. 2002), it has also become an important model system for plant biology research (Sasaki 2005; IRRI 2006). There are two genetically diverged subspecies in cultivated rice, namely *indica* and *japonica*, and each subspecies comprises two or more genetically distinct varietal subgroups within it (Garris et al. 2005; Sweeney and McCouch 2007; Zhao et al. 2011). *Aus* and *indica* are the two major subgroups within the *indica* subspecies, and *aromatic, tropical japonica* and *temperate japonica* are the major subgroups within the *japonica* subspecies. The two subspecies are morphologically distinct and tend to be grown in different geographical regions, with *indica* rice predominating in the tropics and *japonica* varieties cultivated in more temperate climates; they also show partial reproductive isolation (Huang et al. 2012a;

Ouyang and Zhang 2013). Genome-wide sequence divergence between the *indica* and *japonica* subspecies indicates a divergence time of some 200,000–400,000 years ago, which is long before origins of agriculture and thus predates the domestication process; this divergence is also present among individuals within the wild progenitor species, *O. rufipogon* (Ge et al. 1999; Zhu and Ge 2005).

It is widely accepted that *indica* and *japonica* rice were domesticated from *O. rufipogon* in Asia and that this domestication occurred around 10,000 years ago (Oka 1988; Cheng et al. 2003; Fuller et al. 2010). However, there is no consensus on the exact domestication history and the origin(s) of the present-day varietal groups. Two primary alternative hypotheses have been proposed. A single-origin model proposes that rice was domesticated from its wild ancestor only once, and the two cultivated subspecies, *japonica* and *indica*, emerged during the domestication process. Genetic evidence for this model comes from a number of domestication genes where *japonica*-derived domestication alleles are fixed in most or all crop varieties across both subspecies (e.g., *sh4*, controlling reduced seed shattering; *prog1*, controlling erect plant growth; and *rc*, controlling white pericarp/reduced dormancy) (Li et al. 2006; Sweeney et al. 2006; Jin et al. 2008; Tan et al. 2008). Demographic modeling based on SSRs and genome-wide SNPs also supports a single domestication origin (Gao and Innan 2008; Molina et al. 2011).

In contrast, a series of other studies based on phylogenetic and population genetic analyses have proposed that *indica* and *japonica* rice were domesticated independently from different subgroups of *O. rufipogon*. Support is provided by phylogeographic analyses indicating that *indica* and *japonica* are associated with geographically distinct populations of *O. rufipogon*, with *japonica* rice most closely related to populations in southern China and *indica* rice closest to populations in South and Southeast Asia (Cheng et al. 2003; Zhu and Ge 2005; Londo et al. 2006). Population genomic analyses based on ~1,500 wild and cultivated rice accessions similarly indicate that *japonica* was first domesticated from the *O. rufipogon* populations near to the middle area of Pearl River in southern China and that *indica* was subsequently developed from crosses between domesticated *japonica* and local populations of *O. rufipogon* in Southeast and South Asia (Huang et al. 2012a). Consistent with the dual-origin domestication hypothesis, archaeological data provides evidence for early rice cultivation both in the Yangtze Valley (China) and the Ganges Valley (India) beginning at ~8,000–9,000 and ~4,000 years ago, respectively (Fuller et al. 2010; Gross and Zhao 2014; Zuo et al. 2017). In contrast to the demographic models supporting a single origin (Gao and Innan 2008; Molina et al. 2011), gene genealogies across the rice genome indicate that that majority of the *indica* and *japonica* genomes appear to be of independent origin (He et al. 2011). Genomic regions with shared domestication alleles appear to be the rare exception to this pattern and are interpreted under the dual-origin model as evidence of occasional selective introgression during the domestication process (He et al. 2011). Regardless of the number of geographically distinct domestication events, all evidence suggests that *indica* and *japonica* rice are descended from pre-differentiated ancestral *O. rufipogon* populations.

2.2 Evolutionary Roots of Weedy Rice in Different World Regions

As reviewed in the sections below, genetic surveys around the world have revealed multiple independent origins of weedy rice (Reagon et al. 2010; Huang et al. 2017; Li et al. 2017). While different weed strains are all characterized by some weed-adaptive traits (such as seed shattering and dormancy), they also show a broad range of phenotypic variation in both vegetative and reproductive features (Cao et al. 2007; Song et al. 2014; Qiu et al. 2017). This diversity has led to the hypothesis that different weeds may have evolved through different degrees of input from various domesticated and wild *Oryza* populations. Several non-mutually exclusive hypotheses have been proposed, including (1) direct descent from wild rice that adapted to crop fields (De Wet and Harlan 1975; Harlan 1992), (2) intervarietal hybridization within the crop species (Qiu et al. 2014), (3) interspecies hybridization between cultivated and wild rice (Tang and Morishima 1996), and (4) direct de-domestication from cultivated *O. sativa* (including potentially protodomesticates, landraces, or modern varieties) (Li et al. 2017; Qiu et al. 2017).

In considering the evolution of weed strains in different geographical regions of rice cultivation, one of the most important distinctions is whether rice fields occur in proximity to reproductively compatible wild *Oryza* species. If present, wild *Oryza* populations could potentially contribute to the evolution of weed strains – either directly through wild-weed hybridization or indirectly through crop-wild hybridization followed by feralization. In the sections below, we first discuss weedy rice strains in world regions where wild *Oryzas* are absent and then move to the potentially more complex situation of weed strains that co-occur with reproductively compatible wild rice populations. Where data are available, we include comparisons of insights from neutral markers and from candidate genes for weed-associated traits.

2.2.1 No Wild *Oryza* Species Present

United States

While rice of the *indica* subspecies provides over 70% of total rice production and is genetically more diverse (Huang et al. 2010; Zhang et al. 2016), the *japonica* subspecies is the major crop of temperate zones in the northern hemisphere, including in North America, southern Europe, and northeast Asia. In the United States, *tropical japonica* cultivars are grown in the southern Mississippi River flood plain, which is the primary region of US rice farming (Arkansas, Louisiana, Mississippi, southern Missouri, and eastern Texas) (Gealy 2005). *Temperate japonica* varieties are farmed in the secondary rice growing region in northern California. The earliest available record of weedy rice in the United States dates from 1846 (Allston 1846; Craigmiles 1978). Early studies conducted in the 1850s divided US weeds into four different weedy biotypes (Craigmiles 1978).

In the present day, weedy rice is present in both of the US rice cultivation areas. In the southern US rice belt, two major weedy morphotypes are generally recognized based on grain characteristics (Londo and Schaal 2007; Reagon et al. 2010). The strawhull (SH) form shares many domestication traits with cultivated rice and potentially corresponds to the proposed "crop mimic" ecotype described in Sect. 1 above; its defining traits include short stature and straw-colored grain hulls that lack awns. In contrast, the blackhull awned (BHA) form more closely resembles wild *Oryza* species in some traits, including tall stature and seeds that have dark hulls with long awns; however, unlike true wild *Oryza* species, BHA weeds also share a number of crop-like traits with the SH form, including erect growth and reproduction solely by seed with no vegetative propagation (Noldin et al. 1999; Kelly Vaughan et al. 2001; Gealy 2005; Olsen et al. 2007). Both strains are widely distributed through the southern US rice belt, and they are estimated to occur in at least 30% of rice production areas there (Gealy 2005).

In California, a strawhull awned (SHA) morphotype is the major weedy rice type that infests rice fields. This form infested rice fields there from the 1920s to the 1940s (Kennedy 1923; Bellue 1932; Willson 1979), but was completely eliminated in the 1970s. However, new SHA weeds have re-emerged in California since 2003 (Kanapeckas et al. 2016).

Insights from Neutral Markers

As no wild *Oryza* species are native to the United States, hypotheses for the origin of weedy rice strains there have included de-domestication from US cultivars or introduction of an already established weedy rice strain through contamination of seed stock (Londo and Schaal 2007; Reagon et al. 2010; Kanapeckas et al. 2016). Assessments of the evolutionary origin and genetic background of southern US weedy rice have relied primarily on SSRs and DNA sequencing (Estorninos et al. 2000; Kelly Vaughan et al. 2001; Gealy et al. 2002; Londo and Schaal 2007; Reagon et al. 2010). These studies revealed that despite some phenotypic similarity to wild rice, BHA weeds are clearly genetically more closely related to cultivated rice – specifically, *aus* varieties within the *indica* subspecies – than to the wild progenitor. SH weeds are similarly of crop origin; they are genetically most similar to *indica* varieties (sensu stricto). Neither weed strain is genetically similar to the *tropical japonica* cultivars grown in the southern US, although a recent rise in herbicide-resistant rice farming has selectively favored some weeds derived from crop-weed hybridization (Burgos et al. 2014; reviewed in Vigueira et al. 2019) (discussed in Sect. 3).

As neither *aus* nor *indica* varieties have ever been grown commercially in the United States, both weed forms are thought to have originated in Asia and been introduced as weeds through accidental import in contaminated seed stocks. Using genome-wide patterns of single nucleotide polymorphism (SNP) diversity, Reagon et al. (2010) demonstrated that genetic diversity within SH and BHA weeds is very low compared to *aus* and *indica* rice, which is consistent with severe genetic bottlenecks in their establishment in North America. The de-domestication origin of US weeds was recently reconfirmed with whole-genome sequence analyses, which again show that the SH and BHA weedy strains are phylogenetically close to *indica* and *aus* crop ancestors, respectively (Li et al. 2017). With regard to the timing of divergence between weeds and their crop ancestors, BHA weedy rice appears to have diverged from *aus* much earlier than SH from *indica* rice (Reagon et al. 2010; Li et al. 2017 discussed below).

For weedy rice in California, genome-wide SNPs indicate that while the SHA strain shows genetic similarity to the *temperate japonica* rice farmed there, it did not evolve directly from US cultivars. Instead, the California weedy strains appear to have originated from *japonicas* outside of the United States and were thereafter introduced into California as weeds; this weed evolution is estimated to have occurred in the relatively recent past (Kanapeckas et al. 2016).

Insights from Candidate Gene Analyses

Whereas genetic surveys using neutral markers can provide a genome-wide perspective of the evolutionary origin and demographic history of US weedy strains, a complementary source of data can be provided by candidate gene analyses – particularly for genes controlling traits related to domestication or weediness. As weedy rice possesses a combination of wild-like traits (e.g., seed shattering, dormancy) and crop-like traits (e.g., annuality, erect growth), sequence variation in the genes controlling these traits can provide insights into the specific mechanisms by which weed strains have adapted to exploit agroecosystems.

Among agronomic traits, the loss of seed shattering is one of the most characteristic features of rice domestication (Konishi et al. 2006; Li et al. 2006; Li and Olsen 2016). Previous archeological and genetic analyses have already provided a wealth of insights into the timing and locations for the occurrence of non-shattering phenotypes in cultivated rice. For example, archaeobotanical remains indicate that the reduced shattering phenotype was likely fixed in cultivated rice about 6,600–6,900 years ago (Fuller et al. 2009, 2010). Genetic analyses of seed shattering revealed that two quantitative trait loci (QTLs), *sh4* and *qSH1*, contributed to >70% of the phenotypic difference between wild and cultivated rice (Konishi et al. 2006; Li et al. 2006). For the *sh4* gene, a single non-synonymous substitution (G → T) in the Myb3 DNA-binding domain results in the incomplete development of the caryopsis abscission zone (reduced shattering phenotype) in cultivated rice (Li et al. 2006). Further population genetic analyses documented that all cultivated rice accessions carry the non-shattering allele (Zhang et al. 2009).

For the SH and BHA weeds in the United States, even though nearly all weedy strains show a similar degree of shattering to wild rice, both strains carry the *sh4* non-shattering allele fixed in cultivated rice (Thurber et al. 2010) (Fig. 1). This finding demonstrates very clearly that both the BHA and the SH strains are descended from ancestors that had passed through the initial stages of domestication and selection for reduced shattering. Genetic mechanisms underlying the apparent phenotypic reversion to shattering are discussed in Sect. 3.

Similar evidence that the SH and BHA weeds are descended from crop ancestors is provided by the *PROG1* and *OsLG1* genes, which are associated with erect plant architecture and closed-panicle inflorescence architecture, respectively (Fig. 1). As with *sh4* and reduced shattering, both of these genes were targets of selection in the initial stages of rice domestication, and both SH and BHA weedy strains share the same haplotypes as cultivated rice at these loci. These patterns further confirm that both the SH and BHA weeds are derived from cultivated ancestors (Li et al. 2017).

In contrast to these early-selected domestication genes, neither the SH nor BHA strains have been found to carry crop-associated alleles at several varietal-specific improvement genes that were likely targets of selection much later in the history of rice domestication; these include *Phr1* (associated with long-term grain storage), *Waxy* (controlling grain amylose content), and *COLD1* (associated with agroecological adaptation to temperate climates) (Li et al. 2017) (Fig. 1). Thus, while the ancestors of US weed strains were clearly domesticated, they were not the improved varieties of the sort most commonly cultivated around the world today.

Perhaps most interesting are the domestication-related genes where SH strains carry crop alleles but BHA weeds do not (e.g., *Bh4*, controlling hull color; *An-1*, controlling awn production) and where neither weed strain possesses the crop-specific allele (*Rc*, controlling pericarp color/seed dormancy) (Fig. 1). For the genes where the two strains differ, this pattern provides phenotypic confirmation of the neutral marker analyses indicating an earlier divergence of BHA weeds from cultivated rice than SH strains (Reagon et al. 2010; Li et al. 2017 see above). In the case of *Rc*, although the loss-of-function *rc* allele that confers white pericarps and reduced dormancy was strongly favored in domesticated rice and is found in most present-day crop varieties, it is absent or nearly absent in all US weeds. We speculate that there may have been strong selection in the SH and BHA weed strains to maintain the wild-type *Rc* allele because of the importance of seed dormancy in weedy rice adaptation (see also Cui et al. 2016; Li et al. 2017).

Europe

Rice was first introduced into Europe via Italy, where *japonica* varieties began to be grown during the Renaissance (thirteenth to fifteenth centuries) (Faivre-Rampant et al. 2011). From Italy, rice cultivation spread to other regions of the Mediterranean, including Spain and southern France. Weedy rice was described in Italian rice fields as early as the beginning of the nineteenth century (Biroli 1807), but infestations did not become a severe problem until the 1960s with shifts from hand-transplanting to direct seeding (Tarditi and Vercesi 1993; Vidotto and Ferrero 2005). Nowadays weedy rice is estimated to occur in ~70% of the total rice cultivation area in northern Italy (Grimm et al. 2013). Genetic and morphological analyses of Italian cultivated and weedy rice indicate that despite a great morphological diversity among the weed strains occurring there (including strawhull awnless, blackhull awned, and strawhull awned), all weedy ecotypes apparently share a high level of genetic similarity with *japonica* cultivars (Grimm et al. 2013). However, as all rice accessions used in the study were collected in Italy only, it remains unclear whether Italian weedy rice evolved through de-domestication from local *japonica* cultivars or was introduced from elsewhere.

Northeast Asia

In Northeast Asia, another region of *japonica* cultivation, weedy rice infestations have also been a relatively recent phenomenon, dating to widespread mechanized rice farming (Sun et al. 2013; Qiu et al. 2014; Kim et al. 2016; He et al. 2017). In South Korea, *temperate japonica* is the dominant group of cultivars, but *tropical japonica* and *indica* rice cultivation were also introduced recently (Kim et al. 2016). The geographic distribution and genetic composition of Korean weedy rice strains have been characterized in several studies (Heu 1988; Hak-Soo and Mun-Hue 1992; Kim et al. 2016; He et al. 2017; Vigueira et al. 2019). Of these weedy strains, "Aengmi" and "Share" type weeds were described, with the latter mainly occurring on Kanghwa Island and geographically isolated from other weedy strains (Chung and Park 2010). Genetic surveys based on genotyping by sequencing (GBS) and whole-genome data demonstrated that Korean weedy accessions cluster together with local *indica* and *temperate japonica* cultivars, respectively, suggesting the possibility of de-domestication origin of these weeds (Kim et al. 2016; Vigueira et al. 2019). It is interesting that the *indica*-like US and South Korean weeds are largely nonoverlapping in genetic cluster analyses, which potentially suggests independent origins for these temperate-adapted *indica*-derived weed strains (Vigueira et al. 2019). In addition to a de-domestication origin, population genomic analyses have also been used to propose a hybrid origin of Korean weedy rice resulting from crosses between modern cultivars and local landraces (He et al. 2017).

In Japan, *temperate japonica* is the dominant crop in most rice fields. Most varieties are the products of modern breeding, but some traditional landraces still exist, and these may have contributed to the origins of Japanese weedy rice. Weedy rice has been observed in the dry and direct-seeded rice fields since the 1940s but only became a serious problem since the 1980s (Ishii and Akazawa 2003; Ushiki et al. 2005). Similar to South Korean weeds, genetic analyses based on neutral markers revealed two genetically distinct weedy types, *indica*- and *japonica*-like weeds (Akasaka et al. 2009). Of the two weedy types, the *japonica*-like weeds show a close phenotypic and genetic resemblance to the local *japonica* varieties such as "Akebono," "Kibinohana," and "Omachi" (Ushiki et al. 2005; Akasaka et al. 2009). The remaining *indica*-like weedy strains are phylogenetically close to the *indica* rice varieties "Kasalath" and "IR36." These attributes clearly demonstrate the de-domestication origin of Japanese weedy rice strains, with a subset likely descended from local landraces. Additional genetic evidence for this de-domestication scenario also comes from domestication and improvement genes. For example, all Japanese weedy rice strains possess the same non-shattering domestication allele of *sh4* gene to cultivated rice (Akasaka et al. 2011). For the two isozyme loci, *Acp1* and *Sdh1*, *japonica*- and *indica*-like weedy rice strains share the same haplotype with their crop ancestors, respectively, confirming the local de-domestication origin of Japanese weedy rice (Kawasaki et al. 2009).

China

In China, scenarios of weedy rice evolution are more complicated than Korea and Japan, mainly due to the large geographical region and heterogeneous range of environments across which rice cultivation occurs. In general, *japonica* rice is widely cultivated in northern and northeastern China, and *indica* is dominant in southern China. As with other regions of rice cultivation, weedy rice was effectively suppressed in China through traditional hand-transplanted rice farming but has erupted recently because of the adoption of direct-seeding farming (Yu et al. 2005). In northeastern China, cultivated rice was introduced as early as the end of the nineteenth century, but commercial cultivation only occurred there in recent decades (Jin 2010). Although no wild *Oryza* species are naturally distributed in northeastern China, a suite of wild rice traits (e.g., blackhulls and long awns) are found in some weedy rice strains (Qiu et al. 2017). However, population genetic analyses revealed that most of crop- and wild-like weedy rice strains are genetically closer to local *japonica* cultivars rather than *indica* or wild rice, suggesting a role for *japonica* de-domestication in the origin of these Chinese weedy strains (Cao et al. 2007; Sun et al. 2013; Qiu et al. 2017). Similar evidence for de-domestication has been found in other studies in northern and northwestern China, where the weedy rice morphotypes are clustered with either *indica* or *japonica* (Sun et al. 2013; Qiu et al. 2017). Notably, an *indica-japonica* hybridization origin for weedy rice has been described in eastern China where *indica* rice was historically cultivated but has been recently replaced by *japonica* (Qiu et al. 2014). These findings together indicate that de-domestication from different cultivated rice varieties plays the major role in the formation of weedy rice in China, as in other northern temperate rice growing regions.

2.2.2 No *O. rufipogon* but Other Wild *Oryza* Species Present

The genus *Oryza* consists of 23 species that are widely distributed in pantropical areas (Vaughan 1994; Vaughan et al. 2003, 2005). Most members of the genus possess genomes that are diverged from the AA genome that characterizes *O. sativa* and its wild ancestor (*O. rufipogon*) and are thus not interfertile with cultivated rice. However, reproductively compatible wild *Oryza* species besides *O. rufipogon* can be found in several regions where weedy rice occurs, including the neotropics and Africa, and these wild *Oryzas* could in principle contribute to the evolution of local weed strains.

Africa

In Africa, the AA genome-cultivated rice *O. glaberrima* was domesticated from *O. barthii* in the Niger River delta around 3,000 years ago and was disseminated from there to other parts of West Africa (Harlan 1976; Linares 2002; Wang et al. 2014). The concept of "weedy rice" in Africa is different from other world regions, as it collectively refers to the native weedy forms of *O. glaberrima*, *O. barthii*, *O. longistaminata*, and *O. punctata* as well as weedy strains of the introduced Asian *O. sativa* (Delouche et al. 2007). All these weedy forms co-occur in rice fields with cultivated African and Asian rice and cause serious reductions of rice production in West and North Africa, in particular those areas where traditional farming has been replaced by intensification and commercialization of rice agriculture (Delouche et al. 2007). To our knowledge, no detailed genetic characterizations of African weedy rice have been published; the exact origin model and adaptive evolution of these African weedy forms thus remain uninvestigated so far.

Latin America

In Central and South America, the AA genome species *O. glumaepatula* can be found growing in proximity to Asian cultivated and weedy rice. However, studies in Costa Rica based on morphological and genetic analyses do not suggest any direct contribution of this native *Oryza* species to the genetic composition of local weed strains (Arrieta-Espinoza et al. 2005). Similarly, weedy rice strains in South America appear to be descended primarily from domesticated *O. sativa* ancestors (Federici et al. 2001; Andres et al. 2013). In Uruguay and Brazil, different morphotypes of weedy rice have been reported, including blackhull awned and strawhull awnless forms (Federici et al. 2001; Andres et al. 2013). Molecular genetic analyses have suggested that Uruguayan weedy rice is genetically close to local cultivated rice, indicating the possibility of de-domestication (Federici et al. 2001). However, a role for native wild *Oryza* species has not been explicitly assessed in the evolution of weedy rice in South America. Further investigations based on comparative genomics might reveal some role for the native wild species in local adaptation of weedy rice in Latin America.

2.2.3 *O. rufipogon* Present

The wild ancestor of rice (*O. rufipogon*) occurs widely throughout tropical and subtropical regions of southern Asia (Oka 1988; Vaughan 1994; Vaughan et al. 2008), and these wild rice populations are characterized by wide genetic and phenotypic variability (Grillo et al. 2009; Huang et al. 2012b; Civáň et al. 2015). Two wild rice ecomorphs are sometimes recognized based on life history (perennial vs. annual), with the annual form sometimes designated a separate species (*O. nivara*); however, there is no clear evidence for evolutionary divergence between annual and perennial ecotypes, and we refer to both as *O. rufipogon*. As the center of rice domestication, southern Asia is also home to the highest crop varietal diversity, with all five major subgroups cultivated in various parts of this region. This combination of highly diverse crop germplasm and reproductively compatible wild relatives creates an environment in which weedy rice origins could be particularly complex and the genetic diversity of weed strains especially high.

Like other regions of traditional rice cultivation, weedy rice in southern Asia was traditionally kept in check by hand weeding during transplanting of rice seedlings. With shifts to direct seeding, weedy rice infestation has become a serious threat to rice production in South and Southeast Asia in the last several decades (Chauhan 2013; Song et al. 2014). The recent and pervasive emergence of weedy rice is especially well documented in Malaysia and Thailand.

Malaysia

In the last two decades of the twentieth century, 90% of the rice cultivation in Malaysia shifted from traditional transplanting to direct-seeding farming. Weedy rice was first reported in Malaysian rice fields in 1988 (Wahab and Suhaimi 1991), and this was followed by rapid proliferation throughout production areas (Karim et al. 2004). Malaysian weedy rice strains are highly morphologically diverse (Sudianto et al. 2016), with a gradation of grain characters ranging from strains closely resembling cultivated rice (strawhull, awnless) to those showing close similarity to wild rice (brown or blackhulls with awns). Studies using a combination of microsatellites (Song et al. 2014; Neik et al. 2019), genotyping by sequencing (Vigueira et al. 2019), and candidate gene analyses of weed-associated traits (Song et al. 2014; Cui et al. 2016; Neik et al. 2019) have revealed a major role for both domesticated rice, including modern elite *indica* cultivars, and wild rice in the evolution of Malaysian weed strains. Introgression from wild populations is especially evident in candidate gene analyses. For example, whereas the domestication allele of the *sh4* shattering gene is completely fixed in US and northern Asian weedy rice populations (Thurber et al. 2010; Zhu et al. 2012) (Fig. 1), the functional wild-type allele is present at high frequencies in Malaysian weeds, particularly those strains with grain phenotypes similar to wild rice (Song et al. 2014). Interestingly, a high proportion of strains with the wild allele were found to be *sh4* heterozygotes, suggesting recent hybridization between domesticated (or de-domesticated) and wild rice populations. Similar patterns are also found for allelic variation at the hull color gene *Bh4* (Song et al. 2014) and the *An-1* gene controlling presence/absence of awns (Cui et al. 2016; Neik et al. 2019). Malaysian weeds carry disproportionately higher frequencies of the wild-type allele at the *Rc* gene, which controls pericarp color and seed dormancy; this may reflect strong selection for seed dormancy in weedy populations (discussed by Cui et al. 2016).

Thailand

In Thailand, weedy rice also became a serious problem in recent decades, in particular within the commercial *Thai Hom Mali* rice cultivation area (Prathepha 2009a). Genetic investigations based on domestication genes suggested the possibility of gene introgression from both wild and cultivated rice (Prathepha 2009a, b; Wedger et al. 2019). For example, while the Thai weedy rice accessions with red pericarp carry the functional allele of *Rc* gene, the white pericarp accessions share a common genotype with cultivated rice. Similarly, the *badh2* allele (8 bp deletion) of *fgr/BADH2* gene was proven to be responsible for rice fragrance (Bradbury et al. 2005). As expected, three genotypes (*BADH2/BADH2*, *BADH2/badh2*, and *badh2/badh2*) were identified in Thai weedy rice strains (Prathepha 2009a). It should be noted that the domestication allele *badh2* is absent in wild rice but persists at high frequency (54.7%) in the weedy rice population (Prathepha 2009b), indicating the possibility of a wild-crop hybridization origin of weedy rice.

South Asia

In South Asia, weedy rice also exhibits a high level of phenotypic diversity, such as straw and blackhull, white and red pericarp, and awned and awnless (Ishikawa et al. 2005; Rathore et al. 2016; Huang et al. 2017). As in Southeast Asia, independent de-domestication from different cultivated varieties is the major origin model of South Asian weedy rice (Ishikawa et al. 2005; Huang et al. 2017). Interestingly, while the blackhull awned strains there show similarity to the wild-like BHA type found in the United States and South and Southeast Asia, distinct genetic backgrounds have been clearly revealed in several studies (Thurber et al. 2010; Song et al. 2014; Huang et al. 2017). Moreover, as with weed strains in other world regions, the weedy rice strains in this region appear to have evolved independently from crop ancestors, providing further evidence for the parallel evolution of weedy rice worldwide. In addition, however, population genomic analyses have shown that some wild-like weedy strains in South Asia show genetic similarity to wild rice (Huang et al. 2017), suggesting the possibility of weed adaptation through introgression from wild populations as in Southeast Asia. Taken together, these observations suggest that weedy rice can evolve repeatedly worldwide through different genetic mechanisms, which provides an extreme system to address molecular bases underlying recurrent evolution.

3 Insights on Molecular Mechanisms of Weedy Rice Adaptation

Morphological and genetic surveys around the world have clearly established that while weedy rice strains evolved independently in different geographic areas, parallel evolution has resulted in a common suite of weediness traits. These traits collectively consist of adaptations referred to as the "agricultural weed syndrome" (Vigueira et al. 2013; Huang et al. 2017; Li et al. 2017; Qiu et al. 2017). They include features that facilitate escape from dependence on human cultivation for reproduction and propagation, and adaptations to survive, proliferate, and outcompete crops within the agroecosystem. The mechanisms by which weediness traits may emerge could potentially include any of the following: (1) de novo mutations during the process of de-domestication; (2) emergence of weediness traits by recombining standing genetic variation that was already present in crop varieties (e.g., if there is weed evolution through hybridization of genetically distinct crop varieties); (3) adaptive introgression from wild populations (e.g., for traits such as seed shattering and dormancy); and (4) adaptive introgression from crop varieties (e.g., for herbicide resistance). We highlight recent findings for these different mechanisms below.

Among weediness traits, reacquisition of seed shattering is among the most critical steps in the establishment of weedy rice populations, as seed dispersal would be required for persistence in rice fields (De Wet and Harlan 1975; Thurber et al. 2010). In domesticated rice, by contrast, artificial selection for reduced shattering greatly facilitates the harvesting efficiency and renders domesticated rice primarily dependent on humans for survival and propagation (Li and Olsen 2016). Previous studies have identified six shattering-related genes (*sh4*, *qSH1*, *OsSh1*, *SH5*, *OsCPL1*, and *SHAT1*) in cultivated rice, of which *sh4* has been inferred to be a major causative gene that resulted in the phenotypic shift from shattering to non-shattering during rice domestication (Li and Olsen 2016). Consistent with this conclusion, genetic analyses of the *sh4* gene around the world revealed that most of the weedy strains carry the non-shattering *sh4* allele (Thurber et al. 2010; Akasaka et al. 2011; Zhu et al. 2012; Song et al. 2014; Nunes et al. 2015). The fact that weedy rice is highly shattering despite carrying the *sh4* domestication allele strongly suggests a role for either novel mutations or recombined standing variation in the re-emergence of seed shattering in weedy rice. In regions of Southeast Asia where weedy rice occurs in proximity to *O. rufipogon*, the weedy rice shattering phenotype has also been found to be attributable in part to adaptive introgression of the wild-type *sh4* allele from local wild populations (Song et al. 2014; Vigueira et al. 2019).

As a step toward identifying the genetic basis of shattering in US strains of weedy rice, the shattering phenotype was mapped in two recombinant inbred line populations derived from crosses between an *indica* crop cultivar (Dee Geo Woo Gen) and a representative accession of the SH and BHA weed strains (Thurber et al. 2013; Qi et al. 2015). QTL mapping and linkage analyses identified two and five statistically significant QTLs in the SH and BHA mapping populations, respectively, of which *qSH2S* and *qSH3Bb* are estimated to account for 51.7% and 14.1% of the total phenotypic variances in the SH population (Qi et al. 2015). Importantly, these shattering QTLs do not overlap with the genomic locations of previously identified shattering-related genes in cultivated rice, which points to genetic mechanisms other than second-site mutational suppression in the phenotypic reversions to shattering. Equally importantly, the QTL locations do not co-localize between the two US weed-crop mapping populations, indicating different underlying genetic bases for shattering in the two major US weed strains. In a similar study conducted on Chinese weedy rice, Yao et al. (2015) analyzed F_2 offspring derived from the crosses between an *indica* variety Minghui86 and a high shattering Chinese weedy accession. The three shattering-related QTLs (*wd-qsh1*, *wd-qsh3*, and *wd-qsh5*) identified show no overlap with the previously reported shattering-related genes (e.g., *sh4*, *qSH1*, *sh-h*, and *SHAT1*). In addition, eight shattering QTLs were also identified in two RIL populations developed from the crosses between two *japonica* cultivars and a weedy accession (Subudhi et al. 2013). It is interesting to note that although the major-effect QTL identified in that study, *qSH4BR*, overlaps with the *sh4* gene, there is no evidence that mutations within *sh4* itself are responsible for the phenotypic reversion to shattering.

To persist in the agroecosystem, weedy rice must possess sufficient seed dormancy to remain viable in the soil seed bank. Like seed shattering, seed dormancy is a complex trait that is controlled by multiple genes (Gu et al. 2004, 2011; Sugimoto et al. 2010; Ye et al. 2013), and the emergence of shattering in weedy rice remains largely uncharacterized beyond the resolution of QTLs. Among the genetically best-characterized dormancy genes is *Rc*, which encodes a transcription factor that pleiotropically controls both pericarp pigmentation and ABA-mediated seed dormancy. Whereas most modern cultivated rice varieties carry a loss-of-function *rc* allele that confers a white pericarp and reduced dormancy, weedy rice strains worldwide are largely characterized by the presence of functional *Rc* alleles. In the case of US weedy rice, these functional alleles are proposed to be derived from the red-pericarp protodomesticates or landraces that gave rise to the weeds (Gross et al. 2010). In tropical Asia, introgression from hybridizing wild rice populations has also likely played a role in the presence of functional *Rc* alleles in weed strains (Song et al. 2014; Cui et al. 2016; Vigueira et al. 2019). As noted above, in Malaysia, where introgression from both wild and domesticated rice has shaped recent weedy rice evolution, the white pericarp *rc* allele appears to be disproportionately underrepresented in weed strains, which is consistent with selection favoring weed strains that have sufficient seed dormancy to allow seed persistence in crop fields over multiple seasons (Cui et al. 2016).

Over the last two decades, the rice production industry has adopted the widespread use of imidazolinone herbicide-resistant crop varieties as a means of controlling weedy rice and other agricultural weeds (Croughan 2003). Resistance is largely acquired through selection for amino acid variants at the *ALS* (*acetolactate synthase*) gene. As herbicide-resistant rice varieties have increased in use throughout the world, resistant weedy rice strains have begun to proliferate (Olguin et al. 2009;

Azmi et al. 2012; Burgos et al. 2014; Merotto et al. 2016). Crop-weed hybridization appears to be the primary mechanism by which weed populations are acquiring this trait (Roso et al. 2010; Andres et al. 2014; Burgos et al. 2014). This widespread selection for crop introgression will likely have a major impact in reshaping the weedy rice genome in coming decades.

4 Current Advances in Weedy Rice Population Genomics and Future Directions

The availability of an annotated rice reference genome and an abundance of well-characterized domestication-related genes offer an efficient platform to address the genetic mechanisms underlying weedy rice adaptation. However, studies of the functional identification of weediness-related genes lag far behind those of cultivated rice. Fortunately, recent comparative genomics studies have promoted our understanding of the formation of weediness traits in weedy rice. For example, Qi et al. (2015) employed QTL mapping identified a series of QTLs associated with diverse adaptive traits, such as seed shattering, heading date, and emergence data. By comparing the physical positions of these weed-associated QTLs with known functional genes, several domestication-related genes were confirmed to be causally related to the observed phenotypic variation; these include genes for heading date (*Hd1* and *DTH8*), awn length (*An-1*), and pericarp color (*Rc*).

Complementing QTL mapping approaches, population genomics provides an alternative strategy to detect weediness-related genes at the genome-wide level. For example, Li et al. (2017) analyzed the genomes of 183 wild, cultivated, and weedy rice accessions and, through genome scans for signatures of weed-specific adaptation, identified 178 and 307 annotated genes in 2 US weedy strains, SH and BHA, respectively. Interestingly, while some of these genomic regions co-localize to known QTLs for weediness traits (including a major-effect QTL for shattering), many others do not. Likewise, genome-wide scanning based on whole-genome data was also performed in Chinese and Korean weedy rice strains in which a large number of candidate genes were identified (He et al. 2017; Qiu et al. 2017). Collectively these findings suggest that there is a rich pool of candidate genes that can be explored in future studies aimed at characterizing the mechanisms underlying weediness evolution. It seems likely that some key genes underlie phenotypes whose functions have not previously been associated with weediness or competitive ability. Further investigations focusing on functional analyses of weediness candidate genes will play a critical role in advancing our knowledge of weedy rice adaptation.

5 Future Perspectives

With the rapid advance in next-generation sequencing technologies and accumulated genomic resources, pan-genomic analyses may provide an efficient strategy to identify candidate genes related to the weediness traits. Models for such studies can already be found in recent research on the wild-cultivated rice complex, where both single nucleotide polymorphisms (SNPs) and structural variations (SVs) have been shown to underlie important phenotypes (Zhao et al. 2018). In addition, epigenetic control is an important regulatory mechanism that contributes to the phenotypic diversity of plant species. The role of epigenetic regulation in the control of complex phenotypes has been well-documented in cultivated rice (Deng et al., 2016). However, investigations of epigenetic mechanisms underlying weediness traits lag far behind cultivated rice phenotypes. Collectively we propose that future studies should not only focus on the single base-pair DNA sequence variations (e.g., SNPs) and large-scale structural variation (e.g., copy number polymorphisms) but also integrate both genetic and epigenetic regulatory networks to understand the genotype-phenotype connection.

6 Conclusions

Weedy crop relatives have likely been present in and around agricultural fields since the dawn of agriculture. In the case of weedy rice, recent research has revealed that this this major agricultural pest is by no means a single evolutionary entity. Rather, it is a complex assortment of lineages that have arisen multiple times independently and that have evolved and adapted over time with varying levels of genetic input from domesticated and wild rice populations. Population genetics and genomics studies suggest that the adaptations that characterize weedy rice can evolve through a variety of molecular mechanisms; this has almost certainly facilitated the process of convergent weed evolution. It remains to be seen whether these findings for weedy rice are shared with other major weedy crop relatives. While extensive, the research findings reviewed in this chapter are perhaps best considered a good starting point for more detailed characterizations of weediness traits and the mechanisms by which they evolve in weedy rice and other agricultural weed species.

Acknowledgments Weedy rice research in the Olsen and Li labs is supported through the National Science Foundation Plant Genome Research Program (IOS-1032023, IOS-1947609) and Start-up funding at Fudan University (JIH1322105).

References

Akasaka M, Ushiki J, Iwata H, Ishikawa R, Ishii T. Genetic relationships and diversity of weedy rice (*Oryza sativa* L.) and cultivated rice varieties in Okayama Prefecture, Japan. Breeding Sci. 2009;59(4):401–9.

Akasaka M, Konishi S, Izawa T, Ushiki J. Histological and genetic characteristics associated with the seed-shattering habit of weedy rice (*Oryza sativa* L.) from Okayama, Japan. Breeding Sci. 2011;61(2):168–73.

Allaby RG, Fuller DQ, Brown TA. The genetic expectations of a protracted model for the origins of domesticated crops. Proc Natl Acad Sci U S A. 2008;105(37):13982–6.

Allston RFW. The rice plant. DeBow's Review. 1846;1:320–56.

Andres A, Concenço G, Theisen G, Vidotto F, Ferrero A. Selectivity and weed control efficacy of pre-and post-emergence applications of clomazone in Southern Brazil. Crop Prot. 2013;53:103–8.

Andres A, Fogliatto S, Ferrero A, Vidotto F. Susceptibility to imazamox in Italian weedy rice populations and Clearfield® rice varieties. Weed Res. 2014;54(5):492–500.

Arnaud JF, Viard F, Delescluse M, Cuguen J. Evidence for gene flow via seed dispersal from crop to wild relatives in *Beta vulgaris* (Chenopodiaceae): consequences for the release of genetically modified crop species with weedy lineages. Proc Roy Soc Lond B Biol. 2003;270(1524):1565–71.

Arrieta-Espinoza G, Sánchez E, Vargas S, Lobo J, Quesada T, Espinoza AM. The weedy rice complex in Costa Rica. I. Morphological study of relationships between commercial rice varieties, wild *Oryza* relatives and weedy types. Genet Resour Crop Evol. 2005;52(5):575–87.

Azmi M, Azlan S, Yim K, George T, Chew S. Control of weedy rice in direct-seeded rice using the Clearfield production system in Malaysia. Pak J Weed Sci Res. 2012;18:49–53.

Baker HG. Characteristics and modes of origin of weeds. In: Baker HG, Stebbins GL, editors. The genetics of colonizing species. London: Academic Press; 1965. p. 141–72.

Barnaud A, Deu M, Garine E, Chantereau J, Bolteu J, Koïda EO, McKey D, Joly HI. A weed-crop complex in sorghum: the dynamics of genetic diversity in a traditional farming system. Am J Bot. 2009;96(10):1869–79.

Basu C, Halfhill MD, Mueller TC, Stewart CN. Weed genomics: new tools to understand weed biology. Trends Plant Sci. 2004;9(8):391–8.

Bellue MK. Weeds of California seed rice. Calif Dept Agri Bull. 1932;21:290–6.

Biroli G. Del riso. Milan: Tipografia Giovanni Silvestri; 1807. 151pp.

Bradbury LM, Henry RJ, Jin Q, Reinke RF, Waters DL. A perfect marker for fragrance genotyping in rice. Mol Breed. 2005;16(4):279–83.

Burgos NR, Norman RJ, Gealy DR, Black H. Competitive N uptake between rice and weedy rice. Field Crop Res. 2006;99(2):96–105.

Burgos NR, Singh V, Tseng TM, Black H, Young ND, Huang Z, Hyma KE, Gealy DR, Caicedo AL. The impact of herbicide-resistant rice technology on phenotypic diversity and population structure of United States weedy rice. Plant Physiol. 2014;166(3):1208–20.

Cao QJ, Li B, Song ZP, Cai XX, Lu BR. Impact of weedy rice populations on the growth and yield of direct-seeded and transplanted rice. Weed Biol Manage. 2007;7(2):97–104.

Chauhan BS. Strategies to manage weedy rice in Asia. Crop Prot. 2013;48:51–6.

Cheng C, Motohashi R, Tsuchimoto S, Fukuta Y, Ohtsubo H, Ohtsubo E. Polyphyletic origin of cultivated rice: based on the interspersion pattern of SINEs. Mol Biol Evol. 2003;20(1):67–75.

Chung JW, Park YJ. Population structure analysis reveals the maintenance of isolated sub-populations of weedy rice. Weed Res. 2010;50(6):606–20.

Civáň P, Craig H, Cox CJ, Brown TA. Three geographically separate domestications of Asian rice. Nat Plants. 2015;1:15164.

Craigmiles JP. Introduction. Pages 5-6 in red rice research and control. Texas Agric Exp Stn Bul. 1978;1270:46.

Croughan TP. Clearfield rice: it's not a GMO. La Agric. 2003;46(4):24–6.

Cui Y, Song BK, Li LF, Li YL, Huang Z, Caicedo AL, Jia Y, Olsen KM. Little white lies: pericarp color provides insights into the origins and evolution of southeast Asian weedy rice. G3. 2016;6(12):4105–14.

De Wet JM, Harlan JR. Weeds and domesticates: evolution in the man-made habitat. Econ Bot. 1975;29(2):99–108.

Delouche JC, Burgos NR, Gealy DR, de San Martin GZ, Labrada R, Larinde M, Rosell C. Weedy rices: origin, biology, ecology and control, vol. 188. Rome: Food and Agriculture Organization of the United Nations; 2007.

Ellstrand NC, Prentice HC, Hancock JF. Gene flow and introgression from domesticated plants into their wild relatives. Annu Rev Ecol Syst. 1999;30(1):539–63.

Estorninos LE Jr, Gealy DR, Gbur EE, Talbert RE, McClelland MR. Rice and red rice interference. II. Rice response to population densities of three red rice (*Oryza sativa*) ecotypes. Weed Sci. 2005;53(5):683–9.

Estorninos LE, Gealy DR, Talbert RE, Wells BR. Rice research studies 1999. Fayetteville, AR, USA. 2000. p. 463–8.

Faivre-Rampant O, Bruschi G, Abbruscato P, Cavigiolo S, Picco AM, Borgo L, Lupotto E, Piffanelli P. Assessment of genetic diversity in Italian rice germplasm related to agronomic traits and blast resistance (*Magnaporthe oryzae*). Mol Breed. 2011;27(2):233–46.

Federici MT, Vaughan D, Norihiko T, Kaga A, Xin WW, Koji D, Francis M, Zorrilla G, Saldain N. Analysis of Uruguayan weedy rice genetic diversity using AFLP molecular markers. Electron J Biotechnol. 2001;4(3):5–6.

Fuller DQ, Qin L, Zheng Y, Zhao Z, Chen X, Hosoya LA, Sun GP. The domestication process and domestication rate in rice: spikelet bases from the Lower Yangtze. Science. 2009;323(5921):1607–10.

Fuller DQ, Sato Y-I, Castillo C, Qin L, Weisskopf AR, Kingwell-Banham EJ, Song J, Ahn S-M, Van Etten J. Consilience of genetics and archaeobotany in the entangled history of rice. Archaeol Anthrop Sci. 2010;2(2):115–31.

Gao L, Innan H. Nonindependent domestication of the two rice subspecies, *Oryza sativa* ssp. *indica* and ssp. *japonica*, demonstrated by multilocus microsatellites. Genetics. 2008;179(2):965–76.

Garris AJ, Tai TH, Coburn J, Kresovich S, McCouch S. Genetic structure and diversity in *Oryza sativa* L. Genetics. 2005;169(3):1631–8.

Ge S, Sang T, Lu BR, Hong DY. Phylogeny of rice genomes with emphasis on origins of allotetraploid species. Proc Natl Acad Sci U S A. 1999;96(25):14400–5.

Gealy D. Gene movement between rice (*Oryza sativa*) and weedy rice (*Oryza sativa*): a US temperate rice perspective. In: Gressel J, editor. Crop ferality and volunteerism. London: CRC Press; 2005. p. 323–5.

Gealy DR, Tai TH, Sneller CH. Identification of red rice, rice, and hybrid populations using microsatellite markers. Weed Sci. 2002;50(3):333–9.

Gealy DH, Agrama H, Jia MH. Genetic analysis of atypical US red rice phenotypes: indications of prior gene flow in rice fields? Weed Sci. 2012;60(3):451–61.

Goff SA, Ricke D, Lan T-H, Presting G, Wang R, Dunn M, Glazebrook J, Sessions A, Oeller P, Varma H. A draft sequence of the rice genome (*Oryza sativa* L. ssp. *japonica*). Science. 2002;296(5565):92–100.

Gressel J, Valverde BE. A strategy to provide long-term control of weedy rice while mitigating herbicide resistance transgene flow, and its potential use for other crops with related weeds. Pest Manag Sci. 2009;65(7):723–31.

Grillo MA, Li C, Fowlkes AM, Briggeman TM, Zhou A, Schemske DW, Sang T. Genetic architecture for the adaptive origin of annual wild rice, *Oryza nivara*. Evolution. 2009;63(4):870–83.

Grimm A, Fogliatto S, Nick P, Ferrero A, Vidotto F. Microsatellite markers reveal multiple origins for Italian weedy rice. Ecol Evol. 2013;3(14):4786–98.

Gross BL, Zhao Z. Archaeological and genetic insights into the origins of domesticated rice. Proc Natl Acad Sci U S A. 2014;111(17):6190–7.

Gross BL, Steffen FT, Olsen KM. The molecular basis of white pericarps in African domesticated rice: novel mutations at the *Rc* gene. J Evol Biol. 2010;23(12):2747–53.

Gu XY, Kianian SF, Foley ME. Multiple loci and epistases control genetic variation for seed dormancy in weedy rice (*Oryza sativa*). Genetics. 2004;166(3):1503–16.

Gu XY, Foley ME, Horvath DP, Anderson JV, Feng J, Zhang L, Mowry CR, Ye H, Suttle JC, Kadowaki K-I. Association between seed dormancy and pericarp color is controlled by a pleiotropic gene that regulates abscisic acid and flavonoid synthesis in weedy red rice. Genetics. 2011;189(4):1515–24.

Hak-Soo S, Mun-Hue H. Collection and evaluation of Korean red rices I. Regional distribution and seed characteristics. Korean J Crop Sci. 1992;37(5):425–30.

Harlan JR. Origins of African plant domestication. The Hague: Mouton; 1976.

Harlan JR. Crops and man. Madison: American Society of Agronomy; 1992.

He Z, Zhai W, Wen H, Tang T, Wang Y, Lu X, Greenberg AJ, Hudson RR, Wu C-I, Shi S. Two evolutionary histories in the genome of rice: the roles of domestication genes. PLoS Genet. 2011;7(6):e1002100.

He Q, Kim KW, Park YJ. Population genomics identifies the origin and signatures of selection of Korean weedy rice. Plant Biotechnol J. 2017;15(3):357–66.

Heu M. Weed rice "Sharei" showing closer cross-affinity to *japonica* type. Rice Genet Newslett. 1988;5:72–4.

Hilbert L, Neves EG, Pugliese F, Whitney BS, Shock M, Veasey E, Zimpel CA, Iriarte J. Evidence for mid-Holocene rice domestication in the Americas. Nat Ecol Evol. 2017;1:1693–8.

Hill J, Smith RJ, Bayer D. Rice weed control: current technology and emerging issues in temperate rice. Aust J Exp Agr. 1994;34(7):1021–9.

Hillman G, Hedges R, Moore A, Colledge S, Pettitt P. New evidence of Lateglacial cereal cultivation at Abu Hureyra on the Euphrates. Holocene. 2001;11(4):383–93.

Huang X, Sang T, Zhao Q, Feng Q, Zhao Y, Li C, Zhu C, Lu T, Zhang Z, Li M. Genome-wide association studies of 14 agronomic traits in rice landraces. Nat Genet. 2010;42(11):961–7.

Huang X, Kurata N, Wang ZX, Wang A, Zhao Q, Zhao Y, Liu K, Lu H, Li W, Guo Y. A map of rice genome variation reveals the origin of cultivated rice. Nature. 2012a;490(7421):497–501.

Huang P, Molina J, Flowers JM, Rubinstein S, Jackson SA, Purugganan MD, Schaal BA. Phylogeography of Asian wild rice, *Oryza rufipogon*: a genome-wide view. Mol Ecol. 2012b;21(18):4593–604.

Huang Z, Young ND, Reagon M, Hyma KE, Olsen KM, Jia Y, Caicedo AL. All roads lead to weediness: patterns of genomic divergence reveal extensive recurrent weedy rice origins from South Asian *Oryza*. Mol Ecol. 2017;26(12):3151–67.

IRRI. Bring hope, improving lives: strategic plan 2007–2015. Los Baños: IRRI; 2006.

Ishii T, Akazawa M. Direct seeding in dry field and weedy rice in Okayama prefecture. Proceedings of the 18th symposium in Japanese Weedy Association. 2003. p. 7–16.

Ishikawa R, Toki N, Imai K, Sato Y, Yamagishi H, Shimamoto Y, Ueno K, Morishima H, Sato T. Origin of weedy rice grown in Bhutan and the force of genetic diversity. Genet Resour Crop Evol. 2005;52(4):395–403.

Jin Y. A study on modern rice introduction into northeastern China and its impact. Agr Hist China. 2010;3:35–41. In Chinese.

Jin J, Huang W, Gao JP, Yang J, Shi M, Zhu MZ, Luo D, Lin HX. Genetic control of rice plant architecture under domestication. Nat Genet. 2008;40(11):1365–9.

Kanapeckas KL, Vigueira CC, Ortiz A, Gettler KA, Burgos NR, Fischer AJ, Lawton-Rauh AL. Escape to ferality: the endoferal origin of weedy rice from crop rice through de-domestication. PLoS One. 2016;11(9):e0162676.

Karim RS, Man AB, Sahid IB. Weed problems and their management in rice fields of Malaysia: an overview. Weed Biol Manage. 2004;4(4):177–86.

Kawasaki A, Imai K, Ushiki J, Ishii T, Ishikawa R. Molecular constitution of weedy rice (*Oryza sativa* L.) found in Okayama prefecture, Japan. Breeding Sci. 2009;59(3):229–36.

Kelly Vaughan L, Ottis BV, Prazak-Havey AM, Bormans CA, Sneller C, Chandler JM, Park WD. Is all red rice found in commercial rice really *Oryza sativa*? Weed Sci. 2001;49(4):468–76.

Kennedy PB. Observations on some rice weeds in California, vol. 27. Oakland, CA: University of California, California Agricultural Bulletin; 1923. p. 356.

Kim T-S, He Q, Kim K-W, Yoon M-Y, Ra W-H, Li FP, Tong W, Yu J, Oo WH, Choi B. Genome-wide resequencing of KRICE_CORE reveals their potential for future breeding, as well as functional and evolutionary studies in the post-genomic era. BMC Genomics. 2016;17(1):408.

Konishi S, Izawa T, Lin SY, Ebana K, Fukuta Y, Sasaki T, Yano M. An SNP caused loss of seed shattering during rice domestication. Science. 2006;312(5778):1392–6.

Larson G, Piperno DR, Allaby RG, Purugganan MD, Andersson L, Arroyo-Kalin M, Barton L, Vigueira CC, Denham T, Dobney K. Current perspectives and the future of domestication studies. Proc Natl Acad Sci U S A. 2014;111(17):6139–46.

Li LF, Olsen K. Chapter three-to have and to hold: selection for seed and fruit retention during crop domestication. Curr Top Dev Biol. 2016;119:63–109.

Li C, Zhou A, Sang T. Rice domestication by reducing shattering. Science. 2006;311(5769):1936–9.

Li LF, Li YL, Jia Y, Caicedo AL, Olsen KM. Signatures of adaptation in the weedy rice genome. Nat Genet. 2017;49(5):811–4.

Liang D, Qiang S. Current situation and control strategy of weedy rice in China. China Plant Prot. 2011;31:21–4.

Linares OF. African rice (*Oryza glaberrima*): history and future potential. Proc Natl Acad Sci U S A. 2002;99(25):16360–5.

Ling-Hwa T, Morishima H. Genetic characterization of weedy rices and the inference on their origins. Breeding Sci. 1997;47(2):153–60.

Londo J, Schaal B. Origins and population genetics of weedy red rice in the USA. Mol Ecol. 2007;16(21):4523–35.

Londo JP, Chiang Y-C, Hung K-H, Chiang T-Y, Schaal BA. Phylogeography of Asian wild rice, *Oryza rufipogon*, reveals multiple independent domestications of cultivated rice, *Oryza sativa*. Proc Natl Acad Sci U S A. 2006;103(25):9578–83.

Merotto A, Goulart IC, Nunes AL, Kalsing A, Markus C, Menezes VG, Wander AE. Evolutionary and social consequences of introgression of nontransgenic herbicide resistance from rice to weedy rice in Brazil. Evol Appl. 2016;9(7):837–46.

Meyer RS, Purugganan MD. Evolution of crop species: genetics of domestication and diversification. Nat Rev Genet. 2013;14(12):840–52.

Molina J, Sikora M, Garud N, Flowers JM, Rubinstein S, Reynolds A, Huang P, Jackson S, Schaal BA, Bustamante CD. Molecular evidence for a single evolutionary origin of domesticated rice. Proc Natl Acad Sci U S A. 2011;108(20):8351–6.

Neik T-X, Chai J-Y, Tan S-Y, Sudo M, Cui Y, Jayaraj J, Teo S-S, Olsen KM, Song BK. When West meets East: the origins and spread of weedy rice between continental and island Southeast Asia. G3. 2019;9:2941–50.

Noldin JA, Chandler JM, McCauely GN. Red rice (*Oryza sativa*) biology. I. Characterization of red rice ecotypes. Weed Technol. 1999:12–8.

Nunes D, Boa-Sorte N, Grassi MFR, Pimentel K, Teixeira MG, Barreto ML, Dourado I, Galvão-Castro B. Evidence of a predominance of sexual transmission of HTLV-1 in Salvador, the city with the highest prevalence in Brazil. Retrovirology. 2015;12(1):O3.

Oka H. Weedy forms of rice. Origin of cultivated rice. Amsterdam: Elsevier and Japan Society Press; 1988. p. 107–14.

Olguin ERS, Arrieta-Espinoza G, Lobo JA, Espinoza-Esquivel AM. Assessment of gene flow from a herbicide-resistant *indica* rice (*Oryza sativa* L.) to the Costa Rican weedy rice (*Oryza sativa*) in Tropical America: factors affecting hybridization rates and characterization of F1 hybrids. Transgenic Res. 2009;18(4):633–47.

Olsen KM, Caicedo AL, Jia Y. Evolutionary genomics of weedy rice in the USA. J Integr Plant Biol. 2007;49(6):811–6.

Ouyang Y, Zhang Q. Understanding reproductive isolation based on the rice model. Annu Rev Plant Biol. 2013;64:111–35.

Prathepha P. Pericarp color and haplotype diversity in weedy rice (*O. sativa f. spontanea*) from Thailand. Pak J Biol Sci. 2009a;12(15):1075.

Prathepha P. The *badh2* allele of the fragrance (*fgr/BADH2*) gene is present in the gene population of weedy rice (*Oryza sativa f. spontanea*) from Thailand. Am Eurasian J Agric Environ Sci. 2009b;5:603–8.

Qi X, Liu Y, Vigueira CC, Young ND, Caicedo AL, Jia Y, Gealy DR, Olsen KM. More than one way to evolve a weed: parallel evolution of US weedy rice through independent genetic mechanisms. Mol Ecol. 2015;24(13):3329–44.

Qiu J, Zhu J, Fu F, Ye CY, Wang W, Mao L, Lin Z, Chen L, Zhang H, Guo L. Genome re-sequencing suggested a weedy rice origin from domesticated *indica-japonica* hybridization: a case study from southern China. Planta. 2014;240(6):1353–63.

Qiu J, Zhou Y, Mao L, Ye C, Wang W, Zhang J, Yu Y, Fu F, Wang Y, Qian F. Genomic variation associated with local adaptation of weedy rice during de-domestication. Nat Commun. 2017;8:15323.

Rathore M, Singh R, Kumar B, Chauhan B. Characterization of functional trait diversity among Indian cultivated and weedy rice populations. Sci Rep. 2016;6:24176.

Reagon M, Thurber CS, Gross BL, Olsen KM, Jia Y, Caicedo AL. Genomic patterns of nucleotide diversity in divergent populations of US weedy rice. BMC Evol Biol. 2010;10(1):180.

Roso A, Merotto A Jr, Delatorre C, Menezes V. Regional scale distribution of imidazolinone herbicide-resistant alleles in red rice (*Oryza sativa* L.) determined through SNP markers. Field Crop Res. 2010;119(1):175–82.

Sasaki T. The map-based sequence of the rice genome. Nature. 2005;436(7052):793.

Song BK, Chuah TS, Tam SM, Olsen KM. Malaysian weedy rice shows its true stripes: wild *Oryza* and elite rice cultivars shape agricultural weed evolution in Southeast Asia. Mol Ecol. 2014;23(20):5003–17.

Stewart CN. Becoming weeds. Nat Genet. 2017;49(5):654.

Subudhi PK, Singh PK, DeLeon T, Parco A, Karan R, Biradar H, Cohn MA, Sasaki T. Mapping of seed shattering loci provides insights into origin of weedy rice and rice domestication. J Hered. 2013;105(2):276–87.

Sudianto E, Neik T-X, Tam SM, Chuah T-S, Idris AA, Olsen KM, Song B-K. Morphology of Malaysian weedy rice (*Oryza sativa*): diversity, origin and implications for weed management. Weed Sci. 2016;64(3):501–12.

Sugimoto K, Takeuchi Y, Ebana K, Miyao A, Hirochika H, Hara N, Ishiyama K, Kobayashi M, Ban Y, Hattori T. Molecular cloning of *Sdr4*, a regulator involved in seed dormancy and domestication of rice. Proc Natl Acad Sci U S A. 2010;107(13):5792–7.

Suh H, Sato Y, Morishima H. Genetic characterization of weedy rice (*Oryza sativa* L.) based on morpho-physiology, isozymes and RAPD markers. Theor Appl Genet. 1997;94(3–4):316–21.

Sun J, Qian Q, Ma DR, Xu ZJ, Liu D, Du HB, Chen WF. Introgression and selection shaping the genome and adaptive loci of weedy rice in northern China. New Phytol. 2013;197(1):290–9.

Sweeney M, McCouch S. The complex history of the domestication of rice. Ann Bot. 2007;100(5):951–7.

Sweeney MT, Thomson MJ, Pfeil BE, McCouch S. Caught red-handed: *Rc* encodes a basic helix-loop-helix protein conditioning red pericarp in rice. Plant Cell. 2006;18(2):283–94.

Tan L, Li X, Liu F, Sun X, Li C, Zhu Z, Fu Y, Cai H, Wang X, Xie D. Control of a key transition from prostrate to erect growth in rice domestication. Nat Genet. 2008;40(11):1360–4.

Tang L, Morishima H. Genetic characteristics and origin of weedy rice. Beijing: China Agricultural University Press; 1996. p. 211–8.

Tarditi N, Vercesi B. The *Oryza silvatica* as weed: a problem more and more present in rice-growing. Verona: Informatore Agrario; 1993.

Thurber CS, Reagon M, Gross BL, Olsen KM, Jia Y, Caicedo AL. Molecular evolution of shattering loci in US weedy rice. Mol Ecol. 2010;19(16):3271–84.

Thurber CS, Jia MH, Jia Y, Caicedo AL. Similar traits, different genes? Examining convergent evolution in related weedy rice populations. Mol Ecol. 2013;22(3):685–98.

Ushiki J, Ishii T, Ishikawa R. Morpho-physiological characters and geographical distribution of *japonica* and *indica* weedy rice (*Oryza sativa*) in Okayama Prefecture, Japan. Breed Res. 2005;7:179–87.

Vaughan DA. The wild relatives of rice: a genetic resources handbook. Los Banos: International Rice Research Institute; 1994.

Vaughan DA, Morishima H, Kadowaki K. Diversity in the *Oryza* genus. Curr Opin Plant Biol. 2003;6(2):139–46.

Vaughan DA, Kadowaki K-I, Kaga A, Tomooka N. On the phylogeny and biogeography of the genus *Oryza*. Breeding Sci. 2005;55(2):113–22.

Vaughan DA, Lu BR, Tomooka N. The evolving story of rice evolution. Plant Sci. 2008;174(4):394–408.

Vidotto F, Ferrero A. Modelling population dynamics to overcome feral rice in rice. Crop ferality and volunteerism. Boca Raton, FL: CRC Press; 2005. p. 353–68.

Vigueira C, Olsen K, Caicedo A. The red queen in the corn: agricultural weeds as models of rapid adaptive evolution. Heredity. 2013;110(4):303–11.

Vigueira CC, Qi X, Song BK, Li LF, Caicedo AL, Jia Y, Olsen KM. Call of the wild rice: *Oryza rufipogon* shapes weedy rice evolution in Southeast Asia. Evol Appl. 2019;12:93–104. https://doi.org/10.1111/eva.12581.

Wahab A, Suhaimi O. Padi angin characteristics, adverse effects and methods of its eradication. Teknologi Padi. 1991;7:21–31.

Wang M, Yu Y, Haberer G, Marri PR, Fan C, Goicoechea JL, Zuccolo A, Song X, Kudrna D, Ammiraju JS. The genome sequence of African rice (*Oryza glaberrima*) and evidence for independent domestication. Nat Genet. 2014;46(9):982–8.

Warwick SI, Stewart C. Crops come from wild plants: how domestication, transgenes, and linkage together shape ferality. In: Gressel J, editor. Crop ferality and volunteerism. Boca Raton, FL: CRC Press; 2005. p. 9–30.

Wedger MJ, Pusadee T, Wongtamee A, Olsen KM. Discordant patterns of introgression suggest historical gene flow into Thai weedy rice from domesticated and wild relatives. J Hered. 2019;110:601–9.

Weiss E, Kislev ME, Hartmann A. Autonomous cultivation before domestication. Science. 2006;312:1608–10.

Willcox G, Fornite S, Herveux L. Early Holocene cultivation before domestication in northern Syria. Veg Hist Archaeobotany. 2008;17(3):313–25.

Willson JH. Rice in California. Richvale, CA: Butte County Rice Growers Association; 1979. p. 254.

Xia HB, Xia H, Ellstrand NC, Yang C, Lu BR. Rapid evolutionary divergence and ecotypic diversification of germination behavior in weedy rice populations. New Phytol. 2011;191(4):1119–27.

Yao N, Wang L, Yan H, Liu Y, Lu B-R. Mapping quantitative trait loci (QTL) determining seed-shattering in weedy rice: evolution of seed shattering in weedy rice through de-domestication. Euphytica. 2015;204(3):513–22.

Ye H, Beighley DH, Feng J, Gu XY. Genetic and physiological characterization of two clusters of quantitative trait loci associated with seed dormancy and plant height in rice. G3. 2013;3(2):323–31.

Yu J, Hu S, Wang J, Wong GK-S, Li S, Liu B, Deng Y, Dai L, Zhou Y, Zhang X. A draft sequence of the rice genome (*Oryza sativa* L. ssp. *indica*). Science. 2002;296(5565):79–92.

Yu GQ, Bao Y, Shi CH, Dong CQ, Ge S. Genetic diversity and population differentiation of Liaoning weedy rice detected by RAPD and SSR markers. Biochem Genet. 2005;43(5):261–70.

Zhang LB, Zhu Q, Wu ZQ, Ross-Ibarra J, Gaut BS, Ge S, Sang T. Selection on grain shattering genes and rates of rice domestication. New Phytol. 2009;184(3):708–20.

Zhang J, Chen LL, Xing F, Kudrna DA, Yao W, Copetti D, Mu T, Li W, Song JM, Xie W. Extensive sequence divergence between the reference genomes of two elite *indica* rice varieties Zhenshan 97 and Minghui 63. Proc Natl Acad Sci U S A. 2016;113(35):E5163–71.

Zhao K, Tung C-W, Eizenga GC, Wright MH, Ali ML, Price AH, Norton GJ, Islam MR, Reynolds A, Mezey J. Genome-wide association mapping reveals a rich genetic architecture of complex traits in *Oryza sativa*. Nat Commun. 2011;2:467.

Zhao Q, Feng Q, Lu HY, Li Y, Wang AH, Tian QL, Zhan QL, Lu YQ, Zhang L, et al. Pan-genome analysis highlights the extent of genomic variation in cultivated and wild rice. Nat Genet. 2018;50:278–84.

Zhu Q, Ge S. Phylogenetic relationships among A-genome species of the genus *Oryza* revealed by intron sequences of four nuclear genes. New Phytol. 2005;167(1):249–65.

Zhu Y, Ellstrand NC, Lu BR. Sequence polymorphisms in wild, weedy, and cultivated rice suggest seed-shattering locus *sh4* played a minor role in Asian rice domestication. Ecol Evol. 2012;2(9):2106–13.

Ziska LH, Gealy DR, Burgos N, Caicedo AL, Gressel J, Lawton-Rauh AL, Avila LA, Theisen G, Norsworthy J, Ferrero A. Chapter three-weedy (red) rice: an emerging constraint to global rice production. Adv Agron. 2015;129:181–228.

Zuo X, Lu H, Jiang L, Zhang J, Yang X, Huan X, He K, Wang C, Wu N. Dating rice remains through phytolith carbon-14 study reveals domestication at the beginning of the Holocene. Proc Natl Acad Sci U S A. 2017;114(25):6486–91.

Population Genomics of Speciation and Adaptation in Sunflowers

Dan G. Bock, Michael B. Kantar, and Loren H. Rieseberg

Abstract Sunflowers are well-established model organisms in evolutionary biology; studies of them have made important contributions to our understanding of hybridization as an evolutionarily constructive process. Here, after introducing earlier foundational work, we review recent population genomics studies in this group. We discuss the origin of sunflowers, and how genomic data has helped disentangle species relationships. We then review work on past and ongoing speciation, as well as adaptation in natural populations or during domestication and the evolution of invasiveness. Results from these studies have shed light on the nature of sunflower species, revealing that sunflower genomes are mosaics that retain evidence of past and ongoing hybridization with congeners. This occurs even in species for which multiple compounded isolating mechanisms prevent interbreeding. Studies of cultivated sunflowers have similarly clarified that a substantial fraction of the domesticated gene pool is derived from introgressions from as many as half a dozen different species, while also identifying cases of crop-wild gene flow. In invasive species, hybridization may occasionally spur highly competitive genotypes, including in perennial species where the beneficial effects of hybrid vigor can be maintained. Population genomics studies have shown that large chromosomal blocks of high linkage disequilibrium, many of which are chromosomal inversions, facilitate local adaptation of sunflower populations given widespread gene flow. These haploblocks were found to control multiple traits and are often themselves the result of hybridization and introgression. We conclude by considering future research challenges for the sunflower community. These include a thorough characterization of sunflower structural variation and the generation of new reference genomes, revisiting earlier studies based on non-genomic data, and the optimization of

---

D. G. Bock (✉)
Department of Biology, Washington University in St. Louis, St. Louis, MO, USA

M. B. Kantar
Department of Tropical Plant and Soil Sciences, University of Hawaii, Honolulu, HI, USA

L. H. Rieseberg
Department of Botany and Biodiversity Research Centre, University of British Columbia, Vancouver, BC, Canada

transformation methods that can be used to validate the function of ecologically or economically important genes.

Keywords Adaptation · Chromosomal structural evolution · Domestication · Genomics · *Helianthus* · Hybridization · Invasive species · Speciation · Sunflower

1 Introduction

In evolutionary biology, a few species and clades have become sufficiently well studied so as to contribute to the advancement of multiple lines of research on the genetics of speciation and adaptation. Some of these model systems such as the fruit fly *Drosophila melanogaster* or the house mouse *Mus musculus* have been the focus of genetic research since the turn of the twentieth century, shortly after the rediscovery of Mendel's laws of inheritance (Markow 2015; Phifer-Rixey and Nachman 2015). The goal at this stage was to study classical genetics, often in connection with development and physiology (Barr 2003). Since the 1950s, many more model systems have been developed, mainly against a backdrop of molecular and sequencing technology improvements. These advancements have allowed evolutionists to choose study organisms not because they had short generation times and were amendable to laboratory experiments, but because their natural history and ecology permitted new questions to be asked about speciation and adaptation in the wild.

The genus *Helianthus* is a notable example of one such model system. Commonly known as sunflowers, these charismatic plants started to draw the attention of evolutionists beginning in the 1940s and 1950s. This was largely due to experiments performed by botanist and evolutionary biologist Charles B. Heiser Jr. Relying on data from morphology, cytology, and crossing experiments, Heiser meticulously documented the occurrence of hybridization between sunflower taxa (e.g., Heiser 1947, 1951). In a few years, he had amassed evidence for a sufficient number of species to note, in his 1969 monograph of sunflowers, that "the discovery of another interspecific hybrid combination in the genus would scarcely be noteworthy" (Heiser et al. 1969, p. 23). At the time, by combining such experimental evidence with ecological and geographical information, Heiser had contributed some of the most compelling examples of natural introgressive hybridization (Heiser 1949). As such, from the early years of the modern synthesis, sunflowers were part of the debate on whether hybridization is a constructive or destructive force (Dobzhansky 1937; Anderson and Stebbins 1954; Stebbins 1959).

Work performed since the 1990s has expanded Heiser's work significantly. This was achieved using a combination of approaches, including field experiments, genetic and association mapping, and population genetics and genomics in natural hybrid zones (e.g., Rieseberg et al. 1999a, 2003). As a result, sunflower research has considerably added to our understanding of hybrid and non-hybrid speciation, as well as evolution during domestication, evolution of invasiveness, and local adaptation. Here, we review these contributions, focusing on studies that use population genomics.

2 *Helianthus* Diversity and Origin

Helianthus comprises 49 named species, all of which are of North American origin (Heiser et al. 1969; Schilling and Heiser 1981). The diversity captured by the genus is remarkable and can be partitioned along multiple axes. One of these is ecology. Members of the group can be found in environments as disparate as deserts, salt marshes, prairies, rock outcrops, woodland understories, and wetlands (Fig. 1; Heiser et al. 1969; Kantar et al. 2015). For a number of species, local adaptation to these environments has been confirmed by field experiments and population genomic analyses (see Sect. 4.3 below). A second axis of diversity involves reproductive strategy. Specifically, 12 members of the genus are annuals, while the remainder are perennials (Kantar et al. 2014). The annual/perennial distinction can have important bearing on the tempo of speciation (e.g., the speed of hybrid sterility evolution; Owens and Rieseberg 2014) as well as adaptation to local environments or domestication (e.g., Gaut et al. 2015). We will discuss some of these aspects below. Finally, a third major axis of diversity is ploidy level. While most species are diploid, 13 sunflower taxa are polyploids, including tetraploids and hexaploids (Kantar et al. 2014).

Efforts to understand how this diversity originated hinge upon accurate phylogenetic reconstruction. For *Helianthus*, recovering phylogenetic relationships has been notoriously difficult. In retrospect, this is not surprising. Many of the characteristics that make sunflowers an exceptional model system for evolutionary study are also known to negatively impact phylogenetic inference. Such characteristics include rapid speciation and recent origin, very large effective population sizes (Strasburg et al. 2011), a high propensity to hybridize (Sambatti et al. 2012), recent proliferation of repetitive elements (Staton et al. 2012), and multiple rounds of past polyploidization (Barker et al. 2008; Badouin et al. 2017). Under these conditions, it is easy to see why early attempts at phylogenetic reconstruction, drawing from morphology, crossing data, phytochemistry, isozymes, or molecular marker genetic data, were only partially informative.

Schilling and Heiser (1981) used morphology, reproductive strategy, and crossability to infer the first phylogeny for the genus. Subsequent studies provided additional key information, including the identification of *Phoebanthus*, a genus of two perennial species that are narrowly distributed in Florida, as the sister group to *Helianthus* (Schilling 2001). Also notable is the clarification that widespread diploid annual species, including the important oilseed crop *H. annuus*, form a monophyletic clade (Schilling 2001; Timme et al. 2007). Finally, early phylogenetic studies based on genetic data from ribosomal genes were able to identify three putative instances of homoploid hybrid speciation (Rieseberg 1991), which will be discussed below (see Sect. 3.1).

Fig. 1 Representative *Helianthus* diversity. (**a**) *H. anomalus*, (**b**) dune ecotype of *H. petiolaris* ssp. *fallax*. (**c, d**) wild *H. annuus*. (**e**) *H. argophyllus* coastal ecotype. (**f**) cultivar of *H. annuus*. Images courtesy of Nolan C. Kane, Mariana A. Pascual-Robles, and Jason Rick

The use of next-generation sequencing has improved, over the past five years, our understanding of infrageneric species relationships. Stephens et al. (2015) used capture probes and Illumina sequencing to simultaneously isolate 170 phylogenetically informative loci across 37 diploid *Helianthus*. Results showed that most taxa are part of three large clades, one of which contains annuals, and two of which contain perennials. Also, this study provided evidence that the ancestral sunflower was likely a perennial species, with annual life history evolving subsequently three times (Stephens et al. 2015). This is in agreement with observations from other plant taxa, which indicates annual species tend to evolve from perennial ancestors as an adaptation to harsh environmental conditions, such as aridity (Friedman and Rubin 2015). The phylogeny from Stephens et al. (2015) was further used to provide evidence that temperature seasonality constrains genome size expansion in *Helianthus* (Qiu et al. 2019).

Low-coverage whole genome sequencing (i.e., genome skimming) and reduced-representation sequencing have been used to disentangle any remaining ambiguous relationships and revisit previously suspected hybrid species (e.g., Baute et al. 2016; Owens et al. 2016; Zhang et al. 2019). Some of these studies focused on polyploid hybrids. Because these taxa are also perennials, and therefore, fewer generations removed from their progenitors, they have proven particularly difficult to resolve using traditional phylogenetic markers. Information from complete cytoplasmic genomes, as well as much of the rDNA nuclear segment was informative in two cases so far, the hexaploid tuber crop *H. tuberosus* (Bock et al. 2014a), and the critically endangered tetraploid *H. schweinitzii* (Anderson et al. 2019). Lastly, notable among studies using genome skimming is Lee-Yaw et al. (2019). The authors integrated phylogenetic methods and selection analyses to understand drivers of discordance between nuclear and cytoplasmic markers. Cytonuclear discordance has been a common finding in the plant literature and is usually ascribed to incomplete lineage sorting or hybridization. Aside from confirming the occurrence of hybridization and subsequent cytoplasm introgression, results pointed to the contribution of natural selection in driving patterns of plastid DNA variation. This possibility has previously been supported using field experiments in sunflowers (Sambatti et al. 2008) and is being considered increasingly often in other taxa as well (Bock et al. 2014b).

Given what we know so far on sunflower phylogeny, what can we say about the geography, timing, and tempo of *Helianthus* diversification? Considering that *Phoebanthus*, the sister group to *Helianthus*, as well as basal sunflowers such as *H. porteri* have a distribution that is restricted to the South-Eastern US (Schilling 2001; Stephens et al. 2015), it seems likely that the ancestor of *Helianthus* was a perennial species of Central American origin. Further supporting this possibility is the fact that sister-taxa to the clade formed by *Helianthus* and *Phoebanthus* occur in Mexico and South America (Schilling 2001). Molecular clock analyses, while inherently uncertain (Donoghue and Benton 2007), timed the split between the first sunflower species and this ancestor at ~3.6 million years ago (Mason 2018). This places the start of sunflower diversification within the Pliocene, a period characterized by cooling, drying, and considerable vegetation restructuring in

North America, including reductions of closed forest habitat (Mason 2018). This makes sense considering that sunflowers typically occur in open-vegetation environments (Heiser et al. 1969). Subsequent to range expansion from Central America, diversification within North America led to the formation of three major clades. One of these contains species that are more common in dry soils of the South-Western US. Adaptations to these challenging environments likely included an annual reproductive strategy, but also traits that enhance water use efficiency and promote fast growth (Mason and Donovan 2015). Subsequent radiation from the South-Western US occurred later for widespread members of this annual clade, such as *H. petiolaris* (Heiser 1961). The other two ancestral *Helianthus* clades consist of perennials occurring, respectively, in riparian habitats across the Central-Eastern US and water-rich environments of the South-Eastern US. Speciation within all three clades is recent (often <1–2 Myr; Stephens et al. 2015) and occurred in the context of substantial interspecific gene flow (Sambatti et al. 2012; Lee-Yaw et al. 2019) and genome rearrangements (Burke et al. 2004; Barb et al. 2014). We discuss studies investigating these speciation events below.

3 Population Genomics of Sunflower Speciation

Sunflowers are well-known for their contribution to speciation theory. Building on earlier research performed by Heiser (discussed above), species formation has been studied extensively in this group, facilitated in part by its recent origin. This provides the opportunity of studying speciation from taxa that are fully isolated to those that are just transitioning to the status of incipient species (e.g., Andrew and Rieseberg 2013; Ostevik et al. 2016). Here, we follow this speciation continuum. After introducing key previous results, we review population genomics studies of past and ongoing sunflower speciation.

3.1 Sunflowers as Models of Recombinational Speciation

Students of evolutionary biology not familiar with sunflowers are most likely to be introduced to this system in the context of homoploid hybrid or "recombinational" speciation. This is because three sunflower species, *H. anomalus*, *H. deserticola*, and *H. paradoxus* are thought to offer some of the strongest empirical support for this mode of speciation (Yakimowski and Rieseberg 2014). Phylogenetic evidence suggested that these taxa were of hybrid origin and that they likely share the same two progenitors, the widespread annuals *H. annuus* and *H. petiolaris* (Rieseberg 1991). Molecular data further indicate these speciation events occurred rapidly, likely within hundreds of generations (Ungerer et al. 1998; Buerkle and Rieseberg 2008). Despite their presumed hybrid origin, the hybrid neospecies are strongly isolated from their progenitors, as both intrinsic (in the form of almost complete F_1

hybrid sterility; Heiser 1958; Rieseberg 2000) and extrinsic (in the form of ecological differentiation; Rieseberg 1991; Lexer et al. 2003) barriers to gene flow are known. As predicted by the recombinational speciation model (Stebbins 1957; Grant 1958), the hybrids differ from parental taxa in multiple chromosomal rearrangements (Rieseberg et al. 1995a; Lai et al. 2005) that are associated with F_1 sterility (Lai et al. 2005).

Ongoing work is using high-resolution linkage maps based on SNP data to study chromosomal evolution in the genus. Results have so far highlighted that karyotype changes in diploid sunflowers are dominated by inversions and interchromosomal translocations (Ostevik et al. 2020). Among these, inversions predominate (Ostevik et al. 2020). Also, when interpreting results across plant and animal groups, sunflowers appear to have exceptionally high rates of chromosomal evolution (Barb et al. 2014; Ostevik et al. 2020). Non-random patterns have further been identified across the genome, with some chromosomes involved in more translocations than others. Leading explanations for this involve ancestral homology retained after whole genome duplication, as well as repeat element content (Ostevik et al. 2020).

The relevance of chromosomal rearrangements is easy to see when one considers intrinsic isolation during hybrid speciation. Previous studies have shown that hybrid sterility is a result of either direct rearrangement effects or the effects of genic incompatibilities that cluster within rearrangements (Lai et al. 2005). However, this only addresses part of the barriers necessary for hybrid speciation. An important remaining question is how might ecological isolation be achieved at the same time? Theoretical studies indicate that, without niche divergence, hybrids can be outcompeted by numerically superior parental genotypes (Buerkle et al. 2000). As such, hybrid speciation requires the concomitant development of strong ecological divergence (Buerkle et al. 2000).

Luckily, in hybrid sunflower species, ecological isolation can be studied directly. This is because the putative parent taxa and hybrid derivatives are extant. As such, field experiments can be conducted to compare all taxa and/or artificial hybrids, therefore controlling for post-speciation divergence. This was the approach taken by Lexer et al. (2003) to study mechanisms of ecological isolation in *H. paradoxus*. The authors used interspecific BC_2 hybrids between *H. annuus* and *H. petiolaris* and performed QTL mapping for survivorship and elemental uptake traits in salt marshes, the environment typical of *H. paradoxus*. Results provided evidence for strong selection at QTLs with effects in opposing directions. This was early support for the possibility that rapid ecological divergence during hybrid speciation is achieved as a result of selection for recombinants with extreme (transgressive) phenotypes. The mechanistic basis for transgression in this case is complementary gene action (Rieseberg et al. 1999b). Under this model, extreme phenotypes are a product of the "stacking" of alleles from several QTLs that are fixed between parents, and that control the same trait (reviewed in Rieseberg et al. 1999b).

The complementation model provides a convincing explanation for the emergence of transgression and subsequent ecological isolation, but it is not the only mechanism at play. Specifically, gene expression changes likely contribute as well. Lai et al. (2006) used microarray analyses of *H. deserticola* and the two parental taxa. Results showed evidence of transgressive expression in the putative hybrid species as compared to its parents, particularly involving transport-related genes (Lai et al. 2006). These changes are not observed in F_1 *H. annuus/H. petiolaris* hybrids, as estimated using RNA-seq (Rowe and Rieseberg 2013). Thus, current evidence points to gene expression divergence occurring in later generations. Explanations that have been put forth so far include mechanisms independent of the hybridization event such as post-speciation selection, as well as those that are related to the genome merger (Lai et al. 2006). Among this latter category, possibilities include genome rearrangements and transposable element activity (Lai et al. 2006), both of which are elevated in the hybrid sunflower taxa (Renaut et al. 2014a) and have been linked to gene expression divergence (Lai et al. 2006; Dion-Cote et al. 2014; Harewood and Fraser 2014).

3.2 *Non-hybrid Speciation*

While a substantial share of speciation research in sunflowers has focused on understanding mechanisms behind homoploid hybrid species formation, studies of non-hybrid taxa have been extremely insightful as well. These studies sought to understand patterns of introgression along the genome and to document the nature of sunflower species boundaries. These investigations have been facilitated by several characteristics of the sunflower system. For one, detailed information is available on reproductive isolating barriers for some members of the group (e.g., flowering time, pollen competition, hybrid sterility; Rieseberg et al. 1995b; Rieseberg 2000; Sambatti et al. 2012). Using knowledge of barrier strength, estimates can be obtained on the probability of hybridization, which can then be related to genome scans for differentiation between species (e.g., Sambatti et al. 2012). Also, exceptional resources for studying the population genomics of speciation are available in *Helianthus*. These include large EST databases, SNP and expression arrays, high-density genetic maps, and reference genomes (Heesacker et al. 2008; Kane et al. 2013; Badouin et al. 2017; Hübner et al. 2019). Finally, because of the widespread distribution of sunflowers across continental US, contrasts can be made among species pairs that differ in the potential for gene flow. This permits the study of the geographical context of speciation (e.g., Renaut et al. 2013).

An important finding emerging from these studies is that sunflower genomes often are mosaics, with a considerable genomic fraction having an interspecific origin (Kane et al. 2009; Scascitelli et al. 2010; Sambatti et al. 2012; Zhang et al. 2019). This occurs even in species for which interbreeding is rare and prevented by the compounded action of multiple isolating mechanisms. For example, studies of *H. annuus* and *H. petiolaris* have estimated a cumulative barrier close to 1 in both directions (Sambatti et al. 2012), placing these species at the upper end of barrier strength among flowering plants (Lowry et al. 2008). Using widely accepted criteria, these would be considered good species. The intuitive prediction in this case is that extremely rare hybridization translates into strong genetic differentiation between taxa. To the contrary, population genomics has provided evidence of non-independence for the two gene pools. For example, Kane et al. (2009) used 26 microsatellites and 1420 EST-derived orthologs and identified long-term gene flow between nearby populations of *H. annuus* and *H. petiolaris*. This added to previous, more localized examples of genetic exchange between the two species (e.g., Buerkle and Rieseberg 2001; Yatabe et al. 2007).

To reconcile these apparently conflicting results, Sambatti et al. (2012) used coalescent modelling. Results pointed to the large effective population size of these two species as the primary reason for gene flow evidence, even when successful hybridization is extremely rare. Under this scenario, the traces of past introgression are preserved because of very limited genetic drift. Also notable is that the genomic mosaic observed in this case is not merely a result of the recent origin of *H. annuus* and *H. petiolaris*, as revealed by comparisons of even younger speciation events (Sambatti et al. 2012). Beyond documenting the lasting contribution of gene flow to genomic variation in sunflowers, these studies highlighted the important role of ecology in maintaining sunflower species cohesion. In spite of widespread genetic exchange, most species are known to maintain distinctive morphologies and habitat requirements (Kane et al. 2009). Conversely, in the absence of habitat differentiation and under human disturbance or biological invasion, hybridization can be rampant, and even result in the genetic assimilation of species (Kane et al. 2009; Todesco et al. 2016; but see Owens et al. 2016).

If, as discussed above, gene flow has been occurring between *H. annuus* and *H. petiolaris* throughout their evolutionary history, how has this shaped the genomic landscape of divergence? In sympatry and parapatry, the genomic landscape is predicted to be highly heterogenous (Nosil et al. 2009). Peaks of differentiation associated with loci under divergent natural selection are expected to be interspersed by valleys of differentiation corresponding to neutral regions that are being homogenized by gene flow. This contrasts with the expectation in allopatry, where the absence of gene flow should allow neutral and adaptive divergence to accumulate anywhere in the genome (Nosil et al. 2009). Identifying determinants of the genomic landscape of differentiation can contribute to our understanding of speciation. For example, some models of speciation with gene flow predict that initially narrow peaks of differentiation will progressively expand and facilitate divergence of other linked genes (Via 2012). Referred to as "divergence hitchhiking" this process is thought to ultimately lead to the formation of the so-called speciation islands during speciation with gene flow (Via 2012).

To test these predictions in sunflowers, Renaut et al. (2013) used transcriptome data and pairs of taxa that differ in the geography of speciation. These consisted of *H. annuus* and *H. petiolaris* as representatives of sympatric divergence with gene flow. By considering two other taxa, *H. debilis* and *H. argophyllus*, additional comparisons could be made, representing parapatry with and without gene flow, and allopatry (Renaut et al. 2013). Contrary to expectations, the genomic landscape of divergence was not affected by interspecific gene flow. Regardless of sympatric, parapatric, or allopatric categorization, species were found to diverge at numerous independent genomic regions (Renaut et al. 2013). These results notwithstanding,

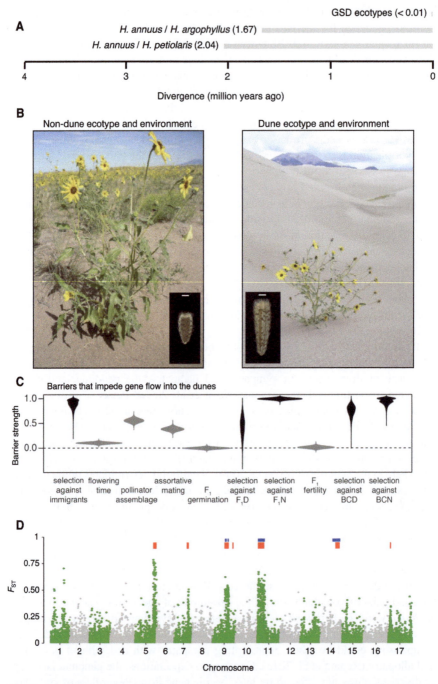

Fig. 2 Population genomics of incipient speciation in Great Sand Dunes (GSD) *H. petiolaris*. (**a**) Divergence time among GSD *H. petiolaris* dune and non-dune ecotypes, as compared to divergence times among two sunflower species pairs for which reproductive isolation has previously been

there was a strong association between islands of divergence and recombination rate. Thus, the study by Renaut et al. (2013) highlighted that, in sunflower species, the functional architecture of the genome, as opposed to gene flow, is a strong predictor of genome divergence (Renaut et al. 2013).

3.3 Sunflower Speciation in Action

Studies of past speciation such as those highlighted above help us understand the process retrospectively: we look back in time and make inferences based on the best available evidence (Via 2009). Ideally, however, we would also observe speciation in action. This is because studies of incipient species can provide valuable information on how gene flow is initially suppressed. Combining the two approaches is possible in sunflowers, since genotypes can be obtained that span the continuum from well-isolated taxa (above) to young ecotypes just transitioning to species status. A well-established system for the study of incipient speciation in *Helianthus* are the dune and non-dune ecotypes of *H. petiolaris* at Great Sand Dunes (GSD) National Park in Colorado (Fig. 2). This is due to a succession of studies that provided detailed information on the evolutionary history of the two ecotypes, reproductive barriers and genomic divergence.

Population genomic data indicate that dune-adapted *H. petiolaris* diverged from an ancestral non-dune population at GSD less than 10,000 years ago (Fig. 2a; Andrew et al. 2013). While gene flow is still occurring (Andrew et al. 2012, 2013), dune and non-dune ecotypes are locally adapted and maintain large divergence at key traits including seed size (Fig. 2b; Ostevik et al. 2016). In line with these observations, analyses of reproductive isolation identified several reproductive barriers that are already active in this system (Fig. 2c; Ostevik et al. 2016). As expected under local adaptation and recent divergence, many barriers were extrinsic. These included both prezygotic (selection against immigrants; divergence in pollinator assemblages) and postzygotic (selection against hybrids; Ostevik et al. 2016) barriers. One intrinsic barrier was identified as well (postpollination assortative mating, Ostevik et al. 2016). Comparisons of barrier strength revealed that, in the

◀─────────────

Fig. 2 (continued) investigated (e.g., using F_1 pollen sterility; see Owens and Rieseberg 2014). Divergence estimates are obtained from Andrew et al. (2013) for GSD and from Mason (2018) for the other two comparisons. (**b**) Representative plants of the non-dune and dune *H. petiolaris* ecotypes, in their respective habitats at GSD, Colorado. Scale bar for seed size indicates 1 mm. Photo credits Kate L. Ostevik (left GSD image), Rose L. Andrew (right GSD image), and Marco Todesco (seed images). (**c**) Barrier strength estimates considering gene flow from non-dune to dune plants, modified from Ostevik et al. (2016). Barriers for which seed size is likely to contribute are highlighted in black. (**d**) F_{ST} comparison between dune and non-dune ecotypes (from Huang et al. 2020). Blue bars are used to indicate the genomic location of seed-size QTLs (as identified in Todesco et al. 2020). Red bars indicate putative inversions

case of *H. petiolaris* at GSD, postzygotic barriers are more effective than prezygotic barriers. As well, extrinsic barriers were shown to be more effective than intrinsic barriers.

How do these barriers impact genome divergence, and how did they evolve? Between-ecotype genome scans identified three major "islands of divergence" against a backdrop of otherwise reduced genomic differentiation (Andrew and Rieseberg 2013). This result further confirmed that the two ecotypes are at the very early stages of speciation, when only a small fraction of the genome is contributing to isolation (Andrew and Rieseberg 2013). While few, the islands of divergence were wide, likely encompassing a large number of genes (Andrew and Rieseberg 2013). Thus, as compared to the substantially older *H. annuus*/ *H. petiolaris* species pair discussed above, the GSD ecotypes better correspond to expectations under "divergence hitchhiking."

An alternative that can explain the large islands of divergence between these ecotypes is the occurrence of chromosomal inversions. Recent studies have increasingly emphasized that structural polymorphism is an important component of adaptive genetic variation (Mérot et al. 2020; Todesco et al. 2020). A follow-up study of the GSD system provided strong evidence for precisely this scenario. Using RADseq and analytical methods that facilitate inversion detection, Huang et al. (2020) demonstrated that islands of divergence between *H. petiolaris* ecotypes do indeed correspond to chromosomal inversions (Fig. 2d). Moreover, these inversions were shown to co-localize with seed-size QTLs (Fig. 2d), as well as environmental variables that differ between dune and non-dune sites (Huang et al. 2020). Thus, in GSD *H. petiolaris*, inversions contribute to divergent adaptation by preventing recombination between co-adapted alleles. Ongoing work is aiming to clarify whether postpollination assortative mating also maps to these inversions (Huang et al. 2020). This is particularly relevant because models of speciation with gene flow emphasize the importance of linkage between loci involved in local adaptation and those involved in assortative mating (Ortiz-Barrientos et al. 2016; Huang and Rieseberg 2020).

4 Population Genomics of Adaptation

The spectacular diversity of *Helianthus* has motivated a number of studies on the genetics of adaptation. Understanding how sunflowers cope with the local environment is relevant for both basic and applied reasons. First, ecological divergence is a major component of sunflower speciation. Thus, understanding local adaptation will provide a window into microevolutionary dynamics at the foundation of species diversity in the genus. Second, *Helianthus* contains two crops, the oilseed *H. annuus* and the tuber crop *H. tuberosus*. Thus, studies can be conducted to understand adaptation during domestication. Even more so, studies of adaptation can help identify genetic resources for breeding stress-resistant cultivars, which is relevant from the perspective of food security under climate change (e.g., Gao et al. 2019).

Lastly, *H. annuus* and *H. tuberosus* contain widespread invasive ecotypes. Thus, these species also allow us to understand evolutionary mechanisms of invasion success.

4.1 Evolution During Domestication

Not only have sunflowers been a model for understanding speciation, they have also been instrumental for understanding evolution under domestication, largely due to work on the oilseed crop *H. annuus*. Key topics that have been addressed include the geographic origin(s) of crops in the Americas, the speed of evolutionary change under domestication, and the genetic mechanisms that drive these changes (Burke et al. 2002; Harter et al. 2004). Despite this, there was still controversy about the origin of the domesticated sunflower as recently as the early twenty-first century (Lentz et al. 2008; Rieseberg and Burke 2008). However, this debate has been settled using analyses of candidate domestication genes (Blackman et al. 2011a). More recently, shotgun sequencing of archeological DNA contributed as well (Wales et al. 2019). These studies have clarified that extant domestic sunflower has a single origin, which occurred around 4,000 years ago in Eastern North America.

Genetics and genomics research over the past two decades has made clear that phenotypic transitions during domestication can be the result of a wide range of mutations (Purugganan 2019). The most common type is a non-synonymous single nucleotide polymorphism. In sunflower, a prime example of this are the flowering time genes. For example, the sunflower *HaFT1* locus, a homolog of a known flowering time regulator, contains a frameshift single nucleotide polymorphism that differentiates wild and domestic *H. annuus* (Blackman et al. 2010). Sequence analyses of the exon containing this frameshift have confirmed that *HaFT1* has been under selection in landrace and elite lines, while evolving neutrally in wild populations (Blackman et al. 2010). Other domestication traits mapped in sunflower include plant architecture and fatty acid synthesis (Wills and Burke 2007; Chapman and Burke 2012). These studies relied on linkage mapping, association mapping, and F_{ST} outlier scans to identify well over 100 candidate domestication genes (Burke et al. 2002; Wills and Burke 2007; Chapman et al. 2008; Mandel et al. 2013; Baute et al. 2015). While this contrasts with observations from other crops in terms of the number of inferred domestication loci, a polygenic signature of domestication is becoming more commonly observed with more widespread use of genomic data (Chen et al. 2020).

Not only were specific genes altered during sunflower domestication, but genome-wide patterns of polymorphism were changed as well. For example, comparisons between wild and domesticated *H. annuus* revealed differences in the content of transposable elements (Mascagni et al. 2015). Also, there have been significant changes in RNA-splicing (Smith et al. 2018). Another transition with genome-wide consequences was the incorporation of a hybrid breeding system, which occurred in the 1970s (Fick and Swallers 1972). This resulted in gene content changes among heterotic groups, including genes involved in pathogen defense (Owens et al. 2019). Heterosis likely resulting from complementation contributed to large yield increases (Owens et al. 2019). This is relevant because, like many other domesticated species (Gaut et al. 2015), sunflower has gone through a bottleneck with genetic diversity being significantly reduced in cultivated lines (Liu and Burke 2006; Wales et al. 2019; Park and Burke 2020). In agreement with this observation, cultivated sunflower genomes were shown to contain more deleterious mutations than wild genomes (Renaut and Rieseberg 2015). However, deleterious polymorphisms currently present in the domestic sunflower gene pool will be challenging to remove because they have accumulated predominately in low-recombination regions (Renaut and Rieseberg 2015).

Following the development of the hybrid breeding system, improvement of domestic sunflower lines also relied on the introduction of genetic material from wild congeners (Baute et al. 2015). Population genomic analyses have helped reveal the genomic extent and the phenotypic consequences of these introgressions. Baute et al. (2015), for example, used transcriptome sequencing to show that introgressed regions account for ~10% of the genome in cultivated sunflower. All modern cultivars examined were found to contain one or more introgression-derived genomic regions (Baute et al. 2015). Hübner et al. (2019) further expanded these results in an analysis of the sunflower pangenome. Results reiterated that approximately 10% of the genome, and 1.5% of genes, originated *via* introgression. Genes involved in biotic resistance were over-represented among those found in introgressed regions. This was confirmed using a GWAS analysis for downy mildew (*Plasmopara halstedii*) resistance, which identified several strong associations that overlap with introgressions from wild species (Hübner et al. 2019).

Understanding the way evolution has changed species under domestication provides new ways to quickly select and develop additional domesticates. For example, known domestication genes represent excellent targets of selection and can help define expectations for breeding programs. A new area of interest in domestication research involves comparative analyses of annual and perennial crops (Gaut et al. 2015). *Helianthus* provides research opportunities in this direction, given that it contains both an annual domesticate (*H. annuus*) and a less well-studied perennial domesticate (*H. tuberosus*; Kantar et al. 2014). Further, there is much interest in domesticating new species to make agricultural systems more sustainable. The *Helianthus* genus provides key species in this endeavor as well (Asselin et al. 2020). We, therefore, anticipate that sunflowers will continue to be important to our understanding of how plants interact with human society, and how this relationship can be improved as more genomic knowledge is gained.

4.2 Evolution of Invasiveness

When considering problematic plants, a distinction can be made between agricultural weeds and invasive species. While admittedly blurry and rarely used, this classification is based on the environment in which the plants typically occur. At one extreme, agricultural weeds are found in highly managed environments, such as cultivated croplands. At the other extreme, invasive species often spread in more natural communities that are minimally altered by human activity (Ellstrand et al. 2010). Aside from this distinction, agricultural weeds and invasive species are similar in many respects. For example, both are known to adapt rapidly in response to novel challenges they encounter. Agricultural weeds rely on adaptations, such as rapid growth, herbicide resistance (Baucom 2019), or crop mimicry (Barrett 1983) to thrive under the unique conditions resulting from cultivation. These include high availability of water and fertilizer, but also targeted removal. Invasive species as well have been shown to respond adaptively to a host of conditions including novel climates (Colautti and Barrett 2013) and reduced competition in empty niches (Dlugosch et al. 2015). Other shared commonalities include the occurrence of hybridization, reduced population size, and serial founder events (Bock et al. 2015; Hodgins et al. 2018). *Helianthus* contains two well-known representatives for both categories, the *H. annuus* agricultural weed and the *H. tuberosus* invasive species. We discuss below how population genomics studies in these taxa have contributed to our understanding of invasiveness in agricultural and natural settings.

The *H. annuus* agricultural weed is common in croplands throughout North America and Europe, where it can have a large economic impact (e.g., Deines et al. 2004). It achieves a patchier distribution in Australia and in Argentina, where it is more often found as a ruderal plant (Presotto et al. 2017). Population genetic and genomic analyses have indicated that weedy *H. annuus* has a diverse origin. This includes multiple derivations from nearby wild populations in North America (Kane and Rieseberg 2008), intraspecific crop-wild hybrid origins in Europe (Muller et al. 2011), and a mixture of wild and interspecific F_1 hybrid origins involving crosses with *H. petiolaris* in Argentina (Mondon et al. 2018). Common garden and drought experiments conducted using North American, Australian, and Argentinian genotypes have highlighted that, relative to wild *H. annuus*, weedy forms have evolved faster growth at the expense of reduced drought tolerance (Mayrose et al. 2011; Koziol et al. 2012; Presotto et al. 2017).

The genetic basis of wild to weedy transition has been investigated for North American *H. annuus*. This was done using genome scans for selection based on 106 EST-derived microsatellites (Kane and Rieseberg 2008) and using microarray-based gene expression analyses (Lai et al. 2008). These studies have highlighted that weediness in this system was likely achieved *via* changes at a small genomic fraction. Kane and Rieseberg (2008) reported evidence of selection at 0.9–5.6% of loci examined. Likewise, Lai et al. (2008) found evidence for significant gene expression differences at 5% of the genes investigated. Moreover, both studies identified limited overlap in outlier loci among weedy populations, thus indicating weediness is easy to evolve in sunflowers, and can have a diverse genetic basis (Kane and Rieseberg 2008). Ongoing work is attempting to expand on these results using whole genome resequencing (Drummond 2018). While comparisons of wild *H. annuus* with independently derived weedy genotypes did reveal some evidence of parallel genetic differentiation, the relative contribution of idiosyncratic changes

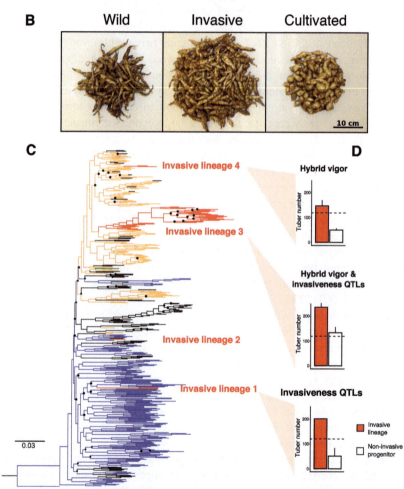

Fig. 3 Population genomics of invasiveness in *H. tuberosus* (modified from Bock et al. 2018). (**a**) Invasive *H. tuberosus* stand (Czech Republic, 2013; photo credit Dan G. Bock). Inset shows the number of sprouts (means ± s.e.m.) produced by invasive and non-invasive genotypes

could not be assessed at this point. Even so, preliminary results have reiterated that the transition to weediness likely involved a small fraction of the genome (<1%; Drummond 2018). Also, outlier regions contained genes associated with a variety of functions, as expected given that wild and weedy sunflowers diverge in multiple traits. These candidate weediness genes are associated with abiotic and biotic stress response, as well as herbicide tolerance (Drummond 2018).

In contrast to *H. annuus*, invasive genotypes of *H. tuberosus* are more often found in natural habitats (Fig. 3a). In Europe, where this perennial sunflower is considered a highly invasive species, populations frequently occur in nutrient- and water-rich soils along river courses (Descombes et al. 2016). A recent study investigated evolution of invasiveness in this system (Bock et al. 2018). Genome-wide SNP data showed that invasive genotypes have a diverse origin: some derive directly from wild accessions, whereas others are wild-domesticated hybrids (Fig. 3c). Common garden results further showed that invasive genotypes invest more in growth. Specifically, invasive plants were found to produce roughly twice the number of vegetative propagules (tubers) that wild and domesticated plants would typically make (Fig. 3b). Moreover, vegetative propagation was shown to be the primary driver of invasive spread in this species. A drought experiment further revealed that increased investment in vegetative propagation is manifested only when water resources are not limiting (Fig. 3a). This result has two implications. First, it shows that invasiveness in *H. tuberosus* evolved by genetic accommodation, *via* adaptive evolution of phenotypic plasticity. Second, it highlights that invasive *H. tuberosus*, similar to weedy *H. annuus*, has adapted to exploit a resource-rich environment in areas of spread. Lastly, association mapping analyses showed that invasive lineages differ in the genetic basis of invasiveness. Namely, two non-exclusive genetic mechanisms were shown to contribute to invasive spread: hybrid vigor and two large-effect QTLs (Fig. 3d; Bock et al. 2018). This result, thus, further strengthens the conclusion that several genetic routes to invasiveness are available in weedy and invasive sunflower populations.

Are results from sunflowers indicative of mechanisms for invasion success likely to be at play in other systems? Recent years have seen an acceleration of research on evolutionary drivers of invasiveness, implementing genomics and field experiments. As such, some preliminary generalizations can be made. For one, it seems likely that increased virulence of introduced genotypes can evolve easily and will have a diverse genetic basis. This is supported by genomic analyses in other taxonomic

Fig. 3 (continued) under water-stress and well-watered conditions. (**b**) Tuber yield from wild, invasive, and cultivated genotypes. (**c**) Maximum-likelihood phylogeny of *H. tuberosus*, including wild (blue), invasive (red), and cultivated (orange) genotypes. Invasive populations have at least four distinct origins. (**d**) For three origins with available phenotype data, means ($\pm$ s.e.m.) are given for tuber number, the main invasiveness trait in this system, in invasive genotypes (red) and closely related non-invasive samples (white). Dotted line shows the mean across the collection. For each origin, inferred genetic mechanisms of invasiveness (i.e., hybrid vigor and/or invasiveness QTLs) are given

groups. For example, Hodgins et al. (2015) traced the signature of natural selection in 35 Asteraceae species, including six major invasive species. Results provided limited evidence for parallel changes in orthologous genes (Hodgins et al. 2015). Concordant results have also been obtained with regard to phenotypic and physiological changes that occur during invasions. Specifically, invasive genotypes have frequently been found to display superior growth under high-resource conditions, but reduced tolerance to abiotic stress. Thus, invasiveness appears to frequently evolve in resource-rich habitats (Hodgins et al. 2018). Such invasion-prone environments may be a result of changes in community structure, including historical declines of competitors (e.g., Dlugosch et al. 2015).

4.3 Local Adaptation with Gene Flow

How does local adaptation originate and persist when gene flow is occurring? Traditionally, gene flow has often been viewed as a disruptor of local adaptation (Tigano and Friesen 2016). This is because, when genetic exchange is rampant, alleles that confer an advantage in the local environment can be swamped by foreign variants. Under this scenario, even if locally suboptimal, alleles with the best fitness across populations will tend to become fixed, and local adaptation will be lost (Lenormand 2002). Modelling studies indicate that the likelihood of such swamping depends on the intensity of gene flow and selection (Tigano and Friesen 2016). Recently, examples of local adaptation with gene flow have become more common, facilitated by genomics tools that allow the topic to be studied in any organism (Tigano and Friesen 2016). As such, there has been a renewed interest in understanding the destructive as well as constructive roles of gene flow during local adaptation. This is particularly relevant in sunflowers, because of the high propensity for gene flow that is characteristic of the genus.

That sunflower populations are locally adapted has been supported by multiple studies using field experiments, population genomics, or a combination of these approaches. Field experiments have used reciprocal transplants, often highlighting photoperiod and soil characteristics such as water and nutrient content as important components of local adaptation in sunflowers (e.g., Sambatti et al. 2008; Whitney et al. 2010; Ostevik et al. 2016). Population genomic screens have searched for the signature of natural selection. The goal in this case was to characterize adaptive evolution from the perspective of types of genomic changes (Moyers and Rieseberg 2013), the number and types of genes (e.g., Kane et al. 2011; Renaut et al. 2012; McAssey et al. 2016), or the contribution of genomic landscape (e.g., Renaut et al. 2014b). Lastly, studies using a combination of the two have aimed to clarify the occurrence of local adaptation using comparisons of quantitative trait vs. neutral genetic differentiation (e.g., Blackman et al. 2011b; Moyers and Rieseberg 2016).

While local adaptation is frequent in sunflowers, how is it achieved given widespread gene flow? A recent study by Todesco et al. (2020) addressed this question. Three species were considered, each containing pairs of locally adapted ecotypes that are well within gene flow contact. For *H. annuus*, sampling covered typical populations found throughout the US, often adapted to mesic soils. Additionally, populations of *H. annuus* subsp. *texanus* were used, which are adapted to sites with increased temperature and pathogen pressure in Texas (Whitney et al. 2010). For *H. petiolaris*, the study included dune and non-dune ecotypes that evolved independently in both Texas and Colorado (Ostevik et al. 2016). Finally, for *H. argophyllus*, sampling covered southern Texas early and late flowering ecotypes. These are differentially adapted to coastal barrier islands (early flowering ecotype) and mainland sites (late flowering ecotype; Fig. 4a; Moyers and Rieseberg 2016). In all, whole genome resequencing data for 1,506 genotypes was used in combination with an extensive developmental, morphological, and environmental dataset.

One of the most striking results was that many of the traits and environmental variables known to be involved in local adaptation in these ecotypes mapped to large chromosomal blocks of high linkage disequilibrium (Todesco et al. 2020). Moreover, many of these chromosomal blocks were found to control multiple traits. For example, in *H. argophyllus*, flowering time as well as leaf nitrogen and carbon content was associated with a 30 Mb region containing two main haplotypes (Fig. 4b–d). Plants with different haplotypes at this region flowered on average 77 days apart (Fig. 4f). Additional analyses confirmed that a large fraction of these plateaus of association correspond to structural variants, represented by inversions and more complex rearrangements. Analyses of the 1,506 genomes further revealed that structural variation is common across the three sunflower species, representing 4–16% of the genome (Todesco et al. 2020). Perhaps equally striking, the divergence times between haplotype blocks was found to exceed the inferred age of the species in which they currently segregate. This suggests some of the haplotype blocks currently contributing to local adaptation in sunflowers may be a result of introgression from older, currently extinct taxa (Todesco et al. 2020).

Aside from clarifying the genetic basis of several adaptive traits, results from this study contributed to our understanding of adaptation with gene flow in at least two important ways. For one, it supports previous theoretical work (Yeaman and Whitlock 2011) indicating that gene flow shapes the genetic architecture of local adaptation. In cases when adaptation occurs with interpopulation genetic exchange, selection will favor modular architectures such as those observed in sunflowers. This is because recombination modifiers including chromosomal inversions prevent the breakup of adaptive allele combinations. At the same time, Todesco et al. (2020) highlight that gene flow can facilitate local adaptation *via* adaptive introgression. This is supported by the finding that haplotype blocks acting as large-effect loci that control multiple adaptive traits can have an interspecific origin (Fig. 4g). This result thus further reinforces the constructive effect of hybridization for biodiversity, a finding that is becoming more and more common as new taxa are being investigated (Tigano and Friesen 2016).

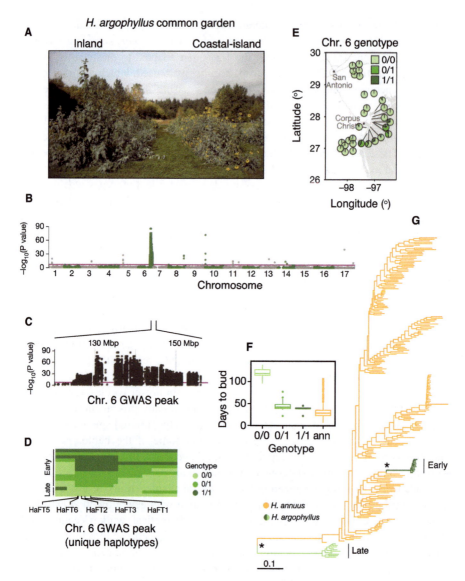

Fig. 4 Population genomics of local adaptation in *H. argophyllus* (modified from Todesco et al. 2020). (**a**) *H. argophyllus* inland and coastal-island plants grown in a common garden (photo credit Brook T. Moyers). (**b**) Flowering time GWAS results. (**c**) Close-up of the chromosome 6 region containing the main association peak. (**d**) Schematic of unique haplotypes contributing to the GWAS flowering time peak, with position of six sunflower homologues of the *FT* flowering-time regulator. (**e**) Geographical distribution of haplotypes identified under the GWAS peak on chromosome 6. Early-flowering genotypes identified using GWAS predominate in coastal populations. (**f**) Flowering time differences among genotypes at the chromosome 6 GWAS peak (ann = *H. annuus*). (**g**) Maximum-likelihood phylogeny of the chromosome 6 region, highlighting the interspecific origin (*H. annuus* – derived) of the early-flowering haplotype. Asterisks indicate well-supported major nodes

4.4 Crop-Wild Gene Flow

In addition to genetic exchange among differentially adapted wild populations, gene flow has been documented between the domesticated sunflower, which is adapted to cultivation, and wild or weedy populations of *H. annuus* and *H. petiolaris* (Linder et al. 1998; Rieseberg et al. 1999c; Muller et al. 2011; Mondon et al. 2018). When the crop is abundant, such gene flow has the potential to reduce the fitness of adjacent wild populations (Ellstrand and Rieseberg 2016). Conversely, some crop traits such as resistance to herbicides or pests could enhance weediness (Snow et al. 2003).

While most studies of crop-wild gene flow in sunflowers have employed a relatively small number of molecular markers, two recent studies provide genome-wide estimates of hybridization and introgression (Corbi et al. 2017; Mondon et al. 2018). In the latter study, Mondon et al. (2018) confirmed the presence of hybridization and introgression between domesticated sunflower (*H. annuus*) and *H. petiolaris* populations in Argentina, despite the strong reproductive barrier between these species (Sambatti et al. 2012). However, it remains unclear whether hybridization is causative or incidental to the weedy life history traits found in Argentinian *H. petiolaris*. Corbi et al. (2017) tracked genome-wide allele frequency shifts in an experimentally synthesized crop × wild sunflower population, which was planted at two natural sites and allowed to evolve for two generations. As expected, most crop alleles were selected against, but a handful were favored. Allelic frequency changes were not closely correlated with shifts in phenotype, possibly suggesting that crop alleles that complemented deleterious alleles were favored rather than a particular phenotype *per se.*

5 Conclusions and Future Directions

As outlined in the introduction, sunflowers have emerged as a useful experimental model for addressing a wide range of evolutionary questions. Most importantly, studies of wild sunflower species have been key to establishing hybridization as a creative force in evolution, facilitating adaptation, and leading to the formation of new species (Rieseberg et al. 2007). Beyond hybridization, sunflowers are unusual in their high rate of chromosomal structural evolution, and studies of wild sunflowers have shown how large structural variants such as chromosomal inversions permit adaptive divergence and speciation in the presence of gene flow (Barb et al. 2014; Huang et al. 2020). Recently, these two previously disparate research themes have been merged with the discovery that polymorphic structural variants segregating in natural populations are often themselves the product of hybridization and introgression (Todesco et al. 2020).

While many of the large structural variants segregating in sunflower populations have been shown to be inversions, others are more complex, possibly representing nested inversions and/or large deletions (Todesco et al. 2020). Also, the comparative genetic mapping and population genomics approaches used to detect structural variation are biased towards detecting large (>1 Mb) structural variants. An important goal of the sunflower community going forward should be to better characterize structural variation in cultivated and wild sunflowers. This will be most efficiently accomplished by generation of reference genomes for all wild sunflower species, focusing initially on the close relatives of the domesticated sunflower and then expanding to the perennials. Such an effort is currently underway, led by the International Consortium on Sunflower Genomics (https://www.heliagene.org/ICSG/). Concomitantly, Hi-C sequencing and Bionano genome mapping of structural variants will be useful to fill in the gaps between reference sequences, as well to infer whether variants represent fixed differences or are segregating within populations or species. Of equal importance will be follow-up studies that attempt to link the structural variants with ecologically important phenotypic traits, climate variables, or soil characteristics.

Beyond generation of new reference sequences, the time has come to revisit well-studied examples of adaptive introgression and hybrid speciation in the sunflower system. Many of these examples were first developed based on cytogenetic and morphological data (e.g., Heiser 1947, 1949, 1951), followed by gene tree analyses (e.g., Rieseberg et al. 1988) and low-resolution comparative genetic mapping (e.g., Rieseberg et al. 1995a). Phylogenomic analyses offer a means to re-examine, enrich, and potentially re-interpret results from these earlier studies (e.g., Owens et al. 2016).

In addition to the wild species, the sunflower genus is blessed with two different crops, one annual (domesticated sunflower) and one perennial (Jerusalem artichoke). Thus, the genus can offer insights into the genetics of domestication and improvement, while permitting a contrast between annual and perennial crops. As with the wild species, arguably the most distinctive feature of the domestication and improvement process in *Helianthus* is hybridization. The cultivated sunflower genome is a mosaic, with introgressions from half a dozen different wild species (Hübner et al. 2019). These introgressions often lack genes that are present in wild *H. annuus*, potentially contributing to linkage drag (Owens et al. 2019). The Jerusalem artichoke is an allohexaploid, with its genome derived via hybridization between diploid Sawtooth Sunflower and tetraploid Hairy Sunflower (Bock et al. 2014a). Hybridization between domesticated and wild Jerusalem artichoke also appears to have played a non-exclusive role in the origin of invasive Jerusalem artichoke (Bock et al. 2018).

As with the wild species, a priority of research going forward should be a reference sequence for Jerusalem artichoke, as well as additional references for the domesticated sunflower. The reference sequences not only provide a means for detecting and characterizing structural variants, but they are also critical for identifying candidate genes underlying domestication or invasiveness traits. Given that >25% of sunflower genes exhibit presence/absence variation among domesticated lines (Hübner et al. 2019), the availability of reference genomes from diverse cultivated lines increases the likelihood that a targeted gene will be present in at least one of them.

Despite successes from population genomic analyses of wild and cultivated *Helianthus*, there are some notable challenges to using sunflowers as an evolutionary or ecological study system. One such challenge is genome size. Sunflower's 3.6 Gb genome is 7.9x the size of the rice genome and 27x the size of the *Arabidopsis* genome. Thus, the recent study by Todesco et al. (2020), which reported on the sequencing of 1,506 sunflower genomes, would be equivalent to sequencing circa 12,000 rice genomes and 41,000 *Arabidopsis* genomes! The differential is even greater for Jerusalem artichoke's 10.9 Gb genome. Because the vast majority of the sunflower genome is made up of highly repetitive transposable elements, Todesco et al. (2020) employed a repeat depletion protocol to reduce the fraction of repetitive sequence by about 3-fold. Such an approach offers the advantage of reducing total sequencing costs while retaining all low copy regions of the genome.

An even greater challenge is sunflower's recalcitrance to transformation, which makes it difficult to validate the function of ecologically and/or agriculturally important candidate genes. Thus, sunflower biologists often rely on heterologous transformation in *Arabidopsis* to explore the phenotypic effects of sunflower alleles (Blackman et al. 2010; Todesco et al. 2020). While the approach works well for genes with phenotypes that can be assayed in *Arabidopsis*, it may not be reliable for traits which do not exist in *Arabidopsis* or whose expression is highly dependent on genetic background. Thus, a clear imperative for the sunflower community is to develop facile transformation and gene editing protocols. Until such methods exist, sunflowers will fail to achieve their potential as an ecological or evolutionary model.

Acknowledgements The authors are grateful to Genome Canada, Genome British Columbia, the Natural Sciences and Engineering Research Council of Canada (NSERC), and the National Science Foundation (NSF) for funding of much of the population genomics research in sunflowers over the past two decades. Also, we thank Rose L. Andrew, Nolan C. Kane, Brook T. Moyers, Kate L. Ostevik, Mariana A. Pascual-Robles, Jason Rick, and Marco Todesco for the sunflower images.

References

Anderson E, Stebbins GL. Hybridization as an evolutionary stimulus. Evolution. 1954;8:378–88.
Anderson J, Kantar M, Bock DG, Grubbs KC, Schilling E, Rieseberg LH. Skim-sequencing reveals the likely origin of the enigmatic endangered sunflower *Helianthus schweinitzii*. Gene. 2019;10:1040.
Andrew RL, Rieseberg LH. Divergence is focused on few genomic regions early in speciation: incipient speciation of sunflower ecotypes. Evolution. 2013;67:2468–82.
Andrew RL, Ostevik KL, Ebert DP, Rieseberg LH. Adaptation with gene flow across the landscape in a dune sunflower. Mol Ecol. 2012;21:2078–91.
Andrew RL, Kane NC, Baute GJ, Grassa CJ, Rieseberg LH. Recent non-hybrid origin of sunflower ecotypes in a novel habitat. Mol Ecol. 2013;22:799–813.

Asselin SR, Brûlé-Babel AL, Van Tassel DL, Cattani DJ. Genetic analysis of domestication parallels in annual and perennial sunflowers (*Helianthus* spp.): routes to crop development. Front Plant Sci. 2020;11:834.

Badouin H, Gouzy J, Grassa CJ, et al. The sunflower genome illuminates the evolutionary history of the Asterids and provides new insights into oil metabolism and flowering time. Nature. 2017;546:148–52.

Barb JG, Bowers JE, Renaut S, Rey JI, Knapp SJ, Rieseberg LH, et al. Chromosomal evolution and patterns of introgression in *Helianthus*. Genetics. 2014;197:969–79.

Barker MS, Kane NC, Matvienko M, Kozik A, Michelmore RW, Knapp SJ, et al. Multiple paleopolyploidizations during the evolution of the compositae reveal parallel patterns of duplicate gene retention after millions of years. Mol Biol Evol. 2008;25:2445–55.

Barr MM. Super models. Physiol Genomics. 2003;13:15–24.

Barrett SH. Crop mimicry in weeds. Econ Bot. 1983;37:255–82.

Baucom RS. Evolutionary and ecological insights from herbicide resistant weeds: what have we learned about plant adaptation, and what is left to uncover? New Phytol. 2019;223:68–82.

Baute GJ, Kane NC, Grassa C, Lai Z, Rieseberg LH. Genome scans reveal candidate domestication and improvement genes in cultivated sunflower, as well as post-domestication introgression with wild relatives. New Phytol. 2015;206:830–8.

Baute GJ, Owens GL, Bock DG, Rieseberg LH. Genome-wide GBS data provide a high-resolution view of wild *Helianthus* diversity, genetic structure and interspecies gene flow. Am J Bot. 2016;103:2170–7.

Blackman BK, Strasburg JL, Raduski AR, Michaels SD, Rieseberg LH. The role of recently derived FT paralogs in sunflower domestication. Curr Biol. 2010;20:629–35.

Blackman BK, Scascitelli M, Kane NC, et al. Sunflower domestication alleles support single domestication center in eastern North America. Proc Natl Acad Sci U S A. 2011a;108:14360–5.

Blackman BK, Michaels SD, Rieseberg LH. Connecting the sun to flowering in sunflower adaptation. Mol Ecol. 2011b;20:3503–12.

Bock DG, Kane NC, Ebert DP, Rieseberg LH. Genome skimming reveals the origin of the Jerusalem Artichoke tuber crop species: neither from Jerusalem nor an Artichoke. New Phytol. 2014a;201:1021–30.

Bock DG, Andrew RL, Rieseberg LH. On the adaptive value of cytoplasmic genomes in plants. Mol Ecol. 2014b;23:4899–911.

Bock DG, Caseys C, Cousens RD, et al. What we still don't know about invasion genetics. Mol Ecol. 2015;24:2277–97.

Bock DG, Kantar M, Caseys C, Matthey-Doret R, Rieseberg LH. Evolution of invasiveness by genetic accommodation. Nat Ecol Evol. 2018;2:991–9.

Buerkle CA, Rieseberg LH. Low intraspecific variation for genomic isolation between hybridizing sunflower species. Evolution. 2001;55:684–91.

Buerkle CA, Rieseberg LH. The rate of genome stabilization in homoploid hybrid species. Evolution. 2008;62:266–75.

Buerkle CA, Morris RJ, Asmussen MA, Rieseberg LH. The likelihood of homoploid hybrid speciation. Heredity. 2000;84:441–51.

Burke JM, Tang S, Knapp SJ, Rieseberg LH. Genetic analysis of sunflower domestication. Genetics. 2002;161:1257–67.

Burke JM, Lai Z, Salmaso M, et al. Comparative mapping and rapid karyotypic evolution in the genus *Helianthus*. Genetics. 2004;167:449–57.

Chapman MA, Burke JM. Evidence of selection on fatty acid biosynthetic genes during the evolution of cultivated sunflower. Theor Appl Genet. 2012;125:897–907.

Chapman MA, Pashley CH, Wenzler J, Hvala J, Tang S, Knapp SJ, et al. A genomic scan for selection reveals candidates for genes involved in the evolution of cultivated sunflower (*Helianthus annuus*). Plant Cell. 2008;20:2931–45.

Chen Q, Samayoa LF, Yang CJ, et al. The genetic architecture of the maize progenitor, teosinte, and how it was altered during maize domestication. PLoS Genet. 2020;16:e1008791.

Colautti RI, Barrett SC. Rapid adaptation to climate facilitates range expansion of an invasive plant. Science. 2013;342:364–6.

Corbi J, Baack EJ, Dechaine JM, Seiler G, Burke JM. Genome-wide analysis of allele frequency change in sunflower crop-wild hybrid populations evolving under natural conditions. Mol Ecol. 2017;27:233–47.

Deines SR, Dille JA, Blinka EL, Regehr DL, Staggenborg SA. Common sunflower (*Helianthus annuus*) and shattercane (*Sorghum bicolor*) interference in corn. Weed Sci. 2004;52:976–83.

Descombes P, Petitpierre B, Morard E, Berthoud M, Guisan A, Vittoz P. Monitoring and distribution modelling of invasive species along riverine habitats at very high resolution. Biol Invasions. 2016;18:1–15.

Dion-Cote AM, Renaut S, Normandeau E, Bernatchez L. RNA-seq reveals transcriptomic shock involving transposable elements reactivation in hybrids of young lake whitefish species. Mol Biol Evol. 2014;31:1188–99.

Dlugosch KM, Cang FA, Barker BS, Andonian K, Swope SM, Rieseberg LH. Evolution of invasiveness through increased resource use in a vacant niche. Nat Plants. 2015;1:1–5.

Dobzhansky TH. Genetics and the origin of species. New York: Columbia University Press; 1937.

Donoghue PCJ, Benton MJ. Rocks and clocks: calibrating the tree of life using fossils and molecules. Trends Ecol Evol. 2007;22:424–31.

Drummond EBM. The role of adaptive evolution in the success of an agricultural weed, *Helianthus annuus*. PhD dissertation. 2018.

Ellstrand NC, Rieseberg LH. When gene flow really matters: gene flow in applied evolutionary biology. Evol Appl. 2016;9:833–6.

Ellstrand NC, Heredia SM, Leak-Garcia JA, Heraty JM, Burger JC, Yao L, et al. Crops gone wild: evolution of weeds and invasives from domesticated ancestors. Evol Appl. 2010;3:494–504.

Fick GN, Swallers CM. Higher yields and greater uniformity with hybrid sunflowers. N Dak Farm Res. 1972;29:7–9.

Friedman J, Rubin MJ. All in good time: understanding annual and perennial strategies in plants. Am J Bot. 2015;102:497–9.

Gao L, Lee JS, Hubner S, Hulke BS, Qu Y, Rieseberg LH. Genotypic and phenotypic analyses indicate that resistance to flooding stress is uncoupled from performance in cultivated sunflower. New Phytol. 2019;223:1657–70.

Gaut BS, Diez CM, Morrell PL. Genomics and the contrasting dynamics of annual and perennial domestication. Trends Genet. 2015;31:709–19.

Grant V. The regulation of recombination in plants. In: Exchange of genetic material: mechanisms and consequences, Cold Spring Harbor Symposium on Quantitative Biology. Cold Spring Harbor, NY; 1958.

Harewood L, Fraser P. The impact of chromosomal rearrangements on regulation of gene expression. Hum Mol Genet. 2014;23:R76–82.

Harter AV, Gardner KA, Falush D, Lentz DL, Bye RA, Rieseberg LH. Origin of extant domesticated sunflowers in eastern North America. Nature. 2004;430:201–5.

Heesacker A, Kishore VK, Gao W, et al. SSRs and INDELs mined from the sunflower EST database: abundance, polymorphisms, and cross-taxa utility. Theor Appl Genet. 2008;117:1021–9.

Heiser CB. Hybridization between the sunflower species *Helianthus annuus* and *H. petiolaris*. Evolution. 1947;1:249–62.

Heiser CB. Natural hybridization with particular reference to introgression. Bot Rev. 1949;15:645–87.

Heiser CB. Hybridization in the annual sunflowers: *Helianthus annuus* X *H. debilis* var. *cucumerifolius*. Evolution. 1951;5:42–51.

Heiser CB. Three new annual sunflowers (*Helianthus*) from the southwestern United States. Rhodora. 1958;60:271–83.

Heiser CB. Morphological and cytological variation in *Helianthus petiolaris* with notes on related species. Evolution. 1961;15:247–58.

Heiser CB Jr. Study in the evolution of the sunflower species *Helianthus annuus* and *H. bolanderi*. Univ Calif Publ Bot. 1949;23:157–208.

Heiser CB, Smith DM, Clevenger SB, Martin WC. The North American sunflowers (*Helianthus*). Mem Torrey Bot Club. 1969;22:1–218.

Hodgins KA, Bock DG, Hahn MA, Heredia SM, Turner KG, Rieseberg LH. Comparative genomics in the Asteraceae reveals little evidence for parallel evolutionary change in invasive taxa. Mol Ecol. 2015;24:2226–40.

Hodgins KA, Bock DG, Rieseberg LH. Trait evolution in invasive species. Annu Plant Rev. 2018;1:1–37.

Huang K, Rieseberg LH. Frequency, origins, and evolutionary role of chromosomal inversions in plants. Front Plant Sci. 2020;11:296.

Huang K, Andrew RL, Owens GL, Ostevik KL, Rieseberg LH. Multiple chromosomal inversions contribute to adaptive divergence of a dune sunflower ecotype. Mol Ecol. 2020;29 (14):2535–49. https://doi.org/10.1111/mec.15428.

Hübner S, Bercovich N, Todesco M, et al. Sunflower pan-genome analysis shows that hybridization altered gene content and disease resistance. Nat Plants. 2019;5:54–62.

Kane NC, Rieseberg LH. Genetics and evolution of weedy *Helianthus annuus* populations: adaptation of an agricultural weed. Mol Ecol. 2008;17:384–94.

Kane NC, King M, Barker MS, Raduski A, Karrenberg S, Yatabe Y, et al. Comparative genomic and population genetic analyses indicate highly porous genomes and high levels of gene flow between divergent *Helianthus* species. Evolution. 2009;63:2061–75.

Kane NC, Barker MS, Zhan S-H, Rieseberg LH. Molecular evolution across the Asteraceae: micro- and macroevolutionary processes. Mol Biol Evol. 2011;28:3225–35.

Kane NC, Burke JM, Marek L, Seiler G, Vear F, Baute G, et al. Sunflower genetic, genomic and ecological resources. Mol Ecol Resour. 2013;13:10–20.

Kantar MB, Baute GJ, Bock DG, Rieseberg LH. Genomic variation in *Helianthus*: learning from the past and looking into the future. Brief Funct Genomics. 2014;3:328–40.

Kantar MB, Sosa CC, Khoury CK, Castaneda-Alvarez NP, Achicanoy HA, Bernau V, et al. Ecogeography and utility to plant breeding of the crop wild relatives of sunflower (*Helianthus annuus* L.). Front Plant Sci. 2015;6:841.

Koziol L, Rieseberg LH, Kane N, Bever JD. Reduced drought tolerance during domestication and the evolution of weediness results from tolerance-growth trade-offs. Evolution. 2012;66:3803–14.

Lai Z, Nakazato T, Salmaso M, Burke JM, Tang S, Knapp SJ, et al. Extensive chromosomal repatterning and the evolution of sterility barriers in hybrid sunflower species. Genetics. 2005;171:291–303.

Lai Z, Gross B, Zou Y, Andrews J, Rieseberg LH. Microarray analysis reveals differential gene expression in hybrid sunflower species. Mol Ecol. 2006;15:1213–27.

Lai Z, Kane NC, Zou Y, Rieseberg LH. Natural variation in gene expression between wild and weedy populations of *Helianthus annuus*. Genetics. 2008;179:1881–90.

Lee-Yaw JA, Grassa CJ, Joly S, Andrew RL, Rieseberg LH. An evaluation of alternative explanations for widespread cytonuclear discordance in annual sunflowers (*Helianthus*). New Phytol. 2019;221:515–26.

Lenormand T. Gene flow and the limits to natural selection. Trends Ecol Evol. 2002;17:183–9.

Lentz DL, Pohl MD, Alvarado JL, Tarighat S, Bye R. Sunflower (*Helianthus annuus* L.) as a pre-Columbian domesticate in Mexico. Proc Natl Acad Sci U S A. 2008;105:6232–7.

Lexer C, Welch ME, Raymond O, Rieseberg LH. The origin of ecological divergence in *Helianthus paradoxus* (Asteraceae): selection on transgressive characters in a novel hybrid habitat. Evolution. 2003;57:1989–2000.

Linder CR, Taha I, Seiler GJ, Snow AA, Rieseberg LH. Long-term introgression of crop genes into wild sunflower populations. Theor Appl Genet. 1998;96:339–47.

Liu A, Burke JM. Patterns of nucleotide diversity in wild and cultivated sunflower. Genetics. 2006;173:321–30.

Lowry DB, Modliszewski JL, Wright KM, Wu CA, Willis JH. The strength and genetic basis of 5 reproductive isolating barriers in flowering plants. Philos Trans R Soc B. 2008;363:3009–21.

Mandel JR, Nambeesan SU, Bowers JE, Marek LF, Ebert D, Rieseberg LH, et al. Association mapping and the genomic consequences of selection in sunflower. PLoS Genet. 2013;9: e1003378.

Markow TA. The secret lives of *Drosophila* flies. Elife. 2015;4:e06793.

Mascagni F, Barghini E, Giordani T, Rieseberg LH, Cavallini A, Natali L. Repetitive DNA and plant domestication: variation in copy number and proximity to genes of LTR-retrotransposons among wild and cultivated sunflower (*Helianthus annuus*) genotypes. Genome Biol Evol. 2015;7:3368–82.

Mason CM. How old are sunflowers? A molecular clock analysis of key divergences in the origin and diversification of *Helianthus* (Asteraceae). Int J Plant Sci. 2018;179:182–91.

Mason CM, Donovan LA. Evolution of the leaf economics spectrum in herbs: evidence from environmental divergences in leaf physiology across *Helianthus* (Asteraceae). Evolution. 2015;69:2705–20.

Mayrose M, Kane NC, Mayrose I, Dlugosch KM, Rieseberg LH. Increased growth in sunflower correlates with reduced defences and altered gene expression in response to biotic and abiotic stress. Mol Ecol. 2011;20:4683–94.

McAssey EV, Corbi J, Burke JM. Range-wide phenotypic and genetic differentiation in wild sunflower. BMC Plant Biol. 2016;16:249.

Mérot C, Oomen RA, Tigano A, Wellenreuther M. A roadmap towards understanding the evolutionary significance of structural genomic variation. Trends Ecol Evol. 2020;35(7):561–72. https://doi.org/10.1016/j.tree.2020.03.002.

Mondon A, Owens GL, Poverene M, Cantamutto M, Rieseberg LH. Gene flow in Argentinian sunflowers as revealed by genotyping by sequencing data. Evol Appl. 2018;11:193–204.

Moyers BT, Rieseberg LH. Divergence in gene expression is uncoupled from divergence in coding sequence in a newly woody sunflower. Int J Plant Sci. 2013;174:1079–89.

Moyers BT, Rieseberg LH. Remarkable life history polymorphism may be evolving under divergent selection in the silverleaf sunflower. Mol Ecol. 2016;25:3817–30.

Muller MH, Latreille M, Tollon C. The origin and evolution of a recent agricultural weed: population genetic diversity of weedy populations of sunflower (*Helianthus annuus* L.) in Spain and France. Evol Appl. 2011;4:499–514.

Nosil P, Funk DJ, Ortiz-Barrientos D. Divergent selection and heterogeneous genomic divergence. Mol Ecol. 2009;18:375–402.

Ortiz-Barrientos D, Engelstadter J, Rieseberg LH. Recombination rate evolution and the origin of species. Trends Ecol Evol. 2016;31:226–36.

Ostevik KL, Andrew RL, Otto SP, Rieseberg LH. Multiple reproductive barriers separate recently diverged sunflower ecotypes. Evolution. 2016;70:2322–35.

Ostevik KL, Samuk K, Rieseberg LH. Ancestral reconstruction of karyotypes reveals an exceptional rate of nonrandom chromosomal evolution in sunflower. Genetics. 2020;214:1031–45.

Owens GL, Rieseberg LH. Hybrid incompatibility is acquired faster in annual than in perennial species of sunflower and tarweed. Evolution. 2014;68:893–900.

Owens GL, Baute GJ, Rieseberg LH. Revisiting a classic case of introgression: hybridization and gene flow in Californian sunflowers. Mol Ecol. 2016;25:2630–43.

Owens GL, Baute GJ, Hubner S, Rieseberg LH. Genomic sequence and copy-number evolution during hybrid crop development in sunflowers. Evol Appl. 2019;12:54–65.

Park B, Burke JM. Phylogeography and the evolutionary history of sunflower (*Helianthus annuus* L.): wild diversity and the dynamics of domestication. Gene. 2020;11:266.

Phifer-Rixey M, Nachman MW. Insights into mammalian biology from the wild house mouse *Mus musculus*. Elife. 2015;4:e05959. https://doi.org/10.7554/eLife.05959.001.

Presotto A, Hernández F, Díaz M, Fernández-Moroni I, Pandolfo C, Basualdo J, et al. Crop-wild sunflower hybridization can mediate weediness throughout growth-stress tolerance trade-offs. Agric Ecosyst Environ. 2017;249:12–21.

Purugganan MD. Evolutionary insights into the nature of plant domestication. Curr Biol. 2019;29: R705–14.

Qiu F, Baack EJ, Whitney KD, Bock DG, Tetreault HM, Rieseberg LH, et al. Phylogenetic trends and environmental correlates of nuclear genome size variation in *Helianthus* sunflowers. New Phytol. 2019;221:1609–18.

Renaut S, Rieseberg LH. The accumulation of deleterious mutations as a consequence of domestication and improvement in sunflowers and other Compositae crops. Mol Biol Evol. 2015;32:2273–83.

Renaut S, Grassa CJ, Moyers BT, Kane NC, Rieseberg LH. The population genomics of sunflowers and genomic determinants of protein evolution revealed by RNAseq. Biology. 2012;1:575–96.

Renaut S, Grassa CJ, Yeaman S, Moyers BT, Lai Z, Kane NC, et al. Genomic islands of divergence are not affected by geography of speciation in sunflowers. Nat Commun. 2013;4:1827.

Renaut S, Rowe HC, Ungerer M, Rieseberg LH. Genomics of homoploid hybrid speciation: diversity and transcriptional activity of LTR retrotransposons in hybrid sunflowers. Philos Trans R Soc B. 2014a;369:20130345.

Renaut S, Owens GL, Rieseberg LH. Shared selection pressure and local genomic landscape lead to repeatable patterns of genomic divergence in sunflowers. Mol Ecol. 2014b;23:311–24.

Rieseberg LH. Homoploid reticulate evolution in *Helianthus* (Asteraceae): evidence from ribosomal genes. Am J Bot. 1991;78:1218–37.

Rieseberg LH. Crossing relationships among ancient and experimental sunflower hybrid lineages. Evolution. 2000;54:859–65.

Rieseberg LH, Burke JM. Molecular evidence and the origin of the domesticated sunflower. Proc Natl Acad Sci U S A. 2008;105:E46.

Rieseberg LH, Soltis DE, Palmer JD. A molecular re-examination of introgression between *Helianthus annuus* and *H. bolanderi* (Compositae). Evolution. 1988;42:227–38.

Rieseberg LH, Fossen CV, Desrochers AM. Hybrid speciation accompanied by genomic reorganization in wild sunflowers. Nature. 1995a;375:313–6.

Rieseberg LH, Desrochers AM, Youn JJ. Interspecific pollen competition as a reproductive barrier between sympatric species of *Helianthus* (Asteraceae). Am J Bot. 1995b;82:515–9.

Rieseberg LH, Whitton J, Gardner K. Hybrid zones and the genetic architecture of a barrier to gene flow between two sunflower species. Genetics. 1999a;152:713–27.

Rieseberg LH, Archer MA, Wayne RK. Transgressive segregation, adaptation, and speciation. Heredity. 1999b;83:363–72.

Rieseberg LH, Kim MJ, Seiler GJ. Introgression between the cultivated sunflower and a sympatric wild relative, *Helianthus petiolaris* (Asteraceae). Int J Plant Sci. 1999c;160:102–8.

Rieseberg LH, Raymond O, Rosenthal DM, et al. Major ecological transitions in wild sunflowers facilitated by hybridization. Science. 2003;301:1211–6.

Rieseberg LH, Kim SC, Randell RA, Whitney KD, Gross BL, Lexer C, et al. Hybridization and the colonization of novel habitats by annual sunflowers. Genetica. 2007;129:149–65.

Rowe HC, Rieseberg LH. Genome-scale transcriptional analyses of first-generation interspecific sunflower hybrids reveals broad regulatory compatibility. BMC Genomics. 2013;14:342.

Sambatti JBM, Ortiz-Barrientos D, Baack EJ, Rieseberg LH. Ecological selection maintains cytonuclear incompatibilities in hybridizing sunflowers. Ecol Lett. 2008;11:1082–91.

Sambatti JBM, Strasburg JL, Ortiz-Barrientos D, Baack EJ, Rieseberg LH. Reconciling extremely strong barriers with high levels of gene exchange in annual sunflowers. Evolution. 2012;21:2078–91.

Scascitelli M, Whitney KD, Randell RA, King M, Buerkle CA, Rieseberg LH. Genome scan of hybridizing sunflowers from Texas (*Helianthus annuus* and *H. debilis*) reveals asymmetric patterns of introgression and small islands of genomic differentiation. Mol Ecol. 2010;19:521–41.

Schilling EE. Phylogeny of *Helianthus* and related genera. Oléagineux Corps Gras Lipides. 2001;8:22–5.

Schilling EE, Heiser CB. An infrageneric classification of *Helianthus* (Compositae). Taxon. 1981;30:393–403.

Smith CCR, Tittes S, Mendieta JP, Collierzans E, Rowe H, Rieseberg LH, et al. Genetics of alternative splicing evolution during sunflower domestication. Proc Natl Acad Sci U S A. 2018;69:789–815.

Snow AA, Pilson D, Rieseberg LH, Paulsen M, Pleskac N, Reagon MR, et al. A Bt transgene reduces herbivory and enhances fecundity in wild sunflowers. Ecol Appl. 2003;13:279–84.

Staton S, Hartman Bakken B, Blackman B, Chapman B, Kane N, Tang S, et al. The sunflower (*Helianthus annuus* L.) genome reflects a recent history of biased accumulation of transposable elements. Plant J. 2012;72:142–53.

Stebbins GL. Self fertilization and population variability in the higher plants. Am Nat. 1957;91:337–54.

Stebbins GL. The role of hybridization in evolution. Proc Am Philos Soc. 1959;103:231–51.

Stephens J, Rogers W, Mason C, Donovan L, Malmberg R. Species tree estimation of diploid *Helianthus* (Asteraceae) using target enrichment. Am J Bot. 2015;102:910–20.

Strasburg JL, Kane NC, Raduski AR, Bonin A, Michelmore R, Rieseberg LH. Effective population size is positively correlated with levels of adaptive divergence among annual sunflowers. Mol Biol Evol. 2011;28:1569–80.

Tigano A, Friesen VL. Genomics of local adaptation with gene flow. Mol Ecol. 2016;25:2144–64.

Timme RE, Simpson BB, Linder CR. High-resolution phylogeny for *Helianthus* (Asteraceae) using the 18S-26S ribosomal DNA external transcribed spacer. Am J Bot. 2007;94:1837–52.

Todesco M, Pascual MA, Owens GL, et al. Hybridization and extinction. Evol Appl. 2016;9:892–908.

Todesco M, Owens GL, Bercovich N, et al. Massive haplotypes underlie ecotypic differentiation in sunflowers. Nature. 2020;584(7822):602–7. https://doi.org/10.1038/s41586-020-2467-6.

Ungerer MC, Baird SJ, Pan J, Rieseberg LH. Rapid hybrid speciation in wild sunflowers. Proc Natl Acad Sci U S A. 1998;95:11757–62.

Via S. Natural selection in action during speciation. Proc Natl Acad Sci U S A. 2009;106:9939–46.

Via S. Divergence hitchhiking and the spread of genomic isolation during ecological speciation-with-gene-flow. Phil Trans R Soc B Biol Sci. 2012;367:451–60.

Wales N, Akman M, Watson RH, et al. Ancient DNA reveals the timing and persistence of organellar genetic bottlenecks over 3,000 years of sunflower domestication and improvement. Evol Appl. 2019;12:38–53.

Whitney KD, Randell RA, Rieseberg LH. Adaptive introgression of abiotic tolerance traits in the sunflower *Helianthus annuus*. New Phytol. 2010;187:230–9.

Wills DM, Burke JM. Quantitative trait locus analysis of the early domestication of sunflower. Genetics. 2007;176:2589–99.

Yakimowski SB, Rieseberg LH. The role of homoploid hybridization in evolution: a century of studies synthesizing genetics and ecology. Am J Bot. 2014;101:1247–58.

Yatabe Y, Kane NC, Scotti-Saintagne C, Rieseberg LH. Rampant gene exchange across a strong reproductive barrier between the annual sunflowers, *Helianthus annuus* and *H. petiolaris*. Genetics. 2007;175:1883–93.

Yeaman S, Whitlock MC. The genetic architecture of adaptation under migration-selection balance. Evolution. 2011;65:1897–911.

Zhang J-Q, Imerovski I, Borkowski K, Huang K, Burge D, Rieseberg LH. Intraspecific genetic divergence within *Helianthus niveus* and the status of two new morphotypes from Mexico. Am J Bot. 2019;106:1229–39.

Genomics of Plant Gene Banks: Prospects for Managing and Delivering Diversity in the Digital Age

Chris Richards ⓘ

Abstract Gene bank collections are at a pivotal point in their history. We review their development and explain how their role is changing with new technologies in computation and analytics and the widespread application of whole genome sequencing for characterizing genetic diversity. Development of genomics technologies is likely to significantly alter the way we search for, use, and valorize plant genetic resources (PGR) in the future. The prospect of searching for alleles of agronomic importance is not far off. We review genomics/ sequencing approaches used to catalog genomic variation and their analytical methods useful in describing diversity in PGR. Importantly, we review how population genomics approaches can be used to characterize diverse collection accessions and the way in which genomic data can help provide information on identity, integrity, and population history for samples maintained in gene banks. We explore sampling approaches used to capture variation within and among accessions. We review case studies of diverse crop systems that provide important examples of how genomic data can be used to both provide critical information for agronomic variation for breeders and provide characterization information integral to managing this diversity. Lastly, we describe how gene banks can be positioned more centrally in the ecosystem of breeding and gene discovery projects acting both as a resource of genetic diversity and a consumer data describing of functional variation. These trends are likely to transform gene banks from merely germplasm service providers to interactive biodiversity informatics research centers.

Keywords Data integration · Domestication biology · Gene banks · Genetic structure · Haplotypes · Molecular evolution · Next-generation sequencing · Population genetics

---

C. Richards (✉)
USDA National Laboratory for Genetic Resources Preservation, Fort Collins, CO, USA
e-mail: chris.richards@usda.gov

1 Introduction

Natural genetic variation, which has evolved over millions of years, provides essential building materials for plant breeding. One key role of a gene bank is to help safeguard natural forms of genetic variation so that even if it is lost from the wild or from farmers' fields, it remains readily available to plant biologists, breeders, and others with interests in plant diversity. Genetic diversity is the foundation of evolutionary change and the source of adaptive variation critical in a changing environment. Plant genetic resources (PGR or germplasm) forms the fundamental building blocks of diversity used in plant breeding and crop improvement programs. Gene banks were established as biorepositories for living seed sources for genetic diversity in cultivated, landrace and wild relatives of crop species. Combined holding of gene banks worldwide now totals some 7.4 million accessions (and growing) covering more than 15,000 genera of crop plants and their wild relative species (FAO 2010; Fu 2017). The imperative was due to a recognition that as new modern agricultural practices were adopted around the world through the efforts of the Green Revolution, the diversity of local varieties, landraces, and wild crop relatives were being displaced and potentially lost to agriculture (Bennett 1968; Khush 2001; Feuillet et al. 2008; van de Wouw et al. 2010; Fu 2017; Ramankutty et al. 2018). There was a realization in the 1960s that the agricultural landscape was changing so rapidly that genetic erosion in diversity would result in irreversible loss and leave agriculture dependent on an ever-narrowing base of germplasm (Brush 1999; Thormann and Engels 2015). This resulted in a worldwide effort to collect and catalog diversity from farmer's fields, from the wild and place them in gene banks in what has been described as a global salvage operation of agricultural diversity (Frankel and Bennett 1970; Scarascia-Mugnozza and Perrino 2002). The objectives of these gene banks were focused on collecting and maintaining living genetic variation not just as preserved samples like herbaria or other biorepositories, but as living samples and used for breeding or research. Genetic resources have primarily served the needs of plant breeders, but their utility was envisioned to provide resource to broad areas of plant sciences including systematics, biogeography, and the molecular evolution of domestication (Khush 2001). The potential these collections represent to agriculture and food security is a resource of global importance (Gepts 2006; Byrne et al. 2018). This chapter discusses the challenges and opportunities that emerge with transforming PGR collections into digital biorepositories where living samples and digital descriptor data are readily accessible. We discuss the technical challenges in characterizing diverse collection structures and techniques emerging that might overcome them. Then, we discuss the impact of these data have on the improved curation of PGR collections and the changing roles these collections might play in the "omics" era.

Decades after their establishment, gene banks continue to struggle with collection characterization and ways to improve access to their diversity (Prada 2009; Glaszmann et al. 2010; Fu 2017). However, the utility and potential of gene banks offer for crop improvement is undergoing a profound change since their development (Walters et al. 2008; Kilian and Graner 2012; Muller et al. 2018; Diez et al. 2018; Belzile et al. 2020). These large and complex ex situ collections are now poised to transition from a salvage operation into a broadly defined phase of intense characterization where tools are emerging for locating novel alleles for agronomically important traits (Hammer 2003).

Gene banks at their very fundamental level provide materials and supporting data to provide vital information needed for researchers and breeders to make choices in requesting. Without detailed descriptive information about the holdings, it is impossible to infer the agronomic value of an accession for breeding or other purposes, its unique identity relative to other accessions and its provenance (Rubenstein et al. 2006; Weise et al. 2020). Data management systems such as GRIN-Global (Kinard et al. 2009) were developed to manage collection inventory, regeneration and viability information, distribution tracking and other curatorial workflow activities. The public facing side of GRIN-Global allows users see what materials are available and allows users to make choices based on relevant characterization information. Gene bank accessions are identified and organized with unique accession numbers which are linked to accompanying passport data – data that describes the location of the collection and its taxonomic identification and other attributes, identifiers and labels intended to facilitate the sharing of standardized information about the collection diversity. The standardization of biological accession records was formalized by the Taxonomic Database Working Group (TDWG) as a set of minimal identifiers codified collectively as the DarwinCore standard and this has been applied globally to biological specimen collections (https://dwc.tdwg.org/). In addition, phenotypic information is typically recorded for each accession on a set of defined descriptors that becomes part of the accession record. Descriptors were established early on as a standardized set of heritable morphological measurements recorded to clarify identity (Gotor et al. 2008). This effort required a methodology for describing germplasm accessions that had international approval and was easy to use. To be effective, the methodology needed to accurately and routinely differentiate between accessions in the same collection and promote collaboration among plant genetic resource (PGR) workers in different countries. Phenotypic descriptors have been applied to ex situ collections with varying degrees of standardization and are subject to environment by genotype interaction. While global efforts to place characterization information into a unified and formal ontology, many gene banks characterize collections without direct coordination. Often the characterization data are applied to subsets of a collection either for a specific study or collected at the time of regeneration. This can often result in data that cannot be combined across an entire collection.

Except for a few well-characterized and valuable crops, most characterization data are sparsely recorded for the same traits across collections. For instance, the US sorghum (*Sorghum bicolor*) collection contains ~42,000 accessions but complete matrix that describes accession by descriptors is only about 2% filled (Reeves et al. 2020). Adding to this challenge, gene bank holdings have continued to rise with little or no additional funding resulting in a decrease in the investment per accession (Fu 2017). The growing disparities in the amount of characterization data collected across accessions in gene banks mean that sectors of the collection have only the basic identity data available which limits the way collection inventory databases can search for holdings with traits other than taxonomy and location. The lack of characterization data limits the utility not only for the user but also for the gene bank manager for curatorial operations.

To address these limitations, over the past decades there have been calls for systematic development of molecular passports using various DNA markers (Bretting and Widrlechner 1995; Karp et al. 1997; van Hintum and van Treuren 2002; van Treuren et al. 2010; McCouch et al. 2012; Mascher et al. 2019). The ability to characterize diversity has changed dramatically with the introduction of high-throughput DNA sequencing methods. The use of molecular markers, revealing polymorphism at the DNA level, has been playing an increasing part in understanding and using genetic diversity in plant germplasm collections for crop improvement (Tanksley and McCouch 1997; Glaszmann et al. 2010; McCouch et al. 2013; Bevan and Uauy 2013; Varshney et al. 2014; Milner et al. 2019; Belzile et al. 2020). To date there have been thousands of projects that have exploited genetic markers for describing the diversity within ex situ collections. These techniques have progressed beyond using a handful of genetic markers to develop whole genome sequences capable of revealing genetic diversity at a remarkably fine scale. These data vary widely in their inherent accuracy and repeatability, in the suitability level (population/phylogenetic) at which they can be most appropriately applied, and in their technical and operational costs.

Whole genome DNA sequence data are driving the development of new approaches for characterizing and searching genetic collections for useful genetic diversity. Approaches that estimate genetic diversity not simply as summary statistics but provide genetic diversity in an explicitly genomic context reveal a more complete understanding of how adaptive differentiation has shaped the genomes of crop plants during domestication. This effort to apply genomic characterization has been applied in part to complement the limitations and gaps of phenotyping collections and also to improve the efficiency of conservation and the improved access this provides to the stakeholder.

2 Genomic Characterization of Plant Genetic Resources in Gene Banks

2.1 Genome Sequencing and Genotyping Approaches

The emergence of low cost, next-generation sequencing technologies has made feasible the systematic genomic characterization of entire collections (Mascher et al. 2019). Large-scale, high-throughput genotyping technologies have greatly facilitated population genomics research and applications (Holliday et al. 2019) and increased the resolution of quantitative trail locus (QTL) mapping and identification of their underlying functional genes. These data have created new opportunities for exploring and utilizing PGR diversity (Kilian and Graner 2012; van Treuren and Van Hintum 2014; Wambugu et al. 2018; Holliday et al. 2019).

Molecular characterization is now the preferred means of estimating genetic diversity within gene bank collections. Two of the most abundant classes of markers (simple sequence repeats (SSRs) and amplified fragment length polymorphisms (AFLP)) have been widely used, were relatively inexpensive to implement and provided a wealth of descriptive data about genetic diversity (e.g., Beckmann and Soller 1990; Rassmann et al. 1991; Powell et al. 1996; Spooner et al. 2005) and have been instrumental in many aspects of conservation such as characterizing plant genetic diversity for purposes of improved acquisition, maintenance, and use. However, as the cost and scalability of obtaining whole genome sequencing data to characterize gene banks genetic diversity drops, these data have become more routinely used in both characterizing genetic mechanisms underlying agronomic traits (gene discovery projects) and providing higher resolution estimates into the population genomics and molecular evolutionary processes shaping diversity in situ and the mode and tempo of selection during domestication of crop plants. These kinds of data are completely changing the way breeders search for and use variation from gene banks (Poland and Rife 2012; Langridge and Waugh 2019; Belzile et al. 2020). In addition, raw sequence data whatever the scale of its application have established metrics for quality and repeatability that have been lacking in SSR and AFLP data sets, allowing more flexibility for these data to be shared, combined and reanalyzed and new analytical approaches evolve (e.g., dePristo et al. 2011). This is especially important when tracking or comparing the identity of germplasm used in different gene banks (e.g., Palme et al. 2020).

There are multiple sequencing platforms for generating high coverage genomic data each with their own capabilities including whole genome sequencing, whole genome exome sequencing, and several forms of reduced representation library construction (Goodwin et al. 2016; Levy and Boone 2019). About a decade ago sequencing options greatly diversified with some notable advancements in price, read lengths, and operational costs. The instrumentation, chemistries, and algorithms for analysis are changing almost monthly and these specific technical considerations are left to other reviews to cover (e.g., Holliday et al. 2019). However, there are broad patterns to the way the technologies are moving. Overall, the axes of variation that help to describe these technologies revolve around DNA sequencing length and methods used for of DNA library preparation.

The dominant platform for short read sequencing uses devices that maximize the parallelization of sample processing, enabling higher throughput in shorter runtimes. Throughput is increased in both the hardware design of the sequencing flow cells (such as Illumina's HiSeq and NovaSeq platforms) or in the way in which samples can be combined (multiplexed) by using DNA barcodes to tag particular sets of samples. While the read lengths for platforms like Illumina have increased, they still remain under 300 base pairs. Compensating for short read length is the high amount of sequence read depth produces in each run, providing increased accuracy. Sequences can also be generated as paired ends of DNA around fragments with a known size which greatly improves efficiency of alignment to a reference genome over individual short reads. The accuracy in base calling has been demonstrated in thousands of studies, however limitations arise when characterizing highly repetitive regions and regions with structural variants like indels, copy number variants, and inversions. Approaches using read depth variation for duplications have been used successfully, but often de novo sequence assembly is the most informative approach (Ho et al. 2020).

Single molecule, long-read sequencing platforms have been developed in part to better assemble these genomic "dark regions," provide greater efficiency in de novo sequence assembly, and enable direct phasing of sequences (Sedlazeck et al. 2018). The two major platforms (PacBio and Oxford-Nonopore) extend sequencing read lengths routinely to 10,000 base pairs. The read length in these platforms has been offset by the lower accuracy rates, but accuracy is improving steadily. The ability to read long fragments as one long continuous series of base calls complements high coverage short read sequencing in several critical areas. First these methods allow direct phasing (especially in repetitive regions or in clustered gene families) of sequence information which can differentiate between the paternal and the maternal genome in diploid organisms (Zhang et al. 2020). Second, the longer read lengths can greatly improve *de novo* genome sequencing of species without a reference genome (Chin et al. 2016). There are numerous examples of long-read *de novo* genome assembly that have re-assembled and improved references developed under other short read methods, fixing errors, filling gaps, and improving resolution (e.g., Zhang et al. 2019a, b). Critically, these methods are more directly able to identify structural variants that have been more difficult to estimate using other methods. These structural variants have often been missed in genomics studies, yet they are relevant and informative characters when estimating diversity of the genomes in populations representing a clade of a species (Gabur et al. 2019; De Coster and Van Broeckhoven 2019).

Linked read (or synthetic long read) methods exploit the accuracy and high coverage of short read sequencing with a novel library preparation (Zheng et al. 2016; Chen et al. 2020). Linked read methods rely on the highly parallelized, high coverage sequencing capacity of short read platforms and focus them on specialized libraries of whole genome DNA preparations that have either been diluted, atomized in micro-droplet aerosols or uniquely labeled using transposon tagged complexes. The general approach is to subdivide and uniquely barcode small sets of high molecular weight molecules which become inserts in a sequencing library. This allows the assembly of the short read sequence data that share a common barcode greatly increasing both the specificity and efficiency in building long contigs for assembly. Linked read sequencing techniques have been especially useful in de novo assembly, generating haplotype information and identifying and cataloging structural variation.

While whole genome sequencing has its advantages in certain applications, a far more common and cost-effective approach for genotyping many samples is to use some form of reduced representation library construction employing a family of protocols under the family of restriction-site associated DNA sequencing or RADseq

(Kilian et al. 2012; Poland and Rife 2012; Davey et al. 2013; Andrews et al. 2016). The libraries typically revolve around fragments delineated by specific DNA endonucleases. There are a number of related DNA library classes that are based on sampling intervals that are bound by restriction sites across the genome. One of the critical hallmarks of these methods is that they are able to identify new variants in samples while the sequencing is done. This combines the discovery and the genotyping phases in one procedure. These methods offer opportunities to develop population scale genotyping at lower costs and have provided the majority of genotyping data in plant genetic resources collections (Scheben et al. 2017). Each method is generic in that the same methodology can be used across different taxa, however comparative studies have indicated there are differences in the way regions are targeted in the genome and the ability to identify rare alleles (Andrews et al. 2016). As genomic regions are more rigorously filtered, sequencing coverage in the resulting fragments tends to increase and missing data decreases. Low coverage RAD sequencing can often result in sparse or missing data. Often missing regions due to allele dropout (technical error due to low coverage sequencing) or actual present/absence variation cannot be directly distinguished in an experiment. Sparse data matrices present challenges for downstream analysis and often these missing loci can be imputed accurate (especially in inbred or pedigreed populations) using a number of algorithms (reviewed in Swarts et al. 2014 but see Brandariz et al. 2016).

Amplicon sequencing using highly multiplexed PCR products has several advantages for targeted genotyping. Both AmpliSeq (Thermo Fisher) and rhAmpSeq (IDT) chemistries represent a specific class of reduced genomic representation used for sequencing that are targeted by primers designed to amplify specific user-defined locations. As genome mapping of functional variation increases, these approaches may prove especially useful in selectively screening gene bank collections for variation at loci of agronomic interest. These methods are scalable to thousands of targeted amplicons and simple and cost-effective enough be suitable for population-level sampling among accessions. By anchoring priming sites to conserved genes, it is possible to design amplicons that sample more diverse flanking regions which can be applied to diverse plant and animal species for phylogenetic diversity studies (Lemmon et al. 2012). As these methods and new technologies become routine, researchers will be able to achieve the accuracy, low cost, and high-throughput necessary for characterizing genetic resources collections. Critical to characterization is being able to process enough samples to adequately represent the diversity of the accession.

2.2 Genomic Data Analytical Approaches

An early first step in the analysis of raw sequence information is some form of alignment and variant calling. Typically, this involves aligning the experimental data to a reference genome and identifying and validating the polymorphic nucleotides. The bioinformatics pipelines for these procedures have been growing for the last decade to accommodate new data forms but the general parameters and approaches have been established (Nielsen et al. 2012). Many of these parameters, such as variant filtering and minor allele frequency tolerances, can have consequences in the read quality and estimates of genetic diversity and differentiation (Linck and Battey 2019; Wu et al. 2019). Critically, the choices made for genome-wide association studies (GWAS) often differ from those focused on estimating population genomic diversity parameters in a collection. This distinction is manifested with the iterative nature of GWAS, often requiring several rounds of validation for accurate mapping, whereas population genomics studies looking for spatial patterns of adaptive differentiation often have only one collection time for samples and genotyping. Alignment accuracy depends critically on the distance (number of substitutions) between the experimental sequence in any one of the alternate alignment sites in the reference. Accuracy in exhaustive alignment searches is critical especially in highly diverse germplasm screening (Marco-Sola et al. 2012). However, a single, arbitrary reference might not adequately capture the genetic diversity in the experimental samples especially when screening more diverse germplasm and crop wild relative species (Clark et al. 2005; Heslot et al. 2013; Brandariz et al. 2016). In these cases, sequences in the experimental sample may not have any corresponding map location in the reference genome. This can influence estimates of genetic diversity in the experimental data by imposing ascertainment bias. One solution to this bias is to develop many independent reference genomes (e.g., Valiente-Mullor et al. 2021) and develop a pangenome approach to quantifying diversity (Nguyen et al. 2014; Golicz et al. 2016; Danilevicz et al. 2020; Terhorst et al. 2017). These approaches are being developed in a number of crop plants for genome/genetic diversity analysis, domestication history and identifying new and useful functional alleles (Hubner et al. 2019; Monat et al. 2019; Scossa et al. 2021). Developing pangenomes within species may provide a more comprehensive understanding of the dynamics of genome size variation (copy number variants and gain and loss of gene sets) and may help to more accurately describe genetic diversity (Bayer et al. 2020). Visualizing and ordering pan genomes as sets of linked sequence blocks (i.e., haplotypes, see below) has been challenging; however, graph theoretical approaches are promising (Franco et al. 2020; Jensen et al. 2020). These methods may also be critical in large-scale genotyping projects, such as genotyping entire gene bank collections, since each haplotype may be tagged with one or more diagnostic SNPs. By using this specific set of SNPs, it might be possible to impute the set of haplotypes and by degree resolve complete genotypes at a cost well below whole genome sequencing.

Pangenomes represent more than just sequence variants since they are able to identify, map, and catalog the structural variation of haplotypes (genomic regions where markers are correlated from strong linkage disequilibrium). The pangenome can identify both allelic variants and set of haplotypes that have dynamically been added or removed from lineages within a species. The size of a haplotype block can be defined arbitrarily in sliding window analysis, estimated computationally using dense whole genome SNPs or it can be defined empirically based on actual read identities (Browning and Browning 2011; Stapleton et al. 2016; Golicz et al. 2016; Zheng et al. 2016; Baetscher et al. 2018; Bansal 2019). It is possible to represent

RAD loci as a set of haplotypes rather than as individual SNPs. Nucleotide polymorphisms are typically binary (two states) and individually they are not as informative as haplotypes are. Within a small genomic region, individual SNPs may convey different evolutionary signals (Li and Jiang 2005; Sehgal et al. 2020). By phasing these SNPs into haplotypes, they can encode multiallelic states which can encode a much larger amount of information regarding the history of the genomic region (Voss-Fels and Snowdon 2016; Ebler et al. 2019). These diverse, physically phased haplotypes produced contain information on molecular evolutionary (coalescence) patterns that can be used to study population structure and history (Malinsky et al. 2018). Mapping studies and GWAS using individual haplotypes vs SNP in a data set have been shown to be efficient and can increase power of resolution for QTL identification (Sehgal et al. 2020).

Sequence information and the variant calls developed in analysis can also provide the basis for fixed SNP arrays which provides high-resolution genotyping at predefined loci. These SNP arrays have been a workhorse for many breeding applications using marker assisted selection (MAS) to track and develop introgression and identify breeding lines (Rasheed et al. 2017). These approaches have also been applied to screening large gene bank collections (Song et al. 2015); however, the accuracy of these genetic diversity studies critically depends on both the ascertainment of the SNP panels being assayed and the range of diversity in the samples.

3 Connecting Gene Bank Curation to Genomic Data

Gene banks were not always established at the outset as structured, well-described collections, they are complex, most certainly redundant and opportunistic in coverage and collected at all stages of crop plant improvement (Walters et al. 2008; Fu 2017). Landraces, traditional varieties, and primitive cultivars represent genetic diversity selected by farmers over 1000s of years and these local lineages contain genetic combinations adapted to particular growing conditions, selected for particular morphologies or flavors. They, therefore, represent variation critical to crop improvement (Hammer and Teklu 2008; Jordan et al. 2011; Lopes et al. 2015). Wild relatives of crop plants represent genetic diversity that predates the domestication process and have been shown to provide additional variation not found in any domesticated lineages, which may have significant agronomic value (Hawkes 1977; Steffenson et al. 2007; Kovach and McCouch 2008; Lam et al. 2010; Warschefsky et al. 2014; Lopes et al. 2015; Brozynska et al. 2016; Mammadov et al. 2018; Zhang et al. 2017, 2019a, b).

Tanksley and McCouch (1997) and more recently the Divseek Initiative (McCouch et al. 2013, 2020) advocate for more effective use of gene bank diversity. However, despite some notable successes in effective introgression of wild allelic variation into modern crop varieties, the genetic resources stored in gene banks are still underutilized in breeding programs (Maxted et al. 2012; Brumlop et al. 2013). This largely stems from the challenges of identifying useful accessions from large and diverse ex situ collections, but it also stems from the lack of characterization and data integration.

3.1 The Accession as a Population

At its most fundamental level, collections are organized into samples called accessions which are given a unique ID that brings together data that describes the identity of the sample (what species or variety? what accession number?), a basic description of its provenance (where was it collected? who was it received from?). These so-called passport data are stored in a central collection database: The Genetic Resources Information Network (GRIN-Global https://www.grin-global.org/). The basic sample unit of a gene bank collection is the accession which is given its own unique identifier. All management practices (inventory, regeneration associated observations and descriptor data) revolve around the individual accession.

While the data structure listing the accession number, plant introduction number or digital object identifiers (DOI) at the collection inventory database might convey the impression that all accessions are uniquely diagnostic and fully resolved as though they were a uniform genetic stock, the biological reality is more complex. Fundamentally, most accessions do not represent uniform set of genotypes but rather they are a heterogenous population of genotypes reflecting the mutational and migration effects that influence all subdivided demes. What distinguishes accessions is simply they are collections that have been sampled in time and maintained in some form of (closed) captive breeding system subject to genetic drift effects. In some instances, such as in many of the small grains like barley (*Hordeum vulgare*), collections are made uniform by recurrent single seed descent. Most crop wild relative accessions have inbreeding and levels of differentiation that reflect the sampling effects of the collector, the inherent breeding system/life history, and the geographical range of the species (Hoban et al. 2018). Any effort to render this genetic diversity, therefore, must take a population genetics/genomics approach for sampling this diversity or to adequately estimate parameters critical to curation such as genetic diversity and differentiation, relatedness and admixture within the collection and analyses that seek to understand the population genomics of domestication history.

Sampling procedures have been developed over the decades for conservation applications (reviewed by Volk et al. 2007). Most sampling strategies are based on a common set of assumptions and extend the probabilistic arguments of the strategies to accommodate rare alleles, multiple loci, and multiple populations. The recommendations of many models are based on assurance with a high probability (>0.95) of the acquisition of alleles at multiple independent loci within a population. Sampling both individuals and loci within individuals represents critical decisions when estimating population genetic parameters, including genetic diversity and differentiation (Bashalkhanov et al. 2009; Manel et al. 2012; Hale et al. 2012). There have been several elegant studies that examine the analytical tradeoffs for different applications, weighing the effects of cost vs data accuracy when estimating processes and patterns in nature; however, in many instances the reference point for many users of plant genetic resources is the ex situ collection itself. Whether the inferring genetic diversity in a finite collection or estimating a more diverse landscape in situ, sampling this variation comes down to the number of individuals that are needed to represent the diversity of the sample. Since many applications in population genetic analysis use estimates of alleles frequencies in differentiation (F_{ST}, D, G_{ST} or AMOVA) (Excoffier et al. 1992) or genetic diversity (allelic richness) and identity, sample considerations have to address accuracy over a range of intra-accession diversity from highly inbred and uniform accessions like inbreeding domesticated soybean (*Glycine max*) to highly diverse population samples like wild outcrossing *Malus* sp. (Hoban et al. 2018). In addition, the information content of the markers is an important component of sampling. For instance, studies that obtain high coverage genomic data of bi-allelic SNPs may require less sampling of individuals in order to achieve sufficient accuracy for intra- and inter-population analyses (Nazareno et al. 2017). However, phased haplotype data take on distinctly more multiallelic character states because they account for linkage disequilibrium that combines multiple SNPs in each haplotype block, and this diversity may influence the way sampling is approached (Reeves et al. 2012).

Sampling the diversity of samples from a large gene bank can be a daunting task when collection sizes are in the tens of thousands accessions and descriptor data is sparse. Defining an entry subset suitable for analysis of genetic diversity is the objective of the core collection concept (Brown 1989). The initial objective of a core collection was to develop a single-entry point into the collection that better served the needs of users by providing a cross-section of collection diversity in a compact set (e.g., Glaszmann et al. 2010). Nominally this approach could be used with whatever data is available in the collection such as locality information or a few common descriptors. The reasoning was that a set comprising of 10% of the entire collection could be selected and distributed for evaluation. However, for many collections even a 10% subset might be in the thousands and put the cost of intensive genotyping out of reach. The approaches themselves have progressed from allocation methods (sampling among defined groups; (Schoen and Brown 1993; van Hintum 1995) to multivariate distance sampling approaches (maximizing distance; Franco et al. 2005) to discrete optimization methods (maximizing diversity and/or distance; Gouesnard et al. 2001; Kim et al. 2007; Thachuk et al. 2009). The algorithms for core set selection have become more efficient, accurate, and scalable. These improvements will be essential as the volume of sequence data increases and could be used for optimizing subset with specific genomic targets (Reeves et al. 2020). Core set algorithms may be particularly useful as integration among genomic, phenotypic, and geospatial data types becomes more important to mine useful variation.

Another method for selecting diverse subsets uses environmental data drawn from collection location coordinates to select accessions based on environmental parameters of the site or origin relating to temperature and water availability. Focused Identification of Germplasm Strategy (FIGS) has been promoted as an

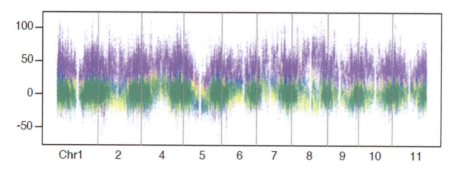

Fig. 1 Relative improvement in haplotype capture in sorghum achieved by using genome-wide single nucleotide polymorphism (SNP) data (purple) instead of geographic (yellow) or environmental(teal) data. Subsets that maximize geographic or environmental diversity capture fewer haplotypes than randomly assembled subsets (y = 0) across much of the genome. Reproduced from Reeves et al. (2020)

approach to identify subsets of accessions with adaptive traits in gene banks or wild populations of plants (Bari et al. 2012). However, a weakness of the FIGS approach is that it looks for adaptive variation that responds to environment (mostly climatic variables). It cannot adequately account for variation that is often influenced by nonequilibrium dynamics and episodes of past gene flow. Not every plant population is adapted to its current location or its current climate, some of the genetic structure in populations can arise from forces other than adaptive differentiation especially in species with recurrent colonization and extinction dynamics. Agronomically useful haplotypes might arise without selection driving them to local high frequencies. Traits that may be useful in agriculture may not always be adaptive in the wild. Interestingly, because of the idiosyncrasy of adaptive evolution, the efficiency of developing subsets maximized for haplotypic variation is not always improved by using surrogate information like environmental data or geographical location (Fig. 1) (Reeves et al. 2012; Reeves and Richards 2018).

Once the accessions are identified, developing an accession level strategy for sampling intra-accession diversity can be challenging. The key to genomic survey information is to not only to identify the sequence variants across the genome, but also to estimate their frequencies in accession populations. This can either be based on probabilistic methods for retaining alleles and a minimum frequency (mentioned above) or based on the logistics that balance cost of genotyping and analytical requirements. Some studies first extract single seed decent lines (inbred) of the selected accessions before beginning genotyping and therefore only sample single individual genotype for each line. This has the advantage in that allows for direct comparison between the accession (or purified stock) and the sequence data (Glaszmann et al. 2010). Critically, it allows the same physical genetic stock to directly connect to derived digital genotypic and phenotypic characterization data. Another method that has gained more traction in studies of ecological genomics is pooled sequencing (Schlotterer et al. 2014). In this case pooled DNA representing scores of individual genotypes sequenced together at high coverage (20×) in the sequencing run and aligned to a reference genome. The aligned reads can be used to catalog the number of microhaplotypes within a genomic position or visualized as a sliding window of haplotype diversity. Methods have been developed to infer the frequency of variants within pools as well as genomic patterns for heterozygosity, and these data can be used to estimate genetic diversity and differentiation measures (Kofler et al. 2011). The advantage to pool sequencing approaches is that the variants are compiled across the genome and the data directly corresponds to an accession unit that can be requested in the gene bank. Finally, another sampling approach draws from information developed from a small diverse panel of genotypes drawn from the collection and sequenced at a high coverage (e.g., Upadhyaya and Ortiz 2001). Variants identified using a diverse set of genotypes will tend to minimize any ascertainment bias. These variant locations can then become the basis for developing a low coverage set (100s) of PCR based KASP markers that can be scaled up to high sample throughput SNP surveys with enough information content to routinely and cost-effectively estimate genetic diversity and structure in the entire collection (Semagn et al. 2013).

3.2 Characterizing Genetic Structure in Gene Bank Collections

Estimating genetic structure and discontinuities between populations is an essential component of intraspecific biodiversity and evolution studies, as well as genome-wide association studies. The range of interrelated analysis possible with data that have a genomic context (polymorphisms within the genome) fundamentally changes the scale of genetic inquiry to include both genome-wide differentiation between populations and individual and also genetic differentiation within and among individual haplotypes and genes (Black et al. 2001; Hamblin et al. 2011; Luikart et al. 2019). The genomic scale of data extends the description of differentiation among groups of individuals to estimating patterns of locus-specific genealogies, linkage relationships and selection at loci underlying phenotypic traits (Josephs et al. 2019). Comparative studies of genome-wide divergence among wild and cultivated species open up a new avenue of understanding domestication history of crops and can help pinpoint the loci that act as drivers for that transition (Ross-Ibarra et al. 2007; Gerbault et al. 2014; Smýkal et al. 2018; Purugganan 2019). Sampling of diversity across the domestication spectrum is tricky because variants that may have been directly selected for in domesticated lineages and are nearly fixed may be quite diverse in wild relatives. Reduction of diversity at specific genomic regions is evidence of selection during domestication and such regions provide loci of key agronomic importance. Signatures of selection during domestication can impart large regions of linkage disequilibrium, where regions around the selected target locus reduce genetic diversity in neighboring loci. This reduction in heterozygosity can help identify both the location and the age of domestication. By combining dense genotypic information in and around candidate genes, it is possible to classify accessions based on their haplotypic diversity and to identify and isolate novel and superior alleles of agronomically important genes from a wide range of crop species, a process defined as allele mining (Leung et al. 2015; Gokidi et al. 2017; Kumari 2018; Belzile et al. 2020). In addition, genomic data can increase the precision of estimates of gene flow, admixture history, relatedness, and kinship structure. Applying these methods to a set of discrete accession samples plays an important role in managing genetic diversity in collections and helps to identify duplications or over representation, and in many instances can lead to identify collection gaps (Bretting and Widrlechner 1995).

The most widely used methods to infer population genetic structure are differentiation measures based on Wright's fixation indices (and their derivative metrics) that measure among *a priori* defined groups, model-based, Bayesian MCMC procedures that minimize Hardy-Weinberg (HWE) and linkage disequilibrium (LD) within subpopulations identified without *a priori* information, and lastly, multivariate ordination analyses that assume no hierarchical structure and operate on only a mathematical distance matrix. The first approach uses individual genotypic frequency information to describe or estimate population genetic parameters, such as deviation from HWE (Wright 1965; Nei 1977; Hedrick 2005). The second class uses Bayesian Markov Chain Monte Carlo (MCMC) methods that have been deployed in the software such as STRUCTURE (Pritchard et al. 2000), which is an objective procedure that identifies groups directly from genetic polymorphism data, rather than relying on subjective, *a priori* notions of existing structure. Groups are assembled with the assumption (predicted by the Wahlund effect) that, if population genetic structure exists, then the mean deviation from Hardy-Weinberg and linkage equilibrium across a sampled population should be less than if the population was treated as a single unit. Accordingly, the procedure maximizes Hardy-Weinberg and linkage equilibrium within K subpopulations by swapping individuals among them during the progression of a Markov chain. Sampling from the chain reveals the posterior probability of assigning individuals to each of the K subpopulations and can be used to estimate admixture (where genotypes have split assignments to different clusters). The last method uses categorical information as the basis for developing interindividual distance matrix. Ordination is most commonly used to decompose complex multi-locus data sets into two- or three-dimensional scatter plots that represent genetic structure spatially, with putative subpopulations forming distinct clusters of points. However, the development of statistically rigorous methods to assign individuals to subpopulations using ordination results has been problematic. The role of ordination has largely been limited to informal visual corroboration of pre-existing hypotheses about population structure. Ordination methods are computationally efficient and recent extensions of these methods have included direct statistical tests of subdivision within multivariate data such as Modal Clustering (Reeves and Richards 2009) or development of discriminant functions based on eigen values derived from principal components such DAPC implemented in the R package Adegenet (Jombart and Ahmed 2011). In many instances, these methods provide a visual inference into structure that cannot otherwise be rendered in two dimensions. Because of the speed of analysis these methods have taken on larger and larger data sets where millions of SNPs are used as character states. While phenetic in nature (as opposed to phylogenetic with an axis of ancestral and derived character states), this visual approach provides hypotheses about the data that are extremely useful.

Estimating genetic diversity in plants, where interspecific hybridization is not uncommon (especially wild and domesticated lineages), can cross the boundaries of population genetic and phylogenetic analysis (Posada and Crandall 2001). There has been a push to press the lower bounds of phylogenetic analysis within the seed plants (e.g., Stuessy et al. 2003) to test population-level evolutionary hypotheses. These methods are useful, but they most often are visualized as tree like relationships (bifurcations) onto systems that often reticulate due to hybridization. Tree structures like neighbor joining methods (Saitou and Nei 1987) offer a hierarchical approach to organizing diversity that may not directly account for conflicts among that arise among markers (where loci might have different patterns of ancestry). When the taxa being structured in this way are outcrossing populations, the topology cannot be fully resolved and may require care in providing a support for each split by using a re-sampling approach, such as bootstrapping the data. Alternative visualization methods are based on graph theoretical properties (split-decomposition approaches) that render tree like networks where splits represent conflicting phylogenetic signals (Huson 1998; Posada and Crandall 2001). Circumscription of natural groups is a critical component of gene bank collection management. Not only is it important to understand the genetic relationships among accessions (especially if they are duplicates) but the structure developed serves as the basis for hypothesis testing for estimating selection for agronomically important traits and reconstructing domestication history.

The variation that could be revealed with genomic data is crucial in developing more robust measurements of accessions distinctiveness or identity using a variety of probabilistic analyses developed for forensic applications DNA profiling and kinship estimations (Lynch and Ritland 1999). The ability to determine with a specified tolerance accession identity help support efforts to identify large duplications and gaps in collections (e.g., Gross et al. 2012; Ellis et al. 2018; Singh et al. 2019). This helps promote the concept of collection rationalization (does the diversity captured in a collection rationalize the size?). The concept of rationalization in gene banks revolves around the idea of efficiency – what is the collection size needed to adequately cover representative diversity of a target crop or species with as little redundancy as possible (van Treuren et al. 2001; Sackville Hamilton et al. 2003; van Treuren et al. 2010). The ability to track collection diversity through unique and identifiable samples is critical to the rationalization of global collections and the most direct way to reduce confusion in synonymy that exists in gene banks collection inventories. The digital Object Identifier approach has resulted in initiatives that seek to coordinate diversity among national gene banks and possibly scale this effort to global information systems (Alercia et al. 2018). Sample tracking in this way might help create linkages between gene bank accessions and the genetic stocks that they contribute to. More challenging is that without modification the precision DOI provides may be at odds with the population's genetic measures of distinctiveness.

4 Selected Empirical Case Studies

The genomic characterization of gene banks has entered a new era where collections of crop plants that were once resource deficient are now able to develop a genomic characterization plan and deliver insights in their collections with unprecedented resolution. The feasibility and technical approach used for genomic characterization at the scale of the entire collection was established by collaborative research group working on several high value commodity crops, such as in the genera *Zea, Oryza, Triticum, Hordeum,* and *Glycine* (Romay et al. 2013; Wang et al. 2018; Sehgal et al. 2015; Milner et al. 2019; Song et al. 2015). These germplasm collection surveys have established not only methodologies and bioinformatics procedures for processing thousands of samples but have also helped developing sample tracking and visualization data management systems that help to connect digital and physical data (e.g., Lee et al. 2005; Konig et al. 2020; Raubach et al. 2020). From the gene bank perspective, these projects provide essential characterization data needed to manage the genetic diversity (e.g., McCouch et al. 2012. Importantly, genomic data may provide more useful descriptive information about the collection identity, diversity, and evolutionary history than any current set of characterization descriptors could provide. In all these crops, sampling was reduced to one individual for each accession by developing inbred lines extracted from the source accession. Accessions were either selected for a core subset or the complete set of gene banks accessions were used (e.g., Milner et al. 2019). In many instances, the species are themselves self-compatible and developing inbred lines was feasible. In every case, the density of SNPs called in the sequencing or SNP array genotyping was sufficient to conduct GWAS for traits of functional importance. In all instances, patterns of linkage disequilibrium in and around genomic positions associated with domestication could reveal patterns of diversity consistent to a selective sweep (where a locus under strong selection goes to high frequency and reduces the diversity in loci nearby) in these loci during domestication.

The barley (*Hordeum vulgare*) project (Milner et al. 2019; Mascher et al. 2019) has opened the door to a system of genomic data integration for use in breeding and curation that influence the way other gene banks characterize their collection and make data available for searching. The cooperative network in the AGENT project (https://agent-project.eu/) was in part inspired by the genotypic data developed around barley and points to new cooperative efforts to coordinate these data in conformation of FAIR data practices (Wilkinson et al. 2016).

In summary, these pioneering genomic surveys of gene bank collections mentioned above affirmed the utility of genomic data to assist in the management of collection by providing high-resolution data on identity, diversity, and history in crops and their wild relatives. The uniformity of the derived samples used for sequencing allowed GWAS to identify quantitative trait loci that underpin important agronomic traits like disease resistance and also traits that were fundamental in the transition from wild species to domesticated crop. Importantly these studies were also able to examine haplotype diversity across a range of phylogenetic lineages which enabled insights into the pan genome and dynamics of haplotype gain and loss among species and during the transition to domestication.

There is now a wave of new collections that have been genomically surveyed altering the way data is used for managing collection and making it available. These projects exploit new sequencing methodologies for improved *de novo* assembly and use new analytical procedures for estimating diversity. Below are project sketches highlighting the most comprehensive project in barley, and newer projects exploring gene bank diversity in collections of lentil (*Lens culinaris*), apple (*Malus* x *domestica*) and lettuce (*Lactuca sativa*).

4.1 Barley

One of the most complete and comprehensive genomic characterizations of a gene bank collection was completed at the Leibniz Institute of Plant Genetics and Crop Plant Research (IPK) in Germany. Barley is a focal study species at the IPK and they coordinated the effort to sequence the barley genome with the International Barley Sequencing Consortium (Schulte et al. 2009) which leads to a high-quality chromosome level assembly of the 5.1 Gbp barley genome (Mascher et al. 2017). The reference was especially notable in that the regions that typically difficult to assemble around the centromeres were resolved at high density providing an unprecedented assessment of the gene space within the genome as well as the regions of highly duplicated transposons. The genotyping of the entire (>22,000 accessions, mostly single plants per accession) of domesticated and wild barley was accomplished using a RADseq approach (GBS). The data produced resulted in identification of over 171,00 SNPs. Notably the number of SNP with minor allele frequency >1% in the samples differed among the wild (~46 K) vs domesticated (~22 K) barley accessions reflecting the large pool of diversity that exists in the wild gene pools that did not enter domestication bottlenecks. The data were used to provide detailed identity measures useful for detecting duplications and variation useful for looking at the geographic structure and domestication history of barley (Milner et al. 2019). Importantly, the data are viewed and managed through a visual analytics web tool that allows users to filter and display these data. The phenotypic descriptor data is directly linked to the genotypic data (there is no intra-accession heterogeneity). The data describing the variation in the collection also have been linked directly to both underlying raw sequencing reads (exportable in Variant Call Format (VCF) and allow users to directly select accessions based on these data and request them through the gene bank through a sample tracking and analytics database called BRIDGE (Konig et al. 2020). The effort to genomically characterize the collection has led to an entirely new way of interaction (visualizing and exploring diversity, selecting subsets and ordering samples) with gene bank collections that have benefitted both curation and users. The BRIDGE system allows more data integration so users can approach the diversity at the level of a particular locus, the complete SNP genotype or at levels such as phenotypic characterization or geography.

4.2 Lentil

Two of the more comprehensive genomic data sets describing cultivated lentil (*Lens culinaris*) accessions and a number of their crop wild relative species, Khazaei et al. (2016) and Dissanayake et al. (2020) used different genotyping approaches to explore genetic diversity, domestication history, and functional allele diversity. Khazaei and colleagues used a fixed SNP array design for genotyping selected set of 352 accessions drawn from 52 countries specifically to represent variation in climatic and agriculture ecological zone including tropical, northern temperate, and Mediterranean growing regions. Filtering for quality score, completeness of data, and minor allele frequencies resulted in a data set of 1194 SNP evenly distributed across the genome. Genetic diversity and structure were examined using model-based Bayesian clustering procedures using STRUCTURE to investigate the number of lineages represented by the accessions. In addition, the authors applied non-hierarchical principal coordinate analysis using inter-individual genetic distances and conducted analysis of molecular variance (AMOVA). A UPGMA tree was also constructed from the genetic distance data. They found that the genetic structure of the collection generally correlated to the eco geographic region of origin and that ongoing distributions of lens into regional breeding programs created several diversity bottlenecks. For example, the genetic clustering data identifies direct genetic relatedness between Mediterranean and Chilean germplasm identifying the origins of introduction into this region from Spain (Fig. 2). A major concern was the narrowness of the germplasm diversity found in north American breeding lines and need to develop breeding strategies that broaden the genetic base of this germplasm. The study is part of a larger effort at the University of Saskatchewan to develop and deploy new improved lentil varieties for northern temperate climates. Projects include exome targeted resequencing (Agile, https://knowpulse.usask.ca/AGILE/1a) and development of a long-read genome assembly of one cultivated and two wild species of lentil. Data management is handled through a database KnowPulse linking gene bank accessions to genomics information and breeding management tools (https://knowpulse.usask.ca/).

Dissanayake and colleagues (2020) sought to refine the estimation of genetic diversity in 467 accessions also selected to represent the range in agro-ecological growing regions. However, their approach was to use transcriptome sequencing methods that focus SNP detection in cDNA transcript pools that are able to detect SNP variation as well as transcript splice variants. There sequencing approach was a RADseq (GBS) reduced representation library preparation. The bioinformatic pipeline aligned transcript reads back to a reference genome, however the reads were

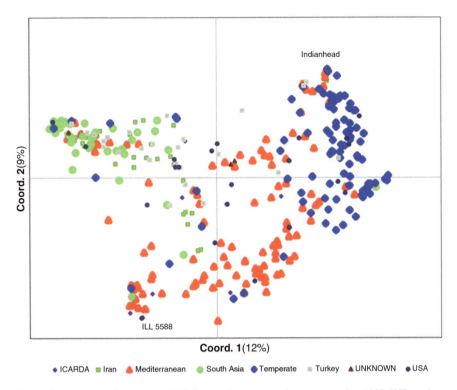

Fig. 2 Principal coordinate plot of 352 *Lens culinaris* accession genotyped at 1192 SNP markers rendered as a bi-plot of the first two principal coordinates. Color reflects region of accession origin. Two distinct control accessions are labeled with their accession name. Reproduced from Khazaei et al. (2016)

aligned separately by species. In total, the procedure produced 422,101 high-quality SNPs. These data allowed the researcher to evaluate genomically mapped allelic variants within transcripts at six chromosomal locations within the set of samples and were able to identify candidate genes and their function and diversity within these regions. The data provided functional diversity and differentiation estimates among the domesticated lineages and among the wild species in *Lens* in the sample panel. The data provided increased resolution for gene bank management and provided options for choosing parents in crossing designs focused on broadening the genetic base of this crop.

4.3 Apple

Apple (*Malus x domestica*) genome characterization has been steadily improving since the first reference quality genome was published in 2010 (Velasco et al. 2010). Recently two papers describing both depth of new sequence characterization and breadth to the collection diversity have been published. Sun and colleague (2020) have assembled a high-quality phased diploid genome of apple (Malus x *domestica*) and its two most prominent wild relative ancestors. Along with the genomic assembly of *Malus sieversii* by Zhang and colleagues (2019a, b) these data represent some of the most detailed genomic assemblies (623–780× coverage) of apple published. The accomplishment is made more impressive because unlike earlier genome assemblies in apple, the genome characterization approach used in the study by Sun et al. (2020) directly sequenced highly heterozygous accessions and characterized the diploid genotypic information derived from over nine million bi-allelic SNPs. In addition to these key reference genomes, the researchers went on to develop pan genome representations for each of the three species using high coverage long read, whole genome sequencing data from 91 samples. Together these data provide detailed views into the domestication history and admixture of apple. From numerous earlier works, the prevailing view is that domesticated apple is a composite lineage derived from *Malus sieversii* in Central Asia and *Malus sylvestris* in Europe. In fact, the admixture history of apple is complex and may have included independent domesticated lineages, each with its own admixture (e.g., Volk et al. 2015). The genomic data in Sun et al. (2020) improves the understanding of the diploid genome and identifies structural variation (such as a 5-Mb inversion on chromosome), transposon distribution, and over 400,000 insertion/deletion polymorphisms. The data gave an unprecedented vie into the genomic history of apple and revealed more precise estimates of admixture proportions derived from wild progenitor species. Genomic structure analysis revealed a remarkable heterogeneity of admixture proportions of wild haplotypes in domesticated apple reflecting the different routes that domestication in this species has taken. Specific genomic positions that have undergone strong selection during domestication have left signatures of diversity in linked neighboring regions (selective sweeps) resulting in the loss of heterozygosity around key loci involved in domestication. The comparative analysis revealed over 1,500 locations under selection between *M. x domestica* and its wild progenitor species. The genomic data supported a coalescent reconstruction of historical divergence and molecular evolution of multiple haplotype blocks. These data were also used to estimate the demographic history of apple during domestication including episodes of reduced effective population size. In total the data provide a high-resolution blueprint for genome enabled breeding in the future and a basis for collection management and rationalization.

Migicovsky et al. (2021) applied genomic characterization more broadly to the USDA collection of over 1,000 apple accessions. The work was based on RADseq genotyping (GBS) of over 30,000 SNPs to reconstruct apple divergence from wild relative species (*M sieversii* and *M. sylvestris*) and also to carefully examine the effects of genetic improvement history of apple breeding on the diversity of *M. x domestica*. The data not only helped to structure the apple collection, but it was also able to estimate and reconstruct pedigree information and pointed out the outsized influence a few popular cultivars have had in the breeding history of the crop (Fig. 3). The consequences are that a few cultivars have more direct clonal relationships, underscoring the relatively narrow genetic base in commercial germplasm

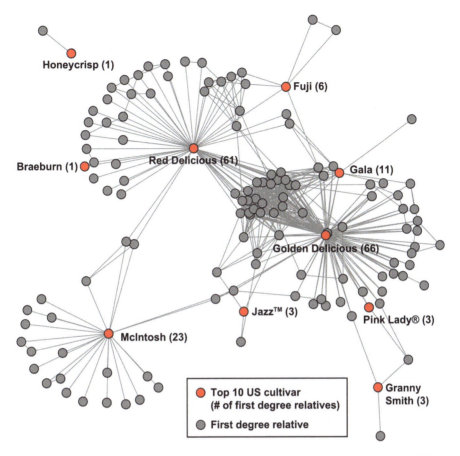

Fig. 3 Network of accessions with a first-degree relationship to one of the top ten apple cultivars sold in the USA. In addition to the top ten apple cultivars, only accessions with at least one first-degree relationships (128) are included. Each accession is represented by a dot. The network of genetic relationships among major cultivars reflects the narrowness of the genetic base for this crop. Reproduced from Migicovsky et al. (2021)

leaving these cultivars more susceptible to emerging disease and a changing climate. These data provide a road map for utilization of gene bank collection in locating novel sources of haplotypic variation useful to crop improvement.

4.4 Lettuce

Genetic diversity of the ex situ collection of lettuce has been an area of focus for collection management approaches at the CGN for decades (Center for Genetic Resource, the Netherlands). Researchers there were early adopters of developing molecular passport information to identify duplicates and develop strategies for maintaining essential diversity in the most efficient way (van Hintum and van Treuren 2002; van Treuren et al. 2010). Recently, the gene bank diversity of single seed descent lines derived from 445 accessions from 47 countries provided a high-resolution view into the domestication history and genetic architecture of domesticated lettuce (Wei et al. 2021). Accessions of cultivated and wild relative species from the CGN were described using whole genome resequencing at high coverage resulting in the development of nearly 180 million SNPs. The resolution enabled the study to describe the origins of domestication using both model-based clustering to infer genetic lineages and admixture and coalescent based phylogenetic analysis to infer divergence times and to locate the likely geographic center of origin in western Turkey. The scale of resolution enables not only reference-based alignment of cultivated lettuce but also enabled *de novo* genome sequence assembly of the wild progenitor species. This data also provides information about the genetic loci involved in diversification of lettuce varieties (grown for both oil and fresh vegetable greens). The study outlined the contribution of wild relative species to several major gene resistance gene families critical to lettuce breeding. These data provide a precise view into a range of previously uncharacterized diversity useful for both crop improvement and conservation.

5 Future Prospects

In 1970 a group of 40 authors (Frankel and Bennett 1970) published a series of articles addressing the benefits of genetic variation to agriculture, forestry, and feeding humanity. They raised concerns that genetic diversity is being lost, is poorly understood and incompletely collected and conserved. They suggested several tactics and solutions to explore, collect, understand, and conserve genetic resources for future generations of plant breeders. Shortly after, a global system of gene banks was established to preserve and use genetic variation to address concerns about "genetic erosion" Harlan (1972). Yet the promise of mobilizing this diversity for crop improvement has stalled, leaving the majority of gene bank collections untapped and uncharacterized.

Currently, gene bank curators act as service providers, making accession diversity (physical specimens) available to the research and breeding communities for evaluation, gene discovery, and applied plant improvement. Information flow about the accessions is largely unidirectional since genomic data is not typically connected back to the gene bank accessions. This structure represents a missed opportunity because gene bank managers would benefit from accessing information about their materials to improve curation efficiency, and researchers and breeders would benefit from having more information about gene bank accessions to enable them to more accurately identify useful diversity for their research and variety development activities.

Genomic characterization of gene banks provides a bridge between breeding and gene discovery projects and PGR management and curation (McCouch et al. 2013, 2020; Varshney et al. 2014; Voss-Fels and Snowdon 2016; Mascher et al. 2019; Belzile et al. 2020). Data standards need to be developed to integrate phenotypic and genomic data with accession characterization data in order to make data sharing and interoperability a reality (e.g., Arnaud et al. 2020). Importantly, with the right capacity development gene bank curation is an integral partner in plant breeding efforts rather than an extractive resource. Integration of candidate genes discovered in breeding programs can serve as the targets for estimating collection diversity and can be used in developing user-defined customized subsets for stakeholder applications (Reeves et al. 2020). High resolution of key functional diversity in gene bank collections provides information critical in developing useful introgression lines, such as nested association mapping populations or multiparent advanced generation inter-cross populations (MAGIC), that help deliver haplotypic diversity from gene bank collection into structured populations critical to breeding and crop improvement (Flint-Garcia et al. 2005; Cavanagh et al. 2008; Langridge and Waugh 2019; Arrones et al. 2020)

Complete genetic characterization of gene bank collection diversity may revolutionize what has been described as seed morgues (Goodman 1990) into living and digital centers of research that might fully realize the impact that the early architects of gene banks had envisioned (Upadhyaya et al. 2007; Kumar et al. 2010; Leung et al. 2015; Ashkani et al. 2015; Mascher et al. 2019). Advances in computational technologies and algorithm development in machine learning represent another advance in our ability to predict phenotypic potential of germplasm based entirely on genotypic information using genomic prediction algorithms (Yu et al. 2016). Ex situ collections represent a remarkable spatial sampling of natural and largely unexplored local agricultural diversity critical for crop improvement to climate change, disease resistance and to adaptive quantitative traits like drought and salinity tolerance (Mascher et al. 2019; McCouch et al. 2020).

The development of these repositories into a mature institutional resource for diversity comes with some district challenges when it comes to both managing and sustaining viable germplasm and also finding relevant agricultural diversity in a sea of accession entries. Genomic collection maintenance, regeneration, documentation, and characterization are areas that virtually all gene banks could improve upon, and all require resources. Identifying gaps in collection coverage, improving collection access through improved data management systems, and clarifying policy governing accession exchange are all vital to improving impact (Fu 2017). In spite of these challenges, collections provide an irreplaceable source of diversity for agricultural crop improvement (Bretting 2018).

As these demands have diversified so to have the responsibilities of curators tasked with maintaining and evaluating these accessions (Byrne et al. 2018; Volk et al. 2019). Curators are currently contending with maintaining the viability of collections, including regeneration viability monitoring and characterization. A high priority in the US National Plant germplasm System (NPGS) are crop wild relatives whose diversity has become critical to allele mining and crop improvement efforts

(Khoury et al. 2020). These species represent complex and challenging biological characteristics that make them especially challenging to maintain in ex situ conditions. Curators faced with this task often have to devote extra resources to their regeneration and evaluation. In addition, curators are responsible for the inventory records and distribution requests for germplasm. In the USA, the NPGS processes on average 250,000 requests a year. These requests are often regulated by material transfer agreements and regulations involving the International Treaty for Plant Genetic Resources for Food and Agriculture (Esquinas-Alcazar 2005). In addition to these skills, it is recognized that the next generation of curators will have to integrate genomics tools into their operations (Volk et al. 2019).

6 Conclusions

Developing a systematic coordination between gene banks, researchers, and breeders is ongoing. There are efforts (e.g., DivSeek International Network, https://divseekintl.org/) leading the way that influence the way the community of gene banks around the world operate. These coordinating efforts (whether focused on best practices or data standards or new analytical approaches) will have a significant impact on the way the diversity in gene banks is characterized and used by stakeholders. We are now entering a phase where the accession remains the organizational unit of the collection, but the unit of interest is a precise combination of genomic haplotypes. Anticipating how to manage the scale of data that is emerging from gene banks around the world will be critical in mobilizing this diversity. Leveraging the ground-breaking advances in "omics" and information technology, computational power of "big data" (e.g., Ramstein et al. 2019) is critically needed to face the agricultural challenges of the future.

References

Alercia A, López FM, Sackville Hamilton NR, Marsella M. Digital object identifiers for food crops – descriptors and guidelines of the global information system. Rome: FAO; 2018.
Andrews KR, Good JM, Miller MR, Luikart G, Hohenlohe PA. Harnessing the power of RADseq for ecological and evolutionary genomics. Nat Rev Genet. 2016;17(2):81–92.
Arnaud E, Laporte M-A, Kim S, Aubert C, Leonelli S, Miro B, et al. The ontologies community of practice: a CGIAR initiative for big data in agrifood systems. Patterns. 2020;1(7):100105.
Arrones A, Vilanova S, Plazas M, Mangino G, Pascual L, Diez MJ, et al. The dawn of the age of multi-parent MAGIC populations in plant breeding: novel powerful next-generation resources for genetic analysis and selection of recombinant elite material. Biology (Basel). 2020;9(8)
Ashkani S, Yusop MR, Shabanimofrad M, Azadi A., Ghasemzadeh A, Azizi P, et al. Allele mining strategies: principles and utilisation for blast resistance genes in rice (Oryza sativa L.). Curr Issues Mol Biol. (2015);17:57–74.

Baetscher DS, Clemento AJ, Ng TC, Anderson EC, Garza JC. Microhaplotypes provide increased power from short-read DNA sequences for relationship inference. Mol Ecol Resour. 2018;18(2):296–305.

Bansal V. Integrating read-based and population-based phasing for dense and accurate haplotyping of individual genomes. Bioinformatics. 2019;35(14):I242–8.

Bari A, Street K, Mackay M, Endresen DTF, De Pauw E, Amri A. Focused identification of germplasm strategy (FIGS) detects wheat stem rust resistance linked to environmental variables. Genet Resour Crop Evol. 2012;59(7):1465–81.

Bashalkhanov S, Pandey M, Rajora OP. A simple method for estimating genetic diversity in large populations from finite sample sizes. BMC Genet. 2009;10:84.

Bayer PE, Golicz AA, Scheben A, Batley J, Edwards D. Plant pan-genomes are the new reference. Nat Plants. 2020;6(8):914–20.

Beckmann JS, Soller M. Toward a unified approach to genetic-mapping of eukaryotes based on sequence tagged microsatellite sites. Bio-Technology. 1990;8(10):930–2.

Belzile F, Abed A, Torkamaneh D. Time for a paradigm shift in the use of plant genetic resources. Genome. 2020;63(3):189–94.

Bennett E. Record of the FAO/IBP technical conference on the exploration, utilization and conservation of plant genetic resources. Rome: FAO; 1968.

Bevan MW, Uauy C. Genomics reveals new landscapes for crop improvement. Genome Biol. 2013;14(6)

Black WC, Baer CF, Antolin MF, DuTeau NM. Population genomics: genome-wide sampling of insect populations. Annu Rev Entomol. 2001;46:441–69.

Brandariz SP, Reymundez AG, Lado B, Malosetti M, Garcia AAF, Quincke M, et al. Ascertainment bias from imputation methods evaluation in wheat. BMC Genomics. 2016;17:13.

Bretting PK. 2017 Frank Meyer Medal for plant genetic resources lecture: stewards of our agricultural future. Crop Sci. 2018;58(6):2233–40.

Bretting PK, Widrlechner M. Genetic markers and plant genetic resource management. Plant Breed Rev. 1995;13:11–86.

Brown A. The case for core collections. In: AHD B, Frankel OH, Marshall DR, Williams JT, editors. Genetic resources. Cambridge: Cambridge University Press; 1989.

Browning SR, Browning BL. Haplotype phasing: existing methods and new developments. Nat Rev Genet. 2011;12(10):703–14.

Brozynska M, Furtado A, Henry RJ. Genomics of crop wild relatives: expanding the gene pool for crop improvement. Plant Biotechnol J. 2016;14(4):1070–85.

Brumlop S, Reichenbecher W, Tappeser B, Finckh MR. What is the SMARTest way to breed plants and increase agrobiodiversity? Euphytica. 2013;194(1):53–66.

Brush SB. Genetic erosion of crop populations in centers of diversity: a revision. Prague: FAO Rome; 1999.

Byrne PF, Volk GM, Gardner C, Gore MA, Simon PW, Smith S. Sustaining the future of plant breeding: the critical role of the USDA-ARS National Plant Germplasm System. Crop Sci. 2018;58(2):451–68.

Cavanagh C, Morell M, Mackay I, Powell W. From mutations to MAGIC: resources for gene discovery, validation and delivery in crop plants. Curr Opin Plant Biol. 2008;11(2):215–21.

Chen ZT, Pham L, Wu TC, Mo GY, Xia Y, Chang PL, et al. Ultralow-input single-tube linked-read library method enables short-read second-generation sequencing systems to routinely generate highly accurate and economical long-range sequencing information. Genome Res. 2020;30(6):898–909.

Chin C-S, Peluso P, Sedlazeck FJ, Nattestad M, Concepcion GT, Clum A, et al. Phased diploid genome assembly with single-molecule real-time sequencing. Nat Methods. 2016;13(12):1050–4.

Clark AG, Hubisz MJ, Bustamante CD, Williamson SH, Nielsen R. Ascertainment bias in studies of human genome-wide polymorphism. Genome Res. 2005;15(11):1496–502.

Danilevicz MF, Fernandez CGT, Marsh JI, Bayer PE, Edwards D. Plant pangenomics: approaches, applications and advancements. Curr Opin Plant Biol. 2020;54:18–25.

Davey JW, Cezard T, Fuentes-Utrilla P, Eland C, Gharbi K, Blaxter ML. Special features of RAD sequencing data: implications for genotyping. Mol Ecol. 2013;22(11):3151–64.

De Coster W, Van Broeckhoven C. Newest methods for detecting structural variations. Trends Biotechnol. 2019;37(9):973–82.

DePristo MA, Banks E, Poplin R, Garimella KV, Maguire JR, Hartl C, et al. A framework for variation discovery and genotyping using next-generation DNA sequencing data. Nat Genet. 2011;43(5):491–8.

Diez MJ, De la Rosa L, Martin I, Guasch L, Cartea ME, Mallor C, et al. Plant genebanks: present situation and proposals for their improvement. The case of the Spanish network. Front Plant Sci. 2018;9:1794.

Dissanayake R, Braich S, Cogan NOI, Smith K, Kaur S. Characterization of genetic and allelic diversity amongst cultivated and wild lentil accessions for germplasm enhancement. Front Genet. 2020;11(546)

Ebler J, Haukness M, Pesout T, Marschall T, Paten B. Haplotype-aware diplotyping from noisy long reads. Genome Biol. 2019;20(1):116.

Ellis D, Chavez O, Coombs J, Soto J, Gomez R, Douches D, et al. Genetic identity in genebanks: application of the SolCAP 12K SNP array in fingerprinting and diversity analysis in the global in trust potato collection. Genome. 2018;61(7):523–37.

Esquinas-Alcazar J. Protecting crop genetic diversity for food security: political, ethical and technical challenges. Nat Rev Genet. 2005;6:946–53.

Excoffier L, Smouse PE, Quattro JM. Analysis of molecular variance inferred from metric distances among DNA haplotypes: application to human mitochondrial DNA restriction data. Genetics. 1992;131(2):479.

FAO. The second report on the state of the world's plant genetic resources. Rome: FAO; 2010.

Feuillet C, Langridge P, Waugh R. Cereal breeding takes a walk on the wild side. Trends Genet. 2008;24(1):24–32.

Flint-Garcia SA, Thuillet AC, Yu J, Pressoir G, Romero SM, Mitchell SE, et al. Maize association population: a high-resolution platform for quantitative trait locus dissection. Plant J. 2005;44(6):1054–64.

Franco J, Crossa J, Taba S, Shands H. A sampling strategy for conserving genetic diversity when forming core subsets. Crop Sci. 2005;45(3):1035–44.

Franco JAV, Gage JL, Bradbury PJ, Johnson LC, Miller ZR, Buckler ES, et al. A maize practical haplotype graph leverages diverse NAM assemblies. bioRxiv. 2020; https://doi.org/10.1101/2020.08.31.268425.

Frankel OH, Bennett E. Genetic resources in plants – their exploration and conservation. Oxford: Blackwell; 1970.

Fu Y-B. The vulnerability of plant genetic resources conserved ex situ. Crop Sci. 2017;57(5):2314–28.

Gabur I, Chawla HS, Snowdon RJ, Parkin IAP. Connecting genome structural variation with complex traits in crop plants. Theor Appl Genet. 2019;132(3):733–50.

Gepts P. Plant genetic resources conservation and utilization: the accomplishments and future of a societal insurance policy. Crop Sci. 2006;46(5):2278–92.

Gerbault P, Allaby RG, Boivin N, Rudzinski A, Grimaldi IM, Pires JC, et al. Storytelling and story testing in domestication. Proc Natl Acad Sci U S A. 2014;111(17):6159–64.

Glaszmann JC, Kilian B, Upadhyaya HD, Varshney RK. Accessing genetic diversity for crop improvement. Curr Opin Plant Biol. 2010;13(2):167–73.

Gokidi Y, Bhanu AN, Chandra K, Singh MN, Hemantaranjan A. Allele mining – an approach to discover allelic variation in crops. J Plant Sci Res. 2017;33(2):167–80.

Golicz AA, Batley J, Edwards D. Towards plant pangenomics. Plant Biotechnol J. 2016;14(4):1099–105.

Goodman MM. Genetic and germplasm stocks worth conserving. J Hered. 1990;81(1):11–6.

Goodwin S, McPherson JD, McCombie WR. Coming of age: ten years of next-generation sequencing technologies. Nat Rev Genet. 2016;17(6):333–51.

Gotor E, Alercia A, Rao VR, Watts J, Caracciolo F. The scientific information activity of bioversity international: the descriptor lists. Genetic Resources and Crop Evolution. 2008;55(5):757–72.

Gouesnard B, Bataillon TM, Decoux G, Rozale C, Schoen DJ, David JL. MSTRAT: an algorithm for building germ plasm core collections by maximizing allelic or phenotypic richness. J Hered. 2001;92(1):93–4.

Gross BL, Henk AD, Forsline PL, Richards CM, Volk GM. Identification of interspecific hybrids among domesticated apple and its wild relatives. Tree Genet Genomes. 2012;8(6):1223–35.

Hale ML, Burg TM, Steeves TE. Sampling for microsatellite-based population genetic studies: 25 to 30 individuals per population is enough to accurately estimate allele frequencies. PLoS One. 2012;7(9):10.

Hamblin MT, Buckler ES, Jannink JL. Population genetics of genomics-based crop improvement methods. Trends Genet. 2011;27(3):98–106.

Hammer K. A paradigm shift in the discipline of plant genetic resources. Genetic Resources and Crop Evolution. 2003;50(1):3–10.

Hammer K, Teklu Y. Plant genetic resources: selected issues from genetic erosion to genetic engineering. J Agric Rural Dev Trop Subtrop. 2008;109(1):15–50.

Harlan JR. Genetics of disaster. J Environ Qual. 1972;1(3):212–5.

Hawkes JG. The importance of wild germplasm in plant breeding. Euphytica. 1977;26:615–21.

Hedrick PW. A standardized genetic differentiation measure. Evolution. 2005;59(8):1633–8.

Heslot N, Rutkoski J, Poland J, Jannink JL, Sorrells ME. Impact of marker ascertainment bias on genomic selection accuracy and estimates of genetic diversity. PLoS One. 2013;8(9):8.

Ho SS, Urban AE, Mills RE. Structural variation in the sequencing era. Nat Rev Genet. 2020;21(3):171–89.

Hoban S, Volk G, Routson KJ, Walters C, Richards C. Sampling wild species to conserve genetic diversity. In: Greene SL, Williams KA, Khoury CK, Kantar MB, Marek LF, editors. North American Crop Wild Relatives, volume 1: conservation strategies. Cham: Springer; 2018. p. 209–28.

Holliday JA, Hallerman EM, Haak DC. Genotyping and sequencing technologies in population genetics and genomics. In: Rajora OP, editor. Population genomics: concepts, approaches and applications. Cham: Springer Nature Switzerland AG; 2019. p. 83–126.

Hubner S, Bercovich N, Todesco M, Mandel JR, Odenheimer J, Ziegler E, et al. Sunflower pan-genome analysis shows that hybridization altered gene content and disease resistance. Nat Plants. 2019;5(1):54–62.

Huson DH. SplitsTree: analyzing and visualizing evolutionary data. Bioinformatics. 1998;14(1):68–73.

Jensen SE, Charles JR, Muleta K, Bradbury PJ, Casstevens T, Deshpande SP, et al. A sorghum practical haplotype graph facilitates genome-wide imputation and cost-effective genomic prediction. Plant Genome. 2020;13(1):e20009.

Jombart T, Ahmed I. Adegenet 1.3-1: new tools for the analysis of genome-wide SNP data. Bioinformatics. 2011;27(21):3070–1.

Jordan DR, Mace ES, Cruickshank AW, Hunt CH, Henzell RG. Exploring and exploiting genetic variation from unadapted sorghum germplasm in a breeding program. Crop Sci. 2011;51(4):1444–57.

Josephs EB, Berg JJ, Ross-Ibarra J, Coop G. Detecting adaptive differentiation in structured populations with genomic data and common gardens. Genetics. 2019;211(3):989–1004.

Karp A, Kresovich S, Bhat K, Ayad W, Hodgkin T. Molecular tools in plant genetic resources conservation: a guide to the technologies. Rome: International Plant Genetic Resources Institute; 1997.

Khazaei H, Caron CT, Fedoruk M, Diapari M, Vandenberg A, Coyne CJ, et al. Genetic diversity of cultivated lentil (Lens culinaris Medik.) and its relation to the world's agro-ecological zones. Front Plant Sci. 2016;7:1093.

Khoury CK, Carver D, Greene SL, Williams KA, Achicanoy HA, Schori M, et al. Crop wild relatives of the United States require urgent conservation action. Proc Natl Acad Sci. 2020;117 (52):33351.

Khush GS. Green revolution: the way forward. Nat Rev Genet. 2001;2(10):815–22.

Kilian B, Graner A. NGS technologies for analyzing germplasm diversity in genebanks. Brief Funct Genomics. 2012;11(1):38–50.

Kilian A, Wenzl P, Huttner E, Carling J, Xia L, Blois H, et al. Diversity arrays technology: a generic genome profiling technology on open platforms. In: Pompanon F, Bonin A, editors. Data production and analysis in population genomics: methods and protocols. Totowa: Humana Press; 2012. p. 67–89.

Kim K-W, Chung H-K, Cho G-T, Ma K-H, Chandrabalan D, Gwag J-G, et al. PowerCore: a program applying the advanced M strategy with a heuristic search for establishing core sets. Bioinformatics. 2007;23(16):2155–62.

Kinard G, Cyr P, Weaver B, Millard M, Gardner C, Bohning M, et al. GRIN-Global: an international project to develop a global plant genebank and information management system. Phytopathology. 2009;99(6):S64.

Kofler R, Pandey RV, Schlötterer C. PoPoolation2: identifying differentiation between populations using sequencing of pooled DNA samples (Pool-Seq). Bioinformatics. 2011;27(24):3435–6.

Konig P, Beier S, Basterrechea M, Schuler D, Arend D, Mascher M, et al. BRIDGE – a visual analytics web tool for barley genebank genomics. Front Plant Sci. 2020;11:701.

Kovach MJ, McCouch SR. Leveraging natural diversity: back through the bottleneck. Curr Opin Plant Biol. 2008;11(2):193–200.

Kumar GR, Sakthivel K, Sundaram RM, Neeraja CN, Balachandran SM, Rani NS, et al. Allele mining in crops: prospects and potentials. Biotechnol Adv. 2010;28(4):451–61.

Kumari R. Allele mining for crop improvement. Int J Pure Appl Biosci. 2018;6(1):1456–65.

Lam HM, Xu X, Liu X, Chen W, Yang G, Wong FL, et al. Resequencing of 31 wild and cultivated soybean genomes identifies patterns of genetic diversity and selection. Nat Genet. 2010;42 (12):1053–9.

Langridge P, Waugh R. Harnessing the potential of germplasm collections. Nat Genet. 2019;51 (2):200–1.

Lee JM, Davenport GF, Marshall D, Ellis THN, Ambrose MJ, Dicks J, et al. GERMINATE. A generic database for integrating genotypic and phenotypic information for plant genetic resource collections. Plant Physiol. 2005;139(2):619–31.

Lemmon AR, Emme SA, Lemmon EM. Anchored hybrid enrichment for massively high-throughput phylogenomics. Syst Biol. 2012;61(5):727–44.

Leung H, Raghavan C, Zhou B, Oliva R, Choi IR, Lacorte V, et al. Allele mining and enhanced genetic recombination for rice breeding. Rice (N Y). 2015;8(1):34.

Levy SE, Boone BE. Next-generation sequencing strategies. Cold Spring Harb Perspect Med. 2019;9(7):11.

Li J, Jiang T. Haplotype-based linkage disequilibrium mapping via direct data mining. Bioinformatics. 2005;21(24):4384–93.

Linck E, Battey CJ. Minor allele frequency thresholds strongly affect population structure inference with genomic data sets. Mol Ecol Resour. 2019;19(3):639–47.

Lopes MS, El-Basyoni I, Baenziger PS, Singh S, Royo C, Ozbek K, et al. Exploiting genetic diversity from landraces in wheat breeding for adaptation to climate change. J Exp Bot. 2015;66 (12):3477–86.

Luikart G, Kardos M, Hand BK, Rajora OP, Aitken SN, Hohenlohe PA. Population genomics: advancing understanding of nature. In: Rajora OP, editor. Population genomics: concepts, approaches and applications. Cham: Springer Nature Switzerland AG; 2019. p. 3–79.

Lynch M, Ritland K. Estimation of pairwise relatedness with molecular markers. Genetics. 1999;152(4):1753–66.

Malinsky M, Trucchi E, Lawson DJ, Falush D. RADpainter and fineRADstructure population inference from RADseq data. Mol Biol Evol. 2018;35(5):1284–90.

Mammadov J, Buyyarapu R, Guttikonda SK, K. Parliament, Abdurakhmonov IY, Kumpatla SP. Wild relatives of maize, rice, cotton, and soybean: treasure troves for tolerance to biotic and abiotic stresses. Front Plant Sci. 2018;9:886.

Manel S, Albert CH, Yoccoz NG. Sampling in landscape genomics. In: Pompanon F, Bonin A, editors. Data production and analysis in population genomics: methods and protocols. Totowa: Humana Press; 2012. p. 3–12.

Marco-Sola S, Sammeth M, Guigo R, Ribeca P. The GEM mapper: fast, accurate and versatile alignment by filtration. Nat Methods. 2012;9(12):1185–8.

Mascher M, Gundlach H, Himmelbach A, Beier S, Twardziok SO, Wicker T, et al. A chromosome conformation capture ordered sequence of the barley genome. Nature. 2017;544(7651):427–33.

Mascher M, Schreiber M, Scholz U, Graner A, Reif JC, Stein N. Genebank genomics bridges the gap between the conservation of crop diversity and plant breeding. Nat Genet. 2019;51(7):1076–81.

Maxted N, Kell S, Ford-Lloyd B, Dulloo E, Toledo Á. Toward the systematic conservation of global crop wild relative diversity. Crop Sci. 2012;52(2):774–85.

McCouch SR, McNally KL, Wang W, Sackville Hamilton R. Genomics of gene banks: a case study in rice. Am J Bot. 2012;99(2):407–23.

McCouch S, Baute GJ, Bradeen J, Bramel P, Bretting PK, Buckler E, et al. Feeding the future. Nature. 2013;499(7456):23–4.

McCouch S, Navabi ZK, Abberton M, Anglin NL, Barbieri RL, Baum M, et al. Mobilizing crop biodiversity. Mol Plant. 2020;13(10):1341–4.

Migicovsky Z, Gardner KM, Richards C, Chao CT, Schwaninger HR, Fazio G, et al. Genomic consequences of apple improvement. Hortic Res. 2021;8(1)

Milner SG, Jost M, Taketa S, Mazon ER, Himmelbach A, Oppermann M, et al. Genebank genomics highlights the diversity of a global barley collection. Nat Genet. 2019;51(2):319–26.

Monat C, Schreiber M, Stein N, Mascher M. Prospects of pan-genomics in barley. Theor Appl Genet. 2019;132(3):785–96.

Muller T, Schierscher-Viret B, Fossati D, Brabant C, Schori A, Keller B, et al. Unlocking the diversity of genebanks: whole-genome marker analysis of Swiss bread wheat and spelt. Theor Appl Genet. 2018;131(2):407–16.

Nazareno AG, Bemmels JB, Dick CW, Lohmann LG. Minimum sample sizes for population genomics: an empirical study from an Amazonian plant species. Mol Ecol Resour. 2017;17(6):1136–47.

Nei M. F-statistics and analysis of gene diversity in subdivided populations. Ann Hum Genet. 1977;41:225–33.

Nguyen N, Hickey G, Zerbino DR, Raney B, Earl D, Armstrong J, et al. Building a pangenome reference for a population. In: Sharan R, editor. Research in computational molecular biology, vol. 8394. Cham: Springer; 2014. p. 207–21.

Nielsen R, Korneliussen T, Albrechtsen A, Li YR, Wang J. SNP calling, genotype calling, and sample allele frequency estimation from new-generation sequencing data. PLoS One. 2012;7(7):10.

Palme AE, Hagenblad J, Solberg SO, Aloisi K, Artemyeva A. SNP markers and evaluation of duplicate holdings of Brassica oleracea in two European genebanks. Plants. 2020;9(8):925.

Poland JA, Rife TW. Genotyping-by-sequencing for plant breeding and genetics. The Plant Genome. 2012;5(3)

Posada D, Crandall KA. Intraspecific gene genealogies: trees grafting into networks. Trends Ecol Evol. 2001;16(1):37–45.

Powell W, Morgante M, Andre C, Hanafey M, Vogel J, Tingey S, et al. The comparison of RFLP, RAPD, AFLP and SSR (microsatellite) markers for germplasm analysis. Molecular Breeding. 1996;2(3):225–38.

Prada D. Molecular population genetics and agronomic alleles in seed banks: searching for a needle in a haystack? J Exp Bot. 2009;60(9):2541–52.

Pritchard JK, Stephens M, Donnelly P. Inference of population structure using multilocus genotype data. Genetics. 2000;155(2):945–59.

Purugganan MD. Evolutionary insights into the nature of plant domestication. Curr Biol. 2019;29(14):R705–14.

Ramankutty N, Mehrabi Z, Waha K, Jarvis L, Kremen C, Herrero M, et al. Trends in global agricultural land use: implications for environmental health and food security. Annu Rev Plant Biol. 2018;69:789–815.

Ramstein GP, Jensen SE, Buckler ES. Breaking the curse of dimensionality to identify causal variants in Breeding 4. Theor Appl Genet. 2019;132(3):559–67.

Rasheed A, Hao Y, Xia X, Khan A, Xu Y, Varshney RK, et al. Crop breeding chips and genotyping platforms: progress, challenges, and perspectives. Mol Plant. 2017;10(8):1047–64.

Rassmann K, Schlötterer C, Tautz D. Isolation of simple-sequence loci for use in polymerase chain reaction-based DNA fingerprinting. Electrophoresis. 1991;12(2-3):113–8.

Raubach S, Kilian B, Dreher K, Amri A, Bassi FM, Boukar O, et al. From bits to bites: advancement of the germinate platform to support prebreeding informatics for crop wild relatives. Crop Sci. 2020;29

Reeves PA, Richards CM. Accurate inference of subtle population structure (and other genetic discontinuities) using principal coordinates. PLoS One. 2009;4(1):e4269.

Reeves PA, Richards CM. Biases induced by using geography and environment to guide ex situ conservation. Conserv Genet. 2018;19(6):1281–93.

Reeves PA, Panella LW, Richards CM. Retention of agronomically important variation in germplasm core collections: implications for allele mining. Theor Appl Genet. 2012;124(6):1155–71.

Reeves PA, Tetreault HM, Richards CM. Bioinformatic extraction of functional genetic diversity from heterogeneous germplasm collections for crop improvement. Agronomy. 2020;10(4)

Romay MC, Millard MJ, Glaubitz JC, Peiffer JA, Swarts KL, Casstevens TM, et al. Comprehensive genotyping of the USA national maize inbred seed bank. Genome Biol. 2013;14(6)

Ross-Ibarra J, Morrell PL, Gaut BS. Plant domestication, a unique opportunity to identify the genetic basis of adaptation. Proc Natl Acad Sci U S A. 2007;104(Suppl 1):8641–8.

Rubenstein KD, Smale M, Widrlechner MP. Demand for genetic resources and the U.S. National Plant Germplasm System. Crop Sci. 2006;46(3):1021–31.

Sackville Hamilton R, Engels J, Van Hintum T. Rationalization of genebank management. A guide to effective management of germplasm collections. In: Engels JMM, Visser L, editors. IPGRI handbooks for genebanks no. 6. Rome: IPGRI; 2003.

Saitou N, Nei M. The neighbor-joining method: a new method for reconstructing phylogenetic trees. Mol Biol Evol. 1987;4(4):406–25.

Scarascia-Mugnozza GT, Perrino P. The history of ex situ conservation and use of genetic resources. In: Engles JMM, Ramanatha RV, Brown AHD, Jackson MT, editors. Managing genetic diversity. Wallingford: CABI; 2002.

Scheben A, Batley J, Edwards D. Genotyping-by-sequencing approaches to characterize crop genomes: choosing the right tool for the right application. Plant Biotechnol J. 2017;15(2):149–61.

Schlotterer C, Tobler R, Kofler R, Nolte V. Sequencing pools of individuals – mining genome-wide polymorphism data without big funding. Nat Rev Genet. 2014;15(11):749–63.

Schoen DJ, Brown AH. Conservation of allelic richness in wild crop relatives is aided by assessment of genetic markers. Proc Natl Acad Sci. 1993;90(22):10623.

Schulte D, Close TJ, Graner A, Langridge P, Matsumoto T, Muehlbauer G, et al. The international barley sequencing consortium—at the threshold of efficient access to the barley genome. Plant Physiol. 2009;149(1):142–7.

Scossa F, Alseekh S, Fernie AR. Integrating multi-omics data for crop improvement. J Plant Physiol. 2021;257:153352.

Sedlazeck FJ, Lee H, Darby CA, Schatz MC. Piercing the dark matter: bioinformatics of long-range sequencing and mapping. Nat Rev Genet. 2018;19(6):329–46.

Sehgal D, Vikram P, Sansaloni CP, Ortiz C, Pierre CS, Payne T, Ellis M, Amri A, Petroli CD, Wenzl P, Singh S. Exploring and mobilizing the gene bank biodiversity for wheat improvement. PLoS One. 2015;10(7):e0132112.

Sehgal D, Mondal S, Crespo-Herrera L, Velu G, Juliana P, Huerta-Espino J, et al. Haplotype-based, genome-wide association study reveals stable genomic regions for grain yield in CIMMYT spring bread wheat. Front Genet. 2020;11(1427)

Semagn K, Babu R, Hearne S, Olsen M. Single nucleotide polymorphism genotyping using kompetitive allele specific PCR (KASP): overview of the technology and its application in crop improvement. Mol Breed. 2013;33(1):1–14.

Singh N, Wu S, Raupp WJ, Sehgal S, Arora S, Tiwari V, et al. Efficient curation of genebanks using next generation sequencing reveals substantial duplication of germplasm accessions. Sci Rep. 2019;9(1):650.

Smýkal P, Nelson MN, Berger JD, Von Wettberg EJB. The impact of genetic changes during crop domestication. Agronomy. 2018;8(7):119.

Song Q, Hyten DL, Jia G, Quigley CV, Fickus EW, Nelson RL, et al. Fingerprinting soybean germplasm and its utility in genomic research. G3 (Bethesda). 2015;5(10):1999–2006.

Spooner D, van Treuren R, de Vicente MC. Molecular markers for genebank management. Rome: PRGRI; 2005. Technical Bulletin #10

Stapleton JA, Kim J, Hamilton JP, Wu M, Irber LC, Maddamsetti R, et al. Haplotype-phased synthetic long reads from short-read sequencing. PLoS One. 2016;11(1):e0147229.

Steffenson BJ, Olivera P, Roy JK, Jin Y, Smith KP, Muehlbauer GJ. A walk on the wild side: mining wild wheat and barley collections for rust resistance genes. Aust J Agr Res. 2007;58(6):532–44.

Stuessy TF, Tremetsberger K, Müllner AN, Jankowicz J, Guo Y-P, Baeza CM, et al. The melding of systematics and biogeography through investigations at the populational level: examples from the genus Hypochaeris (Asteraceae). Basic Appl Ecol. 2003;4(4):287–96.

Sun XP, Jiao C, Schwaninger H, Chao CT, Ma YM, Duan NB, et al. Phased diploid genome assemblies and pan-genomes provide insights into the genetic history of apple domestication. Nat Genet. 2020;52(12):25.

Swarts K, Li H, Romero Navarro JA, An D, Romay MC, Hearne S, et al. Novel methods to optimize genotypic imputation for low-coverage, next-generation sequence data in crop plants. Plant Genome. 2014;7(3) https://doi.org/10.3835/plantgenome2014.2005.0023.

Tanksley SD, McCouch SR. Seed banks and molecular maps: unlocking genetic potential from the wild. Science. 1997;277(5329):1063–6.

Terhorst J, Kamm JA, Song YS. Robust and scalable inference of population history from hundreds of unphased whole genomes. Nat Genet. 2017;49(2):303–9.

Thachuk C, Crossa J, Franco J, Dreisigacker S, Warburton M, Davenport GF. Core Hunter: an algorithm for sampling genetic resources based on multiple genetic measures. BMC Bioinformatics. 2009;10(1):243.

Thormann I, Engels JMM. Genetic diversity and erosion – a global perspective. Genet Divers Erosion Plants. 2015:263–94.

Upadhyaya HD, Ortiz R. A mini core subset for capturing diversity and promoting utilization of chickpea genetic resources in crop improvement. Theor Appl Genet. 2001;102(8):1292–8.

Upadhyaya HD, Furman BJ, Dwivedi SL, Udupa SM, Gowda CLL, Baum M, et al. Development of a composite collection for mining germplasm possessing allelic variation for beneficial traits in chickpea. Plant Genetic Resources. 2007;4(1):13–9.

Valiente-Mullor C, Beamud B, Ansari I, Frances-Cuesta C, Garcia-Gonzalez N, Mejia L, et al. One is not enough: on the effects of reference genome for the mapping and subsequent analyses of short-reads. PLoS Comput Biol. 2021;17(1):29.

van de Wouw M, van Hintum T, Kik C, van Treuren R, Visser B. Genetic diversity trends in twentieth century crop cultivars: a meta analysis. Theor Appl Genet. 2010;120(6):1241–52.

van Hintum T. Hierarchical approaches to the analysis of genetic diversity in crop plants. In: Hodgkin T, Brown AHD, van Hintum TJL, Morales EAV, editors. Core collections of plant genetic resources. Chichester: Wiley; 1995.

van Hintum TJL, van Treuren R. Molecular markers: tools to improve genebank efficiency. Cell Mol Biol Lett. 2002;7(2B):737–44.

van Treuren R, van Hintum TJL. Next-generation genebanking: plant genetic resources management and utilization in the sequencing era. Plant Genet Resour. 2014;12(3):298–307.

van Treuren R, van Soest LJM, van Hintum TJL. Marker-assisted rationalisation of genetic resource collections: a case study in flax using AFLPs. Theor Appl Genet. 2001;103(1):144–52.

van Treuren R, de Groot EC, Boukema IW, de Wiel C, van Hintum TJL. Marker-assisted reduction of redundancy in a genebank collection of cultivated lettuce. Plant Genet Resour Charact Util. 2010;8(2):95–105.

Varshney RK, Terauchi R, McCouch SR. Harvesting the promising fruits of genomics: applying genome sequencing technologies to crop breeding. PLoS Biol. 2014;12(6)

Velasco R, Zharkikh A, Affourtit J, Dhingra A, Cestaro A, Kalyanaraman A, et al. The genome of the domesticated apple (Malus × domestica Borkh). Nat Genet. 2010;42(10):833–9.

Volk GM, Lockwood DR, Richards CM. Wild plant sampling strategies: the roles of ecology and evolution. Plant Breeding Reviews. 2007:285–313.

Volk GM, Namuth-Covert D, Byrne PF. Training in plant genetic resources management: a way forward. Crop Sci. 2019;59(3):853–7.

Volk GM, Henk AD, Baldo A, Fazio G, Chao CT, Richards CM. Chloroplast heterogeneity and historical admixture within the genus Malus. Am J Bot. 2015;102(7):1198–208.

Voss-Fels K, Snowdon RJ. Understanding and utilizing crop genome diversity via high-resolution genotyping. Plant Biotechnol J. 2016;14(4):1086–94.

Walters C, Volk GM, Richards CM. Genebanks in the post-genomic age: emerging roles and anticipated uses. Biodiversity. 2008;9(1-2):68–71.

Wambugu PW, Ndjiondjop MN, Henry RJ. Role of genomics in promoting the utilization of plant genetic resources in genebanks. Brief Funct Genomics. 2018;17(3):198–206.

Wang WS, Mauleon R, Hu ZQ, Chebotarov D, Tai SS, Wu ZC, et al. Genomic variation in 3,010 diverse accessions of Asian cultivated rice. Nature. 2018;557(7703):43–9.

Warschefsky E, Penmetsa RV, Cook DR, von Wettberg EJB. Back to the wilds: tapping evolutionary adaptations for resilient crops through systematic hybridization with crop wild relatives. Am J Bot. 2014;101(10):1791–800.

Wei T, van Treuren R, Liu X, Zhang Z, Chen J, Liu Y, et al. Whole-genome resequencing of 445 Lactuca accessions reveals the domestication history of cultivated lettuce. Nat Genet. 2021;53:752–60.

Weise S, Lohwasser U, Oppermann M. Document or lose it-on the importance of information management for genetic resources conservation in genebanks. Plants (Basel). 2020;9(8):1050.

Wilkinson MD, Dumontier M, Aalbersberg IJ, Appleton G, Axton M, Baak A, et al. The FAIR guiding principles for scientific data management and stewardship. Sci Data. 2016;3(1):160018.

Wright S. The Interpretation of population structure by F-statistics with special regard to systems of mating. Evolution. 1965;19(3):395–420.

Wu X, Heffelfinger C, Zhao H, Dellaporta SL. Benchmarking variant identification tools for plant diversity discovery. BMC Genomics. 2019;20(1):701.

Yu X, Li X, Guo T, Zhu C, Wu Y, Mitchell SE, et al. Genomic prediction contributing to a promising global strategy to turbocharge gene banks. Nat Plants. 2016;2:16150.

Zhang HY, Mittal N, Leamy LJ, Barazani O, Song BH. Back into the wild-apply untapped genetic diversity of wild relatives for crop improvement. Evol Appl. 2017;10(1):5–24.

Zhang H, Yasmin F, Song BH. Neglected treasures in the wild – legume wild relatives in food security and human health. Curr Opin Plant Biol. 2019a;49:17–26.

Zhang LY, Hu J, Han XL, Li JJ, Gao Y, Richards CM, et al. A high-quality apple genome assembly reveals the association of a retrotransposon and red fruit colour. Nat Commun. 2019b;10(1):1494.

Zhang X, Wu R, Wang Y, Yu J, Tang H. Unzipping haplotypes in diploid and polyploid genomes. Comput Struct Biotechnol J. 2020;18:66–72.

Zheng GXY, Lau BT, Schnall-Levin M, Jarosz M, Bell JM, Hindson CM, et al. Haplotyping germline and cancer genomes with high-throughput linked-read sequencing. Nat Biotechnol. 2016;34(3):303–11.

Part II
Population Genomics for Crop Plant Breeding

Part II
Population Genetics for Crop Plant Breeding

Enhancing Crop Breeding Using Population Genomics Approaches

Ryan J. Andres, Jeffrey C. Dunne, Luis Fernando Samayoa, and James B. Holland

Abstract The use of genetic information to predict the value of individuals in plant breeding populations began about 40 years ago. The original paradigm was to identify genomic regions with outsize influence on a trait of economic value and then to use markers in that genomic region to select individuals carrying the desired allelic variants. An explosion of interest in mapping such quantitative trait loci (QTL) followed, with thousands of genomic regions associated with important traits across many species. The practical use of such information lagged well behind the discovery of QTL, however, due mostly to the problem that individual markers were often only associated with a small proportion of genetic variation, such that their value in selection was very small. In a few lucky cases, individual genes with very large effects on important traits were discovered, and these could be more easily turned into useful selection targets. Genome-wide association studies have improved the ability to identify individual variants associated with useful effects in crops, but the fundamental problem of accurately estimating marker effects and using them in selection remains for traits affected by many genes. Genomic selection was proposed by animal breeders as a way to more effectively use the information contained in dense genetic marker sets for the prediction of quantitative traits. Crop breeders subsequently discovered that this approach could be generalized across the diverse population structures and mating systems of plants and have begun implementing genomic selection in crops with success. Here, we first discuss how to identify and select for major genes and QTL in the breeding process. We then discuss the myriad benefits of implementing genomic selection to improve the rate of genetic gain.

R. J. Andres · J. B. Holland (✉)
USDA-ARS Plant Science Research Unit, Raleigh, NC, USA

Department of Crop and Soil Sciences, North Carolina State University, Raleigh, NC, USA
e-mail: Jim.Holland@usda.gov

J. C. Dunne · L. F. Samayoa
Department of Crop and Soil Sciences, North Carolina State University, Raleigh, NC, USA

Keywords Candidate genes · Crop plants · Fine mapping · Genome-wide association studies · Genomic selection · Genotyping · Linkage mapping · Population genomics

1 Introduction

Following the initial demonstrations that specific genomic regions affecting quantitative traits could be identified in crops (Tanksley et al. 1982; Edwards et al. 1987; Paterson et al. 1988), an explosion of interest in mapping such quantitative trait loci (QTL) followed. More recently, a substantial shift in methodology has occurred toward genome-wide association studies (GWAS) to identify QTL or even causal variants underlying quantitative traits. In this chapter, we first review methods used in crop plants to identify major genes via QTL analysis or GWAS. We consider the advantages and disadvantages of each approach, the current state-of-the-art statistical framework generally used, and important features that researchers should consider when designing these studies, such as size of the panel, population structure, and linkage disequilibrium.

Converting QTL and GWAS discoveries into useful breeding tools has remained difficult, in large part because of the mostly polygenic genetic architecture of many important agricultural traits. This type of genetic architecture involves many genes, each with small effects, such that even if truly causal DNA variants can be discovered and their effects accurately estimated, those effects may be too small to justify the effort required for their discovery. Building on developments from animal breeders and quantitative geneticists (Meuwissen et al. 2001; Goddard and Hayes 2007; review in Jonas et al. 2019), crop breeders have more recently adopted genomic selection approaches that provide little biological insight but can be very effective at predicting quantitative trait phenotypes with genome-wide DNA markers. We review important concepts required for effectively integrating genomic selection into practical crop improvement programs, including the size and composition of training data sets, statistical models for genomic prediction, accounting for inbreeding, model updating and retraining, and the implementation of genomic selection on single plants and in off-season nurseries to improve genetic gain per year.

Finally, we consider the possibility that for some traits, some part of the genetic architecture may be truly polygenic, but another part of the genetic variance may be controlled by a small set of genetic variants, each with larger effect than the polygenes. For these traits, a combination of GWAS and genomic selection may usefully target both components of the genetic architecture while still providing some insight into the genes involved in the oligogenic part of the genetic architecture.

2 Identifying Major Genes Causing Large Phenotypic Effects

Genetic dissection of quantitative trait loci (QTL) and Mendelian large-effect genes facilitates the understanding of complex traits by revealing causal genes and associated pathways as targets for genetic manipulation for trait improvement. Major genes or large-effect QTL include loci whose alleles display a large, qualitative effect, segregate by simple Mendelian ratios, and receive relatively little environmental or epistatic influence. Homozygous types are easily distinguished from one another while heterozygotes exhibit either complete dominance or are easily distinguishable as a distinct genotypic group. Such loci should be amenable to mapping in simple, biparental populations (e.g., F_2, backcross (BC) or near-isogenic line (NIL) populations). The cloning of major genes generally proceeds along five major steps as outlined in Fig. 1: (1) initial mapping to determine chromosomal location, (2) fine mapping to establish a small genetic interval, (3) investigation of candidate genes within the interval, (4) confirmation of the causal gene, and (5) additional follow-up studies. Alternatively, candidate gene intervals may be identified through a genome-wide association study (GWAS), followed by steps 3–5 of the same process.

2.1 Initial Linkage Mapping to Determine Chromosomal Location

Mapping of major genes and large-effect QTL begins by crossing inbred lines carrying contrasting alleles at the locus of interest. Using parental lines with diverse ancestry will increase molecular marker polymorphisms, which in turn will facilitate mapping. However, in many cases, parental sources of the rare allele may be limited. Where possible, the use of parents that have already been genotyped will aid in subsequent steps in the process. NIL pairs can also be used in initial mapping populations (Edwards et al. 2017) but can be tedious to develop. The rapid decrease in genotyping costs has favored the use of F_2 populations for initial mapping, since they can be developed rapidly. If scoring individual F_2 plants is difficult or inaccurate, $F_{2:3}$ families can be developed from each F_2 plant to improve the reliability of phenotyping. $F_{2:3}$ families can also be useful for mapping completely dominant traits where heterozygotes cannot be distinguished from both parental types. Backcross populations can also be used, particularly if one allele is transferred from a highly unadapted source. A population of ~180 individuals will generally provide sufficient resolution of the genomic position of the target gene at this stage.

Currently the marker of choice for genotyping is the single-nucleotide polymorphism (SNP) due to its ease of discovery, abundance, and relatively low cost. Simple sequence repeat (SSR) markers may still be valuable for programs lacking either SNP markers or the required genotyping equipment. However, SSRs lack efficient approaches to both identify polymorphisms between parents and then assay those

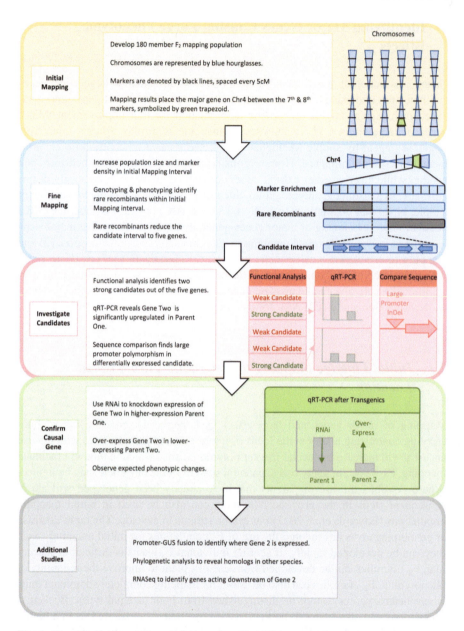

Fig. 1 Generalized scheme for mapping a major effect QTL to a single gene

polymorphisms in the mapping population. Polymorphic SNPs are identified through one of four approaches: whole genome sequencing (WGS), RNA sequencing (RNASeq), genotype-by-sequencing (GBS), or SNP arrays.

WGS involves sequencing both mapping population parents at moderate coverage (10–20X) using next-generation sequencing (NGS) technologies. While this approach is the most expensive, it will produce the highest number of polymorphisms as it should find nearly all that exist between the two parents. Furthermore, once an initial candidate region is identified, WGS should easily reveal genes that are missing, duplicated, or mutated between the two parents, greatly facilitating the investigation of the candidate region. However, the researcher will need to pick a small subset of these SNPs to run on the mapping population. This will involve conversion of these markers to economical, single-plex, PCR-based assays such as PACE (https://3crbio.com/), KASP (He et al. 2014), TaqMan (De La Vega et al. 2005), or rhAmp (Broccanello et al. 2018). These assays can be performed with high throughput by investing in automated equipment, such as the Douglas Scientific Array Tape platform or LGC IntelliQube. Another approach utilizing WGS is that of low-coverage "skim" WGS directly on mapping populations. Since its initial development (Huang et al. 2009), skim sequencing has been used in numerous crops with depth ranging from 0.055 to 4X (Malmberg et al. 2018).

The major advantage of RNASeq is that, if done in the correct tissue, it can also provide expression data that can be used to further investigate candidate genes at latter stages. This expression data can also be useful if little is known about how the major gene influences phenotype. RNASeq is relatively affordable and is less technically demanding in the lab than GBS. The major drawback to using RNASeq is that polymorphic SNPs are only derived from expressed genes. Although RNASeq should generally provide more than enough markers for initial mapping, it is possible mapping resolution may be low if very few expressed genes lie near the gene of interest.

GBS involves whole-genome restriction digests followed by NGS of size-selected fragments that have been multiplexed with ligated adapters. GBS is generally the cheapest of the four strategies, can be run directly on the mapping population without first genotyping the parents, and should provide ample marker density. The major drawbacks to GBS include (1) need to determine an optimal pair of restriction enzymes for each species, (2) procurement of costly adapters, (3) difficulties in reproducing identical SNP sets between different runs, (4) possibility of having large regions of the genome devoid of SNPs, and (5) patent issues prohibiting its use without an acquired license.

All three of the above approaches require robust bioinformatics pipelines tailored to the user's needs that revolve around programs like Bowtie2 (Langmead and Salzberg 2012) and BWA (Li and Durbin 2009) for aligning reads to the reference genome and GATK (McKenna et al. 2010) and SAMtools (Li et al. 2009) for SNP variant calling. Alternatively, SNP arrays can be used, which generally offer higher-quality genotyping data with minimal bioinformatics expertise, but at a higher cost per data point (Barrero et al. 2015; Edwards et al. 2017). While arrays require significant upfront investments of time and resources and lack the ability to easily add new polymorphisms, they reliably provide the same SNP sets between studies, marker positions are known in advance, and marker sequences are available for easy conversion to single-plex PCR marker systems.

Once acquired, phenotypic and genotypic information from the mapping population can then be used for linkage mapping with any of several software programs (e.g., JoinMap (Stam 1993) or QTL Cartographer (Silva et al. 2012)), to generate an initial mapping interval. When working with $F_{2:3}$ populations, genotyping can either be performed on original F_2 plants or bulks of $F_{2:3}$ tissue collected from about eight plants representing each family. A realistic objective for initial mapping is resolution of the gene position to a genetic interval smaller than 10 centimorgans (cM).

Bulked segregant analysis (BSA), in which tissue is pooled among plants of the same phenotype prior to genotyping (Giovannoni et al. 1991; Michelmore et al. 1991), greatly reduces the scale and cost of mapping. BSA is particularly amenable to traits that can be scored accurately prior to tissue collection, such as early season disease resistance and morphological characteristics. It has been successfully used to clone major genes for leaf virescence and seed dormancy using both DNA (Zhu et al. 2017b) and RNA (Barrero et al. 2015) markers, respectively. For well-studied loci, previous cytogenetic or mapping work may have already determined chromosomal location, obviating the need for initial mapping.

2.2 Fine Mapping to Establish a Manageable Candidate Interval

Once a rough chromosomal location has been established, the size of the mapping population can be expanded and the density of markers within that interval increased. Increases in numbers of either individuals or markers ultimately depend on available resources (and luck) but should remain manageable. There is probably little value in going much beyond ~1,200 individuals as the reduction to be gained in the candidate interval is unlikely to justify the costs in time and resources of genotyping so many individuals. The numbers of markers needed for fine mapping depends on the physical size of the interval, but there is often little value to running markers in very close physical proximity to one another as recombination between them is highly unlikely. It is likely that de novo markers will need to be developed at this point with SNP and sequence-tagged site (STS) markers being most common. Sequencing of both parents in the candidate region can greatly assist in marker development (Zuo et al. 2014; Tamura et al. 2015; Ji et al. 2016). This approach has rapidly become more affordable thanks to recently developed targeted re-sequencing strategies such as Illumina AmpliSeq (previously Illumina TruSeq) and Arbor Biosciences' myBaits. Individuals can be genotyped first with the current flanking markers. Only genotypes displaying recombination between those flanking markers need to be genotyped with additional markers in order to pinpoint the recombination breakpoint. However, at a certain point, continuing the search for incredibly rare recombinants is inefficient, and attempting to narrow an interval based on a single recombinant is prone to error.

Moving from a genetic map to a candidate interval requires a high-quality, annotated genome assembly as the physical position of the flanking markers determines the physical candidate region. If necessary, flanking markers can be sequenced and then searched against the genome sequence via BLAST (Altschul et al. 1990) to determine their physical location. The physical order of markers should be in agreement with their genetic order, and programs such as Strudel (Bayer et al. 2011) can assist in visualizing this. All annotated genes between the flanking markers compose the initial candidate list. While results will vary, candidate intervals <100 kb and/or <10 candidate genes should be targeted. It should be noted that genome annotation is an automated, computational process and is not always correct. In certain instances, mutations within the reference genome can cause misannotation of the causal gene (Ji et al. 2016). Furthermore, large deletions may exist within the reference genome that remove the causal gene (Frey et al. 2011). In such scenarios, WGS of the parental lines is likely to be of enormous benefit. Finally, there exists the possibility that the causal variant is a controlling element rather than a change in coding sequence, so there may be no gene within the fine-mapped interval (Salvi et al. 2007; Studer et al. 2011).

2.3 Genome-Wide Association Studies

Genome-wide association study (GWAS) is a powerful approach to genetic analysis that emerged as a complement to linkage analysis. Originally introduced about 15 years ago as a tool to identify genetic variants associated with human diseases (Manolio 2017), GWAS has been used in crop plants to identify genes associated with domestication and agronomic traits (Gupta et al. 2014). GWAS is simply a scaling-up of "association mapping" or "linkage disequilibrium mapping" from testing a few candidate gene markers to testing thousands or millions of markers throughout the genome. Association mapping involves testing the association between a particular sequence polymorphism and trait variation while attempting to avoid false-positive associations that occur due to population structure rather than causality or close linkage between the marker and causal polymorphism (Oraguzie et al. 2007; reviewed in Pino Del Carpio et al. 2019). Genotyping approaches for GWAS are generally similar to the SNP-based methods described above for linkage mapping.

2.3.1 Linkage Disequilibrium and Resolution

Linkage disequilibrium (LD), on which association mapping is based, is a measure of nonrandom association of alleles at different loci (Hartl and Clark 1997). LD is also called "gametic phase disequilibrium," because alleles at unlinked loci may be in nonrandom association (Falconer and Mackay 1996). LD has a critical impact on GWAS because only rarely is the causative variant directly genotyped; instead we rely on LD between causal variants and nearby markers to detect marker-trait associations (Devlin and Risch 1995; Jorde 2000; Flint-Garcia et al. 2003; Clayton 2008). LD for each pair of markers is measured by r^2, the squared correlation between the incidence of alleles at the two loci, with a possible range of 0–1, but values will be less than 1 if allele frequencies differ at the two loci (Flint-Garcia et al. 2003). High values of r^2 imply high levels of association between the alleles at different loci and high levels of LD.

Several factors influence the extent of intrachromosomal LD (Fig. 2), with recombination rate as the most important. Populations with a long history of recombination and high effective population size will tend to have lower levels of LD. In turn, the amount of recombination is influenced by the mating system of the species (Nordborg and Donnelly 1997). LD decays faster in outcrossing species compared to self-fertilizing species, because crossing-over results in recombinations only when both loci are heterozygous (Nordborg 2000). Smaller population sizes or population bottlenecks tend to increase the extent of LD (Ardlie et al. 2002; Rogers 2014) because genetic drift in small populations can increase rare haplotype frequencies by chance. Natural or artificial selection targeted to a gene will decrease sequence diversity and increase LD in the genomic region around that gene (Oraguzie et al. 2007).

Recent intermating between subpopulations with distinct ancestries and allele frequencies (admixture) and migration (gene flow) between subpopulations will also generate LD, even between loci on different chromosomes. This LD will break down in subsequent generations of random matting following the admixture. The LD decay will be slower for pairs of loci that rarely undergo recombination, and even for unlinked loci, linkage equilibrium will not be restored after only one generation of random mating (Pritchard and Przeworski 2001; Ardlie et al. 2002). The linkage disequilibrium due to admixture is a particularly important concern in association mapping since it could give rise to spurious associations at markers unlinked to causal variants. GWAS methods that control for population structure attempt to mitigate this problem.

2.3.2 Population Structure

One of the major concerns in GWAS is minimizing the chance of declaring markers not tightly linked to causal variants as significantly associated with a trait. The "false positive" or "spurious associations" can arise from the use of structured populations. The population structure effect is caused by the unequal distribution of alleles among subpopulations of different ancestries; this is a consequence of historical admixture in the population or the researcher's choice to include individuals from different subpopulations in a common analysis. Association studies often use diverse collections (or "panels") of lines or individuals to sample as much genetic diversity and as many alleles as possible. More diverse panels also tend to have lower LD because they have sampled more historical recombination events, and this can improve the resolution of GWAS. If the trait under study is also correlated with the population

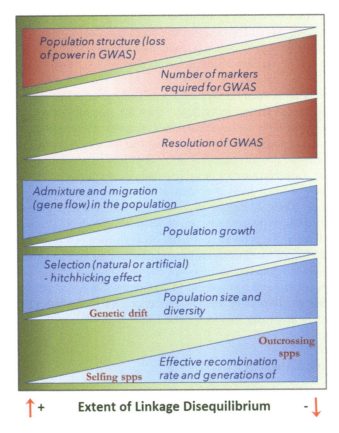

Fig. 2 Primary factors that affect linkage disequilibrium (LD) in a population; higher LD is to the left. Blue triangles represent the effect of each component on increasing or decreasing the extent of LD. Red triangles represent the effect of LD on power or resolution of GWAS conducted in a population

substructure, for example, if different subpopulations have different average trait values, then *any* marker in the genome with sufficient allele frequency differences among the subpopulations will also be associated with trait variation. As an example, consider a panel which comprises two subpopulations: in subpopulation 1, there are two loci with fixed alleles (*A* and *B*), whereas the second subpopulation is fixed for alleles *a* and *b* at the same loci. In the overall panel, these two loci would be in perfect LD even if they are unlinked. Further, consider that the two subpopulations are very distinct for the trait being measured, then both loci could be detected as highly significantly associated with the trait, even if the trait variation is caused by many small-effect polygenes that individually have only very small association with the trait.

Cryptic relatedness refers to the presence of close relatives in a set of individuals that were thought to be unrelated. Population structure and cryptic relatedness are sometimes distinguished as remote common ancestry of a large group of individuals versus recent common ancestry among smaller groups of individuals, respectively (Astle and Balding 2009). The confounding problems caused by cryptic relatedness on GWAS can be even greater than population structure (Devlin and Roeder 1999).

There are several statistical approaches to deal with false-positive associations due to population structure and cryptic relatedness. These methods use a set of random markers distributed throughout the genome to model the population structure and relationships among individuals. Then the quantitative estimates of these relationships can be fit in the association analysis tests to remove trait variation that can be attributed to population structure rather than the specific marker being tested. Population structure can be quantified using Bayesian population substructure analyses that posit a specific number of subpopulations and then identify the proportion of the genome of each individual that is descended from each subpopulation (Pritchard et al. 2000). A simpler approach that does not depend on any genetic models is principal components analysis of the genome-wide SNP genotypes (Reich et al. 2008). Finally, the genomic similarity between each pair of individuals in the population can be estimated using realized genomic relationship matrices (Isik et al. 2017). The pairwise genomic relationships can be combined with population structure estimates or principal components in a common GWAS model to account for both large-scale and finer-scale relationships among the individuals (Yu et al. 2006; Price et al. 2010; Zhou and Stephens 2012; Liu et al. 2014).

2.3.3 Statistical Methods for GWAS

Various statistical methods have been proposed to perform GWAS for dissection of complex traits. For discrete traits, logistic regression models can be used (Zhao and Chen 2012). Crop plant traits sometimes fall into discrete categories and can be analyzed with similar generalized linear models (Andres et al. 2017).

More frequently, crop plant traits are measured as continuous quantitative variables, often with close to normal distributions. Statistical models for GWAS for such traits are usually ordinary least squares/ANOVA or mixed linear models. One of the first approaches used to analyze quantitative traits in plant association studies was the analysis of variance ANOVA (Stich and Melchinger 2010). The null hypothesis to be tested is that there is no difference between the trait mean for the different genotypic classes at the marker being tested (Bush and Moore 2012). This model can be adjusted to include numeric covariates that represent the population structure assignments or principal components from the analysis of random background markers to reduce the chance of spurious marker-phenotype association due to population structure.

The use of pairwise genomic relationships for population structure requires the use of a mixed linear model (MLM). In MLM, the marker being tested is usually considered a fixed effect, and the null hypothesis is that there are no differences among the means of genotypic groups at the marker, just as in the ordinary least squares approach. The important novelty of the MLM is the assumption that each individual or line's genetic background is a random effect whose covariances are proportional to the realized genomic relationship matrix. The random genetic background effect will then be associated with some proportion of the observed variation among lines that matches the pattern of pairwise genomic relationships, and this variation is removed from the single marker being tested. This can help reduce the rate of false-positive associations that occur due to background genomic relationships. For example, the *Dwarf8* gene in maize (*Zea mays*) was reported to be associated with flowering time in a small structured population (Thornsberry et al. 2001). Reanalysis of this association when carefully controlling for population structure and pairwise relationships revealed that the gene-phenotype association was much looser than previously reported (Larsson et al. 2013).

Because each MLM analysis is more computationally intensive than ordinary least squares, it may require a long time to test many thousands or more markers across the whole genome. Variations of the original MLM method proposed by Yu et al. (2006) have been developed to improve computational time and power while still implementing the MLM model in GWAS (Zhang et al. 2010; Listgarten et al. 2012; Zhou and Stephens 2012; Korte et al. 2012; Stram 2014).

In some cases where the quantitative trait is controlled by several QTL of large effect, inflation of single-locus test statistics can be expected. To handle this situation, a multi-locus mixed linear model (MLMM) was proposed (Segura et al. 2012). The MLMM models the effects of multiple markers simultaneously, similar to the idea of composite interval mapping approach in QTL mapping (Jansen 1993; Zeng 1994).

Goddard et al. (2009) proposed an integrated approach of to the estimation of the SNP effects and prediction of trait values that treats the effects of SNP markers as random instead of fixed effects. From this idea, two other variants of MLM and MLMM have been proposed, those are the random mixed linear model (RMLM) and the multi-locus RMLM (MRMLM) approaches (Wang et al. 2016). The FarmCPU approach alternates between treating the test markers as fixed and random to select a set of markers that best represent the observed trait variation (Liu et al. 2016).

2.3.4 Power of GWAS

Power of GWAS represents the ability to detect genetic variants that truly cause phenotypic variation (Teo 2008). Power is influenced by the genetic architecture of the trait of interest, the relationship between population structure and trait variation, marker density, allele frequencies at the marker being tested, and population sample size. Balancing between high power and good control of false positives can be tricky in GWAS. Power to detect significant associations is strongly determined by allele frequency of the functional polymorphism, with rare alleles very difficult to detect (Stich and Melchinger 2010). Power is also influenced by the factors that affect heritability of line mean values: the number of environments and replications per environment used for phenotyping (Stich et al. 2010). The relationship between marker density and power of detection depends on the extent of LD in the target population. By using a family mapping approach to generate extensive LD in a maize association mapping study, Liu et al. (2013) reported that good power can be achieved with low marker density if the QTL have large or medium effect. Higher marker density will increase power, but increasing marker number has a reduced rate of return as the markers saturate the genome.

2.3.5 Detecting Rare Variant Effects in GWAS

There is an active debate about the power of GWAS to detect rare variant associations. Since GWAS was initially designed for common variant identification, typical study designs have low power to detect trait associations with rare alleles unless they have a large effect on the phenotype (Myles et al. 2009). Rare variants may contribute to the so-called missing heritability (variation that cannot be attributed to genetic markers) in complex traits (Manolio et al. 2009; Eichler et al. 2010; Pino Del Carpio et al. 2019). Because of the possibility that GWAS studies tend to "miss" rare variants, new approaches were developed to improve the power of detection of rare variants.

Rare sequence variants are unlikely to be included in preselected SNP arrays, so low-depth whole genome sequencing (WGS) can be used to obtain information on rare variants in a study panel. Power to detect rare variants is greater using larger sample sizes with lower depth sequence information than using small sample sizes with deep sequencing (Li et al. 2011). Information on rare variants detected using high-coverage sequence in an appropriate reference panel of small size can be imputed in a much larger study panel sequenced at low depth, if there are close relationships between the high- and low-sequencing depth groups (Xu et al. 2017).

Exome sequencing is another strategy that has been used to increase the power of GWAS to detect rare variants (Bamshad et al. 2011). It is relatively expensive, since sequencing should be performed at high depth, e.g., 60x–80x (Do et al. 2012), and by design only captures variation in exons. Nevertheless, significant rare variant associations have been reported in humans using this approach (Lange et al. 2014). An alternative approach is to first conduct GWAS with lower density markers, followed by targeted sequencing of specific regions where the initial GWAS detected associations. Rare variants discovered by resequencing in the targeted region can then be tested for association to determine if they cause the observed trait association effect (Johansen et al. 2010). Finally, alternative statistical tests of association of rare variants have been used in human populations, including tests of haplotypes defined by multiple SNPs within small genomic regions, where each haplotype effect is modeled directly, or the number of rare variants carried by each individual within the haplotype region is measured and trait values are regressed on this rare variant "burden" score (Asimit and Zeggini 2010).

2.3.6 Experimental and Population Designs for GWAS

Sample size, extent of replication for phenotyping, genotyping method, and SNP density are important considerations when designing an association mapping study, and all of these factors impact the research budget. One advantage of the use of common diversity panels for association analysis is the availability of lines that have already been genotyped. For example, several densely genotyped diversity panels based on publicly available seed stocks have been developed in maize (Yan et al. 2011). A core diversity panel for association mapping was developed using public maize inbred lines representing diverse geographic origins and including specialty types such as sweet and popcorn (Flint-Garcia et al. 2005). Now almost the entire set of publicly available maize inbreds in the USDA maize seed bank has been densely genotyped, and any subset of lines from it could be used for association analysis (Romay et al. 2013; Zila et al. 2014).

A new generation of association mapping populations has been developed to combine some of the complementary advantages of biparental mapping populations and diverse germplasm panels. In general, these are referred to as multiparental populations. Such populations involve random progeny lines derived from a set of crosses among related parents, so they combine the linkage information within populations with the larger sample of diversity and reduced linkage disequilibrium of multiple parents (Jamann et al. 2015). Nested association mapping populations involve crosses between multiple diverse lines and a common reference parent line (Buckler et al. 2009; McMullen et al. 2009). Multi-parent advanced generation intercross (MAGIC) populations involve intermating multiple founder lines for several generations and then extracting inbred lines from multi-way crosses. In this way, each line contains a mosaic of recombined genomes of numerous founder lines. This approach has been implemented in wheat (*Triticum aestivum*) (Huang et al. 2012; Mackay et al. 2014), rice (*Oryza sativa*) (Bandillo et al. 2013), and maize (Dell'Acqua et al. 2015). More generally, any collection of lines from multiple crosses can be combined into a mapping population even if they do not share a common parent. This approach has been termed as a random-open-parent association mapping (ROAM) population (Xiao et al. 2016). This population structure facilitates the incorporation of new RIL populations into currently existing populations for large-scale genetic analysis.

2.4 Investigation of Candidate Genes Within the Interval

Once a manageable candidate interval has been established through some combination of linkage and association mapping, markers within the region can be employed for marker-assisted selection. The optimal marker to use for selection will either be a causal variant or a "diagnostic" marker in perfect LD with the causal variant across the breeding populations of interest. To ensure that a diagnostic marker is used, it is often helpful to finalize genetic analysis by identifying ("cloning") the causal gene and identifying the critical variants within or nearby the gene that confer the desired phenotypic effect. The majority of gene cloning studies employ some combination of three major techniques to identify the causal gene: functional analysis, sequence comparison, and expression analysis.

2.4.1 Functional Analysis

Functional analysis involves the prediction of a gene's function based on its sequence to identify those that can be plausibly linked to the phenotype of interest. Tools to assists in this process include Pfam (Finn et al. 2016), Gene Ontology (GO) (The Gene Ontology Consortium 2000; Carbon et al. 2017), protein annotation through evolutionary relationship (PANTHER) (Mi et al. 2017), and the Kyoto Encyclopedia of Genes and Genomes (KEGG) (Ogata et al. 1999). Most are already incorporated into publicly available genome browsers. The use of functional analysis requires extensive background knowledge on how the phenotype is determined, likely based on previous work in model species. Even genes with a tentative link cannot be discounted. Genes that can be eliminated include those considered to be highly redundant or housekeeping, as well as those with a clearly defined function highly unlikely to explain the phenotype.

Functional analysis of candidate genes can also be performed by creating gene knockouts in a uniform genetic background via nuclear, sodium azide, or ethyl methanesulfonate (EMS) mutagenesis, followed by screening mutagenized progenies for phenotypic changes (Fu et al. 2009; Krattinger et al. 2009; Moore et al. 2015). Targeting Induced Local Lesions in Genomes (TILLING) can facilitate the identification of polymorphisms introduced by mutagenesis (Hurni et al. 2015).

Transgenic constructs can be developed for all remaining candidates within an interval (Wang et al. 2015). This approach is best suited to situations when only a few candidates remain and to crops highly amenable to transformation. Stable transgenics are considered the gold standard for confirmation that a causal gene underlies a phenotype of interest. This could involve transfer of a functional gene copy (by *Agrobacterium* or biolistics) into a genotype that otherwise lacks a functional gene copy, to see if "transgenic complementation" results in a predicted phenotypic change. A functional gene could also be put under the control of a constitutive or inducible promoter. Or RNA interference (RNAi) constructs could be used to knockdown expression of a functional gene as another means to validate a phenotypic change occurs when that gene is manipulated.

Virus-induced gene silencing (VIGS) is a transient assay that uses a modified virus to temporally knockdown expression of a targeted gene. VIGS has been used successfully as a rapid, efficient confirmation method, particularly in cotton (*Gossypium hirsutum*) (Ma et al. 2016; Wan et al. 2016; Zhu et al. 2017b). Transposable element (TE) populations (e.g., maize Uniform Mu (McCarty et al. 2005), Taiwan Rice Insertional Mutants (TRIM) Database (Hsing et al. 2007)) exist for certain species in order to fast track transgenic confirmation (Yang et al. 2017).

Extending findings regarding the causal gene to orthologs in other species (Chakrabarti et al. 2013; Yang et al. 2017) provides strong confirmation as does transgenically inserting the gene into a different species (Moore et al. 2015). Other methods of confirmation include mutagenesis (Krattinger et al. 2009) and the use of association panels (Barrero et al. 2015; Moore et al. 2015; Wan et al. 2016; Edwards et al. 2017).

Cutting-edge genome editing techniques (e.g., clustered regularly interspaced short palindromic repeats (CRISPR), meganucleases, zinc-finger nucleases (ZFNs), transcription activator-like effector nucleases (TALENs)) offer the potential to massively scale-up the number of candidates that can be altered (Osakabe and Osakabe 2015; Yin et al. 2017; Zhu et al. 2017a, b). While regeneration of whole plants from transformed callus tissue remains a bottleneck facing many crops, innovative approaches for increasing efficiency are an active area of research. One such example is the overexpression of the maize morphogenic genes *WUSCHEL* (*Wus2*) and *BABYBOOM* (*Bbm*) during the transformation process in major monocot crop species (Lowe et al. 2016). Another approach called HI-Edit involves combining the genome editing step with the haploid induction process to enable editing of elite germplasm without regeneration; the loss of paternal chromosomes carrying the editing constructs in early developmental stages results in immediate isolation of edited sequences away from the transgenic editing constructs (Kelliher et al. 2019).

2.4.2 Sequence Comparison

A crop's reference genome sequence can be used to design primers for candidate genes, including promoter and terminator sequences. Longer flanking sequences are always preferred, but at increasing distances from the gene, reliable amplification can become problematic. Amplicons between 400 and 600 bp in length are ideal for Sanger sequencing. Primers should be designed so that amplification products overlap and can be easily assembled into contigs by programs like Sequencher or Geneious. To increase throughput or circumvent amplification problems, longer PCR products can be created, and Sanger sequenced using internal primers. When working with large gene families or polyploids, primers must be gene or genome specific. During primer design, related sequences can be aligned, and areas of disagreement targeted for primer design. Ambiguous base calls in Sanger chromatograms are evidence of non-specific amplification, especially if located at known variable positions. For polyploids, amplification checks in ancestral diploid panels can be performed. Amplification and sequencing of products from parents and other lines of interest can reveal polymorphisms corresponding to phenotype. These polymorphisms can either be directly proposed as causal or serve as novel polymorphic markers for further mapping, association panels, or marker-assisted selection.

Even within coding sequences, the majority of polymorphisms are likely insignificant. Synonymous mutations that do not alter protein sequence can be ignored. The effect of frameshift, nonsense, and nonstop mutations on protein sequence is obvious and has recently been used to identify numerous cloned genes (Ji et al. 2016;

Ma et al. 2016; Edwards et al. 2017). These mutations are generally assumed to dramatically alter protein function, but in the absence of creating or removing domains, effects can be hard to predict and confirm.

Many non-synonymous SNPs and in-frame insertions/deletions (InDels) can be tolerated with no effect on protein function. Only those amino acid changes that alter a conserved functional domain, protein charge, or structure are likely to have a phenotypic effect. Many web-based tools exist to weigh these types of changes in plants including Ensembl Variant Effect Predictor (VEP) (McLaren et al. 2016), Protein Variation Effect Analyzer (PROVEAN) (Choi et al. 2012), and PredictSNP (Bendl et al. 2014). Variant protein sequences can also be entered into InterPro (Finn et al. 2017) to assess changes to predicted functional domains, localization, and taxonomy. However, such effects are difficult to verify experimentally. Available assays depend on the type of protein involved and its predicted function but include yeast one-hybrid (Ma et al. 2016), plant hormone response treatments (Barrero et al. 2015), and analysis of resistance mechanisms (Ji et al. 2016; Yang et al. 2017).

2.4.3 Expression Analysis

Candidate genes can also be investigated for differential expression. Determining appropriate tissue type and development stage can be difficult and require extensive literature review or pilot studies. Particularly for developmental genes, causal changes in expression may occur briefly or only in a small cluster of cells.

Pilot studies can be used to investigate suspected tissue types and can be performed using faster, less expensive techniques such as qRT-PCR or even gel-based semiquantitative approaches. Publicly available expression databases available for many crops (e.g., Rice Expression Database (RED) (Xia et al. 2017), Maize Gene Expression Atlas (Stelpflug et al. 2016)) can also be useful for establishing an expression baseline. qRT-PCR experiments should follow the MIQE (minimum information for publication of quantitative real-time PCR experiments) guidelines to ensure appropriate experimental design and data reporting (Bustin et al. 2009). Primers should be designed to span projected intron gaps. This allows for determination of DNA contamination and confirmation of the coding sequence via cDNA sequencing. 5′ and 3' rapid amplification of cDNA ends (RACE) can be used to confirm transcript ends. Choice of an appropriate reference gene is also critical for qRT-PCR studies and has been established for most major crops (Artico et al. 2010; Manoli et al. 2012).

Although relatively expensive and somewhat more technically demanding, RNASeq is the gold standard for expression studies. Standardized guidelines (similar to MIQE) have not yet been developed, but many useful resources exist to assist in experimental design (Su et al. 2014; Conesa et al. 2016). Isogenic lines are ideal, so as to minimize background expression variation, but are time-consuming to develop. The primary advantage of RNASeq over qRT-PCR is the ability to detect differential expression of genes downstream of the causal locus. Therefore, if the phenotype does not result from differential expression of the causal gene (i.e., an expressed but nonfunctional protein), valuable information will still be produced that can be traced back to the causal gene.

2.5 From Identification of Major Genes or Markers to Selection and Cultivar Development

Ideally, a causal polymorphism can be identified and converted into a genetic marker. This marker can then be used for marker-assisted selection for the desired allele at the target gene. In many instances, however, it is difficult to definitively prove a causal polymorphism and/or easier to convert a nearby polymorphism into a suitable marker for selection. So long as the marker is sufficiently close to the underlying polymorphism, tight linkage should ensure that recombination will not separate the marker from the causal allele. Therefore, an adjoining marker can serve a role functionally identical to that of a marker at the causal locus.

A common use of marker-assisted selection (MAS) for a major gene is marker-assisted backcrossing (MABC) to move a single, desirable gene from an unadapted variety into an elite cultivar needing improvement in the trait controlled by that gene (Fig. 3). The first step is to cross the elite cultivar to the unadapted variety. The resulting F_1 is then backcrossed to the recurrent parent (i.e., the elite variety) to produce the BC_1F_1 generation.

MABC requires a marker (at least) tightly linked to the underlying gene as well as polymorphic flanking markers to select favorable recombinant progenies to guide the incorporation of only a small fragment of the inferior donor genome. Markers evenly distributed throughout the rest of the genome are helpful to promote rapid recovery of the recurrent genome but are not essential. In the BC_1F_1, only individuals heterozygous at the locus of interest are selected for further backcrossing. After each subsequent backcross, progeny are screened for heterozygosity at the locus of interest and recurrent parent homozygosity at as many other loci as possible. The number of backcrosses to be performed will vary, but on average, the donor genome percentage decreases by half each backcross. Using markers distributed throughout the genome allows breeders to select progenies with higher than average proportions of recurrent parent alleles outside of the introgression region, thus reducing the number of generations required to create a highly isogenic introgression line (Frisch et al. 1999). Following the final backcross, selected individuals must be self-pollinated with progeny selected for donor parent homozygosity at the locus of interest and recurrent parent homozygosity elsewhere (leftmost box in Fig. 3).

MAS is particularly helpful when attempting to pyramid multiple major genes affecting the same phenotype into a single line, such as creating durable resistance by stacking multiple major resistance genes to the same pathogen. Markers are needed for each major resistance gene as it is difficult to differentiate the phenotypic effects of each locus when all are segregating in a given population. Major gene markers can also be used as a surrogate or proxy when selecting for traits that are

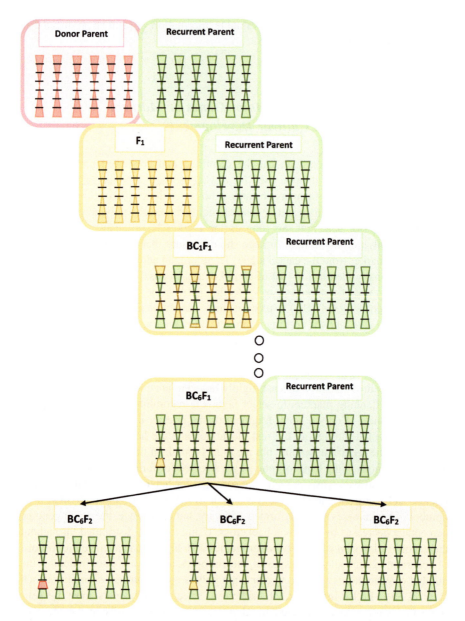

Fig. 3 Simplified hypothetical marker-assisted backcrossing. Red represents donor parent genome, green is recurrent parent genome, and yellow represents heterozygous loci

difficult or expensive to phenotype such as seed or tissue biochemical profiles (e.g., protein content, fatty acid profile). In some cases a rare, beneficial allele may be absent from elite germplasm due to tight repulsion-phase linkage with an important allele at a nearby locus (Fukuoka et al. 2009; Moore et al. 2015). In these situations, major gene markers for both loci are imperative for the identification of rare recombinants that break such linkages to produce the optimal haplotype.

MAS has also been integrated into traditional "forward" breeding methods, such as pedigree selection. The best examples of this have been demonstrated in self-pollinated crops, such as soybean (*Glycine max*), common bean (*Phaseolus vulgaris*), rice, and wheat (Kelly et al. 2003; Cahill and Schmidt 2004; Dubcovsky 2004; Jena and Mackill 2008). Public wheat breeding programs in Australia and the USA have successfully implemented marker-assisted selection for major genes controlling traits such as Fusarium head blight resistance and grain quality for about two decades with the establishment of centralized genotyping facilities to support applied breeding programs (Eagles et al. 2001; Dubcovsky 2004). Straightforward modifications to traditional breeding methods can be used to maximize the probability of recovering inbred lines carrying favorable alleles at a few marker loci without unduly bottlenecking the breeding population, and maintaining variation for traits not tagged by markers (Bonnett et al. 2005). It seems that self-pollinating species tend to have important agronomic traits affected positively by genes with large effects. Conversely, outcrossing and/or more diverse species such as maize are likely to have polygenic inheritance of most agronomically important traits (Buckler et al. 2009; Peiffer et al. 2014).

3 Genomic Selection

A disappointing result of some QTL mapping or GWAS investigations is the inability to validate a useful marker for use in selection programs. This can occur because the marker-trait association is a false positive, its effect is inaccurately estimated, or only small-effect associations collectively explaining only a small part of the genetic variation are discovered. For highly polygenic traits, all of these problems can occur, but the last problem (many genes each with small effects) is a biological issue, not a technical problem that can be overcome with larger sample sizes or more sequencing. Bernardo (2008) reviewed the literature on QTL mapping in plants and commented that "the vast majority of the favorable alleles at these identified QTL reside in journals on library shelves rather than in cultivars that have been improved through the introgression or selection of these favorable QTL alleles." For this reason, alternative methods that can better predict quantitative trait breeding values than QTL or association analysis are needed. Genomic selection provides a way to more accurately predict breeding values from markers than QTL or association analysis, generally at the cost of limited biological interpretability of the models.

Genomic selection uses information from genetic markers dispersed across the entire genome to predict traits of interest (reviewed in Jonas et al. 2019). A set of individuals, families, or lines are genotyped and phenotyped for traits of interest to provide data to build a predictive model that relates genome-wide marker information to trait variation. Then the predictive model allows selection on new materials on which the breeder has only marker information but no trait data. For this reason, genomic selection can accelerate breeding cycles and allow for increased selection intensity, providing breeders the ability to bring novel varieties to market sooner (Schaeffer 2006; Heffner et al. 2009). In this way, implementing genomic selection in a breeding program can significantly reduce input costs or, at least, redirect these resources toward efficient crop improvement.

A number of historical approaches and innovations allowed genomic selection to become feasible. The introduction of allozymes and restriction fragment length polymorphisms (RFLPs) allowed breeders to begin making directional selections at a molecular level (Stuber et al. 1982; Tanksley et al. 1989). Even at that time, breeders could see the potential of using molecular marker technology to impose selection instead of phenotypic information alone. Tanksley et al. (1989) predicted that the use of molecular markers like RFLPs could expedite the movement of desirable genes among varieties, allow the transfer of novel genes from related wild species, and make possible the analysis of complex polygenic characters as ensembles of single Mendelian factors. As envisioned, molecular genetics technologies were widely used to identify quantitative trait loci (QTL) for marker-assisted selection and gene introgression (Hospital and Charcosset 1997; Hospital et al. 1997; Moreau et al. 1998; Bernardo and Charcosset 2006; Xu and Crouch 2008). MAS was shown to be most effective at manipulating major genes and large-effect QTL (Charmet et al. 1999; Hospital et al. 2000; Combs and Bernardo 2013). Traits where markers were predominantly associated with small effects on traits or influenced strongly by environmental factors still required extensive field evaluations during the breeding progress (Moreau et al. 2004), reducing the effectiveness of marker-assisted selection in these cases. The primary limitation to the effectiveness of MAS for small-effect QTL is the relatively poor accuracy of estimates of their effects and positions.

To improve the effectiveness of marker-based selection for quantitative traits, Meuwissen et al. (2001) introduced genomic selection as a method to estimate all of the marker effects simultaneously, rather than select a subset of markers that pass statistical tests of significance. This approach allows markers with small effects to be included in the prediction of breeding values for quantitative traits controlled by many small-effect polygenes (Meuwissen et al. 2001).

3.1 Components of Genomic Selection

Genomic selection requires first the creation of a statistical model that relates marker genotypes to trait values. This prediction model must be built using data on individuals, lines, or families that have both marker and phenotype data. The data used to create the prediction model is referred to as the "training data set" and optimally includes a large sample of lines related to the selection candidates evaluated in replicated, multi-environment trials such that the phenotypic data are reliable. A wide variety of statistical models (described below) can be applied to the training data. These models have in common the use of all markers for prediction, resulting in the simultaneous estimation of allelic effects at each locus. This approach is distinct from QTL mapping and marker-assisted selection, in which each marker is first tested for association with the phenotype, followed by the selection of only "significant" markers to use in breeding value estimation. Generally, there are more markers than lines tested, such that marker effects cannot be estimated using ordinary least squares. Genomic selection prediction models belong to a class of statistical learning models that can use information from more predictors than observations. Although the individual marker effect estimates from these kinds of prediction models are wholly unreliable, the focus of genomic selection is on accurately predicting the genomic estimated breeding values (GEBVs), the combined effects of all markers on the phenotype rather than on individual marker effects.

To evaluate the performance of different statistical models on the available data, the training set can be subdivided into subsets of training data with both marker and trait data and disjoint "test sets," which are subsets of the data where the genotype data are included, but the trait data are "masked" as missing values. Models are fit to the training data subsets, and predictions are made for the "held out" phenotypes of lines in the test set. The correlation between predicted and observed trait values in the test set is an estimation of model "prediction ability." This process can be repeated by partitioning the data into training and test sets multiple times, and the average correlation over training/test set combinations is typically reported as the average prediction ability of the different models. From this preliminary investigation, the breeder may choose which model appears to perform best for use in actual selection.

Once a model to predict GEBVs has been established for the training data, genomic selection can be practiced by applying the model to the selection candidates that truly have only marker information but lack direct phenotypic measurements. A number of simulation and empirical studies suggest that GEBVs based solely on the individual genotype can have a high level of accuracy relative to direct phenotypic selection, although accuracies vary widely among traits and populations (Meuwissen et al. 2001; Habier et al. 2007; Legarra et al. 2008; VanRaden et al. 2009; Zhong et al. 2009; Lorenzana and Bernardo 2009; Jonas et al. 2019).

Since genomic selection (GS) works by associating allelic variation with trait variation, this may prompt reconsideration of how breeding programs are organized. Traditional plant breeding programs are typically organized around selecting on highly heritable traits observed on individual plants or families with limited replication in early generations and concentrating replicated multi-environment evaluations of complex traits like yield on a greatly reduced set of lines that pass the early generation selections. Information from such programs is not easily transferable from generation to generation or across families. With GS, on the other hand, there is the potential to combine information from disparate trials and breeding populations, which can be combined through the common connection of allelic effects at markers. In this scenario, experimental designs can be reconsidered to focus replication on alleles rather than the lines (Knapp and Bridges 1990; Heffner et al. 2009). In other words, it may be more effective to collect phenotypic data on fewer replications of more lines rather than more replications of fewer lines to help build an effective training model.

Breeders should consider modifications to their breeding programs to affect the two main phases of genomic selection: building a prediction model and implementing genomic selection on new selection candidates going forward. Accuracy of prediction models depends on size and composition of the training data set, marker density, reliability of trait data, and statistical model used. GS is optimally applied in situations where phenotypic data are not available or are expensive or difficult to measure; therefore implementation of GS relies on inexpensive and rapid genome-wide marker data collection so that GS can be applied more quickly or easily than phenotypic selection.

3.2 Training Set Composition and Size

Unlike traditional QTL mapping, specific populations do not need to be developed to create GS prediction models. The training set can be sampled directly from the germplasm in the breeding populations of interest, given the breeder has considered the diversity that exists in the population and population size. This opens the possibility of using historical data from a breeding program, if seeds of older materials are available for genotyping (Dawson et al. 2013; Sarinelli et al. 2019).

As a general rule, GS prediction accuracy is improved by using more individuals in the training set. However, there are tradeoffs if the additional individuals added to the training set are not closely related to the selection candidates. For example, increasing training set size by combining lines from different germplasm groups in barley (*Hordeum vulgare*) resulted in decreased prediction accuracy within each group (Lorenz and Smith 2015). Similarly, a training model using lines compiled from numerous distinct breeding programs predicts yield poorly (in some cases negatively) within new breeding crosses (Windhausen et al. 2012). This phenomenon occurs because the marker-trait associations in the training set are not representative of the relationships in the selection candidates. Depending on the diversity of the selection candidate population, training set sizes of a few hundred individuals may be sufficient for prediction. In some cases, even with large training set sizes, a smaller subset of training individuals may give a better prediction model if the individuals are selected to optimize the relationship between training set and selection candidates (Isidro et al. 2015; Akdemir et al. 2015). If genetic data are available before any phenotyping has been performed, the breeder may also select an optimal subset of individuals to phenotype (Rincent et al. 2012).

3.3 Marker Technologies and Marker Density

Implementing genomic selection in a breeding program requires continuous genotyping at a moderate to high density across the genome. Therefore, SNPs are the current marker of choice in modern breeding programs using genomic selection (Rafalski 2002; Elshire et al. 2011). In species where they are available, SNP arrays represent a logical option for GS as they require limited bioinformatics expertise and will reliably reproduce identical SNP sets over multiple runs. However, SNP arrays are resource intensive to develop and are relatively expensive on a per sample basis. Additionally, adding novel SNPs is difficult.

The most inexpensive and perhaps widely used method for SNP genotyping at this time is multiplexed genotype-by-sequencing (GBS). After identifying which sequence reads correspond to which samples, the reads can be aligned to a reference genome, transcriptome, or each other to facilitate identification and calling of SNPs. Publicly available sequence alignment and SNP calling software pipelines have been developed to create de novo assemblies or even pseudo-references based on the reads of a few individuals for the identification of polymorphic loci for all individuals (Lu et al. 2013; Glaubitz et al. 2014; Melo et al. 2016; Manching et al. 2017). The major drawbacks to GBS are the inability to produce identical SNP sets between runs, a requirement for bioinformatics expertise, and issues with freedom to operate due to intellectual property restrictions on the method. Novel NGS-based platforms are also under constant development resulting in decreasing costs for sequencing, while read length, quality, repeatability, and volume are all increasing (Syvänen 2005; Ragoussis 2006; Perkel 2008; Podolak 2010). One such example that might hold promise for genomic selection are amplicon-based targeted resequencing strategies such as LGC's SeqSNP. RNASeq (Azodi et al. 2020) and skim WGS (Malmberg et al. 2018) are also receiving interest as approaches to supply GS models with suitable genotypic data.

Marker-trait associations in both training sets and selection candidates occur due to linkage disequilibrium. Therefore, marker density and average linkage disequilibrium distance in the population need to be scaled appropriately. Meuwissen (2009) stated that for the average linkage disequilibrium between adjacent markers to be considered equal in two populations, the number of markers per centimorgan (cM) divided by the effective population size should be the same for each population. Daetwyler et al. (2008) and Hayes et al. (2009) related the accuracy of a genomic selection model as a function of the training population size, the heritability of the trait of interest, and the number of marker effects estimated; the main consideration of this relationship being the ratio between the training population size and number of marker effects estimated. Since the effective population size in elite crop breeding programs are fairly small (Hamblin et al. 2010), this suggests that a breeder could make accurate predictions with a relatively small training population. Species with more diversity and less overall LD (typically, outcrossing species) will benefit more from increasing marker density, although at some point even in highly diverse populations, gains from increasing marker density diminish.

For example, in a very diverse *Drosophila* population, prediction accuracy increased (although slightly) by increasing marker density up to 200,000 SNPs (Ober et al. 2012).

3.4 Statistical Models for Genomic Prediction

Genomic selection models have in common the assumption that genotypic variation for a trait is conditioned by many QTL, mostly with small (although not necessarily equal) effects, distributed throughout the genome. If the assumption is badly wrong, and the genetic architecture is truly composed of one or a few major effect genes, QTL mapping and marker-assisted selection should be a better approach. For many important agricultural traits, however, the polygenic model of many small-effect QTL appears to be close enough to reality such that the assumptions of GS models tend to predict breeding values with reasonable accuracy (Hill et al. 2008; Gianola et al. 2009).

We consider below a number of variations of genomic selection models that have distinct assumptions about the distribution of QTL effect sizes associated with the many markers in the model (de los Campos et al. 2013). For the most part, these models assume additive genetic control of complex traits, and the models are robust to moderate departures from this assumption. Nonadditive genetic variation can be added to these models by specifically including dominance or epistatic effects (Lee et al. 2008; Lorenzana and Bernardo 2009; Muñoz et al. 2014), although more complex models do not necessarily improve prediction accuracy.

At the core of each model is a basic relationship between phenotype and genotype (Moser et al. 2009).

$$y_i = g(x_i) + e_i$$

where y_i is an observed phenotype of individual i ($i = 1 \ldots n$) and x_i is a $1 \times p$ vector of marker genotypes on individual i, $g(x_i)$ is a function relating genotypes to phenotypes, and e_i is the residual term. The genomic estimated breeding value (GEBV) is, in this case, equal to the $g(x_i)$ term. The statistical models used for genomic selection all attempt to calculate GEBVs by minimizing a cost function. For instance, the cost function in a traditional least squares regression is the sum of squared residuals. The following is a brief overview of commonly used statistical models used in genomic selection.

3.4.1 Ridge Regression Best Linear Unbiased Prediction

Ridge regression best linear unbiased prediction (RR-BLUP), or simply ridge regression, was first proposed by Whittaker et al. (2000) for marker-assisted selection for biparental crosses. The model estimates $g(x_i)$ as

$$g(x_i) = \sum_{k=i}^{p} x_{ik}\beta_k$$

denotes the genotype score for marker k in individual i, β_k is the effect of marker k, and the genetic value is the sum of p marker effects. The normal least squares estimators are modified so that β is estimated using

$$\beta = (\mathbf{X}'\mathbf{X} + \lambda \mathbf{I})^{-1}\mathbf{X}'\mathbf{y}\beta = (\mathbf{X}'\mathbf{X} + \lambda \mathbf{I})^{-1}\mathbf{X}'\mathbf{y}$$

where $\mathbf{X}$ is the matrix of marker data (n rows corresponding to the individuals and m columns corresponding to the numerically coded marker values), $\mathbf{I}$ is an identity matrix, and $\mathbf{y}$ is a vector of trait values. The $\lambda \mathbf{I}$ term differentiates the ordinary least squares and RR-BLUP estimates of β. The $\lambda \mathbf{I}$ term is designed to make $\mathbf{X}'\mathbf{X}$ nonsingular and reduce collinearity between predictors. There are a number of ways to choose a value for λ, including testing a range of values in which the model error is minimized (Whittaker et al. 2000). Another way is to estimate an optimal λ by assuming that marker effects are randomly drawn from a common normal distribution centered at zero and solve the mixed linear model of equations (Henderson 1975; Bernardo and Yu 2007). Then the λ term is equal to var(e)/var(β), where var(e) is the residual variance and var(β) is the common marker effect variance (Piepho 2009). With a small var(β) relative to var(e), the marker effects will be more strongly shrunken toward zero.

This ridge regression model is equivalent to fitting a mixed model $y_i = g_i + e_i$, where the g_i is the GEBV for individual i and the g_i effects are assumed to have a variance-covariance structure equal to $G\widehat{\sigma}_A^2$, where $\mathbf{G}$ is the genomic realized relationship matrix and $\widehat{\sigma}_A^2$ is the estimated additive genetic variance (Habier et al. 2007; Piepho 2009). $\mathbf{G}$ is an $n \times n$ matrix of pairwise relationship estimators; it can be the same as the $\mathbf{K}$ matrix previously described for use in GWAS. This model does not estimate marker effects, but it uses the same marker information as does RR-BLUP to estimate directly the individuals' breeding values. For that reason, this model is referred to as genomic best linear unbiased prediction (GBLUP; Isik et al. 2017).

3.4.2 Least Absolute Shrinkage and Selection Operator (LASSO)

The difference between least absolute shrinkage and selection operator (LASSO) and RR-BLUP is in the cost function estimations (Tibshirani 1996; Bishop 2006). The cost function for RR-BLUP:

$$\sum_i e_i^2 + \frac{\lambda}{2} b'b$$

where **b** is a vector of the marker β estimates. The second term, $\lambda b'b$, is often called the constraint, penalty, or regularizer because a larger λ adds more cost to increasing values of the β, causing them to shrink more back to zero. Since all effects are distributed identically in ridge regression (we assume common variance for all markers), all marker effects are equally shrunken toward zero.

The cost function for least absolute shrinkage and selection operator (LASSO) is:

$$\sum_i e_i^2 + \lambda \sum_k |\beta_k|$$

This constraint makes some predictor coefficients shrink more strongly than RR-BLUP, and some of the coefficients are reduced down to zero (Tibshirani 1996). In this way, the LASSO method creates a more parsimonious model: a subset of the markers has zero effect on the trait values, and the remaining markers have "shrunken" effects to avoid overfitting the model. Unlike in RR-BLUP, the choice of λ is more difficult to evaluate, due to the influence of the predictor subset being selected. As λ approaches zero, the model resembles the solutions derived from ordinary least squares. As λ increases, the absolute value of the regression coefficients gets smaller (de los Campos et al. 2009b). Fortunately, a least angle regression algorithm can help optimize the λ value in LASSO without much computation (Usai et al. 2009; Tibshirani 1996).

3.4.3 Kernel Hilbert Spaces and Support Vector Machine Regression

Gianola et al. (2006) and de los Campos et al. (2009a) introduced the use of reproducing kernel Hilbert spaces (RKHS) regression (Smola and Schölkopf 2004) for genomic selection. The idea was to combine an additive genetic model with a kernel function. The kernel function converted the predictor variables into calculated distances among the observations to produce an $n \times n$ matrix for linear modeling (Smola and Schölkopf 2004). Two terms are introduced to define the spaces in which the individuals reside, the input space and the feature space. The input space is determined to be the multidimensional space in which the location of an individual is determined by the set of marker scores of that individual. Then, the input space is converted to the feature space through the application of a kernel function. Reproducing kernel Hilbert spaces function:

$$g(x) = K_h \alpha$$

where K_h is the matrix of kernel entries quantifying the distances of individuals to each other, similar to the **G** and **K** matrices of realized genomic relationships but representing distances between individuals rather than similarities. More fundamentally, the difference between the genomic relationship matrix and the genetic distance matrix is that genomic relationship matrices assume a linear, polygenic additive model of inheritance, whereas the RKHS model can use a variety of kernel function smoothing parameters (h), including nonlinear relationships. These smoothing parameters dictate the relationship between distances based on markers and distances in trait values. The α vector ($n \times 1$) includes the genotypic effects of individuals within the feature space that can be estimated using mixed model equations (Gianola and de los Campos 2008). Similarly to the RR-BLUP and LASSO methods, h should be optimized for each data set.

3.4.4 Partial Least Squares Regression and Principal Component Regression

Partial least squares (PLS) and principal component (PC) regression are methods of dimension reduction designed to handle a large quantity of independent variables with undefined relationships to the dependent variable (Coxe 2006; Wold 2006). The dimension reduction allows the model to maintain the level of information from the input marker data while handling the large p, small n problem, potentially maintaining a high level of prediction accuracy (Tobias 1995). In both methods, variables extracted as linear combinations of the predictors are used for prediction

$$g(x_i) = \sum_{l=1}^{w} t_{il}\beta_l$$

where t_{il} is the extracted variable as linear combinations of the original predictors and β is the effect associated with the extracted variable. Typically, the number of extracted variables, w, is much smaller than p. Furthermore, the t_i elements are orthogonal, eliminating problems of multicollinearity that exist in the original marker data set. The difference between the PLS and PC resides in the variation in which each are attempting to define. For PC, the extracted variables are ordered by the amount of variation in marker data they explain, but this may not be the same ranking for the proportion of variation they explain in phenotype data. For PLS, the extracted variables are selected based on the amount of variation they can explain in the phenotype data. The appropriate choice of number of components to maintain in the model can be determined through cross-validation (Solberg et al. 2009).

3.4.5 Bayesian Methods

The assumptions of the RR-BLUP model imply that genetic effects are equally distributed across the genome. Meuwissen et al. (2001) suggested that a Bayesian analysis could better represent these genetic effects by relaxing the assumption that all markers have a common variance. One Bayesian model (BayesA) was developed where each marker effect k had an independent variance that followed a normal distribution: $N(0, \mathrm{var}(\beta_k))$. The BayesA method can shrink each marker toward zero at varying degrees. The Bayesian shrinkage regression estimates the variance parameters from a scaled, inverted χ^2 distribution (Xu 2003).

The BayesB method introduces a prior probability (π) that a marker has zero effect on the trait. Meuwissen et al. (2001) suggested that this approach better reflected the underlying genetic architecture, meaning the genetic variance exists at fewer loci and is absent at most. The model is

$$g(x_i) = \sum_{k=1}^{p} x_{ik}\beta_k\Upsilon_k$$

where the only addition to the basic model is the Υ_k term, which simply specifies the presence of marker k in the prediction model. The β_k term follows a normal distribution with a mean of zero and a finite variance. The prior distribution for the variance of β_k follows a mixture distribution

$$\text{var}(\beta_k) = 0, \text{ with probability } \pi$$
$$\text{var}(\beta_k) \sim \chi^{-2}(v, S), \text{ with probability } (1-\pi)$$

Based on the assumptions, BayesB reverts to BayesA if the probability of zero marker effect is $\pi = 0$.

The BayesB model assumes some value for π then obtains the posterior marker effects for that value. A BayesCπ method allows the prior probability π to be estimated from the data. The prior distribution for π is uniform between 0 and 1. This BayesCπ method assumes that the prior variances for the effects of all markers present in the prediction model (Υ_k) are equal. In other words, the effect $\beta_k = 0$ when the marker is excluded from the model, else $\beta_k \sim N\left(0, \sigma_\beta^2\right)$ if the marker is included. The method estimates σ_β^2 equally over all nonzero markers and adds weight over the prior by grouping them (Gianola et al. 2009; Kizilkaya et al. 2010).

Some models are more suitable in different situations. For instance, the use of the RR-BLUP model implies the trait is controlled by many loci with small effects, and all of the marker effects are shrunken to the same degree and included in the model. In contrast, the BayesB assumes most loci have no effect on the trait and are excluded from the model. The remaining markers in the model have effects sampled from distributions with different variances. Using the BayesB model implies there are relatively few causative loci with different effect sizes. Jannink et al. (2010) argued that with greater marker density, QTL in linkage disequilibrium with associated markers and the underlying genetic architecture known, a Bayesian analysis is favored over RR-BLUP, as selection accuracy can be improved under the right conditions (Solberg et al. 2008; Meuwissen 2009). Empirical comparisons from plant breeding studies, however, indicate that the models tend to perform similarly in many cases, and the simpler RR-BLUP models are generally adequate in a wide range of cases (Lorenzana and Bernardo 2009; Heslot et al. 2012).

3.5 Limiting Inbreeding Under Genomic Selection

Populations of the same size can behave differently in terms of their inbreeding rates depending on the individual contributions of gametes to the following generation (Falconer and Mackay 1996). The effective population size influences the level of genetic drift and linkage disequilibrium in the population. The rate of inbreeding is equivalent to the magnitude of random genetic drift, shifting allele frequencies that occur due to the sampling of alleles that are contributing to the progeny. An advantage to having marker data on all of the selection candidates is that the expected inbreeding due to selecting a subset of individuals to form the next breeding generation can be computed. This provides breeders with another tool to monitor inbreeding and to select sets of individuals in a way that balances immediate gain from selection with maintaining genetic diversity to permit long-term genetic gain and avoid inbreeding depression in outcrossing species.

If the breeder specifies a maximum level of inbreeding that will be permitted in one generation of selection, the method of "optimal contribution" chooses the set of individuals with maximum average breeding value whose progenies will not exceed the inbreeding threshold (Sonesson and Meuwissen 2002). Computing the average breeding value for any subset of the individuals is easy, but computing the expected inbreeding coefficients of progenies for all possible subsets is not. Therefore, genetic algorithms can be used to efficiently evaluate a number of possible selection sets and choose the best in practice (Hinrichs et al. 2006). Alternatively, rare alleles can be given extra weight toward the selection criterion to reduce inbreeding under selection (Jannink 2010).

3.6 Retraining the Genomic Selection Model

Selection changes allele frequencies at QTL. Selection is also accompanied by genetic drift due to the limited sampling of parents each generation. Both selection and drift cause LD, which is counteracted by recombination that occurs during the intermating phases of a selection program. The dynamics of allele frequency changes due to selection and drift, and the changes in LD patterns due to selection, drift, and recombination are complex and not easily predicted in a practical breeding program (Jannink 2010). Since LD between markers and QTL impacts the relationship between markers and traits, the changes that occur due to drift and selection in a breeding program will affect the accuracy of the prediction model. In practice, this means that a genomic selection model created from a training data set of individuals will lose effectiveness over generations of selection and recombination (Bernardo and Yu 2007; Combs and Bernardo 2013; Müller et al. 2017). Therefore, breeders need to consider how to balance implementing genomic selection quickly using only marker data for a number of generations versus the need to collect new training data sets with both marker and phenotypic information to retrain and update the prediction model to maintain accuracy (Fig. 4).

3.7 Practical Implementation of Genomic Selection

If breeders have direct phenotypic observations from well-replicated multi-environment trials, these phenotypic data alone are usually sufficient to obtain highly accurate breeding value estimates of the measured lines. Adding marker information to this type of data set can make a small increase in the accuracy of breeding value estimation but is rarely worth the additional cost for that purpose alone. The real power of GS can be realized when reliable phenotype data from extensive field trials is combined with marker information to build a prediction model that is then used on future generations or other sets of lines for which only marker data are available and before investing in costly and time-consuming field trials (Fig. 4). The details of how to execute genomic selection will vary among species, as optimization depends on issues like the ease of making crosses and generating progenies, the availability of greenhouse or off-season nurseries, the speed with which marker data can be collected and processed, and the reliability of phenotypic data at each generation of the breeding process. Detailed protocols for implementing genomic selection have been proposed for wheat (Bassi et al. 2016) and for hybrid crops (Gaynor et al. 2017).

Genomic selection is being rapidly adopted for breeding across a diverse array of plant species, and the accuracy of genomic predictions has been reported for many traits and species (Finn et al. 2016; Grattapaglia 2017; Jain et al. 2017; Rutkoski et al. 2017; Crossa et al. 2017; Annicchiarico et al. 2018; Sarinelli et al. 2019). Head-to-head comparisons of genomic and phenotypic selection are scarce (and difficult to perform), but some empirical results on the effectiveness of genomic selection have been reported. Zhang et al. (2017) reported gains of 0.1 Mg ha^{-1} per year using rapid cycling genomic selection for yield in a tropical maize population, and similar gains from selection have been achieved with GS for maize yield under drought stress (Beyene et al. 2015). GS outperformed MAS over two cycles of selection for improved stover quality in maize (Massman et al. 2013). Yabe et al. (2018) reported that 3 years of multiple cycles of GS for yield in buckwheat (*Fagopyrum esculentum* Moench.) resulted in 21% increase in yield, whereas phenotypic selection achieved 15% increase in the same time. GS was a bit less effective than phenotypic selection for quantitative resistance to stem rust of wheat, however (Rutkoski et al. 2015).

Each trait and population combination may require a unique prediction model to obtain optimal prediction accuracy, but breeders will not have time to explore too many models if they are going to take advantage of the most important aspect of genomic selection for plant breeding, that is, the reduced time needed to turn a cycle of selection. In reality, "wrong" (or "mis-specified") prediction models will often capture a good portion of true breeding values despite being wrong, and there is some robustness to many of the models, with relatively small differences in

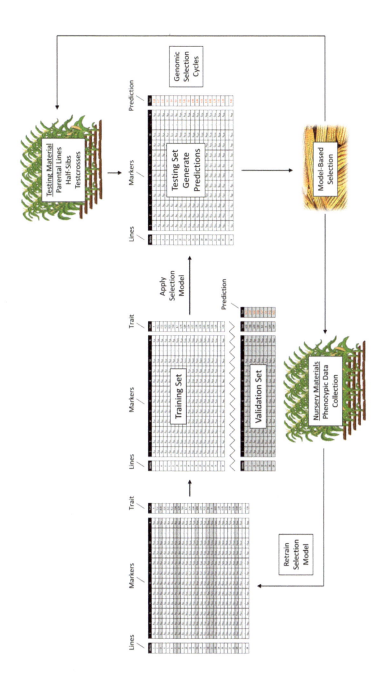

Fig. 4 Application of genomic selection to a crop breeding program. Starting with a training data set with both trait and genotypic data, a selection model is developed to predict values of untested lines to choose breeding parents of new crosses, selection candidates for inbreeding, or lines for hybrid testing. Evaluation of the selected lines in the field provides new data that can be used to update the training data and prediction model

prediction accuracy among them. Therefore, for practical purposes, a small subset of models that encompass a reasonable range of genetic architectures can be fitted to data, with models then selected based on cross-validation (Heslot et al. 2012).

Practical considerations that simplify the collection and processing of genotypic data may have greater impact on the rate of genetic gain over time than small improvements in prediction accuracy that may be achieved by evaluating many different statistical models for prediction. For example, if genomic selection is to be implemented in a generation in which selected individuals are intermated, genetic gain will be doubled if genomic data are available and predictions made before flowering compared to after flowering. Efficient protocols, sample tracking, rapid sequencing library preparation methods, and robust bioinformatics pipelines may allow breeders to collect leaf tissue from small plants soon after planting (or perhaps obtain DNA from seeds) and obtain genomic predictions within the generation time and possibly before flowering of an annual plant species. In this situation, the breeder should choose to use the most easily implemented statistical method for genomic prediction possible if that allows an increase in the number of generations in which genomic selection can be practiced or implementation of selection before flowering.

General rules for choosing training data sets and how and when to update them are hard to provide at this point and will require significantly more experience and empirical results, although effective training set optimization algorithms have been developed (Rincent et al. 2012; Akdemir et al. 2015; Guo et al. 2019). Often, however, prediction models that appear highly effective within a training population of historical data may fail to accurately predict performance of lines within new breeding crosses, particularly segregating early generation lines. The optimal deployment of GS resources among breeding materials is also not yet certain. For example, should prediction aim primarily at selecting sib lines closely related to a training set but which have never been tested in the field or at implementing rapid cycling recurrent selection to increase favorable allele frequencies within broad-based or even biparental populations? How many cycles should a training model be used before it is more efficient to collect new training phenotype data and retrain the model (Combs and Bernardo 2013; Müller et al. 2017)?

4 Future Perspectives

Going forward, the cloning of major genes and QTL will likely focus heavily on major disease resistance genes, particularly novel ones originating in exotic material such as wild species and landraces. As sequencing costs continue to fall, it will become highly practical and efficient to perform WGS on both parents prior to mapping population development. Not only will this allow rapid determination of an abundance of polymorphic markers to use in mapping, but it will also identify polymorphisms within candidate genes and/or large structural rearrangements or insertions/deletions that may underlie the phenotype. In terms of functional validation, the ease and speed of CRISPR will likely make it the dominant fine mapping and functional validation tool, even on larger candidate intervals with 20+ genes. A researcher can generate a CRISPR construct for each candidate gene in the interval to test which gene knockouts cause changes in the target trait.

In terms of genotyping for GS, new approaches capable of repeatedly delivering an identical set of ~5,000 SNPs that are significantly faster and cheaper than either a SNP array or GBS are needed. Ideally, it will be relatively easy to add new markers to the genotyping set as they are discovered and delete others that are fixed in target breeding populations.

5 Conclusions

If a trait is controlled mostly by a single major gene or a few major genes, then targeted marker-assisted selection can be very effective. As described previously, there are numerous important examples of major gene selection with markers in self-pollinated species. In contrast, if the genetic architecture of a trait of interest is highly polygenic, characterized by many genes each with very small effects, then genomic selection will be much more effective (Daetwyler et al. 2010). The true genetic architecture of a trait may be somewhere in between these two extremes, however. Some proportion of the genetic control of a trait may be highly quantitative, influenced by many small-effect genes throughout the genome, but with a handful of genes having outsized influence on variation for the trait. In such cases, the ideas of targeted marker-assisted selection specific for a few key genes or gene regions can be combined with genomic selection for the polygenic component of the genetic architecture.

Some of the Bayesian GS models mentioned previously are capable of capturing the contributions of a polygenic background plus major gene effects (Zhong et al. 2009; de los Campos et al. 2013). Such models are expected to work well in this context; however, there can be a price paid in model complexity and computational demand to execute such models on large (many individuals with many markers) data sets (Heslot et al. 2012).

Explicitly modeling the value of major effect genes may be complicated. For example, a major disease resistance gene may confer immunity to a particular race of pathogen, but the value of this gene to a trait, such as yield, may be difficult to quantify, as the prevalence of the disease in the future may change. In the case of discrete major genes that can be tagged with one or a few markers, the breeder may choose to first conduct truncation selection using a single allele-specific marker for the target gene, to eliminate any lines that do not carry the desired allele. Only lines that pass this initial culling would then be considered for selection on that basis of genomic estimating breeding values for yield and other important traits.

Alternatively, one can collect genotypic and phenotypic data that are appropriate for both genomic selection and genome-wide association study and conduct both types of analyses (Spindel et al. 2016; Bian and Holland 2017; Sarinelli et al. 2019).

This is essentially an extension of the mixed linear model for GWAS, fitting potentially several major effect loci simultaneously. One needs to be careful not to be overly optimistic about the value of GWAS discoveries, however, and proper assessment of the prediction accuracy of this approach requires cross-validation in which GWAS is repeated in every training set subsample separately, which also provides an estimate of the robustness of individual SNP associations to perturbations of the training data sample (Bian and Holland 2017).

As machine learning models are trained on breeding populations, many variations on this theme of combined polygenic plus larger-effect genes can be explored. Gene regions that are regularly identified as having important "signal" in terms of affecting trait variation can be prioritized for more detailed genetic characterization and validation. GWAS discoveries are now being characterized in more detail to understand how they influence complex trait variation (Yang et al. 2017). This provides a purely forward genetics approach to understanding the genetics of complex traits and is sometimes the only reasonable path in cases where the biological pathways underlying variation for a trait are unknown. In addition to providing biological insights, these studies can provide genetic markers useful for targeting specific genes in breeding programs. Finally, looking into the future, gene editing tools like CRISPR-CAS9 hold the promise of revolutionary advances in breeding of crops (Scheben and Edwards 2017), but a prerequisite to the use of gene editing is to know which genetic variations underpin useful trait variation. Therefore, the application of gene editing to complex traits in crops will require continued genetic analysis to identify causal sequence variation, and the challenge of doing this for complex traits and sequence variants with small effects that are hard to measure accurately will remain with us for the foreseeable future.

References

Akdemir D, Sanchez JI, Jannink J-L. Optimization of genomic selection training populations with a genetic algorithm. Genet Sel Evol. 2015;47:38. https://doi.org/10.1186/s12711-015-0116-6.
Altschul SF, Gish W, Miller W, Myers EW, Lipman DJ. Basic local alignment search tool. J Mol Biol. 1990;215:403–10. https://doi.org/10.1016/S0022-2836(05)80360-2.
Andres RJ, Coneva V, Frank MH, Tuttle JR, Samayoa LF, Han S-W, et al. Modifications to a *LATE MERISTEM IDENTITY1* gene are responsible for the major leaf shapes of upland cotton (*Gossypium hirsutum* L.). Proc Natl Acad Sci. 2017;114:E57–66. https://doi.org/10.1073/pnas.1613593114.
Annicchiarico P, Nazzicari N, Pecetti L, Romani M. Genomic selection for biomass yield of perennial and annual legumes. In: Breeding grasses and protein crops in the era of genomics. Cham: Springer; 2018. p. 259–64.
Ardlie KG, Kruglyak L, Seielstad M. Patterns of linkage disequilibrium in the human genome. Natl Rev. 2002;3:299–309. https://doi.org/10.1038/nrg777.
Artico S, Nardeli SM, Brilhante O, Grossi-de-Sa MF, Alves-Ferreira M. Identification and evaluation of new reference genes in *Gossypium hirsutum* for accurate normalization of real-time quantitative RT-PCR data. BMC Plant Biol. 2010;10:49. https://doi.org/10.1186/1471-2229-10-49.

Astle W, Balding DJ. Population structure and cryptic relatedness in genetic association studies. Stat Sci. 2009;24:451–71.

Azodi CB, Pardo J, VanBuren R, de los Campos G, Shiu SH. Transcriptome-based prediction of complex traits in maize. Plant Cell. 2020;32:139–51. https://doi.org/10.1105/tpc.19.00332.

Bamshad MJ, Ng SB, Bigham AW, Tabor HK, Emond MJ, Nickerson DA, et al. Exome sequencing as a tool for Mendelian disease gene discovery. Nat Rev Genet. 2011;12:745–55. https://doi.org/10.1038/nrg3031.

Bandillo N, Raghavan C, Muyco P, Sevilla MAL, Lobina IT, Dilla-Ermita C, et al. Multi-parent advanced generation inter-cross (MAGIC) populations in rice: progress and potential for genetics research and breeding. Rice. 2013;6:11. https://doi.org/10.1186/1939-8433-6-11.

Barrero JM, Cavanagh C, Verbyla KL, Tibbits JFG, Verbyla AP, Huang BE, et al. Transcriptomic analysis of wheat near-isogenic lines identifies PM19-A1 and A2 as candidates for a major dormancy QTL. Genome Biol. 2015;16:93. https://doi.org/10.1186/s13059-015-0665-6.

Bassi FM, Bentley AR, Charmet G, Ortiz R, Crossa J. Breeding schemes for the implementation of genomic selection in wheat (*Triticum* spp.). Plant Sci. 2016;242:23–36. https://doi.org/10.1016/J.PLANTSCI.2015.08.021.

Bayer M, Milne I, Stephen G, Shaw P, Cardle L, Wright F, et al. Comparative visualization of genetic and physical maps with strudel. Bioinformatics. 2011;27:1307–8. https://doi.org/10.1093/bioinformatics/btr111.

Bendl J, Stourac J, Salanda O, Pavelka A, Wieben ED, Zendulka J, et al. PredictSNP: robust and accurate consensus classifier for prediction of disease-related mutations. PLoS Comput Biol. 2014;10:1–11. https://doi.org/10.1371/journal.pcbi.1003440.

Bernardo R. Molecular markers and selection for complex traits in plants: learning from the last 20 years. Crop Sci. 2008;48:1649–64.

Bernardo R, Charcosset A. Usefulness of gene information in marker-assisted recurrent selection: a simulation appraisal. Crop Sci. 2006;46:614. https://doi.org/10.2135/cropsci2005.05-0088.

Bernardo R, Yu J. Prospects for genomewide selection for quantitative traits in maize. Crop Sci. 2007;47:1082–90. https://doi.org/10.2135/cropsci2006.11.0690.

Beyene Y, Semagn K, Mugo S, Tarekegne A, Babu R, Meisel B, et al. Genetic gains in grain yield through genomic selection in eight bi-parental maize populations under drought stress. Crop Sci. 2015;55:154–63. https://doi.org/10.2135/cropsci2014.07.0460.

Bian Y, Holland JB. Enhancing genomic prediction with genome-wide association studies in multiparental maize populations. Heredity (Edinb). 2017;118:585–93. https://doi.org/10.1038/hdy.2017.4.

Bishop CM. Pattern recognition and machine learning. Springer, New York; 2006.

Bonnett DG, Rebetzke GJ, Spielmeyer W. Strategies for efficient implementation of molecular markers in wheat breeding. Mol Breed. 2005;15:75–85. https://doi.org/10.1007/s11032-004-2734-5.

Broccanello C, Chiodi C, Funk A, McGrath JM, Panella L, Stevanato P. Comparison of three PCR-based assays for SNP genotyping in plants. Plant Methods. 2018;14:1–8. https://doi.org/10.1186/s13007-018-0295-6.

Buckler ES, Holland JB, Bradbury P, Acharya C, Brown P, Browne C, et al. The genetic architecture of maize flowering time. Science. 2009;325:714–8.

Bush WS, Moore JH. Chapter 11: genome-wide association studies. PLoS Comput Biol. 2012;8: e1002822. https://doi.org/10.1371/journal.pcbi.1002822.

Bustin SA, Benes V, Garson JA, Hellemans J, Huggett J, Kubista M, et al. The MIQE guidelines: minimum information for publication of quantitative real-time PCR experiments. Clin Chem. 2009;55:611–22. https://doi.org/10.1373/clinchem.2008.112797.

Cahill DJ, Schmidt DH. Use of marker assisted selection in a product development breeding program. In: Fischer T, Turner N, Angus O, et al., editors. 4th international crop science congress. Brisbane, Australia; 2004.

Carbon S, Dietze H, Lewis SE, Mungall CJ, Munoz-Torres MC, Basu S, et al. Expansion of the gene ontology knowledgebase and resources: the gene ontology consortium. Nucleic Acids Res. 2017;45:D331–8. https://doi.org/10.1093/nar/gkw1108.

Chakrabarti M, Zhang N, Sauvage C, Munos S, Blanca J, Canizares J, et al. A cytochrome P450 regulates a domestication trait in cultivated tomato. Proc Natl Acad Sci. 2013;110:17125–30. https://doi.org/10.1073/pnas.1307313110.

Charmet G, Robert N, Perretant MR, Gay G, Sourdille P, Groos C, et al. Marker-assisted recurrent selection for cumulating additive and interactive QTLs in recombinant inbred lines. TAG Theor Appl Genet. 1999;99:1143–8. https://doi.org/10.1007/s001220051318.

Choi Y, Sims GE, Murphy S, Miller JR, Chan AP. Predicting the functional effect of amino acid substitutions and Indels. PLoS One. 2012;7. https://doi.org/10.1371/journal.pone.0046688.

Clayton D. Linkage disequilibrium mapping of disease susceptibility genes in human populations. Int Stat Rev. 2008;68:23–43.

Combs E, Bernardo R. Genomewide selection to introgress semidwarf maize germplasm into U.S. corn belt inbreds. Crop Sci. 2013;53:1427. https://doi.org/10.2135/cropsci2012.11.0666.

Conesa A, Madrigal P, Tarazona S, Gomez-Cabrero D, Cervera A, McPherson A, et al. A survey of best practices for RNA-seq data analysis. Genome Biol. 2016;17:13. https://doi.org/10.1186/s13059-016-0881-8.

Coxe KL. Principal components regression analysis. In: Encyclopedia of statistical sciences. Hoboken, NJ: Wiley; 2006.

Crossa J, Pérez-Rodríguez P, Cuevas J, Montesinos-López O, Jarquín D, de los Campos G, et al. Genomic selection in plant breeding: methods, models, and perspectives. Trends Plant Sci. 2017;22:961–75.

Daetwyler HD, Villanueva B, Woolliams JA. Accuracy of predicting the genetic risk of disease using a genome-wide approach. PLoS One. 2008;3:e3395. https://doi.org/10.1371/journal.pone.0003395.

Daetwyler HD, Pong-Wong R, Villanueva B, Woolliams JA. The impact of genetic architecture on genome-wide evaluation methods. Genetics. 2010;185:1021–31. https://doi.org/10.1534/genetics.110.116855.

Dawson JC, Endelman JB, Heslot N, Crossa J, Poland J, Dreisigacker S, et al. The use of unbalanced historical data for genomic selection in an international wheat breeding program. Field Crop Res. 2013;154:12–22. https://doi.org/10.1016/j.fcr.2013.07.020.

De La Vega FM, Lazaruk KD, Rhodes MD, Wenz MH. Assessment of two flexible and compatible SNP genotyping platforms: TaqMan® SNP genotyping assays and the SNPlex™ genotyping system. Mutat Res. 2005;573:111–35. https://doi.org/10.1016/j.mrfmmm.2005.01.008.

de los Campos G, Gianola D, Rosa GJM. Reproducing kernel Hilbert spaces regression: a general framework for genetic evaluation. J Anim Sci. 2009a;87:1883–7. https://doi.org/10.2527/jas.2008-1259.

de los Campos G, Naya H, Gianola D, Crossa J, Legarra A, Manfredi E, et al. Predicting quantitative traits with regression models for dense molecular markers and pedigree. Genetics. 2009b;182:375–85. https://doi.org/10.1534/genetics.109.101501.

de los Campos G, Hickey JM, Pong-Wong R, Daetwyler HD, Calus MPL. Whole-genome regression and prediction methods applied to plant and animal breeding. Genetics. 2013;193:327–45. https://doi.org/10.1534/genetics.112.143313.

Dell'Acqua M, Gatti DM, Pea G, Cattonaro F, Coppens F, Magris G, et al. Genetic properties of the MAGIC maize population: a new platform for high definition QTL mapping in zea mays. Genome Biol. 2015;16:167. https://doi.org/10.1186/s13059-015-0716-z.

Devlin B, Risch N. A comparison of linkage disequilibrium measures for fine-scale mapping. Genomics. 1995;29:311–22.

Devlin B, Roeder K. Genomic control for association studies. Biometrics. 1999;55:997–1004. https://doi.org/10.1111/j.0006-341X.1999.00997.x.

Do R, Kathiresan S, Abecasis GR. Exome sequencing and complex disease: practical aspects of rare variant association studies. Hum Mol Genet. 2012;21:R1–9. https://doi.org/10.1093/hmg/dds387.

Dubcovsky J. Marker-assisted selection in public breeding programs. The wheat experience. Crop Sci. 2004;44:1895. https://doi.org/10.2135/cropsci2004.1895.

Eagles HA, Bariana HS, Ogbonnaya FC, Rebetzke GJ, Hollamby GJ, Henry RJ, et al. Implementation of markers in Australian wheat breeding. Aust J Agr Res. 2001;52:1349. https://doi.org/10.1071/AR01067.

Edwards MD, Stuber CW, Wendel JF. Molecular-marker-facilitated investigations of quantitative-trait loci in maize. I. Numbers, genomic distribution and types of gene action. Genetics. 1987;116:113–25.

Edwards KD, Fernandez-Pozo N, Drake-Stowe K, Humphry M, Evans AD, Bombarely A, et al. A reference genome for *Nicotiana tabacum* enables map-based cloning of homeologous loci implicated in nitrogen utilization efficiency. BMC Genomics. 2017;18:448. https://doi.org/10.1186/s12864-017-3791-6.

Eichler EE, Flint J, Gibson G, Kong A, Leal SM, Moore JH, et al. Missing heritability and strategies for finding the underlying causes of complex disease. Nat Rev Genet. 2010;11:446–50. https://doi.org/10.1038/nrg2809.

Elshire RJ, Glaubitz JC, Sun Q, Poland JA, Kawamoto K, Buckler ES, et al. A robust, simple genotyping-by-sequencing (GBS) approach for high diversity species. PLoS One. 2011;6:e19379. https://doi.org/10.1371/journal.pone.0019379.

Falconer DS, Mackay TFC. Introduction to quantitative genetics. 4th ed. Harlow: Addison Wesley Longman Limited; 1996.

Finn RD, Coggill P, Eberhardt RY, Eddy SR, Mistry J, Mitchell AL, et al. The Pfam protein families database: towards a more sustainable future. Nucleic Acids Res. 2016;44:D279–85. https://doi.org/10.1093/nar/gkv1344.

Finn RD, Attwood TK, Babbitt PC, Bateman A, Bork P, Bridge AJ, et al. InterPro in 2017-beyond protein family and domain annotations. Nucleic Acids Res. 2017;45:D190–9. https://doi.org/10.1093/nar/gkw1107.

Flint-garcia SA, Thornsberry JM, Buckler ES. Structure of linkage disequilibrium in plants. Annu Rev Plant Biol. 2003;54:357–74. https://doi.org/10.1146/annurev.arplant.54.031902.134907.

Flint-Garcia SA, Thuillet A-C, Yu J, Pressoir G, Romero SM, Mitchell SE, et al. Maize association population: a high-resolution platform for quantitative trait locus dissection. Plant J. 2005;44:1054–64. https://doi.org/10.1111/j.1365-313X.2005.02591.x.

Frey TJ, Weldekidan T, Colbert T, Wolters PJCC, Hawk JA. Fitness evaluation of Rcg1, a locus that confers resistance to colletotrichum graminicola (Ces.) G.W. Wils. Using near-isogenic maize hybrids. Crop Sci. 2011;51:1551–63. https://doi.org/10.2135/cropsci2010.10.0613.

Frisch M, Bohn M, Melchinger AE. Comparison of selection strategies for marker-assisted backcrossing of a gene. Crop Sci. 1999;39:1295–301. https://doi.org/10.2135/cropsci1999.3951295x.

Fu D, Uauy C, Distelfeld A, Blechl A, Epstein L, Chen X, et al. A kinase-START gene confers temperature-dependent resistance to wheat stripe rust. Science. 2009;323:1357–60. https://doi.org/10.1126/science.1166289.

Fukuoka S, Saka N, Koga H, Ono K, Shimizu T, Ebana K, et al. Loss of function of a proline-containing protein confers durable disease resistance in rice. Science. 2009;325:998–1001. https://doi.org/10.1126/science.1175550.

Gaynor RC, Gorjanc G, Bentley AR, Ober ES, Howell P, Jackson R, et al. A two-part strategy for using genomic selection to develop inbred lines. Crop Sci. 2017;57:2372–86. https://doi.org/10.2135/cropsci2016.09.0742.

Gianola D, de los Campos G. Inferring genetic values for quantitative traits non-parametrically. Genet Res (Camb). 2008;90:525. https://doi.org/10.1017/S0016672308009890.

Gianola D, Fernando RL, Stella A. Genomic-assisted prediction of genetic value with semiparametric procedures. Genetics. 2006;173:1761–76. https://doi.org/10.1534/genetics.105.049510.

Gianola D, de los Campos G, Hill WG, Manfredi E, Fernando R. Additive genetic variability and the Bayesian alphabet. Genetics. 2009;183:347–63. https://doi.org/10.1534/genetics.109.103952.

Giovannoni JJ, Wing RA, Ganal MW, Tanksley SD. Isolation of molecular markers from specific chromosomal intervals using DNA pools from existing mapping populations. Nucleic Acids Res. 1991;19:6553–68. https://doi.org/10.1093/nar/19.23.6553.

Glaubitz JC, Casstevens TM, Lu F, Harriman J, Elshire RJ, Sun Q, et al. TASSEL-GBS: a high capacity genotyping by sequencing analysis pipeline. PLoS One. 2014;9:e90346. https://doi.org/10.1371/journal.pone.0090346.

Goddard ME, Hayes BJ. Genomic selection. J Anim Breed Genet. 2007;124:323–30.

Goddard ME, Wray NR, Verbyla K, Visscher PM. Estimating effects and making predictions from genome-wide marker data. Stat Sci. 2009;24:517–29. https://doi.org/10.1214/09-STS306.

Grattapaglia D. Status and perspectives of genomic selection in forest tree breeding. In: Genomic selection for crop improvement: new molecular breeding strategies for crop improvement. Cham: Springer; 2017. p. 199–249.

Guo T, Yu X, Li X, Zhang H, Zhu C, Flint-Garcia S, et al. Optimal designs for genomic selection in hybrid crops. Mol Plant. 2019;12:390–401. https://doi.org/10.1016/j.molp.2018.12.022.

Gupta PK, Kulwal PL, Jaiswal V. Association mapping in crop plants: opportunities and challenges. Adv Genet. 2014;85:109–47. https://doi.org/10.1016/B978-0-12-800271-1.00002-0.

Habier D, Fernando RL, Dekkers JCM. The impact of genetic relationship information on genome-assisted breeding values. Genetics. 2007;177:2389–97. https://doi.org/10.1534/genetics.107.081190.

Hamblin MT, Close TJ, Bhat PR, Chao SM, Kling JG, Abraham KJ, et al. Population structure and linkage disequilibrium in US barley germplasm: implications for association mapping. Crop Sci. 2010;50:556–66.

Hartl DL, Clark AG. Principles of population genetics. Sunderland, MA: Sinauer Associates, Inc.; 1997.

Hayes BJ, Bowman PJ, Chamberlain AJ, Goddard ME. Invited review: genomic selection in dairy cattle: progress and challenges. J Dairy Sci. 2009;92:433–43. https://doi.org/10.3168/jds.2008-1646.

He C, Holme J, Anthony J. Crop Breeding. 2014;1145:75–86. https://doi.org/10.1007/978-1-4939-0446-4.

Heffner EL, Sorrells ME, Jannink JL. Genomic selection for crop improvement. Crop Sci. 2009;49:1–12.

Henderson CR. Best linear unbiased estimation and prediction under a selection model. Biometrics. 1975;31:423. https://doi.org/10.2307/2529430.

Heslot N, Yang H-P, Sorrells ME, Jannink J-L. Genomic selection in plant breeding: a comparison of models. Crop Sci. 2012;52:146. https://doi.org/10.2135/cropsci2011.06.0297.

Hill WG, Goddard ME, Visscher PM. Data and theory point to mainly additive genetic variance for complex traits. PLoS Genet. 2008;4:e1000008.

Hinrichs D, Wetten M, Meuwissen THE. An algorithm to compute optimal genetic contributions in selection programs with large numbers of candidates. J Anim Sci. 2006;84:3212. https://doi.org/10.2527/jas.2006-145.

Hospital F, Charcosset A. Marker-assisted introgression of quantitative trait loci. Genetics. 1997;147:1469–85.

Hospital F, Moreau L, Lacoudre F, Charcosset A, Gallais A. More on the efficiency of marker-assisted selection. TAG Theor Appl Genet. 1997;95:1181–9. https://doi.org/10.1007/s001220050679.

Hospital F, Goldringer I, Openshaw S. Efficient marker-based recurrent selection for multiple quantitative trait loci. Genet Res. 2000;75:357–68. https://doi.org/10.1017/S0016672300004511.

Hsing YI, Chern CG, Fan MJ, Lu PC, Chen KT, Lo SF, et al. A rice gene activation/knockout mutant resource for high throughput functional genomics. Plant Mol Biol. 2007;63:351–64. https://doi.org/10.1007/s11103-006-9093-z.

Huang X, Feng Q, Qian Q, Zhao Q, Wang L, Wang A, et al. High-throughput genotyping by whole-genome resequencing. Genome Res. 2009;19:1068–76. https://doi.org/10.1101/gr.089516.108.

Huang BE, George AW, Forrest KL, Kilian A, Hayden MJ, Morell MK, et al. A multiparent advanced generation inter-cross population for genetic analysis in wheat. Plant Biotechnol J. 2012;10:826–39. https://doi.org/10.1111/j.1467-7652.2012.00702.x.

Hurni S, Scheuermann D, Krattinger SG, Kessel B, Wicker T, Herren G, et al. The maize disease resistance gene *Htn1* against northern corn leaf blight encodes a wall-associated receptor-like kinase. Proc Natl Acad Sci. 2015;112:8780–5. https://doi.org/10.1073/pnas.1502522112.

Isidro J, Jannink J-L, Akdemir D, Poland J, Heslot N, Sorrells ME. Training set optimization under population structure in genomic selection. Theor Appl Genet. 2015;128:145–58. https://doi.org/10.1007/s00122-014-2418-4.

Isik F, Holland J, Maltecca C. Genetic data analysis for plant and animal breeding. New York: Springer; 2017.

Jain A, Roorkiwal M, Pandey MK, Varshney RK. Current status and prospects of genomic selection in legumes. In: Genomic selection for crop improvement: new molecular breeding strategies for crop improvement. Cham: Springer; 2017. p. 131–47.

Jamann TM, Balint-Kurti PJ, Holland JB. QTL mapping using high-throughput sequencing. New York, NY: Humana Press; 2015. p. 257–85.

Jannink JL. Dynamics of long-term genomic selection. Genet Sel Evol. 2010;42:35.

Jannink J-L, Lorenz AJ, Iwata H. Genomic selection in plant breeding: from theory to practice. Brief Funct Genomics. 2010;9:166–77. https://doi.org/10.1093/bfgp/elq001.

Jansen RC. Interval mapping of multiple quantitative trait loci. Genetics. 1993;135:205–11.

Jena KK, Mackill DJ. Molecular markers and their use in marker-assisted selection in rice. Crop Sci. 2008;48:1266. https://doi.org/10.2135/cropsci2008.02.0082.

Ji H, Kim S-R, Kim Y-H, Suh J-P, Park H-M, Sreenivasulu N, et al. Map-based cloning and characterization of the BPH18 gene from wild rice conferring resistance to brown planthopper (BPH) insect pest. Sci Rep. 2016;6:34376. https://doi.org/10.1038/srep34376.

Johansen CT, Wang J, Lanktree MB, Cao H, McIntyre AD, Ban MR, et al. Excess of rare variants in genes identified by genome-wide association study of hypertriglyceridemia. Nat Genet. 2010;42:684–7. https://doi.org/10.1038/ng.628.

Jonas E, Fikse F, Rönnegård L, Mouresan EF. Genomic selection. In: Rajora OP, editor. Population genomics: concepts, approaches and applications. Cham: Springer Nature Switzerland AG; 2019. p. 427–82.

Jorde LB. Linkage disequilibrium and the search for complex disease genes. Genome Res. 2000;10:1435–44. https://doi.org/10.1101/gr.144500.

Kelliher T, Starr D, Su X, Tang G, Chen Z, Carter J, et al. One-step genome editing of elite crop germplasm during haploid induction. Nat Biotechnol. 2019;37:287–92. https://doi.org/10.1038/s41587-019-0038-x.

Kelly JD, Gepts P, Miklas PN, Coyne DP. Tagging and mapping of genes and QTL and molecular marker-assisted selection for traits of economic importance in bean and cowpea. Field Crop Res. 2003;82:135–54.

Kizilkaya K, Fernando RL, Garrick DJ. Genomic prediction of simulated multibreed and purebred performance using observed fifty thousand single nucleotide polymorphism genotypes. J Anim Sci. 2010;88:544–51. https://doi.org/10.2527/jas.2009-2064.

Knapp SJ, Bridges WC. Using molecular markers to estimate quantitative trait locus parameters: power and genetic variances for unreplicated and replicated progeny. Genetics. 1990;126:769–77.

Korte A, Vilhjálmsson BJ, Segura V, Platt A, Long Q, Nordborg M. A mixed-model approach for genome-wide association studies of correlated traits in structured populations. Nat Genet. 2012;44:1066. https://doi.org/10.1038/ng.2376.

Krattinger SG, Lagudah ES, Spielmeyer W, Singh RP, Huerta-espino J, Mcfadden H, et al. Pathogens in wheat. Science. 2009;323:1360–3.

Lange L, Hu Y, Zhang H, Al E. Whole-exome sequencing identifies rare and low-frequency coding variants associated with LDL cholesterol. Am J Hum Genet. 2014;94:233–45. https://doi.org/10.1016/J.AJHG.2014.01.010.

Langmead B, Salzberg SL. Fast gapped-read alignment with Bowtie 2. Nat Methods. 2012;9:357–9. https://doi.org/10.1038/nmeth.1923.

Larsson SJ, Lipka AE, Buckler ES. Lessons from Dwarf8 on the strengths and weaknesses of structured association mapping. PLoS Genet. 2013;9(2):e1003246. https://doi.org/10.1371/journal.pgen.1003246.

Lee SH, van der Werf JHJ, Hayes BJ, Goddard ME, Visscher PM. Predicting unobserved phenotypes for complex traits from whole-genome SNP data. PLoS Genet. 2008;4:e1000231. https://doi.org/10.1371/journal.pgen.1000231.

Legarra A, Robert-Granié C, Manfredi E, Elsen JM. Performance of genomic selection in mice. Genetics. 2008;180:611–8. https://doi.org/10.1534/genetics.108.088575.

Li H, Durbin R. Fast and accurate short read alignment with burrows-wheeler transform. Bioinformatics. 2009;25:1754–60. https://doi.org/10.1093/bioinformatics/btp324.

Li H, Handsaker B, Wysoker A, Fennell T, Ruan J, Homer N, et al. The sequence alignment/map format and SAMtools. Bioinformatics. 2009;25:2078–9. https://doi.org/10.1093/bioinformatics/btp352.

Li Y, Sidore C, Kang HM, Boehnke M, Abecasis GR. Low-coverage sequencing: implications for design of complex trait association studies. Genome Res. 2011;21:940–51. https://doi.org/10.1101/gr.117259.110.

Listgarten J, Lippert C, Kadie CM, Davidson RI, Eskin E, Heckerman D. Improved linear mixed models for genome-wide association studies. Nat Methods. 2012;9:525–6. https://doi.org/10.1038/nmeth.2037.

Liu W, Maurer HP, Reif JC, Melchinger AE, Utz HF, Tucker MR, et al. Optimum design of family structure and allocation of resources in association mapping with lines from multiple crosses. Heredity (Edinb). 2013;110:71–9. https://doi.org/10.1038/hdy.2012.63.

Liu Y, Wu H, Chen H, Liu Y, He J, Kang H, et al. A gene cluster encoding lectin receptor kinases confers broad-spectrum and durable insect resistance in rice. Nat Biotechnol. 2014;33:301–5. https://doi.org/10.1038/nbt.3069.

Liu X, Huang M, Fan B, Buckler ES, Zhang Z. Iterative usage of fixed and random effect models for powerful and efficient genome-wide association studies. PLoS Genet. 2016;12:e1005767. https://doi.org/10.1371/journal.pgen.1005767.

Lorenz AJ, Smith KP. Adding genetically distant individuals to training populations reduces genomic prediction accuracy in barley. Crop Sci. 2015;55:2657. https://doi.org/10.2135/cropsci2014.12.0827.

Lorenzana RE, Bernardo R. Accuracy of genotypic value predictions for marker-based selection in biparental plant populations. Theor Appl Genet. 2009;120:151–61. https://doi.org/10.1007/s00122-009-1166-3.

Lowe K, Wu E, Wang N, Hoerster G, Hastings C, Cho M-J, et al. Morphogenic regulators *Baby boom* and *Wuschel* improve monocot transformation. Plant Cell. 2016;28:1998–2015. https://doi.org/10.1105/tpc.16.00124.

Lu F, Lipka AE, Glaubitz J, Elshire R, Cherney JH, Casler MD, et al. Switchgrass genomic diversity, ploidy, and evolution: novel insights from a network-based SNP discovery protocol. PLoS Genet. 2013;9:e1003215. https://doi.org/10.1371/journal.pgen.1003215.

Ma D, Hu Y, Yang C, Liu B, Fang L, Wan Q, et al. Genetic basis for glandular trichome formation in cotton. Nat Commun. 2016;7:10456. https://doi.org/10.1038/ncomms10456.

Mackay IJ, Bansept-Basler P, Barber T, Bentley AR, Cockram J, Gosman N, et al. An eight-parent multiparent advanced generation inter-cross population for winter-sown wheat: creation, properties, and validation. G3 (Bethesda). 2014;4:1603–10. https://doi.org/10.1534/g3.114.012963.

Malmberg MM, Barbulescu DM, Drayton MC, Shinozuka M, Thakur P, Ogaji YO, et al. Evaluation and recommendations for routine genotyping using skim whole genome re-sequencing in canola. Front Plant Sci. 2018;871:1–15. https://doi.org/10.3389/fpls.2018.01809.

Manching H, Sengupta S, Hopper KR, Polson S, Ji Y, Wisser RJ. Phased genotyping-by-sequencing enhances analysis of genetic diversity and reveals divergent copy number variants in maize. G3 (Bethesda). 2017;7(7):2161–70.

Manoli A, Sturaro A, Trevisan S, Quaggiotti S, Nonis A. Evaluation of candidate reference genes for qPCR in maize. J Plant Physiol. 2012;169:807–15. https://doi.org/10.1016/j.jplph.2012.01.019.

Manolio TA. In retrospect: a decade of shared genomic associations. Nature. 2017;546:360–1. https://doi.org/10.1038/546360a.

Manolio TA, Collins FS, Cox NJ, Goldstein DB, Hindorff LA, Hunter DJ, et al. Finding the missing heritability of complex diseases. Nature. 2009;461:747–53. https://doi.org/10.1038/nature08494.

Massman JM, Jung HJG, Bernardo R. Genomewide selection versus marker-assisted recurrent selection to improve grain yield and Stover-quality traits for cellulosic ethanol in maize. Crop Sci. 2013;53:58–66. https://doi.org/10.2135/cropsci2012.02.0112.

McCarty DR, Mark Settles A, Suzuki M, Tan BC, Latshaw S, Porch T, et al. Steady-state transposon mutagenesis in inbred maize. Plant J. 2005;44:52–61. https://doi.org/10.1111/j.1365-313X.2005.02509.x.

McKenna A, Hannan M, Banks E, Sivachenko A, Cibulskis K, Kernytsky A, et al. The genome analysis toolkit: a MapReduce framework for analyzing next-generation DNA sequencing data. Genome Res. 2010;20:1297–303. https://doi.org/10.1101/gr.107524.110.20.

McLaren W, Gil L, Hunt SE, Riat HS, Ritchie GRS, Thormann A, et al. The Ensembl variant effect predictor. Genome Biol. 2016;17:122. https://doi.org/10.1186/s13059-016-0974-4.

McMullen MD, Kresovich S, Villeda HS, Bradbury P, Li H, Sun Q, et al. Genetic properties of the maize nested association mapping population. Science. 2009;325:737–40. https://doi.org/10.1126/science.1174320.

Melo ATO, Bartaula R, Hale I. GBS-SNP-CROP: a reference-optional pipeline for SNP discovery and plant germplasm characterization using variable length, paired-end genotyping-by-sequencing data. BMC Bioinformatics. 2016;17:29. https://doi.org/10.1186/s12859-016-0879-y.

Meuwissen TH. Accuracy of breeding values of "unrelated" individuals predicted by dense SNP genotyping. Genet Sel Evol. 2009;41:35. https://doi.org/10.1186/1297-9686-41-35.

Meuwissen THE, Hayes BJ, Goddard ME. Prediction of total genetic value using genome-wide dense marker maps. Genetics. 2001;157:1819–29.

Mi H, Huang X, Muruganujan A, Tang H, Mills C, Kang D, et al. PANTHER version 11: expanded annotation data from gene ontology and Reactome pathways, and data analysis tool enhancements. Nucleic Acids Res. 2017;45:D183–9. https://doi.org/10.1093/nar/gkw1138.

Michelmore RW, Paran I, Kesseli RV. Identification of markers linked to disease-resistance genes by bulked segregant analysis: a rapid method to detect markers in specific genomic regions by using segregating populations. Proc Natl Acad Sci. 1991;88:9828–32. https://doi.org/10.1073/pnas.88.21.9828.

Moore JW, Herrera-Foessel S, Lan C, Schnippenkoetter W, Ayliffe M, Huerta-Espino J, et al. A recently evolved hexose transporter variant confers resistance to multiple pathogens in wheat. Nat Genet. 2015;47:1494–8. https://doi.org/10.1038/ng.3439.

Moreau L, Moreau L, Charcosset A, Charcosset A. Marker-assisted selection efficiency in populations of finite size. Genetics. 1998;148(3):1353–65.

Moreau L, Charcosset A, Gallais A. Experimental evaluation of several cycles of marker-assisted selection in maize. Euphytica. 2004;137:111–8.

Müller D, Schopp P, Melchinger AE. Persistency of prediction accuracy and genetic gain in synthetic populations under recurrent genomic selection. G3 (Bethesda). 2017;7:801–11. https://doi.org/10.1534/G3.116.036582.

Muñoz PR, Resende MFR, Gezan SA, Resende MDV, de los campos G, Kirst M, et al. Unraveling additive from nonadditive effects using genomic relationship matrices. Genetics. 2014;198:1759–68. https://doi.org/10.1534/genetics.114.171322.

Myles S, Peiffer J, Brown PJ, Ersoz ES, Zhang Z, Costich DE, et al. Association mapping: critical considerations shift from genotyping to experimental design. Plant Cell. 2009;21:2194–202. https://doi.org/10.1105/tpc.109.068437.

Nordborg M. Linkage disequilibrium, gene trees and selfing: an ancestral recombination. Genetics. 2000;154:923–9.

Nordborg M, Donnelly P. The coalescent process with selfing. Genetics. 1997;146:1185–95.

Ober U, Ayroles JF, Stone EA, Richards S, Zhu D, Gibbs RA, et al. Using whole-genome sequence data to predict quantitative trait phenotypes in *Drosophila melanogaster*. PLoS Genet. 2012;8: e1002685. https://doi.org/10.1371/journal.pgen.1002685.

Ogata H, Goto S, Sato K, Fujibuchi W, Bono H. KEGG: kyoto encyclopedia of genes and genomes. Nucleic Acids Res. 1999;27:29–34.

Oraguzie NC, Wilcox PL, Rikkerink EHA, de Silva HN. Association mapping in plants. New York, NY: Springer; 2007.

Osakabe Y, Osakabe K. Genome editing with engineered nucleases in plants. Plant Cell Physiol. 2015;56:389–400. https://doi.org/10.1093/pcp/pcu170.

Paterson AH, Lander ES, Hewitt JD, Peterson S, Lincoln SE, Tanksley SD. Resolution of quantitative traits into Mendelian factors by using a complete linkage map of restriction fragment length polymorphisms. Nature. 1988;335:721–6.

Peiffer JA, Romay MC, Gore MA, Flint-Garcia SA, Zhang Z, Millard MJ, et al. The genetic architecture of maize height. Genetics. 2014;196:1337–56. https://doi.org/10.1534/genetics.113.159152.

Perkel J. SNP genotyping: six technologies that keyed a revolution. Nat Methods. 2008;5:447–53. https://doi.org/10.1038/nmeth0508-447.

Piepho HP. Ridge regression and extensions for genomewide selection in maize. Crop Sci. 2009;49. https://doi.org/10.2135/cropsci2008.10.0595.

Pino Del Carpio D, Lozano R, Wolfe MD, Jannink J-L. Genome-wide association studies and heritability estimation in the functional genomics era. In: Rajora OP, editor. Population genomics: concepts, approaches and applications. Cham: Springer Nature Switzerland AG; 2019. p. 361–425.

Podolak E. Sequencing's new race. Biotechniques. 2010;48(2):105–11.

Price AL, Zaitlen NA, Reich D, Patterson N. New approaches to population stratification in genome-wide association studies. Nat Rev Genet. 2010;11:459–63. https://doi.org/10.1038/nrg2813.

Pritchard JK, Przeworski M. Linkage disequilibrium in humans: models and data. Am J Hum Genet. 2001;69:1–14.

Pritchard J, Stephens M, Donnelly P. Inference of population structure using multilocus genotype data. Genetics. 2000;155:945–59. https://doi.org/10.1111/j.1471-8286.2007.01758.x.

Rafalski A. Applications of single nucleotide polymorphisms in crop genetics. Curr Opin Plant Biol. 2002;5:94–100.

Ragoussis J. Genotyping technologies for all. Drug Discov Today Technol. 2006;3:115–22.

Reich D, Price AL, Patterson N. Principal component analysis of genetic data from gene expression to disease risk. Nat Genet. 2008;40:491–3.

Rincent R, Laloë D, Nicolas S, Altmann T, Brunel D, Revilla P, et al. Maximizing the reliability of genomic selection by optimizing the calibration set of reference individuals: comparison of methods in two diverse groups of maize inbreds (*Zea mays* L.). Genetics. 2012;192:715–28. https://doi.org/10.1534/genetics.112.141473.

Rogers AR. How population growth affects linkage disequilibrium. Genetics. 2014;197:1329–41. https://doi.org/10.1534/genetics.114.166454.

Romay MC, Millard MJ, Glaubitz JC, Peiffer JA, Swarts KL, Casstevens TM, et al. Comprehensive genotyping of the USA national maize inbred seed bank. Genome Biol. 2013;14:R55. https://doi.org/10.1186/gb-2013-14-6-r55.

Rutkoski J, Singh RP, Huerta-Espino J, Bhavani S, Poland J, Jannink JL, et al. Genetic gain from phenotypic and genomic selection for quantitative resistance to stem rust of wheat. Plant Genome. 2015;8. https://doi.org/10.3835/plantgenome2014.10.0074.

Rutkoski JE, Crain J, Poland J, Sorrells ME. Genomic selection for small grain improvement. In: Genomic selection for crop improvement: new molecular breeding strategies for crop improvement. Cham: Springer; 2017. p. 99–130.

Salvi S, Sponza G, Morgante M, Tomes D, Niu X, Fengler KA, et al. Conserved noncoding genomic sequences associated with a flowering-time quantitative trait locus in maize. Proc Natl Acad Sci U S A. 2007;104:11376–81. https://doi.org/10.1073/pnas.0704145104. https://doi.org/0704145104 [pii].

Sarinelli JM, Murphy JP, Tyagi P, Holland JB, Johnson JW, Mergoum M, et al. Training population selection and use of fixed effects to optimize genomic predictions in a historical USA winter wheat panel. Theor Appl Genet. 2019;132:1247–61. https://doi.org/10.1007/s00122-019-03276-6.

Schaeffer LR. Strategy for applying genome-wide selection in dairy cattle. J Anim Breed Genet. 2006;123:218–23.

Scheben A, Edwards D. Genome editors take on crops. Science. 2017;355:1122–3. https://doi.org/10.1126/science.aal4680.

Segura V, Vilhjálmsson BJ, Platt A, Korte A, Seren Ü, Long Q, et al. An efficient multi-locus mixed-model approach for genome-wide association studies in structured populations. Nat Genet. 2012;44:825–30. https://doi.org/10.1038/ng.2314.

Silva LDCE, Wang S, Zeng Z-B. Composite interval mapping and multiple interval mapping: procedures and guidelines for using windows QTL cartographer. In: Rifkin S, editor. Quantitative trait loci (QTL). Methods in molecular biology (methods and protocols), vol. 871. New York: Humana Press; 2012. p. 75–119.

Smola AJ, Schölkopf B. A tutorial on support vector regression. Stat Comput. 2004;14:199–222.

Solberg TR, Sonesson AK, Woolliams JA, Meuwissen THE. Genomic selection using different marker types and densities. J Anim Sci. 2008;86:2447–54.

Solberg TR, Sonesson AK, Woolliams JA, Meuwissen TH. Reducing dimensionality for prediction of genome-wide breeding values. Genet Sel Evol. 2009;41:29. https://doi.org/10.1186/1297-9686-41-29.

Sonesson AK, Meuwissen TH. Non-random mating for selection with restricted rates of inbreeding and overlapping generations. Genet Sel Evol. 2002;34:23. https://doi.org/10.1186/1297-9686-34-1-23.

Spindel JE, Begum H, Akdemir D, Collard B, Redoña E, Jannink J-L, et al. Genome-wide prediction models that incorporate de novo GWAS are a powerful new tool for tropical rice improvement. Heredity (Edinb). 2016;116:395–408. https://doi.org/10.1038/hdy.2015.113.

Stam P. Construction of integrated genetic linkage maps by means of a new computer package: join map. Plant J. 1993;3:739–44. https://doi.org/10.1111/j.1365-313X.1993.00739.x.

Stelpflug SC, Sekhon RS, Vaillancourt B, Hirsch CN, Buell CR, de Leon N, et al. An expanded maize gene expression atlas based on RNA sequencing and its use to explore root development. Plant Genome. 2016;9. https://doi.org/10.3835/plantgenome2015.04.0025.

Stich B, Melchinger AB. An introduction to association mapping in plants. CAB Rev Perspect Agric Vet Sci Nutr Nat Resour. 2010;5:1–9. https://doi.org/10.1079/PAVSNNR20105039.

Stich B, Utz HF, Piepho H-P, Maurer HP, Melchinger AE. Optimum allocation of resources for QTL detection using a nested association mapping strategy in maize. Theor Appl Genet. 2010;120:553–61. https://doi.org/10.1007/s00122-009-1175-2.

Stram DO. Correcting for hidden population structure in single marker association testing and estimation. In: Design, analysis, and interpretation of genome-wide association scans. New York, NY: Springer; 2014. p. 135–81.

Stuber CW, Goodman MM, Moll RH. Improvement of yield and ear number resulting from selection at Allozyme loci in a maize population1. Crop Sci. 1982;22:737. https://doi.org/10.2135/cropsci1982.0011183X002200040010x.

Studer A, Zhao Q, Ross-Ibarra J, Doebley J. Identification of a functional transposon insertion in the maize domestication gene tb1. Nat Genet. 2011;43:1160–3. https://doi.org/10.1038/ng.942.

Su Z, Łabaj PP, Li S, Thierry-Mieg J, Thierry-Mieg D, Shi W, et al. A comprehensive assessment of RNA-seq accuracy, reproducibility and information content by the sequencing quality control consortium. Nat Biotechnol. 2014;32:903–14. https://doi.org/10.1038/nbt.2957.

Syvänen A-C. Toward genome-wide SNP genotyping. Nat Genet. 2005;37:S5–S10. https://doi.org/10.1038/ng1558.

Tamura Y, Hattori M, Yoshioka H, Yoshioka M, Takahashi A, Wu J, et al. Map-based cloning and characterization of a Brown Planthopper resistance gene BPH26 from *Oryza sativa* L. ssp. indica cultivar ADR52. Sci Rep. 2015;4:5872. https://doi.org/10.1038/srep05872.

Tanksley SD, Rick CM, Medina-Filho H. Use of naturally-occurring enzyme variation to detect and map genes controlling quantitative traits in an interspecific backcross of tomato. Heredity (Edinb). 1982;49:11–25. https://doi.org/10.1038/hdy.1982.61.

Tanksley SD, Young ND, Paterson AH, Bonierbale MW. RFLP mapping in plant breeding: new tools for an old science. Biotechnology. 1989;7:257–64. https://doi.org/10.1038/nbt0389-257.

Teo YY. Common statistical issues in genome-wide association studies: a review on power, data quality control, genotype calling and population structure. Curr Opin Lipidol. 2008;19:133–43. https://doi.org/10.1097/MOL.0b013e3282f5dd77.

The Gene Ontology Consortium. Gene ontology: tool for the unification of biology. Nat Genet. 2000;25:25–9. https://doi.org/10.1038/75556.Gene.

Thornsberry JM, Goodman MM, Doebley J, Kresovich S, Nielsen D, Buckler ES. Dwarf8 polymorphisms associate with variation in flowering time. Nat Genet. 2001;28:286–9. https://doi.org/10.1038/90135.

Tibshirani R. Regression shrinkage and selection via the Lasso. J R Stat Soc Ser B. 1996;58:267–88.

Tobias RD. An introduction to partial least squares regression. In: Proceedings of the 20th annual SAS users group international conference. 1995. p. 1250–7. https://doi.org/http://support.sas.com/techsup/technote/ts509.pdf.

Usai MG, Goodard ME, Hayes BJ. LASSO with cross-validation for genomic selection. Genet Res. 2009;91:427–36.

VanRaden PM, Van Tassell CP, Wiggans GR, Sonstegard TS, Schnabel RD, Taylor JF, et al. Invited review: reliability of genomic predictions for North American Holstein bulls. J Dairy Sci. 2009;92:16–24. https://doi.org/10.3168/jds.2008-1514.

Wan Q, Guan X, Yang N, Wu H, Pan M, Liu B, et al. Small interfering RNAs from bidirectional transcripts of *GhMML3_A12* regulate cotton fiber development. New Phytol. 2016;210:1298–310. https://doi.org/10.1111/nph.13860.

Wang Y, Cao L, Zhang Y, Cao C, Liu F, Huang F, et al. Map-based cloning and characterization of BPH29, a B3 domain-containing recessive gene conferring brown planthopper resistance in rice. J Exp Bot. 2015;66:6035–45. https://doi.org/10.1093/jxb/erv318.

Wang S-B, Feng J-Y, Ren W-L, Huang B, Zhou L, Wen Y-J, et al. Improving power and accuracy of genome-wide association studies via a multi-locus mixed linear model methodology. Sci Rep. 2016;6:19444. https://doi.org/10.1038/srep19444.

Whittaker JC, Thompson R, Denham MC. Marker-assisted selection using ridge regression. Genet Res. 2000;75:249–52.

Windhausen VS, Atlin GN, Hickey JM, Crossa J, Jannink J-L, Sorrells ME, et al. Effectiveness of genomic prediction of maize hybrid performance in different breeding populations and environments. G3 (Bethesda). 2012;2:1427–36.

Wold H. Partial least squares. In: Encyclopedia of statistical sciences. Hoboken, NJ: Wiley; 2006.
Xia L, Zou D, Sang J, Xu X, Yin H, Li M, et al. Rice expression database (RED): an integrated RNA-Seq-derived gene expression database for rice. J Genet Genomics. 2017;44:235–41. https://doi.org/10.1016/j.jgg.2017.05.003.
Xiao Y, Tong H, Yang X, Xu S, Pan Q, Qiao F, et al. Genome-wide dissection of the maize ear genetic architecture using multiple populations. New Phytol. 2016;210:1095–106. https://doi.org/10.1111/nph.13814.
Xu S. Estimating polygenic effects using markers of the entire genome. Genetics. 2003;163:789–801.
Xu Y, Crouch JH. Marker-assisted selection in plant breeding: from publications to practice. Crop Sci. 2008;48:391. https://doi.org/10.2135/cropsci2007.04.0191.
Xu C, Wu K, Zhang J-G, Shen H, Deng H-W. Low-, high-coverage, and two-stage DNA sequencing in the design of the genetic association study. Genet Epidemiol. 2017;41:187–97. https://doi.org/10.1002/gepi.22015.
Yabe S, Hara T, Ueno M, Enoki H, Kimura T, Nishimura S, et al. Potential of genomic selection in mass selection breeding of an Allogamous crop: an empirical study to increase yield of common buckwheat. Front Plant Sci. 2018;9:276. https://doi.org/10.3389/fpls.2018.00276.
Yan J, Warburton M, Crouch J. Association mapping for enhancing maize (L.) genetic improvement. Crop Sci. 2011;51:433. https://doi.org/10.2135/cropsci2010.04.0233.
Yang Q, He Y, Kabahuma M, Chaya T, Kelly A, Borrego E, et al. A gene encoding maize caffeoyl-CoA O-methyltransferase confers quantitative resistance to multiple pathogens. Nat Genet. 2017;49:1364–72. https://doi.org/10.1038/ng.3919.
Yin K, Gao C, Qiu J-L. Progress and prospects in plant genome editing. Nat Plants. 2017;3:17107. https://doi.org/10.1038/nplants.2017.107.
Yu J, Pressoir G, Briggs WH, Vroh Bi I, Yamasaki M, Doebley JF, et al. A unified mixed-model method for association mapping that accounts for multiple levels of relatedness. Nat Genet. 2006;38:203–8. https://doi.org/10.1038/ng1702.
Zeng Z-B. Precision mapping of quantitative trait loci. Genetics. 1994;136:1457–68.
Zhang Z, Ersoz E, Lai C-Q, Todhunter RJ, Tiwari HK, Gore MA, et al. Mixed linear model approach adapted for genome-wide association studies. Nat Genet. 2010;42:355–60. https://doi.org/10.1038/ng.546.
Zhang X, Pérez-Rodríguez P, Burgueño J, Olsen M, Buckler E, Atlin G, et al. Rapid cycling genomic selection in a multiparental tropical maize population. G3 (Bethesda). 2017;7:2315–26. https://doi.org/10.1534/g3.117.043141.
Zhao J, Chen Z. A two-stage penalized logistic regression approach to case-control genome-wide association studies. J Probab Stat. 2012;2012:1–15. https://doi.org/10.1155/2012/642403.
Zhong S, Dekkers JCM, Fernando RL, Jannink J-L. Factors affecting accuracy from genomic selection in populations derived from multiple inbred lines: a barley case study. Genetics. 2009;182:355–64. https://doi.org/10.1534/genetics.108.098277.
Zhou X, Stephens M. Genome-wide efficient mixed-model analysis for association studies. Nat Genet. 2012;44:821–4. https://doi.org/10.1038/ng.2310.
Zhu C, Bortesi L, Baysal C, Twyman RM, Fischer R, Capell T, et al. Characteristics of genome editing mutations in cereal crops. Trends Plant Sci. 2017a;22:38–52. https://doi.org/10.1016/j.tplants.2016.08.009.
Zhu J, Chen J, Gao F, Xu C, Wu H, Chen K, et al. Rapid mapping and cloning of the virescent-1 gene in cotton by bulked segregant analysis-next generation sequencing and virus-induced gene silencing strategies. J Exp Bot. 2017b;68:4125–35. https://doi.org/10.1093/jxb/erx240.
Zila CT, Ogut F, Romay MC, Gardner CA, Buckler ES, Holland JB. Genome-wide association study of Fusarium ear rot disease in the U.S.A. maize inbred line collection. BMC Plant Biol. 2014;14:372. https://doi.org/10.1186/s12870-014-0372-6.
Zuo W, Chao Q, Zhang N, Ye J, Tan G, Li B, et al. A maize wall-associated kinase confers quantitative resistance to head smut. Nat Genet. 2014;47:151–7. https://doi.org/10.1038/ng.3170.

Population Genomics Along With Quantitative Genetics Provides a More Efficient Valorization of Crop Plant Genetic Diversity in Breeding and Pre-breeding Programs

Peter Civan, Renaud Rincent, Alice Danguy-Des-Deserts, Jean-Michel Elsen, and Sophie Bouchet

Abstract The breeding efforts of the twentieth century contributed to large increases in yield but selection may have increased vulnerability to environmental perturbations. In that context, there is a growing demand for methodology to re-introduce useful variation into cultivated germplasm. Such efforts can focus on the introduction of specific traits monitored through diagnostic molecular markers identified by QTL/association mapping or selection signature screening. A combined approach is to increase the global diversity of a crop without targeting any particular trait.

A considerable portion of the genetic diversity is conserved in genebanks. However, benefits of genetic resources (GRs) in terms of favorable alleles have to be weighed against unfavorable traits being introduced along. In order to facilitate utilization of GR, core collections are being identified and progressively characterized at the phenotypic and genomic levels. High-throughput genotyping and sequencing technologies allow to build prediction models that can estimate the

P. Civan
GDEC, INRAE, Université Clermont Auvergne, Clermont-Ferrand, France

R. Rincent
GQE-Le Moulon, INRAE, Univ. Paris-Sud, CNRS, AgroParisTech, Université Paris-Saclay, Clermont-Ferrand, France

A. Danguy-Des-Deserts
GDEC, Université Clermont Auvergne, Clermont-Ferrand, France

GenPhySE (Génétique Physiologie et Systèmes d'Elevage), Université de Toulouse, INRAE, ENVT, Castanet-Tolosan, France

J.-M. Elsen
GenPhySE (Génétique Physiologie et Systèmes d'Elevage), Université de Toulouse, INRAE, ENVT, Castanet-Tolosan, France

S. Bouchet (✉)
INRAE, Université Clermont Auvergne, Clermont-Ferrand, France
e-mail: sophie.bouchet@inrae.fr

genetic value of an entire genotyped collection. In a pre-breeding program, predictions can accelerate recurrent selection using rapid cycles in greenhouses by skipping some phenotyping steps. In a breeding program, reduced phenotyping characterization allows to increase the number of tested parents and crosses (and global genetic variance) for a fixed budget. Finally, the whole cross design can be optimized using progeny variance predictions to maximize short-term genetic gain or long-term genetic gain by constraining a minimum level of diversity in the germplasm. There is also a potential to further increase the accuracy of genomic predictions by taking into account genotype by environment interactions, integrating additional layers of omics and environmental information.

Here, we aim to review some relevant concepts in population genomics together with recent advances in quantitative genetics in order to discuss how the combination of both disciplines can facilitate the use of genetic diversity in plant (pre) breeding programs.

Keywords Genomic predictions · Genotype by environment interaction · Long-term genetic gain · Mating design · Multi-trait · Plant genetic resources · Population genomics · (Pre-)breeding · QTLs · Selection signatures

1 Introduction

The challenge in agriculture today is to produce enough food for an increasing population, using less land, water, fertilizer, and pesticides to limit ecological impacts. Global environmental changes, rainfall variability, nitrogen cycle alteration, higher temperature, and atmospheric CO_2 concentration strongly impact crop plants phenology (Jagadish et al. 2016), resistance to pathogens/insects outbreaks (Deutsch et al. 2018), and yield (Brisson et al. 2010). As a consequence, genetic gain for stress tolerance has become one of the most important targets in plant breeding. In this context, genetic variants present in modern varieties, traditional local varieties (i.e., landraces), and wild relatives may be of interest for crop plant breeders (McCouch et al. 2013).

For about 10,000 years humans have been exerting selection pressure, both consciously (by selecting the "best" seeds or animals to contribute to the next generation) and involuntarily (through farming practices and expansion of the natural distribution range), gradually changing domesticated plants and animals to suit their needs. This piecemeal process of selection gradually morphed into breeding, and first commercially successful plant breeding emerged at the end of the nineteenth century. The "best" (that was selected) has been covering many different criteria (Allard 1999) in different species, times, countries, environments, and now depends on the end-users/markets targeted. Multi-trait indexes have been empirically or economically built from phenotypic observations and expert opinion. This is generally called phenotypic selection (PS). Rapid genetic gain has been secured by breeders by selecting only the highest-performing parental individuals in order to ensure high mean performance of progeny. The classical strategy in crop plants is to cross performant lines for different traits, aiming to obtain some recombinant lines in the progeny that cumulate a maximum of chosen criteria. However, continuous application of truncation selection (selection of the bests) without regular re-introduction of new alleles in the germplasm leads to a rapid loss of genetic diversity around loci under selection by hitch-hiking effects and all along the genome by drift. This can have negative consequences for loci not monitored in the process and reduce the long-term potential of the program. Additionally, truncation selection overlooks favorable alleles that only occur in lines that are not highly performing. The most famous examples of the negative impact of reduced diversity in cultivated plants concern disease resistance. All cultivated potato (*Solanum tuberosum*) varieties cultivated in Ireland were susceptible to late blight, leading to the Great famine in 1845–1849 (Mizubuti and Fry 2006). Similarly, maize (*Zea mays*) varieties that all contained the common male sterile genetic background were susceptible to southern leaf blight that caused 15% losses in 1970–1971 in the USA (Ullstrup 1972).

Through times, cultivated plants have experienced various genetic bottlenecks through selection and drift that accompanied domestication, migrations, and subsequent local adaptation (Spillane and Gepts 2001). These bottlenecks explain the reduction of genetic diversity compared to wild relatives or local traditional varieties referred to as landraces. Only a few studies have focused on long-term changes in genetic diversity in breeding programs, for instance in maize, *Zea mays* (Labate et al. 1999; Feng et al. 2006; van Inghelandt et al. 2010; Gerke et al. 2015; Allier et al. 2019d) and soybean, *Glycine max* (Bruce et al. 2019). There is evidence that modern breeding further reduced genetic diversity (Simmonds 1962; Cooper et al. 2001; Fu 2006, 2015) and changed its geographical distribution because of large open breeding systems. Such impacts of modern breeding are dramatic in the case of bread wheat, *Triticum aestivum*. Although landrace diversity is composed of two major genetic groups, Europe and Asia, Asian alleles are almost absent in worldwide modern lines. Note however that some extrinsic (from related species) DNA segments were introgressed into elite lines by breeders creating neo-diversity in bread wheat, *Triticum aestivum* (Balfourier et al. 2019), maize, *Zea mays* (Hufford et al. 2012), barley, *Hordeum vulgare* (Brown and Clegg 1983), soybean, *Glycine max* (Doyle 1988; Hyten et al. 2006; Han et al. 2016; Sedivy et al. 2017) or peanut, *Arachis hypogaea* (Fonceka et al. 2012), to list a few examples. Fu (2006) showed that genome-wide reduction of crop genetic diversity was minor but allelic reduction at some major QTLs was important. Directional selection actually tends to fix favorable alleles at some QTLs and neighboring regions by linkage drag (Maynard-Smith and Haigh 1974). So, there is an urgent need for efficient methodologies to monitor local and global diversity in breeding programs in order to maintain short-term and long-term genetic gain. It has been actually shown that a large genetic base of elite germplasm not only at known QTLs of agronomic interest but also all along the genome would assure long-term genetic gain and increase resilience of crop plants to biotic and abiotic stresses in unpredictable environmental conditions (Malézieux et al. 2009).

The way breeders rank selection candidates have changed through times. Genome-wide molecular markers and derived tools can now guide their decisions to complement the phenotyping information. With exponential capacities of genotyping and sequencing, improvements in computing and data storage, methodological and statistical developments, genomic selection (GS, Meuwissen et al. 2001) is becoming an essential tool to not only improve accuracy of selection, accelerate genetic gain using rapid cycles, optimize resource allocation, but also to better manage/introduce genetic diversity in breeding programs by optimizing parental contribution and cross design. Recurrent selection schemes (Hallauer and Sears 1972) for population improvement can also be re-visited with the help of GS to cumulate a maximum of favorable alleles in pre-breeding lines that can be integrated in breeding programs. Moreover, any useful information about loci controlling the variation of traits of agronomic interest (allele effects, genomic annotation) or subjected to historical evolutionary constraints (selection by environment or human), about genitors (genetic group, passport, and environmental data), can be used individually or as a covariate in prediction models to optimize selection process or cross designs. Therefore, population genomics in combination with quantitative genetics can provide relevant tools to evaluate, manage, and introduce GR in (pre)-breeding programs.

In this chapter, we first discuss how population genomics helps to assess genetic diversity, identify genes under selection, select candidate genes, and manage genetic diversity in crop plants. Then, we review the methodologies developed in genomic prediction and quantitative genetics to manage long-term genetic gain and genetic diversity in breeding programs. Finally, we present some future perspectives to optimize diversity valorization.

2 Population Genomics of Crop Plants Genetic Resources

2.1 Genetic Diversity in Crop Fields and Genebanks

Only a few crop species are widely cultivated around the world. Four crop species (wheat, maize, rice, and soybean) cover half of all land harvested worldwide – (http://www.fao.org/faostat/en/#data/QC). Moreover, most of the widely-grown species are represented by very few varieties in the fields. Such limited genetic variation in elite germplasm increases vulnerability to market and environmental changes. Mitigation of this situation through introduction of genetic innovations relies on GR that are maintained in around 1700 genebanks worldwide. However, only a small proportion of available GR has been explored and used so far, and it is believed that their comprehensive genomic characterization could enhance their utilization in breeding.

It is estimated that 24% of allelic diversity has been lost in maize compared to teosinte (Vigouroux et al. 2005), 70% in wheat compared to wild emmer (Haudry et al. 2007), and 30% in yam (*Dioscorea alata*) compared to its wild relatives

(Akakpo et al. 2017). This reduction of genetic diversity commonly observed in most crops is attributed to domestication and selection. Domestication corresponds to subsampling of wild progenitor species and results in what is called "the domestication bottleneck" (Goodman 1999, 2005; Meyer and Purugganan 2013; Allaby et al. 2019). Some observations suggest that the domestication bottleneck was not a rapid process associated with the dawn of agriculture, but rather a gradual genetic diversity loss that occurred during millennia (Allaby et al. 2019). The recent selection associated with modern plant breeding had a comparatively smaller impact on genetic diversity (van Heerwaarden et al. 2012), which has been well documented in maize and wheat (Reif et al. 2005; Glémin and Bataillon 2009; Meyer and Purugganan 2013). However, even the fraction of genetic diversity that has been retained in modern crops may not be effectively utilized today (Tenaillon and Charcosset 2011; Balfourier et al. 2019).

While breeding schemes rarely include diverse Genetic Resources (GR), the importance of collecting and characterizing genetic resources is widely recognized (McCouch et al. 2012). Genebank collections are an invaluable reservoir of favorable alleles that are not present in the cultivated gene pool. Examples of traits that have been successfully introgressed from GR into elite cultivars and had significant impact on crop production are numerous. In wheat for instance, dwarfing genes (reduced height loci *Rht-B1* and *Rht-D1*) and genes conferring durable resistance against a wide spectrum of insects and diseases were introgressed by Norman Borlaug during the Green Revolution. The Sorghum Conversion Program in the USA introgressed dwarf and photoperiod-insensitive alleles into African sorghum landraces to adapt them to temperate environments (Klein et al. 2008). The Germplasm Enhancement of Maize project (GEM) (Goodman et al. 2000) enabled massive introgression of GR alleles into the elite germplasm. Introgression lines have been massively produced for peanut as well, using wild relatives (Foncéka et al. 2009). Apart from the genes that control phenology (dwarfing genes, photoperiod insensitivity), great achievements include major genes of disease resistance, such as a resistance gene against grassy stunt virus introgressed from wild rice *Oryza nivara* (Plucknett 1987), leaf rust resistance genes in bread wheat introgressed from *Aegilops* (Kuraparthy et al. 2007) or other relatives (Steffenson et al. 2007; Ellis et al. 2014), and other genes providing resistance to biotic and abiotic stresses (Huang et al. 2016). There are also a few examples proving that wild gene pools contain genetic variants that can improve quality and yield, e.g. in tomato, *Lycopersicum esculentum* (Gur et al. 2004), wheat, *Triticum aestivum* (Uauy et al. 2006), maize, *Zea mays* (Ribaut and Ragot 2007) and rice, *Oryza sativa* (Imai et al. 2013).

Since comprehensive phenotyping and genomic characterization of all the GR is beyond the capacities of genebanks or other interested parties, "core collections" are often identified with the objective to represent most of the genetic diversity according to available information (passport and/or genotypes). These core collections are being intensively phenotyped on national levels (e.g., French initiatives Breedwheat https://breedwheat.fr and Amaizing https://amaizing.fr), or within international initiatives, such as the Seeds of Discovery platform (https://seedsofdiscovery.org) for wheat and maize.

2.2 Detection of Selection Signatures

From the breeders' perspective, the value of genetic resources is given by the presence of agronomically interesting phenotypes that can be introgressed into elite germplasm. However, given the large number of genebank accessions multiplied by the number of potentially useful traits, phenotypic information is rarely available for genetic resources. Moreover, genetic determinants of many important traits are still poorly characterized. Among these knowledge gaps, genomicists explore crop genomes with the "bottom-up" approach (Ross-Ibarra et al. 2007) aiming to identify gene variants beneficial for crop production without phenotyping. This approach assumes that positive selection is the central force in the process of domestication and adaptation, and it is therefore possible to identify domestication and adaptive genes by screening signatures of selection along the genome. The general methodology is based on comparing genomically local diversity measures in the target population to some reference values, which can be modeled under the assumption of neutrality, or estimated from a genome-wide average in the population or orthologous regions of distinct populations. Although results are usually evaluated under some statistical framework to distinguish effects of selection and other evolution forces, specificity and sensitivity of these tests remain problematic: However, identification of genomic regions with limited genetic variability has double utility in the context of breeding. On the one hand, it helps to discover genes responsible for domestication and adaptation traits, and on the other hand, it points out to loci where re-introduction of lost diversity can boost resilience to environmental challenges.

Strong positive selection on a genetic variant with low initial population frequency results in a "selective sweep" (Maynard-Smith and Haigh 1974), the fixation of one haplotype around the selected allele. The following sections contain a brief description of the major genomic signatures of selection, together with a non-exhaustive list of available software tools. It needs to be emphasized that all these tools suffer to various extent from imperfect power (the ability to find real selection signals), specificity (the ability to filter out false positives), and resolution (the ability to identify the causal loci within long sweeps), as observed in association studies. Our ability to detect signatures of positive selection in a sample of genomes depends on the time elapsed since the selection episode, its strength and duration, the mutation and the recombination rates that break up haplotypes, as well as demographic events (intensity of bottlenecks, migration, differentiation, expansion...), which can actually create diversity patterns that resemble selection signatures. The resolution of the methods mostly depends on the extent of LD.

These factors need to be considered when interpreting genome-wide signatures of selection, and robust statistical thresholds are important to identify outlier loci, either with respect to the rest of the genome or to another real or simulated population that was not under selection. In practice, multiple statistics need to be collected, and the more tests converge on the same result, the higher is the confidence that the identified locus is truly under selection. As in association studies, the identified loci need to be treated as "candidates" until phenotypes are established and the role of the genes is confirmed experimentally to avoid false positives (Pavlidis and Alachiotis 2017).

2.2.1 Decrease of Genetic Diversity

The most prominent signature of positive selection is the decrease of genetic diversity. As the frequency of the selected variant increases in the population, linked variation diminishes due to the genetic hitch-hiking effect (Maynard-Smith and Haigh 1974). This decrease of variation is easily detectable by comparing the nucleotide diversity (*Pi; He*) of the studied population (e.g., a population from a specific environment, or a crop as a whole) to a reference population (a population from a different environment, or a wild progenitor). A major difficulty is to distinguish selection-related decrease of diversity from stochastic variation resulting from demographic processes. In practice, several-fold decrease of genetic diversity with respect to the reference population is regarded as a sign of selection. Outlier loci are identified based on a distribution of the values across the genome.

The decrease of nucleotide diversity in the vicinity of a selected variant is mainly due to a change of allelic frequencies on linked sites, rather than to a decrease of the total number of polymorphic sites. This shift in the Site Frequency Spectrum (SFS) toward high- and low-frequency derived variants is another signature of selection (Braverman et al. 1995) and is attributed to the fact that neutral variants that are initially linked with the beneficial allele increase in frequency, while newly-emerging neutral variants hitchhike with the selected allele, and therefore remain in the population. This shift in the SFS can be measured by a summary statistics Tajima's D (1989), where the average number of nucleotide differences between pairs of sequences (*Pi; π*) is compared to the total number of polymorphic sites scaled by the sample size (Watterson's Theta; θ_W). The lack of medium-frequency variants in the vicinity of the selected allele causes a decrease in π while the total number of polymorphic sites may remain unaffected, and this pattern is reflected by negative values of Tajima's D.

The statistical basis for the identification of the SFS shifts as signatures of selection was improved by the introduction of a Composite Likelihood Ratio (CLR) test (Kim and Stephan 2002). The CLR test compares the probability of the observed polymorphism data emerging under a standard neutral model with the probability of the data emerging under a selective sweep model. Nielsen (2005) introduced SweepFinder, a modification of the CLR test where the standard neutral model is replaced with an empirical SFS of the entire data set, which increases the robustness of the test under different demographic scenarios (e.g., mild bottlenecks). SweeD (Pavlidis et al. 2013) is another implementation of the CLR test that is numerically more stable and faster when analyzing large numbers of genomes.

2.2.2 Increase of LD

Local Increase of LD

A variety of tools that detect signatures of selection rely on the observation that haplotypes (stretches of DNA sequence uninterrupted by recombination) of recently selected genes extend much further than expected under neutrality. Extended Haplotype Homozygosity (EHH) (Sabeti et al. 2002), Integrated Haplotype Score (iHS) (Voight et al. 2006), Cross-population Extended Haplotype Homozygosity (XPEHH) which measures the reduction in haplotype diversity in cross-population comparisons (Sabeti et al. 2007), and nSL (Ferrer-Admetlla et al. 2014) are all based on the model of a hard selective sweep, where a de novo adaptive mutation arises on a haplotype that quickly sweeps toward fixation, reducing genetic diversity around the locus. If selection is strong enough, this occurs faster than recombination or mutation can act to break up the haplotype, and thus a signal of high haplotype homozygosity can be observed extending from an adaptive locus. These statistics, nSL in particular, retain some power to detect soft sweeps as well. They are implemented in Selscan that has been optimized for large datasets (Szpiech and Hernandez 2014).

Apart from the extended haplotypes, positive selection creates another specific pattern of LD. As the frequency of a beneficial mutation increases, together with the frequency of linked neutral variants, recombinations sometimes occur on either side of the selected mutation. Since recombinations on the two sides are independent, and double recombinations are much less likely, pairs of variants on each side of the beneficial mutation show elevated LD, but pairs of variants compared across the beneficial mutation show lower LD. This pattern can be measured by the ω-statistics (Kim and Nielsen 2004) that has been implemented in OmegaPlus (Alachiotis et al. 2012). Since the ω-statistics can be assessed at each interval between two SNPs, this method has the potential, at least in theory, to identify the locus under selection very precisely. However, it should also be noted that ω is only applicable when haplotypes are known, either on phased data or inbreds (e.g., in self-pollinating species).

All three aforementioned signatures of selection – decrease in nucleotide diversity, shift in the SFS, and a specific LD pattern – can be assessed simultaneously by RAiSD, a tool introduced by Alachiotis and Pavlidis (2018). On modeled data, this composite evaluation test outperforms tools that measure those signatures of selection individually. However, unlike other methods, RAiSD assumes that polymorphisms are sampled evenly across the genome, and this assumption may be severely violated (e.g., in exome data).

Global Increase of LD

In breeding programs, the detection of inbreeding is also relevant. Runs of homozygosity (ROH) are lengths of contiguous homozygous segments due to transmission of identical haplotypes by parents in heterozygotes. It is the percentage of genome that is identical by descent. Individuals that have undergone recent inbreeding will exhibit long runs of homozygosity (MacLeod et al. 2009; Peripolli et al. 2017). ROH was adapted by Allier et al. (2019d) for inbreds and named ROHe (Runs of Expected Homozygosity).

Some of the causal factors behind the occurrence of ROH are population phenomena, such as genetic drift, population bottlenecks, inbreeding, and intensive artificial selection. Consequently, the identification and characterization of ROH can provide insights into how a population has evolved over time in the past, and additionally, into how a population has to be managed in the future in long-term breeding programs. Intense selection regimes in livestock populations have already alerted the scientific community about the need for strategies to preserve populations, characterize and monitor inbreeding, and manage the genetic diversity by optimizing genetic contributions and mating (See Sects. 3.3.7 and 3.3.8).

2.2.3 Extreme Differentiation

Local adaptation can also be indicated by extreme differentiation of allelic frequencies between genetic groups, especially in contrasted environments. As allelic differentiation can also result from demographic events, it is important to interpret outlier loci cautiously, especially when hierarchical population structure is detected. A relevant strategy to identify significant outliers relies on building a distribution of expected values of the tested statistics in the absence of selection, using neutral coalescent simulations (Bellucci et al. 2014).

A number of statistics are available (Cruickshank and Hahn 2014), with F_{ST} (single site divergence index) (Wright 1931) being the most common. Outlier differentiation methods rely on a hypothesis that under certain conditions (migration-drift equilibrium under a neutral island model with spatially uniform migration and gene flow), population differentiation of allele frequencies across a large number of loci can be used to infer the process of selection acting on a subset of loci (Lewontin and Krakauer 1973). Loci with F_{ST} values (or other genetic distance measure) significantly greater than the genome-wide distribution of the statistics are presumed to be under diversifying selection or linked to those under selection. FDIST(2) (Beaumont and Nichols 1996) implemented in LOSITAN (Antao et al. 2008) assumes a finite island model to generate null F_{ST} distribution and can deal with heterozygosity. ARLEQUIN (Excoffier and Lischer 2009) adds hierarchical genetic structure to the inference. BayeScan (Foll and Gaggiotti 2008) uses a Bayesian method to estimate the relative probability that each locus is under selection. FLK (Bonhomme et al. 2010) uses a modified version of the Lewontin and Krakauer (1973) test for selection by comparing allele frequencies of different populations in a neighbor-joining tree constructed using a matrix of Reynold's genetic distances (Reynolds et al. 1983). Its extension hapFLK (Fariello et al. 2013) calculates haplotype-based frequency differentiation index among hierarchically structured populations. It is robust with respect to bottlenecks and migration and can detect incomplete sweeps. OutFLANK (Whitlock and Lotterhos 2015) does not invoke any specific demographic model and uses a modified version of the Lewontin and Krakauer method to infer a null F_{st} distribution. X^TX (Günther and Coop 2013) implemented in Baypass or Bayenv2 (Coop et al. 2010, p. 2010), employs a Bayesian hierarchical model to test individual SNPs against a null model generated by the covariance in allele frequencies between populations from the entire set of SNPs.

PCAdapt (Luu et al. 2017) assumes that genes under selection are outliers with respect to the prevalent population structure. It calculates z-scores that measure the relatedness of each SNP to the first K principal components of genome-wide variation in a population. The computation uses Mahalanobis distance, which is robust in the presence of hierarchical population structure. Comparisons on simulated data revealed that the false discovery rate of PCAdapt is around 10%, similarly to HapFLK and OutFLANK (Whitlock and Lotterhos 2015), but much lower compared to Bayescan (40%), which is negatively impacted by admixture. PCAdapt and HapFLK are the most powerful tools in scenarios of population divergence and range expansion.

Differentiation among populations can also be detected by comparing site frequency spectra. Selection does not only shift the SFS in the vicinity of the beneficial mutation toward high- and low-frequency variants (see Sect. 2.2.1), but it also causes multilocus allele frequency spectra to differ between two populations. A CLR test can be used to assess whether such local genetic differentiation departs from the expectation under neutrality, as implemented in XP-CLR (Chen et al. 2010) (https://reich.hms.harvard.edu/software).

Additional methods for identification of loci involved in local adaptation exist, but may not be applicable on large data sets (Hoban et al. 2016).

2.2.4 Specific Cases of Genetic Differentiation

Domestication/Selection

Genome scans have been performed in order to detect signatures of selection in most major crops where whole or partial genome sequence data is available for at least 100 accessions. The scans usually aim to distinguish domestication signatures (obtained by comparing traditional landraces to wild progenitors) from genetic improvement signatures (obtained by comparing landraces and modern cultivars).

Hufford et al. (2012) detected 484 loci showing signatures of domestication and 695 loci showing signatures of genetic improvement (with 23% overlap) in maize, using differentiation indexes (F_{ST}, XP-CLR), diversity indexes (π, ρ, Tajima's D and normalized Fay and Wu's H), and haplotype lengths in each genetic group. These results suggest that some traits are of continuous agronomic importance since domestication and additional traits became of interest during the breeding era. In total, 6–11% of the identified loci had no annotation and could correspond to regulatory regions. Some identified candidates of domestication genes controlled flowering time, nitrogen metabolism, thousand kernel weight, phyllotaxy, and seed germination. Some genetic improvement candidates were involved in the biosynthesis of a plant growth hormone gibberellin, or in drought tolerance pathway. According to a gene expression survey, the greatest changes in expression (presence or absence of expression) were observed in the domestication genes, i.e. between wild and cultivated lineages. Expression of the candidate targets of selection in cultivated lines was more homogeneous, with subtle variations, perhaps indicating the importance of fine-tuned expression, as opposed to "on and off" variability. This observation suggests that while the domestication period mostly selected particular gene variants, selection during the improvement period acted predominantly on cis-acting regulatory elements. This information is of interest for private breeding programs (Allier et al. 2019d) that intend to monitor global and local losses and gains of diversity over time in their germplasm using genetic and genomic indicators.

In bread wheat, regions that lost diversity during domestication (Haudry et al. 2007), improvement (Reif et al. 2005; Cavanagh et al. 2013), or both (Pont et al. 2019) have also been investigated. By examining genetic differentiation (PCAdapt) and diversity patterns (Reduction Of Diversity π Index, Tajima's D), Pont et al. (2019) confirmed selection signatures for domestication genes conferring brittle rachis (*Brt*), tenacious glume (*Tg*), homoeologous pairing (*Ph*) and non-free-threshing character (*Q*); improvement genes controlling photoperiod sensitivity (*Ppd*), vernalization (*VRN*), reduced height (*Rht*), glutenins (*Glu*) and gliadins (*Gli*), frizzy panicle (*FSP*), grain number (*GNS*), waxy proteins (*Wx*), and plant architecture (*uniCULm*). Through F_{ST} screening among landraces and cultivars, Cavanagh et al. (2013) found introgression patterns surrounding phenology genes (*Rht-B1*: dwarfing, *Ppd-B1*: photoperiod insensitivity, *Vrn1*: flowering time) and the *Sr36* gene involved in resistance to stem rust.

In tetraploid wheat, Maccaferri et al. (2019) used genetic diversity differentiation indexes (F_{ST}, hapFLK, XP-CLR, and XP-EHH) between wild and domesticated emmer (*T. turgidum* ssp. *dicoccoides*, *T.t.* ssp. *dicoccum*, respectively), durum landraces, and cultivars. They confirmed selection on domestication genes (two brittle rachis regions, a glume QTL controlling threshability) and improvement genes, some of which are associated with disease resistance (e.g., *Sr13* and *Lr14*) and grain yellow pigment content loci (e.g., *Psy-B1*). They also identified *TdHMA3-B1* as the best candidate involved in phenotypic variation of Cd accumulation in the grain. The non-functional *TdHMA3-B1b* allele could be responsible for a reduction in root vacuolar sequestration of Cd and Zn. It was suggested that under Zn-limiting conditions, this allele increases the pool of Zn available for transport to the shoot, thereby sustaining shoot growth.

Genetic diversity and differentiation indexes (Tajima's D and F_{st}, respectively) screened on wild and domesticated yam aided the detection of root development (SCARECRFOW-LIKE gene), starch biosynthesis (*Sucrose Synthase 4 and Sucrose Phosphate Synthase 1*), and photosynthesis related genes (Akakpo et al. 2017) that likely facilitated habitat change during domestication (from cultivation under trees to open field cultivation).

Adaptive Introgression

Screening for past introgressions, i.e. DNA fragments that were absent in the cultivated gene pool until some point in time and appeared in recent material from crosses with related species is another way to identify candidates of agronomic interest (Hufford et al. 2013; Racimo et al. 2015; Schaefer et al. 2016; Rochus et al. 2018). From a basic point of view, detecting gene flow or gene introgression from a distinct population or a different species can help understanding adaptation to various environments and evolution (Anderson 1953; Rieseberg and Wendel 1993). When it increases fitness, it is referred to as "adaptive introgression." It can also help reconstructing speciation processes.

In principle, genetic introgression can be implied when genealogy of a locus in a population or species (i.e., "local ancestry") does not match the "global ancestry" estimated based on genome-wide variation. This approach has been used, for example, to reveal the portion of loci introgressed from *japonica* rice into the *indica* cultivar 93–11 (Yang et al. 2011). Since it is impractical to quantify alternative gene tree topologies on a genome-wide scale using more than a few individuals (the number of possible rooted trees grows exponentially with the increasing number of tips), other methods are necessary to study introgressions on a population level.

Several approaches can be employed to detect past admixtures that concern the whole genome without resolving the ancestry of individual loci. They quantify fractions of genomes associated with distinct populations. These include multivariate analyses, such as PCA (Patterson et al. 2006; Jombart et al. 2009), or model-based clustering algorithms like STRUCTURE, NewHybrids, ADMIXTURE, and sNMF (Pritchard et al. 2000; Anderson and Thompson 2002; Alexander et al. 2009; Frichot et al. 2014).

Different algorithms have been proposed to detect individual introgressed fragments. A model-based inference implemented in HAPMIX (Price et al. 2009) uses phased data (i.e., known haplotypes) and known ancestral populations (assuming only two contributing populations). The central idea of the method is to view haplotypes of each admixed individual as being sampled from the reference populations. At each position in the genome, HAPMIX estimates the likelihood that a haplotype from an admixed individual originated from one reference population or the other. A Hidden Markov Model (HMM) is used to combine these likelihoods with information from neighboring loci, to provide a probabilistic estimate of ancestry at each locus (Fig. 1).

HAPMIX, LAMP-LD, and RFMix packages for local ancestry inference were developed to provide accurate results on human data and recent admixture events but may be difficult to parameterize for crop species. A recently published open-source software Loter does not require any biological parameters and can be applied to a wide range of species (Dias-Alves et al. 2018). Performance testing on simulated datasets revealed that HAPMIX is severely impacted by imperfect haplotype reconstruction, and Loter is the least impacted by increased time since admixture. Loter was used to infer local ancestries in aromatic rices that originated millennia ago through an admixture event between *japonica* rice and Indian *aus*-like rice (Civáň

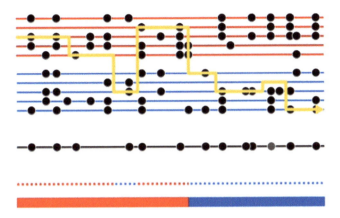

Fig. 1 Schematic representation of the Markov model used for ancestry inference. The black lower line in this figure represents a chromosomal segment from an admixed individual, carrying a number of typed mutations (black circles). The underlying ancestry is shown in the bottom color bar and reveals an ancestry change from the first population (red) to the second population (blue). The admixed chromosome is modeled as a mosaic of segments of DNA from two sets of individuals drawn from different reference populations (red and blue horizontal lines, respectively) closely related to the donors in the admixture event. The yellow line shows how the admixed chromosome is reconstructed with respect to this mosaic. The dotted line above the bottom color bar shows the reference population being copied from along the chromosome – note that at most positions, this is identical to the true underlying ancestry, but with occasional "miscopying" from the other population (blue dotted segment occurring within red ancestry segment). Reproduced from Price et al. (2009)

et al. 2019). However, the authors noted that the local ancestry inference was affected by sample size and missing data in simulations.

Another methodological approach mostly used to provide global estimates of gene flow is based on quantification of shared derived variation among non-sister taxa or populations (Kulathinal et al. 2009; Patterson et al. 2012; Peter 2016). In the absence of gene flow, a correct and rooted four-taxon tree will have the two most-recently diverged taxa sharing statistically equal amounts of derived variants with their non-sister taxon. Deviations from this expectation indicate gene flow. Popular implementations of this concept (often called ABBA-BABA) include the D-statistics (Green et al. 2010; Durand et al. 2011) and f-statistics (Reich et al. 2009, 2012). However, ABBA-BABA is based on a neutral evolution model, and its robustness in the presence of selection has not been tested. Since selection can increase local similarity among non-sister groups similarly to introgression, Čiváň and Brown (2018) argue that only variants demonstrably absent in their ancestral population should be counted toward the introgression signal.

Numerous cases of spontaneous introgression from wild relatives to cultivated species have been described (Hajjar and Hodgkin 2007; Guarino and Lobell 2011; Dempewolf et al. 2017; Burgarella et al. 2019). In maize, adaptive *mexicana* alleles were incorporated into the cultivated gene pool during the expansion of maize agriculture to the highlands of central Mexico (Matsuoka et al. 2002). Some of these introgressed alleles have been functionally validated and shown to provide adaptations to altitude, biotic and abiotic stresses (Hufford et al. 2013; Fustier et al. 2019). In potato (*Solanum tuberosum*), the origin of tuberization under long days was traced to an introgression event from *Solanum microdontum* (Hardigan et al. 2017). Introgression from *Populus balsamifera* (balsam poplar) in *P. trichocarpa* (black cottonwood) was detected through Tajima's D, F_{ST} and LD scans of local admixture, and complemented by analyses of gene expression (Suarez-Gonzalez et al. 2016). The team identified the *Populus PSEUDORESPONSE REGULATOR5* (*PRR5*) to be a strong candidate improving biomass, as well as cold, drought, and salinity tolerance. This gene was shown to work as a transcriptional regulator important for the circadian clock mechanism in Arabidopsis (Nakamichi et al. 2010, 2016, 2020). In poplar, it is upregulated at the onset of short days and it may play a crucial role in the timing of the onset of bud dormancy (Ruttink et al. 2007; Ko et al. 2011). A second candidate gene identified by Suarez-Gonzalez et al. (2016) is *COMT1* (*CAFFEIC ACID 3-O-METHYLTRANSFERASE 1*) that could be involved in lignification and/or pathogen defense (Barakat et al. 2011). In recent polyploids, such as bread wheat, authors consider PAV (presence-absence variation) or CNVs (copy number variation) identified from re-sequencing data as signals of putative introgressions (Balfourier et al. 2019; Cheng et al. 2019). Cheng et al. (2019) measured F_{ST} and π ratio between wild and cultivated lines and found 79 segments supposedly introgressed from wild relatives, co-localizing with 124 QTLs (grain yield, disease resistance, plant height).

The case of aromatic rice offers an example of how disentangling local ancestry and introgression history could aid the breeding process. It has been revealed that Basmati-like aromatic varieties of rice (Glaszmann 1987) originated from hybridizations between cultivated *japonica* rice (29–47% of the genome) introduced to the Indian subcontinent millennia ago, and local wild lineages of the Himalayan foothills similar to the present-day *aus* varieties (Civáň et al. 2019). They possess some characteristics highly valued by consumers (fragrance, texture of cooked rice, grain elongation after cooking). Rice stickiness and texture after cooking is mainly determined by starch synthesis pathways, and particularly, the ratio of amylose and amylopectin. While *japonica* rice is often sticky (or glutinous) due to low amylose content, *aus* and *indica* varieties are generally nonglutinous. Aromatic varieties have intermediate amylose content and medium gel consistency and Civáň et al. (2019) showed that they have mixed ancestry at the two genes, *Waxy* (Olsen et al. 2006) and *ALK* (Gao et al. 2011), responsible for these characteristics. Many aromatic landraces produce grain of superior quality in terms of fragrance and cooking properties, but are tall, lodging susceptible and low-yielding. Unfortunately, breeding efforts to cross aromatic landraces with high-yielding elite cultivars or introduce dwarfing genes were met with limited success. This is mainly attributed to cross incompatibility between aromatic and *indica* varieties, and high sterility in crosses (Singh et al. 2000). Local ancestry inference (Civáň et al. 2019) revealed that most, but not all aromatic varieties carry a *japonica*-derived variant of the *S5* gene responsible for *japonica-indica* hybrid sterility (Chen et al. 2008). Identification of high-quality aromatic landraces carrying non-*japonica* variants of *S5* could therefore be the first step in successful breeding of elite aromatic cultivars.

Environmental Differentiation: Landscape Genomics

Landscape genomics investigates associations of genetic variants with environmental variables, such as temperature, precipitation, altitude, and latitude gradients (Balkenhol et al. 2019). The goal is to identify candidate genes under selection, possibly indicating local adaptation, using outlier differentiation methods (see Sect. 2.2.3) and genetic-environment association (GEA) tests. Landscape genomics should not be confused with landscape genetics (Manel et al. 2003), which focuses on the effects of landscape variables on gene flow and population structure.

GEAs require environmental data such as WorldClim (http://www.worldclim.org, 2015). Bayenv2 tests for large allele frequency differences across environmental gradients by comparing observed allele frequency differences to transformed normal distribution of underlying population allele frequencies. Latent factor mixed models (LFMM) (Frichot et al. 2013) include population structure as latent (or hidden) variables to limit false positive signals. Spatial generalized linear mixed models (SGLMMs) (Guillot et al. 2014) are a computationally more efficient extension. Samβada (Stucki et al. 2017) is a multivariate analysis method that accounts for population structure with estimates of spatial autocorrelation in the data. When the trait of interest or the climatic gradient is correlated to the population structure, PCAdapt can also be used. Some other methods exist and are summarized in Rellstab et al. (2015).

Although most landscape genomics studies concern non-cultivated species, e.g. Arabidopsis (Hancock et al. 2011), associations with climatic data have also been investigated in forest trees (e.g., Sork et al. 2016; Rajora et al. 2016; Collevatti et al. 2020). Landscape genomics has been studied extensively in poplar (Suarez-Gonzalez et al. 2018), and it has been shown that introgressed genomic regions are enriched for disease resistance genes (TIR and LRR domains gene ontology terms) (Suarez-Gonzalez et al. 2016). In common bean (botanical name), 26 loci with selection signatures were found (Rodriguez et al. 2016), some of them involved in responses to environmental stress, such as drought response, cold acclimation or chilling susceptibility, and adaptation to different conditions of light and temperature. Four of these loci were also found to be under selection during domestication.

For sorghum, Lasky et al. (2015) showed that genome-environment associations can predict adaptive traits. Bellis et al. (2020) looked at correlation between allelic frequencies and Striga pressure. They demonstrated that local adaptation to this parasitic plant was partially controlled by the *LOW GERMINATION STIMULANT 1* (*LGS1*) gene. Wang et al. (2020) found some loci that may control seed mass adaptation to precipitation gradients.

In wild pearl millet (*Pennisetum glaucum*), Berthouly-Salazar et al. (2016) focused on climate gradients in Mali and Niger and collected genotype data from 11 populations, together with RNAseq data from a subset of four populations.

Looking at the genetic diversity patterns within populations (Tajima's D), differentiation among populations (F_{ST}, Bayescan), and correlation with environmental variables (centered loadings outliers using a PCA approach for each gradient), they found contigs displaying consistent signatures of selection among populations. Two of these contigs were associated with abiotic and biotic stress responses.

Time series data (that track samples over time) can be very informative for detecting genetic regions under selection. The factors involved in selection may be unknown.

Variety names of pearl millet, their phenotypes and climate data for a period from 1976 to 2003 were collected from a region of Niger that suffered from recurrent drought (Vigouroux et al. 2011). The research showed that an allele of the PHYC locus responsible for earliness increased in frequency over time at a rate that exceeds possible effect of genetic drift and sampling. A correlation between phenology and rainfall suggested that selection of this gene had a direct effect on earliness under shorter rainy seasons. It is noteworthy that this short-term adaptation was not due to introduction of new varieties, but due to within-variety selection, highlighting the importance to conserve within-variety diversity in genebanks.

Time series of a private breeding program (Allier et al. 2019d) can also be of interest to investigate genomic regions under selection or drift. Note that underlying population genetic structure and demographic history, when not properly accounted for, can generate many false positives. For instance, serial population bottlenecks occurring during founder effects of small populations migrating to new areas can result in fixed allelic differences due to genetic drift (Excoffier and Lischer 2009). Also, recent population range expansions from refugia can generate correlations between allele frequencies and environmental variables that are not due to selection.

Genome scan analyses are biased to detect loci with large effects, because power to detect small-effect loci is generally low. Since most phenotypic traits are likely to be polygenic, and thus governed by many loci of small effect, genome scans probably miss most of the loci involved in local adaptation (Stephan 2016; Rajora et al. 2016). Recently, Bayesian and other multilocus approaches have been developed (Rajora et al. 2016; Gompert et al. 2017). Some nonlinear functions have also been proposed to model the importance of environmental variables in explaining turnover of allele frequencies (Fitzpatrick and Keller 2015).

In reality, very few candidates for agronomically important genes revealed through genomic scans have been experimentally validated. Such validation usually requires validation by phenotyping in a controlled experimentation, an association study demonstrating a link between the genotypes and phenotypes (Saïdou et al. 2014) on a large panel of individuals, and a transformation experiment proving the function of the identified gene. All these experiments are costly, laborious, and technically difficult, but essential to convince breeders to monitor those genes in their germplasm. When validated, beneficial alleles can be introgressed into elite germplasm through backcrossing using diagnostic markers (Dempewolf et al. 2017), flanking markers, and sometimes genome-wide markers to minimize linkage drag and introduction of undesirable traits.

3 Population Genomics and Quantitative Genetics Assisted Infusion of Genetic Diversity in Breeding and Pre-breeding Programs

In elite germplasm, genetic diversity is generally limited compared to ancestral diversity. In that context, genebanks are a reservoir of underexploited favorable alleles. In case of one single favorable allele to introgress from genebanks to elites, flanking molecular markers can accelerate the process. But it takes a long time to validate QTLs and allele effects in different genetic backgrounds and design diagnostic markers to monitor the favorable allele in breeding programs. For example, it took 50 years (1960–2010) from the discovery of submergence-tolerant rice landraces to the successful release of submergence-tolerant rice varieties. It necessitated fine mapping and molecular characterization of the *SUBMERGENCE 1* (*SUB1*) locus and an introgression process (Bailey-Serres et al. 2010). This explains why only little use has been made of GR (Goodman 1999; Glaszmann et al. 2010; Wang et al. 2017).

The first reason why introgression process is so long is that elite lines contain much more favorable alleles than GR in general. It takes several generations of recurrent backcrosses with elites and selection to fill up this performance gap between GR and elites. The challenge is to break only unfavorable allelic associations while not breaking the favorable ones when crossing elites and GR. We may find co-adapted alleles in a cluster of genes (tall plant and late flowering alleles, high yield potential, and low protein content for instance) that have been selected by natural selection, creating local epistasy. No recombinants exist even in experimental populations because genes are too close. The recombination that would be desirable for agronomic purposes (small plant and late flowering, high yield potential and high protein content) may be difficult to obtain for ecophysiological or mechanical (low recombining regions, no diversity) reasons (Mayr 1954). The major unfavorable alleles of plant GR to eliminate are involved in phenology and local adaptation (e.g., flowering time, photoperiod sensitivity, height...) because they may not suit the targeted environment.

Genomic predictions (see Sect. 3.3) could help diversity infusion. Predicting GR genetic values using models that are trained on GR core collections is feasible when core collections are phenotyped in (and adapted to) targeted environments. But predicting which elite by GR crosses are of interest remains a statistical challenge because marker effects may depend on genetic backgrounds (Rio et al. 2019). We may need to first produce and evaluate recombinant lines between different genetic groups we want to cross to get an accurate marker effect estimation (GR alleles in elite genetic background in our case).

3.1 Production and Evaluation of Elite × GR Recombinants

Multi-parental crosses between elites and GR may be a good option to combine QTL detection for multiple-trait, identification of favorable GR alleles, selection of pre-breeding lines that could be introduced in breeding programs and training prediction models. Multi-parental Advanced Backcross (AB-QTL) populations (Narasimhamoorthy et al. 2006), pyramidal Multiparent Advanced Generation Inter-Cross (MAGIC) populations (Cavanagh et al. 2008; Leung et al. 2015), Nested Association Mapping (NAM) populations connected by one common parent for US maize (Buckler et al. 2009), European maize (Bauer et al. 2013), US sorghum (Bouchet et al. 2017), Back-Cross-NAM for sorghum (Jordan et al. 2011) have been developed for that purpose. It has been shown that the connection between populations by one or several common parents improves power of QTL detection. According to simulations, Stich (2009) proposed the triple round robin design connected by donors as the most efficient multi-parental design to maximize power of QTL detection as well as maximize genetic gain. But the production of this type of population remains long and laborious. The optimal connection design is not straightforward to predict from a statistical point of view. The choice of parents is often based on empirical information from different sources, the recipient parents being chosen for performance and GR for specific criteria that breeders want to improve, such as disease resistance for instance.

3.2 Maker Assisted Selection

Marker Assisted Selection (MAS) is promising to accelerate and optimize introgression process (Charmet et al. 1999; Servin et al. 2004). It has been successful for the introgression of maize earliness (Simmonds 1979; Smith and Beavis 1996), flowering time and yield under drought (Ribaut and Ragot 2007) as well as many disease resistance genes (Sanz-Alferez et al. 1995; Thabuis et al. 2004). But it becomes very demanding when multiple genes need to be pyramided simultaneously. Very large population sizes are actually necessary to get a reasonable certainty that an individual with the target genotype can be identified. Gene pyramiding strategies using marker-assisted introgression have been proposed (Hospital and Charcosset 1997; Servin et al. 2004; Canzar and El-Kebir 2011; Xu et al. 2011; Beukelaer et al. 2015). If all genes cannot be fixed in a single step of selection, it is necessary to cross again selected individuals with individuals having the favorable alleles that are missing using a marker-based recurrent selection (Charmet et al. 1999; Bernardo and Charcosset 2006). To cumulate more loci in a single genotype, Hospital et al. (2000) proposed a marker-based recurrent selection (MBRS) method using a QTL complementation strategy in a randomly mating population, which is feasible only in open-pollinated species. More recently, Valente et al. (2013) developed the software Optimas and Han et al. (2017) proposed the

Predicted Cross Value (PCV) algorithm to select at each generation crosses that maximize the likelihood of pyramiding desirable alleles in their progeny. A forward variable selection model can be used to select QTLs that explain significant genetic variance (Jansen 1993; Segura et al. 2012) instead of using arbitrary statistical thresholds. Note that Hospital and Charcosset (1997) advised that all QTLs should be given the same weight in the cross molecular score estimation to avoid rapid fixation of main QTLs and loss of small-effect alleles in the process. Control of genetic background with a few molecular markers was proposed in plants by Hospital and Charcosset (1997). With the same idea, the Genotype-Assisted Selection (GAS) concept was introduced by Meuwissen and Sonesson (2004) in animals to control polygenic background genes while selecting favorable alleles at QTLs. They proposed a multi-generation optimization of optimum contribution selection (GAOC: Genotype-Assisted Optimum Contribution) (see Sects. 3.3.7 and 3.3.8) while increasing the frequency of the positive QTL allele to increase genetic gain.

For complex traits that are controlled by a large number of genes, such as yield, MAS is often associated with substantial linkage drag, i.e., introduction of linked unfavorable alleles along with the target favorable allele (Peng et al. 2014) and often was a failure (Simmonds 1993; Hospital and Charcosset 1997; Ribaut and Ragot 2007). An approach using Genomic Selection (GS) addressing this problem in introgression schemes has been proposed (Ødegaard et al. 2009), who demonstrated that backcrosses assisted by genomic selection in fish is the best strategy compared to synthetic production or phenotypic selection to simultaneously select for elite productivity and donor disease resistance for instance. In wheat, Heffner et al. (2010, 2011a, b) came to a similar conclusion by comparing a breeding scheme including MAS with 20 QTLs or MAS followed by GS. Heffner et al. (2010) actually showed that expected annual genetic gain from GS exceeded that of Marker Assisted Recurrent Selection (MARS) for complex traits by about threefold for maize and twofold for winter wheat using analytical simulations of rapid cycles by skipping some phenotyping steps (Bernardo 2009), in a pre-breeding process in particular.

3.3 Genomic Predictions

First predictions in animals, human, and plants were based on pedigrees (Henderson 1975; Falconer et al. 1996; Bijma and Woolliams 1999). Then Lande and Thompson (1990) proposed to estimate the molecular score of an individual by adding its allelic effects at QTLs involved in trait variation. It was later shown that allele effects were overestimated in QTL detection (Beavis et al. 1994; Beavis 1998) and that the significance threshold to select the list of QTLs could be questionable. As the infinitesimal model (Fisher 1918) considering that traits are controlled by many loci of small effects was the best model to explain the variation of many complex traits such as yield, Whittaker et al. (2000) and Meuwissen et al. (2001) proposed to use all available independent markers (hundreds to millions) to build a prediction model that estimates the genetic value of unphenotyped candidates based on a related training population that is phenotyped and genotyped. Considering that genotyping is dense enough, each QTL should be in linkage disequilibrium with at least one marker. The model is thus able to capture more genetic variance than including significant associations only. The principle is to regress phenotypic values on all markers considered as random effects using a linear model. The critical difference with the Lande and Thompson (1990) approach is that we do not set a significance threshold for the loci selected for trait prediction, but we use them all. This molecular score is called Genomic Estimated Breeding Value (GEBV) or genetic value and is an estimation of the additive effects of all loci.

The first to implement GS was the US dairy industry (VanRaden et al. 2007; VanRaden 2008). It doubled genetic gain (Schaeffer 2006; García-Ruiz et al. 2016) for this species particularly well suited for the implementation of GS. It is now applied to many other animal species (Hayes et al. 2009). Daetwyler et al. (2008) showed how to use genomic prediction for analyzing the genetic risk of human diseases. In plants, Bernardo and Yu (2007) and Heffner et al. (2011a) showed the first promising results using simulations, and Lorenzana and Bernardo (2009) using empirical bi-parental data.

More details on genomic selection and prediction models are presented in another chapter of this book (Andres et al. 2020). Here we discuss, how genomic predictions could be used to optimize re-introduction of genetic diversity in plant breeding and pre-breeding programs.

3.3.1 Selection of a Relevant Training Set

Assuming that the number of markers and the training population size is optimal, the accuracy of the calibration model strongly depends on congruence between the allelic composition of the training population (to build the prediction model) and the allelic composition of the candidates whose performance is to be predicted (Habier et al. 2007). When the prediction uses unrelated populations to train the prediction equations, prediction accuracy actually becomes negligible (Crossa et al. 2014). Different ways of estimating prediction accuracy of a training population were developed and have been reviewed (Brard and Ricard 2015). Methods to optimize the composition of the calibration set prior to phenotyping have been proposed based on the Prediction Error Variance or on the Coefficient of Determination (Laloë 1993) of contrasts in unstructured (Rincent et al. 2012) or in structured populations (Rincent et al. 2017b). The algorithm of Rincent et al. (2012) has also been extended to optimize the training population for multiple correlated traits using a criterion called CDmulti (Ben-Sadoun et al. 2020). Other approaches based on spatial sampling (Bustos-Korts et al. 2016), or kinship coefficients (Rincent et al. 2012, 2017b) potentially taking genetic architecture into account (Mangin et al. 2019) were also developed.

3.3.2 Genomic Predictions Assisted Introgression

Using simulations, genomic predictions were shown to be efficient for rapid introgression of GR alleles when implementing 3 cycles per year in maize (Bernardo 2009; Combs and Bernardo 2013). Among 100 simulated QTLs, the adapted inbreds had the favorable allele at 50 or more QTLs and the GR at 50 or less QTLs. They compared 1 year of phenotypic selection versus 3 cycles of genomic selection. The results indicated that a useful strategy should involve 7–8 cycles of genomic selection (2–3 years). They showed that genetic gain was higher when starting from an F_2 population rather than a backcross population, even when the number of favorable alleles was substantially larger in the adapted parent than in the GR parent. Note that they used random mating in their simulations. This procedure would require only 3 years to get some progenies that could be integrated in the breeding program. Allier et al. (2019b) showed, using simulated data and optimal parental contribution method (see Sects. 3.3.6 and 3.3.7), that in a context of multiple allele introgression from a donor into one or several elites, three-way crosses and backcrosses were more adapted than two-way crosses when donors underperform the elite population. They demonstrated that three-way crosses should be preferred because they produce more progeny variance and combine alleles from more parents. This supports the strategy adopted in the Germplasm Enhancement of Maize project (Goodman et al. 2000). Two-way crosses were actually more adapted when donors outperform the elite population for the targeted trait.

3.3.3 Predictions of Accessions' Genetic Values Conserved in Genebanks

Using genotypes and phenotypes of a representative set of genebank accessions, we can build a model to predict the GEBV of the rest of the collection (Yu et al. 2016). As genotyping is less expensive than phenotyping, we can identify this way supplementary GR of interest (Crossa et al. 2016; Brauner et al. 2018, 2019). For instance, in maize, Allier et al. (2019c) calibrated a prediction model on a population, assembling a continuum from old dent accessions to elite iodent material, including founders of breeding pools, elite material released into public domain, and elite material from different private breeders. Yield predictive ability between the calibration population and RAGT2n company germplasm was 0.404 and allowed to detect landraces of agronomic interest to be introduced in the breeding program. But this strategy is possible only if the divergence is not too large between landraces and elite material and the predictive ability is sufficient. It is also necessary that the traits can be evaluated for the landraces in targeted environments. It turned out to be an appropriate approach for biomass sorghum (Yu et al. 2016) and dent maize. But for many species, the presence of some major genes involved in phenology may hinder good quality phenotyping of landraces, because of incapacity to flower, to mature on time or lodging. In that case, unadapted accessions may carry interesting favorable alleles but cannot reveal their potential in the targeted environment. Good quality phenotypes may necessitate to first convert GR by eliminating major phenology unfavorable alleles or to phenotype elite x GR hybrids instead of GR (Longin and Reif 2014). If we consider dominance effects of favorable over deleterious alleles for those major genes involved in phenology, heterozygous hybrids between elites and GR are expected to get favorable major alleles from elites that annul or at least reduce the effects of deleterious alleles from GR. But this assumes that hybrids are technically easy to produce which may be a challenge for autogamous species, at least laborious and expensive.

3.3.4 Optimization of the Allocation of Resources

Thanks to resource allocation optimizations some budget can be saved in evaluation of major traits (yield in general) and be transferred to

1. increase the size of the germplasm (the number of progenitors, crosses, and progenies per cross), leading to an increased genetic variance and a higher chance of creating outstanding individuals.
2. evaluate new traits (such as quality).

Different strategies have been proposed:

1. skip some field evaluation steps, which is relevant in long-lived species such as trees, or when trait values are expensive and/or become available late in the cycle (Hayes et al. 2009),
2. optimize the experimental design, i.e., minimize the number of lines or replicates evaluated in each environment,

Lorenz (2013) showed that it was advantageous to maximize population size at the expense of replication in a breeding program. Endelman et al. (2014), Heslot and Feoktistov (2017), and Akdemir (Akdemir and Isidro-Sánchez 2019) proposed efficient strategies to optimize field evaluation (sparse design) using genomic predictions. Ben-Sadoun et al. (2020) showed that it was possible to reduce budget by 25% for a fixed accuracy of French Bread Making Score by phenotyping it in a reduced number of environments. The idea is to evaluate all alleles in all environments, not all individuals.

3. accelerate cycles: speed (pre)breeding (2 cycles per year for winter wheat, 3 cycles for maize, up to 6 cycles for spring wheat) using adequate growth chambers and greenhouses protocols (Christopher et al. 2015; Hickey et al. 2017; Ghosh et al. 2018) to increase the rate of development,
4. diminish the cost of evaluation of an expensive trait using indirect measurements and optimize phenotyping of both correlated traits (Ben-Sadoun et al. 2020). The strategy is called Trait-Assisted genomic selection (TA) (Fernandes et al. 2018).

It is possible to predict two correlated traits simultaneously using multivariate best linear unbiased prediction (BLUP) (Henderson and Quaas 1976). Those models benefit from information contained in both genetic correlation between traits and genetic relationship among individuals (Calus and Veerkamp 2011). The training population is genotyped and phenotyped for both traits. Each training individual is phenotyped for at least one trait. If the candidate population is genotyped but not phenotyped for any of the traits, the strategy is called Multi-Trait genomic prediction (MT). If some of the candidates are phenotyped for the secondary trait, the strategy is called Trait-Assisted genomic selection (TA) (Fernandes et al. 2018).

As for single trait prediction, under a major QTL genetic architecture, Jia and Jannink (2012) found that Bayesian multivariate models (BayesA or BayesCπ) performed better than multi-trait GBLUP model. But for polygenic genetic architecture, multi-trait GBLUP model was equal to the Bayesian multivariate models. Note that Jiang et al. (2015) developed Bayesian multivariate models that consider correlated SNP effects. Montesinos-López et al. (2016) extended the model to a Bayesian multi-trait and multi-environment genomic prediction model (BMTME) that takes into account the correlation between traits and the three-way interaction term (Trait × Genotype × Environment). More recently, multi-trait deep learning (MTDL) models have been developed to reduce the computational resources (Montesinos-López et al. 2018, 2019). MT models can actually suffer from a high computational demand, time, and some convergence problems (Michel et al. 2018). Obviously, genetic correlation between traits is a key factor determining the MT advantage over single trait (ST) methods (Calus and Veerkamp 2011; Jia and Jannink 2012; Hayashi and Iwata 2013; Guo et al. 2014). Although MT models improve the predictive ability when the targeted trait has a low heritability and the secondary trait has higher heritability, the advantage of MT models to predict high heritability traits is low (Jia and Jannink 2012; Hayashi and Iwata 2013; Iwata et al. 2013; Guo et al. 2014). Studies using experimental data demonstrated that advantage of MT models to predict individuals which have not been phenotyped either for the trait of interest or the correlated trait was small or null in pine tree, *Pinus taeda* (Jia and Jannink 2012), soybean, *Glycine max* (Bao et al. 2015), rye, *Secale cereale* (Schulthess et al. 2016), maize, *Zea mays* (dos Santos et al. 2016), bread wheat, *Triticum aestivum* (Michel et al. 2018; Schulthess et al. 2018; Lado et al. 2018), and sorghum, *Sorghum bicolor* (Fernandes et al. 2018). Several studies using experimental data demonstrated that TA models perform better than ST and MT models in terms of accuracy. The TA models using high-throughput phenotyping, for instance, improved the prediction accuracy of bread wheat grain yield by up to 70% (Rutkoski et al. 2016; Sun et al. 2017; Crain et al. 2018). TA models also improved bread wheat baking quality-related parameters using protein content (Michel et al. 2018) or dough rheological traits (Lado et al. 2018) as correlated traits. Measuring dough strength (W) instead of French Bread Making Score for 75% of the population maintains accuracy by reducing budget of phenotyping by up to 65% (Ben-Sadoun et al. 2020). For a fixed budget, it can increase predictive ability by up to 0.14. Predictive ability of Fusarium head blight severity in hybrid bread wheat was improved using plant height and heading date as correlated traits (Schulthess et al. 2018). Fernandes et al. (2018) showed that TA models increased prediction accuracy by up to 50% when using plant height as correlated trait to predict yield in sorghum.

Robert et al. (2020) proposed a new TA approach, in which the secondary trait is not phenotyped for the selection candidates, but predicted with crop-growth models. The advantage is that it is not necessary to sow the selection candidates, as only the genotypic information is used.

3.3.5 Mating Optimization

The breeder's goal is to obtain "transgressive" individuals (with extreme genetic values) for at least one trait, cumulating as many favorable alleles as possible, putatively coming from different parents. While animal breeders optimize the choice of males, plant breeders may want to optimize mating between two or more parents. Cross design is essential but without accurate tools to guarantee its performance, breeders often select highest-performing parents to ensure high mean performance of progeny, and may focus on one or two traits. The problem is that highest-performing individuals may present similar sets of alleles and may actually produce less genetic variance in progeny than parents that have less but complementary favorable alleles. Because it is not feasible to evaluate all possible crosses in the field, it would be valuable to predict the value of a cross or a global cross design before it is made. Instead of focusing on the performance of parents, the idea is to estimate a proxy of the value of top progenies, i.e., the predicted mean and variance of the progeny. Attempts have been made using distances between parents based on phenotypes (Souza and Sorrells 1991a, b; Utz et al. 2001), genetic distance based on molecular markers (Bohn et al. 1999; Hung et al. 2012), molecular scores (summing QTL effects), or GEBV (summing marker effects estimated by ridge regression) (Tiede et al. 2015), but they were not really successful.

In a pre-breeding program context, it is even more obvious that the interest of a donor for a recipient elite individual depends on its genetic value (which can predict mean performance of progeny) but also its originality at QTLs (which will contribute to increase genetic variance and long-term genetic progress). A first approach was to count the proportion of favorable alleles and complementarity of parents at QTLs (Dudley 1984, 1988; Bernardo 2014). Van Berloo and Stam (1998) discriminated among crosses using a marker score from QTL flanking marker genotypes weighted by their effects.

The idea of Genomic Mating (GM) strategies is to use genomic predictions to optimize complementation of parents to be mated (Akdemir and Isidro-Sánchez 2016). As progeny genetic variance is generated by randomly sampling parental chromosomes during meiotic division, then recombination between those chromosomes, if we can accurately estimate marker effects as well as recombination rates between markers, we can optimize mating such as maximizing the probability to get individuals that cumulate a maximum of favorable alleles. In theory, the value of a cross, or the Usefulness Criterion (UC) of a cross (Schnell and Utz 1975) is the expected genetic value of the selected fraction (the bests) of the progeny

$$UC = \mu + ih\sigma_A$$

with μ the population mean, i the selection intensity, h the square root of heritability of the trait, and σ_A the additive genetic standard deviation among progeny.

Between Two Parents for Biparental Populations

To calculate UC of crosses, we need marker effect and recombination rates estimates. In plants, meiotic recombination maps are usually estimated from bi-parental populations. Note that we can use some other types of populations (F_2, BC, HD), using adapted transformation to get the meiotic recombination rate. Several unconnected or connected populations can also be analyzed together to build consensus or composite maps cumulating more recombination information. A higher resolution method is to infer historical recombination maps from landraces or wild populations (Choi and Henderson 2015; Petit et al. 2017; Danguy des Deserts et al. 2021).

A first strategy is to simulate progeny in silico (stochastic simulations) by randomly producing crossing-overs along parental gametes according to a recombination map (Bernardo and Charcosset 2006). The value of a cross is the mean of the top progeny genetic values, the number of individuals belonging to this top group depending on the intensity of selection (Iwata et al. 2013; Bernardo 2014; Lian et al. 2015; Mohammadi et al. 2015). Note that in plants, we observe a significant negative relationship between parental mean and progeny genetic variance (Mohammadi et al. 2015). This study also showed that mid-parent value explains 99.99% of mid-progeny value and 82–88% of top-progeny value. Mid-parent value and estimated progeny genetic variance explained 99.5 of top-progeny value. This demonstrates the usefulness to estimate genetic variance and not only mean of progeny to estimate cross value. The problem with large breeding programs is that stochastic simulations are compute-intensive. So, attempts are made to predict variance using mathematic formulas (analytically). The mean of a cross is predicted by the mid-parent value in self-pollinated species or the mean of testcross performance in a hybrid crop. Several formulas have been proposed to predict the progeny variance. A first way is to estimate the value of the best possible progeny. We can determine historical haplotype blocks along the genome based on linkage disequilibrium and consider that recombination occurs only between those blocks. The effect of one haplotype block is the sum of its individual allele effects. Daetwyler et al. (2015) defined the Optimal Haploid Value (OHV) of an heterozygous individual as the sum of the effects of the best allele at each haplotypic block, corresponding to the genetic value of the best theoretical gamete to pass on to the next generation. They demonstrated that for a wheat program using DH technology (i.e., getting homozygous lines by gamete cultivation and chromosome doubling using colchicine treatment), genetic gain was improved (up to 0.6 standard deviations) when estimating the value of a cross as the OHV of the corresponding F_1 heterozygous individual

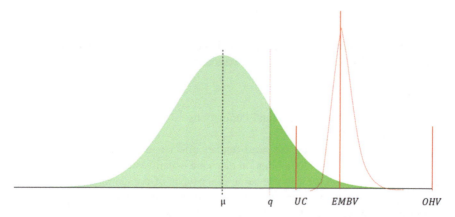

Fig. 2 Different indexes to estimate cross values. Along the distribution of a simulated progeny from one cross is indicated. μ: the mean of the simulated progeny, q: the 10% quantile of the top progenies, UC: the mean of the 10% top progenies, EMBV: the mean of the best progeny for 1,000 different simulations, OHV: the best possible progeny (if all recombinations are possible and population size is unlimited)

compared to standard GS. It also preserved a substantially greater amount of genetic diversity in the population. Müller et al. (2018) proposed the Expected Maximum haploid Breeding Value (EMBV) (Fig. 2). This is the expected GEBV of the best out of N DH lines produced by an F_1 using haplotypic blocks. Compared to OHV, EMBV actually takes into account the fact that the number of progenies produced is limited, the best theoretical progeny being impossible to reach. It can be estimated by stochastic simulations or using an analytical formula. Another analytical way to estimate the value of a bi-parental population explicitly includes the vectors of recombination rate between markers that are polymorphic between parents and marker effects (Zhong and Jannink 2007). This analytical formula estimates the probabilities of transmission of alleles at all QTLs from an F_1 individual (obtained by crossing two parents) to its gametes. In other words, the probability to get an outstanding progeny depends on the distribution of favorable alleles between parents and on the probability to break linkage between loci in repulsion phase and not to break linkage between loci in coupling. If we are interested in two genes that are genetically close to each other, if alleles are in the repulsion phase in the parental genotypes (neither parent has both favorable alleles), recombination widens the variance of the cross by providing extreme genotypes (you can get both favorableand unfavorable alleles in some progeny). On the opposite, if alleles are in the coupling phase in parents (you already observe the best and worse combinations), recombination reduces the variance (Zhong and Jannink 2007; Tiede et al. 2015) by getting combinations of intermediate effects. Formulas considering recombination rate between polymorphic markers were derived to calculate cross values for RILs and DH at generation k (Lehermeier et al. 2017b). The authors confirmed that predicting genetic variance in cross prediction increases genetic gain by 18% in maize compared to predicting the mean only. Formulas were also derived to optimize three- and four-way cross designs (Allier et al. 2019b). The implementation is much faster compared to *in-silico* simulations. But note that in practice, we can only use analytical formulas to predict the next generation variance but not several generations ahead. We actually need to recover the parent genotypes at each cycle to estimate the variance of following generations.

Although Uemoto et al. (2015) suggested to filter MAF ($\geq 5\%$) to improve the prediction of GEBV, markers with low MAF should be kept for cross value prediction as they may be in greater linkage disequilibrium with low MAF QTLs and provide better predictions of the gametic variance (Santos et al. 2019).

Although the superiority of predictions of GEBV using haplotypes instead of single markers has not been demonstrated, Cole and VanRaden (2011) and Bonk et al. (2016) recommended the use of haplotypes to predict cross values in order to limit sampling errors when estimating individual marker effects. Another way to take into account local LD and uncertainty of markers estimates is to use Bayesian estimates of single marker effects (Sorensen et al. 2001; Lehermeier et al. 2017a). The idea is that combinations of alleles in haplotypic blocks may be better estimated (if present in the training population) than individual SNPs.

In a maize pre-breeding context, Allier et al. (2019c) compared different indexes to estimate cross values: the Modified Roger's Distance (MRD) between parents, the proportion of favorable alleles in donors (K) and recipients (J) (Bernardo 2014), OHV, genetic variance VarG in progeny and Lerhermeier's UC (Lehermeier et al. 2017b). They considered different selection rates in the progeny to calculate UC, 5% (UC1) and $10^{-8}\%$ (UC2). The main conclusion was that one might consider UC1 or OHV with large haplotypes for short-term genetic gain prediction, OHV with small haplotypes or UC2 with stringent selection for long-term genetic gain prediction. In other words, complementarity between parents is more important to consider for long-term genetic gain. Another conclusion was that in genetic diversity conservation programs, one might just want to maximize progeny variance (VarG) for the trait of interest, or the MRD between donor and recipient in the absence of trait-specific considerations.

At the Population Level

The long-term potential of a breeding program relies on the efficiency to combine favorable alleles scattered within many individuals (Goddard 2009; Jannink 2010). In a pre-breeding program where we want to increase the number of favorable alleles in a population, this can be optimized using Genotype-Building (GB) strategies. It is the founder population as a whole and not individuals or parents which must cumulate favorable alleles at a maximum number of haplotypic blocks. A parent (founder) is chosen for its complementarity with others. It may carry only a few rare but very favorable alleles and have a low individual genetic value. Considering that the best allele combination (ideotype) is known, there may be many possible cross designs to get there. Because we cannot test all founder populations and cross combinations, the challenge is to build algorithms and solvers so that calculations are feasible and solutions are realistic.

The first proposed strategy was to select a subset of founders that possess altogether the best possible combination of haplotypic alleles along the genome. The Genotype-Building (GB) value of a subpopulation (Kemper et al. 2012) measures the GEBV of an ideal heterozygous progeny that would get the two best haplotype segments from two founders for each block. The Optimal Population Value (OPV) (Goiffon et al. 2017) is an extension of GB for inbreds. It measures the GEBV of the best possible homozygous progeny that can be produced, i.e., the value of the progeny that would get the best allele for each haplotypic segment in the founders. Note that it supposes an unlimited number of generations. The second extension is to consider time and resource constraints. Moeinizade et al. (2019) proposed the LAS (Look-Ahead Selection) algorithm where they improve the population for a few generations, starting with a subset of founders that maximizes OPV, and finally select for the best individuals in the last generation. They also consider a limited budget and vary the numbers of progenies produced from different crosses based on the genetic diversity of the parents: they spend more resources on those crosses that have wider predicted phenotypic distributions and thus higher probabilities of producing outstanding progenies. As for OHV, for GB, OPV, and LAS we assume adjacent markers are likely to segregate together and are grouped into representative haplotype blocks, recombination events occurring only between haplotype blocks.

3.3.6 The Theory of Contributions

According to the "breeder's equation" (Lush 1937), genetic gain is limited by the initial additive genetic variance in breeder's germplasm

$$\Delta\mu = ih\sigma_A$$

with $\Delta\mu$ (genetic gain) the expected change in mean performance per generation, i the selection intensity, h the square root of heritability of the trait, and σ_A the additive genetic standard deviation among progeny.

The level of diversity depends on the effective population size Ne (Fisher 1930; Wright 1931), which refers to the number of breeding individuals in an idealized panmictic population with absence of selection that would show the same amount of genetic diversity than the real population. Genetic diversity is generally measured by the expected heterozygosity He (Nei 1973). While the expected response to selection is proportional to the selection intensity, the number of reproductors and the corresponding effective population size is inversely proportional to the square of selection intensity on major QTLs (Sanchez et al. 2006). Consequently, maximizing selection intensity (using GS for instance) to maximize short-term genetic gain reduces the effective population size and long-term genetic gain.

The genetic gain is also proportional to the product of individuals' contributions (i.e., the number of offspring of each cross) and deviations from population mean (Woolliams and Thompson 1994; Woolliams et al. 1999). The rate of inbreeding, i.e. loss of diversity, is inversely proportional to the square of individuals' contributions (Robertson 1961; Wray and Thompson 1990).

3.3.7 Optimization of Contributions with Diversity Constraints

Based on the theory of contributions, the optimum contribution concept has been developed in animal breeding programs (James and McBride 1958; Wray et al. 1990; Wray and Goddard 1994; Brisbane and Gibson 1995; Meuwissen 1997; Woolliams et al. 2015) and tree breeding (Kerr et al. 1998; Hallander and Waldmann 2009a, b) to limit inbreeding. These methods have been recently adapted in crop breeding (Akdemir and Isidro-Sánchez 2016; Lin et al. 2016; Cowling et al. 2017; De Beukelaer et al. 2017; Akdemir et al. 2018; Gorjanc et al. 2018; Allier et al. 2019a, b, 2020).

The vector of parental contribution to the next generation is chosen at a predefined rate of population inbreeding (Wray and Goddard 1994; Meuwissen 1997), penalizing this way individuals that are too closely related and maintaining genetic diversity. The solution is a compromise between short- and long-term genetic gain, is heuristic, and can be optimized using different types of algorithms such as evolutionary algorithms, genetic algorithms in particular (Holland 1962; Goldberg and Holland 1988). When there is no explicit solution for complex problems (some objectives are not independent), simulated annealing and genetic algorithms are efficient to explore the solution space, obtain a pseudo-optimal solution, and limit the risk to get local minimum solutions. Simulated annealing (Metropolis et al. 1953) uses a Monte Carlo criteria, i.e., a probability of acceptance of a solution. New solutions are proposed until the algorithm converges, i.e., no new solution improves the objective functions. The number of iterations with different starting points for the decision variables and the choice of convergence criteria to decide to stop the algorithm are essential. Genetic algorithms (Holland 1962) work on a population of solutions instead of individual solutions: (1) The algorithm generates a population of possible solutions, each one with defined values for the decision variables. (2) The values defined for decision variables are considered as alleles at different loci (one locus = one decision variable and one allele = one value for the decision variable). (3) The algorithm creates new solutions from existing solutions by "reproduction" with mechanisms similar to genetic evolution (genetic recombination, mutation, selection). Those transition rules are probabilistic. Different reproduction (one point, uniform) and selection (roulette wheel, tournament, rank) operators exist to propose and choose solutions.

Allier et al. (2019b) combined optimum contribution with Usefulness Criterion (UC) (Lehermeier et al. 2017b) strategies in maize. They evaluated the interest of a multi-parental cross implying a donor and one or several elite recipients using the UCPC (Usefulness Criterion Parental Contribution) criterion. They simultaneously predicted the full multivariate progeny distribution (mean, variance, and pairwise covariances) for the agronomic trait, genome-wide contribution of parents, and contributions at favorable alleles. They showed using this strategy that three-way crosses were more efficient for long-term genetic gain when donors are less performant than elites.

In animals, to maintain diversity, Bijma et al. (2020) proposed to produce a number of offspring that is proportional to the gametic variance of the reproductor to accelerate response to recurrent selection.

In addition to contributions, we can optimize mating in Optimal Cross Selection (OCS) approaches. It aims at identifying the optimal set of crosses maximizing the expected genetic value in the progeny under a constraint on genetic diversity in plants. It combines optimal contribution with optimal mating in a multi-objective problem that can be also optimized by heuristic algorithms. The classical OCS approach controls for genetic diversity in the total progeny. Allier et al. (2019a) applied OCS under a constraint on genetic diversity in the selected fraction of the progeny that is used as parents of the next generation accounting for within-family variance. They applied UCPC-based (Usefulness Criterion Parental Contribution) OCS in maize using a differential evolution algorithm (Storn and Price 1997; Kinghorn et al. 2009; Kinghorn 2011). They showed that OCS with constraints on UCPC and He was more efficient than classical OCS for long-term genetic gain with limited reduction of short-term genetic gain. Akdemir et al. (2018) maximized within-cross variance (Shepherd and Kinghorn 1998) and mating for multiple traits. It gives the list of parent mates that maximize gain, maximize cross variance, and minimize inbreeding. It is called Multi-Objective Optimized Breeding (MOOB). Compared to standard multi-trait breeding, the gains from multi-objective optimized parental proportions approaches were about 20–30% higher at the end of long-term simulations of breeding cycles.

The budget and the technical solutions are so different between species, private/public sector, that it is difficult to propose one algorithm that would handle conflicting objectives and satisfy the whole community (Wellmann 2019; Wellmann and Bennewitz 2019).

GS being more efficient at fixing major QTLs, it accelerates the loss of genetic diversity at QTLs according to simulation studies (Jannink 2010; Lin et al. 2016; Ben-Sadoun et al. 2020). Moreover, using RR-BLUP, the rare allele effects are shrunk toward zero, which increases the risk to lose individuals with rare favorable alleles and decreases long-term genetic gain (Goddard 2009; Jannink 2010; Habier et al. 2010; Pszczola et al. 2012). Several authors suggested to up-weight rare favorable alleles (Goddard 2009; Jannink 2010; Sun et al. 2014; Liu et al. 2015a) to select individuals for the next generation. They obtained encouraging results by simulation but did not propose stabilized rules to assign relevant weights to markers.

Considering computation time, several papers concluded the possibility for elite material to pre-select the population of eligible crosses based on parental mean genetic values before optimizing the cross design according to progeny genetic variance estimations (Zhong and Jannink 2007; Lehermeier et al. 2017b). They show that the genetic gain at the following generation is similar when considering all possible crosses or when removing couples with lower mean genetic values. But the conclusion may be different in more diverse materials. In that case, crosses that have high variance but low mean may be interesting for long-term genetic gain, if we wait a sufficient number of generations to give a chance to rare favorable alleles to be selected in a pre-breeding population for instance.

3.3.8 Multiple Traits Optimization

The performance of new varieties often depends on multiple traits and/or constraints. The targeted ideotype can be a compromise between yield and quality for instance, with specific molecule concentrations for the industry. Breeding for multiple traits simultaneously is challenging because some traits are uncorrelated or unfavorably correlated due to linkage or pleiotropy. Bulmer effect (Bulmer 1971) actually mechanically creates negative correlations between traits under selection, yield and protein content in wheat, for instance. Moreover, the economic value of different traits may not be equally important.

In classical multi-trait selection where traits are not correlated or negatively correlated, we have several strategies: (1) tandem selection: we select each trait singly (at different steps or generations), (2) independent culling: we reject individuals that are not meeting required standards for all traits, (3) index selection: traits are combined, using different weights corresponding to economic value, into a score that is considered as a single trait. The problem is that it may exclude the best individuals for each trait and some beneficial alleles. And it does not control for inbreeding.

Although single-objective optimization problems may have a unique optimal solution, the chance to find the best solution to a multi-objective problem is very low. The solution will be a compromise, especially when traits are antagonistic. And there may be several interesting solutions depending on the ranking of objectives. Algorithms propose a multiplicity of compromise solutions called Pareto optimal solutions after judiciously scanning the decision space, i.e., different combinations of equality and inequality constraints. Population of solutions are classified into boundaries according to their level of dominance (see more explanations in Figs. 3 and 4 for two traits, Fig. 5 for three traits).

Note that at the end of a multi-objective optimization, the decision maker still has to select the preferred solution from the Pareto frontier using its own decision rules, i.e., ranking or weighing objective functions like in index strategies.

3.3.9 Production of Varieties Adapted to Local Constraints

The objective of plant breeders is to produce new varieties well adapted to target environments. For this purpose, they evaluate candidate lines for several years in multi-environment trials. Because phenotyping is expensive, only a limited number of lines are evaluated each year in a small number of environments.

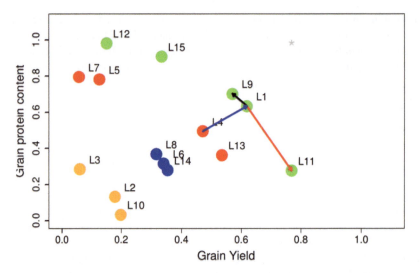

Fig. 3 Genetic values of grain yield and grain protein content for 15 genotypes. Different colors represent different levels of solutions (green: first level non-dominated, red: second level dominated by first level, blue: third level dominated by first and second level, orange: fourth level). Reproduced from Akdemir et al. (2018)

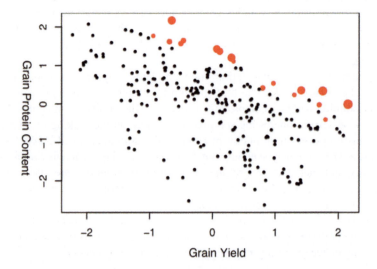

Fig. 4 Optimization of grain yield and grain protein content: Pareto frontier curve. Red points indicate the individuals that are selected by the algorithm. The size of the points is proportional to their contribution to the next generation. Reproduced from Akdemir et al. (2018)

Using genomic predictions accounting for Genotype by Environment Interactions (GEI), we can explore more combinations of genotypes and environments that we cannot afford observing in the field. We can use historical breeding databases

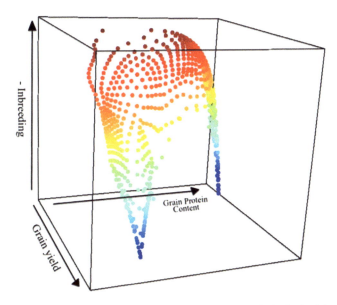

Fig. 5 Pareto optimal solutions for parental contributions (wheat data) obtained by solving the optimization problem for three parameters (three dimensions). The objective was to improve grain yield (GY) and grain protein content (GPC) while controlling group coancestry, i.e., maximizing GY, GPC, and the negative of inbreeding. The redness of the points indicates closeness to ideal solutions. Pareto optimal solutions are represented by a knee. Reproduced from Akdemir et al. (2018)

including numerous years and environment observations to calibrate those models. Different approaches have been proposed in the last decade.

Genotype by Environment Interactions (GEI) Predictions

While classical GS models rely on main effects and are not able to predict GEI (Crossa et al. 2010; Ly et al. 2013), those were adapted to predict environment-specific effects (Schulz-Streeck et al. 2013; Lopez-Cruz et al. 2015; Crossa et al. 2016; Bandeirae Sousa et al. 2017), with possibly a genetic covariance between environments (Burgueno et al. 2012; Lado et al. 2016; Cuevas et al. 2017, 2018). These approaches, similar to multi-trait models, can increase prediction accuracy and can predict missing phenotypes of observed varieties (sparse testing) or unobserved varieties. They are more efficient in the sparse testing scenario in which information on a given variety can be shared between similar environments. Specific R packages were developed to fit simple GEI models with optimized computational properties (De Coninck et al. 2016; Granato et al. 2018). But these models cannot be used to make predictions of a genotype performance in new environments, as they rely on the phenotypic data to estimate the covariance between environments.

To extend the predictions to new environments, Heslot et al. (2014), Jarquín et al. (2014), Malosetti et al. (2016), Millet et al. (2019), and Rincent et al. (2019) proposed to characterize environments with environmental covariates, like molecular markers are used to characterize varieties. These covariates are pedoclimatic characteristics supposed to affect the plants (precipitations, extreme temperature, radiation deficit) at the different developmental stages (Brancourt-Hulmel 1999). A crop model can be used to estimate the timing of the developmental stages, so that the covariates are estimated for a period during which they are supposed to impact plants. This work is inspired from the factorial regression methodology (Brancourt-Hulmel et al. 2000) in which a regression on a covariate explains the variability of the trait in presence of GEI. A generalization of factorial regression on a given covariate to the GBLUP mixed modeling context was proposed (Ly et al. 2017, 2018). In these studies, the covariate has a variety specific random effect with a variance/covariance matrix structured by the kinship. This allows predicting the sensitivity of new varieties to this covariate. It is important to note that the QTLs affecting main effects are not necessarily the same as the QTLs affecting GEI, and this can be taken into account in the statistical models at the marker level (Heslot et al. 2014) or at the kinship level (Rincent et al. 2019). These models involving environmental covariates are particularly useful in the context of climate change, because they can predict the behavior of various varieties in virtual prospective scenarios. If a relevant database exists to calibrate the GS model, it could be used to identify in-silico interesting combination of alleles to face given environmental conditions. If we consider that the genetic diversity available in the elite pool is not sufficient, the prediction models can also be used to screen genebanks for valuable GEI (Crossa et al. 2016; Yu et al. 2016).

Ecophysiological Modeling

The adaptation of plants to their environment has been long studied by ecophysiologists. Their research has allowed developing Crop Growth Models (CGM), which describe plant development using mechanistic relationships with physiological parameters and environmental covariates as inputs. In other words, the CGM simulates GEI by taking into account the specificities of the varieties (genetic parameter) and of the environments (environmental variables). Different ways of using CGM to predict GEI were proposed in the past.

The first application is to predict the developmental stages of the plants to estimate if stress appeared at critical stages. This strategy was applied in wheat and maize (Heslot et al. 2014; Jarquín et al. 2014; Malosetti et al. 2016; Ly et al. 2017; Millet et al. 2019; Rincent et al. 2019). Numerous studies indeed revealed that CGM were efficient to predict phenology even for new varieties (White and Hoogenboom 1996; Nakagawa et al. 2005; Yin et al. 2005; Messina et al. 2006). CGM can also be used to directly derive environmental covariates (Ly et al. 2017; Rincent et al. 2019). In Rincent et al. (2019), CGM SiriusQuality (Martre et al. 2006) was used to estimate dry matter stress index (DMSI) that directly relates to the impacts of temperature, drought, and N deficit, alone or in combination, to daily biomass loss. The idea is to produce stress indexes as close as possible to what the crop experienced in the field. Such variables directly simulated by the CGM were shown to better capture GEI than basic pedoclimatic covariates.

The second application is much more ambitious: the genetic model and the CGM are fully integrated within the Gene-Based Modeling approach (GBM). In GBM, the CGM simulates the development of each variety by using variety specific genetic parameters as input. These genetic parameters (phyllochron, sensitivity to photoperiod) characterize the varieties independently from the environment and are thus supposed to be stable across environments. Once the genetic parameters are estimated for the calibration set, a GS model can be calibrated to predict the genetic parameters of new varieties. These predictions can then be used as input of the CGM to predict the target trait of the new varieties in various environments. The interest and feasibility of this approach coupling CGM and genetics have been validated for leaf elongation rate in maize (Reymond et al. 2003; Chenu et al. 2008), fruit quality (Quilot et al. 2005; Prudent et al. 2011), and phenology of various species (White and Hoogenboom 1996; Nakagawa et al. 2005; Yin et al. 2005; Messina et al. 2006; White et al. 2008; Uptmoor et al. 2012; Zheng et al. 2013; Bogard et al. 2014; Onogi et al. 2016; Rincent et al. 2017a). Recently, Technow et al. (2015), Cooper et al. (2016), and Messina et al. (2018) have illustrated the possibility of coupling CGM and GS models for predicting highly integrated traits such as grain yield. One major advantage of their approach and that of the work of Onogi et al. (2016) is that the genetic parameters and the marker effects are jointly estimated, and so information can be shared between individuals thanks to genotypic data. However, using GBM to predict such complex traits remain challenging, as numerous genetic parameters have to be phenotyped or estimated on the training population. More recently, Robert et al. (2020) proposed to combine GBM with a trait-assisted prediction approach. The GBM is used to predict a secondary trait (heading date) for the test set in all environments. This secondary trait is easy to predict, and its relationship to the target trait (yield) is environment specific and thus allows predicting environment-specific effects in bread wheat.

A last application of CGM is to help clustering environments with similar properties. The objective is to use the CGM to characterize the stressing conditions experienced by the plants in each environment, and then to group environments with similar scenarios. Taking pedoclimatic data and variety characteristics as input, CGM can indeed produce daily stress indexes from sowing to maturity. It has been shown that clustering based on stress scenarios identified by CGM was more relevant than clustering based on the experimental protocols (e.g., non-irrigated vs irrigated) and that it was efficient to capture GEI (Chenu et al. 2011; Touzy et al. 2019). For example, it can happen that in a multi-environment trial, an irrigated trial is more subjected to drought than a non-irrigated trial at another location. In contrast, the CGM is able to finely characterize each environment by taking into account the environmental conditions and the plant development. Once the CGM-based clustering is obtained, reference GS models (or GWAS) can be applied within each cluster, GEI being taken into account by the clustering.

Perspectives in the Field of GEI Prediction

Phenotyping is one of the main bottlenecks in plant breeding. GS models allow predicting new varieties in observed environments or new environments for observed varieties, but large phenotype databases are necessary to calibrate the GS models accurately. High-throughput phenotyping platforms and tools which allow phenotyping at the organ level, at the plant level, or at the plot/field level (Tardieu et al. 2017) constitute a great opportunity to calibrate GEI models. This observation can be used to calibrate CGM (Reymond et al. 2003) or as environment-specific proxies of the target trait (Amani et al. 1996). The systematic and wide use of sensors in the breeding programs will probably allow using deep learning approaches, supposed to be the most efficient when such large datasets are available. Note that in all the approaches described in this section, there were only two kinds of data involved in the model: genomic and phenotypic data. The introduction of other omics data such as transcriptomics, proteomics, and metabolomics in the models will probably allow a better understanding of how a given variety grows in various environments (see Sect. 4.2 below). The introduction of this information in "phenomic" prediction models or in Genomic-like Omics Based prediction models (GLOB) was proven to improve accuracy (Fu et al. 2012; Riedelsheimer et al. 2012; Rincent et al. 2018; Schrag et al. 2018). The combined use of phenomics and genomics is used in pre-breeding for yield potential in stressed environments under the International Wheat Yield Partnership (IWYP, https://iwyp.org/) (Reynolds et al. 2021). Once those tools are cost effective, they could be integrated routinely in breeding programs.

3.3.10 Application to Pre-breeding

When performance gap between donors and elites is too large, it may be judicious to improve a pre-breeding population before introducing GR in a breeding program. For a few generations, starting from relevant founders that bring complementary alleles and mating optimization, we can increase gradually the number of favorable alleles in the population. It is only after a sufficient number of generations that we start selecting individuals based on their genetic value to cross them with elites. Gorjanc et al. (2016) provided guidelines based on stochastic simulations. Starting from 3,000 genotyped maize landraces, they evaluated different pre-breeding programs that differed according to the population to initiate crosses: (1) the best landraces, (2) the best testcrosses, or (3) the best DH seeds derived from testcrosses. They tested different (1) sizes for the pre-breeding program, (2) levels of diversity within the 3,000 landraces, (3) trait heritabilities, (4) number of markers, (5) number of crosses and progeny size per cross, and (6) number of phenotypic observations. The highest genetic gain was achieved by initiation with testcrosses. But it was reconstructing the elite genome and not utilizing the landrace favorable alleles. The best compromise to start a pre-breeding program was to start from landraces. This process can be accelerated by using existing composite or recurrent selection

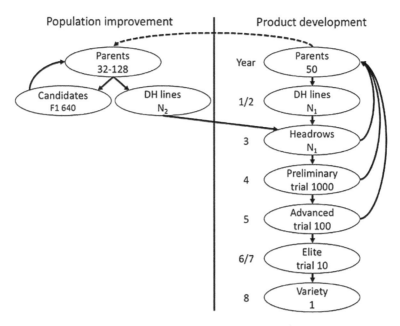

Fig. 6 Two-part breeding program. The population improvement component is based on recurrent selection that brings new genitors to the breeding program. Conventional strategy is based on variety development from elite parents. Reproduced from Gorjanc et al. (2018)

populations or inbred lines derived from local landraces. A recent initiative to characterize and use a part of the untapped variation in maize landraces is the Seeds of Discovery project (SeeD: http://seedofdiscovery.org). SeeD develops germplasm with 75% or more elite and 25% or less landrace genome to provide donors carrying new alleles.

Two-step breeding programs with an integrated pre-breeding program using rapid cycles (recurrent selection) (Gaynor et al. 2017; Gorjanc et al. 2018) is an efficient way to improve long-term genetic gain according to simulations (Fig. 6). An improvement population is produced by recurrent genomic selection with several cycles per year to increase the mean value of GR population in the pre-breeding program. A development population is produced using standard methods to develop new lines in the breeding program. It delivered about 2.5 times larger genetic gain compared to a conventional program for the same investment according to Gaynor et al. (2017) simulations. OCS increased long-term genetic gain by 15–78% depending on the number of parents.

Allier et al. (2020) proposed a strategy in three steps in case of a very large gap between elites and GR. They called base broadening phase (pre-breeding) the recurrent improvement of GR to decrease the performance gap with elites. It is kept independent from breeding programs until performance is satisfying. Best progenies are then crossed with elites to produce a bridging population. And the best bridging progenies can be parents in standard breeding programs. Allier et al.

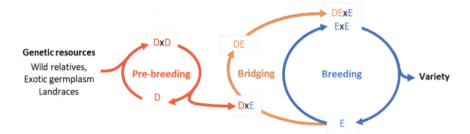

Fig. 7 Diagram illustrating the respective positioning of pre-breeding, bridging and breeding from genetic resources to variety release. Reproduced from Allier et al. (2020)

(2020) compared simulated breeding programs introducing donors with different performance levels. They observed that with recurrent introductions of improved donors, it is possible to maintain the genetic diversity and increase mid- and long-term performances with only limited penalty at short-term. When donors are already high-yielding, the bridging step could be skipped (Fig. 7).

From a practical point of view, several open-source software have been proposed. The R packages Rqtl (Broman et al. 2003), Popvar (Mohammadi et al. 2015), and software Alphasim (Faux et al. 2016) simulate bi-parental populations. The R Package Breeding Scheme Language (Yabe et al. 2017) simulates breeding programs. Multi-stage breeding schemes for hybrids using economic constraints are implemented in the R package Selectiongain (Mi et al. 2014, 2016).

To optimize mating for multiple traits, the R Package Genomic mating (Akdemir and Isidro-Sánchez 2016) and the software Alphamate (Gorjanc and Hickey 2018) have been proposed.

Forward stochastic simulations are proposed in python language in the software SeqBreed (Pérez-Enciso et al. 2020) and MoBPS (Pook et al. 2020), the last one implementing the optimum contribution method in an R environment.

To estimate the probability of getting the best progeny out of N with a specific cross, we can use the R package EMBV (Müller 2017). For qualitative traits controlled by major genes, the probability to cumulate a maximum of favorable alleles can be optimized using the software Optimas (Valente et al. 2013) or PCV (Han et al. 2017).

4 Future Perspectives

4.1 Improvement of Databases

We discussed above how diagnostic markers and genomic predictions can help the introduction of GR beneficial alleles from landraces or wild relatives in breeding populations. Operating procedures for conservation of those accessions have been in place for decades in genebanks, but there is a lack of means and methodological results to optimize the discovery and transfer of beneficial alleles into modern varieties, especially for quantitative traits or multi-trait improvement (Mascher et al. 2019). What is essential to valorize those accessions is the existence of international databases with curated and standardized information (e.g., passport, curated phenotypes, validated GEBV, alleles at validated QTLs, introgressions, cloned genes, and site under ancient or recent selection pressure). There is actually no doubt that the better the database, the better the predictions and the integration of useful information to users. Many initiatives emerged to build national databases (https://www.ars-grin.gov GRIN-Global in the USA). Some national genebanks connect their database to regional (The European Search Catalogue for Plant Genetic Resources: EURISCO, https://eurisco.ipk-gatersleben.de) and international networks, such as the Global Gateway to Genetic Resources (Genesys, https://www.genesys-pgr.org). But not much information is shared beside the passport data. It is not straightforward to standardize experimental protocols, file formats and merge different databases. But this effort would facilitate integration of information and exchange of seeds among genebanks, plant geneticists, and breeders.

For plant phenotypic data management, the number of national initiatives multiplies for many species (Adam-Blondon et al. 2016), in particular in the phenomics context (Neveu et al. 2019). We can also cite the dataverse phenotypic database for CIMMYT wheat and maize trials (www.cimmyt.org/resources/data/). A multispecies integrative information system dedicated to plant and fungi pests called GNPIS has been developed in France, for instance (Pommier et al. 2019). It bridges genetic and genomic data, allowing researchers' access to both genetic information (e.g., genetic maps, quantitative trait loci, association genetics, markers, polymorphisms, germplasms, phenotypes and genotypes) and genomic data (e.g., genome sequences, physical maps, genome annotation and expression data). For genomic data and genome sequences in particular, transplant is an EU-funded project aiming at building hardware, software, and data infrastructure (Spannagl et al. 2016).

On the plant pathogen side, monitoring is generally organized at the national scale. The Australian cereal rust control program is estimated to save the industry $289 million per year from resistance breeding, for instance. The European project Rustwatch (H2020 Sustainable Food Security-2017) tends to gather and standardize information about wheat cultivation surfaces, rust pressure, pathogen races, allelic composition of varieties and their bypass dates, in a standardized database to better understand the dynamic of bypass.

On the breeder side, from a pedigree and phenotype database in the UK, Fradgley et al. (2019) evaluated historical parental contributions in wheat and detected adaptation and selection signatures comparing genetic diversity levels with or without selection (experimental data vs simulated data, respectively) using gene dropping. Similar databases exist for oats, *Avena sativa* (Tinker and Deyl 2005) and rice, for instance (Bruskiewich et al. 2003).

An interesting initiative from NIAB is to propose a Toolbox to wheat breeders including evaluated wheat material introgressed with wild relatives (synthetic lines) (https://triticeaetoolbox.org).

The university of California Davis (UC Davis) proposes a list of public wheat diagnostic markers online (MASwheat https://maswheat.ucdavis.edu).

For genomic selection, a project has started called Genomic Open-source Breeding informatics initiative (GOBii: http://gobiiproject.org/), funded by the Bill & Melinda Gates Foundation. The objective is to develop open-source data management, marker- and genomic-assisted breeding tools (PrAPI), for under-resourced breeding programs in particular, including trainings and workshops around the world (Selby et al. 2019).

The DivSeek project in the USA tends to bridge the gap between information requirements of genebank curators, plant breeders, and more targeted upstream biological researchers. They built a cooperative information platform for phenomics and genomics and gather a collaborative network of genebanks, breeders, scientists, database and computational experts for metadata curation. The objective is to share methodologies, open-source software and best practices related to genetic resources. For maize, the SeeD project established a breeder's core of 4,000 landrace accessions that were genotyped and phenotyped, including testcross performance (http://seedsofdiscovery.org). For wheat, the Heat and Drought Wheat Improvement.

Consortium (HeDWIC, http://www.hedwic.org/) coordinated by CIMMYT aims at boosting heat and drought breeding using genomic and phenomic tools.

Then it is a long-term joint research goal to organize the conversion of information from population genomics and quantitative genetics to the development of some useful material for breeders. And public research may play an essential role in this activity, providing that means and foundings are sufficient.

4.2 Integration of Omics to Better Decipher Genome/Phenome Relationship

Elite varieties have mainly been selected for production and post-harvest qualities with less attention to other features such as drought tolerance, nutrient use efficiency or durable pest and disease resistance. The effects of these factors have been mitigated by the use of treatments such as irrigation, fertilizers, and pesticides. Now that governments promote a more sustainable agriculture, breeding for stress tolerance may become common rules once the tools and methodologies are available. A better understanding of ecophysiolocal and expression determinants is essential to breed for stress tolerance. However large-scale phenotyping of physiological traits and generating data for population genomics and other "omics" aspects, for many varieties in different conditions with biological replicates, is still not affordable. But costs are likely to drop soon (Zivy et al. 2015).

4.2.1 Sequencing Fragments with Known DNA Patterns (Target Candidates)

Instead of sequencing the whole genome of accessions, we can target exome or specific domains such as LRR that are typical of resistance genes. Jupe et al. (2013), using Resistance gene enrichment Sequencing (RenSeq), reannotated the NB-LRR gene family and rapidly mapped resistance loci in segregating populations from hexaploid bread wheat. Arora et al. (2019), using R gene enrichment sequencing, a sequence capture bait library optimized for *Ae. tauschii* NLR domains and k-mer based association genetics (AgRenSeq) on a diverse panel (195 *Ae. tauschii* accessions), rapidly cloned four rust genes (*Sr33*, *Sr46*, *Sr45*, *SrTA1662*). Using mutagenesis coupled with exome capture and NLR-baits (MutRenSeq), Steuernagel et al. (2016) rapidly cloned *Sr22* and *Sr45* genes.

4.2.2 Population Transcriptomics

With the availability of Next Generation Sequencing (NGS) technologies, the possibility to directly sequence mRNA at relatively reduced cost becomes available.

Genomic predictions using whole-genome SNPs or GWAS are limited in capturing epistasis. Because mRNA, small RNA (sRNA) sequences and metabolic data are involved in transcriptional, translational, and post-translational processes, we expect them to provide such information. For instance, GWAS on transcripts allowed detecting candidate genes controlling oil content in maize, and their sequencing to detect polymorphisms and favorable alleles (Li et al. 2013). In grain maize, they evaluated the ability of this kind of data in parental lines to predict the performance of untested hybrids. They found that mRNA data are a superior predictor for grain yield and whole-genome SNP data for grain dry matter content, while sRNA performed relatively poorly for both traits. Combining mRNA and genomic data as predictors resulted in high predictive abilities across both traits and could contribute to more efficient selection of hybrid candidates in maize (Schrag et al. 2018).

RNA sequences can differentiate between isoforms of a gene family, a widespread phenomenon in complex crop genomes, which is difficult using DNA sequences. For example, in wheat, Oono et al. (2013) discovered this way phosphate starvation-responsive genes. Ramírez-González et al. (2018) showed differential expression of homoeolog genes due to epigenetic modifications and variation in transposable elements within promoters. The measurement of tissue and stress-specific co-expression networks throughout the development allows reconstructing regulatory networks. Some kernel component candidates were found using this strategy (Wen et al. 2016).

4.2.3 Population Proteomics

Carpentier et al. (2011) identified protein polymorphisms correlated to drought tolerance using shotgun approaches in banana and Grimaud et al. (2013) found cold-acclimation-related proteins in pea. Virlouvet et al. (2011) identified the *ZmASR1* gene under an abundance proteins QTL (pQTLs), candidate for drought tolerance in maize. The same gene was also associated in tomato, grape, lily, and banana (Maskin et al. 2001; Çakir et al. 2003; Wang et al. 2005; Henry et al. 2011).

4.2.4 Population Metabolomics: Phenotypes Targeting Candidate Metabolic Pathways

Metabolomics can detect targeted primary (sugars, organic- and amino-acids...) and secondary metabolites (photosynthates necessary to biomass formation, flavonoids, sugar-phosphates, phytohormones, phytoalexins) without genome sequence information. But it is not yet possible to work on the entire metabolome. Doerfler et al. (2014) detected 15 metabolites QTLs (mQTLs) of the flavonoid-pathway for cold and light stress in *Arabidopsis thaliana*. Pathogen induced markers were identified for *Rhizoctonia solani* in potatoes (Aliferis and Jabaji 2012), fungal pathogens in soybean (Aliferis et al. 2014), and bacterial blight-resistance in rice (Wu et al. 2012). An aroma (mesifurane) candidate gene was detected in strawberry, *Fragaria* x *ananassa* (Zorrilla-Fontanesi et al. 2012). The use of metabolomics in breeding has been reviewed in Fernandez et al. (2021).

4.2.5 Population Epigenomics

Epigenomic variations are involved in the control of plant developmental processes and shaping phenotypic plasticity to the environment (Gallusci et al. 2017; Moler et al. 2019). The elucidation of epigenetic regulatory networks using DNA methylation information should improve crop models. For instance, we can predict lycopene accumulation during tomato fruit ripening (Liu et al. 2015b), anthocyanin accumulation in apple (El-Sharkawy et al. 2015), energy-use efficiency in canola lines (Hauben et al. 2009).

Concerning histone marks, as they are likely to be erased following meiosis, they are of little interest to breeding applications in sexually propagated crops. But they can be relevant for clonally propagated crops, for pathogen resistance, for instance (Jaskiewicz et al. 2011).

It is well known that DNA mutation, copy number variants or methylation, in genes, promoters or regulatory regions can affect gene expression, which modifies phenotypes in different environmental contexts. Many studies also showed that re-arrangements of loci on chromosomes, inversions, insertions of transposable elements, deletions can also lead to gene silencing. All those types of polymorphisms/annotation could help improving genomic prediction models. Molecular markers at the vicinity of genes actually tend to link more to causal variants in maize (reference). QTL effects are higher in genic regions (Wallace et al. 2014), which is consistent with the fact that a large portion of variability of gene expression is attributed to cis polymorphisms in maize (Schadt et al. 2003). Taking into account the proximity of molecular markers to genes actually improves prediction of agronomic traits in diverse populations of hybrid maize (Ramstein et al. 2020).

To facilitate and optimize those models, we still need the development of generalized methods that integrate multiple data types.

4.2.6 Integration of Different Population "Omics" Information

The long-term objective is to be able to integrate all possible "omics" information on the same samples. We will be able to detect eQTLs, pQTLs, and mQTLs and look for co-localization with molecular marker-based QTLs (cis-QTLs), giving direct access to the genes, favorable alleles, and regulatory factors outside of the gene (trans-QTL). As skills are spread in different groups, a European network named COST project was organized to help building regulation networks from integrated databases. To make it useful to breeders, the first objective is to define traits of interest for specific climatic zones or constraints.

Then, cellular phenotyping (transcriptome/proteome/metabolome) will help building more realistic models to predict phenome in the field. Models taking into account non-additive effects, nonlinear relationships between enzyme concentrations and metabolic fluxes (Fiévet et al. 2010; Vacher and Small 2019) could actually explain even more genetic variance and improve predictions.

5 Conclusion

Integration of concepts and tools of population genomics and quantitative genetics can lead to a better valorization of genetic diversity in crop (pre)breeding programs.

Advances in population genomics offer a new dimension to quantitative genetics in the form of increasing data on genetic diversity and structure, identification of new candidate genes of agronomic interest associated with signatures of selection, associations with environmental covariates and phenotypes, and prediction of genetic values of various plant genetic resources.

Genomic predictions can detect germplasm of interest in genebanks without the need of phenotyping if the calibration population is relevant and the quality of phenotyping is satisfactory. Good quality phenotyping will actually always be a cornerstone to efficient plant breeding and predictions. Genomic predictions can help to optimize the time and cost of the breeding process, allowing a transfer of budget to test a larger number of genitors and crosses. It can accelerate recurrent selection to produce pre-breeding and breeding lines that contain new favorable alleles. It can predict optimum parental contribution and mating in (pre)breeding programs to optimize short-term genetic gain but also assure long-term genetic gain by constraining germplasm diversity. Currently, the main methodological challenge here is a good estimation of marker effects and progeny variance.

Increasingly detailed multi-omic characterization of genetic resources (through genomics, transcriptomics, methylomics, proteomics) is expected to help understand and predict the genome-phenome relationship, and ultimately design ideotypes for particular growth conditions and uses. The hope is that additional layers of omics data will improve estimation of marker by environment effect. Currently, several technical hurdles are preventing industrial implementation of multi-omics approaches in the breeding process. On the fundamental level, effects of epigenetic variation on gene expression – on the background of nucleotide variation – are still difficult to detect, quantify, and generalize. Also, it remains to be seen whether genotype is a good predictor of methylome, transcriptome, and metabolome, i.e. whether training sets characterized with multi-omic data can improve genomic prediction of candidates that have been genotyped with SNPs, giving higher weights in prediction models to QTLs. Moreover, multi-omics approaches in the next generation of genomic prediction can only come with increased analytical complexity and cost. Nonetheless, recent years have witnessed an emergence and proliferation of methods designed for multi-omic data integration and analysis, and with the continuous drop of sequencing costs, multi-omics crop research will attract significant efforts in the immediate future. With a combination of multi-omic, agronomic, phenological and physiological data, supplemented with precise environment characterization (weather, soils, crop management) and targeted trialing, we are set on the path to decipher the complex GxE interactions and predict the performance of existing and new varieties in current and future environments.

For practical applications, it is necessary to integrate population genomics and other "omics" information with phenotypes in common public databases, so that robust methodologies and decision tools could be developed to convert this information into feasible protocols. In that context, one role of public research could be to develop and disseminate databases, new methodologies, and produce decision tools that could be validated by breeders in interactive projects. Public research could also coordinate the design, production, and evaluation of ready-to-use crop plant resources, pre-breeding genitors in particular.

References

Adam-Blondon A-F, Alaux M, Pommier C, Cantu D, Cheng Z-M, Cramer GR, et al. Towards an open grapevine information system. Hortic Res. 2016;3(1):1–8.

Akakpo R, Scarcelli N, Dansi A, Djedatin G, Thuillet A-C, Rhoné B, et al. Molecular basis of African yam domestication: analyses of selection point to root development, starch biosynthesis, and photosynthesis related genes. BMC Genomics. 2017;18(1):782.

Akdemir D, Isidro-Sánchez J. Efficient breeding by genomic mating. Front Genet. 2016;7:210.

Akdemir D, Isidro-Sánchez J. Design of training populations for selective phenotyping in genomic prediction. Sci Rep. 2019;9(1):1–15.

Akdemir D, Beavis W, Fritsche-Neto R, Singh AK, Isidro-Sánchez J. Multi-objective optimized genomic breeding strategies for sustainable food improvement. Heredity. 2018;122(5):672–83.

Alachiotis N, Pavlidis P. RAiSD detects positive selection based on multiple signatures of a selective sweep and SNP vectors. Commun Biol. 2018;1(1):1–11.

Alachiotis N, Stamatakis A, Pavlidis P. OmegaPlus: a scalable tool for rapid detection of selective sweeps in whole-genome datasets. Bioinformatics. 2012;28(17):2274–5.

Alexander DH, Novembre J, Lange K. Fast model-based estimation of ancestry in unrelated individuals. Genome Res. 2009;19(9):1655–64.

Aliferis KA, Jabaji S. FT-ICR/MS and GC-EI/MS metabolomics networking unravels global potato sprout's responses to Rhizoctonia solani infection. PLoS One. 2012;7(8)

Aliferis KA, Faubert D, Jabaji S. A metabolic profiling strategy for the dissection of plant defense against fungal pathogens. PLoS One. 2014;9(11)

Allaby RG, Ware RL, Kistler L. A re-evaluation of the domestication bottleneck from archaeogenomic evidence. Evol Appl. 2019;12(1):29–37.

Allard RW. Principles of plant breeding: Wiley; 1999.

Allier A, Lehermeier C, Charcosset A, Moreau L, Teyssèdre S. Improving short and long term genetic gain by accounting for within family variance in optimal cross selection. Front Genet. 2019a;10:1006.

Allier A, Moreau L, Charcosset A, Teyssèdre S, Lehermeier C. Usefulness criterion and post-selection parental contributions in multi-parental crosses: application to polygenic trait introgression. G3: Genes, Genomes, Genetics. 2019b;9(5):1469–79.

Allier A, Teyssèdre S, Lehermeier C, Charcosset A, Moreau L. Genomic prediction with a maize collaborative panel: identification of genetic resources to enrich elite breeding programs. Theor Appl Genet. 2019c:1–15.

Allier A, Teyssèdre S, Lehermeier C, Claustres B, Maltese S, Melkior S, et al. Assessment of breeding programs sustainability: application of phenotypic and genomic indicators to a North European grain maize program. Theor Appl Genet. 2019d;132(5):1321–34.

Allier A, Teyssèdre S, Lehermeier C, Moreau L, Charcosset A. Optimized breeding strategies to harness genetic resources with different performance levels. BMC Genomics. 2020;21:1–16.

Amani I, Fischer RA, Reynolds MP. Canopy temperature depression association with yield of irrigated spring wheat cultivars in a hot climate. J Agron Crop Sci. 1996;176(2):119–29.

Anderson E. Introgressive hybridization. Biol Rev. 1953;28(3):280–307.

Anderson EC, Thompson EA. A model-based method for identifying species hybrids using multilocus genetic data. Genetics. 2002;160(3):1217–29.

Andres RJ, Dunne JC, Samayoa LF, Holland JB. Enhancing crop breeding using population genomics approaches. In: Population genomics. Cham: Springer; 2020. p. 1–45.

Antao T, Lopes A, Lopes R, Beja-Pereira A, Luikart G. LOSITAN: a workbench to detect molecular adaptation based on a Fst-outlier method. BMC Bioinformatics. 2008;9(1):323.

Arora S, Steuernagel B, Gaurav K, Chandramohan S, Long Y, Matny O, et al. Resistance gene cloning from a wild crop relative by sequence capture and association genetics. Nat Biotechnol. 2019;37(2):139–43.

Bailey-Serres J, Fukao T, Ronald P, Ismail A, Heuer S, Mackill D. Submergence tolerant rice: SUB1's journey from landrace to modern cultivar. Rice. 2010;3(2):138–47.

Balfourier F, Bouchet S, Robert S, De Oliveira R, Rimbert H, Kitt J, et al. Worldwide phylogeography and history of wheat genetic diversity. Sci Adv. 2019;5(5):eaav0536.

Balkenhol N, Dudaniec RY, Krutovsky KV, Johnson JS, Cairns DM, Segelbacher G, et al. Landscape genomics: understanding relationships between environmental heterogeneity and genomic characteristics of populations. In: Rajora OP, editor. Population genomics: concepts, approaches and applications. Cham: Springer Nature Switzerland AG; 2019. p. 261–322.

Bandeirae Sousa M, Cuevas J, de Oliveira Couto EG, Pérez-Rodríguez P, Jarquín D, Fritsche-Neto R, et al. Genomic-enabled prediction in maize using kernel models with genotype x environment interaction. G3: Genes, Genomes, Genetics. 2017;7(6):1995–2014.

Bao Y, Kurle JE, Anderson G, Young ND. Association mapping and genomic prediction for resistance to sudden death syndrome in early maturing soybean germplasm. Mol Breed. 2015;35(6):128.

Barakat A, Yassin NBM, Park JS, Choi A, Herr J, Carlson JE. Comparative and phylogenomic analyses of cinnamoyl-CoA reductase and cinnamoyl-CoA-reductase-like gene family in land plants. Plant Sci. 2011;181(3):249–57.

Bauer E, Falque M, Walter H, Bauland C, Camisan C, Campo L, et al. Intraspecific variation of recombination rate in maize. Genome Biol. 2013;14(9):R103.

Beaumont MA, Nichols RA. Evaluating loci for use in the genetic analysis of population structure. Proc Biol Sci. 1996;263(1377):1619–26.

Beavis WD. QTL analyses: power, precision, and accuracy. Molecular dissection of complex traits. Boca Raton: CRC Press; 1998. p. 145–62.

Beavis W, Smith O, Grant D, Fincher R. Identification of quantitative trait loci using a small sample of topcrossed and F4 progeny from maize. Crop Sci. 1994;34(4):882–96.

Bellis ES, Kelly EA, Lorts CM, Gao H, DeLeo VL, Rouhan G, et al. Genomics of sorghum local adaptation to a parasitic plant. Proc Natl Acad Sci. 2020;117(8):4243–51.

Bellucci E, Bitocchi E, Ferrarini A, Benazzo A, Biagetti E, Klie S, et al. Decreased nucleotide and expression diversity and modified coexpression patterns characterize domestication in the common bean. Plant Cell. 2014;26(5):1901–12.

Ben-Sadoun S, Rincent R, Auzanneau J, Oury FX, Rolland B, Heumez E, et al. Economical optimization of a breeding scheme by selective phenotyping of the calibration set in a multi-trait context: application to bread making quality. Theor Appl Genet. 2020;133:2197–212.

Bernardo R. Genomewide selection for rapid introgression of exotic germplasm in maize. Crop Sci. 2009;49(2):419–25.

Bernardo R. Genomewide selection of parental inbreds: classes of loci and virtual biparental populations. Crop Sci. 2014;54(6):2586–95.

Bernardo R, Charcosset A. Usefulness of gene information in marker-assisted recurrent selection: a simulation appraisal. Crop Sci. 2006;46(2):614–21.

Bernardo R, Yu J. Prospects for genomewide selection for quantitative traits in maize. Crop Sci. 2007;47(3):1082–90.

Berthouly-Salazar C, Thuillet A-C, Rhoné B, Mariac C, Ousseini IS, Couderc M, et al. Genome scan reveals selection acting on genes linked to stress response in wild pearl millet. Mol Ecol. 2016;25(21):5500–12.

Beukelaer HD, Meyer GD, Fack V. Heuristic exploitation of genetic structure in marker-assisted gene pyramiding problems. BMC Genet. 2015;16

Bijma P, Woolliams JA. Prediction of genetic contributions and generation intervals in populations with overlapping generations under selection. Genetics. 1999;151(3):1197–210.

Bijma P, Wientjes YC, Calus MP. Breeding top genotypes and accelerating response to recurrent selection by selecting parents with greater Gametic variance. Genetics. 2020;214(1):91–107.

Bogard M, Ravel C, Paux E, Bordes J, Balfourier F, Chapman SC, et al. Predictions of heading date in bread wheat (Triticum aestivum L.) using QTL-based parameters of an ecophysiological model. J Exp Bot. 2014;65(20):5849–65.

Bohn M, Utz HF, Melchinger AE. Genetic similarities among winter wheat cultivars determined on the basis of RFLPs, AFLPs, and SSRs and their use for predicting progeny variance. Crop Sci. 1999;39(1):228–37.

Bonhomme M, Chevalet C, Servin B, Boitard S, Abdallah J, Blott S, et al. Detecting selection in population trees: the Lewontin and Krakauer test extended. Genetics. 2010;186(1):241–62.

Bonk S, Reichelt M, Teuscher F, Segelke D, Reinsch N. Mendelian sampling covariability of marker effects and genetic values. Genet Sel Evol. 2016;48(1):36.

Bouchet S, Olatoye MO, Marla SR, Perumal R, Tesso T, Yu J, et al. Increased power to dissect adaptive traits in global sorghum diversity using a nested association mapping population. Genetics. 2017;206(2):573–85.

Brancourt-Hulmel M. Crop diagnosis and probe genotypes for interpreting genotype environment interaction in winter wheat trials. Theor Appl Genet. 1999;99(6):1018–30.

Brancourt-Hulmel M, Denis JB, Lecomte C. Determining environmental covariates which explain genotype environment interaction in winter wheat through probe genotypes and biadditive factorial regression. Theor Appl Genet. 2000;100(2):285–98.

Brard S, Ricard A. Is the use of formulae a reliable way to predict the accuracy of genomic selection? J Anim Breed Genet. 2015;132(3):207–17.

Brauner PC, Müller D, Schopp P, Böhm J, Bauer E, Schön C-C, et al. Genomic prediction within and among doubled-haploid libraries from maize landraces. Genetics. 2018;210(4):1185–96.

Brauner PC, Schipprack W, Utz HF, Bauer E, Mayer M, Schön C-C, et al. Testcross performance of doubled haploid lines from European flint maize landraces is promising for broadening the genetic base of elite germplasm. Theor Appl Genet. 2019;132(6):1897–908.

Braverman JM, Hudson RR, Kaplan NL, Langley CH, Stephan W. The hitchhiking effect on the site frequency spectrum of DNA polymorphisms. Genetics. 1995;140(2):783–96.

Brisbane JR, Gibson JP. Balancing selection response and rate of inbreeding by including genetic relationships in selection decisions. Theor Appl Genet. 1995;91(3):421–31.

Brisson N, Gate P, Gouache D, Charmet G, Oury FX, Huard F. Why are wheat yields stagnating in Europe? A comprehensive data analysis for France. Field Crop Res. 2010;119(1):201–12.

Broman KW, Wu H, Sen S, Churchill GA. R/qtl: QTL mapping in experimental crosses. Bioinformatics. 2003;19(7):889–90.

Brown AHD, Clegg MT. Isozyme assessment of plant genetic resources. Isozymes Curr Top Biol Med Res. 1983;11:285–95.

Bruce RW, Torkamaneh D, Grainger C, Belzile F, Eskandari M, Rajcan I. Genome-wide genetic diversity is maintained through decades of soybean breeding in Canada. Theor Appl Genet. 2019;132(11):3089–100.

Bruskiewich RM, Cosico AB, Eusebio W, Portugal AM, Ramos LM, Reyes MT, et al. Linking genotype to phenotype: the international rice information system (IRIS). Bioinformatics. 2003;19(suppl_1):i63–5.

Buckler ES, Holland JB, Bradbury PJ, Acharya CB, Brown PJ, Browne C, et al. The genetic architecture of maize flowering time. Science. 2009;325(5941):714–8.

Bulmer MG. The effect of selection on genetic variability. Am Nat. 1971;105(943):201.

Burgarella C, Barnaud A, Kane NA, Jankowski F, Scarcelli N, Billot C, et al. Adaptive introgression: an untapped evolutionary mechanism for crop adaptation. Front Plant Sci. 2019;10

Burgueno J, de los Campos G, Weigel K, Crossa J. Genomic prediction of breeding values when modeling genotype x environment interaction using pedigree and dense molecular markers. Crop Sci. 2012;52:707.

Bustos-Korts D, Malosetti M, Chapman S, Biddulph B, van Eeuwijk F. Improvement of predictive ability by uniform coverage of the target genetic space. G3: Genes, Genomes, Genetics. 2016;6(11):3733.

Çakir B, Agasse A, Gaillard C, Saumonneau A, Delrot S, Atanassova R. A grape ASR protein involved in sugar and abscisic acid signaling. Plant Cell. 2003;15(9):2165–80.

Calus MP, Veerkamp RF. Accuracy of multi-trait genomic selection using different methods. Genet Sel Evol. 2011;43(1):1.

Canzar S, El-Kebir M. A mathematical programming approach to marker-assisted gene pyramiding. In: International workshop on algorithms in bioinformatics. Berlin: Springer; 2011. p. 26–38.

Carpentier SC, Panis B, Renaut J, Samyn B, Vertommen A, Vanhove A-C, et al. The use of 2D-electrophoresis and de novo sequencing to characterize inter-and intra-cultivar protein polymorphisms in an allopolyploid crop. Phytochemistry. 2011;72(10):1243–50.

Cavanagh C, Morell M, Mackay I, Powell W. From mutations to MAGIC: resources for gene discovery, validation and delivery in crop plants. Curr Opin Plant Biol. 2008;11(2):215–21.

Cavanagh CR, Chao S, Wang S, Huang BE, Stephen S, Kiani S, et al. Genome-wide comparative diversity uncovers multiple targets of selection for improvement in hexaploid wheat landraces and cultivars. Proc Natl Acad Sci. 2013;110(20):8057–62.

Charmet G, Robert N, Perretant M, Gay G, Sourdille P, Groos C, et al. Marker-assisted recurrent selection for cumulating additive and interactive QTLs in recombinant inbred lines. Theor Appl Genet. 1999;99(7):1143–8.

Chen J, Ding J, Ouyang Y, Du H, Yang J, Cheng K, et al. A triallelic system of S5 is a major regulator of the reproductive barrier and compatibility of indica–japonica hybrids in rice. Proc Natl Acad Sci. 2008;105(32):11436–41.

Chen H, Patterson N, Reich D. Population differentiation as a test for selective sweeps. Genome Res. 2010;20(3):393–402.

Cheng H, Liu J, Wen J, Nie X, Xu L, Chen N, et al. Frequent intra- and inter-species introgression shapes the landscape of genetic variation in bread wheat. Genome Biol. 2019;20(1):136.

Chenu K, Chapman SC, Hammer GL, Mclean G, Salah HBH, Tardieu F. Short-term responses of leaf growth rate to water deficit scale up to whole-plant and crop levels: an integrated modelling approach in maize. Plant Cell Environ. 2008;31(3):378–91.

Chenu K, Cooper M, Hammer GL, Mathews KL, Dreccer MF, Chapman SC. Environment characterization as an aid to wheat improvement: interpreting genotype–environment interactions by modelling water-deficit patterns in North-Eastern Australia. J Exp Bot. 2011;62(6):1743–55.

Choi K, Henderson IR. Meiotic recombination hotspots – a comparative view. Plant J. 2015;83(1):52–61.

Christopher J, Richard C, Chenu K, Christopher M, Borrell A, Hickey L. Integrating rapid phenotyping and speed breeding to improve stay-green and root adaptation of wheat in changing, water-limited, Australian environments. Procedia Environ Sci. 2015;29:175–6.

Civáň P, Brown TA. Role of genetic introgression during the evolution of cultivated rice (Oryza sativa L.). BMC Evol Biol. 2018;18(1):57.

Civáň P, Ali S, Batista-Navarro R, Drosou K, Ihejieto C, Chakraborty D, et al. Origin of the aromatic group of cultivated rice (Oryza sativa L.) traced to the Indian subcontinent. Genome Biol Evol. 2019;11(3):832–43.

Cole JB, VanRaden PM. Use of haplotypes to estimate Mendelian sampling effects and selection limits. J Anim Breed Genet. 2011;128(6):446–55.

Collevatti RG, dos Santos JS, Rosa FF, Amaral TS, Chaves LJ, Ribeiro MC. Multi-scale landscape influences on genetic diversity and adaptive traits in a neotropical savanna tree. Front Genet. 2020;11:259.

Combs E, Bernardo R. Genomewide selection to introgress semidwarf maize germplasm into US Corn Belt inbreds. Crop Sci. 2013;53(4):1427–36.

Coop G, Witonsky D, Di Rienzo A, Pritchard JK. Using environmental correlations to identify loci underlying local adaptation. Genetics. 2010;185(4):1411–23.

Cooper HD, Spillane C, Hodgkin T, Cooper H. Broadening the genetic base of crops: an overview. In: Broadening the genetic base of crop production. New York: CABI; 2001. p. 1–23.

Cooper M, Technow F, Messina C, Gho C, Totir LR. Use of crop growth models with whole-genome prediction: application to a maize multienvironment trial. Crop Sci. 2016;56(5):2141–56.

Cowling WA, Li L, Siddique KH, Henryon M, Berg P, Banks RG, et al. Evolving gene banks: improving diverse populations of crop and exotic germplasm with optimal contribution selection. J Exp Bot. 2017;68(8):1927–39.

Crain J, Mondal S, Rutkoski J, Singh RP, Poland J. Combining high-throughput phenotyping and genomic information to increase prediction and selection accuracy in wheat breeding. Plant Genome. 2018;11(1)

Crossa J, Campos GDL, Perez P, Gianola D, Burgueno J, Araus JL, et al. Prediction of genetic values of quantitative traits in plant breeding using pedigree and molecular markers. Genetics. 2010;186(2):713–24.

Crossa J, Perez P, Hickey J, Burgueno J, Ornella L, Cerón-Rojas J, et al. Genomic prediction in CIMMYT maize and wheat breeding programs. Heredity. 2014;112(1):48–60.

Crossa J, Jarquín D, Franco J, Pérez-Rodríguez P, Burgueño J, Saint-Pierre C, et al. Genomic prediction of gene bank wheat landraces. G3: Genes, Genomes, Genetics. 2016;6(7):1819–34.

Cruickshank TE, Hahn MW. Reanalysis suggests that genomic islands of speciation are due to reduced diversity, not reduced gene flow. Mol Ecol. 2014;23(13):3133–57.

Cuevas J, Crossa J, Montesinos-López OA, Burgueño J, Pérez-Rodríguez P, de los Campos G. Bayesian genomic prediction with genotype x environment interaction kernel models. G3: Genes, Genomes, Genetics. 2017;7(1):41–53.

Cuevas J, Granato I, Fritsche-Neto R, Montesinos-Lopez OA, Burgueño J, e Sousa MB, et al. Genomic-enabled prediction Kernel models with random intercepts for multi-environment trials. G3: Genes, Genomes, Genetics. 2018;8(4):1347–65.

Daetwyler HD, Villanueva B, Woolliams JA. Accuracy of predicting the genetic risk of disease using a genome-wide approach. PLoS One. 2008;3(10)

Daetwyler HD, Hayden MJ, Spangenberg GC, Hayes BJ. Selection on optimal haploid value increases genetic gain and preserves more genetic diversity relative to genomic selection. Genetics. 2015;200(4):1341–8.

Danguy des Deserts A, Bouchet S, Sourdille P, Servin B. Evolution of recombination landscapes in diverging populations of bread wheat. bioRxiv. 2021;13(8):evab152.

De Beukelaer H, Badke Y, Fack V, De Meyer G. Moving beyond managing realized genomic relationship in long-term genomic selection. Genetics. 2017;206(2):1127–38.

De Coninck A, De Baets B, Kourounis D, Verbosio F, Schenk O, Maenhout S, et al. Needles: toward large-scale genomic prediction with marker-by-environment interaction. Genetics. 2016;203(1):543–55.

Dempewolf H, Baute G, Anderson J, Kilian B, Smith C, Guarino L. Past and future use of wild relatives in crop breeding. Crop Sci. 2017;57(3):1070–82.

Deutsch CA, Tewksbury JJ, Tigchelaar M, Battisti DS, Merrill SC, Huey RB, et al. Increase in crop losses to insect pests in a warming climate. Science. 2018;361(6405):916–9.

Dias-Alves T, Mairal J, Blum MGB. Loter: a software package to infer local ancestry for a wide range of species. Mol Biol Evol. 2018;35(9):2318–26.

Doerfler H, Sun X, Wang L, Engelmeier D, Lyon D, Weckwerth W. mzGroupAnalyzer-predicting pathways and novel chemical structures from untargeted high-throughput metabolomics data. PLoS One. 2014;9(5)

dos Santos JPR, de Castro Vasconcellos RC, Pires LPM, Balestre M, Von Pinho RG. Inclusion of dominance effects in the multivariate GBLUP model. PLoS One. 2016;11(4)

Doyle JJ. 5S ribosomal gene variation in the soybean and its progenitor. Theor Appl Genet. 1988;75(4):621–4.

Dudley JW. Theory for identification and use of exotic germplasm in maize breeding programs. Maydica. 1984;29:391–407.

Dudley JW. Evaluation of maize populations as sources of favorable alleles. Crop Sci. 1988;28(3):486–91.

Durand EY, Patterson N, Reich D, Slatkin M. Testing for ancient admixture between closely related populations. Mol Biol Evol. 2011;28(8):2239–52.

Ellis JG, Lagudah ES, Spielmeyer W, Dodds PN. The past, present and future of breeding rust resistant wheat. Front Plant Sci. 2014;5:641.

El-Sharkawy I, Liang D, Xu K. Transcriptome analysis of an apple (Malus x domestica) yellow fruit somatic mutation identifies a gene network module highly associated with anthocyanin and epigenetic regulation. J Exp Bot. 2015;66(22):7359–76.

Endelman JB, Atlin GN, Beyene Y, Semagn K, Zhang X, Sorrells ME, et al. Optimal design of preliminary yield trials with genome-wide markers. Crop Sci. 2014;54(1):48–59.

Excoffier L, Lischer H. Arlequin ver 3.5 user manual; an integrated software package for population genetics data analysis. Swiss Institute of Bioinformatics. 2009.

Falconer DS, Mackay TF, Frankham R. Introduction to quantitative genetics (4th edn). Trends Genet. 1996;12(7):280.

Fariello MI, Boitard S, Naya H, San Cristobal M, Servin B. Detecting signatures of selection through haplotype differentiation among hierarchically structured populations. Genetics. 2013;193(3):929–41.

Faux A-M, Gorjanc G, Gaynor RC, Battagin M, Edwards SM, Wilson DL, et al. AlphaSim: software for breeding program simulation. Plant Genome. 2016;9(3):1–14.

Feng L, Sebastian S, Smith S, Cooper M. Temporal trends in SSR allele frequencies associated with long-term selection for yield of maize. Maydica. 2006;51(2):293.

Fernandes SB, Dias KO, Ferreira DF, Brown PJ. Efficiency of multi-trait, indirect, and trait-assisted genomic selection for improvement of biomass sorghum. Theor Appl Genet. 2018;131 (3):747–55.

Fernandez O, Millet EJ, Rincent R, Prigent S, Pétriacq P, Gibon Y. Chapter seven – plant metabolomics and breeding. In: Pétriacq P, Bouchereau A, editors. Plant metabolomics in full swing, Advances in botanical research, vol. 98. Cambridge: Academic Press; 2021. p. 207–35.

Ferrer-Admetlla A, Liang M, Korneliussen T, Nielsen R. On detecting incomplete soft or hard selective sweeps using haplotype structure. Mol Biol Evol. 2014;31(5):1275–91.

Fiévet JB, Dillmann C, de Vienne D. Systemic properties of metabolic networks lead to an epistasis-based model for heterosis. Theor Appl Genet. 2010;120(2):463.

Fisher SRA. The correlation between relatives on the supposition of Mendelian inheritance. Trans R Soc Edinb. 1918;52:399–433.

Fisher RA. The genetical theory of natural selection. Oxford: Clarendon Press; 1930. 272 p.

Fitzpatrick MC, Keller SR. Ecological genomics meets community-level modelling of biodiversity: mapping the genomic landscape of current and future environmental adaptation. Ecol Lett. 2015;18(1):1–16.

Foll M, Gaggiotti O. A genome-scan method to identify selected loci appropriate for both dominant and codominant markers: a Bayesian perspective. Genetics. 2008;180(2):977–93.

Foncéka D, Hodo-Abalo T, Rivallan R, Faye I, Sall MN, Ndoye O, et al. Genetic mapping of wild introgressions into cultivated peanut: a way toward enlarging the genetic basis of a recent allotetraploid. BMC Plant Biol. 2009;9(1):103.

Fonceka D, Tossim H-A, Rivallan R, Vignes H, Lacut E, de Bellis F, et al. Construction of chromosome segment substitution lines in peanut (Arachis hypogaea L.) using a wild synthetic and QTL mapping for plant morphology. PLoS One. 2012;7(11):e48642.

Fradgley N, Gardner KA, Cockram J, Elderfield J, Hickey JM, Howell P, et al. A large-scale pedigree resource of wheat reveals evidence for adaptation and selection by breeders. PLoS Biol. 2019;17(2):e3000071.

Frichot E, Schoville SD, Bouchard G, François O. Testing for associations between loci and environmental gradients using latent factor mixed models. Mol Biol Evol. 2013;30(7):1687–99.

Frichot E, Mathieu F, Trouillon T, Bouchard G, François O. Fast and efficient estimation of individual ancestry coefficients. Genetics. 2014;196(4):973–83.

Fu Y-B. Impact of plant breeding on genetic diversity of agricultural crops: searching for molecular evidence. Plant Genet Resour. 2006;4(1):71–8.

Fu Y-B. Understanding crop genetic diversity under modern plant breeding. Theor Appl Genet. 2015;128(11):2131–42.

Fu J, Falke KC, Thiemann A, Schrag TA, Melchinger AE, Scholten S, et al. Partial least squares regression, support vector machine regression, and transcriptome-based distances for prediction of maize hybrid performance with gene expression data. Theor Appl Genet. 2012;124 (5):825–33.

Fustier M-A, Martínez-Ainsworth NE, Aguirre-Liguori JA, Venon A, Corti H, Rousselet A, et al. Common gardens in teosintes reveal the establishment of a syndrome of adaptation to altitude. PLoS Genet. 2019;15(12):e1008512.

Gallusci P, Dai Z, Génard M, Gauffretau A, Leblanc-Fournier N, Richard-Molard C, et al. Epigenetics for plant improvement: current knowledge and modeling avenues. Trends Plant Sci. 2017;22(7):610–23.

Gao Z, Zeng D, Cheng F, Tian Z, Guo L, Su Y, et al. ALK, the key gene for gelatinization temperature, is a modifier gene for gel consistency in rice. J Integr Plant Biol. 2011;53(9):756–65.

García-Ruiz A, Cole JB, VanRaden PM, Wiggans GR, Ruiz-López FJ, Van Tassell CP. Changes in genetic selection differentials and generation intervals in US Holstein dairy cattle as a result of genomic selection. Proc Natl Acad Sci U S A. 2016;113(28):E3995.

Gaynor RC, Gorjanc G, Bentley AR, Ober ES, Howell P, Jackson R, et al. A two-part strategy for using genomic selection to develop inbred lines. Crop Sci. 2017;57(5):2372–86.

Gerke JP, Edwards JW, Guill KE, Ross-Ibarra J, McMullen MD. The genomic impacts of drift and selection for hybrid performance in maize. Genetics. 2015;201(3):1201–11.

Ghosh S, Watson A, Gonzalez-Navarro OE, Ramirez-Gonzalez RH, Yanes L, Mendoza-Suárez M, et al. Speed breeding in growth chambers and glasshouses for crop breeding and model plant research. Nat Protoc. 2018;13(12):2944–63.

Glaszmann JC. Isozymes and classification of Asian rice varieties. Theoret Appl Genetics. 1987;74(1):21–30.

Glaszmann J, Kilian B, Upadhyaya H, Varshney R. Accessing genetic diversity for crop improvement. Curr Opin Plant Biol. 2010;13(2):167–73.

Glémin S, Bataillon T. A comparative view of the evolution of grasses under domestication. New Phytol. 2009;183(2):273–90.

Goddard M. Genomic selection: prediction of accuracy and maximisation of long term response. Genetica. 2009;136(2):245–57.

Goiffon M, Kusmec A, Wang L, Hu G, Schnable PS. Improving response in genomic selection with a population-based selection strategy: optimal population value selection. Genetics. 2017;206(3):1675.

Goldberg DE, Holland JH. Genetic algorithms and machine learning. Mach Learn. 1988;3(2):95–9.

Gompert Z, Egan SP, Barrett RD, Feder JL, Nosil P. Multilocus approaches for the measurement of selection on correlated genetic loci. Mol Ecol. 2017;26(1):365–82.

Goodman MM. Broadening the genetic diversity in maize breeding by use of exotic germplasm. In: Genetics and exploitation of heterosis in crops, ASA, CSSA, and SSSA books; 1999. p. 139–48.

Goodman MM. Broadening the US maize germplasm base. Maydica. 2005;50(3/4):203.

Goodman MM, Moreno J, Castillo F, Holley RN, Carson ML. Using tropical maize germplasm for temperate breeding. Maydica. 2000;45(3):221–34.

Gorjanc G, Hickey JM. AlphaMate: a program for optimizing selection, maintenance of diversity and mate allocation in breeding programs. Bioinformatics. 2018;34(19):3408–11.

Gorjanc G, Jenko J, Hearne SJ, Hickey JM. Initiating maize pre-breeding programs using genomic selection to harness polygenic variation from landrace populations. BMC Genomics. 2016;17(1):30.

Gorjanc G, Gaynor RC, Hickey JM. Optimal cross selection for long-term genetic gain in two-part programs with rapid recurrent genomic selection. Theor Appl Genet. 2018;131(9):1953–66.

Granato I, Cuevas J, Luna-Vázquez F, Crossa J, Montesinos-López O, Burgueño J, et al. BGGE: a new package for genomic-enabled prediction incorporating genotype × environment interaction models. G3: Genes, Genomes, Genetics. 2018;8(9):3039–47.

Green RE, Krause J, Briggs AW, Maricic T, Stenzel U, Kircher M, et al. A draft sequence of the Neandertal genome. Science. 2010;328(5979):710–22.

Grimaud F, Renaut J, Dumont E, Sergeant K, Lucau-Danila A, Blervacq A-S, et al. Exploring chloroplastic changes related to chilling and freezing tolerance during cold acclimation of pea (Pisum sativum L.). J Proteomics. 2013;80:145–59.

Guarino L, Lobell DB. A walk on the wild side. Nat Clim Change. 2011;1(8):374–5.

Guillot G, Vitalis R, le Rouzic A, Gautier M. Detecting correlation between allele frequencies and environmental variables as a signature of selection. A fast computational approach for genome-wide studies. Spat Stat. 2014;8:145–55.

Günther T, Coop G. Robust identification of local adaptation from allele frequencies. Genetics. 2013;195(1):205–20.

Guo G, Zhao F, Wang Y, Zhang Y, Du L, Su G. Comparison of single-trait and multiple-trait genomic prediction models. BMC Genet. 2014;15(1):30.

Gur A, Semel Y, Cahaner A, Zamir D. Real time QTL of complex phenotypes in tomato interspecific introgression lines. Trends Plant Sci. 2004;9(3):107–9.

Habier D, Fernando R, Dekkers J. The impact of genetic relationship information on genome-assisted breeding values. Genetics. 2007;177(4):2389.

Habier D, Tetens J, Seefried F-R, Lichtner P, Thaller G. The impact of genetic relationship information on genomic breeding values in German Holstein cattle. Genet Sel Evol. 2010;42(1):5.

Hajjar R, Hodgkin T. The use of wild relatives in crop improvement: a survey of developments over the last 20 years. Euphytica. 2007;156(1–2):1–13.

Hallander J, Waldmann P. Optimization of selection contribution and mate allocations in monoecious tree breeding populations. BMC Genet. 2009a;10(1):70.

Hallander J, Waldmann P. Optimum contribution selection in large general tree breeding populations with an application to Scots pine. Theor Appl Genet. 2009b;118(6):1133–42.

Hallauer AR, Sears JH. Integrating exotic germplasm into corn belt maize breeding programs 1. Crop Sci. 1972;12(2):203–6.

Han Y, Zhao X, Liu D, Li Y, Lightfoot DA, Yang Z, et al. Domestication footprints anchor genomic regions of agronomic importance in soybeans. New Phytol. 2016;209(2):871–84.

Han Y, Cameron JN, Wang L, Beavis WD. The predicted cross value for genetic introgression of multiple alleles. Genetics. 2017;205(4):1409–23.

Hancock AM, Brachi B, Faure N, Horton MW, Jarymowycz LB, Sperone FG, et al. Adaptation to climate across the Arabidopsis thaliana genome. Science. 2011;334(6052):83–6.

Hardigan MA, Laimbeer FPE, Newton L, Crisovan E, Hamilton JP, Vaillancourt B, et al. Genome diversity of tuber-bearing Solanum uncovers complex evolutionary history and targets of domestication in the cultivated potato. Proc Natl Acad Sci U S A. 2017;114(46):E9999.

Hauben M, Haesendonckx B, Standaert E, Kelen KVD, Azmi A, Akpo H, et al. Energy use efficiency is characterized by an epigenetic component that can be directed through artificial selection to increase yield. Proc Natl Acad Sci U S A. 2009;106(47):20109–14.

Haudry A, Cenci A, Ravel C, Bataillon T, Brunel D, Poncet C, et al. Grinding up wheat: a massive loss of nucleotide diversity since domestication. Mol Biol Evol. 2007;24(7):1506–17.

Hayashi T, Iwata H. A Bayesian method and its variational approximation for prediction of genomic breeding values in multiple traits. BMC Bioinformatics. 2013;14(1):34.

Hayes B, Bowman P, Chamberlain A, Goddard M. Invited review: genomic selection in dairy cattle: progress and challenges. J Dairy Sci. 2009;92(2):433.

Heffner EL, Lorenz AJ, Jannink J-L, Sorrells ME. Plant breeding with genomic selection: gain per unit time and cost. Crop Sci. 2010;50(5):1681–90.

Heffner EL, Jannink J-L, Iwata H, Souza E, Sorrells ME. Genomic selection accuracy for grain quality traits in biparental wheat populations. Crop Sci. 2011a;51(6):2597–606.

Heffner EL, Jannink J-L, Sorrells ME. Genomic selection accuracy using multifamily prediction models in a wheat breeding program. Plant Genome. 2011b;4(1):65–75.

Henderson CR. Best linear unbiased estimation and prediction under a selection model. Biometrics. 1975:423–47.

Henderson CR, Quaas RL. Multiple trait evaluation using relatives' records. J Anim Sci. 1976;43(6):1188–97.

Henry IM, Carpentier SC, Pampurova S, Van Hoylandt A, Panis B, Swennen R, et al. Structure and regulation of the Asr gene family in banana. Planta. 2011;234(4):785.

Heslot N, Feoktistov V. Optimization of selective phenotyping and population design for genomic prediction. bioRxiv. 2017:172064.

Heslot N, Akdemir D, Sorrells ME, Jannink J-L. Integrating environmental covariates and crop modeling into the genomic selection framework to predict genotype by environment interactions. Theor Appl Genet. 2014;127(2):463–80.

Hickey LT, Germán SE, Pereyra SA, Diaz JE, Ziems LA, Fowler RA, et al. Speed breeding for multiple disease resistance in barley. Euphytica. 2017;213(3):64.

Hoban S, Kelley JL, Lotterhos KE, Antolin MF, Bradburd G, Lowry DB, et al. Finding the genomic basis of local adaptation: pitfalls, practical solutions, and future directions. Am Nat. 2016;188(4):379–97.

Holland JH. Outline for a logical theory of adaptive systems. J ACM. 1962;9(3):297–314.

Hospital F, Charcosset A. Marker-assisted introgression of quantitative trait loci. Genetics. 1997;147(3):1469–85.

Hospital F, Goldringer I, Openshaw S. Efficient marker-based recurrent selection for multiple quantitative trait loci. Genet Res. 2000;75(3):357–68.

Huang L, Raats D, Sela H, Klymiuk V, Lidzbarsky G, Feng L, et al. Evolution and adaptation of wild emmer wheat populations to biotic and abiotic stresses. Annu Rev Phytopathol. 2016;54:279–301.

Hufford MB, Xu X, van Heerwaarden J, Pyhajarvi T, Chia J-M, Cartwright RA, et al. Comparative population genomics of maize domestication and improvement. Nat Genet. 2012;44(7):808–11.

Hufford MB, Lubinksy P, Pyhäjärvi T, Devengenzo MT, Ellstrand NC, Ross-Ibarra J. The genomic signature of crop-wild introgression in maize. PLoS Genet. 2013;9(5):e1003477.

Hung HY, Browne C, Guill K, Coles N, Eller M, Garcia A, et al. The relationship between parental genetic or phenotypic divergence and progeny variation in the maize nested association mapping population. Heredity. 2012;108(5):490–9.

Hyten DL, Song Q, Zhu Y, Choi I-Y, Nelson RL, Costa JM, et al. Impacts of genetic bottlenecks on soybean genome diversity. Proc Natl Acad Sci. 2006;103(45):16666–71.

Imai I, Kimball JA, Conway B, Yeater KM, McCouch SR, McClung A. Validation of yield-enhancing quantitative trait loci from a low-yielding wild ancestor of rice. Mol Breeding. 2013;32(1):101–20.

Iwata H, Hayashi T, Terakami S, Takada N, Saito T, Yamamoto T. Genomic prediction of trait segregation in a progeny population: a case study of Japanese pear (Pyrus pyrifolia). BMC Genet. 2013;14(1):81.

Jagadish SVK, Bahuguna RN, Djanaguiraman M, Gamuyao R, Prasad PVV, Craufurd PQ. Implications of high temperature and elevated CO_2 on flowering time in plants. Front Plant Sci. 2016;7

James JW, McBride G. The spread of genes by natural and artificial selection in closed poultry flock. J Genet. 1958;56(1):55.

Jannink J-L. Dynamics of long-term genomic selection. Genet Sel Evol. 2010;42(1):35.

Jansen RC. Interval mapping of multiple quantitative trait loci. Genetics. 1993;135(1):205–11.

Jarquín D, Crossa J, Lacaze X, Du Cheyron P, Daucourt J, Lorgeou J, et al. A reaction norm model for genomic selection using high-dimensional genomic and environmental data. Theor Appl Genet. 2014;127(3):595–607.

Jaskiewicz M, Conrath U, Peterhänsel C. Chromatin modification acts as a memory for systemic acquired resistance in the plant stress response. EMBO Rep. 2011;12(1):50–5.

Jia Y, Jannink J-L. Multiple-trait genomic selection methods increase genetic value prediction accuracy. Genetics. 2012;192(4):1513–22.

Jiang J, Zhang Q, Ma L, Li J, Wang Z, Liu JF. Joint prediction of multiple quantitative traits using a Bayesian multivariate antedependence model. Heredity. 2015;115(1):29–36.

Jombart T, Pontier D, Dufour A-B. Genetic markers in the playground of multivariate analysis. Heredity. 2009;102(4):330–41.

Jordan D, Mace E, Cruickshank A, Hunt C, Henzell R. Exploring and exploiting genetic variation from unadapted sorghum germplasm in a breeding program. Crop Sci. 2011;51(4):1444–57.

Jupe F, Witek K, Verweij W, Śliwka J, Pritchard L, Etherington GJ, et al. Resistance gene enrichment sequencing (R en S eq) enables reannotation of the NB-LRR gene family from sequenced plant genomes and rapid mapping of resistance loci in segregating populations. Plant J. 2013;76(3):530–44.

Kemper KE, Bowman PJ, Pryce JE, Hayes BJ, Goddard ME. Long-term selection strategies for complex traits using high-density genetic markers. J Dairy Sci. 2012;95(8):4646–56.

Kerr RJ, Goddard ME, Jarvis SF. Maximising genetic response in tree breeding with constraints on group coancestry. Silvae Genet. 1998;47(2):165–73.

Kim Y, Nielsen R. Linkage disequilibrium as a signature of selective sweeps. Genetics. 2004;167 (3):1513–24.

Kim Y, Stephan W. Detecting a local signature of genetic hitchhiking along a recombining chromosome. Genetics. 2002;160(2):765–77.

Kinghorn BP. An algorithm for efficient constrained mate selection. Genet Sel Evol. 2011;43(1):4.

Kinghorn BP, Banks R, Gondro C, Kremer VD, Meszaros SA, Newman S, et al. Strategies to exploit genetic variation while maintaining diversity. In: Adaptation and fitness in animal populations. Springer; 2009. p. 191–200.

Klein RR, Mullet JE, Jordan DR, Miller FR, Rooney WL, Menz MA, et al. The effect of tropical sorghum conversion and inbred development on genome diversity as revealed by high-resolution genotyping. Crop Sci. 2008;48(Supplement_1):S-12-S-26.

Ko J-H, Prassinos C, Keathley D, Han K-H, Li C. Novel aspects of transcriptional regulation in the winter survival and maintenance mechanism of poplar. Tree Physiol. 2011;31(2):208–25.

Kulathinal RJ, Stevison LS, Noor MAF. The genomics of speciation in drosophila: diversity, divergence, and introgression estimated using low-coverage genome sequencing. PLoS Genet. 2009;5(7)

Kuraparthy V, Sood S, Chhuneja P, Dhaliwal HS, Kaur S, Bowden RL, et al. A cryptic wheat–Aegilops triuncialis translocation with leaf rust resistance gene Lr58. Crop Sci. 2007;47 (5):1995–2003.

Labate JA, Lamkey KR, Lee M, Woodman WL. Temporal changes in allele frequencies in two reciprocally selected maize populations. Theor Appl Genet. 1999;99(7-8):1166–78.

Lado B, Barrios PG, Quincke M, Silva P, Gutiérrez L. Modeling genotype × environment interaction for genomic selection with unbalanced data from a wheat breeding program. Crop Sci. 2016;56(5):2165–79.

Lado B, Vázquez D, Quincke M, Silva P, Aguilar I, Gutiérrez L. Resource allocation optimization with multi-trait genomic prediction for bread wheat (Triticum aestivum L.) baking quality. Theor Appl Genet. 2018;131(12):2719–31.

Laloë D. Precision and information in linear models of genetic evaluation. Genet Sel Evol. 1993;25 (6):557–76.

Lande R, Thompson R. Efficiency of marker-assisted selection in the improvement of quantitative traits. Genetics. 1990;124(3):743–56.

Lasky JR, Upadhyaya HD, Ramu P, Deshpande S, Hash CT, Bonnette J, et al. Genome-environment associations in sorghum landraces predict adaptive traits. Sci Adv. 2015;1(6)

Lehermeier C, de Los Campos G, Wimmer V, Schön C-C. Genomic variance estimates: with or without disequilibrium covariances? J Anim Breed Genet. 2017a;134(3):232–41.

Lehermeier C, Teyssèdre S, Schön C-C. Genetic gain increases by applying the usefulness criterion with improved variance prediction in selection of crosses. Genetics. 2017b;207(4):1651–61.

Leung H, Raghavan C, Zhou B, Oliva R, Choi IR, Lacorte V, et al. Allele mining and enhanced genetic recombination for rice breeding. Rice. 2015;8(1):34.

Lewontin RC, Krakauer J. Distribution of gene frequency as a test of the theory of the selective neutrality of polymorphisms. Genetics. 1973;74(1):175–95.

Li H, Peng Z, Yang X, Wang W, Fu J, Wang J, et al. Genome-wide association study dissects the genetic architecture of oil biosynthesis in maize kernels. Nat Genet. 2013;45(1):43.

Lian L, Jacobson A, Zhong S, Bernardo R. Prediction of genetic variance in biparental maize populations: genomewide marker effects versus mean genetic variance in prior populations. Crop Sci. 2015;55(3):1181–8.

Lin Z, Cogan NO, Pembleton LW, Spangenberg GC, Forster JW, Hayes BJ, et al. Genetic gain and inbreeding from genomic selection in a simulated commercial breeding program for perennial ryegrass. Plant Genome. 2016;9(1)

Liu R, How-Kit A, Stammitti L, Teyssier E, Rolin D, Mortain-Bertrand A, et al. A DEMETER-like DNA demethylase governs tomato fruit ripening. Proc Natl Acad Sci U S A. 2015a;112 (34):10804–9.

Liu H, Meuwissen TH, Sørensen AC, Berg P. Upweighting rare favourable alleles increases long-term genetic gain in genomic selection programs. Genet Sel Evol. 2015b;47(1):19.

Longin CFH, Reif JC. Redesigning the exploitation of wheat genetic resources. Trends Plant Sci. 2014;19(10):631–6.

Lopez-Cruz M, Crossa J, Bonnett D, Dreisigacker S, Poland J, Jannink J-L, et al. Increased prediction accuracy in wheat breeding trials using a marker × environment interaction genomic selection model. G3: Genes, Genomes, Genetics. 2015;5(4):569–82.

Lorenz AJ. Resource allocation for maximizing prediction accuracy and genetic gain of genomic selection in plant breeding: a simulation experiment. G3: Genes, Genomes, Genetics. 2013;3 (3):481–91.

Lorenzana RE, Bernardo R. Accuracy of genotypic value predictions for marker-based selection in biparental plant populations. Theor Appl Genet. 2009;120(1):151–61.

Lush JL. Animal breeding plans. Ames: Collegiate Press, Inc; 1937.

Luu K, Bazin E, Blum MG. pcadapt: an R package to perform genome scans for selection based on principal component analysis. Mol Ecol Resour. 2017;17(1):67–77.

Ly D, Hamblin M, Rabbi I, Melaku G, Bakare M, Gauch HG, et al. Relatedness and genotype × environment interaction affect prediction accuracies in genomic selection: a study in cassava. Crop Sci. 2013;53(4):1312.

Ly D, Chenu K, Gauffreteau A, Rincent R, Huet S, Gouache D, et al. Nitrogen nutrition index predicted by a crop model improves the genomic prediction of grain number for a bread wheat core collection. Field Crop Res. 2017;214:331–40.

Ly D, Huet S, Gauffreteau A, Rincent R, Touzy G, Mini A, et al. Whole-genome prediction of reaction norms to environmental stress in bread wheat (Triticum aestivum L.) by genomic random regression. Field Crop Res. 2018;216:32–41.

Maccaferri M, Harris NS, Twardziok SO, Pasam RK, Gundlach H, Spannagl M, et al. Durum wheat genome highlights past domestication signatures and future improvement targets. Nat Genet. 2019;51(5):885–95.

MacLeod IM, Hayes BJ, Goddard ME. A novel predictor of multilocus haplotype homozygosity: comparison with existing predictors. Genet Res. 2009;91(6):413–26.

Malézieux E, Crozat Y, Dupraz C, Laurans M, Makowski D, Ozier-Lafontaine H, et al. Mixing plant species in cropping systems: concepts, tools and models: a review. In: Lichtfouse E, Navarrete M, Debaeke P, Véronique S, Alberola C, editors. Sustainable agriculture. Dordrecht: Springer; 2009. p. 329–53.

Malosetti M, Bustos-Korts D, Boer MP, van Eeuwijk FA. Predicting responses in multiple environments: issues in relation to genotype × environment interactions. Crop Sci. 2016;56 (5):2210–22.

Manel S, Schwartz MK, Luikart G, Taberlet P. Landscape genetics: combining landscape ecology and population genetics. Trends Ecol Evol. 2003;18(4):189–97.

Mangin B, Rincent R, Rabier C-E, Moreau L, Goudemand-Dugue E. Training set optimization of genomic prediction by means of EthAcc. PLoS One. 2019;14(2)

Martre P, Jamieson PD, Semenov MA, Zyskowski RF, Porter JR, Triboi E. Modelling protein content and composition in relation to crop nitrogen dynamics for wheat. Eur J Agron. 2006;25 (2):138–54.

Mascher M, Schreiber M, Scholz U, Graner A, Reif JC, Stein N. Genebank genomics bridges the gap between the conservation of crop diversity and plant breeding. Nat Genet. 2019;51 (7):1076–81.

Maskin L, Gudesblat GE, Moreno JE, Carrari FO, Frankel N, Sambade A, et al. Differential expression of the members of the Asr gene family in tomato (Lycopersicon esculentum). Plant Sci. 2001;161(4):739–46.

Matsuoka Y, Mitchell SE, Kresovich S, Goodman M, Doebley J. Microsatellites in Zea – variability, patterns of mutations, and use for evolutionary studies. Theor Appl Genet. 2002;104 (2-3):436–50.

Maynard-Smith J, Haigh J. The hitch-hiking effect of a favourable gene. Genet Res. 1974;23 (1):23–35.

Mayr E. Change of genetic environment and evolution. London: Allen and Unwin; 1954.

McCouch SR, Kovach MJ, Sweeney M, Jiang H, Semon M. The dynamics of rice domestication: a balance between gene flow and genetic isolation. In: Biodiversity in agriculture: domestication, evolution, and sustainability. Cambridge University Press; 2012. p. 311–29.

McCouch S, Baute GJ, Bradeen J, Bramel P, Bretting PK, Buckler E, et al. Feeding the future. Nature. 2013;499(7456):23–4.

Messina CD, Jones JW, Boote KJ, Vallejos CE. A gene-based model to simulate soybean development and yield responses to environment. Crop Sci. 2006;46(1):456–66.

Messina CD, Technow F, Tang T, Totir R, Gho C, Cooper M. Leveraging biological insight and environmental variation to improve phenotypic prediction: integrating crop growth models (CGM) with whole genome prediction (WGP). Eur J Agron. 2018;100:151–62.

Metropolis N, Rosenbluth A, Rosenbluth M, Teller A, Teller E. Simulated annealing. J Chem Phys. 1953;21(161-162):1087–92.

Meuwissen THE. Maximizing the response of selection with a predefined rate of inbreeding. J Anim Sci. 1997;75(4):934–40.

Meuwissen TH, Sonesson AK. Genotype-assisted optimum contribution selection to maximize selection response over a specified time period. Genet Res. 2004;84(2):109–16.

Meuwissen THE, Hayes BJ, Goddard ME. Prediction of total genetic value using genome-wide dense marker maps. Genetics. 2001;157(4):1819.

Meyer RS, Purugganan MD. Evolution of crop species: genetics of domestication and diversification. Nat Rev Genet. 2013;14(12):840–52.

Mi X, Utz HF, Technow F, Melchinger AE. Optimizing resource allocation for multistage selection in plant breeding with R package. Crop Sci. 2014;54(4):1413–8.

Mi X, Utz HF, Melchinger AE. Selectiongain: an R package for optimizing multi-stage selection. Comput Stat. 2016;31(2):533–43.

Michel S, Kummer C, Gallee M, Hellinger J, Ametz C, Akgöl B, et al. Improving the baking quality of bread wheat by genomic selection in early generations. Theor Appl Genet. 2018;131 (2):477–93.

Millet EJ, Kruijer W, Coupel-Ledru A, Prado SA, Cabrera-Bosquet L, Lacube S, et al. Genomic prediction of maize yield across European environmental conditions. Nat Genet. 2019;51 (6):952–6.

Mizubuti ESG, Fry WE. Potato late blight. In: Cooke BM, Jones DG, Kaye B, editors. The epidemiology of plant diseases. Dordrecht: Springer; 2006. p. 445–71.

Moeinizade S, Hu G, Wang L, Schnable PS. Optimizing selection and mating in genomic selection with a look-ahead approach: an operations research framework. G3: Genes, Genomes, Genetics. 2019;9(7):2123.

Mohammadi M, Tiede T, Smith KP. PopVar: a genome-wide procedure for predicting genetic variance and correlated response in biparental breeding populations. Crop Sci. 2015;55 (5):2068–77.

Moler ERV, Abakir A, Eleftheriou M, Johnson JS, Krutovsky KV, Lewis LC, et al. Population epigenomics: advancing understanding of phenotypic plasticity, acclimation, adaptation and diseases. In: Rajora OP, editor. Population genomics: concepts, approaches and applications [Internet]. Cham: Springer International Publishing AG; 2019.

Montesinos-López OA, Montesinos-López A, Crossa J, Toledo FH, Pérez-Hernández O, Eskridge KM, et al. A genomic Bayesian multi-trait and multi-environment model. G3: Genes, Genomes, Genetics. 2016;6(9):2725–44.

Montesinos-López OA, Montesinos-López A, Crossa J, Gianola D, Hernández-Suárez CM, Martín-Vallejo J. Multi-trait, multi-environment deep learning modeling for genomic-enabled prediction of plant traits. G3: Genes, Genomes, Genetics. 2018;8(12):3829–40.

Montesinos-López OA, Montesinos-López A, Luna-Vázquez FJ, Toledo FH, Pérez-Rodríguez P, Lillemo M, et al. An R package for Bayesian analysis of multi-environment and multi-trait multi-environment data for genome-based prediction. G3: Genes, Genomes, Genetics. 2019;9 (5):1355–69.

Müller D. embvr: computation of expected maximum haploid breeding values. Zenodo. 2017.

Müller D, Schopp P, Melchinger AE. Selection on expected maximum haploid breeding values can increase genetic gain in recurrent genomic selection. G3: Genes, Genomes, Genetics. 2018;8 (4):1173–81.

Nakagawa H, Yamagishi J, Miyamoto N, Motoyama M, Yano M, Nemoto K. Flowering response of rice to photoperiod and temperature: a QTL analysis using a phenological model. Theor Appl Genet. 2005;110(4):778–86.

Nakamichi N, Kiba T, Henriques R, Mizuno T, Chua N-H, Sakakibara H. PSEUDO-RESPONSE REGULATORS 9, 7, and 5 are transcriptional repressors in the arabidopsis circadian clock. Plant Cell. 2010;22(3):594.

Nakamichi N, Takao S, Kudo T, Kiba T, Wang Y, Kinoshita T, et al. Improvement of arabidopsis biomass and cold, drought and salinity stress tolerance by modified circadian clock-associated PSEUDO-RESPONSE REGULATORs. Plant Cell Physiol. 2016;57(5):1085–97.

Nakamichi N, Kudo T, Makita N, Kiba T, Kinoshita T, Sakakibara H. Flowering time control in rice by introducing arabidopsis clock-associated PSEUDO-RESPONSE REGULATOR 5. Biosci Biotechnol Biochem. 2020;84(5):970–9.

Narasimhamoorthy B, Gill BS, Fritz AK, Nelson JC, Brown-Guedira GL. Advanced backcross QTL analysis of a hard winter wheat × synthetic wheat population. Theor Appl Genet. 2006;112(5):787–96.

Nei M. Analysis of gene diversity in subdivided populations. Proc Natl Acad Sci U S A. 1973;70 (12):3321–3.

Neveu P, Tireau A, Hilgert N, Nègre V, Mineau-Cesari J, Brichet N, et al. Dealing with multi-source and multi-scale information in plant phenomics: the ontology-driven Phenotyping Hybrid Information System. New Phytol. 2019;221(1):588–601.

Nielsen R. Molecular signatures of natural selection. Annu Rev Genet. 2005;39(1):197–218.

Ødegaard J, Yazdi MH, Sonesson AK. Incorporating desirable genetic characteristics from an inferior into a superior population using genomic selection. Genetics. 2009;181(2):737–45.

Olsen KM, Caicedo AL, Polato N, McClung A, McCouch S, Purugganan MD. Selection under domestication: evidence for a sweep in the rice waxy genomic region. Genetics. 2006;173 (2):975–83.

Onogi A, Watanabe M, Mochizuki T, Hayashi T, Nakagawa H, Hasegawa T, et al. Toward integration of genomic selection with crop modelling: the development of an integrated approach to predicting rice heading dates. Theor Appl Genet. 2016;129(4):805–17.

Oono Y, Kobayashi F, Kawahara Y, Yazawa T, Handa H, Itoh T, et al. Characterisation of the wheat (Triticum aestivum L.) transcriptome by de novo assembly for the discovery of phosphate starvation-responsive genes: gene expression in Pi-stressed wheat. BMC Genomics. 2013;14 (1):77.

Patterson N, Price AL, Reich D. Population structure and eigenanalysis. PLoS Genet. 2006;2(12)

Patterson N, Moorjani P, Luo Y, Mallick S, Rohland N, Zhan Y, et al. Ancient admixture in human history. Genetics. 2012;192(3):1065.

Pavlidis P, Alachiotis N. A survey of methods and tools to detect recent and strong positive selection. J Biol Res. 2017;24(1):7.

Pavlidis P, Živkovic D, Stamatakis A, Alachiotis N. SweeD: likelihood-based detection of selective sweeps in thousands of genomes. Mol Biol Evol. 2013;30(9):2224–34.

Peng T, Sun X, Mumm RH. Optimized breeding strategies for multiple trait integration: I. Minimizing linkage drag in single event introgression. Mol Breeding. 2014;33(1):89–104.

Pérez-Enciso M, Ramírez-Ayala LC, Zingaretti LM. SeqBreed: a python tool to evaluate genomic prediction in complex scenarios. Genet Sel Evol. 2020;52(1):7.

Peripolli E, Munari DP, Silva M, Lima ALF, Irgang R, Baldi F. Runs of homozygosity: current knowledge and applications in livestock. Anim Genet. 2017;48(3):255–71.

Peter BM. Admixture, population structure, and F-statistics. Genetics. 2016;202(4):1485–501.

Petit M, Astruc J-M, Sarry J, Drouilhet L, Fabre S, Moreno CR, et al. Variation in recombination rate and its genetic determinism in sheep populations. Genetics. 2017;207(2):767.

Plucknett DL. Gene banks and the world's food: Princeton University Press; 1987.

Pommier C, Michotey C, Cornut G, Roumet P, Duch E, et al. Applying FAIR principles to plant phenotypic data management in GnpIS. Plant Phenomics. 2019;2019:1671403.

Pont C, Leroy T, Seidel M, Tondelli A, Duchemin W, Armisen D, et al. Tracing the ancestry of modern bread wheats. Nat Genet. 2019;51(5):905–11.

Pook T, Schlather M, Simianer H. MoBPS-modular breeding program simulator. G3: Genes, Genomes, Genetics. 2020;10(6):1915–8.

Price AL, Tandon A, Patterson N, Barnes KC, Rafaels N, Ruczinski I, et al. Sensitive detection of chromosomal segments of distinct ancestry in admixed populations. PLoS Genet. 2009;5(6): e1000519.

Pritchard JK, Stephens M, Donnelly P. Inference of population structure using multilocus genotype data. Genetics. 2000;155(2):945–59.

Prudent M, Lecomte A, Bouchet J-P, Bertin N, Causse M, Génard M. Combining ecophysiological modelling and quantitative trait locus analysis to identify key elementary processes underlying tomato fruit sugar concentration. J Exp Bot. 2011;62(3):907–19.

Pszczola M, Strabel T, Mulder HA, Calus MPL. Reliability of direct genomic values for animals with different relationships within and to the reference population. J Dairy Sci. 2012;95 (1):389–400.

Quilot B, Génard M, Lescourret F, Kervella J. Simulating genotypic variation of fruit quality in an advanced peach × Prunus davidiana cross. J Exp Bot. 2005;56(422):3071–81.

Racimo F, Sankararaman S, Nielsen R, Huerta-Sánchez E. Evidence for archaic adaptive introgression in humans. Nat Rev Genet. 2015;16(6):359–71.

Rajora OP, Eckert AJ, Zinck JWR. Single-locus versus multilocus patterns of local adaptation to climate in eastern white pine (Pinus strobus, Pinaceae). PLoS One. 2016;11(7):e0158691.

Ramírez-González RH, Borrill P, Lang D, Harrington SA, Brinton J, Venturini L, et al. The transcriptional landscape of polyploid wheat. Science. 2018;361(6403)

Ramstein GP, Larsson SJ, Cook JP, Edwards JW, Ersoz ES, Flint-Garcia S, et al. Dominance effects and functional enrichments improve prediction of agronomic traits in hybrid maize. Genetics. 2020;215(1):215–30.

Reich D, Thangaraj K, Patterson N, Price AL, Singh L. Reconstructing Indian population history. Nature. 2009;461(7263):489–94.

Reich D, Patterson N, Campbell D, Tandon A, Mazieres S, Ray N, et al. Reconstructing Native American population history. Nature. 2012;488(7411):370–4.

Reif JC, Hamrit S, Heckenberger M, Schipprack W, Peter Maurer H, Bohn M, et al. Genetic structure and diversity of European flint maize populations determined with SSR analyses of individuals and bulks. Theor Appl Genet. 2005;111(5):906–13.

Rellstab C, Gugerli F, Eckert AJ, Hancock AM, Holderegger R. A practical guide to environmental association analysis in landscape genomics. Mol Ecol. 2015;24(17):4348–70.

Reymond M, Muller B, Leonardi A, Charcosset A, Tardieu F. Combining quantitative trait loci analysis and an ecophysiological model to analyze the genetic variability of the responses of maize leaf growth to temperature and water deficit. Plant Physiol. 2003;131(2):664–75.

Reynolds J, Weir BS, Cockerham CC. Esimation of the coancestry coefficient: basis for a short-term genetic distance. Genetics. 1983;105(3):767–79.

Reynolds MP, Lewis JM, Ammar K, Basnet BR, Crespo-Herrera L, Crossa J, et al. Harnessing translational research in wheat for climate resilience. J Exp Bot. 2021;72(14):5134–57.

Ribaut J-M, Ragot M. Marker-assisted selection to improve drought adaptation in maize: the backcross approach, perspectives, limitations, and alternatives. J Exp Bot. 2007;58(2):351–60.

Riedelsheimer C, Czedik-Eysenberg A, Grieder C, Lisec J, Technow F, Sulpice R, et al. Genomic and metabolic prediction of complex heterotic traits in hybrid maize. Nat Genet. 2012;44 (2):217–20.

Rieseberg LH, Wendel JF. Introgression and its consequences in plants. Hybrid Zones Evol Process. 1993;70:109.

Rincent R, Laloë D, Nicolas S, Altmann T, Brunel D, Revilla P, et al. Maximizing the reliability of genomic selection by optimizing the calibration set of reference individuals: comparison of methods in two diverse groups of maize inbreds (Zea mays L.). Genetics. 2012;192(2):715–28.

Rincent R, Charcosset A, Moreau L. Predicting genomic selection efficiency to optimize calibration set and to assess prediction accuracy in highly structured populations. Theor Appl Genet. 2017a;130(11):2231–47.

Rincent R, Kuhn E, Monod H, Oury F-X, Rousset M, Allard V, et al. Optimization of multi-environment trials for genomic selection based on crop models. Theor Appl Genet. 2017b;130 (8):1735–52.

Rincent R, Charpentier J-P, Faivre-Rampant P, Paux E, Le Gouis J, Bastien C, et al. Phenomic selection is a low-cost and high-throughput method based on indirect predictions: proof of concept on wheat and poplar. G3: Genes, Genomes, Genetics. 2018;8(12):3961–72.

Rincent R, Malosetti M, Ababaei B, Touzy G, Mini A, Bogard M, et al. Using crop growth model stress covariates and AMMI decomposition to better predict genotype-by-environment interactions. Theor Appl Genet. 2019;132(12):3399–411.

Rio S, Mary-Huard T, Moreau L, Charcosset A. Genomic selection efficiency and a priori estimation of accuracy in a structured dent maize panel. Theor Appl Genet. 2019;132(1):81–96.

Robert P, Le Gouis J, Breadwheat Consortium T, Rincent R. Combining crop growth modelling with trait-assisted prediction improved the prediction of genotype by environment interactions. Front Plant Sci. 2020;11

Robertson A. Inbreeding in artificial selection programmes. Genet Res. 1961;2(2):189–94.

Rochus CM, Tortereau F, Plisson-Petit F, Restoux G, Moreno-Romieux C, Tosser-Klopp G, et al. Revealing the selection history of adaptive loci using genome-wide scans for selection: an example from domestic sheep. BMC Genomics. 2018;19(1):71.

Rodriguez M, Rau D, Bitocchi E, Bellucci E, Biagetti E, Carboni A, et al. Landscape genetics, adaptive diversity and population structure in Phaseolus vulgaris. New Phytol. 2016;209 (4):1781–94.

Ross-Ibarra J, Morrell PL, Gaut BS. Plant domestication, a unique opportunity to identify the genetic basis of adaptation. Proc Natl Acad Sci. 2007;104(suppl 1):8641–8.

Rutkoski J, Poland J, Mondal S, Autrique E, Pérez LG, Crossa J, et al. Canopy temperature and vegetation indices from high-throughput phenotyping improve accuracy of pedigree and genomic selection for grain yield in wheat. G3: Genes, Genomes, Genetics. 2016;6(9):2799–808.

Ruttink T, Arend M, Morreel K, Storme V, Rombauts S, Fromm J, et al. A molecular timetable for apical bud formation and dormancy induction in poplar. Plant Cell. 2007;19(8):2370.

Sabeti PC, Reich DE, Higgins JM, Levine HZ, Richter DJ, Schaffner SF, et al. Detecting recent positive selection in the human genome from haplotype structure. Nature. 2002;419 (6909):832–7.

Sabeti PC, Varilly P, Fry B, Lohmueller J, Hostetter E, Cotsapas C, et al. Genome-wide detection and characterization of positive selection in human populations. Nature. 2007;449(7164):913–8.

Saïdou A-A, Thuillet A-C, Couderc M, Mariac C, Vigouroux Y. Association studies including genotype by environment interactions: prospects and limits. BMC Genet. 2014;15(1):3.

Sanchez L, Caballero A, Santiago E. Palliating the impact of fixation of a major gene on the genetic variation of artificially selected polygenes. Genet Res. 2006;88(2):105–18.

Santos DJA, Cole JB, Lawlor TJ Jr, VanRaden PM, Tonhati H, Ma L. Variance of gametic diversity and its application in selection programs. J Dairy Sci. 2019;102(6):5279–94.

Sanz-Alferez S, Richter TE, Hulbert SH, Bennetzen JL. The Rp3 disease resistance gene of maize: mapping and characterization of introgressed alleles. Theor Appl Genet. 1995;91(1):25–32.

Schadt EE, Monks SA, Drake TA, Lusis AJ, Che N, Colinayo V, et al. Genetics of gene expression surveyed in maize, mouse and man. Nature. 2003;422(6929):297–302.

Schaefer J, Duvernell D, Campbell DC. Hybridization and introgression in two ecologically dissimilar Fundulus hybrid zones. Evolution. 2016;70(5):1051–63.

Schaeffer L. Strategy for applying genome-wide selection in dairy cattle. J Anim Breed Genet. 2006;123(4):218–23.

Schnell FW, Utz HF. Bericht über die Arbeitstagung der Vereinigung österreichischer Pflanzenzüchter. Gumpenstein: BAL Gumpenstein; 1975.

Schrag TA, Westhues M, Schipprack W, Seifert F, Thiemann A, Scholten S, et al. Beyond genomic prediction: combining different types of omics data can improve prediction of hybrid performance in maize. Genetics. 2018;208(4):1373–85.

Schulthess AW, Wang Y, Miedaner T, Wilde P, Reif JC, Zhao Y. Multiple-trait-and selection indices-genomic predictions for grain yield and protein content in rye for feeding purposes. Theor Appl Genet. 2016;129(2):273–87.

Schulthess AW, Zhao Y, Longin CFH, Reif JC. Advantages and limitations of multiple-trait genomic prediction for Fusarium head blight severity in hybrid wheat (Triticum aestivum L.). Theor Appl Genet. 2018;131(3):685–701.

Schulz-Streeck T, Ogutu JO, Gordillo A, Karaman Z, Knaak C, Piepho H-P. Genomic selection allowing for marker-by-environment interaction. Plant Breeding. 2013;132(6):532–8.

Sedivy EJ, Wu F, Hanzawa Y. Soybean domestication: the origin, genetic architecture and molecular bases. New Phytol. 2017;214(2):539–53.

Segura V, Vilhjalmsson BJ, Platt A, Korte A, Seren U, Long Q, et al. An efficient multi-locus mixed-model approach for genome-wide association studies in structured populations. Nat Genet. 2012;44(7):825–30.

Selby P, Abbeloos R, Backlund JE, Basterrechea Salido M, Bauchet G, Benites-Alfaro OE, et al. BrAPI – an application programming interface for plant breeding applications. Bioinformatics. 2019;35(20):4147–55.

Servin B, Martin OC, Mézard M. Toward a theory of marker-assisted gene pyramiding. Genetics. 2004;168(1):513–23.

Shepherd RK, Kinghorn BP. A tactical approach to the design of crossbreeding programs. In: Proceedings of the sixth world congress on genetics applied to livestock production. Armidale: University of New England; 1998. p. 431–8.

Simmonds NW. Variability in crop plants, its use and conservation. Biol Rev. 1962;37(3):422–65.

Simmonds NW. Principles of crop improvement. London: Longman; 1979.

Simmonds NW. Introgression and incorporation. Strategies for the use of crop genetic resources. Biol Rev. 1993;68(4):539–62.

Singh RP, Huerta-Espino J, Rajaram S. Achieving near-immunity to leaf and stripe rusts in wheat by combining slow rusting resistance genes. Acta Phypathol Entomol Hung. 2000;35(1/4):133–40.

Smith S, Beavis W. Molecular marker assisted breeding in a company environment. In: The impact of plant molecular genetics. Springer; 1996. p. 259–72.

Sorensen D, Fernando R, Gianola D. Inferring the trajectory of genetic variance in the course of artificial selection. Genet Res. 2001;77(1):83–94.

Sork VL, Squire K, Gugger PF, Steele SE, Levy ED, Eckert AJ. Landscape genomic analysis of candidate genes for climate adaptation in a California endemic oak, Quercus lobata. Am J Bot. 2016;103(1):33–46.

Souza E, Sorrells ME. Prediction of progeny variation in oat from parental genetic relationships. Theor Appl Genet. 1991a;82(2):233–41.

Souza E, Sorrells ME. Relationships among 70 North American oat germplasms: I. Cluster analysis using quantitative characters. Crop Sci. 1991b;31(3):599–605.

Spannagl M, Alaux M, Lange M, Bolser DM, Bader KC, Letellier T, et al. transPLANT resources for triticeae genomic data. Plant Genome. 2016;9(1).

Spillane C, Gepts P. Evolutionary and genetic perspectives on the dynamics of crop genepools. In: Broadening the genetic base of crop production. CABI; 2001. p. 25–70.

Steffenson BJ, Olivera P, Roy JK, Jin Y, Smith KP, Muehlbauer GJ. A walk on the wild side: mining wild wheat and barley collections for rust resistance genes. Aust J Agr Res. 2007;58(6):532–44.

Stephan W. Signatures of positive selection: from selective sweeps at individual loci to subtle allele frequency changes in polygenic adaptation. Mol Ecol. 2016;25(1):79–88.

Steuernagel B, Periyannan SK, Hernández-Pinzón I, Witek K, Rouse MN, Yu G, et al. Rapid cloning of disease-resistance genes in plants using mutagenesis and sequence capture. Nat Biotechnol. 2016;34(6):652.

Stich B. Comparison of mating designs for establishing nested association mapping populations in maize and Arabidopsis thaliana. Genetics. 2009;183(4):1525–34.

Storn R, Price K. Differential evolution – a simple and efficient heuristic for global optimization over continuous spaces. J Global Optim. 1997;11(4):341–59.

Stucki S, Orozco-terWengel P, Forester BR, Duruz S, Colli L, Masembe C, et al. High performance computation of landscape genomic models including local indicators of spatial association. Mol Ecol Resour. 2017;17(5):1072–89.

Suarez-Gonzalez A, Hefer CA, Christe C, Corea O, Lexer C, Cronk QCB, et al. Genomic and functional approaches reveal a case of adaptive introgression from Populus balsamifera (balsam poplar) in P. trichocarpa (black cottonwood). Mol Ecol. 2016;25(11):2427–42.

Suarez-Gonzalez A, Hefer CA, Lexer C, Cronk QCB, Douglas CJ. Scale and direction of adaptive introgression between black cottonwood (Populus trichocarpa) and balsam poplar (P. balsamifera). Mol Ecol. 2018;27(7):1667–80.

Sun C, VanRaden PM, Cole JB, O'Connell JR. Improvement of prediction ability for genomic selection of dairy cattle by including dominance effects. PLoS One. 2014;9(8)

Sun J, Rutkoski JE, Poland JA, Crossa J, Jannink J-L, Sorrells ME. Multitrait, random regression, or simple repeatability model in high-throughput phenotyping data improve genomic prediction for wheat grain yield. Plant Genome. 2017;10(2).

Szpiech ZA, Hernandez RD. selscan: an efficient multithreaded program to perform EHH-based scans for positive selection. Mol Biol Evol. 2014;31(10):2824–7.

Tajima F. Statistical method for testing the neutral mutation hypothesis by DNA polymorphism. Genetics. 1989;123(3):585–95.

Tardieu F, Cabrera-Bosquet L, Pridmore T, Bennett M. Plant phenomics, from sensors to knowledge. Curr Biol. 2017;27(15):R770–83.

Technow F, Messina CD, Totir LR, Cooper M. Integrating crop growth models with whole genome prediction through approximate Bayesian computation. PLoS One. 2015;10(6):e0130855.

Tenaillon MI, Charcosset A. A European perspective on maize history. C R Biol. 2011;334(3):221–8.

Thabuis A, Palloix A, Servin B, Daubeze AM, Signoret P, Lefebvre V. Marker-assisted introgression of 4 Phytophthora capsici resistance QTL alleles into a bell pepper line: validation of additive and epistatic effects. Mol Breed. 2004;14(1):9–20.

Tiede T, Kumar L, Mohammadi M, Smith KP. Predicting genetic variance in bi-parental breeding populations is more accurate when explicitly modeling the segregation of informative genomewide markers. Mol Breed. 2015;35(10):199.

Tinker NA, Deyl JK. A curated Internet database of oat pedigrees. Crop Sci. 2005;45(6):2269–72.

Touzy G, Rincent R, Bogard M, Lafarge S, Dubreuil P, Mini A, et al. Using environmental clustering to identify specific drought tolerance QTLs in bread wheat (T. aestivum L.). Theor Appl Genet. 2019;132(10):2859–80.

Uauy C, Brevis JC, Dubcovsky J. The high grain protein content gene Gpc-B1 accelerates senescence and has pleiotropic effects on protein content in wheat. J Exp Bot. 2006;57 (11):2785–94.

Uemoto Y, Sasaki S, Kojima T, Sugimoto Y, Watanabe T. Impact of QTL minor allele frequency on genomic evaluation using real genotype data and simulated phenotypes in Japanese Black cattle. BMC Genet. 2015;16(1):134.

Ullstrup AJ. The impacts of the southern corn leaf blight epidemics of 1970-1971. Annu Rev Phytopathol. 1972;10(1):37–50.

Uptmoor R, Li J, Schrag T, Stützel H. Prediction of flowering time in Brassica oleracea using a quantitative trait loci-based phenology model. Plant Biol. 2012;14(1):179–89.

Utz HF, Bohn M, Melchinger AE. Predicting progeny means and variances of winter wheat crosses from phenotypic values of their parents. Crop Sci. 2001;41(5):1470–8.

Vacher M, Small I. Simulation of heterosis in a genome-scale metabolic network provides mechanistic explanations for increased biomass production rates in hybrid plants. NPJ Syst Biol Appl. 2019;5(1):1–10.

Valente F, Gauthier F, Bardol N, Blanc G, Joets J, Charcosset A, et al. OptiMAS: a decision support tool for marker-assisted assembly of diverse alleles. J Hered. 2013;104(4):586–90.

van Berloo R, Stam P. Marker-assisted selection in autogamous RIL populations: a simulation study. TAG Theor Appl Genet. 1998;96(1):147–54.

van Heerwaarden J, Hufford MB, Ross-Ibarra J. Historical genomics of North American maize. Proc Natl Acad Sci. 2012;109(31):12420–5.

van Inghelandt D, Melchinger AE, Lebreton C, Stich B. Population structure and genetic diversity in a commercial maize breeding program assessed with SSR and SNP markers. Theor Appl Genet. 2010;120(7):1289–99.

VanRaden PM. Efficient methods to compute genomic predictions. J Dairy Sci. 2008;91 (11):4414–23.

VanRaden PM, Tooker ME, Cole JB, Wiggans GR, Megonigal JH Jr. Genetic evaluations for mixed-breed populations. J Dairy Sci. 2007;90(5):2434–41.

Vigouroux Y, Mitchell S, Matsuoka Y, Hamblin M, Kresovich S, Smith JSC, et al. An analysis of genetic diversity across the maize genome using microsatellites. Genetics. 2005;169 (3):1617–30.

Vigouroux Y, Barnaud A, Scarcelli N, Thuillet A-C. Biodiversity, evolution and adaptation of cultivated crops. C R Biol. 2011;334(5):450–7.

Virlouvet L, Jacquemot M-P, Gerentes D, Corti H, Bouton S, Gilard F, et al. The ZmASR1 protein influences branched-chain amino acid biosynthesis and maintains kernel yield in maize under water-limited conditions. Plant Physiol. 2011;157(2):917–36.

Voight BF, Kudaravalli S, Wen X, Pritchard JK. A map of recent positive selection in the human genome. PLoS Biol. 2006;4(3):e72.

Wallace JG, Bradbury PJ, Zhang N, Gibon Y, Stitt M, Buckler ES. Association mapping across numerous traits reveals patterns of functional variation in maize. PLoS Genet. 2014;10(12): e1004845.

Wang H-J, Hsu C-M, Jauh GY, Wang C-S. A lily pollen ASR protein localizes to both cytoplasm and nuclei requiring a nuclear localization signal. Physiol Plant. 2005;123(3):314–20.

Wang C, Hu S, Gardner C, Lübberstedt T. Emerging avenues for utilization of exotic germplasm. Trends Plant Sci. 2017;22(7):624–37.

Wang J, Hu Z, Upadhyaya HD, Morris GP. Genomic signatures of seed mass adaptation to global precipitation gradients in sorghum. Heredity. 2020;124(1):108–21.

Wellmann R. Optimum contribution selection for animal breeding and conservation: the R package optiSel. BMC Bioinformatics. 2019;20(1):1–13.

Wellmann R, Bennewitz J. Key genetic parameters for population management. Front Genet. 2019;10:667.

Wen W, Liu H, Zhou Y, Jin M, Yang N, Li D, et al. Combining quantitative genetics approaches with regulatory network analysis to dissect the complex metabolism of the Maize Kernel. Plant Physiol. 2016;170(1):136–46.

White JW, Hoogenboom G. Simulating effects of genes for physiological traits in a process-oriented crop model. Agron J. 1996;88(3):416–22.

White JW, Herndl M, Hunt LA, Payne TS, Hoogenboom G. Simulation-based analysis of effects of Vrn and Ppd loci on flowering in wheat. Crop Sci. 2008;48(2):678–87.

Whitlock MC, Lotterhos KE. Reliable detection of loci responsible for local adaptation: inference of a null model through trimming the distribution of F ST. Am Nat. 2015;186(S1):S24–36.

Whittaker JC, Thompson R, Denham MC. Marker-assisted selection using ridge regression. Genet Res. 2000;75(2):249–52.

Woolliams JA, Thompson R. A theory of genetic contributions. In: Proceedings of the 5th world congress on genetics applied to livestock production. Guelph; 1994. p. 127–34.

Woolliams JA, Bijma P, Villanueva B. Expected genetic contributions and their impact on gene flow and genetic gain. Genetics. 1999;153(2):1009–20.

Woolliams JA, Berg P, Dagnachew BS, Meuwissen THE. Genetic contributions and their optimization. J Anim Breed Genet. 2015;132(2):89–99.

Wray NR, Goddard ME. Increasing long-term response to selection. Genet Sel Evol. 1994;26(5):431.

Wray NR, Thompson R. Prediction of rates of inbreeding in selected populations. Genet Res. 1990;55(1):41–54.

Wray NR, Woolliams JA, Thompson R. Methods for predicting rates of inbreeding in selected populations. Theor Appl Genet. 1990;80(4):503–12.

Wright S. Evolution in Mendelian populations. Genetics. 1931;16(2):97–159.

Wu J, Yu H, Dai H, Mei W, Huang X, Zhu S, et al. Metabolite profiles of rice cultivars containing bacterial blight-resistant genes are distinctive from susceptible rice. Acta Biochim Biophys Sin. 2012;44(8):650–9.

Xu P, Wang L, Beavis WD. An optimization approach to gene stacking. Eur J Oper Res. 2011;214(1):168–78.

Yabe S, Iwata H, Jannink J-L. A Simple package to script and simulate breeding schemes: the breeding scheme language. Crop Sci. 2017;57(3):1347–54.

Yang C, Sakai H, Numa H, Itoh T. Gene tree discordance of wild and cultivated Asian rice deciphered by genome-wide sequence comparison. Gene. 2011;477(1–2):53–60.

Yin X, Struik PC, Tang J, Qi C, Liu T. Model analysis of flowering phenology in recombinant inbred lines of barley. J Exp Bot. 2005;56(413):959–65.

Yu X, Li X, Guo T, Zhu C, Wu Y, Mitchell SE, et al. Genomic prediction contributing to a promising global strategy to turbocharge gene banks. Nat Plants. 2016;2(10):1–7.

Zheng B, Biddulph B, Li D, Kuchel H, Chapman S. Quantification of the effects of VRN1 and Ppd-D1 to predict spring wheat (Triticum aestivum) heading time across diverse environments. J Exp Bot. 2013;64(12):3747–61.

Zhong S, Jannink J-L. Using quantitative trait loci results to discriminate among crosses on the basis of their progeny mean and variance. Genetics. 2007;177(1):567–76.

Zivy M, Wienkoop S, Renaut J, Pinheiro C, Goulas E, Carpentier S. The quest for tolerant varieties: the importance of integrating "omics" techniques to phenotyping. Front Plant Sci. 2015;6:448.

Zorrilla-Fontanesi Y, Rambla J-L, Cabeza A, Medina JJ, Sánchez-Sevilla JF, Valpuesta V, et al. Genetic analysis of strawberry fruit aroma and identification of O-methyltransferase FaOMT as the locus controlling natural variation in mesifurane content. Plant Physiol. 2012;159(2):851–70.

Open Access This chapter is licensed under the terms of the Creative Commons Attribution 4.0 International License (http://creativecommons.org/licenses/by/4.0/), which permits use, sharing, adaptation, distribution and reproduction in any medium or format, as long as you give appropriate credit to the original author(s) and the source, provide a link to the Creative Commons license and indicate if changes were made.

The images or other third party material in this chapter are included in the chapter's Creative Commons license, unless indicated otherwise in a credit line to the material. If material is not included in the chapter's Creative Commons license and your intended use is not permitted by statutory regulation or exceeds the permitted use, you will need to obtain permission directly from the copyright holder.

Population Genomics and Molecular Breeding of Sorghum

Arthur Bernardeli, Cynthia Maria Borges Damasceno,
Jurandir Vieira de Magalhães, Vander Fillipe de Souza,
Janaína de Oliveira Melo, Amanda Avelar de Oliveira,
Maria Lúcia Ferreira Simeone, Aluízio Borém, Robert Eugene Schaffert,
Rafael Augusto da Costa Parrella, and Maria Marta Pastina

Abstract *Sorghum bicolor* (L.) Moench is a multipurpose crop worldwide, and it is used as a source of food, fodder, feed, and fuel. It is currently considered a promising alternative crop for generating bioenergy from sugar and lignocellulosic biomass. Sweet sorghum accumulates soluble sugar in its juicy stalk, being more appropriate for first-generation ethanol production, while biomass sorghum can be used for second-generation ethanol or bioelectricity production. For the development of superior sorghum cultivars dedicated to grain or biomass usage, breeding programs usually focus on several traits related to bioenergy, most of them exhibiting a quantitative inheritance. The genetic architecture and the environmental effects involved in the phenotypic expression of major sorghum traits are key factors to improve sorghum as a grain or bioenergy crop. In this context, understanding the availability of sorghum genetic resources, diversity, and molecular information at

A. Bernardeli
Universidade Federal de Viçosa (UFV), Viçosa, MG, Brazil

University of Nebraska-Lincoln (UNL), Lincoln, NE, USA

C. M. B. Damasceno · J. V. de Magalhães · M. L. F. Simeone · R. E. Schaffert ·
R. A. da Costa Parrella · M. M. Pastina (✉)
Embrapa Milho e Sorgo, Sete Lagoas, MG, Brazil
e-mail: marta.pastina@embrapa.br

V. F. de Souza
Universidade Federal de Lavras (UFLA), Lavras, MG, Brazil

J. de Oliveira Melo
Universidade Federal dos Vales do Jequitinhonha e Mucuri (UFVJM), Diamantina, MG, Brazil

A. A. de Oliveira
Escola Superior de Agricultura Luiz de Queiroz/Universidade de São Paulo (ESALQ/USP), Piracicaba, Brazil

A. Borém
Universidade Federal de Viçosa (UFV), Viçosa, MG, Brazil

the RNA and DNA level is fundamental to perform effective selection strategies in sorghum. Population genomics research is providing key insights into genome structure, genomic variation, genetic diversity and population structure, origin and domestication history, and genetic architecture of adaptation to local climate as well as genomic-assisted selection and breeding in sorghum. In this chapter, we thoroughly discuss the progress made on these population genomics and molecular breeding aspects in sorghum and present future perspectives. Substantial research progress has been made on understanding the genetic basis of economically important sorghum traits through the identification of QTLs via linkage mapping and genome-wide association study (GWAS). Genomic selection or genome-wide selection has a great potential for increasing genetic gains in sorghum breeding programs for bioenergy and other traits.

Keywords Biomass and sweet sorghum · Domestication · Genetic and genomic diversity · Genomic resources · Genomic selection · GWAS · Landscape genomics · Origin · QTL mapping · *Sorghum bicolor*

1 Introduction

Sorghum [*Sorghum bicolor* (L.) Moench] is a C4 plant species native to the dry lands of northern Africa, and its domestication dates to 5,000 years ago (Smith and Frederiksen 2000). It has a wide genetic diversity, which provides a rich source of morphological and phenological traits to the development of cultivars for different purposes and products (Upadhyaya et al. 2009). Worldwide, sorghum is grown for food (grain and syrup), animal feed (pasture, green forage, and silage), and biofuels, showing a broad adaptability even under semiarid and arid regions (Wondimu et al. 2021). According to the end-use, four types of sorghum are commonly cropped: biomass, forage, sweet, and grain sorghums. In 2017, the cultivated global area for sorghum reached 40.7 million hectares (FAO 2017).

Sweet and biomass sorghum types have emerged as alternative raw materials for multiple uses in the generation of bioenergy, such as first- and second-generation ethanol, cogeneration of energy through the combustion of dry biomass, and production of biogas (Amaducci et al. 2000; Almeida et al. 2019). Sorghum features include fast growth, high yield potential, and tolerance to abiotic stresses, such as drought, heat, and aluminum toxicity. Additionally, sorghum is capable of mechanized cropping, including seed sowing, crop management, and harvesting (Bennett and Anex 2009; Zegada-Lizarazu and Monti 2012). Bioenergy sorghum cultivars have specific features according to the biomass-energy conversion process. The most important target traits for biofuel production include crop cycle duration, plant height, total biomass yield, and sugar and fiber contents (Regassa and Wortmann 2014). Such traits are quantitative in nature, controlled by many quantitative trait loci (QTL) (Mace et al. 2019), which are likely under environmental influence.

Grain sorghum is second only to maize (*Zea mays*) as the C4 cereal in terms of grain production. Seed size and number are two components that contribute to sorghum grain yield, although there is limited information available about the genetic control of these traits (Tao et al. 2017). For example, it is well known that greater seed size provides survival competitiveness and rapid initial development of sorghum plant, but also results in smaller number of seeds (Westoby et al. 1992). Over the years of breeding and domestication, the selection process of grain yield in sorghum has resulted in some observable genetic alterations, especially the increased frequency of favorable yield-related alleles, accompanied by a reduced genetic diversity (Doebley et al. 2006).

To cope with challenges of selecting even more productive types of sorghum under a narrowed genetic diversity, studies regarding sorghum genotype by environment interaction and, more recently, genome-based approaches that assist on selection strategies were implemented, such as the one performed by Bernardino et al. (2021) and Mengistu et al. (2020). Those types of research were particularly enabled after the genomic information of sorghum has becoming more available and properly noted (Reddy et al. 2009). Mengistu et al. (2020) assessed the genetic diversity of 342 sorghum lines in Ethiopia, and the 5060 SNP markers revealed the existence of higher genetic variation among and within accessions. These levels of diversity suggest that there is room for selections that might result in positive genetic gain for target traits among and within the genotypes. Furthermore, Bernardino et al. (2021) concluded that highly diverse populations that is generated after crossing numerous founders from a multi-crossing process can emerge as a bridge for gene discovery and to deploy gene-specific markers, useful to the pre-breeding steps for abiotic tolerance.

In this chapter, we discuss the progress made in population genomics research, genetic basis of economically important traits, and genomics-assisted molecular breeding of sorghum. Population genomics research is unraveling key insights into genome structure, genomic variation, genetic diversity and population structure, origin and domestication history, and genetic architecture of adaptation to local climate and assisting genomic-assisted selection and breeding in sorghum.

2 Genomic and Plant Genetic Resources

2.1 Genome and Pan-Genome

A pan-genome corresponds to a species-wide catalogue of genomic variation, or more generally, structural variation, repetitive portions, or chromosomal rearrangements, that affect (potentially non-coding) sequences of 50 or more base pairs (bp) in size (Bayer et al. 2020; Jayakodi et al. 2021). Paterson et al. (2009) reported the first genome assembly of sorghum. Later, an improved assembly occurred (McCormick et al. 2018) and a new reference genome was released (Cooper et al. 2019). However, a single reference genome is insufficient to represent the complete genomic diversity and contents of a species (Bayer et al. 2020; Qamar et al. 2020; Ruperao et al. 2021). Then, the pan-genome study of a species emerges as an important alternative. The main advantage of genome sequences are their availability and accessibility to link genotypes to phenotype. In sorghum, ICRISAT and USDA has been curating both genomic and phenotypic datasets to maintain the repositories (Boyles et al. 2019).

Sorghum is the first C4 grass crop with fully assembled genome using traditional sequencing but is far behind crops such as maize and rice (*Oryza sativa*) in terms of the de novo genome assemblies of more diverse sorghum populations Those new references of divergent sorghum types would bring greater clarity for genome structure and genomic diversity and will provide a resource for population genomics studies as well to identify genes underlying the main sorghum traits (Boyles et al. 2019). The available diversity sequence panels play an important role in sorghum breeding, mainly in the current era of genomics. According to Boyles et al. (2019), each panel has drawbacks regarding the germplasm included and sequencing methodology. Table 1 depicts the available high-density sequenced diversity panels up to date.

Sorghum [*Sorghum bicolor* L. Moench) shows a wide genomic diversity (Kong et al. 2000; Hart et al. 2001; Upadhyaya et al. 2009; Ruperao et al. 2021) which can be explored in pan-genome research aiming at the crop improvement for different traits and conservation of sorghum genetic resources. Pan-genomes correspond to a set of genome sequences and the study of structural variation within species which includes core genes, variable genes, which are present/absent among individuals, as well as structural variants, repetitive portions of the genome and chromosomal rearrangements (Bayer et al. 2020; Jayakodi et al. 2021). The first genome assembly of the sorghum genome was performed by Paterson et al. (2009). Later, an improved assembly was performed (McCormick et al. 2018), and a new reference genome has been made available (Cooper et al. 2019). However, these genome assemblies have been limited to an accession of researchers' choice, a single reference genome is insufficient to represent the complete genomic diversity and contents of a species (Bayer et al. 2020; Qamar et al. 2020; Ruperao et al. 2021). This way, the pan-genome study of a species emerges as an important approach.

Associations between structural variants and multiple types of biotic and abiotic stress resistance have been reported in sorghum. An example of structural variants caused by variation in gene content is the common absence of a sulfotransferase gene (presence/absence variations type) in different lines resulted in resistance to the parasitic weed Striga via reducing its germination stimulant activity in sorghum (Gobena et al. 2017; Lyu 2017). One of the most important causes of structural variation is transposable elements whose activity can be considered as an important contributor to genome size variation among different species (Fedoroff 2012; Tao et al. 2019). Variation in SbMATE expression likely results from changes in tandemly repeated cis sequences flanking a transposable element (a miniature inverted repeat transposable element – MITE) insertion in the SbMATE promoter, which are recognized by the aluminum-responsive transcription factors (Magalhaes et al. 2007; Melo et al. 2019).

The rich diversity in wild relatives can also contribute to identify genes for improving other traits (Tao et al. 2019). In sorghum, a 2.2-kb deletion at sh1 was

Table 1 Established diversity panels with sequence data

Panel	No. of accessions	Original citation	Genomic data	Sequencing method	Repository	Curator
World core collection	2427	Grenier et al. (2001)	Unavailable	Unavailable	Unavailable	ICRISAT
Min-Core collection	242	Upadhyaya et al. (2009)	Morris et al. (2013)	GBS (*ApeKI*)	SRA062716	ICRISAT
SAP	390	Casa et al. (2008)	Morris et al. (2013)	GBS (*ApeKI*)	SRA062716	USDA-GRIN
BAP	390	Brenton et al. (2016)	Brenton et al. (2016)	GBS (*ApeKI*) (WSG in progress)	PRJNA298892	USDA-GRIN
SCP+exotic parents	1160	Thurber et al. (2013)	Thurber et al. (2013)	GBS (PstI-HF + *BfaI* or *HinPII*)	SRP022956	USDA-GRIN
Expanded SCP lines	700	Hayes et al. (2015)	Hayes et al. (2015)	GBS (PstI-HF + *BfaI* or *HinPII*)	Illinois Data Bank	USDA-GRIN
Queensland diversity panel	1033	E. Mace, personal communication, 2018	E. Mace, personal communication, 2018	GBS (TBD)	TBD	University of Queensland/ Australian Grains Genebank
Nigerian diversity panel	516	Borgel and Sequier (1977)	Maina et al. (2018)	GBS (*ApeKI*)	SRP132525	USDA-GRIN

Reproduced from Boyles et al. (2019) with permission from rightsholder

one of three causal mutations that changed the shattering habitat of wild sorghum in domesticated sorghum (Lin et al. 1995). This way, the pan-genome studies of different species for a given genus, as the crop's gene pool comprises many species, especially wild relatives with diverse genetic stock (Khan et al. 2020) can be useful for crop improvement as well as understanding phylogenies and domestication.

In sorghum, a pan-genome was generated from the assembly of 13 genomes (Tao et al. 2021) and the reference genomes previously published (Paterson et al. 2009; McCormick et al. 2018; Cooper et al. 2019) resulting in substantial gene-content variation, with 64% of gene families showing presence/absence variation among the genomes (Tao et al. 2021). Extensive genetic variation was identified within the pan-genome, and the presence/absence variants were shown to be under selection during sorghum domestication (Tao et al. 2021).

Regarding GWAS, the advance in pan-genome research through the genotyping of thousands or millions of SNPs across large sets of diverse individuals may facilitate the gene discovery, because if variants associated with a trait are not present in the reference genome, then QTL mapping or GWAS will not be able to detect them. This way, reference genome assemblies of set of individuals can contribute to identification of novel variants. In the pan-genome context, SNPs can be used to assist and expand genomic prediction studies (Della Coletta et al. 2021). Genomic prediction for biomass trait has been shown accuracy methods in sorghum and evidences a potential for revealing biological processes relevant to the studied quantitative traits (Oliveira et al. 2018). Traditionally, SNPs identified relative to a single reference genome have been used for genomic selection. However, identifying markers within a pan-genome framework have the potential to improve the prediction accuracy (Della Coletta et al. 2021). In face of this and considering the potential of genomic prediction, Jensen et al. (2020) developed a pan-genome database for sorghum with haplotype and variants information from diverse progenies, including founders, which can contribute for increase the accuracy of genomic prediction. This pan-genome database can contribute to maintain variant information, impute SNPs, decrease the cost of genotyping and, consequently, facilitate the genomic predictions studies (Jensen et al. 2020).

Pan-genomic studies can provide useful resources for evolutionary studies, functional genomics, and crop improvement (Jayakodi et al. 2021) for different agronomic characters, such as yield, metabolite biosynthesis, response to biotic and abiotic stresses, and biomass. Different perspectives on the application of pan-genomes are possible, including discovery of genes, polymorphisms, and ncRNAs (Qamar et al. 2020; Jayakodi et al. 2021).

Pan-genome sequences help to understand mutations, function losses, and deletions and insertions of genes that control important roles in sorghum breeding target traits. Recently, Cooper et al. (2019) provided a new reference genome for sorghum and revealed high levels of sequence similarity between sweet and grain types, possibly higher than what has been observed among rice types. Genome-wide prediction and association projects in sorghum breeding can leverage the available sorghum repositories that contain genomic sequences to assemble genotyping chips and perform introgressions of favorable alleles.

2.2 Transcriptome

Recently, substantial transcriptome resources have become available in sorghum that could be used for population genomics studies in this species. These resources are the products of studies carried out for gene expression profiling related to signaling pathways (Baek et al. 2019), phosphorus starvation tolerance (Zhang et al. 2019), and response to drought stress (Varoquaux et al. 2019). The latter two studies are particularly important for sorghum breeding.

Considering that phosphorus is a vital element for sorghum survival and development to deliver optimal grain yield, it is important to better understand the gene expression in the roots through transcriptome profiling. Zhang et al. (2019) provided valuable information on phosphorus starvation tolerance of 29 sorghum genotypes grown under phosphorus deficiencies in pot and hydroponic experiments. The authors evaluated root morphological traits and identified differentially expressed genes though quantitative real-time polymerase chain reaction (qRT-PCR) RNA-seq. The authors identified a total of 2089 differentially expressed genes in response to phosphorus deficiency, suggesting candidate genes that play key roles in complex mechanisms of phosphorus starvation tolerance in sorghum, such as malate influences in root morphology and plant hormone signals, mainly in tolerant accessions. Those accessions can, therefore, be used as donors of favorable alleles of phosphorus deficiency tolerance.

Most transcriptomics studies in plants are conducted under controlled conditions such as greenhouse, potting media, or hydroponic solution. However, Varoquaux et al. (2019) performed transcriptomic analysis of field-drought stressed sorghum from seedling to maturity. Drought can cause leaf rolling in pre-flowering stages, and leaf senescence and stalk lodging in post-flowering stages (Rosenow et al. 1983). Varoquaux et al. (2019) profiled 40 transcriptomes sampled from leaves and roots of RTx430 (drought-tolerant) and BTx642 (stay-green) genotypes under well-watered and pre- and post-flowering water-stressed conditions. The authors described that sorghum could rapidly acclimate to limited water conditions, involving massive changes in the transcriptome, and impacted more than 40% of all expressed genes of several mechanisms in terms of how quickly the plants can recover from the stress.

2.3 Genetic Maps

Genetic maps provide genetically mapped markers for population genomics studies and a framework for mapping QTLs for desired traits. Mapped loci and markers are of great importance for narrow-sense population genomics studies (see Luikart et al. 2019; Rajora 2019). The first studies that addressed genome mapping in sorghum DNA started at the end of the twentieth century using RLFP and DArT markers (Chittenden et al. 1994; Mace et al. (2008). More recently, several maps were

Table 2 Genetic linkage maps in sorghum used for developing a consensus reference genetic map by Mace et al. (2009)

Pedigree	Pedigree	Population size	No. of markers	Reference
BTx623/IS3620C	RIL	137	792	Xu et al. (1994)
R890562/ICSV745	RIL	119	488	Tao et al. (2003)
R931945-22/IS8525	RIL	146	410	Mace et al. (2008), Parh (2005)
B923296/SC170-6-8	RIL	88	189	None, posteriorly used in Mace and Jordan (2011)
BTx642/QL12	RIL	94	117	None, posteriorly used in Jordan et al. (2011)
SAR10/SSM249	RIL	183	807	Trouche (personal communication), Rami (personal communication)

integrated with the sorghum sequenced genome with a number of genetic linkage maps been developed for this crop (Table 2).

The construction of linkage maps is almost always required for map-based gene cloning and for applying marker-assisted selection (MAS) in breeding programs, even for genome-sequenced crop species. To date, several linkage maps constructed from different molecular marker systems have been reported for sorghum; for review see Mace et al. (2009), Mace and Jordan (2010, 2011), Rajendrakumar and Rakshit (2015), Kulwal (2017), Marthur et al. (2017), and Mace et al. (2019). Due to operational limitations and dominant nature of the earlier used markers, mostly restriction fragment length polymorphisms (RFLP), amplified fragment length polymorphisms (AFLP), and random amplified polymorphic DNA (RAPD), several linkage maps based on simple sequence repeat (SSR) markers have later been developed in sorghum (Hulbert et al. 1990; Berhan et al. 1993; Peng et al. 1999; Boivin et al. 1999; Singh and Lohithaswa 2006; Ejeta and Kenoll 2007). This was an important achievement for advancing sorghum genetics, allowing more precise QTL mapping and marker-assisted breeding, and opening the possibility of more effective genome comparison to related crops (Tao et al. 1998; Xu et al. 2000; Moens et al. 2006; Ramu et al. 2009; Yonemaru et al. 2009; Guan et al. 2011; Kong et al. 2013). However, the need for more saturated maps in sorghum led to the use of more high-throughput platforms based on higher number of markers, especially single nucleotide polymorphism (SNP) and, at lesser extent, diversity array technology (DArTs) markers (Table 2; Haussmann et al. 2002; Bowers et al. 2003; Mace et al. 2008, Mace et al. 2009; Li et al. 2009; Kong et al. 2013; Shen et al. 2015).

The published genetic maps of sorghum come from wide crosses, which do not represent the germplasms that are used in commercial sorghum breeding programs (Feltus et al. 2006), and therefore are of limited use in sorghum breeding. Using an innovative approach, Mace et al. (2009) developed a consensus genetic map of sorghum by integrating multiple component maps and high-throughput diversity array technology (DArT) markers. The maps were selected from six multiple segregating populations from distinct genetic pools as described in Table 2. It integrated over 2000 loci using RFLP, SSR, and mostly DarT markers. The marker positions of each component maps were merged, using the Texas AM University-ARS map as the reference map, which had the greatest number of markers and common loci to the other maps. This reference served as the backbone from which the other marker positions were assigned. This consensus map provides a great resource for multiple purposes. More sorghum genetic mapping studies in addition to the consensus one can be found at OZ Sorghum website (Alliance of Australian Sorghum Researchers, https://aussorgm.org.au).

The sorghum genetic maps, with special regard to the consensus one (Mace et al. 2009), can be directly applied in a sorghum breeding scenario and provide resources to identify beneficial alleles and causative markers in candidate genes, to develop SNP markers and chips for marker-assisted selection (marker-QTL associations, GWAS, and GWS), and to better understand the genetic diversity and linkage disequilibrium of sorghum breeding pools.

2.4 Plant Genetic Resources

The main genetic resources used in plant breeding are the germplasm collections, comprised of elite, core, and accessions of genotypes. In the case of sorghum, wild ancestors play advantageous traits such as palatable grains and high yield, but the alleles that provide those traits have been lost through domestication and selection that led to reduced genetic diversity. Mutegi et al. (2012) and Fernandez et al. (2014) noticed that wild sorghum populations harbor a higher genetic diversity in comparison to modern cultivars. Venkateswaran et al. (2019a, b) claimed that although the genetic diversity reduced due to domestication, the genetic variability increased within each sorghum group in terms of sorghum species, which means that new phenotypes are fixed to the subgroups.

Despite the fact that there are abundant sorghum genetic resources, there are still barriers that prevent the use of sorghum wild relatives. The allele introgression process from wild to cultivated sorghum faces issues of genetic incompatibility, that is, the pre- and post-zygotic reproductive sterility barriers due to genome size, chromosome morphology, pollen-pistil incompatibilities, and embryo abortions (Ananda et al. 2020; Venkateswaran et al. 2019a). Some experiments have succeeded to overcome this type of issue, such as in the crossing between *S. bicolor* and *S. macrospermum* (Price et al. 2006). Some other examples of hybridizations between S. *bicolor* and wild relatives are present in Table 3.

In addition, sorghum accessions from diverse gene pools have been listed as sources for introgression of favorable alleles of yield (Bramel-Cox and Cox 1988), resistance to parasitic weeds (Rich et al. 2004), disease resistance (Tsukiboshi et al. 1998; Mojtahedi et al. 1993; Viaene and Abawi 1998), and insect resistance (Nwanze et al. 1995; Dweikat 2005). De novo wild sorghum domestication can provide further positive impacts on cultivated sorghum breeding in a less complicated process of allele introgression. Also, researchers have made significant

Table 3 Experimental details of hybridization between *S. bicolor* and its wild relatives

Taxon	Status	References
S. bicolor and *S. almum*	Successful hybrids	Endrizzi (1957)
S. bicolor and *S. angustum*	Unsuccessful (in vivo rescue of the developing embryos were required)	Price et al. (2006)
S. bicolor and *S. bicolor* subsp. *drummondii*	Unassisted hybridization	Schmidt et al. (2013)
S. bicolor and *S. bicolor* subsp. *drummondii*	Successful hybrids	Werle et al. (2014)
S. bicolor and *S. halepense*	Successful hybrids	Endrizzi (1957), Hadley (1958), Sangduen and Hanna (1984), Piper and Kulakow (1994), Cox et al. (2002), Dweikat (2005), Magomere et al. (2015)
S. bicolor and *S. halepense*	Natural introgression	Morrell et al. (2005)
S. bicolor and *S. macrospermum*	Successful introgression using embryo rescue	Price et al. (2006), Kuhlman et al. (2010)
S. bicolor and *S. nitidum*	Unsuccessful (in vivo rescue of the developing embryos were required)	Price et al. (2006)
S. bicolor and *S. propinquum*	Successful hybrids but no use in sorghum improvement	Paterson et al. (1995), Wooten (2001)
S. bicolor and *S. versicolor*	Successful hybrids	Sun et al. (1991)
S. bicolor and *S. bicolor* subsp. *verticilliflorum*	Successful hybrids	Cox et al. (1984), Jordan et al. (2004)

Reproduced from Ananda et al. (2020) and Ohadi et al. (2017) with permission from rightsholders

progress in the introgression of alleles from wild sorghum to cultivated *S. bicolor* (Table 4).

3 Population Genomics

3.1 Origin and Domestication

Although archaeological findings support the theory of sorghum's origin and domestication in Africa (De Wet and Harlan 1971), it is still widely debated (Fuller and Stevens 2018; Venkateswaran et al. 2019a, b). The earliest evidence suggests the use of wild sorghum as food by the Sahara's hunter-gatherers around 7500 BC (Venkateswaran et al. 2019a, b), while the earliest evidence of domesticated sorghum-based foods refers to the Neolithic population of Sudan around 4000 BC.

Table 4 Introgression of wild sorghum alleles to *S. bicolor*

Taxon	Gene pool	Traits	Status	References
S. propinquum	1	Increase grain yield, increase height, and earliness of development	Successfully introgressed to *S. bicolor*	Wooten (2001)
S. bicolor subsp. *verticilliflorum*	1	Increase grain yield	Successfully introgressed to *S. bicolor*	Cox et al. (1984), Jordan et al. (2004)
S. halepense	2	Perennialism	Successfully introgressed to *S. bicolor*	Cox et al. (2002), Dweikat (2005)
S. bicolor subsp. *verticilliflorum*	1	Ability to grow in drought conditions, seeds with tolerance to high temperatures, high yield, parasite resistance	Candidate for sorghum improvement	Bramel-Cox and Cox (1988), Rich et al. (2004)
S. bicolor nothosubsp. *drummondii*	1	Allelopathic properties, resistance to ergot and nematodes	Candidate for sorghum improvement	Mojtahedi et al. (1993), Tsukiboshi et al. (1998), Viaene and Abawi (1998), Baerson et al. (2008)
S. halepense	2	Resistance to green bug, chinch bug, and sorghum shoot fly	Candidate for sorghum improvement	Nwanze et al. (1995), Dweikat (2005)
S. angustum	3	Resistance to egg laying by sorghum midge	Candidate for sorghum improvement	Sharma and Franzmann (2001)
S. amplum	3	Resistance to egg laying by sorghum midge	Candidate for sorghum improvement	Sharma and Franzmann (2001)
S. bulbosum	3	Resistance to egg laying by sorghum midge	Candidate for sorghum improvement	Sharma and Franzmann (2001)
S. macrospermum	3	Insect and disease resistance, higher growth rate and an insignificant aboveground dhurrin content under drought conditions	Successfully introgressed to *S. bicolor*	Kuhlman et al. (2008), Cowan et al. (2020)
S. brachypodum	3	Higher growth rate and an insignificant aboveground dhurrin content under drought conditions	Candidate for sorghum improvement	Cowan et al. (2020)
S. exstans	3	Resistance to shoot fly	Candidate for sorghum improvement	Kamala et al. (2009)

(continued)

Table 4 (continued)

Taxon	Gene pool	Traits	Status	References
S. stipoideum	3	Resistance to shoot fly	Candidate for sorghum improvement	Kamala et al. (2009)
S. matarankense	3	Resistance to shoot fly	Candidate for sorghum improvement	Kamala et al. (2009)
S. leiocladum	3	Cold tolerance	Candidate for sorghum improvement	Fiedler et al. (2016)

Reproduced from Ananda et al. (2020) with permission from rightsholder

The sorghum's diffusion has occurred to diverse climates across Africa, India, the Middle East, and East Asia (Ananda et al. 2020).

The domestication process of sorghum gave rise to a wide genetic diversity, which provides a rich source of morphological and phenological traits for the development of cultivars for different purposes and products (Upadhyaya et al. 2009). The genus *Sorghum* Moench, which belongs to the *Androgoponeae* tribe within family *Poaceae*, is heterogeneous, including five subgenera (Dahlberg 2000). The subgenera *Sorghum* includes both rhizomatous species (*Sorghum halepense* and *Sorghum propinquum*) and annual types (*Sorghum bicolor*) that comprise cultivated sorghums in addition to wild and weedy sorghums (De Wet 1978).

Cultivated sorghums were classified into five basic morphological races, based on spikelet and inflorescence morphology, namely bicolor, guinea, caudatum, kafir, and durra (Harlan and De Wet 1972). The most primitive grain sorghum is possibly the bicolor race (Dahlberg 2000), which may reflect early domestication events in the Northeastern region of Africa (De Wet 1978; De Wet and Harlan 1971). The guinea race is primarily a West African race, but it is also found in the Malawi region in Southeastern Africa, whereas caudatum sorghums are somewhat limited to the original regions of bicolor domestication (De Wet 1978). In turn, kafir sorghums may have been domesticated from bicolor types east and south of the Savanna belt and subsequently taken to India (Kimber 2000), whereas durra may have been domesticated in India (Harlan and Stemler 1976). Figure 1 provides insights on how sorghum main races have developed throughout the domestication process.

Domestication is the basis for the successful cultivation of sorghum worldwide. According to De Alencar Figueredo et al. (2008), domestication is driven by the selection of favorable alleles in important and visible traits, and it is usually accompanied by a significant loss of diversity due to genetic drift and selection sweeps. To better understand the progress of sorghum domestication, various molecular approaches have been used on diverse panels of representative accessions and resulted in a satisfactory understanding of ecogeographic domestication patterns, mainly in the northeastern part of sub-Saharan Africa. De Alencar Figueredo et al. (2008) used a core sample of 194 accessions of sorghum from International

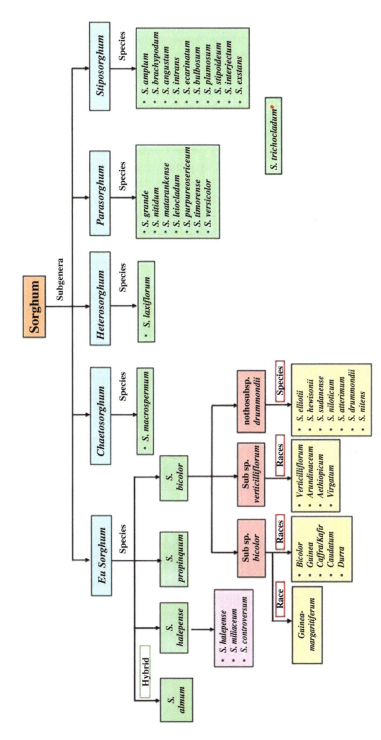

Fig. 1 Classification of sorghum. *The exact position within the phylogeny is still uncertain. (Reproduced from Ananda et al. (2020) with permission from rightsholder)

Crops Research Institute for the Semi-Arid Tropics (ICRISAT) and Centre de Coopération Internationale en Recherche Agronomique pour le Développement (CIRAD) germplasm banks. It was possible to further investigate the sorghum evolution from the described sequences of the six genes under survey, obtained after genomic DNA extraction and sequencing of PCR products. As a result, they recognized patterns of novel diversity, once southern and western Africa genotypes displayed the most frequent and ancient alleles, in contrast to other groups that presented more allele diversity. Their results indicated that both mutation and intragenic recombination generated alleles on the course of domestication, and the positive selection of those led to a high frequency levels of those. The mutation phenomena in sorghum followed positive selection can be considered crop neo diversity or de novo variation as advocated by Rasmusson and Phillips (1997). Gene sequences are not only useful for the identification of diversity patterns in sorghum. When gene sequences and transcriptomics are used jointly, they can lead to interpret functions at gene levels. For example, Francis et al. (2016) described good candidate genes for sorghum target traits that can be upregulated in response to drought stress.

In addition, Smith et al. (2019) provided some valuable insights on how population genomic information can be deployed to assess the genetic pools in sorghum when tracking the trajectory of domesticated sorghum (*Sorghum bicolor* ssp. *bicolor*). These authors compared the archaeological genomes against a global sorghum panel according to a principal coordinate analysis using 1,894 SNPs. As a result, 1,046 sorghum lines were clustered according to 7 main distinct groups, being ancient genomes, Asian durras, caudatum, east African durras, guinea, kafir, and unclassified group. These are the groups that were introgressed over time in the cultivated sorghum, presenting similarities within group, but distinctions among groups (Fig. 2). Also, it was proved that the level of heterozygosity has decreased to a third of its initial values over time in the sorghum domestication. Therefore, population genomics helped to elucidate the pathways of origin, domestication, and selection of sorghum, which are paramount for modern sorghum breeding.

3.2 Genetic Diversity and Population Structure

The patterns of genetic diversity in the species were likely influenced from selection and adaptive events during domestication episodes. For example, some guinea race sorghums, with their large and usually open inflorescences, are adapted for cultivation in high rainfall areas in West Africa (De Wet 1978). Tolerance to aluminum (Al) toxicity in the soil has been found to be much more prevalent in guinea sorghums, and to a lesser extent in caudatum types, compared to other morphological races (Caniato et al. 2011). Subsequently, a haplotype network analysis using SNP markers based on the major Al tolerance locus, AltSB (Magalhaes et al. 2004), suggested a single geographic and racial origin of Al tolerance mutations in primordial guinea domesticates in West Africa (Caniato et al. 2014), which is consistent

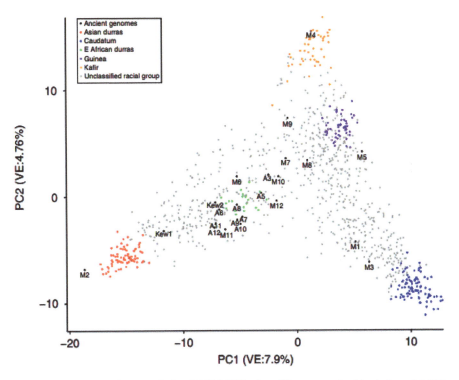

Fig. 2 Principal coordinate analysis of 1,894 SNPs from 23 genomes in this study and 1,046 sorghum lines described in Thurber et al. (2013). PC principal coordinate; VE variance explained. (Reproduced from Smith et al. (2019) with permission from rightsholder)

with prevalence of Al tolerance in guinea sorghums. In addition, guinea-race varieties tend to be adapted to stress factors present in West Africa, such as poor soil fertility, Striga parasitism, low soil pH and lodging, among others (Weltzien et al. 2006). The compact panicle that is typical of durra sorghums suggests adaptation to low-rainfall environments with less grain mold incidence (Kimber 2000). Key haplotypes for important traits may be subpopulation-specific, as reported for maturity loci in Morris et al. (2013). Natural selection for traits of different races has resulted in non-random occurrence of certain phenotypes with respect to the sorghum diversity landscape, key haplotypes for important traits may be subpopulation-specific, as reported for maturity loci in Morris et al. (2013). As such, understanding the patterns of genetic diversity in sorghum becomes fundamental for efficient pre-breeding strategies aimed at developing sorghum cultivars for agricultural or bioenergy purposes.

Brenton et al. (2016) assembled a 390-member biomass and sweet sorghum bioenergy association panel (SBAP) and compared it to the sorghum association panel (SAP) with respect to different aspects including population structure.

Previously, population structure in the SAP, whose members represent all major cultivated races from diverse geographic origins, was shown to be closely related both to racial and geographic origins (Casa et al. 2008). Genomic regions near some plant height and maturity genes had higher degree of polymorphism in the SBAP compared to the SAP, which is in line with selection for early maturity taking place in the SAP but not so much in the high-biomass, late flowering SBAP accessions. Once again in the SBAP, population structure analysis uncovered subpopulations featuring each of the five botanical races in sorghum (Brenton et al. 2016). This study suggests that tall, photoperiod-sensitive accessions suitable for bioenergy purposes are widespread across the sorghum diversity continuum, increasing the opportunities for selection of superior types at least to some extent regardless of racial and geographic origins.

A genetic diversity analysis performed in 125 sweet sorghum genotypes with SSR markers uncovered three main groups, which could be traced to historical and modern syrup, modern sugar/energy types, and amber types (Murray et al. 2009). Accordingly, using SSR markers genotyped in a large grain sorghum panel, these three sweet sorghum groups clustered with kafir/bicolor, caudatum, and bicolor types, respectively. These authors indicated that only a few sweet sorghum cultivars from the sweet sorghum panel represent most of the SNP alleles that were identified, suggesting significant redundancy within the sweet sorghum germplasm. Finally, although clustering patterns were found to be somewhat related to racial origin, sweet sorghums appear to be distinct concerning their origin. Wang et al. (2009) reported consistency between diversity patterns and geographic origins of sweet sorghums, suggesting the selection of accessions from different geographic regions for the development of new cultivars.

Recently, Cooper et al. (2019) presented a new reference genome based on an archetypal sweet sorghum line, which revealed few changes in gene content or overall genome structure, but differential expression of genes related to sugar metabolism, when compared to the current grain sorghum reference. Thus, they concluded that the high level of genomic similarity between sweet and grain sorghum reflects their historical relatedness, rather than their current phenotypic differences. To assess genetic diversity and population structure of sorghum landraces using SNP markers, Mengistu et al. (2020) genotyped 342 sorghum genotypes with 5,060 SNP markers. The genotypes were divided into 10 subgroups according to the collected regions, and the gene diversity, heterozygosity, major allele frequency, and polymorphism were analyzed (Table 5). From these results, it was possible to conclude that the populations are highly homozygous and have approximately the same ratio of MAF, despite different ranges of gene diversity due to unrelatedness between these populations. Therefore, substantial genetic gains would probably be achieved with the selection of favorable alleles within population, and with the adoption of inter-populational recurrent selection for target traits.

Table 5 Estimation of gene diversity, heterozygosity, PIC, and major allele frequency in 342 sorghum accessions based on 5,060 SNP markers

Collection regions	Number of accessions	Gene diversity	Heterozygosity	MAF	PIC
Oromiya	119	0.241	0.084	0.518	0.26
Tigray	53	0.239	0.078	0.518	0.251
Gambella	43	0.242	0.089	0.517	0.255
Benshangul	11	0.261	0.117	0.519	0.241
SNNP	35	0.253	0.066	0.518	0.26
Somali	5	0.179	0.053	0.517	0.184
Amara	47	0.236	0.08	0.518	0.082
Afar	8	0.236	0.043	0.522	0.238
Dire-Dawa	18	0.211	0.041	0.519	0.256
Improved varieties	3	0.237	0.02	0.52	0.195

Reproduced from Mengistu et al. (2020) with permission from rightsholder
MAF major allele frequency; *PIC* polymorphic information content; *SNNP* Southern Nations, Nationalities and Peoples

3.3 Landscape Genomics

Sorghum is a staple crop adapted to multiple environmental conditions worldwide, mainly to the arid and semi-arid areas of the globe. Therefore, it is essential to describe the genetic basis of adaptation to local climate, as well as the genotype by environment interaction, and responses at gene level to contrasting environments, which is particularly important in face to climate change. Several authors performed studies on these aspects, with special attention to the research performed by Wang et al. (2020), Maina et al. (2018), and Lasky et al. (2015). Wang et al. (2020) provided fundamental information that seed mass can be shaped by precipitation levels. For example, a promising hypothesis from this study is that large sorghum seeds can be favored by dry environments due to seedling size effect, which provide the plant competitive and survival advantage in early stages under drought stress. Maina et al. (2018) identified many loci associated with target traits and farmers preferences, such as maturity, flowering time, and kernel color. Landscape genomics in sorghum can be deployed for breeding to climate resilience and crop conservation, as well as phylogeographic studies for both cultivated and wild species, as described by Lasky et al. (2015). These authors also stated that more accurate predictions certain target traits in sorghum can be achieved by the use of low-density SNP arrays, as in the case of sorghum response to aluminum toxicity, suggesting the presence of large effect SNPs. Still, environment stimuli shape a significant proportion of genetic variation, as it can be seen in Fig. 3.

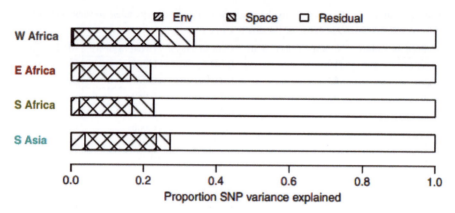

Fig. 3 Proportion of total SNP variation among accessions explained in RDA by environmental variables or spatial structure within each region. (Reproduced from Lasky et al. (2015) with permission from rightsholder)

4 Molecular Breeding for Bioenergy Traits

4.1 Bioenergy Traits

Bioenergy sweet and high-biomass sorghum are annual C4 crops with great potential as feedstock for biofuel production. These bioenergy sorghums show high dry matter yields, sugary stems (in sweet sorghums), and wide adaptability to different climate and soil conditions (Maw et al. 2017; Shukla et al. 2017). At grain maturity, the juice extracted from sweet sorghum stems is rich in fermentable sugars, predominantly sucrose with some glucose and fructose (Regassa and Wortmann 2014; Simeone et al. 2017).

Like sugarcane, sweet sorghum stems can be mechanically harvested and pressed to release juice, which typically contains 12–22% of fermentable sugars. After filtering the juice, the sugars can be directly fermented by yeast to generate ethanol (Wang et al. 2015). Usually, for each 10 tons of crushed sweet sorghum, 5–6 tons of wet bagasse are generated (Negro et al. 1999). Once the juice is removed, the bagasse is mainly composed of lignocellulose and cellulose, which can be converted into ethanol, other industrial biomaterials, valuable chemicals (Wang et al. 2015; Khalil et al. 2015; Antonopoulou et al. 2008), or even for generating electricity through combustion (Bennett and Anex 2009; Mathur et al. 2017).

Stem sugar concentration tends to peak 20 to 30 days after flowering until grain maturity (Briand et al. 2018). Schaffert and Parrella (2012) developed a Period of Industrial Utilization (PIU) protocol establishing minimum production values required for the success of sweet sorghum as a bioenergy crop in Brazil. The implantation of the PIU protocol is important for planning and management of continuous harvest and processing in the industrial plant (Souza et al. 2016).

One of the most important attributes in a plant designed for bioenergy production is biomass yield, which can relate to sugars and/or lignocellulose accumulation. Therefore, plants that flower late in a growing season accumulate the most biomass. The modern grain sorghum has mutations in maturity loci (*Ma1–Ma6*), which confers insensitivity to photoperiod (Marguerat and Bahler 2010). Some of the *Ma* genes code for photoreceptors, flowering repressors (*Ma1, Ma6*) or modulate the expression of other *Ma* genes (Yang et al. 2014; Murphy et al. 2014; Casto et al. 2019). Photoperiod-sensitive materials will have a longer vegetative growth and therefore accumulate more biomass than photoperiod insensitive materials. Time to flowering in sorghum varies approximately from 50 to 150 days after planting depending on field conditions, genotype, latitude, and day-length.

Bioenergy sorghum hybrids sensitive to photoperiod can be obtained by crossing a photoperiod-insensitive male sterile A-line (*ma1ma1*) with a photoperiod-sensitive pollinator fertile restoring R-line (*Ma1Ma1*) (Quinby 1974). The resulting hybrids are photoperiod sensitive and heterozygous for the dominant *Ma1* maturity gene (*Ma1ma1*) (Silva et al. 2018). For the development of sweet sorghum cultivars, breeding programs aim at producing hybrids with high stem juice extraction, high sugar content, and predominance of sucrose in the juice, as well as high purity, that is, high values for the ratio between sucrose and total soluble solids or Brix. The ideal genotypes would have these two traits combined, that is, higher biomass with high sugar yields (Mathur et al. 2017). Other quality attributes such as starch, pH, and phenolic compounds can be highly influenced by environmental and maturity conditions which can adversely affect processing and/or fermentation (Cole et al. 2017; Silva et al. 2017; Costa et al. 2014; Lingle et al. 2013). The number of phenolic compounds varies according to the genotype and negatively affects the ethanol production process (Ravaneli et al. 2011).

The Dry Stalk (*D*) locus in sorghum controls a qualitative difference between dry, pithy-white stems and midribs (genotypes *DD* or *Dd*), and juicy-green stems and midribs (genotype *dd*) (Smith and Frederiksen 2000). Therefore, both parental lines need to be homozygous for the recessive allele *d* to produce high yielding sweet sorghum hybrids (Rooney 2000). On the other hand, for the development of biomass sorghum hybrids, just one homozygous parent for the dominant allele *D* is required to result in a dry stem hybrid (genotype *Dd*) with high fiber content (Silva et al. 2018). Due to the difference in midrib color between juicy stem (genotype *dd*) and dry stem (genotypes *DD* or *Dd*) materials, midrib color can be used as a morphological marker, assisting breeders to select the desired genotypes (Table 6).

Some studies have reported the existence of heterosis in sorghum for sugar content in the juice, and for biomass and sugar production (Pfeiffer et al. 2010; Oliveira et al. 2019). Thus, the development of hybrids is a viable and promising strategy to increase sweet and biomass sorghum productivity. However, for sweet sorghum, it is noteworthy that both parental lines must have a high concentration of sucrose in the juice so that the hybrids have high potential for total sugar production. On the other hand, to develop biomass sorghum hybrids, the two parental lines should not have favorable alleles for genes related to sugar production, since the presence of sugars in the stalk can cause corrosion in the boiler during the biomass

Table 6 Summary of dominant inheritance trait between sweet and biomass sorghum

Crop	Trait	Loci	Feature	Reference
Sweet sorghum	Photoperiod insensitivity	*Ma* recessive alleles	Early flowering in long days	Quinby (1974), Rooney and Aydin (1999)
	Juicy stalk	*D* recessive alleles	Green stems and midribs	Smith and Frederiksen (2000)
Biomass sorghum	Photoperiod sensitivity	*Ma* dominant alleles	High biomass accumulation in long days	Quinby (1974), Rooney and Aydin (1999)
	Dry stalk	*D* dominant alleles	Pithy-white stems and midribs	Smith and Frederiksen (2000)

combustion process for the generation of electricity or high-pressure steam (Mathur et al. 2017).

Focusing on the production of cellulosic ethanol, sorghum breeding programs have also been developing photoperiod-sensitive hybrids with recessive alleles for the brown midrib genes (*bmr* loci), which were generated in sorghum using diethyl sulfate mutagenesis (Porter et al. 1978), attending the demand for high biomass alternative materials for biofuel production. Several mutagenized *bmr* sorghum lines have been characterized and the corresponding genes involved in specific enzymatic steps of monolignol biosynthetic pathway have been identified in recent years (Bout and Vermerris 2003; Saballos et al. 2009; Sattler et al. 2009, 2014). The presence of *bmr-6* recessive alleles reduces the activity of one of the key enzymes, cinnamyl alcohol dehydrogenase (CAD), involved in the synthesis of lignin in sorghum, resulting in lower levels of this component in the plant. Sorghum *bmr* mutants can present up to 50% reduction in lignin, which can improve the conversion efficiency of sorghum biomass to simple sugars and make cellulosic ethanol production more economically feasible (Dien et al. 2009; Godin et al. 2016). The *bmr-6* recessive alleles can be introduced via marker-assisted backcrossing in a set of A, B, and R-lines related to the best conventional (non-*bmr*) biomass sorghum hybrids (Sattler et al. 2009). The *bmr-6* sorghum hybrids are characterized by high dry matter production, around 30 t ha-1, and a greater efficiency in the conversion to cellulosic ethanol (Dien et al. 2009; Silva et al. 2018). Preliminary laboratory analyses showed that *bmr-6* genotypes were approximately 30% more efficient in the enzymatic hydrolysis of the biomass compared to conventional hybrids (Dien et al. 2009; Godin et al. 2016). Thus, considering the productivity levels and the high efficiency in enzymatic hydrolysis, which is a costly process during the cellulosic ethanol production, *bmr-6* hybrids have proven to be an excellent alternative for second-generation fuel production technologies.

4.2 QTL Mapping

Sorghum is currently considered a promising alternative crop for generating bioenergy from sugar and lignocellulosic biomass, especially sweet sorghum, which like sugarcane accumulates soluble sugar in its juicy stalk, and biomass sorghum that can be used for biofuels or bioelectricity production (Zhang et al. 2013). Because of these phenotypic differences between these types of cultivated sorghum, breeding programs must focus on several traits related to bioenergy generation, most of them exhibiting a quantitative (polygenic) inheritance. In this context, understanding the genetic architecture and the environmental effects involved in the phenotypic expression of these traits is a required task to improve sorghum as a bioenergy crop.

Most of the genetic mapping studies related to biofuel production is based on the direct and indirect characterization of biomass yield and composition in sorghum. The traits mainly assessed can be divided basically in nonstructural carbohydrates, that is, stem sugar components (total soluble solids – Brix, sucrose, and total sugars), structural carbohydrates, that is, composition and content of lignocellulosic biomass and its conversion efficiency into biofuel, as well as agronomic traits such as flowering time, plant height, juice weight, and fresh and dry biomass weight. For the purpose of this chapter, plant architecture components, such as panicle length, stem circumference, internode number, and tiller number, will not be included as primary bioenergy traits. For information on their role in contributing to increase biomass energy potential in sorghum, see Anami et al. (2015) and Mathur et al. (2017).

Sorghum breeding programs may benefit greatly from the development of marker-assisted strategies that can increase the selection efficiency and consequently the rate of genetic gain over breeding cycles. Although many QTL studies have been performed in bioenergy sorghum, the use of QTL-based marker-assisted selection (QTL-MAS) is still limited to traits controlled by a few major-effect QTLs, such as the *bmr* genes for lignin content (Silva et al. 2020). Additionally, QTL mapping results are in general specific to the population background and to the environment (Table 3), making it difficult to find constitutive QTLs for using in QTL-MAS strategies.

Quantitative trait loci (QTL) mapping is a useful tool to dissect and to understand the genetic architecture of polygenic traits. Several QTL mapping studies, based on linkage analysis and/or association analysis, have been reported for bioenergy sorghum, using different populations and molecular marker systems (Table 3). The linkage mapping studies have mostly explored recombination events between markers within experimental populations, either F_2 derivatives or recombinant inbred lines (RIL) progenies, obtained from controlled crosses between sweet and non-sweet sorghum lines. They have also used a relatively low number of markers compared to current studies using high-density SNP marker genotyping platforms, such as genotyping-by-sequencing (GBS, Elshire et al. 2011). Additionally, most of these studies provided linkage maps and located QTLs by single marker (SM; Tanksley et al. 1982; Edwards et al. 1987) analysis, simple or composite interval mapping (IM and CIM, respectively; Lander and Botstein 1989; Zeng 1993, 1994), applying univariate models for single environment, or using marginal adjusted means across multiple environments (Table 7). However, strong QTL-by-

Table 7 QTL mapping, genotype-phenotype association, and GWAS studies carried out in bioenergy sorghum

Traits	Reference	QTL mapping approach	Population size	Population origin	Molecular markers	Statistical method	Results
Sugar components (sucrose percentage, sugar yield), dry matter yield, flowering time, and plant height	Natoli et al. (2002)	Linkage analysis	99 F_3	Sweet (LP29/1) × Sweet (LP113A)	144 (SSR and AFLP)	IM / CIM	No QTL for Brix. A single QTL for sucrose was detected on linkage group G (marker interval P12M35_C29 and E32M34_17), explaining 18.8% of the phenotypic variance.
Sugar content (Brix)	Bian et al. (2006)	Linkage analysis	207 $F_{2:3}$	Sweet (Early Folger) × Grain (N32B)	327 (RFLP, AFLP, and SSR)	SM / CIM	Two QTLs showing additive and over-dominance effects for Brix.
Stem sugar and grain nonstructural carbohydrates	Murray et al. (2008a)	Linkage analysis	176 $F_{4:5}$ and 165 $F_{5:6}$	Sweet (Rio) × Grain (BTx623)	300 (AFLP and SSR)	IM / CIM	QTLs for Brix and stem sugar concentration (total juice sugars) were mapped to nearly identical locations on chromosome 3.
Stem and leaf structural carbohydrates	Murray et al. (2008b)	Linkage analysis	176 $F_{4:5}$	Sweet (Rio) × Grain (BTx623)	300 (AFLP and SSR)	IM / CIM	Structural biomass yield "hot spots" for QTL co-localization appeared in similar locations as nonstructural carbohydrate QTL and agronomic traits QTL. Plant height QTLs on chromosomes 7 and 9 were co-localized with increased stem and total biomass QTLs. A delayed flowering time QTL on chromosome

Stem sugar-related and other agronomic traits	Ritter et al. (2008)	Linkage analysis	184 F$_6$	Sweet (R9188) × Grain (R9403463-2-1)	228 (SSR and AFLP)	MPM	6 co-localized with increased stem, leaf, and total biomass. Total structural carbohydrates QTL co-localized with the major Brix QTL on chromosome 3. For sucrose and sugar content, QTLs were generally located to five genomic regions: one on chromosome SBI-01, SBI-06, and SBI-10, and two on chromosome SBI-05. Two QTLs on chromosome SBI-05 and one on SBI-06 were identified for high Brix content.
Plant height and stem sugar (Brix)	Murray et al. (2009)	Association analysis	125 genotypes	Panel (mostly sweet)	369 (SNP and SSR)	GLM / MLM	One association for Brix was detected on chromosome 1, 12kb from a glucose-6-phosphate isomerase homolog.
Sugar components (Brix, glucose, sucrose, and total sugar content) and sugar-related traits	Shiringani et al. (2010)	Linkage analysis	188 F$_{5:6}$	Sweet (SS79) × Grain (M71)	157 (AFLP, SSR, and EST-SSR)	CIM	One QTL was mapped on SBI-02 for stem juice weight and a total of 15 QTLs were located on seven different chromosomes for sugar content in stem juice. QTLs were detected for Brix on SBI-02, -04, and -06.

(continued)

Table 7 (continued)

Traits	Reference	QTL mapping approach	Population size	Population origin	Molecular markers	Statistical method	Results
Plant height, stem and leaf fresh weight, juice weight, and Brix	Guan et al. (2011)	Linkage analysis	186 F_2 and $F_{2:3}$	Sweet (L-Tian) × Grain (Shihong137)	636 (SSR and PIP)	CIM	Six QTLs controlling juice weight were mapped, three on SBI-01 and one on each of the following chromosomes: SBI-04, SBI-07, and SBI-09. QTLs on chromosome SB-01 and SB-04 were mapped in only one environment. Four QTLs related to Brix were detected in two different environments on SBI-01, SBI-02, SBI-03, and SBI-07.
Neutral detergent fiber, acid detergent fiber, acid detergent lignin, cellulose, hemicellulose, fresh leaf mass, stripped stalk mass, dry stalk mass, and biomass	Shiringani and Friedt (2011)	Linkage analysis	188 $F_{5:6}$	Sweet (SS79) × Grain (M71)	215 (SSR and AFLP)	CIM / QTL × E via Anova	A total of 72 additive QTLs associated with fiber quality traits were detected on 10 chromosomes and the highest cluster was identified on chromosome 6.
Sugar components (Brix, juice) and agronomic traits	Burks et al. (2015)	GWAS	252 genotypes	Panel (80 sweet sorghum and 172 landraces)	42.926 (SNP)	MLM / cMLM	A major QTL for midrib color, sugar yield, juice volume, and moisture was identified on chromosome 6.
Brix, stalk diameter, and plant height	Disasa et al. (2018)	Linkage analysis	192 F_2	Sweet (Gambella) × Grain (Sorcoll 163)	76 (SSR)	SM / IM / ICIM	Two major QTLs detected on linkage groups SBI-05 and SBI-06.
Neutral detergent fiber and acid detergent fiber	Li et al. (2015)	Linkage analysis	184 F_2	Sorghum (Tx623A) × Sudangrass	124 (SSR)	ICIM	A total of 12 QTLs detected for forage quality-related traits. They were distributed on chromosomes 1, 2, 3, 6, 8, and 9.

Saccharification yield	Wang et al. (2011)	Association analysis	242 genotypes	Mini core collection (Upadhyaya et al. 2009)	703 (SSR)	GLM / MLM	Two markers on the sorghum chromosomes 2 (23-1062) and 4 (74-508c) were associated with saccharification yield. Localization of these markers based on the whole genome sequence indicates that 23-1062 is 223 kb from a b-glucanase (Bg) gene and 74-508c is 81 kb from a steroid-binding protein (Sbp) gene. These markers were found physically close to genes encoding plant cell wall synthesis enzymes such as xyloglucan fucosyltransferase (149 kb from 74-508c) and UDP-D-glucose 4-epimerase (46 kb from 23-1062).
Agronomic traits, sugar components (Brix) and different calculations for dry biomass and juice yield	Felderhoff et al. (2012)	Linkage analysis	185 $F_{3:4}$	Sweet (Rio) × Grain (BTx3197)	488 (SNP)	SMA / IM / CIM	A QTL for Brix on chromosome 3 was consistently the most significant. Two other QTLs were identified on chromosome 1 and 2. For juice yield and juice by height, QTLs on chromosomes 1, 4 and 7 were detected. However, a QTL on chromosome 1 co-localized with Brix.

(continued)

Table 7 (continued)

Traits	Reference	QTL mapping approach	Population size	Population origin	Molecular markers	Statistical method	Results
Fresh stalk weight, stalk juice weight, and Brix	Lv et al. (2013)	Association analysis	119 genotypes	Panel (43 sweet and 76 grain)	51 (SSR)	GLM	The SSR marker, xtxp340, on chromosome 1 was associated with sugar concentration (Brix) at two distinct environments. Markers on chromosome 10, Undhsbm105 and xtxp141, were also highly associated with stalk juice weight.
Saccharification yield	Wang et al. (2013)	Association analysis	242 genotypes	Mini core collection (Upadhyaya et al. 2009)	14.739 (SNP)	MLM	Seven marker loci were associated with saccharification yield. Candidate genes from the seven loci were identified but must be validated, with the most promising candidates being beta-tubulin, and NST1, a transcription factor for cell wall biosynthesis in fibers.
Fresh stalk weight, stalk dry weight, and percent moisture	Han et al. (2015)	Linkage analysis and deep sequencing technologies	611 F_2	Dry stalk landrace (G21) × Elite cultivar (Jiliang2)	141 (SSR)	ICIM	Detection of the qSW6, a major QTL related to stem water content on chromosome 6, from 48,320,691 bp to 52,840,234 bp.
Biomass fiber composition (acid detergent fiber, neutral detergent fiber, and lignin) and agronomic traits	Brenton et al. (2016)	GWAS	390 genotypes	Two sorghum panels (152 sweet lines and 238 biomass types)	232.303 (SNP)	GLM / MLM / cMLM	Association scans revealed a total of eight significant SNPs, representing five loci, two of them located on chromosome 4 and three on

						chromosome 6, as well as 22 genes. Most notably, one of the loci on chromosome 4 causes an amino acid change in a vacuolar iron transporter family protein; and the other contains a B-box zinc finger protein. Chromosome 6 had two genes coding for cellulase enzymes.
(anthesis, plant height, and dry weight)						
Plant height, biomass, juice weight, and Brix	Wang et al. (2016)	Linkage analysis	181 $F_{2:7}$	Sweet (L-Tian) × Grain (Shihong137)	247 (SSR)	CIM	Two QTLs controlling juice weight were mapped on SBI-01 and SBI-09. Only one QTL for Brix was detected on SBI-02. In addition, a Brix-QTL was detected on chromosome 7, between Xtxp160 and SB4087 markers, in two distinct environments.
Cellulose, hemicellulose, neutral detergent fiber, and acid detergent fiber	Li et al. (2018)	GWAS	245 genotypes	Sorghum diversity panel	85585 (SNP)	MLM (6 methods)	A total of 42 SNPs was identified. One main quantitative trait nucleotide (QTN) was identified for cellulose content on chromosome 2, and was associated to a candidate gene encoding a kinesin-like protein.

AFLP amplified fragment length polymorphism; *CIM* composite interval mapping; *cMLM* compressed mixed linear model; *EST* expressed sequence tags; *GLM* general linear model; *GWAS* genome-wide association study; *ICIM* inclusive composite interval mapping; *IM* interval mapping; *MLM* mixed linear model; *MPM* multipoint map; *PIP* potential intron polymorphism; *QTL* quantitative trait locus; *RFLP* restriction fragment length polymorphism; *SIM* simple interval mapping; *SMA* single marker analysis; *SNP* single nucleotide polymorphisms; *SSR* simple sequence repeats

environment interaction effects were observed for most of the quantitative traits related to bioenergy.

Both linkage and association analyses rely on the existence of linkage disequilibrium (LD), that is, the non-random association of alleles between markers and QTL (Flint-Garcia 2003). However, association analysis has emerged as a powerful approach by exploring evolutionary and historical recombination events between loci within a collection of individuals, allowing a better mapping resolution than linkage analysis (Yu and Buckler 2006; Yu et al. 2006). The QTL mapping resolution depends on the level of LD between markers and causative polymorphisms. Although linkage analysis does not require large numbers of markers, its mapping resolution is low due to the limited number of recombinants within the mapping progeny. By contrast, association analysis can provide high-resolution QTL mapping, but, in general, it requires high-density markers distributed across the genome to detect causal polymorphisms (Mackay et al. 2009; Zhu et al. 2008). Based on Table 7, it is possible to observe the evolution in the application of different molecular marker systems for population genotyping in QTL mapping studies, as well as the increase in genome coverage capacity and mapping resolution in more recent studies for bioenergy sorghum.

In association analysis, genetic diversity panels, composed by accessions of germplasm banks sets of elite breeding lines or cultivars, are often used as mapping populations, which allow maximizing the explored genetic variability. However, in these mapping populations, factors other than linkage may affect LD between markers and QTL (Flint-Garcia 2003; Zhu et al. 2008). For example, the population structure and the degree of genetic relationship between individuals may result in spurious associations between phenotype and genotype, requiring the inclusion of these factors in appropriate statistical models to reduce the number of false positives (Yu et al. 2006). In bioenergy sorghum, association mapping studies were usually performed in panels composed by diverse genotypes with distinct geographic origins, genetic backgrounds, and bioenergy purposes (Table 7). For example, Brenton et al. (2016) performed association analyses in a 390-member sorghum bioenergy association panel (SBAP), composed by worldwide cultivars and germplasm accessions of sweet and biomass sorghum types. However, population structure analysis uncovered SBAP subpopulations featuring each of the five botanical races (bicolor, caudatum, durra, guinea, and kafir) in sorghum. Thus, association scans were performed using a general linear model (GLM), a mixed linear model (MLM), and a compressed MLM (cMLM; Zhang et al. 2010). Model fit was compared by examining the QQ-plots, and the superior fit was observed for the cMLM, which internally controlled for population structure (Q-matrix) and kinship (K-matrix) among individuals and used clustering analysis to assign individuals to groups. In other studies, Q, K and/or Q+K models have already been shown to increase statistical power and reduce false positives when population structure and distinct degrees of genetic relationship between individuals are present in the association mapping population.

Genomic regions and candidate genes identified through linkage and/or association analysis, showing significant effects on bioenergy traits in sorghum, are described in Table 7. Over 850 QTLs have been identified in sorghum for bioenergy-related traits (Anami et al. 2015). Mace and Jordan (2010, 2011) performed a comprehensive meta-analysis using 48 QTL mapping studies conducted in sorghum through linkage analyses from 1995 to 2010 and constructed a consensus genetic map containing 771 QTLs related to 161 traits. Thirty-six QTLs, including eight QTLs for plant height, three for maturity, two related to stem biomass, and five related to sugars, were reported on chromosome 6 in a region of only 5 cM, where genes regulating plant height and maturity were also located, the *Dw2* and *Ma1* genes, respectively. *Dw2* is a dwarfing gene regulating the length of stem internodes and therefore the plant height in sorghum (Quinby et al. 1974). Additionally, Mace and Jordan (2010) mapped the highest number of the targeted genes (seven out of 35) at chromosome 6, including the gene related to stem midrib type, that is, pithy-white or juicy-green stem and midribs (gene *D*) in sorghum.

Sugar composition and content are important traits to consider during the development of potential sweet sorghum cultivars for biofuel production. Not surprisingly, most of the QTL studies for biomass yield and quality, focusing on first-generation ethanol production, were performed for sugar-related traits, and therefore a sweet sorghum line was generally used as one of the mapping population's parents. Many genomic regions have been identified for sugar content in the stalk (Table 7). Murray et al. (2008a) and Ritter et al. (2008) verified that favorable alleles related to sugar content could also come from grain parents and that there is a low negative correlation between sugar content in the stalk and grain yield. However, Murray et al. (2008a), using a RIL population derived from Rio (sweet sorghum) and BTx623 (grain sorghum), identified a major-effect QTL on chromosome 3 explaining 25% of the genetic variability observed for stalk sugar concentration, but not co-localized with any other QTL associated to grain yield.

Other QTL mapping studies (Table 7), using different mapping populations and genetic backgrounds, have also reported a major-effect QTL on chromosome 3, explaining from 18% to 25 % of the total phenotypic variation for sugar concentration (Felderhoff et al. 2012; Murray et al. 2008b; Natoli et al. 2002). Lekgari et al. (2010), using a RIL population evaluated across four environments, detected a QTL on chromosome 1 explaining 33.9 % of the phenotypic variation, but nonsignificant regions were detected in chromosome 3. Murray et al. (2009) also mapped a QTL on chromosome 1 explaining around 9 % of the phenotypic variation for Brix and sugar concentration, and later, via association analysis, this QTL was located 12 kb from a sorghum homolog of glucose-6-phosphate isomerase, an important enzyme in sugar metabolism. In a review, Anami et al. (2015) reported a total of 38 QTLs for Brix, 12 for glucose, 14 for sucrose, 22 for total sugar, and two for fructose content in the stalk that were detected in previous QTL studies in sorghum.

QTLs for sugar concentration were mapped along the sorghum genome on chromosomes 1, 2, 3, 4, 5 and 7, using linkage and/or association mapping strategies. In general, the various results corroborate that most Brix QTLs have additive effects (Felderhoff et al. 2012; Bian et al. 2006; Guan et al. 2011; Lekgari et al. 2010; Lv et al. 2013; Murray et al. 2008a, b, 2009; Natoli et al. 2002; Ritter et al. 2008; Shiringani et al. 2010; Brown et al. 2006; Wang et al. 2016; Disasa et al. 2018).

Thus, to develop high-Brix hybrids, it is necessary to generate male and female lines with high Brix content. However, the soluble solids content (Brix) has a physiological limit of approximately 25% (Felderhoff et al. 2012; Lv et al. 2013). Since stem sugar yield is mainly affected by stalk juice yield and Brix, breeding efforts should also focus on increasing juice yield. In fact, Murray et al. (2008a) reported that stem sugar yield was almost twice more affected by stalk juice yield than by sugar concentration.

Recent studies have reported QTLs for stalk juice yield, using juicy and dry stalk genotypes. In addition to large-effect associations detected on chromosome 6 by GWAS (Burks et al. 2015), Xia et al. (2018) reported the D locus in a region of 36 kb on chromosome 6 using fine-mapping, in which they identified a premature stop codon in a NAC transcription factor as the candidate polymorphism. Although several sugar yield-related QTLs have been identified in a few mapping populations (Table 7), QTLs showed effects of varying significances and magnitudes on the trait, according to the environmental conditions. Therefore, it is still necessary to conduct phenotypic and molecular screening of new sweet sorghum genotypes for sugar related traits, to identify material that can contribute superior alleles related to sugar yield and biofuel production (Silva et al. 2017).

Biomass yield is another target trait in bioenergy sorghum breeding programs and is mainly affected by two factors: plant height and flowering time. Thus, several biomass composition QTLs corresponded to pleiotropic effects of known plant height and maturity-related genes. However, some of them were linked to specific sugar-related genes (Murray et al. 2008a; Ritter et al. 2008; Shiringani et al. 2010; Guan et al. 2011). Over 60 loci have already been identified for plant height in sorghum. For example, dwarfing genes (*Dw1–Dw4*) regulate the length of stem internodes and therefore control plant height (Quinby et al. 1974). QTLs associated to one or more of the *Dw* loci have been identified in several mapping studies (Pereira and Lee 1995; Rami et al. 1998; Hart et al. 2001; Kebede et al. 2001; Murray et al. 2008a; Ritter et al. 2008; Guan et al. 2011; Sabadin et al. 2012; Takai et al. 2012; Reddy et al. 2013, among others). Therefore, biomass sorghum, which is photoperiod sensitive, accumulates more biomass than grain sorghum (Olson et al. 2012). Several QTLs associated with flowering time (such as days to flowering and anthesis date) have been identified (Natoli et al. 2002; Lekgari et al. 2010; Higgins et al. 2014). Furthermore, many studies have reported plant height QTLs co-localized with QTLs for flowering time. For example, QTLs on chromosome 6, where *Dw2* and *Ma1* genes are closely linked, controlling height and flowering time, respectively (Klein et al. 2008). Additionally, flowering time QTLs have not only co-localized with plant height, they also co-localized with biomass and juice yield traits. For example, Shiringani et al. (2010) identified a QTL on chromosome 6 co-localized with QTLs for flowering time, plant height, Brix, sucrose, and sugar content.

Studying the genetic architecture of biomass-related traits is important for the selection and development of biomass sorghum as a bioenergy crop. Plant biomass is mainly composed of cell walls which contain structural carbohydrates (cellulose and hemicellulose) and lignin, a polyphenol (Carpita and McCann 2008). While lignin can negatively affect biomass conversion to cellulosic ethanol, it increases biomass calorific value, an important trait for bioelectricity generation through biomass combustion (Davison et al. 2006; Chen and Dixon 2007; Vermerris et al. 2007). Because of the largely available sorghum variability for lignin content, high and low lignin sorghum can be developed according to their end-use purpose (Mullet et al. 2014). Therefore, manipulation of content and composition of lignin in sorghum can improve biomass conversion efficiency for biofuel production (Zhao et al. 2013; Saballos et al. 2012). In addition, high cellulose levels are also important and desirable for cellulosic ethanol production because of its high energy density nature, compared to nonstructural carbohydrates (sugar and starch). Thus, enzymatic depolymerization of cellulose into glucose molecules allows cellulosic ethanol production through fermentation process (Biswal et al. 2015).

The most relevant findings in mapping studies for structural carbohydrates yield and composition are also summarized in Table 7. Murray et al. (2008a, b) was the first study to address at the same time genetic variation in yield and composition of the whole plant (stem, leaf, and grain) for bioenergy purposes, and identified chromosome *hot spots* for structural carbohydrate-biomass yield in which QTLs were co-localized with nonstructural carbohydrate and agronomic traits-associated loci. The authors also highlighted several lignocellulosic related QTLs, including leaf and stem structural biomass yield, composition, and regrowth QTLs. Not surprisingly, they also reported that QTLs for structural and nonstructural carbohydrate yields co-localized with loci for height, flowering time, as well as stand density. However, QTLs associated to biomass composition traits did not show as much co-localization when comparing tissues and environments evaluated. The authors also calculated theoretical ethanol production as a trait and found that biomass yield contributed more to variation than its composition.

To identify loci linked to biomass conversion efficiency, Vandenbrink et al. (2013) used a RIL population derived from an interspecific cross between *Sorghum bicolor* × *Sorghum propinquum* (wild species). A total of 49 QTLs from leaf and stem biomass were associated with biomass conversion efficiency during enzymatic hydrolysis of cellulose (saccharification). In addition, the authors also identified QTLs related to stem biomass crystallinity index (CI), a feature that can interfere with the biomass conversion process (Vandenbrink et al. 2013). Using a different approach, Wang et al. (2011, 2013) performed an association mapping considering saccharification yield presented as milligrams of glucose per grams of dry biomass in laboratory. Shiringani and Friedt (2011) identified a total of 72 additive QTLs associated with fiber quality traits across all chromosomes and the highest cluster was located on chromosome 6. Various genetic elements involved in fiber-related traits should be considered during biomass sorghum breeding. QTLs associated to biomass content (neutral detergent fiber, acid detergent fiber, acid detergent lignin, cellulose, hemicellulose) and biomass yield traits (total fresh and dry biomass weight) have also been identified (Murray et al. 2008b; Shiringani and Friedt 2011). The combination of favorable alleles for high fiber quality with high biomass in single RILs may represent interesting new sorghum breeding material to produce biomass and bioenergy.

Finally, as summarized in Table 7, QTL mapping studies via linkage analysis for bioenergy traits in sorghum commonly used SM analysis, IM and CIM for single traits, or multi-trait CIM (MT-CIM; Jiang and Zeng 1995). Murray et al. (2008a, b) reported that both IM and CIM consistently identified and mapped QTLs at the same positions when studying nonstructural carbohydrates in sorghum. However, CIM detected more significant QTL effects with smaller confidence intervals (LOD confidence intervals). The power and precision of linkage analysis depends on the genetic contrast of the mapping population, but also its size and the number of marker loci used to cover the genome. In Table 7, it is noticeable that most linkage mapping studies used less than 200 individuals (RILs or F_2 generations) and a low-density marker system, usually SSR markers or a mix with AFLP/RFLP markers, reducing the chance of finding causal polymorphisms or candidate genes. Nonetheless, with advancing of next-generation sequencing technologies, high-throughput genotyping platforms have been widely used in different crop species, making available high-density SNP markers distributed across the whole genome. Additionally, over the years, with the decreasing costs of high-density marker systems, GWAS have become even more common, bringing the possibility to identify some candidate genes in bioenergy sorghum (Table 7).

4.3 GWAS

It is vital to better understand the genetic basis of important traits in sorghum and identify the genomic regions and loci that underly them. Traditionally, structured group of genotypes derived from bi-parental crosses such as F_2, backcrossing, or recombinant inbred line population have been deployed to assess the underlying regions after molecular marker mapping, with the disadvantage of low genomic resolution (Girma et al. 2019). Recently, genome-wide association studies (GWAS) using high-density SNP markers distributed across the whole genome have led to the discovery of candidate genes related to bioenergy traits in sorghum (Table 7). On the other hand, GWAS can overcome the pitfalls of usual QTL mapping, and requires the comprehension of population structure and prior to performing the association analyses to avoid spurious associations (Wu et al. 2011). Although previous association mapping studies (Table 7) have shown success in identifying genomic regions underlying major traits in sorghum, recent GWAS studies on plant height, and disease resistance (Girma et al. 2019; Miao et al. 2020), seed quality traits (Kirmani et al. 2020; Rhodes et al. 2014), lignin content (Niu et al. 2020), plant fresh weight in the presence and absence of herbicide (Baek et al. 2019), and traits related to adaptation to tropical soils (Bernardino et al. 2021) have been addressed in sorghum.

Sorghum is widely grown in tropical and sub-Saharan lands (Bernardino et al. 2021), which accounts for a great part of the acid and arid soils in the globe, and therefore can impact negatively sorghum yield due to limited uptake of water and essential nutrients (Foy et al. 1993; Bernardino et al. 2021). The consequences of growing sorghum under low pH soil conditions not only impact one or few traits, but an interplay of abiotic stress traits. Then, GWAS is an efficient tool in sorghum breeding programs to identify the specific loci in selected lines that could lead to a broad adaptation to tropical soils. This is essentially what Bernardino et al. (2021) performed in their research, using a sorghum multi-parental random mating population (BRP13R), genotyped with 43,825 SNP markers after quality control, and phenotyped for grain yield, plant dry matter, root diameter, total surface area, and aluminum tolerance under low phosphorus (P) and low aluminum (Al) growing conditions. In spite of the reduced population size of BRP13R, its intermediate levels of linkage disequilibrium (LD), low population structure, and multi-allelic nature enabled identification of peaks of association between SNPs and Al tolerance, not necessarily using the gene-specific markers for Al tolerance (*SbMATE*-specific markers). Those factors were advantageous over mapping populations from other association studies such as the ones from Melo et al. (2019). According to Bernardino et al. (2021), P-starvation tolerance SNPs were associated with root morphology; and some SNPs in total LD with P-efficiency genes explained up to 4% of the genetic variance in grain yield (up to 200 kg ha^{-1}). The representation of the associated SNPs to the target traits are presented in Fig. 4. In general, it depicts the relevance of regions within chromosomes 3 and 6, from where gene-specific markers for pre-breeding context and introgression of favorable alleles can be deployed. BRP13R also showed its importance as a source population for breeding purposes and for genetic studies due to its high-recombination aspects. This study could, therefore, detect in fine and efficient way some hotspot regions within chromosomes that cause variation in certain sorghum breeding relevant key traits, and also being a more flexible tool to describe genetic architecture than the usual association mapping that were traditionally performed decades ago (Table 7).

In a population genomics study, McCormick et al. (2018) used the reference genome and re-sequenced genomes for genetically diverse race-specific 176 sorghum accessions. The authors estimated the genome size which was 24.6% (174.5 Mb) higher than the reference genome and identified 398 SNPs significantly associated with important agronomic traits, of which, 92 were in genes. Moreover, drought gene expression analysis identified 1,788 genes that are functionally linked to different conditions, of which 79 were absent from the reference genome (Ruperao et al. 2021). Using the whole genome sequence data from 354 sorghum diverse accessions, the work identified two million SNPs and 3.9 million indel sites, which represented the functional genome diversity (Ruperao et al. 2021). The genome-wide association study (GWAS) was used to understand the functional utility of the pan-genome. Three of the genic SNPs were associated with plant biomass on chromosome 9, while six were associated with plant height trait association on extra-contigs (Ruperao et al. 2021).

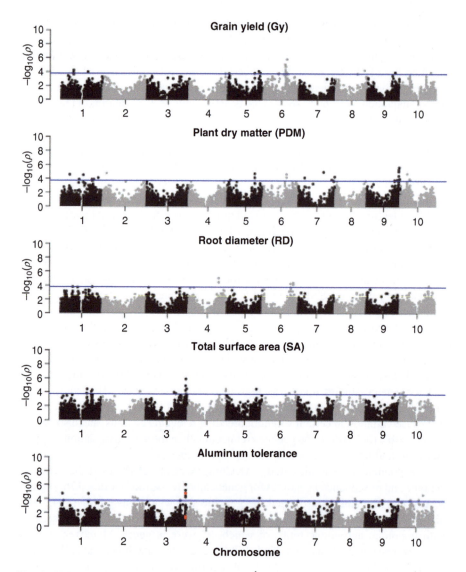

Fig. 4 GWAS profiles for grain yield (GY, ton ha^{-1}), plant dry matter (PDM, ton ha^{-1}), root morphology traits and Al tolerance. The root morphology traits, root diameter (RD, in mm) and total surface area (SA, in cm^2), were assessed after 13 days in nutrient solution with low-P. Al tolerance was measured by relative net root growth after 5 days of ± Al exposure to nutrient solution with an Al^{3+} activity. The horizontal line in blue depicts the significance threshold based on the Bonferroni correction for multiple, independent tests (alpha = 0.05), which was defined according to the extent of LD for each sorghum chromosome. Colored in red are SNPs within the AltSB locus where *SbMATE* is located. (Reproduced from Bernardino et al. (2021) with permission from rightsholder)

4.4 Genomic Selection

Due to the decreasing costs of genotyping thousands or millions of markers, genomic selection (GS; initially proposed by Meuwissen et al. 2001) has emerged as an alternative genome-wide marker-based method to predict breeding values using high-density markers distributed across the genome. GS maximizes the proportion of the genetic variance captured by markers by including both minor and major marker effects in the statistical genetics models, being more advantageous than QTL-MAS for quantitative traits. Thus, accurate predictions can be achieved even for untested genotypes, reducing the number of field-tested materials and, consequently, the phenotyping costs in breeding programs (Krchov and Bernardo 2015). The benefits of GS are even more evident when traits are difficult, time-consuming and/or expensive to measure, or when several environments need to be evaluated.

Although the potential of GS to improve the genetic gains per unit of time in a breeding program, few studies have been conducted in sorghum compared to other crops. Yu et al. (2016) used GS to explore the genetic diversity of sorghum accessions stored in germplasm banks. First, these authors selected a reference set of 962 accessions from 34,844 photoperiod-sensitive sorghum genotypes, collected from 33 countries, covering five sorghum races: durra, guinea, bicolor, caudatum, and kafir. The 962-accession reference set was genotyped using GBS, obtaining 340,496 high-quality SNPs. Then, a set of 299 accessions was selected as the training (TRN) population, representing the overall diversity of the reference set. This TRN set was phenotyped for eight biomass-related traits. Based on the genotype and the phenotype data of the TRN set, extensive cross-validation (CV) schemes were performed to assess the prediction accuracy of five different statistical methods: RR-BLUP, exponential kernel, Gaussian kernel, BayesLASSO, and BayesCπ. A prediction accuracy of 0.69 was found for biomass yield, ranging from 0.35 to 0.78 for the other traits. In addition, the prediction accuracy for all eight traits were relatively stable and generally similar among different statistical methods. Empirical experiments with a 200-accession validation (VLD) set chosen from the reference set confirmed the high prediction accuracy achieved for biomass yield. Additionally, indirect GS using plant height, root lodging, and stalk number provided a slightly lower predictive ability (0.71) than direct GS on biomass yield (0.76). These results suggest that it is possible to design GS strategies to evaluate the genetic diversity of sorghum germplasm banks, allowing the identification of new sources of favorable alleles for bioenergy traits. Figure 5 provides an overview on how genome-based selection can be deployed in a sorghum breeding pipeline, from data acquisition, through genomic selection, ending in cultivar release for each breeding cycle.

Using a breeding population of bioenergy sorghum, Fernandes et al. (2018) evaluated the accuracy of single-trait, multi-trait, and trait-assisted GS strategies to predict biomass yield. In the trait-assisted GS strategy, the use of correlated traits showing higher heritabilities can increase the predictive ability of GS models for

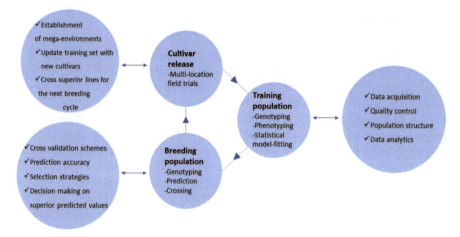

Fig. 5 Sorghum breeding pipeline scheme using genomic selection as an assistive tool

low-heritability traits, such as biomass yield. In this study, correlated traits, such as moisture, plant height, and the area under the growth progress curve, were used to predict biomass yield via indirect and trait-assisted GS. The prediction accuracy was estimated through a fivefold cross-validation (CV) scheme. The trait-assisted GS and the multi-trait GS strategies used 100% and 80% of the phenotypic data of correlated traits to predict biomass yield, respectively. Multi-trait GS did not provide any increase in the prediction accuracy for biomass yield when compared to the single-trait GS model. Probably, because of the limited number of correlated traits measured. By contrast, trait-assisted GS provided substantial improvements in the prediction accuracy for biomass yield when compared to the other GS strategies. The increases in the prediction accuracy ranged from 11.8% for the trait-assisted GS strategy using biomass yield and moisture (YM), to 50% using biomass yield and the area under the growth progress curve (YA), compared to the single-trait GS model. Therefore, the trait-assisted GS is a promising strategy that can be used to predict low-heritability target traits.

The use of multi-trait genomic prediction models applied to grain yield (GY) and stay-green (SG) in grain sorghum was also evaluated by exploiting information from auxiliary traits (Velazco et al. 2019a). For GY, multi-trait models that used SG and PH as auxiliary traits showed an improvement in the predicitive ability of 18% compared to the ST G-BLUP. These results support that multi-trait genomic evaluation combining routinely measured traits may be used to improve prediction of crop productivity and drought adaptability in grain sorghum and can also be extended for practical implementation in other major crops. Habyarimana et al. (2020) applied the multi-trait approach based on the optimum index selection to a panel of 369 biomass sorghum genotypes. They demonstrated for the first time the potential of genomic selection modelling of index selection including biofuel traits such as dry biomass yield, plant height, and dry mass fraction of the fresh material. In all evaluated scenarios, the genomic selection accuracy using all traits simultaneously was higher compared to the ones using single traits. In addition, this work also investigated the use of genomic selection indices applied to regrow from the rhizomes using the population grown from seeds as training set. The results were promising for using the information from the trial grown from seed to predict the performance of the populations regrown from rhizomes even two winter seasons after the original trial was sown. The use of multi-trait index selection improved traits such as plant height and the dry mass fraction of the fresh material that were weakly predicted when the selection target was regrown from the rhizomes.

Oliveira et al. (2018) fitted GS models in a panel of 200 bioenergy sorghum accessions, and then performed functional enrichment analyses using the predicted marker effects. The 200-accession panel was split into two sub-panels, according to the bioenergy breeding purpose, that is, 100 biomass sorghum and 100 sweet sorghum types. The 200 accessions were genotyped via GBS, resulting in approximately 258,000 SNPs. Sub-panel I was evaluated in 2 years (2011 and 2012), while sub-panel II in only 1 year (2011) in Sete Lagoas, State of Minas Gerais, Brazil, for days to flowering, plant height, fresh and dry matter yield, and fiber, cellulose, hemicellulose, and lignin content. Principal component analysis detected two sub-populations equivalent to the two sub-panels. Several GS models, such as BayesA, BayesB, BayesCπ, BayesLasso, Bayes Ridge Regression, and random regression BLUP were compared. In general, all predictive abilities were high, ranging from 0.85 for neutral detergent fiber to 0.66 for days to flowering; all tested GS models provided similar results for each of the nine traits. Prediction accuracy across sub-panels, that is, using the 100 accessions of sub-panel I as TRN set to predict the TST set of 100 genotypes of sub-panel II, was considerably lower when compared to those obtained in the joint analysis of both panels. Finally, the functional enrichment analysis performed with the marker-predicted effects provided different numbers of enriched gene ontology (GO) terms and unveiled several interesting associations for the different biomass traits, with potential for revealing biological processes relevant to the studied quantitative traits.

The use of a combined pedigree, markers and phenotypic data for optimization of genomic prediction in sorghum, was investigated by Velazco et al. (2019b). A dataset from testcross evaluations of sorghum parental lines across multiple environments using several testers and including different production and adaptability traits was used to apply predictive models using a blended kinship matrix construct as a weighted combination of matrices A and G, constructed as $\boldsymbol{K} = w\boldsymbol{A} + (1 - w)\boldsymbol{G}$. The weighting factor w represents the fraction of total additive variance that is not captured by markers, that is, the variance explained by genealogical relationships. A sequence of eight weight factors varying from 0.1 to 0.8 with increments of 0.1 was tested and compared with the A and G kinship matrices. The use of a kinship matrix that integrates both source of information yielded better predictive performance compared to G-BLUP for different traits and prediction scenarios evaluated. The optimal weighting factor was chosen based on the maximization of the predictive abilities and this factor was generally consistent with the weights that optimized model fitting. Traits with lower heritability, as grain yield (GY) and stay-green (SG), showed a higher impact of including genealogy information to improve genomic predictions; this implies that the optimal weighting factor is trait-specific. These results highlighted the potential use of combined pedigree and genomic information and might be relevant for breeding programs with limited genotyping resources, especially for the prediction of low heritability traits.

Single-cross hybrids (Shull 1908) have not only been used to explore heterosis in outcrossing species, but also in self-pollinating species such as grain sorghum (Jordan et al. 2003). In hybrid testing, the main challenge faced by breeders is to find a promising combination among many possible single-cross hybrids between inbred lines (Bernardo 1994; Schrag et al. 2010). Because it is unfeasible to obtain and to evaluate all possible pairwise inbred combinations, predicting the performance of untested single-cross hybrids prior to field trials is highly desirable. Thus, GS models have also been used for hybrid prediction in different crop species, such as maize, rice, wheat (*Triticum* spp.), and grain sorghum (Dias et al. 2019; Guo et al. 2019; Hunt et al. 2018; Hunt et al. 2020; Technow et al. 2012), based on single or multi-environment statistical approaches. Hunt et al. (2018) found high prediction accuracies of hybrid performance for grain yield, using a multi-environment trial data, relatively low marker density, and high average LD. It is well known that in hybrids crops the genetic variance can be partitioned in additive, dominance, and epistatic components. The advantage of including dominance effects in the analysis of multi-environment trials of sorghum hybrids was investigated by Hunt et al. (2020). They showed that partitioning genetic variance into additive, dominance, and residual genetic variances improved the prediction accuracy by an average of 15% and a maximum of 60% compared to models that just considered additive effects and residual genetic effects. In addition, they investigated the genotype by environment interactions (G×E) by comparing the FA1 (single-factor analytic) and FA2 (two-factor analytic) structures. Since an FA2 model fitted better than an FA1 model for both the additive and dominance terms, the role played by the G×E interaction in the data was evident. The G×E caused a change in hybrid rankings between trials with a difference of up to 25% of the hybrids in the top 10% of each trial. Therefore, genomic prediction of hybrid performance, with or without including dominance effects, is a promising strategy to maximize the genetic variability, reduce phenotyping costs, accelerate breeding cycles, and increase the genetic gains, which can also be applied in bioenergy sorghum breeding programs.

5 Future Perspectives

Over the past decades, substantial progress has been made in the development of genetic resources in breeding programs for bioenergy sorghum. Several QTL mapping studies have been performed to understand the genetic architecture and the environmental effects involved in the phenotypic expression of bioenergy-related traits. Linkage and/or association analyses have detected QTLs and/or candidate genes for a wide range of traits related to sugar content, biomass yield, and composition. However, the success of QTL-MAS has been limited to traits controlled by few major-effect QTLs. Recently, GS has emerged as an alternative genome-wide marker-based method to predict yet-to-be seen genetic responses. Including both minor and major marker effects in the genetic-statistical models, GS maximizes the proportion of the genetic variance captured by markers, being more advantageous than QTL-MAS for highly quantitative traits. Although few GS studies have been applied to bioenergy sorghum, all of them showed high prediction accuracy for biomass yield, suggesting the potential utility of this marker-assisted strategy to explore the genetic variability of germplasm banks, to increase the selection efficiency, and to accelerate breeding cycles and the genetic gains in breeding programs. Benefits of integrating marker-assisted strategies into breeding programs for bioenergy sorghum and other crop species may become more evident over the next years, with the decreasing costs and the increasing genome coverage capacity of high-throughput genotyping platforms, and the massive availability of phenotypic data from high-throughput phenotyping platforms. GWS- and GWAS-based selections are not the only primary factors to drive successful results in sorghum breeding nowadays. Population genetics enables sorghum breeding programs to thrive in consolidated areas and in farmlands where sorghum modern cultivars are not yet adapted. Exploiting different sources of adequate genetic variability using information from population genetic studies can shift up selection gains. Diverse sorghum germplasm composed by modern cultivars and wild relatives can form the basis for breeding programs and represent the source of favorable alleles for sorghum key traits.

In addition to breeding, sorghum crop can benefit from population genomics through findings about contrasting genomic patterns among genetic regions where this crop is adapted and provide deeper insights into the levels of connectivity and differentiation between populations that were positively selected in different ways depending on the region. Current advances in computational capacity can provide evidences of sorghum adaption throughout history that enabled this crop to occupy new environments, such as the fixation of certain genes and how they are inherited, the population structure, and genetic architecture of key and adaptive traits. Population genomics can be deployed to preserve sorghum biodiversity and genetic resources for conservational purposes in a long term perspective, once basic features of population size, inbreeding depression, and genetic structure between distinct natural sorghum populations can estimated using population genomic-based approaches.

6 Conclusions

Sorghum is one of the few major crop species that can be grown for multiple purposes according to its subtypes (grain, forage, sweet, and biomass) and that is adapted to the most diverse environments around the globe. The comprehension of population genetics of sorghum was fundamental to segment specific genotypes based on their geographic origins, genetic diversity, genetic pools, which led to establish core germplasms and elite populations in sorghum breeding programs. For the plant breeding perspective, the development of genomic tools and the enhancement of genetic-statistical models, further insights on marker association, gene cloning and validations, and selection strategies for major traits delivered satisfactory genetic gains in sorghum breeding programs. The development of next-generation/whole genome sequencing techniques allowed population genomics to provide even more benefits to major crops, including sorghum, enabling the long-term management of sorghum accessions based on the region of origin, genetic pools, and adaptive and key traits. Therefore, the integration of the discussed topics can enable the rapid development of stable superior sorghum cultivars adapted to specific or broad range of environmental conditions, commercial purposes, and to conserve modern and wild sorghum germplasm that can benefit both breeding and genetic biodiversity.

References

Almeida LGF, Parrella RAC, Simeone MLF, Ribeiro PCO, Santos AS, Costa ASV, et al. Biomass Bioenergy. 2019;122:343–8.

Amaducci S, Amaducci MT, Benati R, Venturi G. Crop yield and quality parameters of four annual fibre crops (hemp, kenaf, maize and sorghum) in the north of Italy. Ind Crop Prod. 2000;11:179–86.

Anami SE, Zhang LM, Xia Y, Zhang YM, Liu ZQ, Jing HC. Sweet sorghum ideotypes: genetic improvement of the biofuel syndrome. Food Energy Secur. 2015;4(3):159–77.

Ananda GK, Myrans H, Norton SL, Gleadow R, Furtado A, Henry RJ. Wild sorghum as a promising resource for crop improvement. Front Plant Sci. 2020;11:1108.

Antonopoulou G, Gavala HN, Skiadas IV, Angelopoulos K, Lyberatos G. Biofuels generation from sweet sorghum: fermentative hydrogen production and anaerobic digestion of the remaining biomass. Bioresour Technol. 2008;99:110–9.

Baek YS, Goodrich LV, Brown PJ, James BT, Moose SP, Lambert KN, et al. Transcriptome profiling and genome-wide association studies reveal GSTs and other defense genes involved in multiple signaling pathways induced by herbicide safener in grain sorghum. Front Plant Sci. 2019;10:192.

Baerson SR, Dayan FE, Rimando AM, Nanayakkara NPD, Liu C, Schröder J, et al. Functional genomics investigation of allelochemical biosynthesis in Sorghum bicolor root hairs. J Biol Chem. 2008;283:3231–47. https://doi.org/10.1074/jbc.M706587200.

Bayer PE, Golicz AA, Scheben A, Batley J, Edwards D. Plant pan-genomes are the new reference. Nat Plants. 2020;6:914–20.

Bennett AS, Anex RP. Production, transportation and milling costs of sweet sorghum as a feedstock for centralized bioethanol production in the upper Midwest. Bioresour Technol. 2009;100(4):1595–067.

Berhan A, Hulbert S, Butler L, Bennetzen J. Structure and evolution of the genomes of Sorghum bicolor and Zea mays. Theo Appl Genet. 1993;86(5):598–604.

Bernardino KC, de Menezes CB, de Sousa SM, Guimarães CT, Carneiro P, Schaffert RE, et al. Association mapping and genomic selection for sorghum adaptation to tropical soils of Brazil in a sorghum multiparental random mating population. Theor Appl Genet. 2021;134(1):295–312.

Bernardo R. Prediction of maize single-cross performance using RFLPs and information from related hybrids. Crop Sci. 1994;34:20–5.

Bian YL, Yazaki S, Inoue M, Cai HW. QTLs for sugar content of stalk in sweet sorghum (Sorghum bicolor L. Moench). Agric Sci China. 2006;5:736–44.

Biswal AK, Hao Z, Pattathil S, Yang X, Winkeler K, Collins C, et al. Downregulation of GAUT12 in Populus deltoides by RNA silencing results in reduced recalcitrance, increased growth and reduced xylan and pectin in a woody biofuel feedstock. Biotechnol Biofuels. 2015;8:41.

Boivin K, Deu M, Rami JF, Trouche G, Hamon P. Towards a saturated sorghum map using RFLP and AFLP markers. Theor Appl Genet. 1999;98(2):320–8.

Borgel A, Sequier J. Prospection de Mils penicillaires et Sor- ghos en Afrique de l'Ouest. Campagne 1976: Niger; 1977.

Bout S, Vermerris W. A candidate-gene approach to clone the sorghum Brown midrib gene encoding caffeic acid O-methyltransferase. Mol Genet Genomics. 2003;269:205–14.

Bowers JE, Abbey C, Anderson S, Chang C, Draye X, Hoppe AH, et al. A high-density genetic recombination map of sequence-tagged sites for sorghum, as a framework for comparative structural and evolutionary genomics of tropical grains and grasses. Genetics. 2003;165:367–86.

Boyles RE, Brenton ZW, Kresovich S. Genetic and genomic resources of sorghum to connect genotype with phenotype in contrasting environments. Plant J. 2019;97(1):19–39.

Bramel-Cox PJ, Cox TS. Use of wild sorghums in sorghum improvement. In: W. D, editor. Proceedings of the 43rd annual corn and Sorghum industry research conference, USA. Washington DC: American Seed Trade Association; 1988. p. 13–26.

Brenton ZW, Cooper EA, Myers MT, Boyles RE, Shakoor N, Zielinski KJ, et al. A Genomic Resource for the Development, Improvement, and Exploitation of Sorghum for Bioenergy. Genetics. 2016;204:21–33.

Briand CH, Geleta SB, Kratochvil RJ. Sweet sorghum (Sorghum bicolor [L.] Moench) a potential biofuel feedstock: Analysis of cultivar performance in the Mid-Atlantic. Renew Energy. 2018;129(A):328–33.

Brown PJ, Klein PE, Bortiri E, Acharya CB, Rooney WL, Kresovich S. Inheritance of inflorescence architecture in sorghum. Theor Appl Genet. 2006;113:931–42.

Burks PS, Kaiser CM, Hawkins EM, Brown PJ. Genomewide association for sugar yield in sweet Sorghum. Crop Sci. 2015;55:2138–48.

Caniato FF, Guimaraes CT, Hamblin M, Billot C, Rami JF, Hufnagel B, et al. The relationship between population structure and aluminum tolerance in cultivated sorghum. PLoS One. 2011;6(6):e20830.

Caniato FF, Hamblin MT, Guimaraes CT, Zhang Z, Schaffert RE, Kochian LV, et al. Association mapping provides insights into the origin and the fine structure of the sorghum aluminum tolerance locus, AltSB. PLoS One. 2014;9(1):e87438.

Carpita NC, McCann MC. Maize and sorghum: genetic resources for bioenergy grasses. Trends Plant Sci. 2008;13:415–20.

Casa AM, Pressoir G, Brown PJ, Mitchell SE, Rooney WL, Tuinstra MR, et al. Community resources and strategies for association mapping in sorghum. Crop Sci. 2008;48:30–40. https://doi.org/10.2135/cropsci2007.02.0080.

Casto AL, Mattison AJ, Olson SN, Thakran M, Rooney WL, Mullet JE. Maturity2, a novel regulator of flowering time in Sorghum bicolor, increases expression of SbPRR37 and SbCO in long days delaying flowering. PLoS One. 2019;14(4):e0212154.

Chen F, Dixon RA. Lignin modification improves fermentable sugar yields for biofuel production. Nat Biotechnol. 2007;25:759–61.

Chittenden LM, Schertz KF, Lin YR, Wing RA, Paterson AH. A detailed RFLP map of Sorghum bicolor × S. propinquum, suitable for high-density mapping, suggests ancestral duplication of Sorghum chromosomes or chromosomal segments. Theor Appl Genet. 1994;87:925–33.

Cole MR, Eggleston G, Petrie E, Uchimiya M, Dalley C. Cultivar and maturity effects on the quality attributes and ethanol potential of sweet sorghum. Biomass Bioenergy. 2017;96:183–92.

Cooper EA, Brenton ZW, Flinn BS, Jenkins J, Shu S, Flowers D, et al. A new reference genome for Sorghum bicolor reveals high levels of sequence similarity between sweet and grain genotypes:

implications for the genetics of sugar metabolism. BMC Genomics. 2019;20:420. https://doi.org/10.1186/s12864-019-5734-x.

Costa GHG, Freita CM, Freita LA, Mutton MJR. Effects of different coagulants on Sweet sorghum juice clarification. Sugar Tech. 2014;16:1–4.

Cowan MF, Blomstedt CK, Norton SL, Henry RJ, Møller BL, Gleadow R. Crop wild relatives as a genetic resource for generating low-cyanide, drought-tolerant Sorghum. Environ Exp Bot. 2020;169:103884. https://doi.org/10.1016/j.envexpbot.2019.103884.

Cox T, House L, Frey KJ. Potential of wild germplasm for increasing yield of grain sorghum. Euphytica. 1984;33(3):673–84. https://doi.org/10.1007/BF00021895.

Cox T, Bender M, Picone C, Tassel DV, Holland J, Brummer E, et al. Breeding perennial grain crops. CRC Crit Rev Plant Sci. 2002;21(2):59–91. https://doi.org/10.1080/0735-260291044188.

Dahlberg JA. Classification and characterization of sorghum. In: Smith CW, Frederiksen RA, editors. Sorghum: origin, history, technology, and production. 1st ed. New York: Wiley; 2000. p. 99–130.

Davison BH, Drescher SR, Tuskan GA, Davis MF, Nghiem NP. Variation of S/G ratio and lignin content in Populus family influences the release of xylose by dilute acid hydrolysis. Appl Biochem Biotechnol. 2006;129-132:427–35.

De Alencar Figueiredo LF, Calatayud C, Dupuits C, Billot C, Rami JF, Brunel D, et al. Phylogeographic evidence of crop neodiversity in sorghum. Genetics. 2008;179(2):997–1008.

De Wet JMJ. Systematics and evolution of sorghum sect. Sorghum (Gramineae). Am J Bot. 1978;65(4):477–84.

De Wet JMJ, Harlan JR. The origin and domestication of Sorghum bicolor. Econ Bot. 1971;25(2):128–35.

Della Coletta R, Qiu Y, Ou S, Hufford MB, Hirsch CN. How the pan-genome is changing crop genomics and improvement. Genome Biol. 2021;22(1):1–19.

Dias KODG, Gezan SA, Guimarães CT, Nazarian A, Silva LC, et al. Improving accuracies of genomic predictions for drought tolerance in maize by joint modeling of additive and dominance effects in multi-environment trials. Heredity. 2019;121:24–37.

Dien BS, Sarath G, Pedersen JF, Sattler SE, Chen H, Funnell-Harris DL, et al. Improved sugar conversion and ethanol yield for forage sorghum (Sorghum bicolor L. Moench) lines with reduced lignin contents. Bioenergy Res. 2009;2:153–64.

Disasa T, Feyissa T, Admassu B, Fetene M, Mendu V. Mapping of QTLs associated with brix and biomass-related traits in Sorghum using SSR markers. Sugar Tech. 2018;20(3):275–85.

Doebley JF, Gaut BS, Smith BD. The molecular genetics of crop domestication. Cell. 2006;127:1309–21. https://doi.org/10.1016/j.cell.2006.12.006.

Dweikat I. A diploid, interspecific, fertile hybrid from cultivated sorghum, Sorghum bicolor and the common Johnsongrass Weed Sorghum halepense. New Strateg Plant Improv. 2005;16:93–101. https://doi.org/10.1007/s11032-005-5021-1.

Edwards MD, Stuber CW, Wendel JF. Molecular-marker-facilitated investigation of quantitative-trait loci in maize. 1. Numbers, genomic distribution and types of gene action. Genetics. 1987;116:113–25.

Ejeta G, Kenoll J. Marker-assisted selection in sorghum. In: Varshney RK, Tuberosa R, editors. Genomics-assisted crop improvement. Netherlands: Springer; 2007. p. 187–205.

Elshire RJ, Glaubitz JC, Sun Q, Poland JA, Kawamoto K, Buckler ES, et al. A Robust, Simple Genotyping-by-Sequencing (GBS) Approach for High Diversity Species. PLoS One. 2011;6(5):e19379.

Endrizzi JE. Cytological studies of some species and hybrids in the Eu-sorghums. Bot Gaz. 1957;119:1–10. https://doi.org/10.1086/335954.

FAO. Food and Agriculture Organization of the United Nations Database of agricultural production. FAO Statistical Databases, 2017. http://faostat.fao.org/site/339/default.aspx. Accessed 03 Nov 2021.

Fedoroff NV. Transposable elements, epigenetics, and genome evolution. Science. 2012;338:758–67.

Felderhoff TJ, Murray SC, Klein PE, Sharma A, Hamblin MT, Kresovich S, et al. QTLs for energy-related traits in a sweet × grain Sorghum (Sorghum bicolor (L.) Moench) mapping population. Crop Sci. 2012;52:2040–9.

Feltus FA, Hart GE, Schertz KF, Casa AM, Kresovich S, Abraham S, et al. Alignment of genetic maps and QTLs between inter- and intraspecific sorghum populations. Theor Appl Genet. 2006;112:1295–305.

Fernandes SB, Dias KOG, Ferreira DF, Brown PJ. Efficiency of multi-trait, indirect, and trait-assisted genomic selection for improvement of biomass sorghum. Theor Appl Genet. 2018;131:747–55.

Fernandez MGS, Okeno JA, Mutegi E, Fessehaie A, Chalfant S. Assessment of genetic diversity among sorghum landraces and their wild/weedy relatives in western Kenya using simple sequence repeat (SSR) markers. Conserv Genet. 2014;15:1269–80. https://doi.org/10.1007/s10592-014-0616-x.

Fiedler K, Bekele WA, Matschegewski C, Snowdon R, Wieckhorst S, Zacharias A, et al. Cold tolerance during juvenile development in sorghum: a comparative analysis by genomewide association and linkage mapping. Plant Breed. 2016;135(5):598–606. https://doi.org/10.1111/pbr.12394.

Flint-Garcia SA. Structure of linkage disequilibrium in plants. Annu Rev Plant Biol. 2003;54:357–74.

Foy CD, Duncan RR, Waskom RM, Miller DR. Tolerance of sorghum genotypes to an acid, aluminum toxic Tatum subsoil. J Plant Nutr. 1993;16:97–127. https://doi.org/10.1080/01904169309364517.

Francis A, Dhaka N, Bakshi M, Jung KH, Sharma MK, Sharma R. Comparative phylogenomic analysis provides insights into TCP gene functions in Sorghum. Sci Rep. 2016;6(1):1–13.

Fuller DQ, Stevens CJ. Sorghum domestication and diversification: a current archaeobotanical perspective. In Plants and people in the African past (pp. 427–452). Springer, Cham; 2018.

Girma G, Nida H, Seyoum A, Mekonen M, Nega A, Lule D, et al. A large-scale genome-wide association analyses of Ethiopian sorghum landrace collection reveal loci associated with important traits. Front Plant Sci. 2019;10:691.

Gobena D, Shimels M, Rich PJ, Ruyter-Spira C, Bouwmeester H, Kanuganti S, et al. Mutation in Sorghum low germination Stimulant 1 alters strigolactones and causes Striga resistance. Proc Natl Acad Sci U S A. 2017;114:4471–6.

Godin B, Nagle N, Sattler S, Agneessens R, Delcarte J, Wolfrum E. Improved sugar yields from biomass sorghum feedstocks: comparing low-lignin mutants and pretreatment chemistries. Biotechnol Biofuels. 2016;9:251.

Grenier C, Hamon P, Bramel-Cox PJ. Core collection of sorghum. Crop Sci. 2001;41:241. https://doi.org/10.2135/cropsci2001.411241x.

Guan Y, Wang H, Qin L, Zhang H, Yang Y, Gao F, et al. QTL mapping of bioenergy related traits in Sorghum. Euphytica. 2011;182(3):431–40.

Guo T, Yu X, Li X, Zhang H, Zhu C, Flint-Garcia S, et al. Optimal designs for genomic selection in hybrid crops. Mol Plant. 2019;12(3):390–401.

Habyarimana E, Lopez-Cruz M, Baloch FS. Genomic selection for optimum index with dry biomass yield, dry mass fraction of fresh material, and plant height in biomass Sorghum. Genes. 2020;11(61):1–16.

Hadley HH. Chromosome numbers, fertility and rhizome expression of hybrids between grain sorghum and johnsongrass. Agron J. 1958;50:278–82. https://doi.org/10.2134/agronj1958.00021962005000050015x.

Han Y, Lv P, Hou S, Li S, Ji G, Ma X, et al. Combining next generation sequencing with bulked segregant analysis to fine map a stem moisture locus in Sorghum (Sorghum bicolor L. Moench). PLoS One. 2015;10(5):e0127065.

Harlan JR, De Wet JMJ. A Simplified classification of cultivated sorghum. Crop Sci. 1972;12(2): 172–6.

Harlan JR, Stemler A. The races of sorghum in Africa. In: Origins of African plant domestication. Paris: Mouton; 1976. p. 465–478.

Hart GE, Schertz KF, Peng Y, Syed NH. Genetic mapping of Sorghum bicolor (L.) Moench QTLs that control variation in tillering and other morphological characters. Theor Appl Genet. 2001;103:1232–42. https://doi.org/10.1007/s001220100582.

Haussmann G, Hess E, Seetharama N, Welz G, Geiger H. Construction of a combined sorghum linkage map from two recombinant inbred populations using AFLP, SSR, RFLP, and RAPD markers, and comparison with other sorghum maps. Theor Appl Genet. 2002;105(4):629–37.

Hayes CM, Burow GB, Brown PJ, Thurber C, Xin Z, Burke JJ. Natural variation in synthesis and catabolism genes influences dhurrin content in sorghum. Plant Genome. 2015;8(2) https://doi.org/10.3835/plantgenome2014.09.0048.

Higgins RH, Thurber CS, Assaranurak I, Brown PJ. Multiparental mapping of plant height and flowering time QTL in partially isogenic sorghum families. G3 Genes Genomes. Genetics. 2014;4:1593–602.

Hulbert S, Richter T, Axtell J, Bennetzen J. Genetic mapping and characterization of sorghum and related crops by means of maize DNA probes. Proc Natl Acad Sci. 1990;87(11):4251–5.

Hunt CH, Eeuwijk FA, Mace ES, Hayes BJ, Jordan DR. Development of genomic prediction in sorghum. Crop Sci. 2018;58:690–700.

Hunt CB, Hayes BJ, Eeuwijk FA, Mace ES, Jordan DR. Multi-environment analysis of sorghum breeding trials using additive and dominance genomic relationships. Theor Appl Genet. 2020;133:1009–18.

Jayakodi M, Schreiber M, Stein N, Mascher M. Building pan-genome infrastructures for crop plants and their use in association genetics. DNA Res. 2021;28:1–9. https://doi.org/10.1093/dnares/dsaa030.

Jensen SE, Charles JR, Muleta K, Bradbury PJ, Casstevens T, Deshpande SP, et al. A sorghum practical haplotype graph facilitates genome-wide imputation and cost-effective genomic prediction. Plant Genome. 2020;13(1687):1–15.

Jiang CJ, Zeng ZB. Multiple-trait analysis of genetic-mapping for quantitative trait loci. Genetics. 1995;140:1111–27.

Jordan D, Tao Y, Godwin I, Henzell R, Cooper M, McIntyre C. Prediction of hybrid performance in grain sorghum using RFLP markers. Theor Appl Genet. 2003;106:559–67.

Jordan J, Butler D, Henzell B, Drenth J, McIntyre L. Diversification of Australian sorghum using wild relatives. Paper presented at the new directions for a diverse planet (Brisbane, Australia: Proceedings of the 4th International Crop Science Congress), 2004.

Jordan DR, Klein RR, Sakrewski KG, Henzell RG, Klein PE, Mace ES. Mapping and characterization of Rf 5: a new gene conditioning pollen fertility restoration in A 1 and A 2 cytoplasm in sorghum (Sorghum bicolor (L.) Moench). Theor Appl Genet. 2011;123(3):383–96.

Kamala V, Sharma H, Manohar Rao D, Varaprasad K, Bramel P. Wild relatives of sorghum as sources of resistance to sorghum shoot fly, Atherigona soccata. Plant Breed. 2009;128(2): 137–42. https://doi.org/10.1111/j.1439-0523.2008.01585.x.

Kebede H, Subudhi PK, Rosenow DT, Nguyen HT. Quantitative trait loci influencing drought tolerance in grain sorghum (Sorghum bicolor L. Moench). Theor Appl Genet. 2001;103:266–76.

Khalil SRA, Abdelhafez AA, Amer EAM. Evaluation of bioethanol production from juice and bagasse of some sweet sorghum varieties. Ann Agric Sci. 2015;60(2):317–24.

Khan AW, Garg V, Roorkiwal M, Golicz AA, Edwards D, Varshney RK. Super-Pangenome by integrating the wild side of a species for accelerated crop improvement. Trends Plant Sci. 2020;25:148–15. https://doi.org/10.1016/j.tplants.2019.10.012.

Kimani W, Zhang LM, Wu XY, Hao HQ, Jing HC. Genome-wide association study reveals that different pathways contribute to grain quality variation in sorghum (Sorghum bicolor). BMC Genomics. 2020;21(1):1–19.

Kimber CT. Origins of domesticated sorghum and its early diffusion to India and China. In: Smity CW, Frederiksen RA, editors. Sorghum: origin, history, technology, and production. 1st ed. New York: Wiley; 2000. p. 3–98.

Klein RR, Mullet JE, Jordan DR, Miller FR, Rooney WL, Menz MA, et al. The effect of tropical sorghum conversion and inbred development on genome diversity as revealed by high-resolution genotyping. Crop Sci. 2008;48:S-12.

Kong L, Dong J, Hart GE. Characteristics, linkage-map positions, and allelic differentiation of Sorghum bicolor (L.) Moench DNA simple-sequence repeats (SSRs). Theor Appl Genet. 2000;101:438–48. https://doi.org/10.1007/s001220051501.

Kong W, Jin H, Franks C, Kim C, Bandopadhyay R, Rana M, et al. Genetic analysis of recombinant inbred lines for Sorghum bicolor × Sorghum propinquum. G3 Genes Genomes. Genetics. 2013;3(1):101–8.

Krchov LM, Bernardo R. Relative efficiency of genome wide selection for testcross performance of doubled haploid lines in a maize breeding program. Crop Sci. 2015;55:2091–9.

Kuhlman LC, Burson BL, Klein PE, Klein RR, Stelly DM, Price HJ, et al. Genetic recombination in Sorghum bicolor× S. macrospermum interspecific hybrids. Genome. 2008;51:749–56. https://doi.org/10.1139/G08-061.

Kuhlman LC, Burson BL, Stelly DM, Klein PE, Klein RR, Price H, et al. Early-generation germplasm introgression from Sorghum macrospermum into sorghum (S. bicolor). Genome. 2010;53(6):419–29. https://doi.org/10.1139/g10-027.

Kulwal PL. Association mapping and genomic selection—where does Sorghum stand? In: Rakshit S, Wang Y-H, editors. The Sorghum genome, compendium of plant genomes; 2017. https://doi.org/10.1007/978-3-319-47789-3_7.

Lander ES, Botstein D. Mapping mendelian factors underlying quantitative traits using RFLP linkage maps. Genetics. 1989;121:185–99.

Lasky JR, Upadhyaya HD, Ramu P, Deshpande S, Hash CT, Bonnette J, et al. Genome-environment associations in sorghum landraces predict adaptive traits. Sci Adv. 2015;1(6):e1400218.

Lekgari AL. Genetic mapping of quantitative trait loci associated with bioenergy traits, and the assessment of genetic variability in sweet sorghum (Sorghum bicolor (L.). Moench). Theses, Dissertations, and Student Research in Agronomy and Horticulture. University of Nebraska, Lincoln, 2010. https://digitalcommons.unl.edu/agronhortdiss/11.

Li M, Yuyama N, Luo L, Hirata M, Cai H. In silico mapping of 1758 new SSR markers developed from public genomic sequences for sorghum. Mol Breed. 2009;24(1):41–7.

Li JQ, Wang LH, Zhan QW, Liu YL, Zhang Q, Li JF, et al. Mapping quantitative trait loci for five forage quality traits in a sorghum-sudangrass hybrid. Genet Mol Res. 2015;14(4):13266–73.

Li J, Tang W, Zhang YW, Chen KN, Wang C, Liu Y, et al. Genome-wide association studies for five forage quality-related traits in Sorghum (Sorghum bicolor L.). Front Plant Sci. 2018;9:1146.

Lin YR, Schertz KF, Paterson AH. Comparative analysis of QTLs affecting plant height and maturity across the Poaceae, in reference to an interspecific sorghum population. Genetics. 1995;141(1):391–411.

Lingle SE, Tew TL, Rukavina H, Boykin DL. Post-harvest changes in sweet Sorghum II: pH, acidity, protein, starch, and mannitol. Bioenergy Res. 2013;6:178–87.

Luikart G, Kardos M, Hand B, Rajora OP, Aitkin S, Hohenlohe PA. Population genomics: advancing understanding of nature. In: Rajora OP, editor. Population genomics: concepts, approaches and applications. Springer International Publishing AG; 2019. p. 3–79.

Lv P, Ji G, Han Y, Hou S, Li S, Ma X, et al. Association analysis of sugar yield-related traits in sorghum (Sorghum bicolor (L.)). Euphytica. 2013;193:419–31.

Lyu J. Striga resistance: Cloak the strigolactone. Nat Plants. 2017;3:17067. https://doi.org/10.1038/nplants.2017.67.

Mace ES, Jordan DR. Location of major effect genes in sorghum (Sorghum bicolor (l.) Moench). Theor Appl Genet. 2010;121:1339–56.

Mace ES, Jordan DR. Integrating sorghum whole genome sequence information with a compendium of sorghum QTL studies reveals uneven distribution of QTL and of gene-rich regions with significant implications for crop improvement. Theor Appl Genet. 2011;123:169–91.

Mace ES, Xia L, Jordan DR, Halloran K, Parh DK, Huttner E, et al. DarT markers: diversity analyses and mapping in Sorghum bicolor. BMC Genomics. 2008;9(1):1–11.

Mace ES, Rami JF, Bouchet S, Klein PE, Klein RR, Kilian A, et al. A consensus genetic map of sorghum that integrates multiple component maps and high-throughput Diversity Array Technology (DarT) markers. BMC Plant Biol. 2009;9(1):1–14.

Mace ES, Innes D, Hunt C, Wang X, Tao Y, Baxter J, et al. The Sorghum QTL Atlas: a powerful tool for trait dissection, comparative genomics and crop improvement. Theor Appl Genet. 2019;132:751–66.

Mackay TFC, Stone EA, Ayroles JF. The genetics of quantitative traits: challenges and prospects. Nat Rev Genet. 2009;10:565–77.

Magalhaes JV, Garvin DF, Wang Y, Sorrells ME, Klein PE, Schaffert RE, et al. Comparative mapping of a major aluminum tolerance gene in sorghum and other species in the Poaceae. Genetics. 2004;167(4):1905–14.

Magalhaes JV, Liu J, Guimarães CT, Lana UGP, Alves VMC, Wang Y-H, et al. A gene in the multidrug and toxic compound extrusion (MATE) family confers aluminum tolerance in sorghum. Nat Genet. 2007;39:1156–61. https://doi.org/10.1038/ng2074.

Magomere TO, Obukosia SD, Shibairo SII, Ngugi EK, Mutitu E. Evaluation of relative competitive ability and fitness of Sorghum bicolor x Sorghum halepense and Sorghum bicolor x Sorghum sudanense F1 hybrids. J Biol Sci. 2015;15:1–15. https://doi.org/10.3923/jbs.2015.1.15.

Maina F, Bouchet S, Marla SR, Hu Z, Wang J, Mamadou A, et al. Population genomics of sorghum (Sorghum bicolor) across diverse agroclimatic zones of Niger. Genome. 2018;61(4):223–32.

Marguerat S, Bahler J. RNA-seq: from technology to biology. Cell Mol Life Sci. 2010;67:569–79.

Mathur S, Umakanth AV, Tonapi VA, Sharma R, Sharma MK. Sweet sorghum as biofuel feedstock: recent advances and available resources. BMC Biotech Biofuels. 2017;10:146.

Maw MJW, Houx JH, Fritschi FB. Nitrogen use efficiency and yield response of high biomass sorghum in the lower Midwest. Agron J. 2017;109:115–21.

McCormick RF, Truong SK, Sreedasyam A, Jenkins J, Shu S, Sims D, et al. The Sorghum bicolor reference genome: improved assembly, gene annotations, a transcriptome atlas, and signatures of genome organization. Plant J. 2018;93:338–54. https://doi.org/10.1111/tpj.13781.

Melo JO, Martins LGC, Barros BA, et al. Repeat variants for the SbMATE transporter protect sorghum roots from aluminum toxicity by transcriptional interplay in cis and trans. Proc Natl Acad Sci. 2019;116:313–8. https://doi.org/10.1073/PNAS.1808400115.

Mengistu G, Shimelis H, Laing M, Lule D, Assefa E, Mathew I. Genetic diversity assessment of sorghum (Sorghum bicolor (L.) Moench) landraces using SNP markers. South Afr J Plant Soil. 2020;37(3):220–6.

Meuwissen T, Hayes BJ, Goddard ME. Prediction of total genetic value using genome-wide dense marker maps. Genetics. 2001;157:1819–29.

Miao C, Xu Y, Liu S, Schnable PS, Schnable JC. Increased power and accuracy of causal locus identification in time series genome-wide association in sorghum. Plant Physiol. 2020;183(4):1898–909.

Moens P, Wu Y, Huang Y. An SSR genetic map of Sorghum bicolor (L.) Moench and its comparison to a published genetic map. Genome. 2006;50(1):84–9.

Mojtahedi H, Santo GS, Ingham RE. Suppression of Meloidogyne chitwoodi with Sudangrass cultivars as green manure. J Nematol. 1993;25:303–11.

Morrell P, Williams-Coplin T, Lattu A, Bowers J, Chandler J, Paterson A. Crop-to-weed introgression has impacted allelic composition of johnsongrass populations with and without recent exposure to cultivated sorghum. Mol Ecol. 2005;14:2143–5. https://doi.org/10.1111/j.1365-294X.2005.02579.x.

Morris GP, Ramu P, Deshpande SP, Hash CT, Shah T, Upadhyaya HD, et al. Population genomic and genome-wide association studies of agroclimatic traits in sorghum. PNAS. 2013;110 (2):453–8.
Mullet J, Morishige D, McCormick R, Truong S, Hilley J, McKinley B, et al. Energy Sorghum—a genetic model for the design of C4 grass bioenergy crops. J Exp Bot. 2014;65(13):3479–89.
Murphy RL, Morishige DT, Brady JA, Rooney WL, Yang S, Klein PE, et al. Ghd7 (Ma 6) represses Sorghum flowering in long days: Ghd7 Alleles enhance biomass accumulation and grain production. Plant Genome. 2014;7(2):1–10.
Murray SC, Rooney WL, Mitchell SE, Sharma A, Klein PE, Mullet JE, et al. Genetic improvement of Sorghum as a biofuel feedstock: I. QTL for stem sugar and grain nonstructural carbohydrates. Crop Sci. 2008a;48:2165–79.
Murray SC, Rooney WL, Mitchell SE, Sharma A, Klein PE, Mullet JE, et al. Genetic improvement of Sorghum as a biofuel feedstock: II. QTL for stem and leaf structural carbohydrates. Crop Sci. 2008b;48:2180–93.
Murray SC, Rooney WL, Hamblin MT, Mitchell SE, Kresovich S. Sweet sorghum genetic diversity and association mapping for Brix and height. Plant Genome. 2009;2(1):48.
Mutegi E, Sagnard F, Labuschagne M, Herselman L, Semagn K, Deu M, et al. Local scale patterns of gene flow and genetic diversity in a crop– wild–weedy complex of sorghum (Sorghum bicolor (L.) Moench) under traditional agricultural field conditions in Kenya. Conserv Genet. 2012;13:1059–71. https://doi.org/10.1007/s10592-012-0353-y.
Natoli A, Gorni C, Chegdani F, Marsan PA, Colombi C, Lorenzoni C. Identification of QTLs associated with sweet sorghum quality. Maydica. 2002;47(3/4):311–22.
Negro MJ, Solano ML, Ciria P, Carrasco J. Composting of sweet sorghum bagasse and other wastes. Bioresour Technol. 1999;67(1):89–92.
Niu H, Ping J, Wang Y, Lv X, Li H, Zhang F, et al. Population genomic and genome-wide association analysis of lignin content in a global collection of 206 forage sorghum accessions. Mol Breed. 2020;40(8):1–13.
Nwanze KF, Seetharama N, Sharma HC, Stenhouse JW. Biotechnology in pest management: Improving resistance in sorghum to insect pests. Afr Crop Sci J. 1995;3:209–15.
Ohadi S, Hodnett G, Rooney W, Bagavathiannan M. Gene flow and its consequences in Sorghum spp. Crit Rev Plant Sci. 2017;36(5-6):367–85.
Oliveira AA, Pastina MM, Souza VF, Parrella RAC, Noda RW, et al. Genomic prediction applied to high-biomass sorghum for bioenergy production. Mol Breed. 2018;38:49.
Oliveira ICM, Marçal TS, Bernardino KC, Ribeiro PCO, Parrella RAC, et al. Combining ability of biomass Sorghum lines for agroindustrial characters and multitrait selection of photosensitive hybrids for energy cogeneration. Crop Sci. 2019;59:1555–66.
Olson S, Ritter K, Rooney W, Kemanian A, McCarl B, Zhang Y, et al. High biomass yield energy sorghum: developing a genetic model for C 4 grass bioenergy crops. Biofuels Bioprod Biorefin. 2012;6:640–55.
Parh DK. DNA-based markers for ergot resistance in sor[1]ghum. In PhD thesis University of Queensland, School of Land and Food Sciences; 2005.
Paterson AH, Schertz KF, Lin YR, Liu SC, Chang YL. The weediness of wild plants: molecular analysis of genes influencing dispersal and persistence of johnsongrass, Sorghum halepense (L.) Pers. Proc Natl Acad Sci U S A. 1995;92:6127–31. https://doi.org/10.1073/pnas.92.13.6127.
Paterson AH, Bowers JE, Bruggmann R, Dubchak I, Grimwood J, Gundlach H, et al. The Sorghum bicolor genome and the diversification of grasses. Nature. 2009;457:551–6. https://doi.org/10.1038/nature07723.
Peng Y, Schertz K, Cartinhour S, Hart G. Comparative genome mapping of Sorghum bicolor (L.) Moench using an RFLP map constructed in a population of recombinant inbred lines. Plant Breed. 1999;118(3):225–35.
Pereira MG, Lee M. Identification of genomic regions affecting plant height in sorghum and maize. Theor Appl Genet. 1995;90:380–8.

Pfeiffer TW, Bitzer MJ, Toy JJ, Pedersen JF. Heterosis in sweet Sorghum and selection of a new sweet Sorghum hybrid for use in syrup production in Appalachia. Crop Sci. 2010;50(5): 1788–94.

Piper JK, Kulakow PA. Seed yield and biomass allocation in Sorghum bicolor and F1 and backcross generations of S. bicolor x S. halepense hybrids. Can J Bot. 1994;72:468–74. https://doi.org/10.1139/b94-062.

Porter KS, Axtell JD, Lechtenberg VL, Colenbrander VF. Phenotype, fiber composition, and in vitro dry matter disappearance of chemically induced brown midrib (bmr) mutants of sorghum. Crop Sci. 1978;18:205–8.

Price HJ, Hodnett GL, Burson BL, Dillon SL, Stelly DM, Rooney WL. Genotype dependent interspecific hybridization of Sorghum bicolor. Crop Sci. 2006;46:2617–22. https://doi.org/10.2135/cropsci2005.09.0295.

Qamar TM, Zhu X, Khan MS, Xing F, Chen LL. Pan-genome: a promising resource for noncoding RNA discovery in plants. Plant Genome. 2020;13:1–16. https://doi.org/10.1002/tpg2.20046.

Quinby JR. The genetic control of flowering and growth in sorghum. Adv Agron. 1974;25:125–62.

Rajendrakumar P, Rakshit S. Genomics and bioinformatics resources. In: Madhusudhana R, Rajendrakumar P, Patil JV, editors. Sorghum molecular breeding; 2015. https://doi.org/10.1007/978-81-322-2422-8_10.

Rajora OP, editor. Population genomics: concepts, approaches and application: Springer Nature; 2019. 823 pp. ISBN 978-3-030-04587-6; ISBN 978-3-030-04589-0 (eBook)

Rami JF, Dufour P, Trouche G, Fliedel G, Mestres C, Davrieux F, et al. Quantitative trait loci for grain quality, productivity, morphological and agronomical traits in sorghum (Sorghum bicolor L. Moench). Theor ApplGenet. 1998;97:605–16.

Ramu P, Kassahun B, Senthilvel S, Kumar C, Jayashree B, Folkertsma R, et al. Exploiting rice–sorghum synteny for targeted development of EST-SSRs to enrich the sorghum genetic linkage map. Theor Appl Genet. 2009;119(7):1193–204.

Rasmusson DC, Phillips RL. Plant breeding progress and genetic diversity from de novo variation and elevated epistasis. Crop Sci. 1997;37:303–8.

Ravaneli GC, Garcia DB, Madaleno LL, Mutton MA, Stupiello PJ, Mutton MJR. Spittlebug impacts on sugarcane quality and ethanol production. Pesquisa Agropecuária Brasileira. 2011;46(2):120–9.

Reddy BVS, Ramesh S, Reddy PS, Kumar AA. Genetic enhancement for drought tolerance in sorghum. Plant Breed Rev. 2009;31:189–222.

Reddy DS, Bhatnagar-Mathur P, Cindhuri KS, Sharma KK. Evaluation and validation of reference genes for normalization of quantitative real-time PCR based gene expression studies in peanut. PLoS One. 2013;8:e78555.

Regassa TH, Wortmann CS. Sweet sorghum as a bioenergy crop: literature review. Biomass Bioenergy. 2014;64:348–55.

Rhodes DH, Hoffmann L Jr, Rooney WL, Ramu P, Morris GP, Kresovich S. Genome-wide association study of grain polyphenol concentrations in global sorghum [Sorghum bicolor (L.) Moench] germplasm. J Agric Food Chem. 2014;62(45):10916–27.

Rich PJ, Grenier C, Ejeta G. Striga resistance in the wild relatives of sorghum. Crop Sci. 2004;44: 2221–9. https://doi.org/10.2135/cropsci2004.2221.

Ritter KB, Jordan DR, Chapman SC, Godwin ID, Mace ES, Mcintyre CL. Identification of QTL for sugar-related traits in a sweet x grain sorghum (Sorghum bicolor L. Moench) recombinant inbred population. Mol Breed. 2008;22:367–84.

Rooney WL. Genetics and cytogenetics. In: Smith CW, Frederiksen RA, editors. Sorghum: origin, history, technology, and production. New York: Wiley; 2000. p. 261–307.

Rooney WL, Aydin S. Genetic control of a photoperiod sensitive response in Sorghum bicolor (L.) Moench. Crop Sci. 1999;39:397–400.

Rosenow DT, Quisenberry JE, Wendt CW, Clark LE. Drought tolerant Sorghum and cotton germplasm, vol. 12. Amsterdam: Elsevier B.V; 1983.

Ruperao P, Thirunavukkarasu N, Gandham P, Selvanayagam S, Govindaraj M, Nebie B, et al. Sorghum pan-genome explores the functional utility for genomic-assisted breeding to accelerate the genetic gain. Front Plant Sci. 2021;12:1–17. https://doi.org/10.3389/fpls.2021.666342.

Sabadin PK, Malosetti M, Boer M, Tardin FD, Santos F, Guimarães CT, et al. Studying the genetic basis of drought tolerance in sorghum by managed stress trials and adjustments for phenological and plant height differences. Theoretical and applied genetics. Theor Appl Genet. 2012;124:1389–402.

Saballos A, Ejeta G, Sanchez E, Kang C, Vermerris W. A genomewide analysis of the cinnamyl alcohol dehydrogenase family in sorghum (Sorghum bicolor (L.) Moench) identifies SbCAD2 as the Brown midrib6 gene. Genetics. 2009;181:783–95.

Saballos A, Sattler SE, Sanchez E, Foster TP, Xin Z, Kang CH, et al. Brown midrib2 (Bmr2) encodes the major 4-coumarate: coenzyme A ligase involved in lignin biosynthesis in sorghum (Sorghum bicolor (L.) Moench). Plant J. 2012;70(5):818–30.

Sangduen N, Hanna W. Chromosome and fertility studies on reciprocal crosses between two species of autotetraploid Sorghum bicolor (L.) Moench and S. halepense (L.) Pers. J Hered. 1984;75:293–6. https://doi.org/10.1093/oxfordjournals.jhered.a109936.

Sattler SE, Saathoff AJ, Haas EJ, Palmer NA, Funnell-Harris DL, Sarath G, et al. A nonsense mutation in a cinnamyl alcohol dehydrogenase gene is responsible for the sorghum brown midrib 6 phenotype. Plant Physiol. 2009;150:584–95.

Sattler SE, Saballos A, Xin Z, Funnell-Harris DL, Vermerris W, Pedersen JF. Characterization of novel Sorghum brown midrib mutants from an EMS-mutagenized population. G3 Genes Genomes. Genetics. 2014;4(11):2115–24.

Schaffert RE, Parrella RDC. Planejamento industrial. 2012.

Schmidt JJ, Pedersen JF, Bernards ML, Lindquist JL. Rate of shattercane x sorghum hybridization in situ. Crop Sci. 2013;53:1677–85. https://doi.org/10.2135/cropsci2012.09.0536.

Schrag TA, Mohring J, Melchinger AE, Kusterer B, Dhillon BS, Piepho HP, et al. Prediction of hybrid performance in maize using molecular markers and joint analyses of hybrids and parental inbreds. Theor Appl Genet. 2010;120(2):451–61.

Sharma HC, Franzmann BA. Host-plant preference and oviposition responses of the sorghum midge, Stenodiplosis sorghicola (Coquillett) (Dipt., Cecidomyiidae) towards wild relatives of sorghum. J Appl Entomol. 2001;125:109–14. https://doi.org/10.1046/j.1439-0418.2001.00524.x.

Shen X, Liu ZQ, Mocoeur A, Xia Y, Jing HC. PAV markers in Sorghum bicolor: genome pattern, affected genes and pathways, and genetic linkage map construction. Theor Appl Genet. 2015;128:623–37.

Shiringani AL, Friedt W. QTL for fibre-related traits in grain x sweet sorghum as a tool for the enhancement of sorghum as a biomass crop. Theor Appl Genet. 2011;123:999–1011.

Shiringani AL, Frisch M, Friedt W. Genetic mapping of QTLs for sugar-related traits in a RIL population of Sorghum bicolor L. Moench. Theor Appl Genet. 2010;121(2):323–36.

Shukla S, Felderhoff TJ, Saballos A, Vermerris W. The relationship between plant height and sugar accumulation in the stems of sweet sorghum (Sorghum bicolor (L.) Moench). Field Crop Res. 2017;203:181–91.

Shull GH. The composition of a field of maize. J Hered. 1908;4:296–301.

Silva MJ, Pastina MM, Souza VF, Schaffert RE, Carneiro PCS, Noda RW, et al. Phenotypic and molecular characterization of sweet sorghum accessions for bioenergy production. PLoS One. 2017;12:1.

Silva MJ, Carneiro PCS, Carneiro JES, Damasceno CMB, Pastina MM, Simeone MLF, et al. Evaluation of the potential of lines and hybrids of biomass sorghum. Ind Crop Prod. 2018;125:379–85.

Silva MJ, Damasceno CMB, Guimarães CT, Pinto MO, Barros BA, Carneiro JES, et al. Introgression of the bmr6 allele in biomass sorghum lines for bioenergy production. Euphytica. 2020;216:1–12.

Simeone MLF, Parrella RAC, Schaffert RE, Damasceno CMB, Leal MCB, Pasquini C. Near infrared spectroscopy determination of sucrose, glucose and fructose in sweet sorghum juice. Microchem J. 2017;134:125–30.

Singh H, Lohithaswa H. Genome mapping and molecular breeding in plants, Cereals and Millets. In: Kole C, editor. Sorghum. Berlin Heidelberg: Springer Verlag; 2006.

Smith CW, Frederiksen RA. Sorghum: origin, history, technology, and production. New York: Wiley; 2000. p. 824.

Smith O, Nicholson WV, Kistler L, Mace E, Clapham A, Rose P, et al. A domestication history of dynamic adaptation and genomic deterioration in Sorghum. Nature plants. 2019;5(4):369–79.

Souza RSE, Parrella RAC, Souza VF, Parrella NNLD. Maturation curves of sweet sorghum genotypes. Ciência e Agrotecnologia (Online). 2016;40:46–56.

Sun Y, Suksayretrup K, Kirkham MB, Liang GH. Pollen tube growth in reciprocal interspecific pollinations of Sorghum bicolor, and S. versicolor. Plant Breed. 1991;107:197–202.

Takai T, Yonemaru J, Kaidai H, Kasuga S. Quantitative trait locus analysis for days-to-heading and morphological traits in an RIL population derived from an extremely late flowering F1 hybrid of sorghum. Euphytica. 2012;187:411–20.

Tanksley SD, Medina-Filho H, Rick CM. Use of naturally-occurring enzyme variation to detect and map genes controlling quantitative traits in an interspecific backcross of tomato. Heredity. 1982;49:11–25.

Tao Y, Jordan D, Henzell R, McIntyre C. Construction of a genetic map in a sorghum recombinant inbred line using probes from different sources and its comparison with other sorghum maps. Aust J Agric Res. 1998;49(5):729–36.

Tao YZ, Hardy A, Drenth J, Henzell RG, Franzmann BA, Jordan DR, et al. Identifications of two different mech[1]anisms for sorghum midge resistance through QTL map[1]ping. Theor Appl Genet. 2003;107:116–22.

Tao Y, Mace ES, Tai S, Cruickshank A, Campbell BC, Zhao X, et al. Whole-genome analysis of candidate genes associated with seed size and weight in Sorghum bicolor reveals signatures of artificial selection and insights into parallel domestication in cereal crops. Front Plant Sci. 2017;8:1237.

Tao Y, Zhao X, Mace E, Henry R, Jordan D. Exploring and exploiting pan-genomics for crop improvement. Mol Plant. 2019;12:156–69. https://doi.org/10.1016/j.molp.2018.12.016.

Tao Y, Luo H, Xu J, Cruickshank A, Zhao X, Teng F, et al. Extensive variation within the pan-genome of cultivated and wild sorghum. Nat Plants. 2021;7(6):766–73. https://doi.org/10.1038/s41477-021-00925-x.

Technow F, Riedelsheimer C, Schrag TA, Melchinger AE. Genomic prediction of hybrid performance in maize with models incorporating dominance and population specific marker effects. Theor Appl Genet. 2012;125:1181–94.

Thurber CS, Ma JM, Higgins RH, Brown PJ. Retrospective genomic analysis of sorghum adaptation to temperate-zone grain production. Genome Biol. 2013;14:R68. https://doi.org/10.1186/gb-2013-14-6-r68.

Tsukiboshi T, Koga H, Uematsu T, Shimanuki T. Resistance of sorghum and sudangrass to ergot caused by Claviceps sp. and the cultural control of the disease. In: Shikenjo NS, editor. Bulletin of the National Grassland Research Institute. Nishianasuno, Japan: National Grassland Research Institute; 1998. p. 28–35.

Upadhyaya HD, Pundir RPS, Dwivedi SL, Gowda CLL, Reddy VG, Singh S. Developing a mini core collection of sorghum for diversified utilization of germplasm. Crop Sci. 2009;49(5):1769–80.

Vandenbrink JP, Goff V, Jin H, Kong W, Paterson AH, Feltus AF. Identification of bioconversion quantitative trait loci in the interspecific cross Sorghum bicolor × Sorghum propinquum. Theor Appl Genet. 2013;126:2367–80.

Varoquaux N, Cole B, Gao C, Pierroz G, Baker CR, Patel D, et al. Transcriptomic analysis of field-droughted sorghum from seedling to maturity reveals biotic and metabolic responses. Proc Natl Acad Sci. 2019;116(52):27124–32.

Velazco JG, Jordan DR, Mace ES, Hunt CH, Malosetti M, Eeuwijk FA. Genomic prediction of grain yield and drought-adaptation capacity in Sorghum is enhanced by multi-trait analysis. Front Plant Sci. 2019a;10:997.

Velazco JG, Malosetti M, Hunt CH, Mace ES, Jordan DR, Eeuwijk FA. Combining pedigree and genomic information to improve prediction quality: an example in Sorghum. Theor Appl Genet. 2019b;132:2055–67.

Venkateswaran K, Elangovan M, Sivaraj N. Origin, domestication and diffusion of Sorghum bicolor. In: Aruna C, Visarada KBRS, Bhat BV, Tonapi VA, editors. Breeding Sorghum for diverse end uses. Cambridge, UK: Woodhead Publishing); 2019a. p. 15–31.

Venkateswaran K, Sivaraj N, Pandravada SR, Reddy MT, Babu BS. Classification, distribution and biology. In: Aruna C, Visarada KBRS, Bhat BV, Tonapi VA, editors. Breeding Sorghum for diverse end uses. Cambridge, UK: Woodhead Publishing; 2019b. p. 33–60.

Vermerris W, Saballos A, Ejeta G, Mosier NS, Ladisch MR, Carpita NC. Molecular breeding to enhance ethanol production from corn and sorghum stover. Crop Sci. 2007;47:S142–53.

Viaene NM, Abawi GS. Management of Meloidogyne hapla on Lettuce in organic soil with Sudangrass as a cover crop. Plant Dis. 1998;82:945–52. https://doi.org/10.1094/PDIS.1998.82.8.945.

Wang ML, Zhu C, Barkley NA, Chen Z, Erpelding JE, Murray SC, et al. Genetic diversity and population structure analysis of accessions in the US historic sweet sorghum collection. Theor Appl Genetics. 2009;120(1):13–23.

Wang YH, Poudel DD, Hasenstein KH. Identification of SSR markers associated with saccharification yield using pool-based genome-wide association mapping in sorghum. Genome. 2011;54:883–9.

Wang YH, Acharya A, Burrell AM, Klein RR, Klein PE, Hasenstein KH. Mapping and candidate genes associated with saccharification yield in sorghum. Genome. 2013;56:659–65.

Wang L, Wang YY, Wang DQ, Xu J, Yang F, Liu G, et al. Dynamic changes in the bacterial community in Moutai liquor fermentation process characterized by deep sequencing. J Inst Brewing. 2015;121:603–8.

Wang HL, Zhang HW, Du RH, Chen GL, Liu B, Yang YB, et al. Identification and validation of QTLs controlling multiple traits in sorghum. Crop Pasture Sci. 2016;67:193–203.

Wang J, Hu Z, Upadhyaya HD, Morris GP. Genomic signatures of seed mass adaptation to global precipitation gradients in sorghum. Heredity. 2020;124(1):108–21.

Weltzien E, Rattunde H, Frederick W, Clerget B, Siart S, Touré A, et al. Sorghum diversity and adaptation to drought in West Africa. In: Enhancing the use of crop genetic diversity to manage abiotic stress in agricultural production systems, Proceedings of the workshop on enhancing the use of crop genetic diversity, Budapest, Hongrie, 23–27 May 2005; 2006.

Werle R, Schmidt JJ, Laborde J, Tran A, Creech CF, Lindquist JL. Shattercane x ALS-tolerant sorghum F1 hybrid and shattercane interference in ALS-tolerant Sorghum. J Agric Sci. 2014;6:159–65.

Westoby M, Jurado E, Leishman M. Comparative evolutionary ecology of seed size. Trends Ecol Evol. 1992;7:368–72. https://doi.org/10.1016/0169-5347(92)90006-W.

Wondimu Z, Dong H, Paterson AH, Worku W, Bantte K. Genetic diversity, population structure and selection signature in Ethiopian Sorghum (Sorghum bicolor L.[Moench]) germplasm. bioRxiv. 2021.

Wooten DR. The use of Sorghum propinquum to enhance agronomic traits in sorghum Ph.D. thesis, [College Station (TX)]: (Texas A&M University), 2001.

Wu C, Dewan A, Hoh J, Wang Z. A comparison of association methods correcting for population stratification in case-control studies. Ann Hum Genet. 2011;75:418–27. https://doi.org/10.1111/j.1469-1809.2010.00639.x.

Xia J, Zhao Y, Burks P, Pauly M, Brown PJ. A sorghum NAC gene is associated with variation in biomass properties and yield potential. Plant Direct. 2018;2:1–11.

Xu GW, Magill CW, Schertz KF. Hart GE: a RFLP linkage map of Sorghum bicolor (L.) Moench. Theor Appl Genet. 1994;89:139–45.

Xu J, Weerasuriya Y, Bennetzen J. Construction of genetic map in sorghum and fine mapping of the germination stimulant production gene response to Striga asiatica. Acta Genet Sin. 2000;28(9): 870–6.

Yang S, Murphy RL, Morishige DT, Klein PE, Rooney WL, Mullet JE. Sorghum phytochrome B inhibits flowering in long days by activating expression of SbPRR37 and SbGHD7, repressors of SbEHD1, SbCN8 and SbCN12. PLoS One. 2014;9:e105352.

Yonemaru J, Ando T, Mizubayashi T, Kasuga S, Matsumoto T, Yano M. Development of genome-wide simple sequence repeat markers using whole-genome shotgun sequences of sorghum (Sorghum bicolor (L.) Moench). DNA Res. 2009;16(3):187–93.

Yu J, Buckler ES. Genetic association mapping and genome organization of maize. Curr Opin Biotechnol. 2006;17:155–60.

Yu J, Pressoir G, Briggs WH, Bi IV, Yamasaki M, Doebley JF, et al. A unified mixed-model method for association mapping that accounts for multiple levels of relatedness. Nat Genet. 2006;38:203–8.

Yu X, Li X, Guo T, Zhu C, Wu Y, et al. Genomic prediction contributing to a promising global strategy to turbocharge gene banks. Nat Plants. 2016;2:16150.

Zegada-Lizarazu W, Monti A. Are we ready to cultivate sweet sorghum as a bioenergy feedstock? A review on field management practices. Biomass Energy. 2012;40:1–12.

Zeng ZB. Theoretical basis for separation of multiple linked gene effects in mapping quantitative trait loci. PNAS. 1993;90(23):10972–6.

Zeng ZB. Precision mapping of quantitative trait loci. Genetics. 1994;136(4):1457–68.

Zhang Z, Ersoz E, Lai CQ, Todhunter RJ, Tiwari HK, Gore MA, et al. Mixed linear model approach adapted for genome-wide association studies. Nat Genet. 2010;42(4):355–60.

Zhang Y, Wang L, Xin H, Li D, Ma C, Ding X, et al. Construction of a high-density genetic map for sesame based on large scale marker development by specific length amplified fragment (SLAF) sequencing. BMC Plant Biol. 2013;13(1):141.

Zhang J, Jiang F, Shen Y, Zhan Q, Bai B, Chen W, et al. Transcriptome analysis reveals candidate genes related to phosphorus starvation tolerance in sorghum. BMC Plant Biol. 2019;19(1):1–18.

Zhao Q, Nakashima J, Chen F, Yin YB, Fu CX, Yun JF, et al. LACCASE is necessary and nonredundant with PEROXIDASE for lignin polymerization during vascular development in Arabidopsis. Plant Cell. 2013;25:3976–87.

Zhu C, Gore M, Buckler ES, Yu J. Status and prospects of association mapping in plants. Plant Genome. 2008;1:5–20.

Population Genomics and Genomics-Assisted Trait Improvement in Tea (*Camellia sinensis* (L.) O. Kuntze)

Tony Maritim, Romit Seth, Ashlesha Holkar, and Ram Kumar Sharma

Abstract "Tea" is undoubtedly one of the most widely consumed beverage processed from young shoots of tea [*Camellia sinensis* (L.) O. Kuntze] plant. The popularity of tea is attributed to its taste, aroma, and multiple health benefits. Developing new tea cultivars with novel characteristics remains the ultimate goal for breeders. Several breeding approaches ranging from conventional techniques, including crossbreeding and mutational breeding, to a series of molecular breeding methods such as marker-assisted selection and genetic modification have been employed in tea. Additionally, advances in high-throughput sequencing technologies, mainly genomics, transcriptomics, metabolomics, and proteomics, have led to the generation of massive dataset, which can be integrated with phenotypic data to identify key genes and pathways controlling important traits in tea. The publication of tea genome has also led to the integration of genotyping-by-sequencing (GBS) approach in genome-wide association studies (GWAS), genomic selection (GS),

---

Ashlesha Holkar died before publication of this work was completed.

T. Maritim
Biotechnology Department, CSIR-Institute of Himalayan Bioresource Technology (CSIR-IHBT), Palampur, Himachal Pradesh, India

Tea Breeding and Genetic Improvement Division, KALRO-Tea Research Institute, Kericho, Kenya

Academy of Scientific and Innovative Research (AcSIR), Ghaziabad, India

R. Seth
Biotechnology Department, CSIR-Institute of Himalayan Bioresource Technology (CSIR-IHBT), Palampur, Himachal Pradesh, India

Plants for Human Health Institute – North Carolina State University, Kannapolis, NC, USA

A. Holkar · R. K. Sharma (✉)
Biotechnology Department, CSIR-Institute of Himalayan Bioresource Technology (CSIR-IHBT), Palampur, Himachal Pradesh, India

Academy of Scientific and Innovative Research (AcSIR), Ghaziabad, India
e-mail: ramsharma@ihbt.res.in; rksharma.ihbt@gmail.com

genomic/genetic diversity, genetic linkage analysis, and ascertainment of trait-specific molecular markers. In spite of the advances made to expedite breeding in tea, success remains elusive due to bottlenecks including high heterozygosity, perennial nature, and self-incompatibility. The field of population genomics, which integrates advances in sequencing approaches, bioinformatics, and statistical analysis into research, has emerged as a novel tool for understanding new and long-standing queries on domestication, evolutionary history, genetic diversity, and population structure of tea plant. Moreover, with the availability of *C. sinensis* reference genome, population genomics allows for physical mapping of adaptive and molecular variants responsible for genotypic and phenotypic variations across the genome. In this chapter, we summarize achievements and future prospects of conventional, molecular breeding and population genomics in assessing/managing genetic diversity, unraveling the origin/evolution and domestication of tea plants, and identification of trait-specific genes in tea plants. Integration of population genomics with proteomics, metabolomics, lipidomics, and epigenomics is also highlighted in this chapter.

Keywords Domestication · Evolution and resequencing · Functional genomics · Genomics · Tea improvement · Transcriptomics

1 Introduction

Tea, *Camellia sinensis* (L.) O. Kuntze, belongs to the family Theaceae and is one of the most economically important perennial crops (Liu et al. 2015). Tea has a long history of cultivation; it was first cultivated in China in approximately 2500 BC as a medicinal herb, after which it became a popular beverage (Liam 2019). Tea consumption in China is woven into the fabric of many cultures and has roots, which date back to the mists of prehistory. The mental clarity and zen-like calm it provides, inspired many emperors to lead a religious following of tea. From China, tea was introduced in Japan, the Netherlands, and later Europe in 1200 AD, 1610 AD, and 1650 AD, respectively. Until the mid-nineteenth century, China was the main supplier of tea to the west, but after the Opium Wars involving Great Qing and the British Government, Britain sought an alternative source by cultivating tea in India, from where tea cultivation spread through the British Empire to the rest of the world (Liam 2019). Current statistics indicate that tea is grown for commercial purposes in more than 60 countries, including 22 in Asia, 18 in Africa, 11 in South and Central America, 9 in Europe and Eurasia, and 2 in North America. China, India, Kenya, and Sri Lanka are the leading producers of tea at 41.9%, 23.0%, 8.24%, and 6.0%, respectively (Table 1, Fig. 1; ITC 2017). It is also estimated that two billion cups of tea are consumed daily around the world at a per capita consumption rate of <120 ml/day (Katiyar and Mukhtar 1996). According to the FAO report of 2018, earnings from tea exports increased by over 75% between 2006 and 2016, contributing significantly to improve rural income and household food security in tea-producing countries (Bramel and Chen 2019). Additionally, tea is rich in

Table 1 Tea production in top ten leading countries in 2017

Country	Area under tea (Ha)	Production (mt)	Total export	Price (USD/kg)
China	2,787,077	2,350,000	328,692	4.52
India	570,267	1,267,366	218,392	2.91
Kenya	210,323	473,011	480,330	2.48
Sri Lanka	192,925	292,574	280,874	4.31
Turkey	77,300	253,312	6,117	4.67
Vietnam	133,000	180,000	142,000	1.66
Indonesia	119,498	125,500	51,464	2.2
Argentina	40,827	84,000	78,177	1.24
Bangladesh	57,282	82,721	539	3.42
Japan	42,400	77,100	4,251	25.58

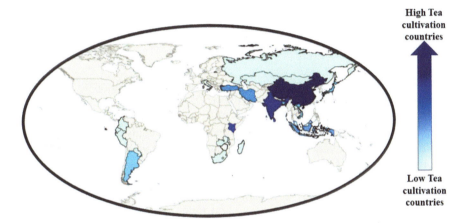

Fig. 1 Top tea-producing nations in the world

nutritive and biologically active compounds that contribute significantly to its taste, flavor, aroma, and medicinal properties (Jin et al. 2018a, b). It also contributes to environmental conservation through enhanced water infiltration, reduced surface erosion, and mitigation of global warming through carbon sequestration (Cheserek 2011).

2 Conventional Tea Genetics and Breeding

2.1 Tea Genetics

The genetic origin of tea, *Camellia sinensis*, is clouded with uncertainty. The plant is thought to have a common evolutionary history with wild species such as *C. irrawadiensis* ($2n = 30$), *C. caudata* ($2n = 30$), and *C. kissi* ($2n = 30$) (Wachira et al. 2013). Cultivated tea is diploid in nature ($2n = 30$), although naturally

Camellia sinensis var *assamica* *Camellia sinensis* var *assamica* ssp. Lasiocalyx *Camellia sinensis* var *sinensis*

Fig. 2 Different varieties of commercially cultivated tea. (Original figures of shoots and seeds captured in our gardens)

occurring polyploids [triploids ($3n = 45$), tetraploids ($4n = 60$), and aneuploids] with important quality-related characteristics such as high polyphenol content have been identified (Wachira and Kiplangat 1991; Magoma et al. 2000). Based on morphological and biochemical attributes, commercially cultivated tea can be classified into three main varieties including *Camellia sinensis* var. *sinensis*, *Camellia sinensis* var. *assamica*, and *Camellia sinensis* var. *assamica* ssp. Lasiocalyx, commonly referred to as chinery, assamica, and cambod varieties, respectively (Fig. 2) (Bezbaruah 1976). However, wild varieties such as *Camellia taliensis*, *Camellia crassicolumna*, *Camellia tachangensis*, *Camellia gymnogyna*, *and Camellia irrawadiensis* are also considered important genetic resources for diversification and trait improvement in tea. *Camellia irrawadiensis* (Wilson's camellia) and *Camellia taliensis* (Forest camellia) are believed to have contributed important quality-related traits, namely, low caffeine and purple pigmentation, to the modern tea cultivars (Ogino et al. 2009; Kamunya et al. 2009).

Although conventional methods (crossbreeding and mutation breeding) have significantly contributed to genetic improvement in tea, these methods have several limitations (Mukhopadhyay et al. 2016). Furthermore, continuous application of these techniques could narrow down the genetic pool from where important traits in tea are drawn, leading to a decline in quality and rendering tea cultivars vulnerable to changing climatic conditions. Therefore, several integrated omics (plant biotechnology, molecular markers, genomics) approaches have been devised for the analysis and manipulation of genetic variability and the development of improved cultivars within a short time. Recent advances in high-throughput sequencing technologies have generated massive "omics" datasets highlighting the expression of genes related to genes and variants linked to important traits in tea. Integration of omics and genetic datasets with phenotypic data can efficiently elucidate molecular mechanisms controlling the highly polygenic traits and help to expedite molecular breeding efforts for trait improvement in tea. This chapter provides an overview of the achievements and limitations of conventional breeding methods and advances in molecular, population genomics and genomic-based approaches.

2.2 Conventional Breeding Strategies for Tea Improvement

2.2.1 Classical Plant Breeding

Classical plant breeding (CPB) is the deliberate intra- or inter-specific crossing of naturally diverse genotypes to create novel lines/varieties with improved agricultural traits such as high yield, high quality, and tolerance to biotic and abiotic stresses (Shehasen 2019). Tea is a highly heterogeneous plant that exhibits inbreeding depression and self-incompatibility (Seth et al. 2019); hence, classical breeding is the most convenient method for developing improved varieties. In CPB, breeders generate genetic variability, select superior progenies, and test in multiple sites (genotype-environment interaction) for stability and adaptability (Kamunya et al. 2012). More than 2,500 tea cultivars currently under commercialization were developed or identified through classical breeding and clonal selection, globally (Bramel and Chen 2019) (Table 2).

In CPB, selection of tea cultivars that are superior is mostly based on phenotypic characteristics such as tolerance to biotic and abiotic stress, leaf pigmentation, leaf pubescence, biochemical components, namely, polyphenols, caffeine, carotenoids, and volatiles and yield potential (Banerjee 1992; Seurei 1996). Examples of commercially available tea cultivars with special characteristics related to quality are presented in Table 3. Although conventional breeding has contributed significantly to varietal improvement in tea, the long juvenile phase (more than 20 years) limits the process to F_1 generation, which may carry undesired traits. This necessitates subsequent generations like F_2 to clearly segregate, which will prolong breeding time further (Mondal et al. 2004). Additionally, selection of superior genotypes based on morphological and biochemical characteristics is not effective due to either environmental influences or individual biases.

Table 2 Global status of tea cultivars/varieties/crosses

Varieties/crosses	No.
Camellia sinensis var. *sinensis*	2,192
Camellia sinensis var. *assamica*	230
Camellia sinensis var. *publimba*	3
Camellia sinensis var. *assamica* ssp. Lasiocalyx	49
Camellia sinensis var. *assamica* × var. *sinensis*	39
Total	*2,513*

Source: Bramel and Chen (2019)

Table 3 Description of tea cultivars with special quality-related attributes

Sn	Attributes	Cultivar	Country	Reference
1.	High polyphenol content	TRFK 6/8	Kenya	Kamunya et al. (2012)
2.	High polyphenol content	GW Ejulu	Kenya	Kamunya et al. (2012)
3.	High quality potential	EPK TN15–23	Kenya	Kamunya et al. (2012)
4.	High quality, early flushing, and high yielding	Him Sphurti	India	Sud et al. (2012)
5.	High quality	DTI 1	Sri Lanka	Kottawa-arachchi et al. (2014)
6.	Caffeine-free	Hongyacha*	China	Jin et al. (2018a, b)
7.	Low caffeine content	Guangdong tea*	China	Mondal et al. (2004)
8.	Anthocyanin-rich	TRFK 306	Kenya	Kerio et al. (2012)
9.	Anthocyanin-rich	Zijuan	China	Bao et al. (2008)
10.	Anthocyanin-rich	Ziyan	China	Lai et al. (2016)
11.	Anthocyanin-rich	Sunrouge	Japan	Nesumi et al. (2012)
12.	Dark purple and densely pubescent shoots	TRI-2043	Sri Lanka	Kottawa-arachchi et al. (2014)
13.	Poor fermenter	TRFK 12/2	Kenya	Kamunya et al. (2012)

*denote naturally caffeine free/less wild tea plants

2.2.2 Mutation Breeding

Mutation breeding is one of the practical methods of genetic improvement in which physical, chemical, and biological mutagens are used to induce genetic variation in plants. As compared to crossbreeding, which involves the production of new genetic combinations or hybrids, mutation breeding can improve one or a few traits in an already outstanding genotype without necessarily affecting other phenotypic traits (Bado et al. 2015). According to the FAO/IAEA Mutant Variety Database (MVD), 3,314 mutant varieties belonging to 180 plant species with improved characteristics such as yield, maturity, quality, and tolerant to biotic and abiotic stress have been registered for commercialization. Among them, two tea varieties, namely, "Fufeng" (high yield and quality) and "Tea Noh PI 2" (self-compatible) are included. In Kenya, mutation breeding is being implemented through in vitro mutagenesis, where chemical mutagens (colchicines, hydroxyquinoline, and sulfanilamide) have been incorporated in culture media with the aim of developing novel tea cultivars that are tolerant to stress factors (Bramel and Chen 2019). The heterozygous nature of tea offers great possibility to increase mutation rate and develop many mutants within a short span of time, however, the feasibility of the technique is affected by several factors, such as the explant used, method of induction, mutagen dosage, and selection of a suitable genotype (Kharkwal et al. 2004).

Although conventional breeding approaches have offered great opportunity to understand the genetic variation in different tea germplasms, the approach has limitations. Biotechnological advances, including the use of genomic resources and molecular breeding approaches, have emerged as key tools that help for scrutinizing all genes and their relationships in order to identify their effect on phenotypic expressions of important traits in tea. Moreover, population genomics, which entails the analysis of genomic data from large population of samples, has also emerged as an extremely powerful approach for studying genetic diversity, population structure, evolution, phylogeny, and domestication (Weigel and Nordborg 2015).

3 Genomic Resources in Tea

Genomics is the study of the structure, function, and inheritance of the entire set of genetic material in an organism. Unlike genetics, which focuses mainly on the function and composition of individual genes, genomics scrutinizes all genes and their relationships in order to identify their combined effect on phenotypic expressions. The emergence of NGS has facilitated genomics research in many agricultural plants. Rice is the first crop to be sequenced and has highly annotated genome (Yu et al. 2002). The availability of genome information allows genome-wide association studies to be performed more precisely in order to dissect complex traits. In tea, two draft genomes belonging to two varieties, *Camellia sinensis* var. *assamica* (CSA) and *Camellia sinensis* var. *sinensis* (CSS), were published in 2017 and 2018, respectively (Xia et al. 2017; Wei et al. 2018). Genome assembly of CSA genome using ~707.88 Gb of Illumina (Hiseq 2000) short reads representing 98% of the estimated genome size revealed 37,618 scaffolds (N50 of 449 kb) with a total size of 3.02 Gb and 258,790 contigs (N50 of 20 kb) with a total size of 2.58 Gb. A total of 36,951 protein-coding genes were predicted out of which, 33,415 (90.43%) could be functionally classified. Homology search of annotated non-coding RNA also revealed 700 transfer RNA (tRNA), 2,860 ribosomal RNA (rRNA), 454 small nucleolar RNA (snoRNA), 223 small nuclear RNA (snRNA), and 233 micro-RNA (miRNA). Additionally, 80.89% of the assembly was annotated as repeat sequences. More recently, genome assembly of CSS genome using ~125.4 Gb of Single Molecule Real Time (PacBio) long read data and ~1,325 Gb Illumina short reads data, representing 93% of genome size, yielded 14,051 scaffolds with a total size of 3.1 Gb and N50 of 67.07 kb. A total of 33,932 protein-coding genes were identified, with at least 64% of the assembly annotated as transposable elements.

The findings from the two genomes suggest that tea may have evolved as a result of whole genome duplication (WGD) events that occurred ~30–100 million years ago. Further identification of 867,339 and 59,765 simple sequence repeats in CSA and CSS genomes, respectively, undoubtedly offers a huge resource for expediting breeding and genetic improvement through implementation of MAS in tea. The advent of tea genome laid foundations for resequencing of diverse lines. Recently, Xia et al. (2020b) reported resequencing of 81 diverse tea accessions along with the phylogenetic tree reporting of their diverse origins and evolution. Genomic resource of 139 tea accessions from different parts of Asia revealed the increased

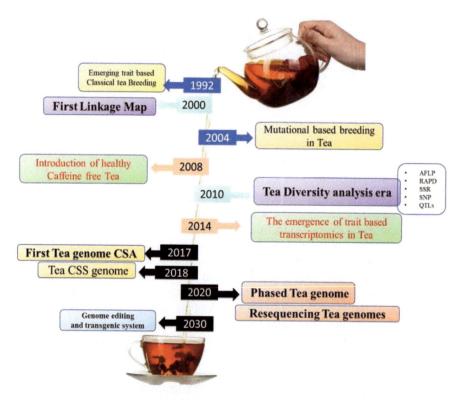

Fig. 3 A timeline illustration of Genetics & Genomics efforts in tea. (The figures is created by authors based on the summary of the public knowledge on Tea breeding and genomics)

hybridization and heterozygosity with increase of tea cultivation around the world (Wang et al. 2020a). Zhang et al. (2021) reported a phased genome based on single sperm sequencing of 135 cells from a single cultivar. Allele-specific expression provides insights into the evolution of tea. A timeline representing genetics and genomics efforts made in tea plant is depicted in Fig. 3.

3.1 Transcriptomics

Transcriptomics is the study of complete set of RNAs (transcriptome) present in a tissue of an organism with the aim of classifying them, determining their transcriptional structure and expression profile under specific condition or developmental stage (Wang et al. 2009). As compared to genome, the expression of transcriptome is dynamic. Rapid advances in RNA-seq technologies with improved sensitivity, accuracy, effective sequencing cost, and increased run-rate have resulted in enormous transcriptome data. In tea, transcriptome sequencing is a comprehensive and effective method for identification of genes, determining their expression patterns, and elucidating regulatory mechanism and networks involved in secondary metabolism (Shi et al. 2011). Tea quality is determined primarily by three classes of secondary metabolites, namely, phenolic compounds (18–36%), amino acid (1–4%), and volatile compounds (0.03%) (Zeng et al. 2019a, b). Therefore, these compounds are major targets in breeding for quality in tea, and their biosynthetic pathways are a subject of study. The first RNA-seq-based metabolite gene analysis in tea was performed by Shi et al. (2011) using a pooled sample comprising seven different tissues. In this study, genes involved in three major pathways (flavonoids, theanine, and caffeine) related to tea quality and tastes were identified. Subsequently, with the advances in technology and reduced cost of sequencing, several transcriptome sequencing projects designed to identify quality (taste and flavor)-related genes and determine their expression dynamics have been implemented in tea (Li et al. 2015b, 2017; Guo et al. 2017a; Wang et al. 2018; Wei et al. 2016). Recently, a comparative analysis integrating both transcriptome and metabolome of three tea cultivars revealed molecular mechanisms responsible for the formation of the unique flavor "Yin Rhyme" in "Tieguanyin" and the relationship between accumulation of secondary metabolites and sensory quality in tea (Guo et al. 2019). Additionally cross-species comparative transcriptome analysis between cultivated tea (*C. sinensis*) and wild relative (*C. taliensis*) has been used to identify genes associated with quality in tea (Zhang et al. 2016). Furthermore, the development of anthocyanin-rich tea cultivars has enthused a lot of interest due to their color, pleasant taste, sweet-woody flavor, and numerous health benefits (Joshi et al. 2017; Lai et al. 2016; Kerio et al. 2013). According to these studies, structural genes [phenylalanine ammonia lyase (PAL), chalcone synthase (CHS), chalcone isomerase (CHI), \trans-cinnamate 4-hydroxylase (C4H), flavanone 3-hydroxylase (F3H), flavonoid 3′,5′-hydroxylase (F3′5′H), dihydroflavonol 4-reductase (DFR), anthocyanidin synthase (ANS), anthocyanidin reductase (ANR) and UDP-glucose: flavonoid 3-O-glucosyltransferase (UFGT)] and transporters [glutathione-S-transferase (GST), the adenosine triphosphate (ATP)-binding cassette (ABC), and multidrug and toxic compound extrusion (MATE)] involved in flavonoid including anthocyanin accumulation were reported in tea (Maritim et al. 2021b). Furthermore, four levels of regulation, namely, regulation at gene expression level, transcription factor level [v-myb avian myeloblastosis viral oncogene homolog (MYB), basic helix-loop-helix (bHLH) and WD40], phenylalanine substrate level, and regulation at pathway branching have been identified in tea (Wang et al. 2017b; Sun et al. 2016). Considering the importance of aroma in quality evaluation and classification of tea, a significant number of studies have been undertaken to understand tea aroma biology (Ho et al. 2015). More than 700 volatile compounds associated with aroma formation have been identified in tea, and classified as fatty acid derived volatiles (VFADs), volatile terpenes (VTs), and glycosidically bound volatiles (GBVs). Additionally, genes involved in the biosynthesis of aroma compounds including two UGTs involved in the glucosylation and xylosylation of volatiles to β-primeveroside and linalool/nerolidol synthase gene have been identified in tea (Ohgami et al. 2015; Liu et al. 2018). Omics analysis using diverse genotypes has

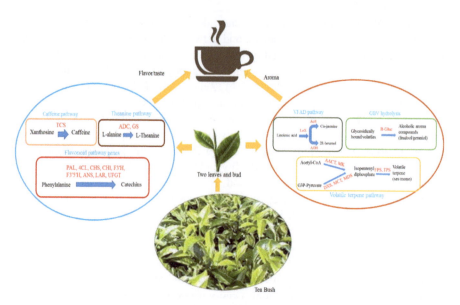

Fig. 4 Differentially expressed pathway genes attributing flavor and aroma to tea. *PAL*: phenylalanine ammonia lyase; *4CL*: 4-coumarate: CoA ligase; *CHS*: chalcone synthase; *CHI*: chalcone isomerase; *F3H*: flavonoid 3-hydroxylase; *F3′H*: flavonoid 3′-hydroxylase; *F3′5′H*: flavonoid 3′5-hydroxylase; *ANS*: anthocyanin synthase; *UFGT*: UDP-glucose flavonoid 3-O-glucosyl transferase; *LAR*: leucoanthocyanidin reductase; *TCS*: tea caffeine synthase; *ADC*: arginine decarboxylase; *GS*: glutamine synthase; *LOX*: lipoxygenase; *ADH*: alcohol dehydrogenase; *AOS*: allene oxide synthase; *β-Glu*: β-glucosidase; *AACT*: acetoacetyl-CoA thiolase; *MK*: mevalonate kinase; *DXS*: 1-deoxy-D-xylulose-5-phosphate synthase; *MCT*: 2-C-methyl-D-erythritol 4-phosphate cytidylyltransferase; *MDS*: 2-C-methyl-d-erythritol 2,4-cyclodiphosphate synthase, *FPS*: farnesyl phosphate synthase; *TPS*: terpene synthase. (Created by authors based of finding of Maritim et al. (2021a))

also helped uncover key candidate genes controlling quality-related pathways in tea (Maritim et al. 2021a). Recently, diverse tea cultivars exhibiting distinct quality-related characteristics were used in transcriptomic analysis to identify enriched pathways (Fig. 4) and candidate genes associated with biosynthesis of aroma and flavor forming compounds in tea (Maritim et al. 2021a). Among them, β-glucosidases, involved in hydrolysis of glycosidically bound volatile (GBV) that emit aroma, allene oxide synthase, allene oxide cyclase, alcohol dehydrogenase, and lipoxygenases involved in biosynthesis of fatty acid derived volatiles (VFADs) were identified. Similarly, the differential expression of volatile terpene, flavonoid, theanine, and caffeine biosynthetic genes revealed their critical role in regulating quality attributes in tea. Moreover, significant enrichment of key genes in the predicted interactome revealed a synchronized mechanism of regulation of quality attributes in tea (Maritim et al. 2021a). Moreover, trait-specific SNP markers present in quality-related genes were identified and successfully validated at the DNA level (Maritim et al. 2021a). Transcriptome data generated from anthocyanin-rich pigmented tea cultivars enrich the genomic resources for ascertainment of trait-specific non-synonymous SNPs that can be utilized for the development of high-

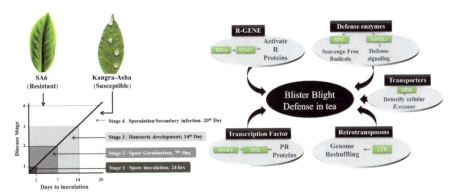

Fig. 5 Blister blight disease response in sensitive and tolerant tea cultivar using transcriptomics approach. (Source: Jayaswall et al. (2016))

density genotyping platform for genome-wide association studies in tea (Maritim et al. 2021b, c).

Likewise, genome-transcriptional analysis of contrasting cultivars identified 149 blister blight defense-related genes corresponding to defense-related enzymes, resistance genes, multidrug-resistant transporters, transcription factors, retrotransposons, metacaspases, and chaperons involved in blister blight resistance (Fig. 5; Jayaswall et al. 2016, Singh et al. 2019).

In one recent study, heat stress-associated transcriptional network with integrated metabolic pathways was successfully predicted using extreme tolerant and sensitive cultivars identified after screening 20 diverse tea cultivars (Seth et al. 2021). Subsequently, using a comprehensive omics approach, key roles of tea-specific heat shock proteins (HSPs) and heat shock transcription factors (HSFs) in transcriptional regulation of starch metabolism, aquaporins, chlorophyll biosynthesis, calcium, and ethylene-mediated plant signaling systems were identified attributing thermotolerance in tea. The identification of tea-specific HSPs (CsHSP90) and its inhibition exhibited enhanced heat shock response in tea (https://www.nature.com/articles/s41438-021-00532-z, Seth et al. 2021).

A similar study was carried out on the drought stress response in tea, which indicates the key role of ABA-dependent/-independent pathways, membrane transporters, and antioxidant defense system in attributing tolerance to the tea plant (Parmar et al. 2019). This information is useful in unraveling the regulatory mechanism involved in response to various biotic and abiotic stresses in tea and other crop plants. Further exploration of mating systems using omics has also helped in elucidation of pollen-pistil interactions underpinning self-incompatibility and cross-compatibility in tea (Fig. 4). By the identification of 42 key genes exhibiting differential expression patterns in stigma-style and ovary, the authors suggested late-acting gametophytic self-incompatibility in tea, which initiates in style and is sustained up to the ovary (Fig. 6; Seth et al. 2019).

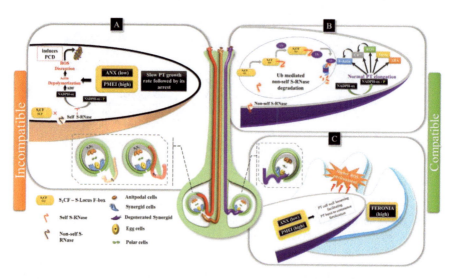

Fig. 6 Pollen-pistil cross-talks during self-incompatible and cross-compatible interactions. (**a**) Ceased and deviated pollen tubes representing incompatible interactions in style; The self S-RNase (csRNS) in SP style inhibits phosphorylation of NADPH-ox, resultantly inducing programmed cell death (PCD) via depolymerization of actin cytoskeleton. (**b**) Normal PT elongation in style in CP as non-self S-RNase undergoes ubiquitin mediated protein degradation. (**c**) Cross PT growth arrest followed by its burst within synergids commencing fertilization. (Source: Seth et al. (2019))

3.2 Functional Genomics

As appropriate methods for genetic transformation in tea are being developed, scientists have resorted to cloning functional genes in the tea plant (Xia et al. 2020b). This has greater significance in elucidating the molecular mechanism controlling important traits in tea. Interestingly, the majority of the genes cloned in tea are associated with the biosynthesis of key quality-related compounds such as polyphenols, anthocyanin, theanine, caffeine, and volatiles (Zhang et al. 2019). Molecular cloning and expression of genes such as aquaporin gene family (Yue et al. 2014), glutathione peroxidase in response to biotic and abiotic stress (Fu 2014), TY1-copia retrotransposons (Yao et al. 2017), and leucoanthocyanidin reductase (Zaman et al. 2018) are few economically important genes characterized in tea. In a recent review, Xia et al. (2020b) summarized some of the genes that have been cloned and functionally characterized in tea.

4 Population Genomics

Population genomics integrates advances in sequencing technologies, bioinformatics tools, and statistical methods to understand complex processes such as domestication, evolution, genetic diversity, and population structure in plants (Rajora

2019). The population genomics approaches have been used to elucidate the genetic diversity, population structure, evolutionary history, and domestication of cultivated tea as highlighted below.

4.1 Genetic Diversity, Population Structure, and Linkage Disequilibrium

Since the discovery of tea plants, its spread across the world's tropics and subtropical regions, and self-incompatibility and out-crossing nature has made it a highly heterogeneous crop plant with great diversity at both the genetic and morphological levels. The use of molecular markers, including RAPD, AFLP, SSR, and SNPs, to estimate genetic diversity and determine the phylogenetic relationships is widely reported in tea (Parmar et al. 2022; Raina et al. 2012; Chen and Yamaguchi 2002; Chen et al. 2005; Wachira et al. 1995, 1997). However, the conventional analysis of genetic diversity is laborious and time-consuming (Xia et al. 2020b).

Recent advances in next-generation sequencing technologies leading to publication of tea genome has led to the integration of genotyping-by-sequencing (GBS) approach in genome-wide association studies (GWAS), genomic selection (GS), genomic/genetic diversity, genetic linkage analysis, and ascertainment of trait-specific molecular markers. Evaluating genetic diversity and population genetic structure is an important aspect for unraveling domestication events and genetic relatedness of tea plants. Additionally, linkage disequilibrium (LD), also defined as the non-random association of alleles at different loci within a given population is a powerful tool for identifying regions of functional importance in a gene, detecting gene conversion and natural selection, and inferring recombination rates among the populations (Gabriel et al. 2002). The use of GBS approach to determine the genetic diversity, population structure, and linkage disequilibrium (LD) pattern is essential for expediting the development on tea breeding strategies.

In 2019, genotyping-by-sequencing (GBS) was first used to determine the genetic diversity, population structure, and LD pattern of 415 tea accessions (Niu et al. 2019). In this study, more than 79,000 high-quality SNPs were ascertained. Their polymorphism information content (PIC) and genetic diversity (GD) index revealed higher level of genetic diversity in the cultivated types than the wild types. Population structure analysis using these markers also clustered the 415 accessions into four populations, with the ancient landraces (tea plants older than 100 years) exhibiting complex genetic structure than both the wild and modern landraces (samples from tea gardens years). Moreover, LD decays of the four different populations varied between 1 kb and ~35 kb for groups GP02 and GP01, respectively (Niu et al. 2019). Later, genome resequencing of 120 ancient Chinese tea plants identified 8,082,370 high-quality SNPs that were used to cluster the 120 ancient tea plants into three distinct groups subdivided into 7 sub-populations based on genetic structure analysis (Lu et al. 2021). Recently, GBS analysis of 253 cultivars comprising modern and ancient landraces revealed varying genetic diversity indicators (Zhao et al. 2022). The ancient cultivars exhibited significantly higher nucleotide diversity (Pi) and

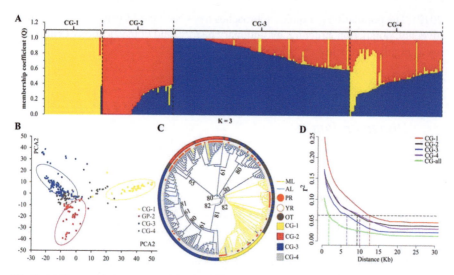

Fig. 7 (**a**) The population structure of 253 tea accessions showing CG-1 (yellow), CG-2 (red) and CG-3 (blue). (**b**) The principal component analysis (PCA) scatter diagram generated using the first and second principal components. The four inferred populations, CG-1 (yellow), CG-2 (red), CG-3 (blue), and CG-4 (gray). (**c**) Neighbor joining tree showing the inferred populations grouped according to their cultivation statuses (*ML* modern landraces, *AL* ancient landraces) and geographical location (*PR* Pearl River Basin, *YR* Yangtze River Basin, and *OT* other). (**d**) The LD decay plot showing the physical distance of four inferred populations groups. (Source: Zhao et al. (2022))

minor allele frequency (*MAF*), while the modern cultivars had significantly higher heterozygosity (*Ho*). Based on the population structure analysis, the 253 cultivars were clustered into four groups, three pure groups harboring 37, 45, and 112 accessions, respectively, and an admixture group containing 59 accessions (Fig. 7) Similarly, the LD decay of the four groups corresponded to physical distance of ~13 kb, ~9 kb, ~6 kb, and ~10 kb for groups CG1–4, respectively, (Fig. 7) (Zhao et al. 2022).

Considering the key role of structural variants (SVs) in the evolution and diversity of plant phenotypic traits, genome resequencing of 107 tea accessions revealed 44,240 high confident SVs comprising deletions, duplications, inversions, insertions, and translocations (Chen et al. 2022). Interestingly, 49.5% of identified protein-coding genes that overlapped with SVs were expressed in the roots, stem, buds, and young and mature leaves (Fig. 8). SV-based analysis of phylogenetic relationships and population structure of the tea plants corroborated with previous findings using SNPs.

4.2 Origin and Evolution of Tea Plants

Although the cultivated tea plants are believed to have originated in China (Chen et al. 2012; Hashimoto 2001), the absence of wild ancestors has rendered this theory

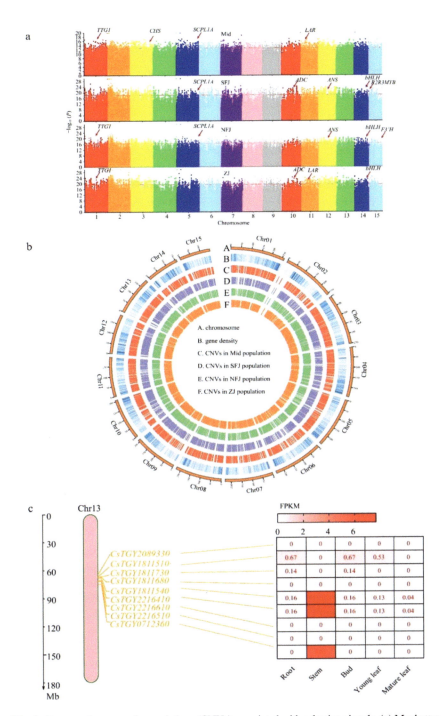

Fig. 8 Feature of copy number variations (CNVs) associated with selection signals. (**a**) Manhattan plots of domesticated CNV sites in four subpopulations (Mid, SFJ, NFJ, and ZJ), genes associated with important functions of tea plant are highlighted in the Manhattan plots. (**b**) CNVs distribution across whole genome, circos plots represent CNV locations among four subpopulations from the outside to the inside. (**c**) Under selected TPS family genes and expression analysis on chromosome 13. (Source: Chen et al. (2022))

vague and called for global attention. Earlier studies using SSR markers resulted in one-sided and rather contentious findings due to poor representation of sample collections and insufficient markers for diversity evaluation (Zhao et al. 2014; Wambulwa et al. 2016; Meegahakumbura et al. 2018). The use of genome-wide SNP markers derived from RAD sequencing revealed distinct genetic divergence between *C. sinensis* and its wild relatives and provided an insight into the artificial selection of tea plants at a genome-wide level (Yang et al. 2016). Moreover, the sequencing and annotation of different tea genomes formed a solid foundation for unraveling the mystery behind the origin and domestication of tea plants and opened up a new chapter in population genomics research (Xia et al. 2017; Wei et al. 2018). Furthermore, integration of population sequencing approach in tea has provided a better understanding of domestication, breeding, and classification together with the point and direction of divergent selection in both *sinensis* and *assamica* type populations. In a recent study, genome resequencing of a global collection of 139 tea accessions showed that hybridization has enhanced the heterozygosity and gene flow among tea populations (Wang et al. 2020a). In-depth analysis showed that selection for disease resistance and flavor formation during domestication was stronger in *C. sinensis* var. *sinensis* than the *C. sinensis* var. *assamica* populations (Fig. 9) (Wang et al. 2020a). The validation of chromosome-level genome of tea also revealed that one Camellia Recent Tetraploidization (CRT) event may have occurred 58.9–61.7 million years ago (Mya) after the core-eudicot common hexaploidization event (146.6–152.7 Mya). Duplication of several genes also occurred after CRT events, resulting in increased functionally divergent genes responsible for stress response and biosynthesis of major metabolites in tea. Interestingly, two-catechin and caffeine-related QTLs are believed to have arisen from the CRT events, suggesting key role of polyploidy in diversification of tea plants (Chen et al. 2020).

4.3 Domestication in Tea

Before the era of genome sequencing, nuclear microsatellites and cpDNA regions were used to identify three independent domestication centers of tea accessions from China and India. Interestingly, the China and Assam type teas are believed to have diverged 22,000 year ago during the last glacial maximum and subsequently split into the Chinese Assam type tea and Indian Assam type teas about 2,770 year ago (Meegahakumbura et al. 2018). This approach however had limitations. Unraveling of chromosome-scale genome of *Camellia oleifera*, plants showed that artificial selection of alleles involved in oil biosynthesis contributed to domestication of oil-Camellia (Lin et al. 2022). Recent analysis of the SVs identified that SVs subjected to artificial selection showed that most genes under domestication were enriched in key quality-related metabolic pathways such as theanine and terpenoid biosynthesis (Fig. 8) (Chen et al. 2022).

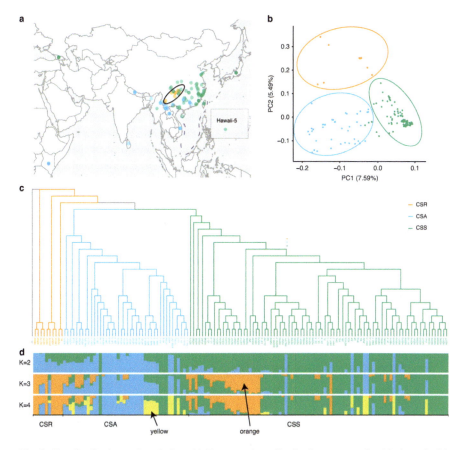

Fig. 9 Tea distribution and evolution. (**a**) Tea accessions distribution representing black oval with highest nucleotide polymorphism. (**b**) Principal component analysis (PC1 & PC2) of the tea populations splitting into three clusters. The *Camellia sinensis* var. *sinensis* (CSS) samples were found to cluster more tightly than the *C. sinensis* var. *assamica* (CSA) samples. (**c**) A phylogenetic tree of tea showing *Camellia sasanqua* Thunb. asan outgroup, and the *C. sinensis*-related species (CSR) as the closest to the outgroup. (**d**) Tea population structure representing CSS (Gree), CSA (blue) and CSR (yellow) populations. Yellow and orange are marked with arrows. (Source: Wang et al. (2020a))

5 Molecular Breeding and Trait Improvement

5.1 Genetic Manipulation

After the green revolution, plant scientists shifted their focus toward gene revolution, which involves modification of qualitative and quantitative traits of an organism by introducing foreign genes that represses or enhances gene expression, a process commonly referred to as the genetic engineering (Ali et al. 2011). As compared to conventional breeding, it is more effective and less time-consuming (Ahmad et al.

2015). With the advances in recombinant-DNA technology and development of gene-transfer protocols, agrobacterium-mediated transformation has been exploited for genetic engineering in woody plants. In tea, consumer demand for caffeine-free/less tea with original tea flavor, health benefits, and improved yield has prompted researchers to explore genetic engineering. Mohanpuria et al. (2008) developed caffeine-less transgenic tea through agrobacterium-mediated transformation. However, commercialization of the technology remains elusive due to cost involved and the bactericidal effect of tea polyphenols, which inhibits regeneration process (Mukhopadhyay et al. 2013). Despite the use of antioxidants or polyphenol adsorbents to ameliorate phenolic inhibition, production of transgenic tea plants remains elusive. Additionally, public antipathy to genetically modified food remains a barrier to the application of genetic engineering. Therefore, instead of manipulating DNA to develop caffeine-free tea, researcher's hopes lie in natural selection and screening of thousands of genotypes. Recently, Chinese scientists identified a low yielding, naturally caffeine-free, and theobromine-rich tea variety "hongyacha" (Jin et al. 2018a, b). Given the potential of the variety and consumer concern, it would be worthwhile to improve the yield component through existing breeding techniques. However, this would take more than 20 years before a high-yielding caffeine-free tea is available for commercialization.

5.2 Mapping of Quantitative Trait Loci

Identification of genetic loci linked to a specific trait is critical to facilitate crop improvement and can only be achieved using established markers. Unlike morphological or some biochemical markers, DNA markers are reproducible, less time-consuming, and not influenced by environmental conditions (Hu et al. 2014). Several DNA-based markers ranging from traditional gel-based AFLP (amplified fragment length polymorphism), RAPD (randomly amplified polymorphic DNA), SSR (simple sequence repeat), ISSR (inter-simple sequence repeat) to modern sequenced-based marker, namely, STS (sequence tagged site), EST-SSR, DArT, and SNP (single nucleotide polymorphism) have been used in genetic studies for assessment of genetic diversity (Wachira et al. 1995, 1997; Paul et al. 1997; Bhardwaj et al. 2013; Sharma et al. 2009, 2010, 2011), assessment of mating systems in tea (Muoki et al. 2007), and identification of quantitative trait loci (QTLs) (Hackett et al. 2000; Kamunya et al. 2010; Koech et al. 2018, 2019). Moreover, markers associated with quality traits in tea (Mphangwe et al. 2013; Elangbam and Misra 2016) have been identified through this approach (Table 4).

Furthermore, SNP markers have been used to map QTLs associated with key quality parameters in tea (Xu et al. 2018; Ma et al. 2018). However, most of these studies are genomic marker-based which limits the identification of expression-based variation (Hansey et al. 2012). Transcriptome sequencing is the most suitable approach to ascertain high quality gene-specific sequence variation and novel transcripts in a cost-effective manner irrespective of genome size (Hansey et al.

Table 4 Molecular markers associated with quality trait in tea

Methodology	Marker	Quality parameter associated	Reference
	RAPD	Black tea quality (3)	Mphangwe et al. (2013)
	CAPS	Catechin content	Elangbam and Misra (2016)
QTL mapping	SSR	Catechin content (25 QTLs)	Ma et al. (2014)
	EST-SSR	Polyphenol (5) Water extract content (5) Free amino acid (6) Caffeine (3)	Hui et al. (2016)
QTL mapping	DArT	Caffeine (6 QTLs) Cetechin content (25 QTLs) Theaflavin (3 QTLs) Tea taster score (liquor color, brightness, astringency, briskness, and aroma) (9)	Koech et al. (2018)
QTL mapping	SNP	27 QTLs (flavonoid/caffeine content)	Xu et al. (2018)
QTL mapping	SSR/SNP	10 QTLs (caffeine, theobromine, caffeine + theobromine and caffeine: theobromine content)	Ma et al. (2018)
Association mapping	SSR/SNP	Caffeine (87)	Jin et al. (2016)
Transcriptional analysis	SNP	Flavonoid (145), caffeine (117), theanine (15), volatile terpenes (23), fatty acid-derived volatiles (19), glycosidically bound volatile biosynthesis (51)	Maritim et al. (2021a)

NB: Values in bracket are the number of markers associated to the trait

2012). This approach has been used to identify functional SNPs in many plant species, including *Persea americana* (Kuhn et al. 2019), *Phoebe chekiangensis* (He et al. 2017), *Populus nigra* (Rogier et al. 2018), *Pinus monticola* (Liu et al. 2014), and *Camellia sinensis* (Parmar et al. 2019).

5.2.1 Linkage-Based QTL Mapping

Linkage mapping (LM) is a powerful DNA marker-based tool that facilitates identification of genomic regions and genes controlling a phenotypic trait in a biparental population. It offers great opportunity to improve breeding speed, precision, and efficiency through marker-assisted selection. The first linkage map in tea was constructed by Hackett et al. (2000) from a biparental cross using a combination of RAPD and AFLP markers. Since then, several studies on linkage and QTL mapping have been performed using various types of markers in tea. Kamunya et al. (2010) used RAPD, AFLP, and SSR to identify QTLs associated with yield. In the last couple of years, several studies have reported on QTL mapping of important traits including quality in tea. Ma et al. (2014) constructed a linkage map from a pseudo test-cross of 406 F_1 individuals using SSR markers spanning 1,143 cM with average distance of 2.9 cM between adjacent markers. In this study, 25 QTLs associated with catechins were identified. Based on the linkage map described in a previous study (Ma et al. 2015), 10 novel QTLs controlling caffeine (CAF), theobromine (TBR), ratio of caffeine and theobromine (CTR), and sum of caffeine and theobromine (SCT) content have been identified (Ma et al. 2018). Additionally, Koech et al. (2018) identified 43 QTLs including 6 associated with caffeine content, 25 for catechins, 3 for theaflavins, and 9 linked to tea taste, from a consensus genetic map constructed based on DArTseq platform. A high-density linkage map constructed from 327 F_1 population derived from two clonal tea cultivars has also been used to identify 27 novel QTLs associated with flavonoid and caffeine content in tea (Xu et al. 2018). In addition, research on QTL linked to other desired phenotypic traits such as early spring bud flush has also received a lot of interest due to their role in quality especially post dormancy. Tan et al. 2018 reported two major QTLs of TBF trait (timing of spring bud flush) on LG01 using sib population of "Longjing 43" * "Baihazao" and also tested the QTL stability in another tea population with early spring cultivar as parents (Tan et al. 2018). However, the tea characteristics, such as self-incompatibility and long juvenile period, limit the creation of a large and suitable population for fine mapping and precise identification of QTLs, which makes linkage-based QTL mapping laborious and time-consuming. However, integration of omics approaches, such as whole genome resequencing in QTL mapping, provides a new insight into tea trait development. A high-quality QTL map generated for leaf area traits of tea by taking advantage of the whole genome resequencing of "Jinxuan," "Yuncha1," and their 96 F1 hybrid offspring identified 8,956 high-quality SNPs in 15 linkage group of tea (An et al. 2021). Timing of spring bud flush, leaf size area, and tea quality traits are directly proportional to the trait improvement in tea thereby increasing its economic value.

5.2.2 Association Mapping

Association mapping (AM), a powerful tool which identifies QTNs by examining trait-marker association in a set of genotypes with unknown ancestry, has been used to overcome limitations associated with linkage-based QTL mapping in plants. It offers higher resolution and high number of alleles and requires less time to establish association populations (Stich and Melchinger 2010). In tea, AM has been used to determine genetic 502 association between caffeine synthase gene (TCS1) and caffeine content in tea (Jin et al. 2016). Furthermore, high resolution AM and high detection power linkage mapping have been integrated and used to identify association between 48 DArTseq markers and quality-related phenotypic traits in tea (Koech et al. 2019). Nonetheless, AM has less power to detect rare alleles and is affected by population structure. To overcome challenges associated with AM and LM, there is need to integrate QTL mapping and RNA-seq and or use next-generation sequencing (NGS)-based bulk segregant analysis to map important

QTLs and genes associated with important traits in tea (Wang et al. 2017a; Tagaki et al. 2013).

Recently, DArTseq marker-based QTL analysis performed, using 3-year yield data, generated from a population of 261 F_1 progeny derived from two biparental reciprocal crosses of a high quality and drought-tolerant tea cultivars, revealed 13 novel QTLs with LOD values ranging from 1.98 to 7.24, and explained 3.4–12% of the phenotypic variation (Malebe et al. 2021). The total length of the parental maps was 1028.1 cM for the high quality and 1026.6 cM for the drought-tolerant cultivar and an average locus spacing of 5.5 cM and 5.4 cM, respectively. Moreover, two segregating populations genotyped using 1,421 DArTseq markers enabled the identification of genome-based prediction models with great potential for expediting selection of parents with desired traits even at the seedling stage (Koech et al. 2020).

5.2.3 Bulk Segregant Analysis

Bulk segregant analysis (BSA) is a simple method for rapid and efficient identification of markers, genes, or genomic regions associated with a specific trait using genetic markers that are polymorphic in a mapping population (Michelmore et al. 1991). In tea, Kamunya et al. (2010) used BSA to identify QTLs associated with yield based on a combination of RAPD, AFLP, and SSR markers. Recent advances in high-throughput sequencing technology have led to an integration of BSA and NGS to expedite identification of genes or markers associated with important traits in plants (Liu et al. 2012). Furthermore, in order to map genes present in populations in which polymorphic markers are present or absent, BSA that combines RNA-seq with BSA has been modified as bulk segregant RNA-seq (BSR-seq) (Liu et al. 2012). Notwithstanding the limited information available on the application of this approach to dissect quality trait in tea, recently published draft tea genomes (Xia et al. 2017; Wei et al. 2018) offer great opportunity for future application of BSR-seq in tea.

6 Emerging Population Genomics Disciplines of Tea

6.1 Population Epigenomics

Epigenetics refers to the study of changes in the regulation of gene activity and expressions that are not dependent on DNA sequence (Jeyaraj et al. 2019). Epigenetic mechanisms involving DNA methylation, chromatin variation, and the synthesis of small non-coding RNAs (miRNA) regulate important biological processes such as stress response, growth, transport, transcription, and oxidation-reduction in plants (Moler et al. 2019; Jeyaraj et al. 2019). MiRNAs are endogenous small non-protein-coding RNA molecules (20–24 nucleotides) that negatively regulate gene expression through targeted mRNA cleavage, translational repression, and silencing of gene transcripts at post-transcriptional level in plants (Kim 2013). Since the identification of the first plant miRNA in *A. thaliana* (Reinhart et al. 2002), several plant miRNAs have been identified using computational/bioinformatics tools, high-throughput sequencing, microarray analysis, small RNA cloning, and degradome sequencing (Zhu and Luo 2013). Currently, a total of 19,029 miRNAs (8,615 hairpin precursors and 10,414 mature miRNAs) from 82 plant species are available in the miRBase database (http://www.mirbase.org/, Release 22.1) (Kozomara et al. 2019). In tea, most studies on regulation of structural genes are focused on transcription regulation (TFs) with limited reports on the role of miRNAs. Most of the miRNAs identified in tea are associated with biotic and abiotic stress response (Wang et al. 2020b; Jeyaraj et al. 2017, 2019; Guo et al. 2017b; Liu et al. 2016). However, they have not been registered or deposited in the public database. Furthermore, characterization of miRNAs associated with biosynthesis of quality-related metabolites may provide deeper insights on regulation of quality formation in tea. Jeyaraj et al. (2017) identified 175 conserved and 83 novel mRNA including those involved in regulation of genes related to phenylpropanoid pathway in two leaves and buds tissue. Sun et al. (2017) identified 55 conserved miRNAs in tea, of which 31 were validated for their presence using qRT-PCR. Furthermore, seven miRNA molecules that regulate key genes (CHI, CHS, DFR, ANR, LAR, and F3H) involved in catechin biosynthesis were identified. Subsequently, Zhao et al. (2019) integrated metabolomics, sRNA sequencing, degradome and transcriptome sequencing to identify four classes of miRNA-TF pairs that might be involved in seasonal regulation of terpenoid biosynthesis in tea. In light of the important roles of miRNAs and the availability of draft tea genomes, further studies on miRNAs present in various tea cultivars need to be done. In future, epigenomics studies should be done at the population level to identify within- and between population epigenomic variation. Development of techniques, such as bisulfite and ChIP sequencing, makes it possible to study epigenome variation at the population level (Moler et al. 2019).

6.2 Population Proteomics

Despite the numerous achievements made through transcriptome sequencing, quantitative mRNA data has a limitation as it cannot always predict protein expression levels due to posttranscriptional modifications/regulations such as RNA splicing, RNA editing, RNA stability, protein phosphorylation, protein degradation, protein ubiquitination, and nuclear export (Wu et al. 2019). Therefore, proteomics (analysis of protein expression profile at specific physiological status or biological stage) complements transcriptome studies by providing a deeper insight into protein functions and metabolisms. With the advances in protein quantification methods [isobaric tags for relative and absolute quantitation (iTRAQ) and mass spectrometry (MS)], coupled with available genome information and modern bioinformatics tools, proteome studies aimed at elucidating molecular mechanism attributed to trait-variation in tea have been implemented. Li et al. (2015a) first used iTRAQ to identify 233 differentially expressed protein (DEPs) [including those involved in polyphenol biosynthesis] between apical bud and young expanding leaves. Additionally, Zhou et al. (2017) used MS (MALDI-TOF/TOF-MS) to identify 46 DEPs including those involved in polyphenol biosynthesis in tender purple and mature green leaves of tea. A comparative proteomic analysis of postharvest leaves of tea plant at different stages of withering [the first step in tea processing] allowed identification of 863 DEPs that may influence tea quality (Wu et al. 2017). Subsequently, iTRAQ was recently used to study regulatory network of flavonoid biosynthesis in tea leaves at different stages of maturity (Wu et al. 2019). Furthermore, Kumari et al. (2020) used both conventional two-dimensional electrophoresis (2-DE) and MS (Q-TOF IMS-LC-MS) to identify 211 DEPs including those involved in anthocyanin biosynthesis during leaf color transition in tea. Although these studies have revealed important information, the number lags behind as compared to transcriptomic studies in tea.

The proteomics studies should be performed at the population level in tea in future to identify within and between population key proteins along with population transcriptomic studies.

6.3 Population Metabolomics

Metabolomics [comprehensive study of metabolites present in a biological sample] is a postgenomic technique that determines the relationship between gene expression and metabolic networks in plants (Escudero-Martinez et al. 2017). Metabolomics provide chemical evidence that allows proper understanding of biological phenomena occurring in tissue at a specific time. There are two main approaches in metabolomics: targeted and untargeted. In targeted metabolomics, specific known metabolites are profiled preferably using mass spectrometry and nuclear magnetic resonance (NMR) because of their specificity and reproducibility (Patti et al. 2012). Untargeted metabolomics, on the other hand, is the global profiling of the entire metabolome using liquid chromatography followed by mass spectrometry (LC/MS) because they permit detection of several metabolites (Patti et al. 2012). In tea, flavonoids, alkaloids, theanine, and volatile compounds such as terpenes are the main subjects of intensive metabolomic studies due to their dominance in tea. Jiang et al. (2019) summarized some of the application of metabolomics in evaluation of quality in tea. Additionally, Li et al. (2020) recently used nontargeted metabolomics to characterize different tea cultivars for suitability processing and determine their influence on sensory quality of finished products. The study identified O-rutinosylation and O-glucosylation as predominant glycosylation process in black tea cultivars, whereas O-glucosyl/galactosyl-rutinosylation were predominant in cultivars suitable for green tea and both green and black tea. Furthermore, a comprehensive profiling of flavor-related metabolites revealed metabolic changes taking place during oolong tea processing and their effect on quality (Chen et al. 2020). Subsequently, targeted and nontargeted metabolomics were recently used successfully to unravel chemical changes in pu-erh tea during pile fermentation (Long et al. 2019; Chen et al. 2019).

Lipids not only play a role in growth, development, and response to environmental cues but also contribute significantly to sensory properties in plants (Liu et al. 2017). In tea, lipids are the de novo precursors for biosynthesis of flavor and aroma compounds, such as hexenal and hexenol, which contribute to the fresh and greenish odor of green tea (Ho et al. 2015). Therefore, quantification, profiling, or identification of novel types is one of the critical ways to characterize and select quality tea cultivars. Till date, limited information is available on lipids profiling in tea. Liu et al. (2017) used UPLC/MS to characterize lipids in fresh shoots of tea grown under different rates of nitrogen fertilizer. In this study, 178 lipids species were identified. Additionally, increased nitrogen fertilization enhanced the ratio of glycolipids [monogalactosyldiacylglycerol (MGDG) and digalactosyldiacylglycerol (DGDG)], which are rich in linolenic acid, a precursor for the formation of aroma compounds in tea.

Population metabolomics studies have not yet been reported in tea and should be undertaken to understand among and within populations' metabolic variation.

7 Conclusion and Future Prospects

Although conventional and biotechnological advances have contributed significantly to varietal improvement in tea, the techniques have limitations as highlighted in this chapter. Therefore, the following multidisciplinary approaches may be embraced in future:

(a) With the release of multiple tea genomes of *C. sinensis* var. *sinensis* and *C. sinensis* var. *assamica*, the foundation for revealing the genetic basis of important traits in tea is established. Moreover, identification and development of sequence-based markers that can be used for the development of high-density genotyping platform for genome-wide association studies and marker-assisted breeding in tea are now possible. Futuristically, resequencing of trait-specific tea cultivars for discovery of single nucleotide polymorphism (SNP), InDels, and structural variants should be the next frontier.

(b) Continuous advancement in NGS technologies and emergence of several omics platforms such as genomics, transcriptomics, proteomics, metabolomics, lipidomics, and microbiomics is leading biology to the "big data" era. Moreover, integration of these approaches often referred to as "*multi-omics*" approach (Bao 2019) will provide a comprehensive understanding of molecular dynamics and regulatory mechanisms associated with tea quality formation.

(c) Despite the promise that genetic modification was a perfect solution to food security and climate change, the application of this technique has been affected by unfounded health and environmental concerns. As a result, scientists have developed less risky biotechnological tools, that is, *genome editing*, that facilitate precise, efficient, and targeted modification (Zhang et al. 2018). Currently, there are four gene editing tools: zinc-finger nucleases (ZFNs), transcription factor-like effector nucleases (TALENs), oligonucleotide-directed mutagenesis (ODM), and the recently developed clustered regularly interspaced short palindromic repeats (CRISPRs) that are available for research and development (Kamburova et al. 2017). All these approaches use sequence-specific nucleases to cleave DNA at specific regions resulting in single or few nucleotide changes that are comparable to naturally induced mutation. Among these approaches, CRISPR/Cas9 is cheap, simple, and highly efficient (Zhang et al. 2018). CRISPR/Cas9 has been applied successfully in many plants; however, there is no report on the application of gene editing in tea plant (*Camellia sinensis*). In woody perennial plant species, gene editing is limited to *Populus*, where CRISPR/Cas9 has been used with 100% efficacy to create knockout mutation in two genes involved in lignin and flavonoid biosynthesis (Zhou et al. 2015). Considering the breakthrough in *Populus*, economically important traits in tea could be modified for site-marching and customer preferences. Similarly, the recent application of CRISPR/Cas9 to overcome self-incompatibility in potato, *Solanum tuberosum* (Ye et al. 2018), may trigger a lot of interest among tea breeders with the hope of breaking the self-incompatibility barrier in tea plants.

Population genomics studies are at an early stage in tea, which should be intensified to construct a pangenome, understand population-level genome-wide genetic, epigenetic, transcriptomic, and proteomic variations in wild and cultivated tea populations; identify the genetic effects of domestication and genes involved; understand the origin, evolution, phylogeny, and phylogeography of tea; and examine speciation and admixture. Furthermore, population genomics will assist in the genetic improvement of tea via genomic selection, conservation, and sustainable management of tea genetic resources.

Funding The financial support was provided by Council of Scientific and Industrial Research, New Delhi, concerning MLP 146, MLP201, GAP 309, Govt. of India.

Acknowledgments The Director, CSIR-IHBT, is gratefully acknowledged for providing necessary facilities for conducting research. This chapter is published in memory of Ms Ashlesha Holkar (third author), who deceased on when this chapter was under revision. This is CSIR-IHBT publication no 5374.

References

Ahmad HM, Rahman MU, Azeem F, Ali Q. QTL mapping for the improvement of drought tolerance in cereal crops: a review. Life Sci J. 2015;12(4s):102–8. https://doi.org/10.7537/marslsj1204s15.14.

Ali Q, Ahsan M, Tahir MHN, Elahi M, Farooq J, Waseem M. Gene expression and functional genomic approach for abiotic stress tolerance in different crop species. IJAVMS. 2011;2:221–48. https://doi.org/10.5455/ijavms.20110521104951.

An Y, Chen L, Tao L, Liu S, Wei C. QTL mapping for leaf area of tea plants (*Camellia sinensis*) based on a high-quality genetic map constructed by whole genome resequencing. Front Plant Sci. 2021;12:705285. https://doi.org/10.3389/fpls.2021.705285.

Bado S, Forster BP, Nielen S, Ali AM, Lagoda PJ, Till BJ, et al. Plant mutation breeding: current progress and future assessment. Plant Breed Rev. 2015;39:23–88. https://doi.org/10.1002/9781119107743.ch02.

Banerjee B. Botanical classification of tea. In: Wilson K, Clifford N, editors. Tea Cultivation and Consumption, Chapman and Hall, London, 1992. p. 25–51.

Bao J. Biotechnology for rice grain quality improvement. In: Jinsong Bao, editor. Rice. St. Paul: AACC International Press; 2019. p. 443–71.

Bao YX, Xia LF, Li YY, Liang MZ. A new tea tree cultivar 'Zijuan'. Acta Hort Sin. 2008;6.

Bezbaruah HP. Tea varieties in cultivation-an appraisal. Two Bud. 1976;23:13–9.

Bhardwaj P, Kumar R, Sharma H, Tewari R, Ahuja PS, Sharma RK. Development and utilization of genomic and genic microsatellite markers in Assam tea (Camellia assamica ssp. assamica) and related Camellia species. Plant Breed. 2013;132:748–63. https://doi.org/10.1111/pbr.12101.

Bramel P, Chen L. A global strategy for the conservation and use of tea genetic resources. Technical report. 2019. https://doi.org/10.13140/RG.2.2.20411.05922.

Chen L, Yamaguchi, S. Genetic diversity and phylogeny of tea plant (*Camellia sinensis*) and its related species and varieties in the section *Thea* genus *Camellia* determined by randomly amplified polymorphic DNA analysis. J Hortic Sci Biotech. 2002;77:729–32.

Chen L, Apostolides Z, Chen ZM. Global Tea Breeding: Achievements, Challenges and Perspectives. Springer, Berlin, Heidelberg, 2012.

Chen J, Wang PS, Xia YM, Xu M, Pei SJ. Genetic diversity and differentiation of *Camellia sinensis* L. (cultivated tea) and its wild relatives in Yunnan province of China, revealed by morphology, biochemistry and allozyme studies. Genet Resour Crop Evol. 2005;52:41–52.

Chen S, Liu H, Zhao X, Li X, Shan W, Wang X, et al. Non-targeted metabolomics analysis reveals dynamic changes of volatile and non-volatile metabolites during oolong tea manufacture. Food Res Int. 2019;128:108778. https://doi.org/10.1016/j.foodres.2019.108778.

Chen JD, Zheng C, Ma JQ, Jiang CK, Ercisli, S, Yao, MZ, Chen L. The chromosome-scale genome reveals the evolution and diversification after the recent tetraploidization event in tea plant. Hort Res. 2020;7.

Chen S, Fang J, Wang Y, Wang P, Zhang S. Evolutionary genomics of structural variation in the tea plant, *Camellia sinensis*. Trop Plants 2022;1:2.

Cheserek B. A report on evidence and impacts of climate change in tea growing areas in Kenya. TRFK Q Bull. 2011;15:2–3.

Elangbam M, Misra AK. Development of CAPS markers to identify Indian tea (*Camellia sinensis*) clones with high catechin content. Genet Mol Res. 2016;15(2):1–13. https://doi.org/10.4238/gmr.15027860.

Escudero-Martinez CM, Morris JA, Hedley PE, Bos JIB. Barley transcriptome analyses upon interaction with different aphid species identify thionins contributing to resistance. Plant Cell Environ. 2017;40(11):2628–43.

Fu JY. Cloning of a new glutathione peroxidase gene from tea plant (*Camellia sinensis*) and expression analysis under biotic and abiotic stresses. Bot Stud. 2014;55(1):1–6.

Gabriel SB, Schaffner SF, Nguyen H, Moore JM, Roy J, Blumenstiel B, et al. The structure of haplotype blocks in the human genome. Sci. 2002;296(5576):2225–9.

Guo F, Guo Y, Wang P, Wang Y, Ni D. Transcriptional profiling of catechins biosynthesis genes during tea plant leaf development. Planta. 2017a;246(6):1139–52. https://doi.org/10.1007/s00425-017-2760-2.

Guo Y, Zhao S, Zhu C, Chang X, Yue C, Wang Z, et al. Identification of drought-responsive miRNAs and physiological characterization of tea plant (*Camellia sinensis* L.) under drought stress. BMC Plant Biol. 2017b;17:211. https://doi.org/10.1186/s12870-017-1172-6.

Guo Y, Zhu C, Zhao S, Zhang S, Wang W, Fu H, et al. De novo transcriptome and phytochemical analyses reveal differentially expressed genes and characteristic secondary metabolites in the original oolong tea (*Camellia sinensis*) cultivar 'Tieguanyin' compared with cultivar 'Benshan'. BMC Gen. 2019;20:265.

Hackett CA, Wachira FN, Paul S, Powell W, Waugh R. Construction of a genetic linkage map for *Camellia sinensis* (tea). Heredity. 2000;85(4):346–55. https://doi.org/10.1046/j.1365-2540.2000.00769.x.

Hansey CN, Vaillancourt B, Sekhon RS, De Leon N, Kaeppler SM, Buell CR. Maize (Zea mays L.) genome diversity as revealed by RNA-sequencing. PLoS One. 2012;7(3):e33071. https://doi.org/10.1371/journal.pone.0033071.

Hashimoto M. The Origin of the Tea Plant, in *Proceedings of 2001 International Conference on O-CHA (tea) Culture and Science (Session II)* (Shizuoka:) 2001.

He B, Li Y, Ni Z, Xu LA. Transcriptome sequencing and SNP detection in Phoebe chekiangensis. PeerJ. 2017;5:e3193. https://doi.org/10.7717/peerj.3193.

Ho CT, Zheng X, Li SM. Tea aroma formation. Food Sci Human Wellness. 2015;4:9–27. https://doi.org/10.1016/j.fshw.2015.04.001.

Hu C-Y, Tsai Y-Z, Lin S-F. Development of STS and CAPS markers for variety identification and genetic diversity analysis in tea germplasm in Taiwan. Bot Stud. 2014;55:12. https://doi.org/10.1186/1999-3110-55-12.

Hui S, Liu JJ, He W, Wen B, Li MF, Feng JC. Association analysis of traits related to tea (*Camellia sinensis*) quality with EST-SSRs in Southern Henan area. J Agric Biotech. 2016;24(9):1328–36. https://www.cabdirect.org/cabdirect/search/?q=au%3a%22Su+Hui%22.

ITC. World tea statistics for year 2016. 2017.

Jayaswall K, Mahajan P, Singh G, Parmar R, Seth R, Raina A, et al. Transcriptome analysis reveals candidate genes involved in blister blight defense in tea (Camellia sinensis (L) Kuntze). Sci Rep. 2016;6:30412. https://doi.org/10.1038/srep30412.

Jeyaraj A, Zhang X, Hou Y, Shangguan M, Gajjeraman P, Li Y, et al. Genome-wide identification of conserved and novel microRNAs in one bud and two tender leaves of tea plant (*Camellia sinensis*) by small RNA sequencing, microarray-based hybridization and genome survey scaffold sequences. BMC Plant Biol. 2017;17(1):212. https://doi.org/10.1186/s12870-017-1169-1.

Jeyaraj A, Wang X, Wang S, Liu S, Zhang R, Wu A, et al. Identification of regulatory networks of microRNAs and their targets in response to *Colletotrichum gloeosporioides* in tea plant (*Camellia sinensis* L.). Front Plant Sci. 2019;10:1096. https://doi.org/10.3389/fpls.2019.01096.

Jiang CK, Ma JQ, Apostolides Z, Chen L. Metabolomics for a Millenniums-Old Crop: tea plant (*Camellia sinensis*). J Agric Food Chem. 2019;67:6445–57. https://doi.org/10.1021/acs.jafc.9b01356.

Jin JQ, Yao MZ, Ma CL, Ma JQ, Chen L. Association mapping of caffeine content with TCS1 in tea plant and its related species. Plant Physiol Biochem. 2016;105:251–9. https://doi.org/10.1016/j.plaphy.2016.04.032.

Jin JQ, Chai YF, Liu YF, Zhang J, Yao MZ, Chen L. Hongyacha, a naturally caffeine-free tea plant from Fujian, China. J Agric Food Chem. 2018a;66(43):11311–9. https://doi.org/10.1021/acs.jafc.8b03433.

Jin JQ, Liu YF, Ma CL, Ma JQ, Hao WJ, Xu YX, et al. A novel F3′ 5′ H allele with 14 bp deletion is associated with high catechin index trait of wild tea plants and has potential use in enhancing tea quality. J Agric Food Chem. 2018b;66(40):10470–8. https://doi.org/10.1021/acs.jafc.8b04504.

Joshi R, Rana A, Kumar V, Kumar D, Padwad YS, Yadav SK, et al. Anthocyanins enriched purple tea exhibits antioxidant, immunostimulatory and anticancer activities. J Food Sci Technol. 2017;54(7):1953–63. https://doi.org/10.1007/s13197-017-2631-7.

Kamburova VS, Nikitina EV, Shermatov SE, Buriev ZT, Kumpatla SP, Emani C, et al. Genome editing in plants: an overview of tools and applications. Int J Agron. 2017.

Kamunya SM, Wachira FN, Nyabundi KW, Kerio L, Chalo RM. The tea research foundation of Kenya pre-releases purple tea variety for processing health tea product. Tea. 2009;30(2):3–10.

Kamunya SM, Wachira FN, Pathak RS, Korir R, Sharma V, Kumar R, et al. Genomic mapping and testing for quantitative trait loci in tea (*Camellia sinensis* (L.) O. Kuntze). Tree Genet Genomes. 2010;6(6):915–29. https://doi.org/10.1007/s11295-010-0301-2.

Kamunya SM, Wachira FN, Pathak RS, Muoki RC, Sharma RK. Tea improvement in Kenya. In: Chen L, Apostolides Z, Chen ZM, editors. Global tea breeding. Advanced topics in science and technology in China. Berlin, Heidelberg: Springer; 2012.

Katiyar SK, Mukhtar H. Tea in chemoprevention of cancer: epidemiologic and experimental studies. Int J Oncol. 1996;8:221–38. https://doi.org/10.3892/ijo.8.2.221.

Kerio LC, Wachira FN, Wanyoko JK, Rotich MK. Characterization of anthocyanins in Kenyan teas: extraction and identification. Food Chem. 2012;131(1):31–8. https://doi.org/10.1016/j.foodchem.2011.08.005.

Kerio LC, Wachira FN, Wanyoko JK, Rotich MK. Total polyphenols, catechin profiles and antioxidant activity of tea products from purple leaf coloured tea cultivars. Food Chem. 2013;136:1405–13. https://doi.org/10.1016/j.foodchem.2012.09.066.

Kharkwal MC, Pandey RN, Pawar SE. Chapter 26: Mutation breeding for crop improvement. In: Jain HK, Kharkwal MC, editors. Plant breeding. Dordrecht: Springer; 2004. p. 601–45.

Kim VN. MicroRNA biogenesis: coordinated cropping and dicing. Nat Rev Mol Cell Bio. 2013;6:376–85.

Koech RK, Malebe PM, Nyarukowa C, Mose R, Kamunya SM, Apostolides Z. Identification of novel QTL for black tea quality traits and drought tolerance in tea plants (*Camellia sinensis*). Tree Genet Genomes. 2018;14:9. https://doi.org/10.1007/s11295-017-1219-8.

Koech RK, Mose R, Kamunya SM, Apostolides Z. Combined linkage and association mapping of putative QTLs controlling black tea quality and drought tolerance traits. Euph. 2019;215:162.

Koech RK, Malebe PM, Nyarukowa C, Mose R, Kamunya SM, Loots T, et al. Genome-enabled prediction models for black tea (*Camellia sinensis*) quality and drought tolerance traits. Plant Breed. 2020;139(5):1003–15. https://doi.org/10.1111/pbr.12813.

Kottawa-Arachchi JD, Gunasekare MTK, Ranatunga MAB, Punyasiri PAN, Jayasinghea L, Karunagodad RP. Biochemical characteristics of tea (Camellia L. spp.) germplasm accessions in Sri Lanka: correlation between black tea quality parameters and organoleptic evaluation. Int J Tea Sci. 2014;10:3–13.

Kozomara A, Birgaoanu M, Griffiths-Jones S. miRBase: from microRNA sequences to function. Nucleic Acids Res. 2019;47:D155–62. https://doi.org/10.1093/nar/gky1141.

Kuhn DN, Livingstone DS III, Richards JH, Manosalva P, Van den Berg N, Chambers AH. Application of genomic tools to avocado (Persea americana) breeding: SNP discovery for genotyping and germplasm characterization. Sci Hortic. 2019;246:1–11. https://doi.org/10.1016/j.scienta.2018.10.011.

Kumari M, Thakur S, Kumar A, Joshi R, Kumar P, Shankar R, et al. Regulation of color transition in purple tea (*Camellia sinensis*). Planta. 2020;251:1–18. https://doi.org/10.1007/s00425-019-03328-7.

Lai YS, Li S, Tang Q, Li HX, Chen SX, Li PW, et al. The dark-purple tea cultivar "Ziyan" accumulates a large amount of delphinidin-related anthocyanins. J Agric Food Chem. 2016;64 (13):2719–26. https://doi.org/10.1021/acs.jafc.5b04036.

Li Q, Li J, Liu S, Huang J, Lin H, Wang K, et al. A comparative proteomic analysis of the buds and the young expanding leaves of the tea plant (*Camellia sinensis* L.). Int J Mol Sci. 2015a;16(6): 14007–38. https://doi.org/10.3390/ijms160614007.

Li CF, Zhu Y, Yu Y, Zhao QY, Wang SJ, Wang XC, et al. Global transcriptome and gene regulation network for secondary metabolite biosynthesis of tea plant (Camellia sinensis). BMC Genomics. 2015b;16(1):560. https://doi.org/10.1186/s12864-015-1773-0.

Li J, Lv X, Wang L, Qiu Z, Song X, Lin J, et al. Transcriptome analysis reveals the accumulation mechanism of anthocyanins in "Zijuan" tea (*Camellia sinensis* var. asssamica (Masters) kitamura) leaves. Plant Growth Regul. 2017;81(1):51–61. https://doi.org/10.1007/s10725-016-0183-x.

Li J, Wang J, Yao Y, Hua J, Zhou Q, Jiang Y, et al. Phytochemical comparison of different tea (*Camellia sinensis*) cultivars and its association with sensory quality of finished tea. LWT. 2020;117:108595. https://doi.org/10.1016/j.lwt.2019.108595.

Liam D. Making tea. Nature. 2019;566:S2–4. https://doi.org/10.1038/d41586-019-00395-4.

Lin P, Wang K, Wang Y, Hu Z, Yan C, Huang H, et al. The genome of oil-Camellia and population genomics analysis provide insights into seed oil domestication. Gen Biol. 2022;23:1–21.

Liu S, Yeh CT, Tang HM, Nettleton D, Schnable PS. Gene mapping via bulked segregant RNA-seq (BSR-seq). PLoS One. 2012;7(5):e36406. https://doi.org/10.1371/journal.pone.0036406.

Liu JJ, Chan D, Sturrock R, Sniezko RA. Genetic variation and population differentiation of the endochitinase gene family in *Pinus monticola*. Plant Syst Evol. 2014;300(6):1313–22. https://doi.org/10.1007/s00606-013-0963-y.

Liu Y, Wang D, Zhang S, Zhao H. Global expansion strategy of Chinese herbal tea beverage. Adv J Food Sci Technol. 2015;7(9):739–45.

Liu SC, Xu YX, Ma JQ, Wang WW, Chen W, Huang DJ, et al. Small RNA and degradome profiling reveals important roles for microRNAs and their targets in tea plant response to drought stress. Physiol Plant. 2016;158(4):435–51. https://doi.org/10.1111/ppl.12477.

Liu M, Burgos A, Ma L, Zhang Q, Tang D, Ruan J. Lipidomics analysis unravels the effect of nitrogen fertilization on lipid metabolism in tea plant (*Camellia sinensis* L.). BMC Plant Biol. 2017;17:165. https://doi.org/10.1186/s12870-017-1111-6.

Liu GF, Liu JJ, He ZR, Wang FM, Yang H, Yan YF, Wei S. Implementation of CsLIS/NES in linalool biosynthesis involves transcript splicing regulation in *Camellia sinensis*. Plant cell environ. 2018;41(1):176–86.

Long P, Wen M, Granato D, Zhou J, Wu Y, Hou Y, et al. Untargeted and targeted metabolomics reveal the chemical characteristic of pu-erh tea (*Camellia assamica*) during pile-fermentation. Food Chem. 2019;311:125895. https://doi.org/10.1016/j.foodchem.2019.125895.

Lu L, Chen H, Wang X, Zhao Y, Yao X, Xiong B, et al. Genome-level diversification of eight ancient tea populations in the Guizhou and Yunnan regions identifies candidate genes for core agronomic traits. Hort Res. 2021;8.

Ma J-Q, Yao M-Z, Ma C-L, Wang X-C, Jin J-Q, Wang X-M, et al. Construction of a SSR-based genetic map and identification of QTLs for catechins content in tea plant (*Camellia sinensis*). PLoS One. 2014;9(3):e93131. https://doi.org/10.1371/journal.pone.0093131.

Ma JQ, Huang L, Ma CL, Jin JQ, Li CF, Wang RK, et al. Large-scale SNP discovery and genotyping for constructing a high-density genetic map of tea plant using specific-locus amplified fragment sequencing (SLAF-seq). PLoS One. 2015;10(6):e0128798. https://doi.org/10.1371/journal.pone.0128798.

Ma JQ, Jin JQ, Yao MZ, Ma CL, Xu YX, Hao WJ, et al. Quantitative trait loci mapping for theobromine and caffeine contents in tea plant (Camellia sinensis). J Agric Food Chem. 2018;66(50):13321–7. https://doi.org/10.1021/acs.jafc.8b05355.

Magoma G, Wachira F, Obanda M, Imbuga M, Agong SG. The use of catechins as biochemical markers in diversity studies of tea (Camellia sinensis). Genet Resour Crop Evol. 2000;47:107–14. https://doi.org/10.1023/A:1008772902917.

Malebe MP, Koech RK, Mbanjo EGN, Kamunya SM, Myburg AA, Apostolides Z. Construction of a DArT-seq marker–based genetic linkage map and identification of QTLs for yield in tea (Camellia sinensis (L.) O. Kuntze). Tree Genet Genomes. 2021;17(1):1–11. https://doi.org/10.1007/s11295-021-01491-1.

Maritim TK, Seth R, Parmar R, Sharma RK. Multiple-genotypes transcriptional analysis revealed candidates genes and nucleotide variants for improvement of quality characteristics in tea (*Camellia sinensis* (L.) O. Kuntze). Genomics. 2021a;113(1):305–16. https://doi.org/10.1016/j.ygeno.2020.12.020.

Maritim T, Masand M, Seth R, Sharma RK. Transcriptional analysis reveals key insights into seasonal induced anthocyanin degradation and leaf color transition in purple tea (*Camellia sinensis* (L.) O. Kuntze). Sci Rep. 2021b;11:1244. https://doi.org/10.1038/s41598-020-80437-4.

Maritim TK, Korir RK, Nyabundi KW, Wachira FN, Kamunya SM, Muoki RC. Molecular regulation of anthocyanin discoloration under water stress and high solar irradiance in pluckable shoots of purple tea cultivar. Planta. 2021c;254(5):1–17.

Meegahakumbura MK, Wambulwa MC, Li MM, Thapa KK, Sun YS, Möller, M, et al. Domestication origin and breeding history of the tea plant (*Camellia sinensis*) in China and India based on nuclear microsatellites and cpDNA sequence data. Front Plant Sci. 2018;8:2270.

Michelmore RW, Paran I, Kesseli RV. Identification of markers linked to disease-resistance genes by bulked segregant analysis: a rapid method to detect markers in specific genomic regions by using segregating populations. Proc Natl Acad Sci. 1991;88(21):9828–32. https://doi.org/10.1073/pnas.88.21.9828.

Mohanpuria P, Rana NK, Yadav SK. Transient RNAi based gene silencing of glutathione synthetase reduces glutathione content in *Camellia sinensis* (L.) O. Kuntze somatic embryos. Biol Plant. 2008;52:381–4. https://doi.org/10.1007/s10535-008-0080-x.

Moler E, Abakir A, Eleftheriou M, Johnson JS, Krutovsky KV, Lewis LC, et al. 2018. Population epigenomics: advancing understanding of phenotypic plasticity, acclimation, adaptation and diseases. In: Rajora OP, editor. Population genomics: concepts, approaches and applications. Springer International Publishing AG; 2019. p. 179–260. https://doi.org/10.1007/13836_2018_59.

Mondal TK, Bhattacharya A, Laxmikumaran M, Ahuja PS. Recent advances in tea biotechnology. Plant Cell Tissue Organ Cult. 2004;75:795–856. https://doi.org/10.1023/B:TICU.0000009254.87882.71.

Mphangwe NI, Vorster J, Steyn JM, Nyirenda HE, Taylor NJ, Apostolides Z. Screening of tea (*Camellia sinensis*) for trait-associated molecular markers. Appl Biochem Biotechnol. 2013;171 (2):437–49. https://doi.org/10.1007/s12010-013-0370-4.

Mukhopadhyay M, Sarkar B, Mondal TK. Omics advances in tea (*Camellia sinensis*). In: Barh D, editor. Omics applications in crop science. Boca Raton: CRC Press, Taylor and Francis Group; 2013. p. 347–66.

Mukhopadhyay M, Mondal TK, Chand PK. Biotechnological advances in tea (*Camellia sinensis* [L.] O. Kuntze): a review. Plant Cell Rep. 2016;35(2):255–87. https://doi.org/10.1007/s00299-015-1884-8.

Muoki RC, Wachira FN, Pathak RS, Kamunya SM. Assessment of the mating system of *Camellia sinensis* in biclonal seed orchards based on PCR markers. J Hortic Sci Biotechnol. 2007;82(5): 733–8. https://doi.org/10.1080/14620316.2007.11512298.

Nesumi A, Ogino A, Yoshida K, Taniguchi F, Maeda Yamamoto M, Tanaka J, et al. "Sunrouge", a new tea cultivar with high anthocyanin. Jpn Agric Res Q. 2012;46:321–8. https://doi.org/10.6090/jarq.46.321.

Niu S, Song Q, Koiwa H, Qiao D, Zhao D, Chen Z, et al. Genetic diversity, linkage disequilibrium, and population structure analysis of the tea plant (*Camellia sinensis*) from an origin center, Guizhou plateau, using genome-wide SNPs developed by genotyping-by-sequencing. BMC Plant Bio. 2019;19(1):1–12.

Ogino A, Tanaka J, Taniguchi F, Yamamoto MP, Yamada K. Detection and characterization of caffeine-less tea plants originating from interspecific hybridization. Breed Sci. 2009;59:277–83. https://doi.org/10.1270/jsbbs.59.277.

Ohgami S, Ono E, Horikawa M, Murata J, Totsuka K, Toyonaga H, et al. Volatile glycosylation in tea plants: sequential glycosylations for the biosynthesis of aroma β-primeverosides are catalyzed by two *Camellia sinensis* glycosyltransferases. Plant Physiol. 2015;168(2):464–77. https://doi.org/10.1104/pp.15.00403.

Parmar R, Seth R, Singh P, Singh G, Kumar S, Sharma RK. Transcriptional profiling of contrasting genotypes revealed key candidates and nucleotide variations for drought dissection in Camellia sinensis (L.) O. Kuntze. Sci Rep. 2019;9:7487. https://doi.org/10.1038/s41598-019-43925-w.

Parmar R, Seth R, Sharma RK. Genome-wide identification and characterization of functionally relevant microsatellite markers from transcription factor genes of tea (Camellia sinensis (L.) O. Kuntze). Sci Rep. 2022;12:1–14. https://doi.org/10.1038/s41598-021-03848-x.

Patti GJ, Yanes O, Siuzdak G. Innovation: metabolomics: the apogee of the omics trilogy. Nat Rev Mol Cell Boil. 2012;13(4):263–9. https://doi.org/10.1038/nrm3314.

Paul S, Wachira FN, Powell W, Waugh R. Diversity and genetic differentiation among populations of Indian and Kenyan tea (Camellia sinensis (L.) O. Kuntze) revealed by AFLP markers. Theor Appl Genet. 1997;94(2):255–63.

Raina SN, Ahuja PS, Sharma RK, Das SC, Bhardwaj P, Negi R, et al. Genetic structure and diversity of India hybrid tea. Genet Resour Crop Evol. 2012;59:1527–41.

Rajora OP. Population genomics: concepts, approaches and application: Springer Nature Switzerland AG; 2019. 823pp. ISBN 978-3-030-04587-6; ISBN 978-3-030-04589-0 (eBook).

Reinhart BJ, Weinstein EG, Rhoades MW, Bartel B, Bartel DP. MicroRNAs in plants. Genes Dev. 2002;16(13):1616–26. https://doi.org/10.1101/gad.1004402.

Rogier O, Chateigner A, Amanzougarene S, Lesage-Descauses MC, Balzergue S, Brunaud V, et al. Accuracy of RNAseq based SNP discovery and genotyping in Populus nigra. BMC Genomics. 2018;19(1):1–12. https://doi.org/10.1186/s12864-018-5239-z.

Seth R, Bhandawat A, Parmar R, Singh P, Kumar S, Sharma RK. Global transcriptional insights of pollen-pistil interactions commencing self-incompatibility and fertilization in tea [Camellia sinensis (L.) O. Kuntze]. Int J Mol Sci. 2019;20:539. https://doi.org/10.3390/ijms20030539.

Seth R, Maritim TK, Parmar R, Sharma RK. Underpinning the molecular programming attributing heat stress associated thermotolerance in tea (Camellia sinensis (L.) O. Kuntze). Hortic Res. 2021;8:99. https://doi.org/10.1038/s41438-021-00532-z.

Seurei P. Tea improvement in Kenya: a review. Tea. 1996;17:76–81.

Sharma R, Bhardwaj P, Negi R, Mohapatra T, Ahuja P. Identification, characterization and utilization of unigene derived microsatellite markers in tea (*Camellia sinensis* L.). BMC Plant Biol. 2009;9:1. https://doi.org/10.1186/1471-2229-9-53.

Sharma RK, Negi MS, Sharma S, Bhardwaj P, Kumar R, Bhattacharya E, et al. AFLP-based genetic diversity assessment of commercially important tea germplasm in India. Biochem Genet. 2010;48(7):549–64. https://doi.org/10.1007/s10528-010-9338-z.

Sharma H, Kumar R, Sharma V, Kumar V, Bhardwaj P, Ahuja PS, et al. Identification and cross-species transferability of 112 novel unigene-derived microsatellite markers in tea (*Camellia sinensis*). Am J Bot. 2011;98(6):e133-8. https://doi.org/10.3732/ajb.1000525.

Shehasen MZ. Tea plant (*Camellia sinensis*) breeding mechanisms role in genetic improvement and production of major producing countries. Int J Res Stud Sci Eng Technol. 2019;6:10–20.

Shi CY, Yang H, Wei CL, Yu O, Zhang ZZ, Jiang CJ, et al. Deep sequencing of the Camellia sinensis transcriptome revealed candidate genes for major metabolic pathways of tea-specific compounds. BMC Genomics. 2011;12(1):131. https://doi.org/10.1186/1471-2164-12-131.

Singh G, Singh G, Seth R, Parmar R, Singh P, Singh V, et al. Functional annotation and characterization of hypothetical protein involved in blister blight tolerance in tea (Camellia sinensis (L) O. Kuntze). J Plant Biochem Biotechnol. 2019;28:447–59. https://doi.org/10.1007/s13562-019-00492-5.

Stich B, Melchinger AE. An introduction to association mapping in plants. CABI Rev. 2010;5:1–9. https://doi.org/10.1079/PAVSNNR20105039.

Sud RK, Gulati A, Singh S, Sharma RK, Singh RD, Dhadwal VS, et al. HIM SPHURTI (CSIR-IHBT-T-01): A cultivar of china hybrid tea (*Camellia sinensis*). CSIR IHBT Technical Bulletin. 2012. https://www.ihbt.res.in/images/TechnicalBulletin/HIMSPHURTI.pdf.

Sun B, Zhu Z, Cao P, Chen H, Chen C, Zhou X, et al. Purple foliage coloration in tea (*Camellia sinensis* L.) arises from activation of the R2R3-MYB transcription factor CsAN1. Sci Rep. 2016;6:3253. https://doi.org/10.1038/srep32534.

Sun P, Cheng C, Lin Y, Zhu Q, Lin J, Lai Z. Combined small RNA and degradome sequencing reveals complex microRNA regulation of catechin biosynthesis in tea (*Camellia sinensis*). PLoS One. 2017;12(2):e0171173. https://doi.org/10.1371/journal.pone.0171173.

Tagaki H, Abe A, Yoshida K, Kosugi S, Natsume S, Mitsouoka C, et al. QTL-seq: rapid mapping of quantitative trait loci in rice by whole genome resequencing of DNA from two bulked populations. The plant J. 2013;74(1):174–83.

Tan LQ, Peng M, Xu LY, Wang LY, Wei K, Zou Y, et al. The validation of two major QTLs related to the timing of spring bud flush in *Camellia sinensis*. Euphytica. 2018;214(1):1–12. https://doi.org/10.1007/s10681-017-2099-6.

Wachira FN, Kiplangat JK. Newly identified Kenyan polyploid tea strains. Tea. 1991;12(1):10–3.

Wachira FN, Waugh R, Powell W, Hackett CA. Detection of genetic diversity in tea (*Camellia sinensis*) using RAPD markers. Genome. 1995;38(2):201–10.

Wachira FN, Powell W, Waugh R. An assessment of genetic diversity among *Camellia sinensis* L (cultivated tea) and its wild relatives based on randomly amplified polymorphic DNA and organelle- specific STS. Heredity. 1997;78:603–11.

Wachira FN, Kamunya S, Karori S. et al. The tea plants: botanical aspects. In: Preedy VR, editor. Tea in health and disease prevention. Elsevier. 2013. pp. 3–17.

Wambulwa MC, Meegahakumbura MK, Chalo R, Kamunya S, Muchugi A, Xu JC, et al. Nuclear microsatellites reveal the genetic architecture and breeding history of tea germplasm of East Africa. Tree Gen & Geno. 2016;12(1):1–10.

Wang Z, Gerstein M, Snyder M. RNA-seq: a revolutionary tool for transcriptomics. Nat Rev Genet. 2009;10:57–63. https://doi.org/10.1038/nrg2484.

Wang SS, Cao M, Ma X, Chen WK, Zhao J, Sun CQ, et al. Integrated RNA sequencing and QTL mapping to identify candidate genes from *Oryza rufipogon* associated with salt tolerance at the seedling stage. Front Plant Sci. 2017a;8:1427. https://doi.org/10.3389/fpls.2017.01427.

Wang L, Pan D, Liang M, Abubakar YS, Li J, Lin J, et al. Regulation of Anthocyanin biosynthesis in purple leaves of Zijuan tea (*Camellia sinensis* var. kitamura). Int J Mol Sci. 2017b;18(4):833. https://doi.org/10.3390/ijms18040833.

Wang W, Zhou Y, Wu Y, Dai X, Liu Y, Qian Y, et al. Insight into catechins metabolic pathways of *Camellia sinensis* based on genome and transcriptome analysis. J Agric Food Chem. 2018;66 (16):4281–93. https://doi.org/10.1021/acs.jafc.8b00946.

Wang X, Feng H, Chang Y, Ma C, Wang L, Hao X, et al. Population sequencing enhances understanding of tea plant evolution. Nat Commun. 2020a;11(1):1–10. https://doi.org/10.1038/s41467-020-18228-8.

Wang S, Liu S, Liu L, Li R, Guo R, Xia X, et al. miR477 targets the phenylalanine ammonia-lyase gene and enhances the susceptibility of the tea plant (*Camellia sinensis*) to disease during Pseudopestalotiopsis species infection. Planta. 2020b;251(3):59. https://doi.org/10.1007/s00425-020-03353-x.

Wei K, Zhang Y, Wu L, Li H, Ruan L, Bai P, et al. Gene expression analysis of bud and leaf color in tea. Plant Physiol Biochem. 2016;107:310–8. https://doi.org/10.1016/j.plaphy.2016.06.022.

Wei C, Yang H, Wang S, Zhao J, Liu C, Gao L, et al. Draft genome sequence of *Camellia sinensis* var. sinensis provides insights into the evolution of the tea genome and tea quality. Proc Natl Acad Sci. 2018;115(18):E4151–8. https://doi.org/10.1073/pnas.1719622115.

Weigel D, Nordborg M. Population genomics for understanding adaptation in wild plant species. Annu Rev Genet. 2015;49:315–38. https://doi.org/10.1146/annurev-genet-120213-092110.

Wu ZJ, Ma HY, Zhuang J. iTRAQ-based proteomics monitors the withering dynamics in postharvest leaves of tea plant (*Camellia sinensis*). Mol Genet Genomics. 2017;293:45–59. https://doi.org/10.1007/s00438-017-1362-9.

Wu LY, Fang ZT, Lin JK, Sun Y, Du ZZ, Guo YL, et al. Complementary iTRAQ proteomic and transcriptomic analyses of leaves in tea plant (*Camellia sinensis* L.) with different maturity and regulatory network of flavonoid biosynthesis. J Proteome Res. 2019;18(1):252–64. https://doi.org/10.1021/acs.jproteome.8b00578.

Xia EH, Zhang HB, Sheng J, Li K, Zhang QJ, Kim C, et al. The tea tree genome provides insights into tea flavor and independent evolution of caffeine biosynthesis. Mol Plant. 2017;10(6):866–77. https://doi.org/10.1016/j.molp.2017.04.002.

Xia E, Tong W, Hou Y, An Y, Chen L, Wu Q, et al. The reference genome of tea plant and resequencing of 81 diverse accessions provide insights into its genome evolution and adaptation. Mol Plant. 2020a;13(7):1013–26. https://doi.org/10.1016/j.molp.2020.04.010.

Xia E, Tong W, Wu Q, Wei S, Zhao J, Zhang Z-Z, et al. Tea plant genomics: achievements, challenges and perspectives. Hortic Res. 2020b;7:7. https://doi.org/10.1038/s41438-019-0225-4.

Xu L, Wang L, Wei K, Tan L-Q, Su J-J, Cheng H. High-density SNP linkage map construction and QTL mapping for flavonoid-related traits in a tea plant (*Camellia sinensis*) using 2b-RAD sequencing. BMC Genomics. 2018;19:955. https://doi.org/10.1186/s12864-018-5291-8.

Yao J, Xiaoyu L, Cheng P, Yeyun L, Jiayue J, Changjun J. Cloning and analysis of reverse transcriptases from Ty1-copia retrotransposons in *Camellia sinensis*. Biotechnol Biotechnol Equip. 2017;31(4):663–9. https://doi.org/10.1080/13102818.2017.1332492.

Yang H, Wei CL, Liu H, Wu JL, Li ZG, Zhang L, et al. Genetic divergence between *Camellia sinensis* and its wild relatives revealed via genome-wide SNPs from RAD sequencing. Plose One. 2016;11(1):e0151424.

Ye M, Peng Z, Tang D, Yang Z, Li D, Xu Y, et al. Generation of self-compatible diploid potato by knockout od S-RNase. Nat Plant 2018:4(9):651–4.

Yu J, Hu S, Wang J, Wong GK-S, Li S, Liu B, et al. A draft sequence of the rice genome (Oryza sativa L. ssp. indica). Science. 2002;296:79–92. https://doi.org/10.1126/science.1068037.

Yue C, Cao H, Wang L, Zhou Y, Hao X, Zeng J, et al. Molecular cloning and expression analysis of tea plant aquaporin (AQP) gene family. Plant Physiol Biochem. 2014;83:65–76. https://doi.org/10.1016/j.plaphy.2014.07.011.

Zaman A, Gogoi M, Kalita MC. Cloning and expression study of Leucoanthocyanidin reductase gene (CaLAR) from *Camellia assamica*. Biotechnol Ind J. 2018;14(1):158.

Zeng C, Lin H, Liu Z, Liu Z. Metabolomics analysis of Camellia sinensis with respect to harvesting time. Food Res Int. 2019a;128:108814. https://doi.org/10.1016/j.foodres.2019.108814.

Zeng L, Watanabe N, Yang Z. Understanding the biosyntheses and stress response mechanisms of aroma compounds in tea (*Camellia sinensis*) to safely and effectively improve tea aroma. Crit Rev Food Sci Nutr. 2019b;59(14):2321–34. https://doi.org/10.1080/10408398.2018.1506907.

Zhang Y, Skaar I, Sulyok M, Liu X, Rao M, Taylor JW. The microbiome and metabolites in fermented pu-erh tea as revealed by high-throughput sequencing and quantitative multiplex metabolite analysis. PLoS One. 2016;11(6):e0157847. https://doi.org/10.1371/journal.pone.0157847.

Zhang Q, Xing HL, Wang ZP, Zhang HY, Yang F, Wang XC, Chen QJ. Potential high-frequency off-target mutagenesis induced by CRISPR/Cas9 in Arabidopsis and its prevention. Plant Mol Biol. 2018;96(4–5):445–56.

Zhang Z, Feng X, Wang Y, Xu W, Hunag K, Hu M, et al. Advances in research on functional genes of tea plant. Gene. 2019;711:143940. https://doi.org/10.1016/j.gene.2019.143940.

Zhang W, Luo C, Scossa F, Zhang Q, Usadel B, Fernie AR, et al. A phased genome based on single sperm sequencing reveals crossover pattern and complex relatedness in tea plants. Plant J. 2021;105(1):197–208. https://doi.org/10.1111/tpj.15051.

Zhao DW, Yang JB, Yang SX, Kato K, Luo, JP. Genetic diversity and domestication origin of tea plant *Camellia taliensis* (Theaceae) as revealed by microsatellite markers. BMC Plant Bio. 2014;14:1–12.

Zhao M, Su XQ, Nian B, Chen LJ, Zhang DL, Duan SM, et al. Integrated meta-omics approaches to understand the microbiome of spontaneous fermentation of traditional Chinese pu-erh tea. mSystems. 2019;4:e00680-19. https://doi.org/10.1128/msystems.00680-19.

Zhao Z, Song Q, Bai D, Niu S, He Y, Qiao D, et al. Population structure analysis to explore genetic diversity and geographical distribution characteristics of cultivated-type tea plant in Guizhou Plateau. BMC Plant Bio. 2022;22(1):1–14.

Zhou X, Jacobs TB, Xue LJ, Harding SA, Tsai CJ. Exploiting SNPs for biallelic CRISPR mutation in the outcrossing woody perennial populous reveals 4-coumarate:CoA ligase specificity and redundancy. New Phyt. 2015;208(2):298–301.

Zhou Q, Chen Z, Lee J, Li X, Sun W. Proteomic analysis of tea plants (*Camellia sinensis*) with purple young shoots during leaf development. PLoS One. 2017;12(5):e0177816. https://doi.org/10.1371/journal.pone.0177816.

Zhu QW, Luo YP. Identification of miRNAs and their targets in tea (*Camellia sinensis*). J Zhejiang Univ Sci. 2013;14(10):916–23. https://doi.org/10.1631/jzus.B1300006.

Part III
Population Genomics of Major Crop Plants

Population Genomics of Maize

**Marcela Pedroso Mendes Resende, Ailton José Crispim Filho,
Adriana Maria Antunes, Bruna Mendes de Oliveira,
and Renato Gonçalves de Oliveira**

Abstract One of the most explored crop plants in genomics studies is maize (*Zea mays* L.). It has served as a model for developing and incorporating biotechnology and genomics approaches in breeding programs of several other crop plants. From this perspective, the genomic information available in the last decades has helped geneticists and breeders better understand maize evolution and diversity, gene flow, admixture, and inbreeding and outbreeding depression, and identify genes and genetic variants related to many traits, increasing genetic gains in breeding programs. In this chapter, we present a review of population genomics aspects related to the genetic history of the origin and domestication of maize, the genes that underlined this process, and the possible introgressions that originated the multiple maize races known. We present the synthesis of information on genetic diversity of maize races and the heterotic groups that have become the genetic base of temperate and tropical commercial germplasms. The genomic diversity of maize, especially after the release of inbred line B73, is presented by focusing on the advances made regarding maize pan-genomes and epigenomics. We also discuss how population genomics helped maize breeding and highlight some results in genome-wide association studies and genomic selection, although these results are not yet fully exploited in maize germplasm conservation. Finally, we briefly present some future perspectives into the application of population genomics in maize conservation, selection, and breeding. Although many advances have already been made in population genomics of maize, the more we elucidate the multiple aspects of maize genetics, the more questions arise, unraveling new key insights into maize population genomics and breeding.

Keywords Epigenomics · Evolution · Genomic diversity · Genomic selection · Germplasm conservation · Pan-genomes · *Zea mays*

M. P. M. Resende (✉) · A. J. C. Filho · A. M. Antunes · B. M. de Oliveira · R. G. de Oliveira
Escola de Agronomia, Universidade Federal de Goiás, Goiânia, GO, Brazil
e-mail: marcelapmr@ufg.br

1 Introduction

Maize (*Zea mays* L. ssp. *mays*) is a diploid, monoecious, allogamous grass that can be grown on virtually all continents, with genotypes adapted to commercial cultivation at latitudes from 58 °N to 40 °S and elevations from sea level up to 3,800 m (Hufford et al. 2021; Ortiz et al. 2010). It is the cereal with the highest production volume worldwide, with more than 1,100 million metric tons (mmt) in the 2020–2021 season. The USA, China, and Brazil are the main maize-producing countries, representing more than 64% of the global production (USDA 2021). Moreover, the maize crop has significant socioeconomic importance and more than 3,500 forms of direct and indirect use such as in human consumption, as animal feed, and as raw material to produce alcohol, oil, starch, dextrose, and syrups (Môro and Fritsche-Neto 2015).

Maize is also one of the most explored crops in genetics, genomics, and breeding studies and has been used to develop genetic designs and breeding methods later applied to several autogamous and allogamous crops. In addition, it is one of the most used species to incorporate biotechnology into breeding programs, with advanced genome sequencing and a large amount of genomic data available in scientific literature and databases.

This chapter presents a review on the progress made on population genomics of maize, including genetic aspects of the origin and domestication of maize, its genetic and genomic diversity, and the application of population genomics in maize conservation, selection, and breeding. Our interest is not to delve into the available population genomics methodologies as this information can be found in books (Rajora 2019) and papers focused on methods (e.g., Holliday et al. 2018). Instead, we have synthesized the information obtained in recent years on various population genomics aspects, which have contributed to a better understanding of maize evolutionary history, origin, domestication, and genomic diversity, and applied in maize selection and breeding.

We would also like to point out two things. First, we do not aim to cover all the theoretical contents available on maize genetics. As already highlighted, maize is a massively studied species. Therefore, we are aware that many of the topics presented here can be complemented, while other important ones can be added, e.g., population transcriptomics, population genetic structure, and several other aspects of the population genomics of wild and cultivated maize. The second thing is that population genomics like any other science area is constantly evolving. Much of what was discussed regarding the origin, evolution, genetics, and genomics of maize was reshaped as new studies and technologies provided novel discoveries. Therefore, many aspects presented here may be revised in the future. However, these studies will still be relevant since the future does not exist without the past, and knowledge is built brick by brick, experiment by experiment, without frontiers or immediate outcomes. We hope that the reviews presented here can contribute to constructing this knowledge in the future and help train students, teachers, and researchers interested in maize.

2 Origin and Domestication

Maize belongs to the family Poaceae, along with other important agronomic crops, e.g., wheat (*Triticum* spp.), rice (*Oryza sativa*), oats (*Avena sativa*), sorghum (*Sorghum bicolor*), barley (*Hordeum vulgare*), and sugarcane (*Saccharum officinarum*). These main grasses are supposed to have arisen from a common ancestor in the last 55–70 million years (Buckler and Stevens 2006). Among cultivated species, maize evolution is particularly obscure due to its complex domestication history compared to other cereals. While wheat, barley, and rice, for example, show small morphological changes compared to their wild ancestors, maize shows substantial changes compared to its closest relatives, the teosintes, especially with regard to the branching pattern and the structure of the female inflorescence (Fig. 1) (Doebley et al. 2006; Stitzer and Ross-Ibarra 2018). These differences are so substantial that, initially, teosintes were considered botanically closer to rice than to maize (Doebley 2004).

Like maize, teosinte has a main stalk terminated by male inflorescences (tassels) and produces primary lateral branches in one or more nodes of the main stalk. In teosinte, however, the lateral branches are long and terminated by male inflorescences, while in maize, these branches are short and terminated by female inflorescences (ears). In teosintes, the female inflorescences are located at the end of the secondary branches. With regard to the structure of the female inflorescence,

Fig. 1 Teosinte compared to maize. (Reproduced with permission from National Science Foundation. Credit: Nicolle Rager Fuller, National Science Foundation. Source: https://www.nsf.gov/news/mmg/media/images/corn-and-teosinte_h1.jpg)

Fig. 2 Teosinte compared to maize. (**a**) A teosinte female inflorescence (left), which arises as a secondary branch from tillers, and tassel (right). (**b**) An ear (left) and tassel (right) of maize. Size bar in A and B is 10 cm. (**c**) Teosinte kernel (left) and maize kernel (right). The teosinte kernel is hidden by hardened glumes (see Glossary). The maize kernel is exposed and reveals the endosperm (En) and embryo (Em). The embryo is surrounded by the scutellum (Sc), the nutritive tissue of the cotyledon. (**d**) A comparison of teosinte on the left, maize on the right, and the F_1 of maize and teosinte in the middle. Image credits: (**d**) John Doebley, Department of Genetics, University of Wisconsin – Madison; all other images, Sarah Hake. (Reproduced from Hake and Ross-Ibarra (2015))

teosintes produce spikes that disarticulate due to an abscission layer, with 5 to 12 grains protected by a glume, an external and rigid covering. In contrast, maize produces one or two ears that lack abscission layers and, therefore, are intact when ripe, usually showing more than 12 rows of exposed grains (Fig. 2) (Sangoi and Bortoluzzi 2018). From this perspective, modern maize is represented by plants with morphological characteristics that limit its occurrence in nature, highlighting the dependence on human cultivation for survival (Stitzer and Ross-Ibarra 2018).

The morphological differences between maize and teosintes represented a great paradox for botanists in the first half of the twentieth century (Buckler and Stevens 2006). However, advances in genetics, biochemistry, and cytology have revealed similarities between the genomes of maize and some types of teosinte. They have the same number of chromosomes and similar chromosomal morphologies and can be crossed to produce fertile hybrids. Thus, although initially classified as a member of the genus *Euchlaena*, teosinte became a part of the genus *Zea*, the same as maize (Doebley and Iltis 1980; Iltis and Doebley 1980).

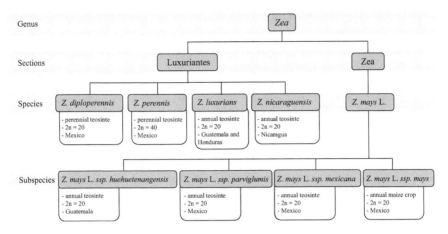

Fig. 3 Sections, species, and subspecies from the *Zea* genus

The genus *Zea* is divided into two sections and five recognized species (Fig. 3). The species *Zea mays* L. includes four subspecies: three wild teosintes and the subspecies *Zea mays* L. ssp. *mays*, which is the cultivated maize we know today (Buckler and Stevens 2006). These species are highly adapted to their local environments and form large populations across Central America (Hufford et al. 2012b).

2.1 Theories on the Origin of Maize

Given the significant taxonomic differences between maize and its closest relatives, especially with regard to plant branching and the morphology of the female inflorescence (Doebley and Iltis 1980; Iltis and Doebley 1980), the explanation for the origin of maize was, for a long time, controversial (Buckler and Stevens 2006), and some theories proposed in the past had flaws. The first theory proposed that maize, teosinte, and *Tripsacum* descended from an extinct common ancestor native to Mexico or Guatemala and evolved individually due to different genetic and environmental mechanisms (Weatherwax 1950). Known as the Common Origin Theory or Divergent Evolution Theory, this hypothesis, in addition to not being testable, did not explain, among other issues, the development of primitive and modern maize races (Sangoi and Bortoluzzi 2018).

Another explanation, known as the Tripartite Theory, proposed that maize could be the ancestor of teosinte (Mangelsdorf 1974; Mangelsdorf and Reeves 1938, 1939). According to this proposal, modern maize evolved from wild maize with similar morphology to the current one. Therefore, this extinct or undiscovered wild maize could be the progenitor of the maize that, after crossing with *Tripsacum*, gave rise to annual teosintes with intermediate morphological characteristics between maize and *Tripsacum*. Thus, the similarities between maize and teosinte would be explained by *Tripsacum*, while their extreme morphological differences would be attributed to an extinct common ancestor (Buckler and Stevens 2006). This theory was widely accepted until the 1960s. However, genetic mapping studies involving these three species showed that *Tripsacum* does not have the genes for the capsule that protects teosinte grains. Therefore, it is genetically impossible for *Tripsacum* to be one of its parents (Sangoi and Bortoluzzi 2018).

The most accepted theory today is that teosinte is the wild ancestor of maize. This hypothesis had already been raised by Beadle (1939), who supported it until the end of his career (Beadle 1980) when it was finally accepted (Doebley 1990). Several genetic and physiological studies carried out at that time showed that, despite showing significant morphological differences, maize and annual teosinte are not only members of the same genus but also belong to the same species (Kato 1984; Kato and Lopes 1990; Kato 1976). The most logical conclusion is that maize originated from conscious or unconscious human selection, gathering beneficial characteristics for human cultivation and consumption and taking advantage of mutations in the loci that determine differences between the subspecies (Buckler and Stevens 2006; Sangoi and Bortoluzzi 2018; Doebley 1990; Lemmon et al. 2014).

2.2 The Genetic History of Maize Origin and Domestication

Phylogenetic analyses suggest a single domestication event, with a single monophyletic maize lineage derived from the teosinte subspecies *Zea mays* L. ssp. *parviglumis* (hereafter called ssp. *parviglumis*) around 9,000 years ago in the narrow lowland region of the Balsas River Valley, Mexico. This teosinte, whose genetic and molecular constitution is similar to cultivated maize, is currently considered its probable direct ancestor (Doebley 1990; Matsuoka et al. 2002; Stitzer and Ross-Ibarra 2018; van Heerwaarden et al. 2011).

The origin of maize from the domestication of teosinte ssp. *parviglumis* goes through the morphological differences between these two subspecies, and five main differences are highlighted in the literature (Doebley 2004; Stitzer and Ross-Ibarra 2018). Teosinte has a cupulate fruitcase formed by an internode and a glume, protecting each kernel; disarticulated ears at maturity such that the individual fruitcases become the seed dispersal units; a single-spikelet (kernel-bearing structure) in each cupulate fruitcase; fruitcases borne in two ranks on opposite sides of the longitudinal axis of the ear; and plants with long lateral branches terminated by male inflorescences. In maize, the undeveloped cupule and glume do not surround the kernel, forming the cob; the ears remain intact at maturity, facilitating harvest; there are two mature spikelets (kernels) on each cupule; cupules are borne in four or more ranks; and plants have short lateral branches terminated by female inflorescences (Figs. 1 and 2). Other traits involved in maize domestication are the loss of seed dormancy (Avendaño López et al. 2011), the loss of photoperiod sensitivity (Huang et al. 2018), and nutritional changes (Hanson et al. 1996; Whitt et al. 2002).

For some time, hypotheses about the genetic architecture of these traits ranged from the belief that they were controlled by a few major-effect loci (Beadle 1939; Mangelsdorf and Reeves 1939) to theories that claimed they were polygenic (Iltis 1983). However, recent studies show a balance between these theories (Stitzer and Ross-Ibarra 2018) as subsequent fine-mapping experiments suggest that these loci are probably just linked collections of genes with larger effects (Quijada et al. 2009). In this scenario, genome-wide analysis has highlighted several thousand genes with evidence of selection during domestication (Wright 2005; Hufford et al. 2012c).

The genes involved in the genetic control of these traits are spread across all ten chromosomes of the species. However, five regions on chromosomes 1, 2, 3, 4, and 5 seem to be responsible for controlling the main phenotypic differences between the subspecies, including about 64–70% of major-effect loci explaining more than 10% of the phenotypic variance of these traits (Briggs et al. 2007; Doebley and Stec 1993, 1991). Although the effects of some genes were essential, different regions across the genome contributed to maize domestication. From this perspective, the origin of maize is due to crucial mutations in these main loci and the movement of mutant genes to a more favorable genetic base controlled by smaller and more numerous loci (Sangoi and Bortoluzzi 2018; Doebley 2004). Dominance, epistasis, and pleiotropy associated with a large proportion (~85%) of transposable elements in the maize genome (Schnable et al. 2009) played a significant role in modulating the phenotypes on which selection acted, which may partially explain phenotypic changes during domestication (Stitzer and Ross-Ibarra 2018; Briggs et al. 2007; Jianming Yu et al. 2008).

Compatible interactions between many loci were necessary to change phenotypes during maize domestication. It is estimated that 52–58% of the genes have overexpressed alleles considering the entire transcriptome of maize compared to the expression pattern of teosinte and that 70–80% of the genes present in maize have allelic dominance over those in teosinte, reflecting the selection to obtain genotypes more suitable for cultivation (Hufford et al. 2012c; Lemmon et al. 2014; Swanson-Wagner et al. 2012; Wang et al. 2018b). As a result, maize experienced a genetic bottleneck that removed a substantial proportion of the diversity found in its ancestor and altered the allele frequencies throughout its genome. Furthermore, the domestication genes underwent a bottleneck ten times as strong as the rest of the maize genome, though an excess of rare alleles indicates that their diversity may be starting to recover (Hufford et al. 2012c).

The estimated size of the maize domestication bottleneck ranges from 5% to 10%, lasting approximately 9,000 generations and only recently starting to recover (Beissinger et al. 2016; Eyre-Walker et al. 1998; Wang et al. 2017b; Wright 2005). Isoenzyme and nucleotide data indicate that maize lost about 25–30% of its genetic diversity compared to teosinte due to the domestication bottleneck (Doebley 2004), resulting in the lack of genetic variability in modern maize for many relevant traits (Briggs et al. 2007; Guanghui Xu et al. 2017; Xue et al. 2016). Approximately 1,200 genes (2–4% of genes) in the entire maize genome have been affected by artificial selection (Wright 2005). Thus, a large part of the maize genome was selected during domestication for a relatively short period, and compatible interactions between many loci probably occurred, explaining why phenotypes changed so much (Stitzer and Ross-Ibarra 2018; Briggs et al. 2007; Hufford et al. 2012c).

2.3 Genes Underlying Domestication

Several studies involving population genomic analyses have confirmed the history of selection in several maize genes. Hufford et al. (2012a, b, c) identified 484 domestication loci, 107 of which were selected during breeding, and 695 improvement loci covering approximately 7.6% of the maize genome. We highlighted some genes that control traits related to plant and inflorescence architecture, plant and kernel color, and kernel composition (Table 1).

One of the first genes identified in the control of domestication traits in maize was *teosint branched*1 (tb1), located on chromosome 1L and associated with the difference in apical dominance between maize and teosinte, changing plant architecture (Doebley et al. 1997). The tb1 maize allele appeared 28,000 years ago, before maize domestication (Studer et al. 2011). It is found at frequencies of up to 44% in teosinte populations, although its presence is not enough to generate a phenotype similar to that of maize, suggesting that other genes also play roles in differentiating the plant architecture of maize and teosinte (Vann et al. 2015). The QTL (quantitative trait locus) region where the tb1 gene is found explains 35.9% of the variation in the number of branches (Doebley and Stec 1991).

The tb1 gene represses the growth of axillary meristems and promotes the elongation of the maize stalk. It is likely to be deactivated or less expressed in teosinte, giving rise to long lateral branches terminated by male tassels. The insertion of a 65-kb-long transposable element near this gene caused its greater expression (Studer et al. 2011), resulting in short lateral branches tipped by female ears and suggesting that long-range chromatin interactions may have played an important role in maize domestication (Li et al. 2019). Tb1 binds to many regions of the genome and is pleiotropic for apical dominance, lateral branch length, leaf growth on lateral branches, development of pedicled spikes, and root architecture, in addition to being closely linked to the cell cycle (Gaudin et al. 2014; Hubbard et al. 2002; Studer et al. 2017).

Like tb1, the *grassy tillers*1 gene (gt1) is associated with a QTL region that controls plant architecture. The maize allele suppresses plant tillering in response to shading as the plant integrates external environmental signals with endogenous ones to control its architecture (Whipple et al. 2011), an essential process in maize domestication. This allele appeared 13,000 years ago and is found segregating at low frequencies in teosinte (Wills et al. 2013).

*Teosinte glume architecture*1 (tga1) is a single gene locus located in a QTL close to the centromere of the 4S chromosome and directly regulated by tb1. It acts in the development of the glume, a protective grain layer in teosinte absent in maize (Studer et al. 2017; Wang et al. 2005). The tga1 is probably the result of the selection

Table 1 Some genes under selection during maize domestication

Gene	Trait	Reference
Genes controlling plant and inflorescence architecture		
tb1 – *teosint branched1*	Apical dominance, tillering, and inflorescence position	Doebley e Hubbard (1997)
gt1 – *grassy tillers1*	Branching architecture	Whipple et al. (2011)
tga1 – *teosinte glume architecture1*	Seed casing	Wang et al. (2005)
ba1 – *barren stalk1*	Transcriptional regulator (bHLH); plant and inflorescence structure	Gallavotti et al. (2004)
ra1 – *ramosa1*	Transcriptional regulator (MYB); inflorescence structure	Vollbrecht et al. (2005)
zfl2 – *Zea Floricaula/Leafy2*	Number of ranks of cupules around the circumference of the ear	Doebley (2004)
tcb1 – *teosinte crossing barrier1*	Pollen-pistil incompatibility	Evans and Kermicle (2001)
tru1 – *tassels replace upper ears1*	Plant architecture	Dong et al. (2017)
krn1 – *kernel row number1*	Kernel row number	Wang et al. (2019)
Genes controlling kernel color		
c1 – *colored aleurone1*	Transcriptional regulator (MYB); kernel color	Hanson et al. (1996)
r1 – *colored1*	Transcriptional regulator (bHLH); kernel color	Hanson et al. (1996)
y1 – *yellow1*	Phytoene synthase; carotenoid content; kernel color	Palaisa et al. (2003)
Genes controlling kernel composition		
su1 – *sugary1*	Isoamylase; sweet corn gene; starch biosynthetic enzymes	Dinges et al. (2001), Whitt et al. (2002)
sh1 e sh2 – *shrunken1* and 2	Pyrophosphorylase; supersweet corn	Bhave et al. (1990)
bt2 – *brittle2*	Starch biosynthetic enzymes	Whitt et al. (2002)
ae1 – *amylose extender1*	Starch biosynthetic enzymes	Whitt et al. (2002)
wx1 – *waxy1*	Starch biosynthetic enzymes	Whitt et al. (2002)

of a new non-synonymous mutation (Wang 2015b) fixed in the last 10,000 years (Wang et al. 2005). The maize allele of tga1 is dominant over the teosinte allele. It is a transcriptional regulator with phenotypic effects on several traits, including cell lignification, silica deposition in cells, three-dimensional growth, and organ size (Dorweiler and Doebley 1997). The rachis internodes of the teosinte female inflorescence are deep, creating an invagination where the grain develops. Longer, thicker, and upward angled glumes form a rigid capsule that protects the grains, making consumption difficult. In maize, the mutation of this gene produces a softer glume and allows these structures to be less developed, being part of the cob where they are inserted and not individually protecting the grains, which facilitates consumption. This was probably one of the first targets of selection performed by Native Americans during maize domestication (Buckler and Stevens 2006). Unlike rice, where tga1 has pleiotropic effects on inflorescences and vegetative structures, the expression of tga1 in maize is restricted to the ear (Preston et al. 2012; Wang et al. 2015a), relieving it of possible restrictions in selection due to pleiotropic effects in other parts of the plant (Stitzer and Ross-Ibarra 2018).

The architecture of the male (tassel) and female (ear) inflorescences reflects the number, arrangement, and activity of the apical meristems of the maize stalk. In this scenario, the *barren stalk*1 gene (ba1) was one of the first to be associated with changes in the pattern of the inflorescence selected for agronomic purposes. It encodes a non-canonical basic helix–loop–helix protein necessary to initiate all lateral shoot meristems in maize, regulating the development of the vegetative lateral meristem together with the bt1 gene (Gallavotti et al. 2004). The development of spikes and branches in the tassel and ear is controlled by the *ramosa*1 gene (ra1), although two others (*ramosa*2 and *ramosa*3) may also participate (Cassani et al. 2006). In addition to these, the *Zea Floricaula/Leafy*2 gene (zlf2), located on the 2S chromosome, seems to have a minor function in increasing the number of rows of cupules in the rachis (Doebley 2004).

Grain color diversity in maize resulted from selection in some genes, among which *yellow*1 (y1), *colored aleurone*1 (c1), and *red color*1 (r1) can be highlighted. While teosinte grains have white endosperm, the yellow endosperm of maize may have originated from a natural mutation in the y1 gene that encodes the phytoene synthase enzyme (PSY), which produces yellow grains with higher carotenoid levels (Palaisa et al. 2003). In the presence of the PSY enzyme, a product of the y1 gene, the carotenoids produced in the endosperm result in a yellow phenotype; in their absence, carotenoids cannot be synthesized, resulting in a colorlessness or white endosperm. The functional difference appears to involve a change in promoter sequences, such that y1, which is normally expressed in leaves (Buckner et al. 1996), is expressed in developing maize kernels (Doebley et al. 2006). Both phenotypes were targets of selection in the past. The yellow phenotype was selected in the 1930s when US farmers quickly adopted maize genotypes with yellow grains after noticing their nutritional advantage since carotenoids are precursors of vitamin A synthesis (Mangelsdorf and Fraps 1931). In contrast, white grains were selected simply due to the preference for white maize products in some cultures (Poneleit 2001).

The color shades of maize kernels result from variants of transcriptional regulators c1 and r1 (Hanson et al. 1996). These genes are part of the anthocyanin biosynthetic pathway, which allows the development of pigmentation in maize kernels. The anthocyanin pathway in maize includes eight enzymatic genes (a1, a2, bz1, bz2, c2, chi, pr, and whp) that catalyze the biosynthesis or transport of anthocyanin, and five regulatory genes (b, GI, pl, r, and vpl) that govern the tissue-specific expression of anthocyanin synthesis. Changes in the regulatory sequences of these genes, probably mediated by transposable elements, are responsible for activating the anthocyanin pathway (Doebley et al. 2006).

Wright (2005) identified 30 candidate genes with putative effects on plant growth and auxin response grouped close to QTL. These genes contribute to phenotypic differences between maize and teosinte. Among them, the author identified a class of genes that participate in the biosynthesis of proteins, which may explain the difference in amino acid composition between maize and teosinte, an important selection target for nutrition.

Maize starch production is one of the most important routes related to grain yield. More than 20 genes are involved in this process, six of which play important roles in this pathway: *amylose extender1* (ae1), *brittle2* (bt2), *shrunken1* (sh1), *shrunken2* (sh2), *sugary1* (su1), and *waxy1* (wx1) (Whitt et al. 2002). The sh1, sh2, and bt2 genes act at the beginning of the pathway by converting sucrose into glucose. Mutant alleles of the ae1 and su1 genes encode enzymes that reduce the amylopectin content, while alleles of the wx1 gene are responsible for reducing amylose production. In general, approximately 75% of the total starch in maize kernels is amylopectin; the remainder is amylose. These proportions range from 100% amylopectin (waxy maize) to 50–45% amylopectin (high-amylose maize). The proportion of amylose and amylopectin influences the appearance, structure, and quality of maize grains. The selection of these genes was essential to increase the yield and quality during maize domestication and breeding (Whitt et al. 2002; Hu et al. 2021).

During its domestication, maize showed a reduction in the genetic diversity of starch synthesis loci. Whitt et al. (2002) reported that the su1, bt2, and ae1 genes had a dramatic reduction (three- to sevenfold) in genetic diversity in the teosinte ssp. *parviglumis*. Most of this genetic diversity is due to rare divergent haplotypes at ae1 and su1. Selection for bt2 and su1 probably occurred before the worldwide spread of maize germplasm; however, the selection is still happening for the ae1 gene.

Mutations in these genes are also responsible for sweet corn varieties, with more sugar and less starch in the endosperm. The mutants of the su1 gene were the first to be commercially explored, but mutations in other genes contributed to the variability of current sweet corn varieties, not only concerning the sugar content but also other traits under selection (Hu et al. 2021). Two independent mutations in the su1 gene suggested at least two independent sweet corn origins (Whitt et al. 2002). However, the presence of three alleles in a conserved haplotype in the recent assembly of the sweet corn genome suggests a potential common origin (Hu et al. 2021).

Buckler and Stevens (2006) argue that, although significant changes occurred in maize due to the effect of a few genes, other modifier genes were needed for the maize plant to evolve into what we know today. Perhaps hundreds or even thousands of genes are involved in increasing ear size, adapting to diverse agricultural environments, and increasing the nutrient content of maize kernels. Many genes involved in the evolutionary process of maize are quantitative and show significant epistatic interactions, highlighting a highly complex process (Sangoi and Bortoluzzi 2018).

2.4 Effects of Crop-Wild Introgression in Maize

After its initial domestication, maize spread rapidly across the Americas, reaching the southwestern USA approximately 4,500 years ago and the coast of South America 6,700 years ago (Kistler et al. 2018; Merrill et al. 2009). Over time, demographic changes in maize and the gene flow between maize and teosintes outside the center of origin caused changes in the maize genome that led to additional losses in genetic diversity during the expansion of the crop, in addition to an increased frequency of deleterious alleles in regions linked to domestication loci in maize populations. The higher mutation rate of maize compared to ssp. *parviglumis* is mainly driven by the domestication bottleneck. This increase is particularly pronounced in Andean maize, which experienced a more significant founder event than other maize populations and underwent more severe mutations (Wang et al. 2017b).

During its diffusion, maize has adapted to several environments that diverge with regard to elevation, latitude, temperature, and precipitation, resulting in a vast differentiation of maize races (Vigouroux et al. 2008). Thus, the maize "born" in lowlands finally reached the Mexican mountains, occupied by the subspecies *Zea mays* L. ssp. *mexicana* (hereafter called ssp. *mexicana*), an upland teosinte that diverged from ssp. *parviglumis* 60,000 years ago, resulting in substantial reciprocal introgression between maize and ssp. *mexicana* (Hufford et al. 2013; Ross-Ibarra et al. 2009; Calfee et al. 2021). This event increased the genetic variability of maize and favored the development of traits that made maize more suitable for cultivation. Little historical gene flow occurred between the teosinte subspecies ssp. *parviglumis* and ssp. *mexicana*, with maize serving as a genetic bridge between them (Hufford et al. 2012b).

The majority of the ancestral traits of ssp. *mexicana* introgressed in maize over 1,000 generations ago, subsequently diverging and being sorted by selection in individual populations (Calfee et al. 2021). The introgression between sympatric populations of maize and ssp. *mexicana* involved traits such as larger spikes and seeds and red hairy leaf sheaths, contrasting with the green and glabrous leaf sheaths of ssp. *parviglumis*. These are essential traits for maize adaptation to the low temperatures of the Mexican highlands (Hufford et al. 2012a; Lauter and Doebley 2002). Approximately 2–10% of the genomes of upland maize populations are estimated to have derived from ssp. *mexicana* (Matsuoka et al. 2002; Warburton et al. 2011), while 4–8% of ssp. *mexicana* derived from maize (Fukunaga et al. 2005), reaching a 20% mixture among the populations of these subspecies (van Heerwaarden et al. 2011).

Hufford et al. (2013) reported a diffuse and asymmetric gene flow in sympatric populations of maize and ssp. *mexicana*. The introgression of ssp. *mexicana* in maize is small or rare in domestication loci (gt1, tb1, tga1, su1, bt2, tcb1) but significant in loci supposedly involved in highland adaptation. Nine introgression regions of ssp. *mexicana* have been identified in maize, three of which span the centromeres of chromosomes 5, 6, and 10, suggesting that Mexican highland maize may harbor centromeric or pericentromeric sequences of ssp. *mexicana*. Two of the introgression regions found are of particular interest: one on chromosome 4, which overlaps a QTL for dark red and highly hairy leaf sheaths, and another on chromosome 9, overlapping a QTL that includes the *macrohairless*1 (mhl1) locus, which promotes hair formation on the leaf and sheath of maize. However, little introgression was observed in the opposite gene flow from maize to ssp. *mexicana* (Hufford et al. 2013).

The introgression of ssp. *mexicana* reduced the prevalence of deleterious alleles in upland maize, probably due to the greater effect size of the ancestor of ssp. *mexicana* and the efficient selection against long-term deleterious alleles (Wang et al. 2017a). The distribution of ssp. *mexicana* and maize in a global diversity panel suggests a broader contribution of ssp. *mexicana* to modern improved maize. Maize colonization of highland environments in Mexico may have been facilitated by adaptive introgression from local ssp. *mexicana* populations (Hufford et al. 2013; Calfee et al. 2021). Thus, Mexican upland maize populations may have given rise to the genotypes currently cultivated in the Americas (Matsuoka et al. 2002; van Heerwaarden et al. 2011).

The genomes of maize varieties found in South America suggest that South American maize was isolated from the wild teosinte gene pool before domestication traits were fixed. Maize may have first arrived in South America as a partially domesticated species, where improved varieties evolved far from the gene flow influence of their wild ancestors. South America may have been a secondary breeding center where new varieties emerged in parallel with the domestication process in Central America (Kistler et al. 2018). Later, hybrids of the South American varieties were probably reintroduced in Central America, increasing the genetic variability associated with favorable agronomic traits in Mexican varieties (Kistler et al. 2020).

3 Genetic Diversity and Population Structure in Maize

Maize is possibly one of the species with the highest genetic diversity among plants and, therefore, one of the most diverse crops in the world (Giordani et al. 2019). Since the beginning of its domestication and dissemination through different regions, new populations have emerged and proposals for classifying these germplasms have been established (Carena et al. 2010). Thus, knowing the population structure and the remaining genetic diversity within maize germplasm pools is crucial for basic research, conservation, and utilization of these genetic resources in the breeding process (Warburton et al. 2011; Adu et al. 2019). This information is directly related to species survival through adaptability to environmental changes and avoidance of genetic vulnerability (Aci et al. 2018; Wen et al. 2012). However, despite the global importance and highly diverse genetic base of maize, the germplasm pool typically used by maize breeders to develop commercial products is still very restricted (Carena et al. 2010).

3.1 Classification of Maize Germplasm

After its domestication, maize expanded across continents, and farmers started to select maize populations in new environments, increasing their yield and adaptability (Prasanna 2012). The new populations, called landraces, have become the basis of maize diversity for many years. Landraces are populations with a historical origin, locally adapted, genetically diverse, lacking formal genetic improvements, and associated with a set of agricultural practices and knowledge base (Perales and Golicher 2014; Villa et al. 2005).

The emergence of landraces was influenced by mutations and geographic or genetic isolation mechanisms, people migration, and germplasm movement to different geographic areas (Carena et al. 2010). Local farmers have historically selected maize landraces mainly due to their tolerance to biotic and abiotic stresses. Some classic examples of landraces are Tuxpeño, La Posta Sequia, Cónica, and Breve de Padilla (drought tolerant), Oloton (tolerant to acidic soils), and the cross between Chalqueño and Ancho de Tehuacán (tolerant to alkaline soils), all originated in Mexico. There are also Reid and Lancaster in the USA, generating the main North American heterotic groups, and the Sikkim population and its variations from India, with great diversity in kernel type and color, popcorn, and high prolificacy (Carena et al. 2010; Prasanna 2012; van Heerwaarden et al. 2011).

Therefore, a continuous gene flow occurred between maize populations with spread worldwide, establishing new gene pools and increasing the genetic and phenotypic variability compared to the original parent germplasms. Several authors have proposed ways to classify this diverse germplasm. Sturtevant (1899) and Kuleshov (1933) classified the maize germplasm into eight groups based on the kernel content, in which the difference in only one gene was necessary to change the classification (Carena et al. 2010):

1. *Z. mays indurata*: flint endosperm, distributed throughout the Western Hemisphere
2. *Z. mays amylacea*: floury endosperm, occurred south of the northern range of flints in North America and the southwestern states of the USA, predominating in the Andean valley of southern Colombia, Peru, and Bolivia; the greatest diversity occurred in Peru
3. *Z. mays indentata*: dent endosperm, occurred in the US Corn Belt and the central and southern states of Mexico
4. *Z. mays everta*: popcorn, collected in multiple localities
5. *Z. mays saccharate*: sweet corn, collected mainly in the central and northeastern regions of the USA and nearly absent in the south and the tropics
6. *Z. mays amylea saccharate*: starchy-sugary endosperm, found in Bolivia and Peru
7. *Z. mays ceratina*: waxy endosperm, restricted to eastern Asia
8. *Z. mays tunicata:* pod, occurred spontaneously in different areas

Subsequently, new classifications were proposed based on many morphological traits and geographical aspects from which these new populations originated or adapted, establishing the term "race" to distinguish populations. The terms "landraces" and "races" are still used as synonyms in maize studies. The emergence of these two terms is a more ideological issue of authors who proposed different classifications of maize germplasm than something more applied. The possible basic difference between them lies in the use of the population by specific production systems. When a farmer establishes a new population adapted to the climate, temperature, elevation, and cultivation conditions of his region, it becomes a landrace. In contrast, a race would be a level above, without so many specifications and establishing a population race based on a sufficient number of common characteristics among individuals to recognize them as a group (Perales and Golicher 2014). However, both races and landraces were the basis for the emergence of what breeders and geneticists know as heterotic groups.

Many catalogs describing maize races from several locations can be found on the USDA website (https://www.ars.usda.gov/midwest-area/ames/plant-introduction-research/home/races-of-maize/). In this scenario, we highlight three classic studies that describe maize races from Mexico (Wellhausen et al. 1952), South America, Brazil (Paterniani and Goodman 1977), and North America (Brown and Goodman 1977).

In Mexico, 59 maize races have already been cataloged (Parra et al. 2006; Sanchez et al. 2000). One of the first studies on the classification of Mexican races was carried out by Wellhausen et al. (1952), who proposed a classification system consisting of 25 Mexican maize races and 7 poorly defined races based on samples collected between 1943 and 1951, a system that is still widely used today (Perales and Golicher 2014). The authors mainly considered not only the geographic distribution but also the morphological, genetic, cytological, physiological, and agronomic characteristics, forming five main groups:

1. Ancient Indigenous Races: Palomero Toluqueño; Arrocillo Amarillo; Chapalote; Nal-Tel
2. Pre-Columbian Exotic Races: Cacahuacintle; Harinoso de Ocho; Olotón; Maíz Dulce
3. Prehistoric Mestizos: Cónico; Reventador; Tabloncillo; Tehua; Tepecintle; Comiteco; Jala; Zapalote Chico; Zapalote Grande; Pepitilla; Olotillo; Tuxpeño; Vandeño
4. Modern Incipient Races: Chalqueño; Celaya; Cónico Norteño; Bolita
5. Poorly Defined Races: Conejo; Mushito; Complejo Serrano de Jalisco; Zamorano Amarillo; Maíz Blando de Sonora; Onaveño; Dulcillo del Noroeste

Paterniani and Goodman (1977) surveyed more than 250 maize races found in Brazil and adjacent areas in their collections and from studies conducted by other authors and reported that about 50% of the races were adapted to low altitudes (0–1,000 m), 10% were adapted to intermediate altitudes (1,000–2,000 m), and 40% were adapted to higher altitudes (greater than 2,000 m). Based on the type of endosperm, approximately 40% of the races had floury endosperm, 30% had flint endosperm, 20% had dent endosperm, 10% were popcorn, and 3% were sweet corn.

Considering their origin, morphology, and geographic dispersion, the maize races were classified into four large groups:

1. Indigenous Races: Pipoca Guarani; Moroti; Caingang; Lenha; Entrelaçado
2. Ancient Commercial Races: Cristal Sulino; Cristal; Canário de Ocho; Cateto Sulino; Precoce; Cateto Sulino; Cateto Sulino Grosso; Cateto; Cateto Nortista
3. Recent Commercial Races: Dente Riograndense; Dente Paulista; Dente Branco; Semi-dentado; Cravo
4. Exotic Commercial Races: Hickory King; Tusón

Brown and Goodman (1977) described some maize races found in three major locations: (1) Mexico and Central America, (2) South America, and (3) the USA. Among the North American races, the authors highlighted nine broad racial complexes:

1. Northern Flints
2. Great Plains Flints and Flours
3. Pima-Papago
4. Southwestern Semidents
5. Southwestern 12 Row
6. Southern Dents
7. Derived Southern Dents
8. Southeastern Flints
9. Corn Belt Dents

Among the racial complexes for the USA, five (Great Plains Flints and Flours, Pima-Papago, Southwestern Semidents, Southwestern 12 Row, and Derived Southern Dents) had little impact on maize breeding in the USA (Carena et al. 2010). Corn Belt Dent is one of these main racial complexes. Its varieties arose by hybridization between the Northern Flints and the Southern Dents. The Northern Flints may have come from the Mexican race Harinoso de Ocho in the southwestern USA, while the San Marcenô and Serrano races came from Guatemala (Brown and Anderson 1947; Galinat and Gunnerson 1963). Southern Dent may have originated from the Mexican Tuxpeño, whereas Derived Southern Dents seem to have arisen from the cross between Southern Dents and Southeastern Flints, Northern Flints and Corn Belt Dents, and played an important role in breeding programs in the southeastern USA and as sources of prolificacy in Corn Belt Dents (Carena et al. 2010).

3.2 Genetic Diversity and Structure of Maize Genetic Resources

Population structure analysis provides a valuable understanding of genetic diversity and helps guide selection. It depends on the study of the occurrence of gene linkage, the mating system, and a series of environmental factors (Carena et al. 2010). Many assessment methods, including pedigree data and morphological and molecular markers, have been used to estimate maize genetic diversity (Adu et al. 2019; Badu-Apraku et al. 2021; Giordani et al. 2019; Mutegi et al. 2015; Stanley et al. 2020).

Many studies have been carried out in different regions and breeding programs to understand how the genetic diversity of maize is structured and its potential use. These approaches mainly considered the types of maize grains (dent and flint) (Dinesh et al. 2016; Adu et al. 2019; Badu-Apraku et al. 2021; Shu et al. 2021; van Inghelandt et al. 2010), landrace populations (Liu et al. 2010; McLean-Rodríguez et al. 2021; Nelimor et al. 2020; Vega-Alvarez et al. 2017), sweet corn (Hu et al. 2021; Mahato et al. 2018), and popcorn (Yu et al. 2021).

Characterizing, cataloging, and conserving these landrace populations have become crucial to maintain the genetic diversity of maize and allow breeding programs for current and future changes in the production systems and edaphoclimatic conditions. Breeding programs explore much of this genetic diversity contained within and among landrace populations. The ability of these populations to withstand drastically extreme environments due to their evolution process and the presence of unique alleles makes them essential genetic resources for germplasm management and to identify useful varieties for the genetic improvement of maize (Aci et al. 2018; Wen et al. 2012). In general, genetic diversity conservation in landrace populations is mainly done by traditional farmers, associations, and public agencies (Vega-Alvarez et al. 2017).

Maize plants can vary from 0.5 to 5 m in height at flowering, reaching maturity between 60 and 330 days after planting, producing from 1 to 4 ears per plant and 10 to 1800 grains per ear, yielding from 0.5 to 23.5 t ha^{-1}, and showing wide variation in characteristics of interest for consumers, e.g., quality and flavor (Ortiz et al. 2010; Wen et al. 2012). According to the latest survey by the Global Crop Diversity Trust (GCDT 2007), the total number of unique maize germplasm accessions exceeds 27,000 in this consortium. Moreover, according to the Maize Genetic Cooperation Stock Center (or USDA-ARS GSZE) of the Department of Plant Sciences at the University of Illinois (USA), nearly 100,000 accessions of maize mutants are conserved and characterized in their facilities, forming an extensive collection available for maize breeders around the world. Data describing this collection of mutants can be accessed at Maize Genetics and Genomics (MaizeGDB; http://www.maizegdb.org; verified 2021). These mutants represent the enormous diversity of maize kernel types, e.g., dent, flint, sweet corn, popcorn, and a massive variety of kernel colors and shapes (Hu et al. 2021; Prasanna et al. 2010).

3.3 Genetic Diversity in Heterotic Groups of Maize

The maize seed market is dominated by large multinational companies that mainly trade hybrids. To increase the probability of obtaining better hybrids, breeders allocate lines into different germplasm groups, called heterotic groups, expecting to explore heterosis or hybrid vigor when crossing lines from different groups (Adu et al. 2019; Badu-Apraku et al. 2021; Shu et al. 2021).

The genetic divergence of heterotic groups is maintained by recycling lines within these groups to produce new elite lines. Open crossing between lines generates new populations from which new lines are selected and crossed with lines from other heterotic groups (testcross). The goal is to generate elite lines with a high frequency of favorable alleles, complementary to lines from other groups. Since lines are judged by their performance in a testcross, greater genetic divergence between heterotic groups is expected as the result of selection over time, improving the performance of the hybrid (Beckett et al. 2017).

The International Maize and Wheat Improvement Center (CIMMYT) first divided their heterotic groups by grain types (dent or flint), maturity (early, intermediate, and late), and kernel colors (white and yellow) (Ortiz et al. 2010). Heterotic group development in tropical maize at CIMMYT began recently in the 1980s. However, most of the CIMMYT lines are derived from broad germplasm pools, populations, and open-pollinated varieties; therefore, heterotic patterns in tropical maize are still not clear (Wu et al. 2016b).

Genetic diversity and heterotic groups can also be assessed using molecular markers, e.g., microsatellites (simple sequence repeats; SSR) and single nucleotide polymorphism (SNP) (Aci et al. 2018; Almeida et al. 2011; Hufford et al. 2021; N'DA et al. 2016; van Inghelandt et al. 2010). The molecular characterization of CIMMYT lines from dent and flint groups showed a clear population structure and three major environmental adaptation groups (lowland tropical, subtropical/mid-altitude, and highland tropical subgroups). However, the GBS SNPs were unable to separate the heterotic groups established based on combining ability tests, probably due to the diverse origin and incomplete pedigree information, the marker system used, the shorter hybrid-oriented breeding history for tropical germplasm, and the use of different testers across breeding programs. The same line can be assigned to the flint or dent group depending on the test used, which may result in the mixing of heterotic groups (Wu et al. 2016b).

A study in Brazil used 81 SSR markers in 90 tropical germplasm lines and revealed three tropical heterotic pools, which have been used by maize seed companies in Brazil and worldwide (Lanes et al. 2014). Another study, using tens of thousands of SNPs markers and aiming to identify potential high-temperature tolerant lines, observed genetic variability between maize lines from Western and Central Africa with the potential to contribute with new beneficial alleles to maize breeding programs in the tropics (Adu et al. 2019).

Some studies established heterotic groups within North American germplasm, but many are repeated or sub-grouped, suggesting a narrow genetic base in the American hybrid production chain (Table 2). The Iowa Stiff Stalk Synthetic or BSSS is an example of the restricted variability explored by breeders. It is the most successful germplasm in history, formed using only sixteen inbred lines. Inbred line B73, used to obtain the first complete assembly of the maize genome, is derived from recurrent selection cycles in BSSS. Thus, landrace populations and exotic and open-pollinated varieties are rapidly disappearing due to the widespread use of germplasms with

Table 2 Heterotic groups proposed for American germplasm

Heterotic groups	Reference
Reid Yellow Dent, Lancaster, and other subgroups	Gethi et al. (2002)
Seven ancient key lines: B73, Mo17, PH207, PHG39 (from B37), LH123Ht, LH82, and PH595	Mikel and Dudley (2006)
B73, Mo17, PH207, A632, Oh43, B37 and their mixture	Nelson et al. (2008)
Iowa Stiff-Stalk Synthetic (BSSS) and non-Iowa Stiff-Stalk Synthetic (non-BSSS)	Lu et al. (2009)
BSSS (B14, B37, B73) and non-BSSS (Iodent, Oh43, Mo17, and other subgroups)	Bernardo (2014)
Stiff Stalk (SS), Non-Stiff Stalk (NSS), Tropical-Subtropical (TST), and MIXED subpopulations	Cui et al. (2020)

narrow genetic bases and the recycling of lines in hybrid breeding programs, resulting in a worrisome genetic vulnerability for maize (Carena et al. 2010).

The reduction of genetic diversity in commercial hybrid programs is quite clear. Commercial hybrids only differ by transgenic events since their parent lines or populations are related (Carena et al. 2010). From this perspective, hundreds of inbred lines from 28 different breeding companies, which dominate the North American seed market, have been structured into just three main heterotic groups: Stiff Stalk, Non-Stiff Stalk, and Iodent (Fig. 4), confirming the previously presented main heterotic groups (Table 2) of North American germplasm (Beckett et al. 2017).

Maize germplasm can also be categorized according to its origin, either temperate or tropical. The essence of US germplasm consists of temperate lines, whereas lines from South America and Africa are mainly tropical. Lines from Mexico and Asia show less differentiation between these groups. However, even tropical lines are related to temperate lines, especially classical inbred lines from the US germplasm, such as B73 and Mo17 (Bernardo 2014). This relationship is explained by maize dissemination to a wide geographic range. Tropical and temperate maize are estimated to have diverged approximately 3,000 to 5,000 years ago, and the effective size of each population decreased after division. Many genomic regions identical by descent are shared by tropical and temperate lines, probably due to gene flow between them or the artificial selection during maize domestication and improvement (Li et al. 2017).

It is eminent how restricted is the exploitation of genetic diversity by maize breeders around the world. This diversity is mainly structured among and within maize landrace populations instead of germplasm banks and heterotic groups of breeding institutions. Therefore, obtaining information about genetic diversity and its structure in breeding populations is essential for any breeding program to develop outstanding marketable products (Adu et al. 2019).

Software, markers, and genotyping platforms are no longer obstacles to understanding the dynamics of genetic diversity in a crop as studied and cultivated as maize. The main limiting factor is to perform an accurate high-throughput phenotyping for the comprehensive and efficient characterization of global maize

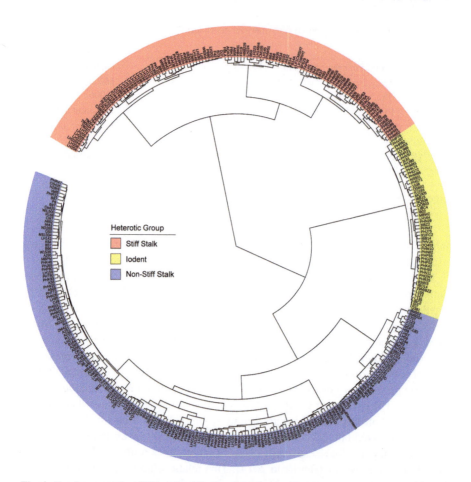

Fig. 4 Dendrogram of ex-PVP and public founder inbreds. Circular dendrogram of ex-PVP and public founder inbreds divided into three heterotic groups. This dendrogram, shown with relative scaled branch lengths and colored according to generally known maize heterotic groups, is based on a cluster analysis using Ward's minimum distance variance method on the matrix of Nei's genetic distance (Nei, 1972; Ward, 1963). Scaled branch lengths allow a visual representation of the relative proportion of genetic difference between the three main heterotic groups. Consultation of available pedigrees confirm the accuracy of heterotic group placement for individual inbreds (Mikel and Dudley 2006; Gerdes et al. 1993; Cross 1989). Note: this tree is presented in a rooted format with the primary purpose of illustrating genetic distance while retaining legible inbred names. While no inference is made about common ancestors, the Stiff Stalk and Iodent/Non-Stiff Stalk portions form an ingroup/outgroup interaction, thus ensuring that the presentation of a tree in rooted format is still an acceptable depiction of the detailed population stratification. https://doi.org/10.1371/journal.pone.0189277.g008. (Reproduced with permission from Travis J. Beckett)

germplasm for key traits, particularly biotic and abiotic stress tolerance and nutritional quality (Prasanna 2012). Some global networks have been created to obtain phenotypic data on a large scale and share them, e.g., the Seeds of Discovery-CIMMYT, the Latin American Maize Program (LAMP), and the US-Germplasm Enhancement of Maize (US-GEM) Project.

Phenotyping and genomic sequencing analyses showed that inbred lines historically used as male and female parents in heterotic groups show both convergent and divergent changes for different sets of agronomic traits. Increased genetic differentiation between female and male lines may have occurred across the breeding eras, showing a positive correlation between increasing heterozygosity levels in the differentiated genes and heterosis in hybrids (Li et al. 2022). This enormous availability of genetic diversity and genomic information is essential to understand the basis of modern hybrid maize breeding and to develop elite maize lines, new varieties, and superior hybrids that can guarantee food security.

4 Maize Genome and Its Structure and Genetic Diversity

The genome sequencing of maize is one of the most advanced among cereal crops, with about 2.3 Gb of nucleotide sequence information assembled and annotated for inbred line B73. Several versions of the B73 genome are available in databases, and the most recent was obtained using modern technologies, such as real-time single-molecule sequencing (PacBio's SMRT) and high-resolution optical mapping (Schnable et al. 2009; Jiao et al. 2017). This genome has an exceptional diversity of indels and related variation, including copy-number variation and presence-absence variation (Wang and Dooner 2006; Swanson-Wagner et al. 2010), which may be a rich source of phenotypic diversity.

Epigenetic studies were also carried out for B73 with emphasis on DNA methylation mechanisms, modification of histone proteins, and the regulatory function of some RNAs. The molecular control of epigenetic mechanisms is still not well understood. However, emerging studies have revealed that epigenetic variation is one of the key factors in biological diversity evolution (Yu et al. 2020). Maize is one of the cultivated species with the highest genetic variability, and pan-genome analysis has revealed extensive structural variations within the species (Hufford et al. 2021).

4.1 The Maize Genome

As mentioned before, the genome sequencing of maize is one of the most advanced among cereal crops, with a large amount of genomic data available in the scientific literature and databases. Maize genome sequences and annotations are available in GenBank (https://www.ncbi.nlm.nih.gov/genome/?term=Zea+mays#) and the Maize Genetics and Genomics Database (MaizeGDB) (http://www.maizegdb.org).

The first complete assembly of the maize genome was published in 2009 using inbred line B73 based on Sanger sequencing data (Fig. 5) (Schnable et al. 2009). The initial genome assembly had 2.3 Gb of nucleotide sequence information, of which 85% were annotated as transposable elements (TE). TEs are DNA sequences capable

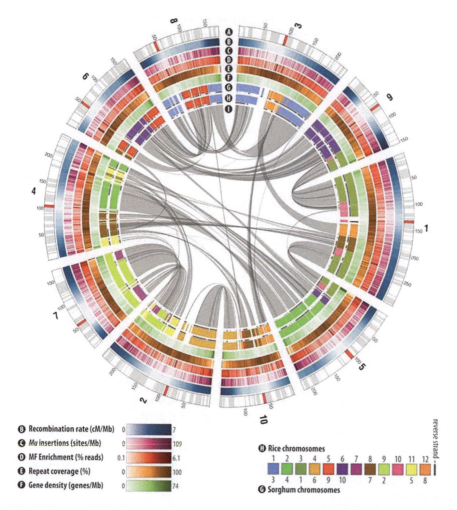

Fig. 5 The maize B73 reference genome (B73 RefGen_v1): Concentric circles show aspects of the genome. Chromosome structure (**a**). Reference chromosomes with physical fingerprint contigs (11) as alternating gray and white bands. Presumed centromeric positions are indicated by red bands (31); enlarged for emphasis. Genetic map (**b**). Genetic linkage across the genome, on the basis of 6,363 genetically and physically mapped markers (14, 19). Mu insertions (**c**). Genome mappings of nonredundant Mu insertion sites (14, 19). Methyl-filtration reads (**d**). Enrichment and depletion of methyl filtration. For each nonoverlapping 1-Mb window, read counts were divided by the total number of mapped reads. Repeats (**e**). Sequence coverage of TEs with RepeatMasker with all identified intact elements in maize. Genes (**f**). Density of genes in the filtered gene set across the genome, from a gene count per 1-Mb sliding window at 200-kb intervals. Sorghum synteny (**g**) and rice synteny (**h**). Syntenic blocks between maize and related cereals on the basis of 27,550 gene orthologs. Underlined blocks indicate alignment in the reverse strand. Homoeology map (**i**). Oriented homoeologous sites of duplicated gene blocks within maize. (Reproduced from Schnable et al. (2009). https://www.science.org/doi/10.1126/science.1178534. Reproduced with permission from Richard K. Wilson)

of transposing, i.e., moving in order to be inserted into new chromosomal locations. Maize TEs have been classified as retrotransposons, DNAs transposons, and rolling circles (RC). Retrotransposons were found to be the most frequent elements, making up 75% of the nucleotide sequences in the maize genome. This class of TE includes long terminal repeat elements (LTR) and the LTR *Gypsy* and LTR *Copia* subfamilies (Schnable et al. 2009).

A total of 406 retrotransposon families with non-uniform distribution in the genome were identified in this assembly, each with at least ten copies. Copia elements were found mainly in gene-rich euchromatic regions, whereas Gypsy elements were overrepresented in gene-poor heterochromatic regions. The centromeric regions showed variable amounts of retrotransposons and satellite repeats (CentC).

There was an intense proliferation of LTR retrotransposons during the evolution of the maize genome, significantly contributing to expanding the genome size of the species. Several TE superfamilies, e.g., CACTA (Spm/En), hAT (Ac), PIF/Harbinger, and Mutator, were discovered for the first time in maize by Schnable et al. (2009).

The high repetitive content of the maize genome, composed mainly of TEs, creates computational challenges to obtain a contiguous and precise genome assembly (Haberer et al. 2005). It makes it harder to annotate other elements that constitute the genome, including gene sequences. Approximately 32,000 protein-coding genes and some microRNA (miRNA) were identified in the first assembled version of the genome. The exon sizes of maize genes were close to those reported for sorghum (*Sorghum bicolor*), which is phylogenetically close to maize; however, the introns were larger. The existence of large introns in maize genes can be explained by the insertion of repetitive elements (Schnable et al. 2009).

The original assembly of the maize genome was the reference for many studies and allowed significant progress in maize genomics. However, this assembly consisted of 100,000 short contigs, many of which were arbitrarily ordered and oriented on the ten chromosomes of the species (Jiao et al. 2017). Advanced sequencing technologies and bioinformatics allowed several other maize genome assemblies and the improvement of the reference genome. In this scenario, 45 assembled versions of the maize genome are currently deposited in the GenBank database, some with a complete representation of the genome. The last two reference sets were released by the NCBI RefSeq in 2017 and 2020, both for inbred line B73.

The maize genome assembly published in 2017 was obtained using single-molecule real-time sequencing (SMRT Sequencing – PacBio) and high-resolution optical mapping. Next-generation sequencing technologies, e.g., PacBio, generate nucleotide sequence data on a fast, low-cost, and large scale, providing long nucleotide sequences that facilitate assembling genomes with large repetitive fractions. The PacBio data obtained for the maize genome allowed assembling 2,958 contigs, half of which were larger than 1.2 Mb. These data were integrated with high-resolution optical mapping to create a hybrid assembly with 625 scaffolds, a highly sensitive methodology used to study repetitive DNA sequences from genomes. Super-scaffolds were built without the help of the BAC (bacterial artificial chromosomes) physical map that guided previous genome assemblies. The final mount had 2.1 Gb and a considerable increase in contig length compared to the previous assemblies. Furthermore, improvements were obtained in the assembly of intergenic spaces and centromeres, and the orientation of pericentromeric regions was corrected. Telomeric sequences were identified on 14 of the 20 chromosomal arms (Jiao et al. 2017).

Substantial improvements were also achieved in the resolution of gaps and correction of gene order and orientation (Jiao et al. 2017). This genome assembly annotated 39,324 protein-coding genes, an average of four exons per gene, and 156 bp in exon length. The 3′UTR and 5′UTR regions were annotated for 25,383 and 26,035 genes, respectively. The cDNA sequencing data improved gene annotation, identifying 131,319 transcripts that correspond to an average of 3.3 transcripts per gene, with 1,281 bp in mean length. The assembly also improved the coverage of regulatory sequences, identifying promoter regions more accurately (Jiao et al. 2017).

TEs corresponded to 64% of the assembled genome, 59.98% of which consisted of LTR retrotransposons. A total of 184,067 TE copies were identified, some exclusive to maize. First, the retrotransposons were sequentially nested in some stretches of the genome. Then, the retrotransposons interrupted by the insertion of other transposable elements, e.g., DNA transposons and helithrons, were noted (Jiao et al. 2017). Helithrons are rolling-circle transposons, and the characterization of TEs present in maize is important since they influence the size, organization, and evolution of genomes.

The genomic information of maize is constantly updated, and a current reference version has been available since 2020 (assembly Zm-B73-REFERENCE-NAM-5.0). Launched by 25 founders of the NAM Consortium, the assembly was produced using long PacBio reads and the mate-pair sequencing strategy. The scaffolds were validated by optical mapping (Bionano) and sorted and oriented using data from the maize pangenome. This assembly had 687 scaffolds and a total string length of 2.1Gb. The scaffolds were anchored to the ten maize chromosomes, with no gaps between them. The DNA sequences of chromosomes have an average size of 213 Mb, a maximum size of 308 Mb, and a minimum size of 152 Mb. The GC content of the entire genome is 46.8%. A total of 47,029 genes and 5,006 pseudogenes were annotated. Due to the complexity of the maize genome, the new assembly obtained still has some gaps to be unraveled in the future, and a complete and accurate reference genome will provide important tools to characterize the structural and functional genetic variations of maize (Hufford et al. 2021).

4.2 The Maize Epigenome

The sequencing and assembly of the maize reference genome also provided advances in epigenomics, a field that studies epigenetic elements, genome modifications that do not involve changes in the DNA sequence but alter gene expression and cell phenotype (Moler et al. 2018). Several studies have been carried out in recent decades to understand the molecular mechanisms of epigenetic regulation in maize. The main epigenetic mechanisms studied were DNA methylation, histone protein modification, and the regulatory function of some RNAs (Yu et al. 2020).

DNA methylation involves the transfer of a methyl group to the 5-position of cytosine. In plants, this covalent modification occurs in the symmetric dinucleotide CG, in CHG, or in the asymmetric CHH, where H can be the nucleotide with the A, T, or C base (Chan et al. 2005). The transfer of the methyl group from S-adenosyl-L-methionine to cytosine bases is catalyzed by methyltransferase enzymes. The Zmet1, Zmet2, Zmet3, Zmet5, and Zmet7 genes from maize encode methyltransferase enzymes (Li et al. 2014b). Some enzymes establish de novo methylations, while others copy methylation patterns during DNA replication. Thus, methylation patterns are inherited after each cell division. There are also demethylase enzymes, which remove methyl groups from DNA. Methylation is a dynamic epigenetic marker that varies spatially and temporally. Modifications in DNA methylation patterns allow the epigenetic regulation of gene expression. In most cases, methylation is associated with gene silencing as it prevents the recognition of methylated regions by the transcription machinery (Yu et al. 2020).

Some studies have produced methylation profiles of the genome of maize inbred lines B73, Mo17, and W22 (Yu et al. 2020). Methodologies to investigate DNA methylation have significantly advanced in recent years, and it is currently possible to study methylation on a genome-wide scale. The technique of DNA sequencing treated with sodium bisulfite was developed in 2008 and is still widely used today. Its principle is to use sodium bisulfite to convert unmethylated cytosines into uracil. Then, high-throughput DNA sequencing is compared to the reference genome of the species. All methylated cytokines are read as cytokines after sequencing since they are not converted into uracil during the reaction with sodium bisulfite (Ji et al. 2015).

Approximately 30% of the cytokines present in the maize genome are methylated. CG and CHG methylation are more abundant in TE sequences and less abundant in the gene context. CHH methylation is found in some TE types and in gene transcription start sites. Methylation is an epigenetic mark that varies according to the stage of development of the maize plant. It is observed less frequently during the embryonic stage, increasing moderately with tissue aging. During early embryonic development, genome demethylation confers totipotency to cells and, as development progresses, epigenetic marks inherited from parents are reestablished to promote cell differentiation. A differentiated cell achieves an epigenetic signature that reflects its phenotype. In this context, DNA methylation was analyzed in the leaf, embryo, and endosperm of the B73 and Mo17 inbred lines, revealing some differences in the methylation levels of the different tissues. Also, the B73 and Mo17 methylation profiles were significantly different (Yu et al. 2020).

Histone modifications are also epigenetic mechanisms that alter gene expression. In chromatin, the histone proteins are organized as an octamer (nucleosome) with two histones of each type, H2A, H2B, H3, and H4, allowing DNA condensation. Each histone has a globular domain and an N-terminal tail that can undergo modifications, e.g., acetylation, phosphorylation, methylation, and glycosylation, among others. These covalent modifications in histones change the charges and physical properties of chromatin, increasing or decreasing DNA accessibility to the transcription machinery. For example, histone acetylation by histone acetyltransferase enzymes activates chromatin for transcription, promoting chromatin remodeling and allowing DNA interactions with the proteins involved in transcription. When transcription of a gene is no longer required, acetylation is reduced by histone deacetylase (HDAC) enzymes. Thus, histones associated with active gene nucleosomes are rich in acetyl groups (so-called hyperacetylated), whereas inactive genes are hypo-acetylated (Dion et al. 2005; Yu et al. 2020).

More than 60 types of histone modifications were detected in maize, including acetylation, phosphorylation, ubiquitination, methylation, proline isomerization, and ADP-ribosylation. Most studies used specific antibodies or mass spectrometry (Yu et al. 2020). Histone modifications are important for normal maize development, and the down-regulation or overexpression of the histone deacetylase gene HDA101, for example, results in morphological and developmental defects. These epigenomic modifications regulate the development of lateral roots, floral meristems, and inflorescence meristems in maize (Li et al. 2011).

Histone modifications are epigenetic mechanisms that play important roles in maize responses to environmental stresses. Soil salinization, for example, affects maize growth and yield, and the adaptive cellular response of this crop to salinity is an increased expression of two histone acetyltransferase genes, ZmHATB and ZmGCN5. The increase in histone acetylation was also observed in maize grown under drought, increasing the expression of the ZmDREB2A gene and leading to tolerance to osmotic stress (Yu et al. 2020).

Histone methylation patterns and modifications are interacting epigenetic mechanisms. The methyl groups in DNA bind to several proteins, including proteins that alter the local conformation of chromatin, e.g., deacetylases. Hypermethylated DNA is usually associated with hypoacetylated chromatin. Transcription is activated when DNA is hypomethylated and histones are hyperacetylated. In contrast, transcription is repressed when DNA is hypermethylated and histones are hypoacetylated. Thus, epigenetic modifications are important mechanisms to control gene expression (Moler et al. 2018; Yu et al. 2020).

Epigenetic regulation is responsible for genetic imprinting, which results in the expression of maternal or paternal alleles only. Imprinted genes are silenced by DNA methylation and local histone modifications. In maize, the overexpression of the *colored*1 (r1) gene promotes pigment accumulation in the aleurone cells of maize endosperm. Thus, when the maternal allele is expressed, the plant produces colored kernels; when the parental allele is expressed, the plant produces colorless or mottled kernels. Hundreds of candidate genes for genetic imprinting have been identified in maize endosperm (Yu et al. 2020).

Some non-coding RNAs also play a role in controlling epigenetic phenomena. They regulate gene expression by interacting with chromatin-modifying complexes. Interference RNAs are double-stranded RNAs that act in post-transcriptional silencing as they bind to complementary nucleotide sequences in target messenger RNAs to inhibit translation and induce messenger RNA degradation. Interference RNAs also recruit enzymes that carry out chromatic modifications, e.g., DNA methylation and histone acetylation. Therefore, they also act in the regulation of epigenetic alterations. In maize, interfering RNAs participate in development regulation and adaptation to biotic and abiotic stress. RNAs also control the activation and repression of TEs (Wang et al. 2017a).

The molecular control of epigenetic mechanisms is still not well understood, but the interest in epigenomics is growing among researchers. Emerging studies have revealed that epigenetic variation is one of the key factors in biological diversity evolution. Epigenetic variation is associated with the regulation of gene expression and contributes to phenotypic diversity. Thus, epigenomic information is useful to improve crop quality and yield, and identifying the maize epigenomic landscape may provide evidence of silent epialleles, which can be reactivated to generate morphological variations. Thus, identifying epigenomic marks and their biological functions in maize can be used in future genetic manipulation to improve several traits of interest (Yu et al. 2020; Ji et al. 2015).

There is also evidence for epigenetic variation between maize populations with relatively stable trans-generational inheritance. Approximately 700 differentially methylated regions (DMRs) were identified between B73 and Mo17, mostly controlled by *cis*-acting differences and exhibit relatively stable inheritance, indicating examples of pure epigenetic variation that is not conditioned by genetic differences (Eichten et al. 2011). Over 300 genes with expression patterns significantly associated with DNA methylation variation were identified among maize genotypes, suggesting that DNA methylation variation is influenced by genetic and epigenetic variation and can influence expression level of genes in the population (Eichten et al. 2013). Comparing whole-genome sequencing (WGS) and whole-genome bisulfite sequencing (WGBS) data on populations of modern maize, landrace, and teosinte (*Zea mays* ssp. parviglumis), Xu et al. (2020) identified correlated DMRs populations with recent selection, suggesting that methylation variation also influence adaptive evolution. The epigenetic variation among individuals of the same or related species may provide important contributions to phenotypic variation even in the absence of genetic differences.

4.3 The Maize Pan-Genome

Most of the genomic information currently available for maize belongs to inbred line B73, chosen to generate the reference genome of the species. However, improvements in sequencing technologies, genome assembly, and annotation methods have allowed the development of multiple genomes from different individuals within a species (Coletta et al. 2021). A single reference genome typically samples only a fraction of the species genome, while high-quality genomic data have been generated for other maize inbred lines. The study of several genomes (pan-genomes) is important to characterize the extensive structural variation within a species, especially in crop plants such as maize (Hufford et al. 2021).

Hufford et al. (2021) described the assembly and annotation of 25 founder genomes of the NAM (Nested Association Mapping) population and an enhanced B73 genome. The 26 genomes were sequenced using PacBio, the reads were assembled into contigs using a temperature-tolerant hybrid approach, and sequencing errors were corrected using Illumina sequencing reads. Scaffolding used Bionano optical maps, and the scaffolds were ordered into pseudomolecules using data from recombinant NAM inbred lines and maize pan-genome markers. The mRNA from ten different tissues for each maize line was also sequenced to support gene annotation (Hufford et al. 2021).

The pipeline revealed an average of 40,621 protein-coding genes and 4,998 non-protein-coding genes per maize genome. A total of 103,538 genes were identified among the 26 genomes. They also compared the gene catalog and pooled genes with high sequence similarity. Of the total genes, 58.39% were present in all 26 genomes, 8.22% were present in 24–25 genomes, 31.75% were present in 2–23 genomes, and 1.64% were present in only one genome. A total of 16,267 genes had tandem duplications in at least one genome. Thus, the maize genomes varied in gene content due to tandem duplicated genes and the presence and absence of specific genes (Hufford et al. 2021).

The 26 maize genomes were also compared for repetitive elements. The authors identified 27,228 transposable element families, of which 59.7% were present in all 26 genomes and 2.5% were unique to one genome. *Gypsy* retrotransposons were more abundant in pericentromeric regions, while *Copia* retrotransposons were more frequent on chromosomal arms. Tropical maize lines had more *Gypsy* retrotransposons than temperate lines. Some maize lines showed more than 15% of tandem repeats. CentC centromere repeats and the telomere TR-1 repeats were also identified (Hufford et al. 2021).

The comparison between the 25 maize lines and the B73 genome revealed 791,101 structural variants larger than 100 bp in size. About 49% of structural variants were larger than 5Kb, and 25% were smaller than 500 bp. Tropical lines had more structural variants than temperate lines, which are more similar to B73. Among structural variants, 35 inversion polymorphisms and five insertion-deletion polymorphisms larger than 1 Mb explained a high percentage of maize phenotypic variance, which is relevant for suitability. These structural variations are associated, for example, with tolerance to abiotic and biotic stresses and the flowering period (Coletta et al. 2021). Phenotypic variation among the 26 maize plants whose genomes were sequenced and assembled can also be explained by the presence of many single nucleotide polymorphisms (SNPs). Thus, characterizing structural variations is important for domestication and can facilitate future crop improvement efforts (Hufford et al. 2021; Coletta et al. 2021).

Another comparative study between maize genomes was carried out by Hu et al. (2021) with sweet corn (Ia453-sh2). The authors compared the assembled genome with six other previously assembled field maize genomes (B73, Mo17, W22, EP1, F7, and DK105). The sweet corn genome was obtained by single-molecule, real-time, long-read sequencing (PacBio SMRT), BioNano optical mapping, and Hi-C Dovetail mapping technologies. The final assembly was 2.29 Gb long and was divided into ten superscaffolds (pseudochromosomes) with 2.11 Gb and 8.440 unassigned contigs with 177.3 Mb. Flow cytometry analysis estimated that the genome size of Ia453-sh2 was 4.8% larger than that of the B73 genome (Hu et al. 2021).

Gene annotation was supported by RNA-seq data. Of the 38,384 genes identified in the sweet corn genome (Ia453-sh2), 22,322 are shared by all seven genomes (sweet corn and field maize genomes), of which 7,864 protein-coding genes are highly conserved. A total of 16,667 genes did not have orthologs in at least one of the studied genomes, and 5,545 genes were unique to only one line. The authors identified an average of 148 genes specific to sweet corn (Ia453-sh2), six of which were found exclusively in the sweet corn genome (Hu et al. 2021).

Transposable elements (TEs) showed a homogeneous distribution on the ten sweet corn pseudochromosomes, slightly increasing near the centromeres. The 2,647,709 TEs belonging to 17 superfamilies occupy 1.69 Gb (82.69%) of the genome. Retrotransposons corresponded to 69.85% and DNAs transposons to 12.84%. The authors also annotated 843,793 LTR *Gypsy* (784 Mb) and 456,321 LTR *Copia* (420 Mb). Sweet corn has fewer TEs than the six field maize genomes. In contrast, some TE superfamilies, e.g., LTR *Copia* and TIR Pif/Harbinger, are more frequent in sweet corn (Hu et al. 2021).

The alignments between the sweet corn genome and the six field maize genomes (pair-by-pair alignment) resulted in an average syntenic match of 68.95%. Each alignment identified, on average, 806,274 small insertions and deletions (indels <100 bp) and 23,664 large structural variations (100–100,000 bp). An important structural variation identified in sweet corn is the mutation of the sh2 allele. This mutation conditioned a loss of gene function, increasing the sugar content in the endosperm. Two structural rearrangements modified the sh2 allele in sweet corn: (1) the first half of the gene was split by the insertion of an LTR *Copia* retrotransposon and (2) the second half of the gene was inverted and separated from the first part by some transposable elements, including an LTR *Copia* and an LTR *Gypsy*. These findings highlight the importance of comparative studies to elucidate the genomic divergence between maize lines (Hu et al. 2021).

Bornowski et al. (2021) sequenced the genomes of six maize inbred lines that represent the diversity and history of the stiff-stalk heterotic group to understand the genomic diversity of this group and support genetic and functional studies. As we discussed before, the stiff-stalk heterotic group is an important source of inbreds used in commercial hybrid production. From this perspective, inbred lines B84, LH145, NKH8431, PHB47, and PHJ40 were sequenced using the Illumina HiSeq-2500 and PacBio Sequel I platforms. The assemblies obtained from the sequencing ranged from 2.13 (NKH8431) to 2.18 Gbp (LH145) and had sizes comparable to the assembly of the reference genome of B73 (v4–2.13Gb). The genomes showed a high proportion of complete orthologs. Transposable elements (TEs) were annotated based on structural features and homology with a pan-stiff-stalk TE library. Approximately 87% of each genome was annotated as TEs using a combination of methodologies. On average, 75.69% of the genomes consisted of LTR retrotransposons, with *Gypsy* and *Copia* elements contributing 46.69% and 25.26%, respectively. In the six studied genomes, the genes were annotated in parallel using ab initio predictions combined with evidence of transcripts from various tissues, e.g., leaves, internodes, roots, stems, and self-pollinated endosperm. A total of 49,986 genes were annotated for B73, and the number of genes ranged from 50,861 (B84) to 52,133 (LH145) for the other five inbred lines.

A cladogram generated by orthologous clustering was used to study the relationship between the stiff-stalk inbred lines, two inbred lines outside the stiff-stalk heterotic group (Mo17 and PH207), and *Sorghum bicolor* (L.). As expected, sorghum was distantly related to maize lines, and Mo17 and PH207 were not grouped with the stiff-stalk inbred lines. B73 and B84 were closely grouped in the cladogram, while PHB47 and PHJ40 were grouped separately from the other inbred lines (Fig. 6). The alignment of genes from each stiff-stalk line with each genome set also highlighted the greater divergence of PHJ40 and PHB47 from the other lines of the stiff-stalk heterotic group. Synteny analysis between the B73 genome and the assemblies obtained for the five stiff-stalk inbred lines highlighted from 1,178 (B84) to 1,737 (PHJ40) collinear blocks among the five inbred lines, revealing extensive collinearity across the five stiff-stalk genomes, even though unique components of the maize pangenome were detected. The five inbred lines differed from B73 due to structural variations, including deletions, insertions, inversions, and

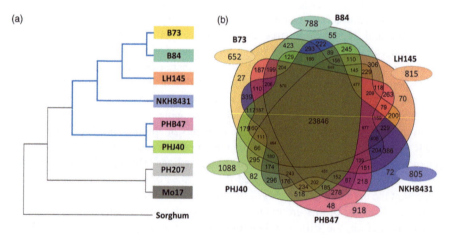

Fig. 6 (**a**) Cladogram showing the relationships among proteomes. The cladogram was constructed and rooted from ancestral orthologous groups with the STAG and STRIDE algorithms (Emms & Kelly, 2017, 2018), respectively. Inbred lines belonging to the stiff-stalk lineage are indicated in blue. All branches had multiple sequence alignment support values of 100%. (**b**) Venn diagram of orthologous and paralogous group occupancy across six stiff-stalk inbreds. Intersections indicate orthologous groups containing at least one gene from a given stiff-stalk inbred. There were 23,846 "core" orthologous groups containing at least one protein from all stiff-stalk inbreds, representing 55.57% of the total orthologous groups assigned (including singletons). Similarly, 31.83% of orthologous groups were missing at least one stiff-stalk inbred, and 12.62% of paralogous groups were unique to a stiff-stalk inbred (i.e., inbred-specific paralogs plus singletons). The number of singletons for each stiff-stalk inbred is shown in an ellipse overlaying the Venn diagram. (Reproduced from Bornowski et al. (2021))

genomic duplications. Despite their highly selected and adapted nature, these results show substantial genetic and genomic diversity within the assembled inbreds available to maize breeders. These lines can be used to study the allelic variation within the stiff-stalk heterotic group, contributing to adaptation, genotype-by-environment interactions, and combining the ability between the stiff-stalk and non-stiff-stalk groups (Bornowski et al. 2021).

5 Population Genomics for Maize Selection and Breeding

Advances in Next-Generation Sequencing (NGS) in the last 20 years have reduced the costs of obtaining molecular markers with genome wide coverage (Bhat et al. 2016). The development of these markers increased the knowledge about the genomic constitution and the genetic architecture of many plant species (Benavente and Giménez 2021), providing large-scale access to genomic information. Therefore, significant advances were possible in the construction of genetic maps (or genetic linkage maps), association studies, and genomic selection in several crops, including maize (Guimarães et al. 2018). Moreover, genomic approaches based on genome-wide marker data led to more robust statistical models to establish marker-trait associations and genomics-assisted selection.

5.1 Mapping Populations and Genetic Maps

Genetic maps are essential to several genetic studies since they provide "reference points" along the genome, indicating the relative position and distances between markers along chromosomes. These maps can be used for gene tagging, map-based cloning of genes, and QTL-mapping, in addition to supporting Marker-Assisted Selection (MAS) in plant breeding programs (Jindal et al. 2020; Semagn et al. 2006), synteny studies (Zhao et al. 2021), associative mapping (Alseekh et al. 2021), and genome assembly (Lin et al. 2021; Zenda et al. 2021). Furthermore, genetically mapped markers are useful in population genomics studies, especially to study adaptive genetic variation (Luikart et al. 2018).

The first genetic map of maize was constructed in 1986 using RFLP (Restriction Fragment Length Polymorphism) (Helentjaris et al. 1986). Since then, increasingly saturated genetic maps with different markers have been published (Guimarães et al. 2018). A significant milestone in the construction of genetic maps in maize was the designation of "bin," intervals along the maps used to simplify the location of genes. "Bin" is a segment of approximately 20 cM flanking two RFLP markers defined as "core" markers (Gardiner et al. 1993), routinely used nowadays to define the location of the loci.

Several maize genetic maps were constructed from different mapping populations and molecular markers. Information on the genetic and physical positions of molecular markers in the maize genome and their sequences are available in the Maize Genetics and Genomics Database (MaizeGDB: www.maizegdb.org). Moreover, information on the maize genome can also be found in the Panzea (www.panzea.org), MaizeGenetics (www.maizegenetics.net), and Gramene (www.gramene.org) databases (Guimarães et al. 2018).

A mapping population is a group of individuals suitable for genetic mapping using Mendelian inheritance principles. Several types of populations have been developed in maize over the years: F_2, backcross, RILs (Recombinant Inbred Lines), DH (Double Haploids) (Benavente and Giménez 2021; Zhao et al. 2018), and the North Carolina design III approach with a RIL population (Samayoa et al. 2017). Regardless of the type of mapping population, the purpose is to obtain a population with genetic variability and linkage disequilibrium (LD) between the alleles of different loci. In maize, the occurrence of LD depends on the type of population studied. Overall, the LD is limited due to the high recombination rates and the presence of transposable elements (mainly retrotransposons), rapidly declining as the physical distance between loci in the genome increases (Flint-Garcia et al. 2003; Dinesh et al. 2016; Olmos et al. 2017).

Many maize mapping populations were developed over time. These populations can be categorized as first, second, and third generation (Gireesh et al. 2021). The first generation of mapping populations, also known as biparental populations, refers to populations obtained from the cross between two parents contrasting for a trait of interest, e.g., F_2, $F_{2:3}$, backcross (BC_1, BC_2), NILs (Near Isogenic Lines), RILs (Recombinant Inbred Lines), and DH (Double Haploids). Biparental populations are used to identify and map qualitative and quantitative traits, e.g., tolerance to drought, salinity, heat (Gao et al. 2019; Gireesh et al. 2021; Luo et al. 2017, 2019).

The limited number of crossover events within biparental populations results in a quite low mapping resolution, whereas the number of alleles at a given locus is limited to those present in the two parents (Gage et al. 2020). In this case, increasing the density of markers (beyond one marker per 15 cM) to increase the resolution of the QTL mapping may not be a useful strategy if the population size is not increased (Kearsey and Farquhar 1998). Furthermore, the population may show a low level of allelic variations (since there are two possible alleles for each locus) and segregate for only a few traits, i.e., those contrasting between the two parents (Rakshit et al. 2012).

The most studied biparental population in maize is IBM (Intermated B73 x Mo17), composed of RILs obtained after five cycles of intercrossing from F_2 (Lee et al. 2002). Many intercrossing cycles allowed extra recombination events after the F_2 generation, increasing the resolution of the genetic maps obtained with this population (Guimarães et al. 2018). The linkage maps of IBM, with many levels of marker saturation, are available at MaizeGDB.

The second generation of mapping populations refers to natural populations, germplasm bank collections, or improved genotype sets used to map complex traits via association mapping. In this case, panels containing a diverse set of genotypes with phenotypic diversity for traits of interest are mapped to establish a marker-trait association using high-density genotyping and multi-environment phenotyping (Gireesh et al. 2021).

Association mapping has advantages compared to linkage analyses, e.g., the detection of valid genetic associations for the entire population and not just a specific cross, increased resolution and mapping accuracy for locating associations, greater allelic variation, and reduction of research costs and time since there is no need to develop experimental populations (Zhu et al. 2008). High allelic diversity and mapping resolution are due to historical recombination events and low LD within a population of diverse individuals (Gage et al. 2020). Despite this, one of the major confounding factors in association studies is the population structure among individuals (Flint-Garcia et al. 2005), resulting in false-positive detection. For this reason, a common practice to control such effects is to use population structure and kinship as factors within an association model (Pritchard et al. 2000a). Furthermore, rare alleles are hardly detected during associative mapping, even if their effects are significant (Cockram and Mackay 2018).

The third generation of mapping populations, also called next-generation mapping populations, was proposed to combine the principles of linkage mapping and association mapping by developing multiparent populations using multiple crossing designs (Gireesh et al. 2021). They seek to solve the weaknesses of the two previous generations, showing higher allelic diversity, more recombination events, more power to detect rare alleles, and lower confounding caused by the population structure (Gage et al. 2020). Three multiparent mapping populations stand out for maize: Nested Association Mapping (NAM), Multiparent Advanced Generation Inter-Cross (MAGIC), and Random-Open-Parent Association Mapping (ROAM).

The NAM population was the first multiparent population for plants, consisting of a set of 25 recombinant inbred lines (RIL) derived from crosses between B73 and 25 other genetically diverse maize inbred lines (McMullen et al. 2009; Yu et al. 2008). The population structure shares a common parent across all populations, the maize line B73, which allows scaling the additive effects of all identified QTL against the common background of B73 alleles, an intuitive application of QTL results in breeding programs. It combines the advantages of linkage analysis, allowing the separate study of each cross by QTL mapping in biparental populations, and association analysis, studying all crosses together. This population is useful to dissect the genetic architecture of complex maize traits, such as flowering time, leaf architecture, stalk strength, and plant height (Buckler et al. 2009; Peiffer et al. 2013, 2014; Tian et al. 2011). However, the use of the NAM population has statistical limitations. The absence of intercrossing between non-B73 parents potentially masks the effects of causal QTL by confounding population structure in cases in which quantitative trait loci segregate among RILs populations but not within them. Furthermore, unbalanced parental composition probably dilutes the GWAS efficiency at any locus with multi-allelic effects (Xiao et al. 2017).

Unlike NAM, the MAGIC population allows crossing all parents to balance their contribution (Cavanagh et al. 2008). These populations were created by intercrossing a moderate number (usually 8 or 16) of inbred parent lines, resulting in maize populations with more recombination events and greater allelic richness than a biparental population and less confounding compared to an association panel. Overall, MAGIC populations have fewer parents than NAM populations but more complex mating schemes (Gage et al. 2020). MAGIC requires the intercrossing of all parents in multiple rounds, resulting in a balanced parental composition and numerous crossover events in the progeny, increasing the statistical power and resolution of the mapping. Moreover, MAGIC populations have the potential for a greater diversity of recombinant haplotypes due to the combinatorial arrangement of donor haplotypes compared to NAM, in which a single common parent is present in all recombinant haplotypes (Ladejobi et al. 2016). However, inferring the origin of identity by descent (IBD) across the genome of MAGIC progenies is mathematically complex, especially with a high number of initial founding parents (Mott et al. 2000). The extensive efforts to develop NAM and MAGIC populations and the impossibility of expanding the number of progenies for traits with low variability limit the use of these populations (Xiao et al. 2017).

The Random-Open-Parent Association Mapping (ROAM) population is a more recent, public multiparent maize population that combines multiple RIL populations to improve the genetic resolution and statistical power and to identify minor effects and low-frequency variants (Pan et al. 2016; Xiao et al. 2017). This population is created from 14 randomly intercrossed parents, from which ten RIL populations are obtained. Since it does not strictly depend on the cross between specific parents, constructing a de novo population mapping allows the direct integration of new populations into currently existing population resources for large-scale genetic analysis whenever necessary. Similar to NAM, the ROAM population allows directly inferring the IBD status within each biparent population. The continued inclusion of new populations into the ROAM population allows achieving a better balance of the parental contribution to increase the QTL detection power. However, it is impossible to guarantee that each new parent contributes one allele for a given QTL, and the robustness of ROAM GWAS models may decline due to the assumption that each parent will independently contribute an allele (Xiao et al. 2017).

The socioeconomic importance of maize contributed to a vast history of studies involving QTL mapping using different segregating populations in different germplasms. Biparental populations are often used to map QTLs in maize, especially IBM. In this case, mapping resolution depends on the number of recombination events that occur during the creation of the population, which are generally limited, and only the loci that show allelic variation between parents can be analyzed (Benavente and Giménez 2021). These disadvantages can be overcome by using multiparent populations, e.g., NAM (Garin et al. 2017) and MAGIC (Dell'Acqua et al. 2015).

RIL and DH populations are composed of inbred lines and can be used to carry out multiple phenotyping over several years and different locations. However, these populations do not allow to study intralocus allelic interactions (dominance). Since maize is an allogamous species with a high degree of heterosis for most quantitative traits, populations that allow estimating dominance effects (e.g., F_2 and those derived from North Carolina design III) are worth mentioning (Guimarães et al. 2018).

5.2 Marker-Assisted Selection

Marker-Assisted Selection (MAS) is mainly used to select parents for hybridization, introgression, pyramiding alleles of interest, and in situations where direct phenotypic selection is more costly/time-consuming or gene expression requires biological or specific environmental conditions (Alzate-Marin et al. 2005; Sakiyama et al. 2014; Singh and Singh 2015; Poland and Rutkoski 2016). The most significant application of MAS is undoubtedly in molecular breeding, in which it is used to obtain disease-resistant genotypes when the genetic control of resistance is controlled by one or a few genes (qualitative trait).

MAS was also proposed for quantitative traits related to the plant morphology of maize, e.g., height, yield, and traits that determine crop responses to abiotic stresses. Both are controlled by multiple genes or QTLs, with minor individual effects on phenotypic variation (Benavente and Giménez 2021). In this scenario, many studies have aimed to identify genes and/or QTLs and validate markers for MAS in maize (Table 3).

Unlike what happened for disease resistance, the use of MAS for quantitative traits was not very successful. The vast number of QTLs controlling quantitative traits provides a poor explanation of the phenotypic variation by the markers associated with them, which does not make up for the cost and time of validating and developing usable tags. Moreover, it is hard to find the same QTLs in different experiments due to interactions between QTLs and different environments (Bernardo 2016).

From another perspective, a trend of studies has involved MAS with polygenic traits in maize, aiming to improve the nutritional quality of the grain. Despite being controlled by several genes, these traits have some genes with more significant effects that can be validated as QTLs and used in MAS to select maize lines or progenies with biofortified grains. Given the massive importance of maize as a food base for several developing countries, especially in Africa, and the future global food security scenario, the nutritional enrichment of grains is a very attractive topic for maize breeders. Therefore, developing maize hybrids and populations with better nutritional quality (e.g., high vitamin A contents) through MAS and QTL mapping has become a trend to provide sustainable and cost-effective solutions for alleviating nutritional deficiency (Duo et al. 2021; Kebede et al. 2021; Singh et al. 2021).

The difference in phenotypic variation explained by the QTLs identified for traits related to disease resistance compared to morphological traits in maize is remarkable. In most cases, plant resistance is conditioned by one or a few genes, allowing the identification of genes or QTLs with more significant effects, which are validated and used in MAS (Singh and Singh 2015). From this perspective, other techniques are more suitable in the molecular breeding of quantitative traits, e.g., Genome-Wide Association Studies (GWAS) and Genome-Wide Selection (GWS).

Table 3 Marker-Assisted Selection (MAS) studies for diseases resistance and morphological traits in maize

R^2 (%)[a]	Chromosome	Traits related to disease resistance	Reference
59.2	5 and 7	Gibberella ear rot – *Fusarium graminearum*	Ali et al. (2005)
41.3	8	Maize rough dwarf disease	Tao et al. (2013)
19–20	5	Northern corn leaf blight – *Exserohilum turcicum*	Chen et al. (2016)
54.0	3, 8 and 9	Southern leaf blight – *Cochliobolus heterostrophus* (Drechs.) Drechs.	Kaur et al. (2019)
70.0	3	Sugarcane mosaic virus	De Souza et al. (2019)
18.9	5	Bacterial leaf streak – *Xanthomonas vasicola* pv. *vasculorum*	Qiu et al. (2020)
48.0–65.0	4	Southern corn rust – *Puccinia polysora*	Deng et al. (2020)
33.2	7	Gibberella ear rot – *Fusarium graminearum*	Yuan et al. (2020)
28.5	2 and 8	Northern corn leaf blight – *Exserohilum turcicum*	Ranganatha et al. (2021)
43.0–78.0	10	Southern corn rust – *Puccinia polysora*	Lv et al. (2021)
55.7	1	Gray leaf spot – *Cercospora zeae-maydis*	Sun et al. (2021)
43.3	1	Fusarium ear rot – *Fusarium verticillioides*	Wen et al. (2021)
R^2 (%)[a]	Chromosome	Morphological traits	Source
20.2	8	Ear height	Wei et al. (2009)
39.0–40.0	10	Anthesis and silking date photoperiod	Coles et al. (2010)
36.7	1	Prolificacy	Wills et al. (2013)
32.3	3	Plant height	Teng et al. (2013)
23.0	3	Kernel width	Wang et al. (2020)
15.3	9	Kernel length	Wang et al. (2020)
18.8	3	Ear length	Yang et al. (2020)
11.1	2	Ear diameter	Yang et al. (2020)
13.7	2	Ear row number	Yang et al. (2020)
17.2	2	Kernel number per row	Yang et al. (2020)
14.6	5	100 Kernel weight	Yang et al. (2020)
11.8	6	Grain weight per plant	Yang et al. (2020)

[a]Percentage of phenotypic variation explained by QTL

5.3 Genome-Wide Association Studies (GWAS)

Genome-Wide Association Study (GWAS) is an association mapping strategy to fine-map QTLs with multiple recombination events that lead to the rapid decay of linkage disequilibrium (LD) (Flint-Garcia et al. 2003). It involves rapidly scanning markers across the genome to discover possible associations between genetic markers and phenotypes of interest, relying upon historical recombination events accumulated over many generations to provide higher resolution power and greater allele frequency (Chen and Lipka 2016; Platt et al. 2010).

Maize is an excellent crop plant for GWAS due to its high genetic variability, rapid decay of linkage disequilibrium, availability of distinct sub-populations, and abundant SNP information (Xiao et al. 2017; Shikha et al. 2021). The first maize GWAS was reported in 2008 and was used to map genes affecting the oleic acid content in maize kernels (Beló et al. 2008). Since the release of the B73 reference genome (Schnable et al. 2009), many studies have promoted maize genetic research into the genomics era. The advances in maize GWAS highlight its potential as a powerful tool to understand the genetic architecture of complex quantitative traits, annotate candidate genes to understand the genome structure and the constitution associated with each trait of interest, and enhance maize breeding by identifying beneficial alleles (Xiao et al. 2017; Shikha et al. 2021).

5.3.1 Advances Using GWAS in Maize

Most maize traits are controlled by many small-effect genes (Wallace et al. 2014), i.e., they are characterized by a continuum of phenotypes and show complex inheritance. However, some loci have greater genetic effects on a trait than others, which can be assessed by GWAS to discover and validate causal variants underlining traits (Xiao et al. 2017). Thus, although GWAS can be (and effectively is) used to find associations for quantitative traits in maize, it seems to be more effective for less complex traits. The largest effect sizes in maize typically explain less than 5% of the total variation, with total identifiable genetic variation spread across up to several dozen genes (Wallace et al. 2014).

A set of studies highlighted the potential of GWAS to identify QTL for multiple traits in maize (Table 4), ranging from molecular (including the transcriptome) to cellular (i.e., metabolites) and from the individual morphological scale (agronomic, yield, or reproductive characteristics) to the interaction with different environmental factors (biotic or abiotic stress tolerance). They vary according to the sample size, number of markers, and populations (Xiao et al. 2017).

GWAS was also applied to detect significant associations for quality traits in maize and understand its genetic basis (Tang et al. 2015; Shikha et al. 2021). For cell quality and kernel composition, for example, few large-effect QTL explain more than 10% of phenotypic variation (Li et al. 2013; Wen et al. 2014). Zheng et al. (2021) identified 3, 7, 21, 8, and 10 SNPs significantly associated with the moisture, protein, oil, starch, and lysine contents, respectively, explaining from 3.77 to 32.79% of the variation in these traits. For SNPs associated with several regions of the maize genome, they classified candidate genes into 11 biological processes, 13 cell components, and 6 functions. The numbers of candidate genes related to moisture, protein, oil, starch, and lysine were 77, 46, 103, 136, and 49, respectively. Candidate genes in biological processes were mainly involved in cellular and metabolic processes; in cellular components, candidate genes are mainly concentrated in the organelle, cell, and part of the cell; finally, candidate genes in molecular functions were mainly involved with catalytic activity and binding (Zheng et al. 2021).

Table 4 Traits studied via GWAS in maize

Trait category	Phenotype	Population	Sample size	No. of marker	Reference
Molecular and cellular	Gene expression	IAP	368	557 K	Fu et al. (2013)
		IAP	368	1.25 M	Liu et al. (2017)
	Secondary metabolome	IAP	368	557 K	Wen et al. (2014)
	Oil concentration	IAP	508	557 K	Li et al. (2013)
		USNAM + IAP	4,699 + 282	1.6 M/52 K	Cook et al. (2012)
	Carotenoid	IAP	380	476 K	Suwarno et al. (2015)
		IAP	201	284 K	Owens et al. (2014)
	Tocopherol	IAP	513	56 K	Li et al. (2012)
		IAP	252	294 K	Lipka et al. (2013)
	Carbon and nitrogen metabolism	IAP	263	56 K	Liu et al. (2016a)
		USNAM	4,699	1.6 M	Zhang et al. (2015)
		USNAM + IAP	4,699 + 282	1.6 M/52 K	Cook et al. (2012)
	Amino acids	IAP	289	56 K	Riedelsheimer et al. (2012)
		USNAM + IAP	4,699 + 282	1.6 M/52 K	Cook et al. (2012)
	Leaf lipidome	IAP	289	56 K	Riedelsheimer et al. (2013)
	Leaf metabolome	IAP	289	56 K	Riedelsheimer et al. (2012)
	Drought-related metabolites	IAP	318	157 K	Zhang et al. (2016)
	Iron homeostasis	IAP	267	438 K	Benke et al. (2015)
	Carotenoid levels	Inbreed lines	416	172 K	Baseggio et al. (2020)
	Nitrogen use efficiency	IMAS	424	955 K	Ertiro et al. (2020)
	Fertility in haploids	DTMA + CML	400	214 K	Chaikam et al. (2019)

Developmental and agronomic	Plant architecture	ROAM	10	56 K	Pan et al. (2017)
	Forage quality	IAP	368	557 K	Wang et al. (2016a)
	Shoot apical meristem	IAP	384	1.2 M	Leiboff et al. (2015)
	Flowering time	IAP	1,487	8.2 K	Van Inghelandt et al. (2012)
		IAP	368	557 K	Yang et al. (2013)
		IAP	513	557 K	Yang et al. (2014)
		IAP	346	60 K	Farfan et al. (2015)
		USNAM	5,000 + 281	1.1 K	Buckler et al. (2009)
		USNAM + IAP	5,000 + 281	1.1 K	Hung et al. (2012)
		USNAM + CNNAM + Ames	4,763 + 1971 + 1745	950 K	Xiaopeng Li et al. (2016)
		MAGIC	529	54 K	Dell'Acqua et al. (2015)
	Plant height related	IAP	284	41 K	Weng et al. (2011)
		IAP	289	56 K	Riedelsheimer et al. (2012)
		IAP	513	557 K	Yang et al. (2014)
		IAP	258	224 K	Xiaopeng Li et al. (2016)
		IAP	346	60 K	Farfan et al. (2015)
		USNAM	4,892	1.6 M	Peiffer et al. (2014)
		MAGIC	529	54 K	Dell'Acqua et al. (2015)
		F_1 population	300	29 K	Zhang et al. (2019b)
	Leaf architecture	USNAM	4,892	1.6 M	Tian et al. (2011)
		IAP500	513	557 K	Yang et al. (2014)
		USNAM + NCRPIS	4,892 + 2,572	1.6 M/405 K	Xue et al. (2016)
	Husk traits	IAP	508	557 K	Cui et al. (2016)
	Tassel architecture	IAP	513	557 K	Yang et al. (2014)
		USNAM	4,892	1.6 M	Brown et al. (2011)
		USNAM + CNNAM + IAP	4,623 + 1972 + 945	500 K/ 500 K/ 44 K	Wu et al. (2016a)

(continued)

Table 4 (continued)

Trait category	Phenotype	Population	Sample size	No. of marker	Reference
	Ear height	IAP	513	557 K	Yang et al. (2014)
		IAP	346	60 K	Farfan et al. (2015)
		USNAM	4,892	1.6 M	Peiffer et al. (2014)
		USNAM + CNNAM + Ames	4,763 + 1971 + 1745	950 K	Xiaopeng Li et al. (2016)
		MAGIC	529	54 K	Dell'Acqua et al. (2015)
	Stalk strength	IAP	368	557 K	Liu et al. (2017)
		USNAM + NCRPIS	4,536 + 2,293	1.6 M/681 K	Peiffer et al. (2013)
	Root related	IAP	267	438 K	Benke et al. (2015)
		Ames	384	681 K	Pace et al. (2015)
	Stover quality	Lines + Inbreed lines + DH lines	424 + 276 + 1,026	181 k	Vinayan et al. (2021)
Yield	Ear architecture	IAP	513	557 K	Yang et al. (2014)
		IAP	513	49 K	Liu et al. (2015a)
		IAP	368	557 K	Liu et al. (2015b)
		USNAM	4,892	1.6 M	Brown et al. (2011)
		USNAM + NCRPIS	4,892 + 2,572	1.6 M/405 K	Xue et al. (2016)
		ROAM	1887	185 K	Xiao et al. (2016)
		XP panel	400	940 K	Yang et al. (2015)
	Grain size	IAP	513	557 K	Yang et al. (2014)
		MAGIC	529	54 K	Dell'Acqua et al. (2015)
	Biomass	IAP	289	56 K	Riedelsheimer et al. (2012)
	Kernel moisture	DH population lines	249	56 K	Yinchao Zhang et al. (2020)

Stress resistance	Disease resistance	IAP	1,487	8.2 K	Van Inghelandt et al. (2012)
		IAP	527	557 K	Chen et al. (2015)
		IAP	1,687	201 K	Zila et al. (2014)
		IAP	999	56 K	Ding et al. (2015)
		IAP	890	56 K	Mahuku et al. (2016)
		IAP	818	43.4 K	Chen et al. (2016)
		IAP	274	246 K	Mammadov et al. (2015)
		IAP	287	261 K	Warburton et al. (2015), Tang et al. (2015)
		IAP	380 + 235	259 K/264 K	Gowda et al. (2015)
		IAP	267	47 K	Zila et al. (2013)
		IAP	346	60 K	Farfan et al. (2015)
		IAP	267	287 K	Horn et al. (2014)
		USNAM	4,892	1.6 M	Kump et al. (2011), Poland et al. (2011)
		Inbreed Lines	380	278 K	Gowda et al. (2021)
		UEM	157	837 K	Kuki et al. (2018)
		DH populations lines + IMAS	558 + 380	293 K	Sitonik et al. (2019)
		Inbred lines	230	226 K	Stagnati et al. (2020)
		CAAM, DTMA, IMAS	419 + 285 + 380	64 K/69 K/69 K	Rashid et al. (2020)
	Insect resistance	IAP	302	246 K	Luis Fernando Samayoa et al. (2015)
		MAGIC	672	224 K	Jiménez-Galindo et al. (2019)
	Low phosphorus tolerance	Inbred lines	356	12 K	Qing-Jun Wang et al. (2019)

(continued)

Table 4 (continued)

Trait category	Phenotype	Population	Sample size	No. of marker	Reference
	Hypersensitive response	IAP	231	47 K	Olukolu et al. (2013)
		USNAM	3,381	26.5 M	Olukolu et al. (2014)
	Drought tolerance	IAP	80	1 K	Hao et al. (2011)
		IAP	368	525 K	Liu et al. (2013)
		IAP	318	157 K	Zhang et al. (2016)
		IAP	346	60 K	Farfan et al. (2015)
		IAP	240	30 K	Thirunavukkarasu et al. (2014)
	Water tolerance	IAP	350	56 K	Xue et al. (2013)
	Salt tolerance	Association populations	445	1.2 M	Luo et al. (2019)
		Inbred lines	420	220 K	Sandhu et al. (2020)
	Cold tolerance	IAP	125	56 K	Juan Huang et al. (2013)
		IAP	375	56 K	Strigens et al. (2013)
		Dent + Flint	306 + 292	50 K	Revilla et al. (2016)
		MAGIC	406	223 K	Yi et al. (2020)
		Inbred lines	900	156 K	Yi et al. (2021)

Xiao et al. (2017) highlighted some successful cases of GWAS applied in maize breeding. One of the most successful molecular breeding projects of CIMMYT identified associations of candidate genes related to the pro-vitamin A biofortification of maize. These alleles were introgressed into elite genotypes to decrease vitamin A deficiency (VAD) in developing countries (Sherwin et al. 2012; Harjes et al. 2008). In another case, the dissection of the ZmCCT gene, which has a significant effect on photoperiod sensitivity, showed that one of its alleles reduces gene expression, promoting early flowering (Hung et al. 2012; Yang et al. 2013). Moreover, two loci (lg1 and lg2) were found to be associated with upper leaf angle, correlating to increased efficiency of solar radiation capture and showing potential to increase grain yield (Tian et al. 2011). Finally, an intergenic sequence (3-Kb within locus KRN4) was found to be responsible for variation in the maize kernel row number in temperate elite inbred lines of maize (Liu et al. 2015a).

All these cases are related to less complex traits. As is the case with any phenotypic or genomics study, GWAS may be more challenging for complex biological phenotypes that result from effects and interactions between multiple phenotypic traits. In this case, an alternative is to measure correlated traits instead of the complex trait itself. The maize inflorescence, for example, reflects an interaction between tassel, ear, and vegetative traits (Bonnett 1954). Several QTLs that contribute to inflorescence and leaf architecture have been identified to date (Brown et al. 2011; Calderón et al. 2016; Li et al. 2015; Pan et al. 2017; Tian et al. 2011; Wu et al. 2016a). Some genetic components underlying these QTL may contribute to two or more traits, a phenomenon called pleiotropy (Stearns 2010). This theory is supported by the QTL regions identified for tassel (Brown et al. 2011; Wu et al. 2016a) and leaf traits (Tian et al. 2011) near genes for ligule absence.

This strategy has also been used for traits related to biotic and abiotic stresses. In this case, GWAS promotes insights into the genetic basis and physiological mechanisms of resistance, in addition to elucidating biological routes and correlated genes that provide resistance (Shikha et al. 2021). Promising results were obtained for biotic stresses, e.g., resistance to maize lethal necrosis (Gowda et al. 2015), resistance to *Gibberella* ear rot (Han et al. 2018), tar spot resistance (Mahuku et al. 2016), and abiotic stresses, e.g., cold, drought, water submergence, salt, and heavy metals (Shikha et al. 2021). For drought tolerance, a complex quantitative trait highly influenced by the environment, some association studies identified favorable natural variants of maize genes by measuring the seedling survival rate under water stress conditions. This strategy reduced trait complexity and enhanced the success of GWAS (Liu et al. 2013; Mao et al. 2015; Wang et al. 2016b).

GWAS applied to transcriptomic variation demonstrated that eQTL (expression QTL) often explains a large proportion of phenotypic expression variation. Since the factors influencing a trait range from gene expression to the level of metabolites in single cell types, the number of QTLs identified may shift from highly quantitative to qualitative or single locus (Xiao et al. 2017).

With regard to understanding the genome structure, GWAS showed that genic and nearly genic regions contribute most to maize trait variation, especially in the 5′UTR (untranslated region) regions. Non-synonymous mutated SNPs and large copy number variants (CNVs) are the most functionally enriched, while intergenic regions show significant depletion for functional SNPs. In addition, non-synonymous SNPs are the most significant drivers of expression regulation, with many SNP-eQTL associations (Xiao et al. 2017; Li et al. 2012; Wallace et al. 2014).

These systematic studies suggest that gene regulation at the expression level could play a key role in phenotypic diversity. Under this hypothesis, the expression landscape of immature maize kernels has been extensively explored (Fu et al. 2013; Liu et al. 2017). Similar conclusions were obtained previously for quantitative traits, i.e., that non-synonymous SNPs are the most significant drivers of expression regulation, showing a higher number of SNP-eQTL associations (Liu et al. 2017).

Lu et al. (2021) found significant associations for 29 SNPs related to dry matter accumulation in stages V3 and V6 of maize. Plant dry matter is an essential index to assess plant growth and development as its accumulation during early development significantly affects crop growth and yield. In maize, dry matter accumulation during early growth stages is positively correlated with the number of flowers per ear (Gonzalez et al. 2019). From this perspective, gene annotation was performed according to the most recent maize B73 reference genome (B73 RefGen_v4), resulting in 224 unique candidate genes for dry matter in the leaf, sheath, and total shoot in stages V3 and V6. These traits did not show overlapped candidate genes and, of the unique candidate genes identified, 18 were associated with at least two traits (Lu et al. 2021).The data revealed a network of relationships between genes and dry matter traits (Fig. 7), containing seven large nodes or hubs representing the dry matter content in plant parts at stage V3, four nodes at stage V6, 1,103 small round nodes (candidate genes), and 1,253 edges (interactions between traits and genes) (Lu et al. 2021). These studies (Table 4) have proven that GWAS is a powerful tool to address the association between genotype and phenotype based on the development of next-generation sequencing technology, achieving significant progress with maize over the past decade (Zheng et al. 2021).

5.3.2 Statistical Concerns in Maize GWAS

GWAS models in maize usually employ statistical approaches that consider the population structure and family kinship (Chen and Lipka 2016; Lipka et al. 2015; Price et al. 2010; Pritchard et al. 2000b). However, population structure or family kinship relationships can lead to false-positive associations if these effects are not accounted for in the GWAS model (Lipka et al. 2015; Zhang et al. 2010). Therefore, the search for models that consider population structure effects and treat them computationally efficiently has been an active research area in maize. One of the most effective strategies to eliminate false positives is to adjust the population structure as covariates in a general linear model (GLM) or adjust the population structure and the total genetic effect of each individual as covariates in a mixed linear model (MLM) to adjust test markers (Price et al. 2006; Jianming Yu et al. 2006).

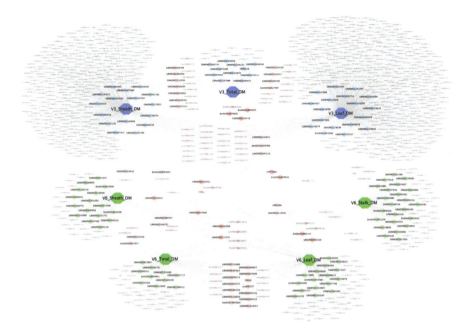

Fig. 7 Trait–gene network of dry matter traits. Of the seven large octagon nodes, the three blue nodes represent stage V3 traits (V3 Leaf DM, V3 Sheath DM, and V3 Total DM), and the four green nodes represent stage V6 traits (V6 Leaf DM, V6 Sheath DM, V6 Stalk DM, and V6 Total DM). Genes are represented by round nodes, and different colors and size indicate different attributes. The 158 genes with detailed functional descriptions were labeled as large round nodes, with colors corresponding to their related traits. The gray small nodes represent genes that correlate with only one specific trait, and the pink and red nodes (both large and small ones) indicate two-trait shared genes and three-trait shared genes, respectively. (Reproduced from Lu et al. (2021))

In order to consider the population effect, a matrix (Q) can be used to adjust the fixed effect of populations. Thus, statistical tests on genetic markers (S) can be performed with a GLM, one marker at a time. The model can be conceptually presented as follows:

$$y = Q + S + e$$

where y corresponds to the observed genotype, Q is the marker vector, and e corresponds to the residue. This model is also known as the Q model.

With regard to the control of false-positive associations between traits and markers, the unified mixed linear model (MLM) is undoubtedly the most suitable for GWAS datasets (Yu and Buckler 2006). In addition to including the tested marker as a fixed effect, this model includes fixed-effect covariates that account for the population structure and a random polygenic effect to control the relationship between individuals (Chen and Lipka 2016).

In order to ensure that these additional terms fit properly with false-positive signals, sets of genome-wide markers are often used to obtain the fixed-effect covariates (called Q) as well as a kinship matrix (K) that estimates the variance-covariance between individuals (i.e., the variance-covariance of the random polygenic effect) (Chen and Lipka 2016; Xiao et al. 2017). This model can be conceptually presented as:

$$y = Q + K + S + e$$

where y is the observed genotype, Q is the marker vector, K is the parentage matrix, and e is the residue. This model is also known as the $Q + K$ model. Previous studies have shown that the $Q + K$ model controls false positives better than naive models such as the t-test, which only fits test markers (Jianming Yu et al. 2006).

Overall, unified MLM successfully identified associations between markers and traits with moderate to large effect sizes (Lipka et al. 2015). However, a limitation of this model is that it often fails to detect small-effect loci underlying complex trait variation (Atwell et al. 2010). Thus, there remains a critical need to modify the traditional unified MLM in order for it to have sufficient statistical power to detect weak signals while adequately controlling false positives (Chen and Lipka 2016). One way to increase the detection power of GWAS via MLM is to increase the sample size. However, the standard MLM method is inefficient for large datasets, with thousands of individuals, due to the considerable computational effort required.

Many algorithms were developed to get around the computational effort problem caused by the MLM model and optimize it, including Efficient Mixed-Model Association (EMMA), EMMA eXpedited (EMMAX), Population Parameters Previously Determined (P3D), Genome-Wide Efficient Mixed-Model Association (GEMMA), Factored Spectrally Transformed Linear Mixed-Model (FaST-LMM), GRAMMAR-Gamma, two-step methods based on fast variance components, and Circulating Probability Unification (FarmCPU) (Kang et al. 2008; Lippert et al. 2011; Liu et al. 2016b; Zhang et al. 2010; Zhou and Stephens 2012).

The EMMA algorithm simplifies matrix operations through spectral decomposition (Kang et al. 2008). However, this "exact method" of solving mixed-model equations with whole-genome markers iteratively has limited value when testing millions of markers to identify all recombination events. A trade-off between computational speed and statistical power can be made by just estimating model parameters once and then testing the markers iteratively using an approximate method, typically including P3D and residual approach. The P3D approach, including P3D and the EMMA algorithm, provides similar benefits to the residual approach of the GRAMMAR algorithm; however, the P3D algorithm and GRAMMAR are technically different (Xiao et al. 2017).

FaST-LMM and GEMMA are two algorithms proposed to improve the speed of the exact method by optimizing the mixed-model equation. In summary, improved methods usually proposes revised mixed-model equations that refactor the conventional likelihood function of the mixed-model to a form analogous to the probability of a linear regression model. This optimization simplifies the estimation of multidimensional parameters in a one-dimensional numerical optimization problem, dramatically reducing the computational load for each iteration. These methods work at even higher speeds than approximate methods in large scale GWAS (Xiao et al. 2017).

From this perspective, two strategies were developed to solve the confounding problem and improve the statistical power of MLM methods. The first is Compressed MLM (CMLM), which groups individuals and adjusts the genetic values of the groups as random effects. This method improves the statistical power compared to traditional MLM methods (Zhang et al. 2010). The second is Enriched CMLM (ECMLM), which continuously improves the statistical power by optimizing the definition of group relatedness rather than constantly using relatedness algorithms (Li et al. 2014a). Both strategies conduct genetic marker tests, one at a time. However, testing multiple markers simultaneously is more advantageous. This can be done using the test in a graded MLM algorithm, called Multi-Locus Mixed-Model (MLMM) (Segura et al. 2012), in which the general relatedness derived from all available markers is used to define the structure of variance and covariance of the genetic effects of individuals.

The FarmCPU algorithm uses the Fixed Effect Model (FEM) and the Random Effect Model (REM) iteratively. FEM is used to test genetic markers, one at a time. Pseudo QTNs are included as covariates to control false positives (Liu et al. 2016b).

Most statistical methods are designed to solve big data challenges focusing on either marker size or sample size. Among MLM methods, CMLM and FaST-LMM work best with large sample sizes, whereas P3D/EMMAX and GRAM-MAR-Gamma work best with high marker densities. However, FarmCPU was developed to handle both large sample sizes and high marker densities (Liu et al. 2016b).

Finally, alternatives have been proposed to deal with pleiotropic traits in maize. Pleiotropy appears to be minimal in maize (Wallace et al. 2014), only occurring in closely related traits, e.g., inflorescence traits (Buckler et al. 2009; Brown et al. 2011) and other traits related to central carbon and nitrogen metabolism (Zhang et al. 2010). However, some alleles considered pleiotropic might be simply tightly linked genes, each affecting different traits (Wallace et al. 2014). This can be dealt with by comparing the results of GWAS studies carried out for a single trait. These results are combined to identify associations between markers and statistically significant traits (Wei and Johnson 1985). This approach consists of performing a GWAS for each trait separately, with the association between marker and trait in many GWAS constituting an indication of pleiotropy, called pleiotropic QTL (pQTL) (Chai et al. 2018; Visscher and Yang 2016). Alternatively, some procedures can be implemented to compile GWAS information for each trait, estimate marker effects and *p-value* for each marker, and then make pleiotropic inferences (Huang et al. 2011; O'Reilly et al. 2012; van der Sluis et al. 2013).

The multivariate statistical methods available to detect pQTLs when several traits are evaluated in the same study can be categorized into two approaches (Galesloot et al. 2014; Solovieff et al. 2013). The first uses a mixed linear model (MLM) with a matrix of correlated traits as the response variable. These multivariate MLMs (mvMLMs) are commonly applied to plants and include covariates to reduce false positives arising from population structure and relatedness (Zhou and Stephens 2014). The second approach uses data reduction methods to create composite features. One of these methods converts t features into t nonlinearly correlated principal components (PC) (Klei et al. 2008). Each of these PCs can then be used in univariate GWAS to identify genomic regions with peak-associated markers. Genomic marker data associated with these PCs are assumed to be associated with traits of interest (W. Zhang et al. 2018). All PCs, including those that explain even a small amount of variation, can help identify pQTL (Aschard et al. 2014; Avery et al. 2011).

Identifying pQTLs is advantageous since it can result in genetic gain for several traits simultaneously instead of one trait at a time. For example, a pQTL associated with the maize inflorescence and related vegetative traits is important as it can result in a simultaneous genetic gain for the maize inflorescence as a whole rather than just an individual trait. The combined use of currently available multivariate GWAS approaches facilitates detecting pQTLs and, therefore, identifying pleiotropic causal mutations. One of the greatest advantages of pQTL analyses is the easier quantification of the genetic architecture of low heritability traits correlated with highly heritable traits (Rice et al. 2020).

GWAS has successfully identified many loci associated with phenotypic, expression, and metabolic traits in maize. However, only some of the genetic factors underlying these associations were explained, while the vast majority remain unexplained. Next-generation sequencing and bioinformatics tools are currently implemented to decipher the genetic diversity of traits of interest. These analyses are upgraded by multi-omics data to improve the understanding of the genetic basis of phenotypic diversity (Alseekh et al. 2021; Scossa et al. 2021).

5.4 Genomic Selection (GS)

Genomic selection (GS) is one of the most powerful tools available for maize breeding, using a genome-wide marker to estimate breeding values and increase genetic gains with fewer breeding cycles (Rice and Lipka 2021). This marker-assisted breeding method, also referred to as Genome-Wide Selection (GWS), was developed in the 1990s based on mixed linear models and later expanded to Bayesian approaches (Bernardo 1994; Meuwissen et al. 2001). It incorporates genome-wide molecular marker effects to obtain genomic estimated breeding values (GEBVs). First, a prediction model is trained to calculate the GEBVs of a test or validation set, considering a training set that includes genotypic and phenotypic data. Then, GEBVs from exclusively genotyped individuals are obtained to enable the selection of those with the desired values, not requiring field evaluation (Meuwissen et al. 2001; Cui et al. 2020). The objective of GS is not to identify specific QTLs as GWAS but rather to determine the genetic potential of an individual. In general, GS models deal better with complex quantitative traits.

Maize was the target crop used to demonstrate the application of GS in 1994, proposed as Genomic Best Linear Unbiased Prediction (GBLUP) (Cui et al. 2020). Since then, GS has become a powerful and attractive tool in breeding programs, especially after the introduction of more accessible molecular markers into the genome based on genotyping chip technology, e.g., Diversity Array Technology (DArT) and single nucleotide polymorphisms (SNP), and the development of high-throughput genome-wide genotyping platforms, known as Next Generation Sequencing (NGS) (Desta and Ortiz 2014; Bhat et al. 2016; Wang et al. 2018a). In maize, GS reduces the selection time by half compared to phenotypic selection (Lorenzana and Bernardo 2009). Although there are several studies with this approach for different crops, the ideal strategy and the steps for implementing GS in a plant breeding program are still debated in plant science (Robertsen et al. 2019).

5.4.1 Advances Using GS in Maize

Bernardo (1994) pioneered the technical application of genomic selection using GBLUP to predict the yield of single-cross maize hybrids from crosses between Iowa Stiff Stalk Synthetic (SSS) and non-SSS inbred lines using RFLP (restriction fragment length polymorphism) markers. The predictive ability obtained in the validation set ranged from 0.65 to 0.80, signaling a powerful emerging tool available to maize breeders. However, GS began to be effectively applied in public and private maize breeding programs only from the mid-2000s with the availability of high-throughput, low-cost, genome-wide distributed markers (SNPs) and prediction models with robust statistical approaches (Mixed Linear Models, Bayesian statistics, and Machine Learning) (Bhat et al. 2016).

Since its conception, GS in maize has mainly focused on predicting the performance of hybrids from a set of inbred lines. This is because many single-cross hybrids can be obtained by crossing a few lines, leading to significant expenses in field evaluations. Since the introduction of doubled haploid technology, the development and evaluation of too many crosses became a significant challenge. Breeders expect GS to allow a more dynamic, faster, and cost-effective selection process. Single-cross hybrids are often used as training populations for comprising elite germplasms in maize breeding programs. In contrast, parental inbred lines are usually employed to create biparental populations and are extensively evaluated in multi-environmental trials, allowing to estimate genotype-environment interactions (Massman et al. 2013).

Therefore, in order to predict single-cross maize hybrids, GS only needs genotyped inbred lines and some hybrids obtained based on a systematic grid of crossings that guarantees the representation of the descendants to compose the training population and calibrate the prediction model. Phenotypic data from the field evaluation of produced hybrids are used to predict the performance of untested hybrids that could be obtained from a set of lines. Since the tested and untested hybrids are half-sibs, prediction accuracies are expected to be high due to close relatedness (Zhao et al. 2015).

Many GS were carried out for multiple traits in maize using all kinds of statistical models, resulting in significant predictive ability (Lorenzana and Bernardo 2009; Massman et al. 2013; Crossa et al. 2014; Jacobson et al. 2014; Lian et al. 2014; Zhao et al. 2015; Cantelmo et al. 2017; Mendes and de Souza 2016; Fritsche-Neto et al. 2021; Rice and Lipka 2021; Zhang et al. 2019a). The results suggest that GS effectively predicts the genomic breeding values of single-cross maize hybrids and identifies superior combinations in different environments for multiple traits, including complex ones.

One of the factors that directly affect the predictive ability of statistical models is the relationship between the training, calibration, and selection of populations. Li et al. (2021) used data from four distinct hybrid maize populations and demonstrated that the prediction abilities of GS between populations were lower than those of GS within populations. Thus, including related individuals from the validation set of one population into the training set of another population (or vice versa) dramatically increases the GS predictive abilities among structured hybrid populations. Also, better prediction accuracies can be achieved by maximizing the number of crosses per parent in the training set (Fristche-Neto et al. 2018).

GS can be used early in maize breeding programs to identify lines with high general combining ability (GCA) and specific combining ability (SCA) (Kadam et al. 2016) or to predict GEBVs in the combination of elite and diverse germplasm when introgressing exotic haplotype segments (Bernardo 2009; Gorjanc et al. 2016). It has been applied in recurrent selection programs, although in a very incipient way and for quantitative traits related to disease resistance. In this case, extensive field trials, laborious inoculations, and a complex evaluation of plants, ears, and grains are necessary to obtain progenies with greater resistance to a given disease (Liu et al. 2021; Holland et al. 2020). Similar to the recurrent selection in autogamous species, partially inbred maize lines ($S_{0:1}$ or $S_{0:2}$) are genotyped with thousands of loci and have their phenotypes evaluated for disease resistance in field trials or greenhouses. With this information, prediction models are trained so that, from a single phenotyping experiment, progenies are selected for the next selection cycle only by their GEBVs. This approach allows increasing the number of inbred lines far beyond the limits supported by a field experiment, effectively increasing selection intensity, saving resources, and conducting recurrent selection cycles faster (Bernardo and Jianming 2007).

Successful cases of GS application in recurrent selection programs for disease resistance in maize are taking place in studies conducted by the US Department of Agriculture (USDA) in partnership with the North Carolina State University (NCSU) to obtain maize lines with greater resistance to Fusarium ear rot (FER), a disease caused by *Fusarium verticillioides* (Sacc.) Nirenberg (Holland et al. 2020; Butoto et al. 2021). In addition to reducing grain weight and affecting the yield, this disease also reduces grain quality by producing fumonisins (FUM), a mycotoxin associated with risks to human and animal health (Guo et al. 2020). Holland et al. (2020) used this approach and found predictive abilities of up to 0.46 for FER and 0.67 for FUM, suggesting that the implementation of GS in this population could result in greater selection gains over time (2 cycles per year) compared to recurrent phenotypic selection, which requires two years to complete a cycle under the climatic conditions of the USA.

Regardless of how GS is applied, the first genomic prediction models were based on the infinitesimal additive model, in which only the additive effects of the inbred lines were considered in the analysis since, in theory, these are transmitted directly from the parents to the progenies (José Crossa et al. 2010, 2011). However, depending on the genetic architecture of the trait, non-additive effects, e.g., dominance and epistasis, can increase the prediction accuracy of genomic models to explore heterosis, especially in maize (Wang et al. 2020; Alves et al. 2021).

Modeling non-additive effects in maize genomics studies can increase the prediction accuracy in GS and help better explore heterosis (Technow et al. 2012; Kadam et al. 2016; Alves et al. 2021; Fritsche-Neto et al. 2021). From this perspective, Santos et al. (2016) simulated 400 crosses in a partial diallel scheme (20 × 20) between two groups of maize inbred lines, one with lines derived from B73 and the other with lines from Mo17. Different prediction models were tested, e.g., GBLUP and Bayesians. Incorporating dominance effects into prediction models made the predictive accuracy jump from 0.70, 0.76, and 0.77 to 0.83, 0.90, and 0.94 for grain yield, with heritability values of 0.30, 0.50, and 0.70, respectively.

Since maize is grown on almost all continents, considering the environmental effects of the interaction between genotypes (GE interaction) in prediction models is fundamental to select superior genotypes in a wide range of environmental conditions or to maximize recommendations for specific environments (Mendes and de Souza 2016; Bandeira e Sousa et al. 2017; Jarquin et al. 2021). In the presence of GE interaction, the predictive accuracy of genomic prediction models significantly decreases when the model is calibrated in one environment and validated in another (Mendes and de Souza 2016).

Acosta-Pech et al. (2017) proposed to test the effect of the environments and the GE interaction using two genomic prediction models: (M1) GBLUP additive-dominant + environment and (M2) GBLUP additive-dominant + environment + GE. They used an extensive dataset including 2,724 maize hybrids derived from 531 lines (507 dents and 24 flints) evaluated for 12 years (2004 to 2015) at 58 different locations for starch content, dry matter content, and silage yield. The predictive abilities for the three traits ranged from 0.42 to 0.50 for M1 and from 0.48 to 0.60 for M2. The authors concluded that the incorporation of the effects of the GE interaction and of any other effect implies more consistent and accurate predictions by adding information to the model.

GE interaction is a source of information for ranking genotypes or recommending hybrids and exploring the genetic correlations between environments through covariates that characterize the environments where the tests were conducted (Basnet et al. 2019). Correlations between environments can influence the ability of predictive models by sharing environmental information.

On the other hand, GBLUP encompasses the correlation between environments and genomic information. In a study with up to eight randomly absent environments (66% missing data for the validation set), untested single hybrid predictions were accurate to 0.40 with an average coincidence index of at least 80 and 50% for selections between and within environments, respectively (Dalsente Krause et al. 2020). Models that estimate the effects of the environment, genotype, markers, and the GE interaction maintain the accuracy of predictive abilities even when all genotypes are overlapped in all test environments or when there is no overlap, i.e., when there are missing genotypes in the test set (Jarquin et al. 2020).

Multi-environment models, also known as reaction norms models, have been proposed in maize to capitalize on the effects of GE interaction. To incorporate the interaction effects into GS models, these models include environmental covariates, e.g., rainfall, temperature, radiation, and humidity. From this perspective, Jarquín et al. (2014) proposed a multi-environmental model based on reaction norms to accommodate DNA markers and high-density environmental covariates as well as their interactions. As a result, they detected a 17% to 34% increase in the predictive accuracy of multi-environment models, including interaction effects, compared to the traditional model using only the main effects (markers and environments).

Reaction norm models allow using historical data accumulated over time, whether phenotypic or environmental, to predict the phenotypic behavior of genotypes in unsampled environments. In the case of hybrid maize breeding, the reaction norm model can be applied for the genomic prediction of unformed (and thus unphenotyped) maize hybrids in sampled and non-sampled environments, which is innovative within the program. This implies savings in labor, financial resources, and time spent to obtain superior combinations with specific adaptation to different regions.

Determining and measuring all environmental factors is called "envirotyping" (Xu 2016). Adding envirotyping effects to GS multi-environment models results in predictions with actual values closer to those observed in the field than when considering only the effects of the GE interaction (Costa-Neto et al. 2021a). However, the difficulty in implementing multi-environmental models lies in obtaining accurate envirotyping data that adequately characterize the environments, i.e., accurate data on environmental covariates. Costa-Neto et al. (2021b) proposed an R software package to integrate environmental information (envirotyping) in various fields of plant and animal research or evolutionary ecology, called EnvRtype. The software collects environmental variable data from public platforms based on a weather system called NASA's Prediction of Worldwide Energy Resources (NASA POWER, https://power.larc.nasa.gov/), which is updated daily and can be accessed anywhere for free. Thus, this software can easily obtain the environmental data required for multi-environment models.

Many studies have investigated the accuracy of genomic predictions in maize breeding. However, there is still little empirical evidence on the functional performance of inbred lines selected based on phenotypic and genomic selection, i.e., untested lines selected exclusively by GEBV. However, the comparison between the performance of hybrids from double-haploid lines (DH lines) of the maize breeding program for South Africa at CIMMYT showed that the selection for grain yield using hybrids subjected to the absence and presence of water stress is equivalent between genomic selection and phenotypic selection. Although the results of these two strategies are equivalent, genomic selection promises less work time and lower costs, reaching about 68% of the cost of phenotypic selection (Beyene et al. 2019).

GS has undoubtedly changed the effectiveness and methodologies used in maize breeding programs (and in plant breeding in general) as it undergoes intense scientific research. Its implementation in maize breeding programs, whether public or private, is essential to achieve better selection efficiency, cost reduction, and spend less time to obtain superior genotypes for traits of interest.

6 Conservation and Sustainable Management of Maize Genetic Resources

The geographic dispersion associated with many adaptive modifications, crosses between different local races, and selection by farmers have made maize one of the crop species with the greatest genetic variability. The secondary maize gene pool is represented by eight wild teosinte taxa (Guzzon et al. 2021). Moreover, there are more than 300 maize races, with many varieties among them (Machado et al. 2011), constituting an immeasurable biological treasure for humanity.

The maintenance of the genetic variability of maize, as in any other crop, is extremely important to ensure its survival and preserve its natural ability to respond to different types of stresses. The greater the variability, the greater the plasticity to adjust to environmental changes. When the germplasm is very uniform, organisms react very similarly to environmental changes. For example, the inexistence of alleles conditioning resistance to drought can lead to the disappearance of the species if, by chance, this phenomenon occurs. Despite the existence of genetic variability in maize conservation, strategies for the genetic resources of this species should be established, especially because its process of domestication and artificial selection imposed by humans tends, over time, to result in genetic erosion, i.e., the loss of genetic variation (Machado et al. 2011).

Genetic erosion in maize is mainly associated with the replacement of local varieties by commercial crops. However, commercial maize breeding has exploited only an elite fraction (5–10%) of all existing variability, and farmer preferences for commercial cultivars over native varieties are due to characteristics such as higher yields, disease resistance, and market factors, e.g., uniform quality (Guzzon et al. 2021). Therefore, to avoid genetic erosion, it is necessary to study and conserve the genetic variability in maize genomes.

The genetic conservation of maize has been carried out in situ and ex situ. In situ maize conservation is accomplished through teosinte conservation in reserves, e.g., the *Reserva de la Biosfera Sierra de Manantlán* in Mexico and the *Reserva de Recursos Genéticos de Apacunca* in Nicaragua. Teosinte populations are threatened due to overgrazing, the use of herbicides, and the cultivation of maize and other crops in their natural habitats (Guzzon et al. 2021).

A complementary strategy to in situ conservation for local maize varieties is on-farm conservation, a form of genetic conservation of agrobiodiversity mainly carried out by small farmers, local communities, and indigenous populations. The conservation of local varieties in farms favors environmental interactions and gene exchange with wild species, allowing local varieties to evolve in their original distribution range and contributing to the maintenance of maize genetic variability (Guzzon et al. 2021). Thus, farming communities and their agricultural practices can significantly contribute to the conservation of genetic resources.

Ex situ maize conservation occurs in germplasm banks and includes collections of maize landraces, improved populations (synthetics, varieties, or segregating genotypes from breeding programs), inbreds, hybrids, genetic stocks (natural genes and transgenes), and wild species (caryopses and clones). Maize seeds are orthodox and tolerate dehydration and storage for long periods in cold chambers (-20 °C), allowing the storage of many accessions in small spaces. Seeds can also be cryopreserved, i.e., kept at negative temperatures (-196 °C).

The number of maize crop accessions conserved worldwide is believed to be higher than 135,000. One of the largest databases is the germplasm bank of the International Maize and Wheat Improvement Center (CIMMYT) in Mexico, which has approximately 26,500 accessions from Latin America and the Caribbean. Other important maize banks are located at the *Instituto Nacional de Investigaciones Forestales, Agrícolas y Pecuarias* (INIFAP) in Mexico, the *Instituto de Ciencia y Tecnologia Agrícola* (ICTA) in Guatemala, the *Instituto Nacional de Innovación Agraria* (INIA) in Peru, the *Instituto Nacional de Tecnología Agropecuaria* (INTA) in Argentina, and the *National University* in Colombia (Guzzon et al. 2021).

In Brazil, the germplasm bank of *Embrapa Milho e Sorgo* has approximately 4,000 accessions, most of which (82%) are from local varieties collected in the Brazilian territory. The accessions are grouped into racial compounds formed by the national collection (3.9%), improved accessions (6.0%), introduced accessions (7.8%), and wild relatives (0.2%). Maize accessions in these collections are described according to many distinguishing characteristics (Teixeira and Costa 2010). CIMMYT published the descriptors for maize as the suggestion of an international pattern to characterize maize germplasm (Teixeira and Costa 2010).

Ex situ maize conservation is also performed through in vivo collections of wild teosinte taxa. These collections are established by collecting and planting individuals in an area different from the in situ condition. There are in vivo collections at CIMMYT, INIFAP, and the University of Guadalajara in Mexico. Ex situ conservation of wild maize is important since populations are threatened, and many have already disappeared (Guzzon et al. 2021).

One of the main activities of germplasm banks is to characterize accessions, expanding the knowledge about their databases in order to make their use viable in breeding programs. Maize genotypes stored in germplasm banks may contain genes for resistance to pests and environmental stresses, high productivity and precocity, and other desirable characteristics that can be explored in new maize cultivars. Furthermore, knowing the genetic variability of maize can foster sustainable use strategies, e.g., strengthening breeding programs for local traditional varieties and increasing the access of local farmers to quality germplasm, ensuring the profitability of the cultivation of local maize varieties (Guzzon et al. 2021).

The refined and complete genetic characterization of maize germplasm accessions is currently based on evaluating morphological, physiological, agronomic, and molecular characteristics. The characterization of variability is performed using genetic markers, i.e., morphological or molecular characteristics that differentiate individuals. Morphological markers are easily identifiable phenotypic characteristics, e.g., kernel color and grain type, plant height, number of leaves above the ear, and leaf shape (de Lima et al. 2020). Molecular markers, in turn, are DNA fragments that reveal polymorphism between individuals. Several DNA-based molecular markers, including microsatellites and SNPs (single nucleotide polymorphism), are used to characterize maize accessions (Adu et al. 2019; Shu et al. 2021). These markers can generate a large amount of information about the genetic identity, diversity, gene frequency, and phylogenetic relationships of maize genetic resources. More than 80,000 maize mutant stocks are conserved and annotated at the Maize Genetic Cooperation Stock Center (USDA-ARS GSZE) of the University of Illinois, USA (Prasanna 2012). The description of all maize mutant stocks of this collection can be accessed at MaizeGDB, the Maize Genetics and Genomics Database (http://www.maizegdb.org).

Although maize is a crop with one of the most significant volumes of published genomics information, these results are not yet exploited to their full potential in maize conservation. Many studies have shown how the genetic history of maize evolution affected its genome and phenotypic diversity, which can be explored in breeding and preserved for future use and food security. Advanced genome sequencing and pan-genome studies involving maize have revealed extensive structural variations within the species (Hufford et al. 2021), and epigenetic variation has stood out as one of the key factors in the evolution of maize biological diversity (Yu et al. 2020). The results obtained through population genomics can help identify the extension and origin of genetic variability, in addition to selecting crucial genes in maize germplasm, estimating individual inbreeding, detecting hybridization, inferring gene flow, and quantifying the population structure to delineate more appropriate stock management scales and conservation units in maize. SNP and haplotype analyses, for example, have revealed the extent of changes in the genetic purity of maize accessions during regeneration in ex situ gene banks and helped recommend the best practices for maintaining the original genetic diversity of gene bank accessions (Wen et al. 2011).

Population genomics can be very useful in maize conservation as it allows the large-scale genotyping of molecular markers. Increasing the number of loci in the analysis also increases the power and accuracy of various important parameters related to genetic diversity assessment, e.g., effective population size, migration rate, and selection coefficient. Moreover, examining multiple loci facilitates identifying and excluding loci under selection (outlier loci) that cause biased parameter estimates, increasing the accuracy of the estimation. Thus, the most direct contribution of genomics to conservation is to vastly increase the precision and accuracy of parameter estimates that require neutral loci by genotyping hundreds to thousands of neutral loci in many individuals (Luikart et al. 2003).

Thus, incorporating population genomics into conservation studies in maize can improve decisions related to how to prioritize landscapes and identify variation in locally adapted populations (Johnson et al. 2018) as well as to estimate whether the amount of genetic variation in a population is closely related to the effective population size (N_e). It can also help understand the evolution of fitness-related phenotypes and the genetic basis of inbreeding depression in maize. This understanding can guide conservation strategies and the management of wild populations in captive breeding programs, avoiding inbreeding depression and invoking genetic rescue by restoring the gene flow (Luikart et al. 2018). Effective conservation of maize genetic variability is crucial to ensure its preservation and availability for future generations.

7 Future Perspectives

Maize is one of the most explored crops in genetics, genomics, and breeding studies. It is used to develop models and methodologies later applied to other plants and incorporate biotechnology into breeding programs. The genome sequencing of maize is one of the most advanced, with a large amount of genomic data available in scientific literature and databases. Moreover, maize is one of the cultivated species with the greatest genetic variability, and pan-genome studies have revealed extensive structural variations within the species (Hufford et al. 2021). This information has helped geneticists and breeders better understand maize evolution and diversity, gene flow, admixture, inbreeding and outbreeding depression, speciation and helped identify important genes and genetic variants related to disease resistance, local adaptation, and improved genetic gains in breeding programs. However, the more we elucidate some aspects of maize genetics, the more questions arise, and a new universe of possibilities is unraveled.

With regard to maize conservation, genomic information (new and existing) should be used to effectively guide decisions related to how to prioritize landscapes, estimate the amount of genetic variation in populations, and propose conservation and management strategies for wild populations in captive breeding programs, avoiding inbreeding depression and invoking genetic rescue by restoring the gene flow (Luikart et al. 2003, 2018).

Since its release, the reference genome sequence B73 has been extensively used for maize functional genomics research, expanding the knowledge about maize domestication and the genetic basis of important traits. From this perspective, the information obtained by the several additional maize genome sequences reported in the last five years will greatly contribute to revealing the genetic diversity of maize, the connection between genetic variation and phenotypic variation, and maize improvement. It is also expected that the graph-reference genome approach and other newly developed methods will soon be adopted in maize research to integrate the increasingly available amounts of variation data and genome sequences. Several gene-expression profiles have already contributed much toward identifying development networks and could also provide the first hints of the functions of genes of interest (Liu et al. 2020).

The availability of high-quality maize genomes and pan-genomes has provided new discoveries in maize domestication. In the next years or decades, these methods may perhaps reveal unexpected aspects of maize evolution and provide insights regarding the genetic architecture of local adaptation based on conserved regions of introgression. However, many aspects of the contribution of spp. *mexicana* to highland adaptation in maize remain to be resolved (Hufford et al. 2013). With pan-genome information, maize breeders can more effectively identify causal genetic variants underlying domestication traits and apply gene-editing tools to rapidly achieve desirable agronomic traits. Together, pan-genome and CRISPR/Cas9 technologies can allow the de novo domestication of wild plants and reduce barriers to the use of genetic variation from wild maize relatives (Coletta et al. 2021).

GWAS has already identified several QTLs of interest in maize, presenting opportunities to enable novel, gene-editing approaches for breeding programs (Coletta et al. 2021). Transgenic drought-tolerant maize was already developed after the detection of ZmVPP1 by GWAS (Wang et al. 2016b). Therefore, as genome-editing technologies continue to improve, the use of GWAS is expected to increase the identification of target genes for editing.

Another significant contribution that population genomics can bring to maize breeding is the reduction of the experimental effort, especially in genomic selection, where the use of information on kinship, prediction histories, and sparse models can help reduce the need for experimental areas and the number of crosses and inputs used in a breeding program (Ames and Bernardo 2020). In the case of using populations with close genetic relationships and limited historical data, maize breeders can adopt approaches such as the selective testing of population subsets in preliminary testing to reduce the cost of early-stage testing. As historical data accumulates, these strategies will converge to eliminate all preliminary production tests and reduce the time needed to develop new varieties and generation intervals, resulting in the early recycling of new inbreds (Atanda et al. 2021).

Unbalanced prediction models can be used to optimize resources, including predicting a particular hybrid in an untested environment or a hybrid combination that has not been tested in any environment. When using models with field tests based on different relationships between observed and unobserved genotypes (O/NO) in test environments, models based on marker effects and including the GE interaction are the ones that capture the greatest phenotypic variability (smaller residual variance). Genomic models that consider the GE interaction provide higher prediction accuracy than main-effect models in different allocation compositions comprising different combinations of O/NO genotypes in environments. Reducing the size of test populations slightly decreases accuracy; however, the predictive ability is regained when we increase the number of common genotypes tested in environments (Jarquin et al. 2020).

Models that estimate the effects of the environment, genotype, markers, and the GE interaction allow maintaining accuracy when two extreme situations occur, when all genotypes are overlapping, or when there is no overlap of genotypes in the test environments, thus reducing the size of the training set. Therefore, it is possible to save resources by optimizing projects using models based on genomic effects and including GE interaction. Due to the number of genotypes tested, it is recommended (but not necessary) to have a small proportion of overlapping genotypes across environments, whereas a large proportion of genotypes should not overlap across environments (Jarquin et al. 2020).

Another aspect to be considered is that maize breeding involves two critical steps. In addition to developing superior inbred lines from improved populations, it is necessary to identify the elite combinations of two inbred lines. With the development of double haploid lines and other technologies, breeders have developed a large number of inbreds that need to be evaluated for their performance in crosses. Thus, the number of potential crosses grows rapidly, making field evaluation of hybrid performance time- and resource-intensive. In this scenario, sparse partial diallel is becoming increasingly common in breeding practice. The general combining ability is primarily a measure of additive effects and indicates which inbreds should be selected. An accurate prediction of the general combining abilities may increase the selection efficiency of inbred lines and then accelerate hybrid crossing in breeding programs, especially in scenarios with datasets based only on sparse partial diallel cross designs (SPDC). Then, the best inbred lines can be selected for different heterotic groups, and some corresponding testers can be used to carry out field validation tests (Wang et al. 2020).

Finally, a significant reduction in the experimental effort can be achieved in maize breeding programs using population genomics. From this perspective, reducing the need for individual phenotyping and the number of crosses required can reduce the experimental areas and resources required to obtain new cultivars.

References

Aci MM, Lupini A, Mauceri A, Morsli A, Khelifi L, Sunseri F. Genetic variation and structure of maize populations from Saoura and Gourara Oasis in Algerian Sahara. BMC Genet. 2018;19(1): 1–10. https://doi.org/10.1186/s12863-018-0655-2.

Ali ML, Taylor JH, Jie L, Sun G, William M, Kasha KJ, et al. Molecular mapping of QTLs for resistance to Gibberella Ear Rot, in Corn, caused by Fusarium Graminearum. Genome. 2005;48 (3):521–33. https://doi.org/10.1139/G05-014.

Acosta-Pech R, Crossa J, de los Campos G, Teyssèdre S, Claustres B, Pérez-Elizalde S, et al. Genomic models with genotype × environment interaction for predicting hybrid performance: an application in maize hybrids. Theor Appl Genet. 2017;130(7):1431–40. https://doi.org/10.1007/s00122-017-2898-0.

Adu GB, Badu-Apraku B, Akromah R, Garcia-Oliveira AL, Awuku FJ, Gedil M. Genetic diversity and population structure of early-maturing tropical maize inbred lines using SNP markers. PLoS One. 2019;14(4):1–12. https://doi.org/10.1371/journal.pone.0214810.

Almeida C, Amorim EP, Neto JFB, Filho JAC, Cruz MJ, de Melo Sereno. Genetic variability in populations of sweet corn, common corn and teosinte. Crop Breeding and Applied Biotechnology. 2011;11(1):64–9. https://doi.org/10.1590/s1984-70332011000100009.

Alseekh S, Kostova D, Bulut M, Fernie AR. Genome-wide association studies: assessing trait characteristics in model and crop plants. Cell Mol Life Sci. 2021;78(15):5743–54. https://doi.org/10.1007/s00018-021-03868-w.

Alves FC, Galli G, Matias FI, Vidotti MS, Morosini JS, Fritsche-Neto R. Impact of the complexity of genotype by environment and dominance modeling on the predictive accuracy of maize hybrids in multi-environment prediction models. Euphytica. 2021;217(3):37. https://doi.org/10.1007/s10681-021-02779-y.

Alzate-Marin AL, Cervigni GDL, Moreira MA, Barros EG. Seleção Assistida Por Marcadores Moleculares Visando Ao Desenvolvimento de Plantas Resistentes a Doenças, Com Ênfase Em Feijoeiro e Soja. Fitopatol Bras. 2005;30(4):333–42. https://doi.org/10.1590/s0100-41582005000400001.

Ames NC, Bernardo R. Genomewide predictions as a substitute for a portion of phenotyping in maize. Crop Sci. 2020;60(1):181–9. https://doi.org/10.1002/csc2.20082.

Aschard H, Vilhjálmsson BJ, Greliche N, Morange PE, Trégouët DA, Kraft P. Maximizing the power of principal-component analysis of correlated phenotypes in genome-wide association studies. Am J Hum Genet. 2014;94(5):662–76. https://doi.org/10.1016/j.ajhg.2014.03.016.

Atanda SA, Olsen M, Burgueño J, Crossa J, Dzidzienyo D, Beyene Y, et al. Maximizing efficiency of genomic selection in CIMMYT's tropical maize breeding program. Theor Appl Genet. 2021;134(1):279–94. https://doi.org/10.1007/s00122-020-03696-9.

Atwell S, Huang YS, Vilhjálmsson BJ, Willems G, Horton M, Li Y, et al. Genome-wide association study of 107 phenotypes in Arabidopsis Thaliana inbred lines. Nature. 2010;465(7298):627–31. https://doi.org/10.1038/nature08800.

Avendaño López, Natividad A, de Jesús Sánchez González J, Corral JAR, Larios LDLC, Santacruz-Ruvalcaba F, et al. Seed dormancy in Mexican teosinte. Crop Sci. 2011;51(5):2056–66. https://doi.org/10.2135/cropsci2010.09.0538.

Avery CL, He Q, North KE, Ambite JL, Boerwinkle E. University of North Carolina at Chapel Hill (N01-HC-55015), Baylor Medical College (N01-HC-55016), University of Mississippi Medical Center (N01-HC-55021). PLoS Genet l WwwPlosgeneticsOrg. 2011;7(10):1. https://doi.org/10.1371/journal.pgen.1002322.

Badu-Apraku B, Garcia-Oliveira AL, Petroli CD, Hearne S, Adewale SA, Gedil M. Genetic diversity and population structure of early and extra-early maturing maize germplasm adapted to sub-Saharan Africa. BMC Plant Biol. 2021;21(1):1–15. https://doi.org/10.1186/s12870-021-02829-6.

Bandeirae Sousa, Massaine JC, de Oliveira Couto EG, Pérez-Rodríguez P, Jarquín D, Fritsche-Neto R, et al. Genomic-enabled prediction in maize using Kernel models with genotype × environment interaction. G3 Genes|Genomes|Genet. 2017;7(6):1995–2014. https://doi.org/10.1534/g3.117.042341.

Baseggio M, Murray M, Magallanes-Lundback M, Kaczmar N, Chamness J, Buckler ES, et al. Natural variation for carotenoids in fresh Kernels is controlled by uncommon variants in sweet corn. Plant Genome. 2020;13(1):e20008. https://doi.org/10.1002/tpg2.20008.

Basnet BR, Crossa J, Dreisigacker S, Pérez-Rodríguez P, Manes Y, Singh RP, et al. Hybrid wheat prediction using genomic, pedigree, and environmental covariables interaction models. Plant Genome. 2019;12(1):180051. https://doi.org/10.3835/plantgenome2018.07.0051.

Beadle GW. Teosinte and the origin of maize. J Hered. 1939;30(6):245–7. https://doi.org/10.1093/oxfordjournals.jhered.a104728.

Beadle GW. The ancestry of corn. Sci Am. 1980;242(1):112–9.

Beckett TJ, Jason Morales A, Koehler KL, Rocheford TR. Genetic relatedness of previously plant-variety-protected commercial maize Inbreds. PLoS One. 2017;12(12):1–23. https://doi.org/10.1371/journal.pone.0189277.

Beissinger TM, Wang L, Crosby K, Durvasula A, Hufford MB, Ross-Ibarra J. Recent demography drives changes in linked selection across the maize genome. Nat Plants. 2016;2(7):16084. https://doi.org/10.1038/nplants.2016.84.

Beló A, Zheng P, Luck S, Shen B, Meyer DJ, Li B, et al. Whole genome scan detects an allelic variant of Fad2 associated with increased oleic acid levels in maize. Mol Gen Genomics. 2008;279(1):1–10. https://doi.org/10.1007/s00438-007-0289-y.

Benavente E, Giménez E. Modern approaches for the genetic improvement of rice, wheat and maize for abiotic constraints-related traits: a comparative overview. Agronomy. 2021;11(2):376. https://doi.org/10.3390/agronomy11020376.

Benke A, Urbany C, Stich B. Genome-wide association mapping of iron homeostasis in the maize association population. BMC Genet. 2015;16(1):1–13. https://doi.org/10.1186/s12863-014-0153-0.

Bernardo R. Prediction of maize single-cross performance using RFLPs and information from related hybrids. Crop Sci. 1994;34(1):20–5. https://doi.org/10.2135/cropsci1994.0011183X003400010003x.

Bernardo R. Genomewide selection for rapid introgression of exotic germplasm in maize. Crop Sci. 2009;49(2):419–25. https://doi.org/10.2135/cropsci2008.08.0452.

Bernardo R. Essentials of plant breeding. Woodbury: Stemma Press; 2014.

Bernardo R. Bandwagons I, too, have known. Theor Appl Genet. 2016;129(12):2323–32. https://doi.org/10.1007/s00122-016-2772-5.

Bernardo R, Jianming Y. Prospects for Genomewide selection for quantitative traits in maize. Crop Sci. 2007;47(3):1082–90. https://doi.org/10.2135/cropsci2006.11.0690.

Beyene Y, Gowda M, Olsen M, Robbins KR, Pérez-Rodríguez P, Alvarado G, et al. Empirical comparison of tropical maize hybrids selected through genomic and phenotypic selections. Front Plant Sci. 2019;10(November) https://doi.org/10.3389/fpls.2019.01502.

Bhat JA, Ali S, Salgotra RK, Mir ZA, Dutta S, Jadon V, et al. Genomic selection in the era of next generation sequencing for complex traits in plant breeding. Front Genet. 2016;7(December) https://doi.org/10.3389/fgene.2016.00221.

Bhave MR, Lawrence S, Barton C, Curtis Hannah L. Identification and molecular characterization of Shrunken-2 CDNA Clones of maize. Plant Cell. 1990;2(6):581–8. https://doi.org/10.1105/tpc.2.6.581.

Bonnett OT. The inflorescences of maize. Science. 1954;120(3107):77–87. https://doi.org/10.1126/science.120.3107.77.

Bornowski N, Michel KJ, Hamilton JP, Shujun O, Seetharam AS, Jenkins J, et al. Genomic variation within the maize stiff-stalk heterotic germplasm Pool. Plant Genome. 2021; https://doi.org/10.1002/tpg2.20114.

Briggs WH, McMullen MD, Gaut BS, Doebley J. Linkage mapping of domestication loci in a large maize–teosinte backcross resource. Genetics. 2007;177(3):1915–28. https://doi.org/10.1534/genetics.107.076497.

Brown WL, Edgar. Anderson. The northern Flint corns. Ann Mo Bot Gard. 1947;34(1):1–28.

Brown WL, Goodman MM. Races of maize. In: Sprague GF, editor. Corn and corn improvement. Madison: American Society of Agronomy; 1977. p. 49–88.

Brown PJ, Upadyayula N, Mahone GS, Tian F, Bradbury PJ, Myles S, et al. Distinct genetic architectures for male and female inflorescence traits of maize. PLoS Genet. 2011;7(11):1002383. https://doi.org/10.1371/journal.pgen.1002383.

Buckler ES, Stevens NM. 4. Maize origins, domestication, and selection. In: Darwin's harvest. Columbia University Press; 2006. p. 67–90. https://doi.org/10.7312/motl13316-005.

Buckler ES, Holland JB, Bradbury PJ, Acharya CB, Brown PJ, Browne C, et al. The genetic architecture of maize flowering time. Science. 2009;325(5941):714–8. https://doi.org/10.1126/science.1174276.

Buckner B, Miguel PS, Janick-Buckner D, Bennetzen JL. The Y1 gene of maize codes for phytoene synthase. Genetics. 1996;143(1):479–88.

Butoto EN, Marino TP, Holland JB. Effects of artificial inoculation on trait correlations with resistance to Fusarium ear rot and Fumonisin contamination in maize. Crop Sci. 2021;61(4): 2522–33. https://doi.org/10.1002/csc2.20551.

Calderón CI, Yandell BS, Doebley JF. Fine mapping of a QTL associated with Kernel row number on chromosome 1 of maize. PLoS One. 2016;11(3) https://doi.org/10.1371/journal.pone.0150276.

Calfee E, Gates D, Anne L, Taylor Perkins M, Coop G, Ross-Ibarra J. Selective sorting of ancestral introgression in maize and teosinte along an elevational cline. BioRxiv. 2021: 2021.03.05.434040. https://doi.org/10.1101/2021.03.05.434040.

Cantelmo NF, Von Pinho RG, Balestre M. Genome-wide prediction for maize single-cross hybrids using the GBLUP model and validation in different crop seasons. Mol Breed. 2017;37(4):51. https://doi.org/10.1007/s11032-017-0651-7.

Carena MJ, Hallauer AR, Miranda Filho JB. Quantitative genetics in maize breeding. Quantitative genetics in maize breeding. 3rd ed. New York: Springer New York; 2010. https://doi.org/10.1007/978-1-4419-0766-0.

Cassani E, Landoni M, Pilu R. Characterization of the Ra1 maize gene involved in inflorescence architecture. Sex Plant Reprod. 2006;19(3):145–50. https://doi.org/10.1007/s00497-006-0031-7.

Cavanagh C, Morell M, Mackay I, Powell W. From mutations to MAGIC: resources for gene discovery, validation and delivery in crop plants. Curr Opin Plant Biol. 2008;11(2):215–21. https://doi.org/10.1016/j.pbi.2008.01.002.

Chai L, Chen Z, Bian R, Zhai H, Cheng X, Peng H, et al. Dissection of two quantitative trait loci with pleiotropic effects on plant height and spike length linked in coupling phase on the short arm of chromosome 2D of common wheat (Triticum Aestivum L.). Theor Appl Genet. 2018;131(12):2621–37. https://doi.org/10.1007/s00122-018-3177-4.

Chaikam V, Gowda M, Nair SK, Melchinger AE, Boddupalli PM. Genome-wide association study to identify genomic regions influencing spontaneous fertility in maize haploids. Euphytica. 2019;215(8):1–14. https://doi.org/10.1007/s10681-019-2459-5.

Chan SWL, Henderson IR, Jacobsen SE. Gardening the genome: DNA methylation in Arabidopsis Thaliana. Nat Rev Genet. 2005;6(5):351–60. https://doi.org/10.1038/nrg1601.

Chen AH, Lipka AE. The use of targeted marker subsets to account for population structure and relatedness in genome-wide association studies of maize (Zea Mays L.). G3: Genes Genomes Genet. 2016;6(8):2365–74. https://doi.org/10.1534/g3.116.029090.

Chen G, Wang X, Hao J, Yan J, Ding J. Genome-wide association implicates candidate genes conferring resistance to maize rough dwarf disease in maize. PLoS One. 2015;10(11):e0142001. https://doi.org/10.1371/journal.pone.0142001.

Chen J, Shrestha R, Ding J, Zheng H, Chunhua M, Wu J, et al. Genome-wide association study and QTL mapping reveal genomic loci associated with Fusarium ear rot resistance in tropical maize germplasm. G3: Genes Genomes Genet. 2016;6(12):3803–15. https://doi.org/10.1534/g3.116.034561.

Cockram J, Mackay I. Genetic mapping populations for conducting high-resolution trait mapping in plants. In: Varshney RK, Pandey MK, Chitikineni A, editors. Plant genetics and molecular biology. Springer; 2018. p. 109–38.

Coles ND, McMullen MD, Balint-Kurti PJ, Pratt RC, Holland JB. Genetic control of photoperiod sensitivity in maize revealed by joint multiple population analysis. Genetics. 2010;184(3):799–812. https://doi.org/10.1534/genetics.109.110304.

Coletta RD, Qiu Y, Shujun O, Hufford MB, Hirsch CN. How the Pan-genome is changing crop genomics and improvement. Genome Biol. 2021;22(1):1–19. https://doi.org/10.1186/s13059-020-02224-8.

Cook JP, McMullen MD, Holland JB, Tian F, Bradbury P, Ross-Ibarra J, et al. Genetic architecture of maize Kernel composition in the nested association mapping and inbred association panels. Plant Physiol. 2012;158(2):824–34. https://doi.org/10.1104/pp.111.185033.

Costa-Neto G, Fritsche-Neto R, Crossa J. Nonlinear Kernels, dominance, and envirotyping data increase the accuracy of genome-based prediction in multi-environment trials. Heredity. 2021a;126(1):92–106. https://doi.org/10.1038/s41437-020-00353-1.

Costa-Neto G, Galli G, Carvalho HF, Crossa J, Fritsche-Neto R. "EnvRtype : a software to interplay enviromics and quantitative genomics in agriculture." Edited by D-J de Koning. G3 Genes|Genomes|Genetics. 2021b;11(4) https://doi.org/10.1093/g3journal/jkab040.

Cross HZ. ND265: a new parental line of early corn. North Dakota Farm Res. 1989;47(3):19–21.

Crossa J, De Los G, Campos PP, Gianola D, Burgueño J, Araus JL, et al. Prediction of genetic values of quantitative traits in plant breeding using pedigree and molecular markers. Genetics. 2010;186(2):713–24. https://doi.org/10.1534/genetics.110.118521.

Crossa J, Pérez P, de los Campos G, Mahuku G, Dreisigacker S, Magorokosho C. Genomic selection and prediction in plant breeding. J Crop Improv. 2011;25(3):239–61. https://doi.org/10.1080/15427528.2011.558767.

Crossa J, Pérez P, Hickey J, Burgueño J, Ornella L, Cerón-Rojas J, et al. Genomic prediction in CIMMYT maize and wheat breeding programs. Heredity. 2014;112(1):48–60. https://doi.org/10.1038/hdy.2013.16.

Cui Z, Luo J, Qi C, Ruan Y, Li J, Zhang A, et al. Genome-Wide Association Study (GWAS) reveals the genetic architecture of four husk traits in maize. BMC Genomics. 2016;17(1):1–14. https://doi.org/10.1186/s12864-016-3229-6.

Cui Z, Dong H, Zhang A, Ruan Y, He Y, Zhang Z. Assessment of the potential for genomic selection to improve husk traits in maize. G3 Genes|Genomes|Genetics. 2020;10(10):3741–9. https://doi.org/10.1534/g3.120.401600.

Dalsente Krause M, das Graças Dias KO, dos Santos JPR, de Oliveira AA, Guimarães LJM, Pastina MM, et al. Boosting predictive ability of tropical maize hybrids via genotype-by-environment interaction under multivariate GBLUP models. Crop Sci. 2020;60(6):3049–65. https://doi.org/10.1002/csc2.20253.

Dell'Acqua M, Gatti DM, Pea G, Cattonaro F, Coppens F, Magris G, et al. Genetic properties of the MAGIC maize population: a new platform for high definition QTL mapping in Zea Mays. Genome Biol. 2015;16(1):1–23. https://doi.org/10.1186/s13059-015-0716-z.

Deng C, Lv M, Li X, Zhao X, Li H, Li Z, et al. Identification and fine mapping of QsCR4.01, a novel major Qtl for resistance to Puccinia Polysora in maize. Plant Dis. 2020;104(7):1944–8. https://doi.org/10.1094/PDIS-11-19-2474-RE.

Desta ZA, Ortiz R. Genomic selection: genome-wide prediction in plant improvement. Trends Plant Sci. 2014;19(9):592–601. https://doi.org/10.1016/j.tplants.2014.05.006.

Dinesh A, Patil A, Zaidi PH, Kuchanur PH, Vinayan MT, Seetharam K. Genetic diversity, linkage disequilibrium and population structure among CIMMYT maize inbred lines, selected for heat tolerance study. Maydica. 2016;61(29):1–7.

Ding J, Ali F, Chen G, Li H, Mahuku G, Yang N, et al. Genome-wide association mapping reveals novel sources of resistance to northern corn leaf blight in maize. BMC Plant Biol. 2015;15(1):1–11. https://doi.org/10.1186/s12870-015-0589-z.

Dinges JR, Colleoni C, Myers AM, James MG. Molecular structure of three mutations at the maize sugary1 locus and their allele-specific phenotypic effects. Plant Physiol. 2001;125:1406–18.

Dion MF, Altschuler SJ, Wu LF, Rando OJ. Genomic characterization reveals a simple histone H4 acetylation code. Proc Natl Acad Sci. 2005;102(15):5501–6. https://doi.org/10.1073/pnas.0500136102.

Doebley J. Molecular evidence and the evolution of maize. Econ Bot. 1990;44(S3):6–27. https://doi.org/10.1007/BF02860472.

Doebley J. The genetics of maize evolution. Annu Rev Genet. 2004;38(1):37–59. https://doi.org/10.1146/annurev.genet.38.072902.092425.

Doebley JF, Iltis HH. Taxonomy of Zea (Gramineae). I. a Subgeneric classification with key to taxa. Am J Bot. 1980;67(6):982–93. https://doi.org/10.1002/j.1537-2197.1980.tb07730.x.

Doebley J, Stec A. Genetic analysis of the morphological differences between maize and teosinte. Genetics. 1991;129(1):285–95.

Doebley J, Stec A. Inheritance of the morphological differences between maize and teosinte: comparison of results for two F2 populations. Genetics. 1993;134(2):559–70. https://doi.org/10.1093/genetics/134.2.559.

Doebley J, Stec A, Hubbard L. The evolution of apical dominance in maize. Nature. 1997;386 (6624):485–8. https://doi.org/10.1038/386485a0.

Doebley JF, Gaut BS, Smith BD. The molecular genetics of crop domestication. Cell. 2006;127(7): 1309–21. https://doi.org/10.1016/j.cell.2006.12.006.

Dong Z, Li W, Unger-Wallace E, Yang J, Vollbrecht E, Chuck G. Ideal crop plant architecture is mediated by Tassels Replace Upper Ears1, a BTB/POZ Ankyrin repeat gene directly targeted by TEOSINTE BRANCHED1. Proc Natl Acad Sci. 2017;114(41):E8656–64. https://doi.org/10.1073/pnas.1714960114.

Dorweiler JE, Doebley J. Developmental analysis of Teosinte Glume Architecture1 : a key locus in the evolution of maize (Poaceae). Am J Bot. 1997;84(10):1313–22. https://doi.org/10.2307/2446130.

Duo H, Hossain F, Muthusamy V, Zunjare RU, Goswami R, Chand G, et al. Development of sub-tropically adapted diverse Provitamin-a rich maize Inbreds through marker-assisted pedigree selection, their characterization and utilization in hybrid breeding. PLoS One. 2021;16 (2 February):1–22. https://doi.org/10.1371/journal.pone.0245497.

Eichten SR, Swanson-Wagner RA, Schnable JC, Waters AJ, Hermanson PJ, Liu S, et al. Heritable epigenetic variation among maize Inbreds. PLoS Genet. 2011;7(11) https://doi.org/10.1371/journal.pgen.1002372.

Eichten SR, Briskine R, Song J, Li Q, Swanson-Wagner R, Hermanson PJ, et al. Epigenetic and genetic influences on DNA methylation variation in maize populations. Plant Cell. 2013;25(8): 2783–97. https://doi.org/10.1105/tpc.113.114793.

Emms DM, Kelly S. STRIDE: species tree root inference from gene duplication events. Mol Biol Evol. 2017;34:3267–78. https://doi.org/10.1093/molbev/msx259.

Emms DM, Kelly S. STAG: species tree inference from all genes. bioRxiv. 2018:267914. https://doi.org/10.1101/267914.

Ertiro BT, Labuschagne M, Olsen M, Das B, Prasanna BM, Gowda M. Genetic dissection of nitrogen use efficiency in tropical maize through genome-wide association and genomic prediction. Front Plant Sci. 2020;11(April):474. https://doi.org/10.3389/fpls.2020.00474.

Evans MMS, Kermicle JL. Teosinte crossing Barrier1, a locus governing hybridization of teosinte with maize. Theor Appl Genet. 2001;103(2–3):259–65. https://doi.org/10.1007/s001220100549.

Eyre-Walker A, Gaut RL, Hilton H, Feldman DL, Gaut BS. Investigation of the bottleneck leading to the domestication of maize. Proc Natl Acad Sci. 1998;95(8):4441–6. https://doi.org/10.1073/pnas.95.8.4441.

Farfan ID, Barrero GN, La Fuente D, Murray SC, Isakeit T, Huang PC, et al. Genome wide association study for drought, Aflatoxin resistance, and important agronomic traits of maize hybrids in the sub-tropics. PLoS One. 2015;10(2):e0117737. https://doi.org/10.1371/journal.pone.0117737.

Flint-Garcia SA, Thornsberry JM, Edward IV SB. Structure of linkage disequilibrium in plants. Annu Rev Plant Biol. 2003;54:357–74. https://doi.org/10.1146/annurev.arplant.54.031902.134907.

Flint-Garcia SA, Thuillet AC, Jianming Y, Pressoir G, Romero SM, Mitchell SE, et al. Maize association population: a high-resolution platform for quantitative trait locus dissection. Plant J. 2005;44(6):1054–64. https://doi.org/10.1111/j.1365-313X.2005.02591.x.

Fristche-Neto R, Akdemir D, Jannink J-L. Accuracy of genomic selection to predict maize single-crosses obtained through different mating designs. Theor Appl Genet. 2018;131(5):1153–62. https://doi.org/10.1007/s00122-018-3068-8.

Fritsche-Neto R, Galli G, Borges KLR, Costa-Neto G, Alves FC, Sabadin F, et al. Optimizing genomic-enabled prediction in small-scale maize hybrid breeding programs: a roadmap review. Front Plant Sci. 2021;12(July) https://doi.org/10.3389/fpls.2021.658267.

Fu J, Cheng Y, Linghu J, Yang X, Kang L, Zhang Z, et al. RNA sequencing reveals the complex regulatory network in the maize Kernel. Nat Commun. 2013;4(1):1–12. https://doi.org/10.1038/ncomms3832.

Fukunaga K, Hill J, Vigouroux Y, Matsuoka Y, Jesus Sanchez G, Liu K, et al. Genetic diversity and population structure of teosinte. Genetics. 2005;169(4):2241–54. https://doi.org/10.1534/genetics.104.031393.

Gage JL, Monier B, Giri A, Buckler ES. Ten years of the maize nested association mapping population: impact, limitations, and future directions. Plant Cell. 2020;32(7):2083–93. https://doi.org/10.1105/tpc.19.00951.

Galesloot TE, Van Steen K, Kiemeney LALM, Janss LL, Vermeulen SH. A comparison of multivariate genome-wide association methods. PLoS One. 2014;9(4):95923. https://doi.org/10.1371/journal.pone.0095923.

Galinat WC, Gunnerson JH. Spread of eight-rowed maize from the prehistoric southwest. Harvard University Botanical Museum Leaflets. 1963;20(5):117–60.

Gallavotti A, Zhao Q, Kyozuka J, Meeley RB, Ritter MK, Doebley JF, et al. The role of Barren Stalk1 in the architecture of maize. Nature. 2004;432(7017):630–5. https://doi.org/10.1038/nature03148.

Gao J, Wang S, Zhou Z, Wang S, Dong C, Cong M, et al. Linkage mapping and genome-wide association reveal candidate genes conferring Thermotolerance of seed-set in maize. J Exp Bot. 2019;70(18):4849–63. https://doi.org/10.1093/jxb/erz171.

Gardiner JM, Coe EH, Melia-Hancock S, Hoisington DA, Chao S. Development of a Core RFLP map in maize using an immortalized F2 population. Genetics. 1993;134:917–30.

Garin V, Wimmer V, Mezmouk S, Malosetti M, van Eeuwijk F. How do the type of QTL effect and the form of the residual term influence QTL detection in multi-parent populations? A case study in the maize EU-NAM population. Theor Appl Genet. 2017;130(8):1753–64. https://doi.org/10.1007/s00122-017-2923-3.

Gaudin ACM, McClymont SA, Soliman SSM, Raizada MN. The effect of altered dosage of a Mutant Allele of Teosinte branched 1 (Tb1-Ref) on the root system of modern maize. BMC Genet. 2014;15(1):23. https://doi.org/10.1186/1471-2156-15-23.

GCDT. Global Strategy for the Ex Situ Conservation and Utilization of Maize Germplasm. Global Crop Diversity Trust, no. September, 2007. https://www.croptrust.org/wp/wp-content/uploads/2014/12/Maize-Strategy-FINAL-18Sept07.pdf.

Gerdes JT, Behr CF, Coors JG, Tracy WF. Compilation of North American maize breeding germplasm. Madison, WI: Crop Science Society of America; 1993. https://doi.org/10.2135/1993.compilationofnorthamerican.

Gethi JG, Labate JA, Lamkey KR, Smith ME, Kresovich S. SSR variation in important U.S. maize inbred lines. Crop Sci. 2002;42(3):951–7. https://doi.org/10.2135/cropsci2002.9510.

Giordani W, Scapim CA, Ruas PM, de Fátima C, Ruas RC-S, Coan M, et al. Genetic diversity, population structure and AFLP markers associated with maize reaction to southern rust. Bragantia. 2019;78(2):183–96. https://doi.org/10.1590/1678-4499.20180180.

Gireesh C, Sundaram RM, Anantha SM, Pandey MK, Madhav MS, Rathod S, et al. Nested Association Mapping (NAM) populations: present status and future prospects in the genomics era. Crit Rev Plant Sci. 2021;40(1):49–67. https://doi.org/10.1080/07352689.2021.1880019.

Gonzalez VH, Lee EA, Lewis Lukens L, Swanton CJ. The relationship between floret number and plant dry matter accumulation varies with early season stress in maize (Zea Mays L.). Field Crop Res. 2019;238(March):129–38. https://doi.org/10.1016/j.fcr.2019.05.003.

Gorjanc G, Jenko J, Hearne SJ, Hickey JM. Initiating maize pre-breeding programs using genomic selection to harness polygenic variation from landrace populations. BMC Genomics. 2016;17(1):30. https://doi.org/10.1186/s12864-015-2345-z.

Gowda M, Das B, Makumbi D, Babu R, Semagn K, Mahuku G, et al. Genome-wide association and genomic prediction of resistance to maize lethal necrosis disease in tropical maize germplasm. Theor Appl Genet. 2015;128(10):1957–68. https://doi.org/10.1007/s00122-015-2559-0.

Gowda M, Makumbi D, Das B, Nyaga C, Kosgei T, Crossa J, et al. Genetic dissection of Striga Hermonthica (Del.) Benth. Resistance via genome-wide association and genomic prediction in tropical maize germplasm. Theor Appl Genet. 2021;134(3):941–58. https://doi.org/10.1007/s00122-020-03744-4.

Guimarães CT, da Costae Silva L, Mendes FF, Pastina MM, de Souza IRP, Damasceno CMB. Mapeamento de QTLs e Seleção Assistida Por Marcadores Moleculares. In: de Milho M, Lima R, Borém A, editors. 1st ed; 2018. p. 307–28. Editora UFV.

Guo Z, Zou C, Liu X, Wang S, Li W-X, Jeffers D, et al. Complex genetic system involved in Fusarium ear rot resistance in maize as revealed by GWAS, bulked sample analysis, and genomic prediction. Plant Dis. 2020;104(6):1725–35. https://doi.org/10.1094/PDIS-07-19-1552-RE.

Guzzon F, Rios LWA, Cepeda GMC, Polo MC, Cabrera AC, Figueroa JM, et al. Conservation and use of Latin American maize diversity: pillar of nutrition security and cultural heritage of humanity. Agronomy. 2021;11(1):172. https://doi.org/10.3390/agronomy11010172.

Haberer G, Young S, Bharti AK, Gundlach H, Raymond C, Fuks G, et al. Structure and architecture of the maize genome. Plant Physiol. 2005;139(4):1612–24. https://doi.org/10.1104/pp.105.068718.

Hake S, Ross-Ibarra J. Genetic, evolutionary and plant breeding insights from the domestication of maize. elife. 2015;4(March) https://doi.org/10.7554/eLife.05861.

Han S, Thomas Miedaner H, Utz F, Schipprack W, Schrag TA, Melchinger AE. Genomic prediction and GWAS of Gibberella ear rot resistance traits in dent and Flint lines of a public maize breeding program. Euphytica. 2018;214(1):6. https://doi.org/10.1007/s10681-017-2090-2.

Hanson MA, Gaut BS, Stec AO, Fuerstenberg SI, Goodman MM, Coe EH, et al. Evolution of anthocyanin biosynthesis in maize Kernels: the role of regulatory and enzymatic loci. Genetics. 1996;143(3):1395–407. https://doi.org/10.1093/genetics/143.3.1395.

Hao Z, Li X, Xie C, Weng J, Li M, Zhang D, et al. Identification of functional genetic variations underlying drought tolerance in maize using Snp markers. J Integr Plant Biol. 2011;53(8):641–52. https://doi.org/10.1111/j.1744-7909.2011.01051.x.

Harjes CE, Rocheford TR, Bai L, Brutnell TP, Kandianis CB, Sowinski SG, et al. Natural genetic variation in lycopene epsilon cyclase tapped for maize biofortification. Science. 2008;319 (5861):330–3. https://doi.org/10.1126/science.1150255.

Heerwaarden JV, Doebley J, Briggs WH, Glaubitz JC, Goodman MM, De Jesus J, et al. Genetic signals of origin, spread, and introgression in a large sample of maize landraces. Proc Natl Acad Sci. 2011;108(3):1088–92. https://doi.org/10.1073/pnas.1013011108.

Helentjaris T, Slocum M, Wright S, Schaefer A, Nienhuis J. Construction of genetic linkage maps in maize and tomato using restriction fragment length polymorphisms. Theor Appl Genet. 1986;72(6):761–9. https://doi.org/10.1007/BF00266542.

Holland JB, Marino TP, Manching HC, Wisser RJ. Genomic prediction for resistance to Fusarium ear rot and Fumonisin contamination in maize. Crop Sci. 2020;60(4):1863–75. https://doi.org/10.1002/csc2.20163.

Holliday JA, Hallerman EM, Haak DC. Genotyping and sequencing technologies in population genetics and genomics. 2018:83–125. https://doi.org/10.1007/13836_2017_5.

Horn F, Habekuß A, Stich B. Genes involved in barley yellow dwarf virus resistance of maize. Theor Appl Genet. 2014;127(12):2575–84. https://doi.org/10.1007/s00122-014-2400-1.

Hu Y, Colantonio V, Müller BSF, Leach KA, Nanni A, Finegan C, et al. Genome assembly and population genomic analysis provide insights into the evolution of modern sweet corn. Nat Commun. 2021;12(1) https://doi.org/10.1038/s41467-021-21380-4.

Huang J, Johnson AD, O'Donnell CJ. PRIMe: a method for characterization and evaluation of pleiotropic regions from multiple genome-wide association studies. Bioinformatics. 2011;27(9):1201–6. https://doi.org/10.1093/bioinformatics/btr116.

Huang J, Zhang J, Li W, Wei H, Duan L, Feng Y, et al. Genome-wide association analysis of ten chilling tolerance indices at the germination and seedling stages in maize. J Integr Plant Biol. 2013;55(8):735–44. https://doi.org/10.1111/jipb.12051.

Huang C, Sun H, Dingyi X, Chen Q, Liang Y, Wang X, et al. ZmCCT9 enhances maize adaptation to higher latitudes. Proc Natl Acad Sci. 2018;115(2):E334–41. https://doi.org/10.1073/pnas.1718058115.

Hubbard L, McSteen P, Doebley J, Hake S. Expression patterns and mutant phenotype of Teosinte Branched1 correlate with growth suppression in maize and teosinte. Genetics. 2002;162(4):1927–35. https://doi.org/10.1093/genetics/162.4.1927.

Hufford MB, Bilinski P, Pyhäjärvi T, Ross-Ibarra J. Teosinte as a model system for population and ecological genomics. Trends Genet. 2012a;28(12):606–15. https://doi.org/10.1016/j.tig.2012.08.004.

Hufford MB, Martínez-Meyer E, Gaut BS, Eguiarte LE, Tenaillon MI. "Inferences from the historical distribution of wild and domesticated maize provide ecological and evolutionary insight." Edited by John P. Hart. PLoS One. 2012b;7(11):e47659. https://doi.org/10.1371/journal.pone.0047659.

Hufford MB, Xun X, van Heerwaarden J, Pyhäjärvi T, Chia J-M, Cartwright RA, et al. Comparative population genomics of maize domestication and improvement. Nat Genet. 2012c;44(7):808–11. https://doi.org/10.1038/ng.2309.

Hufford MB, Lubinksy P, Pyhäjärvi T, Devengenzo MT, Ellstrand NC, Ross-Ibarra J. "The genomic signature of crop-wild introgression in maize". Edited by Rodney Mauricio. PLoS Genet. 2013;9(5):e1003477. https://doi.org/10.1371/journal.pgen.1003477.

Hufford MB, Seetharam AS, Woodhouse MR, Chougule KM, Shujun O, Liu J, et al. De novo assembly, annotation, and comparative analysis of 26 diverse maize genomes. Science. 2021;373(August):655–62. https://www.biorxiv.org/content/10.1101/2021.01.14.426684v1.abstract

Hung HY, Shannon LM, Tian F, Bradbury PJ, Chen C, Flint-Garcia SA, et al. ZmCCT and the genetic basis of day-length adaptation underlying the Postdomestication spread of maize. Proc Natl Acad Sci U S A. 2012;109(28):E1913–21. https://doi.org/10.1073/pnas.1203189109.

Iltis HH. From Teosinte to maize: the catastrophic sexual transmutation. Science. 1983;222(4626):886–94. https://doi.org/10.1126/science.222.4626.886.

Iltis HH, Doebley JF. Taxonomy of Zea (Gramineae). II. Subspecific categories in the Zea Mays complex and a generic synopsis. Am J Bot. 1980;67(6):994. https://doi.org/10.2307/2442442.

Jacobson A, Lian L, Zhong S, Bernardo R. General combining ability model for Genomewide selection in a Biparental cross. Crop Sci. 2014;54(3):895–905. https://doi.org/10.2135/cropsci2013.11.0774.

Jarquín D, Crossa J, Lacaze X, Du Cheyron P, Daucourt J, Lorgeou J, et al. A reaction norm model for genomic selection using high-dimensional genomic and environmental data. Theor Appl Genet. 2014;127(3):595–607. https://doi.org/10.1007/s00122-013-2243-1.

Jarquin D, Howard R, Crossa J, Beyene Y, Gowda M, Martini JWR, et al. Genomic prediction enhanced sparse testing for multi-environment trials. G3: Genes Genomes Genet. 2020;10(8):2725–39. https://doi.org/10.1534/g3.120.401349.

Jarquin D, de Leon N, Romay C, Bohn M, Buckler ES, Ciampitti I, et al. Utility of climatic information via combining ability models to improve genomic prediction for yield within the genomes to fields maize project. Front Genet. 2021;11:592–769. https://doi.org/10.3389/fgene.2020.592769.

Ji L, Neumann DA, Schmitz RJ. Crop epigenomics: identifying, unlocking, and harnessing cryptic variation in crop genomes. Mol Plant. 2015;8(6):860–70. https://doi.org/10.1016/j.molp.2015.01.021.

Jiao Y, Peluso P, Shi J, Liang T, Stitzer MC, Wang B, et al. Improved maize reference genome with single-molecule technologies. Nature. 2017;546(7659):524–7. https://doi.org/10.1038/nature22971.

Jiménez-Galindo JC, Malvar RA, Butrón A, Santiago R, Samayoa LF, Caicedo M, et al. Mapping of resistance to corn borers in a MAGIC population of maize. BMC Plant Biol. 2019;19(1):1–17. https://doi.org/10.1186/s12870-019-2052-z.

Jindal SK, Dhaliwal MS, Meena OP. Molecular advancements in male sterility systems of capsicum: a review. Plant Breed. 2020;139(1):42–64. https://doi.org/10.1111/pbr.12757.

Johnson JS, Krutovsky KV, Rajora OP, Gaddis KD, Cairns DM. Advancing biogeography through population genomics. In: Rajora O, editor. Population genomics. Cham: Population Genomics. Springer; 2018. https://doi.org/10.1007/13836_2018_39.

Kadam DC, Potts SM, Bohn MO, Lipka AE, Lorenz AJ. Genomic prediction of single crosses in the early stages of a maize hybrid breeding pipeline. G3 Genes|Genomes|Genetics. 2016;6(11):3443–53. https://doi.org/10.1534/g3.116.031286.

Kang HM, Zaitlen NA, Wade CM, Kirby A, Heckerman D, Daly MJ, et al. Efficient control of population structure in model organism association mapping. Genetics. 2008;178(3):1709–23. https://doi.org/10.1534/GENETICS.107.080101.

Kaur M, Vikal Y, Kaur H, Pal L, Kaur K, Chawla JS. Mapping quantitative trait loci associated with southern leaf blight resistance in maize (Zea Mays L.). J Phytopathol. 2019;167(10):591–600. https://doi.org/10.1111/jph.12849.

Kato YTA. Cytological studies of maize [Zea Mays L.] and Teosinte [Zea Mexicana Schrader Kuntze] in relation to their origin and evolution. Massachusetts Agricultural Experiment Station, 1976.

Kato YTA. Chromosome morphology and the origin of maize and its races. Evol Biol. 1984;17:219–53.

Kato YTA, Lopes RA. Chromosome knobs of the perennial teosintes. Maydica. 1990;35(2):125–41.

Kearsey MJ, Farquhar AGL. QTL analysis in plants; where are we now? Heredity. 1998;80(2):137–42. https://doi.org/10.1038/sj.hdy.6885001.

Kebede D, Mengesha W, Menkir A, Abe A, Garcia-Oliveira AL, Gedil M. Marker based enrichment of Provitamin a content in two tropical maize synthetics. Sci Rep. 2021;11(1):1–10. https://doi.org/10.1038/s41598-021-94586-7.

Kistler L, Yoshi Maezumi S, Gregorio J, de Souza NAS, Przelomska FM, Costa OS, et al. Multiproxy evidence highlights a complex evolutionary legacy of maize in South America. Science. 2018;362(6420):1309–13. https://doi.org/10.1126/science.aav0207.

Kistler L, Thakar HB, VanDerwarker AM, Domic A, Bergström A, George RJ, et al. Archaeological Central American maize genomes suggest ancient gene flow from South America. Proc Natl Acad Sci. 2020;117(52):33124–9. https://doi.org/10.1073/pnas.2015560117.

Klei L, Luca D, Devlin B, Roeder K. Pleiotropy and principal components of heritability combine to increase power for association analysis. Genet Epidemiol. 2008;32(1):9–19. https://doi.org/10.1002/gepi.20257.

Kuki MC, Scapim CA, Rossi ES, Mangolin CA, Do Amaral AT, Pinto RJB. Genome wide association study for gray leaf spot resistance in tropical maize core. PLoS One. 2018;13(6):e0199539. https://doi.org/10.1371/journal.pone.0199539.

Kuleshov NN. World's diversity of phenotypes of maize. J Am Soc Agron. 1933;25(10):688–700.

Kump KL, Bradbury PJ, Wisser RJ, Buckler ES, Belcher AR, Oropeza-Rosas MA, et al. Genome-wide association study of quantitative resistance to Southern Leaf Blight in the maize nested association mapping population. Nat Genet. Nature Publishing Group. 2011; https://doi.org/10.1038/ng.747.

Ladejobi O, Elderfield J, Gardner KA, Chris Gaynor R, Hickey J, Hibberd JM, et al. Maximizing the potential of multi-parental crop populations. Appl Transl Genom. 2016;11:9–17. https://doi.org/10.1016/j.atg.2016.10.002.

Lanes ECM, Viana JMS, Paes GP, Paula MFB, Maia C, Caixeta ET, et al. Population structure and genetic diversity of maize Inbreds derived from tropical hybrids. Genet Mol Res. 2014;13(3):7365–76. https://doi.org/10.4238/2014.September.12.2.

Lauter N, Doebley J. Genetic variation for phenotypically invariant traits detected in teosinte: implications for the evolution of novel forms. Genetics. 2002;160(1):333–42. https://doi.org/10.1093/genetics/160.1.333.

Lee M, Sharopova N, Beavis WD, Grant D, Katt M, Blair D, et al. Expanding the genetic map of maize with the Intermated B73 x Mo17 (IBM) population. Plant Mol Biol. 2002;48(5–6): 453–61. https://doi.org/10.1023/A:1014893521186.

Leiboff S, Li X, Heng Cheng H, Todt N, Yang J, Li X, et al. Genetic control of morphometric diversity in the maize shoot apical meristem. Nat Commun. 2015;6(1):1–10. https://doi.org/10.1038/ncomms9974.

Lemmon ZH, Bukowski R, Sun Q, Doebley JF. "The role of Cis regulatory evolution in maize domestication". Edited by Hunter Fraser. PLoS Genet. 2014;10(11):e1004745. https://doi.org/10.1371/journal.pgen.1004745.

Li Q, Yang X, Shutu X, Cai Y, Zhang D, Han Y, et al. Genome-wide association studies identified three independent polymorphisms associated with α-tocopherol content in maize Kernels. PLoS One. 2012;7(5):e36807. https://doi.org/10.1371/journal.pone.0036807.

Li K, Wang H, Xiaojiao H, Liu Z, Yujin W, Huang C, et al. Genome-wide association study dissects the genetic architecture of oil biosynthesis in maize kernels. Nat Genet. 2013;45(1):43–50. https://doi.org/10.1038/ng.2484.

Li M, Liu X, Bradbury P, Yu J, Zhang Y-M, Todhunter RJ, et al. Enrichment of statistical power for genome-wide association studies. BMC Biol. 2014a;12(1):1–10. https://doi.org/10.1186/S12915-014-0073-5.

Li Q, Eichten SR, Hermanson PJ, Zaunbrecher VM, Song J, Wendt J, et al. Genetic perturbation of the maize methylome. Plant Cell. 2014b;26(12):4602–16. https://doi.org/10.1105/tpc.114.133140.

Li C, Li YY, Shi Y, Song Y, Zhang D, Buckler ES, et al. Genetic control of the leaf angle and leaf orientation value as revealed by ultra-high density maps in three connected maize populations. PLoS One. 2015;10(3) https://doi.org/10.1371/journal.pone.0121624.

Li X, Zhou Z, Ding J, Yabin W, Zhou B, Wang R, et al. Combined linkage and association mapping reveals QTL and candidate genes for plant and ear height in maize. Front Plant Sci. 2016;7 (JUNE2016):833. https://doi.org/10.3389/fpls.2016.00833.

Li X, Jian Y, Xie C, Jun W, Yunbi X, Zou C. Fast diffusion of domesticated maize to temperate zones. Sci Rep. 2017;7(1):1–11. https://doi.org/10.1038/s41598-017-02125-0.

Li E, Liu H, Huang L, Zhang X, Dong X, Song W, et al. Long-range interactions between proximal and distal regulatory regions in maize. Nat Commun. 2019;10(1):2633. https://doi.org/10.1038/s41467-019-10603-4.

Li D, Zhenxiang X, Riliang G, Wang P, Jialiang X, Dengxiang D, et al. Genomic prediction across structured hybrid populations and environments in maize. Plan Theory. 2021;10(6):1174. https://doi.org/10.3390/plants10061174.

Li C, Guan H, Jing X, Li Y, Wang B, Li Y, et al. Genomic insights into historical improvement of heterotic groups during modern hybrid maize breeding. Nat Plants. 2022;8(7):750–63. https://doi.org/10.1038/s41477-022-01190-2.

Li W, Liu H, Cheng ZJ, Ying Hua S, Han HN, Zhang Y, et al. Dna methylation and histone modifications regulate de Novo shoot regeneration in Arabidopsis by modulating Wuschel expression and auxin signaling. PLoS Genet. 2011;7(8) https://doi.org/10.1371/journal.pgen.1002243.

Lian L, Jacobson A, Zhong S, Bernardo R. Genomewide prediction accuracy within 969 maize biparental populations. Crop Sci. 2014;54(4):1514–22. https://doi.org/10.2135/cropsci2013.12.0856.

Lima BC d, Dudek G, Chaves MHM, Martins AG, Missio VC, Missio RF. DIVERSIDADE GENÉTICA EM ACESSOS DE MILHO CRIOULO / Genetic diversity in landrace maize. Brazil J Dev. 2020;6(10):82712–26. https://doi.org/10.34117/bjdv6n10-631.

Lin G, He C, Zheng J, Koo DH, Le H, Zheng H, et al. Chromosome-level genome assembly of a regenerable maize inbred line A188. Genome Biol. 2021;22(1):1–30. https://doi.org/10.1186/s13059-021-02396-x.

Lipka AE, Gore MA, Magallanes-Lundback M, Mesberg A, Lin H, Tiede T, et al. Genome-wide association study and pathway-level analysis of Tocochromanol levels in maize grain. G3: Genes Genomes Genet. 2013;3(8):1287–99. https://doi.org/10.1534/g3.113.006148.

Lipka AE, Kandianis CB, Hudson ME, Jianming Y, Drnevich J, Bradbury PJ, et al. From association to prediction: statistical methods for the dissection and selection of complex traits in plants. Curr Opin Plant Biol. 2015;24(April):110–8. https://doi.org/10.1016/j.pbi.2015.02.010.

Lippert C, Listgarten J, Liu Y, Kadie CM, Davidson RI, Heckerman D. FaST linear mixed models for genome-wide association studies. Nat Methods. 2011;8(10):833–5. https://doi.org/10.1038/nmeth.1681.

Liu ZZ, Guo RH, Zhao JR, Cai YL, Wang FG, Cao MJ, et al. Analysis of genetic diversity and population structure of maize landraces from the South Maize Region of China. Agric Sci China. 2010;9(9):1251–62. https://doi.org/10.1016/S1671-2927(09)60214-5.

Liu S, Wang X, Wang H, Xin H, Yang X, Yan J, et al. Genome-wide analysis of ZmDREB genes and their association with natural variation in drought tolerance at seedling stage of Zea Mays L. PLoS Genet. 2013;9(9):e1003790. https://doi.org/10.1371/journal.pgen.1003790.

Liu L, Yanfang D, Huo D, Wang M, Shen X, Yue B, et al. Genetic architecture of maize Kernel row number and whole genome prediction. Theor Appl Genet. 2015a;128(11):2243–54. https://doi.org/10.1007/s00122-015-2581-2.

Liu L, Yanfang D, Shen X, Li M, Sun W, Huang J, et al. KRN4 controls quantitative variation in maize Kernel row number. PLoS Genet. 2015b;11(11):e1005670. https://doi.org/10.1371/journal.pgen.1005670.

Liu N, Xue Y, Guo Z, Li W, Tang J. Genome-wide association study identifies candidate genes for starch content regulation in maize kernels. Front Plant Sci. 2016a;7(JULY2016):1046. https://doi.org/10.3389/fpls.2016.01046.

Liu X, Huang M, Fan B, Buckler ES, Zhang Z. Iterative usage of fixed and random effect models for powerful and efficient genome- wide association studies. PLoS Genet. 2016b;12(2):1–24. https://doi.org/10.1186/1471-2156-13-100.

Liu H, Luo X, Niu L, Yingjie Xiao L, Chen JL, Wang X, et al. Distant EQTLs and non-coding sequences play critical roles in regulating gene expression and quantitative trait variation in maize. Mol Plant. 2017;10(3):414–26. https://doi.org/10.1016/j.molp.2016.06.016.

Liu J, Fernie AR, Yan J. The past, present, and future of maize improvement: domestication, genomics, and functional genomic routes toward crop enhancement. Plant Commun. 2020;1(1):100010. https://doi.org/10.1016/j.xplc.2019.100010.

Liu Y, Guanghui H, Zhang A, Loladze A, Yingxiong H, Wang H, et al. Genome-wide association study and genomic prediction of Fusarium ear rot resistance in tropical maize germplasm. Crop Journal. 2021;9(2):325–41. https://doi.org/10.1016/j.cj.2020.08.008.

Lorenzana RE, Bernardo R. Accuracy of genotypic value predictions for marker-based selection in biparental plant populations. Theor Appl Genet. 2009;120(1):151–61. https://doi.org/10.1007/s00122-009-1166-3.

Lu X, Wang J, Wang Y, Wen W, Zhang Y, Jianjun D, et al. Genome-wide association study of maize aboveground dry matter accumulation at seedling stage. Front Genet. 2021;11(January) https://doi.org/10.3389/fgene.2020.571236.

Lu Y, Yan J, Guimarães CT, Taba S, Hao Z, Gao S, et al. Molecular characterization of global maize breeding germplasm based on genome-wide single nucleotide polymorphisms. Theor Appl Genet. 2009;120(1):93–115. https://doi.org/10.1007/s00122-009-1162-7.

Luikart G, England PR, Tallmon D, Jordan S, Taberlet P. The power and promise of population genomics: from genotyping to genome typing. Nat Rev Genet. 2003;4(12):981–94. https://doi.org/10.1038/nrg1226.

Luikart G, Kardos M, Hand BK, Rajora OP, Aitken SN, Hohenlohe PA. Population genomics: advancing understanding of nature. In: Rajora O, editor. Population genomics. Cham: Population Genomics. Springer; 2018. https://doi.org/10.1007/13836_2018_60.

Luo M, Zhao Y, Zhang R, Xing J, Duan M, Li J, et al. Mapping of a major QTL for salt tolerance of mature field-grown maize plants based on SNP markers. BMC Plant Biol. 2017;17(1):1–10. https://doi.org/10.1186/s12870-017-1090-7.

Luo X, Wang B, Gao S, Zhang F, Terzaghi W, Dai M. Genome-wide association study dissects the genetic bases of salt tolerance in maize seedlings. J Integr Plant Biol. 2019;61(6):658–74. https://doi.org/10.1111/jipb.12797.

Lv M, Deng C, Li X, Zhao X, Li H, Li Z, et al. Identification and fine-mapping of RppCML496, a major QTL for resistance to Puccinia Polysora in maize. Plant Genome. 2021;14(1):1–7. https://doi.org/10.1002/tpg2.20062.

Machado AT, Torres C, de Toledo, Lourenço L. Manejo Da Diversidade Genética e Melhoramento Participativo de Milho Em Sistemas Agroecológicos genetic diversity management and maize participatory breeding under agroecological systems. Revista Brasileira de Agroecologia. 2011;6(1):127–36. http://orgprints.org/24142/1/Machado_Manejo.pdf

Mahato A, Shahi JP, Singh PK, Kumar M. Genetic diversity of sweet corn Inbreds using agro-morphological traits and microsatellite markers. 3 Biotech. 2018;8(8):1–9. https://doi.org/10.1007/s13205-018-1353-5.

Mahuku G, Chen J, Shrestha R, Narro LA, Guerrero KVO, Arcos AL, et al. Combined linkage and association mapping identifies a major QTL (QRtsc8-1), conferring tar spot complex resistance in maize. Theor Appl Genet. 2016;129(6):1217–29. https://doi.org/10.1007/s00122-016-2698-y.

Mammadov J, Sun X, Gao Y, Ochsenfeld C, Bakker E, Ren R, et al. Combining powers of linkage and association mapping for precise dissection of QTL controlling resistance to gray leaf spot disease in maize (Zea Mays L.). BMC Genomics. 2015;16(1):1–16. https://doi.org/10.1186/s12864-015-2171-3.

Mangelsdorf PC. Corn: its origin, evolution, and improvement. Cambridge: Belknap Press of Harvard University Press; 1974.

Mangelsdorf PC, Fraps GS. A direct quantitative relationship between vitamin A in corn and the number of genes for yellow pigmentation. Science. 1931;73:241–2.

Mangelsdorf PC, Reeves RG. The origin of maize. Proc Natl Acad Sci. 1938;24(8):303–12. https://doi.org/10.1073/pnas.24.8.303.

Mangelsdorf PC, Reeves RG. Origin of Indian corn and its relatives. Texas: Agricultural and Mechanical College of Texas; 1939.

Mao H, Wang H, Liu S, Li Z, Yang X, Yan J, et al. A transposable element in a NAC gene is associated with drought tolerance in maize seedlings. Nat Commun. 2015;6(1):1–13. https://doi.org/10.1038/ncomms9326.

Massman JM, Gordillo A, Lorenzana RE, Bernardo R. Genomewide predictions from maize single-cross data. Theor Appl Genet. 2013;126(1):13–22. https://doi.org/10.1007/s00122-012-1955-y.

Matsuoka Y, Vigouroux Y, Goodman MM, Sanchez JG, Buckler E, Doebley J. A single domestication for maize shown by multilocus microsatellite genotyping. Proc Natl Acad Sci. 2002;99(9):6080–4. https://doi.org/10.1073/pnas.052125199.

McLean-Rodríguez FD, Costich DE, Camacho-Villa TC, Pè ME, Dell'Acqua M. Genetic diversity and selection signatures in maize landraces compared across 50 years of in situ and ex situ conservation. Heredity. 2021;126(6):913–28. https://doi.org/10.1038/s41437-021-00423-y.

McMullen MD, Kresovich S, Villeda HS, Bradbury P, Li H, Sun Q, et al. Genetic properties of the maize nested association mapping population. Science. 2009;325(5941):737–40. https://doi.org/10.1126/science.1174320.

Mendes MP, Lopes C, de Souza. Genomewide prediction of tropical maize single-crosses. Euphytica. 2016;209(3):651–63. https://doi.org/10.1007/s10681-016-1642-1.

Merrill WL, Hard RJ, Mabry JB, Fritz GJ, Adams KR, Roney JR, et al. The diffusion of maize to the Southwestern United States and its impact. Proc Natl Acad Sci. 2009;106(50):21019–26. https://doi.org/10.1073/pnas.0906075106.

Meuwissen THEE, Hayes BJ, Goddard ME. Prediction of total genetic value using genome-wide dense marker maps. Genetics. 2001;157(4):1819–29. https://doi.org/10.1093/genetics/157.4.1819.

Moler ERV, Abakir A, Eleftheriou M, Johnson JS, Krutovsky KV, Lewis LC, et al. Population epigenomics: advancing understanding of phenotypic plasticity, acclimation, adaptation and diseases. 2018:179–260. https://doi.org/10.1007/13836_2018_59.

Mikel MA, Dudley JW. Evolution of North American Dent Corn from public to proprietary germplasm. Crop Sci. 2006;46(3):1193–205. https://doi.org/10.2135/cropsci2005.10-0371.

Môro GV, Fritsche-Neto R. Importância e Usos Do Milho No Brasil. In: Galvão JCC, Borém A, Pimentel MA, editors. Milho: Do Plantio à Colheita. Editora UFV; 2015. p. 9–25.

Mott R, Talbot CJ, Turri MG, Collins AC, Flint J. A method for fine mapping quantitative trait loci in outbred animal stocks. Proc Natl Acad Sci U S A. 2000;97(23):12649–54. https://doi.org/10.1073/pnas.230304397.

Mutegi E, Snow AA, Rajkumar M, Pasquet R, Ponniah H, Daunay MC, et al. Genetic diversity and population structure of wild/ weedy eggplant (solanum Insanum, Solanaceae) in Southern India: implications for conservation. Am J Bot. 2015;102(1):140–8. https://doi.org/10.3732/ajb.1400403.

N'da HA, Akanvou L, Pokou N'd D, Akanza KP, Kouakou CK, Zoro BIIA. Genetic diversity and population structure of maize landraces from Cte DIvoire. Afr J Biotechnol. 2016;15(44):2507–16. https://doi.org/10.5897/ajb2016.15678.

Nei M. Genetic distance between populations. Am Nat. 1972;1:283–92.

Nelimor C, Badu-Apraku B, Garcia-Oliveira AL, Tetteh A, Paterne A, N'guetta ASP, et al. Genomic analysis of selected maize landraces from Sahel and Coastal West Africa reveals their variability and potential for genetic enhancement. Genes. 2020;11(9):1–14. https://doi.org/10.3390/genes11091054.

Nelson PT, Coles ND, Holland JB, Bubeck DM, Smith S, Goodman MM. Molecular characterization of maize Inbreds with expired U.S. plant variety protection. Crop Sci. 2008;48(5):1673–85. https://doi.org/10.2135/cropsci2008.02.0092.

O'Reilly PF, Hoggart CJ, Pomyen Y, Calboli FCF, Elliott P, Jarvelin MR, et al. MultiPhen: joint model of multiple phenotypes can increase discovery in GWAS. PLoS One. 2012;7(5):34861. https://doi.org/10.1371/journal.pone.0034861.

Olmos SE, Lia VV, Eyhérabide GH. Genetic diversity and linkage disequilibrium in the argentine public maize inbred line collection. 2017;15(6):515–26. https://doi.org/10.1017/S1479262116000228.

Olukolu BA, Negeri A, Dhawan R, Venkata BP, Sharma P, Garg A, et al. A connected set of genes associated with programmed cell death implicated in controlling the hypersensitive response in maize. Genetics. 2013;193(2):609–20. https://doi.org/10.1534/genetics.112.147595.

Olukolu BA, Wang GF, Vontimitta V, Venkata BP, Marla S, Ji J, et al. A genome-wide association study of the maize hypersensitive defense response identifies genes that cluster in related pathways. PLoS Genet. 2014;10(8):e1004562. https://doi.org/10.1371/journal.pgen.1004562.

Ortiz R, Taba S, Chávez VH, Tovar MM, Yunbi X, Yan J, et al. Conserving and enhancing maize genetic resources as global public goods-a perspective from CIMMYT. Crop Sci. 2010;50(1):13–28. https://doi.org/10.2135/cropsci2009.06.0297.

Owens BF, Gore MA, Magallanes-Lundback M, Tiede T, Diepenbrock CH, Kandianis CB, et al. A Foundation for Provitamin a biofortification of maize: genome-wide association and genomic prediction models of carotenoid levels. Genetics. 2014;198(4):1699–716. https://doi.org/10.1534/genetics.114.169979.

Pace J, Gardner C, Romay C, Ganapathysubramanian B, Lübberstedt T. Genome-wide association analysis of seedling root development in maize (Zea Mays L.). BMC Genomics. 2015;16(1):1–12. https://doi.org/10.1186/s12864-015-1226-9.

Palaisa KA, Morgante M, Williams M, Rafalski A. Contrasting effects of selection on sequence diversity and linkage disequilibrium at two phytoene synthase loci[W]. Plant Cell. 2003;15(8):1795–806. https://doi.org/10.1105/tpc.012526.

Pan Q, Li L, Yang X, Tong H, Shutu X, Li Z, et al. Genome-wide recombination dynamics are associated with phenotypic variation in maize. New Phytol. 2016;210(3):1083–94. https://doi.org/10.1111/nph.13810.

Pan Q, Yuancheng X, Li K, Peng Y, Zhan W, Li W, et al. The genetic basis of plant architecture in 10 maize recombinant inbred line populations. Plant Physiol. 2017;175(2):858–73. https://doi.org/10.1104/pp.17.00709.

Parra JR, de Jesús J, González S, Cordero ÁAJ, Valtierra JAC, López JGM, et al. Maíces Nativos Del Occidente de México 1: Colectas 2004. Scientia-CUCBA. 2006;8(1):1–139.

Paterniani E, Goodman MM. Races of maize in Brazil and adjacent areas. Edited by CIMMYT, 1977.

Peiffer JA, Flint-Garcia SA, De Leon N, McMullen MD, Kaeppler SM, Buckler ES. The genetic architecture of maize stalk strength. PLoS One. 2013;8(6):e67066. https://doi.org/10.1371/journal.pone.0067066.

Peiffer JA, Romay MC, Gore MA, Flint-Garcia SA, Zhang Z, Millard MJ, et al. The genetic architecture of maize height. Genetics. 2014;196(4):1337–56. https://doi.org/10.1534/genetics.113.159152.

Perales H, Golicher D. "Mapping the diversity of maize races in Mexico." edited by Hany A. El-Shemy. PLoS One. 2014;9(12):e114657. https://doi.org/10.1371/journal.pone.0114657.

Platt A, Vilhjálmsson BJ, Nordborg M. Conditions under which genome-wide association studies will be positively misleading. Genetics. 2010;186(3):1045–52. https://doi.org/10.1534/genetics.110.121665.

Poland J, Rutkoski J. Advances and challenges in genomic selection for disease resistance. Annu Rev Phytopathol. 2016;54:79–98. https://doi.org/10.1146/annurev-phyto-080615-100056.

Poland JA, Bradbury PJ, Buckler ES, Nelson RJ. Genome-wide nested association mapping of quantitative resistance to northern leaf blight in maize. Proc Natl Acad Sci U S A. 2011;108(17): 6893–8. https://doi.org/10.1073/pnas.1010894108.

Poneleit CG. Breeding white endosperm corn. In: Hallauer AR, editor. Specialty corns. 2nd ed. Boca Raton: CRC Press; 2001. p. 235–73.

Prasanna BM. Diversity in global maize germplasm: characterization and utilization. J Biosci. 2012;37(5):843–55. https://doi.org/10.1007/s12038-012-9227-1.

Prasanna BM, Pixley K, Warburton ML, Xie CX. Molecular marker-assisted breeding options for maize improvement in Asia. Mol Breed. 2010;26(2):339–56. https://doi.org/10.1007/s11032-009-9387-3.

Preston JC, Wang H, Kursel L, Doebley J, Kellogg EA. The role of Teosinte Glume Architecture (Tga1) in coordinated regulation and evolution of grass glumes and inflorescence axes. New Phytol. 2012;193(1):204–15. https://doi.org/10.1111/j.1469-8137.2011.03908.x.

Price AL, Patterson NJ, Plenge RM, Weinblatt ME, Shadick NA, Reich D. Principal components analysis corrects for stratification in genome-wide association studies. Nat Genet. 2006;38(8): 904–9. https://doi.org/10.1038/ng1847.

Price AL, Kryukov GV, de Bakker PIW, Purcell SM, Staples J, Wei LJ, et al. Pooled association tests for rare variants in Exon-resequencing studies. Am J Hum Genet. 2010;86(6):832–8. https://doi.org/10.1016/j.ajhg.2010.04.005.

Pritchard JK, Stephens M, Rosenberg NA, Donnelly P. Association mapping in structured populations. Am J Hum Genet. 2000a;67(1):170–81. https://doi.org/10.1086/302959.

Pritchard JK, Stephens M, Donnelly P. Inference of population structure using multilocus genotype data. Genetics. 2000b;155(2):945–59. https://doi.org/10.1093/GENETICS/155.2.945.

Qiu Y, Kaiser C, Schmidt C, Broders K, Robertson AE, Jamann TM. Identification of quantitative trait loci associated with maize resistance to bacterial leaf streak. Crop Sci. 2020;60(1):226–37. https://doi.org/10.1002/csc2.20099.

Quijada P, Shannon LM, Glaubitz JC, Studer AJ, Doebley J. Characterization of a major maize domestication QTL on the short arm of chromosome 1. Maydica. 2009;54(4):401–8.

Rajora OP, editor. Population genomics. Cham: Springer International Publishing; 2019. https://doi.org/10.1007/978-3-030-04589-0.

Rakshit S, Rakshit A, Patil JV. Multiparent intercross populations in analysis of quantitative traits. J Genet. 2012;91(1):111–7.

Ranganatha HM, Lohithaswa HC, Pandravada A. Mapping and validation of major quantitative trait loci for resistance to northern corn leaf blight along with the determination of the relationship between resistances to multiple Foliar pathogens of maize (Zea Mays L.). Front Genet. 2021;11 (January):1–13. https://doi.org/10.3389/fgene.2020.548407.

Rashid Z, Sofi M, Harlapur SI, Kachapur RM, Dar ZA, Singh PK, et al. Genome-wide association studies in tropical maize germplasm reveal novel and known genomic regions for resistance to northern corn leaf blight. Sci Rep. 2020;10(1):1–16. https://doi.org/10.1038/s41598-020-78928-5.

Revilla P, Rodríguez VM, Ordás A, Rincent R, Charcosset A, Giauffret C, et al. Association mapping for cold tolerance in two large maize inbred panels. BMC Plant Biol. 2016;16(1):1–10. https://doi.org/10.1186/s12870-016-0816-2.

Rice BR, Lipka AE. Diversifying maize genomic selection models. Mol Breed. 2021;41(5):33. https://doi.org/10.1007/s11032-021-01221-4.

Rice BR, Fernandes SB, Lipka AE. Multi-trait genome-wide association studies reveal loci associated with maize inflorescence and leaf architecture. Plant Cell Physiol. Oxford University Press. 2020; https://doi.org/10.1093/pcp/pcaa039.

Riedelsheimer C, Lisec J, Czedik-Eysenberg A, Sulpice R, Flis A, Grieder C, et al. Genome-wide association mapping of leaf metabolic profiles for dissecting complex traits in maize. Proc Natl Acad Sci U S A. 2012;109(23):8872–7. https://doi.org/10.1073/pnas.1120813109.

Riedelsheimer C, Brotman Y, Méret M, Melchinger AE, Willmitzer L. The maize leaf Lipidome shows multilevel genetic control and high predictive value for agronomic traits. Sci Rep. 2013;3 (1):1–7. https://doi.org/10.1038/srep02479.

Robertsen C, Hjortshøj R, Janss L. Genomic selection in cereal breeding. Agronomy. 2019;9(2):95. https://doi.org/10.3390/agronomy9020095.

Ross-Ibarra J, Tenaillon M, Gaut BS. Historical divergence and gene flow in the Genus Zea. Genetics. 2009;181(4):1399–413. https://doi.org/10.1534/genetics.108.097238.

Sakiyama NS, Ramos HCC, Caixeta ET, Pereira MG. Plant breeding with marker-assisted selection in Brazil. Crop Breed Appl Biotechnol. 2014;14(1):54–60. https://doi.org/10.1590/s1984-70332014000100009.

Samayoa LF, Malvar RA, Olukolu BA, Holland JB, Butrón A. Genome-wide association study reveals a set of genes associated with resistance to the Mediterranean Corn Borer (Sesamia Nonagrioides L.) in a maize diversity panel. BMC Plant Biol. 2015;15(1):1–15. https://doi.org/10.1186/s12870-014-0403-3.

Samayoa LF, Malvar RA, Butrón A. QTL for maize Midparent Heterosis in the heterotic pattern American Dent × European Flint under Corn Borer pressure. Front Plant Sci. 2017;8(April):1–8. https://doi.org/10.3389/fpls.2017.00573.

Sanchez GJJ, Goodman MM, Stuber CW. Isozymatic and morphological diversity in the races of maize of Mexico. Econ Bot. 2000;54(1):43–59.

Sandhu D, Pudussery MV, Kumar R, Pallete A, Markley P, Bridges WC, et al. Characterization of natural genetic variation identifies multiple genes involved in salt tolerance in maize. Function Integ Genom. 2020;20(2):261–75. https://doi.org/10.1007/s10142-019-00707-x.

Sangoi L, da Costa Bortoluzzi RL. Botânica, Origem, Evolução e Dispersão. In: de Milho M, Lima R, Borém A, editors. . 396. UFV; 2018.

Santos JP, Dos R, de Castro Vasconcellos RC, Pires LPM, Balestre M, Von Pinho RG. Inclusion of dominance effects in the multivariate GBLUP model. Edited by Qin Zhang. Plos One. 2016;11 (4):e0152045. https://doi.org/10.1371/journal.pone.0152045.

Schnable PS, Ware D, Fulton RS, Stein JC, Wei F, Pasternak S, et al. The B73 maize genome: complexity, diversity, and dynamics. Science. 2009;326(5956):1112–5. https://doi.org/10.1126/science.1178534.

Scossa F, Alseekh S, Fernie AR. Integrating multi-omics data for crop improvement. J Plant Physiol. 2021;257(February):153352. https://doi.org/10.1016/j.jplph.2020.153352.

Segura V, Vilhjálmsson BJ, Platt A, Korte A, Seren Ü, Long Q, et al. An efficient multi-locus mixed-model approach for genome-wide association studies in structured populations. Nat Genet. 2012;44(7):825–30. https://doi.org/10.1038/ng.2314.

Semagn K, Bjørnstad A, Ndjiondjop MN. Principles, requirements and prospects of genetic mapping in plants. Afr J Biotechnol. 2006;5(25):2569–87. https://doi.org/10.4314/ajb.v5i25.56082.

Sherwin JC, Reacher MH, Dean WH, Ngondi J. Epidemiology of vitamin A deficiency and Xerophthalmia in at-risk populations. Trans R Soc Trop Med Hyg. 2012;106(4):205–14. https://doi.org/10.1016/j.trstmh.2012.01.004.

Shikha K, Shahi JP, Vinayan MT, Zaidi PH, Singh AK, Sinha B. Genome-wide association mapping in maize: status and prospects. 3. Biotech. 2021;11(5):244. https://doi.org/10.1007/s13205-021-02799-4.

Shu G, Cao G, Li N, Wang A, Wei F, Li T, et al. Genetic variation and population structure in China summer maize germplasm. Sci Rep. 2021;11(1):1–13. https://doi.org/10.1038/s41598-021-84732-6.

Singh BD, Singh AK. Marker-assisted plant breeding: principles and practices. New Delhi: Springer India; 2015. https://doi.org/10.1007/978-81-322-2316-0.

Singh J, Sharma S, Kaur A, Vikal Y, Cheema AK, Bains BK, et al. Marker-assisted pyramiding of lycopene-ε-cyclase, β-carotene Hydroxylase1 and Opaque2 genes for development of biofortified maize hybrids. Sci Rep. 2021;11(1):1–15. https://doi.org/10.1038/s41598-021-92010-8.

Sitonik C, Suresh LM, Beyene Y, Olsen MS, Makumbi D, Oliver K, et al. Genetic architecture of maize chlorotic mottle virus and maize lethal necrosis through GWAS, linkage analysis and genomic prediction in tropical maize germplasm. Theor Appl Genet. 2019;132(8):2381–99. https://doi.org/10.1007/s00122-019-03360-x.

Solovieff N, Cotsapas C, Lee PH, Purcell SM, Smoller JW. Pleiotropy in complex traits: challenges and strategies. Nat Rev Genet. 2013; https://doi.org/10.1038/nrg3461.

De Souza IRP, Guilhen JHS, De Andrade CDLT, De Pinto M, O, De Lana UG, P, Pastina MM. Major effect Qtl on chromosome 3 conferring maize resistance to sugarcane Mosaic virus. Revista Brasileira de Milho e Sorgo. 2019;18(3):322–39. https://doi.org/10.18512/1980-6477/rbms.v18n3p322-339.

Stagnati L, Rahjoo V, Samayoa LF, Holland JB, Borrelli VMG, Busconi M, et al. A genome-wide association study to understand the effect of Fusarium Verticillioides infection on seedlings of a maize diversity panel. G3: Genes, Genomes, Genetics. 2020;10(4):1685–96. https://doi.org/10.1534/g3.119.400987.

Stanley A, Menkir A, Paterne A, Ifie B, Tongoona P, Unachukwu N, et al. Genetic diversity and population structure of maize inbred lines with varying levels of resistance to. Plan Theory. 2020;9:1223.

Stearns FW. One hundred years of pleiotropy: a retrospective. Genetics. 2010; https://doi.org/10.1534/genetics.110.122549.

Stitzer MC, Ross-Ibarra J. Maize domestication and gene interaction. New Phytol. 2018;220(2):395–408. https://doi.org/10.1111/nph.15350.

Strigens A, Freitag NM, Gilbert X, Grieder C, Riedelsheimer C, Schrag TA, et al. Association mapping for chilling tolerance in Elite Flint and Dent maize inbred lines evaluated in growth chamber and field experiments. Plant Cell Environ. 2013;36(10):1871–87. https://doi.org/10.1111/pce.12096.

Studer A, Zhao Q, Ross-Ibarra J, Doebley J. Identification of a functional transposon insertion in the maize domestication gene Tb1. Nat Genet. 2011;43(11):1160–3. https://doi.org/10.1038/ng.942.

Studer AJ, Wang H, Doebley JF. Selection during maize domestication targeted a gene network controlling plant and inflorescence architecture. Genetics. 2017;207(2):755–65. https://doi.org/10.1534/genetics.117.300071.

Sturtevant EL. Varieties of corn. Washington: USDA Off. Exp. Stn. Bull; 1899.

Sun H, Zhai L, Teng F, Li Z, Zhang Z. QRgls1.06, a major QTL conferring resistance to Gray leaf spot disease in maize. Crop J. 2021;9(2):342–50. https://doi.org/10.1016/j.cj.2020.08.001.

Suwarno WB, Pixley KV, Palacios-Rojas N, Kaeppler SM, Babu R. Genome-wide association analysis reveals new targets for carotenoid biofortification in maize. Theor Appl Genet. 2015;128(5):851–64. https://doi.org/10.1007/s00122-015-2475-3.

Swanson-Wagner RA, Eichten SR, Kumari S, Tiffin P, Stein JC, Ware D, et al. Pervasive gene content variation and copy number variation in maize and its undomesticated progenitor. Genome Res. 2010;20(12):1689–99. https://doi.org/10.1101/gr.109165.110.

Swanson-Wagner R, Briskine R, Schaefer R, Hufford MB, Ross-Ibarra J, Myers CL, et al. Reshaping of the maize transcriptome by domestication. Proc Natl Acad Sci U S A. 2012;109 (29):11878–83. https://doi.org/10.1073/pnas.1201961109.

Tang JD, Andy Perkins W, Williams P, Warburton ML. Using genome-wide associations to identify metabolic pathways involved in maize aflatoxin accumulation resistance. BMC Genomics. 2015;16(1):1–12. https://doi.org/10.1186/s12864-015-1874-9.

Tao Y, Liu Q, Wang H, Zhang Y, Huang X, Wang B, et al. Identification and fine-mapping of a QTL, QMrdd1, that confers recessive resistance to maize rough Dwarf disease. BMC Plant Biol. 2013;13(1):1–13. https://doi.org/10.1186/1471-2229-13-145.

Technow F, Riedelsheimer C, Schrag TA, Melchinger AE. Genomic prediction of hybrid performance in maize with models incorporating dominance and population specific marker effects. Theor Appl Genet. 2012;125(6):1181–94. https://doi.org/10.1007/s00122-012-1905-8.

Teixeira FF, Costa FM. Catacterização de Recursos Genéticos de Milho. Embrapa Milho e Sorgo. 2010;1:10.

Teng F, Zhai L, Liu R, Bai W, Wang L, Huo D, et al. ZmGA3ox2, a candidate gene for a major QTL, QPH3.1, for plant height in maize. Plant J. 2013;73(3):405–16. https://doi.org/10.1111/tpj.12038.

Thirunavukkarasu N, Hossain F, Arora K, Sharma R, Shiriga K, Mittal S, et al. Functional mechanisms of drought tolerance in subtropical maize (Zea Mays L.) identified using genome-wide association mapping. BMC Genomics. 2014;15(1):1–12. https://doi.org/10.1186/1471-2164-15-1182.

Tian F, Bradbury PJ, Brown PJ, Hung H, Sun Q, Flint-Garcia S, et al. Genome-wide association study of leaf architecture in the maize nested association mapping population. Nature Genet. 2011;43. Nature Publishing Group https://doi.org/10.1038/ng.746.

USDA. World Agricultural Production. 2021. https://apps.fas.usda.gov/psdonline/circulars/production.pdf.

van der Sluis S, Posthuma D, Dolan CV. TATES: efficient multivariate genotype-phenotype analysis for genome-wide association studies. PLoS Genet. 2013;9(1):1003235. https://doi.org/10.1371/journal.pgen.1003235.

Van Inghelandt D, Melchinger AE, Lebreton C, Stich B. Population structure and genetic diversity in a commercial maize breeding program assessed with SSR and SNP markers. Theor Appl Genet. 2010;120(7):1289–99. https://doi.org/10.1007/s00122-009-1256-2.

Van Inghelandt D, Melchinger AE, Martinant JP, Stich B. Genome-wide association mapping of flowering time and northern corn leaf blight (Setosphaeria Turcica) resistance in a vast commercial maize germplasm set. BMC Plant Biol. 2012;12(1):1–15. https://doi.org/10.1186/1471-2229-12-56.

Vann L, Kono T, Pyhäjärvi T, Hufford MB, Ross-Ibarra J. Natural variation in Teosinte at the domestication locus Teosinte Branched1 (Tb1). PeerJ. 2015;3(April):e900. https://doi.org/10.7717/peerj.900.

Vega-Alvarez I, Santacruz-Varela A, Rocandio-Rodríguez M, Córdova-Téllez L, López-Sánchez H, Muñoz-Orozco A, et al. Genetic diversity and structure of native maize races from Northwestern Mexico. Pesquisa Agropecuaria Brasileira. 2017;52(11):1023–32. https://doi.org/10.1590/S0100-204X2017001100008.

Vigouroux Y, Glaubitz JC, Matsuoka Y, Goodman MM, Sanchez JG, Doebley J. Population structure and genetic diversity of new world maize races assessed by DNA microsatellites. Am J Bot. 2008;95(10):1240–53. https://doi.org/10.3732/ajb.0800097.

Villa TC, Camacho NM, Scholten M, Ford-Lloyd B. Defining and identifying crop landraces. Plant Genet Resour. 2005;3(3):373–84. https://doi.org/10.1079/pgr200591.

Vinayan MT, Seetharam K, Raman Babu PH, Zaidi MB, Nair SK. Genome wide association study and genomic prediction for Stover quality traits in tropical maize (Zea Mays L.). Sci Rep. 2021;11(1):1–14. https://doi.org/10.1038/s41598-020-80118-2.

Visscher PM, Yang J. A Plethora of pleiotropy across complex traits. Nat Genet. 2016; https://doi.org/10.1038/ng.3604.

Vollbrecht E, Springer PS, Goh L, Buckler IV ES, Martienssen R. Architecture of floral branch systems in maize and related grasses. Nature. 2005;436(7054):1119–26. https://doi.org/10.1038/nature03892.

Wallace JG, Bradbury PJ, Zhang N, Gibon Y, Stitt M, Buckler ES. "Association mapping across numerous traits reveals patterns of functional variation in maize" Edited by Justin O Borevitz. PLoS Genet. 2014;10(12):e1004845. https://doi.org/10.1371/journal.pgen.1004845.

Wang Q, Dooner HK. Remarkable variation in maize genome structure inferred from haplotype diversity at the Bz locus. Proc Natl Acad Sci. 2006;103(47):17644–9. https://doi.org/10.1073/pnas.0603080103.

Wang H, Nussbaum-Wagler T, Li B, Zhao Q, Vigouroux Y, Faller M, et al. The origin of the naked grains of maize. Nature. 2005;436(7051):714–9. https://doi.org/10.1038/nature03863.

Wang H, Studer AJ, Zhao Q, Meeley R, Doebley JF. Evidence that the origin of naked Kernels during maize domestication was caused by a single amino acid substitution in Tga1. Genetics. 2015a;200(3):965–74. https://doi.org/10.1534/genetics.115.175752.

Wang X, Yang Z, Chenwu X. A comparison of genomic selection methods for breeding value prediction. Sci Bullet. 2015b;60(10):925–35. https://doi.org/10.1007/s11434-015-0791-2.

Wang H, Liu R, Yang J, Liu H, Sun Y. Theoretical model for elliptical tube laterally impacted by two parallel rigid plates. Appl Math Mech (English Edition). 2016a;37(2):227–36. https://doi.org/10.1007/s10483-016-2027-8.

Wang X, Wang H, Liu S, Ferjani A, Li J, Yan J, et al. Genetic variation in ZmVPP1 contributes to drought tolerance in maize seedlings. Nat Genet. 2016b;48(10):1233–41. https://doi.org/10.1038/ng.3636.

Wang C, Yang Q, Wang W, Li Y, Guo Y, Zhang D, et al. A transposon-directed epigenetic change in ZmCCT underlies quantitative resistance to Gibberella stalk rot in maize. New Phytol. 2017a;215(4):1503–15. https://doi.org/10.1111/nph.14688.

Wang L, Beissinger TM, Lorant A, Ross-Ibarra C, Ross-Ibarra J, Hufford MB. The interplay of demography and selection during maize domestication and expansion. Genome Biol. 2017b;18 (1):215. https://doi.org/10.1186/s13059-017-1346-4.

Wang X, Yang X, Zhongli H, Chenwu X. Genomic selection methods for crop improvement: current status and prospects. Crop J. Crop Science Society of China/ Institute of Crop Sciences. 2018a; https://doi.org/10.1016/j.cj.2018.03.001.

Wang X, Chen Q, Yaoyao W, Lemmon ZH, Guanghui X, Huang C, et al. Genome-wide analysis of transcriptional variability in a large maize-teosinte population. Mol Plant. 2018b;11(3):443–59. https://doi.org/10.1016/j.molp.2017.12.011.

Wang Q-J, Yuan Y, Liao Z, Jiang Y, Wang Q, Zhang L, et al. Genome-wide association study of 13 traits in maize seedlings under low phosphorus stress. Plant Genome. 2019;12(3):190039. https://doi.org/10.3835/plantgenome2019.06.0039.

Wang N, Wang H, Zhang A, Liu Y, Diansi Y, Hao Z, et al. Genomic prediction across years in a maize doubled haploid breeding program to accelerate early-stage testcross testing. Theor Appl Genet. 2020;133(10):2869–79. https://doi.org/10.1007/s00122-020-03638-5.

Warburton ML, Garrison Wilkes S, Taba AC, Mir C, Dumas F, Madur D, et al. Gene flow among different Teosinte Taxa and into the domesticated maize gene pool. Genet Resour Crop Evol. 2011;58(8):1243–61. https://doi.org/10.1007/s10722-010-9658-1.

Warburton ML, Tang JD, Windham GL, Hawkins LK, Murray SC, Wenwei X, et al. Genome-wide association mapping of aspergillus Flavus and Aflatoxin accumulation resistance in maize. Crop Sci. 2015;55(5):1857–67. https://doi.org/10.2135/cropsci2014.06.0424.

Ward JH. Hierarchical grouping to optimize an objective function. J Am Stat Assoc. 1963;58 (301):236–44.

Weatherwax P. The history of corn. Sci Mon. 1950;71

Wei LJ, Johnson WE. Combining dependent tests with incomplete repeated measurements. Biometrika. 1985;72(2):359–64. https://doi.org/10.1093/biomet/72.2.359.

Wei M, Fu J, Li X, Wang Y, Li Y. Influence of Dent Corn genetic backgrounds on QTL detection for plant-height traits and their relationships in high-oil maize. J Appl Genet. 2009;50(3):225–34. https://doi.org/10.1007/BF03195676.

Wellhausen EJ, Roberts LM, Hernandez EX, Mangelsdorf PC. Races of maize in Mexico. Cambridge: Bussey Inst. Harvard University Press; 1952.

Wen W, Franco J, Chavez-Tovar VH, Yan J, Taba S. Genetic characterization of a core set of a tropical maize race Tuxpeño for further use in maize improvement. PLoS One. 2012;7(3):1–10. https://doi.org/10.1371/journal.pone.0032626.

Wen W, Li D, Li X, Gao Y, Li W, Li H, et al. Metabolome-based genome-wide association study of maize Kernel leads to novel biochemical insights. Nat Commun. 2014;5(1):1–10. https://doi.org/10.1038/ncomms4438.

Wen J, Shen Y, Xing Y, Wang Z, Han S, Li S, et al. QTL mapping of Fusarium ear rot resistance in maize. Plant Dis. 2021;105(3):558–65. https://doi.org/10.1094/PDIS-02-20-0411-RE.

Wen W, Taba S, Shah T, Chavez Tovar VH, Yan J. Detection of genetic integrity of conserved maize (Zea Mays L.) germplasm in Genebanks using SNP markers. Genet Resour Crop Evol. 2011;58(2):189–207. https://doi.org/10.1007/s10722-010-9562-8.

Weng J, Xie C, Hao Z, Wang J, Liu C, Li M, et al. Genome-wide association study identifies candidate genes that affect plant height in Chinese elite maize (Zea Mays l.) inbred lines. PLoS One. 2011;6(12):e29229. https://doi.org/10.1371/journal.pone.0029229.

Whipple CJ, Kebrom TH, Weber AL, Yang F, Hall D, Meeley R, et al. Grassy Tillers1 promotes apical dominance in maize and responds to shade signals in the grasses. Proc Natl Acad Sci. 2011;108(33):E506–12. https://doi.org/10.1073/pnas.1102819108.

Whitt SR, Wilson LM, Tenaillon MI, Gaut BS, Buckler ES. Genetic diversity and selection in the maize starch pathway. Proc Natl Acad Sci. 2002;99(20):12959–62. https://doi.org/10.1073/pnas.202476999.

Wills DM, Whipple CJ, Takuno S, Kursel LE, Shannon LM, Ross-Ibarra J, et al. "From many, one: genetic control of prolificacy during maize domestication." edited by Hopi E. Hoekstra. PLoS Genet. 2013;9(6):e1003604. https://doi.org/10.1371/journal.pgen.1003604.

Wright SI. The effects of artificial selection on the maize genome. Science. 2005;308(5726): 1310–4. https://doi.org/10.1126/science.1107891.

Wu X, Li YY, Shi Y, Song Y, Zhang D, Li C, et al. Joint-linkage mapping and GWAS reveal extensive genetic loci that regulate male inflorescence size in maize. Plant Biotechnol J. 2016a;14(7):1551–62. https://doi.org/10.1111/pbi.12519.

Wu Y, Vicente FS, Huang K, Dhliwayo T, Costich DE, Semagn K, et al. Molecular characterization of CIMMYT maize inbred lines with genotyping-by-sequencing SNPs. Theor Appl Genet. 2016b;129(4):753–65. https://doi.org/10.1007/s00122-016-2664-8.

Xiao Y, Tong H, Yang X, Shizhong X, Pan Q, Qiao F, et al. Genome-wide dissection of the maize ear genetic architecture using multiple populations. New Phytol. 2016;210(3):1095–106. https://doi.org/10.1111/nph.13814.

Xiao Y, Liu H, Liuji W, Warburton M, Yan J. Genome-wide association studies in maize: praise and stargaze. Mol Plant. Cell Press. 2017; https://doi.org/10.1016/j.molp.2016.12.008.

Xu Y. Envirotyping for deciphering environmental impacts on crop plants. Theor Appl Genet. 2016; https://doi.org/10.1007/s00122-016-2691-5.

Xu G, Wang X, Huang C, Dingyi X, Li D, Tian J, et al. Complex genetic architecture underlies maize tassel domestication. New Phytol. 2017;214(2):852–64. https://doi.org/10.1111/nph.14400.

Xu G, Lyu J, Li Q, Liu H, Wang D, Zhang M, et al. Evolutionary and functional genomics of DNA methylation in maize domestication and improvement. Nat Commun. 2020;11(1):5539. https://doi.org/10.1038/s41467-020-19333-4.

Xue Y, Warburton ML, Sawkins M, Zhang X, Setter T, Yunbi X, et al. Genome wide association analysis for nine agronomic traits in maize under well watered and water stressed conditions. Theor Appl Genet. 2013;126(10):2587–96. https://doi.org/10.1007/s00122-013-2158-x.

Xue S, Bradbury PJ, Casstevens T, Holland JB. Genetic architecture of domestication-related traits in maize. Genetics. 2016;204(1):99–113. https://doi.org/10.1534/genetics.116.191106.

Yang Q, Li Z, Li W, Lixia K, Wang C, Ye J, et al. CACTA-like transposable element in ZmCCT attenuated photoperiod sensitivity and accelerated the postdomestication spread of maize. Proc Natl Acad Sci U S A. 2013;110(42):16969–74. https://doi.org/10.1073/pnas.1310949110.

Yang N, Yanli L, Yang X, Huang J, Zhou Y, Ali F, et al. Genome wide association studies using a new nonparametric model reveal the genetic architecture of 17 agronomic traits in an enlarged maize association panel. PLoS Genet. 2014;10(9):e1004573. https://doi.org/10.1371/journal.pgen.1004573.

Yang J, Jiang H, Yeh CT, Jianming Y, Jeddeloh JA, Nettleton D, et al. Extreme-phenotype genome-wide association study (XP-GWAS): a method for identifying trait-associated variants by sequencing pools of individuals selected from a diversity panel. Plant J. 2015;84(3):587–96. https://doi.org/10.1111/tpj.13029.

Yang J, Liu Z, Chen Q, Yanzhi Q, Tang J, Lübberstedt T, et al. Mapping of QTL for grain yield components based on a DH population in maize. Sci Rep. 2020;10(1):1–11. https://doi.org/10.1038/s41598-020-63960-2.

Yi Q, Malvar RA, Álvarez-Iglesias L, Ordás B, Revilla P. Dissecting the genetics of cold tolerance in a multiparental maize population. Theor Appl Genet. 2020;133(2):503–16. https://doi.org/10.1007/s00122-019-03482-2.

Yi Q, Álvarez-Iglesias L, Malvar RA, Romay MC, Revilla P. A worldwide maize panel revealed new genetic variation for cold tolerance. Theor Appl Genet. 2021;134(4):1083–94. https://doi.org/10.1007/s00122-020-03753-3.

Yu J, Buckler ES. Genetic association mapping and genome Organization of Maize. Curr Opin Biotechnol. 2006; https://doi.org/10.1016/j.copbio.2006.02.003.

Yu J, Pressoir G, Briggs WH, Bi IV, Yamasaki M, Doebley JF, et al. A unified mixed-model method for association mapping that accounts for multiple levels of relatedness. Nat Genet. 2006;38(2):203–8. https://doi.org/10.1038/ng1702.

Yu J, Holland JB, McMullen MD, Buckler ES. Genetic design and statistical power of nested association mapping in maize. Genetics. 2008;178(1):539–51. https://doi.org/10.1534/genetics.107.074245.

Yu J, Xu F, Wei Z, Zhang X, Chen T, Pu L. Epigenomic landscape and epigenetic regulation in maize. Theor Appl Genet. 2020;133(5):1467–89. https://doi.org/10.1007/s00122-020-03549-5.

Yu D, Wang H, Wei G, Qin T, Sun P, Youlin L, et al. Genetic diversity and population structure of popcorn germplasm resources using genome-wide SNPs through genotyping-by-sequencing. Genet Resour Crop Evol. 2021;68(6):2379–89. https://doi.org/10.1007/s10722-021-01137-0.

Yuan G, Chen B, Peng H, Zheng Q, Li Y, Xiang K, et al. QTL mapping for resistance to ear Rot caused by Fusarium Graminearum using an IBM Syn10 DH population in maize. Mol Breed. 2020;40(9). https://doi.org/10.1007/s11032-020-01158-0.

Zenda T, Liu S, Dong A, Duan H. Advances in cereal crop genomics for resilience under climate change. Life. 2021;11(6):1–34. https://doi.org/10.3390/life11060502.

Zhang Z, Ersoz E, Lai C-QQ, Todhunter RJ, Tiwari HK, Gore MA, et al. Mixed linear model approach adapted for genome-wide association studies. Nat Genet. 2010;42(4):355–60. https://doi.org/10.1038/ng.546.

Zhang N, Gibon Y, Wallace JG, Lepak N, Li P, Dedow L, et al. Genome-wide association of carbon and nitrogen metabolism in the maize nested association mapping population. Plant Physiol. 2015;168(2):575–83. https://doi.org/10.1104/pp.15.00025.

Zhang X, Warburton ML, Setter T, Liu H, Xue Y, Yang N, et al. Genome-wide association studies of drought-related metabolic changes in maize using an enlarged SNP panel. Theor Appl Genet. 2016;129(8):1449–63. https://doi.org/10.1007/s00122-016-2716-0.

Zhang W, Xue G, Shi X, Zhu B, Wang Z, Gao H, et al. PCA-based multiple-trait GWAS analysis: a powerful model for exploring pleiotropy. Animals. 2018;8(12) https://doi.org/10.3390/ani8120239.

Zhang H, Yin L, Wang M, Yuan X, Liu X. Factors affecting the accuracy of genomic selection for agricultural economic traits in maize, cattle, and pig populations. Front Genet. 2019a;10(MAR) https://doi.org/10.3389/fgene.2019.00189.

Zhang Y, Wan J, He L, Lan H, Li L. Genome-wide association analysis of plant height using the maize F1 population. Plan Theory. 2019b;8(10):432. https://doi.org/10.3390/plants8100432.

Zhang Y, Yu H, Guan Z, Liu P, He Y, Zou C, et al. Combined linkage mapping and association analysis reveals genetic control of maize kernel moisture content. Physiol Plant. 2020;170(4):508–18. https://doi.org/10.1111/ppl.13180.

Zhao Y, Mette MF, Reif JC. "Genomic selection in hybrid breeding" edited by F. Ordon. Plant Breed. 2015;134(1):1–10. https://doi.org/10.1111/pbr.12231.

Zhao J, Li H, Yuhui X, Yin Y, Huang T, Zhang B, et al. A consensus and saturated genetic map provides insight into genome anchoring, Synteny of Solanaceae and leaf- and fruit-related QTLs in Wolfberry (Lycium Linn.). BMC Plant Biol. 2021;21(1):1–13. https://doi.org/10.1186/s12870-021-03115-1.

Zheng Y, Yuan F, Huang Y, Zhao Y, Jia X, Zhu L, et al. Genome-wide association studies of grain quality traits in maize. Sci Rep. 2021;11(1):9797. https://doi.org/10.1038/s41598-021-89276-3.

Zhao X, Luo L, Cao Y, Liu Y, Li Y, Wenmei W, et al. Genome-wide association analysis and QTL mapping reveal the genetic control of cadmium accumulation in maize leaf. BMC Genomics. 2018;19(1):1–13. https://doi.org/10.1186/s12864-017-4395-x.

Zhou X, Stephens M. Genome-wide efficient mixed-model analysis for association studies. Nat Genet. 2012;44(7):821–4. https://doi.org/10.1038/ng.2310.

Zhou X, Stephens M. Efficient multivariate linear mixed model algorithms for genome-wide association studies. Nat Methods. 2014;11(4):407–9. https://doi.org/10.1038/nmeth.2848.

Zhu C, Gore M, Buckler ES, Yu J. Status and prospects of association mapping in plants. Plant Genome. 2008;1(1):5–20. https://doi.org/10.3835/plantgenome2008.02.0089.

Zila CT, Fernando Samayoa L, Santiago R, Butrón A, Holland JB. A genome-wide association study reveals genes associated with Fusarium ear rot resistance in a maize core diversity panel. G3: Genes Genomes Genet. 2013;3(11):2095–104. https://doi.org/10.1534/g3.113.007328.

Zila CT, Ogut F, Romay MC, Gardner CA, Buckler ES, Holland JB. Genome-wide association study of Fusarium ear rot disease in the U.S.A. maize inbred line collection. BMC Plant Biol. 2014;14(1):1–15. https://doi.org/10.1186/s12870-014-0372-6.

Population Genomics of Pearl Millet

Ndjido Ardo Kane and Cécile Berthouly-Salazar

Abstract Pearl millet [*Pennisetum glaucum* (L.) R. Br., syn. *Cenchrus americanus* (L.) Morrone)], one of the major staple food for around 100 million people across Africa and India, is a so-called "orphan" crop. Recently, sequencing of its genome offers numerous possibilities to advance pearl millet conservation, management, and breeding to cope against climate change and for agricultural productivity enhancement. To decipher genetic diversity and genomic patterns of adaptation and regions under selection, genetic differences between and within populations at the genome scale can be analyzed using latest population genomics approaches and sequencing technologies. A conceptual framework of population genomics in pearl millet and a series of case studies that led to the identification of genomic signatures of variation in space and time to evolutionary and adaptive processes are presented. In this chapter, we discuss how population genomics, despite some complexities, paves new paths for understanding genomic diversity and genetic basis of adaptation and facilitating genomic-assisted breeding and genetic resources conservation in pearl millet, a vital crop.

Keywords Breeding · Climate adaptation · Genetic diversity · Pearl millet · Population genomics

---

N. A. Kane (✉)
Institut Sénégalais de Recherches Agricoles (ISRA), Centre d'Étude Régional pour l'amélioration de l'Adaptation à la Sécheresse, Thiès, Senegal

Laboratoire mixte international Adaptation des Plantes et microorganismes associés aux Stress Environnementaux (LAPSE), Dakar, Senegal
e-mail: ndjido.kane@isra.sn

C. Berthouly-Salazar
Laboratoire mixte international Adaptation des Plantes et microorganismes associés aux Stress Environnementaux (LAPSE), Dakar, Senegal

DIADE, Université Montpellier, Institut de Recherche pour le Développement, Montpellier, France

1 Introduction

Pearl millet [*Pennisetum glaucum* (L.) R. Br., syn. *Cenchrus americanus* (L.) Morrone)] (Soreng et al. 2003) is called as an "orphan" crop even though it is a staple food for around 100 million population of Africa and India. It is also used as a feed for birds and livestock and in traditional alcohol brewing. It has an excellent nutritional composition (high protein and fiber content) and is richer in energy and in essential minerals (like iron and zinc) than other major cereals (Malik 2015). Gluten-free with a low glycemic index and hypoallergenic properties, pearl millet could, therefore, be promoted as a nutrient-rich food for health and food security.

Annual C4, short-growing grass, pearl millet is cultivated in the driest environments around the world. It is widely grown across Africa, India, China, Russia and the USA (National Research Council 1996) and being expanded into some non-traditional regions, such as North Africa, Brazil, Mexico, Canada, West and Central Asia (Upadhyaya et al. 2017). Wild progenitor (*P. glaucum* subsp. *monodii*, syn. *Cenchrus americanus* ssp. *monodii*) (Soreng et al. 2003) is distributed in the sub-Sahelian region in West Africa, along a contrasted south-north gradient of humidity. Archeological clues (Manning et al. 2011) date pearl millet domestication around 4,000–5,000 years ago whereas population genomics research estimates the start of its diffusion about 4,900 years ago (Burgarella et al. 2018).

Pearl millet, together with other millets [finger millet (*Eleusine coracana*), proso millet (*Panicum miliaceum*), and foxtail millet (*Panicum italicum*)], produced approximately 30 million tons of grain in 2016 compared to 1 billion ton for maize or 740 million tons of rice or wheat (FAOSTAT 2017). Global production remains low because of many constraints like low soil fertility and biotic and abiotic stresses (Waddington et al. 2010). Quasi-totality of the global millet production and seed demand comes from systems in Africa and Asia characterized by extensive production practices and limited adoption of improved varieties, particularly in West and Central Africa.

Nevertheless, pearl millet has become a favored crop due to its high resilience to dry and poor soils. Highly allogamous, it showed an impressive and untapped genetic diversity (Pucher et al. 2015; Diack et al. 2017) that, if exploited efficiently, could allow to overcome limits enforced by the global climate change. With latest whole genome sequencing technologies and reduced costs, population genomics studies of diversity have become more affordable in such so-called "neglected or underutilized" crops. When combined with precise and accurate phenotyping platforms, these technologies offer a powerful tool for identifying the genetic basis of key agricultural traits and for predicting the value of individuals in a plant breeding population (Varshney et al. 2014; Vadez et al. 2012). Better knowledge of genetic variation implications in a high gene flow species is imperative not only for improving our basic understanding of domestication and adaptive response to global change but also for predicting future trajectories of breeding programs.

Population genomics provides powerful approaches to provide better understanding and new key insights into genetic diversity and population structure, including adaptive genetic variation and differentiation, evolution, and genetic basis of climate adaptation, abiotic and biotic stress tolerance, and domestication, and assist plant breeding and conservation and sustainable management of plant genetic resources (Rajora 2019; Luikart et al. 2019). Therefore, population genomics could help identifying genetic variants underlying domestication and adaptation in pearl millet and could also assist genetic resources management of this crop species. Such approaches could rely on genome information to identify signatures of adaptive genetic variation and link them to evolutionary and adaptive processes. Indeed, genome sequencing data from populations provides a plethora of molecular markers that could be applied in various genetic, plant breeding, and other studies, including studies of local adaptation. With the development and availability of a reference genome, analytical tools, and multiple and more robust genetic resources, population genomics should contribute to advance breeding and development of new improved varieties of pearl millet, contributing toward enhancement of agricultural productivity and germplasm management.

In this chapter, we discuss how population genomics paves new paths for providing better understanding and key insights into genomic diversity and genetic basis of adaptation and facilitating genomics-assisted breeding and genetic resources conservation in pearl millet.

2 Population Genomics Conceptual Framework for Pearl Millet

Genetic variation determines the potential of an organism to evolve and adapt to its environment. The fitness of adapting to changing environments increases the frequency of beneficial alleles as a result of positive selection and phenotypic performances. Population genomics allows the identification of molecular variation. Contrasting patterns of divergence along genomic windows, the so-called genomic scans, allow to investigate the ancestry state of the region (i.e., introgression pattern) and to detect footprints of selection.

For autogamous species, genomic scan approaches to detect the causative allele of adaptation are more difficult to use due to linkage disequilibrium (LD). For allogamous species, a weak population structure and a rapid LD are key attributes to help studying the genetic basis of adaptation to the environment, especially in cereals. Comparative population genomics showed extensive synteny between pearl millet and its panicoid cereal relatives sorghum (*Sorghum bicolor*) and foxtail millet (*Panicum italicum L.*), with multiple rearrangements in the pearl millet lineage (Hu et al. 2015). These features make pearl millet an excellent model for studying the process of domestication and introgression in *Poaceae*.

A conceptual framework for effective use of large populations and genome-wide data set in pearl millet is presented in Fig. 1. It integrates molecular insights from

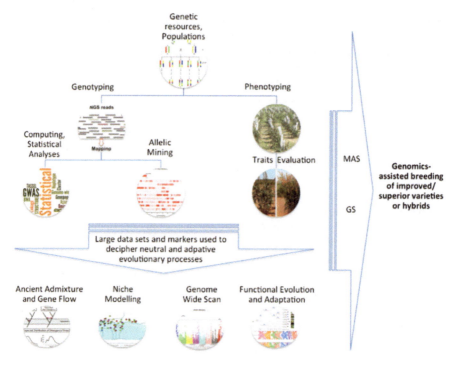

Fig. 1 A conceptual framework of population genomics in pearl millet

high-resolution genotyping of genetic resources, phenotyping evaluation and computational tools used for analyzing large data sets, which are crucial to answer scientific questions of conservation biology, population structural diversity across crop genomes, and adaptive processes. The advent of new genome-wide sequencing generated abundant molecular markers spread across the genome and allows patterns of genomic divergence to be investigated at previously unattainable scales. These approaches and technologies are being used in marker-assisted selection (MAS) and in genomic selection (GS) to predict the trait value of individuals in marker-assisted breeding populations. MAS and GS help identify superior genotypes and breeding lines, which assist to recombine selected genotypes and lines of interest to produce superior genotypes through plant breeding. To implement GS, a genotype-phenotype association model developed from a breeding population is used to predict the performance of related breeding population for various phenotypic traits based only on the marker genotype data (Bassi et al. 2015).

Genetic resources are genotyped using high-resolution genome sequencing technologies and analytical tools to identify genetic variation. Large data sets and markers are generated and combined with phenotyping and trait evaluation to study genome-wide natural selection, ancient admixture, gene flow, and functional and adaptive processes. Marker-assisted selection (MAS) uses molecular markers tightly linked with the QTLs and inherited together in linkage disequilibrium.

Genomic selection (GS) identifies selected promising individuals based on genomics estimated breeding values by combining marker genotype data with phenotype from several environments and pedigree. Both MAS and GS are then implemented in genomics-assisted breeding of new improved varieties/hybrids of pearl millet.

Historically, genetic diversity and evolutionary patterns between populations were investigated using isozymes or dominant markers, such as RAPD. Later, highly polymorphic markers such as SSR, SNPs, DArT, CISP, and SSCP-SNP were developed to better assess genetic diversity and population structure and conduct association and linkage mapping studies. Some of these markers have been successfully used in MAS for identification of genomic regions associated with traits of yield under abiotic stresses (Shivhare and Lata 2017). Yet, the number of markers was still not enough to identify cryptic patterns of genetic differentiation or to pinpoint adaptive processes in highly allogamous crops.

2.1 Applying High-Resolution Sequencing Technologies

Genome-wide sequencing technologies have revolutionized the assessment of allelic variation at the genome- or at chromosome-scale resolution. In pearl millet, whole genome sequencing, reduced representation libraries sequencing [genotyping-by-sequencing (GBS), restriction site-associated DNA sequencing (RAD-seq), and transcriptome sequencing (RNA-seq)], and targeted sequencing (exome capture) have been applied for genomics studies, including detection of signal of selection or gene identification and function.

2.1.1 Whole Genome Sequence

Recent sequencing of nearly one thousand individuals (963 cultivated accessions and 31 wild progenitors) resulted in high-quality assembly of a pearl millet reference genome (Varshney et al. 2017). Genome sequencing confirmed the genome size of pearl millet to be 1.79GB, with a high GC content (49.7%) and repetitive elements (80%), and revealed an enrichment in genes involved for *wax* biosynthesis and transmembrane transporters of secondary metabolites, which may contribute to its adaptation to heat and drought conditions. Genome sequencing revealed both loss of diversity and differentiation occurred as a result of and during domestication (Varshney et al. 2017).

2.1.2 Reduced Representation Sequencing

The application of reduced representation sequencing (RRS) to 500 pearl millet accessions reported a total of 83,875 SNPs, which were used to characterize genomic diversity, population structure, and synteny with other cereals (Hu et al. 2015).

A GBS-based comparative linkage map with the foxtail millet (*Setaria italica*), maize (*Zea mays*), rice (*Oryza sativa*), sorghum (*Sorghum bicolor*), soybean (*Glycine max*), and wheat (*Triticum aestivum*) genomes indicated extensive regions of synteny, as well as some large-scale rearrangements in the pearl millet lineage (Hu et al. 2015; Varshney et al. 2017). Additionally, RRS approach has been used to investigate geographic distribution of rare alleles in pearl millet in Western and Eastern Sahel (Berthouly-Salazar et al. 2016). Using spatially explicit simulations based on the intensity of gene flow between sympatric wild and cultivated populations, the origin of pearl millet domestication in Africa was inferred, revealing a geographic clustering in three major groups: West, Central, and East Africa. This supports the feasibility of combining RRS with an assembled reference genome to improve the reliability of genotype calls and subsequently downstream inferences in pearl millet.

2.1.3 Targeted Sequencing

In targeted sequencing, only a part of the genome is sequenced that focuses on all coding regions but excludes less well-conserved intronic regions. Depending on the species, coding sequences represent 1–2% of the genome (Warr et al. 2015) and comprise high level of functional variants and low repeat content (Kaur and Gaikwad 2017). This approach requires at least partial knowledge about the transcriptome, making it more difficult to predict which variants might be relevant to a trait. However, it allows to examine genome synteny, identification of QTLs and candidate regulatory genes as well as to assist in genome assembly of species without reference genome (Varshney et al. 2017; Kaur and Gaikwad 2017). Likewise, targeting the chloroplast genome generates intraindividual polymorphic SNPs useful to study gene flow in plants as well as evolutionary and phylogenetic processes and relationships. A good example is the exome capture protocol via hybridization validated on chloroplast genomes of African rice (*Oryza glaberrima* spp.), pearl millet (*Pennisetum glaucum* spp.), fonio (*Digitaria exilis* spp.), and African yam (*Dioscorea* spp.) species (Mariac et al. 2014). This approach was effective for assembling the chloroplast genome in silico and to detect SNPs which can be applied to population genetics and phylogeography studies of this kind (Mariac et al. 2014).

2.2 Plant Resources and Populations

Population genomics relies on gene pool of the species and their wild relatives to support the selection of ideal parental lines as allele donors for the desired traits in breeding schemes. Even though less genetic resources were developed in pearl millet compared with other major crops, there are reference sets of lines, inbred population mapping, and mutants that could be used as good starting point for population genomics studies.

2.2.1 Germplasms

Landraces, wild species, elite improved open-pollinated cultivars, hybrids, and inbred lines can be used to address in-depth characterization of adaptive signatures to local climatic conditions, cultural practices, and breeding traits. A pearl millet inbred germplasm association panel (PMiGAP) of 346 lines was established, which is assembled from a large set of 1,000 diverse breeding lines and inbred mapping population parents and representative of cultivated germplasm from Africa and Asia (Shivhare and Lata 2017; Sehgal et al. 2015). This panel represents an excellent gene pool for population genomics studies to determine genetic variation leading to pearl millet evolution and adaptation to harsh environments as well as to select parental or heterotic pool for breeding improvement.

Landraces are dynamic populations that show highly variable phenotypes and are often genetically diverse in equilibrium with environment, pests, and diseases. Therefore, they could be used for the identification and the characterization of adaptation linked to local environmental conditions as they provide a snapshot of phenotypic distribution and genetic diversity structure across traditional production systems. Evidence of this comes from Upadhyaya et al. (Upadhyaya et al. 2017) who used 15,979 landraces of pearl millet originating in 34 countries and conserved in the ICRISAT Genebank. They have revealed how pearl millet landraces are phenotypically diversified in space and in time. Indeed, phenotypic diversity can be very localized; the observation of long panicles of pearl millet only in Central West Africa suggests a strong local adaptation to farmers' preference and/or environment. On the other hand, correlations between genetic basis of flowering time control and photoperiod sensitivity response of pearl millet landraces from West Africa were found with spatial genetic structure and *PgPhyC* and *PgMADS11* allele frequencies (Diack et al. 2017; Saïdou et al. 2009; Mariac et al. 2011). Likewise, landraces could be used to address temporal evolution and adaptive response to environmental changes. A study showed that polymorphism within the *PgPhyC* gene from landraces of Niger was associated with earliness in flowering time to cope against recurrent drought periods that happened in West Sahel over three decades (Vigouroux et al. 2011).

Domesticated crops are often not genetically distinct from their wild ancestors. While the diversity of the cultivated population can be limited, wild relative populations encompass an interesting reservoir of genetic diversity and adaptation capacity even though genetic loss between cultivated pearl millet and wild progenitor has been estimated to be 20% (Mariac et al. 2006; Oumar et al. 2008). Living in more extreme conditions, wild relatives are subjected to strong natural selection; thus, a rapid and contrasted response to environmental changes is expected in wild relative species. Population genomics could greatly assist in exploiting wild genetic diversity either by introducing artificially adaptive genes through selection program or by characterizing large collections of diverse wild germplasm in order to reduce phenotyping bottleneck.

2.2.2 Second-Generation Biparental Populations

Genetic linkage mapping and QTL mapping studies use biparental populations such as F_2, recombinant inbred lines (RILs), or double haploids (DH). Association mapping studies using backcross nested association mapping (BCNAM) or multi-parent advanced generation intercross (MAGIC) approach use multi-parent allelic diversity to a maximum extent in fine mapping of QTL for desirable traits using an integrated mapping population (Serba and Yadav 2016). A crossing scheme for the development of multi-parent BCNAM or MAGIC recombinant inbred lines for pearl millet was proposed that involves half-diallel mating of eight molecularly diverse inbred germplasm, intercrossing of the recombinants, and repeated selfing (Serba and Yadav 2016). These second-generation BCNAM or MAGIC populations are based on high-density map to identify potentially novel genes or QTLs associated with agronomic performances.

2.2.3 Mutant Populations

Limited studies have been conducted using chemical or physical mutagenesis induction on plant and seed of pearl millet (Serba et al. 2017). Induced mutations in pearl millet are used to identify new alleles and for practical application in breeding rather than a way to create additional variability, which is already high within this crop. Indeed, the induced mutation constitutes a way of generating an array of genotypes, especially for genetic studies and trait mapping in pearl millet (Serba et al. 2017). Homozygous lines treated with mutagenic agents create allelic changes leading to a spectrum of mutants with novel traits that cannot be generated with normal genetic recombination. Half a century ago, the importance of using EMS-induced mutation to identify targeted traits like dwarfing genes for reducing plant stature has been reported (Burton and Forston 1966). More recently, teams in India, the USA, and Senegal are developing new mutant populations for functional population genomics and trait mapping. One advantage of these new resources will be the possibility offered by high-resolution genotyping approaches to scan a large panel of individuals, to identify allelic variation, and to link them to genes/QTLs within mutant populations.

3 Progress in Population Genomics of Pearl Millet

Because pearl millet is well adapted to harsh climate conditions and susceptible to various biotic and abiotic stresses like other crops, it provides an excellent model to decipher genetic architecture underlying local adaptation and stress responses. To link DNA polymorphisms with trait variation in order to identify genomic regions where genes governing traits of interest are located, statistical models are often used.

Genome-wide association studies (GWAS) analyze the DNA variation among individuals of a population having varying phenotypes for a particular trait and then attempt to associate genetic variants with phenotypes. To provide coverage of the entire genome by genotyping a subset of variants, GWAS relies on linkage disequilibrium (LD) decay, a nonrandom association of alleles at different loci used to understand past evolutionary and demographic events, to map genes that are associated with quantitative or inherited traits. LD throughout the genome reveals population history, breeding system, and geographic pattern whereas LD in each genomic region can reveal the history of natural selection, gene conversion, mutation, and other forces that cause gene frequency evolution (Slatkin 2008).

3.1 Domestication

Despite low genetic structure observed in pearl millet populations, studying broader geographic sampling can provide interesting features on local adaptation and domestication origins. The whole genome sequencing of 192 geo-referenced pearl millet accessions suggested Western Sahara as a center of pearl millet domestication (Burgarella et al. 2018). In addition, 290 genomic regions were detected under selection for each cultivated group of pearl millet; however, only four were common to all geographical groups, and nearly 50% were specific to one geographical region (Burgarella et al. 2018). Further, these authors provided evidence that wild-to-crop gene flow increased genetic diversity in cultivated pearl millet leading to diversity hotspots in Western and Eastern Sahel and potential adaptive introgression of 15 genomic regions.

3.2 Genetic Diversity and Population Structure

Until now, low genetic differentiation between cultivated landraces of pearl millet from West Africa and across the world has been observed based on a limited number of genetic markers (Oumar et al. 2008; Dussert et al. 2015). Only phenotypic variation and differentiation along the latitudinal gradient (Upadhyaya et al. 2017), with a fine sampling strategy at a national scale (Diack et al. 2017), were able to provide an evidence for a subtle differentiation between early- and late-flowering landraces (Fst = 0.004). Further, the population structure of landraces from multiple agroecological zones in Senegal highlighted a high genetic diversity and an untapped potential of this germplasm. However, a study employing genome-wide SNP data has revealed high genomic diversity, a clear population structure between accessions and Senegalese landraces, and a geographic structure among countries but not within Senegal due to the small number of late-flowering landraces in their data set (Hu et al. 2015). This exemplifies the power of population genomics approaches to unravel key information on genetic diversity and population structure that was not possible earlier using a limited number of molecular markers. Therefore, population genomics could help in characterizing the genetic diversity or deciphering fine-scale population genetic structure for a better management or breeding of pearl millet germplasms.

3.3 Adaptation to Biotic and Abiotic Stress Responses

Populations diverge in response to biotic and abiotic stresses. Diseases, parasitic weeds, drought, or high temperatures mainly limit pearl millet genotypes' performance. Investigating adaptation and identifying the causative alleles underlying adaptation are quite challenging. However, population genomics approaches by combining several molecular methods, sampling approaches, and statistical methodologies can help to meet this challenge. Using gene markers (a 3 bp indel of *PHYTOCHROME C* gene and a 24 bp indel of *MADS 11*) in pearl millet and genotyping more than 7000 individuals in 17 populations of cultivated pearl millet over a period of 2 years, allele frequencies have been estimated, and evidence for temporal and spatial selection acting on the flowering cycle has been found (Mariac et al. 2016). Variation in adaptive traits also involved *cis*-regulatory mutations that might have fewer deleterious pleiotropic effects than protein-coding mutations. There is the hypothesis that enrichment of adaptive *cis*-regulatory divergence builds up over time and their mutations are not predominantly involved in rapid adaptive evolution (Rhoné et al. 2017). This has been tested in pearl millet by analyzing SNPs identified as under selection and showing no excess of *cis*-acting mutations for allele-specific expression. In this study, the authors were able to demonstrate that the intraspecific-level *cis*-regulatory variations depend on the divergence time between genotypes and are not predominantly involved in adaptive evolution.

3.3.1 Resistance to Biotic Stress

Downy mildew, caused by fungi *Sclerospora graminicola*, is one of the most devastating diseases of pearl millet (Fig. 2). However, little attention has been given toward identification and characterization of the genetic basis of mechanisms underlying pearl millet's resistance to downy mildew. A screening of 48 pearl millet inbred lines against 9 diverse pathotypes of *S. graminicola* from 5 major pearl millet-growing states in India has identified 17 lines that provided strongly differential reactions to the 9 pathogen isolates (Hash et al. 2006). Eight of these lines have been used as parents of pearl millet mapping population progeny sets used to map host-plant resistance to *S. graminicola*. QTLs for downy mildew resistance have been mapped to linkage groups associated with the presence of genes that spanned 10–40 cM to the closest markers (Dwivedi et al. 2011). De novo transcriptome analysis revealed upregulation of phenylpropanoid pathway genes in pearl millet genotypes resistant to downy mildew, and an inference was made for potential

Fig. 2 Effect of downy mildew on pearl millet. Severe infection of downy mildew causes malformed panicles and grain production losses in pearl millet. Picture taken by Ndjido A. Kane/ISRA

hypersensitive response and systemic acquired resistance as possible defense mechanisms operating against downy mildew infection in pearl millet (Kulkarni et al. 2016). Breeding for resistance against downy mildew disease and incorporating this trait into high-yielding cultivars of pearl millet were attempted among parental genotypes ICMR-01007 (P1) and ICMR-01004 (P2) and their progenies (F_1 and F_2 populations). With the whole genome of one of the most virulent pathotypes of *S. graminicola* sequenced, not only advanced breakthroughs into co-evolutionary patterns of the pathogen with the crop and the environment could be made, but also it helped in management and breeding for resistance to downy mildew disease into high-yielding cultivars of pearl millet. Indeed, population genomics could be applied to identify fraction of genomes or traits that determine the variability of plant-pathogen interaction to be used as markers for plant disease control.

Striga hermonthica (Del.) Benth. is the most widespread and destructive parasitic weed affecting cereals, and pearl millet is not an exception (Fig. 3). Genetic basis of resistance to *Striga* appears to be complex, and its nature could be either qualitative controlled by one or two major genes (monogenic or oligogenic) or quantitative governed by several alleles at multiple loci (polygenic). However, several QTLs for *Striga* resistance in pearl millet with large effects against the pathogen population from India and West African countries have been identified (Kountche et al. 2013). They were successfully transferred to the background of elite parents to improve their resistance to the pearl millet *Striga* through MAS. A *Striga*-resistant gene pool was developed in West African pearl millet, and the response was assessed to five

Fig. 3 Infected pearl millet plant by *Striga* (*Striga hermonthica* (*Del.*) *Benth.*). *Striga*, so-called witchweed, extracts nutrients from host plant, induces senescence of leaves, and causes nutrient deficiency. *Striga* infection mimics drought symptoms and leads to grain losses (empty grain panicle). Picture taken by Ndjido A. Kane/ISRA

cycles of recurrent selection targeting resistance. Assessment of this gene pool together with parental landraces, experimental varieties derived from previous cycles, and local checks in *Striga*-infested fields (Kountche et al. 2013) has shown genetic variation and heritability that enable further genetic gains through selection. Introgression of pearl millet *Striga* resistance into the background of the male parent of the hybrid HHB 67 has shown limited resolution through linkage mapping due to the relatively few recombinants generated from the parents (Hash et al. 2006). Varying heritability, scarcity of donor sources, limited resolution for QTL identification, and large $G \times E$ interactions underline the need for population genomics studies as well as a multilocation assessment.

3.3.2 Tolerance to Abiotic Stress

Drought is the major constraint to pearl millet, and the mechanism of response can take various forms: high plant water status (dehydration avoidance), low plant water potential (drought avoidance), water use efficiency (water lost by evapotranspiration), or drought escape (rapid development to complete a life cycle before drought) (Fang and Xiong 2015). Drought tolerance is at the same time a complex polygenic trait, influenced significantly by its environment and controlled by hundreds of genes (Hu and Xiong 2014). Population genome scans and association genetics framework along the environmental gradient have revealed allelic variation correlated with climatic metrics and putative adaptive traits (Berthouly-Salazar et al. 2016; Ousseini et al. 2017a). Using an association mapping strategy with a mixed model in 11 populations of wild pearl millet across two north-south gradients, one in Niger and one in Mali, Ousseini et al. (2017b) found two SNPs in myoxin-like gene, a target of selection associated with fitness and adaptation to aridity in wild pearl millet. GWAS identified new QTLs for biomass production in early drought stress conditions and for stay-green characters (Debieu et al. 2018). The authors reported genes involved in the siroheme, and wax biosynthesis pathways were found to co-locate with associated loci, as highlighted from the genome sequencing (Varshney et al. 2017). Additionally, high-resolution sequencing of multi-parent populations captures allelic variations underlying not only stress-related QTLs but also key agronomic traits or adaptive signatures. Numerous other types of populations (F_2, RILs, DH, etc.) have been developed and used to evaluate drought-related traits, such as leaf traits, osmotic adjustment capabilities, water potential, ABA content, stability of the cell membrane, or root system. The latter plays a crucial role in the drought avoidance mechanisms, as root architecture, anatomy, depth, and density are important indicators of drought response. Genetic variability of root exudation (Ndour et al. 2017) and root growth (Passot et al. 2016) in pearl millet were characterized and led to a genetic association study on a panel of 188 RILs to identify genomic regions associated with these traits linked to water and soil nutrients uptake (Ndour et al. 2017). This provides evidence that root characterization might have a significant contribution in pearl millet breeding for drought tolerance.

4 Future Prospects for Pearl Millet Population Genomics

The availability of pearl millet genome offers (1) a reference sequence to pinpoint allelic variations at the genome-wide level, (2) an annotation to identify and characterize genes and their roles in influencing particular traits, and (3) the prospects to match the potential candidate genes or loci to linked regions under selection. This can also improve the information about either specific genomic regions or specific genes, define heterotic pools, establish marker-trait associations, and use genomic selection strategy to breed new variety or hybrid.

One challenge was to decrypt letter-by-letter pearl millet genome information, and another is to link millions of SNPs and other genetic variants with highly diverse phenotypes. In addition to that, the paradigm stating that phenotypes are driven by genetic and environmental variations is nowadays extended. It is well established that some phenotypic variation is explained by epigenetic factors. As new technologies sequence DNA at high-resolution, methylation state could be scrutinized by identifying 5-methylcytosine at single base-pair level. Such approach, either bisulfite genomic sequencing (Frommer et al. 1992) or CHIP-seq (Barski et al. 2017), is qualitative and quantitative and allows considering other factors underlying phenotypes rather than limiting them to genetic and environmental variations. This, moreover, shows some complexities to consider when conducting population genomics studies, particularly in highly allogamous crop, such as pearl millet. First, population size can have very important effects on patterns of molecular variation. One may argue that the larger the population, the more powerful are the statistics needed to establish the robust genotype-phenotype associations. Relatively, in smaller populations, natural selection is less effective, i.e., they would be less able to either eliminate deleterious mutations or fix advantageous ones. Such pattern of polymorphism or variance has been found in model species *Arabidopsis thaliana* (Weinreich and Rand 2000) and *Saccharomyces cerevisiae* (Gu et al. 2005). In pearl millet, however, with a set of 90 RILs, GWAS based on patterns of LD and genotype by environment interactions have allowed to identify tightly linked SNPs (and other variants) in genes associated in space and time with climate variability and agronomic performance of pearl millet (Saïdou et al. 2009, 2014a, b; Vigouroux et al. 2011). New directions are to genotype and phenotype large populations from intercrossing, backcrossing and conduct association and linkage mapping of key agronomic traits.

Second, a high-quality reference genome is desirable as a resource to determine polymorphism levels and density of variants (SNPs) among individuals. Within a crop having a relatively large genome size, such as pearl millet, if the number of SNPs is not sufficiently large, adaptive signatures might not be detected. On the other hand, when the number of SNPs is large and spread across the genome, chances to score variation in specific traits are high, and rapid LD decay is expected in sets of lines that represent the variation. Populations with both rapid LD decay and high genetic diversity constitute good gene pool to perform genome-wide association studies for genetic dissection of complex traits.

Third, most traits are polygenic; consequently, the effect of one particular gene may be minimal. Therefore, it is imperative to understand the multilocus architecture underlying polygenic traits, such as local adaptation (Rajora et al. 2016). There is also a potential for precise genome editing for breeding and toward development of new superior pearl millet varieties or hybrids. Candidate gene approach in pearl millet could be used to detect selection footprints associated with environmental and/or stress responses or to identify the key agronomic traits in the background of cross-incompatible wild species. This approach, using linkage maps based on functional markers, was proven practical to identify numerous genes or loci known to potentially play an important role in local adaptation, or that are involved in a biological process related to environmental factors. Such genes are markers to be used for assisted breeding or genetic engineering. Another trend gaining attention is the CRISPR technology as a genome editing method, and many modifications of CRISPR/Cas9 technology are being successfully used in several crops with more complex genomes than pearl millet (Song et al. 2016). Potential applications of CRISPR in pearl millet would be to enhance through gene editing/pyramiding yields (quality and quantity) and performance under driest environments. Despite complexities/difficulties to regenerate viable, non-sterile individuals to advance in generation, biolistic method for gene delivery has been lately used to transform pearl millet (Lata 2015). Hence, it can be inferred that CRISPR technology offers new paths for pearl millet genetic improvement.

Crop improvement has historically been done through phenotypic selection, and recent advances in genomics come to reinforce it through genomics-assisted breeding through either MAS or GS. Where MAS uses molecular markers tightly linked with the QTLs and inherited together in LD, GS identifies selected promising individuals based on genomics estimated breeding values by combining marker genotype data with phenotype from several environments and pedigree. But both MAS and GS can increase the efficiency of incorporating desirable traits into elite cultivars or hybrids. Under a climatic change context, intraspecific breeding can be effective to transfer back adaptive genes present in wild relatives into domesticated cultivars because intraspecific progeny lines show more fitness than interspecific ones under harsh conditions. Accessions of wild pearl millet identified showing tolerance to environmental stress or resistance to fungal agent of downy mildew or to parasitic weed *Striga* in sub-Saharan West Africa constitute parental lines to introduce genetic traits and values to progenies. Use of genetically heterogeneous open-pollinated varieties with pyramided alleles of resistance or intraspecific crosses with susceptible cultivated varieties could lead to stability or introgression of resistance in new variety or hybrid of pearl millet.

Genomic selection (GS) in plant breeding programs helps select superior lines among experimental cultivars by reducing the breeding cycles. It is an applicable methodology to pearl millet since its genome has been sequenced and assembled and high-density markers are available. Recently, GS models were applied using grain yield and dense molecular marker information of pearl millet obtained from conventional RAD-seq and tunable GBS (tGBS) (Jarquin et al. 2020). Based on genotyping platforms and markers, population phenotypes used in calibration data sets, and genotype-by-environment (GxE) interaction information, the authors found that the performance of hypothetical genotypes among experimental cultivars could be predicted and improved by GS. This finding supports that GS is applicable for estimating breeding values and offers the opportunity to increase the productivity of experimental cultivars by assisting with increased genetic gains.

More species have their genome sequenced. However, there is a need for assembly and annotation of a reference transcriptome. Transcriptome assembly from high-throughput sequencing of mRNA (RNA-seq) is a powerful tool for detecting variations in sequences and gene expression between individuals, tissues, or conditions. Therefore, it becomes feasible to compare transcripts between individuals within a population and detect transcriptomic diversity patterns and outlier regions where the positive selection may have impacted genetic diversity. Population-level gene expression could be performed to estimate the average level and the magnitude of the variance of gene expression within the population sequenced. Additionally, gene expression diversity could also be examined to study the correlation between phenotypic variation and gene expression patterns if pearl millet varieties or landraces are transplanted from native to harsher habitats. For pearl millet genome annotation, RNA-seq data improve the quality of the read mapping and, by extension, the assembly. This accuracy could help genetic diversity and population structure characterization and deepen our understanding of how pearl millet adapted to harsh and driest areas.

Other approaches as phylogenomics, population epigenomics, or niche modeling offer promising perspectives regarding genome-wide population analyses of orphan crops. For example, niche modeling could provide news insights into how pearl millet has conquered harsh dry regions around West Africa or will be subjected to selection in predicted future environments.

5 Conclusions

Application of population genomics approaches in pearl millet has enabled us to understand genome-wide genetic diversity, population genetic structure, nucleotide diversity, linkage disequilibrium, and recombination rates across the genome. This has helped identifying genetic variants that shaped domestication and adaptation of pearl millet. Due to the inherent high and unpredictable climate variability, it is imperative to better use extant genetic diversity since it can potentially reduce the vulnerability of the production systems and enhance production stability. New allelic combinations from various sources would allow overcoming the major constraints of pearl millet production, where, until now, a little progress has been made. For example, application of population genomics to breeding process called biofortification could help release varieties with higher nutritional quality, making available nutrient-rich pearl millet varieties to provide nutritional security to small farmers and raising its marketability. Therefore, it is obvious that population genomics will reshape pearl millet breeding and genetic resource management program as it offers new paths and advantages. Longtime lagged behind other major cereals, evidence is that pearl millet should no more be considered as "orphan," and population genomics should help harness its potentials to fight malnutrition, food insecurity, and agricultural losses under an unpredictable climate.

References

Barski A, Cuddapah S, Cui K, Roh T-Y, Schones DE, Wang Z, et al. High-resolution profiling of histone methylations in the human genome. Cell. 2017;129(4):823–37. https://doi.org/10.1016/j.cell.2007.05.009.

Bassi FM, Bentley AR, Charmet G, Ortiz R, Crossa J. Breeding schemes for the implementation of genomic selection in wheat (*Triticum* spp.). Plant Sci. 2015;242:23–36.

Berthouly-Salazar C, Thuillet AC, Rhoné B, Mariac C, Ousseini IS, Couderc M, et al. Genome scan reveals selection acting on genes linked to stress response in wild pearl millet. Mol Ecol. 2016;25:5500–12.

Burgarella C, Cubry P, Kane NA, Varshney RK, Mariac C, Liu X, et al. A western Sahara centre of domestication inferred from pearl millet genomes. Nat Ecol Evol. 2018;2(9):1377–80.

Burton GW, Forston JC. Inheritance and utilization of five dwarfs in pearl millet (*Pennisetum typhoides*) breeding. Crop Sci. 1966;6:69–70. https://doi.org/10.2135/cropsci1966.0011183X000100010022x.

Debieu M, Sine B, Passot S, Grondin A, Akata E, Gangashetty P, et al. Response to early drought stress and identification of QTLs controlling biomass production under drought in pearl millet. PLoS One. 2018;13(10):e0201635.

Diack O, Kane NA, Berthouly-Salazar C, Gueye MC, Diop BM, Fofana A, et al. New genetic insights into pearl millet diversity as revealed by characterization of early- and late-flowering landraces from Senegal. Front Plant Sci. 2017;8:1–9. http://journal.frontiersin.org/article/10.3389/fpls.2017.00818/full.

Dussert Y, Snirc A, Robert T. Inference of domestication history and differentiation between early- and late-flowering varieties in pearl millet. Mol Ecol. 2015;24(7):1387–402.

Dwivedi S, Upadhyaya H, Senthilvel S, Hash C, Fukunaga K, Diao X, et al. Millets: genetic and genomic resources. In: Plant breeding reviews. Hoboken, NJ: Wiley; 2011. p. 247–375. https://doi.org/10.1002/9781118100509.ch5.

Fang Y, Xiong L. General mechanisms of drought response and their application in drought resistance improvement in plants. Cell Mol Life Sci. 2015;72:673–89.

Frommer M, McDonald LE, Millar DS, Collis CM, Watt F, Grigg GW, et al. A genomic sequencing protocol that yields a positive display of 5-methylcytosine residues in individual DNA strands. Proc Natl Acad Sci U S A. 1992;89(5):1827–31. http://www.ncbi.nlm.nih.gov/pmc/articles/PMC48546/.

Gu Z, David L, Petrov D, Jones T, Davis RW, Steinmetz LM. Elevated evolutionary rates in the laboratory strain of *Saccharomyces cerevisiae*. Proc Natl Acad Sci U S A. 2005;102(4):1092–7. http://www.ncbi.nlm.nih.gov/pmc/articles/PMC545845/.

Hash CT, Sharma A, Kolesnikova-Allen MA, Singh SD, Thakur RP, Raj AB, et al. Teamwork delivers biotechnology products to Indian small-holder crop-livestock producers: pearl millet hybrid "HHB 67 Improved" enters seed delivery pipeline. J SAT Agric Res. 2006;2(1):1–3.

Hu H, Xiong L. Genetic engineering and breeding of drought-resistant crops. Annu Rev Plant Biol. 2014;65(1):715–41. https://doi.org/10.1146/annurev-arplant-050213-040000.

Hu Z, Mbacké B, Perumal R, Guèye MC, Sy O, Bouchet S, et al. Population genomics of pearl millet (*Pennisetum glaucum* (L.) R. Br.): comparative analysis of global accessions and Senegalese landraces. BMC Genomics. 2015;16(1):1048. http://www.biomedcentral.com/1471-2164/16/1048.

Jarquin D, Howard R, Liang Z, Gupta SK, Schnable JC, Crossa J. Enhancing hybrid prediction in pearl millet using genomic and/or multi-environment phenotypic information of inbreds. Front Genet. 2020;10:1294.

Kaur P, Gaikwad K. From genomes to GENE-omes: exome sequencing concept and applications in crop improvement. Front Plant Sci. 2017;8:2164. https://www.frontiersin.org/article/10.3389/fpls.2017.02164.

Kountche BA, Hash CT, Dodo H, Laoualy O, Sanogo MD, Timbeli A, et al. Development of a pearl millet Striga-resistant genepool: response to five cycles of recurrent selection under Striga-infested field conditions in West Africa. Field Crop Res. 2013;154:82–90.

Kulkarni KS, Zala HN, Bosamia TC, Shukla YM, Kumar S, Fougat RS, et al. De novo transcriptome sequencing to dissect candidate genes associated with pearl millet-downy mildew (*Sclerospora graminicola* Sacc.) interaction. Front Plant Sci. 2016;7:847. http://www.ncbi.nlm.nih.gov/pmc/articles/PMC4916200/.

Lata C. Advances in omics for enhancing abiotic stress tolerance in millets. Proc Indian Natl Sci Acad. 2015;81:397–417.

Luikart G, Kardos M, Hand BK, Rajora OP, Aitken SN, Hohenlohe PA. Population genomics: advancing understanding of nature. In: Rajora OP, editor. Population genomics: concepts, approaches and applications. Cham: Springer Nature Switzerland AG; 2019. p. 3–79.

Malik S. Pearl millet-nutritional value and medicinal uses! Int J Adv Res Innov Ideas Educ. 2015;3:414–8.

Manning K, Pelling R, Higham T, Schwenniger JL, Fuller DQ. 4500-Year old domesticated pearl millet (*Pennisetum glaucum*) from the Tilemsi Valley, Mali: new insights into an alternative cereal domestication pathway. J Archaeol Sci. 2011;38:312–22.

Mariac C, Robert T, Allinne C, Remigereau MS, Luxereau A, Tidjani M, et al. Genetic diversity and gene flow among pearl millet crop/weed complex: a case study. Theor Appl Genet. 2006;113(6):1003–14.

Mariac C, Jehin L, Saïdou AA, Thuillet AC, Couderc M, Sire P, et al. Genetic basis of pearl millet adaptation along an environmental gradient investigated by a combination of genome scan and association mapping. Mol Ecol. 2011;20:80–91.

Mariac C, Scarcelli N, Pouzadou J, Barnaud A, Billot C, Faye A, et al. Cost-effective enrichment hybridization capture of chloroplast genomes at deep multiplexing levels for population genetics and phylogeography studies. Mol Ecol Resour. 2014;14:1103–13.

Mariac C, Ousseini IS, Alio A-K, Jugdé H, Pham J-L, Bezançon G, et al. Spatial and temporal variation in selection of genes associated with pearl millet varietal quantitative traits in situ. Front Genet. 2016;7:130. http://journal.frontiersin.org/article/10.3389/fgene.2016.00130.

National Research Council. Lost crops of Africa: volume I: grains. Washington, DC: The National Academies Press; 1996. https://www.nap.edu/catalog/2305/lost-crops-of-africa-volume-i-grains.

Ndour PMS, Gueye M, Barakat M, Ortet P, Bertrand-Huleux M, Pablo A-L, et al. Pearl millet genetic traits shape rhizobacterial diversity and modulate rhizosphere aggregation. Front Plant Sci. 2017;8:1288. http://www.ncbi.nlm.nih.gov/pmc/articles/PMC5529415/.

Oumar I, Mariac C, Pham J-L, Vigouroux Y. Phylogeny and origin of pearl millet (*Pennisetum glaucum* [L.] R. Br) as revealed by microsatellite loci. Theor Appl Genet. 2008;117:489–97.

Ousseini IS, Bakasso Y, Kane NA, Couderc M, Zekraoui L, Mariac C, et al. Myosin XI is associated with fitness and adaptation to aridity in wild pearl millet. Heredity (Edinb). 2017a;119(2):88–94.

Ousseini I, Bakasso Y, Kane N, Couderc M, Zekraoui L, Mariac C, et al. Myosin XI is associated with fitness and adaptation to aridity in wild pearl millet. Heredity (Edinb). 2017b:1–7. http://www.nature.com/doifinder/10.1038/hdy.2017.13.

Passot S, Gnacko F, Moukouanga D, Lucas M, Guyomarc'h S, Ortega BM, et al. Characterization of pearl millet root architecture and anatomy reveals three types of lateral roots. Front Plant Sci. 2016;7:829. http://journal.frontiersin.org/Article/10.3389/fpls.2016.00829/abstract.

Pucher A, Sy O, Angarawai II, Gondah J, Zangre R, Ouedraogo M, et al. Agro-morphological characterization of west and central African pearl millet accessions. Crop Sci. 2015;55(2):737–48.

Rajora OP, editor. Population genomics: concepts, approaches and applications. Cham: Springer Nature Switzerland AG; 2019.

Rajora OP, Eckert AJ, Zinck JWR. Single-locus versus multilocus patterns of local adaptation to climate in eastern white pine (*Pinus strobus*, Pinaceae). PLoS One. 2016;11(7):e0158691.

Rhoné B, Mariac C, Couderc M, Berthouly-Salazar C, Ousseini IS, Vigouroux Y. No excess of CIS-regulatory variation associated with intraspecific selection in wild pearl millet (*Cenchrus Americanus*). Genome Biol Evol. 2017;9:388–97.

Saïdou AA, Mariac C, Luong V, Pham JL, Bezançon G, Vigouroux Y. Association studies identify natural variation at PHYC linked to flowering time and morphological variation in pearl millet. Genetics. 2009;182:899–910.

Saïdou AA, Clotault J, Couderc M, Mariac C, Devos KM, Thuillet AC, et al. Association mapping, patterns of linkage disequilibrium and selection in the vicinity of the PHYTOCHROME C gene in pearl millet. Theor Appl Genet. 2014a;127:19–32.

Saïdou A-A, Thuillet A-C, Couderc M, Mariac C, Vigouroux Y. Association studies including genotype by environment interactions: prospects and limits. BMC Genet. 2014b;15:3. http://bmcgenet.biomedcentral.com/articles/10.1186/1471-2156-15-3.

Sehgal D, Skot L, Singh R, Srivastava RK, Das SP, Taunk J, et al. Exploring potential of pearl millet germplasm association panel for association mapping of drought tolerance traits. PLoS One. 2015;10(5):e0122165.

Serba DD, Yadav RS. Genomic tools in pearl millet breeding for drought tolerance: status and prospects. Front Plant Sci. 2016;7:1724. http://www.ncbi.nlm.nih.gov/pmc/articles/PMC5118443/.

Serba DD, Perumal R, Tesso TT, Min D. Status of global pearl millet breeding programs and the way forward. Crop Sci. 2017;57:2891–905. https://doi.org/10.2135/cropsci2016.11.0936.

Shivhare R, Lata C. Exploration of genetic and genomic resources for abiotic and biotic stress tolerance in pearl millet. Front Plant Sci. 2017;7:2069. http://journal.frontiersin.org/article/10.3389/fpls.2016.02069/full.

Slatkin M. Linkage disequilibrium – understanding the evolutionary past and mapping the medical future. Nat Rev Genet. 2008;9(6):477–85. http://www.ncbi.nlm.nih.gov/pmc/articles/PMC5124487/.

Song G, Jia M, Chen K, Kong X, Khattak B, Xie C, et al. CRISPR/Cas9: a powerful tool for crop genome editing. Crop J. 2016;4(2):75–82. http://www.sciencedirect.com/science/article/pii/S2214514116000192.

Soreng R, Davidse G, Peterson P, Zuloaga F, Judziewicz E, Filgueiras T, et al. On-line taxonomic novelties and updates, distributional additions and corrections, and editorial changes since the four published volumes of the catalogue of New World Grasses (Poaceae) published in contributions from the United States National Herbarium, vols. 39, 41, 46, and 48. In: Soreng RJ, Davidse G, Peterson PM, Zuloaga FO, Filgueiras TS, Judziewicz EJ, Morrone O, editors. Catalogue of New World Grasses. St. Louis: Missouri Botanical Garden; 2003.

Upadhyaya HD, Reddy KN, Ahmed MI, Kumar V, Gumma MK, Ramachandran S. Geographical distribution of traits and diversity in the world collection of pearl millet [*Pennisetum glaucum* (L.) R. Br., synonym: *Cenchrus americanus* (L.) Morrone] landraces conserved at the ICRISAT genebank. Genet Resour Crop Evol. 2017;64(6):1365–81.

Vadez V, Hash T, Bidinger FR, Kholova J. II.1.5 Phenotyping pearl millet for adaptation to drought. Front Physiol. 2012;3:386.

Varshney RK, Terauchi R, McCouch SR. Harvesting the promising fruits of genomics: applying genome sequencing technologies to crop breeding. PLoS Biol. 2014;12(6):1–8.

Varshney RK, Shi C, Thudi M, Mariac C, Wallace J, Qi P, et al. Pearl millet genome sequence provides a resource to improve agronomic traits in arid environments. Nat Biotechnol. 2017;35(10):969–76. http://www.nature.com/doifinder/10.1038/nbt.3943.

Vigouroux Y, Mariac C, de Mita S, Pham JL, Gérard B, Kapran I, et al. Selection for earlier flowering crop associated with climatic variations in the Sahel. PLoS One. 2011;6(5):e19563. http://www.ncbi.nlm.nih.gov/pmc/articles/PMC3087796/.

Waddington SR, Li X, Dixon J, Hyman G, de Vicente MC. Getting the focus right: production constraints for six major food crops in Asian and African farming systems. Food Secur. 2010;2(1):27–48. https://doi.org/10.1007/s12571-010-0053-8.

Warr A, Robert C, Hume D, Archibald A, Deeb N, Watson M. Exome sequencing: current and future perspectives. G3. 2015;5(8):1543–50. http://www.g3journal.org/content/5/8/1543.abstract.

Weinreich DM, Rand DM. Contrasting patterns of nonneutral evolution in proteins encoded in nuclear and mitochondrial genomes. Genetics. 2000;156(1):385–99.

Potato Population Genomics

Xiaoxi Meng, Heather Tuttle, and Laura M. Shannon

Abstract Potatoes are the third most widely grown food crop in the world. Due to their ability to grow on marginal land and produce large amounts of nutritious food with relatively few inputs, they have shaped human history. They are part of a wide range of international cuisines and growing in popularity. However, due to the complexity of autotetraploidy, clonal growth, and a wide range of wild relatives, we know less about potato genomics, diversity, and evolution than we do about comparable grain crops. Population genomics questions such as the relationship between potatoes and their numerous wild relatives, the timing and location of domestication and range expansion, and the patterns of selection and variation left in the potato genome by these histories remain open. Although there is a long history of potato genetics research, the current moment is particularly exciting. Potatoes are being reinvented as a diploid inbred-hybrid crop by scientists in the public and private sector around the world. New tools are under development for quantitative genetics and breeding in autopolyploid crops, for which potato serves as a model species. The availability of sequence data for potato and its relatives is growing rapidly. In this chapter, we review the currently available population genetics and genomics data in potato and the inferences that have been drawn from it as well as the questions it presents.

Keywords Admixture · Autotetraploid and polyploid · Domestication · Genetic resource conservation · Genetic variation and structure deleterious variation · Genome and transcriptome resources · Introgression · Population epigenomics · *Solanum tuberosum*

X. Meng · H. Tuttle · L. M. Shannon (✉)
Department of Horticultural Science, University of Minnesota, St. Paul, MN, USA
e-mail: lmshannon@umn.edu

1 Introduction

Potatoes (*Solanum tuberosum* L.) are vital to human food security. They are the most widely grown vegetable crop in the world; only grains are grown more commonly (FAO 2019). In 2019, 111 billion dollars of potatoes were produced internationally, with the leading producer being China, which grew 91 million tons of potatoes (FAO 2019). This is reflective of rapidly expanding markets. Until the 1990s, the majority of potatoes were grown in the USA and Europe. More recently, production has been expanding in Asia, Africa, and Latin America (FAO 2008). In 2019, India, Bangladesh, Peru, Egypt, Pakistan, and Kazakhstan were among the top-twenty potato producing nations. Particularly in India and China, this represents a specific national effort to encourage adoption of potatoes into the national diet and increase cultivation to feed expanding populations (FAO 2019). In 2008, the United Nations declared the international year of the potato in an effort to support potato production and trade as a practical solution to growing populations in changing climates (FAO 2008). While incorporation of potatoes into cuisines in India and China is an ongoing project, potatoes are traditional dietary staples in a variety of American and European cuisines. Potatoes played a major role in European history providing a cheap, easy-to-produce, nutritious food source that allowed for the urban population growth crucial for the industrial revolution (Pollan 2001). As predicted by the novel and film *The Martian* (Weir 2014), NASA is exploring potato as a food source that could be produced in space (Mansfield 2014).

Potatoes are central to our past, present, and future for two reasons: they are extraordinarily nutritious and can be grown on marginal land. Although potatoes have been erroneously villainized by people promoting low carb diets, they are high in fiber and protein and are satiating. They are a major source of potassium, vitamin C, and antioxidants (Beals 2019; Navarre et al. 2019). They have the potential to improve human health outcomes (Ezekiel et al. 2013; McGill et al. 2013; Charepalli et al. 2015; Bibi et al. 2018; Reddivari et al. 2019; Vanamala 2019). An exclusively milk-and-potato diet provides all the nutrients needed for human life (Pollan 2001). Furthermore, potatoes historically have been grown in difficult conditions from mountain sides in the Andes to lands rejected for export farming in Ireland. Potatoes require less water and space per calorie or unit protein than any other staple crop (Renault and Wallender 2000). Traditionally, in times of limited food availability, humans have eaten potatoes. It seems inevitable that as populations expand and climates change, potatoes will continue to be a major source of sustenance.

Despite the clear importance of potato, we know less about their genetics, especially population genetics and genomics, than comparably important grain crops. The primary reason for this is autopolyploidy. Commercial potatoes in the USA and Europe are autotetraploid. Genetic analysis of autopolyploid species presents technical and theoretical challenges. In particular, the three heterozygous genotypic classes in tetraploids (ABBB, AABB, AAAB) complicate genotype calling from array data (Schmitz Carley et al. 2017; Voorrips et al. 2011), create a need for increased read depth when sequencing (Uitdewilligen et al. 2013), and muddle attempts at phasing and haplotype construction (Bourke et al. 2016). Simply collapsing the heterozygotes into a single genotypic class reduces the accuracy of quantitative (Endelman et al. 2018) and population (Meirmans et al. 2018) genetic models. Many of the basic principles and mathematical models used in genomic and evolutionary research must be adjusted for autotetraploids. For example, although autopolyploid chromosomes often form random bivalents during meiosis, they can also form multivalent pairing structures which can result in gametes containing pairs of sister chromatids rather than pairs of homologous chromosomes (Gallais 2003). This presents challenges for mapping (Hackett et al. 2014). Autotetraploidy similarly diverges from the models for linkage disequilibrium, dominance relationships, and Hardy-Weinberg equilibrium (Meirmans et al. 2018; Gallais 2003). It is not only more difficult to generate genomic data about potatoes than about comparable diploids or allopolyploids that data is also challenging to interpret in absence of autopolyploid-specific models.

An additional challenge is presented by the fact that potato primarily reproduces clonally. This means that potatoes have gone through fewer generations since domestication than sexually reproducing species. Furthermore, in a clonal species, fitness does not necessarily depend on ability to reproduce sexually. Limited selective pressure maintaining sexual reproduction pathways means potatoes are often difficult to cross. This is particularly true of commercial tetraploid varieties which are often male sterile and sometimes completely infertile (Bethke and Jansky 2021). Conversely, species barriers among potatoes and their wild relatives seem relatively weak. Potato was domesticated from a species complex, and its post domestication history has been shaped by introgression from a wide variety of sympatric species (Hardigan et al. 2017; Vos et al. 2015; Hoopes et al. 2022). The variety of contributors to the potato genome introduces complexity which further complicates the study of potato origins and genetic diversity. The combination of these factors has resulted in extremely heterozygous individuals. Potato exhibits higher heterozygosity than any of the other major domesticated crop species (Hardigan et al. 2017). These levels of heterozygosity pose technical challenges for sequencing and genome assembly.

Although the biology and history of potato resulted in a difficult-to-study crop with a complex genome, the study of potato is a necessity. It is the kind of resource-efficient, resilient-to-environmental-stress, highly nutritious crop that humans rely on in difficult times. Potato is the kind of crop we need as our environment and agricultural spaces change. Understanding potato genomes, evolution, and population genetics and genomics will support efforts to develop productive and widely adapted potato varieties to meet the challenges ahead. In this chapter, we review the progress made thus far in population genomics of potato and highlight some of the new methods for genetic analysis in autopolyploids which will facilitate further progress.

2 Key Questions for Population Genomics Research in Potato

2.1 Potato Domestication: Where? When? How? From What?

The origins of potato remain relatively mysterious. We know potato was domesticated at least once in the Andes on the border of Peru and Bolivia near Lake Titicaca (Spooner et al. 2005a). Additionally, a secondary southern domestication event in Chile has been hypothesized (Sukhotu et al. 2004). Alternatively, potato may have followed a similar pattern to rice (*Oryza sativa*) with a single domestication event followed by adaptation to new environments facilitated by introgression from local species. In favor of this hypothesis, the history of cultivated potato is characterized by extensive introgression (Hardigan et al. 2017) and it is notoriously difficult to distinguish between de novo domestication and high rates of local introgression (Choi et al. 2017; Choi and Purugganan 2018). It is hypothesized that, like most wild potatoes, the original northern domesticated potatoes were diploid (Spooner et al. 2005a). Potatoes in southern South America tend to be tetraploid. Therefore, if there was a second southern domestication event, the original southern domesticated potatoes are likely to have been tetraploid (Sukhotu et al. 2004). Polyploidization often coincides with or rapidly follows domestication events (Salman-Minkov et al. 2016) which suggests this is also a plausible scenario.

The putative wild progenitor is alternatively called *Solanum candoelleanum* (Vos et al. 2015) or *S. bukasovii* (Spooner et al. 2005a), and it is referred to as both a species (Spooner et al. 2005a) and a species complex (Sukhotu et al. 2004). Some of the confusion is due to the USDA taxonomy which names *S. candoelleanum* as a single species which includes 48 previously described taxa (Spooner et al. 2018). However, recent trees built on sequence data found that *S. medians*, *S. megistracrolobum*, and *S. raphanifolium* were most closely related to cultivated lineages, suggesting alternate potential wild progenitors (Hardigan et al. 2017). Introgression from wild species has been shown to cloud origin relationships (Vonholdt et al. 2010; Freedman et al. 2014; Shannon et al. 2015), and *S. candoelleanum* is the wild species exhibiting the most introgression with cultivated potato (Hardigan et al. 2017). Quantifying wild species introgression into potato is a crucial step to clarifying origin relationships.

The timing and size of the domestication and improvement bottlenecks are unknown. It has long been orthodoxy among potato breeders that the migration bottlenecks into Europe and the USA were narrow, with small founding populations, leading to little diversity (Hirsch et al. 2013), but recent genotyping and sequencing studies demonstrate extensive diversity in the US germplasm suggesting a much wider bottleneck (Hardigan et al. 2016; Pham et al. 2017).

Understanding the origins and evolution of staple crops tells us about human history and provides insights which can be used in crop improvement and breeding. Furthermore, since crop domestication is evolution on a shortened time scale with increased selective pressure leaving more identifiable signatures in the genome, it is often used as a model for evolution (Doebley et al. 2006). Developing a more comprehensive picture of potato domestication will not only inform breeding and improvement efforts by describing the potato genome, but will also provide a model for autopolyploid evolution.

2.2 What Is the Relationship Between Wild and Cultivated Potato Diversity?

Not only is our understanding of potato history limited, but also the dynamics of current populations are not well understood. Potato is highly heterozygous; according to some reports, significantly more so than other domesticated species or its own wild relatives (Hardigan et al. 2017). This was an unexpected result because of the hypothesized narrow bottle neck, and because in general domestication creates a reduction in diversity (Doebley et al. 2006). Furthermore, other analyses have arrived at different conclusions with respect to the relative genetic diversity in cultivated and wild potato (Li et al. 2018b). The role ploidy plays in these calculations and population-level diversity remain underexplored.

2.3 What Role Do Deleterious Variants Play in the Potato Genome?

The potato genome is rife with deleterious alleles, which have implications for breeding. However, these deleterious alleles have primarily been explored in diploid landrace germplasm (Zhang et al. 2019, 2021a) or diploidized potatoes from few tetraploid cultivars (Manrique-Carpintero et al. 2018; Achakkagari et al. 2022). Predictions of deleterious alleles in tetraploids (Hoopes et al. 2022) have relied on protein effects rather than evolutionary approaches (Kono et al. 2018). Increased ploidy has the effect of reducing selective pressure and hiding recessive alleles (Monnahan and Brandvain 2020). Therefore, we would expect even more deleterious alleles in the cultivated tetraploid germplasm. Uncovering these alleles will allow breeders to purge them from the genome (Zhang et al. 2021a).

2.4 What Is the Genetic Basis of Commercial Traits?

Potatoes are a horticultural crop which means that a series of quality traits are essential to their marketability. The largest market class in the USA is processing 'Russet' for French fries, hash browns, and similar purposes. Russets must have an oblong shape, a within range specific gravity, russeted skin, and white flesh. Chipping potatoes must be round, with thin white skin, and white flesh. They must have easily detached stolons, a specific gravity within a particular range, and be approximately the size and shape of a baseball. Both must have shallow eyes and no eyebrows, a regular shape, and no internal defects. They must be able to be stored at cold temperatures without converting starches to sugars. Deviation from these standards makes a variety unsalable. However, for the vast majority of these traits, we do not know the underlying genetic mechanism. Color in potato is reasonably well understood (De Jong et al. 2003, 2004; Jung et al. 2005, 2009; Zhang et al. 2009a, b), although the picture is more complex than the major effect alleles identified (Caraza-Harter and Endelman 2020). A round- versus oblong-shaped locus has been identified (Endelman and Jansky 2016). The genetics of cold sweetening are sufficiently understood to facilitate the development of GMO varieties that can be stored at extremely cold temperatures (Ellis et al. 2020a). However, the genetic basis of traits as basic as russeting remain unexplained.

2.5 How Does Dosage at Different Ploidy Levels Affect These Traits?

Potato breeding is undergoing a community-wide change, as public sector and private sector breeders around the world explore converting potato to a diploid inbred crop (Jansky et al. 2016; Lindhout et al. 2011). Thus far, breeders have shown great success replicating tetraploid phenotypes at the diploid level (Zhang et al. 2021a; Manrique-Carpintero et al. 2018; Alsahlany et al. 2021). However, it seems likely that ploidy interacts with some crucial commercial traits in potato. For example, tetraploid potato is self-fertile while diploid potato is not (Clot et al. 2020). As diploid potatoes are developed, it will be crucial to understand the genetic mechanism behind traits at both diploid and tetraploid levels, providing the opportunity to explore the role of dosage effects in the control of complex traits in autopolyploids.

2.6 How Can We Best Use Population Genomic Insights to Support Breeding?

The development of new tools and genetic resources for potato in recent years has led to great leaps in potato breeding techniques, including a focus on genomic selection. Understanding population genomics of a crop can inform breeding strategies, because demography defines genetic architecture of crucial traits and partitioning of variation, deleterious and otherwise, among populations (Turner-Hissong et al. 2020). Developing a more thorough understanding of the history and population genetic patterns of potato will provide crucial information as we continue to improve potato breeding methodology. In particular, an understanding of how individual heterozygosity translates to population-level diversity within market class, identification of deleterious alleles, and clarifying the relationship between wild species with biotic and abiotic stress resistance and cultivated potato, would allow breeders to make more informed decisions regarding crossing and selection.

3 Genomic, Transcriptomic, and Other Resources for Population Genomics Research

Population genetics is the study of genetic variation within and between populations and the forces that shape that variation. Power, accuracy, and applicability in population genetics and genomics studies depend in part on the datasets used for these studies (Luikart et al. 2019; Rajora 2019). In this section, we provide an overview of the available resources including genomes, transcriptomes, genetic maps, and marker technologies.

3.1 Genomes

The most complete reference genome for potato is DM1-3 516 R44 (Hardigan et al. 2016; Potato Genome Sequencing Consortium et al. 2011; Sharma et al. 2013; Pham et al. 2020), henceforth referred to as DM. DM is a doubled monoploid of a diploid *S. tuberosum group phureja*, which was created to eliminate the problems with genome assembly of highly heterozygous individuals. The original assembly was created with a whole-genome shotgun sequencing approach scaffolded with fosmids and BACs (Potato Genome Sequencing Consortium et al. 2011). The genome was revised using genetic maps and synteny with tomato, *Solanum lycopersicum* (Sharma et al. 2013) and unscaffolded contigs were assembled and aligned (Hardigan et al. 2016). The latest version relies on Oxford Nanopore Sequencing and Hi-C to create the most complete reference genome sequence for potato to date (Pham et al. 2020). The primary limitation of DM as a reference genome is that it is not representative of the majority of potatoes grown and consumed which are highly heterozygous autotetraploids from *S. tuberosum group tuberosum* (Uitdewilligen et al. 2013).

In 2022, eight tetraploid *S. tuberosum group tuberosum* genomes were published, five of these genomes are phased. The largest sequencing project was the potato pan genome project which produced genomes for six cultivars: Atlantic a US chipping clone, 'Castle Russet' a US cultivar for French fry processing, two European fresh market cultivars 'Columba' and 'Spunta', and two European starch cultivars 'Altus' and 'Avenger' (Hoopes et al. 2022). Unphased genome assemblies were created for all six clones using NRgene's Denovo magic technology. Hi-C and Oxford Nanopore data was combined with mapping populations to phase 'Atlantic' and 'Castle Russet' (Hoopes et al. 2022). Meanwhile, 'Altus' was phased using HiFi reads and a mapping population (Mari et al. 2022). A different approach was taken in sequencing 'Otava,' a tetraploid *S. tuberosum* cultivar from the Czech Republic (Sun et al. 2022). There, haplotypes were phased by combining PacBio long-read sequencing and 10X genomics pollen sequencing. The only group to create a phased tetraploid genome without a mapping population sequenced 'Qingshu 9' and used a combination of HiFi and HiC reads to deliver phased chromosomes (Wang et al. 2022).

Sequencing and phasing an autotetraploid remains a long, complicated, and expensive process. An alternate strategy is to sequence diploids. There are 12 *S. tuberosum group tuberosum* diploid sequences. First, Solyntus is a diploid F_9 inbred line which its creators describe as *S. tuberosum* (van Lieshout et al. 2020). Unfortunately, due to its industry origin, the pedigree of Solyntus is unclear beyond the use of *S. chacoense* as a self-compatibility donor in the original cross. Sequencing was done using Oxford Nanopore long reads and Illumina short reads resulting in a genome consisting of 116 contigs assembled into pseudo-molecules. Second, RH089-039-16 (commonly referred to as RH), is a heterozygous diploid with a pedigree consisting of dihaploidized tetraploid commercial varieties including Katahdin, Chippewa, and Primura (Zhou et al. 2020). 10X genomics and Illumina whole-genome sequencing were combined to create scaffolds, and a mapping population was used to assign scaffolds to linkage groups. The process was repeated with circular consensus sequencing. The two genomes were merged, and Oxford Nanopore and Hi-C were used to improve and clean up the assembly. A related RH line and another *S. tuberosum tuberosum* were sequenced as part of a larger diversity sequencing effort using circular consensus sequencing and Hi-C (Tang et al. 2022). Finally, nine legacy clones from Agriculture Agri-Food Canada's historical efforts at diploid breeding were sequenced using 10X genomics (Achakkagari et al. 2022). These clones were developed through a combination of diploidizing tetraploid material, introgression breeding, and inbreeding. Eight of the nine clones are primarily tuberosum.

There have also been sequencing efforts for landraces. Specifically, there is a *Solanum tuberosum stenotonum* assembly which was created using a combination of short reads, Hi-C, and PacBio long reads (Yan et al. 2021). *Stenotonum* is hypothesized to have an ancestral relationship with *tuberosum tuberosum*. Seventeen additional de novo assemblies of *S. tuberosum group stenotomum, ajanhuri, goniocalyx, and phureja* developed through circular consensus sequencing and Hi-C are available (Tang et al. 2022). Kyriakidou et al. (Kyriakidou et al. 2020a) sequenced 12 potato landraces, ranging from diploid to pentaploid, to compare structural variation by aligning them to the DM and M6 references. Plastome (Achakkagari et al. 2020) and mitogenome (Achakkagari et al. 2021) assemblies are also available for these clones. The authors created de novo assemblies for six polyploid landraces: *S. chaucha, S. x juzepczukii, S. tuberosum ssp. tuberosum, S. x curtilobum*, and two examples of *S. tuberosum ssp. andigenum* (Kyriakidou et al.

2020b), using Illumina short reads for all clones and 10X genomics and PacBio long reads for *S. tuberosum ssp. andigenum*.

Wild potato is an essential source of stress resistance genes for potato breeding (Jansky et al. 2013) and played an essential role in the evolution of cultivated potato (Hardigan et al. 2017; Hoopes et al. 2022). One important wild potato genome assembly is M6, a *Solanum chacoense* clone notable for its self-compatibility, which was assembled from short-read Illumina sequencing (Leisner et al. 2018). Additionally, de novo assemblies of 22 wild species from section *Petoa* and two from the neighboring section *Etuberosum* are available (Tang et al. 2022; Hosaka et al. 2022). Particular attention was paid to *S. candolleanum*, cultivated potato's closest relative, for which there are four unique de novo assemblies (Tang et al. 2022). Reference-based assemblies are also available for *S. commersonii* (Aversano et al. 2015) and *S. pinnatisctum* (Tiwari et al. 2021), and there is a chloroplast genome for *S. hjertingii* (Park 2022).

3.2 Transcriptomes

Creating a potato transcriptome does not carry with it all the challenges of autotetraploid genome assembly, and therefore while sequencing data is limited there are a plethora of transcriptomes. There are two distinct sets of gene models based on DM used in these transcriptomes: one is from the Potato Genome Sequencing Consortium (Potato Genome Sequencing Consortium et al. 2011), and the other is from the Tomato Genome Consortium (Sato et al. 2012). These two gene models have been reconciled (Petek et al. 2020). A 44K expression array has been developed by the Potato Oligo Chip Initiative (Kloosterman et al. 2008), which has been used to examine quality (Ducreux et al. 2008) and disease traits (Ali et al. 2014; Burra et al. 2014; Stare et al. 2017).

Transcriptome sequencing approaches provide a widely used alternative to the array. The most comprehensive of these being the transcriptome for DM which includes 32 tissues and growth conditions (Massa et al. 2011). As with the genomes described above, there is extensive variation between cultivars, which can be captured with a pan-transcriptome approach (Petek et al. 2020). Comparative approaches are valuable not only because they are more representative of the potato transcriptome as a whole, but also because they can be used to identify pathways which differ between clones (Liu et al. 2015). Transcriptomic analyses are available for a variety of abiotic stress conditions including drought (Gong et al. 2015; Moon et al. 2018; Yang et al. 2019a), heat (Liu et al. 2021; Zhang et al. 2021b), salinity (Li et al. 2020), freezing (Carvallo et al. 2011), nitrogen limitation (Gálvez et al. 2016; Tiwari et al. 2020a; Zhang et al. 2020), and damp storage conditions (Peivastegan et al. 2019). Additionally, transcriptomes have been used to study resistance to a plethora of diseases in potato including scab (Kaiser et al. 2020; Fofana et al. 2020), root knot nematode (Macharia et al. 2020), potato viruses A (Li et al. 2017) and X (Herath and Verchot 2021), and potato wart disease (Li et al.

2021). Transcriptomics has also been used to explore plant response to possible management practices used to mitigate disease (Lemke et al. 2020; Yang et al. 2020). There are mitochondrial (Varré et al. 2019) and microRNA (miRNA) (Lakhotia et al. 2014; Zhang et al. 2013) transcriptomes available, as well as *S. commersonii* transcripts from multiple growth conditions (Carvallo et al. 2011; Zuluaga et al. 2015).

3.3 Genetic Maps

Developing linkage maps for potatoes is made challenging by their autotetraploid nature and their tendency to be primarily outcrossers. The first genetic map in tetraploid potato was created from an F_1 population in 1998 using Amplified Fragment Length Polymorphisms (AFLPs) (Meyer et al. 1998). This work was based on a simulation study demonstrating it was possible to combine simplex (AAAB) and duplex (AABB) markers into a single autotetraploid map (Hackett et al. 1998). This was followed with the release of the autotetraploid mapping software TetraploidMap (Hackett and Luo 2003; Hackett et al. 2007) and TetraploidSNPMap (Hackett et al. 2017), specifically for high density markers. The R package PERGOLA allows for mapping in F_2 and backcross populations created from homozygous autotetraploid parents (Grandke et al. 2017). Of course, homozygous autotetraploids are difficult to generate and so F_1 mapping is generally preferred (Gallais 2003). Three R packages allow for genetic map creation from F_1 polyploid populations, PolymapR (Bourke et al. 2018a), MapPoly (Mollinari and Garcia 2019), and NetGWAS (Behrouzi and Wit 2019). PolymapR is distinguished by its ability to estimate recombination fractions and parental linkage phases. MapPoly uses a hidden Markov model (HMM) and produces genotype probability estimates from that HMM. NetGWAS is uniquely designed for data sets with highly uneven or missing data and its graph-based approach is often faster. In the case of multiple F_1 populations which share parents, haplotypes can be inferred using PolyOrigin (Zheng et al. 2021).

Before the advent of advanced mapping techniques, linkage maps were primarily created in diploid relatives of potato, as linkage mapping is easier and more flexible for diploids (Felcher et al. 2012). As the potato community works to convert potato to a diploid inbred crop (Jansky et al. 2016; Lindhout et al. 2011), diploidized versions of tetraploid potatoes can be used to create mapping populations (Manrique-Carpintero et al. 2018; Clot et al. 2020). This approach takes advantage of the ease of linkage mapping in a diploid while still reflecting haplotypes from their respective tetraploid parents.

3.4 QTL Mapping and GWAS

In addition to the challenges of creating a linkage map, specifically the potential for four haplotypes, autotetraploidy poses difficulties for mapping phenotypes onto genomes. One such challenge is that the three classes of heterozygosity open up the possibility for multiple types of dominance (Rosyara et al. 2016). A second is the necessity of large populations, because the probability of observing a given genotype is lower than in a diploid due to tetraploid segregation ratios (Gallais 2003). Regardless, the phenotype mapping techniques used in diploids are available and widely used in autotetraploid potato.

TetraploidSNPmap (Hackett et al. 2017) and PERGOLA (Grandke et al. 2017), described above, can be used for QTL mapping from cross populations. PolyqtlR (Bourke et al. 2018a) and QTLpoly (da Silva et al. 2020) are two additional R packages which can be used for QTL mapping in F_1 populations. QTLpoly takes a multiple interval mapping approach, which allows users to fit multiple QTLs at once (da Silva et al. 2020), while PolyqtlR relies on identical by descent (IBD) probabilities and uses an HMM algorithm. Chen et al. (Chen et al. 2021) present a method for QTL mapping allowing for quadrivalent pairing and double reduction, but they have not released software. QTL mapping has been used extensively to identify causative loci for disease resistance (Santa et al. 2018), maturity (Li et al. 2018a), and agronomic traits (Massa et al. 2018; da Pereira et al. 2021). As with linkage mapping, QTL mapping with diploid relatives to get around the challenges posed by tetraploids is a commonly used strategy (Endelman and Jansky 2016; Braun et al. 2017; Manrique-Carpintero et al. 2015; Kloosterman et al. 2012).

Genome-wide association mapping (GWAS) has also been implemented in potato, most notably using the GWASpoly package (Rosyara et al. 2016). GWASpoly allows for scans assuming three dominance models: additive, simplex dominance, and duplex dominance. These authors' simulations suggest that using the model that corresponds to true gene action provides the most power. Similarly, the appropriateness of using a kinship (K) or population structure (Q) based model to correct for relatedness depends on the trait being mapped (Sharma et al. 2018). GWAS has been used to identify loci controlling morphological traits (Yousaf et al. 2021; Zia et al. 2020), tuber quality (Byrne et al. 2020; Klaassen et al. 2019), scab tolerance (Kaiser et al. 2020; Koizumi et al. 2021; Yuan et al. 2020), and drought resistance (Toubiana et al. 2020; Aliche et al. 2019; Tagliotti et al. 2021).

Both QTL mapping and GWAS of tetraploids require large populations as compared to diploids. One way to circumvent the need for a large population is using mapping by sequencing approaches. One such approach is Comparative Subsequent Sets Analysis (CoSSA) which has been used to map resistance to potato wart disease (Prodhomme et al. 2019). This method is based on bulk segregant analysis in an F_1 population. K-mers found only in bulked resistant germplasm are distinguished from those found only in bulked susceptible germplasm. The k-mers identified as resistance related are filtered for read depth, based on the assumption that all resistant material will have the resistance k-mers. Finally, if the parents are sequenced, k-mers can be filtered for IBD with the resistant parent. Another mapping by sequencing approach is RenSeq which uses capture sequencing for R-genes to facilitate cloning of resistance genes (Jupe et al. 2013). RenSeq has been used to identify late blight (Jupe et al. 2013; Witek et al. 2016; Witek et al. 2021; Chen et al. 2018), potato cyst nematode (Strachan et al. 2019), and PVY (Grech-Baran et al. 2020; Torrance et al. 2020) resistance.

3.5 Genetic Markers

DNA polymorphism analysis has been a powerful tool in potato research for the last three decades. Available conventional marker types include the following: AFLP, Random Amplification of Polymorphic DNA (RAPD), Restriction Fragment Length Polymorphism (RFLP), Simple Sequence Repeat (SSR), Single-Stranded Conformation Polymorphism (SSCP), Sequence Tagged Site (STS), Cleaved Amplified Polymorphic Sequence (CAPS), Sequence Characterized Amplified Region (SCAR), and Kompetitive Allele Specific PCR (KASP) markers. These have been used for linkage maps and QTL mapping (Li et al. 2018a; Collins et al. 1999; Ewing et al. 2000; Ghislain et al. 2001; Huang et al. 2004; Kuhl et al. 2001; Meijer et al. 2018; Meyer et al. 1998; Naess et al. 2000; Oberhagemann et al. 1999; Sandbrink et al. 2000; Song et al. 2003; Sørensen et al. 2008; Van Der Vossen et al. 2003; Visker et al. 2003; Bormann n.d.; Li et al. 2018b; Sliwka 2004), marker assisted selection (MAS) (Clot et al. 2020; Totsky et al. 2020; Fulladolsa et al. 2015; Kasai et al. 2000), association mapping, assessment of population structure, and linkage disequilibrium analyses (D'hoop et al. 2014; D'hoop et al. 2010; D'hoop et al. 2008; Li et al. 2013). Moreover, AFLP, RAPD, RFLP, and SSR markers have been used to investigate taxonomic classification, origin, genetic diversity, varietal identification, and domestication (Spooner et al. 2005a; Sukhotu et al. 2004; Spooner et al. 2010; Gavrilenko et al. 2010; Gebhardt et al. 1991; Sukhotu et al. 2005; Anoumaa et al. 2017; Caputo et al. 2013; Collares and Choer 2004; Ghislain et al. 2009; Hoque et al. 2013; Rocha et al. 2010; Salimi et al. 2016; Song et al. 2016; Lee et al. 2021a). Researchers often combined more than one marker type in order to improve density and coverage (Li et al. 2018a; D'hoop et al. 2014; Bradshaw et al. 2008; Khu et al. 2008). Although these sorts of markers have been useful historically, they have several drawbacks including uneven distribution and difficulty in covering the entire genome which can lead to low resolution of QTL mapping and long QTL interval lengths (Sliwka 2004). In addition, dominant markers, such as AFLP and RAPD, do not differentiate homozygous and heterozygous alleles (Boopathi 2012).

Next-generation sequencing data has enabled the development of high-throughput genotyping techniques capable of generating large numbers of markers for higher coverage and density. The co-dominant nature of these makers also enables estimation of allele dosage in polyploids (Rickert et al. 2003). The most widely used of these marker platforms are single nucleotide polymorphism (SNP) arrays. There were originally two versions of the SNP array, the US-based SolCAP

Infinium array and the European SolSTW array (Uitdewilligen et al. 2013; Vos et al. 2015; Felcher et al. 2012; Hamilton et al. 2011). Both sets of SNPs are now available on the latest SolCAP array which contains over 50,000 SNPs. These arrays are a powerful tool for GWAS (Rosyara et al. 2016; Sharma et al. 2018; Lindqvist-Kreuze et al. 2020; Stich et al. 2013; Vos et al. 2017; Mosquera et al. 2016), linkage analysis and QTL mapping (Felcher et al. 2012; Massa et al. 2018; Odilbekov et al. 2020; Hackett et al. 2013; Douches et al. 2014; Schönhals et al. 2017), measuring genetic diversity and population structure (Hirsch et al. 2013; Pandey et al. 2021; Berdugo-Cely et al. 2017; Kolech et al. 2016), genetic identity determination (Ellis et al. 2018), phylogenetic relationship inference (Hardigan et al. 2015), understanding breeding history (Vos et al. 2015; Hirsch et al. 2013), allelic variation analysis (Manrique-Carpintero et al. 2015), and identifying pathways under selection (Hirsch et al. 2013).

Array genotyping depends on determining dosage based on signal intensity. In a diploid, this results in three classes of intensity, and the boundaries are relatively easy to determine. As the classes of heterozygosity and therefore the classes of intensity increase, forming discrete borders and therefore assigning genotypes to an individual becomes more difficult (Schmitz Carley et al. 2017). Multiple software solutions have been presented for calling marker allele dosage in tetraploid potato including GenomeStudio, ClusterCall, and FitPoly (Schmitz Carley et al. 2017; Voorrips et al. 2011; Zych et al. 2019; Staaf et al. 2008). Of the options currently available, we recommend FitTetra. GenomeStudio relies on the threshold approach most appropriate to diploids, and FitPoly has recently implemented a correction based on prediction of genotype distributions similar to ClusterCall combining both polyploid-specific approaches into a single package.

The arrays were developed using North American and European potato germplasm and although landraces are added as the array increases in size, ascertainment bias and absence of rare alleles remain potential limitations of the array (Hirsch et al. 2013; Sharma et al. 2018). Reduced-representation sequencing methods, such as genotyping-by-sequencing (GBS), restriction-site-associated DNA sequencing (RADseq), diversity arrays technology (DArTseq), and specific length amplified fragment sequencing (SLAF-Seq), coupled with the assembled reference genome, provide cost-effective approaches for obtaining genome-wide high-throughput SNP markers while minimizing ascertainment bias. Polyploidy also presents difficulties for GBS and related technologies, for instance, 60X coverage is required to confidently call tetraploid genotypes as compared to 5X for diploids (Uitdewilligen et al. 2013). Nevertheless GBS has been used for association mapping (Uitdewilligen et al. 2013; Byrne et al. 2020) and genomic prediction (Byrne et al. 2020; Sverrisdóttir et al. 2017) in potato. RADseq has been applied to association mapping (Schönhals et al. 2017), and SLAF-Seq was used for genetic linkage map construction (Yu et al. 2020). DArT array has been successfully employed in genetic linkage map and QTL analysis (Śliwka et al. 2012; Iorizzo et al. 2014; Lebecka et al. 2021; Plich et al. 2018), genetic diversity analysis (Rungis et al. 2017), and microscale genome sequence heterogeneity evaluation (Traini et al. 2013). Whole-genome resequencing with desirable sequencing depth could provide millions of genome-wide reliable SNP, insertion/deletion (INDEL), and structural variation (SV) makers simultaneously. Genetic markers from whole-genome resequencing are powerful tools in studying copy number variations among potato varieties (Hardigan et al. 2016; Achakkagari et al. 2020), wild *Solanum* introgression (Hardigan et al. 2017), selection sweeps (Hardigan et al. 2017; Li et al. 2018b), linkage map and quantitative trait locus analysis (Manrique-Carpintero et al. 2018), genome instability (Amundson et al. 2021), and evolutionary history (Gutaker et al. 2019).

3.6 Plant Resources

Potato is a staple crop, and therefore it and its wild crop relatives are priorities for conservation. Ex situ conservation in gene banks is particularly effective at making potato germplasm widely available to breeders and researchers (Ellis et al. 2020a). Approximately 98,000 potato accessions are housed in 147 gene banks, with an estimated 24,000–29,000 of those being unique (Migdadi et al. 2007). The majority of these accessions (67%) are housed at the largest 20 gene banks (Machida-Hirano 2015). The largest collection is housed at The French National Institute for Agricultural Research (INRA) where potato cultivars are characterized for morphological, agronomic, and disease resistance traits (Esnault et al. 2016). The Vavilov Institute in Russia contains almost 9,000 accessions and is one of the world's oldest gene banks (Dzyubenko 2018). The global in-trust collection, held in the International Potato Center in Lima Peru, contains 8% of the world's accessions and features cryopreservation for long-term storage (Vollmer et al. 2016) as well as a large genotyped collection of landraces (Ellis et al. 2018). Germany and the USA (Bamberg and del Rio 2021; Bamberg and del Rio 2016) each hold 5% of the world's collection, while other genebanks hold less than 3% (Machida-Hirano 2015). Even at smaller gene banks, screening the entire collection for traits of interest is difficult. Therefore, core collections have been identified for tetraploid *S. tuberosum tuberosum* (Esnault et al. 2016), *S. tuberosum andigena* (Huamán et al. 2000; Chandra et al. 2002; Gopal et al. 2013; Tiwari et al. 2013), and diploid species including: *S. tuberosum phureja* (Ghislain et al. 1999), *S. microdontum* (Bamberg and del Rio 2014), *S. jamesii* (Bamberg et al. 2016), *S. fendleri* (Bamberg and del Rio 2016), and *S. demissum* (Bamberg and del Rio 2021).

4 Progress in the Population Genomics of Potatoes

4.1 Understanding Potato Origin, Domestication, and Evolution

Potato has a complex history and phylogeny which we are only beginning to untangle. The advances in sequencing and genotyping technologies described in the last section have facilitated our understanding of potato evolution. Application and development of these tools to increase the breadth and depth of genomic data from potatoes and their wild relatives will yield new insights.

4.1.1 Wild Potatoes

Solanum is one of the largest angiosperm genera (Frodin 2004) and includes important vegetable crops such as tomato, potato, eggplant (*S. melongena*), and pepper (*Capsicum annum*). Potato-tomato divergence occurred 6–8 million years ago (MYA), and the divergence of eggplant and tomato/potato is estimated at 14 MYA (Consortium TG 2012; Wang et al. 2008; Rodriguez et al. 2009; Särkinen et al. 2013). The tomato clade started diversifying only 2 MYA, while the potato clade did so soon after separating from tomato (Särkinen et al. 2013). Wild potato (*Solanum* L. section *Petota* Dumort.) contains about 100 species of varying ploidy levels (Spooner 2009; Simon et al. 2010). This clade is native to the Americas – distributed from the southwestern United States to the southern cone of South America – covering a great ecogeographical range (Spooner and Hijmans 2001; Bradshaw et al. 2006) and featuring extensive diversity (Spooner et al. 2014).

Due to the size and complexity of section *Petoa*, clarifying the phylogeny is challenging. The section is characterized by multiple hybrid origins, introgression, a mixture of sexual and asexual reproduction, polyploidy, possible recent species divergence, phenotypic plasticity, morphological similarity, and allele loss (Zhou et al. 2020; Spooner 2009; Huang et al. 2019; Ames and Spooner 2010; Cai et al. 2012; Spooner and van den Berg 1992). However, understanding the phylogenetic relationship among wild potatoes is crucial for effective germplasm conservation and utilization in breeding programs (Spooner and Salas 2006). To that end, species boundaries, relationships, and origins have been investigated using morphological analyses, crossability (Spooner et al. 2010; Vandenberg et al. 1998; Lara-Cabrera and Spooner 2005; Lara-Cabrera and Spooner 2004), ploidy level, geography, nuclear DNA microsatellites (Lara-Cabrera and Spooner 2005; Raker and Spooner 2002; Ghislain et al. 2006; Ríos et al. 2007; Spooner et al. 2007a), plastid DNA restriction sites (Ríos et al. 2007; Spooner et al. 2007a; Spooner et al. 1993; Castillo and Spooner 1997), plastid DNA deletions (Ríos et al. 2007; Spooner et al. 2007a), AFLPs (Spooner et al. 2005a; Lara-Cabrera and Spooner 2004; Spooner et al. 2007a; Berg et al. 2000; Jiménez et al. 2008; Jacobs et al. 2008), RFLPs/RAPDs (Berg et al. 2000; Miller and Spooner 1999; Bamberg and del Rio 2008), and nuclear orthologs (Spooner et al. 2018; Rodriguez et al. 2009; Cai et al. 2012; Spooner et al. 2008; Rodríguez et al. 2010; Fajardo and Spooner 2011) sometimes in combination. More recently, phylogenomic approaches have been applied using plastid genomes (Huang et al. 2019; Gagnon et al. 2022), and genome-wide nuclear SNPs (Hardigan et al. 2017; Li et al. 2018b; Hardigan et al. 2015). These methods have produced conflicting hypotheses for species number, interrelationships, and hybrid origins (Li et al. 2018b; Huang et al. 2019; Ames and Spooner 2010; Spooner et al. 2007a; Berg et al. 2000; van den Berg et al. 2002; Spooner et al. 2001). The most recent molecular and morphological data led to the lumping of taxa together and a reduction in species number (Spooner et al. 2014). An illustration of the logic behind lumping species comes from the *S. brevicaule* complex, which had been previously described as consisting of approximately 20 morphologically similar wild taxa (Spooner et al. 2005a). Morphological, RAPD/RFLP (Miller and Spooner 1999), AFLP (Spooner et al. 2005a), whole-genome sequencing data (Li et al. 2018b), and plastid DNA sequence (Huang et al. 2019) failed to distinguish between species in this complex. Instead, most studies found two primary groupings defined by geography. The northern group located in Peru was renamed as a single species *S. candolleanum,* while the southern group located in Bolivia and Argentina was renamed as the species *S. brevicaule* (Spooner 2016).

Currently, 107 wild species (Spooner et al. 2014) and 4 cultivated species (Spooner et al. 2007b) are recognized in *Solanum L. sect. Petota*, a reduction from the 235 previously recognized species (Hawkes 1990). The section has historically been divided into 4 clades (1 to 4) according to plastid restriction site data (Spooner and van den Berg 1992; Bamberg and del Rio 2008) or 3 clades (1+2, 3, and 4) based on nuclear orthologs, whole-genome DNA sequence data, and full plastid DNA sequence data (Li et al. 2018b; Rodriguez et al. 2009; Huang et al. 2019; Cai et al. 2012; Spooner et al. 2001). In 2018, a new clade, *Neocardenasii* was added as a sister clade to 1+2, bringing the total clade number back to 4 (Spooner et al. 2018; Huang et al. 2019).

Wild species have played a crucial role in shaping the potato genome (Hardigan et al. 2017) and provide a vital source for disease and stress resistance genes (Jansky et al. 2013). Improving our understanding of this diverse and complicated section will provide vital insight into potato history and novel alleles for potato improvement.

4.1.2 Domestication

Newly available sequencing data will be an essential tool in reconstructing the history of domestication events (Arnoux et al. 2021; Freedman et al. 2014; Wang et al. 2017; Wang et al. 2021). However, at the time of publication, the full potential of this data has not been explored. From the available studies, it is clear that cultivated potatoes were domesticated from wild *Solanum* species native to the Andes Mountains, most likely *S. candolleanum* (Spooner et al. 2005a; Li et al. 2018b; Spooner et al. 2014), about 10,000 years ago and spread globally after their migration to Europe during colonization of South America (Gutaker et al. 2019; Spooner et al. 2014). In general, it is thought that the group of diploids previously known *S. stenotomum subsp. stenotomum* which has been lumped into *S. tuberosum* were the first domesticated potatoes, and other species and subspecies arose from them (Hawkes 1990; Grun 1990; Sukhotu and Hosaka 2006). Historically, there have been two competing hypotheses of *stenotomum* domestication – multiple origin and single origin.

The multiple origin hypothesis suggests that the distribution and diversity (both morphological and in terms of chloroplast DNA) of cultivated potatoes indicate origin from several distinct wild gene pools (Sukhotu et al. 2004; Miller and Spooner 1999; Grun 1990; Sukhotu and Hosaka 2006; Hosaka and Hanneman 1988; Hosaka 1995; Huamán and Spooner 2002; Gavrilenko et al. 2013). Chloroplast DNA genotypes from cultivated species are often shared with *S. brevicaule* sharing a similar range (Hosaka 1995). This suggests successive domestication events of the Andean potatoes and parallel differentiation of wild species from the ancestral species complex (Hosaka 1995). Evidence from cpDNA markers and nuclear DNA RFLP markers from Andigena accessions and closely related cultivated and wild species revealed a progressive and continuous variation of cpDNA from Peruvain wild species to cultivated diploid potatoes and then to cultivated tetraploids (Sukhotu and Hosaka 2006). However, multiple origins are notoriously difficult to differentiate from a single origin with extensive introgression and hybridization (Choi et al. 2017; Choi and Purugganan 2018; Vonholdt et al. 2010; Freedman et al. 2014; Shannon et al. 2015).

The alternative hypothesis, then, is that the group formerly known as *S. stenotomum* have a single origin, and all other species and subspecies are a result of introgression and polyploidization (Spooner et al. 2005a; Grun 1990; Gavrilenko et al. 2013; Kardolus 1998; Ugent 1970). Domestication occurred in a relatively restricted geographical region in the Andes including Southern Peru and Northern Bolivia. This was suggested by Hosaka and Hanneman (Hosaka and Hanneman 1988) after observing the polymorphic nature of the chloroplast DNA of *S. tuberosum subsp. andigena*. Similarly, data from high-resolution cpDNA markers and nuclear DNA RFLPs support the frequent genetic exchange occurring among cultivated and wild species (Sukhotu and Hosaka 2006). Spooner et al. (Spooner et al. 2005a) conducted phylogenetic analyses based on the representative cladistic diversity of 362 individual, wild, and landrace tuber-bearing potatoes with AFLP marker analysis and concluded that the cultivated species are in the same clade as the northern Brevicaule complex, aka *S. candolleanum*. Phylogenetic analyses that used plastid DNA SSR markers, full plastid DNA sequences, and resequencing data also revealed cultivated potatoes have a closer relationship to the northern members of the *S. brevicaule* complex and supported the monophyletic origin of the landrace cultivars (Sukhotu et al. 2004; Li et al. 2018b; Huang et al. 2019; Sukhotu and Hosaka 2006; Gavrilenko et al. 2013). Introgression analysis has demonstrated that all the studied cultivated groups harbored a significant contribution from *Solanum candolleanum* (Hardigan et al. 2017; Hoopes et al. 2022; Li et al. 2018b). One explanation for the prevalence of *S. candolleanum* introgressions is an origin relationship (Li et al. 2018b). However, as discussed above, extensive introgression can and often does occur from sympatric plants which are not the progenitor.

4.1.3 Cultivated Potato

As with wild potatoes, contrasting taxonomic systems for cultivated potatoes have been reported. Prior taxonomic treatments recognized 1 to 21 distinct Linnean species or various cultivar groups (Hawkes 1990; Huamán and Spooner 2002;

Dodds 1962; Bukasov 1971; Lechnovich 1971; Ochoa 1990, 1999). Hawkes' (Hawkes 1990) treatment of the 7 cultivated species and 7 subspecies was, until recently, the most widely used taxonomic system (Spooner et al. 2010). These systems relied on intuitive judgments based on distinct ploidy levels, morphology, and ecogeographical criteria. Recent studies fail to support the phylogenetic classification by Hawkes (1990), Ochoa (1990, 1999), and Bukasov (1978). In combination with the known crossing data and morphological results of Huamán and Spooner (2002), Spooner et al. (2007a) examined 742 cultivated accessions with 50 nuclear SSR primer pairs and a plastid DNA deletion marker to provide a reclassification of the cultivated potatoes into four species: *S. ajanhuiri* (diploid), *S. juzepczukii* (triploid), *S. curtilobum* (pentaploid), and *S. tuberosum*. This taxonomy is the result of lumping several taxa from previous phylogenies together. Specifically, *S. tuberosum* includes diploids: *S. stenotomum subsp. stenotomum, S. stenotomum subsp. goniocalyx, S. phureja;* triploid *S. x chaucha;* and tetraploids: *S. tuberosum subsp. andigenum* and *S. tuberosum subsp. tuberosum* (Table 1). The plants formerly known as *S. tuberosum subsp. tuberosum* form the Chilotantum group of lowland Chilean landraces, while the rest of *S. tuberosum* forms the Andigenum group in the high Andes from Western Venezuela to Northern Argentina (Spooner et al. 2012). The two groups of *S. tuberosum* can be distinguished by their cytoplasmic sterility factors, morphological traits, daylength adaptation, microsatellite markers, plastid differences of a 241-bp deletion in the chloroplast (cp) DNA molecule, and a co-evolved mitochondrial (mt) DNA type (Spooner et al. 2005b).

The Spooner taxonomy was supported by other studies using nuclear microsatellites (Spooner et al. 2010; Raker and Spooner 2002; Ghislain et al. 2006), DNA sequence data of nuclear orthologs (waxy gene) (Rodríguez et al. 2010), plastid microsatellites (Gavrilenko et al. 2013), and plastid DNA deletion data (Gavrilenko et al. 2013; Ames and Spooner 2008). SNP array data from a sample of the global intrust collection of potatoes suggests groupings between the Hawkes and Spooner taxonomies (Ellis et al. 2018). The best STRUCTURE (Pritchard et al. 2000) result was $k = 6$, with clear groups for *S. stenotomum subsp. stenotomum, S. tuberosum subsp. tuberosum, S. phureja*, and *S. juzepczukii*. The two remaining populations do not seem to correspond to any of the populations included in the study; instead, they seem to be primarily contributors of admixture. Perhaps they represent wild germplasm which has contributed to cultivated potato or cultivated potatoes not represented in this sample. The other germplasm included in this study is heavily admixed, although all groups have substantial admixture except *S. tuberosum subsp. tuberosum* and *S. juzepczukii* which may appear artificially distinct due to comparatively narrow genetic bases. *S. stenotomum subsp. goniocalyx* appears to be a hybrid of *S. stenotomum subsp. stenotomum* and *S. phureja. S. tuberosum subsp. andigenum* and to a lesser extent *S. x chaucha* are a mix of everything. This supports *S. juzepczukii* as the only distinct species and the separation between the Chilotantum and Andigenum groups. However, it also indicates that *S. phureja* represents distinct germplasm.

Discordant phylogenetic relationship both within and between plastid and nuclear datasets or between gene tree and species is common in potato (Huang et al. 2019;

Table 1 Two proposed potato taxonomies

Spooner	Hawkes	Ploidy	Image
S. anjanhuiri	S. x anjanhuiri	2x	
S. tuberosum andigenum group	S. stenotomum subsp. stenotomum	2X	
	S. stenotomum subsp. goniocalyx	2X	
	S. phureja	2X	
	S. x chaucha	3X	
	S. tuberosum subsp. andigenum	4X	
S. tuberosum chilotanum group	S. tuberosum subsp. tuberosum	4X	
S. juzepczukii	S. x juzepczukii	3X	
S. curtilobum	S. x curtilobum	5X	

Listed ploidies are the most common for each taxon. Images are of an example variety from each species, taken by the author at the International Potato Center in Lima Peru

Ghislain et al. 2006; Jacobs et al. 2008; Spooner et al. 2008; Gagnon et al. 2022). Different germplasm bases, different evaluation environments, different trait scoring methods, and number of replications of measurements can lead to technical bias on taxonomic results. Other possible processes that can lead to such incongruence are plastid or chloroplast capture, gene duplication, historical hybridization, introgression, horizontal gene transfer, and incomplete lineage sorting (Huang et al. 2019; Gagnon et al. 2022). Phylogenomics using massive multilocus datasets boosted by next-generation sequencing techniques can facilitate the resolution of complex phylogenetic problems, reduce systematic biases, and recognize processes like incomplete lineage sorting, gene duplication, and horizontal transfer (Posada 2016). In addition, mapping against multiple genomes (pan-genome) can minimize phylogenetic discovery bias and mapping bias for phylogenetic relationship construction (Patané et al. 2018).

4.1.4 Evolutionary Forces Shaping the Potato Genome

The potato genome has been shaped by selection and admixture. Potato domestication involved natural selection and breeding by humans of wild species for shorter stolons, larger tubers, lower glycoalkaloid content, attractive tuber skin and flesh color, reduced sexual fertility, vine vigor, and extensive segregation for flower and foliage traits (Spooner et al. 2005a; Machida-Hirano 2015; Spooner et al. 2014; Gavrilenko et al. 2013). Detection of genomic regions that have been targeted by selection can contribute to understanding the process of domestication and facilitate the identification of genes which control important agronomic traits. Hardigan and colleagues (Hardigan et al. 2015) evaluated allelic variations within a diversity panel of germplasm from *Solanum* section *Petota* using the 8K SolCAP array. They identified several carbohydrate metabolism genes containing SNPs with highly divergent allele frequencies between wild diploids and cultivated tetraploids. The carbohydrate pathway plays a key role in tuber development, and this pattern of divergence is consistent with the expectation for selected genes. A genome-wide scan for signatures of selection (F_{ST} and Tajima's D) comparing 67 sequenced potato relatives, including South American landraces, North American cultivars and wild-diploid species, identified 2,622 highly differentiated genes (Hardigan et al. 2017). The highly differentiated genes included genes controlling glycoalkaloid biosynthesis, carbohydrate metabolism, the shikimate pathway, cell cycling/endoreduplication, circadian rhythm, and reduced pollen fertility. Less than 16% of the identified genes were shared by the North American modern potato cultivars and the Andigenum landraces, as we would expect for domestication genes. The observation of the relatively limited overlapping gene set indicated a small number of genes involved in the initial domestication, while adaptation of upland and lowland populations and improvement in the USA and Europe targeted a wider variety of distinct loci (Hardigan et al. 2017; Ghislain and Douches 2020). A similar study examining differences in levels of polymorphism between 21 diploid landraces and 178 diploid wild potatoes identified 609 putatively selected genes (Li et al. 2018b). These genes were involved in glycoalkaloid biosynthesis, tuberization, tuber size, tuber number, and disease resistance. No selection signals in genes involved in the starch biosynthesis pathway were found. When considering these results, it must be remembered that many evolutionary forces create similar genetic signatures to selection (Stephan 2019). High differentiation or low polymorphism is not conclusive evidence of selection, and so these lists should be treated as candidate loci rather than definitively identified regions.

Selection acts on genetic variation; either standing genetic variation, mutation, or as seems to be the case in potato introgressed variation from sympatric species. Selection on variation resulting from admixture is known as adaptive introgression (Suarez-Gonzalez et al. 2018; Burgarella et al. 2019). Admixture and adaptive introgression have been hypothesized to be a primary evolutionary mechanism in section *Petota* (Hoopes et al. 2022; Hawkes 1990; Ugent 1970). Phenotypic studies showed that 9.5% of wild potatoes in Argentina, Brazil, Paraguay, and Uruguay are interspecific hybrids (Hawkes and Hjerting 1969). This was confirmed for *S. berthaultii* and *S. tarijense* using genetic data (Spooner et al. 2007b). D statistics from whole-genome sequencing suggest that there has been extensive introgression from section *Etuberosum* throughout section *Petoa* (Tang et al. 2022). Furthermore, the variation in ploidy among domesticated potato may have arisen from hybridization. Triploid *S. juzepczukii* originated via natural hybridization between diploid accessions of *S. stenotomum* and frost-resistant tetraploid wild species *S. acaule* (Gavrilenko et al. 2010; Rodríguez et al. 2010; Gavrilenko et al. 2013; Kardolus 1998). Subsequent natural crosses between *S. juzepczukii* and Andean tetraploid landraces led to the formation of pentaploid *S. curtilobum* (Rodríguez et al. 2010; Gavrilenko et al. 2013). *S. ajanhuiri* originated through interspecific hybridization between Andean diploid cultivar *S. stenotomum* and wild species *S. megistacrolobum* (Rodríguez et al. 2010).

Admixture is also evident in tetraploid potato. For example, Chilean potatoes (*S. tuberosum chilotantum*) evolved from the *S. tuberosum* Andigenum group by adapting to the long-day conditions in Southern Chile. Restriction enzyme analysis of chloroplast DNA shows a geographical cline from the Andean region to coastal Chile (Hosaka and Hanneman 1988; Hosaka 2003), indicating an original Andean origin for *chilotantum*. Two potential sources of admixture have been identified. *S. berthaultii* is the likely source of T-type chloroplast DNA in Chilean landraces (Spooner et al. 2007b; Gavrilenko et al. 2013; Hosaka 2003), while nuclear and plastid SSRs as well as single-copy nuclear gene waxy markers indicate *S. maglia* as a potential paternal contributor (Rodríguez et al. 2010; Gavrilenko et al. 2013; Spooner et al. 2012; Ugent et al. 1987).

Even after its hypothesized role in speciation, admixture has continued to shape the potato genome (Ugent 1970; Debener et al. 1991; Celis et al. 2004). Admixture both among cultivated species (Spooner et al. 2018; Ellis et al. 2018) and market classes (Hirsch et al. 2013; Pandey et al. 2021) is evident. Sequence data from cultivated potato shows at least 20 wild species have contributed to both US-improved potato and South American landraces (Fig. 1) (Hardigan et al. 2017; Hoopes et al. 2022). The limitation on this finding appears to be the number of species sequenced rather than the contributors to *S. tuberosum* germplasm, as nearly all examined wild species have contributed to cultivated potato. In one study, 73% of the alleles identified in the wild germplasm were also found in tetraploid North American cultivars (Hardigan et al. 2017). Although North American cultivars showed the highest level of introgression, all species represented in the North American germplasm were also represented in South American tetraploid landraces. Introgressions were least prevalent in the diploid landraces but nevertheless present.

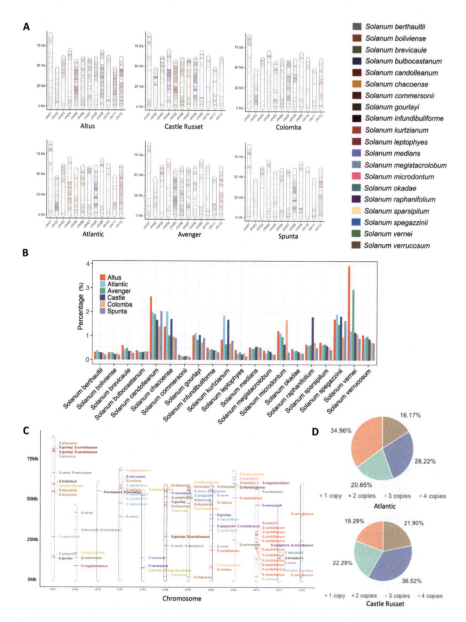

Fig. 1 (**a**) The physical distribution of introgressions at least 10 kb in length from 20 wild species in the six tetraploid cultivars in the pan genome. (**b**) The distribution of those introgressions by wild species in each of the six tetraploid cultivars. (**c**) The introgressions shared by all 6 cultivars. (**d**) The dosage of introgressions shared by all six cultivars in each of the fully phased cultivars. (These figures are reproduced with permission from Hoopes et al. 2022)

This may be due to a real difference in levels or introgression or simply a by-product of the smaller number of alleles per plant in a diploid resulting in a lower chance of observing any given allele in a diploid population.

The crossing of wild species and cultivated varieties with the intent of increasing disease resistance has been one of the major innovations in the history of modern potato breeding. This is an ongoing essential tool for breeders (Li et al. 2018b; Machida-Hirano 2015; McCann et al. 2010) who have targeted cyst nematodes, late blight, and other viruses using *S. demissum*, *S. stoloniferum*, and *S. tuberosum* Group Andigena clone CPC-1673 and *S. vernei* (Bradshaw and Ramsay 2005). Recent efforts focus on the identification of molecular markers for these introgressed R-genes (Harris and Nielsen 2013). The introduction of wild species or landrace germplasm can be difficult because it often comes with unintended wild or landrace alleles as with the case of diploid self-compatibility segregating in the tetraploid germplasm (Clot et al. 2020).

The development of introgression breeding as a modern strategy in approximately 1945 provides another method for the identification of introgressed SNPs. Vos et al. (2015) evaluated changes in the composition of the European potato gene pool over time using the Infinium potato 12K SNP array and dated the point at which SNPs were introduced to the genome. SNPs which were monomorphic in clones released before 1945 and polymorphic in one or more of the genotypes released after 1945 are assumed to be the result of introgression breeding. Many of these alleles are also found in regions identified as introgressions from sequencing data (Hoopes et al. 2022).

While we have records of recent introgression breeding, the patterns of introgression in the potato genome suggest that introgression is an ongoing historical phenomenon (Hoopes et al. 2022). Introgressions were found not only in cultivars known to have wild species in their pedigrees like 'Castle Russet,' but also in cultivars with no record of introgression breeding like 'Spunta' (Fig. 1). Similarly, wild species for which we have records of their use in introgression breeding are represented in all six cultivars, but so are wild species for which there is no evidence of their use in introgression breeding. Furthermore, introgressions tend to be shared. While in 'Atlantic' and 'Castle Russet' 88% of loci exhibited more than four haplotypes pointing to minimal allele sharing, 50.5% of introgressed regions were shared by more than three cultivars. All six cultivars shared 6.8% of the introgressed regions. These shared introgressions are old. The length of introgressions is generally correlated with their age (Harris and Nielsen 2013), and this is consistent with the Hoopes et al. (2022) data where the introgressions identified as introduced post 1945 were longer than introgressions overall. On the other hand, the 6.8% of introgressions shared by all six cultivars were shorter than expected ($p = 0.008$). Additionally, these introgressions are more homozygous than expected. While cultivars in the pan genome had on average three haplotypes per loci, only approximately 25% of introgressions shared in all six cultivars were present in simplex (Fig. 1). The length and ubiquity of these introgressions point to a historical pattern of admixture, while their comparatively high allele frequency suggests selection.

Evidence suggests these introgressions are functional. Genes found within introgressions are more likely to be expressed (FPKM $>=$ 1) and highly expressed (FPKM $>$10) than genes in the genome overall ($p <$ 0.00001) (Hoopes et al. 2022). The shared introgression regions are enriched for stress resistance genes, a finding consistent with previous results (Hardigan et al. 2017). Several of the shared introgressed regions show introgressions from multiple sources, which could be the result of selection. Taken together, these results suggest that adaptive introgression may have played an important role throughout potato's domestication and range expansion.

4.1.5 Introduction to Europe

Cultivated potato was first recorded in Europe in the Canary Islands about 100 km west of Morocco, 1,200 km southwest of mainland Spain, and 5,000 km northeast of northern South America, in the mid-1500s (Ríos et al. 2007; Hawkes and Francisco-Ortega 1993). European potatoes then spread throughout the world and became the foundation for all modern cultivated potatoes (Hawkes 1990). The origin of European potato is controversial with some studies pointing toward an Andean origin (Gutaker et al. 2019; Hosaka and Hanneman 1988; Ames and Spooner 2008), others pointing toward a lowland Chilean origin (Ghislain et al. 2009; Juzepchuk 1929), and still others indicating multiple origins (Ríos et al. 2007; Spooner et al. 2005b). Data from historical herbarium specimens from 1660 to 1896 suggests that the first potatoes imported into Europe were Andean (Gutaker et al. 2019). However, these imports were followed by an influx of Chilean potatoes which quickly became the predominant potato germplasm in Europe (Spooner et al. 2005a; Gutaker et al. 2019; Ames and Spooner 2008; Yoshida et al. 2013). Surveys of chlorotype frequency within the potato population in Europe from 1650 to 2000 indicate an increase of Chilean-related chlorotypes emerging in the late eighteenth century (Gutaker et al. 2019), while the Andean chlorotypes were diminished by half in Europe between seventeenth and twentieth centuries. There was a small resurgence of Andean chlorotypes in the years 1846–1891, coinciding with the Irish Famine (Gutaker et al. 2019). This pattern of a decrease in Andean chlorotypes coupled with an increase in Chilean chlorotypes was repeatedly observed in herbarium specimens of *Solanum tuberosum* collected in Europe from the early 1700s to 1910 (Ames and Spooner 2008). Gene flow between Europe and South America has been bidirectional. Contemporary Chilean potatoes contain 83% European ancestry and differ strongly from historical Chilean samples, suggesting that a reintroduction of European potatoes to Chile occurred in the nineteenth century (Gutaker et al. 2019).

4.1.6 Adaptation for Range Expansion

Potatoes were domesticated near the equator, and therefore the first domesticated potatoes were obligate short-day plants which could not produce tubers under the long-day conditions typical of the growing season in southern Chile, Europe, and North America (Kloosterman et al. 2013). Thus, acquiring the capacity for tuberization under long-day conditions was a prerequisite for introduction to temperate zones. Fine mapping in wild and cultivated diploids identified *StCDF1* a transcriptional regulator which controls maturity and tuber size (Kloosterman et al. 2013). A truncated allele of *StCDF1* allows for tuberization under long-day conditions, thus facilitating range expansion outside equatorial zones for potato. There are at least three truncated alleles which differ in effect and appear to act additively (Hoopes et al. 2022). The most likely source of these truncated alleles is wild species in particular *S. microdontum* (Hardigan et al. 2017) or *S. maglia* (Gavrilenko et al. 2013; Spooner et al. 2012). Although *StCDF1* is the major driver of maturity in contemporary European and US potato, the truncated *StCDF1* allele is not present in early herbarium samples from Europe or those from lowland Chile (Gutaker et al. 2019). The first European sample with the long-day allele of *StCDF1* is from 1810. Potatoes grown prior to 1810 in Europe exhibit a reduction in diversity on chromosome 10 in a region containing photoperiod response genes associated with the gibberellic acid synthesis when compared with short-day Andean potatoes. The target of selection appears to be *SP6A* florigen. This suggests that the first potatoes introduced to Europe were preadapted to long-day conditions through the gibberellin pathway rather than of carrying the gain-of-function alleles in the *StCDF1* gene that confer adaptation to a European climate. The herbarium data supports the hypothesis that *StCDF1* variants in modern European potatoes arose de novo rather than through introgression after the mid-1700s and rapidly fixed due to their dominance and potential breeding advantage (Gutaker et al. 2019).

Day length is not the only crucial difference between environments, genes which allow potatoes to respond to biotic and abiotic stress as their range expands have been under selection in potato throughout its history (Bradshaw et al. 2006). Many of the genes found in introgressed regions in potato play a role in environmental stress response, and similarly stress response genes are often part of copy number variants (Hardigan et al. 2017; Hoopes et al. 2022; Hardigan et al. 2016). These stress response genes, while crucial to the evolution of potato, are less well described than *StCDF1*.

4.2 Potato Genetic Diversity and Population Structure

4.2.1 Genetic Diversity

There are disparate reports of potato genetic diversity. Historically it has been suggested that US- and European-cultivated potato has limited genetic diversity

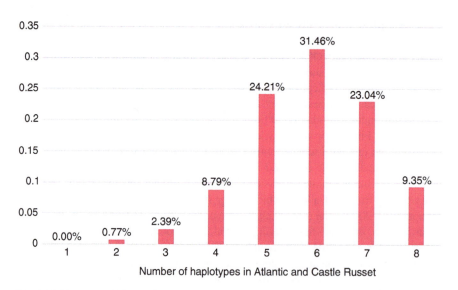

Fig. 2 A histogram of the number of haplotypes per CDS for Atlantic and Castle Russet where a haplotype is defined as 100% identity in a coding sequence (CDS). (This figure was made with data from the pan genome (Hoopes et al. 2022))

due to a small number of founders resulting in a narrow genetic base (Hirsch et al. 2013). Similarly, SSR marker analysis suggests that most cultivated species share four haplotypes, and 96% of the diploid *S. andigenum* share two haplotypes (Gavrilenko et al. 2013). However, sequenced cultivated potato is highly heterozygous with between 0.05 and 0.08 SNPs per base pair in tetraploids, much higher than comparable crops (Hardigan et al. 2017; Hoopes et al. 2022). On average all six cultivars in the pan genome have three haplotypes per loci, and when two cultivars are compared there are generally more than four haplotypes (Fig. 2) (Hoopes et al. 2022). Of course, this depends on the history of the cultivar, 'Otava' exhibited almost 50% identity by descent presumably due to recent inbreeding (Mari et al. 2022). However, the other available sequence points to this inbreeding being an aberration (Hoopes et al. 2022; Sun et al. 2022). All data points to extensive structural variation in potato affecting up to 30% of genes (Hoopes et al. 2022; Hardigan et al. 2016; Mari et al. 2022) and directly affecting expression (Pham et al. 2017) both in US tetraploid *S. tuberosum* and in the landraces (Jansky et al. 2013). SNP array studies also suggest high levels of heterozygosity, although SNP array markers are preselected for heterozygosity; hence, ascertainment bias is an issue (Hirsch et al. 2013; Pandey et al. 2021; Hardigan et al. 2015). Taken together, recent data implies that older genotyping technologies are underestimating genetic diversity in potato.

Genetic diversity in wild potato is also a subject of debate. Observation of phenotypes suggests tremendous diversity in wild species (Bamberg and del Rio 2021). The SSR data which indicated cultivated potato only has 4 haplotypes indicated 69 haplotypes in *S. brevicaule* and *S. candollonlleanum* (Gavrilenko et al. 2013). However, AFLPs (Bryan et al. 2017) and SNP array data (Hardigan et al. 2015) suggest low nucleotide diversity in wild potato especially in comparison with cultivated potato, and resequencing data suggests low relative heterozygosity in wild species (Hardigan et al. 2017). This runs counter to our expectations for the relationship between wild and domesticated species. Domestication and improvement bottlenecks generally create reduced diversity in domesticated species as compared to wild species (Doebley et al. 2006). Recorded history (Hirsch et al. 2013) and Tajima's D estimates in US breeding programs (Pandey et al. 2021) indicate a history of bottlenecks that would create the expected reduction in genetic diversity and yet the most complete sequence data on domesticated potato suggests the existence of extremely high heterozygosity.

We see five possible explanations for the discrepancy between recent observations and expectation. First, genetic diversity in wild species may be underestimated. Ascertainment bias is an issue when using the SNP array, as evidenced by the species most distantly related to the *S. tuberosum* the SNP array was based on appearing the least diverse in analyses. Specifically, *S. jamesii*, *S. bulbocastanum*, and *S. pinnatisectum* were all reported to have less than 1% heterozygosity as compared to 30% in the landraces (Hardigan et al. 2015). These species are also the most diverged from *S. tuberosum* with an estimated divergence 7.51 million years ago. Several attempts to compensate for biases in SNP detection support the hypothesis that wild species genetic diversity is being underestimated (Bamberg and del Rio 2021; Bamberg and del Rio 2004; Bamberg et al. 2015; Haynes et al. 2017). Furthermore, the sequenced genomes of *S. commersonii* (Aversano et al. 2015) and *S. chacoense* (Leisner et al. 2018) both have SNP frequencies of approximately 1.5% which is similar to if not higher than the published estimates for cultivated potato.

Second, the comparisons between wild and cultivated species may not be fair comparisons. While many cultivated potatoes are tetraploid, the majority of wild species are diploid. Polyploids are inherently more heterozygous than diploids simply due to the number of observations per locus; see Meirmans et al. (2018) for an in-depth explanation. When resequencing data from wild and cultivated diploids were compared, the wild diploids were more diverse (Li et al. 2018b). Similarly, while Hardigan et al. (2017) found comparatively high heterozygosity in tetraploid cultivars they found higher nucleotide diversity in wild diploids, suggesting that while there is more genetic diversity at the population level for wild species, individual cultivated tetraploids represent more of that diversity simply because they have more alleles (Hardigan et al. 2017). Additionally, SNP filtering could account for the difference between diploids and tetraploids reported in Hardigan et al. (2017). With a more stringent filter the SNP frequency in cultivated potato decreases approximately 20-fold (Huang et al. 2018; Hardigan et al. 2018), such that it is in line with the resequencing data from Li et al. (2018b). The ploidy effect may also explain the observed difference in heterozygosity between diploid and tetraploid Andean landraces (Sukhotu et al. 2005; Sukhotu and Hosaka 2006).

Third, there may actually be a very small number of haplotypes in cultivated potato, but those haplotypes could be maintained in heterozygous states due to balancing selection. Breeders have long focused on enhancing heterozygosity (Hirsch et al. 2013) which could create this effect. Furthermore, the Tajima's D values reported within breeding programs consistent with a bottleneck are also consistent with balancing selection (Pandey et al. 2021).

Fourth, population decline in wild populations and expansion in domesticated populations could produce the reported pattern of comparatively less diversity in wild species (Shannon et al. 2015). This seems highly unlikely for potato because it primarily reproduces clonally, and the estimated generation time for cultivated potato is long.

Finally, introgression breeding has played a major role in shaping the potato genome, with introgressions coming from a plethora of wild species (Hardigan et al. 2017). It is possible that the genetic diversity reported in cultivated potato reflects the accumulated genetic diversity from multiple species in section *Petoa*. While wild species primarily reproduce with conspecifics or at least sympatric species, cultivated potato has had the opportunity to be more promiscuous as it has traveled, and breeders have used a variety of methods to break crossing barriers (Bradshaw and Ramsay 2005; Jansky and Peloquin 2006; Plaisted and Hoopes 2008). Cultivated potato then would carry not only the alleles which passed through domestication, migration, and improvement bottlenecks or arose thereafter, but also many alleles from the wild species to which we are comparing them. The difference in hybridization rates between diploids and tetraploids and the role of hybridization in polyploid formation has also been proposed as an explanation for disparities in heterozygosity (Gavrilenko et al. 2013).

These five hypotheses are nonexclusive, and any combination of them may turn out to explain the observed patterns of genetic diversity. In order to evaluate the majority of these hypotheses, additional sequence data from wild and cultivated species is needed. Additionally, methods of diversity comparison which compensate for differences in ploidy will facilitate a more accurate assessment of relative species diversity. Understanding this genetic diversity will shed light on the history of potato and the relationship between species.

4.2.2 Population Structure in Breeding Germplasm

In cultivated potato, population structure is primarily defined by market class. In North America, the largest market class is russet potatoes for processing; these are consumed as French fries, tater tots, frozen hash browns, etc. The second largest market class is chipping potatoes which are round white potatoes used for creating potato chips. Finally, there are fresh market/specialty potatoes intended for culinary use, and these include yellow skinned yellow fleshed potatoes, red skinned white fleshed potatoes, and a variety of pigmented potatoes. Based on SolCAP array genotype data from a genetic diversity panel, russets form a distinct population, as do yellow and white fresh market potatoes. All pigmented potatoes, red and purple, group together. There are three distinct populations of chipping potatoes (Hirsch et al. 2013). Reports from the breeding program at Texas A&M indicate a single germplasm pool for chips and russets and a separate pool for reds (Pandey et al. 2021), although the inclusion of multiple 'Russet Norkotah' sports may result in artificial groupings of populations that would otherwise appear distinct. When US russet potatoes are assessed across breeding programs, they form three populations based on pedigree (Bali et al. 2018).

Population structure in Chinese potatoes appears to depend on the date/period of introduction (Wang et al. 2019). There are two primary germplasm groups, a low diversity group of landraces derived from historical introductions of potato to China and a higher diversity group resulting from potatoes brought into China since 1995. Within these groups, there are smaller subpopulations defined by origin and geographic location. This is consistent with previous reports of weak population structure dependent on location within China (Duan et al. 2019). There is also minimal evidence for structure within the tetraploid germplasm in Korea (Lee et al. 2021b) and Europe (Stich et al. 2013; Gebhardt et al. 2004; Malosetti et al. 2007; Selga et al. 2022). In general, tetraploid potato breeding populations outside of South America exhibit minimal population structure. This is reasonably expected given their shared origin and the tendency of breeders to cross for maximum heterozygosity.

4.3 Population Epigenomics

Epigenetic variation, including variance in DNA methylation pattern, histone modification, and chromatin state play important roles in phenotypic plasticity, acclimation, and adaptation in plants (Tiwari et al. 2015; Moler et al. 2019). Epigenetic studies in potato have focused on DNA methylation. In plants, the main form of DNA methylation occurs at the fifth position of cytosine yielding three cytosine contexts (CG sites, CHG sites, and CHH sites, where H represents A, C, or T) (Smyda-Dajmund et al. 2021; Alonso et al. 2015; Pikaard and Mittelsten Scheid 2014). DNA methylation is stably inherited and regulates gene expression and transposon silencing (Finnegan et al. 2000; Gehring and Henikoff 2007). In potato DNA methylation has been implicated in regulation of cold-induced sweetening (Shumbe et al. 2020), somatic hybridization (Tiwari et al. 2015; Smyda-Dajmund et al. 2021), gene regulation under low-temperature stress (Xin et al. 2015), resistance to late blight (Meller et al. 2018), divergent evolution of duplicated genes (Wang et al. 2018), autopolyploidization (Aversano et al. 2013), and anthocyanin biosynthesis (D'Amelia et al. 2020). Whole-genome potato DNA methylation status can be assessed by high-performance liquid chromatography (HPLC) (D'Amelia et al. 2020), Methylation-Sensitive Amplification Polymorphism (MSAP) markers (Tiwari et al. 2015; Smyda-Dajmund et al. 2021; Xin et al. 2015; Aversano et al. 2013; Cara et al. 2019; Marfil et al. 2009; Harding 1994; Ibañez et al. 2021; Joyce and Cassells 2004), and whole-genome bisulfite sequencing (Wang et al. 2018).

MSAP markers use methylation-sensitive restriction enzymes to characterize the methylation patterns of anonymous 5'-CCGG sequences from random genomic DNA (Cara et al. 2020). This approach is an extension of the original AFLP technique and can be adapted to evaluate population-level methylation because it is relatively inexpensive and simple to perform (Cara et al. 2020). MASP markers have been used to survey DNA methylation status during the creation of diploid inbred lines (Nakamura and Hosaka 2010). Methylation was measured in parents and progeny for each selfing generation. This study provided evidence that the homozygosity/heterozygosity of methylated DNA may contribute to inbreeding depression and heterosis through the regulation of transcription factors (Nakamura and Hosaka 2010). MASP markers were also used to demonstrate that populations of synthetic, natural, and somatic hybrids exhibited different methylation patterns than their parental genotypes, implying epigenetic changes occur during interspecific hybridization (Tiwari et al. 2015; Smyda-Dajmund et al. 2021; Cara et al. 2019). Similarly, autopolyploidization alters methylation status (D'Amelia et al. 2020). Other MASP studies have found that methylation can explain plant morphology (Marfil et al. 2009; Joyce and Cassells 2004) and contribute to stress response and environmental adaptation (Xin et al. 2015; Harding 1994; Ibañez et al. 2021). In many of these studies, both epigenetic and genetic variation was evident; however, in several cases, genetics remained constant and the only changes were in methylation status (Tiwari et al. 2015; Aversano et al. 2013; Marfil et al. 2009).

Histone modification, including acetylation, methylation, and Ser/Thr phosphorylation, has been investigated to a lesser extent in potato. These modifications have been implicated in resistance to late blight, dormancy break, protein attributes in tubers, and cold stress response (Meller et al. 2018; Salvato et al. 2014; Law and Suttle 2004; Zeng et al. 2019; Sun et al. 2021). Furthermore, H3K4me3-modified nucleosomes in combination with accessible chromatin and stowaway transposons determine the location of meiotic crossovers in potato (Marand et al. 2017). Despite the crucial role histone modification plays in potato biology, most studies have focused on individual genomes rather than reaching the population level.

Micro RNA (miRNA) plays an important role in defining the epigenetic status of a plant. These are a family of small single-stranded noncoding RNA molecules in plants and animals, many of which are conserved across organisms (Djami-Tchatchou et al. 2017). MiRNA is involved in signal transduction (Deng et al. 2021) and regulates gene expression by degrading targeted mRNA to create translational repression at the post-transcriptional level (Deng et al. 2021; Bartel 2004). Additionally, it has been shown to determine vine architecture and tuberization (Bhogale et al. 2014), possibly through controlling the stolon-to-tuber transition (Kondhare et al. 2018). Tuber skin and flesh color are also influenced by miRNA (Bonar et al. 2018). Finally, miRNA contributes to disease resistance including verticillium wilt (Yang et al. 2015), late blight (Natarajan et al. 2018), PVA (Li et al. 2017), and PVY (Szajko et al. 2019); and to abiotic stress response including cold (Esposito et al. 2020), drought (Yang et al. 2013; Yang et al. 2016), nitrogen deficiency (Tiwari et al. 2020b), and salinity (Kitazumi et al. 2015; Zhang et al. 2018). Over the past decade, numerous groups have contributed to the genome-wide identification and characterization of potato and wild relative miRNAs through comparative genomics, computational prediction, high-throughput sequencing, and bioinformatic analysis in a variety of tissues including leaf, stolon,

root, stem, and tuber (Lakhotia et al. 2014; Esposito et al. 2020; Yang et al. 2016; Tiwari et al. 2020b; Zhang et al. 2013; Zhang et al. 2009c; Yang et al. 2010; Xie et al. 2011). The primary database for miRNAs, miRBase (http://www.mirbase.org/index.shtml) contains 575 potato sequences, including 228 miRNA primary transcripts (miRbase, release 22.1). Despite evidence of a crucial functional role for miRNA, we do not yet have a picture of how they differ within and across populations.

Approximately 18% of the potato genome is made up of transposons (Zavallo et al. 2020). These transposable elements (TEs) impact gene expression (Zollinger 2010), tuber skin color (Momose et al. 2010), and stress response (Tüfekçi et al. 2019; Esposito et al. 2019) as well as play an important role in miRNA evolution, exonization, and cis-regulation of associated genes (Mehra et al. 2015). Although early studies of TEs in potato focused on specific transposon families (Leisner et al. 2018; Momose et al. 2010; Esposito et al. 2019; Mehra et al. 2015; Kuang et al. 2009; Gaiero et al. 2019; Tang et al. 2014), a comprehensive profile of the TEs found in the DM reference genome is now available (Zavallo et al. 2020). The most common TEs in cultivated and wild potatoes are long terminal repeat (LTR) retrotransposons (Potato Genome Sequencing Consortium et al. 2011; Zavallo et al. 2020; Gaiero et al. 2019). Although the distribution, transcriptional activity, and insertion age depend on the TE superfamily (Zavallo et al. 2020; Esposito et al. 2019; Mehra et al. 2015), in general the repeat profiles of tomato and potato are very similar suggesting many of the transposons found in potato are ancient (Gaiero et al. 2019). TEs have provided useful genetic markers for population-level studies. For example, inter-primer-binding sites (iPBS) retrotransposon marker system (Demirel et al. 2018), inter-Retrotransposon Amplified Polymorphism (IRAP) primers (Novakova et al. 2009), and inter-SINE amplified polymorphism (ISAP) (Wenke et al. 2015) were employed for genotype-specific high-resolution fingerprinting and distinguishing large numbers of *Solanum tuberosum* genotypes.

Variation in transposable elements, miRNA, methylation, histone modification, and chromosome state clearly play a role in potato functional diversity. As we develop a more complete understanding of the potato genome and scalable analysis tools, population-level study of epigenetic variation will become more tractable. Elucidating the distribution of these variants within and among populations will no doubt increase our understanding of potato history and evolution and the dynamics of important phenotypes within potato populations.

4.4 Genotype-Phenotype Associations

4.4.1 GWAS and Linkage Disequilibrium

GWAS, also called linkage disequilibrium mapping, is a powerful approach to associate marker variants with complex traits (Berdugo-Cely et al. 2017; Korontzis et al. 2020; Zhang et al. 2010). GWAS makes use of diverse genetic material and historic recombination events to increase mapping resolution and identify markers likely to be tightly linked to causative alleles in relevant germplasm pools (Sharma et al. 2018; D'hoop et al. 2014; Stich et al. 2013; Ersoz et al. 2007; Gaut and Long 2003). GWAS populations are also generally useful for a wide variety of traits unlike biparental crosses which are often created to target a specific trait. In potato, GWAS has been used to examine various agro-morphological and quality traits including yield, starch content, chip quality, maturity, tuber skin/flesh color, russeting, tuber shape, tuber size, tuber number, tuber weight, tuber-specific gravity, plant height, eye depth, fry color, after cooking/baking darkening, cooking type, enzymatic browning, common scab resistance, wart disease resistance, late blight resistance, root and stolon traits, drought response, cold-induced sweetening, phosphorus content of starch, and metabolites (Kaiser et al. 2020; Felcher et al. 2012; Sharma et al. 2018; Yousaf et al. 2021; Zia et al. 2020; Byrne et al. 2020; Aliche et al. 2019; Yuan et al. 2020; D'hoop et al. 2014; Hamilton et al. 2011; Stich et al. 2013; Kloosterman et al. 2013; Mosquera et al. 2016; Schönhals et al. 2017; Berdugo-Cely et al. 2017; Sverrisdóttir et al. 2017; Schönhals et al. 2016; Levina et al. 2021; Prodhomme et al. 2020; Khlestkin et al. 2019; Pandey et al. 2022; Diaz et al. 2021; Zhang et al. 2022; Tagliotti et al. 2021).

Linkage disequilibrium (LD) is the base for association mapping, and the rate of LD decay between loci determines the power to assess marker-trait associations. Faster LD decay improves mapping resolution (Zia et al. 2020; D'hoop et al. 2010; Mackay and Powell 2007). Reports of LD decay in potato vary widely, due to differences in population type and origin, molecular marker type and number, and threshold used for LD determination. While some authors have reported LD decay at 270-280 bp (Stich et al. 2013; Mosquera et al. 2016), others have reported distances as long as 3.27 Mb (Sharma et al. 2018). Vos et al. (2017) suggested $LD_{1/2,90}$ (the distances at which LD equals one-half of its maximum fitted of the 90th percentile of the r^2 value) as an unbiased estimator for comparing the extent of LD decay across different studies. The author reported a decline in LD over the last century from an $LD_{1/2,90}$ of 1.5 Mb in cultivars released before 1945 to 0.6 Mb in the recent cultivars released after 2005 (Vos et al. 2017). Using the same LD estimator, decay at 0.55 Mb (Bourke et al. 2016) and 0.9 Mb (Sharma et al. 2018) were observed in CIP advanced tetraploid potato clones and European tetraploid cultivars and breeding lines, respectively.

Although the goal of GWAS is generally to identify causative alleles, the precision of the genomic region identified depends on the number of historic recombination events (i.e., the population size and diversity) and our ability to detect those events (i.e., the number of markers). GWAS has been successfully carried out in potato with as few as 100 individuals and as many as 500 including wild species, genetic stocks, cultivars, breeding lines, bi-parietal crosses, and/or landrace materials using SNP arrays ranging from 8K to 25K (Kaiser et al. 2020; Felcher et al. 2012; Sharma et al. 2018; Yousaf et al. 2021; Zia et al. 2020; Aliche et al. 2019; Yuan et al. 2020; D'hoop et al. 2014; Mosquera et al. 2016; Berdugo-Cely et al. 2017; Gaut and Long 2003; Schönhals et al. 2016). Although the SNP arrays provide sufficient markers, GBS data can be preferable for tagging rare variants especially in highly diverse populations and is often cheaper (Byrne et al. 2020; Stich et al. 2013; Sverrisdóttir et al. 2017). Whole-genome resequencing ensures that even low frequency variants are detected; however, it is comparatively expensive and unless a huge population is used, there is rarely enough power to make use of all the variants (Inostroza et al. 2018). Standard GWAS software including TASSEL and GAPIT has been used in GWAS analysis (Zia et al. 2020; Mosquera et al. 2016; Schönhals et al. 2017). However, ability to carry out GWAS studies in potato has been greatly improved by the development of GWASpoly, an R package specifically designed for GWAS in autopolyploids (Rosyara et al. 2016). This software uses a Q+K model, takes into account polyploid gene dosage, and allows for the multiple types of dominance possible in polyploids. These enhanced tools not only facilitate our ability to conceptualize genetic architecture in potato but also have facilitated study in other autopolyploid organisms demonstrating the value of potato as a model for autopolyploidy (Ferrão et al. 2018; Bourke et al. 2018b; Yang et al. 2019b; Matias et al. 2019; Inostroza et al. 2018).

4.4.2 Genomic Selection

While genomic selection has a long history of success in other crops, it is relatively new in potato. However, since 2017 genomic selection studies have been carried out for agronomic traits such as yield, yield components (Endelman et al. 2018; Gemenet et al. 2020; Habyarimana et al. 2017; Selga et al. 2021; Sood et al. 2020; Stich and Van Inghelandt 2018) and tuber size distribution (Habyarimana et al. 2017); tuber quality traits including tuber dry matter (Endelman et al. 2018; Sverrisdóttir et al. 2017; Habyarimana et al. 2017; Sood et al. 2020; Stich and Van Inghelandt 2018; Caruana et al. 2019; Sverrisdóttir et al. 2018), flesh color (Habyarimana et al. 2017; Sood et al. 2020; Caruana et al. 2019), skin color (Caruana et al. 2019), eye depth (Sood et al. 2020; Caruana et al. 2019), and tuber shape (Sood et al. 2020); cooking quality traits such as chip color (Endelman et al. 2018; Sverrisdóttir et al. 2017; Sood et al. 2020; Caruana et al. 2019; Sverrisdóttir et al. 2018), color when boiled (Sood et al. 2020; Caruana et al. 2019), and skin loss when boiled (Sood et al. 2020); vine traits including number of stems per plant (Habyarimana et al. 2017) and maturity (Gemenet et al. 2020; Sood et al. 2020; Stich and Van Inghelandt 2018; Caruana et al. 2019); and resistance to late blight (Gemenet et al. 2020; Selga et al. 2021; Stich and Van Inghelandt 2018; Sverrisdóttir et al. 2018), resistance to scab (Enciso-Rodriguez et al. 2018), PVY (Gemenet et al. 2020), and early blight (Sood et al. 2020). GBS (Sverrisdóttir et al. 2017; Sood et al. 2020; Sverrisdóttir et al. 2018), the SolCAP array (Endelman et al. 2018; Stich and Van Inghelandt 2018; Enciso-Rodriguez et al. 2018), SilicoDArT (Habyarimana et al. 2017), and transcriptome-based markers (Caruana et al. 2019) have all been shown to provide sufficient data for accurate prediction of breeding values.

As with many of the analyses described above, genomic selection is complicated by the autotetraploid nature of potato. In particular, the challenges of describing relatedness (Gallais 2003) and the multiple kinds of dominance possible in an autotetraploid (Rosyara et al. 2016). Kerr et al. (2012) described methods for determining pedigree-based relatedness matrixes for autopolyploids, which Slater et al. (2016) implemented in potatoes. In general, calculated genetic relationships based on marker data are more informative because they allow for actual rather than estimated covariance between relatives. Additive marker-based relatedness matrices were used by Slater et al. (2016), Sverrisdóttir et al. (2017), and Habyarimana et al. (2017). Endelman et al. (2018) introduced a nonadditive genetic covariance matrix which allows for the estimation of dominance variance; however, none of the traits he examined exhibited a large amount of dominance variance. This result has been confirmed by others examining a variety of traits (Gemenet et al. 2020; Amadeu et al. 2020). Generally, dominance estimations improve goodness of fit but not prediction accuracy in potato (Amadeu et al. 2020). The use of Bayesian models as an alternative to GBLUP and modeling known diagnostic markers as fixed effects may result in improved detection of dominance variance (Stich and Van Inghelandt 2018). Alternately, combining genetic and pedigree data into a single matrix improves prediction accuracy as compared to GBLUP for many traits (Sood et al. 2020). Potato breeders are increasingly implementing genomic selection in their breeding programs as tools continue to develop, and we can expect to see developments in this area in the near future.

4.5 *Conservation and Sustainable Management of Genetic Resources*

As a clonal crop, cultivated potato presents challenges for ex situ collection management. Stocks cannot be maintained as botanical seed and tubers generally last under a year in storage. Tissue culture presents an alternative, although maintenance of tissue culture stock even on long-term media is time intensive. Cryopreservation is the most effective long-term storage method for potato (Vollmer et al. 2016). Regardless of the challenges, numerous genebanks employ a combination of these approaches to maintain over 98,000 accessions of cultivated and wild potato. Ex situ conservation is a powerful approach for conserving potato diversity, particularly when the primary objective is making that diversity available to breeders and researchers (Ellis et al. 2020b). However, due to a long history of exploitation of Indigenous resources by the Global North, local laws and treaty obligations limit collection in many of the centers of potato biodiversity. Therefore, these ex situ collections are mostly static.

In order to maintain the full range of potato diversity in situ preservation is essential. A promising strategy for this is to support indigenous communities in their conservation efforts and traditional practices, which in turn maintain biodiversity in situ. A particularly successful example of this strategy in potatoes is Parque de la Papa which consists of five indigenous communities in the Andes near Pisac, Cusco, Peru. These communities have banded together to form a biocultural park centered on preserving nature and cultural practices with the support of a local NGO, Asociación ANDES. At the founding of the park, potatoes were repatriated from the International Potato Center in Lima to augment existing populations. The potato park preserves not only potato varieties but also indigenous knowledge about these varieties and their properties (IPCCA 2021).

Potato breeding relies heavily on introgressions from wild species to confer biotic and abiotic stress resistance. Therefore, it is essential to conserve wild relatives as well as cultivated potato (Jansky et al. 2013). Protecting crop wild relatives requires protecting their environment and cataloging their dispersal. Efforts have been made to describe and preserve S. *kurtzianum* in Argentina (Marfil et al. 2015) and 21 endemic species in Bolivia (Cadima et al. 2013). In situ and ex situ conservation of potatoes and their wild relatives will allow us to continue researching, breeding, and eating potatoes well into the future.

5 Conclusion

Potato genetics is undergoing a renaissance. A wealth of new genomic data has started to become available, including sequences for tetraploid individuals and wild relatives. This sort of data is necessary to answer questions about potato origins and genetic diversity. Simultaneously, new methods for autotetraploid genetics are being developed, which improve our ability to genotype, phase, map, and carry out genomic selection in tetraploids. These methods facilitate new insight into potato phenotypes and genomes, allowing us to identify causal variants and haplotypes and trace those in breeding populations. At the same time, potato is being reinvented as a diploid inbred crop which will allow breeders to take advantage of tools like hybrid breeding and introgression which have been used successfully in diploids. We expect this to shorten the breeding cycle, increase the rate of genetic gain, and allow breeders to respond more nimbly to changes in climate and consumer preferences. It is an exciting time to be a potato geneticist.

These innovations come to us not a moment too soon. Global climates are changing and the predictions for agriculture are often bleak. Historically, people in challenging climates, natural and political, have eaten potatoes because they are nutritious and grow well on marginal lands. These are the kind of crops that will take us into the future. Understanding the evolutionary history and extant diversity of potato will facilitate the improvements to the potato genome that will let us continue humanity's relationship with the spud well into the next century.

References

About: Parque de la Papa - Program of IPCCA. In: IPCCA | Indigenous Peoples' Biocultural Climate Change Assessment Initiative. https://www.ipcca.info/about-parque-de-la-papa. Accessed 1 Dec 2021

Achakkagari SR, Bozan I, Anglin NL, et al. Complete mitogenome assemblies from a panel of 13 diverse potato taxa. Mitochondrial DNA B Resour. 2021;6:894–7. https://doi.org/10.1080/23802359.2021.1886016.

Achakkagari SR, Kyriakidou M, Gardner KM, et al. Genome sequencing of adapted diploid potato clones. Front Plant Sci. 2022;13 https://doi.org/10.3389/fpls.2022.954933.

Achakkagari SR, Kyriakidou M, Tai HH, et al. Complete plastome assemblies from a panel of 13 diverse potato taxa. PLoS One. 2020;15:e0240124. https://doi.org/10.1371/journal.pone.0240124.

Ali A, Alexandersson E, Sandin M, et al. Quantitative proteomics and transcriptomics of potato in response to Phytophthora infestans in compatible and incompatible interactions. BMC Genomics. 2014;15 https://doi.org/10.1186/1471-2164-15-497.

Aliche EB, Oortwijn M, Theeuwen TPJM, et al. Genetic mapping of tuber size distribution and marketable tuber yield under drought stress in potatoes. Euphytica. 2019;215:186. https://doi.org/10.1007/s10681-019-2508-0.

Alonso C, Pérez R, Bazaga P, Herrera CM. Global DNA cytosine methylation as an evolving trait: phylogenetic signal and correlated evolution with genome size in angiosperms. Front Genet. 2015;6 https://doi.org/10.3389/fgene.2015.00004.

Alsahlany M, Enciso-Rodriguez F, Lopez-Cruz M, et al. Developing self-compatible diploid potato germplasm through recurrent selection. Euphytica. 2021;217:47. https://doi.org/10.1007/s10681-021-02785-0.

Amadeu RR, Ferrão LFV, de Oliveira IB, et al. Impact of dominance effects on autotetraploid genomic prediction. Crop Sci. 2020;60:656–65. https://doi.org/10.1002/csc2.20075.

Ames M, Spooner D. Phylogeny of Solanum series Piurana and related species in Solanum section Petota based on five conserved ortholog sequences. 2010. https://doi.org/10.1002/TAX.594009

Ames M, Spooner DM. DNA from herbarium specimens settles a controversy about origins of the European potato. Am J Bot. 2008;95:252–7.

Amundson KR, Ordoñez B, Santayana M, et al. Rare instances of haploid inducer DNA in potato dihaploids and ploidy-dependent genome instability. Plant Cell. 2021; https://doi.org/10.1093/plcell/koab100.

Anoumaa M, Yao N, Kouam E, et al. Genetic diversity and core collection for potato (Solanum tuberosum L.) cultivars from Cameroon as revealed by SSR markers. Am J Potato Res. 2017;94: 10.1007/s12230-017-9584-2.

Arnoux S, Fraïsse C, Sauvage C. Genomic inference of complex domestication histories in three Solanaceae species. J Evol Biol. 2021;34:270–83. https://doi.org/10.1111/jeb.13723.

Aversano R, Caruso I, Aronne G, et al. Stochastic changes affect Solanum wild species following autopolyploidization. J Exp Bot. 2013;64:625–35. https://doi.org/10.1093/jxb/ers357.

Aversano R, Contaldi F, Ercolano MR, et al. The Solanum commersonii genome sequence provides insights into adaptation to stress conditions and genome evolution of wild potato relatives. Plant Cell. 2015;27:954–68. https://doi.org/10.1105/tpc.114.135954.

Bali S, Patel G, Novy R, et al. Evaluation of genetic diversity among russet potato clones and varieties from breeding programs across the United States. PLoS One. 2018;13:e0201415. https://doi.org/10.1371/journal.pone.0201415.

Bamberg J, del Rio A. Selection and validation of an AFLP marker core collection for the wild potato Solanum microdontum. Am J Potato Res. 2014;91:368–75. https://doi.org/10.1007/s12230-013-9357-5.

Bamberg J, del Rio A. Accumulation of genetic diversity in the US potato Genebank. Am J Potato Res. 2016;93:430–5. https://doi.org/10.1007/s12230-016-9519-3.

Bamberg J, del Rio A. A metric for species representation in the US potato Genebank. Am J Potato Res. 2021; https://doi.org/10.1007/s12230-021-09833-4.

Bamberg J, del Rio A, Coombs J, Douches D. Assessing SNPs versus RAPDs for predicting heterogeneity and screening efficiency in wild potato (Solanum) species. Am J Potato Res. 2015;92:276–83. https://doi.org/10.1007/s12230-014-9428-2.

Bamberg J, del Rio A, Kinder D, et al. Core collections of potato (Solanum) species native to the USA. Am J Potato Res. 2016;93:564–71. https://doi.org/10.1007/s12230-016-9536-2.

Bamberg J, del Rio AH. Proximity and introgression of other potato species does not explain genetic dissimilarity between Solanum verrucosum populations of Northern and Southern Mexico. Am J Potato Res. 2008;85:Article number: 232.

Bamberg JB, del Rio AH. Genetic heterogeneity estimated by RAPD polymorphism of four tuber-bearing potato species differing by breeding system. Am J Pot Res. 2004;81:377–83. https://doi.org/10.1007/BF02870198.

Bartel DP. MicroRNAs: genomics, biogenesis, mechanism, and function. Cell. 2004;116:281–97. https://doi.org/10.1016/s0092-8674(04)00045-5.

Beals KA. Potatoes, nutrition and health. Am J Potato Res. 2019;96:102–10. https://doi.org/10.1007/s12230-018-09705-4.

Behrouzi P, Wit EC. De novo construction of polyploid linkage maps using discrete graphical models. Bioinformatics. 2019;35:1083–93. https://doi.org/10.1093/bioinformatics/bty777.

Berdugo-Cely J, Valbuena RI, Sánchez-Betancourt E, et al. Genetic diversity and association mapping in the Colombian Central Collection of Solanum tuberosum L. Andigenum group using SNPs markers. PLoS One. 2017;12:e0173039.

Berg RG, Groendijk-Wilders N, Zevenbergen M, Spooner D. Molecular systematics of Solanum series Circaeifolia (Solanum section Petota) based on AFLP and RAPD markers: Undefined; 2000.

Bethke PC, Jansky SH. Genetic and environmental factors contributing to reproductive success and failure in potato. Am J Potato Res. 2021;98:24–41. https://doi.org/10.1007/s12230-020-09810-3.

Bhogale S, Mahajan AS, Natarajan B, et al. MicroRNA156: a potential graft-transmissible microRNA that modulates plant architecture and tuberization in Solanum tuberosum ssp. andigena. Plant Physiol. 2014;164:1011–27. https://doi.org/10.1104/pp.113.230714.

Bibi S, Navarre D, Sun X, et al. Beneficial effect of potato consumption on gut microbiota and intestinal epithelial health. Am J Potato Res. 2018; https://doi.org/10.1007/s12230-018-09706-3.

Bonar N, Liney M, Zhang R, et al. Potato miR828 is associated with purple tuber skin and flesh color. Front Plant Sci. 2018;9 https://doi.org/10.3389/fpls.2018.01742.

Boopathi NM. Genetic mapping and marker assisted selection: basics, practice and benefits: Springer Science & Business Media; 2012.

Bormann CA. Genetic and molecular analysis of quantitative and qualitative late blight resistance in tetraploid potato. 4

Bourke PM, Gitonga VW, Voorrips RE, et al. Multi-environment QTL analysis of plant and flower morphological traits in tetraploid rose. Theor Appl Genet. 2018b;131:2055–69. https://doi.org/10.1007/s00122-018-3132-4.

Bourke PM, van Geest G, Voorrips RE, et al. polymapR—linkage analysis and genetic map construction from F1 populations of outcrossing polyploids. Bioinformatics. 2018a;34:3496–502. https://doi.org/10.1093/bioinformatics/bty371.

Bourke PM, Voorrips RE, Kranenburg T, et al. Integrating haplotype-specific linkage maps in tetraploid species using SNP markers. Theor Appl Genet. 2016;129:2211–26. https://doi.org/10.1007/s00122-016-2768-1.

Bradshaw JE, Bryan GJ, Ramsay G. Genetic resources (including wild and cultivated Solanum species) and progress in their utilisation in potato breeding. Potato Res. 2006;49:49–65. https://doi.org/10.1007/s11540-006-9002-5.

Bradshaw JE, Hackett CA, Pande B, et al. QTL mapping of yield, agronomic and quality traits in tetraploid potato (Solanum tuberosum subsp. tuberosum). Theor Appl Genet. 2008;116:193–211. https://doi.org/10.1007/s00122-007-0659-1.

Bradshaw JE, Ramsay G. Utilisation of the commonwealth potato collection in potato breeding. Euphytica. 2005;146:9–19.

Braun SR, Endelman JB, Haynes KG, Jansky SH. Quantitative trait loci for resistance to common scab and cold-induced sweetening in diploid potato. Plant Genome. 2017;10 https://doi.org/10.3835/plantgenome2016.10.0110.

Bryan GJ, McLean K, Waugh R, Spooner DM. Levels of intra-specific AFLP diversity in tuber-bearing potato species with different breeding systems and ploidy levels. Front Genet. 2017;8:119. https://doi.org/10.3389/fgene.2017.00119.

Bukasov SM. Flora of cultivated plants in the USSR. Volume 9. Potatoes. Flora of cultivated plants in the USSR Volume 9 Potatoes, 1971.

Bukasov SM. Principles of the systematics of potatoes. Trudy po prikladnoi botanike, genetike i selektsii, 1978.

Burgarella C, Barnaud A, Kane NA, et al. Adaptive introgression: an untapped evolutionary mechanism for crop adaptation. Front Plant Sci. 2019;10:4. https://doi.org/10.3389/fpls.2019.00004.

Burra DD, Berkowitz O, Hedley PE, et al. Phosphite-induced changes of the transcriptome and secretome in Solanum tuberosum leading to resistance against Phytophthora infestans. BMC Plant Biol. 2014;14 https://doi.org/10.1186/s12870-014-0254-y.

Byrne S, Meade F, Mesiti F, et al. Genome-wide association and genomic prediction for fry color in potato. Agronomy. 2020;10:90. https://doi.org/10.3390/agronomy10010090.

Cadima X, van Zonneveld M, Scheldeman X, et al. Endemic wild potato (Solanum spp.) biodiversity status in Bolivia: reasons for conservation concerns. J Nat Conserv. 2013;22:10.1016/j.jnc.2013.09.007.

Cai D, Rodríguez F, Teng Y, et al. Single copy nuclear gene analysis of polyploidy in wild potatoes (Solanum section Petota). BMC Evol Biol. 2012;12:70. https://doi.org/10.1186/1471-2148-12-70.

Cara N, Ferrer MS, Masuelli RW, et al. Epigenetic consequences of interploidal hybridisation in synthetic and natural interspecific potato hybrids. New Phytol. 2019;222:1981–93. https://doi.org/10.1111/nph.15706.

Cara N, Marfil CF, Bertoldi MV, Masuelli RW. Methylation-sensitive amplified polymorphism as a tool to analyze wild potato hybrids. Bio-protocol. 2020;10:e3671.

Caraza-Harter MV, Endelman JB. Image-based phenotyping and genetic analysis of potato skin set and color. Crop Sci. 2020;60:202–10. https://doi.org/10.1002/csc2.20093.

Carputo D, Alioto D, Aversano R, et al. Genetic diversity among potato species as revealed by phenotypic resistances and SSR markers. Plant Genet Resour. 2013;11:131–9. https://doi.org/10.1017/S1479262112000500.

Caruana BM, Pembleton LW, Constable F, et al. Validation of genotyping by sequencing using Transcriptomics for diversity and application of genomic selection in tetraploid potato. Front Plant Sci. 2019;10 https://doi.org/10.3389/fpls.2019.00670.

Carvallo MA, Pino M-T, Jeknić Z, et al. A comparison of the low temperature transcriptomes and CBF regulons of three plant species that differ in freezing tolerance: Solanum commersonii, Solanum tuberosum, and Arabidopsis thaliana. J Exp Bot. 2011;62:3807–19. https://doi.org/10.1093/jxb/err066.

Castillo RO, Spooner DM. Phylogenetic relationships of wild potatoes, Solanum series Conicibaccata (Sect. Petota). Syst Bot. 1997;22:45–83. https://doi.org/10.2307/2419677.

Celis C, Scurrah M, Cowgill S, et al. Environmental biosafety and transgenic potato in a centre of diversity for this crop. Nature. 2004;432:222–5.

Chandra S, Huaman Z, Hari Krishna S, Ortiz R. Optimal sampling strategy and core collection size of Andean tetraploid potato based on isozyme data – a simulation study. Theor Appl Genet. 2002;104:1325–34. https://doi.org/10.1007/s00122-001-0854-4.

Charepalli V, Reddivari L, Radhakrishnan S, et al. Anthocyanin-containing purple-fleshed potatoes suppress colon tumorigenesis via elimination of colon cancer stem cells. J Nutr Biochem. 2015;26:1641–9. https://doi.org/10.1016/j.jnutbio.2015.08.005.

Chen J, Leach L, Yang J, et al. A tetrasomic inheritance model and likelihood-based method for mapping quantitative trait loci in autotetraploid species. New Phytol. 2021;230:387–98. https://doi.org/10.1111/nph.16413.

Chen X, Lewandowska D, Armstrong MR, et al. Identification and rapid mapping of a gene conferring broad-spectrum late blight resistance in the diploid potato species Solanum verrucosum through DNA capture technologies. Theor Appl Genet. 2018;131:1287–97. https://doi.org/10.1007/s00122-018-3078-6.

Choi JY, Platts AE, Fuller DQ, et al. The rice paradox: multiple origins but single domestication in Asian rice. Mol Biol Evol. 2017;34:969–79. https://doi.org/10.1093/molbev/msx049.

Choi JY, Purugganan MD. Multiple origin but single domestication led to Oryza sativa. G3 (Bethesda). 2018;8:797–803. https://doi.org/10.1534/g3.117.300334.

Clot CR, Polzer C, Prodhomme C, et al. The origin and widespread occurrence of Sli-based self-compatibility in potato. Theor Appl Genet. 2020;133:2713–28. https://doi.org/10.1007/s00122-020-03627-8.

Collares EAS, Choer E, da Pereira S. Characterization of potato genotypes using molecular markers. Pesq agropec bras. 2004;39:871–8. https://doi.org/10.1590/S0100-204X2004000900006.

Collins A, Milbourne D, Ramsay L, et al. QTL for field resistance to late blight in potato are strongly correlated with maturity and vigour. Mol Breed. 1999;5:387–98. https://doi.org/10.1023/A:1009601427062.

Consortium TG. The tomato genome sequence provides insights into fleshy fruit evolution. Nature. 2012;485:635.

D'Amelia V, Villano C, Batelli G, et al. Genetic and epigenetic dynamics affecting anthocyanin biosynthesis in potato cell culture. Plant Sci. 2020;298:110597. https://doi.org/10.1016/j.plantsci.2020.110597.

D'hoop B, Keizer PL, Paulo MJ, et al. Identification of agronomically important QTL in tetraploid potato cultivars using a marker–trait association analysis. Theor Appl Genet. 2014;127:731–48.

D'hoop BB, Paulo MJ, Kowitwanich K, et al. Population structure and linkage disequilibrium unravelled in tetraploid potato. Theor Appl Genet. 2010;121:1151–70. https://doi.org/10.1007/s00122-010-1379-5.

D'hoop BB, Paulo MJ, Mank RA, et al. Association mapping of quality traits in potato (Solanum tuberosum L.). Euphytica. 2008;161:47–60. https://doi.org/10.1007/s10681-007-9565-5.

da Pereira GS, Mollinari M, Schumann MJ, et al. The recombination landscape and multiple QTL mapping in a Solanum tuberosum cv. 'Atlantic'-derived F1 population. Heredity. 2021;126:817–30. https://doi.org/10.1038/s41437-021-00416-x.

da Silva PG, Gemenet DC, Mollinari M, et al. Multiple QTL mapping in Autopolyploids: a random-effect model approach with application in a Hexaploid Sweetpotato full-sib population. Genetics. 2020;215:579–95. https://doi.org/10.1534/genetics.120.303080.

De Jong WS, De Jong DM, De Jong H, et al. An allele of dihydroflavonol 4-reductase associated with the ability to produce red anthocyanin pigments in potato (Solanum tuberosum L.). Theor Appl Genet. 2003;107:1375–83. https://doi.org/10.1007/s00122-003-1395-9.

De Jong WS, Eannetta NT, De Jong DM, Bodis M. Candidate gene analysis of anthocyanin pigmentation loci in the Solanaceae. Theor Appl Genet. 2004;108:423–32. https://doi.org/10.1007/s00122-003-1455-1.

Debener T, Salamini F, Gebhardt C. The use of RFLPs (restriction fragment length polymorphisms) detects germplasm introgressions from wild species into potato (Solanum tuberosum ssp. tuberosum) breeding lines. Plant Breed. 1991;106:173–81.

Demirel U, Tındaş İ, Yavuz C, et al. Assessing genetic diversity of potato genotypes using inter-PBS retrotransposon marker system. Plant Genet Resour. 2018;16:137–45. https://doi.org/10.1017/S1479262117000041.

Deng K, Yin H, Xiong F, et al. Genome-wide miRNA expression profiling in potato (Solanum tuberosum L.) reveals TOR-dependent post-transcriptional gene regulatory networks in diverse metabolic pathway. PeerJ. 2021;9:10.7717/peerj.10704.

Diaz P, et al. Genomic regions associated with physiological, biochemical and yield-related responses under water deficit in diploid potato at the tuber initiation stage revealed by GWAS. PLoS One. 2021;16(11) Public Library Science:e0259690. https://doi.org/10.1371/journal.pone.0259690.

Djami-Tchatchou AT, Sanan-Mishra N, Ntushelo K, Dubery IA. Functional roles of microRNAs in Agronomically important plants—potential as targets for crop improvement and protection. Front Plant Sci. 2017;8 https://doi.org/10.3389/fpls.2017.00378.

Dodds KS. The classification of the cultivated potatoes. In: Correll DS, editor. The potato and its wild relatives. Texas Res. Found; 1962. p. 517–39.

Doebley JF, Gaut BS, Smith BD. The molecular genetics of crop domestication. Cell. 2006;127:1309–21. https://doi.org/10.1016/j.cell.2006.12.006.

Douches D, Hirsch CN, Manrique-Carpintero NC, et al. The contribution of the Solanaceae coordinated agricultural project to potato breeding. Potato Res. 2014;57:215–24. https://doi.org/10.1007/s11540-014-9267-z.

Duan Y, Liu J, Xu J, et al. DNA fingerprinting and genetic diversity analysis with simple sequence repeat markers of 217 potato cultivars (Solanum tuberosum L.) in China. Am J Potato Res. 2019;96:21–32. https://doi.org/10.1007/s12230-018-9685-6.

Ducreux LJM, Morris WL, Prosser IM, et al. Expression profiling of potato germplasm differentiated in quality traits leads to the identification of candidate flavour and texture genes. J Exp Bot. 2008;59:4219–31. https://doi.org/10.1093/jxb/ern264.

Dzyubenko NI. Vavilov's collection of worldwide crop genetic resources in the 21st century. Biopreserv Biobank. 2018;16:377–83. https://doi.org/10.1089/bio.2018.0045.

Ellis D, Chavez O, Coombs J, et al. Genetic identity in genebanks: application of the SolCAP 12K SNP array in fingerprinting and diversity analysis in the global in trust potato collection. Genome. 2018;61:523–37.

Ellis D, Salas A, Chavez O, et al. Ex situ conservation of potato [Solanum section Petota (Solanaceae)] genetic resources in Genebanks. In: Campos H, Ortiz O, editors. The potato crop: its agricultural, nutritional and social contribution to humankind. Cham: Springer International Publishing; 2020a. p. 109–38.

Ellis GD, Knowles LO, Knowles NR. Developmental and postharvest physiological phenotypes of engineered potatoes (Solanum tuberosum L.) grown in the Columbia Basin. Field Crop Res. 2020b;250:107775. https://doi.org/10.1016/j.fcr.2020.107775.

Enciso-Rodriguez F, Douches D, Lopez-Cruz M, et al. Genomic selection for late blight and common scab resistance in tetraploid potato (Solanum tuberosum). G3 Genes|Genomes|Genetics. 2018;8:2471–81. https://doi.org/10.1534/g3.118.200273.

Endelman JB, Carley CAS, Bethke PC, et al. Genetic variance partitioning and genome-wide prediction with Allele dosage information in autotetraploid potato. Genetics. 2018;209:77–87. https://doi.org/10.1534/genetics.118.30068.

Endelman JB, Jansky SH. Genetic mapping with an inbred line-derived F2 population in potato. Theor Appl Genet. 2016;129:935–43. https://doi.org/10.1007/s00122-016-2673-7.

Ersoz ES, Yu J, Buckler ES. Applications of linkage disequilibrium and association mapping in crop plants. In: Genomics-assisted crop improvement. Springer; 2007. pp. 97–119.

Esnault F, Pellé R, Dantec J-P, et al. Development of a potato cultivar (Solanum tuberosum L.) Core collection, a valuable tool to prospect genetic variation for novel traits. Potato Res. 2016;59:329–43. https://doi.org/10.1007/s11540-016-9332-x.

Esposito S, Aversano R, Bradeen JM, et al. Deep-sequencing of Solanum commersonii small RNA libraries reveals riboregulators involved in cold stress response. Plant Biol. 2020;22:133–42. https://doi.org/10.1111/plb.12955.

Esposito S, Barteri F, Casacuberta J, et al. LTR-TEs abundance, timing and mobility in Solanum commersonii and S. tuberosum genomes following cold-stress conditions. Planta. 2019; https://doi.org/10.1007/s00425-019-03283-3.

Ewing EE, Šimko I, Smart CD, et al. Genetic mapping from field tests of qualitative and quantitative resistance to Phytophthora infestans in a population derived from Solanum tuberosum and Solanum berthaultii. Mol Breed. 2000;6:25–36. https://doi.org/10.1023/A:1009648408198.

Ezekiel R, Singh N, Sharma S, Kaur A. Beneficial phytochemicals in potato — a review. Food Res Int. 2013;2:487–96. https://doi.org/10.1016/j.foodres.2011.04.025.

Fajardo D, Spooner DM. Phylogenetic relationships of Solanum series Conicibaccata and related species in Solanum section Petota inferred from five conserved Ortholog sequences. sbot. 2011;36:163–170. https://doi.org/10.1600/036364411x553252.

FAO. International year of the potato. 2008. https://www.fao.org/potato-2008/en/aboutiyp/index.html. Verified 11/23/201

FAO. Visualize data crops. 2019. http://www.fao.org/faostat/en/#data/QC/visualize. Verified 11/23/2021.

Felcher KJ, Coombs JJ, Massa AN, et al. Integration of two diploid potato linkage maps with the potato genome sequence. PLoS One. 2012;7:e36347. https://doi.org/10.1371/journal.pone.0036347.

Ferrão LFV, Benevenuto J, de Oliveira IB, et al. Insights into the genetic basis of blueberry fruit-related traits using diploid and polyploid models in a GWAS context. Front Ecol Evol. 2018;6:107. https://doi.org/10.3389/fevo.2018.00107.

Finnegan EJ, Peacock WJ, Dennis ES. DNA methylation, a key regulator of plant development and other processes. Curr Opin Genet Dev. 2000;10:217–23. https://doi.org/10.1016/s0959-437x(00)00061-7.

Fofana B, Somalraju A, Fillmore S, et al. Comparative transcriptome expression analysis in susceptible and resistant potato (Solanum tuberosum) cultivars to common scab (Streptomyces scabies) revealed immune priming responses in the incompatible interaction. PLoS One. 2020;15 https://doi.org/10.1371/journal.pone.0235018.

Freedman AH, Gronau I, Schweizer RM, et al. Genome sequencing highlights the dynamic early history of dogs. PLoS Genet. 2014;10:e1004016. https://doi.org/10.1371/journal.pgen.1004016.

Frodin DG. History and concepts of big plant genera. Taxon. 2004;53:753–76.

Fulladolsa AC, Navarro FM, Kota R, et al. Application of marker assisted selection for Potato virus Y resistance in the University of Wisconsin Potato Breeding Program. Am J Potato Res. 2015;92:444–50.

Gagnon E, Hilgenhof R, Orejuela A, et al. Phylogenomic data reveal hard polytomies across the backbone of the large genus Solanum (Solanaceae). Am J Bot. 2022;109(4):580–601. https://doi.org/10.1002/ajb2.1827.

Gaiero P, Vaio M, Peters SA, et al. Comparative analysis of repetitive sequences among species from the potato and the tomato clades. Ann Bot. 2019;123:521–32. https://doi.org/10.1093/aob/mcy186.

Gallais A. Quantitative genetics and breeding methods in autopolyploid plants: Editions Quae; 2003.

Gálvez JH, Tai HH, Lagüe M, et al. The nitrogen responsive transcriptome in potato (Solanum tuberosum L.) reveals significant gene regulatory motifs. Sci Rep. 2016;6:10.1038/srep26090.

Gaut BS, Long AD. The lowdown on linkage disequilibrium. Plant Cell. 2003;15:1502–6.

Gavrilenko T, Antonova O, Ovchinnikova A, et al. A microsatellite and morphological assessment of the Russian National Potato Collection. Genet Resour Crop Evol. 2010;57 https://doi.org/10.1007/s10722-010-9554-8.

Gavrilenko T, Antonova O, Shuvalova A, et al. Genetic diversity and origin of cultivated potatoes based on plastid microsatellite polymorphism. Genet Resour Crop Evol. 2013;60:1997–2015.

Gebhardt C, Ballvora A, Walkemeier B, et al. Assessing genetic potential in germplasm collections of crop plants by marker-trait association: a case study for potatoes with quantitative variation of resistance to late blight and maturity type. Mol Breed. 2004;13:93–102. https://doi.org/10.1023/B:MOLB.0000012878.89855.df.

Gebhardt C, Ritter E, Barone A, et al. RFLP maps of potato and their alignment with the homoeologous tomato genome. Theoret Appl Genet. 1991;83:49–57. https://doi.org/10.1007/BF00229225.

Gehring M, Henikoff S. DNA methylation dynamics in plant genomes. Biochim Biophys Acta. 2007;1769:276–86. https://doi.org/10.1016/j.bbaexp.2007.01.009.

Gemenet DC, Lindqvist-Kreuze H, De Boeck B, et al. Sequencing depth and genotype quality: accuracy and breeding operation considerations for genomic selection applications in autopolyploid crops. Theor Appl Genet. 2020;133:3345–63. https://doi.org/10.1007/s00122-020-03673-2.

Ghislain M, Andrade D, Rodríguez F, et al. Genetic analysis of the cultivated potato Solanum tuberosum L. Phureja Group using RAPDs and nuclear SSRs. Theor Appl Genet. 2006;113: 1515–27. https://doi.org/10.1007/s00122-006-0399-7.

Ghislain M, Douches DS. The genes and genomes of the potato. In: The potato crop. Cham: Springer; 2020. p. 139–162.

Ghislain M, Núñez J, del Rosario HM, Spooner DM. The single Andigenum origin of Neo-Tuberosum potato materials is not supported by microsatellite and plastid marker analyses. Theor Appl Genet. 2009;118:963–9.

Ghislain M, Trognitz B, del Herrera Ma R, et al. Genetic loci associated with field resistance to late blight in offspring of Solanum phureja and S.tuberosum grown under short-day conditions. Theor Appl Genet. 2001;103:433–42. https://doi.org/10.1007/s00122-001-0545-1.

Ghislain M, Zhang D, Fajardo D, et al. Marker-assisted sampling of the cultivated Andean potato Solanum phureja collection using RAPD markers. Genet Resour Crop Evol. 1999;46:547–55. https://doi.org/10.1023/A:1008724007888.

Gong L, Zhang H, Gan X, et al. Transcriptome profiling of the potato (Solanum tuberosum L.) plant under drought stress and water-stimulus conditions. PLoS One. 2015;10:10.1371/journal.pone.0128041.

Gopal J, Kumar V, Kumar R, Mathur P. Comparison of different approaches to establish a Core collection of Andigena (Solanum tuberosum Group Andigena) potatoes. Potato Res. 2013;56: 85–98. https://doi.org/10.1007/s11540-013-9232-2.

Grandke F, Ranganathan S, van Bers N, et al. PERGOLA: fast and deterministic linkage mapping of polyploids. BMC Bioinform. 2017;18:12. https://doi.org/10.1186/s12859-016-1416-8.

Grech-Baran M, Witek K, Szajko K, et al. Extreme resistance to Potato virus Y in potato carrying the Rysto gene is mediated by a TIR-NLR immune receptor. Plant Biotechnol J. 2020;18:655–67. https://doi.org/10.1111/pbi.13230.

Grun P. The evolution of cultivated potatoes. Econ Bot. 1990;44:39–55.

Gutaker RM, Weiß CL, Ellis D, et al. The origins and adaptation of European potatoes reconstructed from historical genomes. Nat Ecol Evol. 2019;3:1093–101.

Habyarimana E, Parisi B, Mandolino G. Genomic prediction for yields, processing and nutritional quality traits in cultivated potato (Solanum tuberosum L.). Plant Breed. 2017;136:245–52. https://doi.org/10.1111/pbr.12461.

Hackett CA, Boskamp B, Vogogias A, et al. TetraploidSNPMap: software for linkage analysis and QTL mapping in autotetraploid populations using SNP dosage data. J Hered. 2017;108:438–42. https://doi.org/10.1093/jhered/esx022.

Hackett CA, Bradshaw JE, Bryan GJ. QTL mapping in autotetraploids using SNP dosage information. Theor Appl Genet. 2014;127:1885–904. https://doi.org/10.1007/s00122-014-2347-2.

Hackett CA, Bradshaw JE, Meyer RC, et al. Linkage analysis in tetraploid species: a simulation study. Genet Res. 1998;71:143–53. https://doi.org/10.1017/S0016672398003188.

Hackett CA, Luo ZW. TetraploidMap: construction of a linkage map in autotetraploid species. J Hered. 2003;94:358–9. https://doi.org/10.1093/jhered/esg066.

Hackett CA, McLean K, Bryan GJ. Linkage analysis and QTL mapping using SNP dosage data in a tetraploid potato mapping population. PLoS One. 2013;8:e63939. https://doi.org/10.1371/journal.pone.0063939.

Hackett CA, Milne I, Bradshaw JE, Luo Z. TetraploidMap for windows: linkage map construction and QTL mapping in autotetraploid species. J Hered. 2007;98:727–9. https://doi.org/10.1093/jhered/esm086.

Hamilton JP, Hansey CN, Whitty BR, et al. Single nucleotide polymorphism discovery in elite North American potato germplasm. BMC Genomics. 2011;12:302. https://doi.org/10.1186/1471-2164-12-302.

Hardigan MA, Bamberg J, Buell CR, Douches DS. Taxonomy and genetic differentiation among wild and cultivated germplasm of Solanum sect. Petota. Plant Genome. 2015;8:plant genome 2014-06.

Hardigan MA, Crisovan E, Hamilton JP, et al. Genome reduction uncovers a large dispensable genome and adaptive role for copy number variation in asexually propagated Solanum tuberosum. Plant Cell. 2016;28:388–405. https://doi.org/10.1105/tpc.15.00538.

Hardigan MA, Laimbeer FPE, Hamilton JP, et al. Reply to Huang et al.: avoiding "one-size-fits-all" approaches to variant discovery. PNAS. 2018;115:E6394–5. https://doi.org/10.1073/pnas.1807622115.

Hardigan MA, Laimbeer FPE, Newton L, et al. Genome diversity of tuber-bearing Solanum uncovers complex evolutionary history and targets of domestication in the cultivated potato. Proc Natl Acad Sci U S A. 2017;114:E9999–E10008. https://doi.org/10.1073/pnas.1714380114.

Harding K. The methylation status of DNA derived from potato plants recovered from slow growth. Plant Cell Tissue Organ Cult. 1994;37:31–8. https://doi.org/10.1007/BF00048114.

Harris K, Nielsen R. Inferring demographic history from a Spectrum of shared haplotype lengths. PLoS Genet. 2013;9:e1003521. https://doi.org/10.1371/journal.pgen.1003521.

Hawkes JG. The potato: evolution, biodiversity and genetic resources: Belhaven Press; 1990.

Hawkes JG, Francisco-Ortega J. The early history of the potato in Europe. Euphytica. 1993;70:1–7.

Hawkes JG, Hjerting JP. The potatoes of Argentina, Brazil, Paraguay and Uruguay. A biosystematic study. The potatoes of Argentina, Brazil, Paraguay and Uruguay A biosystematic study, 1969.

Haynes KG, Zaki HEM, Christensen CT, et al. High levels of heterozygosity found for 15 SSR loci in Solanum chacoense. Am J Potato Res. 2017;94:638–46. https://doi.org/10.1007/s12230-017-9602-4.

Herath V, Verchot J. Transcriptional regulatory networks associate with early stages of potato virus X infection of Solanum tuberosum. Int J Mol Sci. 2021;22 https://doi.org/10.3390/ijms22062837.

Hirsch CN, Hirsch CD, Felcher K, et al. Retrospective view of North American potato (Solanum tuberosum L.) breeding in the 20th and 21st centuries. G3: genes. Genome Genet. 2013;3:1003–13. https://doi.org/10.1534/g3.113.005595.

Hoopes G, Meng X, Hamilton JP, et al. Phased chromosome-scale genome assemblies of tetraploid potato reveal a complex genome, transcriptome, and predicted proteome landscape underpinning genetic diversity. Mol Plant. 2022;15(3):520–36. https://doi.org/10.1016/j.molp.2022.01.003.

Hoque ME, Huq H, Moon NJ. Molecular diversity analysis in potato (Solanum tuberosum L.) through RAPD markers. SAARC J Agric. 2013;11:95–102. https://doi.org/10.3329/sja.v11i2.18405.

Hosaka AJ, Sanetomo R, Hosaka K. A de novo genome assembly of Solanum verrucosum Schlechtendal, a Mexican diploid species geographically isolated from other diploid A-genome species of potato relatives. G3 Genes|Genomes|Genetics. 2022;12:jkac166. https://doi.org/10.1093/g3journal/jkac166.

Hosaka K. Successive domestication and evolution of the Andean potatoes as revealed by chloroplast DNA restriction endonuclease analysis. Theor Appl Genet. 1995;90:356–63.

Hosaka K. T-type chloroplast DNA in Solarium tuberosum L. ssp. tuberosum was conferred from some populations of S. tarijense Hawkes. Am J Potato Res. 2003;80:21–32.

Hosaka K, Hanneman RE. Origin of chloroplast DNA diversity in the Andean potatoes. Theor Appl Genet. 1988;76:333–40.

Huamán Z, Ortiz R, Gómez R. Selecting a Solanum tuberosum subsp. andigena core collection using morphological, geographical, disease and pest descriptors. Am J Pot Res. 2000;77:183–90. https://doi.org/10.1007/BF02853943.

Huamán Z, Spooner DM. Reclassification of landrace populations of cultivated potatoes (Solanum sect. Petota). Am J Bot. 2002;89:947–65.

Huang B, Ruess H, Liang Q, et al. Analyses of 202 plastid genomes elucidate the phylogeny of Solanum section Petota. Sci Rep. 2019;9:1–7.

Huang B, Spooner DM, Liang Q. Genome diversity of the potato. PNAS. 2018;115:E6392–3. https://doi.org/10.1073/pnas.1805917115.

Huang S, Vleeshouwers VGAA, Werij JS, et al. The R3 resistance to Phytophthora infestans in potato is conferred by two closely linked R genes with distinct specificities. MPMI. 2004;17:428–35. https://doi.org/10.1094/MPMI.2004.17.4.428.

Ibañez VN, Masuelli RW, Marfil CF. Environmentally induced phenotypic plasticity and DNA methylation changes in a wild potato growing in two contrasting Andean experimental gardens. Heredity. 2021;126:50–62. https://doi.org/10.1038/s41437-020-00355-z.

Inostroza L, Bhakta M, Acuña H, et al. Understanding the complexity of cold tolerance in white clover using temperature gradient locations and a GWAS approach. Plant Genome. 2018;11:170096. https://doi.org/10.3835/plantgenome2017.11.0096.

Iorizzo M, Gao L, Mann H, et al. A DArT marker-based linkage map for wild potato Solanum bulbocastanum facilitates structural comparisons between SolanumA and B genomes. BMC Genet. 2014;15:123. https://doi.org/10.1186/s12863-014-0123-6.

Jacobs MMJ, van den Berg RG, Vleeshouwers VGAA, et al. AFLP analysis reveals a lack of phylogenetic structure within Solanum section Petota. BMC Evol Biol. 2008;8:145. https://doi.org/10.1186/1471-2148-8-145.

Jansky SH, Charkowski AO, Douches DS, et al. Reinventing potato as a diploid inbred line–based crop. Crop Sci. 2016;56:1412–22. https://doi.org/10.2135/cropsci2015.12.0740.

Jansky SH, Dempewolf H, Camadro EL, et al. A case for crop wild relative preservation and use in potato. Crop Sci. 2013;53:746–54. https://doi.org/10.2135/cropsci2012.11.0627.

Jansky SH, Peloquin SJ. Advantages of wild diploid Solanum species over cultivated diploid relatives in potato breeding programs. Genet Resour Crop Evol. 2006;53:669–74. https://doi.org/10.1007/s10722-004-2949-7.

Jiménez JP, Brenes A, Fajardo D, et al. The use and limits of AFLP data in the taxonomy of polyploid wild potato species in Solanum series Conicibaccata. Conserv Genet. 2008;9:381–7. https://doi.org/10.1007/s10592-007-9350-y.

Joyce SM, Cassells A. Variation in potato microplant morphology in vitro and DNA methylation. Plant Cell Tissue Organ Cult. 2004; https://doi.org/10.1023/A:1016312303320.

Jung CS, Griffiths HM, De Jong DM, et al. The potato P locus codes for flavonoid 3′,5′--hydroxylase. Theor Appl Genet. 2005;110:269–75. https://doi.org/10.1007/s00122-004-1829-z.

Jung CS, Griffiths HM, De Jong DM, et al. The potato developer (D) locus encodes an R2R3 MYB transcription factor that regulates expression of multiple anthocyanin structural genes in tuber skin. Theor Appl Genet. 2009;120:45–57. https://doi.org/10.1007/s00122-009-1158-3.

Jupe F, Witek K, Verweij W, et al. Resistance gene enrichment sequencing (RenSeq) enables reannotation of the NB-LRR gene family from sequenced plant genomes and rapid mapping of resistance loci in segregating populations. Plant J. 2013;76:530–44. https://doi.org/10.1111/tpj.12307.

Juzepchuk SV. A contribution to the question of the origin of the potato, 1929.

Kaiser NR, Coombs JJ, Felcher KJ, et al. Genome-wide association analysis of common scab resistance and expression profiling of tubers in response to Thaxtomin A treatment underscore the complexity of common scab resistance in tetraploid potato. Am J Potato Res. 2020;97:513–22. https://doi.org/10.1007/s12230-020-09800-5.

Kardolus JP. A biosystematic analysis of Solanum acaule: Kardolus; 1998.

Kasai K, Morikawa Y, Sorri VA, et al. Development of SCAR markers to the PVY resistance gene Ry adg based on a common feature of plant disease resistance genes. Genome. 2000;43:1–8.

Kerr RJ, Li L, Tier B, et al. Use of the numerator relationship matrix in genetic analysis of autopolyploid species. Theor Appl Genet. 2012;124:1271–82. https://doi.org/10.1007/s00122-012-1785-y.

Khlestkin VK, Rozanova IV, Efimov VM, Khlestkina EK. Starch phosphorylation associated SNPs found by genome-wide association studies in the potato (Solanum tuberosum L.). BMC Genet. 2019;20:29. https://doi.org/10.1186/s12863-019-0729-9.

Khu D-M, Lorenzen J, Hackett CA, Love SL. Interval mapping of quantitative trait loci for Corky Ringspot disease resistance in a tetraploid population of potato (Solanum tuberosum subsp. tuberosum). Am J Pot Res. 2008;85:129–39. https://doi.org/10.1007/s12230-008-9016-4.

Kitazumi A, Kawahara Y, Onda TS, et al. Implications of miR166 and miR159 induction to the basal response mechanisms of an andigena potato (Solanum tuberosum subsp. andigena) to salinity stress, predicted from network models in Arabidopsis. Genome. 2015;58:13–24. https://doi.org/10.1139/gen-2015-0011.

Klaassen MT, Willemsen JH, Vos PG, et al. Genome-wide association analysis in tetraploid potato reveals four QTLs for protein content. Mol Breed. 2019;39:151. https://doi.org/10.1007/s11032-019-1070-8.

Kloosterman B, Abelenda JA, del Gomez MMC, et al. Naturally occurring allele diversity allows potato cultivation in northern latitudes. Nature. 2013;495:246–50. https://doi.org/10.1038/nature11912.

Kloosterman B, Anithakumari AM, Chibon P-Y, et al. Organ specificity and transcriptional control of metabolic routes revealed by expression QTL profiling of source-sink tissues in a segregating potato population. BMC Plant Biol. 2012;12:1–12. https://doi.org/10.1186/1471-2229-12-17.

Kloosterman B, De Koeyer D, Griffiths R, et al. Genes driving potato tuber initiation and growth: identification based on transcriptional changes using the POCI array. Funct Integr Genomics. 2008;8:329–40. https://doi.org/10.1007/s10142-008-0083-x.

Koizumi E, Igarashi T, Tsuyama M, et al. Association of Genome-Wide SNP markers with resistance to common scab of potato. Am J Potato Res. 2021;98:149–56. https://doi.org/10.1007/s12230-021-09827-2.

Kolech SA, Halseth D, Perry K, et al. Genetic diversity and relationship of Ethiopian potato varieties to germplasm from North America, Europe and the International Potato Center. Am J Potato Res. 2016;93:609–19. https://doi.org/10.1007/s12230-016-9543-3.

Kondhare KR, Malankar NN, Devani RS, Banerjee AK. Genome-wide transcriptome analysis reveals small RNA profiles involved in early stages of stolon-to-tuber transitions in potato under photoperiodic conditions. BMC Plant Biol. 2018;18:284. https://doi.org/10.1186/s12870-018-1501-4.

Kono TJY, Lei L, Shih C-H, et al. Comparative genomics approaches accurately predict deleterious variants in plants. G3 Genes|Genomes|Genetics. 2018;8:3321–9. https://doi.org/10.1534/g3.118.200563.

Korontzis G, Malosetti M, Zheng C, et al. QTL detection in a pedigreed breeding population of diploid potato. Euphytica. 2020;216:1–14.

Kuang H, Padmanabhan C, Li F, et al. Identification of miniature inverted-repeat transposable elements (MITEs) and biogenesis of their siRNAs in the Solanaceae: new functional implications for MITEs. Genome Res. 2009;19:42–56. https://doi.org/10.1101/gr.078196.108.

Kuhl J, Hanneman R, Havey M. Characterization and mapping of Rpi1, a late-blight resistance locus from diploid (1EBN) Mexican Solanum pinnatisectum. Mol Gen Genomics. 2001;265:977–85. https://doi.org/10.1007/s004380100490.

Kyriakidou M, Achakkagari SR, Gálvez López JH, et al. Structural genome analysis in cultivated potato taxa. Theor Appl Genet. 2020a;133:951–66. https://doi.org/10.1007/s00122-019-03519-6.

Kyriakidou M, Anglin NL, Ellis D, et al. Genome assembly of six polyploid potato genomes. Sci Data. 2020b;7:88. https://doi.org/10.1038/s41597-020-0428-4.

Lakhotia N, Joshi G, Bhardwaj AR, et al. Identification and characterization of miRNAome in root, stem, leaf and tuber developmental stages of potato (Solanum tuberosum L.) by high-throughput sequencing. BMC Plant Biol. 2014;14:6. https://doi.org/10.1186/1471-2229-14-6.

Lara-Cabrera SI, Spooner DM. Taxonomy of North and Central American diploid wild potato (Solanum sect. Petota) species: AFLP data. Plant Syst Evol. 2004;248:129–42.

Lara-Cabrera SI, Spooner DM. In: Keating RC, Hollowell VC, Croat TB, editors. Taxonomy of Mexican diploid wild potatoes (Solanum sect. Petota): morphological and microsatellite data. A Festschrift for William G D'arcy : the legacy of a taxonomist; 2005.

Law RD, Suttle JC. Changes in histone H3 and H4 multi-acetylation during natural and forced dormancy break in potato tubers. Physiol Plant. 2004;120:642–9. https://doi.org/10.1111/j.0031-9317.2004.0273.x.

Lebecka R, Śliwka J, Grupa-Urbańska A, et al. QTLs for potato tuber resistance to Dickeya solani are located on chromosomes II and IV. Plant Pathol. 2021;70(7):1745–56. https://doi.org/10.1101/2021.02.19.432067.

Lechnovich VS. Cultivated potato species. In: Bukasov SM, editor. Flora of cultivated plants, chapter 2, vol. IX. Leningrad, Russia: Kolos; 1971. p. 41–304.

Lee KJ, Sebastin R, Cho GT, Yoon M, Lee GA, Hyun DY. Genetic diversity and population structure of potato germplasm in RDA-Genebank: utilization for breeding and conservation. Plants (Basel, Switzerland). 2021b;10(4):752. https://doi.org/10.3390/plants10040752.

Lee K-J, Sebastin R, Cho G-T, et al. Genetic diversity and population structure of potato germplasm in RDA-Genebank: utilization for breeding and conservation. Plan Theory. 2021a;10:752.

Leisner CP, Hamilton JP, Crisovan E, et al. Genome sequence of M6, a diploid inbred clone of the high-glycoalkaloid-producing tuber-bearing potato species Solanum chacoense, reveals residual heterozygosity. Plant J. 2018;94:562–70. https://doi.org/10.1111/tpj.13857.

Lemke P, Moerschbacher BM, Singh R. Transcriptome analysis of Solanum Tuberosum genotype RH89-039-16 in response to Chitosan. Front Plant Sci. 2020;11 https://doi.org/10.3389/fpls.2020.01193.

Levina AV, Hoekenga O, Gordin M, et al. Genetic analysis of potato tuber metabolite composition: genome-wide association studies applied to a nontargeted metabolome. Crop Sci. 2021;61:591–603.

Li L, Tacke E, Hofferbert H-R, et al. Validation of candidate gene markers for marker-assisted selection of potato cultivars with improved tuber quality. Theor Appl Genet. 2013;126:1039–52.

Li P, Fan R, Peng Z, et al. Transcriptome analysis of resistance mechanism to potato wart disease. Open Life Sci. 2021;16:475–81. https://doi.org/10.1515/biol-2021-0045.

Li Q, Qin Y, Hu X, et al. Transcriptome analysis uncovers the gene expression profile of salt-stressed potato (Solanum tuberosum L.). Sci Rep. 2020;10:10.1038/s41598-020-62057-0.

Li X, Xu J, Duan S, et al. Mapping and QTL analysis of early-maturity traits in tetraploid potato (Solanum tuberosum L.). Int J Mol Sci. 2018a;19 https://doi.org/10.3390/ijms19103065.

Li Y, Colleoni C, Zhang J, et al. Genomic analyses yield markers for identifying Agronomically important genes in potato. Mol Plant. 2018b;11:473–84. https://doi.org/10.1016/j.molp.2018.01.009.

Li Y, Hu X, Chen J, et al. Integrated mRNA and microRNA transcriptome analysis reveals miRNA regulation in response to PVA in potato. Sci Rep. 2017;7 https://doi.org/10.1038/s41598-017-17059-w.

Lindhout P, Meijer D, Schotte T, et al. Towards F1 hybrid seed potato breeding. Potato Res. 2011;54:301–12. https://doi.org/10.1007/s11540-011-9196-z.

Lindqvist-Kreuze H, de Boeck B, Unger P, et al. Global multi-environment resistance QTL for foliar late blight resistance in tetraploid potato with tropical adaptation. G3 (Bethesda). 2020;11: jkab251. https://doi.org/10.1093/g3journal/jkab251.

Liu B, Kong L, Zhang Y, Liao Y. Gene and Metabolite integration analysis through transcriptome and metabolome brings new insight into heat stress tolerance in potato (Solanum tuberosum L.). Plants (Basel). 2021;10:10.3390/plants10010103.

Liu Y, Lin-Wang K, Deng C, et al. Comparative transcriptome analysis of white and purple potato to identify genes involved in anthocyanin biosynthesis. PLoS One. 2015;10:e0129148. https://doi.org/10.1371/journal.pone.0129148.

Luikart G, Kardos M, Hand B, Rajora OP, Aitkin S, Hohenlohe PA. 2018. Population genomics: advancing understanding of nature. In: Rajora OP, editor. Population genomics: concepts, approaches and applications. Cham: Springer International Publishing AG; 2019. p. 3–79.

Macharia TN, Bellieny-Rabelo D, Moleleki LN. Transcriptome profiling of potato (Solanum tuberosum L.) responses to Root-Knot Nematode (Meloidogyne javanica) infestation during a compatible interaction. Microorganisms. 2020;8:10.3390/microorganisms8091443.

Machida-Hirano R. Diversity of potato genetic resources. Breed Sci. 2015;65:26–40.

Mackay I, Powell W. Methods for linkage disequilibrium mapping in crops. Trends Plant Sci. 2007;12:57–63. https://doi.org/10.1016/j.tplants.2006.12.001.

Malosetti M, van der Linden CG, Vosman B, van Eeuwijk FA. A Mixed-Model approach to association mapping using pedigree information with an illustration of resistance to Phytophthora infestans in potato. Genetics. 2007;175:879–89. https://doi.org/10.1534/genetics.105.054932.

Manrique-Carpintero NC, Coombs JJ, Cui Y, et al. Genetic map and QTL analysis of agronomic traits in a diploid potato population using single nucleotide polymorphism markers. Crop Sci. 2015;55:2566–79. https://doi.org/10.2135/cropsci2014.10.0745.

Manrique-Carpintero NC, Coombs JJ, Pham GM, et al. Genome reduction in tetraploid potato reveals genetic load, haplotype variation, and loci associated with agronomic traits. Front Plant Sci. 2018;9:944. https://doi.org/10.3389/fpls.2018.00944.

Mansfield CL. Space Spuds to the rescue. https://www.nasa.gov/vision/earth/everydaylife/spacespuds.html. 2014. Verified 11/23/21

Marand AP, Jansky SH, Zhao H, et al. Meiotic crossovers are associated with open chromatin and enriched with stowaway transposons in potato. Genome Biol. 2017;18:203. https://doi.org/10.1186/s13059-017-1326-8.

Marfil CF, Camadro EL, Masuelli RW. Phenotypic instability and epigenetic variability in a diploid potato of hybrid origin, Solanum ruiz-lealii. BMC Plant Biol. 2009;9:21. https://doi.org/10.1186/1471-2229-9-21.

Marfil CF, Hidalgo V, Masuelli RW. In situ conservation of wild potato germplasm in Argentina: example and possibilities. Global Ecol Conserv. 2015;3:461–76. https://doi.org/10.1016/j.gecco.2015.01.009.

Mari RS, Schrinner S, Finkers R, et al. Haplotype-resolved assembly of a tetraploid potato genome using long reads and low-depth offspring data, 2022. https://doi.org/2022.05.10.491293.

Massa AN, Childs KL, Lin H, et al. The transcriptome of the reference potato genome Solanum tuberosum group Phureja Clone DM1-3 516R44. PLoS One. 2011;6:e26801. https://doi.org/10.1371/journal.pone.0026801.

Massa AN, Manrique-Carpintero NC, Coombs J, et al. Linkage analysis and QTL mapping in a tetraploid russet mapping population of potato. BMC Genet. 2018;19:87. https://doi.org/10.1186/s12863-018-0672-1.

Matias FI, Alves FC, Meireles KGX, et al. On the accuracy of genomic prediction models considering multi-trait and allele dosage in Urochloa spp. interspecific tetraploid hybrids. Mol Breed. 2019;39:100. https://doi.org/10.1007/s11032-019-1002-7.

McCann LC, Bethke PC, Simon PW. Extensive variation in fried chip color and tuber composition in cold-stored tubers of wild potato (Solanum) germplasm. J Agric Food Chem. 2010;58:2368–76.

McGill CR, Kurilich AC, Davignon J. The role of potatoes and potato components in cardiometabolic health: a review. Ann Med. 2013;45:467–73. https://doi.org/10.3109/07853890.2013.813633.

Mehra M, Gangwar I, Shankar R. A Deluge of complex repeats: the Solanum genome. PLoS One. 2015;10:e0133962. https://doi.org/10.1371/journal.pone.0133962.

Meijer D, Viquez-Zamora M, van Eck HJ, et al. QTL mapping in diploid potato by using selfed progenies of the cross S. tuberosum × S. chacoense. Euphytica. 2018;214:121. https://doi.org/10.1007/s10681-018-2191-6.

Meirmans PG, Liu S, van Tienderen PH. The analysis of polyploid genetic data. J Hered. 2018;109:283–96. https://doi.org/10.1093/jhered/esy006.

Meller B, Kuźnicki D, Arasimowicz-Jelonek M, et al. BABA-primed histone modifications in potato for intergenerational resistance to Phytophthora infestans. Front Plant Sci. 2018;9 https://doi.org/10.3389/fpls.2018.01228.

Meyer RC, Milbourne D, Hackett CA, et al. Linkage analysis in tetraploid potato and association of markers with quantitative resistance to late blight (Phytophthora infestans). Mol Gen Genet. 1998;259:150–60. https://doi.org/10.1007/s004380050800.

Migdadi H, Fayad M, Ajloni M, et al The second report on the State of the World's plant genetic resources for food and agriculture, 2007.

Miller JT, Spooner DM. Collapse of species boundaries in the wild potato Solanum brevicaule complex (Solanaceae, S. sect. Petota): molecular data. Plant Syst Evol. 1999;214:103–30.

Moler E, Abakir A, Eleftheriou M, Johnson JS, Krutovsky KV, Lewis LC, et al. Population epigenomics: advancing understanding of phenotypic plasticity, acclimation, adaptation and diseases. In: Rajora OP, editor. Population genomics: concepts, approaches and applications. Cham: Springer International Publishing AG; 2019. p. 179–260.

Mollinari M, Garcia AAF. Linkage analysis and haplotype phasing in experimental autopolyploid populations with high ploidy level using Hidden Markov Models. G3 Genes|Genomes|Genetics. 2019;9:3297–314. https://doi.org/10.1534/g3.119.400378.

Momose M, Abe Y, Ozeki Y. Miniature inverted-repeat transposable elements of Stowaway are active in potato. Genetics. 2010;186:59–66. https://doi.org/10.1534/genetics.110.117606.

Monnahan P, Brandvain Y. The effect of autopolyploidy on population genetic signals of hard sweeps. Biol Lett. 2020;16:20190796. https://doi.org/10.1098/rsbl.2019.0796.

Moon K-B, Ahn D-J, Park J-S, et al. Transcriptome profiling and characterization of drought-tolerant potato plant (Solanum tuberosum L.). Mol Cells. 2018;41:979–92. https://doi.org/10.14348/molcells.2018.0312.

Mosquera T, Alvarez MF, Jiménez-Gómez JM, et al. Targeted and untargeted approaches unravel novel candidate genes and diagnostic SNPs for quantitative resistance of the potato (Solanum tuberosum L.) to Phytophthora infestans causing the late blight disease. PLoS One. 2016;11:e0156254.

Naess SK, Bradeen JM, Wielgus SM, et al. Resistance to late blight in Solanum bulbocastanum is mapped to chromosome 8. Theor Appl Genet. 2000;101:697–704. https://doi.org/10.1007/s001220051533.

Nakamura S, Hosaka K. DNA methylation in diploid inbred lines of potatoes and its possible role in the regulation of heterosis. Theor Appl Genet. 2010;120:205–14. https://doi.org/10.1007/s00122-009-1058-6.

Natarajan B, Kalsi HS, Godbole P, et al. MiRNA160 is associated with local defense and systemic acquired resistance against Phytophthora infestans infection in potato. J Exp Bot. 2018;69:2023–36. https://doi.org/10.1093/jxb/ery025.

Navarre DA, Brown CR, Sathuvalli VR. Potato vitamins, minerals and phytonutrients from a plant biology perspective. Am J Potato Res. 2019;96:111–26. https://doi.org/10.1007/s12230-018-09703-6.

Novakova A, Šimáčková K, Barta J, Čurn V. Potato variety identification by molecular markers based on retrotransposon analyses. Czech J Genet Plant Breed. 2009;45(1):1–10.

Oberhagemann P, Chatot-Balandras C, Schäfer-Pregl R, et al. A genetic analysis of quantitative resistance to late blight in potato: towards marker-assisted selection. Mol Breed. 1999;5:399–415. https://doi.org/10.1023/A:1009623212180.

Ochoa CM. The potatoes of South America: Bolivia: Cambridge University Press; 1990.

Ochoa CM. Las papas de Sudamérica: Perú: International Potato Center; 1999.

Odilbekov F, Selga C, Ortiz R, et al. QTL mapping for resistance to early blight in a tetraploid potato population. Agronomy. 2020;10:728. https://doi.org/10.3390/agronomy10050728.

Pandey J, Scheuring DC, Koym JW, et al. Genetic diversity and population structure of advanced clones selected over forty years by a potato breeding program in the USA. Sci Rep. 2021;11:1–18.

Pandey J, et al. Genomic regions associated with tuber traits in tetraploid potatoes and identification of superior clones for breeding purposes. Front Plant Sci. 2022;13, Frontiers Media S.A https://doi.org/10.3389/fpls.2022.952263.

Park T-H. Complete chloroplast genome sequence of Solanum hjertingii, one of the wild potato relatives. Mitochondrial DNA Part B. 2022;7:715–7. https://doi.org/10.1080/23802359.2022.2068983.

Patané JSL, Martins J, Setubal JC. Phylogenomics. Methods Mol Biol. 2018;1704:103–87. https://doi.org/10.1007/978-1-4939-7463-4_5.

Peivastegan B, Hadizadeh I, Nykyri J, et al. Effect of wet storage conditions on potato tuber transcriptome, phytohormones and growth. BMC Plant Biol. 2019;19 https://doi.org/10.1186/s12870-019-1875-y.

Petek M, Zagorščak M, Ramšak Ž, et al. Cultivar-specific transcriptome and pan-transcriptome reconstruction of tetraploid potato. Sci Data. 2020;7:249. https://doi.org/10.1038/s41597-020-00581-4.

Pham GM, Hamilton JP, Wood JC, et al. Construction of a chromosome-scale long-read reference genome assembly for potato. GigaScience. 2020;9 https://doi.org/10.1093/gigascience/giaa100.

Pham GM, Newton L, Wiegert-Rininger K, et al. Extensive genome heterogeneity leads to preferential allele expression and copy number-dependent expression in cultivated potato. Plant J. 2017;92:624–37. https://doi.org/10.1111/tpj.13706.

Pikaard CS, Mittelsten Scheid O. Epigenetic regulation in plants. Cold Spring Harb Perspect Biol. 2014;6 https://doi.org/10.1101/cshperspect.a019315.

Plaisted R, Hoopes R. The past record and future prospects for the use of exotic potato germplasm. Am Potato J. 2008; https://doi.org/10.1007/BF02853982.

Plich J, Przetakiewicz J, Śliwka J, et al. Novel gene Sen2 conferring broad-spectrum resistance to Synchytrium endobioticum mapped to potato chromosome XI. Theor Appl Genet. 2018;131:2321–31. https://doi.org/10.1007/s00122-018-3154-y.

Pollan M. The botany of desire: a plant's-eye view of the world. 1st ed: Random House; 2001.

Posada D. Phylogenomics for systematic biology. Syst Biol. 2016;65:353–6. https://doi.org/10.1093/sysbio/syw027.

Potato Genome Sequencing Consortium, Xu X, Pan S, et al. Genome sequence and analysis of the tuber crop potato. Nature. 2011;475:189–95. https://doi.org/10.1038/nature10158.

Pritchard JK, Stephens M, Donnelly P. Inference of population structure using multilocus genotype data. Genetics. 2000;155:945–59.

Prodhomme C, Esselink D, Borm T, et al. Comparative Subsequence Sets Analysis (CoSSA) is a robust approach to identify haplotype specific SNPs; mapping and pedigree analysis of a potato wart disease resistance gene Sen3. Plant Methods. 2019;15:60. https://doi.org/10.1186/s13007-019-0445-5.

Prodhomme C, Vos PG, Paulo MJ, et al. Distribution of P1 (D1) wart disease resistance in potato germplasm and GWAS identification of haplotype-specific SNP markers. Theor Appl Genet. 2020;133:1–13.

Rajora OP, editor. Population genomics: concepts, approaches and application. Cham: Springer Nature; 2019. p. 823. ISBN 978-3-030-04587-6; ISBN 978-3-030-04589-0 (eBook)

Raker CM, Spooner DM. Chilean tetraploid cultivated potato, Solanum tuberosum, is distinct from the Andean populations. Crop Sci. 2002;42:1451–8. https://doi.org/10.2135/cropsci2002.1451.

Reddivari L, Wang T, Wu B, Li S. Potato: an anti-inflammatory food. Am J Potato Res. 2019;96: 164–9. https://doi.org/10.1007/s12230-018-09699-z.

Renault D, Wallender WW. Nutritional water productivity and diets. Agric Water Manag. 2000;45: 275–96. https://doi.org/10.1016/S0378-3774(99)00107-9.

Rickert AM, Kim JH, Meyer S, et al. First-generation SNP/InDel markers tagging loci for pathogen resistance in the potato genome. Plant Biotechnol J. 2003;1:399–410. https://doi.org/10.1046/j.1467-7652.2003.00036.x.

Ríos D, Ghislain M, Rodríguez F, Spooner DM. What is the origin of the European potato? Evidence from Canary Island landraces. Crop Sci. 2007;47:1271–80.

Rocha EA, Paiva LV, de Carvalho HH, Guimarães CT. Molecular characterization and genetic diversity of potato cultivars using SSR and RAPD markers. Crop Breed Appl Biotechnol. 2010;10:204–10. https://doi.org/10.1590/S1984-70332010000300004.

Rodríguez F, Ghislain M, Clausen AM, et al. Hybrid origins of cultivated potatoes. Theor Appl Genet. 2010;121:1187–98.

Rodriguez F, Wu F, Ané C, et al. Do potatoes and tomatoes have a single evolutionary history, and what proportion of the genome supports this history? BMC Evol Biol. 2009;9:1–16.

Rosyara UR, Jong WSD, Douches DS, Endelman JB. Software for genome-wide association studies in Autopolyploids and its application to potato. Plant Genome. 2016;9: plantgenome2015.08.0073. https://doi.org/10.3835/plantgenome2015.08.0073.

Rungis DE, Voronova A, Kokina A, et al. Assessment of genetic diversity and relatedness in the Latvian potato genetic resources collection by DArT genotyping. Plant Genet Resour. 2017;15: 72–8. https://doi.org/10.1017/S1479262115000398.

Salimi H, Bahar M, Mirlohi A, Talebi M. Assessment of the genetic diversity among potato cultivars from different geographical areas using the genomic and EST microsatellites. Iran. J Biotechnol. 2016;14(270–277):10.15171/ijb.1280.

Salman-Minkov A, Sabath N, Mayrose I. Whole-genome duplication as a key factor in crop domestication. Nat Plants. 2016;2:16115. https://doi.org/10.1038/nplants.2016.115.

Salvato F, Havelund JF, Chen M, et al. The potato tuber mitochondrial proteome. Plant Physiol. 2014;164:637–53. https://doi.org/10.1104/pp.113.229054.

Sandbrink JM, Colon LT, Wolters PJCC, Stiekema WJ. Two related genotypes of Solanum microdontum carry different segregating alleles for field resistance to Phytophthora infestans. Mol Breed. 2000;6:215–25. https://doi.org/10.1023/A:1009697318518.

Santa JD, Berdugo-Cely J, Cely-Pardo L, et al. QTL analysis reveals quantitative resistant loci for Phytophthora infestans and Tecia solanivora in tetraploid potato (Solanum tuberosum L.). PLoS One. 2018;13:e0199716. https://doi.org/10.1371/journal.pone.0199716.

Särkinen T, Bohs L, Olmstead RG, Knapp S. A phylogenetic framework for evolutionary study of the nightshades (Solanaceae): a dated 1000-tip tree. BMC Evol Biol. 2013;13:1–15.

Sato S, Tabata S, Hirakawa H, et al. The tomato genome sequence provides insights into fleshy fruit evolution. Nature. 2012;485:635–41. https://doi.org/10.1038/nature11119.

Schmitz Carley CA, Coombs JJ, Douches DS, et al. Automated tetraploid genotype calling by hierarchical clustering. Theor Appl Genet. 2017;130:717–26. https://doi.org/10.1007/s00122-016-2845-5.

Schönhals EM, Ding J, Ritter E, et al. Physical mapping of QTL for tuber yield, starch content and starch yield in tetraploid potato (Solanum tuberosum L.) by means of genome wide genotyping by sequencing and the 8.3 K SolCAP SNP array. BMC Genomics. 2017;18:1–20.

Schönhals EM, Ortega F, Barandalla L, et al. Identification and reproducibility of diagnostic DNA markers for tuber starch and yield optimization in a novel association mapping population of potato (Solanum tuberosum L.). Theor Appl Genet. 2016;129:767–85.

Selga C, Chrominski P, Carlson-Nilsson U, Andersson M, Chawade A, Ortiz R. Diversity and population structure of Nordic potato cultivars and breeding clones. BMC Plant Biol. 2022;22 (1):1–12.

Selga C, Koc A, Chawade A, Ortiz R. A bioinformatics pipeline to identify a subset of SNPs for genomics-assisted potato breeding. Plan Theory. 2021;10:30. https://doi.org/10.3390/plants10010030.

Shannon LM, Boyko RH, Castelhano M, et al. Genetic structure in village dogs reveals a Central Asian domestication origin. Proc Natl Acad Sci. 2015;112:13639–44.

Sharma SK, Bolser D, de Boer J, et al. Construction of reference chromosome-scale pseudomolecules for potato: integrating the potato genome with genetic and physical maps. G3 Genes|Genomes|Genetics. 2013;3:2031–47. https://doi.org/10.1534/g3.113.007153.

Sharma SK, MacKenzie K, McLean K, et al. Linkage disequilibrium and evaluation of genome-wide association mapping models in tetraploid potato. G3 Genes|Genomes|Genetics. 2018;8: 3185–202. https://doi.org/10.1534/g3.118.200377.

Shumbe L, Visse M, Soares E, et al. Differential DNA methylation in the Vinv promoter region controls Cold Induced Sweetening in potato. bioRxiv. 2020:2020.04.26.062562. https://doi.org/10.1101/2020.04.26.062562.

Simon R, Xie CH, Clausen A, et al. Wild and cultivated potato (Solanum sect. Petota) escaped and persistent outside of its natural range. Invas Plant Sci Manag. 2010;3:286–93.

Slater AT, Cogan NOI, Forster JW, et al. Improving genetic gain with genomic selection in autotetraploid potato. Plant Genome. 2016;9:plantgenome2016.02.0021. https://doi.org/10.3835/plantgenome2016.02.0021.

Sliwka J. Genetic factors encoding resistance to late blight caused by Phytophthora infestans (Mont.) de Bary on the potato genetic map. Cell Mol Biol Lett. 2004;9:855–67.

Śliwka J, Jakuczun H, Chmielarz M, et al. A resistance gene against potato late blight originating from Solanum × michoacanum maps to potato chromosome VII. Theor Appl Genet. 2012;124: 397–406. https://doi.org/10.1007/s00122-011-1715-4.

Smyda-Dajmund P, Śliwka J, Villano C, et al. Analysis of cytosine methylation in genomic DNA of Solanum × michoacanum (+) S. tuberosum somatic hybrids. Agronomy. 2021;11:845. https://doi.org/10.3390/agronomy11050845.

Song J, Bradeen JM, Naess SK, et al. Gene RB cloned from Solanum bulbocastanum confers broad spectrum resistance to potato late blight. PNAS. 2003;100:9128–33. https://doi.org/10.1073/pnas.1533501100.

Song X, Zhang C, Li Y, et al. SSR analysis of genetic diversity among 192 diploid potato cultivars. Hortic Plant J. 2016;2:163–71. https://doi.org/10.1016/j.hpj.2016.08.006.

Sood S, Lin Z, Caruana B, et al. Making the most of all data: combining non-genotyped and genotyped potato individuals with HBLUP. The Plant Genome. 2020;13:e20056. https://doi.org/10.1002/tpg2.20056.

Sørensen KK, Kirk HG, Olsson K, et al. A major QTL and an SSR marker associated with glycoalkaloid content in potato tubers from Solanum tuberosum x S. sparsipilum located on chromosome I. Theor Appl Genet. 2008;117:1–9. https://doi.org/10.1007/s00122-008-0745-z.

Spooner D, Jansky S, Clausen A, et al. The enigma of Solanum maglia in the origin of the Chilean cultivated potato, Solanum tuberosum Chilotanum Group 1. Econ Bot. 2012;66:12–21.

Spooner DM. DNA barcoding will frequently fail in complicated groups: an example in wild potatoes. Am J Bot. 2009;96:1177–89.

Spooner DM. Taxonomy of wild potatoes and their relatives in southern South America (Solanum sects. Petota and Etuberosum). The American Society of Plant Taxonomists, 2016. ©2016, [Ann Arbor, Michigan]

Spooner DM, Anderson GJ, Jansen RK. Chloroplast DNA evidence for the interrelationships of tomtoes, potatoes, and Pepinos (Solanaceae). Am J Bot. 1993;80:676–88. https://doi.org/10.2307/2445438.

Spooner DM, Fajardo D, Bryan GJ. Species limits of Solanum berthaultii Hawkes and S. tarijense Hawkes and the implications for species boundaries in Solanum sect. Petota Taxon. 2007b;56: 987–99.

Spooner DM, Gavrilenko T, Jansky SH, et al. Ecogeography of ploidy variation in cultivated potato (Solanum sect. Petota). Am J Bot. 2010;97:2049–60. https://doi.org/10.3732/ajb.1000277.

Spooner DM, Ghislain M, Simon R, et al. Systematics, diversity, genetics, and evolution of wild and cultivated potatoes. Bot Rev. 2014;80:283–383.

Spooner DM, Hijmans RJ. Potato systematics and germplasm collecting, 1989–2000. Am J Potato Res. 2001;78:237–68.

Spooner DM, Hoekstra R, Vilchez B. Solanum SectionPetota in Costa Rica: taxonomy and genetic resources. Am J Pot Res. 2001;78:91–8. https://doi.org/10.1007/BF02874764.

Spooner DM, McLean K, Ramsay G, et al. A single domestication for potato based on multilocus amplified fragment length polymorphism genotyping. PNAS. 2005b;102:14694–9. https://doi.org/10.1073/pnas.0507400102.

Spooner DM, Nunez J, Rodriguez F, et al. Nuclear and chloroplast DNA reassessment of the origin of Indian potato varieties and its implications for the origin of the early European potato. Theor Appl Genet. 2005a;110:1020–6.

Spooner DM, Núñez J, Trujillo G, et al. Extensive simple sequence repeat genotyping of potato landraces supports a major reevaluation of their gene pool structure and classification. Proc Natl Acad Sci. 2007a;104:19398–403.

Spooner DM, Rodríguez F, Polgár Z, et al. Genomic origins of potato Polyploids: GBSSI gene sequencing data. Crop Sci. 2008;48:S-27-S-36. https://doi.org/10.2135/cropsci2007.09.0504tpg.

Spooner DM, Ruess H, Arbizu CI, et al. Greatly reduced phylogenetic structure in the cultivated potato clade (Solanum section Petota pro parte). Am J Bot. 2018;105:60–70. https://doi.org/10.1002/ajb2.1008.

Spooner DM, Salas A. Structure, biosystematics, and genetic resources. In: Handbook of potato production, improvement, and post-harvest management; 2006.

Spooner DM, van den Berg RG. An analysis of recent taxonomic concepts in wild potatoes (Solarium sect. Petota). Genet Resour Crop Evol. 1992;39:23–37. https://doi.org/10.1007/BF00052651.

Staaf J, Vallon-Christersson J, Lindgren D, et al. Normalization of Illumina Infinium whole-genome SNP data improves copy number estimates and allelic intensity ratios. BMC Bioinform. 2008;9:409. https://doi.org/10.1186/1471-2105-9-409.

Stare T, Stare K, Weckwerth W, et al. Comparison between proteome and transcriptome response in potato (Solanum tuberosum L.) leaves following potato virus Y (PVY) infection. Proteomes. 2017;5:10.3390/proteomes5030014.

Stephan W. Selective sweeps. Genetics. 2019;211:5–13. https://doi.org/10.1534/genetics.118.301319.

Stich B, Urbany C, Hoffmann P, Gebhardt C. Population structure and linkage disequilibrium in diploid and tetraploid potato revealed by genome-wide high-density genotyping using the SolCAP SNP array. Plant Breed. 2013;132:718–24.

Stich B, Van Inghelandt D. Prospects and potential uses of genomic prediction of key performance traits in tetraploid potato. Front Plant Sci. 2018;9 https://doi.org/10.3389/fpls.2018.00159.

Strachan SM, Armstrong MR, Kaur A, et al. Mapping the H2 resistance effective against Globodera pallida pathotype Pa1 in tetraploid potato. Theor Appl Genet. 2019;132:1283–94. https://doi.org/10.1007/s00122-019-03278-4.

Suarez-Gonzalez A, Lexer C, Cronk QCB. Adaptive introgression: a plant perspective. Biol Lett. 2018;14:20170688. https://doi.org/10.1098/rsbl.2017.0688.

Sukhotu T, Hosaka K. Origin and evolution of Andigena potatoes revealed by chloroplast and nuclear DNA markers. Genome. 2006;49:636–47.

Sukhotu T, Kamijima O, Hosaka K. Nuclear and chloroplast DNA differentiation in Andean potatoes. Genome. 2004;47:46–56. https://doi.org/10.1139/g03-105.

Sukhotu T, Kamijima O, Hosaka K. Genetic diversity of the Andean tetraploid cultivated potato (Solanum tuberosum L. subsp. andigena Hawkes) evaluated by chloroplast and nuclear DNA markers. Genome. 2005;48:55–64.

Sun H, Jiao WB, Krause K, et al. Chromosome-scale and haplotype-resolved genome assembly of a tetraploid potato cultivar. Nat Genet. 2022;54:342–8. https://doi.org/10.1038/s41588-022-01015-0.

Sun W, Wang Y, Zhang F. Phosphoproteomic analysis of potato tuber reveals a possible correlation between phosphorylation site occupancy and protein attributes. Plant Mol Biol Rep. 2021;39: 163–78. https://doi.org/10.1007/s11105-020-01243-w.

Sverrisdóttir E, Byrne S, Sundmark EHR, et al. Genomic prediction of starch content and chipping quality in tetraploid potato using genotyping-by-sequencing. Theor Appl Genet. 2017;130: 2091–108. https://doi.org/10.1007/s00122-017-2944-y.

Sverrisdóttir E, Sundmark EHR, Johnsen HØ, et al. The value of expanding the training population to improve genomic selection models in tetraploid potato. Front Plant Sci. 2018;9 https://doi.org/10.3389/fpls.2018.01118.

Szajko K, Yin Z, Marczewski W. Accumulation of miRNA and mRNA targets in potato leaves displaying temperature-dependent responses to Potato virus Y. Potato Res. 2019;62:379–92. https://doi.org/10.1007/s11540-019-9417-4.

Tagliotti ME, Deperi SI, Bedogni MC, Huarte MA. Genome-wide association analysis of agronomical and physiological traits linked to drought tolerance in a diverse potatoes (Solanum tuberosum) panel. Plant Breed. 2021;140(4) Wiley:654–64. https://doi.org/10.1111/pbr.12938.

Tang D, Jia Y, Zhang J, et al. Genome evolution and diversity of wild and cultivated potatoes. Nature. 2022;606:535–41. https://doi.org/10.1038/s41586-022-04822-x.

Tang X, Datema E, Guzman MO, et al. Chromosomal organizations of major repeat families on potato (Solanum tuberosum) and further exploring in its sequenced genome. Mol Gen Genomics. 2014;289:1307–19. https://doi.org/10.1007/s00438-014-0891-8.

Tiwari JK, Buckseth T, Zinta R, et al. Transcriptome analysis of potato shoots, roots and stolons under nitrogen stress. Sci Rep. 2020a;10 https://doi.org/10.1038/s41598-020-58167-4.

Tiwari JK, Buckseth T, Zinta R, et al. Genome-wide identification and characterization of microRNAs by small RNA sequencing for low nitrogen stress in potato. PLoS One. 2020b;15:e0233076. https://doi.org/10.1371/journal.pone.0233076.

Tiwari JK, Rawat S, Luthra SK, et al. Genome sequence analysis provides insights on genomic variation and late blight resistance genes in potato somatic hybrid (parents and progeny). Mol Biol Rep. 2021;48:623–35. https://doi.org/10.1007/s11033-020-06106-x.

Tiwari JK, Saurabh S, Chandel P, et al. Analysis of genetic and epigenetic changes in potato somatic hybrids between Solanum tuberosum and S. etuberosum by AFLP and MSAP markers. Agric Res. 2015;4:339–46. https://doi.org/10.1007/s40003-015-0185-3.

Tiwari JK, Singh BP, Gopal J, et al. Molecular characterization of the Indian Andigena potato core collection using microsatellite markers. Afr J Biotechnol. 2013;12 https://doi.org/10.4314/ajb.v12i10.

Torrance L, Cowan GH, McLean K, et al. Natural resistance to Potato virus Y in Solanum tuberosum Group Phureja. Theor Appl Genet. 2020;133:967–80. https://doi.org/10.1007/s00122-019-03521-y.

Totsky IV, Rozanova IV, Safonova AD, et al. Genomic regions of Solanum tuberosum L. associated with the tuber eye depth. Vestn VOGiS. 2020;24:465–73. https://doi.org/10.18699/VJ20.638.

Toubiana D, Cabrera R, Salas E, et al. Morphological and metabolic profiling of a tropical-adapted potato association panel subjected to water recovery treatment reveals new insights into plant vigor. Plant J. 2020;103:2193–210. https://doi.org/10.1111/tpj.14892.

Traini A, Iorizzo M, Mann H, et al. Genome microscale heterogeneity among wild potatoes revealed by diversity arrays technology marker sequences. Int J Genom. 2013;2013: undefined-undefined. https://doi.org/10.1155/2013/257218.

Tüfekçi ED, İnal B. In silico analysis of drought responsive transposons and transcription factors in Solanum tuberosum L. Harran Tarım ve Gıda Bilimleri Dergisi. 2019;23(189–195):10.29050/harranziraat.439682.

Turner-Hissong SD, Mabry ME, Beissinger TM, et al. Evolutionary insights into plant breeding. Curr Opin Plant Biol. 2020;54:93–100. https://doi.org/10.1016/j.pbi.2020.03.003.

Ugent D. The potato. Science. 1970;170:1161–6.

Ugent D, Dillehay T, Ramirez C. Potato remains from a late Pleistocene settlement in south central Chile. Econ Bot. 1987;41:17–27.

Uitdewilligen JGAML, Wolters A-MA, D'hoop BB, et al. A next-generation sequencing method for genotyping-by-sequencing of highly heterozygous autotetraploid potato. PLoS One. 2013;8: e62355. https://doi.org/10.1371/journal.pone.0062355.

van den Berg RG, Bryan GJ, Del Rio A, Spooner DM. Reduction of species in the wild potato Solanum section Petota series Longipedicellata: AFLP, RAPD and chloroplast SSR data. Theor Appl Genet. 2002;105:1109–14. https://doi.org/10.1007/s00122-002-1054-6.

Van Der Vossen E, Sikkema A, te Hekkert BL, et al. An ancient R gene from the wild potato species Solanum bulbocastanum confers broad-spectrum resistance to Phytophthora infestans in cultivated potato and tomato. Plant J. 2003;36:867–82. https://doi.org/10.1046/j.1365-313X.2003.01934.x.

van Lieshout N, van der Burgt A, de Vries ME, et al. Solyntus, the new highly contiguous reference genome for potato (Solanum tuberosum). G3 (Bethesda). 2020;10:3489–95. https://doi.org/10.1534/g3.120.401550.

Vanamala JKP. Potatoes for targeting Colon cancer stem cells. Am J Potato Res. 2019;96:177–82. https://doi.org/10.1007/s12230-018-09700-9.

Vandenberg R, Miller J, Ugarte M, et al. Collapse of morphological species in the wild potato Solanum brevicaule complex (Solanaceae: sect. Petota). Am J Bot. 1998;85:92.

Varré J-S, D'Agostino N, Touzet P, et al. Complete sequence, multichromosomal architecture and transcriptome analysis of the Solanum tuberosum mitochondrial genome. Int J Mol Sci. 2019;20 https://doi.org/10.3390/ijms20194788.

Visker M, Keizer L, Van Eck H, et al. Can the QTL for late blight resistance on potato chromosome 5 be attributed to foliage maturity type? Theor Appl Genet. 2003;106:317–25. https://doi.org/10.1007/s00122-002-1021-2.

Vollmer R, Villagaray R, Egusquiza V, et al. The potato cryobank at the International Potato Center (CIP): a model for long term conservation of clonal plant genetic resources collections of the future. Cryo Letters. 2016;37:318.

Vonholdt BM, Pollinger JP, Lohmueller KE, et al. Genome-wide SNP and haplotype analyses reveal a rich history underlying dog domestication. Nature. 2010;464:898–902. https://doi.org/10.1038/nature08837.

Voorrips RE, Gort G, Vosman B. Genotype calling in tetraploid species from bi-allelic marker data using mixture models. BMC Bioinf. 2011;12:172. https://doi.org/10.1186/1471-2105-12-172.

Vos PG, Paulo MJ, Voorrips RE, et al. Evaluation of LD decay and various LD-decay estimators in simulated and SNP-array data of tetraploid potato. Theor Appl Genet. 2017;130:123–35. https://doi.org/10.1007/s00122-016-2798-8.

Vos PG, Uitdewilligen JGAML, Voorrips RE, et al. Development and analysis of a 20K SNP array for potato (Solanum tuberosum): an insight into the breeding history. Theor Appl Genet. 2015;128:2387–401. https://doi.org/10.1007/s00122-015-2593-y.

Wang F, Xia Z, Zou M, et al. The autotetraploid potato genome provides insights into highly heterozygous species. Plant Biotechnol J. 2022. https://doi.org/10.1111/pbi.13883

Wang L, Beissinger TM, Lorant A, et al. The interplay of demography and selection during maize domestication and expansion. Genome Biol. 2017;18:215. https://doi.org/10.1186/s13059-017-1346-4.

Wang L, Xie J, Hu J, et al. Comparative epigenomics reveals evolution of duplicated genes in potato and tomato. Plant J. 2018;93:460–71. https://doi.org/10.1111/tpj.13790.

Wang M-S, Zhang J-J, Guo X, et al. Large-scale genomic analysis reveals the genetic cost of chicken domestication. BMC Biol. 2021;19:118. https://doi.org/10.1186/s12915-021-01052-x.

Wang Y, Diehl A, Wu F, et al. Sequencing and comparative analysis of a conserved syntenic segment in the Solanaceae. Genetics. 2008;180:391–408.

Wang Y, Rashid MAR, Li X, et al. Collection and evaluation of genetic diversity and population structure of potato landraces and varieties in China. Front Plant Sci. 2019;0 https://doi.org/10.3389/fpls.2019.00139.

Weir A. The Martian. Reprint ed: Ballantine Books; 2014.

Wenke T, Seibt KM, Döbel T, et al. Inter-SINE Amplified Polymorphism (ISAP) for rapid and robust plant genotyping. Methods Mol Biol. 2015;1245:183–92. https://doi.org/10.1007/978-1-4939-1966-6_14.

Witek K, Jupe F, Witek AI, et al. Accelerated cloning of a potato late blight–resistance gene using RenSeq and SMRT sequencing. Nat Biotechnol. 2016;34:656–60. https://doi.org/10.1038/nbt.3540.

Witek K, Lin X, Karki HS, et al. A complex resistance locus in Solanum americanum recognizes a conserved Phytophthora effector. Nat Plants. 2021;7:198–208. https://doi.org/10.1038/s41477-021-00854-9.

Xie F, Frazier TP, Zhang B. Identification, characterization and expression analysis of MicroRNAs and their targets in the potato (Solanum tuberosum). Gene. 2011;473:8–22. https://doi.org/10.1016/j.gene.2010.09.007.

Xin C, Hou R, Wu F, et al. Analysis of cytosine methylation status in potato by methylation-sensitive amplified polymorphisms under low-temperature stress. J Plant Biol. 2015;58:383–90. https://doi.org/10.1007/s12374-015-0316-1.

Yan L, Zhang Y, Cai G, et al. Genome assembly of primitive cultivated potato Solanum stenotomum provides insights into potato evolution. G3 Genes|Genomes|Genetics. 2021;11: 10.1093/g3journal/jkab262.

Yang J, Zhang N, Ma C, et al. Prediction and verification of microRNAs related to proline accumulation under drought stress in potato. Comput Biol Chem. 2013;46:48–54. https://doi.org/10.1016/j.compbiolchem.2013.04.006.

Yang J, Zhang N, Zhou X, et al. Identification of four novel stu-miR169s and their target genes in Solanum tuberosum and expression profiles response to drought stress. Plant Syst Evol. 2016;302:55–66. https://doi.org/10.1007/s00606-015-1242-x.

Yang L, Mu X, Liu C, et al. Overexpression of potato miR482e enhanced plant sensitivity to Verticillium dahliae infection. J Integr Plant Biol. 2015;57:1078–88. https://doi.org/10.1111/jipb.12348.

Yang W, Liu X, Zhang J, et al. Prediction and validation of conservative microRNAs of Solanum tuberosum L. Mol Biol Rep. 2010;37:3081–7. https://doi.org/10.1007/s11033-009-9881-z.

Yang X, Chen L, Yang Y, et al. Transcriptome analysis reveals that exogenous ethylene activates immune and defense responses in a high late blight resistant potato genotype. Sci Rep. 2020;10 https://doi.org/10.1038/s41598-020-78027-5.

Yang X, Liu J, Xu J, et al. Transcriptome profiling reveals effects of drought stress on gene expression in diploid potato genotype P3-198. Int J Mol Sci. 2019b;20 https://doi.org/10.3390/ijms20040852.

Yang X, Sood S, Luo Z, et al. Genome-wide association studies identified resistance loci to Orange rust and yellow leaf virus diseases in sugarcane (Saccharum spp.). Phytopathology. 2019a;109: 623–31. https://doi.org/10.1094/PHYTO-08-18-0282-R.

Yoshida K, Schuenemann VJ, Cano LM, et al. The rise and fall of the Phytophthora infestans lineage that triggered the Irish potato famine. elife. 2013;2:e00731.

Yousaf MF, Demirel U, Naeem M, Çalışkan ME. Association mapping reveals novel genomic regions controlling some root and stolon traits in tetraploid potato (Solanum tuberosum L.). 3 Biotech. 2021;11:174. https://doi.org/10.1007/s13205-021-02727-6.

Yu X, Zhang M, Yu Z, et al. An SNP-based high-density genetic linkage map for tetraploid potato using specific length amplified fragment sequencing (SLAF-Seq) technology. Agronomy. 2020;10:114. https://doi.org/10.3390/agronomy10010114.

Yuan J, Bizimungu B, De Koeyer D, et al. Genome-Wide Association study of resistance to potato common scab. Potato Res. 2020;63:253–66. https://doi.org/10.1007/s11540-019-09437-w.

Zavallo D, Crescente J, Gantuz M, et al. Genomic re-assessment of the transposable element landscape of the potato genome. Plant Cell Rep. 2020;39 https://doi.org/10.1007/s00299-020-02554-8.

Zeng Z, Zhang W, Marand AP, et al. Cold stress induces enhanced chromatin accessibility and bivalent histone modifications H3K4me3 and H3K27me3 of active genes in potato. Genome Biol. 2019;20:123. https://doi.org/10.1186/s13059-019-1731-2.

Zhang C, Wang P, Tang D, et al. The genetic basis of inbreeding depression in potato. Nat Genet. 2019;51:374–8. https://doi.org/10.1038/s41588-018-0319-1.

Zhang C, Yang Z, Tang D, et al. Genome design of hybrid potato. Cell. 2021a;184:3873–3883.e12. https://doi.org/10.1016/j.cell.2021.06.006.

Zhang F, et al. Resequencing and genome-wide association studies of autotetraploid potato. Mol Hortic. 2022;2(1), BMC:1–18. https://doi.org/10.1186/s43897-022-00027-y.

Zhang G, Tang R, Niu S, Si H, Yang Q, Rajora OP, et al. Heat-stress-induced sprouting and differential gene expression in growing potato tubers: comparative transcriptomics with that induced by postharvest sprouting. Hortic Res. 2021b;8:226. https://doi.org/10.1038/s41438-021-00680-2.

Zhang J, Wang Y, Zhao Y, et al. Transcriptome analysis reveals Nitrogen deficiency induced alterations in leaf and root of three cultivars of potato (Solanum tuberosum L.). PLoS One. 2020;15:10.1371/journal.pone.0240662.

Zhang L, Yao L, Zhang N, et al. Lateral root development in potato is mediated by Stu-mi164 regulation of NAC transcription factor. Front Plant Sci. 2018;9:383. https://doi.org/10.3389/fpls.2018.00383.

Zhang R, Marshall D, Bryan GJ, Hornyik C. Identification and characterization of miRNA transcriptome in potato by high-throughput sequencing. PLoS One. 2013;8 https://doi.org/10.1371/journal.pone.0057233.

Zhang W, Luo Y, Gong X, et al. Computational identification of 48 potato microRNAs and their targets. Comput Biol Chem. 2009b;33:84–93. https://doi.org/10.1016/j.compbiolchem.2008.07.006.

Zhang Y, Cheng S, De Jong D, et al. The potato R locus codes for dihydroflavonol 4-reductase. Theor Appl Genet. 2009a;119:931–7. https://doi.org/10.1007/s00122-009-1100-8.

Zhang Y, Jung CS, De Jong WS. Genetic analysis of pigmented tuber flesh in potato. Theor Appl Genet. 2009c;119:143–50. https://doi.org/10.1007/s00122-009-1024-3.

Zhang Z, Ersoz E, Lai C-Q, et al. Mixed linear model approach adapted for genome-wide association studies. Nat Genet. 2010;42:355–60.

Zheng C, Amadeu RR, Munoz PR, Endelman JB. Haplotype reconstruction in connected tetraploid F1 populations. Genetics. 2021;219(2):iyab106. https://doi.org/10.1101/2020.12.18.423519.

Zhou Q, Tang D, Huang W, et al. Haplotype-resolved genome analyses of a heterozygous diploid potato. Nat Genet. 2020;52:1018–23. https://doi.org/10.1038/s41588-020-0699-x.

Zia MAB, Demirel U, Nadeem MA, Çaliskan ME. Genome-wide association study identifies various loci underlying agronomic and morphological traits in diversified potato panel. Physiol Mol Biol Plants. 2020;26:1003–20. https://doi.org/10.1007/s12298-020-00785-3.

Zollinger T. Epigenetic regulation of Gene Associated Transposable Elements in potato: effect of DNA methylation in the promoter region of Wound induced 1. Thesis Wageningen University. 1940588; 2010.

Zuluaga AP, Solé M, Lu H, et al. Transcriptome responses to Ralstonia solanacearum infection in the roots of the wild potato Solanum commersonii. BMC Genomics. 2015;16 https://doi.org/10.1186/s12864-015-1460-1.

Zych K, Gort G, Maliepaard CA, et al. FitTetra 2.0 – improved genotype calling for tetraploids with multiple population and parental data support. BMC Bioinform. 2019;20:148. https://doi.org/10.1186/s12859-019-2703-y.

Population Genomics of Tomato

Christopher Sauvage, Stéphanie Arnoux, and Mathilde Causse

Abstract Tomato (*Solanum lycopersicum* L.) is an acknowledged model species for research in genetics and genomics, on fruit development and disease resistance, but it also deserves to be a model species for population genomics studies due to the availability of large genetic and genomic resources. In breeding, tomato breeding and genetic improvement largely depends on introgression of beneficial alleles from wild relative species. Since the first release of its high-quality genome sequence in 2012, the genomes of several hundreds of cultivated accessions and chosen wild relatives have been re-sequenced, allowing the discovery of millions single nucleotide polymorphisms (SNP). The study of these genomes confirmed the new phylogenetic organization of the tomato clade, *Solanum* section *Lycopersicon*, composed of 13 species. Recent population and ecological genomics approaches, notably using RNAseq approach, provided new results on speciation and interspecific reproductive barriers. The molecular mechanisms of adaptation to abiotic stress in cultivated and wild tomato species were also analysed and their roles underlined as factors of speciation and diversification. The diversity of ecological conditions of the wild relative species allowed the study of evolutionary and molecular mechanisms of adaptation to abiotic stress in crop and wild tomatoes.

Population genomics studies provided key insights into the two steps of tomato domestication, the intensity of bottlenecks resulting from domestication and further modern breeding, and selection footprints and large genomic regions introgressed from the wild relative species. At the transcriptome level, it was also shown that domestication and modern breeding rewired global patterns of genome expression, notably for stress-related genes. Finally, the availability of genome sequences and SNP markers allowed studying large collections of varieties, developing genome-wide association studies (GWAS) and advancing our knowledge about the genome structure and dynamic (linkage disequilibrium decay, distribution of recombination), but also allowing genes and QTL involved in many traits to be mapped and using the information for breeding new varieties. In this chapter we describe the tomato

C. Sauvage · S. Arnoux · M. Causse (✉)
INRAE, GAFL, Avignon, France
e-mail: mathilde.causse@inrae.fr

history, its domestication and the diversity and phylogeny of its wild relative species. We then present the genomic resources and new insights into the evolution and diversity of tomato accessions due to the impact of domestication and breeding. Finally, some major prospects are discussed.

Keywords Domestication · Ecological adaptation and speciation · Genetic architecture of agronomic traits · Genetic diversity and demography · Genome and transcriptome · Pan-genome · Phylogeny · Population genomics · *Solanum lycopersicum* · Tomato

1 Introduction

Tomato (*Solanum lycopersicum* L.) is the first vegetable grown over the world. It accounts for more than 15% of the world vegetable production (over 177 million metric tons in 2016; Food and Agriculture Organisation [faostat 2016]). Half of the world production is produced in four countries (China 56 MT, India 18 MT, USA 13 MT and Turkey 12 MT). Tomato is a rich source of micronutrients in human diet and is grown for two main usages: processing and fresh market. The major goals of tomato breeding (high productivity, tolerance to biotic and abiotic stresses and high sensory and health value of the fruit) require a good knowledge and management of tomato genetic resources and diversity. Tomato is also an acknowledged model species for research in genetics, fruit development and disease resistance. It has a short life cycle, is easy to cross and self-pollinate in its cultivated form and has a genome size of ~900 Mb and many genetic and genomic resources. Furthermore, the tomato scientific community has access to several databases archiving most of these important data.

Tomato and its 12 closely related species belong to the genus *Solanum* in the large Solanaceae family. All of the species come from the Andean region of South America (Rick 1988). Explorations of tomato centre of origin permitted major advances in the characterization of its genetic and phenotypic diversity. In parallel, ex situ conservation of genetic resources in large national collections ensured the conservation of landraces and wild species. Thus, the genetic potential of tomato's wild relatives for breeding purpose emerged. In parallel, the ecological and taxonomic diversity of tomato turned it into a model species for evolutionary studies. Since the mid-twentieth century, mastering controlled hybridization allowed crosses between wild and cultivated tomato to be performed. Modern genetics and breeding methods contributed to understand the genetic control of agronomical traits but also accentuated the progress and the development of thousands of new cultivars. The value of crop wild relatives in breeding and development of new cultivars was recognized.

The advent of molecular genetic markers in the 1980s raised great hopes for characterization of the genetic diversity in both wild and cultivated crop plants. Great expectations also emerged since the development of molecular techniques to pinpoint genomic regions involved in targeted traits. Dissection of the genetic control of complex traits, using ad hoc techniques from quantitative and population genetics, was possible, leading to the identification of key alleles involved in many traits, originating from several wild relatives. Presently, the tomato genome is nearing high quality, and the genomes of many wild and cultivated accessions have been re-sequenced (Lin et al. 2014; Aflitos et al. 2014; Bolger et al. 2014; Gao et al. 2019) using high-throughput sequencing (HTS) techniques. Large datasets describing the genome expression (transcriptome, proteome and metabolome) are also available providing an overview of the (post-)transcriptional landscape (http://solgenomics.net). Quantitative trait locus (QTL) mapping techniques or genome-wide association studies (GWAS) also facilitated the understanding of the genetic architecture of complex traits and germplasm management of both wild and cultivated tomatoes.

In this chapter we first describe the tomato history, its domestication and the diversity and phylogeny of its wild relative species. We then present the genomic resources available for the tomato clade and how they have provided new insights into the evolution and diversity of tomato accessions. We then focus on the impact of domestication and breeding, before showing how crop wild relatives were used to introgress beneficial alleles and traits and identify important loci for the crop. Finally, some major prospects are discussed.

2 Tomato History, from Past to Modern Era: A Model Plant for Vegetables

Tomato (*Solanum lycopersicum* L.) and its 12 wild relative species originated from the Andean region of South America (de Candolle 1886; Jenkins 1948; Rick and Fobes 1975; Spooner et al. 2005; Peralta et al. 2008; Zuriaga et al. 2009). Its common name "tomato" originated from the Nahuatl (Aztec language) word "tomatl". The origin of domestication was debated over the years, but recent studies untangled this mystery. Briefly, it was first domesticated from the wild species *S. pimpinellifolium* by ancestors of the Inca human population in Ecuadorian and Peruvian regions. The beginning of trade between human populations from South America and Mesoamerica later introduced few individuals in the Mexican region leading to a founder's effect (Blanca et al. 2012). A second strong bottleneck occurred with the Spanish colonization of the American continent when Mesoamerican tomato seeds were brought to Europe. Tomato started to be consumed in Europe as food during the seventeenth and eighteenth centuries, and in 1869 Henry John Heinz founded the first tomato-linked company (Ray 1673; Labate et al. 2007).

Since then, the tomato spread worldwide, and in the early 1920s, genetic improvement of tomato began to obtain the first disease-tolerant cultivars from hybridization with wild progenitors. The first cultivars resistant to *Cladosporium fulvum* and *Fusarium oxysporum* were developed in the 1930s and 1940s with the discovery of resistance genes in the closely related wild tomato species (Langford 1937; Stevens and Rick 1986). Thereafter, tomato genetic improvement has largely been dependent on introgressions of beneficial alleles from wild germplasm (Atherton and Harris 1986), which increased the interest for the knowledge and conservation of crop's wild relative species. After the pioneer Nikolai Vavilov (Kurlovich et al. 2000), the main protagonist in the development of a seed bank for cultivated and wild tomato was Charles M. Rick, who dedicated his life in field trips to South America and established the Tomato Genetics Resource Center (Rick 1990, https://tgrc.ucdavis.edu/). His effort to prospect large number of crop and wild tomato samples and make his collection available to the scientific community is probably a main reason of the position of tomato as a "model species". The other reason to deepen the research on tomato is its economic importance as one of the leading vegetable crops worldwide. As a reference over the past 30 years (1984–2014), the global yield of tomato has increased from 83 to 170 million tons, and the area harvested has increased from three to five million hectares (United Nations Food and Agriculture Organization (FAO) statistics; Food and Agricultural Organization of the United Nations [faostat 2016]). The scientific and agricultural community considerably improved the tomato varieties and growth conditions over the last 50 years, notably for its yield, stress tolerance, fruit properties and pathogen resistances (Bauchet and Causse 2012).

3 Genomic, Transcriptomic and Other Resources for Population Genomics Research

3.1 The Reference Genome of Tomato and Databases

In the early 2000s, the Tomato Genome Consortium was set up from an international consortium of scientists from 14 countries, pooling their funds to sequence the first tomato (*Solanum lycopersicum*) genome (among other Solanaceae species) and provide a publicly available resource. Following the first objective to sequence the 220 Mb of tomato euchromatin, predicted to contain the majority of genes (Mueller et al. 2005), the next-generation sequencing methodologies offered the opportunity to produce a mostly complete high-quality reference genome that finally covered 742 Mb (i.e. 83% of the 900 Mb genome, Sato et al. 2012). This work is part of a larger initiative called the "International Solanaceae Genome Project (SOL): Systems Approach to Diversity and Adaptation". The SOL community aims to help in understanding the genetic basis of plant diversity by offering a big clade-oriented database within the SOL Genomic Network website (SGN, http://solgenomics.net/) that collects and stores all the genome sequences and phenotypic and genomic data available for the Solanaceae and related species (Mueller 2005; Menda et al. 2008; Bombarely et al. 2011). This database, available to all researchers, also implemented supplementary tools, such as solQTL or SGN VIGS (Virus-Induced Gene Silencing)

(Tecle et al. 2010; Fernandez-Pozo et al. 2015). With the same intent to create a tomato gene expression database, the Tomato Expression Atlas (Fernandez-Pozo et al. 2015, 2017), the Tomato EFP Browser, TomPLEX (Winter et al. 2007) and TomExpress (Zouine et al. 2017) are now allowing browsing the transcriptional landscape of each annotated gene from different plant tissues, genotypes and conditions, also displaying the results with graphical outputs.

Following the release of the first reference genome sequence of cultivated tomato (cultivar Heinz1706), the genome sequence and its annotation have been regularly updated from an initial version to the current one, the third (SL4.0, by September 2019), which integrated new whole genome shotgun, full-length BAC and optical sequencing and reduced the number of contig gaps and benefited from the combination of the latest technologies such as long single molecule sequencing, Hi-C proximity ligation and optical maps. This reflects the continuous efforts to offer a high-quality tomato genome to reach the gold standard that is available since many years now in the model plant *Arabidopsis thaliana*.

3.2 Genome and Transcriptome Sequencing of Crop Wild Relative Species

Crop wild relative species are particularly useful in population genomics notably to identify SNP markers and determine the derived/ancestral state to unfold site frequency spectrum, track introgression events for adaptive traits or help phylogeny to be rooted to understand the evolution of traits along (Farris 1982). The advent of the second- and third-generation sequencing technologies (i.e. Hiseq Illumina, long-read technologies, respectively) allowed reaching these objectives by first providing the complete genome sequences of several wild relative species of the cultivated tomato. Indeed, these technologies are more adapted for the outcrossing wild relative species to manage properly their higher level of heterozygosity compared to the cultivated tomato which is self-pollinated and highly homozygous. Among these crop wild relative species, the first genome of *Solanum pennellii* was sequenced using Illumina technology (LA0716 accession, Bolger et al. 2014) and was updated using a de novo assembly based on the nanopore technology (LYC 1722 accession, Schmidt et al. 2017). The main objective was notably to foster our knowledge of traits related to stress tolerance and the evolutionary role of transposable elements on these traits, as *S. pennellii*, endemic to Andean regions in South America, has evolved to thrive in arid habitats. The genome completeness obtained from de novo approach applied to the novel self-compatible (SC) *S. pennellii* accession LYC 1722, compared to the reference sequence of the *S. pennellii* accession LA0716, illustrated the gain obtained from the use of long-read sequencing (i.e. Oxford Nanopore). Schmidt et al. (2017) were able to achieve assemblies for which the N50 contig length was 2.45 Mb (i.e. half of the assembly was in contigs of 2.45 Mb or longer) and the complete genome sequence was assembled in only

899 contigs. While being error prone, the estimated error rate, when using polishing software, was similar to the Illumina technology, down to 0.025%. A complete reference genome was also released for the wild relative species *Solanum lycopersicoides*. Using the PacBio sequencing, with a coverage of 90x, the N50 and genome coverage were estimated to 139 kb and 89.7% (see https://solgenomics.net/organism/Solanum_lycopersicoides/genome). It should be noted that additional genomes are available for *S. pimpinellifolium* (LA1589 accession) and *S. galapagense* (LA0436 accession) but the assembled sequences are highly fragmented, limiting their use (ftp://ftp.solgenomics.net/genomes/Solanum_galapagense/).

In addition, re-sequencing efforts have been conducted to complement large-scale genomic panel studies for tomato mainly dedicated to population structure and GWAS (Aflitos et al. 2014; Lin et al. 2014) or to investigate species barriers (Labate et al. 2014), providing data in crop wild relatives species. Besides complete and re-sequenced genomes, transcriptomic data produced at the genome-wide scale from crop wild relatives have been released in the last years. This approach is relatively powerful to reduce the complexity of the analysis by reducing the genome representation and cope with the higher level of polymorphism in these species. We can briefly mention RNAseq data from Pease et al. (2016) that produced reads across four subgroups (Esculentum, Arcanum, Peruvianum and Hirsutum), from Sauvage et al. (2017) in *S. pimpinellifolium* and from Florez-Rueda et al. (2016) and Bauchet et al. (2017a, b) in *S. peruvianum* and *S. chilense*. The main scientific results obtained from these data are detailed in the next sections of this chapter.

3.3 Community Tomato Genome and Transcriptome Data Bases

The past decade has been really fruitful in producing genomic data such as genome sequences, transcriptomes and metabolomes. The type and quality of data may vary according to their generation technology (e.g. Hiseq vs. PacBio or RNAseq vs. genome sequencing), and, therefore, it might be difficult to compare them within the same analysis. The real challenge is thus to develop databases that are user-friendly and help scientist in handling the amount of data available. The tomato community with the creation of databases like Sol Genomics Network (SGN – https://solgenomics.net/), TOMATOMICS (Kudo et al. 2017), the Tomato Expression Atlas (Fernandez-Pozo et al. 2017) and TomExpress (Zouine et al. 2017) has managed to acquire, collect and share most of the data available. It remains essential for researchers to make the best use of accumulated biological knowledge on tomato. In this context, the SGN database initiated a collection of QTL analyses, but the discrepancy of alignment made it nearly unusable. Using methods developed in human in 2008 (Allen et al. 2008; Zeggini et al. 2008) and later applied in *A. thaliana* (Grimm et al. 2012), GWAS results were aggregated onto a cross-species platform to replicate results and share data. In tomato, many GWA studies have been conducted, especially on traits related to fruit quality, offering the opportunity to understand the genetic architecture of this trait through a GWA meta-analysis and consequently discover new candidate loci and reducing the proportion of uncovered heritability (Zhao et al. 2019).

4 Progress in Population Genomics of Tomato and Key Insights Provided

4.1 Phylogeny, Evolution and Taxonomy

The first botanist to study domesticated tomato was de Tournefort (1694), who recognized its close relationship with the genus *Solanum* but named the tomato genus *Lycopersicon* ("Wolf peach" in Greek). For a better nomenclature, Linnaeus (1753) intended to use consistently Latin binomials. He placed tomato in the *Solanum* genus and named the domesticated tomato as *S. lycopersicum* and its wild relative *S. peruvianum*. The Gardener's and Botanist's Dictionary (Miller and Miller 1768) started using the Linnaeus' binomial system but kept the *Lycopersicon* genus, and it is only in the 1807s edition that the tomato was classified in the *Solanum* genus. After these feeble taxonomic beginnings, most of the taxonomists and gardeners kept the *Lycopersicon esculentum* name until the 1980s when the first phylogenetic studies started confirming the *Solanum* affiliations (Rick and Tanksley 1981; Spooner et al. 1993). The *Linnaeus* nomenclature has gained wide acceptance, but *Lycopersicon* might remain present in the common language. The first phylogenetic studies brought a new growing interest in deciphering the crop and wild tomato evolutionary trees. The 12 wild relative species also followed several nomenclature changes. One recent nomenclature with the ecological characteristics of the species was presented in Bauchet and Causse (2012) that compiled data from Peralta et al. (2008), Moyle (2008) and Grandillo et al. (2011).

Over the last 30 years, numerous studies performed marker-assisted analyses using different types of molecular markers to uncover the phylogenetic organization of the genus *Solanum*. The first genetic marker study led by Palmer focused on chloroplast DNA in 1982 (Palmer and Zamir 1982) and managed to separate the Peruvianum group from the Esculentum group and revealed *S. lycopersicoides* and *S. juglandifolium* as outgroup species. Following this example, a few studies improved the genus phylogeny using chloroplast DNA (Bohs and Olmstead 1997; Olmstead and Palmer 1997; Olmstead et al. 1999), mtDNA (McClean and Hanson 1986), nuclear RFLPs (Miller and Tanksley 1990) and AFLPs (Spooner et al. 2005; Zuriaga et al. 2009). These studies could already untangle the phylogeny of most of the genus *Solanum*, separate and order the current species groups in the *Solanum* section *Lycopersicon* (namely *Hirsutum*, *Peruvianum*, *Arcanum* and *Esculentum*). The sequence data of internal transcribed spacer region of rDNA (Marshall et al.

2001), the Granule-Bound Starch Synthase (GBSSI) genes (Peralta and Spooner 2001), and the two nuclear genes from Zuriega and colleagues (Zuriaga et al. 2009) brought confidence into the main species classification and confirmed the tomato species phylogeny within the genus *Solanum*. Using 14 expressed sequence tags (ESTs), Roselius et al. (2005) completed the marker analyses on wild tomato accessions by estimating population genetic parameters, such as nucleotide polymorphism or recombination rate. The pioneer work of Peralta, Spooner and Knapp refined the taxonomy in the genera, notably by the combined use of morphologic data and molecular marker genotyping (Peralta and Spooner 2000; Spooner et al. 2005; and see Peralta et al. 2008 for the taxonomic monograph). From the many studies they conducted, the topology demonstrated the monophyletic origin of the clade *Solanum*, section *Lycopersicon*, composed of 13 species.

The reference genome availability unlocked high-throughput genomics studies focusing on the wild species speciation event and on the whole tomato genus phylogeny. For the sake of genus phylogeny clarification, the whole transcriptomes of 12 wild tomato species revealed evidences of diversification fuelled by at least 3 sources of adaptive genetic variation being "post-speciation hybridization, rapid accumulation of new mutations, and recruitment from ancestral variation" (Fig. 1;

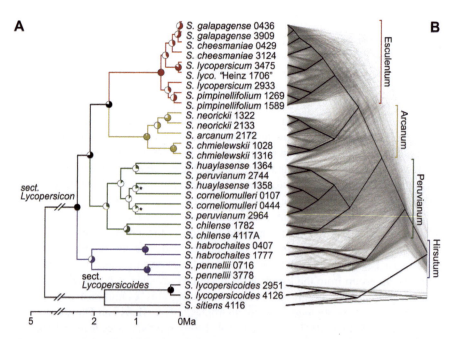

Fig. 1 The phylogeny of *Solanum* sect. *Lycopersicon*. (**a**) A whole-transcriptome concatenated molecular clock phylogeny with section Lycopersicoides as the outgroup. Branch colours indicate the four major subgroups (labels on right). Pie charts on each node indicate majority rule extended bipartition support scores (out of 100) using trees from 100-kb genomic windows. All nodes are supported by 100 bootstrap replicates, except "*" which denotes bootstrap support score of 68. (**b**) A "cladogram" of 2,745 trees (grey) inferred from nonoverlapping 100-kb genomic windows. For contrast, the consensus phylogeny is shown in black (from Pease et al. 2016)

Pease et al. 2016). Multi-locus sequences of two wild species (*S. peruvianum* and *S. chilense*) were implemented in coalescent-based models to infer the evolutionary processes of speciation (Stadler 2008). Two additional wild species *S. arcanum and S. habrochaites* added power to Bayesian methods to decipher their speciation (Tellier et al. 2011; Böndel et al. 2015). The population genomics approaches using 14,043 SNPs on 46 samples of *S. peruvianum* untangled the species complex into 4 separate species, *S. peruvianum* sensu stricto, *S. corneliomuelleri*, *S. huaylasense* and *S. arcanum* (Labate et al. 2014), clarifying the organization of the clade. However, the real number of species of the wild species of tomato remains debated according to the criteria being used.

The ease of hybridization between closely related species within the tomato clade is the foundation of improvement of domesticated cultivars. Such hybridization is not quite current between *S. peruvianum* and *S. chilense* that are two distinct tomato species (Rick and Fobes 1975). The controlled hybridization with crop cultivars is a great opportunity to improve the domesticated tomato varieties, and the wild species will remain the source of adaptative potential to introgress alleles from for tomato in the future.

4.2 Pan-genomics, Genetic Diversity and Demography

4.2.1 Pan-genome

Large-scale genomic characterization of genetic diversity in plants is already going on, especially with the re-sequencing of large sets of accessions. In 2018, over 500 genome sequences were publicly available in tomato, and future projects aim to sequence up to thousands of accessions. For example, these data allowed the identification of domestication footprints and track hybridization events (Lin et al. 2014). Mining and leveraging the sequence data in such large-scale projects require a pan-genomic approach. A pan-genome structure that describes the full complement of genes in a single species has multiple advantages over a single, linear reference genome sequence for population genomics and plant breeding applications. The approach was applied in crop and wild accessions of rice (*Oryza sativa*). Identifying conserved and variable regions allowed to pinpoint new causal variants that underlie complex evolutionary traits (Zhao et al. 2018). In tomato, a pan-genome that includes its wild relative species would provide a single coordinate system to anchor known nucleotide variation (SNP, InDels and CNV, for examples) with phenotypic data. The tomato genome reference was obtained from the Heinz1706 accession that experienced breeding during its history, leading to the fixation or depletion of nucleotide variation. Thus, using a single reference genome is limiting the identification of novel genes from the available germplasm that are not present in this reference genome, especially genes of agronomical interest. Rare CNV were already detected in the tomato genome demonstrating that structural variation exists in this species (Causse et al. 2013). In this context, it makes sense to re-think the idea of a "reference" genome. The pan-genome is also an opportunity to track chromosomal rearrangements between genotypes that may have occurred over micro (i.e. domestication) and macro (i.e. species divergence) timescales. While being computationally challenging, methodological approaches are available to construct, use and visualize pan-genome (The Computational Pan-Genomics Consortium 2016). While seeking for core and dispensable genome fractions to catalogue the gene repertoire of the genus, Gao et al. (2019) obtained a tomato pan-genome constructed using genome sequences of 725 phylogenetically and geographically representative accessions. They discovered 4,873 genes absent from the reference genome. Presence/absence variation analyses revealed substantial gene loss and intense negative selection of genes and promoters during tomato domestication and improvement. Lost or negatively selected genes were particularly enriched for disease resistance genes. The tomato pan-genome completes the reference genome and will be useful for future biological discovery and breeding.

4.2.2 Demographic History and Ecological Niche

The genomes of contemporary crops contain considerable information about their history. Although the general contour of tomato history has been defined with the increasing amount of available data (both SNP genotyping and sequencing) and sampling sizes, its resolution remains partly elusive. Statistical inference methods, adopted from human genomics and based on coalescent theory, have been developed to leverage information contained in these genome-wide data sets and have proven their power to refine parameters of the species history. More precisely, from observed footprints in DNA sequence variation, these methods aim at reconstructing the evolutionary history and providing precise estimates of selective and demographic events (i.e. population effective size growth or decline) that the species of interest experienced. Population genetic summary statistics (i.e. Watterson's Theta, Tajima's D) provide such data to test for demographic events. Numerous methods and models have been developed for demographic inferences (see Schraiber and Akey (2015) for a review) with the most popular ones being the principal component analysis (PCA), Structure (Falush et al. 2003) and Treemix software (Pickrell and Pritchard 2012) that are very powerful towards identifying population structure and mixture. In tomato, these methods have been largely applied and are the basis for further explorations of more complex demographic models that describe events like population divergence, migration and changes in demographic sizes. Towards this objective, methods based on site frequency spectrum (SFS) modelling have been applied in both the cultivated tomato (Lin et al. 2014) and its CWR species to unravel timings of population divergence, for example (Beddows et al. 2017), or investigate the complex domestication history of this crop plant (Razifard et al. 2020). Furthermore, until now, despite the large amount of genomic data, no haplotype-based method has been used to precisely measure coalescence between haplotype in a population to infer changes in its effective size, for example. The sequentially Markov coalescent (SMC) method and its extensions (PSMC (Li and Durbin

2011) and MSMC (Schiffels and Durbin 2014)), operating on full genome sequences, would be a precious tool to precisely decipher this species evolutionary history.

In parallel, past climate change may have contributed significantly to population dynamics and shaped patterns of nucleotide variation. Ecological niche modelling (ENM) building from current bioclimate variables are projected to paleoclimates to predict the variation in geographical population distribution over large timescales. In tomato, the role of geography and ecology in species divergence has been investigated using a combination of climatic, geographic and biological data from nine wild Andean tomato species to describe each species' ecological niche and to evaluate the likely ecological and geographical modes of speciation in this clade (Nakazato et al. 2008, 2010). Both studies demonstrated that the nine studied species experienced an ecological adaptation that drove genetic and phenotypic divergence in association with one or more environmental variables, leading to specific ecological niches following a recent divergence. All of these features could provide a major source of biotic and abiotic stress-responsive genes and genetic mechanisms of adaptation to climate change. Those genetic resources can directly sustain breeding efforts for elite germplasm that would grow under stressful or changing conditions without being detrimental to traits of economic interest, such as yield or fruit quality.

4.3 Ecological, Speciation and Landscape Genomics of the Cultivated Tomato and Its Wild Relatives

Ecological genomics aims at understanding the origin, history and function of the observed natural biological variation, from nucleotide to community levels (Seehausen et al. 2014). In this context, the approach relies on ecological and genomic resources and provides an opportunity to precisely dissect genetic and developmental mechanisms, and to connect a genetic polymorphism to a phenotypic variation, as well as to directly demonstrate the ecological and evolutionary relevance of this phenotypic variation. Many of these studies have been performed in the wild tomato clade (*Solanum* section *Lycopersicon*), a genetic group that has both exceptional diversity and genomic tools (see Haak et al. 2014 for a complete review). Within this section, we will focus on two major processes that are speciation and adaptation and report how much population genomics has addressed these questions in the *Solanum* genus.

4.3.1 Speciation Mechanism and Reproduction Barriers

In tomato, the timing of speciation and the underlying molecular mechanisms of wild species divergence remained unresolved. Other nebulous scientific questions are still not resolved in this complex of species. The transition from self-incompatible (SI) to self-compatible (SC) reproduction system was partly induced by the domestication process, but the main molecular consequences remain elusive. However, *S. habrochaites* is a wild partially self-compatible species revealing that the transition was independent. For example, genes involved in self-incompatibility are poorly characterized at the molecular levels (nucleotide diversity, gene expression levels).

Strong reproductive barriers have been established between some of the species of the genus. Charles Rick's extensive work tested for these barriers by crossing all the tomato wild relative species together during the 1970 and 1980s (Rick 1988; Rick and Chetelat 1995). As several speciation mechanisms seem to be at the origin of wild tomato diversification, there is a current debate on their respective roles/preponderance. On the one side, there are traits responsible for prezygotic isolation (conferring ecological differentiation) that are suspected to be the most important isolation barriers and most efficient in preventing gene flows between the species (Kirkpatrick and Ravigné 2002; Ramsey et al. 2003). On the other side, postzygotic barriers leading to hybrid unviability and sterility, which are more likely permanent and irreversible barriers to gene flow between species (Muller 1942; Coyne and Orr 2004). Moyle (2007) conducted a QTL mapping experiment to decipher the contribution of the pre- and postzygotic isolation barriers between *S. lycopersicum* and *S. habrochaites* using a set of near-isogenic lines. They compared floral morphology between species and investigated sterility traits in hybrid crosses. However, the outcomes of this study remain limited as genome-wide associations were not evident: these traits showed a complex genetic architecture and association with centromeric regions, which warrant further fine-scale investigation, due to limited recombination. More recently, the role of the interspecific reproductive barriers (IRBs) in limiting sympatric hybridization between closely related species was evaluated at three stages, prezygotic (floral morphology), post-mating prezygotic (pollen tube growth) and postzygotic barriers (fruit and seed development), and were measured in situ in Peru by Baek et al. (2016). This study, based on 11 interspecific crosses, demonstrated multiple IRBs with 3 types of post-mating prezygotic IRBs and strong postzygotic IRBs that prevented normal seed development by resulting from aborted endosperm and overgrown endothelium. However, hybridization was possible in some cases, notably from the pair *S. pennellii* and *S. corneliomulleri* with nearly fully developed seeds that produced viable F_1 hybrids. In this latter case, molecular markers confirmed hybridity, which underlies the significant role that genomics tools can play for the study of this process. Thus, current studies on speciation mechanisms in wild tomatoes confirm the intricate role of pre- and postzygotic isolation and suggest that several scenarios underlie the speciation between two sister species.

From then on, population genomics revealed its potential to elucidate the reproductive barriers among species using RNAseq. Following Rick's investigations (Rick 1988), extensive work focusing on postzygotic barriers has also been conducted in the species pair *S. peruvianum* × *S. chilense*. These two species are closely related with partly overlapping geographic ranges in northern Chile and southwestern Peru but are morphologically dissimilar. Roth (2017) demonstrated that crosses between these two species led to high proportions of non-viable seeds due to endosperm failure and arrested embryo development. On the basis of seed size differences in reciprocal hybrid crosses and developmental evidence implicating endosperm failure, they hypothesized that perturbed parental effects (e.g. genomic imprinting or parent-specific allelic expression) were involved in the strong postzygotic barrier. Florez-Rueda et al. (2016) conducted a transcriptomic screen in developing endosperms within intra- and interspecific crosses and estimated the influence of the parent of origin on expression profiles using both homozygous and heterozygous nucleotide differences between parental individuals to identify candidate imprinted genes. As a result, they uncovered systematic shifts of "normal" (intraspecific) maternal/paternal transcript proportions in hybrid endosperms. They showed different behaviours among species. For instance, in *S. peruvianum* hybrid triploid endosperm, the genome-wide increase in maternal proportion almost entirely eliminated paternally expressed imprinted genes. Thus, they demonstrated that changes in parental expression proportions may be the underlying core process at play, leading to transcriptional regulation of the processes compromising the hybrid endosperm development and contributing to hybrid seed failure. However, on the other hand, they cannot reject that the transcriptional rewiring of the imprinted genes was the main source of perturbation of the essential developmental genes. Following this initial study, Roth (2017) extended this work with two additional species pairs and supported the common role of the genomic imprinting between nuclear and cellular endosperm types but also provided evidence for the genome-wide rewiring of gene expression and parental dosage in wild tomato hybrid endosperm as a major postzygotic barrier. These results are very interesting to reinvestigate the Endosperm Balance Number (EBN) hypothesis developed in the early 1980s in *Capsella* (Lagriffol and Monnier 1985). This hypothesis was proposed to explain the basis of normal seed development after intra- and interspecific crosses, through a 2:1 maternal-to-paternal ratio in the hybrid endosperm. Up to now, it was mostly not possible to properly test for how much EBN may act as a powerful isolating mechanism (Carputo et al. 1999).

The release of recombinant inbred lines (RILs), linkage maps and the genome sequence of the domesticated tomato (*Solanum lycopersicum*) were valuable tools for the genetic analysis of IRBs. It provided the basis for QTL detection, read mapping and gene annotation and shed light on the underlying mechanisms involved in reproductive barriers in the tomato genus. However, transgenic methodologies are new tools that are providing opportunities to test the candidate loci involved in these barriers, while the complementation of proteomics and transcriptomics offers insights into the molecular regulation of gene expression to provide a more clear picture of the interspecific reproductive barriers present in wild tomato relatives through the identification of new candidate genes or proteins (Bedinger et al. 2011). Finally, Li and Chetelat (2010, 2015) deciphered the unilateral interspecific incompatibility (UI) system and identified two genes that block cross-hybridization between related species, typically when the pollen donor is self-compatible and the pistil parent is self-incompatible (SI): *ui1.1*, a pollen UI factor in tomato, which encodes an S-locus F-box protein and *ui6.1*, which encodes a Cullin1 protein that functions in both UI and SI.

4.3.2 Ecological and Climate Adaptation

Darwin proposed that phenotypic differentiation among populations resulted from differential adaptation in response to environmental heterogeneity (Kawecki and Ebert 2004). This mechanism is relatively frequent and has been proven experimentally by connecting physiological, genetic and ecological data to measure the fitness over evolutionary timescales. Two main factors have been proposed to influence ecological adaptation, abiotic and biotic stresses, which may also interact together. The advent of high-throughput genomics allowed refining our knowledge of the

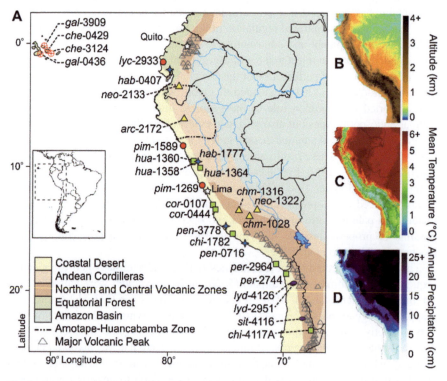

Fig. 2 Geographic distribution and ecological diversity of sampled populations of wild tomato. (**a**) Wild tomato species inhabit diverse ecological zones (shaded regions) along the western coast of South America and the Galápagos Islands. For each sample location, labels indicate species and accession number, and symbols denote major phylogenetic groupings (circle = Esculentum, triangle = Arcanum, square = Peruvianum, star = Hirsutum, oval = outgroup; base map modified from the original obtained from http://www.freevectormaps.com). High variation of (**b**) altitude, (**c**) mean annual temperature and (**d**) annual precipitation across the habitat range of wild tomato species (data from http://www.worldclim.org; plotted using GRASS GIS http://grass.osgeo.org/) (from Pease et al. 2016)

adaptation mechanisms. The tomato clade, along with the two allied species in the outgroup *Solanum* sect. Lycopersicoides, was extensively used towards this objective notably due to its large geographic range (Fig. 2). These contrasting environments are characterized by different stress conditions such as drought, salt, cold and heat. Hereafter, we describe a limited number of applications of population genomics approaches to document the molecular mechanisms of adaptation to abiotic stress in crop and wild tomatoes. For a complete review, including adaptation to biotic stress, see Haak et al. (2014).

Among wild tomato species, it has been demonstrated that the greatest axes of differentiation between species are average annual rainfall and temperature (Nakazato et al. 2010). QTL mapping experiments reported that both *S. chilense* and *S. pennellii* developed distinct strategies to adapt to drought stress. In addition, the comparison with the domesticated tomato identified QTL associated with eco-physiological trait variation and identified a polygenic and complex genetic architecture of drought response based on both main effect and transgressive QTL (Muir and Moyle 2009). Contrasted nucleotide diversity patterns of local adaptation at drought-related candidate genes in wild tomatoes (*S. peruvianum* and *S. chilense*), identified at two major loci in the abscisic acid signalling pathway, were observed (Xia et al. 2010). On the one side, *LeNCED1* exhibited very low nucleotide diversity relative to the eight neutral reference loci that were surveyed in populations of these two species. This suggested that strong purifying selection has been acting on this gene. On the other side, *pLC30-15* exhibited higher levels of nucleotide diversity. Additionally, for these two loci, particularly in *S. chilense*, higher genetic differentiation (F_{ST}) between populations than for the reference loci indicated signatures of selection in response to local adaptation at these loci. In the more drought-tolerant species *S. chilense*, one population (from Quicacha) showed a significant haplotype structure, which appeared to be the result of positive (diversifying) selection (Xia et al. 2010).

Local adaptation is crucial when a species colonizes new habitats. The tomato wild relative species *S. chilense* is an example of native range expansion in southern America from North to South. It provides a strong experimental framework to test for differential hypothesis underlying the mechanisms of local adaptation through colonization. Böndel et al. (2015) tested whether local adaptation occurred more frequently in large ancestral populations or in small derived populations using a population genomics approach. They conducted a population genetic analysis and inferred the past demography of *S. chilense* populations on pooled-sequencing data from 30 genes (8,080 SNPs). Across Chile and Peru, 23 *S. chilense* populations were sampled according to the north to south colonization. Along this cline, a decrease of genetic variation was associated with a relaxed purifying selection and an increasing proportion of non-synonymous polymorphisms from the study of the distribution of fitness effect and by population substructure with at least four genetic groups. In other words, the north to south cline is associated with an increase in deleterious mutations, potentially conferring a decreased adaptive potential to southern populations. Patterns of population structure, natural selection and linkage disequilibrium within these *S. chilense* populations confirmed previously inferred population-specific demographic histories (Arunyawat et al. 2007).

Similarly, spatial genetic analyses revealed a clinal patterns of nucleotide diversity from North to South in other wild tomato species such as the wild relatives *S. peruvianum* and *S. pimpinellifolium* and the cultivated *S. lycopersicum* (Nakazato and Housworth 2011; Nakazato et al. 2012) and in the related Solanaceae species *S. lycopersicoides* and *S. sitiens* (Albrecht et al. 2010), which occur in sympatry with *S. chilense* in northern Chile (Peralta et al. 2008). The combination of ecological and genomic data provided evidence and putative parameters for conservation of both *S. chilense* and *S. peruvianum* (Tellier et al. 2011). In this study, the inferred difference in germination rate between these two species reflected divergent strategy of adaptation for seed dormancy that agreed with previous population genetic analyses and the ecology of these two-sister species. Overall, the "seeds" strategy relied on spending on average, a shorter time in the soil in the specialist species (*S. chilense*) than in the generalist species (*S. peruvianum*) which stayed in dormancy for a longer period.

Using whole transcriptomes from the 13 species of the *Solanum* clade, Pease et al. (2016) not only identified the ecological and genetic factors that promoted the species radiations and inferred the species phylogeny (see Sect. 4.1) but also found evidence for at least three sources of adaptive genetic variation that fuel species radiations. First, they detected introgression events between the early-branching lineages and more recently between individual populations. This supported the hypothesis of adaptive benefits through hybridization. Second, they provided evidence for lineage-specific de novo evolution for loci involved in the production of red fruit colour. Third, they detected environment-specific sorting of ancestral variation among populations that come from different species that shared common environmental conditions. Overall, these results indicated that multiple genetic sources could promote a rapid diversification and endow the speciation mechanism in response to ecological adaptation. Last but not least, this study highlighted the complexity of both ancient and recent species radiations, using a combination of ecological and genomic data.

4.3.3 Adaptation to Climate Change

Food security may be threatened by a combination of events, such as increasing human population and needs, climate change and by the lack of sustainable development. Evolutionary adaptation has been proposed as a tool to understand how some species, such as the tomato, overcome environmental changes by the understanding of local adaptation mechanisms (Mousavi-Derazmahalleh et al. 2018). These changes act as selective pressures and are driven by climate change. However, the success of evolutionary adaptation depends on various factors, one of which being the extent of genetic variation available within the crop species. Many QTL studies have involved crop wild relatives, but just a few wild accessions were used (<10 *S. pimpinellifolium* and *S. habrochaites* and one or two of the other species, as reviewed by Grandillo and Cammareri (2016)). Thus, a large natural diversity, including important alleles for the crop, remains to be discovered and used to improve tomato adaptation. The population genomics approaches provide a unique opportunity to identify genetic variation that can be employed for breeding programmes.

5 Genomic Footprints of Domestication and Modern Breeding Stages

5.1 Deciphering the Domestication and Breeding History

Comparative genomics has proven to be a valuable tool to decipher evolutionary mechanisms and forces that occurred over macro and micro timescales. Comparing nucleotide patterns is the basic idea behind this approach to highlight loci constrained by evolutionary forces. Both domestication and modern breeding stages (also called "improvement") are crucial phases for studying adaptation, genome evolution, and the genetics and evolution of complex traits. Domestication is characterized by the accumulation of non-synonymous variants, the so-called genetic cost of domestication, as proposed by Lu et al. (2006) and reviewed by Moyers et al. (2018), and by selective sweeps (stretch of homozygosity due to specific selection). These processes were evident in the comparisons of crop and wild accessions in many crop species such as soybean (*Glycine max*, Lam et al. 2010), maize (*Zea mays*, Hufford et al. 2013) or rice (*Oryza sativa*, Xu et al. 2012).

Comparative expression profiling extended the approach of comparative genomics in a few crops, such as maize (Swanson-Wagner et al. 2012; Lemmon et al. 2014), cotton (Rapp et al. 2010) or common bean, botanical name (Bellucci et al. 2014). In tomato, the consequences of the domestication syndrome have been deeply studied for phenotypic traits, such as growth habit (plant vigour and flowering time) and fruit traits (set, size, shape, colour and morphology), and many major genes and QTLs underlying these traits have been identified during the last decades (Grandillo and Tanksley 1996; Doganlar et al. 2000; Tanksley 2004; Bai and Lindhout 2007; Chakrabarti et al. 2013). The use of "omics" (i.e. HTS) also shed light into the genomic footprints of domestication and modern breeding in tomato.

Using population genomics approaches, tomato domestication was clarified, notably by delineating the position of *S.l. cerasiforme* and its role in this process (Blanca et al. 2015). To do so, a very large collection of >1,000 accessions was screened using the SOLCAP SNP array (>8,000 SNPs). Tomato domestication seems to have followed a two-step process: a first domestication in South America and a second step in Mesoamerica (Blanca et al. 2015). The distribution of fruit weight and shape alleles supported that domestication of *S. cerasiforme* occurred in the Andean region and clarified the biological status of this genetic group as a true phylogenetic group within tomato.

By re-sequencing the genomes of 166 accessions (SP, SLC and SLL) from South America and Mesoamerica, Razifard et al. (2020) used population genomic inference methods to reconstruct tomato domestication history, focusing on the evolutionary changes that occurred in the intermediate stages of this process. They showed that the origin of SLC may predate domestication and that many traits considered typical of cultivated tomatoes arose in South American SLC accession, but were lost or reduced once these partially domesticated forms spread northwards. These traits were then likely reselected in a convergent fashion in the common cultivated tomato, prior to its expansion around the world.

5.2 Effects on Nucleotide Diversity Patterns

The strong human selection induced by domestication and later by crop improvement left footprints on the plant genome that can be tracked through the genome-wide study of nucleotide diversity with summary statistics such as π and Tajima's D. From these summary statistics, selective sweeps or genetic bottlenecks can be inferred. In tomato, the genome-wide reduction in nucleotide diversity has been one of the most obvious genetic marks of such bottlenecks during the domestication of *S. lycopersicum* from its closest wild relative species *S. pimpinellifolium*. Miller and Tanksley (1990) reported that the amount of genetic variation in the SI species (i.e. *S. peruvianum*) far exceeded (−95%) that found in SC species (*S. lycopersicum*) from the analysis of RFLP markers. More recently, this genetic diversity loss has been supported with revised estimates by many studies (Sato et al. 2012; Koenig et al. 2013; Lin et al. 2014; Blanca et al. 2015; Sauvage et al. 2017; Sahu and Chattopadhyay 2017, Razifard et al. 2020). We observed variable but drastic reduction of the total nucleotide diversity $\left(\frac{\pi \text{CROP}}{\pi \text{WILD}} = 0.37\right)$ reported in Lin et al. (2014) from the comparison between *S. lycopersicum* and *S. pimpinellifolium* at the genome-wide scale and $\frac{\pi \text{CROP}}{\pi \text{WILD}} = 0.65$ reported in Sauvage et al. (2017) from the comparison between *S. lycopersicum* and *S. pimpinellifolium* at the transcriptome-wide scale. However, this drastic reduction has to be cautiously interpreted because these average estimates across the genome may not reflect nucleotide diversity reduction in specific genomic regions.

The extensive use of wild germplasm for breeding purposes was a common practice during the genetic improvement stage in tomato. This had an impact on genome structure/architecture as shown by the extensive work conducted by Labate et al. (2009). When examining genome-wide patterns of nucleotide diversity, small chromosomal regions show non-randomly distributed regions of higher nucleotide diversity in cultivated compared to wild accessions. In *S. lycopersicum*, these regions showed increased allele sharing with *S. pimpinellifolium*, indicating recent introgressions from this species or a closely related other species. Koenig et al. (2013) identified 550 candidate introgressed genes in the reference genome of Heinz1706 and 2,479 in the cultivated accession M82. The large number of candidate loci introgressed in M82 highlights the challenge of linkage drag during breeding using wild accessions and may contribute to reduce genome-wide divergence in nucleotide sequence between cultivated and wild accessions. Similar observations were reported in Blanca et al. (2015) when comparing contemporary *S. lycopersicum* to vintage accessions and in Sauvage et al. (2017) when comparing *S. lycopersicum* to *S. pimpinellifolium*, especially on chromosome 9. Additionally, evidence of a strong genetic bottleneck and relaxation of purifying selection was reported by these authors. Estimates of d_N/d_S in *S. lycopersicum* supported the accumulation of potentially deleterious mutations during its cultivation (Koenig et al. 2013). In contrast, Sahu and Chattopadhyay (2017) identified a continuous and strong purifying selection in the cultivated tomato which may be required to maintain some favoured agronomic traits. In their study, nearly about 1% (8.76 Mb) of the tomato genome (distributed across seven chromosomes) showed very strong purifying selection with Tajima's D estimates lower than -3.0. Breeding may have also contributed to fix haplotypes and reduce nucleotide diversity by favouring one allele of interest (i.e. hard selective sweep). A total of 186 domestication sweeps $\left(\frac{\pi\ S.cerasiforme}{\pi\ S.pimpinellifolium}\right)$ and 133 improvement sweeps $\left(\frac{\pi\ S.cerasiforme}{\pi\ S.lycopersicum}\right)$ covering nearly 8.3% (64.6 Mb) and 7.0% (54.5 Mb) of the species genome were identified, witnessing the frequency of allele fixation during the history of tomato breeding (Lin et al. 2014). Overall, both domestication and improvement sweeps overlapped with known QTL, notably related to fruit weight, a major trait affected during these two stages of the tomato history (i.e. loci *fw2.2*, *fw3.2*, etc.).

5.3 Domestication and Modern Breeding Induced a Transcriptome Rewiring

The comparative genomics approach was extended by using gene expression levels to decipher the genome-wide transcriptional changes induced during the domestication and improvement stages of the tomato history. Expression and co-expression patterns were investigated and showed that specialized as well as general pathways have been affected during both stages. Itkin et al. (2013) showed how tomato turned from "nasty to tasty". More precisely, metabolic pathways and genes directing the synthesis of some anti-nutritional compounds (i.e. steroidal glycoalkaloids (SGAs)) in potato and tomato were elucidated by these authors. Comparative co-expression analyses between tomato and potato coupled with chemical profiling revealed ten genes partaking in SGA biosynthesis. Six of them form a cluster on chromosome 7, whereas an additional two are adjacent in a duplicated genomic region on chromosome 12. Silencing *GLYCOALKALOID METABOLISM 4* pathway prevented accumulation of SGAs in tomato fruit and in potato tubers. This demonstrated that domestication downregulated entire specialized metabolic pathways, locking the production of anti-nutritional compounds.

Patterns of differential expression and co-expression between cultivated and wild tomato species showed major transcriptional changes in genes related to stress response, defence response, photosynthesis, response to high light and redox pathways (Koenig et al. 2013). These molecular functions partly overlapped with genes related to response to stress, the generation of precursor metabolites and energy, metabolic processes, the epigenetic regulation of gene expression and carbohydrate metabolism additionally identified by Sauvage et al. (2017). Enrichment for these categories indicated that abiotic and biotic stresses have played a major role driving transcriptional variation along tomato history. In addition, the comparison of genomic and transcriptomic patterns (Sauvage et al. 2017) showed that both synonymous and non-synonymous polymorphism rates tended to be higher in the wild group than in the cultivated group. This trend was significantly more pronounced for differentially expressed genes (DEG) between crop and wild tomato accessions, than for the non-differentially expressed ones, indicating that purifying selection was significantly weaker in DEG compared with non-DEG. Altogether, this suggests that purifying selection tends to be stronger among DEGs in the wild group.

6 Population Genomics for Modern Breeding

There are two strong interests in studying the crop wild relative species such as wild tomatoes: (1) use the wild relative species to better understand processes and modification triggered by domestication of crop plants (Abbo 2014) and (2) identify and introgress wild relative genes of interest to introduce new genetic diversity lost following the strong diversity bottlenecks and, thus, increasing the crop fitness (Ohmori et al. 1995, 1998). Since the pioneer work of Steve Tanksley's research group, molecular markers were used to construct a high-density genetic map of the tomato genome (Tanksley et al. 1992) and dissect quantitative traits into simple Mendelian factors or QTL (quantitative trait loci) (Paterson et al. 1988, 1991; Tanksley et al. 1992). This also allowed to positionally clone the genetic factors underlying major mutations or quantitative traits (Frary et al. 2000). The low polymorphism detected by RFLP and PCR markers compelled geneticists to study interspecific segregating populations, which were more polymorphic. This also underlined the interest of the wild relative species as a source of new genetic diversity. With the availability of SNP markers, it became possible to study large collections of varieties; develop GWAS and advance our knowledge about the genome structure, such as linkage disequilibrium decay, the structure of haplotypes and the distribution of recombination; and identify early introgressions from wild species (Sim et al. 2012; Lin et al. 2014; Sauvage et al. 2014; Zhao et al. 2019).

6.1 Introgressions from Crop Wild Relative Species Improved the Crop Tomato

The crucial role of crop wild relatives has been identified for many crops (Vincent et al. 2013; Brozynska et al. 2015), but it is particularly pronounced for tomato breeding. This was already suggested by the pioneer work of Charles Rick (1990) who showed the existence of several disease resistance traits in wild tomato species. More than 200 pathogens infect the crop tomato (Bai and Lindhout 2007), and Heirloom varieties are usually susceptible to all of them. Thus, wild relatives were first screened for disease resistance, and many monogenic dominant genes were discovered, first for *Cladosporium fulvum* and *Fusarium oxysporum*. They were subsequently introgressed into cultivars, and nowadays modern hybrids carry up to eight disease resistance genes against fungi (*Cladosporium, Fusarium, Verticillium*), viruses (tobacco mosaic virus, tomato spotted wilt virus, bacteria bacterial wilt) and nematodes. The introgression required the identification of molecular markers linked to these genes, and the genome locations of many of them are now known (Causse and Grandillo 2016). Following the mapping effort, tomato was used as a model species to clone these genes and decipher their structure and their molecular organization (Martin et al. 1993). Wild germplasm has played a crucial role in the modern breeding of cultivated tomato (Stevens and Rick 1986), triggering interest for wild tomatoes species and for the evolution of the group as a whole (Labate et al. 2007).

During the sequencing of the tomato reference genome, the introgression of several chromosomal segments related to *S. pimpinellifolium* was shown (Sato et al. 2012). These introgressions, probably due to the first introgressions of disease resistance genes, were detectable on several chromosomes, suggesting several rounds of introgression.

The large size of introgressions from wild relative species was first shown by Young and Tanksley (1989). This was confirmed at the genome-scale by Lin et al. (2014) who detected, in a set of modern F_1 hybrids, a large exotic fragment on chromosome 9 (more than 50 Mb in length) carrying the tobacco mosaic virus resistance gene $Tm\text{-}2^a$ derived from *S. peruvianum*. In addition, they detected two other major introgressions on chromosome 6: one (>25 Mb in length) carrying the root knot nematode resistance gene *Mi-1* introgressed from *S. peruvianum* and the other (>30 Mb in length) carrying the tomato yellow leaf curl virus resistance gene *Ty-1* from *S. chilense*. Even after many generations of backcrossing, these introgressed fragments remain intact, possibly due to chromosomal rearrangements or a centromeric location that would inhibit recombination, as in the case of *Ty-1* and *Mi-1* (Seah et al. 2004; Verlaan et al. 2013).

6.2 Dissecting the Genetic Architecture of Agronomical Traits

Quantitative trait mapping revealed the potential importance of crop wild relatives even for un-targeted traits. Due to the low genetic diversity within the cultivated compartment (Miller and Tanksley 1990), most of the first mapping populations were based on interspecific crosses between a cultivar and a related wild accession from the section *Lycopersicon* (as reviewed by Foolad (2007); Labate et al. (2007); Grandillo et al. (2011)) or from the *Lycopersicoides* (Pertuzé et al. 2002) and *Juglandifolia* sections (Albrecht et al. 2010). However, intraspecific crosses, notably with cherry tomatoes, have proved their interest notably on fruit quality aspects (Saliba-Colombani et al. 2001). All those populations allowed discovering and/or characterizing a myriad of major genes and QTLs involved in various traits (recent synthesis in Grandillo and Cammareri 2016).

Introgression lines (ILs) derived from interspecific crosses allowed dissecting the effect of unique chromosome fragments from a donor (usually a wild relative species) introgressed into a recurrent elite line. ILs were used for fine mapping and positional cloning of several genes and QTL of interest (*Lin5* for sugar content, *fw2.2* for fruit weight) as reviewed in Rothan et al. (2019). The first IL library was developed from the introgression of an accession of *S. pennellii* (LA716) into M82, a cultivated accession (Eshed and Zamir 1995; Zamir 2001). This progeny was used to identify hundreds of QTLs for fruit traits (Causse et al. 2004), anti-oxidants (Rousseaux et al. 2005), vitamin C (Stevens et al. 2007), fruit metabolomics composition (Alseekh et al. 2013) and volatile aromas (Tadmor et al. 2002). QTL mapping power was increased in IL compared to classical biallelic QTL mapping population. It was further improved by the constitution of a sub-IL set with smaller introgressed fragments (Ofner et al. 2016). Such libraries combining small fragments of wild genomes into the cultivated one were then designed with several species, involving *S. pimpinellifolium* (Doganlar et al. 2002), *S. habrochaites* (Monforte and Tanksley 2000; Finkers et al. 2007) and *S. lycopersicoides* (Canady et al. 2005). ILs were also used to dissect the genetic basis of heterosis (Eshed and Zamir 1995). Heterosis refers to a phenomenon where hybrids between distant varieties or crosses between related species exhibit greater biomass, speed of development and fertility than both parents (Birchler et al. 2010). Heterosis involves genome-wide dominance complementation and inheritance model such as locus-specific overdominance (Lippman et al. 2007). The potential of related wild species even for improving unexpected traits was shown, for instance, some QTL alleles increasing the red colour of the fruit were discovered in *S. pennellii*, a green-fruited species (Causse et al. 2004). Interesting alleles at QTL for fruit volatiles were also detected in several interspecific progenies (Klee 2010).

6.3 Molecular Bases of Trait Diversification

Tomato domestication and later diversification of fruit types led to a large morphological diversity in tomato fruit (with small to large, round, blocky, elongated, pear-shaped fruits, with colour ranging from red to green, white, black, pink, orange or yellow). On the contrary, wild tomato species carry small, round red, orange or green fruits, with a limited intraspecific phenotypic diversity. Using molecular markers, the genetic control of fruit traits has been widely dissected (Grandillo et al. 1999; Lippman and Tanksley 2001; Barrero and Tanksley 2004; Alseekh et al. 2013). The first QTL controlling fruit weight variation, *fw2.2*, was cloned (Frary et al. 2000) followed by several mutations/QTL involved in fruit shape: *LC* and *FAS* which increase locule number (*LC*) and fruit size (*FAS*) (Cong et al. 2008; Muños et al. 2011), OVATE which gives ovoid fruit shape (Liu et al. 2002) and *SUN* which gives an elongated fruit shape or the oxheart shape when associated to *LC* and *FAS* (van der Knaap et al. 2002; Xiao et al. 2008). It was then shown that the combination of alleles at these four genes were responsible for most of the diversity of fruit shape present in the cultivated germplasm (Rodriguez et al. 2011). The allelic distribution of the four genes was then associated with morphological, geographical and historical data in a collection of diverse cultivated accessions, and a model for fruit shape evolution in tomato was established suggesting that selection occurred in distinct chronologic and historic periods: *LC* arose first, followed by *OVATE*, both in *S.l. cerasiforme* background and in distinct populations. *FAS* arose later in a *LC* background. The presence of these three mutations in Latin American germplasm suggested pre-Columbian mutations. Combined with *fw2.2*, they must have strongly contributed to the increase in fruit size during tomato domestication. On the contrary, SUN mutation is not carried by any Latin American material tested, suggesting that SUN mutation appeared post domestication in European material (probably in Italy). This study also showed that the selection for fruit shape is strongly responsible for the underlying genetic structure in tomato cultivars (Rodriguez et al. 2011).

6.4 Breeding Shaped the Genetic Structure of Modern Cultivars

As previously stated above, selection during domestication and subsequent breeding considerably reduced the genetic diversity of the cultivated tomato. To further improve the cultivated tomato, segments of wild tomato genomes were introgressed into modern cultivars (Rick 1960). To better understand how modern breeding had changed the tomato genome, Sim et al. (2011) studied the population structure of 70 tomato lines (with 173 markers) and found clusters that separated the cultivated tomato into processing, fresh market, vintage and landrace varieties. A similar study detected a longer linkage disequilibrium decay in processing tomatoes (7–14 cM) and in fresh market tomato (3–16 cM) than in vintage cultivars (6–8 cM), which

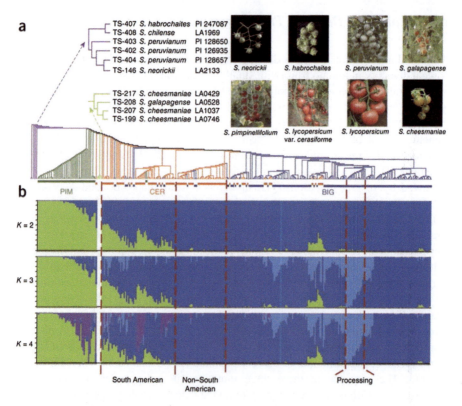

Fig. 3 Genome-wide relationship and fruit morphology in cultivated tomato and its wild relatives. (**a**) The neighbour-joining tree of the population (331 accessions from the red-fruited clade and 10 wild accessions) was generated using 20,111 SNPs at fourfold-degenerate sites. The bars indicate the PIM (green), CER (orange) and BIG (blue) lines. The two branches containing wild accessions are enlarged for visualization. Typical fruits of the species studied are shown. (**b**) Model-based clustering analysis with different numbers of clusters ($K = 2$, 3 and 4). The y-axis quantifies cluster membership, and the x-axis lists the different accessions. The orders and positions of these accessions on the x-axis are consistent with those for the neighbour-joining tree. South American CER, non-South American CER and processing tomato clusters are separated by dashed red lines. (Reproduced from Lin et al. 2014)

revealed the strong selection cost that modern breeding induced in processing and fresh market tomato varieties (Robbins et al. 2011).

More recently, Lin et al. (2014), sequenced genomes from 360 tomato accessions and varieties: 333 representing the diversity of types and varieties from the red-fruited clade (*S. pimpinellifolium*, *S. lycopersicum* var. *cerasiforme* and *S. lycopersicum*) including 166 big-fruited *S. lycopersicum*. They detected two main groups in *S. lycopersicum*: the first including accessions of *S. lycopersicum* with big fruits paired with the non-South American *S. lycopersicum* var. *cerasiforme* and the second composed of the *S. lycopersicum* var. *cerasiforme* that were originated from South America. With a higher resolution ($K = 4$), they could as well detect the processing tomato cluster. They focused on PIM (*S. pimpinellifolium*), CER (*S. lycopersicum* var. *cerasiforme*) and BIG (*S. lycopersicum*) clusters and observed domestication-induced diversity decrease by measuring the number of sites that were polymorphic in each group, from the close wild relative PIM (30.4% of the total 3.5 million SNPs) to the BIG (2.8%) with the intermediate group of CER (6.6%). Following this polymorphism detection, they showed a strong difference in linkage disequilibrium decay occurring between SNPs at physical distance of 8.8 kb in PIM, 256.8 kb in CER and 865.7 kb in BIG (Fig. 3).

The 1,008 tomato accessions that were genotyped using 7,720 SNPs by Blanca et al. (2015) completed the analyses of Lin et al. (2014). In this study, the expected and observed heterozygosity values were higher in PIM (He $=$ 0.21/ Ho $=$ 0.042) than in CER (He $=$ 0.17/ Ho $=$ 0.023) and in BIG (He $=$ 0.12/Ho $=$ 0.012). Known introgressions were detected in modern cultivar compared to so-called vintage ones, by measuring a higher heterozygosity due to the wild introgressions (He $=$ 0.12 vs. 0.09). In a recent paper, Sahu and Chattopadhyay (2017) detected 2,439 SNPs that were only polymorphic in wild accessions; these wild variants were part of 1,594 genes (868 SNPs were located up- and downstream of these genes). With this study they confirmed that chromosomes 1, 2, 3, 8 and 10 were the most affected by domestication and presented high diversity loss. These chromosomes are including the chromosome 2 that is known since 1964 (Kerr and Bailey 1964) as bearing three genes responsible for the change in fruit shape and size (*LC, fw2.2* and *Ovate*).

6.5 Genome-Wide Association Approach Extended the Knowledge of the Genetic Architecture of Agronomical Traits

In plants, the QTL approach has been largely used in biparental and multi-parental crosses (i.e. multi-allelic genetic intercross (MAGIC) – Pascual et al. (2014) – or nested association mapping (NAM) populations). However, this approach is restricted in allelic diversity limiting the genomic resolution to map genetic determinants (Borevitz and Nordborg 2003). The genome-wide association study (GWAS) approach was proposed to overcome the main limitations of traditional gene mapping by (1) providing higher resolution using ancestral polymorphism at the population level and (2) using panels of individuals from populations in which commonly occurring genetic variants can be associated with phenotypic variation. The availability of high-density SNP arrays (Sim et al. 2012; Víquez-Zamora et al. 2013) and sequencing data allowed genome-wide scans to test for significant associations between molecular markers and the quantitative trait variation. GWAS was first applied in large studies of human disease that successfully identified candidate loci (Hindorff et al. 2009). It was then adopted in plants a decade ago.

Overall, these successful studies identified loci that explain large portions of phenotypic variation in phenotypic traits (Brachi et al. 2011).

In major crop species, GWAS was applied to decipher the genetic architecture of complex quantitative traits and benefited from statistical and technical developments. More precisely, the implementation of mixed linear models (MLM) to account for population structure and kinship (or relatedness), estimated in the studied panel, allowed detecting associations with a higher accuracy. Similarly, correction for multiple testing (i.e. FDR or Bonferroni corrections) removed false-positive associations sorting out the most promising candidate loci. Additionally, the size of the GWAS datasets in major crops followed the trend of the power of high-throughput genotyping and sequencing technologies. From a few SSR or SNP makers, a decade ago, actual genomic datasets rely on full-length genome sequence for hundreds of individuals. Tomato was not an exception with numerous GWA studies conducted during the last decade, notably for agronomic traits, such as fruit morphology, metabolomic content or genotype by environment interactions (GxE). In more details, the first association study investigated fruit quality using limited sets of SNPs (<100) spread over the chromosome 2 (Ranc et al. 2012). Then, rapidly, with the development of the SOLCAP SNP genotyping array (nearly 8,000 SNPs), genome-wide-level GWA experiments were conducted to decipher the genetic basis of agronomical traits such as fruit morphology or fruit metabolite contents (Sauvage et al. 2014; Ruggieri et al. 2014; Sacco et al. 2015; Zhang et al. 2015; Bauchet et al. 2017a, b). Then, low coverage sequencing and full genome sequencing provided a broader coverage of the tomato genome, increasing the power to detect new associations notably for traits related to fruit colour (Lin et al. 2014), agronomical traits (Shirasawa et al. 2013; Ye et al. 2017), flavour components (Tieman et al. 2017) or extensive sets of primary and secondary metabolites (Zhu et al. 2018). However, within these latter studies, the interactions between the genotype and its surrounding environment were not considered until Albert et al. (2016) provided a GWAS study of the impact of drought stress onto agronomical and fruit quality traits in tomato, opening the door to further GxE experiments, notably related to biotic and abiotic stress tolerance. Overall, these studies made an extensive use of the combination of population genomics and germplasm collection. They deepened our knowledge of tomato genome dynamics in terms of recombination patterns (through the study of the LD decay) and identified candidate loci that were functionally validated, proving the validity of the approach in this species and more generally in plants.

However, in tomato, as in major crop species, there are limitations to sustain the discovery of new candidate loci underlying complex traits. Breakthroughs have been made in the field of high-throughput phenotyping, such as nano-sensor-assisted phenotyping (Dalal et al. 2017), which complement the production of population genomics data in this field of research. These latter technologies would be easily transferred to decipher the genetic architecture of local adaptation processes, for example, by providing large amounts of data for a reasonable cost. Another limitation is the statistical correction applied for multiple testing that inherently lowers the power of the association approach towards small to medium effect loci. At this stage, population genomics will be of great help to tackle this limitation. Haplotype determination methods are more mature procedures, as a result of the HapMap human project. However, these procedures have been sparsely applied in crop species (Wang et al. 2013). Reports in maize, rice or soybean demonstrated the power of the haplotype approach for adaptive traits such as flowering time (Van Inghelandt et al. 2012), sugar metabolism (Lestari et al. 2011) or salinity resistance (Patil et al. 2016), respectively. Besides the identification of promising candidate loci, the use of haplotypes provided further knowledge of the demographic or selective history of these loci. The same approach can be applied in tomato that experienced drastic changes in nucleotide diversity patterns along its domestication and modern breeding phases. Thus, haplotype makers will strengthen biological interpretations obtained from quantitative genetics and population genomics, offering a broader view of selective forces that acted on loci related to traits of agronomical interest for which the molecular determinants have been identified by GWAS. Dealing with the missing heritability is another limitation of the GWA approach (Brachi et al. 2011) that could be unveiled by population genomics based on the analysis of epi-markers. The approach was successfully applied, notably in human for common diseases (Rakyan et al. 2011) as epigenetic variation affects genes function and can contribute to common disease and in *Arabidopsis thaliana*, for local adaptation (Dubin et al. 2015). Overall, there is a unique opportunity to merge population genetics and population genomics to get the best of both worlds in sustaining breeding efforts while deciphering the selective history of agronomical loci in crops, such as tomato.

6.6 Genomic Selection

Genomic selection (GS) is a promising approach exploiting the density of molecular markers across genomes to offer advanced breeding designs (Goddard and Hayes 2007). More precisely, GS refers to selection decisions based on genome estimated breeding values (GEBV, Hayes et al. (2009)). This approach has the potential to be cost-effective (in both time and money) by reducing generation time or phenotyping effort through its prediction. While being successful in dairy cow breeding, its application in crops remains to be fully implemented in major species such as maize (Crossa et al. 2013). This methodological approach benefits from the availability of large genotyping or sequencing datasets, mainly obtained from GWAS panels, to test its feasibility. The initial step, as described in Heffner et al. (2009), relies on performing a cross-validation (or model training cycle) step where the effect of parameters such as LD decay, size of the training population and density of markers on the correlation between the predicted phenotype and the measured phenotype are evaluated (the so-called "r^2" estimation). Using this knowledge, the most accurate prediction parameters and models can be determined. The cross-validation step offers the best framework to start with and run a first round of GEBV to select the best individual to reproduce.

In tomato, cross-validation studies have already been conducted providing an appreciation about the potential of GS in this species. The studied phenotypic traits were mainly related to fruit quality and showed a high predictability from a medium size GWAS panel of nearly 160 individuals (accuracy up to 0.89 for fruit weight, Duangjit et al. (2016)) but were variable according to the trait heritability: as expected, a low heritability trait was less predictable than high heritability trait. Additionally, the potential of GS was evaluated and showed that (1) reliable phenotype prediction models were constructed from simulated data leading to confident prediction for both yield and flavour, with, for example, an r^2 of 0.807 for solid soluble content (Yamamoto et al. 2016) and (2) quality traits improvement through GS can be reached for F_1 hybrid genotypes (Yamamoto et al. 2017). However, these studies also revealed that GS will be difficult to apply in a breeding context in tomato because of the number of traits to consider and the antagonism between fruit yield and quality traits (sugar content vs. fruit size for example) combined with the high level of LD in modern varieties or the bottleneck of high-throughput phenotyping. GS is also a promising approach towards selecting the most promising lines to produce hybrid progenies by simulating crosses between an almost unlimited number of lines. But overall, tomato germplasm collections remain precious material that should be maintained, deeply characterized and enriched (notably with the addition of crop wild relative species) to support GS and GWA approaches.

7 Future Perspectives

The tomato genus that includes 12 wild species covering a large geoclimatic range provides an excellent framework to investigate the origin and history of the biological variation that occurs at the phenotypic and genomic levels (Haak et al. 2014). High-throughput genomics extends our understanding of these past processes. The combination of the "omics" approaches, notably genomics and transcriptomics with metabolomics and proteomics, provides an exceptional opportunity to get clearer interpretation of the forces at play in the processes of speciation and adaptation within the *Solanum* genus. The availability of a high-quality genome sequence of cultivated tomato was key towards the production of large amounts of results, as presented in this chapter, but efforts should be brought towards a high-quality reference genome for each of the 12 wild species. The use of third-generation sequencing technologies (i.e. Oxford Nanopore sequencing) is about to deliver such promises.

As we previously reported in this chapter, the higher nucleotide diversity from wild relative species will continue to supply breeding improvement. The data from crop and wild tomato species also represent an opportunity to expand scientific studies on plant biotic and abiotic stress responses. Indeed, wild tomato species, being locally well adapted to all kind of extreme environments (from high altitude to arid areas), are a crucial resource for breeders to retrieve traits for future cultivars retaining high quality and performance despite environmental changes. Furthermore, high synteny revealed the common structure within families of plants such as for the Solanaceae (Wang et al. 2008; Rodriguez et al. 2009; Peters et al. 2012; Rinaldi et al. 2016) opening the possible diversity sources to the entire family. Therefore, useful discoveries in species like eggplant, *S. melongena*, or potato, *S. tuberosum*, could be translated to the tomato crop genome. Such translational approach was recently demonstrated by the successful transfer of natural resistance from *Pisum sativum* to *A. thaliana* using new gene editing methods, such as CrispR-cas9 (Bastet et al. 2018). On the opposite, another potentially successful approach to sustain the development of high-yielding crops was recently proposed and could be applied in tomato: This approach, called the "rewilding", consists in improving crops that lost properties, thanks to their ancestors that once possessed such properties to tolerate variable environmental conditions (Palmgren et al. 2015). Finally, new discovery should also come from population epigenomics studies as illustrated in human and yeast (Taudt et al. 2016) towards the understanding of this molecular mechanisms notably in the regulation of trait of interest such as flowering or fruit maturation (Zhong et al. 2013).

The routinely use of genomics-based selection methods is a recent breakthrough facilitating the assessment of genetic variation and discovery of adaptive genes in this species. While additional information is needed, the current utility of selection tools indicates a robust ability to utilize existing variation in the tomato to address the challenges of climate uncertainty. Thus the objective is to properly use population genomics to increase tomato yield, quality and stability of production through advanced breeding strategies, enhancing the resilience of this crop species to climate variability as proposed in Abberton et al. (2016).

8 Conclusions

With the advance in population genomics, our knowledge of tomato evolution history and the genetic bases of important traits for this species is growing fast. With almost L1066 and genomic re-sequenced, a large number of transcriptome data for many tissues and conditions and genomic resources available to the scientific community are also very abundant. These resources were obtained, thanks to the availability of large collections of genetic resources gathered in the past and shared by/to the community. All these data must continue to be managed in a single database in open access to the community for the continuous discovery in tomato population genomics. Population genetics and genomics have advanced our knowledge of tomato domestication events and phylogenomics and helped in the breeding and conservation of genetic resources.

Acknowledgements We would like to warmly thank Dr. Silvana Grandillo, Dr. Morgane Roth, Dr. Margot Paris and Dr. Om P. Rajora for their critical reading of this chapter and helpful insights and comments.

References

Abberton M, Batley J, Bentley A, Bryant J, Cai H, Cockram J, et al. Global agricultural intensification during climate change: a role for genomics. Plant Biotechnol J. 2016;14:1095–8. https://doi.org/10.1111/pbi.12467.

Abbo S, Pinhasi van-Oss R, Gopher A, Saranga Y, Ofner I, Peleg Z. Plant domestication versus crop evolution: a conceptual framework for cereals and grain legumes. Trends Plant Sci. 2014;19:351–60. https://doi.org/10.1016/j.tplants.2013.12.002.

Aflitos S, Schijlen E, de Jong H, de Ridder D, Smit S, Finkers R, et al. Exploring genetic variation in the tomato (*Solanum* section *Lycopersicon*) clade by whole-genome sequencing. Plant J. 2014;80:136–48. https://doi.org/10.1111/tpj.12616.

Albert E, Segura V, Gricourt J, Bonnefoi J, Derivot L, Causse M. Association mapping reveals the genetic architecture of tomato response to water deficit: focus on major fruit quality traits. J Exp Bot. 2016;67:6413–30. https://doi.org/10.1093/jxb/erw411.

Albrecht E, Escobar M, Chetelat RT. Genetic diversity and population structure in the tomato-like nightshades *Solanum lycopersicoides* and *S. sitiens*. Ann Bot. 2010;105:535–54. https://doi.org/10.1093/aob/mcq009.

Allen NC, Bagade S, McQueen MB, Ioannidis JP, Kavvoura FK, Khoury MJ, et al. Systematic meta-analyses and field synopsis of genetic association studies in schizophrenia: the SzGene database. Nat Genet. 2008;40:827–34. https://doi.org/10.1038/ng.171.

Alseekh S, Ofner I, Pleban T, Tripodi P, Di Dato F, Cammareri M, et al. Resolution by recombination: breaking up *Solanum pennellii* introgressions. Trends Plant Sci. 2013;18(10):536–8. https://doi.org/10.1016/j.tplants.2013.08.003.

Arunyawat U, Stephan W, Städler T. Using multilocus sequence data to assess population structure, natural selection, and linkage disequilibrium in wild tomatoes. Mol Biol Evol. 2007;24:2310–22. https://doi.org/10.1093/molbev/msm162.

Atherton JG, Harris GP. Flowering. In: The tomato crop. Dordrecht: Springer; 1986. p. 167–200.

Baek YS, Royer SM, Broz AK, Covey PA, López-Casado G, Nuñez R, et al. Interspecific reproductive barriers between sympatric populations of wild tomato species (*Solanum* section *Lycopersicon*). Am J Bot. 2016;103:1964–78. https://doi.org/10.3732/ajb.1600356.

Bai Y, Lindhout P. Domestication and breeding of tomatoes: what have we gained and what can we gain in the future? Ann Bot. 2007;100:1085–94. https://doi.org/10.1093/aob/mcm150.

Barrero LS, Tanksley SD. Evaluating the genetic basis of multiple-locule fruit in a broad cross section of tomato cultivars. Theor Appl Genet. 2004;109:669–79. https://doi.org/10.1007/s00122-004-1676-y.

Bastet A, Lederer B, Giovinazzo N, Arnoux X, German-Retana S, Reinbold C, et al. Trans-species synthetic gene design allows resistance pyramiding and broad-spectrum engineering of virus resistance in plants. Plant Biotechnol J. 2018:1–13. https://doi.org/10.1111/pbi.12896.

Bauchet G, Causse M. Genetic diversity in tomato (*Solanum lycopersicum*) and its wild relatives. In: Genetic diversity in plants. Rijeka: InTech; 2012.

Bauchet G, Grenier S, Samson N, Bonnet J, Grivet L, Causse M. Use of modern tomato breeding germplasm for deciphering the genetic control of agronomical traits by genome wide association study. Theor Appl Genet. 2017a;130:875–89. https://doi.org/10.1007/s00122-017-2857-9.

Bauchet G, Grenier S, Samson N, Segura V, Kende A, Beekwilder J, et al. Identification of major loci and genomic regions controlling acid and volatile content in tomato fruit: implications for flavor improvement. New Phytol. 2017b;215:624–41. https://doi.org/10.1111/nph.14615.

Beddows I, Reddy A, Kloesges T, Rose LE. Population genomics in wild tomatoes – the interplay of divergence and admixture. Genome Biol Evol. 2017;9:3023–38. https://doi.org/10.1093/gbe/evx224.

Bedinger PA, Chetelat RT, McClure B, Moyle LC, Rose JKC, Stack SM, et al. Interspecific reproductive barriers in the tomato clade: opportunities to decipher mechanisms of reproductive isolation. Sex Plant Reprod. 2011;24:171–87. https://doi.org/10.1007/s00497-010-0155-7.

Bellucci E, Bitocchi E, Ferrarini A, Benazzo A, Biagetti E, Klie S, et al. Decreased nucleotide and expression diversity and modified coexpression patterns characterize domestication in the common bean. Plant Cell. 2014;26:1901–12. https://doi.org/10.1105/tpc.114.124040.

Birchler JA, Yao H, Chudalayandi S, Vaiman D, Veitia RA. Heterosis. Plant Cell Online. 2010;22:2105–12. https://doi.org/10.1105/tpc.110.076133.

Blanca J, Cañizares J, Cordero L, Pascual L, Diez MJ, Nuez F. Variation revealed by SNP genotyping and morphology provides insight into the origin of the tomato. PLoS One. 2012;7:e48198. https://doi.org/10.1371/journal.pone.0048198.

Blanca J, Montero-Pau J, Sauvage C, Bauchet G, Illa E, Díez MJ, et al. Genomic variation in tomato, from wild ancestors to contemporary breeding accessions. BMC Genomics. 2015;16:257. https://doi.org/10.1186/s12864-015-1444-1.

Bohs L, Olmstead RG. Phylogenetic relationships in *Solanum* (Solanaceae) based on *ndhf* sequences. Syst Bot. 1997;22:5. https://doi.org/10.2307/2419674.

Bolger A, Scossa F, Bolger ME, Lanz C, Maumus F, Tohge T, et al. The genome of the stress-tolerant wild tomato species *Solanum pennellii*. Nat Genet. 2014;46:1034–8. https://doi.org/10.1038/ng.3046.

Bombarely A, Menda N, Tecle IY, Buels RM, Strickler S, Fischer-York T, et al. The Sol Genomics Network (solgenomics.net): growing tomatoes using Perl. Nucleic Acids Res. 2011;39:1149–55. https://doi.org/10.1093/nar/gkq866.

Böndel KB, Lainer H, Nosenko T, Mboup M, Tellier A, Stephan W. North-south colonization associated with local adaptation of the wild tomato species *Solanum chilense*. Mol Biol Evol. 2015;32:2932–43. https://doi.org/10.1093/molbev/msv166.

Borevitz JO, Nordborg M. The impact of genomics on the study of natural variation in Arabidopsis. Plant Physiol. 2003;132:718–25. https://doi.org/10.1104/pp.103.023549.

Brachi B, Morris GP, Borevitz JO. Genome-wide association studies in plants: the missing heritability is in the field. Genome Biol. 2011;12:232. https://doi.org/10.1186/gb-2011-12-10-232.

Brozynska M, Furtado A, Henry RJ. Genomics of crop wild relatives: expanding the gene pool for crop improvement. Plant Biotechnol J. 2015; https://doi.org/10.1111/pbi.12454.

Canady MA, Meglic V, Chetelat RT. A library of *Solanum lycopersicoides* introgression lines in cultivated tomato. Genome. 2005;48:685–97. https://doi.org/10.1139/g05-032.

Carputo D, Monti L, Werner JE, Frusciante L. Uses and usefulness of endosperm balance number. TAG Theor Appl Genet. 1999;98:478–84. https://doi.org/10.1007/s001220051095.

Causse M, Grandillo S. Gene mapping in tomato. In: Causse M, Giovannoni J, Bouzayen M, Zouine M, editors. The tomato genome. Berlin: Springer; 2016. p. 23–37.

Causse M, Duffe P, Gomez MC, Buret M, Damidaux R, Zamir D, et al. A genetic map of candidate genes and QTLs involved in tomato fruit size and composition. J Exp Bot. 2004;55:1671–85.

Causse M, Desplat N, Pascual L, Le Paslier MC, Sauvage C, Bauchet G, et al. Whole genome resequencing in tomato reveals variation associated with introgression and breeding events. BMC Genomics. 2013;14:791. https://doi.org/10.1186/1471-2164-14-791.

Chakrabarti M, Zhang N, Sauvage C, Munos S, Blanca J, Canizares J, et al. A cytochrome P450 regulates a domestication trait in cultivated tomato. Proc Natl Acad Sci. 2013;110:17125–30. https://doi.org/10.1073/pnas.1307313110.

Cong B, Barrero LS, Tanksley SD. Regulatory change in YABBY-like transcription factor led to evolution of extreme fruit size during tomato domestication. Nat Genet. 2008;40:800–4. https://doi.org/10.1038/ng.144.

Coyne JA, Orr HA (2004) Speciation. Sinauer.

Crossa J, Pérez P, Hickey J, Burgueño J, Ornella L, Cerón-Rojas J, et al. Genomic prediction in CIMMYT maize and wheat breeding programs. Heredity (Edinb). 2013;112:48.

Dalal A, Rana JS, Kumar A. Ultrasensitive nanosensor for detection of malic acid in tomato as fruit ripening indicator. Food Anal Methods. 2017;10:3680–6. https://doi.org/10.1007/s12161-017-0919-x.

de Candolle A. The origin of cultivated plants. Cambridge: Cambridge University Press; 1886.

de Tournefort JP. Élemens de Botanique, ou Méthode pour connoître les Plantes. de l'Imprimerie royale, Paris. 1694.

Doganlar S, Frary A, Tanksley SD. The genetic basis of seed-weight variation: tomato as a model system. TAG Theor Appl Genet. 2000;100:1267–73. https://doi.org/10.1007/s001220051433.

Doganlar S, Frary A, Daunay MC, Lester RN, Tanksley SD. A comparative genetic linkage map of eggplant (*Solanum melongena*) and its implications for genome evolution in the Solanaceae. Genetics. 2002;161:1697–711.

Duangjit J, Causse M, Sauvage C. Efficiency of genomic selection for tomato fruit quality. Mol Breed. 2016;36:29. https://doi.org/10.1007/s11032-016-0453-3.

Dubin MJ, Zhang P, Meng D, Remigereau M-S, Osborne EJ, Paolo Casale F, et al. DNA methylation in Arabidopsis has a genetic basis and shows evidence of local adaptation. Elife. 2015;4:e05255. https://doi.org/10.7554/eLife.05255.

Eshed Y, Zamir D. An introgression line population of *Lycopersicon pennellii* in the cultivated tomato enables the identification and fine mapping of yield-associated QTL. Genetics. 1995;141:1147–62.

Falush D, Stephens M, Pritchard JK. Inference of population structure using multilocus genotype data: linked loci and correlated allele frequencies. Genetics. 2003;164:1567–87.

Farris JS. Outgroups and parsimony. Syst Biol. 1982;31:328–34. https://doi.org/10.1093/sysbio/31.3.328.

Fernandez-Pozo N, Menda N, Edwards JD, Saha S, Tecle IY, Strickler SR, et al. The Sol Genomics Network (SGN) – from genotype to phenotype to breeding. Nucleic Acids Res. 2015;43: D1036–41. https://doi.org/10.1093/nar/gku1195.

Fernandez-Pozo N, Zheng Y, Snyder SI, Nicolas P, Shinozaki Y, Fei Z, et al. The tomato expression atlas. Bioinformatics. 2017;33:2397–8. https://doi.org/10.1093/bioinformatics/btx190.

Finkers R, van Heusden AW, Meijer-Dekens F, van Kan JAL, Maris P, Lindhout P. The construction of a *Solanum habrochaites LYC4* introgression line population and the identification of QTLs for resistance to Botrytis *cinerea*. Theor Appl Genet. 2007;114:1071–80. https://doi.org/10.1007/s00122-006-0500-2.

Florez-Rueda AM, Paris M, Schmidt A, Widmer A, Grossniklaus U, Städler T. Genomic imprinting in the endosperm is systematically perturbed in abortive hybrid tomato seeds. Mol Biol Evol. 2016;33:2935–46. https://doi.org/10.1093/molbev/msw175.

Foolad MR. Genome mapping and molecular breeding of tomato. Int J Plant Genomics. 2007;2007:1–52. https://doi.org/10.1155/2007/64358.

Frary A, Nesbitt TC, Grandillo S, Knaap E, Cong B, Liu J, et al. fw2.2: a quantitative trait locus key to the evolution of tomato fruit size. Science. 2000;289:85–8. https://doi.org/10.1126/science.289.5476.85.

Gao L, Gonda I, Sun H, Ma Q, Bao K, Tieman DM, et al. The tomato pan-genome uncovers new genes and a rare allele regulating fruit flavor. Nat Genet. 2019;51(6):1044–51. https://doi.org/10.1038/s41588-019-0410-2.

Goddard ME, Hayes BJ. Genomic selection. J Anim Breed Genet. 2007;124:323–30. https://doi.org/10.1111/j.1439-0388.2007.00702.x.

Grandillo S, Cammareri M. Molecular mapping of quantitative trait loci in tomato. In The tomato genome (M. Causse, J. Giovannoni, M. Bouzayen and M. Zouine eds). Berlin, Heidelberg: Springer, 2016. pp. 39–73.

Grandillo S, Tanksley SD. QTL analysis of horticultural traits differentiating the cultivated tomato from the closely related species *Lycopersicon pimpinellifolium*. Theor Appl Genet. 1996;92:935–51. https://doi.org/10.1007/BF00224033.

Grandillo S, Ku HM, Tanksley SD. Identifying the loci responsible for natural variation in fruit size and shape in tomato. Theor Appl Genet. 1999;99:978–87. https://doi.org/10.1007/s001220051405.

Grandillo S, Chetelat R, Knapp S, Spooner D, Peralta I, Cammareri M, et al. Solanum sect. Lycopersicon. In: Wild crop relatives: genomic and breeding resources. Berlin: Springer; 2011. p. 129–215.

Grimm D, Greshake B, Kleeberger S, Lippert C, Stegle O, Scholkopf B, et al. easyGWAS: an integrated interspecies platform for performing genome-wide association studies. 2012. arXiv:1212.4788.

Haak DC, Kostyun JL, Moyle LC. Merging ecology and genomics to dissect diversity in wild tomatoes and their relatives. Adv Exp Med Biol. 2014;781:273–98.

Hayes BJ, Bowman PJ, Chamberlain AJ, Goddard ME. Genomic selection in dairy cattle: progress and challenges. J Dairy Sci. 2009;92:433–43. https://doi.org/10.3168/jds.2008-1646.

Heffner EL, Sorrells ME, Jannink J-L. Genomic selection for crop improvement. Crop Sci. 2009;49:1. https://doi.org/10.2135/cropsci2008.08.0512.

Hindorff LA, Sethupathy P, Junkins HA, Ramos EM, Mehta JP, Collins FS, et al. Potential etiologic and functional implications of genome-wide association loci for human diseases and traits. Proc Natl Acad Sci. 2009;106:9362–7. https://doi.org/10.1073/pnas.0903103106.

Hufford MB, Lubinksy P, Pyhäjärvi T, Devengenzo MT, Ellstrand NC, Ross-Ibarra J. The genomic signature of crop-wild introgression in maize. PLoS Genet. 2013;9:e1003477. https://doi.org/10.1371/journal.pgen.1003477.

Itkin M, Heinig U, Tzfadia O, Bhide AJ, Shinde B, Cardenas PD, et al. Biosynthesis of antinutritional alkaloids in Solanaceous crops is mediated by clustered genes. Science. 2013;341:175–9. https://doi.org/10.1126/science.1240230.

Jenkins JA. The origin of the cultivated tomato. Econ Bot. 1948;2:379–92. https://doi.org/10.1007/BF02859492.

Kawecki TJ, Ebert D. Conceptual issues in local adaptation. Ecol Lett. 2004;7:1225–41. https://doi.org/10.1111/j.1461-0248.2004.00684.x.

Kerr EA, Bailey DL. Resistance to *Cladosporium fulvum* cke. obtained from wild species of tomato. Can J Bot. 1964;42:1541–54. https://doi.org/10.1139/b64-153.

Kirkpatrick M, Ravigné V. Speciation by natural and sexual selection: models and experiments. Am Nat. 2002;159:S22–35. https://doi.org/10.1086/338370.

Klee HJ. Improving the flavor of fresh fruits: genomics, biochemistry, and biotechnology. New Phytol. 2010;187:44–56. https://doi.org/10.1111/j.1469-8137.2010.03281.x.

Koenig D, Jiménez-Gómez JM, Kimura S, Fulop D, Chitwood DH, Headland LR, et al. Comparative transcriptomics reveals patterns of selection in domesticated and wild tomato. Proc Natl Acad Sci. 2013;110:E2655–62. https://doi.org/10.1073/pnas.1309606110.

Kudo T, Kobayashi M, Terashima S, Katayama M, Ozaki S, Kanno M, et al. TOMATOMICS: a web database for integrated omics information in tomato. Plant Cell Physiol. 2017;58:e8. https://doi.org/10.1093/pcp/pcw207.

Kurlovich BS, Rep'ev SI, Petrova MV, Buravtseva TV, Kartuzova LT, Volumeva TA. The significance of Vavilov's scientific expeditions and ideas for development and use of legume genetic resources. Plant Genet Resour Newsl. 2000;124:23–32.

Labate JA, Grandillo S, Fulton T, Muños S, Caicedo AL, Peralta I, et al. Vegetables. Berlin: Springer; 2007.

Labate JA, Robertson LD, Baldo AM. Multilocus sequence data reveal extensive departures from equilibrium in domesticated tomato (*Solanum lycopersicum* L.). Heredity (Edinb). 2009;103:257–67. https://doi.org/10.1038/hdy.2009.58.

Labate JA, Robertson LD, Strickler SR, Mueller LA. Genetic structure of the four wild tomato species in the *Solanum peruvianum* s.l. species complex. Genome. 2014;57:169–80. https://doi.org/10.1139/gen-2014-0003.

Lagriffol J, Monnier M. Effects of endosperm and placenta on development of Capsella embryos in ovules cultivated in vitro. J Plant Physiol. 1985;118:127–37. https://doi.org/10.1016/S0176-1617(85)80141-3.

Lam HM, Xu X, Liu X, Chen W, Yang G, Wong FL, et al. Resequencing of 31 wild and cultivated soybean genomes identifies patterns of genetic diversity and selection. Nat Genet. 2010;42:1053–9. https://doi.org/10.1038/ng.715.

Langford AN. The parasitism of *Cladosporium fulvum* cooke and the genetics of resistance to it. Can J Res. 1937;15c:108–28. https://doi.org/10.1139/cjr37c-008.

Lemmon ZH, Bukowski R, Sun Q, Doebley JF. The role of cis regulatory evolution in maize domestication. PLoS Genet. 2014;10:e1004745. https://doi.org/10.1371/journal.pgen.1004745.

Lestari P, Lee G, Ham T-H, Reflinur, Woo M-O, Piao R, et al. Single nucleotide polymorphisms and haplotype diversity in rice sucrose synthase 3. J Hered. 2011;102:735–46. https://doi.org/10.1093/jhered/esr094.

Li W, Chetelat RT. A pollen factor linking inter- and intraspecific pollen rejection in t tomato. Science. 2010;330:1827–30. https://doi.org/10.1126/science.1197908.

Li W, Chetelat RT. Unilateral incompatibility gene *ui1.1* encodes an *S-locus F-box* protein expressed in pollen of Solanum species. Proc Natl Acad Sci. 2015;112:4417–22. https://doi.org/10.1073/pnas.1423301112.

Li H, Durbin R. Inference of human population history from individual whole-genome sequences. Nature. 2011;475:493.

Lin T, Zhu G, Zhang J, Xu X, Yu Q, Zheng Z, et al. Genomic analyses provide insights into the history of tomato breeding. Nat Genet. 2014;46:1220–6. https://doi.org/10.1038/ng.3117.

Linnaeus C. Species plantarum. Stockholm: Impensis G. C. Nauk; 1753.

Lippman Z, Tanksley SD. Dissecting the genetic pathway to extreme fruit size in tomato using a cross between the small-fruited wild species *Lycopersicon pimpinellifolium* and *L. esculentum* var. Giant heirloom. Genetics. 2001;158:413–22.

Lippman ZB, Semel Y, Zamir D. An integrated view of quantitative trait variation using tomato interspecific introgression lines. Curr Opin Genet Dev. 2007;17:545–52. https://doi.org/10.1016/j.gde.2007.07.007.

Liu J, Van Eck J, Cong B, Tanksley SD. A new class of regulatory genes underlying the cause of pear-shaped tomato fruit. Proc Natl Acad Sci. 2002;99:13302–6.

Lu J, Tang T, Tang H, Huang J, Shi S, Wu CI. The accumulation of deleterious mutations in rice genomes: a hypothesis on the cost of domestication. Trends Genet. 2006;22:126–31. https://doi.org/10.1016/j.tig.2006.01.004.

Marshall JA, Knapp S, Davey MR, Power JB, Cocking EC, Bennett MD, et al. Molecular systematics of *Solanum* section *lycopersicum* (Lycopersicon) using the nuclear ITS rDNA region. Theor Appl Genet. 2001;103:1216–22. https://doi.org/10.1007/s001220100671.

Martin GB, Brommonschenkel SH, Chunwongse J, Frary A, Ganal MW, Spivey R, et al. Map-based cloning of a protein kinase gene conferring disease resistance in tomato. Science. 1993;262:1432–6.

McClean PE, Hanson MR. A community-based annotation framework for linking solanaceae genomes with phenomes. Genetics. 1986;112:649–67.

Menda N, Buels RM, Tecle I, Mueller LA. A community-based annotation framework for linking solanaceae genomes with phenomes. Plant Physiol. 2008;147:1788–99. https://doi.org/10.1104/pp.108.119560.

Miller J, Miller P. The gardeners dictionary. London: John and Francis Rivington as well as 23 others; 1768.

Miller JC, Tanksley SD. RFLP analysis of phylogenetic relationships and genetic variation in the genus *Lycopersicon*. Theor Appl Genet. 1990;80:437–48. https://doi.org/10.1007/BF00226743.

Monforte AJ, Tanksley SD. Development of a set of near isogenic and backcross recombinant inbred lines containing most of the *Lycopersicon hirsutum* genome in a *L. esculentum* genetic background: a tool for gene mapping and gene discovery. Genome. 2000;43:803–13. https://doi.org/10.1139/g00-043.

Mousavi-Derazmahalleh M, Bayer PE, Nevado B, Hurgobin B, Filatov D, Kilian A, et al. Exploring the genetic and adaptive diversity of a pan-Mediterranean crop wild relative: narrow-leafed lupin. Theor Appl Genet. 2018;131:887–901. https://doi.org/10.1007/s00122-017-3045-7.

Moyers BT, Morrell PL, McKay JK. Genetic costs of domestication and improvement. J Hered. 2018;109:103–16. https://doi.org/10.1093/jhered/esx069.

Moyle LC. Comparative genetics of potential prezygotic and postzygotic isolating barriers in a *Lycopersicon* species cross. J Hered. 2007;98:123–35. https://doi.org/10.1093/jhered/esl062.

Moyle LC. Ecological and evolutionary genomics in the wild tomatoes (*Solanum* sect. Lycopersicon). Evolution. 2008;62:2995–3013. https://doi.org/10.1111/j.1558-5646.2008.00487.x.

Mueller LA. The SOL Genomics Network. A comparative resource for solanaceae biology and beyond. Plant Physiol. 2005;138:1310–7. https://doi.org/10.1104/pp.105.060707.

Mueller LA, Tanskley SD, Giovannoni JJ, van Eck J, Stack S, Choi D, et al. The tomato sequencing project, the first cornerstone of the International Solanaceae Project (SOL). Comp Funct Genomics. 2005;6:153–8. https://doi.org/10.1002/cfg.468.

Muir CD, Moyle LC. Antagonistic epistasis for ecophysiological trait differences between Solanum species. New Phytol. 2009;183:789–802. https://doi.org/10.1111/j.1469-8137.2009.02949.x.

Muller CH. Notes on the American flora, chiefly Mexican. Am Midl Nat. 1942;27:470. https://doi.org/10.2307/2421014.

Muños S, Ranc N, Botton E, Bérard A, Rolland S, Duffé P, et al. Increase in tomato locule number is controlled by two single-nucleotide polymorphisms located near *WUSCHEL*. Plant Physiol. 2011;156:2244–54. https://doi.org/10.1104/pp.111.173997.

Nakazato T, Housworth EA. Spatial genetics of wild tomato species reveals roles of the Andean geography on demographic history. Am J Bot. 2011;98:88–98. https://doi.org/10.3732/ajb.1000272.

Nakazato T, Bogonovich M, Moyle LC. Environmental factors predict adaptive phenotypic differentiation within and between two wild andean tomatoes. Evolution. 2008;62:774–92. https://doi.org/10.1111/j.1558-5646.2008.00332.x.

Nakazato T, Warren DL, Moyle LC. Ecological and geographic modes of species divergence in wild tomatoes. Am J Bot. 2010;97:680–93. https://doi.org/10.3732/ajb.0900216.

Nakazato T, Franklin RA, Kirk BC, Housworth EA. Population structure, demographic history, and evolutionary patterns of a green-fruited tomato, *Solanum peruvianum* (Solanaceae), revealed by spatial genetics analyses. Am J Bot. 2012;99:1207–16. https://doi.org/10.3732/ajb.1100210.

Ofner I, Lashbrooke J, Pleban T, Aharoni A, Zamir D. *Solanum pennellii* backcross inbred lines (BILs) link small genomic bins with tomato traits. Plant J. 2016;87:151–60. https://doi.org/10.1111/tpj.13194.

Ohmori T, Murata M, Motoyoshi F. Identification of RAPD markers linked to the *Tm-2* locus in tomato. Theor Appl Genet. 1995;90(3–4):307–11.

Ohmori T, Murata M, Motoyoshi F. Characterization of disease resistance gene-like sequences in near-isogenic lines of tomato. Theor Appl Genet. 1998;96:331–8. https://doi.org/10.1007/s001220050745.

Olmstead RG, Palmer JD. Implications for the phylogeny, classification, and biogeography of solanaceae from cpDNA site variation. Syst Bot. 1997;22:19–29.

Olmstead RG, Sweere JA, Spangler RE, Bohs L, Palmer J. Phylogeny and provisional classification of the Solanaceae based on chloroplast DNA. In: Nee M, Symon D, Lester RN, Jessop JP, editors. Solanaceae IV advances in biology and utilization. Kew: Royal Botanic Gardens; 1999. p. 111–37.

Palmer JD, Zamir D. Chloroplast DNA evolution and phylogenetic relationships in Lycopersicon. Proc Natl Acad Sci U S A. 1982;79:5006–10.

Palmgren MG, Edenbrandt AK, Vedel SE, Andersen MM, Landes X, Osterberg JT, et al. Are we ready for back-to-nature crop breeding? Trends Plant Sci. 2015;20:155–64. https://doi.org/10.1016/j.tplants.2014.11.003.

Pascual L, Desplat N, Huang BE, Desgroux A, Bruguier L, Bouchet JP, et al. Potential of a tomato MAGIC population to decipher the genetic control of quantitative traits and detect causal variants in the resequencing era. Plant Biotechnol J. 2014;13:565–77. https://doi.org/10.1111/pbi.12282.

Paterson AH, Lander ES, Hewitt JD, Peterson S, Lincoln SE, Tanksley SD. Resolution of quantitative traits into Mendelian factors by using a complete linkage map of restriction fragment length polymorphisms. Nature. 1988;335:721.

Paterson AH, Damon S, Hewitt JD, Zamir D, Rabinowitch HD, Lincoln SE, et al. Mendelian factors underlying quantitative traits in tomato: comparison across species, generations, and environments. Genetics. 1991;127:181–97.

Patil G, Do T, Vuong TD, Valliyodan B, Lee J-D, Chaudhary J, et al. Genomic-assisted haplotype analysis and the development of high-throughput SNP markers for salinity tolerance in soybean. Sci Rep. 2016;6:19199.

Pease JB, Haak DC, Hahn MW, Moyle LC. Phylogenomics reveals three sources of adaptive variation during a rapid radiation. PLoS Biol. 2016;14:e1002379. https://doi.org/10.1371/journal.pbio.1002379.

Peralta IE, Spooner DM. Classification of wild tomatoes: a review. Kurtziana. 2000;28:45–54.

Peralta IE, Spooner DM. Granule-bound starch synthase (*GBSSI*) gene phylogeny of wild tomatoes (*Solanum* L. section *Lycopersicon* [Mill.] Wettst. subsection *Lycopersicon*). Am J Bot. 2001;88:1888–902.

Peralta IE, Spooner DM, Knapp S. Taxonomy of wild tomatoes and their relatives (*Solanum* sect. *Lycopersicoides*, sect. *Juglandifolia*, sect. *Lycopersicon*; Solanaceae). Ann Arbor: American Society of Plant Taxonomists; 2008.

Pertuzé RA, Ji Y, Chetelat RT. Comparative linkage map of the *Solanum lycopersicoides* and *S. sitiens* genomes and their differentiation from tomato. Genome. 2002;45:1003–12. https://doi.org/10.1139/g02-066.

Peters SA, Bargsten JW, Szinay D, van de Belt J, Visser RG, Bai Y, et al. Structural homology in the Solanaceae: analysis of genomic regions in support of synteny studies in tomato, potato and pepper. Plant J. 2012;71:602–14. https://doi.org/10.1111/j.1365-313X.2012.05012.x.

Pickrell JK, Pritchard JK. Inference of population splits and mixtures from genome-wide allele frequency data. PLoS Genet. 2012;8. https://doi.org/10.1371/journal.pgen.1002967.

Rakyan VK, Down TA, Balding DJ, Beck S. Epigenome-wide association studies for common human diseases. Nat Rev Genet. 2011;12:529.

Ramsey J, Bradshaw HD, Schemske DW. Components of reproductive isolation between the monkeyflowers *Mimulus lewisii* and *M. cardinalis* (Phrymaceae). Evolution. 2003;57:1520–34.

Ranc N, Muños S, Xu J, Le Paslier M-C, Chauveau A, Bounon R, et al. Genome-wide association mapping in tomato (*Solanum lycopersicum*) is possible using genome admixture of *Solanum lycopersicum* var. *cerasiforme*. G3. 2012;2:853–64. https://doi.org/10.1534/g3.112.002667.

Rapp RA, Haigler CH, Flagel L, Hovav RH, Udall JA, Wendel JF. Gene expression in developing fibres of upland cotton (*Gossypium hirsutum L.*) was massively altered by domestication. BMC Biol. 2010;8:1–15. https://doi.org/10.1186/1741-7007-8-139.

Ray J. Observations topographical, moral and physiological, made in a journey through part of low-countries, Germany, Italy, and France. London: John Martyn; 1673.

Razifard H, Ramos A, Della Valle AL, Bodary C, Goetz E, Manser EJ, et al. Genomic evidence for complex domestication history of the cultivated tomato in Latin America. Mol Biol Evol. pii: msz297. 2020; https://doi.org/10.1093/molbev/msz297.

Rick CM. Hybridization between *Lycopersicon esculentum* and *Solanum pennellii*: phylogenetic and cytogenetic significance. Proc Natl Acad Sci. 1960;46:78–82. https://doi.org/10.1073/pnas.46.1.78.

Rick CM. Tomato-like nightshades: affinities, autoecology, and breeders' opportunities. Econ Bot. 1988;42:145–54.

Rick CM. Perspectives from plant genetics: the tomato genetics stock center. Genet Resour Risk. 1990:11–9.

Rick CM, Chetelat RT. Utilization of related wild species for tomato improvement. Acta Hortic. 1995;412:21–38. https://doi.org/10.17660/ActaHortic.1995.412.1.

Rick CM, Fobes JF. Allozyme variation in the cultivated tomato and closely related species. Bull Torrey Bot Club. 1975;102:376. https://doi.org/10.2307/2484764.

Rick CM, Tanksley SD. Genetic variation in *Solanum pennellii*: comparisons with two other sympatric tomato species. Plant Syst Evol. 1981;139:11–45. https://doi.org/10.1007/BF00983920.

Rinaldi R, Van Deynze A, Portis E, Rotino G, Toppino L, Hill T, et al. New insights on eggplant/tomato/pepper synteny and identification of eggplant and pepper orthologous QTL. Front Plant Sci. 2016;7 https://doi.org/10.3389/fpls.2016.01031.

Robbins MD, Sim S-C, Yang W, Van Deynze A, van der Knaap E, Joobeur T, et al. Mapping and linkage disequilibrium analysis with a genome-wide collection of SNPs that detect polymorphism in cultivated tomato. J Exp Bot. 2011;62:1831–45. https://doi.org/10.1093/jxb/erq367.

Rodriguez F, Wu F, Ané C, Tanksley S, Spooner DM. Do potatoes and tomatoes have a single evolutionary history, and what proportion of the genome supports this history? BMC Evol Biol. 2009;9:191. https://doi.org/10.1186/1471-2148-9-191.

Rodriguez GR, Munos S, Anderson C, Sim S-C, Michel A, Causse M, et al. Distribution of *SUN*, *OVATE*, *LC*, and *FAS* in the tomato germplasm and the relationship to fruit shape diversity. Plant Physiol. 2011;156:275–85. https://doi.org/10.1104/pp.110.167577.

Roselius K, Stephan W, Städler T. The relationship of nucleotide polymorphism, recombination rate and selection in wild tomato species. Genetics. 2005;171:753–63. https://doi.org/10.1534/genetics.105.043877.

Roth MM. Variability of hybrid seed failure in wild tomatoes (Solanum sect. Lycopersicon): phenotypic and molecular signatures in the developing endosperm. PhD thesis 24694, ETH Zurich, Switzerland. 2017.

Rothan C, Diouf I, Causse M. Trait discovery and editing in tomato. Plant J. 2019;97(1):73–90. https://doi.org/10.1111/tpj.14152.

Rousseaux MC, Jones CM, Adams D, Chetelat R, Bennett A, Powell A. QTL analysis of fruit antioxidants in tomato using *Lycopersicon pennellii* introgression lines. Theor Appl Genet. 2005;111:1396–408. https://doi.org/10.1007/s00122-005-0071-7.

Ruggieri V, Francese G, Sacco A, D'Alessandro A, Rigano MM, Parisi M, et al. An association mapping approach to identify favourable alleles for tomato fruit quality breeding. BMC Plant Biol. 2014;14:337. https://doi.org/10.1186/s12870-014-0337-9.

Sacco A, Ruggieri V, Parisi M, Festa G, Rigano MM, Picarella ME, et al. Exploring a tomato landraces collection for fruit-related traits by the aid of a high-throughput genomic platform. PLoS One. 2015;10:e0137139. https://doi.org/10.1371/journal.pone.0137139.

Sahu KK, Chattopadhyay D. Genome-wide sequence variations between wild and cultivated tomato species revisited by whole genome sequence mapping. BMC Genomics. 2017;18:430. https://doi.org/10.1186/s12864-017-3822-3.

Saliba-Colombani V, Causse M, Langlois D, Philouze J, Buret M. Genetic analysis of organoleptic quality in fresh market tomato. 1. Mapping QTLs for physical and chemical traits. Theor Appl Genet. 2001;102:259–72. https://doi.org/10.1007/s001220051643.

Sato S, Tabata S, Hirakawa H, Asamizu E, Shirasawa K, Isobe S, et al. The tomato genome sequence provides insights into fleshy fruit evolution. Nature. 2012;485:635–41. https://doi.org/10.1038/nature11119.

Sauvage C, Segura V, Bauchet G, Stevens R, Do PT, Nikoloski Z, et al. Genome-wide association in tomato reveals 44 candidate loci for fruit metabolic traits. Plant Physiol. 2014;165:1120–32. https://doi.org/10.1104/pp.114.241521.

Sauvage C, Rau A, Aichholz C, Chadoeuf J, Sarah G, Ruiz M, et al. Domestication rewired gene expression and nucleotide diversity patterns in tomato. Plant J. 2017;91:631–45. https://doi.org/10.1111/tpj.13592.

Schiffels S, Durbin R. Inferring human population size and separation history from multiple genome sequences. Nat Genet. 2014;46:919.

Schmidt MH-W, Vogel A, Denton AK, Istace B, Wormit A, van de Geest H, et al. De novo assembly of a new *Solanum pennellii* accession using nanopore sequencing. Plant Cell. 2017;29:2336–48. https://doi.org/10.1105/tpc.17.00521.

Schraiber JG, Akey JM. Methods and models for unravelling human evolutionary history. Nat Rev Genet. 2015;16:727.

Seah S, Yaghoobi J, Rossi M, Gleason CA, Williamson VM. The nematode-resistance gene, *Mi-1*, is associated with an inverted chromosomal segment in susceptible compared to resistant tomato. Theor Appl Genet. 2004;108:1635–42. https://doi.org/10.1007/s00122-004-1594-z.

Seehausen O, Butlin RK, Keller I, Wagner CE, Boughman JW, Hohenlohe PA, et al. Genomics and the origin of species. Nat Rev Genet. 2014;15:176.

Shirasawa K, Fukuoka H, Matsunaga H, Kobayashi Y, Kobayashi I, Hirakawa H, et al. Genome-wide association studies using single nucleotide polymorphism markers developed by re-sequencing of the genomes of cultivated tomato. DNA Res. 2013;20:593–603. https://doi.org/10.1093/dnares/dst033.

Sim SC, Robbins MD, Van Deynze A, Michel AP, Francis DM. Population structure and genetic differentiation associated with breeding history and selection in tomato (*Solanum lycopersicum* L.). Heredity (Edinb). 2011;106:927–35. https://doi.org/10.1038/hdy.2010.139.

Sim SC, Durstewitz G, Plieske J, Wieseke R, Ganal MW, van Deynze A, et al. Development of a large SNP genotyping array and generation of high-density genetic maps in tomato. PLoS One. 2012;7 https://doi.org/10.1371/journal.pone.0040563.

Spooner DM, Anderson GJ, Jansen RK. Chloroplast DNA evidence for the interrelationships of tomatoes, potatoes, and pepinos (Solanaceae). Am J Bot. 1993;80:676. https://doi.org/10.2307/2445438.

Spooner D, Peralta IE, Knapp S. Comparison of AFLPs with other markers for phylogenetic inference in wild tomatoes. Taxon. 2005;54:43–61. https://doi.org/10.2307/25065301.

Stadler T. Lineages-through-time plots of neutral models for speciation. Math Biosci. 2008;216:163–71. https://doi.org/10.1016/j.mbs.2008.09.006.

Stevens MA, Rick CM. Genetics and breeding. In: The tomato crop. Dordrecht: Springer; 1986. p. 35–109.

Stevens R, Buret M, Duffé P, Garchery C, Baldet P, Rothan C, et al. Candidate genes and quantitative trait loci affecting fruit ascorbic acid content in three tomato populations. Plant Physiol. 2007;143:1943–53.

Swanson-Wagner R, Briskine R, Schaefer R, Hufford MB, Ross-Ibarra J, Myers CL, et al. Reshaping of the maize transcriptome by domestication. Proc Natl Acad Sci. 2012;109:11878–83. https://doi.org/10.1073/pnas.1201961109.

Tadmor Y, Fridman E, Gur A, Larkov O, Lastochkin E, Ravid U, et al. Identification of malodorous, a wild species allele affecting tomato aroma that was selected against during domestication. J Agric Food Chem. 2002;50:2005–9. https://doi.org/10.1021/jf011237x.

Tanksley SD. The genetic, developmental, and molecular bases of fruit size and shape variation in tomato. Plant Cell. 2004;16:S181–9. https://doi.org/10.1105/tpc.018119.

Tanksley SD, Ganal MW, Prince JP, de Vicente MC, Bonierbale MW, Broun P, et al. High density molecular linkage maps of the tomato and potato genomes. Genetics. 1992;132:1141–60.

Taudt A, Colomé-Tatché M, Johannes F. Genetic sources of population epigenomic variation. Nat Rev Genet. 2016;17(6):319–32. https://doi.org/10.1038/nrg.2016.45.

Tecle IY, Menda N, Buels RM, van der Knaap E, Mueller L. solQTL: a tool for QTL analysis, visualization and linking to genomes at SGN database. BMC Bioinformatics. 2010;11:525. https://doi.org/10.1186/1471-2105-11-525.

Tellier A, Laurent SJY, Lainer H, Pavlidis P, Stephan W. Inference of seed bank parameters in two wild tomato species using ecological and genetic data. Proc Natl Acad Sci. 2011;108:17052–7. https://doi.org/10.1073/pnas.1111266108.

The Computational Pan-Genomics Consortium. Computational pan-genomics: status, promises and challenges. Brief Bioinform. 2016; https://doi.org/10.1093/bib/bbw089.

Tieman D, Zhu G, Resende MFR, Lin T, Nguyen C, Bies D, et al. A chemical genetic roadmap to improved tomato flavor. Science. 2017;355:391–4. https://doi.org/10.1126/science.aal1556.

van der Knaap E, Lippman ZB, Tanksley SD. Extremely elongated tomato fruit controlled by four quantitative trait loci with epistatic interactions. Theor Appl Genet. 2002;104:241–7. https://doi.org/10.1007/s00122-001-0776-1.

Van Inghelandt D, Melchinger AE, Martinant J-P, Stich B. Genome-wide association mapping of flowering time and northern corn leaf blight (*Setosphaeria turcica*) resistance in a vast commercial maize germplasm set. BMC Plant Biol. 2012;12:56. https://doi.org/10.1186/1471-2229-12-56.

Verlaan MG, Hutton SF, Ibrahem RM, Kormelink R, Visser RGF, Scott JW, et al. The tomato yellow leaf curl virus resistance genes *Ty-1* and *Ty-3* are allelic and code for DFDGD-class RNA-dependent RNA polymerases. PLoS Genet. 2013;9 https://doi.org/10.1371/journal.pgen.1003399.

Vincent H, Wiersema J, Kell S, Fielder H, Dobbie S, Castañeda-Álvarez NP, et al. A prioritized crop wild relative inventory to help underpin global food security. Biol Conserv. 2013;167:265–75. https://doi.org/10.1016/j.biocon.2013.08.011.

Víquez-Zamora M, Vosman B, van de Geest H, Bovy A, Visser RGF, Finkers R, et al. Tomato breeding in the genomics era: insights from a SNP array. BMC Genomics. 2013;14:354. https://doi.org/10.1186/1471-2164-14-354.

Wang Y, Diehl A, Wu F, Vrebalov J, Giovannoni J, Siepel A, et al. Sequencing and comparative analysis of a conserved syntenic segment in the solanaceae. Genetics. 2008;180:391–408. https://doi.org/10.1534/genetics.108.087981.

Wang Z, Cao H, Sun Y, Li X, Chen F, Carles A, et al. Arabidopsis paired amphipathic helix proteins SNL1 and SNL2 redundantly regulate primary seed dormancy via abscisic acid-ethylene antagonism mediated by histone deacetylation. Plant Cell. 2013;25:149–66. https://doi.org/10.1105/tpc.112.108191.

Winter D, Vinegar B, Nahal H, Ammar R, Wilson G V, Provart NJ, et al. An "Electronic Fluorescent Pictograph" browser for exploring and analyzing large-scale biological data sets. PLoS One. 2007;2:e718. https://doi.org/10.1371/journal.pone.0000718.

Xia H, Camus-Kulandaivelu L, Stephan W, Téllier A, Zhang Z. Nucleotide diversity patterns of local adaptation at drought-related candidate genes in wild tomatoes. Mol Ecol. 2010;19:4144–54. https://doi.org/10.1111/j.1365-294X.2010.04762.x.

Xiao H, Jiang N, Schaffner E, Stockinger EJ, van der Knaap E. A retrotransposon-mediated gene duplication underlies morphological variation of tomato fruit. Science. 2008;319:1527–30.

Xu X, Liu X, Ge S, Jensen JD, Hu F, Li X, et al. Resequencing 50 accessions of cultivated and wild rice yields markers for identifying agronomically important genes. Nat Biotechnol. 2012;30:105–11. https://doi.org/10.1038/nbt.2050.

Yamamoto E, Matsunaga H, Onogi A, Kajiya-Kanegae H, Minamikawa M, Suzuki A, et al. A simulation-based breeding design that uses whole-genome prediction in tomato. Sci Rep. 2016;6:1–11. https://doi.org/10.1038/srep19454.

Yamamoto E, Matsunaga H, Onogi A, Ohyama A, Miyatake K, Yamaguchi H, et al. Efficiency of genomic selection for breeding population design and phenotype prediction in tomato. Heredity (Edinb). 2017;118:202–9. https://doi.org/10.1038/hdy.2016.84.

Ye J, Wang X, Hu T, Zhang F, Wang B, Li C, et al. An InDel in the promoter of *Al-ACTIVATED MALATE TRANSPORTER9* selected during tomato domestication determines fruit malate contents and aluminum tolerance. Plant Cell. 2017;29:2249–68. https://doi.org/10.1105/tpc.17.00211.

Young ND, Tanksley SD. RFLP analysis of the size of chromosomal segments retained around the *Tm-2* locus of tomato during backcross breeding. Theor Appl Genet. 1989;77:353–9. https://doi.org/10.1007/BF00305828.

Zamir D. Improving plant breeding with exotic genetic libraries. Nat Rev Genet. 2001;2:983.

Zeggini E, Scott LJ, Saxena R, Voight BF, Marchini JL, Hu T, et al. Meta-analysis of genome-wide association data and large-scale replication identifies additional susceptibility loci for type 2 diabetes. Nat Genet. 2008;40:638–45. https://doi.org/10.1038/ng.120.

Zhang J, Zhao J, Xu Y, Liang J, Chang P, Yan F, et al. Genome-wide association mapping for tomato volatiles positively contributing to tomato flavor. Front Plant Sci. 2015;6:1042. https://doi.org/10.3389/fpls.2015.01042.

Zhao Q, Feng Q, Lu H, Li Y, Wang A, Tian Q, et al. Pan-genome analysis highlights the extent of genomic variation in cultivated and wild rice. Nat Genet. 2018;50:278–84. https://doi.org/10.1038/s41588-018-0041-z.

Zhao J, Sauvage C, Bitton F, Bauchet G, Liu D, Huang S, et al. (2019) Meta-analysis of genome-wide association studies provides insights into genetic control of tomato flavor. Nat Commun. 10(1):1534. doi: https://doi.org/10.1038/s41467-019-09462-w.

Zhong S, Fei Z, Chen YR, Zheng Y, Huang M, Vrebalov J, et al. Single-base resolution methylomes of tomato fruit development reveal epigenome modifications associated with ripening. Nat Biotechnol. 2013;31:154–9.

Zhu G, Wang S, Huang Z, Zhang S, Liao Q, Zhang C, et al. Rewiring of the fruit metabolome in tomato breeding. Cell. 2018;172:249–255.e12. https://doi.org/10.1016/j.cell.2017.12.019.

Zouine M, Maza E, Djari A, Lauvernier M, Frasse P, Smouni A, et al. TomExpress, a unified tomato RNA-Seq platform for visualization of expression data, clustering and correlation networks. Plant J. 2017;92:727–35. https://doi.org/10.1111/tpj.13711.

Zuriaga E, Blanca J, Nuez F. Classification and phylogenetic relationships in Solanum section Lycopersicon based on AFLP and two nuclear gene sequences. Genet Resour Crop Evol. 2009;56:663–78. https://doi.org/10.1007/s10722-008-9392-0.

Population Genomics of Soybean

Milind B. Ratnaparkhe, Rishiraj Raghuvanshi, Vennampally Nataraj,
Shivakumar Maranna, Subhash Chandra, Giriraj Kumawat,
Rucha Kavishwar, Prashant Suravajhala, Shri Hari Prasad,
Dalia Vishnudasan, Subulakshmi Subramanian, Pranita Bhatele,
Supriya M. Ratnaparkhe, Ajay K. Singh, Gyanesh K. Satpute,
Sanjay Gupta, Kunwar Harendra Singh, and Om P. Rajora

Abstract Soybean (*Glycine max* (L.) Merr) is a major oilseed crop globally with major production in the USA, Brazil, Argentina, China, and India. Significant progress has been made in soybean research for increasing yield and in the improvement of other agronomic and physiological traits. During the last three decades, there has been tremendous progress in soybean genetics and genomics research. Population genomics studies can assist understanding genetic diversity, population structure, evolution, and domestication and facilitating genomics-assisted breeding of soybean. In this chapter, we discuss recent progress made in population genomics to assess genetic diversity, population structure, and evolution of cultivated and wild species of soybean; develop pan-genomes; and identify genes and genomic regions under selection from domestication. We also review the application of population genomics for assisting soybean breeding via genome-wide association and genomic

---

M. B. Ratnaparkhe (✉) · R. Raghuvanshi · V. Nataraj · S. Maranna · S. Chandra · G. Kumawat ·
R. Kavishwar · G. K. Satpute · S. Gupta · K. H. Singh
ICAR-Indian Institute of Soybean Research, Indore, Madhya Pradesh, India
e-mail: milind.ratnaparkhe@icar.gov.in

P. Suravajhala · D. Vishnudasan · S. Subramanian
Amrita School of Biotechnology, Amrita University, Kollam, Kerala, India

S. H. Prasad
Centre for Plant Biotechnology and Molecular Biology, Kerala Agricultural University, Thrissur, Kerala, India

P. Bhatele
M. H. College of Home Science & Science for Women, Autonomous, Jabalpur, Madhya Pradesh, India

S. M. Ratnaparkhe
Indore Biotech Inputs and Research (P) Ltd., Indore, Madhya Pradesh, India

A. K. Singh
ICAR-National Institute of Abiotic Stress Management, Khurd, Baramati, Maharashtra, India

O. P. Rajora
Faculty of Forestry and Environmental Management, University of New Brunswick, Fredericton, NB, Canada

selection studies and highlight recent studies that illustrate a range of discoveries enabled by sequencing of populations of soybean genomes. We also present the genome, transcriptome, epigenome, genetic maps, and plant resources available for population genomics studies in soybean.

Keywords Domestication · Evolution · Genetic diversity and population structure · Genomes · Genome-wide association studies · Genomic selection · Pan-genomes · Transcriptomes

1 Introduction

Soybean [*Glycine max* (L.) Merr] is a major oilseed crop in the world and an important source of protein and oil for both humans and animals. It is also used as a raw product for many human health and industrial applications. Other than edible oil (18–22%), it contains protein (38–45%), ash, carbohydrate, minerals, nutritional elements, and antioxidants largely beneficial for humans. Therefore, it has gained popularity in the food, feed, health, and pharmaceutical industries worldwide. Soybean also improves soil fertility by fixing free atmospheric nitrogen through symbiotic association with soil microorganism. Soybean also contains several minerals, and useful nutraceuticals like isoflavones and tocopherols having several health benefits. Therefore, increasing soybean production and conservation and sustainable management of soybean genetic resources remains crucial for global food security. Population genomics studies can help in the understanding of the genetic/genomic diversity, origin and evolution, domestication syndrome, including selective sweeps in domestication and genes involved in domestication, acclimation, adaptation, and facilitate genomics-assisted breeding and conservation of genetic resources (Rajora 2019; Luikart et al. 2019).

In this chapter, we discuss the progress made in population genomics of soybean for understanding the origin, genetic diversity, population structure, evolution, and domestication as well as assisting genomics-assisted breeding. We also provide an overview of genetic, genomic, transcriptomic, epigenomic, and plant resources available to facilitate population genomics studies in soybean.

2 Soybean (*Glycine*) Genus and Species

The genus *Glycine* is divided into two subgenera, *Glycine* Willd. (perennial) and *Soja* (Moench) F.J. Herm (annual). The subgenus *Soja* includes two species: soybean [(*G. max* (L.) Merr.)] and its wild annual progenitor *G. soja* Sieb. & Zucc. (Fig. 1) (Ratnaparkhe et al. 2010). The subgenus *Glycine* contains ~30 wild perennial species. Soybean genetic resources may be categorized into four plausible gene pools (GP). Soybean gene pool-1 (GP-1) consists of biological species that can be crossed to produce vigorous hybrids that exhibit normal meiotic chromosome

Fig. 1 (a) *Glycine max*, (b) *Glycine soja*. The major differences can be seen in terms of height, leaf morphology, and stem characteristics. (Picture taken by Milind Ratnaparkhe)

pairing and possess total seed fertility. Based on this definition, all soybean (*G. max*) germplasm and the wild soybean, *G. soja*, are included in GP-1 with the qualification that seed sterility can be associated with chromosomal structural changes such as inversions and translocations. Gene segregation is normal and gene exchange is generally easy. GP-2 species can hybridize with GP-1 easily, and F_1 plants exhibit at least some seed fertility (Harlan and de Wet 1971). *Glycine max* is without GP-2 because no known species has such a relationship with soybean. It is possible that species in the soybean GP-2 do exist in Southeast Asia where the *Glycine* genus may have originated. However, extensive plant exploration in this part of the world is required to validate this assumption. GP-3 is the third outer limit of potential genetic resources. Hybrids between GP-1 and GP-3 are lethal, and gene transfer is not possible or requires radical techniques, such as embryo rescue (Harlan and de Wet 1971). Based on this definition, GP-3 includes the 26 wild perennial species of the subgenus *Glycine*. These species are indigenous to Australia and are geographically isolated from *G. max* and *G. soja*. Only three species (*G. argyrea*, *G. canescens*, and *G. tomentella*) have been successfully hybridized with soybean; the F_1 hybrids were rescued by embryo culture and were sterile, and most researchers could not proceed beyond the amphidiploid stage, with the exception of Singh et al. (1998a, b). This suggests that only three species belong to GP-3. GP-4 is the extremely outer limit of potential genetic resources. Pre- and post-hybridization barriers inhibit embryo development and premature embryo abortion occurs. Rarely can hybrid seedling lethality, hybrid seed inviability, and inviable F_1 plants can be circumvented by bridge crosses within the genus *Glycine* (Singh et al. 2007a, b). Only a few wild perennial *Glycine* species have been hybridized with soybean. Thus, a majority of species belong to soybean GP-4 as they have not been hybridized with GP-1, and when hybridized did not produce viable F_1 plants (Singh and Hymowitz 1987). Wild species are extremely diverse morphologically, cytologically, and genetically; grow in very diverse climates; and have a wide geographical distribution. They are a rich source of agronomically useful genes and alleles for biotic and abiotic stress tolerance (Ratnaparkhe et al. 2013; Maranna et al. 2023). The annual and perennial soybean species are significantly distantly related.

3 Genomic, Transcriptomic, and Other Resources for Population Genomics Research

A substantial amount of genomic, transcriptomic, proteomic, and metabolomic resources and database have been developed in soybean (Table 1), which are of importance for population genomics studies.

3.1 Genomes

Reference genomes provide a good resource, especially genomic markers, such as SNPs, for population genomics studies (Rajora 2019; Luikart et al. 2019). The soybean genome sequencing project was accomplished by DOE-JGI-Community Sequencing Program. The genome sequence assembly was termed as Glyma-1.0. This revealed approximately 950 Mb genome of expected 1,115 Mb assembled on to 20 chromosomes (Schmutz et al. 2010). The protein-coding regions were predicted to be 66,153, of which over 46,000 genes were predicted with a high confidence level. The genome sequence data and gene annotation of soybean are housed in The Phytozome database (http://www.phytozome.net/). Subsequently, several other projects were conducted on genome assembly of a variety of soybean accessions. Kim et al. (2010) assembled a genome sequence of a wild soybean (*G. soja* var. IT182932) using Illumina-GA and GS-FLX. Shen et al. (2018a) de novo assembled a high-quality genome for cultivar "Zhonghuang 13" (Gmax_ZH13) using single-molecule real-time (SMRT) sequencing, optical mapping, chromosome conformation capture sequencing (Hi-C), and next-generation sequencing (HiSeq). Xie et al. (2019) then assembled a high-quality reference genome for wild soybean W05 in 2019. Valliyodan et al. (2019) de novo assembled references for another two cultivars and one wild soybean using a combination of short- and long-read technologies. Also, several research projects have provided large-scale datasets for population, comparative, and functional genomics studies (Lam et al. 2010;

Population Genomics of Soybean

Table 1 List of online databases useful for population genomics studies in soybean

Databases	Features	Tools	Website
SoyBase and the SoybeanBr-eeder's Toolbox	Genetic and physical maps, QTL, genome sequence, transposable elements, annotations, graphical chromosome visualizer	BLAST search, ESTs search, SoyChip Annotation Search, Potential Haplotype (pHap) and Contig Search, Soybean Metabolic Pathways, Fast Neutron Mutants Search, RNA-Seq Atlas	http://soybase.org/
SGMD The Soybean Genomics and Microarray Database	Integrated view genomic, EST and microarray data	Analytical tools allowing correlation of soybean ESTs with their gene expression profiles	http://bioinformatics.towson.edu/SGMD/
SoyKB-Soybean Knowledge Base	Multi-omics datasets, genes/proteins, miRNAs/sRNAs metabolite profiling, molecular markers, information about plant introduction lines and traits	Germplasm browser, QTL and Trait browser, Fast neutron mutant data, Differential expression analysis, Phosphorylation data, Phylogeny	http://soykb.org/
SoyDB-Soybean transcription factors database	Protein sequences, predicted tertiary structures, putative DNA binding sites, Protein Data Bank (PDB), protein family classifications	PSI-BLAST, Browse database, Family Prediction by HMM, FTP data retrieve	http://casp.rnet.mis souri.edu/soydb/
SoyMetDB-The soybean metabolome database	Soybean metabolomic data	Pathway Viewer	http://soymetdb.org
soyTEdb-Soybean transposable elements database	Williams 82 transposable element database	Browse for Repetitive elements, Transposable Element and Map position, Data retrieval tools	www.soybase.org/soytedb/
SoyProDB-Soybean proteins database,	Several 2D Gel images showing isolated soybean seed proteins	Search tool for 2D spots, Navigation tools for protein data	http://bioinformatics.towson.edu
DaizuBase-An integrated soybean genome database including BAC-based physical maps	BAC-based physical map, linkage map and DNA markers, BACend, BAC contigs, ESTs, full-length cDNAs	Gbrowse, Unified Map, Gene viewer, BLAST	http://daizu.dna.affrc.go.jp/

(continued)

Table 1 (continued)

Databases	Features	Tools	Website
Soybean network (SoyNet)			www.inetbio.org/soynet
A knowledge database of soybean functional networks(SoyFN)		Functional gene network, microRNA functional network, gene annotation, genome browser	http://nclab.hit.edu.cn/SoyFN
SoyXpress-Soybean transcriptome database	Soybean ESTs, Metabolic pathways, Gene Ontology terms, SwissProt identifiers, and Affymetrix gene expression data	BLAST search, Microarray experiments, Pathway search	http://soyxpress2.agrenv.mcgill.ca
SoyGD-The Soybean GBrowse Database, Southern Illinois	Physical map and genetic map, Bacterial artificial chromosome (BAC) fingerprint database, Associated genomic data	Sequence data retrieval tools, Navigation tool for sequence information of different builds	http://soybeangenome.siu.edu/
Deltasoy-An Internet-Based Soybean Database for Official Variety Trials	Official variety trial (OVT) information in soybean, Mississippi OVT data, including yield, location, and disease information	Comparison tools for variety trail data, phenotypic data and disease-related data	http://msucares.com/deltasoy/testlocationmap.htm
Soybean Cyst Nematode proteins database (SCNProDB)		SCN protein identification, 2D gel images data	http://bioinformatics.towson.edu/
Soybean Functional Genomics Database (SFGD)		Gbrowse, microarray expression profiling, transcriptome data, gene co-expression regulatory network, acyl-lipid metabolism pathways, cis-element significance analysis	http://bioinformatics.cau.edu.cn/SFGD/
Soybean Proteome Database		Proteome, Metabolome, Transcriptome datasets, 2D-PAGE and proteomics information, comparative proteomics under flooding, drought, and salt stress	http://proteome.dc.affrc.go.jp/Soybean/
Soybean-VCF2Genomes	To map single sample variant call format (VCF) file against known soybean germplasm collection for identification of the closest soybean accession		http://pgl.gnu.ac.kr/soy_vcf2genome/

Source: Kavishwar et al. (2021)

Ma et al. 2010; Ha et al. 2012; Ashfield et al. 2012; Valliyodan et al. 2016, 2019; Chaudhary et al. 2019; Kim et al. 2019; Ratnaparkhe et al. 2020; Kajiya-Kanegae et al. 2021; Chu et al. 2021; Yang et al. 2022; Liu et al. 2023, 2024; Huang et al. 2024).

3.2 Transcriptomes

Whole transcriptome sequencing and differentially expressed genes provide functional genomic makers for population genomics and population transcriptomics studies. The gene expression patterns in soybean have been investigated using global expression analysis techniques like high-density expression arrays, Serial Analysis of Gene Expression (SAGE), and other functional genomics approaches. Microarray studies for soybean gene expression were conducted for functional studies of key genes associated with biotic and abiotic stresses (Maguire et al. 2002; Thibaud-Nissen et al. 2003; Vodkin et al. 2004). In soybean characterization of genetic elements defining tolerance to biotic and abiotic stresses, seed composition and increasing yield have gained greater interests (Thao et al. 2013; Ramesh et al. 2014; Ratnaparkhe et al. 2022b). Initial exploration for drought tolerance in soybean showed a strong upregulation of root-derived genes and metabolite coumestrol (Tripathi et al. 2016). Additionally, early transcriptional response of soybean roots to drought stress have been studied in great details by Neto et al. (2013). Also, differential expression of genes involved in osmo-protectant biosynthesis also conferred drought tolerance (Ha et al. 2015). Comparative expression and protein-protein interaction analysis of AQPs in cultivated and wild soybean helped in identifying *GmTIP2;1* as a novel candidate gene that would confer salt and water stress tolerance to the plants (Zhang et al. 2016). Functional genomics studies were also conducted to identify the role of microRNAs. MicroRNAs (miRNAs) are key regulators of gene expression and play important roles in many aspects of plant biology. Turner et al. (2012) identified a number of novel miRNAs and previously unknown family members for conserved miRNAs in soybean genome sequence. They classified all known soybean miRNAs based on their phylogenetic conservation and examined their genome organization, family characteristics, and target diversity. Currently, there are several ongoing projects on whole transcriptome sequencing and functional genomics in soybean (Du et al. 2023; Sun et al. 2024).

3.3 Epigenomes

Epigenomic variation can contribute significantly to phenotypic plasticity, stress responses, disease conditions, and acclimation and adaptation (Moler et al. 2019). Epigenome sequencing can provide epigenetic markers for population genomics and epigenomics studies. To understand the impact of epigenetics on crop domestication,

Shen et al. (2018b) investigated the variation of DNA methylation during soybean domestication by whole-genome bisulfite sequencing of 45 soybean accessions, including wild soybeans, landraces, and cultivars. Through methylomic analysis, they identified 5,412 differentially methylated regions (DMRs). These DMRs exhibit characters distinct from those of genetically selected regions. In particular, they have significantly higher genetic diversity. Association analyses suggested that only 22.54% of DMRs could be explained by local genetic variations. Intriguingly, genes in the DMRs that were not associated with any genetic variation were found to be enriched in carbohydrate metabolism pathways (Shen et al. 2018b).

DNA methylation profiling analyses in soybean revealed that hypomethylation could affect the expression of neighboring genes (Song et al. 2013). Kim et al. (2015) found that CG body-methylated genes were abundant in duplicated genes that exhibited higher expression level than single copy genes. It was found that DNA demethylation/methylation also plays critical roles in stress responses, such as continuous cropping stress adaptability (Liang et al. 2019), salinity stress (Song et al. 2013), and cyst nematode infection (Rambani et al. 2015). In addition, hundreds of small RNAs had been identified in soybean (Arikit et al. 2014; Kulcheski et al. 2011; Zhou et al. 2013), some of which showed tissue-specific or time-specific transcriptional patterns, indicating their biological relevance (Arikit et al. 2014). Population genetic analyses suggested a coevolution of MIRNA and miRNA targets during soybean domestication (Zhao et al. 2015).

3.4 Genetic Markers and Genetic and QTL Maps

Highly informative molecular genetic markers are required for population genomics studies, and genetically mapped markers have advantages over unmapped markers (Luikart et al. 2019). Many types of DNA markers, such as RFLP, AFLP, RAPD, SSR, and SNP, have been developed in soybean. The first report of restriction fragment length polymorphism (RFLP) in soybean was on assessment of molecular genetic diversity of the nuclear genome (Apuya et al. 1988). Subsequently, RFLP markers were used extensively for genetic diversity analysis for soybean cultivars, landraces, and germplasm lines and for genome mapping (Keim et al. 1989, 1990, 1997; Skorupska et al. 1989, 1993; Lorenzen et al. 1995; Diers et al. 1992; Lark et al. 1993; Shoemaker and Specht 1995; Mansur et al. 1996; Cregan et al. 1999; Ferreira et al. 2000; Yamanaka et al. 2001; Lightfoot et al. 2005) until SSR and SNP markers became popular (Kumawat et al. 2020).

High-density linkage maps were developed using a combination of SSR and/or SNP markers (Li et al. 2019). Hyten et al. (2008) developed a multiplex assay of 384 SNPs designated as soybean oligo pool all-1 (SoyOPA-1), for genotyping the complex genome of soybean. In order to develop more SNPs, Hyten et al. (2010a) sequenced a total of 3,268 SNP-containing robust STS in six diverse genotypes, resulting in identification of 13,042 SNPs with an average of 3.5 SNP per polymorphic STS. These SNPs along with 5,551 SNPs discovered by Choi et al. (2007) were used to design two Illumina custom 1,536 SNP GoldenGate assays designated as SoyOPA-2 and SoyOPA-3. A set of 1,536 SNPs from the 3,456 SNPs present in three SoyOPAs was selected to include sufficient polymorphic SNP markers distributed throughout the genome that could be used for genetic mapping applications. This set of 1,536 SNPs GoldenGate assay was designated as Universal Soy Linkage Panel 1.0 (USLP1.0). Hyten et al. (2010b) sequenced a reduced representation library of soybean to identify SNPs using high-throughput sequencing methods. A total of 1,536 SNPs were selected from this pool of 7,108 SNPs to create an Illumina GoldenGate assay (SoyOPA-4). The SoyOPA-4 produced 1,254 successful GoldenGate assays indicating a validation and assay conversion rate of 81.6% for the predicted SNPs. Chaisan et al. (2010) used 335,857 publicly available ESTs derived from 18 genotypes for EST clustering and in silico SNP identification. A total of 3,219 EST contigs were established based on three to nine genotypes and a total of 26,735 SNPs were identified. The confirmation of in silico identified SNPs by Sanger sequencing yielded a 15.7% accuracy rate between two cultivars "Williams 82" and "Harosoy". These studies resulted in the development of a large number of SNP markers in soybean, which could be used for population genomics studies and mapping of complex traits.

QTLs underlying more than 100 agronomically important traits have been mapped in soybean. QTL mapping has also progressed for physiological traits, insects and pest resistance, and tolerance to several climatic stresses with improved yield. Current information on all mapped QTLs in soybean is available on the USDA-ARS soybean genetic database *SoyBase* (http://soybase.org). Since SNP markers are abundant, they have become more popular for QTL analysis of various agronomic traits in soybean (https://soybase.org, http://soykb.org). With the availability of whole-genome sequence, Gene/QTL mapping in soybean has become more common (Schmutz et al. 2010). Genome sequencing has greatly assisted in the development of thousands of molecular markers for genetic mapping studies. QTL analysis played a significant role in identifying genomic regions associated with various traits (Chen et al. 2021; Kumar et al. 2023; Tripathi et al. 2021; Chandra et al. 2022; Gao et al. 2024).

While the existing research on wild soybean primarily centers around biotic and abiotic stress factors, a limited number of studies have explored its potential to enhance soybean yield traits. These investigations have uncovered promising alleles within wild soybean germplasm that could contribute to improved soybean yields. Concibido et al. (2003), for instance, successfully mapped a QTL originating from wild soybean PI407305 through the use of BC_2, a breeding population derived from a cross between cultivated soybean (HS-1) and wild soybean (PI407305). Similarly, Wang et al. (2004) identified eight yield-related QTLs in BC2F4, a population created by crossing cultivated soybean (IA2008) with wild soybean (PI468916), revealing the presence of four advantageous alleles in wild soybean. In another study, Li et al. (2008a) mapped a QTL closely linked to the SSR marker satt511 from wild soybean in three different environments, using BC_2F_4, derived from a cross between cultivated soybean (7,499) and wild soybean (PI245331). This research indicated that the additive gene effect could increase yields from 191 to 235 kg ha^{-1}

in wild soybean. Furthermore, Wen et al. (2008) conducted association mapping for agronomic and quality traits in both wild and cultivated soybean populations, uncovering associations that were exclusively present in the wild soybean group. Kan et al. (2012) successfully mapped two QTLs for pod number per plant and one QTL for yield per plant from wild soybean over a span of 2 years. Collectively, these findings strongly suggest that wild soybean harbors alleles that favor increased yield potential. This underscores the feasibility of identifying yield-enhancing alleles in wild soybean germplasm. The exploration of wild soybean's genetic diversity continues to yield numerous novel alleles associated with yield-related traits. As more QTLs related to yield are discovered in various wild accessions, it will become clearer whether these alleles are widespread among all accessions or are concentrated in specific wild variants that are more distant from cultivated soybean varieties, as determined through linkage or association mapping.

3.5 Plant Resources

Soybean germplasm, landraces, and wild species have been collected extensively and maintained at various institutes. Currently, more than 170,000 *G. max* accessions are maintained by more than 160 institutions in nearly 70 countries. The USDA Soybean Germplasm Collection is one of the most intensely used germplasm collections in the world, and the most intensely used in the National Plant Germplasm System (NPGS). The collection includes more than 1,100 wild soybeans from China, Korea, Japan, and Russia, and more than 18,000 cultivated soybeans from China, Korea, Japan, and 84 other countries (Song et al. 2015).

4 Progress in Population Genomics of soybean

4.1 Pan-genomes

A pan-genome is one such strategy that appears to have the potential to capture a species' whole genetic repertoire. The use of a single reference genome has several limitations to investigate genetic diversity across diverse accessions within the species. Mapping of sequencing reads on the single reference genome frequently misses highly polymorphic regions and regions that are absent from the reference genome and therefore a more robust and thorough methodology is necessary.

The first plant pan-genome was developed in 2014 in wild soybean *Glycine soja* (Li et al. 2014). Subsequently, numerous pan-genomes were developed for various crop species and their wild relatives. Later, Liu et al. (2020) constructed a soybean pan-genome by de novo genome assembly of 26 representative wild and cultivated soybeans using long-read sequencing. This assembly produced not only golden-grade genomes for each accession, but also for the first time reported a graph-based genome in plants, which provides a promising platform for future in-depth soybean functional genomic studies (Liu et al. 2020). The initial pan-genome assemblies largely relied on short-read sequencing and only comprised a small number of accessions. Nevertheless, long-read sequencing and hybrid approaches, based on the combination of long- and short-read sequences, have made it possible to reconstruct pan-genome assemblies from a much larger set of accessions recently. Pan-genome from seven wild soybeans has been constructed using second-generation sequencing technology (Li et al. 2014). Liu et al. (2020) discovered sequence variation by using 26 genomes plus three previously reported genome (Wm82, ZH13, and W05). A total of 14,604,953 SNPs and 12,716,823 small insertions and indels ($\leq$50 bp) were discovered (Fig. 2). Furthermore, a total of 723,862 PAVs (>50 bp insertion or deletion), 27,531 CNVs, and 21,886 translocation events were detected. This suggests that soybean has very high genome diversity. Liu et al. (2020) found that >90% of the length variation of the assembled genomes resulted from PAVs, indicating that PAV was a major contributor to the genome size variation in soybean. This study introduced a dataset that includes 27 different wild and cultivated soybean accessions, setting the stage for future research into soybean genomics. Using advanced genome analysis techniques, the study identifies a wide range of genetic variations that were previously undetectable with conventional methods. From the total gene sets, they reported the genome organization as follows: core (35.87%) present in all 27 accessions, soft core (14.2%) present in 25–26 accessions (>90% of the collection), dispensable (49.88%) present in 2–24 accessions, and private (0.05%) present only in one accession (Liu et al. 2020; Fig. 2). The findings from this study are shared with the scientific community through databases like the Genome Sequence Archive (GSA) and Figshare. Additionally, the study employed a graph-based genome approach, which improves the accuracy of identifying variations, especially around structural variations at the pan-genome level. This approach allowed for a more thorough reanalysis of previously collected data. Moreover, by integrating RNA-seq and small RNA sequencing data from individual accessions, the study enables researchers to explore the relationship between genetic variations and gene expression, thus enhancing our understanding of soybean genetics.

Zhuang et al. (2022) assembled chromosome-level genomes of representative perennial species across the genus *Glycine* including five diploids and a young allopolyploid and constructed a *Glycine* super-pan-genome framework by integrating 26 annual soybean genomes. These perennial diploids exhibit greater genome stability and possess fewer centromere repeats than the annuals. Biased subgenomic fractionation occurred in the allopolyploid, primarily by accumulation of small deletions in gene clusters through illegitimate recombination, which was associated with preexisting local subgenomic differentiation. The super-pan-genome framework of the *Glycine* genus includes 109,827 nonredundant protein-coding genes from the perennials. The study indicated that ~70% protein-coding genes were absent in the annual soybean pan-genome, representing a huge repertoire of genetic potential for improvement of the annual crop. This study also unveiled the propensities and consequences of polyploid genome evolution, genetic determinants of the life-history strategy transition, and the causes and mechanisms for subgenome fractionation.

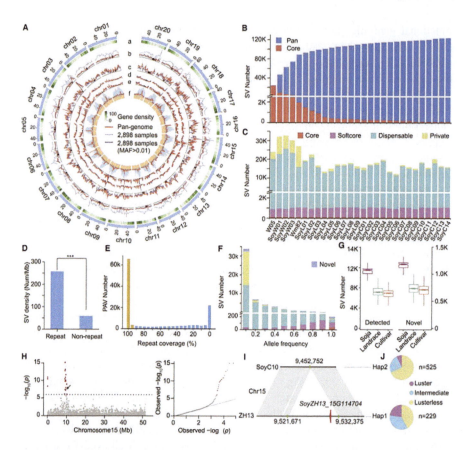

Fig. 2 Distribution of genetic variations from 29 genomes and 2,898 resequenced soybean accessions. (**a**) Analysis of genetic diversity was conducted using data from 29 soybean genomes and 2,898 resequenced soybean accessions. Key findings include the distribution of gene density and various measures of genetic variation such as single nucleotide polymorphism (SNP) density, nucleotide diversity (π), nonsynonymous (dN), and synonymous (dS) mutations. Additionally, larger structural variations and repeat sequences across the soybean genome were highlighted differences among accessions. (**b**) Variants were consolidated across samples in a non-repetitive manner, starting with one accession and progressively incorporating unique variants from others. (**c**) The distribution of variants across different classes was analyzed for each accession. (**d**) Structural variation density was assessed in both repeat and non-repeat regions of the genome using 500-kb windows, with statistical significance determined using Fisher's exact test. (**e**) The relationship between structural variation count and repetitive DNA coverage was explored. (**f**) Structural variations identified across the accessions were plotted against their discovery frequency, utilizing short-read mapping against a graph-based genome. (**g**) The numbers of both previously detected and novel structural variations were compared among wild soybeans, landraces, and cultivars, with boxplots illustrating the distribution. (**h**) Genome-wide association study (GWAS) was conducted to investigate the association between seed luster and presence-absence variants (PAVs) genotyped based on a graph-based genome. (**i**) An example of a 10-kb PAV was provided, highlighting its impact on the presence or absence of an HPS encoding gene, SoyZH13_15G114704. (**j**) The variation in seed luster between two haplotypes of the 10 kb PAV was compared. (Reproduced from Liu et al. (2020))

4.2 Genetic Diversity, Population Structure, and Evolution

Molecular genetic diversity in soybean and its wild species has been examined using various markers, including random amplified polymorphic DNA (RAPD) markers, amplified fragment length polymorphisms (AFLPs), simple-sequence repeats (SSRs), and single nucleotide polymorphisms (SNPs). The level of genetic diversity and geographic differentiation in Chinese cultivated soybean has been extensively studied using the coefficient of parentage (Cui et al. 2000a, b), morphological traits (Dong et al. 2004), and SSR markers (Li et al. 2008b), showing a clear geographic effect on genetic structure. Comparisons between the diversity of different samples of Asian soybean landraces and that of North American cultivars have demonstrated a lower level of diversity in the American pools than in the Asian pools, using either phenotypic characterization (Cui et al. 2000a, 2001) or the coefficient of parentage (Cui et al. 2000b). This reduced diversity was confirmed using DNA sequence analyses to show successive genetic bottlenecks between wild and cultivated soybeans and between Asian landraces and North American cultivars (Hyten et al. 2006).

Typically, genetic diversity clusters by taxon, with a clear differentiation between wild and domesticated taxa (Powell et al. 1996). The genetic structure of *G. max* and *G. soja* typically agree with their geographic locations (Dong et al. 2004; Abe et al. 2003; Xu and Gai 2003; Li et al. 2008b). Molecular and phylogenetic studies have indicated that Chinese and Japanese *G. soja* populations form distinct germplasm pools (Kuroda et al. 2006, 2010), and Asian accessions of *G. max* group generally according to their planting region and also the sowing season (Abe et al. 2003). There has been comprehensive study of genetic relationships of all species in the genus *Glycine*. The annual (subgenus *Soja*) and perennial (subgenus *Glycine*) soybean species are significantly distantly related and have diverged from a common ancestor around 5 MYA (Innes et al. 2008; Wawrzynski et al. 2008; Ashfield et al. 2012).

SNP markers have been used to understand the genetic diversity and population structure in soybean and wild species (Haun et al. 2011; Li et al. 2013). Wang et al. (2018) conducted SNP genotyping in soybean to characterize the genetic structure of 235 cultivars and examined their relationship to geographic origins. The cultivars were obtained from various locations, including different latitudinal regions of China, Japan, and the USA. The majority of the cultivars represented modern varieties, while five were landraces. The genotyping was performed using the SoySNP8k iSelect BeadChip, which targeted 7,189 single nucleotide polymorphisms (SNPs). The analysis of the population structure revealed the presence of seven subgroups as the most likely number of divisions (K). This clustering was supported by a significant delta K value, as indicated in Fig. 3a. Additionally, the neighbor-joining tree depicted in Fig. 3b provided further support for the identified subgroups. Consequently, the cultivars were classified into seven subgroups that generally corresponded to their geographic origins, including Japan, Northern America, central China, Huang-huai region China, Northern area China, and

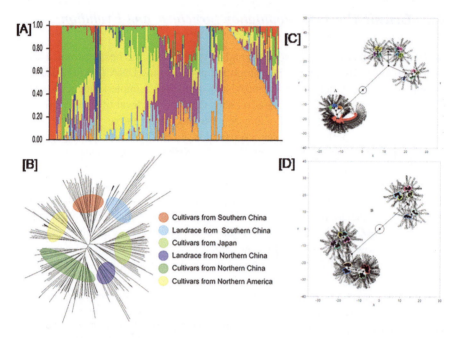

Fig. 3 Genetic diversity and population structure of soybean cultivars. Wang et al. (2018) characterized the genetic structure of the 235 cultivars using SNP genotyping and demonstrated the association between their classification into subgroups and their geographic origins. (**a**) Population structure of 235 cultivars at $K = 7$. Each cultivar is represented by a single vertical line, and color represents one cluster. (**b**) Phylogenetic tree showing into seven subgroups. Another study based on cluster analysis conducted by Shaibu et al. (2021) with SNP and DArT markers revealed two groups with each group having subgroups (**c, d**). (Reproduced from Wang et al. (2018) and Shaibu et al. (2021))

landraces (wild soybean). Another study conducted by Shaibu et al. (2021) employed cluster analysis using SNP and DArT markers. The results of this study revealed the presence of two main groups, each consisting of multiple subgroups (Fig. 3c, d). For the SNP data, Group I comprised four subgroups, with US accessions distributed across all subgroups, while Group II comprised ten subgroups containing accessions from Taiwan and China. The DArT cluster analysis identified nine subgroups within Group I and five subgroups within Group II.

Valliyodan et al. (2021) conducted a thorough investigation into soybean genetic diversity using resequencing data from 481 accessions, which included 52 wildtypes (*Glycine soja*) (Figs. 4 and 5). They employed 25,496 SNPs shared with the SoySNP50K iSelect BeadChip-derived data to construct a phylogenetic tree, revealing distinct clustering by country of origin. This indicates the emergence of phylogenetically distinct lineages within specific geographic regions, with limited genetic exchange between certain locales. Noteworthy findings include the predominant presence of Chinese accessions in the upper portion of the tree and the concentration of US elite lines in the lower middle. Additionally, key accessions significant in US

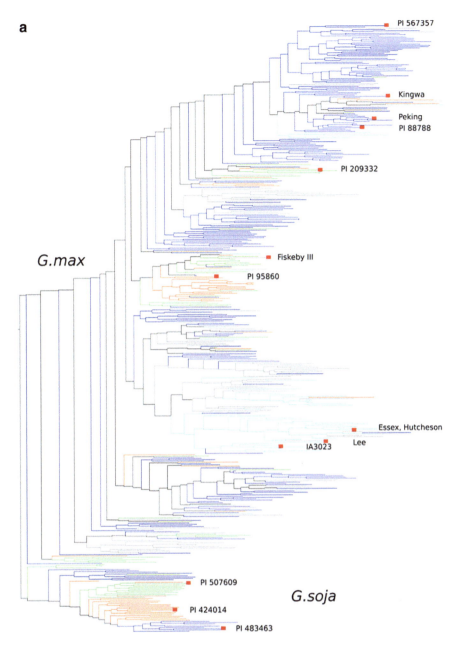

Fig. 4 Phylogenetic analysis of 481 soybean lines based on resequence (Valliyodan et al. 2021). The study used 25,496 SNPs shared with SoySNP50K data to create a phylogenetic tree of soybean accessions, rooted between *G. max* and *G. soja*. In both the *G. max* and *G. soja* clades, the accessions generally cluster by country of origin, as indicated by groupings of colors (countries): Japanese (green), Korean (orange), Chinese (blue), US (light blue), and all other countries in gray. There were distinct lineages observed in specific regions, indicating limited genetic exchange between them. Chinese accessions were predominant in the upper clade, while US elite lines were more prevalent in the middle. Certain accessions, such as PI 88788 and Peking, were highlighted for their importance in disease resistance, while others like Fiskeby III and Lee exhibited salt tolerance. *G. soja* PI 483463 was notable for its recently sequenced genome. (Figure and its legend reproduced from Valliyodan et al. (2021))

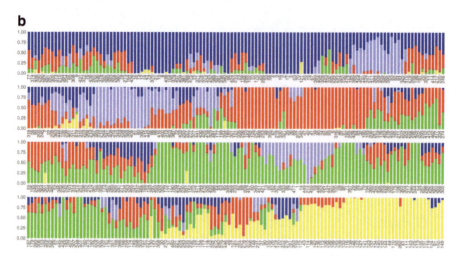

Fig. 5 Genetic structure plot for the 481 resequenced accessions indicating genetic diversity in the soybean germplasm and accessions. A structure plot was generated for 481 accessions using a cluster size of $K = 5$. The order of accessions in the structure plot mirrored that of the phylogenetic tree (Fig. 4a). This structure aligned well with the phylogenetic tree (Fig. 4a), for instance, *G. soja* lines (yellow) clustered together, consistent with their genetic distance from *G. max*. This depiction of structure analyses sheds light on the history of soybean domestication, showcasing independent selection in various locations across Asia. These findings contribute to a more comprehensive understanding of US soybean breeding programs' history, informing future breeding endeavors. (Figure and its legend reproduced from Valliyodan et al. (2021))

breeding programs were highlighted, along with *G. soja* PI 483463, singled out due to its recently sequenced genome. Furthermore, the study generated a genomic structure plot for all 481 accessions, corroborating the earlier phylogenetic findings by grouping *G. soja* lines together, consistent with their genetic distinction from *G. max*. These analyses shed light on the history of soybean domestication, providing a deeper understanding of genetic variation and valuable resources for future research and breeding efforts in soybeans.

Liu et al. (2022) sequenced a total of 2,898 soybean accessions by Illumina technology with an average coverage depth of more than 13× for each accession. These accessions included 103 wild soybeans, 1,048 landraces, and 1,747 cultivars, which represented a full range of soybean geographic distributions. Liu et al. (2022) resequenced 60 vegetable soybean accessions and performed comparative analysis with other resequenced genomes. Other 103 previously resequenced wild soybean and 155 previously resequenced grain soybean accessions were included for comparative genomics studies, population structure analysis, and selective sweep of vegetable, grain, and wild soybean. Results indicated that a total of 1,112 and 1,047 genes are under selection in the vegetable soybean and grain soybean populations as compared with the wild soybean population, respectively. Of these, 134 genes found under selection were shared between vegetable soybean and grain soybean populations. Additionally, four sucrose synthase genes, one sucrose-phosphate synthase gene, and four sugar transport genes were identified as candidate genes related to important traits such as seed sweetness and seed size in vegetable soybean (Liu et al. 2022).

4.3 Domestication Genomics

Population genomics approaches allow to understand the origin of crop plants' genetic basis of the domestication syndrome, including genetic diversity reduction, selection sweeps, and genomic regions and genes involved in domestication (Rajora 2019). Studies have indicated that the cultivated soybean was domesticated from its wild relative [*Glycine soja* (Sieb. and Zucc.)] approximately 5,000 years ago in temperate regions of China (Zhou et al. 2015). Chung et al. (2014) catalogued genome variation in an annual soybean population by high-depth resequencing of ten cultivated and six wild accessions and obtained 3.87 million high-quality single nucleotide polymorphisms (SNPs). Nuclear genome phylogeny supported a single origin for the cultivated soybeans. In this study, tenfold higher linkage disequilibrium (LD) in the wild soybean relatives was identified as compared to wild maize and rice. Despite the small population size, the high LD and large SNP data allowed identification of 206 candidate domestication regions with significantly lower diversity in the cultivated, but not in the wild, soybeans. Valliyodan et al. (2016) reported genomic sequences of 106 soybean accessions representing a wide variety of geographic origins, and with desirable traits used in the soybean research program for crop improvement. Results indicated that US soybean cultivars were grouped into the elite group, and most of Chinese cultivars were grouped into the landrace group. Individuals from the same geographical region tended to cluster together, which reflected isolation by distance during evolution and/or parallel selections in similar ecological habitats accompanied by gene flow. The genetic diversity analysis using the whole-genome SNPs showed a lower level of genetic diversity in cultivated soybeans as compared to wild soybeans, and the genetic diversity decreased from wild to landraces, to elite cultivars. The total number of SNPs and the number of nonsynonymous SNPs was 15% higher in wild soybeans than in the landraces and elite groups. Also, the number of large effect SNPs was nearly the same between landraces and elite groups. Kim et al. (2021) analyzed the genomic variation of 781 soybean individuals consisting of *G. max*, *G. soja*, and hybrid (*G. max* × *G. soja*) accessions using whole-genome sequence data and identified putative deleterious mutations in soybean populations. The results indicated that there is 7.1% reduction of overall deleterious mutations in domesticated soybean relative to wild soybean and a further 1.4% reduction from landrace to improved accessions. The detected domestication-selective sweeps also show reduced levels of deleterious alleles.

A single origin of domesticated soybean has been suggested by several researchers and also supported by recent genome resequencing studies (Zhou et al. 2015; Han et al. 2016). Resequencing of 302 wild, landrace, or improved soybeans suggests that all domesticated soybeans derived from a single cluster of *G. soja* wild soybeans. This supports the single origin hypothesis that all cultivated soybeans originated from a single domestication event (Zhou et al. 2015). Several genetic bottlenecks occurred during soybean domestication and diversification, mainly in the domestication of Asian landraces and in the introduction of few landraces to North America (Hyten et al. 2006). The resequencing of wild and cultivated soybean accessions revealed the consequences of the artificial selection accompanying domestication and showed that genetic diversity was significantly decreased after domestication (Lam et al. 2010). However, naturally occurring introgression was widespread and counteracted genetic bottlenecks during soybean domestication (Wang et al. 2019).

Genomic regions associated with soybean domestication are sought based on test statistics using the levels and pattern of nucleotide diversity, and the extent of LD and haplotype extension for complete or partial selective sweeps. Population differentiation analysis has also been employed in identification of genomic regions that potentially underwent selection in geographic differentiation or modern breeding programs. Loss of pod shattering is a key agronomic trait that was targeted by human selection and is regarded as a milestone of crop domestication. Loss of pod shattering in soybean lines is promoted by a NAC transcription factor, SHAT1–5 (Dong et al. 2014). The domesticated allele of this gene is expressed 15-fold higher than the wild allele. *Pdh1*, encoding a dirigent-like protein involved in lignification, is another gene that affects soybean's pod shattering phenotype (Funatsuki et al. 2014). Shattering-resistant varieties carry a single nucleotide substitution at the beginning of the coding sequence that produces a stop codon. The shattering-resistant allele of *Pdh1* is observed at low frequencies in Japanese and Korean landraces and cultivars and at moderate frequency in China, while about 75% of South Asian landraces carry the resistant allele. Most of the modern North American cultivars possess the resistant allele, indicating that the *Pdh1* gene was utilized as an additional shattering-resistance locus in the modern breeding programs in North America.

Seed hardness is another important trait for soybean domestication and improvement. The gene associated with seed harness has been identified as *GmHs1-1*, which encodes a calcineurin-like metallo-phosphoesterase transmembrane protein (Sun et al. 2015). Determinacy is another important trait associated with the domestication process of soybean. Results demonstrated that soybean stem growth habit is regulated by an epistatic interaction between two major loci, *Dt1* and *Dt2* (Bernard 1972). A major focus of the soybean domestication and diversification process was selection for adaptation to a particular latitudinal photoperiod (Cober and Morrison 2010; Kim et al. 2012). As a short-day flowering plant, its latitudinal expansion requires loss of photoperiod sensitivities. Photoinsensitivity has played a great role in adaptation of soybean by helping it enter into the reproductive phase in long day conditions of higher latitudes and to rapidly complete the life cycle in frost-free period (Saindon et al. 1989). When genotypes adapted to higher latitudes are introduced to lower latitudes, they enter into the reproductive phase without attaining sufficient biomass (Islam et al. 2018). Several genes underlying soybean's latitudinal adaptation have been identified. Analyses of genetic regions that control photoperiod sensitivities of cultivated soybeans, known as maturity loci (*E* loci), indicated variation in DNA sequence of flowering-associated genes. The maturity of plants depends upon the complex and coordinated regulation of photoreceptors, floral meristem, and flowering time genes as well as geographical distribution. Valliyodan et al. (2016) examined the natural variation in 106 PI lines for known maturity (E1–E4) and plant architecture (*Dt*) genes. The functional analysis of mutant alleles for maturity genes showed an early flowering time phenotype. The haplogroup analysis of maturity and plant architecture genes showed a strong correlation, and distinct clusters were associated with growth habits and geographic origin. The analysis of E1–E4 genes (gene plus 3.5 kb upstream/downstream regions) found that the lower maturity group (MG) lines (MG-0, −I, −II) retain one or more mutant allele. Soybean accession, Fiskeby III from Sweden, belonged to MG-0, and showed an entire gene deletion for the E1 locus, leading to the e1-nl allele. In addition, this line showed all of the major mutant alleles for other maturity (E2 and E3) and *Dt* genes.

Among 180 cultivated soybeans surveyed, the percentages of recessive alleles at the major maturity loci *E1*, *E2*, *E3*, and *E4* were found to be 38.3, 84.5, 36.3, and 7.2%, respectively (Zhai et al. 2014), These maturity loci have significantly contributed to diversification or local adaptation. Among these *E* loci, *E1* shows a predominant effect on photoperiodic control of flowering and maturation. Another maturity locus *E2* encodes GmGIa, a homolog of *Arabidopsis* GIGANTEA (GI) that is a component of the circadian clock and a regulator of photoperiodic flowering (Watanabe et al. 2011). The dominant *E2* allele delays flowering and maturity, while the homozygous *e2* alleles elevate expression of *GmFT2a*, leading to early flowering. *E2* allele is more prevalent in wild soybeans and the *e2* allele in cultivated soybeans (Langewisch et al. 2014). The recessive *e2* haplotypes, H1, H2, and H3, display unique geographic patterns (Wang et al. 2016). H1 is widely distributed among cultivated soybeans, while H2 is present in Southern China. H3 is assumed to have been later introgressed from wild soybean independently and is restricted to the Northeast region of China. Several loci that are associated with the photoperiod response in soybean have been identified (Cober et al. 1996, 2010; Watanabe et al. 2009, 2011; Cober and Voldeng 2001a, b; Kong et al. 2010; Zhai et al. 2014; Lu et al. 2020; Lin et al. 2021; Dong et al. 2022; Kou et al. 2022; Hou et al. 2023). Three loci, namely, E2, Tof5, and Tof12 were mainly associated with domestication (Wang et al. 2016; Lu et al. 2020; Kou et al. 2022) and show sequence diversity between cultivated soybean and its annual relative *G. soja*. Parallel selection was observed for Tof5 loci in cultivated soybean and its annual wild relatives, both leading to adaptation to high latitudes (Dong et al. 2022).

4.4 Genome-Wide Association Studies

Genome-wide association studies are based on linkage disequilibrium (LD) between a marker or markers and a phenotypic trait of interest. Variation in LD at a particular-genomic region is affected by mutation, domestication, population admixture, population substructure, level of inbreeding, and selection and confounding effects (Rafalski and Morgante 2004). The extent of LD is also reliant on the recombination rate. Soybean usually shows less decay of LD (longer region is in LD) because the recombination is ineffective to cause LD decay in a homozygous genetic background, whereas high LD decay (shorter region in LD) is common in an outcrossing crop species. SNPs provide an abundant source of DNA polymorphism compared to SSRs, thereby improving the success rate in a variety of applications and determination of genetic relatedness among individuals. Zhu et al. (2003) sequenced 25 diverse soybean genotypes to determine SNP frequency in coding and noncoding regions of soybean genome. The nucleotide diversity (θ) observed was 0.00053 and 0.00111 in coding and in noncoding perigenic DNA, respectively, whereas combined nucleotide diversity of the whole sequence analyzed was 0.00097. Squared allele frequency correlations (r^2) among haplotypes at 54 loci with two or more SNPs indicated low genome-wide LD. A worldwide haplotype map for soybean (GmHapMap) was constructed using whole-genome sequence data for 1,007 *Glycine max* accessions (Torkamaneh et al. 2021). A lower level of genome-wide genetic diversity was observed in soybean as compared to other major crops. Genome-wide LD studies in soybean have facilitated identification of markers and key genes associated with various abiotic stresses and other traits of interest.

The genome-wide association study (GWAS) approach provides opportunities to explore the tremendous allelic diversity existing in natural soybean accessions (Lee et al. 2023). The recent advances in genome sequencing have played a key role in the genome-wide association studies (Abdurakhmonov and Abdukarimov 2008). GWAS is routinely being used in soybean and other plant species, for identifying markers associated with biotic and abiotic stresses and other agronomics traits.

4.4.1 Abiotic Stress Tolerance

Drought stress is a complex polygenic trait comprising several morpho-physiological sub-traits (Singh et al. 2022; Satpute et al. 2022; Ratnaparkhe et al. 2022a; Chaudhary et al. 2019). GWAS for quantitative traits like drought tolerance are predictable to be affected by a confounding population structure. Dhanapal et al. (2015) analyzed carbon isotope ratio ($\delta^{13}C$) in a population of 373 soybean genotypes and found association of 39 SNPs, which were tagged to 21 different loci. Similarly, Kaler et al. (2017) reported 54 SNPs associated with $\delta^{13}C$ and 47 SNPs associated with $\delta^{18}O$. These SNPs were tagged with 46 putative loci and 21 putative loci for $\delta^{13}C$, and $\delta^{18}O$, respectively. Several markers and loci have been reported to be associated with various drought-related traits, namely, chlorophyll fluorescence

(Hao et al. 2012; Herritt et al. 2018), canopy temperature (Kaler et al. 2017), delayed canopy wilting (Steketee et al. 2020), and drought susceptibility index (Chen et al. 2020). GWAS studies for drought tolerance were reported in a germplasm association panel containing 259 Chinese cultivars. The investigation was based on a total of 4,616 SNPs, and 15 SNP trait associations were identified, among which three SNPs were associated with two of the drought-tolerance indices (Liu et al. 2020). Zhang et al. (2022) identified several QTLs for drought conditions by using a RIL population comprising 234 F6:10 lines and a GWAS panel of 259 soybean accessions. They constructed a genetic map by using SNP markers and 18 QTLs were identified on 7 soybean chromosomes in two environments. Besides this, 53 QTLs were also identified in the GWAS panel on 19 soybean chromosomes. A combination of the two populations showed that two SNPs fell within two of the QTL (qPH7-4 and qPH7-6) confidence intervals. They relocated several previously reported drought-tolerance genes in soybean and some other crops but also identified number of nonsynonymous stress-related mutation site differences between the two parents, involving Glyma.07g093000, Glyma.07g093200, Glyma.07g094100, and Glyma.07g094200. One previously unreported new gene related to drought stress, Glyma.07g094200, was found by regional association analysis. The significant SNP CHR7-17619 (G/T) was within an exon of the Glyma.07g094200 gene. In the RIL population, the DSP value of the "T" allele of CHR7-17619 was significantly ($P < 0.05$) larger than the "G" allele in different environments.

Canopy wilting is one of the most important traits that directly justify drought stress. Steketee et al. (2020) conducted a GWAS study on the canopy wilting trait in 162 diverse genotypes and identified 45 SNPs tagging 44 loci, which are significantly associated with the canopy wilting trait, among them, some are found near the previously reported QTLs. The genomic regions discovered across environments can be exploited by breeders to improve soybean drought tolerance.

Yu et al. (2019) conducted GWAS in a panel of 347 soybean genotypes to identify SNPs associated with seed-flooding tolerance-related traits. In this study, three major QTNs, namely, QTN13, qNSR-10, and qEC-7-2 were identified. Further, QTN13 was consistently identified in all three studied traits and in multiple environments. Wu et al. (2020) conducted GWAS in a panel of 384 soybean lines for flooding tolerance, and a total of 14 SNPs were identified across all environments and models. Another GWAS study was conducted with Mixed Linear Model (MLM) and Multi-Locus Random-SNP-Effect Mixed Linear Model (mrMLM) for seed-flooding tolerance (Sharmin et al. 2020). Several common SNPs in between these two models were identified, among these SNPs, two key SNPs, Gm_08_11971416 and Gm_08_46239716, were consistently connected with seed-flooding tolerance-related traits such as electrical conductivity and germination rate.

Kan et al. (2015) conducted association mapping study in soybean by analyzing seed germination under salt stress and identified three loci on chromosome Gm08, Gm09, and Gm18. Using 283 diverse lines of soybean, Zeng et al. (2017a, b) identified eight loci on chromosome Gm02, Gm07, Gm08, Gm10, Gm13, Gm14, Gm16, and Gm20, which are significantly associated with leaf chloride concentrations and leaf chlorophyll concentrations and also confirmed a major locus on Gm03.

Huang et al. (2018) used 192 soybean germplasm lines and identified six genomic regions associated with salt tolerance on soybean chromosome Gm02, Gm03, Gm05, Gm06, Gm08, and Gm18. Do et al. (2019) used two GWAS populations for GWAS studies and identified a major locus on chromosome Gm03 and three new loci on Gm01, Gm08, and Gm18 associated with salt tolerance. Several SNPs were found to be significantly associated with four traits, leaf scorching score, chlorophyll content ratio, leaf sodium content, and leaf chloride content on chromosome Gm08 and Gm18, of which two adjacent peak SNPs were present in the significant region on Gm08 (Do et al. 2019). Zhang et al. (2019) identified 18 SNPs located on chromosome Gm08 and Gm18 associated with salt tolerance at germination stage. Seventeen of the eighteen significant SNPs were located in a major QTL*qST-8*, which was identified by linkage mapping in recombinant inbred lines (RILs) (Zhang et al. 2019).

4.4.2 Root Systems Architecture

Genomics-assisted breeding for new varieties with efficient root system architecture has great potential in increasing nutrient use efficiency and adaptation for challenging climates. Root systems architecture (RSA) is the most important trait that influences several biotic and abiotic stress as well as nutrient use efficiency. An important GWAS study was carried out for RSA by Mandozai et al. (2021) with the help of specific locus amplified fragment sequencing (SLAF-seq). They efficiently applied 5 different GWAS models in 260 spring soybean genotypes and identified 27 significantly associated SNPs distributed on chromosomes 2, 6, 8, 9, 13, 16, and 18 and two of them were found to be associated with multiple root-related traits. Eleven promising root and shoot regulating genes, namely, *Glyma.13G303800*, *Glyma.08G060600*, *Glyma.16G173300*, *Glyma.02G113900*, *Glyma.06G148800*, *Glyma.09G179600*, *Glyma.08G060300*, *Glyma.16G138900*, *Glyma.16G208400*, *Glyma.09G153400*, and *Glyma.18G094200* overlapping with significant SNPs were identified and verified by expression analysis. These genes showed role in positive regulation of root branching number in the spring soybean panel, which could be useful in genomics-assisted breeding programs.

4.4.3 Seed Protein and Oil Contents

Over the past two decades, several studies have revealed numerous seed protein and oil QTLs (SoyBase, the USDA, ARS Soybean Genetics and Genomics Database). Hwang et al. (2014) carried out a GWAS study for seed protein and oil content, using 298 genetically diverse soybean genotypes. A significant association of 25 SNPs with oil content was found, and these were physically located in 13 regions on 12 of the 20 chromosomes. The seed-protein QTL on chromosome 20 has been reported in several studies (Bolon et al. 2010; Silva et al. 2024). A key locus of 8.4 Mb region on chromosome 20 for seed protein content was discussed extensively that depicts

12 potential candidate genes. Further, it was narrowed to a 2.4 Mb region based on GWAS study (Hwang et al. 2014) that harbors only 6 genes, namely, *Glyma20g19620*, *Glyma20g19630*, *Glyma20g19680*, *Glyma20g21030*, *Glyma20g21040*, and *Glyma20g21080*. These genes were also confirmed to be associated with seed protein by expression analysis. The narrower GWAS-defined locus will help to design more precise genomics-assisted allele selection and expedite positional cloning of the genes associated with protein content.

4.5 Genomic Selection

GWAS and genome-wide prediction are key tools in genomics-assisted varietal development (Fig. 2). GWAS is used to assist molecular breeding program by identifying markers significantly associated with a trait of interest. Genomic prediction estimates breeding values in order to rank the selection candidates in molecular breeding. Genomic selection (GS) can potentially accelerate genetic improvement of soybean by reducing the time for selection (Silva et al. 2021; Matei et al. 2018). The US soybean germplasm collection has proven to be a valuable resource for creating GS models (Jarquin et al. 2016). Promising results have displayed successful prediction of grain yield, protein and oil content, plant height, maturity, seed weight (Duhnen et al. 2017; Stewart-Brown et al. 2019; Ravelombola et al. 2021) as well as soybean cyst nematode resistance (Ravelombola et al. 2019, 2020).

With the advent of new high-throughput platforms, genotyping has made GS more affordable and efficient. Jarquin et al. (2016) was one of the first studies examining the potential for GS in soybean for seed yield prediction. They reported a prediction accuracy of 0.64 for seed yield across 301 experimental lines from the soybean breeding program and found little improvement in accuracy when training set size (N_P) exceeded 100 breeding lines. Predicted success when performing GS tends to be higher in studies reporting results with prediction accuracy vs. predictive ability, especially for lower heritability traits. The BARCSoySNP6K iSelect BeadChip was used to genotype a mixed population of 235 soybean cultivars by Ma et al. (2016), and potential for GS was examined for plant height and seed yield. The potential to use GS has also been investigated within larger populations such as the SoyNAM population, which is composed of over 5,500 lines across 40 biparental populations (Xavier et al. 2016). Traits investigated were seed yield, days to maturity, plant height, pod number, node number, and pods per node.

5 Conclusions

Population genomics studies and applications have provided key insights into origin, evolution, genetic diversity, population structure, domestication syndrome, and pan-genomes of soybean. Population genomics has also assisted soybean breeding via genome-wide association studies and genomic selection. These studies and phylogeography studies (see Johnson et al. 2019) should be taken on a larger and higher scale using resequencing of many individuals from many wild and domesticated populations in the future. Populations at natural range margins should be included as these populations may be genetically distinct and harbor genetic variants of adaptation values (Chhatre and Rajora 2014; Pandey and Rajora 2012), especially under climate change conditions. Furthermore, population epigenomics and population transcriptomics studies will benefit understanding genetic and epigenetic basis of acclimation and adaptation of soybean. In future, population proteomics and population metabolomics studies should be undertaken to complement population genomics, population epigenomics, and population transcriptomics studies, and landscape genomics studies to understand genetic architecture of local adaptation (Rajora et al. 2016) in soybean. Population genomics information should be applied for conservation and sustainable management of soybean genetic resources.

References

Abdurakhmonov IY, Abdukarimov A. Application of association mapping to understanding the genetic diversity of plant germplasm resources. Int J Plant Genomics. 2008;2008:574927.

Abe J, Xu DH, Suzuki Y, Kanazawa A, Shimamoto Y. Soybean germplasm pools in Asia revealed by nuclear SSRs. Theor Appl Genet. 2003;106:445–53.

Apuya NR, Frazier BL, Keim P, Roth EJ, Lark KG. Restriction fragment length polymorphisms as genetic markers in soybean, *Glycine max* (L.) Merrill. Theor Appl Genet. 1988;75:889–901.

Arikit S, Xia R, Kakrana A, Huang K, Zhai J, Yan Z, et al. An atlas of soybean small RNAs identifies phased siRNAs from hundreds of coding genes. Plant Cell. 2014;26:4584–601.

Ashfield T, Egan A, Pfeil BE, Chen NW, Podicheti R, Ratnaparkhe MB, et al. Evolution of a complex disease resistance gene cluster in diploid *Phaseolu*s and tetraploid *Glycine*. Plant Physiol. 2012;159:336–54.

Bernard RL. Two genes affecting stem termination in soybeans. Crop Sci. 1972;12:235–9.

Bolon YT, Joseph B, Cannon SB, Graham MA, Diers BW, Farmer AD, et al. Complementary genetic and genomic approaches help characterize the linkage group I seed protein QTL. BMC Plant Biol. 2010;10:41.

Chaisan T, Van K, Kim MY, Kim KD, Choi BS, Lee SH. *In silico* single nucleotide polymorphism discovery and application to marker-assisted selection in soybean. Mol Breed. 2010;29:221–33.

Chandra S, Choudhary M, Bagaria PK, Nataraj V, Kumawat G, Choudhary JR, et al. Progress and prospectus in genetics and genomics of Phytophthora root and stem rot resistance in soybean (Glycine max L.). Front Genet. 2022;13:939182.

Chaudhary J, Shivaraj S, Khatri P, Ye H, Zhou L, Klepadlo M, et al. Approaches, applicability, and challenges for development of climate-smart soybean. In: Kole C, editor. Genomic designing of climate-smart oilseed crops. Berlin: Springer; 2019. p. 1–74.

Chen L, Fang Y, Li X, Zeng K, Chen H, Zhang H, et al. Identification of soybean drought-tolerant genotypes and loci correlated with agronomic traits contributes new candidate genes for breeding. Plant Mol Biol. 2020;102(1–2):109–22.

Chen H, Pan X, Wang F, Liu C, Wang X, Li Y, et al. Novel QTL and meta-QTL mapping for major quality traits in soybean. Front Plant Sci. 2021;8(12):774270.

Chhatre VE, Rajora OP. Genetic divergence and signatures of natural selection in marginal populations of a keystone, long-lived conifer, eastern white pine (Pinus strobus) from Northern Ontario. PLoS One. 2014;9:e97291.

Choi IY, Hyten DL, Lakshmi KM, Qijian S, Julian MC, Charles VQ, et al. A soybean transcript map: gene distribution, haplotype and SNP analysis. Genetics. 2007;176:685–96.

Chu JS-C, Peng B, Tang K, Yi X, Zhou H, Wang H, et al. Eight soybean reference genome resources from varying latitudes and agronomic traits. Sci Data. 2021;8:164.

Chung WH, Jeong N, Kim J, Lee WK, Lee YG, Lee SH, et al. Population structure and domestication revealed by high-depth resequencing of Korean cultivated and wild soybean genomes. DNA Res. 2014;21(2):153–67.

Cober ER, Morrison MJ. Regulation of seed yield and agronomic characters by photoperiod sensitivity and growth habit genes in soybean. Theor Appl Genet. 2010;120:1005–12.

Cober ER, Voldeng HD. Low R: FR light quality delays flowering of E7E7 soybean lines. Crop Sci. 2001a;41:1823–6.

Cober ER, Voldeng HD. A new soybean and photoperiod-sensitivity locus linked to E1 and T. Crop Sci. 2001b;41:698–701.

Cober ER, Tanner JW, Voldeng HD. Genetic control of photoperiod response in early-maturing near-isogenic soybean lines. Crop Sci. 1996;36:601–5.

Cober ER, Molnar SJ, Charette M, Voldeng HD. A new locus for early in soybean. Crop Sci. 2010;50:524–7.

Concibido VC, La Vallee B, McLaird P, Pineda N, Meyer J, Hummel L, et al. Introgression of a quantitative trait locus for yield from Glycine soja into commercial soybean cultivars. Theor Appl Genet. 2003;106:575–86.

Cregan PB, Jarvik T, Bush AL, Shoemaker RC, Lark KG, Kahler AL, et al. An integrated genetic linkage map of the soybean. Crop Sci. 1999;39:1464–90.

Cui Z, Carter TE Jr, Burton JW. Genetic base of 651 Chinese soybean cultivars released during 1923 to 1995. Crop Sci. 2000a;40:1470–81.

Cui Z, Carter TE Jr, Burton JW. Genetic diversity patterns in Chinese soybean cultivars based on coefficient of parentage. Crop Sci. 2000b;40:1780–93.

Cui Z, Carter TE Jr, Burton JW, Wells R. Phenotypic diversity of modern Chinese and North American soybean cultivars. Crop Sci. 2001;41:1954–67.

Dhanapal AP, Ray JD, Singh SK, Hoyos-Villegas V, Smith JR, Purcell LC, et al. Genome-wide association study (GWAS) of carbon isotope ratio ($\delta^{13}C$) in diverse soybean [*Glycine max* (L.) Merr.] genotypes. Theor Appl Genet. 2015;128:73–91.

Diers BW, Keim P, Fehr WR, Shoemaker RC. RFLP analysis of soybean seed protein and oil content. Theor Appl Genet. 1992;83:608–12.

Do TD, Vuong TD, Dunn D, Clubb M, Valliyodan B, Patil G, et al. Identification of new loci for salt tolerance in soybean by high-resolution genome-wide association mapping. BMC Genomics. 2019;20:318.

Dong YS, Zhao LM, Liu B, Wang ZW, Jin ZQ, Sun H. The genetic diversity of cultivated soybean grown in China. Theor Appl Genet. 2004;108:931–6.

Dong Y, Yang X, Liu J, Wang B-H, Liu B-L, Wang Y-Z. Pod shattering resistance associated with domestication is mediated by a NAC gene in soybean. Nat Commun. 2014;5:3352.

Dong LD, Cheng Q, Fang C, Kong LP, Yang H, Hou Z, et al. Parallel selection of distinct Tof5 alleles drove the adaptation of cultivated and wild soybean to high latitudes. Mol Plant. 2022;15:308–21.

Du H, Fang C, Li Y, Kong F, Liu B. Understandings and future challenges in soybean functional genomics and molecular breeding. J Integr Plant Biol. 2023;65:468–95.

Duhnen A, Gras A, Teyssèdre S, Romestant M, Claustres B. Genomic selection for yield and seed protein content in soybean: a study of breeding program data and assessment of prediction accuracy. Crop Sci. 2017;57:1325–37.

Ferreira AR, Foutz KR, Keim P. Soybean genetic map of RAPD markers assigned to an existing scaffold RFLP map. J Hered. 2000;91:392–6.

Funatsuki H, Suzuki M, Hirose A, Inaba H, Yamada T, Hajika M, et al. Molecular basis of a shattering resistance boosting global dissemination of soybean. Proc Natl Acad Sci USA. 2014;111:17797–802.

Gao W, Ma R, Li X, Liu J, Jiang A, Tan P, et al. Construction of genetic map and QTL mapping for seed size and quality traits in soybean (*Glycine max* L.). Int J Mol Sci. 2024;25:2857.

Ha J, Abernathy B, Nelson W, Grant D, Wu X, Nguyen HT, et al. Integration of the draft sequence and physical map as a framework for genomic research in soybean (*Glycine max* (L) Merr) and wild soybean (Glycine soja Sieb and Zucc). G3: Genes Genom Genet. 2012;2(3):321–9.

Ha CV, Watanabe Y, Tran UT, Le DT, Tanaka M, Nguyen KH, et al. Comparative analysis of root transcriptomes from two contrasting drought-responsive Williams 82 and DT2008 soybean cultivars under normal and dehydration conditions. Front Plant Sci. 2015;6:551.

Han Y, Zhao X, Liu D, Li Y, Lightfoot DA, Yang Z, et al. Domestication footprints anchor genomic regions of agronomic importance in soybeans. New Phytol. 2016;209:871–84.

Hao D, Chao M, Yin Z, Yu D. Genome-wide association analysis detecting significant single nucleotide polymorphisms for chlorophyll and chlorophyll fluorescence parameters in soybean (*Glycine max*) landraces. Euphytica. 2012;186:919–31.

Harlan JR, de Wet JMJ. Towards a rational classification of cultivated plants. Taxon. 1971;20:509–17.

Haun WJ, Hyten DL, Xu WW, Gerhardt DJ, Albert TJ, Richmond T, et al. The composition and origins of genomic variation among individuals of the soybean reference cultivar Williams 82. Plant Physiol. 2011;155:645–55.

Herritt M, Dhanapal AP, Purcell LC, Fritschi FB. Identification of genomic loci associated with 21 chlorophyll fluorescence phenotypes by genome-wide association analysis in soybean. BMC Plant Biol. 2018;18(1):312.

Hou Z, Fang C, Liu B, Yang H, Kong F. Origin, variation, and selection of natural alleles controlling flowering and adaptation in wild and cultivated soybean. Mol Breed. 2023;43:36.

Huang L, Zeng A, Chen P, Wu C, Wang D, Wen Z. Genome-wide association analysis of salt tolerance in soybean [*Glycine max* (L) Merr]. Plant Breed. 2018;137:714–20.

Huang Y, Koo DH, Mao Y, Herman EM, Zhang J, Schmidt MA. A complete reference genome for the soybean cv. Jack. Plant Commun. 2024;5(2):100765.

Hwang E-Y, Song Q, Jia G, Specht JE, Hyten D, Costa J, et al. A genome-wide association study of seed protein and oil content in soybean. BMC Genomics. 2014;15:1.

Hyten DL, Song Q, Zhu Y, Choi I-Y, Nelson RL, Costa JM, et al. Impacts of genetic bottlenecks on soybean genome diversity. Proc Natl Acad Sci USA. 2006;103:16666–71.

Hyten DL, Song Q, Choi IY, Yoon MS, Specht JE, Matukumalli LK, et al. High-throughput genotyping with the GoldenGate assay in the complex genome of soybean. Theor Appl Genet. 2008;116:945–52.

Hyten DL, Choi I-Y, Song Q, Specht JE, Carter TE, Shoemaker RC, et al. A high density integrated genetic linkage map of soybean and the development of a 1,536 Universal Soy Linkage Panel for QTL mapping. Crop Sci. 2010a;50:960–8.

Hyten DL, Cannon SB, Song Q, Weeks N, Fickus EW, Shoemaker RC, et al. High-throughput SNP discovery through deep resequencing of a reduced representation library to anchor and orient scaffolds in the soybean whole genome sequence. BMC Genomics. 2010b;11:38.

Innes RW, Ameline-Torregrosa C, Ashfield T, Cannon E, Cannon SB, Chacko B, et al. Differential accumulation of retroelements and diversification of NB-LRR disease resistance genes in duplicated regions following polyploidy in the ancestor of soybean. Plant Physiol. 2008;148:1740–59.

Islam MR, Fujita D, Watanabe S, Zheng SH. Variation in photosensitivity of flowering in the world soybean minicore collections (GmWMC). Plant Prod Sci. 2018;22:220. https://doi.org/10.1080/1343943X.2018.1561197.

Jarquin D, Specht J, Lorenz A. Prospects of genomic prediction in the USDA soybean germplasm collection: historical data creates robust models for enhancing selection of accessions. G3 (Bethesda). 2016;6:2329–41.

Johnson JS, Krutovsky KV, Rajora OP, Gaddis KD, Cairns DM. Advancing biogeography through population genomics. In: Rajora OP, editor. Population genomics: concepts, approaches and applications. Springer International Publishing AG; 2019. p. 539–86. https://doi.org/10.1007/13836_2018_39.

Kajiya-Kanegae H, Nagasaki H, Kaga A, Hirano K, Ogiso-Tanaka E, Matsuoka M, et al. Whole-genome sequence diversity and association analysis of 198 soybean accessions in mini-core collections. DNA Res. 2021;28(1):dsaa032.

Kaler AS, Ray JD, Schapaugh WT, King CA, Purcell LC. Genome-wide association mapping of canopy wilting in diverse soybean genotypes. Theor Appl Genet. 2017;130:2203–17.

Kan GZ, Tong ZF, Hu ZB, Zhang D, Zhang GZ, Yu DY. Mapping QTLs for yield-related traits in wild soybean (*Glycine soja* Sieb. and Zucc.). Soybean Sci. 2012;31:333–40.

Kan G, Zhang W, Yang W, Ma D, Zhang D, Hao D, et al. Association mapping of soybean seed germination under salt stress. Mol Genet Genomics. 2015;290(6):2147–62.

Kavishwar R, Chandra S, Satpute G, Kumawat G, Kamble V, Singh D, et al. Role of omics technologies in understanding drought stress tolerance in soybean. Soybean Res. 2021;19(1):90–7.

Keim P, Shoemaker RC, Palmer RG. Restriction fragment length polymorphism diversity in soybean. Theor Appl Genet. 1989;77:786–92.

Keim P, Diers BW, Olson TC, Shoemaker RC. RFLP mapping in soybean: association between marker loci and variation in quantitative traits. Genetics. 1990;126:735–42.

Keim P, Schupp JM, Travis SE, Clayton K, Zhu T, Shi L, et al. A high-density soybean genetic map based on AFLP markers. Crop Sci. 1997;37:537–43.

Kim MY, Lee S, Van K, Kim T-H, Jeong S-C, Choi I-Y, et al. Whole-genome sequencing and intensive analysis of the undomesticated soybean (*Glycine soja* Sieb. and Zucc.) genome. Proc Natl Acad Sci USA. 2010;107:22032–7.

Kim MS, Lozano R, Kim JH, Bae DN, Kim ST, Park JH, et al. The patterns of deleterious mutations during the domestication of soybean. Nat Commun. 2012;12(1):97.

Kim KD, El Baidouri M, Abernathy B, Iwata-Otsubo A, Chavarro C, Gonzales M, et al. A comparative epigenomic analysis of polyploidy-derived genes in soybean and common bean. Plant Physiol. 2015;168:1433–47.

Kim JY, Jeong S, Kim KH, Lim WJ, Lee HY, Jeong N, et al. Dissection of soybean populations according to selection signatures based on whole-genome sequences. Gigascience. 2019;8(12):giz151.

Kim MS, Lozano R, Kim JH, Bae DN, Kim ST, Park JH, et al. The patterns of deleterious mutations during the domestication of soybean. Nat Commun. 2021;12(1):97.

Kong F, Liu B, Xia Z, Sato S, Kim BM. Two coordinately regulated homologs of FLOWERING LOCUS T are involved in the control of photoperiodic flowering in soybean. Plant Physiol. 2010;154:1220–31.

Kou K, Yang H, Li HY, Fang C, Chen LY, Yue L, et al. A functionally divergent SOC1 homolog improves soybean yield and latitudinal adaptation. Curr Biol. 2022;32:1–15.

Kulcheski FR, de Oliveira LFV, Molina LG, Almerão MP, Rodrigues FA, Marcolino J, et al. Identification of novel soybean microRNAs involved in abiotic and biotic stresses. BMC Genomics. 2011;12:307.

Kumar R, Saini M, Taku M, Debbarma P, Mahto RK, Ramlal A, et al. Identification of quantitative trait loci (QTLs) and candidate genes for seed shape and 100-seed weight in soybean [*Glycine max* (L) Merr]. Front Plant Sci. 2023;4:1074245.

Kumawat G, Maranna S, Gupta S, Tripathi R, Agarwal N, Singh V, et al. Identification of novel genetic sources for agronomic and quality traits in soybean using multi-trait allele specific genic marker assays. J Plant Biochem Biotechnol. 2020;30:160–71.

Kuroda Y, Kaga A, Tomooka N, Vaughan D. Population genetic structure of Japanese wild soybean (*Glycine soja*) based on microsatellite variation. Mol Ecol. 2006;15:959–74.

Kuroda Y, Kaga A, Tomooka N, Vaughan D. The origin and fate of morphological intermediates between wild and cultivated soybeans in their natural habitats in Japan. Mol Ecol. 2010;19:2346–60.

Lam HM, Xu X, Liu X, Chen W, Yang G, Wong FL, et al. Resequencing of 31 wild and cultivated soybean genomes identifies patterns of genetic diversity and selection (2010). Nat Genet. 2010;42(12):1053–9.

Langewisch T, Zhang H, Vincent R, Joshi T, Xu D, Bilyeu K. Major soybean maturity gene haplotypes revealed by SNPViz analysis of 72 sequenced soybean genomes. PLoS One. 2014;9(4):e94150.

Lark KG, Weisemann JM, Matthews BF, Palmer R, Chase K, Macalma T. A genetic map of soybean (*Glycine max* L) using an intraspecific cross of two cultivars: "Minsoy" and "Noir 1". Theor Appl Genet. 1993;86:901–6.

Lee D, Lara L, Moseley D, Vuong TD, Shannon G, Xu D, et al. Novel genetic resources associated with sucrose and stachyose content through genome-wide association study in soybean (*Glycine max* (L.) Merr.). Front Plant Sci. 2023;14:1294659.

Li DD, Pfeiffer TW, Cornelius PL. Soybean QTL for yield and yield components associated with *Glycine soja* alleles. Crop Sci. 2008a;48:571–81.

Li W, Han Y, Zhang D, Yang M, Teng W, Jiang Z, et al. Genetic diversity in soybean genotypes from north-eastern China and identification of candidate markers associated with maturity rating. Plant Breed. 2008b;127:56–61.

Li YH, Zhao SC, Ma JX, Li D, Yan L, Li J, et al. Molecular footprints of domestication and improvement in soybean revealed by whole genome re-sequencing. BMC Genomics. 2013;28(14):579.

Li YH, Zhou G, Ma J, Jiang W, Jin LG. De novo assembly of soybean wild relatives for pan-genome analysis of diversity and agronomic traits. Nat Biotechnol. 2014;32(10):1045–52.

Li R, Jiang H, Zhang Z, Zhao Y, Xie J, Wang Q, et al. Combined linkage mapping and BSA to identify QTL and candidate genes for plant height and the number of nodes on the main stem in soybean. Int J Mol Sci. 2019;21(1):42.

Liang X, Hou X, Li J, Han Y, Zhang Y, Feng N, et al. High-resolution DNA methylome reveals that demethylation enhances adaptability to continuous cropping comprehensive stress in soybean. BMC Plant Biol. 2019;19:79.

Lightfoot DA, Njiti VN, Gibson PT, Kassem MA, Iqbal JM, Meksem K. Registration of the Essex by Forrest recombinant inbred line mapping population. Crop Sci. 2005;45:1678–81.

Lin XY, Liu BH, Weller JL, Abe J, Kong FJ. Molecular mechanisms for the photoperiodic regulation of flowering in soybean. J Integr Plant Biol. 2021;63:981–94.

Liu Y, Du H, Li P, Shen Y, Peng H, Liu S, et al. Pan-genome of wild and cultivated soybeans. Cell. 2020;182(1):162–76.

Liu N, Niu Y, Zhang G, Feng Z, Bo Y, Lian J, et al. Genome sequencing and population resequencing provide insights into the genetic basis of domestication and diversity of vegetable soybean. Hortic Res. 2022;9:uhab052.

Liu S, Liu Z, Hou X, Li X. Genetic mapping and functional genomics of soybean seed protein. Mol Breed. 2023;43:29.

Liu Z, Yang Q, Liu B, Li C, Shi X, Wei Y, et al. De novo genome assembly of a high-protein soybean variety HJ117. BMC Genom Data. 2024;25(1):25.

Lorenzen L, Boutin S, Young N, Specht JE, Shoemaker RC. Soybean pedigree analysis using map-based molecular markers: I. Tracking RFLP markers in cultivars. Crop Sci. 1995;35:1326–36.

Lu SJ, Dong LD, Fang C, Liu SL, Kong LP, Cheng Q, et al. Stepwise selection on homeologous PRR genes controlling flowering and maturity during soybean domestication. Nat Genet. 2020;52:428–36.

Luikart G, Kardos M, Hand B, Rajora OP, Aitkin S, Hohenlohe PA. Population genomics: advancing understanding of nature. In: Rajora OP, editor. Population genomics: concepts, approaches and applications. Springer International Publishing AG; 2019. p. 3–79. https://doi.org/10.1007/13836_2018_60.

Ma J, Shoemaker R, Jackson S, Cannon S. Comparative genomics. In: Bilyeu K, Ratnaparkhe MB, Kole C, editors. Genetics, genomics, and breeding of soybean. New Hampshire: CRC; 2010. p. 245–62.

Ma Y, Reif JC, Jiang Y, Wen Z, Wang D, Liu Z, et al. Potential of marker selection to increase prediction accuracy of genomic selection in soybean (*Glycine max L.*). Mol Breed. 2016;36:113. https://doi.org/10.1007/s11032-016-0504-9.

Maguire TL, Grimmond S, Forrest A, Iturbe-Ormaetxe I, Meksem K, Gresshoff P. Tissue-specific gene expression in soybean (*Glycine max*) detected by cDNA microarray analysis. J. Plant Physiol. 2002;159:1361–4.

Mandozai A, Moussa AA, Zhang Q, Qu J, Du Y, Anwari G, et al. Genome-wide association study of root and shoot related traits in spring soybean (*Glycine max* L.) at seedling stages using SLAF-seq. Front Plant Sci. 2021;12:568995.

Mansur LM, Orf JH, Chase K, Jarvik T, Cregan PB, Lark KG. Genetic mapping of agronomic traits using recombinant inbred lines of soybean. Crop Sci. 1996;36:1327–36.

Maranna S, Kumawat G, Nataraj V, Gill B, Nargund R, Ratnaparkhe MB, et al. Development of improved genotypes for extra early maturity, higher yield and Mungbean Yellow Mosaic India Virus (MYMIV) resistance in soybean (*Glycine max* L.). Crop Pasture Sci. 2023;11(1):22853.

Matei G, Woyann LG, Milioli AS, de Bem Oliveira I, Zdziarski AD, Zanella R, et al. Genomic selection in soybean: accuracy and time gain in relation to phenotypic selection. Mol Breed. 2018;38:117.

Moler E, Abakir A, Eleftheriou M, Johnson JS, Krutovsky KV, Lewis LC, et al. Population epigenomics: advancing understanding of phenotypic plasticity, acclimation, adaptation and diseases. In: Rajora OP, editor. Population genomics: concepts, approaches and applications. Springer International Publishing AG; 2019. p. 179–260. https://doi.org/10.1007/13836_2018_59.

Neto JRCF, Pandolfi V, Guimaraes FCM, Benko-Iseppon AM, Romero C, de Oliveira Silva RL, et al. Early transcriptional response of soybean contrasting accessions to root dehydration. PLoS One. 2013;8(12):83466.

Pandey M, Rajora OP. Genetic diversity and differentiation of core versus peripheral populations of eastern white cedar, *Thuja occidentalis* L. (Cupressaceae). Am J Bot. 2012;99:690–9.

Powell W, Morgante M, Doyle JJ, McNicol JW, Tingey SV, Rafalske AJ. Gene pool variation in genus *Glycine* subgenus soja revealed by polymorphic nuclear and chloroplast micro-satellites. Genetics. 1996;144:793–803.

Rafalski A, Morgante M. Corn and humans: recombination and linkage disequilibrium in two genomes of similar size. Trends Genet. 2004;20:103–11.

Rajora OP. Population genomics: concepts, approaches and application: Springer International Publishing AG; 2019. 823pp. ISBN 978-3-030-04587-6; ISBN 978-3-030-04589-0 (eBook).

Rajora OP, Eckert AJ, Zinck JWR. Single-locus versus multilocus patterns of local adaptation to climate in eastern white pine (*Pinus strobus*, Pinaceae). PLoS One. 2016;11(7):e0158691. https://doi.org/10.1371/journal.pone.0158691.

Rambani A, Rice JH, Liu J, Lane T, Ranjan P, Mazarei M, et al. The methylome of soybean roots during the compatible interaction with the soybean cyst nematode. Plant Physiol. 2015;168:1364–77.

Ramesh SV, Ratnaparkhe MB, Kumawat G, Gupta GK, Husain SM. Plant miRNAome and antiviral resistance: a retrospective view and prospective challenges. Virus Genes. 2014;48:1–14.

Ratnaparkhe MB, Singh RJ, Doyle JJ. Glycine. In: Kole C, editor. Wild crop relatives: genomic and breeding resources. Berlin Heidelberg: Springer; 2010. p. 83–116.

Ratnaparkhe MB, Ramesh SV, Giriraj K, Husain SM, Gupta S. Advances in soybean genomics. In: Gupta S, Nadarajan N, Gupta D, editors. Legumes in the Omic Era. New York: Springer; 2013. p. 41–72.

Ratnaparkhe MB, Marmat N, Kumawat G, Shivakumar M, Kamble VG, Nataraj V, et al. Whole genome re-sequencing of soybean accession EC241780 providing genomic landscape of candidate genes involved in rust resistance. Curr Genomics. 2020;21(7):504–11.

Ratnaparkhe MB, Satpute GK, Kumawat G, Chandra S, Kamble VG, Kavishwar R, et al. Genomic designing for abiotic stress tolerant soybean. In: Kole C, editor. Genomic designing for abiotic stress resistant oilseed crops. Cham: Springer; 2022a. p. 1–73.

Ratnaparkhe MB, Nataraj V, Shivakumar M, Chandra S. Genomic design for biotic stresses in soybean. In: Kole C, editor. Genomic designing for biotic stress resistant oilseed. Cham: Springer; 2022b. p. 1–54.

Ravelombola WS, Qin J, Shi A, Nice L, Bao Y. Genome-wide association study and genomic selection for soybean chlorophyll content associated with soybean cyst nematode tolerance. BMC Genomics. 2019;20:904. https://doi.org/10.1186/s12864-019-6275-z.

Ravelombola WS, Qin J, Shi A, Nice L, Bao Y. Genome-wide association study and genomic selection for tolerance of soybean biomass to soybean cyst nematode infestation. PLoS One. 2020;15:e0235089.

Ravelombola W, Qin J, Shi A, Song Q, Yuan J. Genome-wide association study and genomic selection for yield and related traits in soybean. PLoS One. 2021;16:e0255761. https://doi.org/10.1371/journal.pone.0255761.

Saindon G, Voldeng HD, Beversdorf WD, Buzzell RI. Genetic control of long daylength response in soybean. Crop Sci. 1989;29:1436–9.

Satpute GK, Ratnaparkhe MB, Chandra S, Kamble VG, Kavishwar R, Singh AK, et al. Breeding and molecular approaches for evolving drought-tolerant soybeans. In: Giri B, Sharma MP, editors. Plant stress biology. Singapore: Springer; 2022. p. 83–130.

Schmutz J, Cannon SB, Schlueter J, Ma J, Mitros T, Nelson W, et al. Genome sequence of the paleopolyploid soybean. Nature. 2010;463:178–83.

Shaibu AS, Ibrahin H, Zainab LM, Mohammed IB, Mohammed SG, Yusuf HL, et al. Assessment of the genetic structure and diversity of soybean (*Glycine max* L.) germplasm using diversity array technology and single nucleotide polymorphism markers. Plants. 2021;11(1):68. https://doi.org/10.3390/plants11010068.

Sharmin RA, Bhuiyan MR, Lv W, Yu Z, Chang F, Kong J, et al. RNA-Seq based transcriptomic analysis revealed genes associated with seed-flooding tolerance in wild soybean (*Glycine soja* Sieb & Zucc). Environ Exp Bot. 2020;171:103906.

Shen Y, Liu J, Geng H, Zhang J, Liu Y, Zhang H, et al. De novo assembly of a Chinese soybean genome. Sci China Life Sci. 2018a;61:871. https://doi.org/10.1007/s11427-018-9360-0.

Shen Y, Zhang J, Liu Y, Liu S, Liu Z, Duan Z, et al. DNA methylation footprints during soybean domestication and improvement. Genome Biol. 2018b;19(1):128.

Shoemaker RC, Specht JE. Integration of the soybean molecular and classical genetic linkage groups. Crop Sci. 1995;35:436–46.

Silva ÉDB da, Xavier A, Faria MV. Impact of genomic prediction model, selection intensity, and breeding strategy on the long-term genetic gain and genetic erosion in soybean breeding. Front Genet. 2021;12:637133. https://doi.org/10.3389/fgene.2021.637133.

Silva JNB, Bueno RD, de Sousa TdJF, Xavier YPM, Silva LCC, Piovesan ND, et al. Exploring SoySNP50K and USDA germplasm collection data to find new QTLs associated with protein and oil content in Brazilian genotypes. Biochem Genet. 2024; https://doi.org/10.1007/s10528-024-10698-5.

Singh RJ, Hymowitz T. Intersubgeneric crossability in the genus Glycine Willd. Plant Breed. 1987;98:171–3.

Singh RJ, Klein TM, Mauvais CJ, Knowlton S, Hymowitz T, Kostow CM. Cytological characterization of the transgenic soybean. Theor Appl Genet. 1998a;96:319–24.

Singh RJ, Kollipara KP, Hymowitz T. The genomes of *Glycine canescens* F J Herm and *G. tomentella* Hayata of Western Australia and their phylogenetic relationships in the genus *Glycine* Willd. Genome. 1998b;41:669–79.

Singh RJ, Chung GH, Nelson RL. Landmark research in legumes. Genome. 2007a;50:525–37.

Singh RJ, Nelson RL, Chung GH. Soybean (*Glycine max* (L) Merr). In: Singh RJ, editor. Genetic resources, chromosome engineering and crop improvement, Oilseed crops, vol. 4. Boca Raton: CRC; 2007b. p. 13–50.

Singh AK, Raina SK, Kumar M, Aher L, Ratnaparkhe MB, Rane J, et al. Modulation of GmFAD3 expression alters abiotic stress responses in soybean. Plant Mol Biol. 2022;110(1–2):199–218.

Skorupska H, Albertsen MC, Langholz KD, Palmer RG. Detection of ribosomal RNA genes in soybean, *Glycine max* (L.) Merr., by *in situ* hybridization. Genome. 1989;32:1091–5.

Skorupska HT, Shoemaker RC, Warner A, Shipe ER, Bridges WC. Restriction fragment length polymorphism in soybean germplasm of the southern USA. Crop Sci. 1993;33:1169–76.

Song QX, Lu X, Li QT, Chen H, Hu XY, Ma B, et al. Genome-wide analysis of DNA methylation in soybean. Mol Plant. 2013;6(6):1961–74.

Song Q, Hyten DL, Jia G, Quigley CV, Fickus EW, Nelson RL, et al. Fingerprinting soybean germplasm and its utility in genomic research. G3 (Bethesda). 2015;5(10):1999–2006.

Steketee CJ, Schapaugh WT, Carter TE, Li Z. Genome-wide association analyses reveal genomic regions controlling canopy wilting in soybean. G3 (Bethesda). 2020;10:1413–25.

Stewart-Brown BB, Song Q, Vaughn JN, Li Z. Genomic selection for yield and seed composition traits within an applied soybean breeding program. G3 (Bethesda). 2019;9:2253–65.

Sun L, Miao Z, Cai C, Zhang D, Zhao M, Wu Y, et al. *GmHs1-1*, encoding a calcineurin-like protein, controls hard-seededness in soybean. Nat Genet. 2015;47:939–43.

Sun Z, Lam HM, Lee SH, Li X, Kong F. Soybean functional genomics: bridging theory and application. Mol Breed. 2024;44:2.

Thao NP, Thu NB, Hoang XL, Van Ha C, Tran LS. Differential expression analysis of a subset of drought-responsive GmNAC genes in two soybean cultivars differing in drought tolerance. Int J Mol Sci. 2013;14:23828–41.

Thibaud-Nissen F, Shealy RT, Khanna A, Vodkin LO. Clustering of microarray data reveals transcript patterns associated with somatic embryogenesis in soybean. Plant Physiol. 2003;132(1):118–36.

Torkamaneh D, Laroche J, Valliyodan B, O'Donoughue L, Cober E, Rajcan I, et al. Soybean (*Glycine max*) Haplotype Map (GmHapMap): a universal resource for soybean translational and functional genomics. Plant Biotechnol J. 2021;19(2):324–34.

Tripathi P, Rabara RC, Reese RN, Miller MA, Rohila JS, Subramanian S, et al. A toolbox of genes, proteins, metabolites and promoters for improving drought tolerance in soybean includes the metabolite coumestrol and stomatal development genes. BMC Genomics. 2016;17:102.

Tripathi R, Agrawal N, Kumawat G, Gupta S, Varghese P, Ratnaparkhe MB, et al. QTL mapping for long juvenile trait in soybean accession AGS 25 identifies association between a functional allele of *FT2a* and delayed flowering. Euphytica. 2021;217:36.

Turner M, Yu O, Subramanian S. Genome organization and characteristics of soybean microRNAs. BMC Genomics. 2012;13:169.

Valliyodan B, Qiu D, Patil G, Zeng P, Huang J, Dai L, et al. Landscape of genomic diversity and trait discovery in soybean. Sci Rep. 2016;31(6):23598.

Valliyodan B, Cannon SB, Bayer PE, Shu S, Brown AV, Ren L, et al. Construction and comparison of three reference-quality genome assemblies for soybean. Plant J. 2019;100(5):1066–82.

Valliyodan B, Brown AV, Wang J, Patil G, Liu Y, Otyama PI, et al. Genetic variation among 481 diverse soybean accessions, inferred from genomic re-sequencing. Sci Data. 2021;8:50.

Vodkin LO, Khanna A, Shealy R, Clough SJ, Gonzalez DO, Philip R, et al. Microarray analysis for global expression constructed with a low redundancy set of 27,500 sequenced cDNAs representing an array of developmental stages and physiological conditions of the soybean plant. BMC Genomics. 2004;5:73.

Wang D, Graef GL, Procopiuk AM, Diers BW. Identification of putative QTL that underlie yield in interspecific soybean backcross populations. Theor Appl Genet. 2004;108:458–67.

Wang Y, Gu YZ, Gao HH, Qiu LJ, Chang RZ, Chen SY, et al. Molecular and geographic evolutionary support for the essential role of GIGANTEAa in soybean domestication of flowering time. BMC Evol Biol. 2016;16:79.

Wang YY, Li YQ, Wu HY, Hu B, Zheng JJ, Zhai H, et al. Genotyping of soybean cultivars with medium-density array reveals the population structure and QTNs underlying maturity and seed traits. Front Plant Sci. 2018;9(9):610.

Wang X, Chen L, Ma J. Genomic introgression through interspecific hybridization counteracts genetic bottleneck during soybean domestication. Genome Biol. 2019;20:22.
Watanabe S, Hideshima R, Xia Z, Tsubokura Y, Sato S, Nakamoto Y, et al. Map-based cloning of the gene associated with the soybean maturity locus E3. Genetics. 2009;182:1251–62.
Watanabe S, Xia Z, Hideshima R, Tsubokura Y, Sato S, Harada K. A map-based cloning strategy employing a residual heterozygous line reveals that the GIGANTEA gene is involved in soybean maturity and flowering. Genetics. 2011;188:395–407.
Wawrzynski A, Ashfield T, Chen NWG, Mammadov J, Nguyen A, Podicheti R, et al. Replication of nonautonomous retroelements in soybean appears to be both recent and common. Plant Physiol. 2008;148:1760–71.
Wen ZX, Zhao TJ, Zheng YZ, Liu SH, Wang CE, Wang F, et al. Association analysis of agronomic and quality traits with SSR markers in *Glycine max* and *Glycine soja* in China: I. Population structure and associated markers. Acta Agron Sin. 2008;34:1169–78.
Wu C, Mozzoni LA, Moseley D, Hummer W, Ye H, Chen P, et al. Genome-wide association mapping of flooding tolerance in soybean. Mol Breed. 2020;40(1):1–4.
Xavier A, Muir WM, Rainey KM. Assessing predictive properties of genome-wide selection in soybeans. G3 (Bethesda). 2016;6:2611–6.
Xie M, Chung CY-L, Li MW, Wong F-L, Wang X, Liu A, et al. A reference-grade wild soybean genome. Nat Commun. 2019;10:1216. https://doi.org/10.1038/s41467-019-09142-9.
Xu DH, Gai JY. Genetic diversity of wild and cultivated soybeans growing in China revealed by RAPD analysis. Plant Breed. 2003;122:503–6.
Yamanaka N, Ninomiya S, Hoshi M, Tsubokura Y, Yano M, Nagamura Y, et al. An informative linkage map of soybean reveals QTLs for flowering time, leaflet morphology and regions of segregation distortion. DNA Res. 2001;8:61–72.
Yang C, Yan J, Jiang S, Li X, Min H, Wang X, et al. Resequencing 250 soybean accessions: new insights into genes associated with agronomic traits and genetic networks. Genomics Proteomics Bioinformatics. 2022;20(1):29–41.
Yu Z, Chang F, Lv W, Sharmin RA, Wang Z, Kong J, et al. Identification of QTN and candidate gene for seed-flooding tolerance in soybean [*Glycine max* (L) Merr] using genome-wide association study (GWAS). Genes (Basel). 2019;10(12):957.
Zeng A, Chen P, Korth K, Hancock F, Pereira A, Brye K, et al. Genome-wide association study (GWAS) of salt tolerance in worldwide soybean germplasm lines. Mol Breed. 2017a;37(3):30.
Zeng A, Lara L, Chen P, Luan X, Hancock F, Korth K, et al. Quantitative trait loci for chloride tolerance in "Osage" soybean. Crop Sci. 2017b;57:2345–53.
Zhai H, Lü S, Liang S, Wu H, Zhang X, Liu B, et al. GmFT4, a homolog of flowering locus T, Is positively regulated by E1 and functions as a flowering repressor in soybean. PLoS One. 2014;9(2):e89030.
Zhang J, Wang J, Jiang W, Liu J, Yang S, Gai J, et al. Identification and analysis of NaHCO3 stress responsive genes in wild soybean (*Glycine soja*) roots by RNA-seq. Front Plant Sci. 2016;7:1842.
Zhang W, Liao X, Cui Y, Ma W, Zhang X, Du H, et al. A cation diffusion facilitator, GmCDF1, negatively regulates salt tolerance in soybean. PLoS Genet. 2019;15(1):e1007798.
Zhang Y, Liu Z, Wang X, Li Y, Li Y, Gou Z, et al. Identification of genes for drought resistance and prediction of gene candidates in soybean seedlings based on linkage and association mapping. Crop J. 2022;10:830–9.
Zhao M, Meyers BC, Cai C, Xu W, Ma J. Evolutionary patterns and coevolutionary consequences of MIRNA genes and microRNA targets triggered by multiple mechanisms of genomic duplications in soybean. Plant Cell. 2015;27(3):546–62.
Zhou Z, Wang Z, Li W, Fang C, Shen Y, Li C, et al. Comprehensive analyses of microRNA gene evolution in paleopolyploid soybean genome. Plant J. 2013;76:332–44.

Zhou Z, Jiang Y, Wang Z, Gou Z, Lyu J, Li W. Resequencing 302 wild and cultivated accessions identifies genes related to domestication and improvement in soybean. Nat Biotechnol. 2015;33(4):408–14.

Zhu YL, Song QJ, Hyten DL, Van Tassell CP, Matukumalli LK, Grimm DR, et al. Single-nucleotide polymorphisms in soybean. Genetics. 2003;163:1123–34.

Zhuang Y, Wang X, Li X, Hu J, Fan L, Landis JB, et al. Phylogenomics of the genus *Glycine* sheds light on polyploid evolution and life-strategy transition. Nat Plants. 2022;8:233–44.

Population Genomics of *Phaseolus* spp.: A Domestication Hotspot

Travis A. Parker and Paul Gepts

Abstract The genus *Phaseolus* includes five domesticated species. Collectively, *Phaseolus* beans are produced across all inhabited continents under a variety of environments, mainly for dry beans but also green pods. Building on the development of genetic tools, the last two decades have seen an active development of genomic resources, comprising several whole-genome reference sequences (in common, lima, and tepary beans), gene expressions databases, and genotyped diversity and MAGIC populations. The genus is a domestication hotspot with seven independent domestication events, including two species (common and lima beans) that were each domesticated twice. The five domesticated species represent diverse ecological niches and form a gradient in the intensity of domestication, perenniality, and rate of outcrossing. In common bean, genomic regions and candidate genes have been correlated with different climatic variables to which these wild populations are adapted. The different domestications have induced – to varying extents – a reduction in molecular diversity. Bottom-up (or reverse genetic) approaches have identified regions with reduced genetic diversity (potential selective sweeps) and high differentiation in ~15% of the genome. Although such a large proportion is probably an overestimate due to linkage disequilibrium and background selection, some genome regions coincide with domestication quantitative loci identified through top-down (or direct genetic) approaches, and contain candidate genes. Genes have been identified that control several domestication syndrome traits. These often are loss-of-function mutations, underscoring the importance of this mutation type in adaptation. Traits such as photoperiod insensitivity, determinacy and pigmentation changes reflect convergent evolution intra-specifically and possibly inter-specifically. Further developments in *Phaseolus* population genomics should involve a more detailed analysis of the domestication process across the five domesticated species, a better understanding of differential gene expression networks, application to marker-assisted and genomic selections, and transfer of traits from unadapted germplasm into advanced cultivars.

T. A. Parker · P. Gepts (✉)
Department of Plant Sciences/MS1, Section of Crop and Ecosystem Sciences, University of California, Davis, CA, USA
e-mail: plgepts@ucdavis.edu

Keywords Adaptation · Breeding · Convergent evolution · Domestication syndrome · Domestication triangle · Farmer selection · Gene flow · Growth habit · Long-distance dispersal · Multiple domestications · Photoperiod sensitivity · Pod dehiscence · Syngameon

1 Introduction

Domestication is a crucial component of the transition from foraging to agriculture, which happened in several regions of the world some 10,000 years ago (Larson et al. 2014). The active planting, cultivation, and harvest steps inherent in the agricultural process, combined with the preferences of humans as cultivators and consumers, set the stage for a selective process that resulted in the development of crop plants, each with its characteristic domestication syndrome (Gepts 2004). The details of the domestication process and the types of plants subject to domestication vary among the different centers of agricultural origins and the specific crops (Harlan 1992; Gepts 2014a, b).

Domestication and crop evolution are subject of active research for several reasons. The transition from hunting-gathering to agriculture is one of the most significant milestones in human evolution. First, this transition has had major effects on multiple facets of human life on Earth, including increasing human population levels and altering human well-being (Harper and Armelagos 2013), markedly changing land use and habitat modifications or even destructions (Stephens et al. 2019), the development of cities and city-states and increased urbanization (e.g., Scott 2018), and more hierarchical, less equal societies (e.g., Dong et al. 2017).

Second, crop evolution, in general, and domestication, in particular, are experimental models to study the phenomenon of evolution in general (Bolnick et al. 2018). In his seminal volume, Darwin (1859) devoted the first chapter to the evolution of crop plants and domestic animals to illustrate the potent effects of selection on phenotypic evolution. He would pursue this theme in a later, more detailed two-volume contribution (Darwin 1868). The advantages in using crop plants (and domestic animals) for the study of evolution include the general availability of crop wild relatives in natural populations or as gene bank accessions (especially the descendants of their immediate wild ancestors) (e.g., Warschefsky et al. 2014; Castañeda-Álvarez et al. 2016), knowledge about time frames of domestication and human dispersal (e.g., Larson et al. 2014; Carney and Rosomoff 2009; Spengler 2020), and availability of biological knowledge and genetics and genomics tools for crop species (e.g., Casacuberta et al. 2016; De Filippis 2018).

In contrast, however, with natural evolution, domestication involves human factors as well, leading to a "domestication triangle" in which each apex represents one of the three major categories of factors influencing the process of domestication, including biological, environmental, and human factors (Gepts 2004; Hufford et al. 2019). The human component – often neglected in crop evolution and domestication studies – includes both archaeological data, such as archaeobotanical information, and ethnobiological data, such as contemporary human agency affecting crop biodiversity (Brush 2004; Worthington et al. 2012; Wilkus et al. 2018; Hufford et al. 2019; Zimmerer 2020).

Third, the genetic relationships and genomic structure and function of today's crops, whether they are major, staple crops or underutilized, regional or endemic crops, are dependent on their phylogenetic derivation and recent evolution. In turn, an understanding of the resulting organization of crop genetic diversity and the availability of well-studied genetic resources is a condition *sine qua non* for the conservation of this genetic diversity, both *in situ* and *ex situ* (Gepts 2006; Motley et al. 2006; Hunter et al. 2017a), the development of the improved cultivars and crops (e.g., Hübner et al. 2019; Varshney et al. 2019; Fu 2019), and general improvements in agriculture needed to achieve production and sustainability goals (Brummer et al. 2011; Ray et al. 2013; Hunter et al. 2017b).

In this chapter, we review the evolution of genetic and genomic diversity in the genus *Phaseolus* (Fabaceae; with a few exceptions, $2n = 2x = 22$), especially from the standpoint of evolutionary factors operating before, during, and after domestication. Unlike many other genera of agricultural importance, *Phaseolus* spp. (Fig. 1) has been subjected to multiple domestications: no less than five species were domesticated, two of which were domesticated twice. The degree to which these species were domesticated and the outcome of these domestications differs among the seven domestications, thus, making *Phaseolus* a fascinating experimental model to study the effect and interactions among biological, environmental, and anthropic factors affecting the domestication phenomenon.

Fig. 1 Seeds of domesticated *Phaseolus*. The nutritional and agronomic utility of the genus led to seven independent domestications. Today, it is an important source of nutrition globally. Domesticated *Phaseolus* species are highly diverse in environmental adaptation, resistance to biotic and abiotic stresses, perenniality, pollination type, extent of domestication, seed type, and many other characteristics (Photo: T. Parker)

The five domesticated species of *Phaseolus* differ in numerous capacities, such as their life histories, levels of outcrossing, ecological adaptations, genetic diversity, and degree of domestication. Despite these differences, they have been selected for many of the same properties, such as a reduction in pod shattering, loss of seed dormancy, increased seed size and phenotypic variation, and an emphasis on reproductive investment over vegetative growth. The use of population genomics methods to study these properties allows for a detailed evaluation of fundamental evolutionary properties, including the degree of molecular parallelism and predictability, the role of structural genes versus transcription factors, changes in genetic diversity and linkage disequilibrium, patterns of population differentiation and gene flow, and the parallel evolution of associated organisms, including pathogens (Gepts 2014b, c; Hufford et al. 2019). These relationships have major economic value in conservation of genetic resources, which are valuable for breeding for crop resilience (Gepts 2006; Acosta-Gallegos et al. 2007). Production of *Phaseolus* beans, including *P. vulgaris* in particular, is threatened by the effects of climate change (Ramirez-Cabral et al. 2016). The increasing understanding of population genomics in the genus will be important for the development of new varieties which maintain productivity in inclement ecological and biological conditions. Continued improvements in *Phaseolus* population genomics will facilitate the movement of useful alleles between gene pools and will ultimately be critical to meet production targets for crops in this important genus.

2 *Phaseolus* spp. Beans as a Biocultural Asset: Their Importance in Global Agriculture and Nutrition

The *Phaseolus* domesticates, ranked by production, include common bean (*P. vulgaris* L.), lima bean (*P. lunatus* L.), runner bean (*P. coccineus* L.), tepary bean (*P. acutifolius* A. Gray), and year bean (*P. dumosus* Macfady). Beans of this genus are grown for several purposes, mainly dry seeds used as grain legumes and pods and immature seeds prepared as vegetables. *Phaseolus* beans are produced and consumed on a greater scale than any other non-oilseed pulse (Broughton et al. 2003).

Phaseolus beans are a major part of the culture of regions where they are a staple, such as Latin America and Africa. In Mesoamerica, beans are traditionally grown with maize (*Zea mays* L.) in the milpa cropping system, which provides a complementary combination of crops from an agronomic (Zhang et al. 2014) and dietary standpoint (Lopez-Ridaura et al. (2021). In Rwanda, dry beans are responsible for 65% of daily protein intake for the population, compared to only 4% derived from animal products (Larochelle and Alwang 2014). These patterns highlight the nutritional importance of *Phaseolus* dry beans for hundreds of millions of people globally, particularly in less food-secure regions. The important nutritional role of *Phaseolus* beans has existed in the Americas since ancient times (Zizumbo-Villareal and Colunga GarcíaMarín 2010; Zizumbo-Villarreal et al. 2012, 2014). Ancient farmers selected for a variety of cultural uses for *Phaseolus* beans, and they were foundational for the development of all the major pre-Columbian civilizations in the Americas.

Selections also occurred as *Phaseolus* beans were dispersed beyond their ancestral homelands in the tropical and subtropical Americas. These include modifications to the photoperiod sensitivity, either requiring no environmental cues or requiring new combinations of photoperiod, temperature, and/or precipitation (White and Laing 1989). They also were selected for new cultural uses, such as the loss of pod strings and wall fiber in snap beans (Parker et al. 2021a). Beans have taken on cultural value around the world, with many dishes considered typical of the local cuisine and some rising to the status of national dishes, like "Feijoada" in Brazil, "Gallo Pinto" of Central America, "Cassoulet" of France, "Lobio" of Georgia, "Jajangmyeon" in Korea, "Ugali na maharage ya nazi" in Tanzania, and "Fagioli all'uccelletto" of Italy. This involved immense human-mediated diversifying selection for seed and pod traits (color, size, shape, and texture), leading to variability seen in few other crops. In contrast with many other crops, these consumer selections have left a major impact on the genomes of *Phaseolus* species.

3 Genomic Resources in *Phaseolus* spp.

The genus *Phaseolus* has benefited from an expanding number of genomics tools (Cortinovis et al. 2020; Pérez de la Vega et al. 2017). These include high-quality reference genomes, an extensive set of genetic markers, assays, and databases, RNA-seq based gene expression data, a variety of genotyped diversity panels and biparental populations, and a long history of genetic study (Gepts et al. 2008; Myers and Kmiecik 2017). Among the most important tools for the genus are multiple high-quality reference genomes. The first of these was conducted on the Andean landrace G19833 ("Chaucha Chuga") of common bean (Schmutz et al. 2014; Phytozome id 442: https://phytozome-next.jgi.doe.gov/info/Pvulgaris_v2_1). Additional reference genomes have subsequently been developed for Middle American common bean breeding line BAT93 (Race Mesoamerica: Vlasova et al. 2016; Rendón-Anaya et al. 2017a) and cultivar UI111 (Race Durango: P. McClean, pers. comm.: Phytozome id 734, https://phytozome-next.jgi.doe.gov/info/PvulgarisUI111_v1_1). A reference genome has also been developed for lima bean (Garcia et al. 2021; Phytozome id 563: https://phytozome-next.jgi.doe.gov/info/Plunatus_V1), as well as whole-genome sequences for wild (PI 638833; Phytozome id 581, https://phytozome-next.jgi.doe.gov/info/PacutifoliusWLD_v2_0) and domesticated (G40001-Seq or PI 692269, Frijol Bayo; Phytozome id 580, https://phytozome-next.jgi.doe.gov/info/Pacutifolius_v1_0).

Members of *Phaseolus* sp. have fairly small and consistent genome sizes of approximately 500–600 Mb (2n = 2x = 22), without natural variation in ploidy (Lobaton et al. 2018). Except for the species group Leptostachyus, characterized by an aneuploid reduction (2n = 2x = 20; Delgado-Salinas et al. 2006), a high degree of genetic similarity and genomic collinearity exists among *Phaseolus* species (Almeida and Pedrosa-Harand 2013; Gujaria-Verma et al. 2016; Mina-Vargas et al. 2016; Cortés et al. 2018). Occasional inversions exist between more distantly related species, such as common bean, tepary bean, and lima bean (Gujaria-Verma et al. 2016; Cortés et al. 2018; Garcia et al. 2021) and may be responsible for some of the difficulties in forming stable interspecific hybrids (Mendel 1865; Lamprecht 1941, 1948; Gepts 1981). Collinearity also exists with other major legume genera such as *Vigna* (Lonardi et al. 2019). In contrast, the ancestors of soybean underwent a whole-genome duplication and major reorganization since their divergence with *Phaseolus* and other genera (Schmutz et al. 2010, 2014; Delgado-Salinas et al. 2011). These genomic resources have been used to conduct genotyping-by-sequencing (Ariani et al. 2016) and resequencing of numerous genomes, including *P. acutifolius* and *P. coccineus* (Lobaton et al. 2018; Wu et al. 2020).

Numerous genotyping resources also exist for the genus. These include the comprehensive marker database PhaseolusGenes, which includes SCAR, AFLP, microsatellite, SNP, and other markers (Gepts and Lin 2010; Miller et al. 2018). Several genotyping assays have been developed for the genus, including the common bean Illumina BarCBean6K_3 BeadChip with 5,398 SNPs (Song et al. 2015; now increased to some ~12,000 SNPs, Q. Song, pers. comm.) and an Illumina GoldenGate assay with 768 SNPs, applicable for common and tepary bean (Gujaria-Verma et al. 2016). *Phaseolus* also benefits from the existence of gene expression databases based on RNA sequencing of the Middle American common beans Negro Jamapa and BAT 93 (O'Rourke et al. 2013; Vlasova et al. 2016).

Numerous mapping populations and diversity panels exist in the genus. These include nearly 38,000 accessions in the gene bank of the International Center for Tropical Agriculture (CIAT, https://genebank.ciat.cgiar.org/genebank/beancollection.do), plus tens of thousands of others in other gene banks worldwide (Assefa et al. 2019). Several dozen recombinant inbred and introgression line populations have been developed (Gepts et al. 2008; González et al. 2017), as well as genotyped diversity panels of Andean, Middle American, and wild common bean (Cichy et al. 2015a; Moghaddam et al. 2016; Ariani et al. 2018). These include populations developed specifically to evaluate traits of interest in green bean (e.g., Wallace et al. 2018; Hagerty et al. 2016; Myers et al. 2004). Although these resources are extensive, gaps in sampling are known to exist. Comparisons between ecological niche modeling and the known distribution of populations are useful for identifying where sampling gaps may be found (Ramirez-Villegas et al. 2010, 2020). Even in Mexico, which has been sampled actively, several populations were not represented in gene banks until recently, e.g., in Colima and Jalisco (Zizumbo et al. 2009) or remain to be collected (e.g., Sierra de Pénjamo, Guanajuato; state of

Oaxaca; P. Gepts, unpubl. results). These genotyped populations serve as important resources in the studying of population genomics in the species, ultimately elucidating a variety of demographic patterns and marker-trait associations.

4 Origin and Diversification of *Phaseolus* spp.

4.1 Relationship Between **Phaseolus** *and Other Taxa*

Phaseolus is a member of the order Fabales, within the family Fabaceae (Leguminosae) and the subfamily Papilionoideae. *Phaseolus* spp. belongs to the Phaseoleae tribe of the phaseoloid/millettioid clade, which is generally adapted to warmer growing conditions than the Hologalegina clade (Fig. 2; Gepts et al. 2005). The order Fabales is part of a taxonomic group known as the Rosid I clade, which also includes the evolutionary successful Rosales, Fagales, and Cucurbitales. All of these form nitrogen-fixing relationships to varying degrees. The most recent common ancestor of the Rosid I clade is predicted to have formed symbiotic relationships with nitrogen-fixing *Frankia* bacteria 100 million years ago (Griesmann et al. 2018). During the evolution of the Fabaceae, *Rhizobia* became the dominant bacteria in the interaction, and a highly efficient mechanism of nodulation and nitrogen fixation

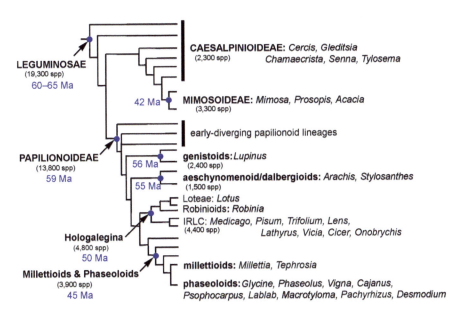

Fig. 2 Schematic phylogeny of the Leguminosae family (from Gepts et al. 2005). The three subfamilies (Caesalpinioideae, Mimosoideae, and Papilionoideae) and major subclades within the Papilionoideae are shown in boldface. Estimated ages in million of years from Lavin et al. (2005)

developed (Masson-Boivin and Sachs 2018). The Fabaceae today is found globally from the arctic to the tropics and encompass species from herbaceous annuals to large trees. The nitrogen-fixing capacity of the family makes them particularly well suited to nutrient-poor soils. The family has 770 genera and 19,500 species, making it the third largest plant family after the Asteraceae and Orchidaceae (LPWG 2017). Economically, it is second in importance after the Poaceae (LPWG 2017). The family underwent approximately 40 (Hammer and Khoshbakht 2015) or 41 (Harlan 1992) documented domestication events, the most of any plant family. Of these, approximately 35 were domesticated for edible seeds, known as pulses, while the rest were domesticated for tubers, as fodder, or for other uses.

The Fabaceae is divided into several subfamilies, the most speciose of which is the Papilionoideae (LPWG 2017). The Papilionoideae contains several early diverging lineages, including peanut (*Arachis* spp.) and lupin (*Lupinus* spp.), both with economically important domesticates. The core Papilionoideae is divided into two major groups, the Galegoid and Phaseolid clades. The Galegoids include numerous cool-season species, including pea (*Pisum sativum*), lentil (*Lens culinaris*), chickpea (*Cicer arietinum*), fava bean (*Vicia faba*), alfalfa (*Medicago sativa*), and clover (*Trifolium* spp.) (Smýkal et al. 2015). The Phaseolids are a warm-season group and include 114 genera with at least 2,000 species (Lewis et al. 2005). This group is itself a hotspot of domestication, with no fewer than 24 strong domestication events, relative to 11 found in all other legumes. Domesticates in this group include pulses of the genera *Phaseolus, Glycine, Vigna, Cajanus, Lablab, Macrotyloma, Sphenostylis,* and *Canavalia,* as well as tubers of *Pachyrhizus* and several forages (Table 1). Two genera, *Vigna* and *Phaseolus,* each have at least seven domestications, with possibly as many as ten in *Vigna* (Tomooka et al. 2014). *Vigna* contains approximately 100 species (Tomooka et al. 2014) while *Phaseolus* contains at least 70 (Maréchal et al. 1978; Freytag and Debouck 2002), meaning that nearly 10% of the species in each genus were domesticated by humans. This process continues to the present day (e.g., Takahashi et al. 2019). *Phaseolus* is native only to the Americas, while the closely related *Vigna* fills similar ecological niches in Africa and Asia.

The lineages that would give rise to *Vigna* and *Phaseolus* diverged approximately 7–9 million years ago (mya, Lavin et al. 2005; Delgado-Salinas et al. 2006, 2011). This is believed to have involved the trans-oceanic dispersal of the progenitor of the Phaseolinae (*Phaseolus* and its closest relatives in the Americas) from Africa to the New World, possibly by migratory birds.

4.2 Relationships Among Phaseolus *Species*

The first divergence within the genus *Phaseolus* occurred approximately 4.6 mya. This led to two major taxonomic groups, known as the A and B clades of the genus. The two categories contain similar numbers of species but vary strongly in their adaptability (Delgado-Salinas et al. 2006). The A clade is found almost exclusively

Table 1 Domesticated species in the Phaseoleae tribe (Fabaceae)

Subtribes	Uses or status	Species
Glycininae	Food	Soybean, *Glycine max*
		Talet bean, *Amphicarpaea bracteata*
		Jícama or yam bean, *Pachyrhizus* sp.
	Forage, erosion control	*Calopogonium* sp.
Phaseolinae	Food	Beans, *Phaseolus* sp.
		• 5 species, 7 domestications: common bean, *P. vulgaris*; lima bean, *P. lunatus*; runner bean, *P. coccineus*; tepary bean, *P. acutifolius*; year bean, *P. dumosus*
		Vigna sp.
		• African: subgenus *Vigna*: 2 species: cowpea, *V. unguiculata*; Bambara groundnut, *V. subterránea*
		• Asia: subgenus *Ceratotropis*: 7+ species: mung bean, *V. radiata*; black gram, *V. mungo*; moth bean, *V. aconitifolia*; rice bean, *V. umbellata*; creole bean, *V. reflexo-pilosa*; azuki bean, *V. angularis*; Zombi pea, *V. vexillata*; Minni payaru (neodomesticate), *V. stipulacea*
		Hyacinth bean, *Lablab purpureus*
		Winged bean, *Psophocarpus tetragonolobus*;
		Ground bean or Hausa bean, *Macrotyloma geocarpa*
	Forage	*Macroptilium atropurpureum*
Cajaninae	Food	Pigeon pea, *Cajanus cajan*
Clitoriinae	Forage	Butterfly pea, *Centrosema* sp.
Diocleinae	Food & forage	Sword bean and jack bean, *Canavalia* sp.

in Mexico, with a small number of species extending into the Southwestern United States and Central American countries to Panama. They are not found in oceanic islands and are confined to a relatively narrow elevation range. In contrast, the clade B species are distributed from the United States-Canada border to Argentina and inhabit much wider elevational ranges, with numerous species found at low elevations and on oceanic islands. Examples of the latter include *P. lignosus* on the Bahamas islands (Debouck 2015) and *P. mollis* on the Galápagos islands (Delgado-Salinas et al. 2006). They also show much greater tolerance to environmental variability and stresses, such as frost and disturbed soils (Delgado-Salinas et al. 2006). This broad adaptability may have served as a pre-adaptation to the complex and variable conditions of the cultivated environment. Of the seven domestication events in the genus, all of them occurred in the B clade. Of the eight subgroups of the genus, domestication occurred in only two, the Lunatus group (with two domestications) and the Vulgaris group. Five domestications were concentrated within just seven species of the Vulgaris group. The species with the widest ranges of the genus, *P. lunatus* and *P. vulgaris*, were each domesticated twice. *P. acutifolius, P. coccineus*, and *P. dumosus* were each domesticated once. Each domesticated species shows characteristics, which distinguish them from the other domesticates (Table 2).

Table 2 Comparative domestication in the Genus *Phaseolus*

Seeds wild: domesticated	Degree of domestication	Species	No. of domestications	Presumed domestication locations	Reproductive systems	Life history	Wild elevation range (masl)	Adaptation	Growth habit	Rhizobial symbiont
	Most intensive	Common bean (*P. vulgaris*)	2	Mexican Pacific region & Southern Andes	Predominantly autogamous	Annual (medium)	600–3,000	Mesic	Bush determinate to climbing	*Rhizobium*
		Lima bean (*P. lunatus*)	2	Western Mexico & Ecuador & N. Peru	Mixed auto- and allogamous	Annual (long)	0–2,100	Hot, dry to humid	Bush determinate to climbing	*Bradyrhizobium*
		Runner bean (*P. coccineus*)	1	Central Mexico	Predominantly allogamous	Perennial	1,100–2,900	Cool and moist	Climbing (some bush)	*Rhizobium*
		Tepary bean (*P. acutifolius*)	1	N.W. Mexico	Auto- to cleistogamous	Annual (short)	0–2,300	Hot and dry	Bush prostrate	*Rhizobium*
	Least intensive	Year bean (*P. dumosus*)	1	Western Guatemalan Highlands	Leaning to allogamous	Pluriannual	1,400–2,000	Intermediate between runner and common bean	Climbing	*Rhizobium*

The family Fabaceae, the subfamily Papilionoideae, the tribe Phaseoleae tribe, and the genus *Phaseolus* are all domestication hotspots with disproportionate numbers of species used as crops (Table 1). These patterns demonstrate that domestication is highly concentrated taxonomically. The ~40–41 clear domestications in the Fabaceae are much more numerous than that found in any other plant family. By comparison, in the four other most speciose plant families, seven domesticated species occur in the Asteraceae, one in the Orchidaceae, one in the Rubiaceae, and 26 in the Poaceae (Hammer and Khishbakht 2015). Despite this, the percent of species that were domesticated in some families, including the Poaceae (0.26%) or Cucurbitaceae (3.67%), is higher than that of the legumes (0.21%). These domestication hotspots contrast strongly with the non-flowering land plants, which include approximately 34,477 species and none are strongly domesticated (Nic Lughadha et al. 2016). These data demonstrate the highly non-random taxonomic placement of plant domestication. Of the 40–41 *Phaseolus* domesticates, approximately two thirds are found in the tribe Phaseoleae alone. The repeated domestications within *Phaseolus* and their close relatives are evidence of their utility and pre-adaptation to domestication.

Members of *Phaseolus* can be classified into a series of gene pools based on their abilities to hybridize (Harlan and de Wet 1971; Fig. 3). For example, all members of *P. vulgaris* are in one another's primary gene pool, as alleles can typically be moved easily between them. *P. dumosus* and *P. coccineus* are in the secondary gene pool of *P. vulgaris*, as there is a notable reduction in fertility between the groups, although hybridization is possible (e.g., Mendel 1865; Gepts 1981; Butare et al. 2012). The tertiary gene pool of *P. vulgaris* includes *P. acutifolius* (e.g., Thomas and Waines 1984), as advanced techniques such as embryo rescue are required for hybridization. Species that are more distantly related and cannot be hybridized are in one another's quaternary gene pools, and for *P. vulgaris* this includes *P. lunatus*. Alleles from four domesticated *Phaseolus* species can be exchanged through hybridization, making genetic progress in most species applicable for the improvement of others (Smartt 1981; Singh 2001; Mina-Vargas et al. 2016; Rendón-Anaya et al. 2017a). All domesticates can be hybridized with wild relatives, including various other species (Fig. 3).

4.3 Diversity and Ecological Niche of Wild P. lunatus

Lima bean (*P. lunatus* L.) is the most phylogenetically distinct of the *Phaseolus* domesticates. Its wild forms inhabit a range much larger than that of any other wild *Phaseolus* species. It is widespread from northern Mexico to both South America and covers a large elevation range (Fig. 4; 0–2,100 m, CIAT 2020). Lima bean is the only domesticated *Phaseolus* species native to the lowland tropics, this allows it to inhabit oceanic islands near the American continent, such as Cuba. It is also very widely distributed in Mexico and is found in all of the country's 14 vegetation types (Rzedowski 1990), compared to smaller numbers of vegetation types for runner bean

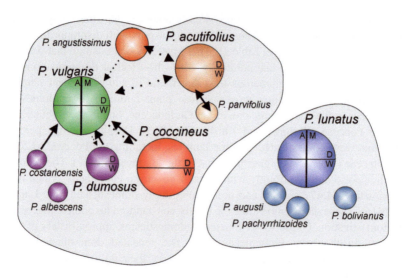

Fig. 3 Crossability relationships among the different domesticated *Phaseolus* species and some of their closest wild relatives. The shaded areas represent two crossability groups or syngameons in which gene flow can take place with varying degrees of difficulty, i.e. *P. lunatus* cannot be crossed with the other domesticated species and related wild species. Continuous arrows indicate (relatively) easy gene introgression. Stippled arrows represent more difficult introgression due to partial lack of viability and/or fertility. Absence of arrows indicates lack of information. *A* Andean, *M* Middle American, *D* Domesticated, *W*: Wild. Multiple sources were used for this diagram, including but not limited to: Belivanis and Doré (1986), Blair et al. (2003), Delgado-Salinas et al. (2006), Gepts (1981), Haghighi and Ascher (1988), Hervieu et al. (1993, 1994), Mejía-Jiménez et al. (1994), Mendel (1865), Mok et al. (1978), Muñoz et al. (2004), Shii et al. (1982), Singh et al. (1998), Souter et al. (2017), Thomas and Waines (1984), Valarmathi et al. (2006), Wall (1970), Wall and Wall (1975)

(10), common bean (9), tepary bean (8), and year bean (1) (Delgado and Gama López 2015). Wild populations of lima bean are also found throughout Central America (Guatemala, El Salvador, Belize, Honduras, Nicaragua, Costa Rica, and Panama) and in several locations in South America, particularly in and near the Andes (Colombia, Ecuador, Peru, and Argentina) (CIAT 2020). These wild populations are structured into four gene pools, two predominantly Andean (AI and AII) and two mainly Mesoamerican (MI and MII) (Serrano-Serrano et al. 2010, 2012; Garcia et al. 2021). The MI gene pool predominates in drier environments of west-central Mexico, while the MII gene pool is most common in the comparatively humid areas of southern Mexico, Central America, and Colombia. The AI gene pool is common from Colombia to Argentina, while the AII gene pool is strictly Colombian (Chacón-Sánchez and Martínez-Castillo 2017; Garcia et al. 2021).

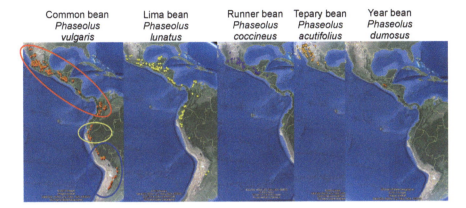

Fig. 4 Distribution of the wild ancestral relatives of the five domesticated *Phaseolus* species. For common bean, the red ellipse includes the accessions of the Middle American gene pool, which includes accessions in Central America and Colombia. The yellow ellipse includes the PhI group in Ecuador and northern Peru, resulting from a long-distance dispersal some 400,000 years ago, and the blue ellipse includes the accessions from the southern Andean gene pool, which includes accessions from southern Peru, Bolivia, and northwestern Argentina. Based on data available from the Genetic Resources Unit of the Centro Internacional de Agricultura Tropical (CIAT), Cali, Colombia: https://genebank.ciat.cgiar.org/genebank/language.do?collection=bean (accessed 2 Oct., 2020)

4.4 Diversity and Ecological Niche of Wild P. acutifolius

The wild range of tepary bean (*P. acutifolius*) spans from the western Mexican states of Jalisco and Colima to the southwestern United States (Fig. 4). Several wild taxa are closely related to domesticated teparies. *P. acutifolius* is divided into vars. *acutifolius*, *latifolius* (including domesticates), and *tenuifolius* (Freytag and Debouck 2002; Blair et al. 2012). *P. parvifolius* is a closely related species, which ranges from northern Mexico to Guatemala (Freytag and Debouck 2002). The *latifolius* and *acutifolius* groups are highly similar genetically and are now generally considered indistinct (Pratt and Nabhan 1988; Blair et al. 2012; Gujaria-Verma et al. 2016). Blair et al. (2012) determined that a genetic continuum exists between the core *P. acutifolius* var. *acutifolius* types and the *P. parvifolius* types. They supported the placement of *P. parvifolius* as a separate species due to its genetic distance from other populations of *P. acutifolius*. Intermediate between these are the *tenuifolius* types, which had previously been considered genetically indistinguishable from the core *acutifolius* group (Muñoz et al. 2006). Gujaria-Verma et al. (2016) identified similar patterns using SNP data. Tepary beans are unique among *Phaseolus* domesticates in their extreme resistance to heat and aridity. This makes them particularly well suited to resist changes associated with climate change (Ramirez-Cabral et al. 2016). Tepary beans also inhabit a particularly wide elevation range, from 0 to 2,300 m (CIAT 2020) which may be related to their resilience to other stresses, including light frost (Souter et al. 2017). The species is also a source of resistance alleles to pathogens such as common bacterial blight (Drijfhout and Blok 1987; Singh and Muñoz 1999). *P. vulgaris, P. coccineus, and P. dumosus* are all in the tertiary gene pool of tepary beans, so these useful alleles need to be moved between species using advanced breeding techniques (Haghighi and Ascher 1988; Muñoz et al. 2004; Mejía-Jiménez et al. 1994), embryo rescue (Thomas and Waines 1984), or bridging lines (Barrera et al. 2020).

4.5 Diversity and Ecological Niches of Wild P. coccineus and P. dumosus

Wild runner bean (*P. coccineus*) is distributed from the state of Chihuahua in Northern Mexico to Guatemala and Honduras (Fig. 4; Delgado Salinas 1988; Freytag and Debouck 2002; Chacón-Sánchez 2018). It is adapted to the highest average elevations of *Phaseolus* domesticates (from 1,100 to 2,900 m; CIAT 2020). These wild types develop large tuberous taproots and are perennial (Freytag and Debouck 2002; Bitocchi et al. 2017). The species is the only *Phaseolus* domesticate to have scarlet flowers and has an extrorse stigma type that prevents self-pollination, leading to extremely high levels of outcrossing and no pollination in the absence of pollinators, like carpenter bees and hummingbirds. The species is also unique in its hypogeal germination, in which cotyledons remain underground after germination (Freytag and Debouck 2002; Bitocchi et al. 2017).

Year bean (*Phaseolus dumosus*) is found only in a small wild range, in the highlands of western Guatemala and Chiapas state, Mexico (Fig. 4; Freytag and Debouck 2002; CIAT 2020). In this range, it is found from 1,400 to 2,000 m. The species' wild range is the narrowest, both in area and elevation, of any domesticated *Phaseolus* species. *P. dumosus* arose through a reticulated network of hybridization between the wild gene pools of *P. vulgaris* and *P. coccineus* (Llaca et al. 1994; Mina-Vargas et al. 2016). The compatibility of hybridization between these species is highly dependent on parent choice, in a direction-dependent manner (Fig. 4, Gepts 1981; Singh 2001; Schwember et al. 2017) as *P. vulgaris* must be the maternal parent for successful hybridization with *P. coccineus* (Wall 1970; Shii et al. 1982). The results of this partial incompatibility have left strong genomic signatures on *P. dumosus*. The maternally inherited plastid genomes of *P. dumosus* are highly similar to *P. vulgaris*, while the nuclear genome is generally more similar to *P. coccineus* (Llaca et al. 1994; Mina-Vargas et al. 2016). *P. dumosus* combines certain properties of *P. vulgaris* and *P. coccineus*. Its epigeal germination and mode of outcrossing are more similar to *P. vulgaris*, but like *P. coccineus*, it has relatively large seeds and adaptation to cool highland environments (Mina-Vargas et al. 2016; Bitocchi et al. 2017). The root structure and life history of *P. dumosus* are intermediate between *P. vulgaris* and *P. coccineus*, as it has strong, thickened storage roots which allow some perenniality, but which are not strongly tuberous like *P. coccineus* (Schmit and Debouck 1991).

P. dumosus and *P. coccineus* are in the secondary gene pool of common bean, and several disease resistances have been introgressed into common bean (e.g., Mahuku et al. 2002a, b; Zapata et al. 2004; Beaver 2020). Nevertheless, three traits with potential agronomic importance – hypogeal germination, scarlet flower color, and extrorse stigma shape could not be introgressed in viable and fertile lines (Lamprecht 1941; Wall 1970; Gepts 1981).

4.6 Diversity and Ecological Niche of Wild P. vulgaris

Common bean (*Phaseolus vulgaris*) has a large wild range that extends from Chihuahua in northern Mexico to Córdoba in Argentina, which is punctuated by several gaps (Fig. 4; Freytag and Debouck 2002; Sirolli et al. 2015). This range is second only to lima bean in terms of distribution area among *Phaseolus* wild relatives. Wild *P. vulgaris* is adapted to middle-elevation regions, from 600 to 3,000 m, although most varieties are not found at the extremes of this range (CIAT 2020). It also requires a dry period in its natural habitat, which promotes pod shattering and prevents vivipary of the seeds inside the pods; this dry period may also hinder competition from larger perennials. This dry period lasts a minimum of one month in the equatorial distribution of the species and up to six months at higher latitudes. The middle-elevation adaptation of common bean is associated with a preference for moderate temperature and precipitation. These ecological factors are not found in many regions of the Neotropics, such as tropical lowlands and mountain highlands, and are responsible for the gaps and population differentiation seen in the species. Three low-elevation gaps include the isthmus of Tehuantepec, the country of Nicaragua, and a region spanning Panama and the Chocó region in northwest Colombia. Three highland gaps exist in the Andes of South America, in northern Ecuador (representing a transition in the distribution from the eastern to the western slope of the Andes, central Peru (representing a transition in opposite direction), and near the Peru-Bolivia border (CIAT 2020; Bitocchi et al. 2017). Dispersals between these may be due to zoochory, the transmission of seeds by animals. In the case of *Phaseolus* beans, birds of the dove family may be responsible, as they are known to regularly eat wild seeds of the genus and have a migratory distribution encompassing Mesoamerica and part of the Andes. This habit is reflected in local names for wild beans, such as "frijol de paloma" ("pigeon bean" or "dove bean"; Debouck et al. 1993; Ariani et al. 2018). Seed dispersal in *Phaseolus* is a continuum ranging between rare, long-distance dispersal, and far more frequent short-distance dispersal by pod shattering leading to ballistic seed scattering (Ariani et al. 2018).

The population of wild common beans from Ecuador and northern Peru is highly genetically differentiated from all other populations of the species. This was first recognized based on its ancestral type I ("Inca") phaseolin (hence, PhI group), which does not display tandem direct repeats of 15 bp in the fourth exon and/or 27 bp in the sixth exon (Slightom et al. 1985; Kami et al. 1995). The lack of these tandem direct repeats is evidence of an ancestral state, as new repeats are more likely to develop than deletions that excise precisely one of the members of the duplication. This phaseolin type is also found in the closest relatives of *P. vulgaris*, such as *P. coccineus* and *P. dumosus*. The early divergence of this population has been supported by detailed nucleotide-based analyses and metabolomic studies (e.g., Rendón-Anaya et al. 2017a; Ariani et al. 2018). The divergence between the Inca phaseolin types and all other beans occurred roughly 260,000 years ago (Rendón-Anaya et al. 2017a) to 373,000 (Ariani et al. 2018).

The next major split in the wild *P. vulgaris* gene pool separates the populations from Middle America, from northern Mexico to Colombia, from those of the southern Andes. This divergence occurred approximately 87,000 (Ariani et al. 2018) to 110,000 (Mamidi et al. 2013), to 165,000 years ago (Schmutz et al. 2014, Table 3). These populations are strongly differentiated, with F_{st} values estimated repeatedly at 0.34 (Schmutz et al. 2014; Ariani et al. 2018). The southern Andean cluster is subdivided into two geographically isolated groups, one found in central and southern Peru and the other from Bolivia and Argentina. These two southern Andean populations are highly genetically similar, suggesting a recent divergence, although they can be distinguished genetically (Koenig and Gepts 1989; Kwak and Gepts 2009; Bitocchi et al. 2012; Rodriguez et al. 2016; Ariani et al. 2018). Population genetics data indicate that a major pre-domestication bottleneck occurred in this southern Andean population, greatly reducing its genetic diversity. This is consistent with a hypothesis of recent migration via long-distance dispersal from the Middle American population, found several thousand kilometers away (Ariani et al. 2018). Since the bottleneck of the southern Andean population, it has undergone a major population expansion which continues to the present day (Rossi et al. 2009; Bitocchi et al. 2012; Mamidi et al. 2013; Schmutz et al. 2014; Ariani et al. 2018). In contrast, no sign of a population bottleneck has been found in the wild Middle American population (Schmutz et al. 2014; Ariani et al. 2018). The allopatric divergence between the Middle American and Andean gene pools has led to Dobzhansky-Muller incompatibility (Orr and Turelli 2001) between certain gene combinations. The *Dominant Lethal-1* (*Dl-1*) allele of certain Middle American types and the *Dl-2* allele of specific Andean accessions interact to cause autoimmune-mediated hybrid lethality, which is a barrier to fertility for these particular genotypes (Shii et al. 1980; Gepts and Bliss 1985; Koinange and Gepts 1992; Hannah et al. 2007).

The present distribution of phaseolin types, especially the presence of the ancestral I phaseolin outside the ancestral region in Middle America, and the genetic diversity of the different gene pools in common bean is intriguing. Three demographic models could explain this pattern (Fig. 5; Ariani et al. 2018). These include (a) a Middle American model where the Middle American evolutionary lineage did not experience a bottleneck, and rather was the source of two offshoot lineages through long-distance dispersal, one leading to the PhI group with ancestral phaseolin (Kami et al. 1995), and a second one giving rise to wild southern Andean *P. vulgaris*; (b) the Northern Peru-Ecuador model, where the PhI group, based on the presence of these ancestral sequences, is the ancestor of both the Middle American and Andean gene pools; and (c) the Protovulgaris model where the Middle American

Table 3 Genetic diversity and divergence statistics for *Phaseolus vulgaris*

Comparison	Marker	Metric	Value	Reference
Genetic diversity, Andean vs. Middle American wild				
Diversity AW/MW	SSR	H_e	0.34	Kwak and Gepts (2009)
Diversity AW/MW	AFLP	H_e	0.55	Rossi et al. (2009)
Diversity AW/MW	Nucleotide	Pi	0.094	Bitocchi et al. (2012)
Diversity AW/MW	Nucleotide	Pi	0.096	Bitocchi et al. (2013)
Diversity AW/MW	Nucleotide	Hd	0.685	Mamidi et al. (2013)
Diversity AW/MW	Nucleotide	Pi	0.23	Schmutz et al. (2014)
Diversity AW/MW	Nucleotide	H_e	0.46	Rodriguez et al. (2016)
Diversity AW/MW	Nucleotide	Pi	0.45	Ariani et al. (2018)
Genetic diversity, PhI vs. Middle American wild				
Diversity PhI/MW	SSR	H_e	0.55	Kwak and Gepts (2009)
Diversity PhI/MW	AFLP	H_e	0.59	Rossi et al. (2009)
Diversity PhI/MW	Nucleotide	Pi	0.25	Bitocchi et al. (2012)
Diversity PhI/MW	Nucleotide	Pi	0.95	Schmutz et al. (2014)
Diversity PhI/MW	Nucleotide	H_e	0.28	Rodriguez et al. (2016)
Diversity PhI/MW	Nucleotide	Pi	0.73	Ariani et al. (2018)
Divergence time, Andean vs. Middle American wild				
Divergence time AW/MW	Nucleotide	Age (years)	110,706	Mamidi et al. (2013)
Divergence time AW/MW	Nucleotide	Age (years)	165,000	Schmutz et al. (2014)
Divergence time AW/MW	Nucleotide	Age (years)	87,410	Ariani et al. (2018)
Divergence time, PhI vs. Middle American wild				
Divergence time PhI vs. AW+MW	Nucleotide	Age (years)	260,000	Rendón-Anaya (2017a)
Divergence time PhI vs. AW+MW	Nucleotide	Age (years)	373,060	Ariani et al. (2018)
Genetic diversity, Middle American domesticate vs. wild				
Diversity MD/MW	AFLP	H_e	0.68	Rossi et al. (2009)
Diversity MD/MW	Nucleotide	Pi	0.28	Bitocchi et al. (2013)
Diversity MD/MW	Nucleotide	Pi	0.83	Schmutz et al. (2014)
Genetic diversity, Andean domesticate vs. wild				
Diversity AD/AW	AFLP	H_e	1.00	Rossi et al. (2009)
Diversity AD/AW	Nucleotide	Pi	0.73	Bitocchi et al. (2013)
Diversity AD/AW	Nucleotide	Pi	1.21	Schmutz et al. (2014)
Diversity AD/AW	Nucleotide	H_e	0.74	Rodriguez et al. (2016)

(continued)

Table 3 (continued)

Comparison	Marker	Metric	Value	Reference
Genetic diversity, Andean vs. Middle American domesticate				
Diversity AD/MD	AFLP	H_e	0.81	Rossi et al. (2009)
Diversity AD/MD	Nucleotide	Pi	0.25	Bitocchi et al. (2013)
Diversity AD/MD	Nucleotide	Pi	0.34	Schmutz et al. (2014)
Diversity AD/MD	Nucleotide	H_e	0.56	Rodriguez et al. (2016)

AW Andean Wild, *MW* Middle American Wild, *PhI* Inca Phaseolin (*P. debouckii*), *MD* Middle American Domesticated, *AD* Andean Domesticated, H_e Expected heterozygosity, *Hd* Haplotype diversity, *Pi* Nucleotide diversity

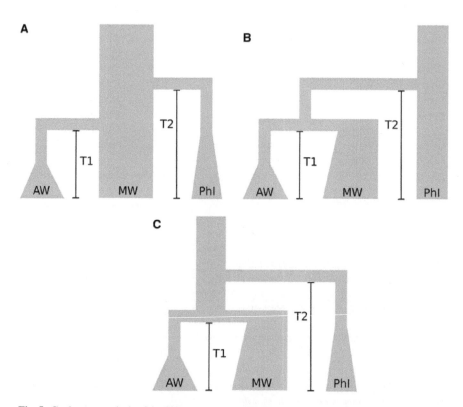

Fig. 5 Coalescent analysis of the SNP genetic diversity of wild common bean, *Phaseolus vulgaris*. (**a**) Mesoamerican demographic model where the Mesoamerican wild (MW) population did not experience any population bottleneck; (**b**) the Northern Peru-Ecuador model where the Northern Peru-Ecuador (PhI) gene pool did not experience any population bottleneck; and (**c**) the Protovulgaris model where the ancestral population went extinct after the Mesoamerican and Andean Differentiation. For further explanations, see text and Ariani et al. (2018). Copyright: CC BY 4.0: https://creativecommons.org/licenses/by/4.0/; no changes made

lineage gave rise to the PhI group through long-distance dispersal and subsequently went extinct after the Middle American and Andean offshoots became differentiated. Multiple lines of evidence favor the first hypothesis (Ariani et al. 2018), including a coalescent analysis (Fig. 5), the topology of trees in which the PhI group is an outgroup to the Andean and Middle American branches (Bitocchi et al. 2012; Ariani et al. 2018), the existence of additional ancestral sequences, such as cpDNA (Chacón S et al. 2007), and the high level of diversity of mtDNA (Khairallah et al. 1992), the metabolomic data showing divergence from the Andean and Middle American metabolomes (Rendón-Anaya et al. 2017a), and the older age of divergence of the PhI group, compared to the southern Andean group (Rendón-Anaya et al. 2017a; Ariani et al. 2018). These numerous molecular and biochemical distinction as well as the ecological differences on the western slope of the Andes, contrasting with the eastern slope distribution of the other Andean wild *P. vulgaris* led Rendón-Anaya et al. (2017b) to classify the PhI group as a distinct species, *P. debouckii* Delgado-Salinas, sister species to *P. vulgaris*.

Wild Middle American common bean populations are further subdivided into several distinct subpopulations (e.g., Bitocchi et al. 2012; Rodriguez et al. 2016; Ariani et al. 2018). While the exact clustering of these varies with sampling, in general three main clusters have been recognized. One of these is found in and around the Trans-Mexican Volcanic Belt, in the central Mexican states of Morelos, Puebla, Guerrero, and the state of Mexico. This group is lower in genetic diversity than the other Middle American wild subpopulations, possibly due to the region's unique volcanic soils, climate, or history of intensive agriculture (Rodriguez et al. 2016). A second subpopulation predominates in the surrounding region to the northwest and southeast, ranging from the isthmus of Tehuantepec to Chihuahua, excluding the central region around Mexico City. This pattern closely resembles spatial and genetic relationships of teosinte, from which maize was domesticated (Moreno-Letelier et al. 2020). A third population is found southwest of the isthmus of Tehuantepec, throughout middle-elevation regions from southern Mexico to Colombia. This extensive distribution encompasses greater ecological diversity than the other wild Middle American subpopulations, although it is not correspondingly higher in genetic diversity (Rodriguez et al. 2016).

The major wild gene pools of common bean vary extensively in genetic diversity. The Middle American population is more geographically expansive and more genetically diverse than all other wild or domesticated gene pools (Chacón S et al. 2007; Kwak and Gepts 2009; Schmutz et al. 2014; Rodriguez et al. 2016; Ariani et al. 2018). The wide geographic range of the Middle American population, from Chihuahua to Colombia, may contribute to this pattern. The wild population of the southern Andes has approximately half the level of genetic diversity of the Middle American population (Table 3), while the PhI population of Ecuador and northern Peru (*P. debouckii*) is roughly intermediate between these Andean and Middle American types (Kwak and Gepts 2009; Bitocchi et al. 2012; Schmutz et al. 2014; Ariani et al. 2018, Table 3). The reduction in genetic diversity of the Andean wild gene pools of common bean is consistent with the long-distance dispersal events that founded these gene pools and their reduced ecological amplitude (Debouck et al.

1993; Freyre et al. 1996). The Andes mountains include all three major populations of common bean, including the southern Andean population, the ancestral phaseolin population of Ecuador and northern Peru, and the Middle American population in the Colombian and Venezuelan Andes. This implies that long-distance dispersal over the isthmus of Panama occurred as least three separate times in the species.

5 Spatial and Temporal Patterns of Domestication

A domestication pattern can be defined by the number of domestication processes or occurrences, their respective locations, the level of gene flow between sympatric wild and domesticated populations, and the type and number of traits ("domestication syndrome") selected for during the respective domestication processes. Determining a crop's region of origin typically depends on several forms of evidence, each with complementary strengths and limitations. Among these are archaeological, historical, linguistic, and botanical/genetic data (de Candolle 1882). Since de Candolle's time, the types of evidence used in domesticated studies have vastly increased (Harlan and de Wet 1973; Gepts 2014a, c; Larson et al. 2014). Advances in scientific approaches have been applied to the study of crop evolution and shed light on the process of crop evolution. For *Phaseolus*, these advances include the use of accelerator mass spectrometry (Kaplan and Lynch 1999), micro-remains like starch grains (Piperno and Dillehay 2008), and genomics of the crop and its wild relatives (Rendón-Anaya et al. 2017a). In this section, we address the putative numbers and locations of domestication of the five domesticated *Phaseolus* species (Table 3). Among several centers of agricultural origins (Harlan 1992; Larson et al. 2014), at least three have been recognized in the Americas, two of which include *Phaseolus* domestications (Pickersgill 2007).

5.1 Archaeological and Linguistic Data

Archaeological evidence (Table 4; Kaplan and Lynch 1999; Piperno and Dillehay 2008; Brown et al. 2014; Gepts 2014a) suggests that *Phaseolus* beans were domesticated possibly as long ago as 8,000–7,000 years ago (*Phaseolus* sp.) and 5,000–4,300 years ago (*P. vulgaris*) in the Andes and no later than 2,300 B. P. (*P. vulgaris*) in Middle America (Kaplan and Lynch 1999). The earliest archaeological evidence of tepary domestication is similar to that of common bean in Middle America, with archaeological evidence dating to at least 2,500 B. P. in the Tehuacán valley (Kaplan and Lynch 1999). Archaeological remains of lima beans have been found in coastal Peru dating back to at least 5600 B. P.

Determining timing of domestication by archaeological methods poses several major hurdles. First, there is a relative lack of archaeological sites in the Americas, compared to, for example, the Near Eastern center of domestication. Thus, the

Table 4 Oldest archaeobotanical remains of *Phaseolus* spp. (from Gepts 2014a)

Location	Taxon (status)	Type	^{14}C age (year B.P.)	Age (year cal. B.P.)	Source
Andes					
Ñanchoc Valley, Peru	*Phaseolus sp.* (domesticated)	Starch grains from teeth calculus	8,210–6,970	8,600–7,000	Piperno and Dillehay (2008)
Chilca, Peru	*P. lunatus* (domesticated)	Pod, non-carbonized	5,600	6,400	Kaplan and Lynch (1999)
Guitarrero Cave, Peru	*P. vulgaris* (domesticated)	Seed, non-carbonized	4,300	5,000	Kaplan and Lynch (1999)
Mesoamerica					
Oaxaca Valley, Mexico	Phaseolinae (wild)	Seed and pod, non-carbonized	7,600	8,300	Kaplan and Lynch (1999)
Tehuacán Valley, Mexico	*P. vulgaris* (domesticated)	Pod, non-carbonized	2,285	2,300	Kaplan and Lynch (1999)
Tehuacán Valley, Mexico	*P. acutifolius* (domesticated)	Seed, non-carbonized	2,360	2,400	Kaplan and Lynch (1999)
Tehuacán Valley, Mexico	*P. coccineus* (domesticated)	Seed, non-carbonized	410	500	Kaplan and Lynch (1999)

absence of earlier archaeobotanical data should not be taken as the evidence for a late domestication of *Phaseolus* spp., relative to other crops in the Americas. Second, there is a preservation bias of archaeobotanical remains, based on the relative adaptation of the different domesticated species and the "survivability" or taphonomy of different plant parts in the archaeobotanical record (e.g., Banning 2020). Species such as *P. coccineus* and *P. dumosus* thrive in humid conditions, which are poorly suited to long-term preservation in the archaeological record, and the earliest archaeological evidence of these domesticates is only 500 years old (Kaplan and Lynch 1999). In contrast, arid conditions are better suited to archaeological preservation, and remains of *P. vulgaris* have been found in the American southwest only 100 years after their appearance in their native homeland in Mexico (Kaplan and Lynch 1999). Thus, a potential preservation bias may affect the relative abundance of archaeobotanical remains among *Phaseolus* species and between *Phaseolus* and other crop plants.

Different plant parts may also have different survival abilities in the archaeobotanical record. The macro-remains such as seeds usually found for *Phaseolus* spp. may have a lower survival rate compared to, for example, a heavily lignified cob of maize (Benz 2001). It is not a coincidence that some of the oldest maize remains are micro-remains identified on grinding stones in Mexico by Piperno et al. (2009). Currently, the oldest *Phaseolus* remains are actually starch grains that survived in the teeth calculus of human remains in Peru (Piperno and Dillehay 2008).

Thus, identification of additional archaeological sites is necessary, especially in geographically relevant areas like river basins in western Mexico (Kwak et al. 2009; Zizumbo-Villareal and Colunga GarcíaMarín 2010; Zizumbo-Villarreal et al. 2012, 2014). A similar need exists for the analysis of existing and additional *Phaseolus* archaeobotanical remains in the Andean domestication regions of *P. vulgaris* and *P. lunatus*, especially in Ecuador and northern Peru (lima bean domestication) and from southern Peru to northwest Argentina for both common bean and lima bean.

Historic evidence is also limited, as it depends on written sources, which only emerged long after the origins of agriculture. In common bean, historic data are relevant mostly in the description of beans post-1492, either to describe the crop in its centers of origin (e.g., Codex Magliabechiano in Mexico; Velasco 1789 description of nuña beans in the Central Andes, cited by Tohme et al. 1995) or in regions of introduction like Europe (sixteenth century herbals by Gerard, Dodoens, and de L'Ecluse; von Martens 1860). Linguistic data does not rely on written records and has been applied to assess the origins of *Phaseolus* beans (Schmit and Debouck 1991; Rodriguez et al. 2016; Brown et al. 2014). Using a paleobiolinguistic approach, Brown et al. (2014) attempted to reconstruct bean terms in proto-languages of the Americas. For Mesoamerica, the oldest reconstructable name for bean was in the Otopamean language (around 3,700 BP) in central Mexico, close to the eastern end of a putative domestication region for common bean, proposed by Kwak et al. (2009). For the Andean domestication region of common bean in the southern Andes (ranging from Southern Peru to northwest Argentina), the oldest reconstructable bean word was in Matacoan, at the confines of the borders between Argentina, Bolivia, and Paraguay, in the eastern piedmont of the Andes. Some disadvantages of paleobiolinguistic data include limitations in sampling of the number of languages and language families in a region, determining subtle changes in the meaning of words, and determining the relationship between societies and the botanical items involved, for example, whether the crop referred to was still wild or had been domesticated at least incipiently (Brown et al. 2014). Botanical and genetic methods overcome several of these constraints and are therefore valuable assets in determining crop origins.

5.2 Botanical, Genetic, and Genomic Data

The last decades have seen a considerable amount of progress in delineating putative domestication areas for the five domesticated *Phaseolus* species (Fig. 6). This progress is attributable to additional botanical explorations that have refined the distribution area of the wild ancestral populations of the *Phaseolus* domesticates (e.g., Debouck et al. 1993; Freyre et al. 1996; Freytag and Debouck 2002). Research on electrophoretic diversity of phaseolin seed protein provided the first unequivocal evidence for multiple domestications in a crop. This evidence obtained was based on a comparison of phaseolin diversity in both wild and domesticated types distributed

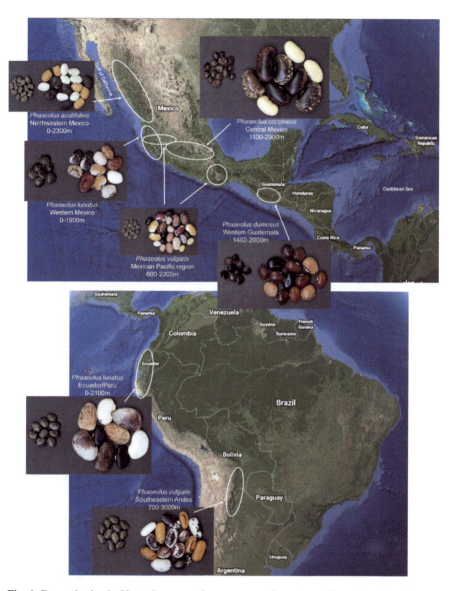

Fig. 6 Domestication in *Phaseolus* occurred over a range of ecogeographic settings. In each case, wild types (shown at left) were selected for characteristics useful to humans, leading to a breadth of phenotypic diversity in domesticated populations of each gene pool (at right). In many cases, this led to highly parallel transitions, such as white seed color across numerous populations. Population genomic approaches have been critical to understanding when, where, and how domestication proceeded. All seeds shown to scale with one another. Seed photos: T. Parker

in both Mesoamerica and the Andes to cover the entire distribution of wild *P. vulgaris* (Gepts et al. 1986). A parallel geographic distribution in phaseolin diversity between wild and domesticated common bean strongly suggested at least two major domestications, in Mesoamerican and in the Andes. Subsequent analyses with additional wild populations that had been obtained with additional explorations and other markers, like allozymes (Koenig and Gepts 1989), restriction fragment length polymorphisms (RFLPs) of nuclear sequences (Becerra Velásquez and Gepts 1994), RFLPs of cytoplasmic sequences (chloroplast: Fofana et al. 1999; Chacón S et al. 2005, 2007; mitochondria: Khairallah et al. 1992), amplified fragment length polymorphisms (Pallottini et al. 2004; Rossi et al. 2009), microsatellites (Kwak and Gepts 2009), and sequencing (Bitocchi et al. 2013; Schmutz et al. 2014; Rodriguez et al. 2016; Rendón-Anaya et al. 2017a; Ariani et al. 2016; Lobaton et al. 2018), have repeatedly confirmed the existence of the two divergent domesticated gene pools.

Additional research has attempted to identify a more specific domestication location for the two domestications in their respective geographic areas. Gepts (1988) identified a region in west-central Mexico, located in the states of Jalisco and neighboring Guanajuato, that represents a putative Mesoamerican domestication center based on morphological introgression information and phaseolin seed protein diversity (Fig. 7a). In this region, several wild Mesoamerican common bean populations did not show any morphological signs of introgression with domesticated types and displayed the Mesoamerican domesticated phaseolin type ("S" type) (Gepts et al. 1986). Gepts (1988) argued that the complexity of the sequence of molecular events leading to the different phaseolin electrophoretic types suggested that each phaseolin type probably had a unique origin. These molecular events include sequence diversification [gene duplications, nucleotide substitutions, tandem direct repeats (α- or ß-phaseolins)], post-translational glycosylations, etc. (e.g., Lioi and Bollini 1984). A consequence of the uniqueness of each phaseolin type is that accessions sharing a phaseolin type have a common ancestor (Gepts 1988). This characteristic of the phaseolin diversity dynamics, combined with the high levels of linkage disequilibrium observed in common bean (Koenig and Gepts 1989; Kwak and Gepts 2009) explains why phaseolin has been so useful as a diagnostic marker to identify the domestication gene pools (Andean vs. Mesoamerican) of common bean and is systematically screened for in the World Collection of *P. vulgaris* (CIAT 2020).

Further evidence for a domestication center in west-central Mexico, more specifically in the Lerma-Santiago basin, was provided by Kwak et al. (2009) based on microsatellite diversity (Fig. 7b, c). Strikingly, this is the same region as the one identified originally by Gepts (Fig. 7a; 1988). This region extends along an east-west axis from Pénjamo in the state of Guanajuato to Mascota in the state of Jalisco. It is a mid-altitude area (1,400–2,100 m) situated between the southern reaches of the Western Sierra Madre and the western end of the Transverse Volcanic Axis to the south (Fig. 7c); it has a subtropical, subhumid, and semi-warm climate (López Soto et al. 2005), a monsoon-influenced humid subtropical climate (Cwa), or a temperate highland tropical climate with dry winters (Cwb).

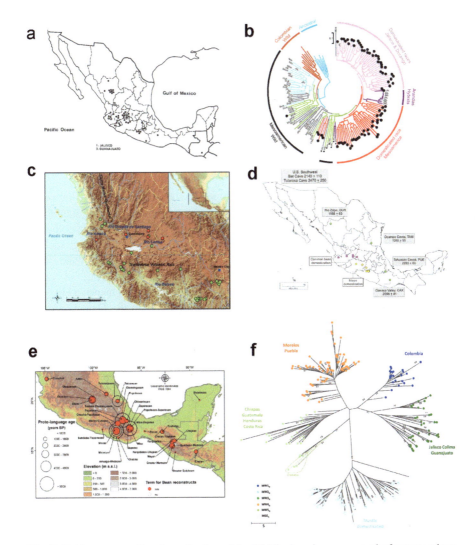

Fig. 7 Evidence supporting domestication of the Middle American gene pool of common bean (*Phaseolus vulgaris*) in the Pacific states of Mexico. From Gepts (1988), Fig. 9: (**a**) asterisks in Jalisco (1) and Guanajuato (2) indicate putative wild populations involved in Middle American based on "S" phaseolin type. From Kwak et al. (2009), Figs. 2 and 4a, b: (**b**) Unrooted neighbor-joining tree of microsatellite diversity showing wild populations from Jalisco and Guanajuato (*) closest to Middle American domesticates; (**c**) Geographic distribution of wild bean accessions: red: wild beans closed to Middle American domesticates, green: other wild bean populations; (**d**) Geographic distribution of putative common-bean (purple symbols) and maize (yellow) domestications in the Pacific West region in Mexico, compared with the disjunct, scarce archaeological remains (green). From Brown et al. (2014), Fig. 2 (Creative Commons Attribution-NonCommercial 4.0 International Public License (CC BY-NC 4.0): (**e**) Bean-term reconstruction for ancient Mesoamerican languages (in red): the larger the circle, the older the reconstructed term. The oldest term is in the Otopamean language (~3,700 BP) at the confluence of the states of Guanajuato, Michoacán, and Querétaro. From Rodriguez et al. (2016), Fig. 6: (**f**) an alternative proposal for a putative domestication center in the Mexican Pacific West (state of Oaxaca)

It should be noted that the putative maize domestication region is also located geographically in the Pacific region of Mexico, namely in the upper Balsas river basin in the state of Guerrero (Fig. 7d). This larger region consisting of river basins flowing from the highlands of central Mexico into the Pacific Ocean may be relatively homogeneous ecologically and culturally (Zizumbo-Villareal and Colunga GarcíaMarín 2010). Recently, Moreno-Letelier et al. (2020) have suggested that Jalisco might also be an alternative domestication region for maize. Further support for a putative domestication region of common bean in the Jalisco-Guanajuato region may be provided by palaeobiolinguistics, which identified the oldest reconstructed word for bean in the Otopamean language (dated at some 3,700 years B.P.) (Fig. 7e; Brown et al. 2014).

Bitocchi et al. (2013) and Rodriguez et al. (2016) suggested an alternative domestication region for the Middle American common bean gene pool, namely in Oaxaca (Fig. 7f). These results are based primarily on genetic and archaeological data. Currently, these forms of evidence suffer from inconsistent sampling, such as the presence of archaeobotanical bean remains in neighboring Tehuacán Valley with its arid environment very conducive to conservation of these remains (Fig. 7d), and a history of limited genetic sampling of wild accessions from Oaxaca (Fig. 7f). Improved genetic sampling of wild accessions, particularly in the Oaxaca region, will be critical to pinpoint where in the Pacific region of Mexico domestication occurred.

An alternative way to time the domestication process in the two gene pools, as a complement to archaeobotanical data, is through genetic evidence. Mamidi et al. (2011) showed that bottleneck events occurred 8,200–6,300 years before present (BP) in the Middle American population and 8,500–7,000 years BP in the Andean population (Mamidi et al. 2011), suggesting when the two domestications of common bean may have occurred.

Identifying a domestication region in the southern Andes for common bean has been difficult because of the low levels of polymorphisms and, concomitantly, the limited geographic differentiation – compared to the wild Mesoamerican gene pool – of the wild common-bean gene pool of the southern Andes, including southern Peru, Bolivia, and northwestern Argentina (e.g., Gepts et al. 1986; Koenig and Gepts 1989; Koenig et al. 1990; Debouck et al. 1993; Kami et al. 1995; Kwak and Gepts 2009). Nevertheless, with an increasing number of more polymorphic markers, it has been possible to hypothesize an Andean domestication region. For example, Beebe et al. (2001) used AFLP markers to suggest a region encompassing Bolivia and possibly northwest Argentina, as a domestication region, as did Rodriguez et al. (2016), relying on SNPs in some 100 genes, and Bitocchi et al. (2013) based on sequence variation in five gene fragments. The result is consistent with linguistic evidence (Brown et al. 2014) and is also supported by the domestication of other food crops in the region, such as peanut (Leal-Bertioli et al. 2017) and cassava (*Manihot esculenta*, Olsen 2004). A genetic bottleneck occurred in Andean common bean 8,500–7,000 years BP, possibly related to domestication (Mamidi et al. 2011). Macro-archaeological remains of domesticated *Phaseolus* occur only several thousand years later (Gepts 2014a).

In tepary bean, several forms of evidence converge on a domestication in or near the Mexican state of Sinaloa. Schinkel and Gepts (1988) determined that wild *P. acutifolius* types with the domesticated-type phaseolin seed protein were found in the states of Sinaloa and Sonora. With allozyme data of the same population, they determined that the neighboring state of Durango was a likely center of domestication (Schinkel and Gepts 1989). Garvin and Weeden (1994) evaluated AFLPs across a large 263-member population, representing wild and domesticated types. Their results suggested that domestication most likely occurred in the states of Sinaloa or Jalisco. Muñoz et al. (2006) agreed that tepary bean was likely domesticated a single time in one of these same two Mexican states. Despite this, a second domestication could not be ruled out. Using microsatellite markers, Blair et al. (2012) determined that tepary bean was most likely domesticated in Sinaloa or Sonora. A subsequent analysis with genome-wide SNP data also indicated that tepary bean was domesticated only once (Gujaria-Verma et al. 2016), confirming the data of Schinkel and Gepts (1988).

Spataro et al. (2011) and Rodriguez et al. (2013) suggested that *P. coccineus* was independently domesticated in Mexico and Central America based on SSR data. Using genomic data, Guerra-García et al. (2017) found evidence of only a single domestication in the Trans-Mexican Volcanic Belt of central Mexico. Despite this, a lack of sampling of Guatemalan and Honduran wild types by Guerra-García et al. makes it difficult to rule out the possibility of a second domestication in those regions. The extremely limited wild range of *P. dumosus* in the highlands of western Guatemala (Fig. 4) and phaseolin data (Schmit and Debouck 1991) suggest that this is likely the site of its domestication. This has also been supported by linguistic and biochemical evidence, and the greatest phaseolin diversity of domesticated types is found in the region (Schmit and Debouck 1991).

Genetic data indicate that lima bean was domesticated twice, occurring in the MI gene pool of Middle America and the AI gene pool of South America (explained in Sect. 4.3; Gutiérrez Salgado et al. 1995; Motta-Aldana et al. 2010; Andueza-Noh et al. 2015; Chacón-Sánchez and Martínez-Castillo 2017; Garcia et al. 2021): Middle American domesticates gave rise to cv.-group "Potato" (small, rounded seeds) and cv.-group "Sieva" (medium-sized, flat seeds), while AI led to Andean domesticates with large seeds (cv.-group "Big Lima"; Mackie 1943; Baudet 1977; Lioi 1994). In Middle America, genetic evidence has supported a domestication in Western Mexico from the MI gene pool (Motta-Aldana et al. 2010). Genetic evidence suggests a bottleneck in the Middle American domesticated gene pool starting 7,700 years ago and lasting until no later than 6,500 years ago, when outcrossing with wild members of MII began (Chacón-Sánchez and Martínez-Castillo 2017). Domestication in the Andes occurred in the AI gene pool in Ecuador and northwestern Peru (Gutiérrez Salgado et al. 1995; Motta-Aldana et al. 2010; Chacón-Sánchez 2018).

Thus, across species, genetic data has been a major asset in understanding the domestication of *Phaseolus* spp. Despite this, much remains to be resolved. The reduced costs of genotyping will be useful for more thorough and comprehensive analyses of crop origins across the genus, with greater representation of both accessions and genetic markers. Beyond this, further archaeological explorations will be useful for better understanding the early phases of crop dispersal. This will be particularly true of micro-remains, which are the source of the earliest archaeological evidence of domesticated *Phaseolus*, but which have not yet been thoroughly explored. A combination of genetic, genomic, archaeological, and linguistic evidence will be critical for understanding the domestication of the genus.

6 Domestication-Related Effects on Genetic Diversity and Gene Flow Between Wild and Domesticated Types

Plant domestication is associated with demographic processes that generally reduce genetic diversity. Many species, including several *Phaseolus* spp., tend to undergo major bottleneck events early in the domestication process, which leads to the fixation of alleles due to selection or genetic drift. Subsequent selection and bottleneck events often occur as crops expand beyond the center of domestication and fill new ecological or cultural niches, further limiting genetic diversity. These processes are associated with increased linkage disequilibrium and linkage drag in the resulting populations. Across populations of *Phaseolus* spp., population structure tends to be strongly related to geographic origin. This is particularly true in the Americas, where the species have been established for the longest period.

Genetic diversity of domesticated populations can be increased through outcrossing with wild relatives and by *de novo* mutations. As such, species with higher levels of outcrossing often undergo less severe domestication bottlenecks, as continuous outcrossing maintains higher levels of genetic diversity and promotes recombination. Similarly, large historic population sizes reduce the degree of the bottleneck effect and provide more opportunity for de novo variation in secondary centers of diversity. Domesticated species of the genus *Phaseolus* vary in several respects relevant to this diversity, such as the rate of outcrossing and degree of use in secondary centers. The genus is therefore an excellent system to evaluate the role of these factors in affecting diversity (Table 1).

6.1 Molecular Marker-Based Approaches to Assess Levels of Genetic and Genomic Diversity in Wild and Domesticated Populations

Domestication of common bean has led to major changes in genetic diversity, relatively high F_{st} values between populations, and increased linkage disequilibrium relative to wild populations (Kwak and Gepts 2009; Bitocchi et al. 2013; Schmutz et al. 2014; Rodriguez et al. 2016). The Middle American gene pool, in particular, underwent a major domestication-related bottleneck relative to its abundant genetic diversity in the wild. These studies have typically found a genetic diversity loss of approximately 50% in the group (Table 3). By comparison, reductions in genetic diversity in the Andean domesticated gene pool were much lower, at approximately 20% (Table 3), but starting from a much lower level of genetic diversity in the southern Andean wild relative. While wild Middle American common beans are much more diverse than their wild Andean counterparts, the disparity in bottleneck strength has led to relatively similar levels of genetic diversity in the domesticated gene pools. Several earlier studies indicated that equal or greater genetic diversity might be present in the Andean gene pool (Gepts et al. 1986; Blair et al. 2006a, 2009; Becerra et al. 2011; McClean et al. 2012; Kwak et al. 2009), while more recent analyses have shown that Middle American domesticates are more diverse (Bitocchi et al. 2013; Schmutz et al. 2014; Rodriguez et al. 2016).

The domesticated gene pool of common bean is divided into several major ecogeographic races (Singh et al. 1991a, b, c). Each of these consists of a unique set of genetic and/or phenotypic traits, including seed type, growth habit, shape and structure of vegetative organs, and photoperiod sensitivity (Singh et al. 1991a). The original classification of these consisted of six total races: race Mesoamerica, race Durango, and race Jalisco of the Middle American gene pool; and race Nueva Granada, race Peru, and race Chile of the Andean gene pool. Subsequent genetic analyses added a fourth Middle American race "Guatemala" (Beebe et al. 2000; Díaz and Blair 2006; Tobar Piñón et al. 2020). Several studies have failed to identify clear differentiation between races Durango and Jalisco at a molecular marker level (Díaz and Blair 2006; Kwak and Gepts 2009), although they are differentiated phenotypically. This has led them to often be described at the molecular level as the combined race Durango-Jalisco. Yet, as the next paragraph shows, their phenotypic differences are important (Singh et al. 1991a).

Among Middle American races, race Mesoamerica is generally adapted to humid conditions of the lowland tropics, often from the south and east of the area of domestication (Kwak and Gepts 2009). Within race Mesoamerica, several studies have determined that the group is subdivided into two major sections (Beebe et al. 2000; Blair et al. 2006a). The first of these consists primarily of upright small, black-seeded bean types from Mexico with a compact and upright growth habit. The second group includes Central American types of numerous colors. These typically have a climbing growth habit. The groups also vary in resistance to several diseases (Beebe et al. 2000; Blair et al. 2006a). In contrast, race Durango is adapted to semi-arid highlands generally to the north of the center of domestication. These are also subdivided into groups with specific adaptations, such as soil types (Beebe et al. 2000). Race Jalisco is also adapted to highland regions, but more humid areas, located in the southern part of Mexico, Central America, and even possibly in Colombia. Race Guatemala combines elements of several races, in that it is found in highland environments to the southeast of the center of domestication and includes mostly climbing types. Race Guatemala is genetically distinct from other Middle American and Andean races (Chacón S et al. 2005; Díaz and Blair 2006; Blair et al. 2009). To what extent race Guatemala is distinct from race Jalisco remains to be determined (Singh et al. 1991b).

In the Andean gene pool, race Nueva Granada also is divided into two genetically differentiated subgroups (Blair et al. 2009; Galeano et al. 2012). Race Peru may also be divided into two subgroups (Blair et al. 2009). The subgroups of races Nueva Granada and Peru show some differences in seed type and growth habit, with some overlap in phenotypic properties. Race Chile tends to cluster with race Nueva Granada, while race Peru is more distantly related to the other ecogeographic clusters (Becerra et al. 2011). This parallels the loss of photoperiod sensitivity found in both groups, leading to adaptation to diverse ecological and geographic environments.

Race Chile has a limited distribution, mainly in the country of the same name. In addition to endemic Chilean varietal types (e.g., Tórtola, Coscorrón, Cuyano), it also includes naturalized types (e.g., Frutilla, Bayo, Manteca, Blanco grande) (Becerra et al. 2011). An analysis of common-bean germplasm collected in the country using phaseolin and allozyme (Paredes and Gepts 1995b) detected a high frequency of introgression from the Middle American gene pool. The introgression from Middle American sources may have been facilitated because of segregations skewing towards Middle American alleles observed by Paredes and Gepts (1995a). The authors suggested that the source of Middle American introgression was race Durango, because of similar latitudes (in the two hemispheres) and adaptation to arid environments, correlated with similar prostrate growth habits and medium seed weights in the two races.

Recently, Trucchi et al. (2021) analyzed by DNA sequencing 30 Argentinean archaeological common-bean seeds, dated by Accelerator Mass Spectrometry to the period spanning 2,500–600 years BP. They observed that the two contemporary race Chile landraces included in their study belonged to the same clade as the archaeological sample, in contrast with contemporary landraces from Argentina, Bolivia, and Peru. Their data suggests that ancient common bean grown in Argentina belonged actually to race Chile and was subsequently replaced by other Andean cultivars belonging to other races like race Peru. Further in-depth analysis is necessary to confirm these preliminary observations with a larger, more representative sample of both archaeological and contemporary bean materials.

The latitudinal distribution of some races is generally highly restricted, particularly for races Peru, Jalisco, and Guatemala. Because of their geographic distribution closer to the Equator, they tend to be more sensitive to photoperiod-temperature interactions and flower mainly in short-day/long night (12 h/12 h) environments. As a consequence, they are underrepresented in gene banks and diversity panels located or based in temperate regions. Races Durango and Chile show some insensitivity to photoperiod and are suited to a wider range of environments. Despite this, they remain comparatively less common outside of the Americas. Races Nueva Granada and Mesoamerica, in contrast, show strong photoperiod insensitivity, broad adaptability, and are globally distributed (Zeven et al. 1999; Cichy et al. 2015a; Moghaddam et al. 2016; Wilkus et al. 2018; Kuzay et al. 2020).

Snap beans show similar levels of expected heterozygosity to dry beans, indicating that there was not a major reduction in genetic diversity during the selection for snap pod traits (Wallace et al. 2018). Both Andean and Middle American snap bean accessions exist, indicating that they were independently selected in both gene pools

(Wallace et al. 2018). The frequency of Middle American and Andean ancestry varies by country. Chinese snap beans, for example, are primarily Middle American in origin while North American snap beans are primarily Andean. Many modern snap bean varieties are the result of hybridization between both common bean gene pools (Wallace et al. 2018).

Hybridization has also occurred between Andean and Middle American dry bean gene pools (Lobaton et al. 2018), despite lethality in certain parental combinations caused by a Dobzhansky-Muller incompatibility (Shii et al. 1980; Gepts and Bliss 1985; Koinange and Gepts 1992; Hannah et al. 2007). In general, gene flow has moved more strongly from Middle American to Andean populations than vice versa. The introgressions are non-random, with specific regions often transferred in one direction, possibly related to introgression of disease resistance specificities related to the co-evolution process between the bean host and associated pathogens (Lobaton et al. 2018). Gepts and Bliss (1985) proposed that microorganisms associated with common bean, whether pathogens or beneficial organisms like *Rhizobium* would show a parallel distribution of genetic diversity into two major geographic gene pools, namely Andean and Middle American. This hypothesis has been confirmed for several pathogens and *Rhizobium*. Notably, resistance against Mesoamerican strains of a pathogen can be identified in Andean genotypes, and vice versa, as shown for angular leaf spot (*Pseudocercospora griseola*: Guzmán et al. 1995; Pastor-Corrales et al. 1998), anthracnose (*Colletotrichum lindemuthianum*: Sicard et al. 1997; Geffroy et al. 1999), and rust (*Uromyces appendiculatus*: Araya et al. 2004). Although exceptions have been observed to this pattern, the existence of host plant-pathogen co-evolution in common bean provides a useful tool in breeding for the identification of additional host plant resistances. It also constitutes a guide to ascertain the representation of a diverse source of strains of the pathogen (e.g., Arunga et al. 2012).

Domestication-related reductions in lima bean genetic diversity vary considerably by population. Small-seeded types were domesticated in Mexico and underwent much greater reductions in genetic diversity than large-seeded types of South America (Gutiérrez Salgado et al. 1995). The MI population of the Middle American gene pool was the only group that underwent a major domestication bottleneck and maintained only 45% (Garcia et al. 2021) to 69% (Chacón-Sánchez and Martínez-Castillo 2017) of the variation found in the wild MI gene pool. As domesticates of the MI gene pool spread into new habitats, they hybridized extensively with Middle American MII wild populations, to the extent that many domesticates are more genetically similar to this wild group. The MII domesticates have 19% greater genetic diversity than the wild MII population, due in part to their hybrid origin. In the independently domesticated Andean gene pool of lima bean, little reduction in genetic diversity was observed, with nearly identical levels of genetic diversity in domesticated and wild types (Gutiérrez Salgado et al. 1995; Chacón-Sánchez and Martínez-Castillo 2017; Garcia et al. 2021). The relatively high levels of genetic diversity in two of the three major groups of domesticated lima bean may be partly due to its relatively high levels of outcrossing relative to other species, with outcrossing rates averaging 38% and sometimes as high as 84% (Penha et al. 2017; Waines and Barnhart 1997).

In tepary bean, several methodologies have independently determined that a major reduction in genetic diversity occurred during domestication. For example, wild tepary beans show 15 unique phaseolin patterns, while domesticated types have only one (Schinkel and Gepts 1988). Similar results have been found using AFLPs (Muñoz et al. 2006), SSRs (Blair et al. 2012), and SNP data (Gujaria-Verma et al. 2016), with extremely low levels of genetic diversity in domesticated populations relative to what exists in the wild. The highly self-pollinating nature of tepary bean is more extreme than that of any other *Phaseolus* domesticate, leading to low levels of outcrossing with wild relatives (Blair et al. 2012) and a major reduction in genetic diversity during domestication.

The single phaseolin type observed by Schinkel and Gepts (1988) suggests a single domestication. Nevertheless, the domesticated gene pool of tepary bean is divided into two major subpopulations, one found in Mexico and the southwestern United States, and the other in Central America. The greatest genetic diversity is found in the Central American group, outside its region of origin, in contrast to the pattern of most species. Most of the genetic diversity of the Central American group is due to a small number of highly divergent accessions, while the majority of the population's members are highly similar (Gujaria-Verma et al. 2016). Tepary beans also show less phenotypic diversity than many other *Phaseolus*, possibly due to the smaller scale of its production and difficulty in recombining useful alleles.

In contrast, almost no reduction in genetic diversity was seen during the domestication of *P. coccineus* (Guerra-García et al. 2017) based on SNP data. This analysis identified four geographically differentiated groups of wild types: Trans-Mexican Volcanic Belt, Sierra Madre del Sur and Chiapas Highlands, Sierra Madre Occidental, and *P. coccineus* subsp. *striatus*. It also identified four domesticated groups: Trans-Mexican Volcanic Belt, Sierra Madre del Sur and Chiapas Highlands, Sierra Madre Occidental, and the Oaxaca Valley, which represented a monophyletic clade, suggesting a single domestication, although the authors could not determine a more specific location (except to exclude the Sierra Madre del Sur-Chiapas Highlands group). The group with the highest genetic diversity was the wild group from the Trans-Mexican Volcanic Belt, in contrast with the wild group from the Sierra Madre Occidental. Two domesticated groups showed high genetic variance, namely the Trans-Mexican Volcanic Belt and the Sierra Madre del Sur-Chiapas Highlands groups.

Early forms introduced to Europe underwent a bottleneck and subsequent radiation after their introduction in the sixteenth century (Spataro et al. 2011; Rodriguez et al. 2013). This radiation may be a result of the runner bean's adaptation to cool and humid conditions, which are also found in many temperate regions of Europe. *P. coccineus* has a secondary center of diversity in Europe today and these forms are genetically distinct from those of the Americas (Spataro et al. 2011; Rodriguez et al. 2013).

In *Phaseolus dumosus*, the greatest genetic diversity is found near the center of origin in the highlands of western Guatemala. Domesticates from Central America show eight seed protein banding patterns, whereas South American accessions show only two (Schmit and Debouck 1991). Despite this, the extremely limited range, the hybrid origin, and limited population size of *P. dumosus* in the wild means that the species as a whole has very low genetic diversity compared to the more widely grown *Phaseolus* species (Schmit and Debouck 1991; Llaca et al. 1994).

6.2 Population Genomics Approaches to Assess Signatures of Domestication

Bottom-up approaches have been used in *Phaseolus* to determine historical patterns of selection based on genetic diversity of modern populations. These seek to identify loci for domestication of improvement based on genomic signatures of selection. Domestication left major genome-wide impacts on *Phaseolus* species, including common bean. Papa et al. (2007) contrasted the genetic diversity of sympatric wild and domesticated common-bean populations in four regions of Mexico (Chiapas, Jalisco, Oaxaca, and Puebla) using AFLP markers distributed on average at 250 kbp. They showed that some 16% of the bean genome showed effects of selection during domestication as reductions in genetic diversity. Strikingly, most of the markers showing significant effects were linked to QTLs identified earlier by Koinange et al. (1996) in the recombinant inbred population Midas x G12873 as controlling the domestication syndrome. The large size of the genome affected by domestication may be an overestimate due to hitchhiking and background selection, especially given the high values for linkage disequilibrium in this predominantly autogamous species and the low density of markers.

Schmutz et al. (2014) identified – in the Middle American gene pool – sliding windows with both low genetic diversity and high differentiation totaling 74 Mbp of sequence (13% of the 587 Mbp genome) as having putatively selected during domestication, similar to the Papa et al. (2007) result. The two domestications appear to have affected different regions of the common-bean genome. Within the Mesoamerican landrace population, chromosomes Pv02, Pv07, and Pv09 were primarily affected, whereas in the Andean population, chromosomes Pv01, Pv02, and Pv10 were involved.

Using a similar approach based on allelic differentiation (F_{st}) and nucleotide diversity (π) of wild and domesticated gene pools of common bean, Schmutz et al. (2014) identified 1,835 genes involved in selective sweeps in the Middle American population and 748 comparable genes in the Andean gene pool. Of these, only 59 gene models showed signs of selection in both gene pools. Although the floral integrator genes *SOC1* and *FT* appear not to be domestication genes in common bean, 25 Middle American and 13 Andean genes controlling these two genes were candidate genes for domestication. The low overlap in selected genes between the two gene pools indicated that different loci may have been involved in domestication of the two gene pools. Thus, selection for similar traits affected different sets of genes or alleles affecting these traits. Further examples of these differences are provided in Sect. 10 discussing convergent evolution.

Demographic modeling was not used to control for false positives in the Schmutz et al. (2014) study, so it is difficult to determine how many of the genes that show selection-like patterns had reduced diversity due to genetic drift during the domestication bottleneck, linkage disequilibrium with selected genes, or other processes. Sampling effects, for example of the Mesoamerican and Andean wild and landrace samples, may also have played a role in the actual identification (or the lack of it) of candidate or causal genes. In this regard, it should be noted that the determinacy gene (*PvTFL1y*; Repinski et al. 2012; Kwak et al. 2012) was not identified as a candidate gene for flowering by Schmutz et al. (2014), in spite of the fact that it was identified in a different sample of the Andean gene pool (Cichy et al. 2015a), nor did the role of the phaseolin seed protein locus as one of the most important seed weight loci (Delaney and Bliss 1991a, b; Johnson et al. 1996).

In *P. coccineus*, Guerra-García et al. (2017) used two approaches to screen for genomic signatures of domestication, cultivar diversification, and natural selection. The two methods agreed on 24 SNP sites related to domestication, 13 to cultivar diversification, and eight related to natural selection; these are considered major candidates for future study. This small set of SNPs recapitulated the genetic and geographic structure of wild and domesticated groups identified with the larger (~12,000) set of SNPs, especially a clear separation of wild and domesticated types.

Similarly, 150 genes have been identified in lima bean with signatures of domestication-related selection, and of these, five are known to have functions related to seed formation, pod development, and growth-related traits important for domestication (Chacón-Sánchez and Martínez-Castillo 2017). Similar approaches have been used to characterize genomic regions, which strongly diverged during and after speciation (in the absence of gene flow: see Fig. 3) and domestication of *P. vulgaris* and *P. lunatus*, each species showing double, geographically independent domestications (Cortés et al. 2018). The species show several regions of divergence, particularly on chromosomes 10 and 11. Chromosome 10 is known to harbor a pericentric inversion between the species. Chromosome 11 was found to harbor numerous signals of divergence related to domestication and speciation. The authors concluded that neighboring signatures of speciation and domestication could be influenced by similar genomic constraints and perhaps incidentally affect genomic differentiation at other scales of divergence (Cortés et al. 2018).

6.3 The Role of Gene Flow in Sympatric Wild and Domestication Populations of Domesticated Species

Outcrossing with wild relatives has played an important role in the genetic diversity of several *Phaseolus* gene pools. For example, some studies have found evidence of multiple domestications within the Middle American gene pool of common bean (e.g., Chacón S et al. 2005), but this has been disputed by more recent work (e.g., Kwak and Gepts 2009; McClean et al. 2012; Bitocchi et al. 2013). The pattern may be due to frequent outcrossing with wild ancestors observed genetically in regions where they are sympatric (Chacón S et al. 2005). This gene flow is primarily one-directional, with three- to fourfold higher levels of outcrossing from domesticates into wild populations than vice versa (Papa and Gepts 2003; Papa et al. 2005, 2007). The predominant direction can be explained by several, not mutually excluding factors, including the dominance of wild traits, the stronger selection by farmers against wild traits compared to the recessive domesticated traits in wild environments, and the larger pollen mass contributed by cultivated populations, generally planted at higher density compared to wild populations.

Wild-weedy-domesticated populations have been identified in several locations of the distribution of wild common bean (Freyre et al. 1996; Beebe et al. 1997; Payró de la Cruz et al. 2005; Zizumbo-Villarreal et al. 2005). Nevertheless, the levels of outcrossing remain rather low such that they do not overcome the selective effects operating in the wild and domesticated gene pools; hence, the wild and domesticated types generally maintain the essential aspects of their respective phenotypes even in sympatry (Fig. 8a). Molecular markers linked to genes differentiating wild and domesticated types (Fig. 8b) show a significant reduction in genetic diversity in domesticated populations (Fig. 8c), consistent with other observations (e.g., Gepts et al. 1986; Sonnante et al. 1994; Table 3). Genetic segregation analyses have shown that these differentiating markers are often, but not always, linked to QTLs or major genes controlling domestication traits and show the highest F_{st} values (e.g., Fig. 8d; Papa et al. 2005, 2007). In contrast, markers unlinked to domestication genes show limited differentiation between sympatric wild and domesticated types, but differentiate different geographic areas harboring these sympatric types (Fig. 8b). These observations suggest that over the years and cumulatively, gene flow plays an important role in shaping the diversity of sympatric wild and domesticated types. Further evidence of the role of gene flow is provided by localized sympatric wild-domesticated complexes, in which shorter spatial distances are correlated with higher levels of gene flow (Fig. 8e, f).

Because of the directionality of this gene flow mainly, but not exclusively, from domesticated to wild populations, and the generally lower levels of molecular diversity in domesticated types (Table 3), this gene flow leads to a displacement of molecular diversity and *in situ* genetic erosion in wild types. In turn, this observation provides an argument for the active collection of additional wild types of common bean, in contrast to the affirmation of Ramirez-Villegas et al. (2010) who considered them to be of low priority.

Middle American lima beans show similar patterns of introgressions as Middle American common beans. Soon after domestication of the MI population, these domesticates spread to environments with MII wild types, and experienced strong outcrossing with this gene pool, to the extent that they are sometimes more similar to those wild populations (Chacón-Sánchez and Martínez-Castillo 2017; Chacón-Sánchez 2018; Garcia et al. 2021). Like common bean, this was at one time

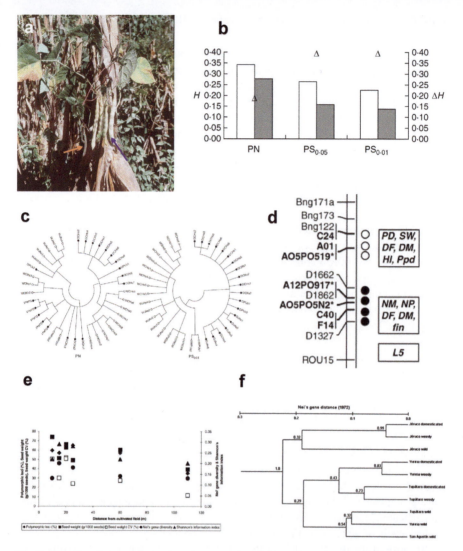

Fig. 8 Gene flow between domesticated common bean (*Phaseolus vulgaris*) and its wild progenitor populations in Mexico. (**a**) Sympatric wild and domesticated populations of common bean in Chiapas, Mexico: orange arrow: pods of wild plant; purple arrow: pods of domesticated plant (photo: P. Gepts). From Papa et al. (2007) (Creative Commons Attribution Non-Commercial License (http://creativecommons.org/licenses/by-nc/2.0/uk/): (**b**) Fig. 1: Genetic diversity (H) in wild (open bars) and domesticated (shaded bars) populations using putatively neutral AFLP markers (PN) and those putatively under selection $P < 0.05$ ($PS_{0.05}$) and $P < 0.01$ ($PS_{0.01}$) between wild and domesticated types; triangles are reductions in genetic diversity (ΔH) from wild to domesticated. (**c**) Fig. 2: Neighbor-joining consensus tree among individual wild and domesticated types for PN (left) and $PS_{0.01}$ (right) AFLP markers; (**d**) Fig. 3 (partial representation, focused on part of chromosome 1: to the left of the linkage group are markers (polymorphic in that study in bold), to the right are symbols indicating significant (black) and non-significant (white) differentiation between wild and domesticated types, while the boxes represent genes or QTLs mapped in this region, notably *fin* for determinate growth habit and *Ppd* for photoperiod sensitivity (for other symbols, consult the original article); (**e**) From Payró de la Cruz et al. (2005): Fig. 2: Relationship between distance of wild bean individuals to cultivated fields (independent variable) and phenotypic and genetic diversity variables. From Zizumbo-Villarreal et al. (2005): (**f**) Fig. 4: Unweighted-Pair-Group-Method-with-Arithmetic-means dendrogram based on Nei's genetic distances using ISSR markers populations of wild-weedy-domesticated complexes

considered evidence of multiple Middle American domestications, before being dismissed by new genetic evidence (Andueza-Noh et al. 2015).

Martínez-Castillo et al. (2007) described the existence of wild-weedy-domesticated lima bean complexes in the Yucatan Peninsula. The gene flow level was low ($Nm < 1$) and, like in common bean (Papa and Gepts 2003), was three times higher from domesticated to wild types than in the opposite direction. Like common bean, Martínez-Castillo et al. (2007) also suggest there may be a risk of genetic assimilation of wild populations.

In runner bean, Guerra-García et al. (2017) identified a basal reticulate pattern, also involving *P. dumosus* and *P. vulgaris*, suggesting ancestral gene flow among the three species, which formed a syngameon (Suarez-Gonzalez et al. 2018), perhaps also including other related wild species like *P. costaricensis* and *P. albescens* (Freytag and Debouck 2002). They did not identify, however, recent gene flow between wild and domesticated groups, despite the predominantly outcrossing nature of the first two species and the documented existence of gene flow in the predominantly selfing *P. vulgaris* (Papa and Gepts 2003; Papa et al. 2005, 2007; Beebe et al. 2001).

6.4 Correlations Between Genomic Diversity and Environmental Variables

In common bean, Rodriguez et al. (2016) used 131 SNP to characterize genetic diversity in a large sample including 417 wild and 160 domesticated accessions. They determined that 20% of the loci showed signatures of selection, with 5% showing strong support. Some of these genes are putatively involved in responses to environmental factors such as cold, light, and drought or they mark genome regions under selection to these same or other factors, taking into account the possible effects of LD. These authors did point out, however, the low levels of LD ($r^2 = 0.04$) observed in Middle American wild bean by Rossi et al. (2009).

An independent study by Ariani et al. (2018) included 246 wild accessions, among which 157 Middle American, 12 Ecuador-Northern Peru (PhI group), and 77 southern Andean entries. Using ~20,000 SNPs obtained by Genotyping-By-Sequencing based on the *Cvi*AII restriction enzyme (Ariani et al. 2016), they identified five geographical subpopulations or gene pools, each distributed in a distinct floristic province of the seasonally dry, Neotropical forest, one of the most threatened biomes on Earth (Banda-R et al. 2016).

Correlations between the distribution of the five subpopulations and climate variables showed that the southernmost Middle American group (Southern Mexico, Central America, and Colombia) was subjected to the highest level of rainfall, whereas the Andean group received the least precipitation. Potential evapotranspiration (PET) was the highest in the two northern Middle American subpopulations. Thus, in the search for additional genetic diversity for heat and drought tolerance in wild common bean, accessions from the northern Mexico may be the most likely candidate. This hypothesis has been verified by Berny Mier y Teran et al. (2019, 2020). Berny Mier y Teran et al. (2019), for example, showed that deeper-rooted and more productive wild common beans are associated with the driest environments of origin, in particular in the northern half of Mexico. In this greenhouse study, genomic regions marked by SNPs distributed on all 11 chromosomes were identified that were correlated with phenotypic traits (e.g., biomass, root depth and plant height, specific leaf area) and environmental variables of the locations of origin (evapotranspiration but – surprisingly – not temperature, soil bulk density). Berny Mier y Teran et al. (2020) compared the yield-enhancing capability of three wild common bean accessions, two originating in arid regions of the northern half of Mexico and one (as a control) from a high-rainfall of the Guatemalan highlands, crossed to the same domesticated, drought-tolerant breeding line (nested design). A QTL analysis conducted on the basis of a joint SNP linkage map revealed factors affecting the number of days to flowering, seed weight, and yield. Generally, the additive effects on yield were larger when they originated in the two arid-region accessions compared to the humid-region accession and led to yield increases under drought stress, suggesting that wild beans from arid areas can provide a valuable source of additional genetic diversity for drought stress tolerance. Further research is needed to introgress this genetic diversity into commercial dry bean classes.

6.5 The Role of Contemporary Farmers in Maintaining and Creating Bean Genetic Diversity

As illustrated by the domestication triangle (Gepts 2004; Hufford et al. 2019), humans do play an important role in the domestication and subsequent dissemination and diversification (or genetic erosion) processes. While archaeobotany plays an obvious role in the first stages of domestication (e.g., Piperno 2012), domestication can also be considered a continuing process that extends to today's era. This agency of contemporary farmers should not come as a surprise given the abundant ethnobotanic record documenting the intricate and intimate knowledge of plants in indigenous societies (e.g., Lira et al. 2016; Rivera Núñez et al. 2012; Métailié 2008). In the context of subsistence agriculture in which part of the harvest is saved to plant the next crop, today's farmers play a role in the maintenance and even the generation of new genetic diversity.

Studies in common bean provide several examples of how farmers influence the genetic diversity in their fields. Farmers can counteract the reduction in diversity that generally characterizes crop domestication, including that of *Phaseolus* beans (see Sect. 6.1, Table 3). Farmers maintain genetic diversity in their fields at a level comparable to those observed in wild beans by including both existing landraces and an improved variety, as well as tolerating wild beans in their fields. There is a distinction, nevertheless, between wild populations and cultivated fields. In the former, molecular diversity is distributed primarily among populations, whereas in the latter, molecular diversity is found mostly predominantly within populations (Zizumbo-Villarreal et al. 2005).

Bean farmers in subsistence agriculture often plant crop mixtures as a risk mitigation strategy and to satisfy diverse culinary needs. Mulumba et al. (2012) observed that both bananas and beans in smallholder farms of Uganda showed high richness and evenness in frequency of traditional varieties. In locations with higher disease incidence of the bean pathogens anthracnose (*Colletotrichum lindemuthianum*) and angular leaf spot (*Pseudocercospora griseola*), a negative correlation between richness and the disease index was observed, suggesting that increased within-field bean diversity can act as a partial disease control measure. In this study, diversity was based on the number of landraces. However, Wilkus et al. (2018) demonstrated that traditional bean varieties grown by households in Uganda include both Andean and Mesoamerican domesticates of *P. vulgaris*. For both diseases, there is a parallel structure of genetic diversity with Andean and Mesoamerican strains, which are more virulent on Andean and Mesoamerican hosts, respectively (Guzmán et al. 1995; Sicard et al. 1997). A household or community varietal mixture consisting of both Andean and Mesoamerican varieties will inherently be more resistant to diseases than varietal portfolios consisting of one or the other gene pool.

Farmers in a traditional village in Oaxaca (Mexico) use three taxa of *Phaseolus* beans in their plantings, including two ecogeographic races of common bean (*P. vulgaris*; races Jalisco and Mesoamerica; Singh et al. 1991a) and runner bean (*P. coccineus*), with distinct altitudinal adaptation based on seed types and nuclear microsatellite markers (Worthington et al. 2012; Soleri et al. 2013). Farmers are aware of these taxonomic subdivisions and their differential adaptation; although their fields are mixtures of beans, they take into account the altitude of each field to adjust the mixtures planted to increase the proportion of the most adapted component. For example, the most recent introduction of beans – race Mesoamerica represented by small, black beans – constituted an attempt to introduce marketable beans; it required opening new fields at lower altitudes, consistent with their adaptation. Analysis of chloroplast microsatellite markers also revealed that the marketable traits of the recently introduced beans – mainly small, black-colored grains – were being introgressed from race Mesoamerican into race Jalisco, i.e., the traditional bean landraces of this village. This would facilitate cultivation of marketable beans at the more convenient altitude of the village. This is a clear case of "creolization," as described also, for example, by Zizumbo-Villarreal et al. (2005) in beans and maize (Bellon and Risopoulos 2001), in which native landraces are improved with traits from introduced varieties.

Based on these examples showing purposeful mixing and gene flow of planting materials, it is clear that farmers are not passive recipients of technology and knowledge but, rather, they have to make difficult decisions that integrate a complex series of biological, climatic, and socio-economic variables, over some of which they have little control. Farmers are definitely an important factor in determining patterns and level of crop diversity.

7 Comparison of the Intensity of Domestication Among the Five Domesticated *Phaseolus* Species

The intensity of domestication can be defined by the degree to which a crop differs from its wild ancestors, as defined by its domestication syndrome. The main factors which determine domestication intensity include the number of traits involved in domestication, the magnitude of the phenotypic change brought about domestication, and the number of distinct domestication events. Darwin (1859) had already remarked that fully domesticated plants had lost the ability to survive on their own and were dependent on humans for their continued propagation. Most crops are only partially domesticated: their domestication syndrome is more limited, and they conserve part of their reproductive ability.

Different crops show different intensities of domestication, as illustrated by the genus *Phaseolus* (Table 5). Of the five domesticated *Phaseolus* species, common bean is the most intensely domesticated species. Not only was the species domesticated twice, it shows the broadest range of domesticated phenotypic variation, affecting growth habit, pod dehiscence, seed dormancy, photoperiod sensitivity, seed weight average, flower and seed color, and diversity of seed colors. In addition, it is used both for its grains as dry or shell beans and for its immature pods, as green beans. Lima bean is in many ways similar to common bean in that it was domesticated twice and shows substantial domesticated phenotypic variation. In contrast with common bean, however, it is not grown for its green, immature pods. Runner bean was domesticated once and is grown for its dry grain and sometimes also for its pods. Interestingly, its flowers are consumed as well, because of their sweet taste, associated with nectar production in this species cross-pollinated by insects and hummingbirds (Burquez and Sarukhán 1980). Its plants are also used for ornamental purposes with its attractive clusters of scarlet flowers.

The two least domesticated species – tepary bean and year bean – have gained interest recently primarily as sources of genetic diversity for disease resistance, introgressed into common bean. They show limited variation for growth habit and seed types. Whereas determinacy has arisen repeatedly across *P. vulgaris*, *P. lunatus*, and *P. coccineus*, the characteristic is unknown in tepary bean or year bean (Gepts 2014a). Similarly, their range in seed color, shapes, and sizes is comparatively low, especially in tepary bean (Freytag and Debouck 2002; Blair et al. 2012). Tepary bean is being studied as a crop with potential in hot, arid areas of the Americas and Africa, for example. Year bean is the most closely related domesticated species to common bean, but it shows very limited variation under domestication, except for seed types, a reduction of pod shattering, and possibly the loss of seed dormancy.

A comparison of these different domesticated *Phaseolus* species provides a view of the potential for domestication once a species is subjected to increasingly diverse selection pressures. It also allows us to explore the gene space affecting specific traits, i.e., different genes and alleles that can play a role in the expression of domestication or other traits. For example, all *Phaseolus* domesticated species

Table 5 Domesticated *Phaseolus* species in decreasing order of domestication intensity from top to bottom of the table, relative to wild common bean, *Phaseolus vulgaris*

	Growth habits[a]	Pod dehiscence	Dormancy	Photoperiod sensitivity	Seed weight average (range; g/100 seed)	White flowers and seeds	Diversity of seed colors, relative to W *P. vulgaris*
Common bean (*P. vulgaris*)	*Mesoamerican*						
	IV, III, II	Reduced: dry bean types (QTLs on Pv03, 08, etc.); Absent: green, snap bean types (?)	Absent (?)	Part of germplasm is photoperiod-insensitive (*ppd*, hr)	W: 8 (2–25) D: 28 (9–100)	Present	++++
	PhI group (Synonym: *P. debouckii*)						
	IV	Fully dehiscent	Present	Photoperiod-sensitive	W: 10 (7–18)	Absent	+
	Andean						
	IV, III, I (*fin*)	Reduced: dry bean types (QTLs on Pv05, etc.); Absent: green, snap bean types (*St*)	Absent (major QTL on Pv03)	Part of germplasm is photoperiod-insensitive	W: 11 (2.8–26) D: 42 (5–128)	Present	++++
Lima bean (*P. lunatus*)	*Mesoamerican*						
	IV, I	Reduced: dry bean types	Absent	Part of germplasm is photoperiod-insensitive	W: 10 (4–25) D: 43 (8–190)	Present	+++
	Andean						
	IV, I	Reduced: dry bean types	Absent	Part of germplasm is photoperiod-insensitive	W: 18 (4–33) D: 95 (28–249)	Present	+++
Runner bean (*P. coccineus*)	IV, I	Reduced: dry bean types	Absent	Part of germplasm is photoperiod-insensitive	W: 14 (0.6–38) D: 97 (20–249)	Present	+++
Tepary bean (*P. acutifolius*)	III only	Reduced: dry bean types	Absent	Part of germplasm is photoperiod-insensitive	W: 3 (2–8) D: 15 (9–27)	Present	++

(continued)

Table 5 (continued)

	Growth habits[a]	Pod dehiscence	Dormancy	Photoperiod sensitivity	Seed weight average (range; g/100 seed)	White flowers and seeds	Diversity of seed colors, relative to W P. vulgaris
Year bean (P. dumosus)	IV only	Reduced: dry bean types	Absent	Photoperiod-sensitive	W: 29 (19–43) D: 73 (26–116)	Present	++
Wild P. vulgaris	IV only	Fully dehiscent	Present	Photoperiod-sensitive (Ppd, Hr)	See above	Absent	+

Based on CIAT URG data (downloaded 06/19/2019)
[a]Growth habits (Singh 1982): IV: Climbing; III: Bush prostrate; II: Bush erect indeterminate; I: Bush erect determinate

show a reduction in pod shattering (Table 5), but a more intensely domesticated species like common bean shows additional factors for pod indehiscence in certain gene pools or ecogeographic races, depending on local adaptation (see next section: Parker et al. 2020a). These additional factors can hide basic genes for adaptation in the early stages of domestication. Thus, by comparing the genetic and molecular basis of a trait, one gets a more complete picture of genes involved in different stages and locations of the domestication processes. Further research is needed to identify these basic genes for adaptation in the different *Phaseolus* bean domestications.

8 Molecular Genetic and Genomic Basis of the Domestication Syndrome in *Phaseolus*

8.1 Gigantism

Gigantism is one of the most conspicuous components of the domestication syndrome in *Phaseolus*. Each domesticated gene pool of *Phaseolus* has much higher average seed weights than its wild relatives (Fig. 9d), often with little to no overlap in seed size (Chacón-Sánchez 2018). The archaeological record also indicates that gigantism occurred at least 7,000 years ago (Piperno and Dillehay 2008). Seed gigantism is therefore arguably a core component of the domestication syndrome of *Phaseolus* (Chacón-Sánchez 2018). In addition to aesthetic and cultural value, these changes could also provide utilitarian value. Genetic variation for seed weight has been mapped to QTLs that also influence harvest index and yield in wild x domesticated populations (Koinange et al. 1996; Blair et al. 2006b), indicating that in some cases seed weight or number may be related to changes in sink strength and resource partitioning. Varieties that produce smaller numbers of larger seeds are also easier to winnow and clean after harvest.

Populations of *Phaseolus* express considerable variation in seed size within and among species (Table 5). In the wild, species living in closed and shaded environments tend to have larger seeds than closely related species of open environments, and variations in wild *Phaseolus* reflect this (Silvertown 1982; Koenig et al. 1990). In the arid-adapted *P. acutifolius*, 100-seed weights of wild types average only 3 g, while the montane *P. coccineus* and *P. dumosus* have 100-seed-weight averages of 14–29 g (Chacón-Sánchez 2018). The smallest increase in size occurred in the weakly domesticated *P. dumosus*, with two to three times larger seeds in domesticated populations. The largest increases occurred in *P. coccineus* and Andean *P. lunatus*, each with five to six times larger seeds in domesticated populations on average.

Variation in seed size and weight of common bean was the first fully quantitative trait to be studied genetically in any species and is linked to white seed color (Sax 1923). These two traits were later mapped to chromosome Pv07 of the common bean genome (Koinange et al. 1996) and were specifically related to the complex

Fig. 9 The domestication syndrome in *Phaseolus* is highly parallel. The process typically involved (**a**) major reductions in pod shattering, (**b**) loss of seed dormancy, (**c**) changes in growth habit, (**d**) gigantism of seeds and other organs, (**e**) diversification of seed color and shape, and (**f**) loss of photoperiod sensitivity, most often conditioned by recessive, loss-of-function alleles. Some domesticated accessions show wild-type characteristics for nearly all these characteristics. Pod shattering and loss of seed dormancy are part of the core domestication process, while an increase in seed size is also nearly universal in domesticates. Other traits such as seed type, growth habit, and photoperiod sensitivity are secondary components of the domestication syndrome and are much more likely to be found in their wild-type state in domesticates. Population genomic analyses have shed light on the genetic control of these traits. In many cases, selection on key genes has been highly parallel even within species. In *P. vulgaris*, for example, at least 10 independent, recessive mutations led to white seed color (*P*, McClean et al. 2018), five recessive mutations led to complete loss of photoperiod sensitivity (*Ppd*, Weller et al. 2019), and 8 recessive mutations led to loss of determinacy (*fin*, Kwak et al. 2012). "W" indicates wild character state, "D" indicates domesticated state. Species names indicate species shown in each photo. Photos: T. Parker

phaseolin locus (Kami et al. 1995). Numerous later studies validated the highly polygenic, quantitative nature of seed weight. Koinange et al. (1996) identified three significant loci, while others identified ten (Blair et al. 2006b), eight (Blair et al. 2009), or three (Pérez-Vega et al. 2010). González et al. (2010) determined that, by that time point, at least 13 different loci had been associated with seed weight in common bean, spanning all chromosomes except Pv10. Ironically, GWAS of common bean seed weight later found the most significant SNPs on chromosome Pv10, indicating that seed weight may be associated with all 11 chromosomes (Schmutz et al. 2014; Moghaddam et al. 2016). These studies supported the complex polygenic nature of seed weight. Many of the identified alleles are important regulators within specific ecogeographic races and others are important among races and gene pools (Blair et al. 2006b, 2009; Moghaddam et al. 2016). In *P. lunatus*, seed weight is also known to be polygenically controlled by loci across at least four chromosomes, Pl03, Pl04, Pl09, and Pl10 (Garcia et al. 2021). While transgressive segregation occurred below the weight of the small-seeded parent, no recombinant inbred line equaled or exceeded the weight of the large-seeded parent, further demonstrating the complex genetic basis of seed size in *Phaseolus*.

Gigantism of other plant structures has also been mapped and is of similar complexity to that of seed size. Koinange et al. (1996) identified three QTLs for seed weight and three QTLs for pod size in their wild x domesticated RIL population. Similarly, in a cross between Andean common beans, Yuste-Lisbona et al. (2014) identified QTLs for pod size parameters on four chromosomes. In crosses between Middle American and Andean domesticated beans, Murube et al. (2020) identified QTLs with an influence on pod size on nine chromosomes, similar to their identification of seed weight QTLs on ten chromosomes. Complex polygenic changes in root architecture also occurred during domestication and are partly related to seed gigantism (Singh et al. 2019). The genetic inheritance of gigantism in common bean serves as a prime example of a highly complex, quantitative domestication trait.

8.2 Growth Habit

In the wild, *Phaseolus* spp. are indeterminate twining and circumnutating vines, which spread and climb over neighboring vegetation. This ancestral habit continues in many domesticated members of *Phaseolus*, particularly those grown in polycultures (Fig. 9d). In the wild, mutations leading to compact growth habit would be highly deleterious, as affected individuals would not be able to compete effectively for light. In many cultivated environments, however, especially those without maize or other support, a compact growth habit can facilitate harvest, improve harvest index, and reduce days to maturity. Growth habits in *Phaseolus* can take numerous forms. These are most often described on a scale ranging from I to IV (Singh 1982) with increasing climbing ability. These range from type I (determinate bush), type II (indeterminate upright), and type III (indeterminate prostrate, non-climbing, to type IV (indeterminate, strongly climbing). Today types I and II make up most commercially produced varieties in many regions, despite their two- to threefold lower yields compared to climbing varieties. Selection for improvements in bush growth habit continues to be a major focus for breeders in the present day (Kelly 2001, 2004; Singh 1999), even though climbing beans tend to be two to three times higher yielding and continue to be grown in countries with high demand for beans and abundant labor, such as the Great Lakes region in Africa (Beebe et al. 2013).

The spectrum of growth habits in indeterminate common beans, ranging from compact bush type II types to strongly climbing type IV types, is regulated by several loci. These typically include photoperiod insensitivity regulated by *Ppd* and *Hr*, as well as other QTLs (White and Laing 1989; Koinange et al. 1996; Moghaddam et al. 2016; MacQueen et al. 2020). Among the most important of these is a major locus on chromosome Pv07, which controls lodging, canopy size, and canopy growth rate. The locus may be the result of a single gene with strong pleiotropy, as would be expected for major changes in stem structure, or the result of several linked genes. Moghaddam et al. (2016) and Parker et al. (2020a) independently identified SNPs strongly related to growth habit in the same single gene model, Phvul.007G221800. Both approaches took advantage of the 280-member Middle American Diversity panel but were grown in different environments and used a variety of contrasting ground- and drone-based approaches to phenotype the population. The effects of this locus are highly stable across environments and across mapping methodologies in the population (Moghaddam et al. 2016).

The gene model Phvul.007G221800 is predicted to encode a leucine-rich repeat receptor-like protein kinase important for plant growth regulation. When close relatives of this gene are mutated in *Arabidopsis*, plants display stunted and remodeled growth (Clouse 2002). The same major Pv07 locus also includes another gene, Phvul.007G218900, which encodes a gene related to *Arabidopsis SUPPRESSOR OF AUXIN RESISTANCE 1 (SAR1)*, may be involved in defining stem thickness and was also identified by MacQueen et al. (2020). This trait has previously been linked to lodging resistance in common bean (Singh 1999).

Beyond the major locus on chromosome Pv07, growth habit in indeterminate varieties is influenced by numerous other loci across chromosomes, including Pv01, Pv04, Pv06, Pv08, Pv09, Pv10, and Pv11 (Moghaddam et al. 2016; Resende et al. 2018; MacQueen et al. 2020), some of which may have comparatively minor effects. Together, these genes lead to the more compact type II and type III growth habits. Type II growth habits primarily originated in members of the ecogeographic races Mesoamerica and Nueva Granada (Singh et al. 1991a). In recent times, these alleles have been moved through phenotypic selection into other races, such as race Durango to obtain more upright pinto or great northern beans (Kelly et al. 1990, 1999).

The major transition to type I (determinate) growth habit occurred independently in *P. vulgaris*, *P. lunatus*, and *P. coccineus*. The difference between these forms is so great that the two classes in common bean were originally classified by Linnaeus as distinct species, *P. vulgaris* (climbing) vs. "*P. nanus*" for determinate, bush beans (Mendel 1865; von Martens 1860; Brücher 1988). The determinate growth habit facilitates harvest, improves harvest index, reduces disease, and leads to more uniform maturity of pods, but is correlated with heat sensitivity (Shonnard and Gepts 1994). The simple genetic basis of this trait was identified early (Norton

Table 6 Domestication-related genes identified based on whole-plant phenotypes

Whole-plant trait (gene symbol if available)	Gene (chromosome)	Gene model[a]	Reference
Common bean (Phaseolus vulgaris)			
Plant habit: Determinacy (*fin*)	*PvTFL1y* (Pv01)	Phvul.001G189200	Repinski et al. (2012), Kwak et al. (2012)
Pigmentation: White seededness & white flowers (*p*)	*PvTT8*, IIIf plant basic Helix-Loop-Helix, clade B protein (Pv07)	Phvul.007G171333	McClean et al. (2018)
Photoperiod response: Insensitivity (*ppd*)	*PvPHYA3* (Pv01)	Phvul.001G221100	Weller et al. (2019)
Pod indehiscence	*PvPDH1* (Pv03)	Phvul.003G252100	Parker et al. (2020a)
	PvMYB26 (Pv05)	Phvul.005G157600	Rau et al. (2019), Parker et al. (2020a)
Seed dormancy	*Pectin acetylesterase 8* (Pv03)	Phvul.003G277600	Soltani et al. (2021)
Lima bean (Phaseolus lunatus)			
Cyanogenesis	β-glucosidase (Pl05)	To be determined	Garcia et al. (2021)
Determinacy	*PlTFL1* (Pl01)		

[a]Phaseolus vulgaris genome v2.1; Schmutz et al. (2014)

1915). The main gene responsible for this in the common bean, called *fin*, maps to chromosome Pv01 (Koinange et al. 1996). In *Arabidopsis*, *TERMINAL FLOWER 1* (*TFL1*) is an important repressor of flowering and regulator of determinacy, therefore making it a major candidate for *fin* (Bradley et al. 1997). Three homologs of *TFL1* have been identified in common bean, called *PvTFL1x*, *PvTFL1y*, and *PvTFL1z* (Kwak et al. 2008). Of these, *PvTFL1y* maps to Pv01 and shows co-segregation with *fin* in large segregating populations (Kwak et al. 2008; Repinski et al. 2012; Table 6). *Agrobacterium*-mediated transformations have shown that the gene is the functional homolog of *TFL1* in *Arabidopsis*, as the common bean is able to restore wild-type phenotypes (Repinski et al. 2012).

Like other domestication-related processes, the transition to determinate growth habit evolved several times in parallel, even within species. In a sample of 47 accessions, Kwak et al. (2012) identified eight mutations at the *PvTFL1y* locus with a predicted loss-of-function effect. These included multiple independent origins in the Andean gene pool and a single origin in the Mesoamerican gene pool. Despite this, 73% of sampled determinate accessions had a single mutation, involving a retrotransposon insertion into exon 4 of the gene. This mutation is also the only one identified in accessions from outside of the Americas. The other mutations of *PvTFL1y* include non-synonymous substitutions, frameshift mutations, and a deletion of the entire coding sequence of the gene. The diversity of mutation types associated with *PvTFL1y* parallels what is seen for other traits, such as white seed color and photoperiod insensitivity. While all varieties with major mutations in

PvTFL1y display the determinate growth habit, various determinate accessions have no known mutation in the gene. In some of these accessions, variation in regulatory sequence around *PvTFL1y* may be responsible (Kwak et al. 2012). In other determinate types, particularly those descended from X-ray mutagenesis programs in Michigan in the mid-twentieth century, this determinacy has been mapped to chromosome Pv07 (Kolkman and Kelly 2003). This indicates that *PvTFL1z* may also be capable of causing the determinate habit. The predominance of a single mutant allele of *PvTFL1y* across much of the world is not yet fully understood but may depend on the genetic background around the *PvTFL1y* locus (Kwak et al. 2012).

Determinacy in Andean lima bean has also been mapped to a region near *PlTFL1y*. The independently domesticated Middle American gene pool of lima bean also includes determinate types, and whether these evolved independently or through introgression from Andean types is uncertain (Garcia et al. 2021). Hybrids between determinate common beans and determinate runner beans are also determinate, indicating that the same gene was likely selected in both species (J. Berny Mier y Teran, pers. comm.). The role of *TFL1* homologs in the control of determinacy has been identified in a very broad range of species, including dicots [Fabaceae: cowpea (*Vigna unguiculata*, Dhanasekar and Reddy 2015), mung bean (*Vigna radiata*, Li et al. 2018), soybean (*Glycine max*, Liu et al. 2010; Tian et al. 2010), pea (Foucher et al. 2003); Solanaceae: tomato (*Solanum lycopersicum*, Pnueli et al. 1998), pepper (*Capsicum spp.*, Elitzur et al. 2009); and Vitaceae: Boss et al. 2006] and monocots (Poaceae: barley (*Hordeum vulgare*, Comadran et al. 2012)]. This indicates that the gene family may be involved in determinacy in other domesticated *Phaseolus* species as well.

8.3 Seed Pigmentation

Variation in seed pigmentation is a primary example of diversifying selection during domestication. The tendency to increase phenotypic diversity during domestication was recognized by Darwin (1859), who attributed it to the human interest in propagating any unique variation held solely by the discoverer. Like seed size, seed color changes have strong cultural value for consumers (see Sect. 2) and many have been mapped in *P. vulgaris* (Bassett 2007). Many highly specific seed coat color phenotypes have independently been selected among species, indicating that orthologous genes may be involved.

Mutations leading to solid white seed color (Fig. 9e) have been identified in all domesticated species of *Phaseolus* (Table 5; CIAT 2020), and therefore serve as a useful example in evaluating seed color diversification. In common bean, white-seeded accessions are found in all ecogeographic races (Singh et al. 1991a) as a result of loss-of-function mutations at the *Pigment* (*P*) locus (Shull 1907; Bassett 2007). *P* pleiotropically eliminates pigmentation across other plant structures, including flowers and stems. The *P* locus in common bean is attributed to the gene model Phvul.007G171333 on Pv07 (McClean et al. 2018). Phvul.007G171333 encodes a basic helix-loop-helix protein that is an ortholog of *Arabidopsis TRANS PARENT TESTA 8* (*TT8*). In *Arabidopsis*, *TT8* forms complexes with proteins of the MYB and WD-repeat families to specify the production of anthocyanidins, proanthocyanidins, and seed coat mucilage (Gonzalez et al. 2008; Zhang and Schrader 2017). In common bean, 10 unique alleles specifying a total loss-of-function have arisen independently (McClean et al. 2018, Table 6). These loss-of-function mutations include amino acid substitutions, frameshift mutations with premature stop codons, and deletions of important protein domains. An eleventh loss-of-function allele is believed to be based on variation in the promoter in types descended from Common Great Northern, but this has not yet been confirmed. In addition, at least four partial loss-of-function mutations of the gene have also been described, leading to several mottling patterns (Bassett 2007). Sequence data for these is not yet available. The 10 mutant haplotypes of the Phvul.007G171333 CDS identified at the sequence level notably outnumber the four known wild-type haplotypes of the gene in the species. These four functional haplotypes vary in nucleotide sequence but have identical amino acid sequences, including between the Andean and Middle American gene pools. These results demonstrate the importance of a functional *P* locus in wild types, as it conditions camouflaged seeds (Table 2), which are protected from seed predators and illustrate that diversity of the gene has increased strongly during domestication, in contrast to the general pattern across most of the genome.

8.4 Pod Shattering

A reduction in wild mechanisms of seed dispersal is a central part of the domestication syndrome in plants, including common bean (Fig. 9; Hammer 1984; Koinange et al. 1996). In the wild, pod shattering serves an important role in seed dispersal for the legume family (Di Vittori et al. 2019; Parker et al. 2021a, b, c, d, e, f). As legume pods dry, differential contraction between pod wall layers leads to helical warping in each pod valve. This creates tension that pulls at a weak layer of cells, called the dehiscence zone, at the suture. In wild-type individuals, the tension produced by the pod walls overcomes the structural integrity of the dehiscence zone, leading to shattering. While this is important for short-range dispersal in wild legumes, seeds dispersed in this way cannot be harvested efficiently, so non-shattering mutants were selected across the legume family. These mutations have reduced the tension of pod walls, increased the strength of the dehiscence zone, or both (Parker et al. 2021a, b, c, d, e, f). Pod twisting force is intrinsically linked to pod drying, so pod shattering is most problematic in arid environments. Varieties developed in humid climates may not have strong resistance to shattering when grown in drier conditions.

In *P. vulgaris*, variation in pod shattering is based primarily on non-homologous mutations between gene pools (Parker et al. 2020a, b; Rau et al. 2019). *PvPdh1* is a major regulator of pod shattering in Middle American common beans (Parker et al. 2020a, b; Table 6). The gene is an ortholog of *POD DEHISCENCE 1* of soybean, which provides shattering resistance that is highly correlated with aridity in the region of origin of landraces (Funatsuki et al. 2014). In common bean, *PvPdh1* was originally identified in a recombinant inbred population derived from Middle American parents and was validated in the BeanCAP Middle American Diversity Panel (MDP) and 108 additional accessions (Parker et al. 2020a). Non-shattering types have a threonine/asparagine polymorphism in a highly conserved amino acid. The wild-type threonine at this position has been strongly conserved since the origin of lignification in plants and is a component of the protein's active site (Kim et al. 2015). Its substitution leads to a complete loss of protein function (Gasper et al. 2016). The mutant allele of *PvPdh1* is at high frequency in types of the ecogeographic race Durango, which is adapted to semi-arid conditions of northern Mexico and the southwestern United States (see Sect. 6.1). The allele is absent in types with entirely race Mesoamerica ancestry. Diversity analyses in the region surrounding *PvPdh1* are consistent with selection on the race Durango variant, which achieves its effect by reducing pod twisting force (Parker et al. 2020b). Whether other loci were responsible for reducing shattering in the initial domestication of Middle American common beans is not yet known.

Several loci affect pod shattering in the Andean gene pool of common bean (Rau et al. 2019; Parker et al. 2020a). Rau et al. (2019) identified a locus on chromosome Pv05 related to a loss of pod shattering in a backcrossed population descended from a snap bean (Andean gene pool) and a wild parent (Middle American gene pool). The role of this locus was also identified in the Andean Diversity Panel (Parker et al. 2020a); it has been mapped to *PvMYB26* (Table 6), and orthologs of this gene have been implicated in massively parallel variation in pod shattering across five species of the closely related *Vigna* (Lo et al. 2018; Takahashi et al. 2020; Watcharatpong et al. 2020). *PvMYB26* is believed to achieve its effect in common bean through differential expression during pod development, leading to differences in fiber deposition (Di Vittori et al. 2020). Several other loci on chromosomes Pv03, Pv04, Pv06, Pv08, and Pv09 are strongly associated with pod shattering in Andean beans, which may be the result of variation in several transcription factors and/or biosynthetic genes (Rau et al. 2019; Parker et al. 2020a).

Several more genes are involved in the extreme loss of pod fiber found in snap bean, which leads to extreme resistance to pod shattering. This involves a reduction in suture fiber, and a total loss of pod wall fiber. The fiber bundle at the sutures is well developed in wild-type and dry bean and can be removed after pods reach full size. In the nineteenth century, mutant screens by Calvin Keeney led to the identification that lacked this pod string. At least two interacting loci are believed to control the loss of pod strings, possibly with a temperature effect (Currence 1930; Drijfhout 1978). One of these, *St*, has been mapped to chromosome Pv02 (Koinange et al. 1996; Hagerty et al. 2016), near the common bean ortholog of *INDEHISCENT*, which regulates pod patterning and fiber development in *Arabidopsis* (Liljegren et al. 2004). Despite this, stringless varieties showed no fixed sequence differences in *PvIND*, and the gene model is separated from the phenotypic marker by 7.8 cM

(Gioia et al. 2013). Davis et al. (2006) mapped a pod fiber QTL to chromosome Pv06, consistent with previous multilocus genetic models for pod string formation. Inheritance of pod wall fiber has typically been attributed to one major locus (Emerson 1904; Koinange et al. 1996) sometimes with secondary modifiers (e.g., Lamprecht 1932). The multiplicity of loci, responsible for complementation in hybrids of parents with different pod shattering loci, and the lack of penetrance of the pod shattering trait in moist environments, account for the persistence of the atavistic pod shattering trait in the domesticated gene pool (Parker et al. 2020a).

PlPdh1 is a lima bean ortholog of the common bean *PvPdh1*. RT-qPCR experiments have indicated that the gene is weakly expressed in domesticated plants of accession G27455 relative to wild types of accession G25230 (Garcia et al. 2021). This may be the result of parallel molecular evolution in the legume family, with alternate forms of causal mutations. An ortholog of this gene was also recently described as a shattering regulator in chickpea (Aguilar-Benitez et al. 2020), extending the degree of molecular parallelism to the more distantly related cool-season legumes. The frequent identification of *MYB26* and *PDH1* orthologs as regulators of pod shattering makes these excellent candidates for regulation of the trait in other domesticated species of *Phaseolus*.

8.5 Photoperiod Sensitivity and Flowering Time

In the wild, almost all species of *Phaseolus* are short-day plants and long nights are critical for the transition to flowering (Gepts and Debouck 1991, p. 467). Deviations from this are rare but have been recorded in temperate species such as *P. polystachios*, in which flowering is inhibited by excessively long or short days (Allard 1947). Mutations for reduced photoperiod sensitivity or neutrality (Fig. 9f) have been selected in at least four of five domesticated species, with the sole possible exception of *P. dumosus* (Table 5). At higher latitudes, photoperiod-sensitive accessions flower late in the season and are often not ready for harvest by the onset of inclement weather in autumn and winter. Photoperiod insensitivity is useful in these conditions, and wild-type varieties are extremely rare. Across environments, photoperiod insensitivity may be related to early maturity, variation in growth habit, and improved harvest index, which may explain why it is found even in tropical environments near the centers of origins for the species (Wallace et al. 1993). The evolution of photoperiod insensitivity was critical for niche and range expansion of *Phaseolus* beans to fit a variety of human demands.

Photoperiod sensitivity is controlled by at least two major genes in common bean, with a significant effect of temperature (Coyne 1967; Leyna et al. 1982; Kornegay et al. 1993b; Koinange et al. 1996; Gu et al. 1998). The interaction of these variables is often adaptive, allowing for consistent harvest times despite variable onset of summer rains (Acosta-Gallegos and White 1995). White and Laing (1989) screened 3,060 globally sourced common bean varieties under variable photoperiod conditions. As expected, photoperiod sensitivity was far more common in varieties bred in low-latitude regions, and was relatively rare in types from temperate climates. Despite this, full sensitivity to photoperiod was found in some varieties from as far north as the Netherlands and Canada. The authors also identified strong patterns between photoperiod neutrality and other traits, such as seed size and growth habit. Photoperiod insensitivity was most common in small-seeded types and those with a bush indeterminate growth habit, attributes later ascribed to race Mesoamerica (Singh et al. 1991a). The distribution of photoperiod sensitivity scores showed three distinct peaks, supporting a two-gene model for the variation in photoperiod seen in common bean.

In crosses between common bean gene pools, White and Laing (1989) and Kornegay et al. (1993b) did not find genetic complementation for photoperiod sensitivity, indicating that selection for photoperiod insensitivity occurred in parallel on the same genes. The major-effect locus controlling photoperiod sensitivity, called *Ppd*, was first named by Wallace et al. (1993) and mapped by Koinange et al. (1996) to Pv01. The locus is loosely linked with the determinacy locus *Fin* of common bean, by a distance of approximately 20 cM (Koinange et al. 1996; Weller et al. 2019). In addition to this major locus, Koinange et al. (1996) mapped a second locus related to photoperiod, also found on chromosome Pv01, but at a distance of more than 50 cM from *Ppd*. The locus, described as *Hr* for *Higher response* (Gu et al. 1998), may be within 50 cm of *Fin*, but in the opposite direction of *Ppd*. It is hypostatic to *Ppd* and strengthens the sensitivity to daylength.

Using the Middle American Diversity Panel, Moghaddam et al. (2016) identified associations between flowering time and two major loci on Pv01, separated by approximately 7 Mb. The two major Pv01 loci identified in this study parallel the results of Koinange et al. (1996). They proposed Phvul.001G064600 (homologous to *MYB56*) and Phvul.001G087900 [homologous to *KNUCKLES (KNU)*] as being major candidate genes that may underlie the QTLs. In *Arabidopsis*, *MYB56* and *KNU* are known to be involved in the transition to flowering.

MacQueen et al. (2020) evaluated a rich phenotypic dataset spanning nearly 40 years of annual trials, averaging up to 20–50 mostly different varieties, at each of 10–20 locations across the USA and Canada (as part of the Cooperative Dry Bean Nursery). They then combined these phenotypic data with millions of SNP data points spanning the 327 total evaluated varieties. The results of their flowering time analysis also supported the role of the common bean KNU homolog Phvul.001G087900, which may underlie *Hr*. Nevertheless, the results of MacQueen et al. (2020) and Moghaddam et al. (2016) were based largely on elite breeding materials from the United States and Canada and may have few strongly photoperiod-sensitive varieties. Wallach et al. (2018) developed a QTL-based model for predicting flowering time in common bean, which supported the major effects of several loci on chromosome Pv01. They also identified other alleles associated with flowering time on chromosomes Pv03, Pv04, and Pv11.

Using an F_2 population descended from Midas and G12873 (used first by Koinange et al. 1996), Weller et al. (2019) identified four groups of flowering times under long days, with at least two genes controlling variation in the trait. Subsequent progeny tests confirmed this prediction. One of the genes, *Ppd*, led to the strongest reduction in photoperiod sensitivity and was epistatic to all others. Sequencing of candidate genes on chromosome Pv01 indicated that all individuals with the Midas allele of Phvul.001G221100 flowered in 38 days or less. Phvul.001G221100 is predicted to encode an ortholog of *PHYTOCHROME A* (*PHYA*) of *Arabidopsis*, which is important in light signaling. This was not detected as a major candidate in evaluations of North American breeding lines by Moghaddam et al. (2016) or MacQueen et al. (2020). Despite this, the gene was independently identified for the regulation of flowering time in the Andean Diversity Panel by Kamfwa et al. (2015) and in a panel of primarily European accessions by Raggi et al. (2019). Another *PHYA* homolog, *E3*, is an important flowering regulator in soybean. Phvul.001G221100 (*PvPHYA3*) was therefore predicted to be the gene model underlying *Ppd* (Table 6).

Subsequent sequencing of *PhPHYA3* by Weller et al. (2019) identified an indel in the first exon of the Midas sequence, leading to premature truncation of the protein. Expression analyses indicated that in its wild-type form, *PvPHYA3* represses transcription of several flowering time-related *FT* genes. Sequencing of *PvPHYA3* in 54 wild and domesticated accessions identified no fewer than five independent mutations in *Ppd,* and the total number could be considerably greater. The strong molecular parallelism found in the photoperiod sensitivity trait of common bean is an indication that the *PHYA* orthologs may also be responsible for controlling the characteristic in other *Phaseolus* species.

8.6 Seed Dormancy and Hard-Seededness

In wild species, seed dormancy (Fig. 9b) prevents germination until conditions are favorable for seedling growth and serves as a form of bet-hedging against environmental stochasticity. In wild legumes, seed dormancy is mainly regulated by hard-seededness. Hard-seeded legumes possess a thick seed coat that is impermeable to water, which may resist imbibition for a year or more (Kaplan 1965). Using CT scan imaging, Soltani et al. (2021) observed that water uptake was initiated from the bean seed lens; using scanning electron microscopy, they further identified cracks on the lens surface, presumably facilitating water entry into non-dormant beans. In the cultivated environment, humans control the timing of planting and harvest and have selected strongly for seeds that emerge uniformly. Beyond this, seeds that are impermeable to water have far greater cooking times (Castellanos et al. 1995; Cichy et al. 2015b). The cost of cooking fuel is a major expense in many traditional societies, so selection against hard-seededness may have served both agronomic and economic needs.

Koinange et al. (1996) identified four QTLs with an effect on seed dormancy. Of these, the most important was a locus on chromosome Pv03 which explained 52% of the phenotypic variation. Secondary loci were identified on chromosomes Pv02, D14 (now part of Pv04), and D15, and these controlled 18%, 19%, and 12% of the variation in dormancy, respectively. Similarly, Cichy et al. (2015b) mapped variation in cooking time to factors on Pv02, Pv03, and Pv06 in a genome-wide association study of the Andean Diversity Panel. Midas, the domesticated parent used by Koinange et al. (1996), is an Andean type and the QTLs identified by the contrasting methods could be related to the same genes.

Recent genetic analysis showed that seed water uptake was associated with a major QTL on Pv03 (Soltani et al. 2021) in the same chromosome region identified by Koinange et al. (1996). A candidate gene was identified, namely a *pectin acetylesterase 8* gene. This gene showed a 5-bp insertion causing a loss-of-function mutation and a 21-fold reduction in expression in non-dormant genotypes. Furthermore, wild beans all showed the functional allele, whereas 77% of the domesticated beans showed the non-functional allele (Soltani et al. 2021).

The *Asper* (*Asp*) and Joker (*J*) loci of common bean are known to influence seed coat properties and permeability to water during soaking (Castellanos et al. 1995). This is partly due to deposition of epicuticular wax surrounding the seed coat (Konzen and Tsai 2018). *Asp* is known to influence the thickness of the palisade epidermis cell layer of the seed coat wall, leading to thinner seed coats and dull texture when mutated. It has been mapped to chromosome Pv07 (Pérez-Vega et al. 2010; Cichy et al. 2014). *Joker* (*J*) allows for the full expression of shininess when found with *Asp*, and pleiotropically modifies seed coat pigmentation. *J* has been mapped to Pv10. Konzen and Tsai (2018) determined that in nearly-isogenic lines varying for only *J* and *Asp*, mutations in either gene significantly increased water absorption relative to fully shiny types. Rate of water uptake was not significantly different based on which gene was in its mutant form. Despite this, the small-seeded line 5–593, possessing wild-type *Asp* and *J*, imbibed water at a far greater rate than several landraces evaluated. This indicates that other genes are also necessary for strong hard-seededness, supporting the role of QTLs identified by previous authors on other chromosomes.

Sandhu et al. (2018) mapped hard-seededness in a recombinant inbred population that segregated for the trait. In addition to the major-effect locus on Pv07 associated with *Asp*, they also identified loci controlling the trait on chromosomes Pv01 and Pv02. It is not clear whether the chromosome Pv02 locus is the same as that identified in Andean domesticates by Koinange et al. (1996) and Cichy et al. (2015a).

Multiple lines of evidence indicate that seed dormancy in common bean is oligogenic. The identification of a candidate gene for the dormancy QTL on Pv03 by Soltani et al. (2021) is a major step towards a better understanding of the genetic control of this trait in *Phaseolus*. The trait is likely to be an area for further consideration in the future, in particular the identification of additional candidate genes and their distribution across gene pools and ecogeographic races.

9 Unresolved Questions in Domestication Studies: A *Phaseolus* Perspective

Many questions regarding the process of domestication remain unresolved. Among these, different approaches have disputed the number of genes and the degree of human intentionality involved in the process.

The number of genes involved is highly controversial and is a central paradox of domestication genomics. Many phenotypes associated with the domestication syndrome are found in a minority of domesticated lines, and few are universal in domesticated populations. In common bean, for example, many landraces show phenotypes similar to wild types, such as photoperiod sensitivity, viny growth habits, and cryptic seed pigmentations. Pod indehiscence, rachis non-brittleness, and seed size have been increased in most domesticates, although overlap continues to exist between most wild and domesticated gene pools. Top-down approaches typically find a relatively small number of genes that control these traits. Even for core pulse domestication traits such as pod shattering, thorough evaluations have found no alleles that are unique to either wild or domesticated populations, and even putative domestication-related alleles are at low frequencies in many domesticated populations (Zhang and Singh 2020). In common bean, the major-effect pod shattering genes, *PvPdh1* and *PvMYB26*, both segregate extensively in domesticated gene pools (Parker et al. 2020a, b). In contrast, a very large number of genes conditioning individual domestication traits are identified by bottom-up approaches, such as 1,835 genes in the Middle American population of common bean (Schmutz et al. 2014). Top-down approaches may underestimate the number of loci regulating a trait due to limitations in population size and sampling. Similarly, the number of genes identified by bottom-up approaches may be overestimated due to effects of population demography, epistatic effects on downstream developmental or metabolic genes, selection for traits unrelated to domestication itself, or linkage disequilibrium (Kantar et al. 2017). The true number and roles of genes involved in domestication is likely intermediate and remains to be determined. Overall, moderate changes in allele frequency during domestication were likely far more common than qualitative fixation of domestication-related alleles.

There has long been a debate on the degree of consciousness during selection of domesticated species (Ross-Ibarra et al. 2007; Larson et al. 2014; Abbo and Gopher 2020). Indeed, much of the dispute regarding intentionality of selection in domestication may be attributable to Darwin himself, who concluded of domestication that "as far as the final result is concerned, [selection] has been followed almost unconsciously. It has consisted in always cultivating the best-known variety, sowing its seeds, and, when a slightly better variety has a chance to appear, selecting it, and so onwards" (Darwin 1859). The statement is somewhat self-contradictory, as the selection described is clearly intentional. Some of the debate regarding consciousness during domestication may be the result of this contradiction, which applies the term consciousness to the overall process of domestication process, rather than selection for any useful traits.

Little evidence supports unconscious selection during the domestication of *Phaseolus*, whereas conscious selection has occurred across a variety of traits. *Phaseolus* includes numerous examples of mutations that were strongly and repeatedly selected, despite having no effect or a negative effect on agronomic fitness. For example, at least eight independent mutations at *PvTFL1y (Fin)*, which controls determinacy, have been selected in *P. vulgaris* alone (Kwak et al. 2012). The mutation leads to smaller plants and generally reduces total seed yield (Kornegay et al. 1992), but greatly simplifies and accelerates harvest, a desirable trait in food-insecure, subsistence agroecosystems. The repeated transition to the determinate type is extremely difficult to explain without invoking conscious human selection. Pod non-shattering and uniform germination also facilitate harvest, and there is little reason to conclude they would not have also been selected consciously. Similarly, *PvPHYA3 (Ppd)* for photoperiod insensitivity was strongly selected in the race Mesoamerica types of the lowland tropics, rather than in race Durango of northern latitudes (Singh et al. 1991a; Weller et al. 2019). Photoperiod insensitive varieties mature earlier, which is highly desirable to subsistence farmers (see above), but typically have lower seed yield (Scully and Wallace 1990). This indicates that they were likely selected intentionally for early harvest, at the expense of total seed yield.

Finally, the selection of dozens of seed color alleles (Bassett 2007) is further evidence of the intentionality of selection. The functional value of many of these, outside of human curiosity, is trivial. This can be compared with increased seed size, which nearly universally distinguishes domesticated *Phaseolus* and wild relatives of the same gene pool and may be considered a core domestication trait (Chacón-Sánchez 2018). Given the conscious selection of numerous traits in *Phaseolus*, some of which are of minimal utilitarian value, there is little reason to believe that ancient peoples were unaware of or unconcerned with other important traits such as pod shattering, germination rate, or seed size. One exception to the rule of conscious selection may exist for domestication-related root traits, which could not have been selected directly, but may have been correlated with improvements in productivity (Singh et al. 2019; Berny Mier y Teran et al. 2019). Foragers and farmers in traditional societies were known to select *Phaseolus* beans, and all other New World crops, on a plant-by-plant basis rather than by mass selection (Sauer 1965; Larson et al. 2014). Mutations involving all the elements of the domestication syndrome would be easily recognizable on a plant-by-plant basis. It is only in the small-seeded cereal grains of the Old World where mass selection and unconscious selection have received widespread support (Ross-Ibarra et al. 2007; Larson et al. 2014). However, these may be the exceptions to the general rule of conscious selection in plant domestication, which has been noted for core domestication traits in other taxa (Abbo et al. 2011). Widespread evidence supports the conscious selection of several domestication-related traits in *Phaseolus*, and there is little reason to conclude that early food-producing cultures were unaware of or unconcerned with the process.

Numerous unanswered questions remain related to the process of domestication. Population genomics methods have shed light on several interrelated patterns related to domestication, such as the degree of molecular parallelism, the number of genes involved, and even the degree of intentionality in the process. Continued improvements in population methods are likely to further clarify remaining questions in these areas.

10 Patterns of Convergent Evolution and Domestication in *Phaseolus*

Many evolutionary lessons can be drawn from the domestication and improvement of *Phaseolus*. Across all seven domestication events, a highly parallel domestication syndrome included a reduction in shattering, a reduction in seed dormancy, gigantism, and seed type diversification. Other traits evolved independently in most of the domesticates, such as photoperiod insensitivity arising in all domestication events except possibly *P. dumosus*, and the bush growth habit, which arose in at least four of seven domesticates. Even within individual species, where it has been studied, repeated mutations within individual genes have often been the rule rather than the exception. This is highlighted by the five known independent loss-of-function alleles at *Ppd* (*PvPHYA3*) for photoperiod insensitivity (Weller et al. 2019) and eight at *fin* (*PvTFL1y*) for determinacy (Kwak et al. 2012). In the greatest display of molecular parallelism yet described in the genus, at least 14 independent, full- or partial-loss-of-function mutations have been described in *P*, a master regulator of seed color (McClean et al. 2018). This intra-specific parallelism suggests that orthologous genes may be responsible for regulating the traits in closely related species. Moreover, this indicates that in many cases evolution may proceed by highly predictable molecular mechanisms (Lenser and Theißen 2013). A further similarity is that selection across several traits of recessive, loss-of-function alleles at a relative small or a single locus conditioning these domestication traits, highlighting the importance of this type of mutational change in adaptation, whether during and subsequent domestication (Gepts 2004, 2014b) or in natural evolution (Monroe et al. 2021).

The major common bean genes for photoperiod insensitivity (*Ppd/PvPHA3*), determinacy (*fin/PvTFL1y*), and absence of seed pigmentation (*P/PvTT8*) fit three of the four criteria proposed by Lenser and Theißen to promote parallel molecular evolution. The first of these factors is a nodal position in gene regulation. In each case, the mutated common bean gene is a transcription factor and works upstream of numerous other transcription factors and structural genes. Mutations in these master regulatory genes may have effects that would require numerous parallel mutations in downstream targets. Further, modifications in master regulator genes can modify how and when the downstream genes are activated. This can initiate developmental programs in ways that might be impossible through mutations of structural genes alone, such as in the case of determinacy. The second of Lenser and Theißen's criteria is that the genes are involved in simple pathways. Indeed, each of the individual common bean genes acts as a molecular switch which controls the presence or absence of a clear qualitative trait. This is in contrast to other traits, such as gigantism, which is a highly complex and quantitatively inherited trait, involving numerous means to the same end. The third major criterion is that the genes operate with little negative pleiotropy. In each of these cases, few if any pleiotropic deleterious side-effects occur due to the mutations. In the case of *P*, for example, white seed color is pleiotropically regulated along with the loss of flavonoid pigments in the stems and flowers, with few negative consequences. In *Arabidopsis*, several genes are known to control seed coat color, but these often have other pleiotropic effects (Buer and Djordjevic 2009). *TTG1*, for example, forms a protein complex with the *P* ortholog *TT8*. When mutated, *TTG1* leads to loss of seed color, but also causes serious defects in the development of root hairs, trichomes, and other epidermal structures (Zhang and Schrader 2017; McClean et al. 2018). The repeated selection for *P* mutants is likely the result of their few negative pleiotropic effects. Thus, our results confirm several of the factors proposed to lead to parallel molecular evolution, including the nodal positioning of the genes, their involvement in simple pathways, and minimal pleiotropic effects (Lenser and Theißen 2013).

In many cases, individual selected mutations achieve transformative useful effects without any known deleterious pleiotropy. In contrast, we have found little evidence thus far to suggest that selection of standing mutations was important in *Phaseolus* domestication. This factor is not relevant to determinacy, white seededness, and photoperiod insensitivity in *Phaseolus*, which would be highly deleterious to wild forms. In soybean, several pod-shattering related genes are found in both their functional and mutant forms in wild populations, although the largest of these, *PDH1*, is nearly absent in the wild (Zhang and Singh 2020). In *Phaseolus*, sequencing of *PvPdh1* in wild accessions failed to identify the shattering-resistance allele (Parker et al. 2020a), supporting the possibility that it may have been a *de novo* mutation in the domesticated gene pool.

Other factors may also be related to molecular parallelism during domestication. For example, most genes with strong parallelism involve loss-of-function mutations. This applies to the major alleles related to seed color, determinacy, photoperiod sensitivity, and pod shattering. Gain-of-function mutations are known to exist in some species, such as *Teosinte branched 1* (*tb1*) of maize but are comparatively rare. While only a single *tb1* gain-of-function mutation is known to control growth architecture in maize (Studer et al. 2011), loss-of-function mutations at *PvTFL1y* have transformed growth architecture at least eight times independently in common bean. If the effects of *PvTFL1y* required a gain-of-function mutation, it is likely that it would have been selected in a much less parallel manner.

Evolution during *Phaseolus* domestication and early crop improvement was often highly parallel. Human-mediated selection often targeted similar genes repeatedly, leading to similar phenotypic results. This is seen in a range of phenotypic traits. The highly parallel nature of domestication indicates that molecular evolution may lead to a small set of predictable outcomes given a specific change in selective pressure.

Novel independent evolution of character states between populations is problematic for GWAS. This is the result of independent mutations across diverse genetic backgrounds, which strongly reduces the correlation between phenotypes and markers near the gene of interest (Bolnick et al. 2018). In common bean, for example, genome-wide mapping of flowering time has not always led to the identification of *Ppd* (Moghaddam et al. 2016; MacQueen et al. 2020), which is known to segregate in the populations studied and has a major effect on flowering time. Because multiple independent mutations exist in the populations, each in diverse genetic backgrounds (Weller et al. 2019), the relationship between the phenotype and linked markers is less clear. This pattern highlights the role of combined approaches, which take advantage of the complementary benefits of GWAS, QTL mapping, and other methods.

11 The Future of Breeding in *Phaseolus*

Major improvements in agricultural productivity are needed to achieve production and sustainability goals (Ray et al. 2013; Ramirez-Cabral et al. 2016; Hunter et al. 2017a, b). Many countries in which *Phaseolus* beans are most widely consumed have rapidly expanding populations, highlighting the need for improved agricultural production efficiency. These advances will require a combination of improved agronomic practices and genetic improvement of crop varieties. A principal advantage of genetic improvement is that it does not require increased use of fertilizer, pesticides, or other inputs. Instead, these new varieties provide higher yields due to novel resistances to biotic or abiotic stresses or improved resource use efficiency. Improved varieties may also have better growth habits, harvest properties postharvest storage quality, or nutritional properties than previous varieties. In all cases, the new types must adhere to consumer quality demands. Today, there is increasing interest in crop quality traits in *Phaseolus*, such as cooking time (Cichy et al. 2015b), nutritional content (Porch et al. 2017; Katuuramu et al. 2018), and culinary value (Walters et al. 2011; Brouwer et al. 2016; Swegarden et al. 2016; Parker et al. 2021b, c, d, e, f).

Marker-assisted and genomic selection will be crucial for plant breeding in *Phaseolus* (Beaver and Osorno 2009; Assefa et al. 2019; Miller et al. 2018; Mukankusi et al. 2019; Keller et al. 2020). These technologies rely on precise mapping of genomic regions that control variation for a trait. Plant breeders can then use this genetic information to identify promising progeny lines with greater selection accuracy, at shorter intervals, and/or at lower cost than using traditional phenotype-based methods. This leads to greater rates of improvement, which is critical for providing a reliable supply of nutritious food to the global population. The use of these methods is likely to greatly increase in coming years (Assefa et al. 2019).

Phaseolus crop wild relatives have large untapped potential. Crop gene pools are typically limited in genetic diversity compared to wild relatives due to historical selection and drift (Hufford et al. 2019). Wild gene pools therefore maintain many useful alleles that are not found in domesticates. Of five studies conducted with wild x domesticated mapping populations of common bean (Berny Mier y Teran et al.

2020), each identified alleles from the wild type that improved domesticated varieties of the crop. In most cases, wild varieties carry alleles that provide resistance to biotic or abiotic stresses. Recent studies have indicated, for example, that yields of the drought-resistant common bean variety SEA5 can be increased by 20% by introgressing alleles from wild accessions collected in the driest region of the species' natural range (Berny Mier y Teran et al. 2020). Wild accessions are sources of resistance to numerous pest and pathogens, such as bruchids and white mold (Debouck 2016; Dohle et al. 2019, Table 7). Hybridization between wild and domesticated accessions has led to the release of several improved cultivars and germplasm lines, with pest and pathogen resistance beyond what has traditionally existed in the domesticated gene pool (e.g., Osborn et al. 2003; Beaver et al. 2012; Kusolwa et al. 2016). Crop wild relatives are also a source for numerous other plant traits, such as increasing seed weight, increasing pod size, reducing seed dormancy (Koinange et al. 1996), improving yield (Blair et al. 2006b; Berny Mier y Teran et al. 2020), enhancing mineral accumulation in seeds (Blair et al. 2013), and improving root architecture (Singh et al. 2019).

Unfortunately, the useful alleles of wild germplasm are tied up in genetic backgrounds that are poorly suited for commercial production. Breeding programs must therefore recombine the beneficial wild alleles with those of the domesticated gene pools, requiring extensive pre-breeding (Singh 2001; Acosta-Gallegos et al. 2007). This often entails backcrossing into domesticated germplasm, with subsequent expansion into numerous locally adapted varieties. Each of these processes requires recurrent selection, which is particularly challenging for alleles with a recessive effect or minor quantitative effect. These processes will be greatly facilitated by a clear population genomic understanding of the relevant variation. Like wild relatives, landraces also harbor abundant genetic diversity that could be applied towards plant breeding, and therefore these types also warrant evaluation and pre-breeding (Gioia et al. 2019).

Genomics-informed breeding strategies will be critical for *Phaseolus* improvement. The wide ecological diversity of domesticated *Phaseolus* species is one of the greatest strengths of the genus. These species range from highly arid environments to cool, humid highlands, and experience a wide range of biotic and abiotic conditions. In the context of climate change, which is predicted to significantly affect the production of common bean (Porch et al. 2013; Ramirez-Cabral et al. 2016), the heat-adapted species *P. lunatus* and *P. acutifolius* will be of great importance (e.g., Medina et al. 2017). Several qualities of *P. acutifolius* will make it of particular interest in coming decades, such as its extreme tolerance for drought and heat, resistance to common bacterial blight, and ability to hybridize with *P. vulgaris*. This hybridization has traditionally required embryo-rescue, limiting its use, but bridge parents have recently been developed which can hybridize without the need for embryo rescue (Barrera et al. 2020). This will be a major step towards employing the value of tepary beans on a widespread level. The outcrossing habit of *P. coccineus* could also be of great value to other species, such as common bean, as a path towards F_1 hybrid seed production (Bannerot 1988; Hervieu et al. 1994;

Table 7 Wild *Phaseolus* germplasm as a source of plant improvement. Wild accessions carry useful alleles that can be incorporated into cultivated germplasm. Modified from Debouck (2016), Dohle et al. (2019)

Category	Trait	Species	Accession	Reference
Pest resistance	Bruchids	*P. vulgaris* L.	G12882, G12866, G12952, G02771, G24582	Schoonhoven et al. (1983), Osborn et al. (1988, 2003), Kornegay et al. (1993a), Acosta-Gallegos et al. (1998)
	Bruchids	*P. acutifolius* A. Gray	G40199	Singh et al. (1998), Kusolwa et al. (2016)
Disease resistance	Bacterial wilt	*P. vulgaris* L.	G12883	Urrea and Harveson (2014)
	Common bacterial blight	*P. vulgaris* L.	PI417662	Beaver et al. (2012)
	Fusarium	*P. vulgaris* L.	G12947	Acosta-Gallegos et al. (2007)
	Web blight	*P. vulgaris* L.	PI417662	Beaver et al. (2012)
	White mold	*P. vulgaris* L.	PI318695	Mkwaila et al. (2011)
	White mold	*P. costaricensis* Freytag & Debouck	G40604	Singh et al. (2009), Schwartz and Singh (2013)
	White mold	*P. polystachios* (L.) Britton et al.	Multiple	Dohle et al. (2019)
Abiotic stress	Cold and drought	*P. acutifolius* A. Gray	PI 638833	Souter et al. (2017)
	Salinity	*P. filiformis*, *P. lunatus*, *P. macvaughii*	Multiple	Bayuelo-Jimenez et al. (2002)
General plant traits	Amino acid content	*P. vulgaris* L.	Multiple	Montoya et al. (2008)
	Increased pod size	*P. vulgaris* L.	G12873	Koinange et al. (1996)
	Increased seed weight	*P. vulgaris* L.	G12873	Koinange et al. (1996)
	Polyphenols	*P. vulgaris* L.	G11025	Espinosa-Alonso et al. (2006)
	Reduced seed dormancy	*P. vulgaris* L.	G12873	Koinange et al. (1996)
	Root traits	*P. vulgaris* L.	G23419	Singh et al. (2019)
	Seed iron, zinc	*P. vulgaris* L.	G10022	Blair et al. (2013)
	Yield improvement	*P. vulgaris* L.	G24404	Blair et al. (2006b)
	Yield improvement	*P. vulgaris* L.	G24423	Acosta-Gallegos et al. (2007)
	Yield improvement	*P. vulgaris* L.	PI319441, PI417653	Berny Mier y Teran et al. (2020)

Gepts 1998). Population genomics advances will facilitate the identification and movement of useful alleles between *Phaseolus* species.

Useful alleles from four domesticated *Phaseolus* species can be readily exchanged, forming a single extended gene pool or syngameon. *Phaseolus lunatus* cannot be successfully hybridized with other domesticates, but it is closely related to several wild species, including *P. polystachios*. *P. polystachios* displays hypogeal germination, which is useful for resistance to early-season frost in cool climates. *P. polystachios* also possesses near-total immunity to white mold, which is a major constraint for lima bean production. The two species have been successfully hybridized, although with a reduction in pollen fertility in the F_1 (Dhaliwal and Pollard 1962). Modern techniques would likely be able to overcome these, leading to major improvements for lima bean (Dohle et al. 2019). Lima bean also faces numerous other challenges, such as a lack of reliable resistance to *Lygus*, which could potentially be introduced from other species (Dohle 2017; Dashner 2018).

12 Conclusions and Future Perspectives

The genus *Phaseolus* is a hotspot for domestication, with seven independent domestications across five species. Today, these domesticates are globally important sources of nutrition, but their production faces a variety of obstacles. Identifying and recombining useful alleles between species and gene pools of the genus, in both and domesticated wild populations, will be critical for addressing these challenges. A thorough genetic understanding of these useful variants, including those involved in domestication, will be critical for maintaining production of a globally important source of nutrition. These explorations will also shed light on the fundamental processes of crop evolution. Population genomics methods, coupled with careful and precise phenotyping, will be of fundamental importance for resolving these major outstanding questions related to *Phaseolus* evolution. Further, marker-assisted selection and genomic selection are certain to be of increasing importance for *Phaseolus* breeding, and these have been made possible by improvements in population genomics methods.

The genus *Phaseolus* sits at the core of a domestication hotspot. Its family, Fabaceae, includes more domesticated species than any other plant family. Among these, approximately two thirds of the domesticates occur within the single tribe Phaseoleae. The repeated domestication of *Phaseolus*, including two species that were domesticated twice, highlights its role as a major domestication hotspot. These domestications serve as repeated natural experiments in evolution. In the process of domestication, members of *Phaseolus* underwent major genetic bottlenecks, with corresponding reductions in genetic diversity. This was most severe in *P. acutifolius*, with a highly self-pollinating habit, and non-existent in *P. coccineus*, an obligate outcrosser. This reinforces the role of outcrossing in preserving genetic diversity during domestication. Further, it indicates that the diverse gene pools of wild beans are likely a reservoir of useful alleles for crop improvement. Population genomics has been and will be useful to detect outcrossing and areas of the genome differentially affected by gene flow.

Domestication involved the selection of several highly parallel traits between and within species. Reductions in pod shattering, a core domestication trait, were largely mediated by mutations in a relatively small number of key genes throughout the Phaseoleae. In contrast, the loss of seed dormancy and particularly increased seed size are mediated by somewhat larger numbers. Secondary elements of the domestication syndrome, such as changes in seed color, growth habit, and photoperiod sensitivity, have been studied in more detail. In these traits, selection on the same genes has operated in extreme examples of parallelism. For each, between five and ten independent mutations with a predicted loss-of-function effect have been identified in *P. vulgaris*. This extreme molecular genetic and genomic parallelism supports the notion that directional selection often proceeds based on molecular changes at predictable loci. This also makes the orthologs of genes influencing a process in any *Phaseolus* species strong candidates for regulating similar roles across other members of the genus. Many of these parallel changes, such as determinate growth habit and white seed color, have cultural value but otherwise deleterious evolutionary effects. Their repeated selection supports the role of conscious selection during crop evolution.

The expanding genomic understanding of agronomic and domestication-related characteristics will be valuable for *Phaseolus* crop improvement. Much of the requisite diversity of the genus is found in germplasm that is poorly adapted for commercial production. In many cases, this extends to minor species, including those which have not been domesticated. Traditional population genetics assays are often not applicable to more distantly related germplasm, as they may differ significantly in SNPs, repeats or indels, and chromosomal rearrangements. This demonstrates the need for expanded whole-genome resources across the genus. The important role of diversity found in non-commercial germplasm also highlights the need to actively preserve and maintain these materials for future pre-breeding and cultivar development. The genetic basis of useful characteristics will be greatly clarified by the versatile and data-rich outputs of population genomics methods. In turn, this knowledge will facilitate the movement of useful alleles into commercial varieties. These genomics-informed breeding methods will be crucial for the production of a globally important source of nutrition.

References

Abbo S, Gopher A. Plant domestication in the neolithic near east: the humans-plants liaison. Quaternary Sci Rev. 2020;242:106412. https://doi.org/10.1016/j.quascirev.2020.106412.

Abbo S, Lev-Yadun S, Gopher A. Origin of near eastern plant domestication: homage to Claude Levi-Strauss and "La Pensée Sauvage". Genet Resour Crop Evol. 2011;58:175–9. https://doi.org/10.1007/s10722-010-9630-0.

Acosta-Gallegos J, White J. Phenological plasticity as an adaptation by common bean to rainfed environments. Crop Sci. 1995;35:199–204. https://doi.org/10.2135/cropsci1995.0011183X003500010037x.

Acosta-Gallegos J, Quintero C, Vargas J, Toro O, Tohme J, Cardona C. A new variant of arcelin in wild common bean, *Phaseolus vulgaris* L., from southern Mexico. Genet Resour Crop Evol. 1998;45:235–42.

Acosta-Gallegos JA, Kelly JD, Gepts P. Prebreeding in common bean and use of genetic diversity from wild germplasm. Crop Sci. 2007;47:S44–59. https://doi.org/10.2135/cropsci2007.04.0008IPBS.

Aguilar-Benitez D, Casimiro-Soriguer I, Torres AM. First approach to pod dehiscence in faba bean: genetic and histological analyses. Sci Rep. 2020;10:17678. https://doi.org/10.1038/s41598-020-74750-1.

Allard HA. The ecology of the wild kidney bean *Phaseolus polystachyos* (L.) BSP. J Wash Acad Sci. 1947;37:306–9.

Almeida C, Pedrosa-Harand A. High macro-collinearity between lima bean (*Phaseolus lunatus* L.) and the common bean (*P. vulgaris* L.) as revealed by comparative cytogenetic mapping. Theor Appl Genet. 2013;126:1909–16. https://doi.org/10.1007/s00122-013-2106-9.

Andueza-Noh RH, Martínez-Castillo J, Chacón-Sánchez M. Domestication of small-seeded lima bean (*Phaseolus lunatus* L.) landraces in Mesoamerica: evidence from microsatellite markers. Genetica. 2015;143:657–69. https://doi.org/10.1007/s10709-015-9863-0.

Araya CM, Alleyne AT, Steadman JR, Eskridge KM, Coyne DP. Phenotypic and genotypic characterization of *Uromyces appendiculatus* from *Phaseolus vulgaris* in the Americas. Plant Disease. 2004;88:830–6. https://doi.org/10.1094/pdis.2004.88.8.830.

Ariani A, Berny Mier y Teran JC, Gepts P. Genome-wide identification of SNPs and copy number variation in common bean (*Phaseolus vulgaris* L.) using genotyping-by-sequencing (GBS). Molecular Breeding. 2016;36:87. https://doi.org/10.1007/s11032-016-0512-9.

Ariani A, Berny Mier y Teran J, Gepts P. Spatial and temporal scales of range expansion in wild *Phaseolus vulgaris*. Mol Biol Evol. 2018;35:119–31. https://doi.org/10.1093/molbev/msx273.

Arunga EE, Ochuodho JO, Kinyua MG, Owuoche JO. Characterization of *Uromyces appendiculatus* isolates collected from snap bean growing areas in Kenya African. J Agric Res. 2012;7:5685–91. https://doi.org/10.5897/AJAR12.1826.

Assefa T, Assibi Mahama A, Brown AV, Cannon EKS, Rubyogo JC, Rao IM, et al. A review of breeding objectives, genomic resources, and marker-assisted methods in common bean (*Phaseolus vulgaris* L.). Molecular Breeding. 2019;39:20. https://doi.org/10.1007/s11032-018-0920-0.

Banda-R K, Delgado-Salinas A, Dexter KG, Linares-Palomino R, Oliveira-Filho A, Prado D, et al. Plant diversity patterns in neotropical dry forests and their conservation implications. Science. 2016;353:1383–7. https://doi.org/10.1126/science.aaf5080.

Bannerot H. The potential of hybrid beans. In: Beebe S, editor. Current topics in breeding of common bean. Cali: Centro Internacional de Agricultura Tropical; 1988. p. 111–34.

Banning EB. Archaeological plant remains. In: Banning EB, editor. The archaeologist's laboratory: the analysis of archaeological evidence. Cham: Springer Nature Switzerland AG; 2020. p. 267–91.

Barrera S, Berny Mier y Teran JC, Diaz J, Leon R, Beebe S, Urrea CA. Identification and introgression of drought and heat adaptation from tepary beans to improve elite common bean backgrounds. Annu Rep Bean Improv Coop. 2020;63:21–2.

Bassett MJ. Genetics of seed coat color and pattern in common bean. Plant Breed Rev. 2007;28:239–315. https://doi.org/10.1002/9780470168028.ch8.

Baudet JC. The taxonomic status of the cultivated types of lima bean (*Phaseolus lunatus* L.). Tropical Grain Legume Bull. 1977;7:29–30.

Bayuelo-Jimenez JS, Debouck DG, Lynch JP. Salinity tolerance in *Phaseolus* species during early vegetative growth. Crop Sci. 2002;42:2184–92. https://doi.org/10.2135/cropsci2002.2184.

Beaver JS. The production and genetic improvement of beans in the Caribbean. Annu Rep Bean Improv Coop. 2020;63:7–12.

Beaver J, Osorno J. Achievements and limitations of contemporary common bean breeding using conventional and molecular approaches. Euphytica. 2009;168:145–75. https://doi.org/10.1007/s10681-009-9911-x.

Beaver JS, Zapata M, Alameda M, Porch TG, Rosas JC. Registration of PR0401-259 and PR0650-31 dry bean germplasm lines. J Plant Regist. 2012;6:81–4. https://doi.org/10.3198/jpr2011.05.0283crg.

Becerra Velásquez VL, Gepts P. RFLP diversity in common bean (*Phaseolus vulgaris* L.). Genome. 1994;37:256–63. https://doi.org/10.1139/g94-036.

Becerra VV, Paredes CM, Debouck D. Genetic relationships of common bean (*Phaseolus vulgaris* L.) race Chile with wild Andean and Mesoamerican germplasm. Chilean J Agric Res. 2011;71:3–15. https://doi.org/10.4067/S0718-58392011000100001.

Beebe S, Toro O, González A, Chacón M, Debouck D. Wild-weed-crop complexes of common bean (*Phaseolus vulgaris* L., Fabaceae) in the Andes of Peru and Colombia, and their implications for conservation and breeding. Genet Resour Crop Evol. 1997;44:73–91. https://doi.org/10.1023/A:1008621632680.

Beebe S, Skroch PW, Tohme J, Duque MC, Pedraza F, Nienhuis J. Structure of genetic diversity among common bean landraces of Middle American origin based on correspondence analysis of RAPD. Crop Sci. 2000;40:264–73. https://doi.org/10.2135/cropsci2000.401264x.

Beebe S, Rengifo J, Gaitan E, Duque MC, Tohme J. Diversity and origin of Andean landraces of common bean. Crop Sci. 2001;41:854–62. https://doi.org/10.2135/cropsci2001.413854x.

Beebe S, Rao I, Mukankusi C, Buruchara R. Improving resource use efficiency and reducing risk of common bean production in Africa, Latin America, and the Caribbean. In: Hershey CH, editor. Eco-efficiency: from vision to reality. Cali: Centro Internacional de Agricultura Tropical (CIAT); 2013. p. 117–34.

Belivanis T, Doré C. Interspecific hybridization of *Phaseolus vulgaris* L. and *Phaseolus angustissimus* A. Gray using in vitro embryo culture. Plant Cell Rep. 1986;5:329–31. https://doi.org/10.1007/bf00268593.

Bellon MR, Risopoulos J. Small-scale farmers expand the benefits of improved maize germplasm: a case study from Chiapas, Mexico. World Dev. 2001;29:799–811. https://doi.org/10.1016/S0305-750X(01)00013-4.

Benz B. Archaeological evidence of teosinte domestication from Guilá Naquitz. Proc Natl Acad Sci. 2001;98:2104–6. https://doi.org/10.1073/pnas.98.4.2104.

Berny Mier y Teran JC, Konzen ER, Medina V, Palkovic A, Ariani A, Tsai SM, et al. Root and shoot variation in relation to potential intermittent drought adaptation of Mesoamerican wild common bean (*Phaseolus vulgaris* L.). Ann Bot. 2019;124:917–32. https://doi.org/10.1093/aob/mcy221.

Berny Mier y Teran J, Konzen E, Palkovic A, Tsai S, Gepts P. Exploration of the yield potential of Mesoamerican wild common beans from contrasting eco-geographic regions by nested recombinant inbred populations. Front Plant Sci. 2020;11:346. https://doi.org/10.3389/fpls.2020.00346.

Bitocchi E, Nanni L, Bellucci E, Rossi M, Giardini A, Zeuli PS, et al. Mesoamerican origin of the common bean (*Phaseolus vulgaris* L.) is revealed by sequence data. Proc Natl Acad Sci. 2012;109:E788–96. https://doi.org/10.1073/pnas.1108973109.

Bitocchi E, Bellucci E, Giardini A, Rau D, Rodriguez M, Biagetti E, et al. Molecular analysis of the parallel domestication of the common bean (*Phaseolus vulgaris*) in Mesoamerica and the Andes. New Phytol. 2013;197:300–13. https://doi.org/10.1111/j.1469-8137.2012.04377.x.

Bitocchi E, Rau D, Bellucci E, Rodriguez M, Murgia ML, Gioia T, et al. Beans (*Phaseolus* ssp.) as a model for understanding crop evolution. Front Plant Sci. 2017;8:722. https://doi.org/10.3389/fpls.2017.00722.

Blair MW, Pantoja W, Muñoz LC, Hincapie A. Genetic analysis of crosses between cultivated tepary bean and wild *Phaseolus acutifolius* and *P. parvifolius*. Annu Rep Bean Improv Coop. 2003;46:27–8.

Blair MW, Giraldo MC, Buendia HF, Tovar E, Duque MC, Beebe SE. Microsatellite marker diversity in common bean (*Phaseolus vulgaris* L.). Theor Appl Genet. 2006a;113:100–9. https://doi.org/10.1007/s00122-006-0276-4.

Blair MW, Iriarte G, Beebe S. QTL analysis of yield traits in an advanced backcross population derived from a cultivated Andean x wild common bean (*Phaseolus vulgaris* L.) cross. Theor Appl Genet. 2006b;112:1149–63. https://doi.org/10.1007/s00122-006-0217-2.

Blair MW, Díaz L, Buendía H, Duque M. Genetic diversity, seed size associations and population structure of a core collection of common beans (*Phaseolus vulgaris* L.). Theor Appl Genet. 2009;119:955–72. https://doi.org/10.1007/s00122-009-1064-8.

Blair M, Pantoja W, Muñoz LC. First use of microsatellite markers in a large collection of cultivated and wild accessions of tepary bean (*Phaseolus acutifolius* A. Gray). Theor Appl Genet. 2012;125:1137–47. https://doi.org/10.1007/s00122-012-1900-0.

Blair M, Izquierdo P, Astudillo C, Grusak M. A legume biofortification quandary: variability and genetic control of seed coat micronutrient accumulation in common beans. Front Plant Sci. 2013;4:275. https://doi.org/10.3389/fpls.2013.00275.

Bolnick DI, Barrett RDH, Oke KB, Rennison DJ, Stuart YE. (Non)parallel evolution. Annu Rev Ecol Evol Syst. 2018;49:303–30. https://doi.org/10.1146/annurev-ecolsys-110617-062240.

Boss PK, Sreekantan L, Thomas MR. A grapevine TFL1 homologue can delay flowering and alter floral development when overexpressed in heterologous species. Funct Plant Biol. 2006;33(1):31–41.

Bradley D, Ratcliffe O, Vincent C, Carpenter R, Coen E. Inflorescence commitment and architecture in Arabidopsis. Science. 1997;275:80–3. https://doi.org/10.1126/science.275.5296.80.

Broughton WJ, Hernandez G, Blair M, Beebe S, Gepts P, Vanderleyden J. Beans (*Phaseolus* spp.) – model food legumes. Plant Soil. 2003;252:55–128. https://doi.org/10.1023/A:1024146710611.

Brouwer BO, Murphy KM, Jones SS. Plant breeding for local food systems: a contextual review of end-use selection for small grains and dry beans in Western Washington. Renew Agric Food Syst. 2016;31:172–84. https://doi.org/10.1017/s1742170515000198.

Brown CH, Clement C, Epps P, Luedeling E, Wichmann S. The paleobiolinguistics of the common bean (*Phaseolus vulgaris* L.). Ethnobiol Lett. 2014;5:104–15. https://doi.org/10.14237/ebl.5.2014.203.

Brücher H. The wild ancestor of *Phaseolus vulgaris* in South America. In: Gepts P, editor. Genetic resources of *Phaseolus* beans. Dordrecht: Kluwer; 1988. p. 185–214.

Brummer EC, Barber WT, Collier SM, Cox TS, Johnson R, Murray SC, et al. Plant breeding for harmony between agriculture and the environment. Front Ecol Environ. 2011;9:561–8. https://doi.org/10.1890/100225.

Brush SB. Farmers' bounty. New Haven, CT: Yale University Press; 2004.

Buer CS, Djordjevic MA. Architectural phenotypes in the transparent testa mutants of Arabidopsis thaliana. J Exp Bot. 2009;60:751–63. https://doi.org/10.1093/jxb/ern323.

Burquez A, Sarukhán KJ. Biología floral de poblaciones silvestres de *Phaseolus coccineus* L. I. Relaciones planta-polinizador. Bol Soc Bot México. 1980;39:5–25.

Butare L, Rao I, Lepoivre P, Cajiao C, Polania J, Cuasquer J, et al. Phenotypic evaluation of interspecific recombinant inbred lines (RILs) of *Phaseolus* species for aluminium resistance and shoot and root growth response to aluminium–toxic acid soil. Euphytica. 2012;186:715–30. https://doi.org/10.1007/s10681-011-0564-1.

Carney JA, Rosomoff RN. In the shadow of slavery: Africa's botanical legacy in the Atlantic world. Berkeley, CA: University of California Press; 2009.

Casacuberta JM, Jackson S, Panaud O, Purugganan M, Wendel J. Evolution of plant phenotypes, from genomes to traits. G3. 2016;6:775–8. https://doi.org/10.1534/g3.115.025502.

Castañeda-Álvarez NP, Khoury CK, Achicanoy HA, Bernau V, Dempewolf H, Eastwood RJ, et al. Global conservation priorities for crop wild relatives. Nat Plants. 2016;2:16022. https://doi.org/10.1038/nplants.2016.22.

Castellanos JZ, Guzmán-Maldonado H, Acosta-Gallegos JA, Kelly JD. Effects of hardshell character on cooking time of common beans grown in the semiarid highlands of Mexico. J Sci Food Agric. 1995;69:437–43. https://doi.org/10.1002/jsfa.2740690406.

Chacón S. MI, Pickersgill B, Debouck DG. Domestication patterns in common bean (*Phaseolus vulgaris* L.) and the origin of the Mesoamerican and Andean cultivated races. Theor Appl Genet. 2005;110:432–44. https://doi.org/10.1007/s00122-004-1842-2.

Chacón S. MI, Pickersgill B, Debouck DG, Arias JS. Phylogeographic analysis of the chloroplast DNA variation in wild common bean (*Phaseolus vulgaris* L.) in the Americas. Plant Systemat Evol. 2007;266:175–95. https://doi.org/10.1007/s00606-007-0536-z.

Chacón-Sánchez MI. The domestication syndrome in *Phaseolus* crop plants: a review of two key domestication traits. In: Pontarotti P, editor. Origin and evolution of biodiversity. Cham: Springer; 2018. p. 37–59.

Chacón-Sánchez MI, Martínez-Castillo J. Testing domestication scenarios of lima bean (*Phaseolus lunatus* L.) in Mesoamerica: insights from genome-wide genetic markers. Front Plant Sci. 2017;8:1551. https://doi.org/10.3389/fpls.2017.01551.

CIAT. Bean germplasm collection and database. Cali: CIAT; 2020. https://genebank.ciat.cgiar.org/genebank/beancollection.do. Accessed 15 Sept 2020

Cichy KA, Fernandez A, Kilian A, Kelly JD, Galeano CH, Shaw S, et al. QTL analysis of canning quality and color retention in black beans (*Phaseolus vulgaris* L.). Molecular Breeding. 2014;33:139–54. https://doi.org/10.1007/s11032-013-9940-y.

Cichy KA, Porch TG, Beaver JS, Cregan P, Fourie D, Glahn RP, et al. A *Phaseolus vulgaris* diversity panel for Andean bean improvement. Crop Sci. 2015a;55:2149–60. https://doi.org/10.2135/cropsci2014.09.0653.

Cichy KA, Wiesinger JA, Mendoza FA. Genetic diversity and genome-wide association analysis of cooking time in dry bean (*Phaseolus vulgaris* L.). Theor Appl Genet. 2015b;128:1555–67. https://doi.org/10.1007/s00122-015-2531-z.

Clouse SD. Brassinosteroid signal transduction: clarifying the pathway from ligand perception to gene expression. Mol Cell. 2002;10:973–82. https://doi.org/10.1016/S1097-2765(02)00744-X.

Comadran J, Kilian B, Russell J, Ramsay L, Stein N, Ganal M, et al. Natural variation in a homolog of Antirrhinum *CENTRORADIALIS* contributed to spring growth habit and environmental adaptation in cultivated barley. Nat Genet. 2012;44:1388–92. https://doi.org/10.1038/ng.2447.

Cortés AJ, Skeen P, Blair MW, Chacón-Sánchez MI. Does the genomic landscape of species divergence in *Phaseolus* beans coerce parallel signatures of adaptation and domestication? Front Plant Sci. 2018;9:1816. https://doi.org/10.3389/fpls.2018.01816.

Cortinovis G, Frascarelli G, Di Vittori V, Papa R. Current state and perspectives in population genomics of the common bean. Plants. 2020;9:330. https://doi.org/10.3390/plants9030330.

Coyne DP. Photoperiodism: inheritance and linkage studies in *Phaseolus vulgaris*. J Hered. 1967;58:313–4. https://doi.org/10.1093/oxfordjournals.jhered.a107628.

Currence TM. Inheritance studies in *Phaseolus vulgaris*. Tech Bull Minn Agric Exp Stat. 1930;68:3–28.

Darwin C. On the origin of species by means of natural selection. London: J. Murray; 1859.

Darwin C. The variation of plants and animals under domestication. London: J. Murray; 1868.

Dashner ZS. Examination of *Lygus* bug resistance in lima bean: polygalacturonase inhibiting proteins as candidate traits, MS thesis. Davis, CA: University of California; 2018.

Davis JW, Kean D, Yorgey B, Fourie D, Miklas PN, Myers JR. A molecular marker linkage map of snap bean (*Phaseolus vulgaris*). Annu Rep Bean Improv Coop. 2006;49:73–4.

de Candolle A. L'origine des plantes cultivées. English translation: the origin of cultivated plants. New York: Appleton; 1882.

De Filippis LF. Underutilised and neglected crops: next generation sequencing approaches for crop improvement and better food security. In: Ozturk M, Hakeem KR, Ashraf M, Ahmad MSA, editors. Global perspectives on underutilized crops. Cham: Springer; 2018. p. 287–380.

Debouck DG. Observations about *Phaseolus lignosus* (Leguminosae: Papilionoideae: Phaseoleae), a bean species from the Bermuda Islands. J Bot Res Inst Texas. 2015;9:107–19.

Debouck DG. Your beans of the last harvest and the possible adoption of bright ideas. In: Lira R, Casas A, Blancas J, editors. Ethnobotany of Mexico: interactions of people and plants in mesoamerica. New York, NY: Springer; 2016. p. 367–87.

Debouck DG, Toro O, Paredes OM, Johnson WC, Gepts P. Genetic diversity and ecological distribution of *Phaseolus vulgaris* in northwestern South America. Econ Bot. 1993;47:408–23. https://doi.org/10.1007/BF02907356.

Delaney DE, Bliss FA. Selection for increased percentage phaseolin in common bean. 1. Comparison of selection for seed protein alleles and S_1 family recurrent selection. Theor Appl Genet. 1991a;81:301–5. https://doi.org/10.1007/bf00228667.

Delaney DE, Bliss FA. Selection for increased percentage phaseolin in common bean. 2. Changes in frequency of seed protein alleles with S_1 family recurrent selection. Theor Appl Genet. 1991b;81:306–11. https://doi.org/10.1007/bf00228668.

Delgado A, Gama López S. Diversidad y distribución de los frijoles silvestres en México, vol. 16. México: Universidad Nacional Autónoma de México Revista Digital Universitaria; 2015. http://www.revista.unam.mx/vol.16/num12/art10/

Delgado Salinas A. Variation, taxonomy, domestication, and germplasm potentialities in *Phaseolus coccineus*. In: Gepts P, editor. Genetic resources of *Phaseolus* beans. Dordrecht: Kluwer; 1988. p. 441–63.

Delgado-Salinas A, Bibler R, Lavin M. Phylogeny of the genus *Phaseolus* (Leguminosae): a recent diversification in an ancient landscape. Syst Bot. 2006;31:779–91. https://doi.org/10.1600/036364406779695960.

Delgado-Salinas A, Thulin M, Pasquet R, Weeden N, Lavin M. *Vigna* (Leguminosae) sensu lato: the names and identities of the American segregate genera. Am J Bot. 2011;98:1694–715. https://doi.org/10.3732/ajb.1100069.

Dhaliwal AS, Pollard LH. Cytological behavior of an F_1 species cross (*Phaseolus lunatus* L. var. Fordhook x *Phaseolus polystachios* L.). Cytologia. 1962;27:369–74. https://doi.org/10.1508/cytologia.27.369.

Dhanasekar P, Reddy KS. A novel mutation in TFL1 homolog affecting determinacy in cowpea (Vigna unguiculata). Mol Genet Genomics. 2015;290:55–65. https://doi.org/10.1007/s00438-014-0899-0.

Di Vittori V, Gioia T, Rodriguez M, Bellucci E, Bitocchi E, Nanni L, et al. Convergent evolution of the seed shattering trait. Genes. 2019;10:68. https://doi.org/10.3390/genes10010068.

Di Vittori V, Bitocchi E, Rodriguez M, Alseekh S, Bellucci E, Nanni L, et al. Pod indehiscence in common bean is associated with the fine regulation of PvMYB26. J Exp Bot. 2020; https://doi.org/10.1093/jxb/eraa553.

Díaz LM, Blair MW. Race structure within the Mesoamerican gene pool of common bean (*Phaseolus vulgaris* L.) as determined by microsatellite markers. Theor Appl Genet. 2006;114:143–54. https://doi.org/10.1007/s00122-006-0417-9.

Dohle S. Development of resources for lima bean (*Phaseolus lunatus*) breeding and genetics research, Plant sciences. Davis, CA: University of California; 2017. p. 108.

Dohle S, Berny Mier y Teran JC, Egan A, Kisha T, Khoury CK. Wild beans (Phaseolus L.) of North America. In: Greene SL, Williams KA, Khoury CK, Kantar MB, Marek LF, editors. North American crop wild relatives, volume 2: important species. Cham: Springer; 2019. p. 99–127.

Dong Y, Morgan C, Chinenov Y, Zhou L, Fan W, Ma X, et al. Shifting diets and the rise of male-biased inequality on the Central Plains of China during Eastern Zhou. Proc Natl Acad Sci. 2017;114:932–7. https://doi.org/10.1073/pnas.1611742114.

Drijfhout E. Influence of temperature-dependent string formation in common bean (*Phaseolus vulgaris* L.). Netherlands J Agric Sci. 1978;26:99–105.

Drijfhout E, Blok WJ. Inheritance of resistance to *Xanthomonas campestris* pv. *phaseoli* in tepary bean (*Phaseolus acutifolius*). Euphytica. 1987;36:803–8. https://doi.org/10.1007/BF00051863.

Elitzur T, Nahum H, Borovsky Y, Pekker I, Eshed Y, Paran I. Co-ordinated regulation of flowering time, plant architecture and growth by FASCICULATE: the pepper orthologue of SELF PRUNING. J Exp Bot. 2009;60:869–80. https://doi.org/10.1093/jxb/ern334.

Emerson RA. Heredity in bean hybrids (*Phaseolus vulgaris*). Agricultural experiment station of Nebraska, Seventeenth annual report. 1904. p. 33–68.

Espinosa-Alonso LG, Lygin A, Widholm JM, Valverde ME, Paredes-Lopez O. Polyphenols in wild and weedy Mexican common beans (*Phaseolus vulgaris* L.). J Agric Food Chem. 2006;54:4436–44. https://doi.org/10.1021/jf060185e.

Fofana B, Baudoin JP, Vekemans X, Debouck DG, Du Jardin P. Molecular evidence for an Andean origin and a secondary gene pool for the Lima bean (Phaseolus lunatus L.) using chloroplast DNA. Theor Appl Genet. 1999;98:202–12. https://doi.org/10.1007/s001220051059.

Foucher F, Morin J, Courtiade J, Cadioux S, Ellis N, Banfield Mark J, et al. DETERMINATE and LATE FLOWERING are two TERMINAL FLOWER1/CENTRORADIALIS homologs that control two distinct phases of flowering initiation and development in pea. Plant Cell. 2003;15:2742–54. https://doi.org/10.1105/tpc.015701.

Freyre R, Ríos R, Guzmán L, Debouck D, Gepts P. Ecogeographic distribution of *Phaseolus* spp. (Fabaceae) in Bolivia. Econ Bot. 1996;50:195–215. https://doi.org/10.1007/BF02861451.

Freytag GF, Debouck DG. Taxonomy, distribution, and ecology of the genus *Phaseolus* (Leguminosae – Papilionoideae) in North America, Mexico and Central America. Fort Worth, TX: Botanical Research Institute of Texas; 2002.

Fu Y-B. A molecular view of flax gene pool. In: Cullis CA, editor. Genetics and genomics of *Linum*, Plant genetics and genomics: crops and models, vol. 23. Cham: Springer; 2019. p. 17–37.

Funatsuki H, Suzuki M, Hirose A, Inaba H, Yamada T, Hajika M, et al. Molecular basis of a shattering resistance boosting global dissemination of soybean. Proc Natl Acad Sci. 2014;111:17797–802. https://doi.org/10.1073/pnas.1417282111.

Galeano C, Cortes A, Fernandez A, Soler A, Franco-Herrera N, Makunde G, et al. Gene-based single nucleotide polymorphism markers for genetic and association mapping in common bean. BMC Genet. 2012;13:48. https://doi.org/10.1186/1471-2156-13-48.

Garcia T, Duitama J, Smolenski Zullo S, Gil J, Ariani A, Dohle S, et al. Comprehensive genomic resources related to domestication and crop improvement traits in Lima bean. Nat Commun. 2021;12(1):702. https://doi.org/10.1038/s41467-021-20921-1.

Garvin DF, Weeden NF. Isozyme evidence supporting a single geographic origin for domesticated tepary bean. Crop Sci. 1994;34:1390–5.

Gasper R, Effenberger I, Kolesinski P, Terlecka B, Hofmann E, Schaller A. Dirigent protein mode of action revealed by the crystal structure of AtDIR6. Plant Physiol. 2016;172:2165–75. https://doi.org/10.1104/pp.16.01281.

Geffroy V, Sicard D, de Oliveira J, Sévignac M, Cohen S, Gepts P, et al. Identification of an ancestral resistance gene cluster involved in the coevolution process between *Phaseolus vulgaris* and its fungal pathogen *Colletotrichum lindemuthianum*. Mol Plant Microbe Interact. 1999;12:774–84. https://doi.org/10.1094/MPMI.1999.12.9.774.

Gepts P. Hibridaciones interespecíficas para el mejoramiento de *Phaseolus vulgaris*. Internal seminar, SE-10-81. Cali: Centro Internacional de Agricultura Tropical; 1981. 17 p.

Gepts P. Phaseolin as an evolutionary marker. In: Gepts P, editor. Genetic resources of *Phaseolus* beans. Dordrecht: Kluwer; 1988. p. 215–41.

Gepts P. Origin and evolution of common bean: past events and recent trends. HortScience. 1998;33:1124–30. https://doi.org/10.21273/HORTSCI.33.7.1124.

Gepts P. Crop domestication as a long-term selection experiment. Plant Breed Rev. 2004;24(Part 2):1–44. https://doi.org/10.1002/9780470650288.ch1.

Gepts P. Plant genetic resources conservation and utilization: the accomplishments and future of a societal insurance policy. Crop Sci. 2006;46:2278–92. https://doi.org/10.2135/cropsci2006.03.0169gas.

Gepts P. Beans, origin and development. In: Smith C, editor. Encyclopedia of global archaeology. New York: Springer; 2014a. p. 822–7.

Gepts P. The contribution of genetic and genomic approaches to plant domestication studies. Curr Opin Plant Biol. 2014b;18:51–9. https://doi.org/10.1016/j.pbi.2014.02.001.

Gepts P. Domestication of plants. In: Leakey RRB, editor. Encyclopedia of agriculture and food systems. San Diego: Elsevier; 2014c. p. 474–86.

Gepts P, Bliss FA. F_1 hybrid weakness in the common bean: differential geographic origin suggests two gene pools in cultivated bean germplasm. J Hered. 1985;76:447–50. https://doi.org/10.1093/oxfordjournals.jhered.a110142.

Gepts P, Debouck DG. Origin, domestication, and evolution of the common bean, *Phaseolus vulgaris*. In: Voysest O, Van Schoonhoven A, editors. Common beans: research for crop improvement. Oxon: CAB; 1991. p. 7–53.

Gepts P, Lin D. Development of PhaseolusGenes, a genome database for marker discovery and candidate gene identification in common bean. Annu Rep Bean Improv Coop. 2010;53:30–1.

Gepts P, Osborn TC, Rashka K, Bliss FA. Phaseolin-protein variability in wild forms and landraces of the common bean (*Phaseolus vulgaris*): evidence for multiple centers of domestication. Econ Bot. 1986;40:451–68. https://doi.org/10.1007/BF02859659.

Gepts P, Beavis WD, Brummer EC, Shoemaker RC, Stalker HT, Weeden NF, et al. Legumes as a model plant family. Genomics for food and feed report of the cross-legume advances through genomics conference. Plant Physiol. 2005;137:1228–35. https://doi.org/10.1104/pp.105.060871.

Gepts P, Aragão FJL, Ed B, Blair MW, Brondani R, Broughton W, et al. Genomics of *Phaseolus* beans, a major source of dietary protein and micronutrients in the tropics. In: Moore PH, Ming R, editors. Genomics of tropical crop plants. Berlin: Springer; 2008. p. 113–43.

Gioia T, Logozzo G, Kami J, Spagnoletti Zeuli P, Gepts P. Identification and characterization of a homologue to the Arabidopsis *INDEHISCENT* gene in common bean. J Hered. 2013;104:273–86. https://doi.org/10.1093/jhered/ess102.

Gioia T, Logozzo G, Marzario S, Spagnoletti Zeuli P, Gepts P. Evolution of SSR diversity from wild types to U.S. advanced cultivars in the Andean and Mesoamerican domestications of common bean (Phaseolus vulgaris). PLoS One. 2019;14:e0211342. https://doi.org/10.1371/journal.pone.0211342.

Gonzalez A, Zhao M, Leavitt JM, Lloyd AM. Regulation of the anthocyanin biosynthetic pathway by the TTG1/bHLH/Myb transcriptional complex in Arabidopsis seedlings. Plant J. 2008;53:814–27. https://doi.org/10.1111/j.1365-313x.2007.03373.x.

González AM, De La Fuente M, De Ron AM, Santalla M. Protein markers and seed size variation in common bean segregating populations. Mol Breed. 2010;25:723–40. https://doi.org/10.1007/s11032-009-9370-z.

González AM, Yuste-Lisbona FJ, Fernández-Lozano A, Lozano R, Santalla M. Genetic mapping and QTL analysis in common bean. In: Pérez de la Vega M, Santalla M, Marsolais F, editors. The common bean genome, Compendium of plant genomes. Cham: Springer; 2017. p. 69–107.

Griesmann M, Chang Y, Liu X, Song Y, Haberer G, Crook MB, et al. Phylogenomics reveals multiple losses of nitrogen-fixing root nodule symbiosis. Science. 2018:eaat1743. https://doi.org/10.1126/science.aat1743.

Gu W, Zhu J, Wallace D, Singh S, Weeden N. Analysis of genes controlling photoperiod sensitivity in common bean using DNA markers. Euphytica. 1998;102:125–32. https://doi.org/10.1023/A:1018340514388.

Guerra-García A, Suárez-Atilano M, Mastretta-Yanes A, Delgado-Salinas A, Piñero D. Domestication genomics of the open-pollinated scarlet runner bean (*Phaseolus coccineus* L.). Front Plant Sci. 2017;8:1891. https://doi.org/10.3389/fpls.2017.01891.

Gujaria-Verma N, Ramsay L, Sharpe AG, Sanderson L-A, Debouck DG, Tar'an B, et al. Gene-based SNP discovery in tepary bean (*Phaseolus acutifolius*) and common bean (*P. vulgaris*) for diversity analysis and comparative mapping. BMC Genomics. 2016;17:1–16. https://doi.org/10.1186/s12864-016-2499-3.

Gutiérrez Salgado A, Gepts P, Debouck D. Evidence for two gene pools of the lima bean, *Phaseolus lunatus* L., in the Americas. Genet Res Crop Evol. 1995;42:15–22. https://doi.org/10.1007/BF02310680.

Guzmán P, Gilbertson RL, Nodari R, Johnson WC, Temple SR, Mandala D, et al. Characterization of variability in the fungus *Phaeoisariopsis griseola* suggests coevolution with the common bean (*Phaseolus vulgaris*). Phytopathology. 1995;85:600–7. https://doi.org/10.1094/Phyto-85-600.

Hagerty CH, Cuesta-Marcos A, Cregan P, Song Q, McClean P, Myers JR. Mapping snap bean pod and color traits, in a dry bean × snap bean recombinant inbred population. J Am Soc Hort Sci. 2016;141:131–8. https://doi.org/10.21273/JASHS.141.2.131.

Haghighi KR, Ascher PD. Fertile, intermediate hybrids between *Phaseolus vulgaris* and *P. acutifolius* from congruity backcrossing. Sex Plant Reprod. 1988;1:51–8. https://doi.org/10.1007/BF00227023.

Hammer K. Das Domestikationssyndrom. Die Kulturpflanze. 1984;32:11–34. https://doi.org/10.1007/bf02098682.

Hammer K, Khishbakht K. A domestication assessment of the big five plant families. Genet Resour Crop Evol. 2015;62:665–89. https://doi.org/10.1007/s10722-014-0186-2.

Hannah MA, Kramer KM, Geffroy V, Kopka J, Blair MW, Erban A, et al. Hybrid weakness controlled by the dosage-dependent lethal (DL) gene system in common bean (*Phaseolus vulgaris*) is caused by a shoot-derived inhibitory signal leading to salicylic acid-associated root death. New Phytol. 2007;176:537–49. https://doi.org/10.1111/j.1469-8137.2007.02215.x.

Harlan JR. Crops and Man. Madison, WI: Crop Science Society of Amera; 1992.

Harlan JR, de Wet JMJ. Towards a rational classification of cultivated plants. Taxon. 1971;20:509–17. https://doi.org/10.2307/1218252.

Harlan JR, de Wet JMJ. On the quality of evidence for origin and dispersal of cultivated plants. Curr Anthropol. 1973;14:51–62. https://doi.org/10.1086/201406.

Harper KN, Armelagos GJ. Genomics, the origins of agriculture, and our changing microbe-scape: time to revisit some old tales and tell some new ones. Am J Phys Anthropol. 2013;152:135–52. https://doi.org/10.1002/ajpa.22396.

Hervieu F, Charbonnier L, Bannerot H, Pelletier G. The cytoplasmic male-sterility (CMS) determinant of common bean is widespread in *Phaseolus coccineus* L. and *Phaseolus vulgaris* L. Curr Genet. 1993;24:149–55. https://doi.org/10.1007/bf00324679.

Hervieu F, Bannerot H, Pelletier G. A unique cytoplasmic male sterility (CMS) determinant is present in three *Phaseolus* species characterized by different mitochondrial genomes. Theor Appl Genet. 1994;88:314–20. https://doi.org/10.1007/BF00223638.

Hübner S, Bercovich N, Todesco M, Mandel JR, Odenheimer J, Ziegler E, et al. Sunflower pan-genome analysis shows that hybridization altered gene content and disease resistance. Nature Plants. 2019;5:54–62. https://doi.org/10.1038/s41477-018-0329-0.

Hufford MB, Berny Mier y Teran JC, Gepts P. Crop biodiversity: an unfinished magnum opus of nature. Annu Rev Plant Biol. 2019;70:727–51. https://doi.org/10.1146/annurev-arplant-042817-040240.

Hunter D, Guarino L, Spillane C, McKeown PC, editors. Routledge handbook of agricultural biodiversity. New York: Taylor & Francis; 2017a.

Hunter MC, Smith RG, Schipanski ME, Atwood LW, Mortensen DA. Agriculture in 2050: recalibrating targets for sustainable intensification. BioScience. 2017b;67:386–91. https://doi.org/10.1093/biosci/bix010.

Johnson WC, Menéndez C, Nodari RO, Koinange EMK, Magnusson S, Singh SP, et al. Association of a seed weight factor with the phaseolin seed storage protein locus across genotypes, environments, and genomes in *Phaseolus* – *Vigna* spp.: Sax (1923) revisited. J Agric Genom. 1996;2. Article 5, http://wheat.pw.usda.gov/jag/papers96/paper596/indexp596.html or http://www.plantsciences.ucdavis.edu/gepts/Sax.htm

Kamfwa K, Cichy KA, Kelly JD. Genome-wide association study of agronomic traits in common bean. Plant Genome. 2015;8 https://doi.org/10.3835/plantgenome2014.09.0059.

Kami J, Becerra Velásquez B, Debouck DG, Gepts P. Identification of presumed ancestral DNA sequences of phaseolin in *Phaseolus vulgaris*. Proc Natl Acad Sci U S A. 1995;92:1101–4. https://doi.org/10.1073/pnas.92.4.1101.

Kantar MB, Nashoba AR, Anderson JE, Blackman BK, Rieseberg LH. The genetics and genomics of plant domestication. BioScience. 2017;67:971–82. https://doi.org/10.1093/biosci/bix114.

Kaplan L. Archaeology and domestication in American *Phaseolus*. Econ Bot. 1965;19:358–68. https://doi.org/10.1007/BF02904806.

Kaplan L, Lynch T. *Phaseolus* (Fabaceae) in archaeology: AMS radiocarbon dates and their significance for pre-Columbian agriculture. Econ Bot. 1999;53:261–72. https://doi.org/10.1007/BF02866636.

Katuuramu DN, Hart JP, Porch TG, Grusak MA, Glahn RP, Cichy KA. Genome-wide association analysis of nutritional composition-related traits and iron bioavailability in cooked dry beans (*Phaseolus vulgaris* L.). Mol Breed. 2018;38:44. https://doi.org/10.1007/s11032-018-0798-x.

Keller B, Ariza-Suarez D, de la Hoz J, Aparicio JS, Portilla-Benavides AE, Buendia HF, Mayor VM, Studer B, Raatz B. Genomic prediction of agronomic traits in common bean (*Phaseolus vulgaris* L.) under environmental stress. Front Plant Sci. 2020;11:1001. https://doi.org/10.3389/fpls.2020.01001

Kelly JD. Remaking bean plant architecture for efficient production. Adv Agron. 2001;71:109–43. https://doi.org/10.1016/S0065-2113(01)71013-9.

Kelly JD. Advances in common bean improvement: some case histories with broader applications. Acta Horticulturae. 2004;637:99–122. https://doi.org/10.17660/ActaHortic.2004.637.11.

Kelly JD, Adams MW, Saettler AW, Hosfield GL, Varner GV, Uebersax MA, et al. Registration of 'Sierra' pinto bean. Crop Sci. 1990;30:745–6. https://doi.org/10.2135/cropsci1990.0011183X003000030062x.

Kelly JD, Hosfield GL, Varner GV, Uebersax MA, Taylor J. Registration of 'Matterhorn' great northern bean. Crop Sci. 1999;39:589–90. https://doi.org/10.2135/cropsci1999.0011183X003900020058x.

Khairallah MM, Sears BB, Adams MW. Mitochondrial restriction fragment polymorphisms in wild *Phaseolus vulgaris* – insights in the domestication of common bean. Theor Appl Genet. 1992;84:915–22. https://doi.org/10.1007/BF00227404.

Kim K-W, Smith CA, Daily MD, Cort JR, Davin LB, Lewis NG. Trimeric structure of (+)-pinoresinol-forming dirigent protein at 1.95 Å resolution with three isolated active sites. J Biol Chem. 2015;290:1308–18. https://doi.org/10.1074/jbc.M114.611780.

Koenig R, Gepts P. Allozyme diversity in wild *Phaseolus vulgaris*: further evidence for two major centers of diversity. Theor Appl Genet. 1989;78:809–17. https://doi.org/10.1007/BF00266663.

Koenig R, Singh SP, Gepts P. Novel phaseolin types in wild and cultivated common bean (*Phaseolus vulgaris*, Fabaceae). Econ Bot. 1990;44:50–60. https://doi.org/10.1007/BF02861066.

Koinange EMK, Gepts P. Hybrid weakness in wild *Phaseolus vulgaris* L. J Hered. 1992;83:135–9. https://doi.org/10.1093/oxfordjournals.jhered.a111173.

Koinange EMK, Singh SP, Gepts P. Genetic control of the domestication syndrome in common-bean. Crop Sci. 1996;36:1037–45. https://doi.org/10.2135/cropsci1996.0011183X003600040037x.

Kolkman JM, Kelly JD. QTL conferring resistance and avoidance to white mold in common bean. Crop Sci. 2003;43:539–48. https://doi.org/10.2135/cropsci2003.5390.

Konzen ER, Tsai SM. Genetic variation of landraces of common bean varying for seed coat glossiness and disease resistance: valuable resources for conservation and breeding. In: Grillo O, editor. Rediscovery of landraces as a resource for the future. London: IntechOpen; 2018. p. 177–93.

Kornegay J, White JW, Ortiz de la Cruz O. Growth habit and gene pool effects on inheritance of yield in common bean. Euphytica. 1992;62:171–80. https://doi.org/10.1007/BF00041751.

Kornegay J, Cardona C, Posso CE. Inheritance of resistance to Mexican bean weevil in common bean, determined by bioassay and biochemical tests. Crop Sci. 1993a;33:589–94. https://doi.org/10.2135/cropsci1993.0011183X003300030034x.

Kornegay J, White JW, Dominguez JR, Tejada G, Cajiao C. Inheritance of photoperiod response in Andean and Mesoamerican common bean. Crop Sci. 1993b;33:977–84. https://doi.org/10.2135/cropsci1993.0011183X003300050021x.

Kusolwa PM, Myers JR, Porch TG, Trukhina Y, González-Vélez A, Beaver JS. Registration of AO-1012-29-3-3A red kidney bean germplasm line with bean weevil, BCMV and BCMNV resistance. J Plant Regist. 2016;10:149–53. https://doi.org/10.3198/jpr2015.10.0064crg.

Kuzay S, Hamilton-Conaty PA, Palkovic A, Gepts P. Is the USDA core collection of common bean representative of genetic diversity of the species, as assessed by SNP diversity? Crop Sci. 2020;60:1398–414. https://doi.org/10.2135/cropsci2019.08.0497.

Kwak M, Gepts P. Structure of genetic diversity in the two major gene pools of common bean (*Phaseolus vulgaris* L., Fabaceae). Theor Appl Genet. 2009;118:979–92. https://doi.org/10.1007/s00122-008-0955-4.

Kwak M, Velasco DM, Gepts P. Mapping homologous sequences for determinacy and photoperiod sensitivity in common bean (*Phaseolus vulgaris*). J Hered. 2008;99:283–91. https://doi.org/10.1093/jhered/esn005.

Kwak M, Kami JA, Gepts P. The putative Mesoamerican domestication center of *Phaseolus vulgaris* is located in the Lerma-Santiago basin of Mexico. Crop Sci. 2009;49:554–63. https://doi.org/10.2135/cropsci2008.07.0421.

Kwak M, Toro O, Debouck D, Gepts P. Multiple origins of the determinate growth habit in domesticated common bean (*Phaseolus vulgaris* L.). Ann Bot. 2012;110:1573–80. https://doi.org/10.1093/aob/mcs207.

Lamprecht H. Beitrage zur Genetik von *Phaseolus vulgaris*. II. Über Vererbung von Hülsenfarbe und Hülsenform. Hereditas. 1932;16:295–340.

Lamprecht H. Die Artgrenze zwischen *Phaseolus vulgaris* L. und *multiflorus* Lam. Hereditas. 1941;27:51–175.

Lamprecht H. Zur Lösung des Artproblems: Neue und bisher bekannte Ergebnisse der Kreuzung *Phaseolus vulgaris* L. x *coccineus* L. und reziprok. Agri Hortique Genet. 1948;6:87–141.

Larochelle C, Alwang JR. Impacts of improved bean varieties on food security in Rwanda, 2014 annual meeting, 27–29 July 2014. Minneapolis, MN: Agricultural and Applied Economics Association; 2014. https://doi.org/10.22004/ag.econ.170567.

Larson G, Piperno DR, Allaby RG, Purugganan MD, Andersson L, Arroyo-Kalin M, et al. Current perspectives and the future of domestication studies. Proc Natl Acad Sci. 2014;111:6139–46. https://doi.org/10.1073/pnas.1323964111.

Lavin M, Herendeen PS, Wojciechowski MF. Evolutionary rates analysis of Leguminosae implicates a rapid diversification of the major family lineages immediately following an Early Tertiary emergence. Syst Biol. 2005;54:575–94. https://doi.org/10.1080/10635150590947131.

Leal-Bertioli SCM, Moretzsohn MC, Santos SP, Brasileiro ACM, Guimarães PM, Bertioli DJ, et al. Phenotypic effects of allotetraploidization of wild *Arachis* and their implications for peanut domestication. Am J Bot. 2017;104:379–88. https://doi.org/10.3732/ajb.1600402.

Lenser T, Theißen G. Molecular mechanisms involved in convergent crop domestication. Trends Plant Sci. 2013;18:704–14. https://doi.org/10.1016/j.tplants.2013.08.007.

Lewis G, Schrire B, Mackinder B, Lock M. Legumes of the world. London: Royal Botanic Garden, Kew Publishing; 2005.

Leyna H, Korban SS, Coyne DP. Changes in patterns of inheritance of flowering time of dry beans in different environments. J Hered. 1982;73:306–8. https://doi.org/10.1093/oxfordjournals.jhered.a109653.

Li S, Ding Y, Zhang D, Wang X, Tang X, Dai D, et al. Parallel domestication with a broad mutational spectrum of determinate stem growth habit in leguminous crops. Plant J. 2018;96:761–71. https://doi.org/10.1111/tpj.14066.

Liljegren SJ, Roeder AHK, Kempin SA, Gremski K, Østergaard L, Guimil S, et al. Control of fruit patterning in Arabidopsis by INDEHISCENT. Cell. 2004;116:843–53. https://doi.org/10.1016/S0092-8674(04)00217-X.

Lioi L. Morphotype relationships in Lima bean (*Phaseolus lunatus* L.) deduced from variation of the evolutionary marker phaseolin. Genet Resour Crop Evol. 1994;41:81–5. https://doi.org/10.1007/BF00053052.

Lioi L, Bollini R. Contribution of processing events to the molecular heterogeneity of four banding types of phaseolin, the major storage protein of *Phaseolus vulgaris* L. Plant Mol Biol. 1984;3:345–53. https://doi.org/10.1007/BF00033381.

Lira R, Casas A, Blancas J. Ethnobotany of Mexico. New York: Springer; 2016.

Liu B, Watanabe S, Uchiyama T, Kong F, Kanazawa A, Xia Z, et al. The soybean stem growth habit gene Dt1 is an ortholog of Arabidopsis TERMINAL FLOWER1. Plant Physiol. 2010;153:198–210. https://doi.org/10.1104/pp.109.150607.

Llaca V, Delgado Salinas A, Gepts P. Chloroplast DNA as an evolutionary marker in the *Phaseolus vulgaris* complex. Theor Appl Genet. 1994;88:646–52. https://doi.org/10.1007/BF01253966.

Lo S, Muñoz-Amatriaín M, Boukar O, Herniter I, Cisse N, Guo Y-N, et al. Identification of QTL controlling domestication-related traits in cowpea (*Vigna unguiculata* L. Walp). Sci Rep. 2018;8:6261. https://doi.org/10.1038/s41598-018-24349-4.

Lobaton J, Miller T, Gil J, Ariza D, de la Hoz J, Soler A, et al. Re-sequencing of common bean identifies regions of inter-gene pool introgression and provides comprehensive resources for molecular breeding. Plant Genome. 2018;11:170068. https://doi.org/10.3835/plantgenome2017.08.0068.

Lonardi S, Muñoz-Amatriaín M, Liang Q, Shu S, Wanamaker SI, Lo S, et al. The genome of cowpea (*Vigna unguiculata* [L.] Walp.). Plant J. 2019;98:767–82. https://doi.org/10.1111/tpj.14349.

López Soto JL, Ruiz Corral JA, Sánchez González JJ, Lépiz Ildefonso R. Adaptación climática de 25 especies de frijol silvestre (*Phaseolus* spp.) en la república mexicana. Rev Fitotec Mex. 2005;28:221–30.

Lopez-Ridaura S, Barba-Escoto L, Reyna-Ramirez CA, Sum C, Palacios-Rojas N, Gerard B. Maize intercropping in the milpa system. Diversity, extent and importance for nutritional security in the Western Highlands of Guatemala. Sci Rep. 2021;11 https://doi.org/10.1038/s41598-021-82784-2.

LPWG. A new subfamily classification of the Leguminosae based on a taxonomically comprehensive phylogeny – The Legume Phylogeny Working Group (LPWG)3. Taxon. 2017;66:44–77. https://doi.org/10.12705/661.

Mackie WW. Origin, dispersal and variability of the lima bean, *Phaseolus lunatus*. Hilgardia. 1943;15:1–29. https://doi.org/10.3733/hilg.v15n01p001.

MacQueen AH, White JW, Lee R, Osorno JM, Schmutz J, Miklas PN, et al. Genetic associations in four decades of multi-environment trials reveal agronomic trait evolution in common bean. Genetics. 2020;215:267–84. https://doi.org/10.1534/genetics.120.303038.

Mahuku GS, Jara C, Cuasquer JB, Castellanos G. Genetic variability within *Phaeoisariopsis griseola* from Central America and its implications for resistance breeding of common bean. Plant Pathol. 2002a;51:594–604. https://doi.org/10.1046/j.1365-3059.2002.00742.x.

Mahuku GS, Jara CE, Cajiao C, Beebe S. Sources of resistance to *Colletotrichum lindemuthianum* in the secondary gene pool of *Phaseolus vulgaris* and in crosses of primary and secondary gene pools. Plant Dis. 2002b;86:1383–7. https://doi.org/10.1094/pdis.2002.86.12.1383.

Mamidi S, Rossi M, Annam D, Moghaddam S, Lee R, Papa R, et al. Investigation of the domestication of common bean (*Phaseolus vulgaris*) using multilocus sequence data. Funct Plant Biol. 2011;38:953–67. https://doi.org/10.1071/fp11124.

Mamidi S, Rossi M, Moghaddam SM, Annam D, Lee R, Papa R, et al. Demographic factors shaped diversity in the two gene pools of wild common bean *Phaseolus vulgaris* L. Heredity. 2013;110:267–76. https://doi.org/10.1038/hdy.2012.82.

Maréchal R, Mascherpa J-M, Stainier F. Etude taxonomique d'un groupe complexe d'espèces des genres *Phaseolus* et *Vigna* (Papilionaceae) sur la base de données morphologiques et polliniques, traitées par l'analyse informatique. Boissiera. 1978;28:1–273. Genève: Conservatoire et Jardin Botaniques

Martínez-Castillo J, Zizumbo-Villareal D, Gepts P, Colunga GarcíaMarín P. Gene flow and genetic structure in the wild-weedy-domesticated complex of *Phaseolus lunatus* L. in its Mesoamerican center of domestication and diversity. Crop Sci. 2007;47:58–66. https://doi.org/10.2135/cropsci2006.04.0241.

Masson-Boivin C, Sachs JL. Symbiotic nitrogen fixation by rhizobia – the roots of a success story. Curr Opin Plant Biol. 2018;44:7–15. https://doi.org/10.1016/j.pbi.2017.12.001.

McClean PE, Terpstra J, McConnell M, White C, Lee R, Mamidi S. Population structure and genetic differentiation among the USDA common bean (*Phaseolus vulgaris* L.) core collection. Genet Resour Crop Evol. 2012;59:499–515. https://doi.org/10.1007/s10722-011-9699-0.

McClean PE, Bett KE, Stonehouse R, Lee R, Pflieger S, Moghaddam SM, et al. White seed color in common bean (*Phaseolus vulgaris*) results from convergent evolution in the *P* (pigment) gene. New Phytol. 2018;219:1112–23. https://doi.org/10.1111/nph.15259.

Medina V, Berny-Mier y Teran JC, Gepts P, Gilbert ME. Low stomatal sensitivity to vapor pressure deficit in irrigated common, lima and tepary beans. Field Crop Res. 2017;206:128–37. https://doi.org/10.1016/j.fcr.2017.02.010.

Mejía-Jiménez A, Muñoz C, Jacobsen HJ, Roca WM, Singh SP. Interspecific hybridization between common and tepary beans: increased hybrid embryo growth, fertility, and efficiency of hybridization through recurrent and congruity backcrossing. Theor Appl Genet. 1994;88:324–31. https://doi.org/10.1007/BF00223640.

Mendel G. Versuche über Pflanzenhybriden. Verhandlungen des naturforschenden Vereines Brünn, 1901. Leipzig: Verlag von Wilhelm Engelmann; 1865. p. 3–47.

Métailié G. Ethnobotany in China. In: Selin H, editor. Encyclopaedia of the history of science, technology, and medicine in non-western cultures. Dordrecht: Springer; 2008. p. 833–5.

Miller T, Gepts P, Kimmo S, Arunga E, Chilagane LA, Nchimbi-Msolla S, et al. Alternative markers linked to the *Phg-2* angular leaf spot resistance locus in common bean using the PhaseolusGenes marker database. Afr J Biotechnol. 2018;17:818–28. https://doi.org/10.5897/AJB2018.16493.

Mina-Vargas AM, McKeown PC, Flanagan NS, Debouck DG, Kilian A, Hodkinson TR, et al. Origin of year-long bean (*Phaseolus dumosus* Macfady, Fabaceae) from reticulated hybridization events between multiple *Phaseolus* species. Ann Bot. 2016;118:957–69. https://doi.org/10.1093/aob/mcw138.

Mkwaila W, Terpstra KA, Ender M, Kelly JD. Identification of QTL for agronomic traits and resistance to white mold in wild and landrace germplasm of common bean. Plant Breed. 2011;130:665–72. https://doi.org/10.1111/j.1439-0523.2011.01876.x.

Moghaddam SM, Mamidi S, Osorno JM, Lee R, Brick M, Kelly J, et al. Genome-wide association study identifies candidate loci underlying agronomic traits in a Middle American diversity panel of common bean. Plant Genome. 2016;9:3. https://doi.org/10.3835/plantgenome2016.02.0012.

Mok DWS, Mok MC, Rabakoarihanta A. Interspecific hybridization of phaseolus vulgaris with *P. lunatus* and *P. acutifolius*. Theor Appl Genet. 1978;52:209–15. https://doi.org/10.1007/BF00273891.

Monroe JG, McKay JK, Weigel D, Flood PJ. The population genomics of adaptive loss of function. Heredity. 2021; https://doi.org/10.1038/s41437-021-00403-2.

Montoya CA, Leterme P, Victoria NF, Toro O, Souffrant WB, Beebe S, et al. Susceptibility of phaseolin to in vitro proteolysis is highly variable across common bean varieties (*Phaseolus vulgaris*). J Agric Food Chem. 2008;56:2183–91. https://doi.org/10.1021/jf072576e.

Moreno-Letelier A, Aguirre-Liguori JA, Piñero D, Vázquez-Lobo A, Eguiarte LE. The relevance of gene flow with wild relatives in understanding the domestication process. R Soc Open Sci. 2020;7:191545. https://doi.org/10.1098/rsos.191545.

Motley TJ, Zerega N, Cross H. Darwin's harvest: new approaches to the origins, evolution, and conservation of crops. New York: Columbia University Press; 2006.

Motta-Aldana JR, Serrano-Serrano ML, Hernández-Torres J, Castillo-Villamizar G, Debouck DG, Chacón MI. Multiple origins of lima bean landraces in the Americas: evidence from chloroplast and nuclear DNA polymorphisms. Crop Sci. 2010;50:1773–87. https://doi.org/10.2135/cropsci2009.12.0706.

Mukankusi C, Raatz B, Nkalubo S, Berhanu F, Binagwa P, Kilango M, et al. Genomics, genetics and breeding of common bean in Africa: a review of tropical legume project. Plant Breed. 2019;138:401–14. https://doi.org/10.1111/pbr.12573.

Mulumba JW, Nankya R, Adokorach J, Kiwuka C, Fadda C, De Santis P, et al. A risk-minimizing argument for traditional crop varietal diversity use to reduce pest and disease damage in agricultural ecosystems of Uganda. Agr Ecosyst Environ. 2012;157:70–86. https://doi.org/10.1016/j.agee.2012.02.012.

Muñoz LC, Blair MW, Duque MC, Tohme J, Roca W. Introgression in common bean × tepary bean interspecific congruity-backcross lines as measured by AFLP Markers. Crop Sci. 2004;44:637–45. https://doi.org/10.2135/cropsci2004.6370.

Muñoz LC, Duque MC, Debouck DG, Blair MW. Taxonomy of tepary bean and wild relatives as determined by amplified fragment length polymorphism (AFLP) markers. Crop Sci. 2006;46:1744–54. https://doi.org/10.2135/cropsci2005-12-0475.

Murube E, Campa A, Song Q, McClean P, Ferreira JJ. Toward validation of QTLs associated with pod and seed size in common bean using two nested recombinant inbred line populations. Mol Breed. 2020;40:7. https://doi.org/10.1007/s11032-019-1085-1.

Myers JR, Kmiecik K. Common bean: economic importance and relevance to biological science research. In: Pérez de la Vega M, Santalla M, Marsolais F, editors. The common bean genome, Compendium of plant genomes. Cham: Springer; 2017. p. 1–20.

Myers JR, Davis JW, Jorgey B, Kean D. Genetic analysis of processing traits in green bean (*Phaseolus vulgaris*). Acta Horticulturae. 2004;637:369–75. https://doi.org/10.17660/ActaHortic.2004.637.46.

Nic Lughadha E, Govaerts R, Belyaeva I, Black N, Lindon H, Allkin R, et al. Counting counts: revised estimates of numbers of accepted species of flowering plants, seed plants, vascular plants and land plants with a review of other recent estimates. Phytotaxa. 2016;272:082–8. https://doi.org/10.11646/phytotaxa.272.1.5.

Norton JB. Inheritance of habit in the common bean. The American Naturalist 1915;49 (585):547–61

O'Rourke JA, Iniguez LP, Bucciarelli B, Roessler J, Schmutz J, McClean PE, et al. A re-sequencing based assessment of genomic heterogeneity and fast neutron-induced deletions in a common bean cultivar. Front Plant Genet Genom. 2013;4:00210. https://doi.org/10.3389/fpls.2013.00210.

Olsen KM. SNPs, SSRs and inferences on cassava's origin. Plant Mol Biol. 2004;56:517–26. https://doi.org/10.1007/s11103-004-5043-9.

Orr HA, Turelli M. the evolution of postzygotic isolation: accumulating Dobzhansky-Muller incompatibilities. Evolution. 2001;55:1085–94. https://doi.org/10.1111/j.0014-3820.2001.tb00628.x.

Osborn TC, Alexander DC, Sun SSM, Cardona C, Bliss FA. Insecticidal activity and lectin homology of arcelin seed protein. Science. 1988;240:207–10. https://doi.org/10.1007/BF00223640.

Osborn TC, Hartweck LM, Harmsen RH, Vogelzang RD, Kmiecik KA, Bliss FA. Registration of *Phaseolus vulgaris* genetic stocks with altered seed protein compositions. Crop Sci. 2003;43:1570–1. https://doi.org/10.2135/cropsci2003.1570.

Pallottini L, Garcia E, Kami J, Barcaccia G, Gepts P. The genetic anatomy of a patented yellow bean. Crop Sci. 2004;44:968–77. https://doi.org/10.2135/cropsci2004.9680.

Papa R, Gepts P. Asymmetry of gene flow and differential geographical structure of molecular diversity in wild and domesticated common bean (*Phaseolus vulgaris* L.) from Mesoamerica. Theor Appl Genet. 2003;106:239–50. https://doi.org/10.1007/s00122-002-1085-z.

Papa R, Acosta J, Delgado-Salinas A, Gepts P. A genome-wide analysis of differentiation between wild and domesticated *Phaseolus vulgaris* from Mesoamerica. Theor Appl Genet. 2005;111:1147–58. https://doi.org/10.1007/s00122-005-0045-9.

Papa R, Bellucci E, Rossi M, Leonardi S, Rau D, Gepts P, et al. Tagging the signatures of domestication in common bean (*Phaseolus vulgaris*) by means of pooled DNA samples. Ann Bot. 2007;100:1039–51. https://doi.org/10.1093/aob/mcm151.

Paredes O, Gepts P. Segregation and recombination in inter-gene pool crosses of *Phaseolus vulgaris* L. J Hered. 1995a;86:98–106. https://doi.org/10.1093/oxfordjournals.jhered.a111556.

Paredes OM, Gepts P. Extensive introgression of Middle American germplasm into Chilean common bean cultivars. Genet Resour Crop Evol. 1995b;42:29–41. https://doi.org/10.1007/BF02310681.

Parker TA, Berny Mier y Teran JC, Palkovic A, Jernstedt J, Gepts P. Pod indehiscence is a domestication and aridity resilience trait in common bean. New Phytol. 2020a;225:558–70. https://doi.org/10.1111/nph.16164.

Parker TA, De Sousa LL, De Oliveira FT, Palkovic A, Gepts P. Toward the introgression of *PvPdh1* for increased resistance to pod shattering in common bean. Theor Appl Genet. 2020b;134:313–25. https://doi.org/10.1007/s00122-020-03698-7.

Parker TA, Palkovic A, Gepts P. Determining the genetic control of common bean early-growth rate using unmanned aerial vehicles. Remote Sens (Basel). 2020c;12:1748. https://doi.org/10.3390/rs12111748.

Parker TA, Lo S, Gepts P. Pod shattering in grain legumes: emerging genetic and environment-related patterns. Plant Cell. 2021a:koaa025. https://doi.org/10.1093/plcell/koaa025.

Parker T, Palkovic A, Brummer E, Gepts P. Registration of 'UC Tiger's Eye' heirloom-like dry bean. J Plant Regist. 2021b;15:16–20. https://doi.org/10.1002/plr2.20084.

Parker T, Palkovic A, Brummer EC, Gepts P. Registration of 'UC Rio Zape' heirloom-like dry bean. J Plant Regist. 2021c;15:37–42. https://doi.org/10.1002/plr2.20095.

Parker T, Palkovic A, Brummer EC, Gepts P. Registration of 'UC Southwest Gold' heirloom-like gold and white mottled bean. J Plant Regist. 2021d;15:48–52. https://doi.org/10.1002/plr2.20117.

Parker T, Palkovic A, Brummer EC, Gepts P. Registration of 'UC Southwest Red' heirloom-like red and white mottled bean. J Plant Regist. 2021e;15:21–7. https://doi.org/10.1002/plr2.20092.

Parker T, Palkovic A, Brummer EC, Gepts P. Registration of 'UC Sunrise' heirloom-like orange and white mottled bean. J Plant Regist. 2021f;15:43–7. https://doi.org/10.1002/plr2.20096.

Pastor-Corrales MA, Jara C, Singh SP. Pathogenic variation in, sources of, and breeding for resistance to *Phaeoisariopsis griseola* causing angular leaf spot in common bean. Euphytica. 1998;103:161–71. https://doi.org/10.1023/a:1018350826591.

Payró de la Cruz E, Gepts P, Colunga GarciaMarín P, Zizumbo Villareal D. Spatial distribution of genetic diversity in wild populations of *Phaseolus vulgaris* L. from Guanajuato and Michoacán, México. Genet Resour Crop Evol. 2005;52:589–99. https://doi.org/10.1007/s10722-004-6125-x.

Penha JS, Lopes ACA, Gomes RLF, Pinheiro JB, Assunção Filho JR, Silvestre EA, et al. Estimation of natural outcrossing rate and genetic diversity in Lima bean (*Phaseolus lunatus* L. var. lunatus) from Brazil using SSR markers: implications for conservation and breeding. Genet Resour Crop Evol. 2017;64:1355–64. https://doi.org/10.1007/s10722-016-0441-9.

Pérez de la Vega M, Santalla M, Marsolais F, editors. The common bean genome. Cham: Springer; 2017.

Pérez-Vega E, Pañeda A, Rodríguez-Suárez C, Campa A, Giraldez R, Ferreira J. Mapping of QTLs for morpho-agronomic and seed quality traits in a RIL population of common bean (*Phaseolus vulgaris* L.). Theor Appl Genet. 2010;120:1367–80. https://doi.org/10.1007/s00122-010-1261-5.

Pickersgill B. Domestication of plants in the Americas: insights from Mendelian and molecular genetics. Ann Bot. 2007;100:925–40. https://doi.org/10.1093/aob/mcm193. ISSN 0305-7364

Piperno DR. New archaeobotanical information on early cultivation and plant domestication involving microplant (phytolith and starch grain) remains. In: Gepts P, Famula T, Bettinger R, Brush S, Damania A, McGuire P, Qualset C, editors. Biodiversity in agriculture – domestication, evolution, and sustainability. Cambridge: Cambridge University Press; 2012. p. 136–59.

Piperno DR, Dillehay TD. Starch grains on human teeth reveal early broad crop diet in northern Peru. Proc Natl Acad Sci. 2008;105:19622–7. https://doi.org/10.1073/pnas.0808752105.

Piperno DR, Ranere AJ, Holst I, Iriarte J, Dickau R. Starch grain and phytolith evidence for early ninth millennium B.P. maize from the Central Balsas River Valley, Mexico. Proc Natl Acad Sci. 2009;106:5019–24. https://doi.org/10.1073/pnas.0812525106.

Pnueli L, Carmel-Goren L, Hareven D, Gutfinger T, Alvarez J, Ganal M, et al. The *SELF-PRUNING* gene of tomato regulates vegetative to reproductive switching of sympodial meristems and is the ortholog of *CEN* and *TFL1*. Development. 1998;125:1979–89.

Porch T, Beaver J, Debouck D, Jackson S, Kelly J, Dempewolf H. Use of wild relatives and closely related species to adapt common bean to climate change. Agronomy. 2013;3:433–61. https://doi.org/10.3390/agronomy3020433.

Porch TG, Cichy K, Wang W, Brick M, Beaver JS, Santana-Morant D, et al. Nutritional composition and cooking characteristics of tepary bean (*Phaseolus acutifolius* Gray) in comparison with common bean (*Phaseolus vulgaris* L.). Genet Resour Crop Evol. 2017;64:935–53. https://doi.org/10.1007/s10722-016-0413-0.

Pratt RC, Nabhan GP. Evolution and diversity of *Phaseolus acutifolius* genetic resources. In: Gepts P, editor. Genetic resources of *Phaseolus* beans. Dordrecht: Kluwer; 1988. p. 409–40.

Raggi L, Caproni L, Carboni A, Negri V. Genome-wide association study reveals candidate genes for flowering time variation in common bean (*Phaseolus vulgaris* L.). Front Plant Sci. 2019;10 https://doi.org/10.3389/fpls.2019.00962.

Ramirez-Cabral NYZ, Kumar L, Taylor S. Crop niche modeling projects major shifts in common bean growing areas. Agric For Meteorol. 2016:218, 102–9, 113. https://doi.org/10.1016/j.agrformet.2015.12.002.

Ramirez-Villegas J, Khoury C, Jarvis A, Debouck DG, Guarino L. A gap analysis methodology for collecting crop genepools: a case study with *Phaseolus* beans. PLoS One. 2010;5:e13497. https://doi.org/10.1371/journal.pone.0013497.

Ramirez-Villegas J, Khoury CK, Achicanoy HA, Mendez AC, Diaz MV, Sosa CC, et al. A gap analysis modelling framework to prioritize collecting for *ex situ* conservation of crop landraces. Divers Distrib. 2020;26(6):730–42. https://doi.org/10.1111/ddi.13046.

Rau D, Murgia ML, Rodriguez M, Bitocchi E, Bellucci E, Fois D, et al. Genomic dissection of pod shattering in common bean: mutations at nonorthologous loci at the basis of convergent phenotypic evolution under domestication of leguminous species. Plant J. 2019;97:693–714. https://doi.org/10.1111/tpj.14155.

Ray DK, Mueller ND, West PC, Foley JA. Yield trends are insufficient to double global crop production by 2050. PLoS One. 2013;8:e66428. https://doi.org/10.1371/journal.pone.0066428.

Rendón-Anaya M, Herrera-Estrella A, Gepts P, Delgado-Salinas A. A new species of *Phaseolus* (Leguminosae, Papilionoideae) sister to *Phaseolus vulgaris*, the common bean. Phytotaxa. 2017a;313:259–66. https://doi.org/10.11646/phytotaxa.313.3.3.

Rendón-Anaya M, Montero-Vargas JM, Saburido-Alvarez S, Vlasova A, Capella-Gutiérrez S, Ordaz-Ortiz JJ, et al. Genomic history of the origin and domestications of common bean in the Americas unveils its closest sister species. Genome Biol. 2017b;18:60. https://doi.org/10.1186/s13059-017-1190-6.

Repinski SL, Kwak M, Gepts P. The common bean growth habit gene PvTFL1y is a functional homolog of Arabidopsis TFL1. Theor Appl Genet. 2012;124:1539–47. https://doi.org/10.1007/s00122-012-1808-8.

Resende RT, de Resende MDV, Azevedo CF, Fonseca e Silva F, Melo LC, Pereira HS, et al. Genome-wide association and regional heritability mapping of plant architecture, lodging and productivity in *Phaseolus vulgaris*. G3. 2018;8:2841–54. https://doi.org/10.1534/g3.118.200493.

Rivera Núñez D, Matilla Séiquer G, Obón de Castro C, Alcaraz Ariza FJ. Plants and humans in the Near East and the Caucasus: ancient and traditional uses of plants as food and medicine, a diachronic ethnobotanical review: (Armenia, Azerbaijan, Georgia, Iran, Iraq, Lebanon, Syria, and Turkey). Murcia: Editum; 2012.

Rodriguez M, Rau D, Angioi SA, Bellucci E, Bitocchi E, Nanni L, et al. European *Phaseolus coccineus* L. landraces: population structure and adaptation, as revealed by cpSSRs and phenotypic analyses. PLoS One. 2013;8:e57337. https://doi.org/10.1371/journal.pone.0057337.

Rodriguez M, Rau D, Bitocchi E, Bellucci E, Biagetti E, Carboni A, et al. Landscape genetics, adaptive diversity and population structure in *Phaseolus vulgaris*. New Phytol. 2016;209:1781–94. https://doi.org/10.1111/nph.13713.

Rossi M, Bitocchi E, Bellucci E, Nanni L, Rau D, Attene G, et al. Linkage disequilibrium and population structure in wild and domesticated populations of *Phaseolus vulgaris* L. Evol Appl. 2009;2:504–22. https://doi.org/10.1111/j.1752-4571.2009.00082.x.

Ross-Ibarra J, Morrell PL, Gaut BS. Plant domestication, a unique opportunity to identify the genetic basis of adaptation. Proc Natl Acad Sci. 2007;104:8641–8. https://doi.org/10.1073/pnas.0700643104.

Rzedowski J. Vegetación Potencial 1:4000 000. IV.8.2. Atlas Nacional de México [en línea]. 1990, vol. II, http://www.conabio.gob.mx/informacion/gis/. Ciudad de México: Instituto de Geografía, Universidad Nacional Autónoma de México; 1990.

Sandhu KS, You FM, Conner RL, Balasubramanian PM, Hou A. Genetic analysis and QTL mapping of the seed hardness trait in a black common bean (Phaseolus vulgaris) recombinant inbred line (RIL) population. Mol Breed. 2018;38(3):34. https://doi.org/10.1007/s11032-018-0789-y.

Sauer CE. American agricultural origins: a consideration of nature and culture. In: Leighly J, editor. Land and life: a selection of the writing of Carl Ortwin Sauer. Berkeley, CA: University of California Press; 1965.

Sax K. The association of size differences with seed coat pattern and pigmentation in *Phaseolus vulgaris*. Genetics. 1923;8:552–60.

Schinkel C, Gepts P. Phaseolin diversity in the tepary bean, *Phaseolus acutifolius* A. Gray. Plant Breed. 1988;101:292–301. https://doi.org/10.1111/j.1439-0523.1988.tb00301.x.

Schinkel C, Gepts P. Allozyme variability in the tepary bean, *Phaseolus acutifolius* A. Gray. Plant Breed. 1989;102:182–95. https://doi.org/10.1111/j.1439-0523.1989.tb00336.x.

Schmit V, Debouck DG. Observations on the origin of *Phaseolus polyanthus* Greenman. Econ Bot. 1991;45:345–64. https://doi.org/10.1007/BF02887077.

Schmutz J, Cannon SB, Schlueter J, Ma JX, Mitros T, Nelson W, Hyten DL, Song QJ, Thelen JJ, Cheng JL, Xu D, Hellsten U, May GD, Yu Y, Sakurai T, Umezawa T, Bhattacharyya MK, Sandhu D, Valliyodan B, Lindquist E, Peto M, Grant D, Shu SQ, Goodstein D, Barry K, Futrell-Griggs M, Abernathy B, Du JC, Tian ZX, Zhu LC, Gill N, Joshi T, Libault M, Sethuraman A, Zhang XC, Shinozaki K, Nguyen HT, Wing RA, Cregan P, Specht J, Grimwood J, Rokhsar D, Stacey G, Shoemaker RC, Jackson SA. Genome sequence of the palaeopolyploid soybean. Nature. 2010;463:178–83. https://doi.org/10.1038/nature08670.

Schmutz J, McClean P, Mamidi S, Wu G, Cannon S, Grimwood J, et al. A reference genome for common bean and genome-wide analysis of dual domestications. Nat Genet. 2014;46:707–13. https://doi.org/10.1038/ng.3008.

Schoonhoven AV, Cardona C, Valor J. Resistance to the bean weevil and the Mexican bean weevil (Coleoptera: Bruchidae) in noncultivated common bean accessions. J Econ Entomol. 1983;76:1255–9. https://doi.org/10.1093/jee/76.6.1255.

Schwartz HF, Singh SP. Breeding common bean for resistance to white mold: a review. Crop Sci. 2013;53:1832–44. https://doi.org/10.2135/cropsci2013.02.0081.

Schwember AR, Carrasco B, Gepts P. Unraveling agronomic and genetic aspects of runner bean (*Phaseolus coccineus* L). Field Crop Res. 2017;206:86–94. https://doi.org/10.1016/j.fcr.2017.02.020.

Scott JC. Against the grain: a deep history of the earliest states. New Haven, CT: Yale University Press; 2018.

Scully BT, Wallace DH. Variation in and relationship of biomass, growth rate, harvest index, and phenology to yield of common bean. J Am Soc Hort Sci. 1990;115:218–25. https://doi.org/10.21273/jashs.115.2.218.

Serrano-Serrano ML, Hernández-Torres J, Castillo-Villamizar G, Debouck DG, Chacón Sánchez MI. Gene pools in wild Lima bean (*Phaseolus lunatus* L.) from the Americas: evidences for an Andean origin and past migrations. Mol Phylogenet Evol. 2010;54:76–87. https://doi.org/10.1016/j.ympev.2009.08.028.

Serrano-Serrano ML, Andueza-Noh RH, Martínez-Castillo J, Debouck DG, Chacón SMI. Evolution and domestication of lima bean in Mexico: evidence from ribosomal DNA. Crop Sci. 2012;52:1698–712. https://doi.org/10.2135/cropsci2011.12.0642.

Shii CT, Mok MC, Temple SR, Mok DWS. Expression of developmental abnormalities in hybrids of *Phaseolus vulgaris* L.: interaction between temperature and allelic dosage. J Hered. 1980;71:218–22. https://doi.org/10.1093/oxfordjournals.jhered.a109353.

Shii CT, Rabakoarihanta A, Mok MC, Mok DWS. Embryo development in reciprocal crosses of *Phaseolus vulgaris* and *Phaseolus coccineus*. Theor Appl Genet. 1982;62:59–64.

Shonnard GC, Gepts P. Genetics of heat tolerance during reproductive development in common bean. Crop Sci. 1994;34:1168–75. https://doi.org/10.2135/cropsci1994.0011183X003400050005x.

Shull GH. Some latent characters of a white bean. Science. 1907;25:828–32. https://doi.org/10.1126/science.25.647.828-b.

Sicard D, Michalakis Y, Dron M, Neema C. Genetic diversity and pathogenic variation of *Colletotrichum lindemuthianum* in the three centers of diversity of its host, *Phaseolus vulgaris*. Phytopathology. 1997;87:807–13. https://doi.org/10.1094/phyto.1997.87.8.807.

Silvertown JW. Introduction to plant population ecology. London: Longman; 1982.

Singh SP. A key for identification of different growth habits of *Phaseolus vulgaris* L. Annu Rep Bean Improv Coop. 1982;25:92–5.

Singh SP, editor. Common bean improvement in the twenty-first century. Dordrecht: Springer-Science+Business Media; 1999.

Singh SP. Broadening the genetic base of common bean cultivars: a review. Crop Sci. 2001;41:1659–75. https://doi.org/10.2135/cropsci2001.1659.

Singh SP, Muñoz CG. Resistance to common bacterial blight among *Phaseolus* species and common bean improvement. Crop Sci. 1999;39:80–9. https://doi.org/10.2135/cropsci1999.0011183X003900010013x.

Singh SP, Gepts P, Debouck DG. Races of common bean (*Phaseolus vulgaris* L., Fabaceae). Econ Bot. 1991a;45:379–96. https://doi.org/10.1007/BF02887079.

Singh SP, Gutiérrez JA, Molina A, Urrea C, Gepts P. Genetic diversity in cultivated common bean: II. Marker-based analysis of morphological and agronomic traits. Crop Sci. 1991b;31:23–9. https://doi.org/10.2135/cropsci1991.0011183X003100010005x.

Singh SP, Nodari R, Gepts P. Genetic diversity in cultivated common bean. I. Allozymes. Crop Sci. 1991c;31:19–23. https://doi.org/10.2135/cropsci1991.0011183X003100010004x.

Singh SP, Debouck DG, Roca WM. Interspecific hybridization between *Phaseolus vulgaris* L. and *P. parvifolius Freytag*. Annu Rep Bean Improv Coop. 1998;41:7–8.

Singh SP, Teran H, Schwartz HF, Otto K, Lema M. Introgressing white mold resistance from *Phaseolus* species of the secondary gene pool into common bean. Crop Sci. 2009;49:1629–37. https://doi.org/10.2135/cropsci2008.08.0508.

Singh J, Gezan S, Vallejos CE. Developmental pleiotropy shaped the roots of the domesticated common bean *Phaseolus vulgaris* L. Plant Physiol. 2019;180(3):1467–79. https://doi.org/10.1104/pp.18.01509.

Sirolli H, Drewes SI, Picca PI, Kalesnik FA. In situ conservation of the wild relative of the common bean (*Phaseolus vulgaris* var. *aborigineus*) at the south of its neotropical distribution: environmental characterization of a population in central Argentina. Genet Resour Crop Evol. 2015;62:115–29. https://doi.org/10.1007/s10722-014-0139-9.

Slightom JL, Drong RF, Klassy RC, Hoffman LM. Nucleotide sequences from phaseolin cDNA clones: the major storage proteins from *Phaseolus vulgaris* are encoded by two unique gene families. Nucleic Acids Res. 1985;13:6483–98. https://doi.org/10.1093/nar/13.18.6483.

Smartt J. Gene pools in *Phaseolus* and *Vigna* cultigens. Euphytica. 1981;30:445–9. https://doi.org/10.1007/bf00034009.

Smýkal P, Coyne CJ, Ambrose MJ, Maxted N, Schaefer H, Blair MW, et al. Legume crops phylogeny and genetic diversity for science and breeding. Crit Rev Plant Sci. 2015;34:43–104. https://doi.org/10.1080/07352689.2014.897904.

Soleri D, Worthington M, Aragón-Cuevas F, Smith SE, Gepts P. Farmers' varietal identification in a reference sample of local *Phaseolus* species in the Sierra Juárez, Oaxaca, Mexico. Econ Bot. 2013;67:283–98. https://doi.org/10.1007/s12231-013-9248-1.

Soltani A, Walter KA, Wiersma AT, Santiago JP, Quiqley M, Chitwood D, et al. The genetics and physiology of seed dormancy, a crucial trait in common bean domestication. BMC Plant Biol. 2021;21(1):58. https://doi.org/10.1186/s12870-021-02837-6.

Song Q, Jia G, Hyten DL, Jenkins J, Hwang E-Y, Schroeder SG, et al. SNP assay development for linkage map construction, anchoring whole-genome sequence, and other genetic and genomic applications in common bean. G3. 2015;5:2285–90. https://doi.org/10.1534/g3.115.020594.

Sonnante G, Stockton T, Nodari RO, Becerra Velásquez VL, Gepts P. Evolution of genetic diversity during the domestication of common-bean (*Phaseolus vulgaris* L). Theor Appl Genet. 1994;89:629–35.

Souter JR, Gurusamy V, Porch TG, Bett KE. Successful introgression of abiotic stress tolerance from wild tepary bean to common bean. Crop Sci. 2017;57:1160–71. https://doi.org/10.2135/cropsci2016.10.0851.

Spataro G, Tiranti B, Arcaleni P, Bellucci E, Attene G, Papa R, et al. Genetic diversity and structure of a worldwide collection of *Phaseolus coccineus* L. Theor Appl Genet. 2011;122:1281–91. https://doi.org/10.1007/s00122-011-1530-y.

Spengler RN III. Anthropogenic seed dispersal: rethinking the origins of plant domestication. Trends Plant Sci. 2020;25:340–8. https://doi.org/10.1016/j.tplants.2020.01.005.

Stephens L, Fuller D, Boivin N, Rick T, Gauthier N, Kay A, et al. Archaeological assessment reveals Earth's early transformation through land use. Science. 2019;365:897–902. https://doi.org/10.1126/science.aax1192.

Studer A, Zhao Q, Ross-Ibarra J, Doebley J. Identification of a functional transposon insertion in the maize domestication gene tb1. Nat Genet. 2011;43:1160–3. https://doi.org/10.1038/ng.942.

Suarez-Gonzalez A, Lexer C, Cronk QCB. Adaptive introgression: a plant perspective. Biol Lett. 2018;14:20170688. https://doi.org/10.1098/rsbl.2017.0688.

Swegarden HR, Sheaffer CC, Michaels TE. Yield stability of heirloom dry bean (*Phaseolus vulgaris* L.) cultivars in Midwest organic production. J Am Soc Hort Sci. 2016;51:8–14. https://doi.org/10.21273/hortsci.51.1.8.

Takahashi Y, Sakai H, Yoshitsu Y, Muto C, Anai T, Pandiyan M, et al. Domesticating *Vigna stipulacea*: a potential legume crop with broad resistance to biotic stresses. Front Plant Sci. 2019;10:1607. https://doi.org/10.3389/fpls.2019.01607.

Takahashi Y, Kongjaimun A, Muto C, Kobayashi Y, Kumagai M, Sakai H, et al. Same locus for non-shattering seed pod in two independently domesticated legumes, *Vigna angularis* and *Vigna unguiculata*. Front Genet. 2020;11:748. https://doi.org/10.3389/fgene.2020.00748.

Thomas CV, Waines JG. Fertile backcross and allotetraploid plants from crosses between tepary beans and common beans. J Hered. 1984;75:93–8. https://doi.org/10.1093/oxfordjournals.jhered.a109901.

Tian Z, Wang X, Lee R, Li Y, Specht JE, Nelson RL, et al. Artificial selection for determinate growth habit in soybean. Proc Natl Acad Sci. 2010;107:8563–8. https://doi.org/10.1073/pnas.1000088107.

Tobar Piñón MG, Mafi Moghaddam S, Lee RK, Villatoro Mérida JC, DeYoung DJ, Reyes BA, et al. Genetic diversity of Guatemalan climbing bean collections. Genet Resour Crop Evol. 2020; https://doi.org/10.1007/s10722-020-01013-3.

Tohme J, Toro Ch O, Vargas J, Debouck DG. Variability in Andean nuña common beans (*Phaseolus vulgaris*, Fabaceae). Econ Bot. 1995;49:78–95. https://doi.org/10.2307/4255694.

Tomooka N, Naito K, Kaga A, Sakai H, Isemura T, Ogiso-Tanaka E, et al. Evolution, domestication and neo-domestication of the genus *Vigna*. Plant Genet Resour. 2014;12:S168–71. https://doi.org/10.1017/S1479262114000483.

Trucchi E, Benazzo A, Lari M, Iob A, Vai S, Nanni L, et al. Ancient genomes reveal early Andean farmers selected common beans while preserving diversity. Nat Plants. 2021; https://doi.org/10.1038/s41477-021-00848-7.

Urrea CA, Harveson RM. Identification of sources of bacterial wilt resistance in common bean (*Phaseolus vulgaris*). Plant Dis. 2014;98:973–6. https://doi.org/10.1094/pdis-04-13-0391-re.

Valarmathi G, Martinez Rojo J, Bett K, Vandenberg A. *Phaseolus acutifolius* as a potential bridge species in the hybridization of *P. vulgaris* and *P. angustissimus*. Annu Rep Bean Improv Coop. 2006;49:125–6.

Varshney RK, Thudi M, Roorkiwal M, He W, Upadhyaya HD, Yang W, et al. Resequencing of 429 chickpea accessions from 45 countries provides insights into genome diversity, domestication and agronomic traits. Nat Genet. 2019;51:857–64. https://doi.org/10.1038/s41588-019-0401-3.

Vlasova A, Capella-Gutiérrez S, Rendón-Anaya M, Hernández-Oñate M, Minoche AE, Erb I, et al. Genome and transcriptome analysis of the Mesoamerican common bean and the role of gene duplications in establishing tissue and temporal specialization of genes. Genome Biol. 2016;17:1–18. https://doi.org/10.1186/s13059-016-0883-6.

von Martens G. Die Gartenbohnen: ihre Verbreitung, Cultur und Benützung. Stuttgart: Verlag von Ebner & Seubert; 1860. https://play.google.com/store/books/details?id=VyhAAAAAcAAJ

Waines JG, Barnhart DR. Outcrossing rates and multiple paternity of common and lima beans. Annu Rep Bean Improv Coop. 1997;40:44–5.

Wall JR. Experimental introgression in the genus *Phaseolus*. 1. Effect of mating systems on interspecific gene flow. Evolution. 1970;24:356–66. https://doi.org/10.2307/2406810.

Wall JR, Wall SW. Isozyme polymorphisms in the study of evolution in the *Phaseolus vulgaris-P. coccineus* complex of Mexico. In: Markert CL, editor. Isozymes IV. New York: Academic Press; 1975.

Wallace DH, Yourstone KS, Masaya PN, Zobel RW. Photoperiod gene control over partitioning between reproductive and vegetative growth. Theor Appl Genet. 1993;86:6–16. https://doi.org/10.1007/BF00223803.

Wallace L, Arkwazee H, Vining K, Myers J. Genetic diversity within snap beans and their relation to dry beans. Genes. 2018;9:587. https://doi.org/10.3390/genes9120587.

Wallach D, Hwang C, Correll MJ, Jones JW, Boote K, Hoogenboom G, et al. A dynamic model with QTL covariables for predicting flowering time of common bean (*Phaseolus vulgaris*) genotypes. Eur J Agron. 2018;101:200–9. https://doi.org/10.1016/j.eja.2018.10.003.

Walters H, Brick MA, Ogg JB. Evaluation of heirloom beans for production in northern Colorado. Annu Rep Bean Improv Coop. 2011;54:56–7.

Warschefsky E, Penmetsa RV, Cook DR, von Wettberg EJB. Back to the wilds: tapping evolutionary adaptations for resilient crops through systematic hybridization with crop wild relatives. Am J Bot. 2014;101:1791–800. https://doi.org/10.3732/ajb.1400116.

Watcharatpong P, Kaga A, Chen X, Somta P. Narrowing down a major QTL region conferring pod fiber contents in yardlong bean (*Vigna unguiculata*), a vegetable cowpea. Genes. 2020;11:363. https://doi.org/10.3390/genes11040363.

Weller JL, Vander Schoor JK, Perez-Wright EC, Hecht V, González AM, Capel C, et al. Parallel origins of photoperiod adaptation following dual domestications of common bean. J Exp Bot. 2019;70:1209–19. https://doi.org/10.1093/jxb/ery455.

White JW, Laing DR. Photoperiod response of flowering in diverse genotypes of common bean (*Phaseolus vulgaris*). Field Crop Res. 1989;22:113–28. https://doi.org/10.1016/0378-4290(89)90062-2.

Wilkus EL, Berny Mier y Teran JC, Mukankusi CM, Gepts P. Genetic patterns of common-bean seed acquisition and early-stage adoption among farmer groups in western Uganda. Front Plant Sci. 2018;9:586. https://doi.org/10.3389/fpls.2018.00586.

Worthington M, Soleri D, Aragón-Cuevas F, Gepts P. Genetic composition and spatial distribution of farmer-managed bean plantings: an example from a village in Oaxaca, Mexico. Crop Sci. 2012;52:1721–35. https://doi.org/10.2135/cropsci2011.09.0518.

Wu J, Wang L, Fu J, Chen J, Wei S, Zhang S, Zhang J, Tang Y, Chen M, Zhu J, Lei L, Geng Q, Liu C, Wu L, Li X, Wang X, Wang Q, Wang Z, Xing S, Zhang H, Blair MW, Wang S. Resequencing of 683 common bean genotypes identifies yield component trait associations across a north–south cline. Nature Genetics. 2020;52:118–25. https://doi.org/10.1038/s41588-019-0546-05.

Yuste-Lisbona F, González A, Capel C, García-Alcázar M, Capel J, Ron A, et al. Genetic variation underlying pod size and color traits of common bean depends on quantitative trait loci with epistatic effects. Mol Breed. 2014;33:1–14. https://doi.org/10.1007/s11032-013-0008-9.

Zapata M, Freytag G, Wilkinson R. Release of five common bean germplasm lines resistant to common bacterial blight: W-BB-11, W-BB-20-1, W-BB-35, W-BB-52 and W-BB-11-56. J Agric Univ Puerto Rico. 2004;88:91–5.

Zeven AC, Waninge J, van Hintum T, Singh SP. Phenotypic variation in a core collection of common bean (*Phaseolus vulgaris* L.) in the Netherlands. Euphytica. 1999;109:93–106. https://doi.org/10.1023/A:1003665408567.

Zhang B, Schrader A. TRANSPARENT TESTA GLABRA 1-dependent regulation of flavonoid biosynthesis. Plants. 2017;6:65. https://doi.org/10.3390/plants6040065.

Zhang J, Singh AK. Genetic control and geo-climate adaptation of pod dehiscence provide novel insights into soybean domestication. G3. 2020;10:545–54. https://doi.org/10.1534/g3.119.400876.

Zhang C, Postma JA, York LM, Lynch JP. Root foraging elicits niche complementarity-dependent yield advantage in the ancient 'three sisters' (maize/bean/squash) polyculture. Ann Bot. 2014;114:1719–33. https://doi.org/10.1093/aob/mcu191.

Zimmerer K. The paradox of chocolate agrobiodiversity: Humboldt, Raimondi, and indigenous smallholders in South America. ReVista. 2020;20. https://revista.drclas.harvard.edu/book/paradox-chocolate-agrobiodiversity/

Zizumbo D, Papa R, Hufford M, Repinski S, Gepts P. Identification of new wild populations of *Phaseolus vulgaris* in western Jalisco, Mexico, near the Mesoamerican domestication center of common bean. Annu Rep Bean Improv Coop. 2009;52:24–5.

Zizumbo-Villareal D, Colunga GarcíaMarín P. Origin of agriculture and plant domestication in West Mesoamerica. Genet Resour Crop Evol. 2010;57:813–25. https://doi.org/10.1007/s10722-009-9521-4.

Zizumbo-Villarreal D, Colunga-GarcíaMarín P, Payró de la Cruz E, Delgado-Valerio P, Gepts P. Population structure and evolutionary dynamics of wild–weedy–domesticated complexes of common bean in a Mesoamerican region. Crop Sci. 2005;35:1073–83. https://doi.org/10.2135/cropsci2004.0340.

Zizumbo-Villarreal D, Flores-Silva A, Colunga-García Marín P. The archaic diet in Mesoamerica: incentive for milpa development and species domestication. Econ Bot. 2012;66:328–43. https://doi.org/10.1007/s12231-012-9212-5.

Zizumbo-Villarreal D, Flores-Silva A, Colunga-GarcíaMarín P. The food system during the formative period in West Mesoamerica. Econ Bot. 2014;68:67–84. https://doi.org/10.1007/s12231-014-9262-y.

Population Genomics of Cotton

Lavanya Mendu, Kaushik Ghose, and Venugopal Mendu

Abstract Cotton (*Gossypium*) is a fiber-producing oil seed crop with significant economic and scientific importance in natural textile fiber and edible oil industries. Cotton is an ideal crop model system for studying polyploidy, evolution, population genomics, and domestication, due to the presence of several species, wild relatives, and progenitors. Population genomics studies can provide key insights into genetic diversity, population structure, evolution, and domestication as well as genotype-phenotype associations and genomics-assisted breeding of *Gossypium* species. Population genomics research in cotton is lagging behind when compared to other crops with simple genomes due to its large complex genome size and polyploidy. However, modern genomics technologies such as sequencing, resequencing, comparative genomics, and whole-genome sequencing (WGS) have generated reference genomes of diploid and tetraploid cotton, which have addressed key questions about the role of polyploidy in the evolution of cotton. Population genomics studies and reference genomes have advanced understanding of genomic variation across species, gene expression changes, domestication, and evolutionary history of several species of *Gossypium*, and genetic variation underlying different traits of cotton such as fiber yield, fiber quality, biotic, and abiotic stress resistance/tolerance. These advances are facilitating selection and breeding of cotton for improved traits through genome-wide association studies (GWAS) and genomic selection. GWAS has discovered candidate genes for different quantitative trait loci (QTLs). Hence, population genomics approach will be a key component in studying genetic history of cotton domestication, improvement of different agronomic traits such as fiber quality, oil yield, disease resistance, and tolerance to extreme environmental conditions. This chapter discusses the recent advances made in cotton population genomics and its future perspectives.

L. Mendu · V. Mendu (✉)
Department of Plant Sciences and Plant Pathology, Montana State University, Bozeman, MT, USA
e-mail: venugopal.mendu@montana.edu

K. Ghose
Department of Plant and Soil Science, Texas Tech University, Lubbock, TX, USA

Keywords Diversification · Domestication · Evolution · Genomic selection · Genomics · *Gossypium* · GWAS · Pangenome · Polyploidy · Population genetic variation

1 Introduction

Cotton (*Gossypium*) is an oil and fiber crop that provides an excellent model system to study domestication, polyploidy, evolution, and single-celled fiber development. Cotton is cultivated worldwide primarily for natural textile fiber production. In addition, cottonseed is used for food, fodder, and oil with an overall annual economic impact of $500 billion and with a supply of 115 million bales for textile production (Chen et al. 2007). There are approximately 54 species (Fryxell 1992; Wang et al. 2018; Yin et al. 2020) in the *Gossypium* genus with 46 diploids and 7 allotetraploids; however, only four species have been domesticated (Shan et al. 2014; Bourke et al. 2018; Wendel et al. 2018). Cultivated cotton is an allotetraploid (AADD) species originated by natural hybridization and genome doubling of two diploid species (AA and DD) (Jiang et al. 1998; Page et al. 2013). Parental diploids of A genome and D genome species by hybridization and polyploidy generated 7 allotetraploid species. The possible progenitors of AD genome (*G. barbadense* and *G. hirsutum*) are D genome (*G. raimondii*) and A genome (*G. herbaceum* and *G. arboreum*) (Yang et al. 2020). Approximately 5–10 million years ago, A diploid genome diverged from D diploid genome which again 1–2 million years ago formed a tetraploid cotton through hybridization of A (*G. arboreum*) and D (*G. raimondii*) genomes (Wendel 1989). Cotton domestication began 4,000 to 5,000 years ago, in which, human artificial selection played a major role. Such domestication and selection pressures resulted in few cultivars and less genetic diversity in the cultivated species. Integrated multi-omics, homology-based cloning, and reverse genetics approaches were successful in identifying useful genes targeted for breeding. Successful breeding programs depend on higher genetic diversity that is available in the germplasm resources. Cotton genomics research progress has been slow due to its large and complex genome made up of two subgenomes A and D in single nucleus. Thus far, genetic improvement of cotton was primarily based on breeding and selection followed by the application of common genomics tools such as physical mapping, QTL mapping, Expressed Sequence Tag (EST) analysis, gene expression profiling, genome-wide association studies (GWAS), and association mapping.

Cotton is an important natural textile fiber crop produced throughout the world, and efforts to improve cotton yield and resistance resulted in narrow genetic diversity (Gross and Strasburg 2010). More than 90% of the cotton production is from the cultivated upland cotton, *G. hirsutum* (Niles and Feaster 1984; Hu et al. 2019). This *Gossypium* genus (diploids and polyploids) with 54 species (Wang et al. 2012) is classified into three diploid lineages such as New World clade (D genome, America) (Pan et al. 2020), Africa-Asian clade (A, B, E, F genomes) (Cronn et al. 2003), and Australian clade (C, G, K genomes) (Chen et al. 2007) based on their place of origin. *Gossypium* species provide a good model system for studying origin, evolution, and domestication (Hu et al. 2019). *Gossypium* is a close family member of African-Madagascan genus *Gossypioides* and Hawaiian genus *Kokia* in which the divergence occurred 12.5 million years ago in Miocene times (Cronn et al. 2002). Based on several phylogenetic studies, the origin of *Gossypium* genus was reported to be 5–10 million years ago (Wendel et al. 2009). Availability of different cotton genome data sets revealed the evolutionary history of cotton as a result of polyploidy and suggested a decaplodization in *Gossypium* genus origin and neo-tetraploidization behind the current cultivated lines (Pan et al. 2020). Thus, polyploidy induced genomic instability in the form of gene loss, inversions, translocations, recombination, increase in repetitive sequences, and higher rate of gene evolution was observed (Pan et al. 2020). Four domesticated species, namely, New World allopolyploids *G. hirsutum*, *G. barbadense* ($2n = 52$) and Old World diploids *G. arboretum* (India) and *G. herbaceum* ($2n = 26$) (Africa) have extensive taxonomy history with 8 genome groups of A (2 species of *G. arboreum* and *G. herbaceum*), B (3-4 species of *G. anomalum*, *G. triphyllum*, *G. capitis-viridis*, and *G. trifurcatum*), C (2 species of *G. sturtianum* and *G. robinsonii*), D (13 to 14 species of *G. thurberi*, *G. armourianum*, *G. harknessii*, *G. davidsonii*, *G. Klotschianum*, *G. aridum*, *G. raimondii*, *G. gossypiodes*, *G. lobatum*, *G. trilobatum*, *G. laxum*, *G. turneri*, *G. schwendimanii*, *G.sp.nov*), E (5 to 9 species, *G. stocksii*, *G. somalense*, *G. areysianum*, *G. incanum*, *G. trifurcatum*, *G. benidirense*, *G. bricchetttii*, *G. vollesenii*, and *G. trifurcatum*), F (1 species of *G. longicalyx*), G (3 species of *G. bickii*, *G. australe*, *G. nelsonii*), K (12 species such as *G. anapoides*, *G. costulatum*, *G. cunninghamii*, *G. enthyle*, *G. exiguum*, *G. londonderriense*, *G. marchantiid*, *G. nobile*, *G. pilosum*, *G. populifolium*, *G. pulchellum*, and *G. rotundifolium*), AD (5 species of *G. hirsutum* , *G. barbadense*, *G. tomentosum*, *G. mustelinum*, and *G. darwinii*), and evolution (Wendel et al. 2009). These 8 genomes with 54 species belong to *Gossypieae* taxonomic tribe of cotton genus (Fryxell 1968, 1979). These genomes were spread geographically to small regions (four genomes: *Lebronnecia*, *Cephaalohibiscus*, *Gossypiodes*, and *Kokia)*, and widely spread regions (four genomes: *Hampea* with 21 species, *Cienfuegosia* with 25 species, *Thespesia* with 17 species, and *Gossypium* with 54 species) (Fryxell 1992; Wendel et al. 2009). As a result of speciation and diversification in *Gossypium*, cotton was spread to Australia, Africa, Peninsula, and Mexico. Structurally this genome diverged from herbaceous perennial in Australia to trees in Mexico. Further, genus diversification caused higher chromosomal evolution. Domesticated species have lower diversity compared to 54 species that existed in the genus because of their geographical origins of the world's tropical and subtropical regions. Genomic analyses of assembled genomes and their comparison revealed differences in gene expression, structural variations, and expanded gene families responsible for speciation and species evolution (Hu et al. 2019). The presence of two genomes A and D in allopolyploids suggests presence of duplicated copies, and reports suggest that there is 3–4% sequence variation between these duplicated genes (Senchina et al. 2003). Additionally, several transoceanic dispersals were identified in cotton genus, and changes associated with structure, ecology, and chromosomal differentiation were reported. In *Gossypium* genus, 25% of genome evolution was known to be driven by interspecific hybridization. Allopolyploid cottons (Upland and Pima) originated 1–2 million years ago (Pleistocene) by trans-oceanic dispersal of A genome African species (diploid) and D genome American species (diploid) hybridization and doubling of chromosomes (Paterson et al. 2012). New World *Gossypium barbadense* (Pima / Egyptian Cotton) was domesticated 4,000 to 5,000 years ago in Peruvian Andes and spread to pre-Columbian domain (Westengen et al. 2005). *G. hirsutum* commonly known as upland cotton was domesticated in the Yucatan peninsula during the same time (Brubaker and Wendel 1994). Wild populations of *G. barbadense* are from Guayas and Los Rios of Ecuador, Tumbes, and Peru. To connect the genetic diversity between Andes and Columbian germplasm of *G. barbadense*, studies reported the usage of amplified fragment length polymorphism (AFLP) fingerprinting (Westengen et al. 2005). During domestication, several phenotypic changes occurred in both Pima and upland cotton, but a prominent change occurred in flowering times not affected by photoperiodism and seed trichomes (Gross and Strasburg 2010). After polyploidization, *G. hirsutum* showed better yield and tolerance traits than *G. barbadense* that produces superior quality fibers which led to the domestication of *G. hirsutum* worldwide. Parallel and convergent domestication was reported in cotton based on the origin of cultivated and wild polyploids (Pan et al. 2020).

Population genomics studies provide a path to understand genetic history behind domestication process (Nazir et al. 2020). Studies using high throughput genome sequencing of *G. hirsutum* landraces, old cultivars, and modern cultivars revealed 93 differential regions and 311 selection sweeps involved in domestication and improvement (Nazir et al. 2020). Various techniques were employed to study genetic diversity in upland cotton. Genetic diversity in cotton has been studied using several molecular markers such as restriction fragment length polymorphism (RFLP), random amplified polymorphic DNA (RAPD) (Iqbal et al. 1997), amplified fragment length polymorphism (AFLP) (Rana et al. 2005), single nucleotide polymorphism (SNP) (Wang et al. 2022), and simple sequence repeats (SSRs) (Noormohammadi et al. 2013). These techniques included use of genome-wide association studies (GWAS). GWAS were used to study the genetic basis of cotton fiber traits (Liu et al. 2018). GWAS help in associating genotypes with phenotypes by identifying candidate genes (Nuzhdin et al. 2012) and subsequent evolutionary changes in wild type and cultivated crops (Naoumkina et al. 2019). In cotton, QTLs and candidate genes associated with cotton fiber quality trait were investigated and identified in upland cotton (Sun et al. 2017; Ma et al. 2018; Naoumkina et al. 2019). Other studies in cotton revealed 12 fiber length genes with SNPs (ns SNP) (Naoumkina et al. 2019).

SSRs (198) were used to analyze genetic diversity and population structure in upland cotton germplasm (302 elite upland accessions) and revealed that there was moderate genetic diversity at DNA level and recognized three subpopulations with future application of these markers in GWAS for useful alleles (Seyoum et al. 2018). SSRs were also employed to find differences in phylogenetic relationship of different accessions of different ecotypes and population structures of New World cottons

(Ulloa et al. 2013). Pangenome, which is the collection of all individuals DNA sequences of a species (Tettelin et al. 2005), was constructed for both *G. hirsutum* and *G. barbadense* in cotton and identified more variable genes in *G. hirsutum* (39,278 genes) than in *G. barbadense* (11,359 genes) (Li et al. 2021a). Pangenome analysis was used for detecting genetic diversity across different species in cotton. Different cotton genotypes and tissues were tested to understand the role of DNA methylation in gene regulation and phenotypic diversity in cotton (Osabe et al. 2014). Using HPLC, the total DNA methylation was reported to be greater than the genetic diversity suggesting the key role of epigenetic regulation in different developmental stages, and different tissues of cotton leading to diversity in the phenotype. Such methylation changes were reported to be involved in salt (NaCl) stress (Zhao et al. 2010), alkali stress (Cao et al. 2011), heterosis (Zhao et al. 2008), and quality of light (Li et al. 2011), in cotton. Population genomics studies helps in the understanding of the genetic/genomic diversity, origin and evolution, domestication syndrome including selective sweeps in domestication and genes involved in domestication, acclimation, adaptation, and facilitate genomics-assisted breeding and conservation of genetic resources (Rajora 2019; Luikart et al. 2020). In this chapter, we discuss the advances made in population genomics of cotton, especially in understanding origin, genome evolution, domestication, conservation of genetic resources, and genomics-assisted breeding. We also provide an overview of genomic and transcriptomic resources of population genomics relevance available in cotton. Finally, we present future perspectives in cotton population genomics.

2 Origin, Polyploidy, and Genome Evolution

The complex genome with polyploidy nature in cotton complicates the domestication and evolution studies due to the duplicated genomes within a single nucleus. Domesticated cotton has double the number of chromosomes resulting in presence of multiple gene copies in a single nucleus. Such doubling of genome results in molecular genetic interactions, genomic evolution, genetic transfer between genomes, gene conversion, and changes in gene expression. The presence of two subgenomes in the same nucleus causes altered expression of certain genes due to the transcription factors, post-transcriptional gene silencing, and translational and post-translational cross talks leading to complex gene regulation. Reports show that *Gossypium* species originated 5–10 million years ago resulting in the diversification of major genomes (Wendel and Cronn 2003). Studies reported that all A genome accessions, such as *G. herbaceum* (A1), *G. arboreum* (A2), and *G. hirsutum* [(AD)1] might have evolved from common ancestor (A0). *G. hirsutum*, the commonly cultivated cotton, originated from natural hybridization of two genomes, A and D genomes (Li et al. 2015). It was originally thought that A genome might have been originated from *G. herbaceum* (A_1 genome) or *G. arboreum* (A_2 genome), while D genome was originated from *G. raimondii* (D_5 genome) (Wendel 1989). Results from other studies question the origin of A genome from A_1 or A_2 genomes (Zahn

2012). The data from studies of whole-genome phylogenetic relationships, molecular tree, and population analysis have provided evidence for the origin of A_1 and A_2 genomes from a common ancestor A_0 in which A_1 is closely related to A_0 while D genome originated from D_5 genomes in the tetraploids (Ma 2020). Genome analysis revealed that A_0 and D_5 combined in the tetraploid approximately 1.6 million years ago. But the speciation of A_1 and A_2 occurred 0.7 million years ago suggesting that A genome is not from A_1 or A_2 genomes as the AD genome hybrid has formed much earlier. Genome sequencing and resequencing studies of three diploids and two tetraploids (A_1, A, D_5, and AD) also suggested the same origin of A and D parental genomes in the tetraploid (Paterson et al. 2012; Wang et al. 2012; Li et al. 2014; Liu et al. 2015; Yuan et al. 2015; Zhang et al. 2015b; Du et al. 2018; Hu et al. 2019; Udall et al. 2019; Wang et al. 2019; Yang et al. 2019; Huang et al. 2020). Sequencing studies also suggested that A_1 and A_2 genome size increased because of transposon insertion and movements. Evolutionary history using genome data sets indicated that decaploidization event in the origin of *Gossypium* genus and tetraploidization event in cultivated cotton origin (Pan et al. 2020). 'A' genome expansion, speciation, and evolution were the results of long terminal repeat bursts that happened million years (5.7 to 0.6) ago as depicted by Gaussian probability density function (Concia et al. 2020). Phylogenomics methods were used to understand molecular evolution and phylogeny of diploid cottons and to support the phylogenomics. The results of whole-genome sequencing (WGS) data suggested that subgenus was formed 6.6 million years ago by transoceanic dispersal in Africa with diversification 0.5 to 2 million years ago in mid Pleistocene followed by multiple events of dispersals spreading to Arizona, Galapagos Islands, and Peru (Grover et al. 2018). Relative comparison of chloroplast DNA and nuclear DNA indicated multiple events of interspecific hybridization.

Gossypium species are from arid and semiarid regions of tropics and subtropics. There are 54 species comprised of eight different genomes (Wendel and Cronn 2003). *Gossypium* species (diploids) were divided into three major clades based on their continental origin into African-Asian clade (A, B, E, F), New World diploids clade (D genome), and Australia group clade (C, G, K genomes) (Wendel and Albert 1992). Cladogram area analysis suggested that Australia or Africa might be the origins of *Gossypium* species. *Gossypium* arose from *Gossypioides* (African Madagascan) and *Kokia* (Hawaiian endemic). Paleo continental construction studies, palynology studies evidence, and chloroplast DNA divergent estimate times together suggested that first clade divergence occurred in Oligocene by an intercontinental dispersal. Transoceanic dispersal (Paterson et al. 2012) occurred in A genome diploid to America's resulted in allopolyploids, which then hybridized with diploid, D genome (one allotetraploid AD genome clade) as suggested by maternal phylogeny studies and area cladogram. The probable period of origin for polyploids is Pleistocene. After polyploidization, there was genome instability mediated by asymmetric changes of inversions, gene loss, translocation, heterologous recombination, repetitive sequences, and evolution rate (Pan et al. 2020).

Four species are domesticated, which are from Africa, Asia (*G. arboreum* and *G. herbaceum*), and America (*G. hirsutum* and *G. barbadense*). *G. hirsutum* originated from Mesoamerica, which was spread to 50 other countries by domestication. Genome doubling mediated by polyploidy played a key role in the evolution of cotton plants. To understand the phylogeny of the *Gossypium* species, studies of 19 chloroplast genomes were compared for divergence within and between species (Chen et al. 2016). This study revealed that cultivated cotton contains large chloroplast genomes, and divergence time of major diploid clade using cp DNA was depicted to be 10–11 million years ago. Additionally, genome origin and phylogeny of polyploids were studied in cotton using restriction fragment length polymorphisms by measuring diploid-polyploids RFC values (restriction fragment correspondence) (Zhang et al. 2015a). In this study, *G. lambatum* with highest RGC was shown as an ancestor to all five D genomes, while A genome of polyploids is suggested as a progenitor for both *G. herbaceum* and *G. arboreum*. In the cultivated tetraploid cotton, there are two genomes, namely, A genome (*G. arboreum*, spinnable fiber-producing species) (Wendel 1989) and D genome (*G. raimondii*, non-spinnable fiber-producing species) (Wendel et al. 1995). Studies revealed that D genome contributed more to the fiber development when compared to A genome in tetraploids rather than equal contribution of both A and D genomes (Xu et al. 2015).

Reference genomes of three diploids, namely, *G. arboreum* (Li et al. 2014), *G. herbaceum* (Huang et al. 2020), *G. raimondii* (Udall et al. 2019) and two tetraploids, namely, *G. hirsutum* (Li et al. 2015) *G. barbadense* (Yuan et al. 2015) were available. Conserved gene order was observed when *G. hirsutum* (AtDt) sequence was compared with diploid *G. arboreum* (AA) and *G. raimondii* (DD) genomes. Common ancestor origin for *G. barbadense* and *G. hirsutum* was revealed by gene collinearity analysis. Comparison of sequence information showed that the progenitor allotetraploid existed 1 to 1.5 million years ago (Li et al. 2015), while divergence of two allotetraploids occurred 1 million years ago (Liu et al. 2015). Recent studies showed that tetraploidization occurred between 1.7 and 1.9 million years ago and divergence time to be 0.4 and 0.6 million years ago (Hu et al. 2019). DD genome has 57% of TE elements, AA genome has 68.5% of AA genome, and the genome AtAtDtDt has 67.2% TE elements (Wang et al. 2016c). Genomes have large proportion of TE elements that are responsible for mutation, duplication, transfer, and new gene generation using genetic and epigenetic mechanisms. The predominant mode of speciation is allopolyploidy. Changes in genome occur because of genome duplication (or polyploidy) which in turn rearranged chromosomes and changed gene expression (Adams 2007) and predominant method for evolution (Wendel et al. 1995). These gene expression changes were studied and indicated the role of methylation changes, modification of histones, and antisense RNA in gene expression changes. More structural rearrangements were observed in A genome than D genome (Fang et al. 2017b).

Cotton population genomics studies gained significant importance recently, particularly after sequencing the tetraploid and parental diploid species. Population genomics studies in cotton identified the role of transposable elements in the evolution of the genomes and domestication history of upland cotton (Wu et al. 2017). The TE element proportion varies among different species of *Gossypium*, and changes in the composition of TE elements in different *Gossypium* species might cause gene expression changes. For example, TE composition of diploid *G. raimondii* and *G. arboreum* are 68.5%, and 57%, respectively, while tetraploid *G. hirsutum* has 67.2% TE composition (Wang et al. 2016c). These genomic regions with TE elements are the highly variable regions leading to gene mutations, duplications, and gene transfer. Transposable elements are known to generate new genes by both genetic and epigenetic mechanisms in cotton. The composition of long terminal repeat (LTR) gypsy retrotransposons was also higher in AA and AADD genomes than in DD genomes (Wendel and Albert 1992). Hence, suggesting the role of TE elements in chromosomal size variation during evolution, and gene expression changes. Several sequencing tools were recently employed in cotton to understand the role of genomic variation in polyploidy, domestication history, and evolutionary processes particularly by sequencing cultivated lines, wild types, and landraces. The availability of sequenced cotton genomes (*G. hirsutum*, *G. barbadense*, *G. raimondii*, *G. arboreum*) advanced cotton genomics in different areas of genome evolution, gene expression, transcriptomics, epigenomics, and proteomics (Wu et al. 2017). Several research studies identified genomic variation linked to important traits in cotton. DNA-based differences were associated with gene expression changes and subsequent phenotype changes. Variation in 93 regions related to domestication in upland cotton was identified by employing different strategies such as high throughput/ resolution chromosome conformation capture (Hi-C), chromatin interaction analysis by paired end tag sequencing (ChIA-PET), DNase hypersensitive site (DHS), and histone markers linked to the chromatin structure of the genome. Whole-genome sequencing is an advanced technology used to generate cotton reference genome sequence, which helps in studying the evolutionary, functional, quantitative trait loci mapping and population genomics (Wang et al. 2006; Paterson et al. 2012; Tang et al. 2015). Initially, WGS was applied to a few model species with small genomes and has been recently extended to many crops including polyploid species, such as cotton. Highly repetitive and polyploid nature of the cotton genome makes genome assembly difficult with the short-read sequencing technologies. A combination of short- and long-read technologies will help in high quality genome sequencing of polyploid species.

The Progenitor D diploids contain the smallest genome with simple features relative to other cotton species, hence, they became the first species to be sequenced in 2012 (Wang et al. 2012). Genome sequence studies showed that the D genome contains approximately 40,976 protein-coding genes, and the D genome has gone through significant chromosomal rearrangements in the evolution process. This study has also suggested eudicots hexaploidization event and cotton-specific whole-genome duplication in D genome. Sequencing of progenitor A genomes (*G. arboretum*-A_1 and *G. herbaceum*-A_2) showed that genome variation in A genome ancestor is equally divergent from D genome ancestor *G. raimondii* (Li et al. 2014; Du et al. 2018). Further, comparison of A genomes (A_1 and A_2) showed that these two species evolved separately as allotetraploid formation occurred before the A_1 and A_2 speciation. Comparison of progenitor diploids with tetraploid genome showed that the gene order is conserved in both diploids and tetraploids (Brubaker et al. 1999; Rong et al. 2004; Desai et al. 2006; Huang et al.

2020). Studies showed that before speciation, *G. arboreum* and *G. raimondii* share two whole-genome duplications (Li et al. 2014). Comparison of species at diploid and tetraploid level shows how interaction exists between domestication and ploidy. Similarly, comparisons of domesticated and nondomesticated forms of cotton elucidate differences in natural selection and artificial selection (Chen et al. 2007). Sequence data is currently available for three diploids and two tetraploids (*G. arboreum, G. herbaceum, G. raimondii, G. hirsutum,* and *G. barbadense*), and there are several collections within the species, which need to be sequenced. Completion of multiple accessions within the species will help in comprehensive understanding of population genetics, domestication, and evolutionary relationships.

Gossypium genus has 45 diploid and 7 tetraploid species, and the cultivated cotton is an allotetraploid (AADD) species originated from two diploid (AA and DD) species. Allotetraploid cotton is a model crop used for studying polyploidy changes in gene expression by polyploidy, and genome evolution. Genome variation across *Gossypium* species showed that genetic diversity is affected by genome duplications and mobile elements transfer in the cultivated allotetraploid. Research data confirms that gene expression changes occur because of polyploidy (Strygina et al. 2020). Several cotton draft genomes are available; however, the intergenic DNA, such as highly repetitive regions, telomeres, and centromeres, were poorly represented (Yang et al. 2019). The *G. raimondii* (D genome) with small, simple genome, also known as a possible ancestor of *G. hirsutum* and *G. barbadense* was the first sequenced cotton species (Paterson et al. 2012; Wang et al. 2022). In 2012, the genome assembly of *G. raimondii* (DD) was reported, which showed the genome size as 740 Mb, but several gaps were found in this assembly. Later, the draft genomes of *G. arboreum* was also reported which was more than double the genome size of *G. raimondii* (Li et al. 2014). Availability of two draft genomes of diploid species facilitated the comparative genomic analysis as agronomically *G. arboreum* produces lint fibers, while *G. raimondii* does not. Lint fiber production is the desired trait in cotton and has scientific and commercial significance. Comparative genomics studies of *G. raimondii* and *G. arboreum* revealed conservation of gene order despite substantial agronomic differences (Li et al. 2015). Later, two independent research groups released the draft genome of tetraploid *G. hirsutum* (TM-1) (Li et al. 2015; Liu et al. 2015; Yuan et al. 2015). *Gossypium australe*, a diploid wild cotton genome, was also sequenced by combining Pac Bio, illumina short read, Bio Nano (DLS), and Hi-C technologies (Cai et al. 2020). This species contains disease-resistant traits and late gland morphogenesis that were used in breeding programs to transfer these selective traits in cultivated cotton. Comparative studies of sequence showed that 73.5% of *G. australe* genome is made up of repeat sequences different from other *Gossypium* species (85.3% in *G. arboreum*, 69.86% in *G. hirsutum*, and 69.83% in *G. barbadense*) (Cai et al. 2020).

With the availability of reference genomes of D genome (Gr), A genome (Ga) diploids, two cultivated tetraploids facilitated the genome-wide analysis of genetic variation during evolution. Comparative analysis of reference genomes showed structural, genetic, and gene expression variations in fiber trait and stress responses (Li et al. 2015; Liu et al. 2015; Zhang et al. 2015b; Hu et al. 2019; Wang et al. 2019). However, these studies could not clearly explain the effect of polyploidy on domestication and selection of wild and cultivated cotton species (Chen et al. 2020). Asymmetric evolution was implicated in the genomes as A genome shows more structural rearrangements, loss of genes, presence of disrupted genes, the divergence of sequence than in D genome (Zhang et al. 2015b). Further, population genomics studies are required to completely understand the mechanism behind the cotton domestication and evolution, particularly fiber-related traits. The genome of *Gossypium* is large, complex, and diverse ranging from diploid A to G (A, B, C, D, E, F, G), and K, in addition to four AD allotetraploids (Wendel et al. 1989; Li et al. 2014). The diploids ($2n = 2x = 26$ chromosomes) have a total of 26 chromosomes, while allotetraploids ($2n = 4x = 52$ chromosomes) have 52 chromosomes (Ulloa et al. 2017). The simple smallest genome belongs to the D genome (Wang et al. 2012). The A and D diploids were considered as common ancestors of the AD genome (Wendel et al. 1989). The closest relative of the D genome is *G. raimondii*, while the A genome's closest relatives are *Gossypium arboretum* and *G. herbaceum*.

Earlier studies on cotton polyploidy pointed out that the transposons spread across the genomes produced polyploidy and might have led to diversification. Availability of genome sequences of upland cotton (*G. hirsutum* L.), Egyptian cotton (*G. barbadense* L.), *G. arboreum* L., and *G. raimondii* has transformed the genomics study in cotton. The first selected species for WGS (whole-genome shotgun) was *G. raimondii* which was reported in 2012 with 740 Mb DNA (Wang et al. 2012); however, this assembly contains contigs N50 (44.9 kb and 135.5 kb) with several thousands of gaps and 60% repeated sequences. There are 13 chromosomes in *G. raimondii* (D diploid); however, the genome size might vary among the diploids with up to threefold higher compared to *G. raimondii* genome (Chen et al. 2007; Abdurakhmonov et al. 2008). The evidence for significant chromosomal rearrangements in *G. raimondii* came from the identification of 2,355 syntenic blocks with approximately 40% paralogs in each block. Further, in polyploid species, availability of ancestral parental and polyploid species facilitates the investigation of independent evolution of genomes as independent species (parental) as well as part of the subgenome (polyploids). Availability of genomes sequences of *G. hirsutum* L., and *G. barbadense* L. along with draft genomes of *G. arboreum* L., *G. herbaceum* L., and *G. raimondii* has transformed the omics studies in cotton leading to the discovery of PAVs, evolution, chromosomal variation, gene expression studies, and genetic variation related to agronomic traits. Nevertheless, poor representation of telomeres, centromeres, and repeat-rich regions in the sequenced draft cotton genomes is a major drawback in the understanding of functionally important genomic regions. A deeper understanding of genomic organization is necessary for understanding the genetic and molecular basis of origin, speciation, and diversification of cotton.

3 Genomic Resources

The major limitation for breeding progress was the lack of genomic information because of highly diverse and large number of unsequenced *Gossypium* species (Yang et al. 2020). Population genomics studies using various available genomic tools revealed the fact that there is a narrow genetic base in upland cotton. Hence, such restricted genetic bases of cotton (*G. hirsutum*) warrant need for polymorphism-based marker breeding in cotton to exploit the available genetic diversity. These molecular markers are highly useful tools to identify the genetic diversity, marker-assisted selection (MAS), linkage mapping, and genomic finger printing (Kalia et al. 2011). Molecular markers reveal existing genetic variation in the population, which can be used in the development of new cultivars (Yang et al. 2020). Further, advances in the identification of different markers, promoters, mapping of useful genes in different species of *Gossypium* revealed candidate genes in the selection of various traits of cotton related to development, disease resistance, oil content, fiber traits, and abiotic stress tolerance. Hence, population genomics studies ease the trait transfer process without the employment of laborious process of transformations in transgenic plant generation. The availability of genome sequences of parental diploid and tetraploid cotton species together with other advances in cotton genomics makes cotton as a model system for studying hybridization, chromosome duplication events, altered gene expression, and independent evolution of diploids and tetraploid genomes. Thus, the cotton population genomics is rapidly advancing with the availability of ancestor diploid (Wang et al. 2012; Li et al. 2014) and cultivated tetraploid (Li et al. 2015; Zhang et al. 2015b) reference genome sequences coupled with other genomic and molecular breeding technologies. Sequencing and resequencing of related diploid and tetraploid species aid population genomics studies to understand the history of cotton domestication, genetic/genomic variation determining cotton fiber quality, biotic and abiotic stress tolerance/ resistance, seed, and fiber yield (Yang et al. 2020; He et al. 2021). In addition, the availability of genome sequences is useful for developing pangenomes to understand the structural variation among individuals within a species that will ultimately help in crop trait improvement (Della Coletta et al. 2021). Pan-genome data studies has advanced cotton breeding by comparing entire genome collections of wild type and cultivated lines, understanding the presence/absence of variation (PAV), role in domestication, and identifying genomic differences behind different agronomic traits (Li et al. 2021a). However, inadequate representation of telomeres, centromeres, and repeat-rich regions in the sequenced draft cotton genomes are major drawbacks in understanding of functionally critical genomic regions at this point. High quality genome sequencing with properly aligned chromosomes is necessary for understanding the genetic and molecular basis of origin, speciation, and diversification of cotton. Further, linking DNA sequence information to the agriculturally important traits was exploited to design such sequence changes in other cotton cultivars for similar trait development using genome-editing technologies. Because of the economic importance of cotton, both genomes of diploid and tetraploids were sequenced repeatedly to develop high quality sequence data to reveal the evolutionary history of cotton (Pan et al. 2020). In conclusion, availability of reference genomes of *Gossypium* species made the sequence information available for DNA markers identification, analysis of germplasm, genotyping, genetic mapping, genomic comparisons, breeding advances, marker-assisted selection, and discovery of QTLs of selective traits such as fiber quality, yield, stress tolerance, and disease resistance in cotton (Bardak et al. 2018).

3.1 Transcriptomes

Available genome sequence information in cotton population is a useful tool to study the genome differences among different species, genome size changes mediated by polyploidy, evolutionary changes during domestication of cultivated cotton, possible genetic diversity, and specific gene functions in driving different agricultural traits. RNA-Seq analysis will capture the genetic variation related to a specific function such as biotic or abiotic stress or fiber development. In addition to capturing the variation arising from genome difference, RNA-Seq captures the variation in expressed genes such as spice variant and alternative polyadenylation. Cotton transcriptome studies were performed for understanding genetic basis of various traits such as stress and fiber development. RNA sequencing was used for low temperature–induced genome-wide gene expression changes for the first time in *G. hirsutum*, a cold-sensitive crop when subjected to low temperature for transcriptomic profiling (Li et al. 2019b). Earlier reports in cotton revealed the role of induced transcription factors, responsive element binding proteins (GhDREB1), and repression of gibberellic acid in cold stress tolerance (Shan et al. 2007). Additionally, peroxidase enzyme, superoxide dismutase, cinnamyl dehydrogenase, and ascorbic acid peroxidase (Wu et al. 2013) and dehydrin GhDhn1 (Wang et al. 2016b) were reported in cold tolerance in *G. hirsutum*. Therefore, genome-wide changes in response to cold stress will reveal molecular mechanism in abiotic stress tolerance mechanisms. Transcriptomic analysis comparison of *G. thurberi*, *G. klotzschianum, G. raimondii,* and *G. trilobum* genomes to the reference genome suggested that there were more differentially expressed genes (DEGs) in cold stress relative to salt stress (Xu et al. 2020). Results also show that there is evolutionary divergence and 34% DEGs overlap between cold and salt stress in all four species. Weighted gene co-expression network and RNA-Seq data supported that *G. hirsutum* and *G. barbadense* fiber quality differences were related to secondary cell wall synthesis and phytohormones candidate genes (Zhang et al. 2022). Comparative RNA-Seq studies showed that 1,530 transcripts were differentially expressed in cotton plants subjected to water-deficit conditions (Bowman et al. 2013). Transcriptome analysis of heavy metal Cadmium stress identified several genes that are differentially expressed in cotton (Han et al. 2019).

Similarly, transcriptomics studies were performed to study differential gene expression during white fly infestation that causes leaf curl disease in cotton

(Naqvi et al. 2019), which showed differential expression of 468 genes in which 220 were upregulated and 248 downregulated. Cotton fiber yield and quality are important traits that determine the cotton market value and downstream processing. Cotton RNA-Seq studies were performed to identify the genes involved in fiber development, fiber-related traits, and fiber quality (Wang et al. 2020; Prasad et al. 2021). To discover the candidate genes involved in cotton fiber length, RNA-Seq was performed using long fiber and short fiber lines (Qin et al. 2019). RNA-Seq has been used in cotton for developmental as well as stress responses; however, most of the predicted gene-models were derived from computational prediction and assembled short-read RNA sequencing data and hence generated inaccurate / incomplete data. Cottongen and CGP gene annotations for the A-diploid (AA) genome produced only coding sequences data with no information on 5' and 3' UTRs. Overall, the RNA-Seq provides genetic variation associated with specific trait, which can be used to population genomics studies as well as trait improvement in cotton. Next-generation sequencing (NGS)-based RNA-Seq was used in gene profiling; however, it cannot generate full-length transcripts and results in possible amplification errors in the library construction because of short reads. Pacific Biosciences (Pacbio) single-molecule long-read Iso-Seq (Isoform sequencing) overcomes this limitation and has been used for transcriptome profiling of tetraploid cotton (*G. hirsutum*) (Abdel-Ghany et al. 2016; Wang et al. 2016a, 2018). Because of some disadvantages such as high error rate (up to 15%), high cost, and lower throughput of transcriptomes in PacBio sequencing, this technology is used in combination with second-generation and third-generation sequencing by Sebastiano et al. to develop isoform data set in human embryonic stem cells (hybrid sequencing: NGS and Pacbio) to improve the sequence quality and assembly (Au et al. 2013). Other technologies such as NGS-based Cap Analysis Gene Expression (CAGE-Seq) and poly A seq were developed for accurate identification of 5' and 3' ends (Batut and Gingeras 2013; Hoque et al. 2013). This helps in capturing the variation in the 3', and 5' ends accurately, which is highly useful for various population genomics studies.

Cotton transcriptome studies have been extensively conducted before and after the genome sequence was released due to its significance in textile industry (Flagel et al. 2012; Bowman et al. 2013; Suzuki et al. 2013; Yoo and Wendel 2014; Hovav et al. 2015; Hasan et al. 2019; Naqvi et al. 2019). Transcriptome studies have focused on salt and drought stress, evolutionary studies, Verticillium resistance, fiber developmental biology, leaf senescence, aphid, and white fly resistance. Drought stress is the most common abiotic factor that affects cotton productivity, and molecular mechanism of drought stress was investigated in three different cotton species (*G. hirsutum, G. barbadense,* and *G. arboreum*) using high-throughput next-generation sequencing (Hasan et al. 2019). Approximately 6,986 genes were identified as differentially expressed genes simultaneously revealing the role of signal transduction of plant hormones and metabolic pathways in drought tolerance. Transcription factors of ethylene-responsive genes, ABA-responsive genes, reactive oxygen species, and heat shock proteins were identified as key players involved in conferring drought tolerance in cotton. On the other hand, salt stress is extreme stress that also affects cotton plants growth, productivity, and fiber production. Transcriptomics studies using RNA-Seq on Illumina Solexa platform suggested that there are at least 6,732 differentially expressed genes under salt stress in *G. klotzschianum* root and leaf samples. Similar to drought stress, salt stress studies also indicated the role of hormones and signal transduction pathways (Wei et al. 2017). In addition to abiotic stresses, biotic stresses also cause a reduction in cotton yields and fiber quality. Verticillium wilt, a vascular disease caused by soil-borne fungus *Verticillium dahlia* kleb, causes severe yield and fiber quality loss in cotton (Zhang et al. 2020).

White fly causes leaf curl disease in cotton resulting in loss of yield and quality of cotton fiber. Comparative RNA-Seq transcriptome studies using white fly infected susceptible *G. hirsutum* line showed 220 upregulated and 248 downregulated genes (Naqvi et al. 2019). Further, qPCR analysis of two susceptible lines (Karishma and MNH 786) showed a similar gene expression pattern confirming the validity of transcriptome data. Transcriptomics studies of the cotton seedlings treated with the *V. dahlia* showed 5,638 differentially expressed genes (Zhang et al. 2020). KEGG analysis showed that the differentially expressed genes (DEGs) belong to plant-pathogen interaction, MAPK signaling pathway, flavonoid biosynthesis, and phenylpropanoid biosynthesis pathways (Zhang et al. 2020). RNA-Seq analysis in cotton plants exposed to excess cadmium showed DEGs in root (4627), stem (3022), and leaves (3854) which identified ABC, CDF, HMA, annexin, and heat shock genes mainly related to oxidation reduction and metal binding functions (Han et al. 2019). Characterization of one of the candidate genes, *GhHMAD5* in upland cotton, showed that silencing of *GhHMAD5* caused the plants sensitive to Cd stress. GO (Gene ontology) analysis of RNA-Seq data revealed the DEGS in *V. dahlia* resistance related to ROS metabolism, hydrogen peroxide metabolism, defense response, and superoxide dismutase and antioxidant activity along with key plant defense genes ERF, CNGC, FLS2, MYB, GST, and CML (Zhang et al. 2017, 2020). Other RNA-Seq studies have shown differential expression of several genes related to hormones, signal transduction, plant-pathogen interaction, phenylpropanoid pathway, and ubiquitin-mediated signals (Zhang et al. 2017; Li et al. 2019a). Comparative RNA-Seq analysis showed that cotton white fly–borne leaf curl disease (ClcuD) resulted in differential expression of 464 genes (Naqvi et al. 2019). Transcriptomic analysis was also performed on cotton xylem, pith, and developing seed fibers, which revealed the tissue-specific expression of several NAC transcription factors and highlighted its role in secondary cell walls regulation of xylem and seed fibers (Li et al. 2017a; MacMillan et al. 2017). RNA-Seq analysis was also performed in different wild type and domesticated lines for cotton fiber development (Yoo and Wendel 2014). The comparative transcriptomic analysis study showed 5,000 differentially expressed genes in primary and secondary cell wall synthesis between wild type and cultivated cotton lines. These studies opened new doors to understand genetic basis of cotton fiber evolution and role of domestication. Population genomics analysis of different cotton lines for gene expression using transcriptome analysis will help in revealing gene expression differences, which could assist selecting the lines with higher expression of useful genes. Further, it captures the variation in the expression of different genes, for example, splice variants and alternative polyadenylation, which helps in studying the population genomics in a particular species.

4 Pangenome

Advances in genome sequencing technologies not only allowed us to compare multiple related plant species but also aided in discovering a high level of genomic polymorphisms (Tao et al. 2019; Bayer et al. 2020; Della Coletta et al. 2021). However, a single reference genome will not represent all the existing genetic variations in a particular species, which has led to the idea of pangenome analysis. Pangenome helps to study structural genomic variation, copy number variation, and chromosomal rearrangement of the genome by extending to core genes that are common to all individuals and varied genes that are present in some individuals within a species. Such structural variants are under less selection pressure but play a significant role in the process of evolution of domesticated genes from their wild relatives (Lye and Purugganan 2019). Complete understanding of the structural variation needs pangenome analyses representing the DNA sequences from all individuals in a species (Li et al. 2021a). Pangenomes were constructed for *G. hirsutum* and *G. barbadense* using reference guided assembly (Li et al. 2021a). *G. hirsutum* pangenome has 102,768 genes with genome size 3388 Mb, and *G. barbadense* pangenome has 80,148 genes with genome size 2575 Mb (Fig. 1). *G. hirsutum* pangenome mapping against resequencing reads showed 561 accessions with 17,100 genes and 1,020 accessions with 85,667 gene. These accessions were further divided into core (63489), soft core (5941), shell (3803), and cloud (12,434) genes in the *G. hirsutum* pangenome. *G. barbadense* pangenomes showed 68,789 core genes, 1,796 soft core genes, 5,867 shell genes, and 2,160 cloud genes. Core genes were shown to be associated with development and cellular metabolism, while variable genes were shown to be associated with defense, stress, and signal transduction responses by GO analysis. Comparison of core and variable genes suggested that core genes express highly relative to variable genes. Further, A and D genome comparison indicated more variable gene expression in A subgenome than in D subgenome (Fig. 1). Data showed that evolutionary rate in D subgenome is greater than A subgenome. This study suggests that pangenomes help in the identification of genes lost during domestication and plant improvement and identified 6,753 genes lost in domestication and 3,866 genes lost in cotton genetic improvement. Construction of variome based on variation and divergence during domestication process using 1,961 cotton accessions discovered 456 Mb of domestication signals (Li et al. 2021a). Variome has identified 84 novel loci out of total 162 loci, 47 of which were found to be associated with 16 agronomic traits, and pangenome analyses showed a loss of 32,569 *Gossypium hirsutum* genes and G. *barbadense* lost 8,851 genes during domestication (Fig. 2). Comparative genome analysis showed 38.2% (39,278) and 14.2% (11,359) of these *Gossypium hirsutum* and *Gossypium*

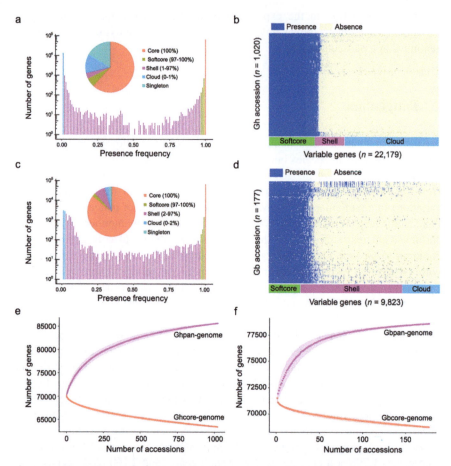

Fig. 1 Pangenomes of *G. hirsutum* and *G. barbadense* species. (**a**) Gene number and presence frequency in *G. hirsutum* pan genes. The pie chart corresponds to the core (present in all accessions), softcore, shell, and cloud genes. The singleton genes in low-depth (<5) accessions were excluded for further PAV analysis. The variable genes are divided into reference and non-reference genes. (**b**) 1020 *G. hirsutum* accessions heatmap showed presence and absence of variable PAVs. (**c**) Gene number and presence frequency in *G. barbadense* pan genes. (**d**) 177 *G. barbadense* accessions heatmap showed presence and absence of variable PAVs. (**e, f**) Saturation curve modeling the increase of pangenome size and decrease of core-genome size in 1020 *G. hirsutum* (**e**) and 177 *G. barbadense* (**f**). The error bar was calculated based on 1000 random combinations with five replicates of cotton genomes. The top and bottom edges in purple and red represent the maximum and minimum gene number. The solid lines represent the number of pan genes and core genes. (Figure reproduced from Li et al. (2021a))

barbadense genes showed presence/absence variation (PAV), and 124 PAVs are associated with important traits such as yield and fiber quality.

In any species, the genome varies in gene content and repetitive regions of the genome. Several mechanisms cause structural variations such as transposable elements that cause many mutations in plant genomes by disrupting the coding

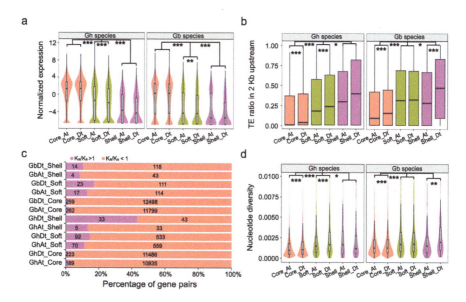

Fig. 2 Comparison of core and variable genes in A and D subgenomes. (**a**) Expression levels of core and variable genes in *G. hirsutum* and *G. barbadense*. The softcore genes are represented by "Soft." (**b**) Ratio of transposable element (TE) insertion frequency in upstream 2 kb of core and variable genes in the A and D subgenomes. (**c**) Ratio of nonsynonymous/synonymous (K_a/K_s) mutations of core and variable genes. (**d**) SNP diversity of core and variable genes. The comparison of gene expression, TE, and SNP diversity between core and variable genes were carried out using a two-sided Kolmogorov-Smirnov test (*$P < 0.05$, **$P < 0.01$, ***$P < 0.001$). (Figure reproduced from Li et al. (2021a))

sequences and errors during meiotic recombination (Jiang et al. 2004; Fedoroff 2012; Zhao et al. 2016; Allendorf 2017; Della Coletta et al. 2021). Further, markers for complex traits were developed and efficiently applied in breeding with the help of pangenome analyses.

5 Genetic Maps and QTL Mapping

Genetic mapping by DNA markers helps in identifying genomic differences among populations and can be used as tools for crop improvement. Mapped markers are of great importance in population genomics studies based on adaptive genetic variation (Luikart et al. 2018). Improving fiber traits is a constant effort for cotton crop, and genetic mapping acts as a tool for the identification of genetic variability at nucleotide level in such selective traits and advances breeding efforts further. Several research teams in cotton genetic mapping used different DNA markers such as random fragment length polymorphism (RFLP) (Reinisch et al. 1994), amplified fragment length polymorphism (AFLP) (Abdalla et al. 2001), random amplified polymorphic DNA (RAPD) (Tatineni et al. 1996), simple sequence repeats (SSRs) markers (Liu et al. 2000), and single nucleotide polymorphism (SNP) markers (Byers et al. 2012). Genetic maps are generated from identified DNA markers with desirable traits in cotton (Hulse-Kemp et al. 2015; Bourke et al. 2018). Fiber quality traits were found to be controlled by multiple loci, and genetic maps were constructed using single nucleotide polymorphism (SNP) assays in cotton. SNP63K was developed for mapping both intraspecific and interspecific populations of *Gossypium* species (Hulse-Kemp et al. 2015). The SNP63K technology was applied in studying genetic diversity, population structure, genotype differences (Hinze et al. 2017), genetic map construction (Hulse-Kemp et al. 2015), genome-wide association studies for fiber quality traits (Huang et al. 2017; Sun et al. 2017), and QTLs investigation (Li et al. 2016; Palanga et al. 2017; Zhang et al. 2017). Another SNP-based technology; SNP80K, was used to study genetic variation in cotton genotypes (Cai et al. 2017). Approximately, 1,506 QTLs regulating fiber quality traits were identified in cotton using Cotton QTLdb (Said et al. 2013, 2015) and are influenced by environmental changes (Li et al. 2016; Tan et al. 2018). Mapping studies identified transcription factor (TFGhi005602) associated with fiber length trait on chromosome 19 (Chen et al. 2015). A high-density map was constructed for tetraploid cotton, which provides deep insights into comparative QTL mapping: map-based cloning, genome structure, and evolution (Blenda et al. 2012). Another mapping approach in cotton uses Whole-Genome DNA Marker Map (WGMM) that acts as a tool for marker-assisted breeding, fine mapping, candidate genes cloning, and quantitative trait locus (QTL), markers, maps, genome-wide association studies (GWAS), and evolution studies (Wang et al. 2013). RFLP markers have also been associated with agronomic traits such as stem and leaf trichome, high gossypol, low seed gossypol, leaf chlorophyll content, and fiber-related traits (Jiang et al. 1998; Vroh Bi et al. 1999; Wright et al. 1999; Saranga et al. 2001; Chee et al. 2005). RFLP markers were used in upland cotton for the study of heterosis and genesis of varieties, and linkage maps were also generated using RFLP markers in cotton by different research groups (Jiang et al. 1998; Shappley et al. 1998).

6 Genetic Diversity and Population Structure

Genetic differentiation between domestic cultivars and their wild relatives is of utmost importance to understand the effect of historical events that determine the diversity in a population (Linck and Battey 2019). The cotton genome is highly diverse with eight diploid (A-G, K) and one allotetraploid genomes (Appels et al. 2018). Studies showed that genetic diversity is higher in *G. hirsutum* species compared to *G. barbadense*, which can be efficiently used for population genomics studies (Hinze et al. 2016; Hinze et al. 2017) (Fig. 3). Sources such as the US National Cotton Germplasm collection and USDA preserve different *Gossypium* accessions for evaluation and efficient use for crop improvement. Studies using SNP

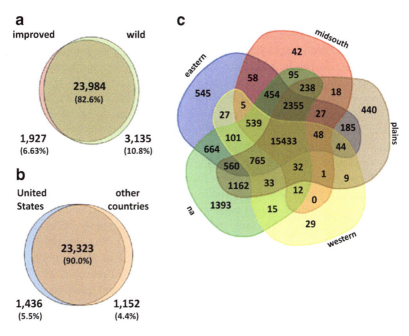

Fig. 3 SNPs unique and common to different sets of *G. hirsutum* germplasm. (**a**) 292 improved and 71 wild samples. (**b**) Improved samples from the United States (185) and from other countries (107). (**c**) Improved types from breeding regions within the United States (eastern, 48 samples; mid-south, 48; plains, 43; western, 12; n/a (unclassified breeding region), 34). (Figure reproduced from Hinze et al. (2017))

markers and CottonSNP63K arrays showed genetic variation associated with lint and fuzz cotton fiber traits (Hinze et al. 2017). Hulse-Kemp et al. (2015) developed the CottonSNP63K, an Illumina array having 45,104 putative intraspecific SNP markers for use within the cultivated cotton species *Gossypium hirsutum* L. and 17,954 putative interspecific SNP markers for use with crosses of other cotton species with *G. hirsutum*. They have used SNPs from 13 different discovery sets to develop a diverse range of *G. hirsutum* germplasm and five other species. They have produced first saturated intraspecific genetic map that form association between 26 linkage groups thus creating a linkage map with high levels of collinearity to the JGI *G. raimondii* Ulbrich reference genome (Hulse-Kemp et al. 2015). High-density SNP 80K arrays was developed in cotton for analysis of genomes and genotyping accessions in *G. hirsutum* as single nucleotide polymorphisms (SNPs) show highest variation across the genome (Cai et al. 2017). Hence, this method with high polymorphism can be used for intraspecific genotyping and analysis of genomes in many cotton accessions. In this study, 100 cotton cultivars were used for resequencing using 82,259 selected SNP markers. Out of the total 777,774 SNP loci, 76.51% showed polymorphism. Results suggested that cotton SNP 80K is highly applicable; produces genotyping results that are economical, repeatable, with highly accurate results; and provides means in cotton for germplasm genotyping, selection, functional genomics research, and molecular breeding.

Cotton fiber length is regulated by multiple genes making it a complex trait. Fiber length was identified to be associated with cluster of SNPs on chromosome D11 (Naoumkina et al. 2019); however, a single dominant mutation of N_1 results in fuzz less phenotype (no fuzz and variable amount of lint), while recessive n_2 mutation results in fuzz phenotype with uniform lint on seeds (Kearney and Harrison 1927; Ware et al. 1947). Seven genes were identified to be differentially expressed in fuzz less mutant when compared with wild type (Naoumkina et al. 2021). Similarly, SSR markers were also employed for studying the genetic diversity and population structure through GWAS and to discover the conserved alleles in cotton germplasm (Seyoum et al. 2018). To assess the genetic diversity and population structure at the DNA level, Seyoum et al. (2018) genotyped 253 Chinese and 49 different exotic elite upland cotton germplasm using 198 simple sequence repeats (SSRs) markers. This led to the identification of 897 alleles, of which 77.7% were polymorphic. Studies in upland cotton discovered core sets involved in allelic richness and demonstrated genetic diversity and population structure. Approximately 22% of alleles were specific to each accession in upland cotton, and these core sets will help in the generation of association mapping (Tyagi et al. 2014). Several factors, such as number of generations before reaching the domestication, selection pressure, effect of a genetic bottleneck, and the rate of gene flow between wild and domesticated cultivars, play a significant role in differentiation of population structure between wild and domesticated cultivars (Meyer and Purugganan 2013).

High-density SNP (single nucleotide polymorphism) 80K array was used in germplasm genotyping, selection verification, genomic variation research, and molecular breeding. Greater diversity was observed in *G. hirsutum* than in *G. barbadense*, and genetic diversity was high in wild cultivars than in improved cotton cultivars. US national cotton germplasm collection provides a reference set for different accessions of *Gossypium*. Improved and wild type accession of *G. hirsutum* and *G. barbadense* were evaluated using 105 mapped SSR markers and found that there were distinct differences between both species (Hinze et al. 2016); hence, these serve as effective tools for use of cotton genetic resources. Genotype by sequencing (GBS) provides a low-cost platform for SNP marker identification, genetic variation, genetic recombination, and genome-wide association studies (Zhang et al. 2019). About 146588 SNPs were detected using stacking tool and mapped to 26 chromosomes through the available reference genomes (Zhang et al. 2019). The highest SNP density was observed on chromosome 7 (8419 SNPs) and lowest SNP density was associated with chromosome 2 (1754). Overall, in the total SNPs identified, 62.9% were transitions (A/G or T/C), while 37.1% account for transversions (A/C or A/T or C/G or G/T). Nazir et al. (2020) used principal component analysis for analyzing genetic relationships of all accessions, and a phylogenetic tree was constructed by 4,329, 838 SNPs (Nazir et al. 2020) (Fig. 4). Both PCA and phylogenetic tree results showed that there are three groups with group 1 containing modern cultivars, group 2 with obsolete cultivars, and group 3 with geographical landraces of *G. hirsutum*. Linkage

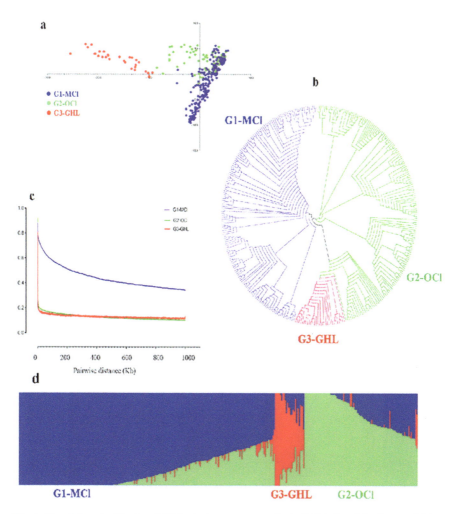

Fig. 4 Population stratification. (**a**) Principal component analysis (PCA) plot of the first two PCAs, i.e., PCA1 (21.898%) and PCA2 (2.988%), for all accessions. Dot color scheme is as G1-MCl = Modern cultivars G2-OCl = Obsolete Cultivars collected from south China, G3-GHL = Geographical landraces of *G. hirsutum*. (**b**) Phylogenetic tree constructed using whole-genome data, distributing genotypes into three clades as per original classification. (**c**) Pairwise linkage disequilibrium (LD) decay in each group. (**d**) Structure results for k-3. (Figure reproduced from Nazir et al. (2020))

disequilibrium decay was more in modern cultivars when compared to obsolete cultivars and geographical landraces. SNP data analysis by Li Yuan et al. (2021) explored population structure of *G. hirsutum* and *G. barbadense* using core genome SNP data (Fig. 5). A total of 1,913 accessions with 256 *G. hirsutum* landraces, 438 *G. hirsutum* cultivars from other countries and the United States, 929 *G. hirsutum* China cultivars, 261 *G. barbadense* accessions, and outgroup of 29 *Gossypium* species were divided into 12 clades using neighbor-joining tree analysis 8 clades constitute *G. hirsutum* accessions, 3 clades form *G. barbadense* accessions, and one clade form rest of the species (Fig. 5).

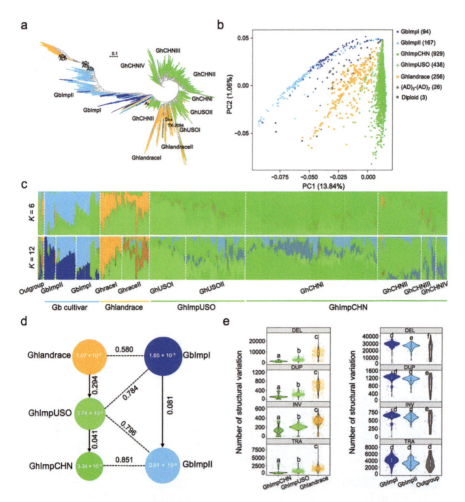

Fig. 5 Population structure and genetic diversity in *G. hirsutum* and *G. barbadense* accessions. (**a**) The unweighted neighbor-joining tree of 1,913 cotton accessions was constructed based on 20,000 random SNPs from core SNPs. The *G. tomentosum* (AD3), *G. mustelinum* (AD4), *G. darwinii* (AD5), *G. ekmanianum* (AD6), *G. stephensii* (AD7) of tetraploid species, *G. arboreum* (A2) and *G. davidsonii* (D3-d) of diploid species serve as outgroup. (**b**) Principal component analysis (PCA) plot of the first two components for all accessions. (**c**) STRUCTURE analysis of all cotton accessions with different numbers of clusters $K = 6$ and $K = 12$ ($K = 12$ is optimal value). The x-axis lists the outgroup species (gray), *G. barbadense* (blue), *G. hirsutum* landrace accessions (orange), and *G. hirsutum* improved accessions (green), respectively, and the y-axis quantifies genetic diversity in each accession. (**d**) Nucleotide diversity (π) and fixation index divergence (F_{ST}) across the five groups. (**e**) The number of deletions, duplications, inversions, and translocations in five populations (two-sided Wilcoxon rank-sum test for adjacent groups, $P < 0.001$). Each node represents one accession. In this analysis, the number of SVs was shown with the TM-1 reference genome. (Figure reproduced from Li et al. (2021a))

Further, population structure showed that *G. barbadense* accessions were branched from *G. hirsutum* landraces GhImpUSO and GhImpCHN. Gh landraces showed higher structural variations than in GhImpUSO and GhImpCHN with nucleotide diversity of 3.74×10^{-4} in GhImpUSO, 3.34×10^{-4} in GhImpCHN, 1.07×10^{-3} in landraces and 1.01×10^{-3} in *G. barbadense*. Further, domestication-related traits driven by genetic variation were investigated by allele frequency difference of nucleotide diversity by comparing each cultivar with its wild progenitor. Additionally, 31 new domestication sweep regions (DSRs) of 43.6 Mb were identified in this study by analysis of domestication selection (Li et al. 2021a, b). Differential expression was observed between wild type and cultivated lines in stress response, cell wall regulation, jasmonic acid, ethylene, and circadian clocks. Selected genes in 120 Mb with 1,006 genes in A subgenome and 2,369 genes in D subgenome of 353 gene pairs were also revealed that were not reported before. D subgenome (441) showed more genes than A subgenome (50) in domestication and improvement suggested more SNP selection signals in D subgenome. Copy number variations (CNVs) allele frequency changes with domesticated; 286 CNV regions revealed domestication selection signals with 297 Mb in A subgenome and 105 Mb in D subgenome (Fig. 6). GWAS was performed on 890 accessions of *G. hirsutum* to recognize QTLs associated with different agronomic traits. As a result, 125 QTLs of 15 agronomic traits with 4,751 candidate genes were revealed in this study in which 47 were novel QTLs and 14 were involved in domestication and improvement.

7 Domestication and Selection

Human selection has a significant role in the cotton domestication due to constant selection pressure for high yield, fiber quality, disease resistance, and abiotic stress tolerance (Nazir et al. 2020). Comparative population genomics studies suggested less variation among the cultivated genomes as human selection played a critical role in driving genetic diversity. The reason behind less genetic variation among cultivated genomes is the selection of the same cultivars as parental lines in modern breeding programs (Abdurakhmonov et al. 2008). Individual farmers started progressing in cotton breeding in the 1800s advancing to pollen manipulations and gene transformations in the 1990s. Public breeders use parental material from commercial cultivars and available public germplasm, while private breeders use in-house lines as parental lines in breeding programs for introgression of useful genetic material into elite lines. Such examples of domesticated species include two common cotton varieties of *G. hirsutum* and *G. barbadense,* which are grown worldwide and have great commercial importance (Fang et al. 2017a). *G. barbadense* and *G. hirsutum* were domesticated in Peruvian Andes and Yucantan peninsula 4,000 to 5,000 years ago (Brubaker and Wendel 1994). Studies of phylogenetic analysis using SNP data revealed *G. hirsutum* and *G. barbadense* into two divergent groups implying dual domestication process (Fang et al. 2017a) (Fig. 7). Interspecific introgression events were observed between these two species.

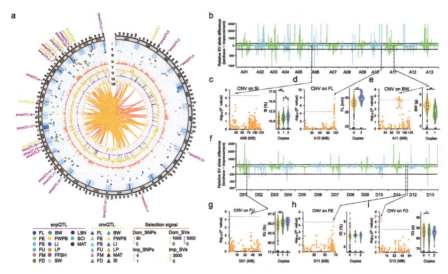

Fig. 6 Multiple-scale variation for subgenomic divergence and GWAS on agronomic traits during cotton domestication. (**a**) Circos plot showing the SNP- and SV-based selection signals and QTLs during cotton domestication and improvement. The selection region was calculated in a 1-Mb sliding window with a step size of 200 kb. I–VIII, Circos plot from outer to inter tracks showing gene density (I), snpQTLs (II), cnvQTLs (III), the ratio of nucleotide diversity (π) based on SNPs between 256 landraces and 1364 improved accessions for domestication (IV), the ratio of nucleotide diversity (π) based on SNPs between 438 GhImpUSO accessions and 929 GhImpCHN accessions for improvement (V), the relative SV allele difference in the comparisons between landrace and improved accessions (VI), and between GhImpUSO and GhImpCHN (VII). The track (VIII) represents the domesticated homologous. Upper and lower panels (VI) represent deletion and duplication variation allele difference, respectively. The snpQTLs were identified using the meta-GWAS analysis of 890 cotton accessions. The outermost circle of the circos plot purple and yellow font shows pleiotropic snpQTLs (psnpQTLs) and pleiotropic cnvQTLs (pcnvQTLs), respectively. (**b–i**) Selective signals of copy number variations (CNVs) between the A (**b**) and D (**f**) subgenome during domestication. The horizontal gray dashed lines show the domestication signal threshold with the ratio of nucleotide diversity between wild/landrace and improved cotton accessions (πlandrace/πImproved >200). (**c–e**) and (**g–i**) Six CNV-based GWAS hits that overlapped with domestication selection signals are shown for seed index (SI) (**c**), fiber length (FL) (**d**), boll weight (BW) (**e**), fiber uniformity (FU) (**g**), fiber elongation (FE) (**h**), and flowering date (FD) (**i**). The threshold of cnvQTL line was −log10P = 4.4. The violin plot showed phenotypic variation with the lead CNV genotype. The numbers in the violin plot show the number of accessions for each copy. The significance difference was calculated with two-sided Wilcoxon rank-sum test (**$P < 0.01$, *$P < 0.05$). (Figure reproduced from Li et al. (2021a))

There is evidence for parallel independent domestication in *G. barbadense* and *G. hirsutum* (Yuan et al. 2021). Other domestication and differentiation studies compared *G. hirsutum* landraces, modern cultivars, and obsolete accessions. The comparison suggested that landraces have higher differentiation than modern cultivars and obsolete accessions. More differentiation was observed in At subgenome than D subgenome (Fig. 8). Cultivated cotton species (*G. hirsutum* and *G. barbadense*) differ from wild progenitors in several traits of seed dormancy,

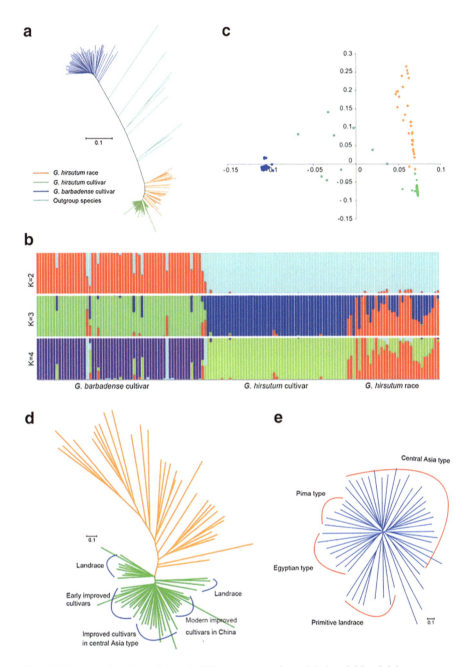

Fig. 7 Phylogenetic relationships of 147 cotton accessions. (**a**) A neighbor-joining tree was constructed using whole-genome SNP data. The cotton samples were divided into *G. hirsutum* races (*orange*), *G. hirsutum* cultivars (*green*), *G. barbadense* cultivars (*dark blue*), and outgroup species (*light blue*). (**b**) Population structure of cotton accessions determined using STRUCTURE. The accessions were divided into three groups when $K = 3$. (**c**) Principal component analysis of all cotton accessions using whole-genome SNP data. (**d**) Phylogenetic relationships between *G. hirsutum* cultivars and races. (**e**) Phylogenetic relationships between *G. barbadense* landraces and cultivars. The *scale bar* indicates the simple matching distance. (Figure reproduced from Fang et al. (2017a))

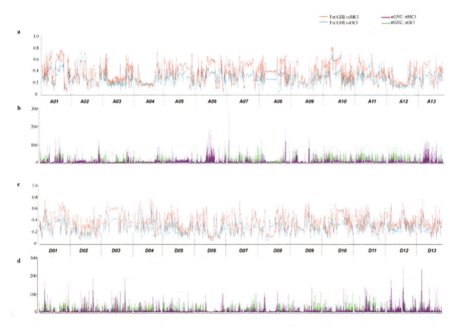

Fig. 8 Genetic differentiation and selection signals among *G. hirsutum* landraces (GHL), modern cultivars (MCl), and obsolete cultivars (OCl) (**a**) Population divergence (F_{ST}) for At subgenome (Chr A01–A13), (**b**) π_{GHl}/π_{MCl} (Purple columns) and π_{GHl}/π_{OCl} (Green columns) values (genetic diversity in *G. hirsutum* landraces (GHL) as compared to MCl and OCl for At subgenome (Chr A01–A13), (**c**) Population divergence (F_{ST}) for Dt subgenome (Chr D01–D13), (**d**) π_{GHl}/π_{MCl} (Purple columns) and π_{GHl}/π_{OCl} (Green columns) values (genetic diversity in *G. hirsutum* landraces (GHL) as compared to MCl and OCl for Dt subgenome (Chr D01–D13). (Figure reproduced from Nazir et al. (2020))

yield, plant architecture, and fiber quality (Applequist et al. 2001; Gross and Strasburg 2010). Two main clades were generated by phylogenetic relationships using maximum likelihood (ML) analysis (Yuan et al. 2021) (Fig. 9). *G. barbadense* and *G. darwinii* formed one clade, while *G. hirsutum*, *G. tomentosum*, *G. ekmanianum*, and *G. stephensii* formed the other clade with *G. mustelinum* outgroup, which was consistent with studies reported earlier for *G. hirsutum* and *G. barbadense* groups (Wendel et al. 1989; Grover et al. 2015a, b; Gallagher et al. 2017; Chen et al. 2020). Highest genetic diversity was observed in *G. mustelinum*, while lowest genetic diversity is observed in *G. stephensii*, and the cultivated species *G. barbadense* showed higher genetic diversity than *G. hirsutum*. The cultivated species genetic diversity was confined to intergenic regions.

Several technologies were used to study the genetic variation in predominant upland cotton varieties, which includes pedigree breeding, morphological markers, biochemical markers, and molecular markers (Wendel et al. 1992; May et al. 1995; Yu et al. 2012; Tyagi et al. 2014; Kuang et al. 2016; Su et al. 2016; He et al. 2019). Both Old World and New World cotton species were primarily cultivated for textile industry. Two Old World cultivated cottons (*G. arboretum* and *G. herbaceum*) are

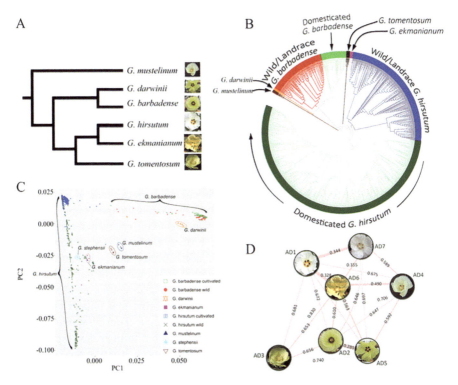

Fig. 9 (a) Expected species relationships based on previous research (b) Phylogenetic relationships among resequenced accessions (*G. stephensii* not shown). (c) PCA of SNP diversity within and among species. As expected, *G. ekmaninum* and *G. stephensii* are close to but not included within wild *G. hirsutum*. Individual species are circumscribed/bracketed; however, small numbers of disjunct accessions are found for *G. ekmaninum* (2), *G. mustelinum* (1), *G. darwinii* (1), *G. barbadense* (1), and *G. hirsutum* (2). (d) Divergence among the 7 allopolyploid cotton species. Weighted F_{st} is depicted as lines among species; species are abbreviated as follows: AD_1 (*G. hirsutum*), AD_2 (*G. barbadense*), AD_3 (*G. tomentosum*), AD_4 (*G. mustelinum*), AD_5 (*G. darwinii*), AD_6 (*G. ekmanianum*), and AD_7 (*G. stephensii*). The width of the lines scale with $1-F_{st}$. (Figures reproduced from Yuan et al. (2021))

known to under cultivation for more than 5,000 years (Chowdhury and Buth 1971; Brubaker et al. 1999). Old World cotton showed two different domestication events from two wild progenitors. Phylogenetic analysis of orthologs and divergence time studies suggested that *G. arboretum* and *G. herbaceum* shared a common ancestor 0.4 to 2.5 million years ago (Renny-Byfield et al. 2016). Structural genomics studies have demonstrated that these Old World cotton species differ by a reciprocal translocation (Gerstel 1953). The two New World domesticated cottons *G. hirsutum* (Upland from Yucatan peninsula) (Brubaker and Wendel 1994) and *G. barbadense* (Pima from Peruvian Andes) (Westengen et al. 2005) were domesticated 4,000 to 5,000 years ago (Westengen et al. 2005). These two New World tetraploid cottons evolved from Old World cotton varieties by a single hybridization event. Allopolyploid cotton species (*G. hirsutum* and *G. barbadense*) have undergone two separate domestication events (Fang et al. 2017a). Phylogenetic analysis showed that *G. hirsutum* cultivars were genetically more related to Latifolium, Richmondi, and Marie-Galante accessions, which was elucidated by GBS (genotype by sequencing) methods and SNP signatures (Zhang et al. 2019). Yucatanense, Morrilli, and Punctatum races were also known to be genetically related to *G. hirsutum* by origin in addition to above races. MethylC-seq between domesticated cotton and its wild relatives showed 519 differentially methylated genes associated with flowering time and seed dormancy (Song et al. 2017). An independent study using the same MethylC-seq technology found 4,248 and 1,164 differentially methylated regions between wild and domesticated soybean (Shen et al. 2018).

Upland cotton, *G. hirsutum*, is grown worldwide with an estimate of 95% of the world's cotton production (Nazir et al. 2020). New World upland cotton cultivars have originated from *G. hirsutum var. yucatanense* (wild type) and experienced thousands of years of artificial selection. *G. hirsutum* species has both wild and domesticated lines. Upland cotton has both domesticated and wild populations within the species. Cultivated/domesticated cotton is known to be self-pollinated, but pollen dispersal studies supported that domesticated cotton is outcrossing (Velázquez-López et al. 2018). Hence, this study reported that upland cotton as mixed mating system using classic reproductive biology. Transcriptome studies revealed that domestication has changed the genes related to fiber development of upland cotton when compared to wild type cotton (Rapp et al. 2010). A total of 9,465 genes were differentially expressed in cultivated cotton relative to wild type (Gross and Strasburg 2010). High yields with good cotton fiber quality of modern upland cultivar were mediated by human selection by relative comparison to wild cultivars. A study on transcriptomes of wild and domesticated cotton cultivars reported that there are gene expression changes at different developmental stages of wild and domesticated cotton fiber leading to difference in cotton fiber development (Rapp et al. 2010). The gene expression changes were observed in approximately 12,626 genes in domesticated cultivar as opposed to 5,273 genes in wild cultivar by comparing cumulative data of all developmental stages and suggestive role of these genes in superior cotton fiber quality in cultivated cotton. Research advances on genetics of artificial selection were made mostly by artificial domestication methods, population genetics, archaeobotany, quantitative genetics, and molecular genetics (Gross and Strasburg 2010). Comparative population genomics studies provided novel tools for genetic dissection of domestication and role of human selection. Studies in these lines have discovered genome-wide associations for fiber yield, quality, and other structural traits (Nazir et al. 2020). Additionally, this study discovered that domestication and improvement were related to 93 distinct regions and 311 selection sweeps. This study also revealed that domestication improved the fiber quality in modern cultivars by increasing gene expression of SCW (secondary cell wall development) genes, but compromised the stress pathway by the downregulation of stress response genes. Studies in cotton employed SNP microarray technologies for three stages of seed trichome (fiber) development in *G. hirsutum* and *G. barbadense* to understand their genetic differences (Hovav et al.

2008). This study suggested that human selection made these species more similar instead of parallel genetic and phenotypic changes. GWAS in cotton identified more loci related to lint yield than fiber quality suggesting stronger selection pressure for lint traits.

8 Cotton Population Epigenomics

Epigenetics is a heritable mechanism by DNA methylation, RNA methylation, histone modifications, and chromatin assembly with no change in the DNA sequence (Moler et al. 2019). Epialleles or alleles created by epigenetic modification are localized variations in methylation pattern between individuals in a population, resulting in a phenotype change (Weigel and Colot 2012; Guo et al. 2015). The role of DNA methylation changes in cotton phenotype differences and development were analyzed in different cotton genotypes and tissues (Osabe et al. 2014). Methylation differences were higher than the genetic differences. The methylation differences in the tissues also suggested significantly greater methylation rates especially in fiber tissue suggesting epigenetic regulation in cotton and fiber development. Epigenetic and functional analysis of allotetraploid, tetraploid, and diploid cotton methylated cytosine's revealed that some of these were related to domestication traits such as flowering time and seed dormancy (Song et al. 2017). *COL2* (CONSTANS(CO)- LIKE) regulates photoperiodicity in allotetraploid cotton. Absence of methylation of *COL2* promotes its expression and flowering while methylation silences *COL2* and delays flowering in cultivated cotton. They noticed the fact that the rate of evolution for methylated genes is higher than the non-methylated ones and identified that the cotton homolog of the *Arabidopsis* photoperiod regulating gene CO-LIKE (COL) known as COL2 is an epiallele in allotetraploid cottons. It was discovered that loss of hypermethylation of COL2D causes significant increase in expression in cotton domesticated varieties compared to its wild relatives, where it is either silenced or repressed (Song et al. 2017). Polyploidy is a common mechanism of evolution in flowering plants that result in genetic/ epigenetic changes including DNA methylation that in turn affects gene expression and phenotype (Chen 2007; Otto 2007). Hybridization and domestication were showed to be responsible for gene expression changes by epigenomic profiling studies of diploid, allopolyploid, and domesticated cotton lines (Jackson 2017; Song et al. 2017). Polyploid evolution preserved the hybridization-mediated methylation changes in cotton. Studies using different cotton genotypes and different tissues suggest that CHG methylation was more divergent than CG methylation (Osabe et al. 2014). Such methylation changes in fiber and other tissue suggest epigenetic regulation of cotton fiber development. Methylome was generated to understand the cotton epigenetic changes for wild and domesticated lines to provide tools for improving cotton by breeding (Song et al. 2017). Several studies have shown differential methylation of DNA is associated with the domestication-related traits (Song et al. 2017; Shen et al. 2018).

9 Genome-Wide Association Studies

GWAS have discovered genetic variation and genes involved in plant structural traits (Su et al. 2018). GWAS facilitate QTL detection for complex traits of plants with advantages of high resolution, low-cost method with no requirement for mapping population as opposed to bi-parental linkage mapping (Saïdou et al. 2014). GWAS approach overcomes the drawbacks of genetic mapping and has been used to identify several genes and QTLs by reaching out to the level of genes (Brachi et al. 2011). Extensive usage of genome-wide association studies (GWAS) in different cotton populations has identified various quantitative trait loci (QTL) associated with seed quality, morphology, fiber quality, yield, abiotic stress, and disease resistance traits (Sun et al. 2012; Said et al. 2013, 2015; Nie et al. 2016; Baytar et al. 2018; Sun et al. 2019; Li et al. 2020b). Till 2015, a minimum of 1,000 fiber quality QTLs spread across 26 chromosomes were identified in cotton using mapping (Said et al. 2013, 2015). In upland cotton, GWAS revealed genetic diversity and identified candidate genes involved in drought stress traits (Hou et al. 2018). In cotton, GWAS identified 9 QTLs related to salt and drought tolerance suggesting a common genetic mechanism or cross link for both ST (salt tolerance) and DT (drought tolerance). This study identified 4 QTLs on chromosome A13, 3 QTL on A01 for DT while 4 QTL on D08 and 3 QTL on A11 for ST (Abdelraheem et al. 2021). Other GWAS in cotton also revealed 163 QTLs for 15 agronomic traits and early maturity candidate gene (Shen et al. 2019). GWAS in upland cotton identified genes associated with early maturity (Li et al. 2021b). Genetic variation in lint yield traits associated with salt stress was also revealed by GWAS (Zhu et al. 2020). GWAS in combination with phenomics in cotton identified 390 genetic loci for drought tolerance (Li et al. 2020a). These GWAS were also reported in verticillium wilt resistance (Li et al. 2017b). These studies revealed the relationship of genomics with phenomics which can be used for trait improvement in cotton by genomic and molecular breeding.

SNP markers were used as a primary source for genetic dissection of traits through QTL mapping and GWAS. The presence of less genetic diversity or narrow genetic base composition of cotton genome made single nucleotide polymorphism (SNP) studies difficult (Byers et al. 2012). Advances in high throughput sequencing provided means to identify SNPs in the cotton genome (Udall et al. 2006). SNPs were used to study diversity, characterization, and mapping for several traits in cotton (An et al. 2008; Van Deynze et al. 2009). Large SNP data sets were developed in cotton using available reference genomes of *G. raimondii* (Paterson et al. 2012; Wang et al. 2012) and *G. arboreum* (Li et al. 2014). Cotton SNP63K array was developed to identify SNPs in cultivated species of *Gossypium hirsutum* and crosses of *G. hirsutum* (Hulse-Kemp et al. 2015). The 80K SNP array was used for genotyping different accessions of cotton (Cai et al. 2017). Several QTLs related to yield, fiber quality, boll size, biotic and abiotic stress-related traits were discovered using SNPs in cotton (Majeed et al. 2019). In cotton, SNPs were used for genetic map construction (Byers et al. 2012), genome-wide association studies

(GWAS) for genome-wide haplotypes (Yano et al. 2016), evolution studies, and exploring genetic diversity (Morin et al. 2004; Huang et al. 2017). Further, SNP markers used in phylogenetic study of cotton diploid genomes revealed that the A_1 genome and A_2 genome are 98% similar (Shaheen et al. 2016). One such example is use of CottonSNP63K array that was successful in identifying SNP markers in large size genome variation of cultivated species. Use of SNP markers reported 35 drought-associated QTLs, 150 QTLs associated for cotton productivity and fiber quality trait in cotton (Zheng et al. 2016; Ulloa et al. 2020). Further, 20 SNPs associated with drought tolerance have been identified using GWAS and RNA-Seq integrated data in cotton (Hou et al. 2018). Three SNPs located on Chromosome D11 tightly associated with fiber length. When compared to the reference TM-1 genome, they identified a unique and shorter fiber associated haplotype (D11-alt) in 10% of RILs. They have also identified 949 differentially expressed genes between these two classes of haplotype. Ultimately, they discovered 12 genes with nonsynonymous SNPs (nsSNPs) significantly associated with long fiber trait (Naoumkina et al. 2019). In multiple other studies done in last few years, scientists have discovered an astonishing number of SNP markers, haplotypes, QTLs, and candidate genes associated with yield and fiber quality in cotton cultivars (Liu et al. 2018; Naoumkina et al. 2019).

Earlier quantitative trait studies in cotton were mainly focused on traditional molecular markers such as RFLP, AFLP, and SSRs (Huang et al. 2017), and the discovery of SSRs/microsatellites helped in the final scoring of heterozygous and homozygous loci (Morgante and Olivieri 1993). Genetic diversity and relatedness extent in cotton was studied using AFLP markers (Pillay and Myers 1999; Murtaza 2006). This method is well known for differentiation of inter- and intraspecific cotton cultivars with potential application in marker-assisted selection. AFLP-based genetic diversity estimates were used for 20 cultivars of *G. hirsutum. L.,* and *G. arboreum.* L. (Murtaza 2006). AFLP markers in cotton were shown to link fiber-related and agronomic traits in which fewer traits were associated to fiber than agronomic traits. The study revealed 50 AFLP markers associated with seven traits which can be used in marker-assisted selection (MAS) to develop cotton cultivars with improved traits (Wu et al. 2007). Several modifications of AFLP were reported in cotton in which AFLP is anchored to ATG for genome-wide analysis (Lu et al. 2008), use of restriction enzymes in cleavage AFLP (cAFLp) to understand genetic relationships among four genotypes from two cultivated cotton species *G. hirsutum* and *G. barbadense* (Zhang et al. 2005a). cAFLP overcomes the limitation of low level of AFLP markers in cotton that hinders use of AFLP in linkage mapping. cAFLP incorporates restriction enzymes for cutting AFLPs, and cAFLP generated 64% more polymorphism in upland cotton and 132% more in Pima cotton relative to AFLPs. Sixty upland cotton genotypes were tested using AFLP markers for studying molecular diversity (Badigannavar et al. 2012), RGA-AFLPs for mapping disease resistance markers in cotton (Niu et al. 2011). AFLP markers were also marked to different chromosomes in upland cotton. Additionally, AFLP markers were used to study root knot nematode (RKN)-resistant cotton cultivars (Niu et al. 2007). AFLP markers were used in combination with RAPD markers in cotton mapping studies

(Malik et al. 2014). Genetic diversity (Abdalla et al. 2001; Rana and Bhat 2004) and map saturation (Lacape et al. 2003; Zhang et al. 2005b) in cotton were also studied using AFLP technology.

Recent advancement in capillary electrophoresis and peak identification software technologies also made it suitable for high-throughput SSR-based genotyping applications (Liang et al. 2018). Hence, these markers were chosen as common markers in the population genomics studies. Recent usage of SNP arrays and sequencing technologies promote QTL mapping precision and genetic mapping resolution for fine mapping. Genome resequencing and use of other technologies will identify novel SNPs in cotton, which will help in advancing cotton genomic technologies by identifying genetic variation within the species and improving molecular breeding. Resequencing studies using upland and Pima cotton revealed structural variations of genotype associated with phenotype with possible future application in crop improvement (Ma et al. 2021). Good fiber quality is one of the most desirable and marketable traits in cotton, and improving fiber quality without compromising yield is one of the major goals for cotton breeders and farmers. But this has been a challenge for the breeders due loss of genetic diversity in current cotton cultivars during the process of domestication and the linkage drag between yield and fiber quality traits.

GWAS can significantly improve the chances of identifying pleiotropic loci to improve both the yield and fiber quality together. In 2017, Sun et al. developed a Cotton 63K Illumina Infinium SNP array for identification of SNPs associated with fiber quality traits (Sun et al. 2017). In a population of 719 upland cotton accessions, they have discovered 10511 SNPs covering 26 chromosomes, among which 46 were found to strongly associated with fiber quality. These 46 SNPs were distributed over 15 chromosomes involving 612 candidate genes involved in signal transduction, polysaccharide biosynthesis, and protein translocation. The SNPs were grouped into two main fiber length- and strength-related haplotypes Dt11 and At07 (Sun et al. 2017). Dong et al. in 2019 identified (Dong et al. 2019) 23 fiber quality associated identified single nucleotide polymorphisms (SNPs) using high-density CottonSNP80K array based GWAS in a population of 408 cotton accessions. They were able to predict 128 candidate genes including two major fiber quality–associated loci (GR1 and GR2) on chromosomes A07 and A13. They were able to put functional annotation to 22 out of these 128 genes, and 11 of these genes showed differential expression during fiber development process. In another study done in the same year, Thyssen et al. in 2021 used a multi-parent advanced generation intercross (MAGIC) population created by random mating to generate a cotton genetic pool that can be used QTL mapping and candidate gene identification (Abdelraheem et al. 2021). From 11 different cultivars, they were able to create a population of 550 recombinant inbred lines (RILs) and used it to identify QTLs and candidate genes associated with fiber length. Their GWAS analysis led to the discovery of single nucleotide polymorphisms (Wang et al. 2017). One of the studies using genome-wide association studies in cotton revealed genetic variation in lint yield under salty conditions (Zhu et al. 2020). Further, studies also revealed that the number of genomic loci linked to the lint yield were greater than the genomic loci determining the fiber quality and also showed the involvement of ethylene genes in lint yield (Fang et al. 2017c). Before the genomes of *Gossypium* species were sequenced, linkage mapping studies identified about 1000 QTLS associated 26 chromosomes, which were found to be fiber related (Liu et al. 2020; Zhu et al. 2020). One of the African wild cotton species *G. longicalyx* has reniform nematode immunity which was targeted to transfer to the cultivated cotton by breeding; however, it was not successful (Grover et al. 2020). Further, genomics studies are required to understand the mechanisms governing this trait.

With the discovery and use of clustered regularly interspaced short palindromic repeats (CRISPER)/cas9 technology which can be applied for DNA deletions / insertions of specific DNA loci on the genome for trait improvement, application of this technology for improving both yield and fiber quality together in upland cotton (*Gossypium hirsutum* L.) is of utmost interest. However, the idea is hampered due to the presence of linkage drag between markers associated with two traits. To avoid this, an alternative approach of identifying pleotropic loci using GWAS has been adopted. In a recent study (Wang et al. 2021), 316 cotton accessions *were genotyped by* restriction-site-associated DNA sequencing (RAD-Seq). Linkage disequilibrium-based analysis using 231 SNP markers with eight traits under nine environments has identified pleotropic loci containing six candidate genes, which can potentially improve yield and fiber quality together. In upland cotton, RAD technology is applied to identify 21,109 SNPs present in both parents that can be used for recombinant line genotyping (Wang et al. 2015). RAD-Seq technology identified distinct genomic regions on chromosome A06 and A08 among different subgroups of in upland cotton lines (He et al. 2020). RAD sequencing of parental lines generated markers for the genetic map that can be targeted for molecular marker-assisted selection (Wang et al. 2015). Double digest restriction-site-associated DNA sequencing (ddRADseq) is applied to address major cotton pest issues such as boll weevil and flea hopper using population genomics (Raszick et al. 2021).

10 Genomic Selection

Plant breeding uses genomic selection (GS) for efficient selection, to reduce cost and breeding time (Gapare et al. 2018). In GS, a completely phenotyped and genotyped training population and another population that was genotyped but not phenotyped are used to estimate genomic estimate breeding value (GEBV) (Meuwissen et al. 2001). Limited availability of phenotypic data promoted the use of genomic selection by efficient use of markers of complex traits in breeding (Islam et al. 2020). Five different statistical methods, namely, genomic BLUP (GBLUP), ridge regression BLUP (rrBLUP), BayesB, Bayesian LASSO, reproducing kernel Hilbert spaces (RKHS), were five GS methods employed to estimate GEBV for fiber quality traits in upland cotton breeding. BayesB method estimated the most accurate GEBV value and suggested to be an accurate GS method. The BayesB method was also used for GS methods of traits of fiber length and strength in upland cotton (Gapare et al.

2018). Genomic selection was used in cotton for fiber quality and yield (Katageri et al. 2020) and has huge potential in cotton crop improvement. For several years, cotton research was using quantitative genetics approaches for determining genetic parameters for different agronomic traits including cotton fiber trait, which are useful for pedigree selection and inbreeding based system. Billings et al. (2022) tested phenotypic and genotypic data in upland cotton for genomic selection in cotton breeding using GWAS for 21 agronomic traits of fiber, seed composition, and yield (Billings et al. 2022). Results showed that genomic selection prediction ability is lower for other traits but higher for fiber quality.

11 Conservation and Sustainable Management and Use of Genetic Resources

Collection and utilization of plant genetic resources is as old as human civilization with the selection of wild cultivars to grow for specific purposes, also known as domestication. With the continuous migration, evolution, and adaptation of the humans across the globe, the plants domesticated by humans have adapted to the new environments by accumulating traits such as pathogen resistance and tolerance to drought, cold, and flooding. The broad and well-organized collection of plant genetic resources of agriculturally important germplasm preservation started at the beginning of the twentieth century to preserve the genetic resources. Preservation of plant germplasm has become increasingly important due to endangerment of genetic resources from rapidly growing human/animal population and adverse climatic conditions. Germplasm collection stores genes and genotypes which forms a critical part of natural resources (Campbell et al. 2010). The first known cotton collection was established by J.P.B. von Rohr in the eighteenth century who maintained Caribbean and South American cotton varieties (Rohr 1791). Efforts to preserve world's cotton genetic resources increased in the last 100 years. Available cotton genetic resources were divided into three pools, namely, primary, secondary, and tertiary pools. The first pool contains 5 tetraploids, the second pool contains 20 diploids, and the third pool contains 25 diploids. This report shows that most of the classified *Gossypium* species were not available in the eight germplasm collections in the world and hence suggested that they are not conserved or close to extinction (Campbell et al. 2010). In the nineteenth century, Parlatore and Todaro collected cotton germplasm (Todaro 1877), and a British company, previously called Empire Cotton Growing Company, was the first known major contributor of cotton germplasm collection. This company has grown cotton varieties exterior to the United States (Hutchinson et al. 1948). In the twentieth century, the collection efforts have increased in Central America and northern South America to find resistant germplasm against boll weevil crisis. Major cotton germplasm collection centers are available in Australia, Brazil, China, France, India, Russia, the United States, and Uzbekistan. In Australia, CSIRO (Commonwealth Scientific and Industrial Research

Organization) and ATGGC (Australian Tropical Grain Germplasm Center) maintain cotton germplasm. Seed lots were collected for each accession by CSIRO, which were renewed every 10 years. Different US resistant germplasms were crossed with Australian cultivars to generate bacterial blight–resistant cultivars. The Australian cultivars were also crossed with Indian and China cotton accession to generate *Fusarium* wilt resistance in Australian accessions. ATGGC maintains native *Gossypium* collections, and 90% collection is *G. hirsutum* and 10% comprises of 4 tetraploids and other diploid species. In Brazil, Brazil Agricultural Research Company maintains the germplasm collections at National Center for Genetic Resources and Biotechnology. Consultative Group on International Agriculture Research (CGIAR) developed International Plant Genetic Resource Institute (IPGRI) in 1974 and established a National Center for Genetic Resources and Biotechnology for cotton germplasm collections in Brazil. This collection contains several cotton accessions covering 39% *G. hirsutum* and 35% *G. barbadense* with rest of the collection of 28 diploids and 3 tetraploids.

In China, Chinese Crop Germplasm Information System (CGRIS) is a freely available largest resource of cotton germplasm collection, and CIRAD, French Agricultural Research Center for International Development, collects the germplasm collections of cotton (5 tetraploids and 27 diploid species) and involves research in France. In India, Central Institute for Cotton Research (CICR) and National Bureau for Plant Genetic Resources (NBPGR) supported by Indian Council of Agricultural Research (ICAR) maintain cotton germplasm collections. These accessions include *G. hirsutum, G. barbadense, G. herbaceum, G. arboretum, and G. arboretum* along with 26 wild species and 32 introgressed cultivars. In Russia, Union of Soviet Socialist Republics (USSR) collected and maintained cotton germplasm resources, and N.I. Vavilov Research Institute of Plant Industry (VIR) collection is the current resource for cotton germplasm in Russia with several accessions of 24 diploids, 3 tetraploids, other diploids, and tetraploid species. In the United States, several Cotton Germplasm Collection Centers were established, namely, National Seed storage currently called as National Center for Genetic Resources and Preservation (NCGRP) for germplasm storage of cotton. National collection of *Gossypium* contains 10,000 accessions covering 45 *Gossypium* species. Further, the germplasm is divided into 7 sections, namely, variety collection, primitive landrace collection, *G. barbadense* collection, A genome species collection, wild species collection, genetic marker collection, and base collection, and USDA-ARS maintains first 5 pools. Cotton accessions data, kept in National Germplasm Resources Information Network (GRIN), is available online and accessible to worldwide scientists. In Uzbekistan, Cotton Breeding Institute of Agriculture Ministry of Uzbekistan, Institute of Genetics and Plant Experimental Biology at the Academy of Sciences, and National University of Uzbekistan are three centers for cotton germplasm collection. Therefore, several national collection centers exist to maintain cotton germplasm around the world for preserving different cotton accessions.

There was no storage of original cotton cultivars until 1880 when there was a boll weevil epidemic. Two cotton cultivars Acala and Kekchi were introduced from germplasm collection in years 1902 and 1906. The National Seed Storage Lab, now known as NCGRP, was established in 1958 for long-term access for germplasm collection and preservation (Percy et al. 2014). Diverse germplasm collections were initiated in 1950 by S1, the first regional research project. In 1960, these germplasm collections were extended to National Center for Genetic Resources and Preservation (NGCRP), which were integrated at College Station, Texas, in 1980. International Board of Plant Genetic Resources currently known as Bioversity International provided a guide for cotton germplasm collections from 1980. Germplasm Resources and Information Network (GRIN) manages the databases of National Plant Germplasm System (NPGS) online. National Cotton Germplasm Collection (NCGC) is part of US National Plant Germplasm Collection that collects accessions of cultivated and wild lines. The data is collected by Cottongen.org. GRIN-global and Cottongen are the online databases for several accessions information in NCGC. For research investigations by scientists, US National Cotton Germplasm Collection forms a platform to access these genetic resources for genetic improvement studies.

In cotton, availability of integrated web database, microsatellite database, and comparative QTL resources helped in advancing *Gossypium* selection for several gene pools (Mehboob ur et al. 2012). Absence of genetic diversity is the major limitation in cotton domestication in which *G. hirsutum* cultivars showed less genetic diversity than *G. barbadense* cultivars because of the introgression of several alleles of *G. hirsutum* into *G. barbadense*. Approximately 26 linkage maps were generated using genomic resources such as 16,162 SSRs, 312 mapped RFLPs with SSRs, RFLP, AFLP, and RAPD markers. Association of DNA markers with 29 QTLs were reported in cotton related to fiber quality, disease resistance, and other plant development traits. About 432 QTLs were mapped to high-density map of 3475 loci in studies using comparative mapping. Transformation methods were also employed in cotton and contains genes such as cry genes, namely, cry1Ac and cry1Ab in addition to herbicide resistance genes. Crispr cas9 studies were emerging in cotton to correct disease-susceptible cultivars and yield-reducing traits (Mubarik et al. 2020). As higher genetic diversity helps in surviving in challenging environments, lower genetic diversity in cotton is leading to genetic erosion. Identification, collection, and use of wild and land races of *Gossypium* into the breeding program will lead the cotton future (Boopathi and Hoffmann 2016). Available genomic sequence resources and functional genomics tools combined with evolution and population structure will help in effective use of natural variation in cotton and storage of such germplasm resources. PGRs were known for such storage of plant germplasms (Wambugu et al. 2018). Hence, population genomics studies can help in investigating genetic diversity in wild progenitors, landraces, and cultivated cultivars in *Gossypium* which then can be collected, stored, and preserved for targeting future use in cotton genetic breeding and crop improvement.

12 Future Perspectives

Comprehensive genome information of cotton diploid and tetraploid species is available through whole-genome sequencing and resequencing which is advancing cotton genomics (Paterson et al. 2012; Wang et al. 2012; Li et al. 2014, 2015; Liu et al. 2015; Yuan et al. 2015; Zhang et al. 2015b; Fang et al. 2017a, c). Availability of transcriptome and molecular markers has facilitated the study of cotton evolutionary history and domestication at the molecular level. In addition, these genome biology studies enabled us to investigate agronomically essential traits in cotton. As cotton crop is a natural textile fiber-producing crop, fiber-related traits such as development and quality were explored in depth using genomic and transcriptomic analysis (Hu et al. 2019; Wang et al. 2019; Yang et al. 2019). However, advancing the use of cotton genomics for breeding is currently hampered due to the presence of gaps in the available genomes, which need to be improved for practical applications. In addition, several studies have focused on genetic diversity; however, integration of genetic/genomic data with phenotypic data is a limiting factor for integration of genomics with phenomics in cotton. Hence, there is a critical need to focus on filling the gap between genotype and phenotype under various environmental conditions. New phenomic platforms must be developed to meet these needs for cotton phenotyping due to its uniqueness of canopy production and boll positioning.

Recent advances in genomics, transcriptomics, and epigenetics tools can help to discover ancient events in the structuration of cotton domestication (Kantar et al. 2017). Technological advancement can also help us to discover gene and protein expression changes as a result of domestication (Hekman et al. 2015; Jiang et al. 2019), validate the domestication-associated candidate genes using functional genomics (Zhou et al. 2020), assess the effect of epialleles (Jensen 2015), and study the still unexplored area of chromatin structural alterations (Concia et al. 2020). The challenges in analyzing domestication genomics of large polyploid species can be addressed with the latest cost-effective next-generation sequencing technologies. Thus, it can open the door to the still undiscovered genetic variation and selection sweeps associated with domestication. In addition to the omics technologies system, biology-based approach can also shed light on the structure of evolutionary events (Piperno 2017). Integration of genomics, transcriptomics, proteomics, genotyping, phenotyping, genome editing, domestication, genomic selection will help in both fundamental and applied cotton research for crop improvement and genetic resource management and conservation.

13 Conclusions

Cotton is a unique crop, which has tremendous scientific and economic significance in agricultural industry. Cotton can be used to study the domestication, polyploidy, and phylogenetic analyses because of the diversity and availability of large number of genomes in *Gossypium* species. The genome of *Gossypium* is complex, and advances in genomic research were slow because of large polyploid genome. Advances in new technologies such as next-generation sequencing, high throughput genotyping, whole-genome/transcriptome assembly, comparative genomics, genome-wide association studies and sequence-based markers provided insights into the genome-wide variation related to cotton evolution and domestication. Cotton genome sequencing laid the foundation for population genomics studies; however, complex nature of the cultivated cotton genome impeded the study of centromeres, highly repetitive regions, 5'end and 3' ends of chromosomes. Thus, there will be misinterpretation on the data at this point. Advances in the long-read sequencing technologies and algorithms will improve the assembly and quality of the genomes. Sequencing and resequencing of more cotton lines coupled with structural/functional genomics studies will improve our understanding of fiber- and oil-related traits, disease/pest resistance, plant structure, stomatal conductance, and abiotic stress tolerance. Despite the commercial and scientific significance, cotton functional genomics is lagging. However, the genome sequence is available, and the difficulty and length involved in plant transformation and long-life cycle are hampering the functional characterization of individual genes. In addition, the cotton commercial lines are not transformable; hence, it is important to develop transformation technologies that are genotype independent, which can advance comparative as well as population genomics. Available information on the genome changes, sequences, and marker differences in traits can be studied in detail in response to different environments to fill such gaps and improve our understanding of economically important traits associated with cotton. Further, the commercial traits such as fiber development and oil biosynthesis need to be investigated using population genomics under different environments to understand the improvement caused by domestication and gene evolution. Overall, cotton is a promising crop with significant commercial and scientific importance, and population genomics techniques will help in making significant progress in fundamental and applied research in cotton.

References

Abdalla AM, Reddy OUK, El-Zik KM, Pepper AE. Genetic diversity and relationships of diploid and tetraploid cottons revealed using AFLP. Theor Appl Genet. 2001;102(2):222–9.

Abdel-Ghany SE, Hamilton M, Jacobi JL, Ngam P, Devitt N, Schilkey F, et al. A survey of the sorghum transcriptome using single-molecule long reads. Nat Commun. 2016;7:11706.

Abdelraheem A, Thyssen GN, Fang DD, Jenkins JN, McCarty JC, Wedegaertner T, et al. GWAS reveals consistent QTL for drought and salt tolerance in a MAGIC population of 550 lines derived from intermating of 11 Upland cotton (Gossypium hirsutum) parents. Mol Genet Genomics. 2021;296(1):119–29.

Abdurakhmonov IY, Kohel RJ, Yu JZ, Pepper AE, Abdullaev AA, Kushanov FN, et al. Molecular diversity and association mapping of fiber quality traits in exotic G. hirsutum L. germplasm. Genomics. 2008;92(6):478–87.

Adams KL. Evolution of duplicate gene expression in polyploid and hybrid plants. J Hered. 2007;98(2):136–41.

Allendorf FW. Genetics and the conservation of natural populations: allozymes to genomes. Mol Ecol. 2017;26(2):420–30.

An C, Saha S, Jenkins JN, Ma D-P, Scheffler BE, Kohel RJ, et al. Cotton (Gossypium spp.) R2R3-MYB transcription factors SNP identification, phylogenomic characterization, chromosome localization, and linkage mapping. Theor Appl Genet. 2008;116(7):1015–26.

Appels R, Eversole K, Stein N, Feuillet C, Keller B, Rogers J, et al. Shifting the limits in wheat research and breeding using a fully annotated reference genome. Science. 2018;361(6403): eaar7191.

Applequist WL, Cronn R, Wendel JF. Comparative development of fiber in wild and cultivated cotton. Evol Dev. 2001;3(1):3–17.

Au KF, Sebastiano V, Afshar PT, Durruthy JD, Lee L, Williams BA, et al. Characterization of the human ESC transcriptome by hybrid sequencing. Proc Natl Acad Sci U S A. 2013;110(50): E4821–30.

Badigannavar A, Myers GO, Jones DC. Molecular diversity revealed by AFLP markers in upland cotton genotypes. J Crop Improv. 2012;26(5):627–40.

Bardak A, Bhatti K, Erdogan O, Mahmood Z, Khan N-U-I, Iqbal M, et al. Genetic mapping in cotton. In: Past, present and future trends in cotton breeding. IntechOpen, London; 2018.

Batut P, Gingeras TR. RAMPAGE: promoter activity profiling by paired-end sequencing of 5′-complete cDNAs. Curr Protoc Mol Biol. 2013;104(1):25B.11.21–25B.11.16.

Bayer PE, Golicz AA, Scheben A, Batley J, Edwards D. Plant pan-genomes are the new reference. Nat Plants. 2020;6(8):914–20.

Baytar AA, Peynircioğlu C, Sezener V, Basal H, Frary A, Frary A, et al. Genome-wide association mapping of yield components and drought tolerance-related traits in cotton. Mol Breed. 2018;38 (6):74.

Billings GT, Jones MA, Rustgi S, Bridges WC Jr, Holland JB, Hulse-Kemp AM, et al. Outlook for implementation of genomics-based selection in public cotton breeding programs. Plants (Basel). 2022;11(11):1446.

Blenda A, Fang DD, Rami J-F, Garsmeur O, Luo F, Lacape J-M. A high density consensus genetic map of tetraploid cotton that integrates multiple component maps through molecular marker redundancy check. PLoS One. 2012;7(9):e45739.

Boopathi NM, Hoffmann LV. Genetic diversity, erosion, and population structure in cotton genetic resources. In: Ahuja MR, Jain SM, editors. Genetic diversity and erosion in plants: case histories. Cham: Springer International Publishing; 2016. p. 409–38.

Bourke PM, Voorrips RE, Visser RGF, Maliepaard C. Tools for genetic studies in experimental populations of polyploids. Front Plant Sci. 2018;9:513.

Bowman MJ, Park W, Bauer PJ, Udall JA, Page JT, Raney J, et al. RNA-Seq transcriptome profiling of upland cotton (Gossypium hirsutum L.) root tissue under water-deficit stress. PLoS One. 2013;8(12):e82634.

Brachi B, Morris GP, Borevitz JO. Genome-wide association studies in plants: the missing heritability is in the field. Genome Biol. 2011;12(10):232.

Brubaker CL, Wendel JF. Reevaluating the origin of domesticated cotton (Gossypium hirsutum; Malvaceae) using nuclear restriction fragment length polymorphisms (RFLPs). Am J Bot. 1994;81(10):1309–26.

Brubaker CL, Paterson AH, Wendel JF. Comparative genetic mapping of allotetraploid cotton and its diploid progenitors. Genome. 1999;42(2):184–203.

Byers RL, Harker DB, Yourstone SM, Maughan PJ, Udall JA. Development and mapping of SNP assays in allotetraploid cotton. Theor Appl Genet. 2012;124(7):1201–14.

Cai C, Zhu G, Zhang T, Guo W. High-density 80 K SNP array is a powerful tool for genotyping G. hirsutum accessions and genome analysis. BMC Genomics. 2017;18(1):654.

Cai Y, Cai X, Wang Q, Wang P, Zhang Y, Cai C, et al. Genome sequencing of the Australian wild diploid species Gossypium australe highlights disease resistance and delayed gland morphogenesis. Plant Biotechnol J. 2020;18(3):814–28.

Campbell T, Saha S, Percy R, Frelichowski J, Jenkins J, Park W, et al. Status of the global cotton germplasm resources. Crop Sci. 2010;50:1161.

Cao D, Gao X, Liu J, Kimatu JN, Geng S, Wang X, et al. Methylation sensitive amplified polymorphism (MSAP) reveals that alkali stress triggers more DNA hypomethylation levels in cotton (Gossypium hirsutum L.) roots than salt stress. Afr J Biotechnol. 2011;10(82): 18971–80.

Chee PW, Draye X, Jiang CX, Decanini L, Delmonte TA, Bredhauer R, et al. Molecular dissection of phenotypic variation between Gossypium hirsutum and Gossypium barbadense (cotton) by a backcross-self approach: III. Fiber length. Theor Appl Genet. 2005;111(4):772–81.

Chen ZJ. Genetic and epigenetic mechanisms for gene expression and phenotypic variation in plant polyploids. Annu Rev Plant Biol. 2007;58(1):377–406.

Chen ZJ, Scheffler BE, Dennis E, Triplett BA, Zhang T, Guo W, et al. Toward sequencing cotton (Gossypium) genomes. Plant Physiol. 2007;145(4):1303–10.

Chen X, Jin X, Li X, Lin Z. Genetic mapping and comparative expression analysis of transcription factors in cotton. PLoS One. 2015;10(5):e0126150.

Chen Z, Feng K, Grover CE, Li P, Liu F, Wang Y, et al. Chloroplast DNA structural variation, phylogeny, and age of divergence among diploid cotton species. PLoS One. 2016;11(6): e0157183.

Chen ZJ, Sreedasyam A, Ando A, Song Q, De Santiago LM, Hulse-Kemp AM, et al. Genomic diversifications of five Gossypium allopolyploid species and their impact on cotton improvement. Nat Genet. 2020;52(5):525–33.

Chowdhury KA, Buth GM. Cotton seeds from the Neolithic in Egyptian Nubia and the origin of Old World cotton. Biol J Linn Soc. 1971;3(4):303–12.

Concia L, Veluchamy A, Ramirez-Prado JS, Martin-Ramirez A, Huang Y, Perez M, et al. Wheat chromatin architecture is organized in genome territories and transcription factories. Genome Biol. 2020;21(1):104.

Cronn RC, Small RL, Haselkorn T, Wendel JF. Rapid diversification of the cotton genus (Gossypium: Malvaceae) revealed by analysis of sixteen nuclear and chloroplast genes. Am J Bot. 2002;89(4):707–25.

Cronn R, Small RL, Haselkorn T, Wendel JF. Cryptic repeated genomic recombination during speciation in gossypium gossypioides. Evolution. 2003;57(11):2475–89.

Della Coletta R, Qiu Y, Ou S, Hufford MB, Hirsch CN. How the pan-genome is changing crop genomics and improvement. Genome Biol. 2021;22(1):3.

Desai A, Chee P, Rong J, May O, Paterson A. Desai A, Chee PW, Rong JK, May OL, Paterson AH. Chromosome structural changes in diploid and tetraploid A genomes of Gossypium. Genome 49: 336-345. Genome / National Research Council Canada = Génome / Conseil national de recherches Canada. 2006;49:336–45.

Dong C, Wang J, Yu Y, Ju L, Zhou X, Ma X, et al. Identifying functional genes influencing Gossypium hirsutum fiber quality. Front Plant Sci. 2019;9:1968.

Du X, Huang G, He S, Yang Z, Sun G, Ma X, et al. Resequencing of 243 diploid cotton accessions based on an updated A genome identifies the genetic basis of key agronomic traits. Nat Genet. 2018;50(6):796–802.

Fang L, Gong H, Hu Y, Liu C, Zhou B, Huang T, et al. Genomic insights into divergence and dual domestication of cultivated allotetraploid cottons. Genome Biol. 2017a;18(1):33.

Fang L, Guan X, Zhang T. Asymmetric evolution and domestication in allotetraploid cotton (Gossypium hirsutum L.). Crop J. 2017b;5(2):159–65.

Fang X, Liu X, Wang X, Wang W, Liu D, Zhang J, et al. Fine-mapping qFS07. 1 controlling fiber strength in upland cotton (Gossypium hirsutum L.). Theor Appl Genet. 2017c;130(4):795–806.

Fedoroff NV. Presidential address. Transposable elements, epigenetics, and genome evolution. Science. 2012;338(6108):758–67.

Flagel LE, Wendel JF, Udall JA. Duplicate gene evolution, homoeologous recombination, and transcriptome characterization in allopolyploid cotton. BMC Genomics. 2012;13:302.

Fryxell PA. A redefinition of the tribe Gossypieae. Bot Gaz. 1968;129(4):296–308.

Fryxell PA. The natural history of the cotton tribe (Malvaceae, tribe Gossypieae): Texas A & M University Press, Scientific publishers, Wuhan, China; 1979.

Fryxell PA. A revised taxonomic interpretation of Gossypium L. (Malvaceae). Rheedea. 1992;2(2): 108–65.

Gallagher JP, Grover CE, Rex K, Moran M, Wendel JF. A New Species of Cotton from Wake Atoll, Gossypium stephensii (Malvaceae). Syst Bot. 2017;42(1):115–23.

Gapare W, Liu S, Conaty W, Zhu Q-H, Gillespie V, Llewellyn D, et al. Historical datasets support genomic selection models for the prediction of cotton fiber quality phenotypes across multiple environments. G3 (Bethesda, Md). 2018;8(5):1721–32.

Gerstel D. Chromosomal translocations in interspecific hybrids of the genus Gossypium. Evolution. 1953;7:234–44.

Gross BL, Strasburg JL. Cotton domestication: dramatic changes in a single cell. BMC Biol. 2010;8 (1):137.

Grover CE, Gallagher JP, Jareczek JJ, Page JT, Udall JA, Gore MA, et al. Re-evaluating the phylogeny of allopolyploid Gossypium L. Mol Phylogenet Evol. 2015a;92:45–52.

Grover CE, Zhu X, Grupp KK, Jareczek JJ, Gallagher JP, Szadkowski E, et al. Molecular confirmation of species status for the allopolyploid cotton species, Gossypium ekmanianum Wittmack. Genetic Resour Crop Evol. 2015b;62(1):103–14.

Grover CE, Arick MA II, Thrash A, Conover JL, Sanders WS, Peterson DG, et al. Insights into the evolution of the New World diploid cottons (Gossypium, Subgenus Houzingenia) based on genome sequencing. Genome Biol Evol. 2018;11(1):53–71.

Grover CE, Pan M, Yuan D, Arick MA, Hu G, Brase L, et al. The Gossypium longicalyx genome as a resource for cotton breeding and evolution. G3: Genes|Genomes|Genetics. 2020;10(5): 1457–67.

Guo Z, Song G, Liu Z, Qu X, Chen R, Jiang D, et al. Global epigenomic analysis indicates that Epialleles contribute to Allele-specific expression via Allele-specific histone modifications in hybrid rice. BMC Genomics. 2015;16(1):232.

Han M, Lu X, Yu J, Chen X, Wang X, Malik WA, et al. Transcriptome analysis reveals cotton (Gossypium hirsutum) genes that are differentially expressed in cadmium stress tolerance. Int J Mol Sci. 2019;20(6):1479.

Hasan MM-U, Ma F, Islam F, Sajid M, Prodhan ZH, Li F, et al. Comparative transcriptomic analysis of biological process and key pathway in three cotton (Gossypium spp.) species under drought stress. Int J Mol Sci. 2019;20(9):2076.

He S, Sun G, Huang L, Yang D, Dai P, Zhou D, et al. Genomic divergence in cotton germplasm related to maturity and heterosis. J Integr Plant Biol. 2019;61(8):929–42.

He S, Wang P, Zhang Y-M, Dai P, Nazir MF, Jia Y, et al. Introgression leads to genomic divergence and responsible for important traits in upland cotton. Front Plant Sci. 2020;11:929.

He S, Sun G, Geng X, Gong W, Dai P, Jia Y, et al. The genomic basis of geographic differentiation and fiber improvement in cultivated cotton. Nat Genet. 2021;53(6):916–24.

Hekman JP, Johnson JL, Kukekova AV. Transcriptome analysis in domesticated species: challenges and strategies. Bioinform Biol Insights. 2015;9(Suppl 4):21–31.

Hinze LL, Gazave E, Gore MA, Fang DD, Scheffler BE, Yu JZ, et al. Genetic diversity of the two commercial tetraploid cotton species in the Gossypium diversity reference set. J Hered. 2016;107(3):274–86.

Hinze LL, Hulse-Kemp AM, Wilson IW, Zhu Q-H, Llewellyn DJ, Taylor JM, et al. Diversity analysis of cotton (Gossypium hirsutum L.) germplasm using the CottonSNP63K Array. BMC Plant Biol. 2017;17(1):37.

Hoque M, Ji Z, Zheng D, Luo W, Li W, You B, et al. Analysis of alternative cleavage and polyadenylation by 3′ region extraction and deep sequencing. Nat Methods. 2013;10(2):133–9.

Hou S, Zhu G, Li Y, Li W, Fu J, Niu E, et al. Genome-Wide Association studies reveal genetic variation and candidate genes of drought stress related traits in cotton (Gossypium hirsutum L.). Front Plant Sci. 2018;9:1276.

Hovav R, Chaudhary B, Udall JA, Flagel L, Wendel JF. Parallel domestication, convergent evolution and duplicated gene recruitment in allopolyploid cotton. Genetics. 2008;179(3):1725.

Hovav R, Faigenboim-Doron A, Kadmon N, Hu G, Zhang X, Gallagher JP, et al. A transcriptome profile for developing seed of polyploid cotton. Plant Genome. 2015;8(1): eplantgenome2014.2008.0041.

Hu Y, Chen J, Fang L, Zhang Z, Ma W, Niu Y, et al. Gossypium barbadense and Gossypium hirsutum genomes provide insights into the origin and evolution of allotetraploid cotton. Nat Genet. 2019;51(4):739–48.

Huang C, Nie X, Shen C, You C, Li W, Zhao W, et al. Population structure and genetic basis of the agronomic traits of upland cotton in China revealed by a genome-wide association study using high-density SNPs. Plant Biotechnol J. 2017;15(11):1374–86.

Huang G, Wu Z, Percy RG, Bai M, Li Y, Frelichowski JE, et al. Genome sequence of Gossypium herbaceum and genome updates of Gossypium arboreum and Gossypium hirsutum provide insights into cotton A-genome evolution. Nat Genet. 2020;52(5):516–24.

Hulse-Kemp AM, Lemm J, Plieske J, Ashrafi H, Buyyarapu R, Fang DD, et al. Development of a 63K SNP array for cotton and high-density mapping of intraspecific and interspecific populations of Gossypium spp. G3 (Bethesda, Md). 2015;5(6):1187–209.

Hutchinson JB, Silow RA, Stephens SG. The evolution of Gossypium and the differentiation of the cultivated cottons. New York: Oxford University Press, 1947. 160 p. $4.25. Sci Educ. 1948;32(3):225–6.

Iqbal MJ, Aziz N, Saeed NA, Zafar Y, Malik KA. Genetic diversity evaluation of some elite cotton varieties by RAPD analysis. Theor Appl Genet. 1997;94(1):139–44.

Islam MS, Fang DD, Jenkins JN, Guo J, McCarty JC, Jones DC. Evaluation of genomic selection methods for predicting fiber quality traits in Upland cotton. Mol Genet Genomics. 2020;295(1):67–79.

Jackson SA. Epigenomics: dissecting hybridization and polyploidization. Genome Biol. 2017;18(1):117.

Jensen P. Adding 'epi-' to behaviour genetics: implications for animal domestication. J Exp Biol. 2015;218(Pt 1):32–40.

Jiang C, Wright RJ, El-Zik KM, Paterson AH. Polyploid formation created unique avenues for response to selection in Gossypium (cotton). Proc Natl Acad Sci U S A. 1998;95(8):4419–24.

Jiang N, Bao Z, Zhang X, Eddy SR, Wessler SR. Pack-MULE transposable elements mediate gene evolution in plants. Nature. 2004;431(7008):569–73.

Jiang Y, Jiang Y, Wang S, Zhang Q, Ding X. Optimal sequencing depth design for whole genome re-sequencing in pigs. BMC Bioinf. 2019;20(1):556.

Kalia R, Rai M, Kalia S, Singh R, Dhawan A. Microsatellite markers: an overview of the recent progress in plants. Euphytica. 2011;177:309–34.

Kantar MB, Nashoba AR, Anderson JE, Blackman BK, Rieseberg LH. The genetics and genomics of plant domestication. Bioscience. 2017;67(11):971–82.

Katageri I, Gowda A, Biradar M, Patil R, RM. Prospects for molecular breeding in cotton, Gossypium spp. In: Plant breeding - current and future views. IntechOpen, London; 2020.

Kearney TH, Harrison GJ. Inheritance of smooth seeds in cotton. J Agric Res. 1927;35:193–217.

Kuang M, Wei S-J, Wang Y-Q, Zhou D-Y, Ma L, Fang D, et al. Development of a core set of SNP markers for the identification of upland cotton cultivars in China. J Integr Agric. 2016;15(5):954–62.

Lacape J-M, Nguyen T-B, Thibivilliers S, Bojinov B, Courtois B, Cantrell RG, et al. A combined RFLP–SSR–AFLP map of tetraploid cotton based on a Gossypium hirsutum × Gossypium barbadense backcross population. Genome. 2003;46(4):612–26.

Li T, Fan H, Li Z, Wei J, Cai Y, Lin Y. Effect of different light quality on DNA methylation variation for brown cotton (Gossypium hirstum). Afr J Biotechnol. 2011;10(33):6220–6.

Li F, Fan G, Wang K, Sun F, Yuan Y, Song G, et al. Genome sequence of the cultivated cotton Gossypium arboreum. Nat Genet. 2014;46(6):567–72.

Li F, Fan G, Lu C, Xiao G, Zou C, Kohel RJ, et al. Genome sequence of cultivated Upland cotton (Gossypium hirsutum TM-1) provides insights into genome evolution. Nat Biotechnol. 2015;33 (5):524–30.

Li C, Dong Y, Zhao T, Li L, Li C, Yu E, et al. Genome-wide SNP linkage mapping and QTL analysis for fiber quality and yield traits in the upland cotton recombinant inbred lines population. Front Plant Sci. 2016;7:1356.

Li P-T, Wang M, Lu Q-W, Ge Q, Rashid MHO, Liu A-Y, et al. Comparative transcriptome analysis of cotton fiber development of Upland cotton (Gossypium hirsutum) and Chromosome Segment Substitution Lines from G. hirsutum × G. barbadense. BMC Genomics. 2017a;18(1):705.

Li T, Ma X, Li N, Zhou L, Liu Z, Han H, et al. Genome-wide association study discovered candidate genes of Verticillium wilt resistance in upland cotton (Gossypium hirsutum L.). Plant Biotechnol J. 2017b;15(12):1520–32.

Li P-T, Rashid MHO, Chen T-T, Lu Q-W, Ge Q, Gong W-K, et al. Transcriptomic and biochemical analysis of upland cotton (Gossypium hirsutum) and a chromosome segment substitution line from G. hirsutum × G. barbadense in response to Verticillium dahliae infection. BMC Plant Biol. 2019a;19(1):19.

Li Z-B, Zeng X-Y, Xu J-W, Zhao R-H, Wei Y-N. Transcriptomic profiling of cotton Gossypium hirsutum challenged with low-temperature gradients stress. Sci Data. 2019b;6(1):197.

Li B, Chen L, Sun W, Wu D, Wang M, Yu Y, et al. Phenomics-based GWAS analysis reveals the genetic architecture for drought resistance in cotton. Plant Biotechnol J. 2020a;18(12):2533–44.

Li B, Tian Q, Wang X, Han B, Liu L, Kong X, et al. Phenotypic plasticity and genetic variation of cotton yield and its related traits under water-limited conditions. Crop J. 2020b;8:966.

Li J, Yuan D, Wang P, Wang Q, Sun M, Liu Z, et al. Cotton pan-genome retrieves the lost sequences and genes during domestication and selection. Genome Biol. 2021a;22(1):119.

Li L, Zhang C, Huang J, Liu Q,Wei H,Wang H, et al. Genomic analyses reveal the genetic basis of early maturity and identification of loci and candidate genes in upland cotton (Gossypium hirsutum L.). Plant Biotechnol J. 2021b;19(1):109–23.

Liang C, Wan T, Xu S, Li B, Li X, Feng Y, et al. Molecular identification and genetic analysis of cherry cultivars using capillary electrophoresis with fluorescence-labeled SSR markers. 3 Biotech. 2018;8(1):16.

Linck E, Battey CJ. Minor allele frequency thresholds strongly affect population structure inference with genomic data sets. Mol Ecol Resour. 2019;19(3):639–47.

Liu S, Cantrell RG, McCarty JC Jr, Stewart JM. Simple sequence repeat–based assessment of genetic diversity in cotton race stock accessions. Crop Sci. 2000;40(5):1459–69.

Liu X, Zhao B, Zheng H-J, Hu Y, Lu G, Yang C-Q, et al. Gossypium barbadense genome sequence provides insight into the evolution of extra-long staple fiber and specialized metabolites. Sci Rep. 2015;5(1):1–14.

Liu R, Gong J, Xiao X, Zhang Z, Li J, Liu A, et al. GWAS analysis and QTL identification of fiber quality traits and yield components in upland cotton using enriched high-density SNP markers. Front Plant Sci. 2018;9:1067.

Liu W, Song C, Ren Z, Zhang Z, Pei X, Liu Y, et al. Genome-wide association study reveals the genetic basis of fiber quality traits in upland cotton (Gossypium hirsutum L.). BMC Plant Biol. 2020;20(1):395.

Lu Y, Curtiss J, Miranda D, Hughs E, Zhang J. ATG-anchored AFLP (ATG-AFLP) analysis in cotton. Plant Cell Rep. 2008;27(10):1645–53.

Luikart G, Kardos M, Hand B, Rajora O, Aitken S, Hohenlohe P. Population genomics: advancing understanding of nature; 2018.

Luikart G, Kardos M, Hand B, Aitken S, Luikart G, Kardos M, et al. Population genomics: advancing understanding of nature: Springer, New York; 2020.

Lye ZN, Purugganan MD. Copy number variation in domestication. Trends Plant Sci. 2019;24(4): 352–65.

Ma Z. Unraveling the puzzle of the origin and evolution of cotton A-genome. J Cotton Res. 2020;3 (1):17.

Ma Z, He S, Wang X, Sun J, Zhang Y, Zhang G, et al. Resequencing a core collection of upland cotton identifies genomic variation and loci influencing fiber quality and yield. Nat Genet. 2018;50(6):803–13.

Ma Z, Zhang Y, Wu L, Zhang G, Sun Z, Li Z, et al. High-quality genome assembly and resequencing of modern cotton cultivars provide resources for crop improvement. Nat Genet. 2021;53(9):1385–91.

MacMillan CP, Birke H, Chuah A, Brill E, Tsuji Y, Ralph J, et al. Tissue and cell-specific transcriptomes in cotton reveal the subtleties of gene regulation underlying the diversity of plant secondary cell walls. BMC Genomics. 2017;18(1):539.

Majeed S, Rana IA, Atif RM, Ali Z, Hinze L, Azhar MT. Role of SNPs in determining QTLs for major traits in cotton. J Cotton Res. 2019;2(1):5.

Malik W, Ashraf J, Iqbal MZ, Khan AA, Qayyum A, Ali Abid M, et al. Molecular markers and cotton genetic improvement: current status and future prospects. Sci World J. 2014;2014:607091.

May OL, Bowman DT, Calhoun DS. Genetic diversity of U.S. upland cotton cultivars released between 1980 and 1990. Crop Sci. 1995;35(6):cropsci1995.0011183X003500060009x.

Mehboob Ur R, Shaheen T, Tabbasam N, Iqbal MA, Ashraf M, Zafar Y, et al. Cotton genetic resources. A review. Agron Sustain Dev. 2012;32(2):419–32.

Meuwissen TH, Hayes BJ, Goddard ME. Prediction of total genetic value using genome-wide dense marker maps. Genetics. 2001;157(4):1819–29.

Meyer RS, Purugganan MD. Evolution of crop species: genetics of domestication and diversification. Nat Rev Genet. 2013;14(12):840–52.

Moler ERV, Abakir A, Eleftheriou M, Johnson JS, Krutovsky KV, Lewis LC, Ruzov A, Whipple AV, Rajora OP. Population epigenomics: Advancing understanding of phenotypic plasticity, acclimation, adaptation and diseases. In: Rajora OP, editor. Population Genomics: Concepts, Approaches and Applications. Cham: Springer International Publishing; 2019. p. 179–260.

Morgante M, Olivieri AM. PCR-amplified microsatellites as markers in plant genetics. Plant J. 1993;3(1):175–82.

Morin PA, Luikart G, Wayne RK, S. N. P. w. g. the. SNPs in ecology, evolution and conservation. Trends Ecol Evol. 2004;19(4):208–16.

Mubarik MS, Ma C, Majeed S, Du X, Azhar MT. Revamping of cotton breeding programs for efficient use of genetic resources under changing climate. Agronomy. 2020;10(8):1190.

Murtaza N. Cotton genetic diversity study by AFLP markers. Electron J Biotechnol (ISSN: 0717-3458). 2006;9(4):9.

Naoumkina M, Thyssen GN, Fang DD, Jenkins JN, McCarty JC, Florane CB. Genetic and transcriptomic dissection of the fiber length trait from a cotton (Gossypium hirsutum L.) MAGIC population. BMC Genomics. 2019;20(1):112.

Naoumkina M, Thyssen GN, Fang DD, Bechere E, Li P, Florane CB. Mapping-by-sequencing the locus of EMS-induced mutation responsible for tufted-fuzzless seed phenotype in cotton. Mol Genet Genomics. 2021;296:1041.

Naqvi RZ, Zaidi SS-E-A, Mukhtar MS, Amin I, Mishra B, Strickler S, et al. Transcriptomic analysis of cultivated cotton Gossypium hirsutum provides insights into host responses upon whitefly-mediated transmission of cotton leaf curl disease. PLoS One. 2019;14(2):e0210011.

Nazir MF, Jia Y, Ahmed H, He S, Iqbal MS, Sarfraz Z, et al. Genomic insight into differentiation and selection sweeps in the improvement of upland cotton. Plants. 2020;9(6):711.

Nie X, Huang C, You C, Li W, Zhao W, Shen C, et al. Genome-wide SSR-based association mapping for fiber quality in nation-wide upland cotton inbreed cultivars in China. BMC Genomics. 2016;17(1):352.

Niles GA, Feaster CV. Breeding. In: Cotton. Madison: American Society of Agronomy; 1984. p. 201–31.

Niu C, Hinchliff D, Cantrell R, Wang C, Roberts P, Zhang J. Identification of molecular markers associated with Root-Knot Nematode resistance in upland cotton. Crop Sci. 2007;47:951–60.

Niu C, Lu Y, Yuan Y, Percy R, Ulloa M, Zhang J. Mapping resistance gene analogs (RGAs) in cultivated tetraploid cotton using RGA-AFLP analysis. Euphytica. 2011;181:65–76.

Noormohammadi Z, Rahnama A, Sheidai M. EST-SSR and SSR analyses of genetic diversity in diploid cotton genotypes from Iran. Nucleus. 2013;56(3):171–8.

Nuzhdin SV, Friesen ML, McIntyre LM. Genotype–phenotype mapping in a post-GWAS world. Trends Genet. 2012;28(9):421–6.

Osabe K, Clement JD, Bedon F, Pettolino FA, Ziolkowski L, Llewellyn DJ, et al. Genetic and DNA methylation changes in cotton (Gossypium) genotypes and tissues. PLoS One. 2014;9(1): e86049.

Otto SP. The evolutionary consequences of polyploidy. Cell. 2007;131(3):452–62.

Page JT, Huynh MD, Liechty ZS, Grupp K, Stelly D, Hulse AM, et al. Insights into the evolution of cotton diploids and polyploids from whole-genome re-sequencing. G3 (Bethesda, Md). 2013;3 (10):1809–18.

Palanga KK, Jamshed M, Rashid MHO, Gong J, Li J, Iqbal MS, et al. Quantitative trait locus mapping for Verticillium wilt resistance in an upland cotton recombinant inbred line using SNP-based high density genetic map. Front Plant Sci. 2017;8:382.

Pan Y, Meng F, Wang X. Sequencing multiple cotton genomes reveals complex structures and lays foundation for breeding. Front Plant Sci. 2020;11(1377):560096.

Paterson AH, Wendel JF, Gundlach H, Guo H, Jenkins J, Jin D, et al. Repeated polyploidization of Gossypium genomes and the evolution of spinnable cotton fibres. Nature. 2012;492(7429): 423–7.

Percy RG, Frelichowski JE, Arnold MD, Campbell TB, Dever JK, Fang DD, et al. The U.-S. National Cotton Germplasm Collection – its contents, preservation, characterization, and evaluation. In: World cotton germplasm resources; 2014.

Pillay M, Myers GO. Genetic diversity in cotton assessed by variation in ribosomal RNA genes and AFLP markers. Crop Sci. 1999;39(6):1881–6.

Piperno DR. Assessing elements of an extended evolutionary synthesis for plant domestication and agricultural origin research. Proc Natl Acad Sci. 2017;114(25):6429.

Prasad P, Khatoon U, Verma RK, Kumar A, Mohapatra D, Bhattacharya P, et al. Unravelling cotton RNAseq repositories to the fiber development specific modules and their alliance with the fiber-related traits. bioRxiv. 2021; 2021.2002.2013.431059

Qin Y, Sun H, Hao P, Wang H, Wang C, Ma L, et al. Transcriptome analysis reveals differences in the mechanisms of fiber initiation and elongation between long- and short-fiber cotton (Gossypium hirsutum L.) lines. BMC Genomics. 2019;20(1):633.

Rajora OPE. Population genomics: concepts, approaches and application: ebook. Springer Link, New York; 2019.

Rana MK, Bhat KV. A comparison of AFLP and RAPD markers for genetic diversity and cultivar identification in cotton. J Plant Biochem Biotechnol. 2004;13(1):19–24.

Rana MK, Singh VP, Bhat KV. Assessment of genetic diversity in upland cotton (Gossypium hirsutum L.) breeding lines by using amplified fragment length polymorphism (AFLP) markers and morphological characteristics. Genetic Resour Crop Evol. 2005;52(8):989–97.

Rapp RA, Haigler CH, Flagel L, Hovav RH, Udall JA, Wendel JF. Gene expression in developing fibres of Upland cotton (Gossypium hirsutum L.) was massively altered by domestication. BMC Biol. 2010;8(1):1–15.

Raszick TJ, Dickens CM, Perkin LC, Tessnow AE, Suh CP-C, Ruiz-Arce R, et al. Population genomics and phylogeography of the boll weevil, Anthonomus grandis Boheman (Coleoptera: Curculionidae), in the United States, northern Mexico, and Argentina. Evol Appl. 2021;14(7): 1778–93.

Reinisch AJ, Dong JM, Brubaker CL, Stelly DM, Wendel JF, Paterson AH. A detailed RFLP map of cotton, Gossypium hirsutum x Gossypium barbadense: chromosome organization and evolution in a disomic polyploid genome. Genetics. 1994;138(3):829–47.

Renny-Byfield S, Page JT, Udall JA, Sanders WS, Peterson DG, Arick MA 2nd, et al. Independent domestication of two Old World cotton species. Genome Biol Evol. 2016;8(6):1940–7.

Rohr JPBV. Anmerkungen über den Cattunbau [electronic resource] : zum Nuzen [sic] der dänischen westindischen Colonien auf allerhöchsten königlichen Befehl geschrieben / von Julius Philip Benjamin von Rohr ; mit einer Vorrede von Herrn D. Philipp Gabriel Hensler. Altona ; Leipzig, Bey Johann Friedrich Hammerich, 1791.

Rong J, Abbey C, Bowers JE, Brubaker CL, Chang C, Chee PW, et al. A 3347-locus genetic recombination map of sequence-tagged sites reveals features of genome organization, transmission and evolution of cotton (Gossypium). Genetics. 2004;166(1):389–417.

Said J, Lin ZX, Zhang X, Song M, Zhang J. Comprehensive meta QTL analysis for fiber quality, yield, yield related and morphological traits, drought tolerance, and disease resistance in tetraploid cotton. BMC Genomics. 2013;14:776.

Said JI, Knapka JA, Song M, Zhang J. Cotton QTLdb: a cotton QTL database for QTL analysis, visualization, and comparison between Gossypium hirsutum and G. hirsutum × G. barbadense populations. Mol Genet Genomics. 2015;290(4):1615–25.

Saïdou A-A, Thuillet A-C, Couderc M, Mariac C, Vigouroux Y. Association studies including genotype by environment interactions: prospects and limits. BMC Genet. 2014;15(1):3.

Saranga Y, Menz M, Jiang CX, Wright RJ, Yakir D, Paterson AH. Genomic dissection of genotype x environment interactions conferring adaptation of cotton to arid conditions. Genome Res. 2001;11(12):1988–95.

Senchina DS, Alvarez I, Cronn RC, Liu B, Rong J, Noyes RD, et al. Rate variation among nuclear genes and the age of polyploidy in gossypium. Mol Biol Evol. 2003;20(4):633–43.

Seyoum M, Du XM, He SP, Jia YH, Pan Z, Sun JL. Analysis of genetic diversity and population structure in upland cotton (Gossypium hirsutum L.) germplasm using simple sequence repeats. J Genet. 2018;97(2):513–22.

Shaheen T, Zafar Y, Rahman M-U. Phylogenetic analysis of cotton species (Diploid genomes) using single nucleotide polymorphisms (SNPs) markers. Pakistan J Agric Sci. 2016;53:283–90.

Shan D-P, Huang J-G, Yang Y-T, Guo Y-H, Wu C-A, Yang G-D, et al. Cotton GhDREB1 increases plant tolerance to low temperature and is negatively regulated by gibberellic acid. New Phytol. 2007;176(1):70–81.

Shan C-M, Shangguan X-X, Zhao B, Zhang X-F, Chao L-M, Yang C-Q, et al. Control of cotton fibre elongation by a homeodomain transcription factor GhHOX3. Nat Commun. 2014;5(1): 5519.

Shappley ZW, Jenkins JN, Zhu J, McCarty JC. Quantitative trait loci associated with agronomic and fiber traits of upland cotton. J Cotton. 1998;2:153.

Shen Y, Zhang J, Liu Y, Liu S, Liu Z, Duan Z, et al. DNA methylation footprints during soybean domestication and improvement. Genome Biol. 2018;19(1):128.

Shen C, Wang N, Huang C, Wang M, Zhang X, Lin Z. Population genomics reveals a fine-scale recombination landscape for genetic improvement of cotton. Plant J. 2019;99(3):494–505.

Song Q, Zhang T, Stelly DM, Chen ZJ. Epigenomic and functional analyses reveal roles of epialleles in the loss of photoperiod sensitivity during domestication of allotetraploid cottons. Genome Biol. 2017;18(1):99.

Strygina K, Khlestkina E, Podolnaya L. Cotton genome evolution and features of its structural and functional organization. Biol Commun. 2020;65(1):15–27.

Su J, Li L, Pang C, Wei H, Wang C, Song M, et al. Two genomic regions associated with fiber quality traits in Chinese upland cotton under apparent breeding selection. Sci Rep. 2016;6(1): 38496.

Su J, Li L, Zhang C, Wang C, Gu L, Wang H, et al. Genome-wide association study identified genetic variations and candidate genes for plant architecture component traits in Chinese upland cotton. Theor Appl Genet. 2018;131(6):1299–314.

Sun F-D, Zhang J-H, Wang S-F, Gong W-K, Shi Y-Z, Liu A-Y, et al. QTL mapping for fiber quality traits across multiple generations and environments in upland cotton. Mol Breed. 2012;30(1): 569–82.

Sun Z, Wang X, Liu Z, Gu Q, Zhang Y, Li Z, et al. Genome-wide association study discovered genetic variation and candidate genes of fibre quality traits in Gossypium hirsutum L. Plant Biotechnol J. 2017;15(8):982–96.

Sun H, Meng M, Yan Z, Lin Z, Nie X, Yang X. Genome-wide association mapping of stress-tolerance traits in cotton. Crop J. 2019;7(1):77–88.

Suzuki H, Rodriguez-Uribe L, Xu J, Zhang J. Transcriptome analysis of cytoplasmic male sterility and restoration in CMS-D8 cotton. Plant Cell Rep. 2013;32(10):1531–42.

Tan Z, Zhang Z, Sun X, Li Q, Sun Y, Yang P, et al. Genetic map construction and fiber quality QTL mapping using the cottonSNP80K array in upland cotton. Front Plant Sci. 2018;9:225.

Tang S, Teng Z, Zhai T, Fang X, Liu F, Liu D, et al. Construction of genetic map and QTL analysis of fiber quality traits for Upland cotton (Gossypium hirsutum L.). Euphytica. 2015;201(2): 195–213.

Tao Y, Zhao X, Mace E, Henry R, Jordan D. Exploring and exploiting pan-genomics for crop improvement. Mol Plant. 2019;12(2):156–69.

Tatineni V, Cantrell RG, Davis DD. Genetic diversity in elite cotton germplasm determined by morphological characteristics and RAPDs. Crop Sci. 1996;36(1): cropsci1996.0011183X003600010033x.

Tettelin H, Masignani V, Cieslewicz MJ, Donati C, Medini D, Ward NL, et al. Genome analysis of multiple pathogenic isolates of Streptococcus agalactiae: Implications for the microbial pan-genome. Proc Natl Acad Sci. 2005;102(39):13950–5.

Todaro DG. Relazione Sulla Cultura Dei Cotoni in Italia, Seguita Da Una Monografia Del Genere Gossypium. Roma Palermo: Stamperia Reale Ditta P. A. Molina Cromo-litografia Visconti; 1877.

Tyagi P, Gore MA, Bowman DT, Campbell BT, Udall JA, Kuraparthy V. Genetic diversity and population structure in the US Upland cotton (Gossypium hirsutum L.). Theor Appl Genet. 2014;127(2):283–95.

Udall JA, Swanson JM, Haller K, Rapp RA, Sparks ME, Hatfield J, et al. A global assembly of cotton ESTs. Genome Res. 2006;16(3):441–50.

Udall JA, Long E, Hanson C, Yuan D, Ramaraj T, Conover JL, et al. De Novo genome sequence assemblies of Gossypium raimondii and Gossypium turneri. G3 Genes|Genomes|Genetics. 2019;9(10):3079–85.

Ulloa M, Abdurakhmonov IY, Perez-M C, Percy R, Stewart JM. Genetic diversity and population structure of cotton (Gossypium spp.) of the New World assessed by SSR markers. Botany. 2013;91(4):251–9.

Ulloa M, Hulse-Kemp AM, De Santiago LM, Stelly DM, Burke JJ. Insights into upland cotton (Gossypium hirsutum L.) genetic recombination based on 3 high-density single-nucleotide polymorphism and a consensus map developed independently with common parents. Genomics Insights. 2017;10:1178631017735104.

Ulloa M, De Santiago LM, Hulse-Kemp AM, Stelly DM, Burke JJ. Enhancing Upland cotton for drought resilience, productivity, and fiber quality: comparative evaluation and genetic dissection. Mol Genet Genomics. 2020;295(1):155–76.

Van Deynze A, Stoffel K, Lee M, Wilkins TA, Kozik A, Cantrell RG, et al. Sampling nucleotide diversity in cotton. BMC Plant Biol. 2009;9(1):125.

Velázquez-López R, Wegier A, Alavez V, Pérez-López J, Vázquez-Barrios V, Arroyo-Lambaer D, et al. The mating system of the wild-to-domesticated complex of Gossypium hirsutum L. Is mixed. Front Plant Sci. 2018;9:945.

Vroh Bi I, Maquet A, Baudoin JP, du Jardin P, Jacquemin JM, Mergeai G. Breeding for "low-gossypol seed and high-gossypol plants" in upland cotton. Analysis of tri-species hybrids and backcross progenies using AFLPs and mapped RFLPs. Theor Appl Genet. 1999;99(7):1233–44.

Wambugu PW, Ndjiondjop M-N, Henry RJ. Role of genomics in promoting the utilization of plant genetic resources in genebanks. Brief Funct Genomics. 2018;17(3):198–206.

Wang B-H, Wu Y-T, Huang N-T, Zhu X-F, Guo W-Z, Zhang T-Z. QTL mapping for plant architecture traits in upland cotton using RILs and SSR markers. Acta Genetica Sinica. 2006;33(2):161–70.

Wang K, Wang Z, Li F, Ye W, Wang J, Song G, et al. The draft genome of a diploid cotton Gossypium raimondii. Nat Genet. 2012;44(10):1098–103.

Wang K, Wendel JF, Hua J. Designations for individual genomes and chromosomes in Gossypium. J Cotton Res. 2018;1(1):1–5.

Wang Z, Zhang D, Wang X, Tan X, Guo H, Paterson AH. A whole-genome DNA marker map for cotton based on the D-genome sequence of Gossypium raimondii L. G3 (Bethesda, Md). 2013;3(10):1759–67.

Wang H, Jin X, Zhang B, Shen C, Lin Z. Enrichment of an intraspecific genetic map of upland cotton by developing markers using parental RAD sequencing. DNA Res. 2015;22(2):147–60.

Wang B, Tseng E, Regulski M, Clark TA, Hon T, Jiao Y, et al. Unveiling the complexity of the maize transcriptome by single-molecule long-read sequencing. Nat Commun. 2016a;7:11708.

Wang J, Mu M, Wang S, Lu X, Chen X, Wang D, et al. Molecular clone and expression of GhDHN1 gene in cotton (Gossypium hirsutum L.). Scientia Agricultura Sinica. 2016b;49(15): 2867–78.

Wang K, Huang G, Zhu Y. Transposable elements play an important role during cotton genome evolution and fiber cell development. Sci China Life Sci. 2016c;59(2):112–21.

Wang M, Tu L, Lin M, Lin Z, Wang P, Yang Q, et al. Asymmetric subgenome selection and cis-regulatory divergence during cotton domestication. Nat Genet. 2017;49(4):579–87.

Wang M, Wang P, Liang F, Ye Z, Li J, Shen C, et al. A global survey of alternative splicing in allopolyploid cotton: landscape, complexity and regulation. New Phytol. 2018;217(1):163–78.

Wang M, Tu L, Yuan D, Zhu D, Shen C, Li J, et al. Reference genome sequences of two cultivated allotetraploid cottons, Gossypium hirsutum and Gossypium barbadense. Nat Genet. 2019;51(2): 224–9.

Wang H, Zhang R, Shen C, Li X, Zhu D, Lin Z. Transcriptome and QTL analyses reveal candidate genes for fiber quality in Upland cotton. Crop J. 2020;8(1):98–106.

Wang P, He S, Sun G, Pan Z, Sun J, Geng X, et al. Favorable pleiotropic loci for fiber yield and quality in upland cotton (Gossypium hirsutum). Sci Rep. 2021;11(1):15935.

Wang J, Zhang Z, Gong Z, Liang Y, Ai X, Sang Z, et al. Analysis of the genetic structure and diversity of upland cotton groups in different planting areas based on SNP markers. Gene. 2022;809:146042.

Ware J, Benedict L, Rolfe W. A recessive naked-seed character in upland cotton. J Hered. 1947;38 (10):313–20.

Wei Y, Xu Y, Lu P, Wang X, Li Z, Cai X, et al. Salt stress responsiveness of a wild cotton species (Gossypium klotzschianum) based on transcriptomic analysis. PLoS One. 2017;12(5): e0178313.

Weigel D, Colot V. Epialleles in plant evolution. Genome Biol. 2012;13(10):249.

Wendel JF. New World tetraploid cottons contain Old World cytoplasm. Proc Natl Acad Sci. 1989;86(11):4132.

Wendel JF, Albert VA. Phylogenetics of the cotton genus (Gossypium): character-state weighted parsimony analysis of chloroplast-DNA restriction site data and its systematic and biogeographic implications. Syst Botany. 1992;17(1):115–43.

Wendel JF, Cronn RC. Polyploidy and the evolutionary history of cotton. Adv Agronom, Academic Press. 2003;78:139–86.

Wendel JF, Olson PD, Stewart JM. Genetic diversity, introgression, and independent domestication of Old World cultivated cottons. Am J Bot. 1989;76(12):1795–806.

Wendel JF, Brubaker CL, Percival AE. Genetic diversity in Gossypium Hirsutum and the origin of upland cotton. Am J Bot. 1992;79(11):1291–310.

Wendel JF, Schnabel A, Seelanan T. Bidirectional interlocus concerted evolution following allopolyploid speciation in cotton (Gossypium). Proc Natl Acad Sci U S A. 1995;92(1):280–4.

Wendel JF, Brubaker C, Alvarez I, Cronn R, Stewart JM. Evolution and natural history of the cotton genus. In: Genetics and genomics of cotton. Springer, New York; 2009. p. 3–22.

Wendel JF, Lisch D, Hu G, Mason AS. The long and short of doubling down: polyploidy, epigenetics, and the temporal dynamics of genome fractionation. Curr Opin Genet Dev. 2018;49:1–7.

Westengen OT, Huamán Z, Heun M. Genetic diversity and geographic pattern in early South American cotton domestication. Theor Appl Genet. 2005;110(2):392–402.

Wright R, Thaxton P, El-Zik K, Paterson A. Molecular mapping of genes affecting pubescence of cotton. J Hered. 1999;90(1):215–9.

Wu J, Jenkins JN, McCarty JC, Zhong M, Swindle M. AFLP marker associations with agronomic and fiber traits in cotton. Euphytica. 2007;153(1):153–63.

Wu H, Zhang J, Shi J, Fan Z, Aliyan R, Zhang P, et al. Physiological responses of cotton seedlings under low temperature stress. Acta Botanica Boreali-Occidentalia Sinica. 2013;33(1):74–82.

Wu Z, Yang Y, Huang G, Lin J, Xia Y, Zhu Y. Cotton functional genomics reveals global insight into genome evolution and fiber development. J Genet Genomics. 2017;44(11):511–8.

Xu Z, Yu J, Kohel RJ, Percy RG, Beavis WD, Main D, et al. Distribution and evolution of cotton fiber development genes in the fibreless Gossypium raimondii genome. Genomics. 2015;106(1):61–9.

Xu Y, Magwanga RO, Jin D, Cai X, Hou Y, Juyun Z, et al. Comparative transcriptome analysis reveals evolutionary divergence and shared network of cold and salt stress response in diploid D-genome cotton. BMC Plant Biol. 2020;20(1):518.

Yang Z, Ge X, Yang Z, Qin W, Sun G, Wang Z, et al. Extensive intraspecific gene order and gene structural variations in upland cotton cultivars. Nat Commun. 2019;10(1):1–13.

Yang Z, Qanmber G, Wang Z, Yang Z, Li F. Gossypium genomics: trends, scope, and utilization for cotton improvement. Trends Plant Sci. 2020;25(5):488–500.

Yano K, Yamamoto E, Aya K, Takeuchi H, Lo P-C, Hu L, et al. Genome-wide association study using whole-genome sequencing rapidly identifies new genes influencing agronomic traits in rice. Nat Genet. 2016;48(8):927–34.

Yin X, Zhan R, He Y, Song S, Wang L, Ge Y, Chen D. Morphological description of a novel synthetic allotetraploid (A1A1G3G3) of Gossypium herbaceum L. and G. nelsonii Fryx. suitable for disease-resistant breeding applications. PLoS One. 2020;15(12): e0242620.

Yoo MJ, Wendel JF. Comparative evolutionary and developmental dynamics of the cotton (Gossypium hirsutum) fiber transcriptome. PLoS Genet. 2014;10(1):e1004073.

Yu JZ, Fang DD, Kohel RJ, Ulloa M, Hinze LL, Percy RG, et al. Development of a core set of SSR markers for the characterization of Gossypium germplasm. Euphytica. 2012;187(2):203–13.

Yuan D, Tang Z, Wang M, Gao W, Tu L, Jin X, et al. The genome sequence of Sea-Island cotton (Gossypium barbadense) provides insights into the allopolyploidization and development of superior spinnable fibres. Sci Rep. 2015;5:17662.

Yuan D, Grover CE, Hu G, Pan M, Miller ER, Conover JL, et al. Parallel and intertwining threads of domestication in allopolyploid cotton. Adv Sci. 2021;8(10):2003634.

Zahn LM. Unraveling the origin of cotton. Science. 2012;335(6073):1148.

Zhang J, Lu Y, Yu S. Cleaved AFLP (cAFLP), a modified amplified fragment length polymorphism analysis for cotton. Theor Appl Genet. 2005a;111(7):1385–95.

Zhang Z-S, Xiao Y-H, Luo M, Li X-B, Luo X-Y, Hou L, et al. Construction of a genetic linkage map and QTL analysis of fiber-related traits in upland cotton (Gossypium hirsutum L.). Euphytica. 2005b;144(1):91–9.

Zhang M, Rong Y, Lee MK, Zhang Y, Stelly DM, Zhang HB. Phylogenetic analysis of Gossypium L. using restriction fragment length polymorphism of repeated sequences. Mol Genet Genomics. 2015a;290(5):1859–72.

Zhang T, Hu Y, Jiang W, Fang L, Guan X, Chen J, et al. Sequencing of allotetraploid cotton (Gossypium hirsutum L. acc. TM-1) provides a resource for fiber improvement. Nat Biotechnol. 2015b;33(5):531–7.

Zhang W, Zhang H, Liu K, Jian G, Qi F, Si N. Large-scale identification of Gossypium hirsutum genes associated with Verticillium dahliae by comparative transcriptomic and reverse genetics analysis. PLoS One. 2017;12(8):e0181609.

Zhang S, Cai Y, Guo J, Li K, Peng R, Liu F, et al. Genotyping-by-sequencing of Gossypium hirsutum races and cultivars uncovers novel patterns of genetic relationships and domestication footprints. Evol Bioinform Online. 2019;15:1176934319889948.

Zhang Y, Yang N, Zhao L, Zhu H, Tang C. Transcriptome analysis reveals the defense mechanism of cotton against Verticillium dahliae in the presence of the biocontrol fungus Chaetomium globosum CEF-082. BMC Plant Biol. 2020;20(1):89.

Zhang J, Mei H, Lu H, Chen R, Hu Y, Zhang T. Transcriptome time-course analysis in the whole period of cotton fiber development. Front Plant Sci. 2022;13:864529.

Zhao Y, Yu S, Xing C, Fan S, Song M. Analysis of DNA methylation in cotton hybrids and their parents. Mol Biol. 2008;42(2):169.

Zhao Y-L, Yu S-X, Ye W-W, Wang H-M, Wang J-J, Fang B-X. Study on DNA cytosine methylation of cotton (Gossypium hirsutum L.) genome and its implication for salt tolerance. Agric Sci China. 2010;9(6):783–91.

Zhao D, Ferguson AA, Jiang N. What makes up plant genomes: the vanishing line between transposable elements and genes. Biochimica et Biophysica Acta (BBA) - Gene Regulatory Mechanisms. 2016;1859(2):366–80.

Zheng JY, Oluoch G, Riaz Khan MK, Wang XX, Cai XY, Zhou ZL, et al. Mapping QTLs for drought tolerance in an F2:3 population from an inter-specific cross between Gossypium tomentosum and Gossypium hirsutum. Genet Mol Res. 2016;15 https://doi.org/10.4238/gmr.15038477.

Zhou J, Li D, Wang G, Wang F, Kunjal M, Joldersma D, et al. Application and future perspective of CRISPR/Cas9 genome editing in fruit crops. J Integr Plant Biol. 2020;62(3):269–86.

Zhu G, Gao W, Song X, Sun F, Hou S, Liu N, et al. Genome-wide association reveals genetic variation of lint yield components under salty field conditions in cotton (Gossypium hirsutum L.). BMC Plant Biol. 2020;20(1):23.

Population Genomics of *Brassica* Species

Yonghai Fan, Yue Niu, Xiaodong Li, Shengting Li, Cunmin Qu, Jiana Li, and Kun Lu

Abstract The genus *Brassica* contains the most economically valuable cultivated dicotyledonous plants in the world. They provide edible oil, protein, and vegetables for human consumption, as well as fodder for livestock. Extensive researches have been performed with the aim of unraveling the complex genomes of the *Brassica* species. Of the six *Brassica* species within the "triangle of U," the genomes of all six *Brassica* species have been sequenced and assembled. The analyses of these genomes have revealed the genetic variation, genomic structure, biogeographical origin, and population evolution of the *Brassica* species, and when combined with large-population resequencing, these data were used to propose the history and genetic effects of domestication and adaptive mechanism of the *Brassica* species. Advances in resequencing technology have enabled the application of high-efficiency breeding strategies in these crop species, involving the identification of genetic variation and genetic loci underlying a trait, genome-wide association studies, and genomic selection. Moreover, population genomics approaches, including population transcriptomics, population epigenomics, and genomic selection studies, have contributed to enhancing our understanding of acclimation, adaptation and disease and insect resistance for populations in the *Brassica* species. Population genomics therefore provides new insights and facilitates the deciphering of the secrets of the evolution, domestication, and adaptation of *Brassica* species.

Yonghai Fan and Yue Niu contributed equally to this work.

Y. Fan · Y. Niu · X. Li · S. Li
College of Agronomy and Biotechnology, Southwest University, Chongqing, China

C. Qu · J. Li · K. Lu (✉)
College of Agronomy and Biotechnology, Southwest University, Chongqing, China

Academy of Agricultural Sciences, Southwest University, Chongqing, China

Engineering Research Center of South Upland Agriculture, Ministry of Education, Chongqing, China
e-mail: drlukun@swu.edu.cn

Keywords Brassica · Population genomics · High-throughput sequencing · GWAS · Domestication · Evolution · Origin and phylogenomics · Genetic diversity and population structure · Genomic selection

1 Introduction

The genus *Brassica* belongs to the tribe Brassicaceae and comprises approximately 40 species, including many crop plants, such as cabbage (*Brassica oleracea*), turnip (*Brassica rapa*), mustard (*Brassica juncea*), and rapeseed (*Brassica napus*) (Warwick et al. 2006, 2010; Franzke et al. 2011). The morphological characteristics of these species include glabrous or single hairs, thin or lumpy roots, rosettes of basal leaves, stalked or clasped stems, corymbose racemes, yellow or white flowers, and long siliques. The *Brassica* genus has a global distribution, although most species are found in temperate regions of the Northern Hemisphere, especially in western Europe, North America, the Mediterranean, and central Asia (Al-Shehbaz 2012). *Brassica oleracea*, for example, is found along the Atlantic coasts of Europe although wild populations are endemic to the Mediterranean basin (Chiang et al. 1993; Sauer 1993; Herve 2003). *Brassica rapa* is extensively distributed across the Mediterranean region, Afghanistan, central Asia, and northwest India (Denford and Vaughan 1977; Song et al. 1988; Zhao et al. 2004).

Brassica species are extensively cultivated around the world for vegetables, condiments (mustard), seed oil, and forage crops. The most commonly grown *Brassica* vegetables are the varieties of *B. oleracea*, particularly cabbage (*Brassica oleracea* var. *capitata*), cauliflower (*Brassica oleracea* var. *botrytis*), and broccoli (*Brassica oleracea* var. *italica*). Cabbage is the most commonly used *Brassica* vegetable, because the production of cauliflower is lower, and broccoli is a relatively newcomer and emerging vegetable. The appearance of "double low" (erucic acid <2% in the oil and glucosinolates <30 mg/g in the meal) rapeseed lines during the 1970s resulted in a *Brassica* oil shown to be suitable for human and animal consumption (Shahidi 1990; Buzza 1995). The lipid profile of CanOLA (Canadian Oil Low Acid) is extremely well balanced for a vegetable oil (low in saturated fats, high in monosaturated fats, and rich in omega-3 fatty acids) (Iniguez-Luy and Federico 2011). Over the past 40 years, the cultivation and processing of rapeseed has grown rapidly, making it the second most abundant oil crop after soybean (*Glycine max*). Many other *Brassica* species are used as vegetables (e.g., *B. rapa*) and oil crops (e.g., *Brassica juncea*, used as an oilseed and for mustard). The emergence and cultivation of these species are therefore important for human food security.

The evolutionary relationships within the *Brassica* genus have been described by the theory of the "triangle of U" (Nagaharu 1935), which was developed based on the analysis of the chromosomes of six closely related *Brassica* species (Fig. 1a): three diploid species, *B. rapa* (AA genome; turnip and Asian cabbages), *B. nigra*

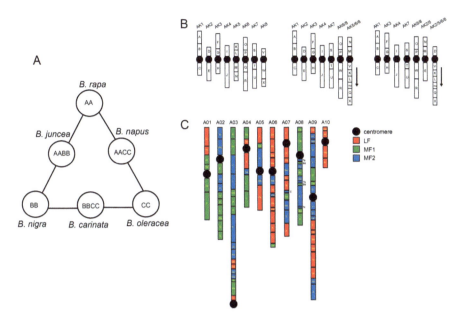

Fig. 1 The relationships and ancestral Brassicaceae genomes of *Brassica* crop plants. (**a**) The theory of "triangle of U," which is used to describe the relationships of *Brassica* crop plants (Nagaharu 1935). (**b**) The ancestral genomes ACK, PCK, and tPCK, each comprising 24 ancestral genome blocks (GBs), chromosome labels in the PCK and tPCK origin from two karyotypes in ancestral ACK genome (Schranz et al. 2006; Mandakova and Lysak 2008; Cheng et al. 2013). (**c**) The chromosomes of *B. rapa* consist of 24 triplicated genome blocks which were assigned to three subgenomes, LF, MF1, and MF2 (Wang et al. 2011c; Cheng et al. 2013). The segments of a single block are labeled in lowercase letters, such as a, b, etc. This figure is modified from Cheng et al. (2013)

(BB genome; black mustard), and *B. oleracea* (CC genome; European cabbages, cauliflower, and broccoli), and three autotetraploid species resulting from pairwise hybridization, *B. napus* (AACC genome; rapeseed, canola, and swede), *B. juncea* (AABB genome; Indian mustard, leaf mustard, and gai choy), and *B. carinata* (BBCC genome; Ethiopian mustard). The theory states that the genomes of *B. rapa*, *B. nigra*, and *B. oleracea* represent the three ancestral *Brassica* species that combined together to generate the allotetraploid species, which, in addition to some cultivars of *B. rapa*, are primarily used as the oil and condiment crops. The allopolyploid *B. juncea* was formed by the hybridization of the diploid ancestors of *B. rapa* and *B. nigra*, after which the vegetable- and oil-use varieties were developed under diversifying selection resulting in vegetable and oilseed mustard in China, oilseed crops in India, canola crops in Canada and Australia, and condiment crops in Europe and other regions (Chen et al. 2013).

Before the twenty-first century, the origin and domestication history of *Brassica* corps were highly debated and remain unclear. At present, these outstanding questions have been gradually resolved by whole-genome sequencing and assemblies and population genomics studies in *Brassica* species. The rapid development of sequencing technology has enabled the generation and analysis of large-scale population genomics data. We can now explore the genetic variation in genomes and domesticated and wild populations, understand evolution and diversification of species and varieties, study the effects of natural/artificial selection to domestication, and explore the history of population structures. To date, the genomes of six species on which the "triangle of U" theory was based, have been sequenced and assembled (Table 1). This has widely expanded our knowledge of the genome structures and evolutionary history of the *Brassica* species. The sequencing of growing numbers of cultivated accessions has provided extensive genomic data and enabled the elucidation of genetic diversity in *Brassica* crop plants. The abundant sub-varieties represent good resources for population genomics studies, which is helpful for researchers to explore the origin and diffusion, population dynamics, and adaptive differentiation of *Brassica* crops in evolutionary history. In particular, genetic loci and variation detected from population genomics approaches could provide valuable resource for the breeding of *Brassica* crops. In this chapter, we review the progress made on population genomics of *Brassica* crop species, focusing on genomes, evolution and phylogenomics in *Brassica* species, and genetic diversity, structure, domestication, agronomic traits associated loci, population epigenomics, and population transcriptomics and genomic selection in *Brassica* populations. Then we present future prospects and further directions and hotspots for population genomics studies in *Brassica* species.

2 Genomic Resources for Population Genomics Research

2.1 Genomes

As described above, the genomes of the "triangle of U" theory *Brassica* species have been sequenced, assembled, and annotated to date. In this section, we summarize their genomic features and the sequencing strategy used (Table 1).

B. rapa was the first *Brassica* species of which the genome was sequenced. Wang et al. (2011c) sequenced the genome of the Chinese cabbage cultivar chiifu401/42 using NGS (Next-generation-sequencing) and then assembled and annotated it. This first reference genome served as a valuable resource for the assembly and annotation of other *Brassica* genomes in subsequent years, such as those of *B. oleracea* and *B. napus*. This first reference genome was only about 283.8 Mb, 58.52% of the estimated size of the *B. rapa* genome; therefore, Cai et al. (2017) de novo assembled the *B. rapa* genome and obtained a new and improved genome assembly for this species, which was 389.2 Mb (version 2) (Cai et al. 2017). Next, Zhang et al. (2018) assembled version 3 of the *B. rapa* genome with a contig N50 size of 1.45 Mb using a combination of sequencing and chromosome conformation capture technologies (Hi-C). This latest version allowed the relationship among genome blocks to be refined and facilitated the accurate identification of the centromere locations in *B. rapa* (Zhang et al. 2018).

Population Genomics of *Brassica* Species

Table 1 The summary of all published reference genomes of *Brassica* species from 2011 to september 2021

Species	Strategy	Sample name	Estimated genome size (Mb)	Genome size (Mb)	Reference
B. nigra	Mate pairs + Illumina	YZ12151	591	396.9	Yang et al. (2016)
B. nigra	Nanopore sequencing	Ni100	570	506	Perumal et al. (2020)
B. nigra	Nanopore sequencing	CN115125	608	537	Perumal et al. (2020)
B. nigra	Illumina Hiseq platform+ Platanus	CGN7651	632	512	Wang et al. (2019)
B. napus	BAC + whole-genome-sequence	Zhongshuang11	1,132	976	Sun et al. (2017)
B. napus (pangenome)	PacBio + Illumina	Westar/No2127/ Zheyou7/Gangan/ Shengli/Tapidor/ Quinta/ZS11	~1,132 Mb	~961	Song et al. (2020)
B. napus	454 + Sanger + Illumina	Darmor-bz	1,130	849.7	Chalhoub et al. (2014)
B. napus	De novo assembly	Tapidor	1,335	635	Bayer and King (2017)
B. juncea	Illumina + PacBio	variety T84–66	922	784	Yang et al. (2016)
B. rapa	BAC + Illumina	Chiifu-401-42	485	283.8	Wang et al. (2011a)
B. rapa	PacBio	Chiifu-401-42	485	389.3	Cai et al. (2017)
B. rapa	Single-molecule sequencing + optical mapping + Hic	Chiifu-401-42	485	455 Mb	Zhang et al. (2018)
B. oleracea	454+Sanger + Illumina	Var.capitataline02–12	630	539.5	Liu et al. (2014)
B. oleracea	454+Sanger + Illumina	Var.alboglabraline TO1000DH	648	488.6	Parkin et al. (2014)
B. oleracea	PacBio + Illumina	var. botrytis (C-8)	603.04	584.6	Sun et al. (2019)
B. oleracea (pangenome)	Illumina + 454 + Sanger	To1000, Cabbage 1/ Cabbage 2/kale/ Brussels sprout/ Kohlrabi/Cauli-flower2/Broccoli/ Macrocarpa	-	587	Golicz et al. (2016)
B. carinata	Nanopore+ Hi-C + Illumina	zd-1	1,150.76	1,086.8	Song et al. (2021)

The reference genome of *B. oleracea* was assembled using two varieties, var. *capitata* line 02–12 (Liu et al. 2014) and var. *alboglabra* line TO1000DH (Parkin et al. 2014). In Liu et al. study, about 85% of the estimated genome was assembled, while only about 75% was assembled by Parkin and colleagues (Liu et al. 2014; Parkin et al. 2014). In 2018, a high-quality reference genome of a broccoli using HDEM with contig N50, more than 9 Mb was assembled by combined long reads and optical maps technology (Belser et al. 2018). In 2019, Sun et al. assembled *B. oleracea* genome using cauliflower var. *botrytis* (C-8). The resulting genome was 584.6 Mb, and numerous terminal repeats and transposable elements were also identified in this variety (Sun et al. 2019). Recently, a genome of the JZS (also named 02–12), a heading cabbage belonging to *B. oleracea* ssp. *capitata*, was assembled by three technologies (Illumina, PacBio, and Hi-C). Comparative analysis with the previous genome sequences of two other genomes, assembled by long-read technology TO1000 and HDEM, identified extensive gene orders and gene structural variations (SV), including 5,270 JZS v2 specific genomic segments (5.00 Mb in total), 6,438 HDEM specific genomic segments (7.11 Mb in total), and 5,307 TO1000 specific genomic segments (4.78 Mb in total). In addition, they found that the genome-specific amplification of Gypsy-like LTR-RTs occurred around 0–1 MYA (million years ago) (Cai et al. 2020).

In 2016, the first *B. nigra* and *B. juncea* genomes were sequenced and assembled by Yang et al. (2016), and the assembled genome sizes were 396.9 and 784 Mb, respectively. In 2019, Wang et al. produced a new *B. nigra* genome assembly comprising 484 Mb using the accession CGN7651 (Wang et al. 2019). Recently, the genomes of two other *B. nigra* varieties, CN115125 and Ni100, were assembled, resulting in genome sizes of 537 and 506 Mb, respectively (Perumal et al. 2020).

The *B. napus* genome was first assembled using a European winter oilseed cultivar, Darmor-*bzh* (Chalhoub et al. 2014). In 2017, Bayer and King assembled a new *B. napus* genome using the Tapidor cultivar, but fewer genes were found in this genome than the Darmor-*bzh* genome (Bayer and King 2017). The same year, the genome of an Asian semi-winter *B. napus* cultivar, Zhongshuang 11 (ZS11), was also sequenced and integrated into the ongoing analyses (Sun et al. 2017). The semi-winter type ZS11 and the winter-type Darmor-*bzh* maintained high levels of genomic collinearity, and the ZS11 genome harbored several cultivars-specific segmental homeologous exchanges (HEs). Moreover, the ZS11 genome displayed potential genomic introgressions with *B. rapa*. In 2020, the genome of *B. napus* cultivar, ZS11, was assembled again by Chen et al. (2020a).

Recently, the first *B. carinata* genome has been successfully sequenced and assembled using the accession of "zd-1," and the assembled genome size was 1,086.8 Mb, accounting for 94.44% of the estimated genome (Song et al. 2021). At the present, the genomes of all six *Brassica* species in "triangle of U's" have been resolved, which provide valuable insights into the genome evolution of the six *Brassica* species in "triangle of U's".

In addition, the pangenomes of *B. oleracea* and *B. napus* were assembled, which contributed to the progress of understanding genome variation at the population level in *Brassica* genus (Golicz et al. 2016; Song et al. 2020). We describe these pangenomes below in Sect. 2.2.

2.2 Pangenomes

A large number of genes, which control major agronomic traits, have presence/absence variation (PAV, a major class of genome structural variations) and copy number variation in plant species (Tao et al. 2019). However, the reference genome of a single individual is unable to capture/identify the whole gene content because of the PAV. Therefore, there is an increasing awareness that it is essential to construct pangenomes, which integrate the genomes of many accessions from different ecotypes and genotypes (Song et al. 2020).

In *Brassica* crop species, Golicz et al. (2016) constructed and analyzed a 587 Mb *B. oleracea* pangenome sequence using nine morphologically diverse *B. oleracea* varieties and a wild relative, *Brassica macrocarpa*. A large number of genes containing PAVs were found to influence phenotypic and agronomic traits, including auxin function, flowering time, and glucosinolate metabolism (Golicz et al. 2016). To better understand the genetic diversity of the *B. napus* morphotypes and their cultivation history, Song et al. (2020) reported on the sequencing, *de novo* assembly, and annotation of the genomes of eight *B. napus* accessions, including four semi-winter types, two winter types, and two spring types. The pan-reference genome size was estimated to be ~1.8 Gb and it contains 152,185 genes (Fig. 2). The genome size and gene number increased as the number of genomes increased but the number of orthologous gene clusters did not increase after combining six genomes, indicating that the gene families in the *B. napus* pan-reference genome tend to be saturated when integrating six genomes representing the three ecotypes. On the base of the pangenome, the authors explored the evolution and intraspecific genomic variations of *B. napus*. Millions of small variations and 77.2–149.6 Mb of PAVs were identified. Besides, researchers found that PAV-based genome-wide association study (PAV-GWAS) is complementary to SNP-GWAS in identifying associations with phenotypes caused by SVs (Song et al. 2020). The pangenome analysis provides important resources for understanding the genome architecture and population genomics studies and accelerating the genetic improvement of *B. napus*.

2.3 Genetic Maps and Genetic Markers

Genetic marker technology has advanced rapidly and has been widely used in plant breeding and genetic studies. Moreover, genome-wide and genetically mapped genetic markers provide an excellent resource for population genomics studies (Luikart et al. 2019). The first molecular markers developed and used in *Brassica* were restriction fragment length polymorphism (RFLP) markers, which were mainly used in the 1990s (Song et al. 1991; Chyi et al. 1992; Teutonico and Osborn 1994; Lagercrantz and Lydiate 1996). Thereafter, amplified fragment length polymorphism and random amplified polymorphic DNA (RAPD) markers became the main types of molecular markers used by geneticists and breeders between 1995

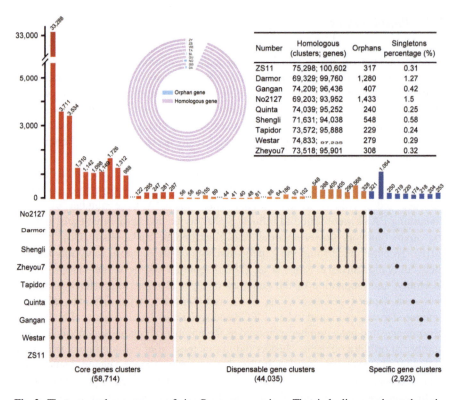

Fig. 2 The core- and pan-genome of nine *B. napus* accessions. The circle diagram shows the ratio of homologous genes to orphan genes and the table lists the detailed number. The histograms below show the core-gene clusters, dispensable gene clusters, and specific gene clusters. This figure is reproduced from Song et al. (2020)

and 2006 (Jung et al. 2006). Finally, from 2005 to the present, sequence-tagged markers, such as simple sequence repeats (SSR; Choi et al. 2007; Kim et al. 2009; Wang et al. 2011d; Kebede et al. 2012; Suwabe et al. 2012; Yu et al. 2013), single nucleotide polymorphism(s) (SNPS/SNPs) (Li et al. 2011; Wang et al. 2011b; Paritosh et al. 2013; Chung et al. 2014), and insertion–deletions (InDels; Choi et al. 2007; Kim et al. 2009; Wang et al. 2011b; Suwabe et al. 2012; Yu et al. 2013; Liu et al. 2013) gradually became commonly used.

Genetic linkage maps are highly effective tools for the primary and fine-mapping of the genomic regions carrying major genes, quantitative trait loci (QTLs) associated with desired agronomic traits and population genomics studies. In traditional linkage maps, the orientation of linkage groups was different in different maps, and there was a need to ensure that linkage group number and orientation were correct. Now, with the development of high-throughput sequencing technologies and the availability of the *Brassica* reference genomes, sequence-tagged markers are typically used in genetic linkage mapping. These markers are transferable across mapping populations, enabling them to serve as anchors to link mapping populations and determining their physical positions in the reference genome. Genetic maps with dense markers allow for finer comparative genome analyses with related genome sequences and the faster map-based cloning of major genes and QTLs. Genetically mapped markers are highly suitable for various population genomics studies (Luikart et al. 2019), including genetic diversity and population genetic structure of natural populations.

In recent years, many genetic maps of *Brassica* species with sequence-tagged markers have been built (Bonnema 2015). A high-density linkage map for *B. rapa* containing 5392 high-quality SNP markers was constructed by analyzing a recombinant inbred line population using specific-locus amplified fragment sequencing (SLAF-seq), which is a relatively new low-cost and highly efficient high-throughput sequencing technique (Liu et al. 2019). Yu et al. (2019) constructed a high-density genetic map with 6694 SLAF markers and 12,980 SNPs for *B. oleracea* var. *italica*. A genetic linkage map of an F_2 population from two *B. carinata* accessions with contrasting morphological characteristics was constructed using a set of 6464 high-quality DArTseq markers and subsequently used for a QTL analysis (Raman et al. 2017). A 60,000 (60K) SNP array for *B. napus* (A and C subgenomes) was published in 2012 (Snowdon and Luy 2012; Edwards et al. 2013). This SNP array enables the efficient generation of high-density, sequence-based, genome-wide polymorphism screens and genetic maps. This array has also been tested on both *B. rapa* and *B. oleracea* doubled haploid (DH) populations and many accessions (Bonnema 2015). In Sect. 3.3, we discuss the uses of molecular makers in *Brassica* populations for genetic diversity and population structure analysis in detail.

3 Progress in Population Genomics of *Brassica*

3.1 Taxonomy and Genetic Relationships of Brassica Species

The genus *Brassica* is one of the most economically important genera in the family Brassicaceae (Cruciferae). In this family, Heywood (1993) recorded 380 genera and 3,000 species, whereas 419 genera and 4,130 species were recorded in the study of Judd et al. (1999). After that, Mabberley (1997) found 365 genera and 3,250 species, and Warwick et al. (2006) recorded 338 genera and 3,709 species belonging to the family Brassicaceae. The *Brassica* genus contains numerous species, including major vegetable and oilseed crops with a lot of agronomic traits. This genus contains six economically important species, including three diploid species (*B. oleracea*, 2n = 18; *B. rapa*, 2n = 20; *B. nigra*, 2n = 16) and three allotetraploid species (*B. napus*, 2n = 38; *B. juncea*, 2n = 36; *B. carinata*, 2n = 34), with a wide range of genetic and morphological diversity. The whole-genome triplication (WGT) has been an important event in the speciation and the expansion of numerous morphotypes in the *Brassica* genus. The subsequent genomic rearrangement and gene evolution initiated by WGT promoted the appearance of a variety of *Brassica* plants (Cheng et al. 2014). Many traits are shared but evolved independently and in parallel, such as heading leaves and enlarged roots.

B. oleracea, as an important vegetable crop species, includes several cole crops: cabbage (*B. oleracea* subspecies *capitata*), cauliflower (*B. oleracea* subspecies *botrytis*), brussels sprout (*B. oleracea* subspecies *gemmifera*), broccoli (*B. oleracea* subspecies *italica*), Kale and collards (*B. oleracea* subspecies *acephala*), and kohlrabi (*B. oleracea* subspecies *gongylodes*) (El-Esawi 2017). The cole crops are categorized by extreme morphological characteristics, such as enlarged inflorescence (cauliflower and broccoli), enlarged stems and marrowstem (kohlrabi and kale), enlarged single apical bud (cabbage), and axillary buds (brussels sprout) (Hong et al. 2008). In general, *B. oleracea* crop species grow slowly and present a large storage capacity for nutrients.

B. rapa is wildly cultivated for leaf and root vegetables and oilseed as well. Different morphotypes have different usages (Cheng et al. 2014), and three groups well-defined based on the morphological characters of *B. rapa* (CFIA 1999): (1) The oil-type *B. rapa*, such as Polish rape or summer turnip rape, produce large, full seeds for oil extraction, and have a specific form with "double low," i.e., low erucic acid in its oil and low glucosinolate content in its meal protein; (2) The leaf type *B. rapa*, which develops beautiful leaf patterns and colors, thus used as ornamental plants, such as chinensis group (pak-choi, celery mustard), the pekinensis group (Chinese cabbage), and the perviridis group (tendergreen); (3) The rapiferous type *B. rapa*, the plants usually present large leaves and are used as the important vegetable sources, including the rapifera group (turnip, rapini) and the ruvo group (turnip broccoli, Italian turnip) (CFIA 1999). The seven known varieties of vegetable *B. rapa* types, var. *campestris*, var. *pekinensis*, var. *parachinensis*, var. *japonica*, var. *narinosa*, var. *chinensis* and var. *rapa*, the var. *campestris* is known as the crudest leaf vegetable, var. *pekinensis* and var. *narinosa* have an ability to cold tolerance. The var. *chinensis* is differentiated from oilseed rape types of China and var. *parachinensis* is derived from var. *chinensis*. The var. *japonica* is a leaf vegetable, which comes from Japan, and var. *rapa* (turnip) is a vegetable, which is widely cultivated worldwide (Rakow 2004)

B. nigra, also known as black mustard, is an annual plant and its seeds are often used as a condiment. *B. nigra* is always planted in the tropical northern Africa, temperate parts of Europe and parts of Asia. The plant height of *B. nigra* can reach up to 2 m and this species does not require the induction of vernalization for flowering (Rakow 2004).

B. napus is cultivated mainly as oilseed or canola rape in many countries, due to its highly productive yield. *B. napus* could be divided into three ecotypes: the winter, semi-winter, and spring ecotypes. Its high-yield potentiality might be related to the high photosynthetic rate per unit leaf area which is positively related to chloroplast number per unit leaf area and to chloroplast volume (El-Esawi 2017). Interestingly, the special root-forming *B. napus* types, known as tuber-bearing swede or rutabaga, are cultivated as vegetables and fodder for animals (Rakow 2004).

B. juncea is suitable to semi-arid conditions, and it has a great potential of seed productions. *B. juncea* has many branching and tall plants. *B. juncea* contains enormous morphotype diversity, and five subspecies are recognized, including ssp. *juncea, crispifolia, foliosa, integrifolia*, and *napiformis*. The seed size of *B. juncea* is small with yellow and black-red coat. *B. carinata*, also named as Ethiopian mustard, is taller than other *Brassica* species with the pale purple stem. *B. carinata* as a winter crop contains many desirable agronomic characteristics, including heat, drought, lodging tolerance and disease resistance (Ojiewo et al. 2014; Odongo et al. 2017; Raman et al. 2017). Comparing with other *Brassica* species, *B. carinata* can adapt to a much wider range of environmental conditions and grow in extremely unfavorable environments (Ban et al. 2017).

The genetic relationship of the six main species of the *Brassica* genus was described as the U triangle (Nagaharu 1935). Pachytene chromosome morphology studies have identified the basic genomes of *Brassica* crops, *B. rapa* is the AA genome ($2n = 20$), *B. nigra* is the BB genome ($2n = 16$) and CC genome ($2n = 18$) is for *B. oleracea* (Branca and Cartea 2011). Comparing genomic libraries between *B. napus* and *B. oleracea*, studies identified numerous shared fragments among A, B, and C-genomes, suggesting the origin of allotetraploid species (*B. napus, B. carinata,* and *B. juncea*) from its parental diploid species (Hosaka et al. 1990; Branca and Cartea 2011). These three allotetraploid species, *B. napus* is the allotetraploid species derived from the crosses between *B. rapa* and *B. oleracea*; *B. juncea* is an allotetraploid species formed by two progenitor species, *B. rapa* and *B. nigra*; *B. carinata* is an allotetraploid species originated from the crosses between *B. nigra* and *B. oleracea* (Nagaharu 1935; Tsunoda 1980; Rakow 2004; Chen et al. 2013). Moreover, phylogenetic studies illustrated a common ancestor with $n = 6$ in the evolution of *Brassica* species (Song et al. 1990; Branca and Cartea 2011), and 36 cytodemes for *Brassica* crops, such as *Enarthrocarpus, Hirschfeldia, Eruca, Erucastrum, Diplotaxis, Sinapis,* and *Trachystoma* genera (Harbered 1976; Branca and Cartea 2011). However, comparing with AB and BC amphihaploids, the close genetic relationship between the A and C genome has been illustrated through the meaningfully higher amounts of chromosome pairing in AC amphihaploids (Attia and Röbbelen 1986; Attia et al. 1987; Navabi 2009). On the other side, there is few pairings between the B-genome chromosomes with the AC chromosomes, suggesting that the B-genome is more distantly related to the A and C genome (Navabi 2009).

3.2 Evolution and Phylogenomics of Brassica Genus

Brassica genomes have experienced extensive chromosome reshuffling during their evolutionary history. Chromosome variation following polyploidization has been prevalent and played an important role in the emergence of novel functions and in speciation (Wendel 2000; Gaeta and Chris 2010; Soltis et al. 2016). The ancestral *Brassica* karyotype has been under debate for more than half a century, although the sequencing of the *B. rapa* genome provided the first opportunity to investigate the ancestral genomes of this genus. Based on deciphering of the *Brassica* genomes, the diploid ancestral *Brassica* genomes were speculated to resemble the translocated Proto Calepineae Karyotype before the WGT (Fig. 1b) (Mandakova and Lysak 2008; Cheng et al. 2013). A comparative analysis of *B. rapa* and *A. thaliana* revealed that 71 of the 72 (3 × 24) expected genome blocks (GBs) in the *B. rapa* genome comprised ancestral GBs from the Ancestral Crucifer Karyotype ($n = 8$; Fig. 1c; Schranz et al. 2006), and the genomic synteny comparison between them suggested three syntenic copies for each GB. According to the rate of gene loss (fractionation) in the *Brassica* genomes, the GBs were classified into three subgenomes, LF (the least fractionated), MF1 (the medium fractionated), and MF2 (the most fractionated) (Wang et al. 2011c). After the WGT, the genomic reshuffling occurred independently in the subgenomes during rediploidization, and subgenomic points with similar breakages or fusions should be the chromosomal boundaries of their diploid ancestor. With the assembly of the genome of the Chinese cabbage cultivar chiifu401/42 at 2011, the specific WGT event in *Brassica* genus was studied in detail (Lysak et al. 2005; Wang et al. 2011c; Kagale et al. 2014). The WGT of *Brassica* species occurred ~11 MYA (Song et al. 2020). Following the deciphering of the *Brassica* genome data, the evolutionary relationships of the *Brassica* species have been deeply elucidated. The *Brassica* lineages diverged from the Arabidopsis lineages ~20–35 MYA (Yang et al. 1999; Town et al. 2006). *B. nigra* separated from the ancestor of *B. oleracea* and *B. rapa* about 6.0–14.6 MYA (Lysak et al. 2005; Navabi et al. 2013), while the separation of *B. oleracea* and *B. rapa* occurred approximately 4.0 MYA (Lysak et al. 2005; Navabi et al. 2013), although one study suggested this latter divergence occurred 6.8 MYA (Sun et al. 2019).

Furthermore, the version 3 reference genome of *B. rapa* revealed a new duplication event that occurred in the *B. rapa* genome ~1.2 MYA, which was accompanied by an expansion of a long terminal repeat retrotransposon (Zhang et al. 2018). Asymmetrical gene loss, transposable elements, and gene expression were widely examined, which revealed that the genome duplication and gene divergence that occurred in *B. oleracea* may have influenced its biochemical and morphological variation (Liu et al. 2014; Parkin et al. 2014). A comparative genomic analysis revealed that cauliflower diverged from the ancestral *B. oleracea* ~3.0 MYA, which was later than the divergence of *B. oleracea* var. *capitata* (~ 2.6 MYA) and other *Brassica* species (over 2.0MYA) (Sun et al. 2019). By comparing the chromosomes of *B. nigra* with that of *B. rapa* and *B. oleracea*, it was revealed that most of the variation between the B and A/C subgenomes had occurred before the A/C divergence, although more variation was detected on the C chromosomes following their divergence (Wang et al. 2019). By comparing two genomes of *B. nigra* accessions, a novel centromere-associated ALE class I element was identified, which appears to have proliferated through relatively recent nested transposition events (<1 MYA) (Perumal et al. 2020).

The allotetraploid species of *Brassica*, *B. napus* (2n = AACC), *B. juncea* (2n = AABB), and *B. carinata* (2n = BBCC), were derived from diploid species *B. rapa* (2n= AA), *B. oleracea* (2n = CC), and *B. nigra* (2n = BB) (Nagaharu 1935; Tsunoda 1980; Rakow 2004; Chen et al. 2013). *B. napus*, diverged from its progenitors ~7,500 years ago or less (Chalhoub et al. 2014; Sun et al. 2017). Further research revealed that A subgenome may have evolved from the ancestor of European turnip (~106–1,170 years ago) and the C subgenome may have evolved from the common ancestor of kohlrabi, cauliflower, broccoli, and Chinese kale (~108–898 years ago) (Lu et al. 2019). The semi-winter type *B. napus* (from China) were found to be closer to turnip than to spring type and winter type (from Europe, Canada, Australia and other countries) based on population structure analysis (Lu et al. 2019; Song et al. 2020). Other researchers found that the A and C subgenomes were engaged in subtle structural, functional, and epigenetic crosstalk (Chalhoub et al. 2014). The genomic analysis suggested that the divergence of many duplicate genes contributed to the diversification of the *B. napus* varieties, resulting in the evolution of various agronomic traits, such as oil biosynthesis, seed glucosinolate contents, and flowering (Chalhoub et al. 2014). A comparative genomic analysis revealed that *B. juncea* evolved from its progenitors ~39,000–55,000 years ago (Yang et al. 2016). The *B. juncea* genome was compared with that of *B. napus*, revealing that the A subgenomes of *B. juncea* and *B. napus* each had independent origins (Yang et al. 2016). The study showed that the A subgenome of *B. napus* originated from the *B. rapa* subspecies European turnip (*B. rapa* subsp. *rapa*), while the A subgenome of *B. juncea* is closely related to the *B. rapa* subgroup oilseed (*B. rapa* subsp. *tricolaris*). The genome of *B. napus* was used in a comparative genomic analysis along with that of *B. juncea*, which revealed subgenomic features, including biased gene retention and expression in the B subgenome. Recently, the genomic analysis revealed that *B. carinata* evolved from its progenitors ~47,000 years ago, and the relationship between BcaB subgenome and BcaC subgenome is greater than that between BjuB subgenomes and BnaC subgenomes, and also from their diploid parents (Song et al. 2021). Overall, the *Brassica* species containing the B subgenome are more ancient than those with only the A or C subgenome, and the *Brassica* species with the A subgenome appeared earlier than those with the C subgenome (Yang et al. 2016).

3.3 Genetic Diversity and Population Structure

Genetic diversity represents the variation of individual genotypes within and among species and provides the basis for crop survival, adaptation, evolution, domestication, and genetic improvement (El-Esawi 2017). The genetic diversity allows individuals or populations to have the capacity to adjust to a changing environment, whether from human or natural factors. Within a species range, the genetic composition of whole populations varies from place to place (El-Esawi 2015), and these differences provide an opportunity for dispersing individuals to create a new population or change allele frequencies from mating in small populations (Meffe and Carroll 1994; Husband and Schemske 1996; Falk et al. 2001). Differences among populations may be derived from locations, conditions, survival and reproduction, and gene flow, resulting in the accumulation of genetic differences and furtherly developing a new species (Falk et al. 2001, El-Esawi 2015). During the past three decades, genetic diversity is estimated based on morphological, cytological, biochemical, and molecular markers. The molecular markers used include isozymes, RAPD, inter-simple sequence ISSR, AFLPs, RFLPs and SSRs, and genome-wide or large-scale makers SNPs, INDELs. Comparing with earlier molecular markers, genome-wide or large-scale markers based on high-throughput sequencing can reveal adaptive or functionally important loci, and these makers have become the ideal makers for genetic diversity analyses and molecular breeding. SNP or INDELs are widely spread across genomes, enabling us to explore genetic diversity among individuals and populations, as well as detecting mutations, genetic drift, selection, and population structure using these markers. Besides their abundance in genomes, SNP or INDELs have the advantages of being codominant and amenable to high-throughput automation (El-Esawi et al. 2016). Moreover, the high-quality SNP array not only provides new key insights into various population genomics aspects, such as genetic diversity and population structure, but also is suitable for genetic fine-mapping, genome-wide association studies (GWAS), and genomic selection.

Genetic diversity in *Brassica* genus has been studied based on earlier to modern molecular markers, including isozymes (Sekhon and Gupta 1995), RAPD (Jain et al. 1994), inter-simple sequence ISSR (Bornet and Branchard 2004), AFLPs (Negi et al. 2004), RFLPs (Song et al. 1995; Cavell et al. 1998), SSRs (Tautz 1989; Zhou et al. 2006; Chen et al. 2008), and SNP (Bird et al. 2017). Studies on genetic diversity and population structure of *Brassica* species have allowed us to understand their survival, adaptation, evolution, and origin and facilitated genetic improvement, and the classification of the *Brassica* populations according to their genotypes, morphotype, and geography. Genetic diversity analyses have also facilitated the identification and elucidating evolutionary histories of wild, cultivated, and landraces of *Brassica* (Persson et al. 2001; Pu et al. 2007; Christensen et al. 2011; El-Esawi et al. 2016). Using SNPs or INDELs makers, researchers have developed valuable material or methods to evaluate the multinational *Brassica* diversity, which provided improved resolution of genetic relationships of subspecies within *Brassica* species (Pang et al. 2015; Bird et al. 2017; Zhang et al. 2017; Lu et al. 2019; Wu et al. 2019; Khedikar et al. 2020). Here, we review a few case studies for the "triangle of U" theory species in *Brassica* genus.

3.3.1 *B. oleracea*

The genetic diversity and population structure *B. oleracea* have been well studied in the past three decades. According to the morphotype and geography, *B. oleracea* accessions are divided into numerous groups, such as west Mediterranean group, central Mediterranean group, Aegean group, and Atlantic group (Lázaro and Aguinagalde 1998a). Through the analyses of genetic diversity in populations, researchers studied the genetic variation in *B. oleracea* from different eco-geographies, and divided cultivated, wild, and feral populations. Besides, the important genetic resources were identified in some regions that could be used for further research, such as wild cabbages in the Atlantic region (Maggioni et al. 2020,

Mittell et al. 2020). Here, we summarize several studies that focused on *B. oleracea* genetic diversity and population structure.

Lázaro and Aguinagalde (1998a) analyzed the genetic diversity of *B. oleracea* among 36 populations of wild taxa and two cultivated forms using isozyme variation at 11 loci and found that 67% of the genetic diversity resided within populations 33% among populations. Using genetic distances among taxa, these populations could be clustered into three different groups, including west Mediterranean group (e.g., *B. oleracea*, *B. alboglabra,* and *B. bourgeaui*), the central Mediterranean area group (e.g., *B. villosa*, *B. villosa subsp. Drepanensis* and *B. rupestris*), and Aegean group (e.g., *B. cretica*) (Lázaro and Aguinagalde 1998a). At the same time, they evaluated the genetic diversity of 29 wild populations of *B. oleracea* and several wild taxa and two cultivars using RAPD markers and found that *B. oleracea* populations were clustered with those of *B. bourgeaui*, *B. alboglabra*, *B. montana* and *B. incana* in the first cluster, *B. insularis*, *B. macrocarpa*, *B. villosa,* and *B. rupestris* populations in the second cluster, and *B. cretica* and *B. hilarionis* populations formed the third cluster. Further genetic diversity analysis suggested that *B. cretica* subspecies was the most diverse (Lázaro and Aguinagalde 1998b). Using AFLP markers, Christensen et al. (2011) analyzed the genetic diversity and structure of *B. oleracea* among 17 accessions of kale landraces, cultivars, and wild populations from Europe and found that the average gene diversity (based on allele frequencies) ranged from 0.11 to 0.27; some landraces presented higher levels of diversity than the wild populations, and the most genetically diverse landraces were those collected from areas where kales are known to be extensively grown (Christensen et al. 2011). Across Ireland, El-Esawi et al. (2016) assessed the genetic diversity and population structure of *B. oleracea* germplasm using microsatellite markers and found that all of the studied accessions have an excess of heterozygotes (observed heterozygosity, 0.699; expected heterozygosity, 0.417). They showed that 27.1% of the total genetic variation was among accessions, and 72.9% of the variation resided within accessions (El-Esawi et al. 2016). Recently, Maggioni et al. (2020) surveyed the genetic diversity and structure of nine wild populations and five locally cultivated accessions of *B. oleracea* in the northwestern French coastal areas using AFLP markers. The population structure analysis showed a low level of interpopulation genetic differentiation, which could be explained by a relatively recent origin of all populations from a common source. The genetic diversity of wild populations was similar to, or lower than that of cultivated accessions and the wild Penly/Petites Dalles cabbage; and the authors concluded that the Saint-Saëns cabbage could be used for conservation and breeding due to its high levels of genetic diversity (Maggioni et al. 2020). The genetic diversity and structure of feral populations of *B. oleracea* along the Atlantic coast in western Europe were assessed by Mittell et al. (2020), and no evidence of isolation by distance was found. Several geographically related populations were found to be more genetically distinct from each other than distant populations, and some distant populations showed shared genetic ancestry. Their results suggested that wild cabbages are an important genetic resource in the Atlantic region, and that their relationship with current crop varieties should be examined (Mittell et al. 2020).

3.3.2 B. rapa

The genetic diversity of *B. rapa* has also been estimated based on morphological, cytological, biochemical, and molecular markers. Through the genetic diversity, population structure and phylogenetic analyses, geographic origin, genetic variation, interpopulation, and morphotypes have been well understood in this species. Generally, the *B. rapa* accession could be grouped according to their morphotype and geography, including European turnips, Asian turnips, yellow/brown sarson, Chinese cabbage and bok choy, choy sum, and tatsoi (Bird et al. 2017). Moreover, the genetic diversity of *B. rapa* deepens our understanding of the evolutionary history among wild, cultivated, and landraces taxa, also improves resolution of genetic relationships of subspecies within *B. rapa* (Pang et al. 2015; Bird et al. 2017; Aissiou et al. 2018). Here, we summarize several studies conducted on *B. rapa* genetic diversity and population structure.

The genetic diversity of European and Chinese *B. rapa* was first studied using isozyme markers by Zhao and Becker (1998), who found two distinct groups corresponding to European and Chinese origins. Later, Persson et al. (2001) studied genetic diversity and relationships of 31 accessions of turnip (*B. rapa* ssp. *rapa*) from the Nordic area, also using isozymes, and reported five clusters among the 31 accessions, with one cluster including 25 of the 31 accessions. The level of heterozygosity in the landraces was relatively low but it was higher than that of variety "Ostersundom" (Persson et al. 2001). Using RAPD markers, He et al. (2002) examined the genetic diversity in vegetable *B. rapa* originated from China. The results showed that the vegetable and oil-yielding *B. rapa* formed subgroups according to their geographical origin of collection, and the genetic diversity level in the spring type was higher than that in the winter type (He et al. 2002). Using SSR markers, Soengas et al. (2011) assessed the genetic diversity and structure of 80 populations of *B. rapa subsp. rapa* from northwestern Spain. The population genetic analysis suggested a broad range of genetic diversity, with most of the genetic variation residing within populations and low level of interpopulation genetic differentiation – all populations clustered in one group. Based on these results, Soengas et al. (2011) suggested existence of highly variable metapopulation in the studied samples. Pang et al. (2015) investigated the genetic diversity and population structure of 238 fixed lines of leafy *B. rapa* with SSR markers and SNP markers. The population structure analysis revealed four subpopulations in accordance with the geographical origins and morphological traits. Particularly, the Chinese cabbage group was subdivided into three subgroups and presented obvious correlation with leaf- and heading-related traits. Their work developed a valuable material for evaluating the multinational *B. rapa* diversity (Pang et al. 2015). Zhang et al. (2017) evaluated the genetic diversity and relationship of 54 *B. rapa* accessions (42 Chinese accessions and 12 exotic accessions) using SSR and SRAP markers. The PCA and population structure analyses separated the accessions into five major clusters. The exotic *B. rapa* accessions grouped in Cluster I with the exception of yellow sarson from India, Cluster II was the Chinese *B. rapa*, and other *Brassicas* formed the Cluster III, Cluster IV, and Cluster V. The proportion of the genetic variation was 26.1% among populations and 73.90% within populations (Zhang et al. 2017). Aissiou et al. (2018) used SSR markers and morphological traits to explore genetic diversity among 18 Algerian *B. rapa* accessions and compared with that previously reported in other *Brassica* species, including *B. napus* varieties used. Their results showed that the wild and cultivated Algerian accessions had the highest allelic richness as compared to the worldwide *B. rapa* diversity. Population structure analysis revealed that wild and cultivated *B. rapa* accessions from Algeria formed two clusters regardless of their local geographic origin and showed their distinction from other *B. rapa* groups described previously (Aissiou et al. 2018). In a global genetic diversity analysis, Bird et al. (2017) assessed a panel of 333 *B. rapa* accessions from National Plant Germplasm System using 18,272 SNP markers. The population genetic and phylogenetic analyses clustered these accessions into five groups, corresponding to the morphotype and geography, including European turnips, Asian turnips, yellow/brown sarson (*B. rapa* ssp. *trilocularis* and ssp. *dichotoma*), Chinese cabbage (*B. rapa* ssp. *pekinensis*), and bok choy, choy sum, and tatsoi (*B. rapa* ssp. *chinensis*, ssp. *parachinensis*, ssp. *narinosa*). Moreover, evidence for polyphyly and/or paraphyly was found, especially for oilseed morphotypes (*B. rapa* ssp. *oleifera* and ssp. *dichotoma*) and turnips. Their results provided improved resolution of genetic relationships of subspecies within *B. rapa* (Bird et al. 2017).

3.3.3 *B. napus*

Compared to its parental species (*B. oleracea* and *B. rapa*), although *B. napus* (rapeseed, canola) has a shorter evolutionary history, it has adapted to diverse climate and latitudes and formed three main ecotype groups: winter, semi-winter, and spring types. Wang et al. (2014) investigated genome-wide genetic changes, including population structure, genetic relatedness, the extent of linkage disequilibrium, nucleotide diversity, and genetic differentiation based on F_{ST} outlier detection, for a panel of 472 *B. napus* inbred accessions (including inbred lines of commercial cultivars and landraces produced from 1950s to 2010s) using a 60K *Brassica* Infinium® SNP array. They found that the genetic diversity increased meaningfully in both China and Europe during the periods 1950–1970 and 1971–1980, while it maintained at a stable level in recent years compared with cultivars produced in previous decades. Besides, comparing to European samples, Chinese germplasm presented a lower level of genetic diversity (Wang et al. 2014). Using SNPs, Gazave et al. (2016) assessed genetic diversity of 782 accessions, including winter and spring types, collected from 33 countries across Europe, Asia, and America. The strong population structure was observed corresponding broadly with the accessions' growth habit and geography; the accessions clustered into three major genetic groups: including spring, winter European, and winter Asian groups. Compared with the winter European group using subpopulation-specific polymorphism patterns, winter Asia group showed enriched genetic diversity, while a smaller effective breeding population for spring group compared to winter Europe. Their results suggested different geographic origins for the two subgenomes of *B. napus*, with phylogenetic analysis placing winter Europe and winter Asia as basal clades for the other subpopulations in the C and A subgenomes, respectively (Gazave et al. 2016). Subsequently, using genotyping by sequencing transcriptomics and/or whole-genome resequencing, Malmberg et al. (2018) analyzed the global genetic diversity and population structure of 633 *B. napus* samples, representing 627 canola varieties from 27 countries, including 258 Australian samples, 271 European samples, 69 Asian samples, 8 North American samples, 2 New Zealand samples, 1 African sample, and 24 samples of unknown. The samples clustered into spring, winter, and semi-winter groups. The cluster results were also consistent with the geographic origin of the samples, because the majority of Australian samples were of spring types, European accessions winter types, and most of Asian varieties, Chinese semi-winter types, and Asian winter types (Malmberg et al. 2018). Lu et al. (2019) resequenced 588 diverse *B. napus* accessions from 21 countries and also studied 199 *B. rapa* and 119 *B. oleracea* accessions. They suggested that the A subgenome of *B. napus* may have evolved from the ancestor of European turnip and that C subgenome may have evolved from the common ancestor of kohlrabi, cauliflower, broccoli, and Chinese kale, and the original form of *B. napus* may be the winter oilseed (Lu et al. 2019). About the same time, Wu et al. (2019) also resequenced 991 rapeseed germplasm accessions, including 658 winter, 145 semi-winter, and 188 spring types from 39 countries (Fig. 3a). The population genetic structure and principal component analysis (PCA) clustered these rapeseed accessions into three groups (group 1-3), roughly corresponding to their ecotype groups revealing that the winter, semi-winter, and spring ecotypes are genetically different from each other (Fig. 3b, c). The Principal component (PC) 1 accounted for 11.19% of the total variation and distinguished the winter-type accessions from the semi-winter and spring types, whereas PC2 accounted for 6.90% of the total variation and distinguished the semi-winter types from the spring types (Fig. 3d). However, 8.9% accessions did not group with their ecotypic group, suggesting that genomic modifications may have occurred during their adaptation to different environments (Wu et al. 2019). Using microsatellite markers, Chen et al. (2020b) elucidated the genetic diversity of feral rapeseeds in Japan by genotyping 537 accessions from various regions in Japan. The results suggested moderate genetic diversity and high inbreeding levels within the feral populations. Analysis of molecular variance revealed a greater proportion of genetic diversity among individuals than between populations (Chen et al. 2020b). In the recent pangenome studies of *B. napus*, Song et al. (2020) assessed 210 *B. napus* accessions, included semi-winter oilseed rapes from China and winter-type oilseed rapes and spring-type oilseed rapes from Europe, Canada, Australia and other countries. The *accessions* were clustered into different groups roughly corresponding to three ecotypes. The phylogenetic tree and PCA results showed that semi-winter oilseed rapes were closer to turnip than to spring-type oilseed rapes and winter-type oilseed rapes, corresponding to the PCA results. Their result suggested that the genetic diversity of these selected accessions

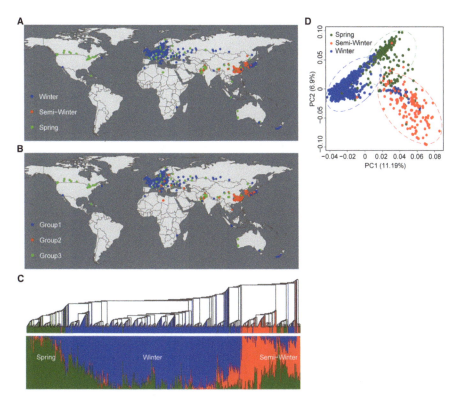

Fig. 3 Distribution, population structure, and PCA of 991 rapeseed germplasm accessions. (**a**) Geographic distribution based on country of origin and ecotype information. (**b**) Geographic distribution based on population structure and PCA. (**c**) Population structure analysis. (**d**) PCA plot of the first two components (PC1 and PC2). This figure is reproduced with permission from Wu et al. (2019)

typically represented the genetic diversity of natural rapeseed populations (Song et al. 2020).

Through various genetic diversity and population structure analyses, the three main ecotype groups, winter, semi-winter, and spring types could be easily genetically distinguished. This genetic differentiation among different ecotypes could be due to natural selection for adaptation to different environments. Similar with its parental species, the *B. napus* genetic diversity is also consistent with its geography, and the A and C subgenomes have different geographic origins. Increasing number of studies have revealed the origin of A and C subgenomes based on SNP makers, while no true wild *B. napus* is found and the original form of *B. napus* still remains debatable.

3.3.4 B. juncea

Like *B. oleracea*, *B. rapa,* and *B. napus*, the genetic diversity of *B. juncea* has been examined based on genetic markers, and its genotypes also have shown genetic differentiation according to their geographical origin. Here, we briefly review key studies reporting genetic diversity and population structure in *B. juncea*. Jain et al. (1994) documented genetic variation among 12 Indian and 11 East European *B. juncea* accessions using RAPD markers. These populations clustered into two groups, one group contained only East European accessions while another group comprised all Indian and four exotic genotypes (Jain et al. 1994). Using AFLP markers, Burton et al. (2004) assessed the genetic diversity of 92 *B. juncea* accessions, including 77 canola-quality lines from Canada and the Australia, and 15 non-canola-quality lines from India, Russia, Canada, and China. These authors found that the 92 *B. juncea* breeding lines formed two groups, with Indian genotypes forming a distinct group and the rest forming the second group (Burton et al. 2004). Similarly, Srivastava et al. (2001) assessed 21 natural *B. juncea* accessions originating from India, Russia, Australia, and Canada and nine resynthesized lines using AFLP markers, and divided them into three diversity groups. Among them, all Indian types clustered into one group; the resynthesized lines formed the second cluster and all East European types (from Australia, Canada, and Russia) formed the third group (Srivastava et al. 2001). Singh et al. (2013) established the genetic diversity of 50 genotypes of Indian mustard using RAPD markers and biological yield. High variation was observed for the yield. The molecular markers and agronomic traits suggested division of genotypes into different clusters, but no association was found between clusters and geographic origin (Singh et al. 2013). Vinu et al. (2013) examined the genetic diversity of 44 mustard genotypes from India and abroad using SSR markers and yield and component traits data. Based on genetic distances, these tested genotypes could be separated into four groups corresponding with their origin and pedigree records (Vinu et al. 2013). In southwest and western China, studies also identified high RAPD diversity in *B. juncea* (Li et al. 1997; An et al. 1999), and the *B. juncea* landraces were found to be closely related to agro ecological adaptations in southwest China (Pu et al. 2007). Similarity, the molecular genetic diversity associated with geological and biological conditions was also reported for 101 *B. juncea* accessions from western China (Xu et al. 2008). Meanwhile, these authors found that the genetic diversity of *B. juncea* winter types was higher than that of the spring types, and *B. juncea* in Shaanxi and Xinjiang provinces presented higher genetic diversity than those observed in Tibet (Xu et al. 2008). With further research, Chen et al. (2013) reported two distinct groups from 119 oilseed forms of *B. juncea* using SSR markers. One group contains lines from central and western India and eastern China. Another included forms from northern and eastern India, central and western China, as well as European and Australian accessions (Chen et al. 2013). These two distinct groups were also confirmed from a population structure analysis of 122 *B. juncea* accessions with nuclear markers, one group included accessions from East Europe and India and another group consisted of genotypes from India, China, and Australia (Kaur et al. 2014).

3.3.5 *B. nigra* and *B. carinata*

Comparing with *B. oleracea*, *B. rapa*, *B. napus,* and *B. juncea*, relatively less work has been conducted on genetic diversity and population structure in *B. nigra* and *B. carinata*. Negi et al. (2004) reported high levels of AFLP diversity in 18 *B. nigra* accessions and all *B. nigra* accessions were found to cluster with *B. juncea* accessions. Pradhan et al. (2011) examined microsatellite genetic diversity of 180 *B. nigra* Koch genotypes from 60 accessions, and one line each of *B. juncea* (Czern) and *B. carinata* (Braun). Most of the *B. nigra* accessions were highly heterozygous and the accessions clustered along their geographic origin. The authors (Pradhan et al. 2011) identified 44 accessions (73%) as truly *B. nigra*, which formed morphologically and genetically distinct groups in accordance with their geographic origin, notably in Ethiopia, Israel, India, and Europe. Oduor et al. (2015) sequenced chloroplast DNA intron (trnF–trnL) from 284 individuals of *B. nigra* from North America, which included 36 native and 15 invasive populations. The genetic diversity was found to be similar between invasive range and native range populations, and the authors (Oduor et al. 2015) suggested that invasive *B. nigra* populations may have been originated from multiple native North American populations. Recently, Liu et al. (2021) reported high microsatellite genetic diversity in 83 *B. nigra*, 16 *B. juncea*, and other *Brassica* accessions, and that *B. nigra* accessions clustered in four groups. The subgroup B-III presented the closest genetic relationship with *B. juncea*, while subgroup B-IV was the most distant group (Liu et al. 2021).

Warwick et al. (2006) evaluated the genetic diversity of 66 *B. carinata* accessions from western Canada based on AFLP markers and compared it with that of 20 *B. juncea* and 7 *B. nigra* accessions. The results showed that *B. carinata* was less genetically diverse than *B. juncea* and *B. nigra*. Recently, Khedikar et al. (2020) discovered over 10,000 genome-wide SNPs using genotype by sequencing of 620 *B. carinata* accessions and using the genomes of *B. nigra* and *B. oleracea* as reference. They reported low levels of genetic diversity among the *B. carinata* accessions. The *B. carinata* accessions grouped into two distinct subpopulations, with 88% accessions clustering with the population from Ethiopia, and the collection of breeding accessions forming the other cluster (Khedikar et al. 2020).

3.4 *The Origin and Domestication of* **Brassica** *Species*

Crop domestication is an artificial selection process that transforms wild plants into cultivated crops to meet the demands of humans (Doebley et al. 2006). Understanding the genetic/genomic basis of crop domestication can provide an understanding of the origin of domesticated crops and a foundation for further genetic improvement of current crops, especially to address environmental and climate challenges. Domestication of *Brassica* species has resulted in several agricultural crops. Most of the domestication studies have focused on *B. rapa*, *B. oleracea,* and *B. napus*, due to these three species serve as major oil and vegetable-type crops and have been widely cultivated throughout the world.

The domestication of the diverse *B. rapa* crops most likely occurred about 5,500 years ago (3500 BCE). Qi et al. (2017) have provided strong evidence to support a European–Central Asian origin of *B. rapa* crops by combining demographic inferences and written records of *B. rapa* domestication events with high-throughput transcriptome data of 126 accessions representing the geographic and crop diversity. Previous archeological and linguistic lines of evidence suggest that turnip is likely the first domesticated *B. rapa* in the European-Central Asian region (Ignatov et al. 2010; Reiner et al. 1995). Unlike *B. rapa*, the domestication process of *B. oleracea* has not been fully clarified, either regarding its initial location or the progenitor species involved. *B. oleracea* is an economically important and outcrossing species domesticated as early as 2000 BCE (4,021 years ago) and has been diverged into many unique botanical types such as broccoli, cauliflower, cabbage, kale, Chinese kale, and Brussels sprouts. According to the ancient Greek and Latin written records, the domestication of *B. oleracea* presumably occurred in the Mediterranean region. To clarify the domestication footprints of landrace and improved *B. oleracea* broccoli, Stansell et al. (2018) employed genotyping-by-sequencing of 85 landrace and improved *B. oleracea* accessions (Fig. 4). Their population structure and phylogenetic analyses (Fig. 4) added weight to the broccoli-first domestication model (Stansell et al. 2018).

B. napus was domesticated from its diploid progenitors *B. rapa* and *B. oleracea* in the Mediterranean region ~7,500 years ago (Chalhoub et al. 2014). The literature recorded that winter *B. napus* was first cultivated in Europe. Around the year 1,700, spring *B. napus* was developed and spread to England in the late eighteenth century (Bonnema 2012). The semi-winter ecotype was mainly cultivated in China, which was introduced from Europe in the 1930–1940s (Qian et al. 2006). However, no truly wild *B. napus* populations were known. To further understand the origin and domestication process of *B. napus*, a population genomics study was conducted based on resequencing data of 588 diverse *B. napus* accessions from 21 countries (Lu et al. 2019). The study uncovered that the A subgenome may have evolved from the ancestor of European turnip and the C subgenome may have evolved from the common ancestor of kohlrabi, cauliflower, broccoli, and Chinese kale. Additionally, they proposed a model for origin and evolutionary history of *B. napus* (Fig. 5) and posited that *B. napus* was originated from the hybridization between domesticated *B. rapa* and *B. oleracea* ~1,910–7,180 years ago. The LD (linkage disequilibrium) and demographic analyses also support that the original *B. napus* is winter oilseed, and the spring and semi-winter *B. napus* developed ~416 and ~60 years ago, and non-oilseed *B. napus* developed ~277 years ago (Fig. 5) (Lu et al. 2019).

B. juncea, an allotetraploid derived from interspecific hybridization between *B. rapa* and *B. nigra*, has arisen about 10,000 years ago, mainly cultivated in China and India. Evidence suggests that India and China are secondary centers of diversity of *B. juncea* (Vavilov 1951). In 2013, a study consisting of 123 *B. juncea* accessions from four countries including China, India, Australia, and Europe was

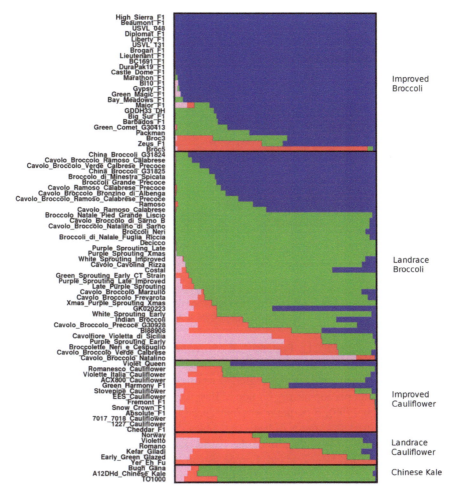

Fig. 4 Population structure ($K = 4$) of 85 unique *B. oleracea* entries using 21,680 SNPs. This figure is reproduced from Stansell et al. (2018)

conducted (Chen et al. 2013). Their results support polyphyletic origin of *B. juncea*, such as West Asia as the primary center of diversity, and India and China as the secondary centers of diversity.

B. nigra or black mustard, probably originated in central and south Europe and has been collected and recorded by the ancients for its medicinal value (Prakash et al. 2011). The domestication of *B. nigra* may have been performed by the Mediterranean civilizations, such as Mesopotamia, Egyptian, Greek, and Roman, in which it was used not only as a medicinal plant, but also as an important condiment (Prakash et al. 2011). Likewise, *B. carinata* is also believed to have arisen about 47,000 years ago as an important allopolyploid crop derived from interspecific crosses between *B. nigra* and *B. oleracea* in the Mediterranean region where both species were

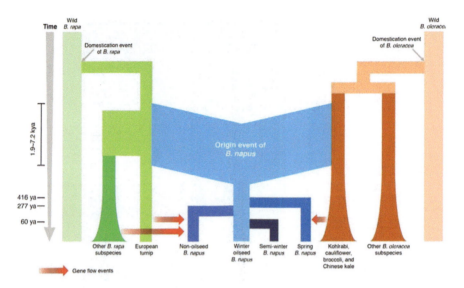

Fig. 5 Proposed model for origin and evolutionary history of *B. napus*. The origin time of *B. napus* and divergence times of spring, semi-winter, and non-oilseed *B. napus* from winter oilseed *B. napus* were estimated using SMC^{++4}. This figure is reproduced from Lu et al. (2019)

present (Song et al. 2021). Presently, *B. carinata* is grown mainly on the East African Plateau, particularly in Ethiopia and in parts of the east and west coasts of the African continent, whereas the region of its origin is thought to be located in the highlands of Ethiopia and nearby portion of East Africa and the Mediterranean coast (Gómez-Campo 1999). *B carinata* was introduced to North America from Ethiopia in 1957 to be used as a source of leafy vegetables (Stephens 2009). Recently, it has been cultivated in different parts of the world, including South Asia (India and Pakistan), Australia, New Zealand, Europe (Spain, Italy, Greece, and the UK), and South America (Chile and Uruguay) as an oil crop with desirable energy potential (Seepaul et al. 2016; Zada et al. 2013; Prakash et al. 2011).

3.4.1 Domesticated Traits and Underlying Genes

Interestingly, although originated from different geographical locations, similar morphotypes are often selected for in *Brassica* crops, clearly illustrating convergent crop domestication (Lenser and Theissen. 2013). In recent years, the knowledge of the genes involved in *Brassica* crop domestication has increased dramatically, which could provide key insights into how molecular convergence contributed to *Brassica* crops domestication. Here, we summarize two crucial morphological transitions and the major genes underlying *Brassica* crops domestication.

Inflorescence Architecture

The inflorescence architecture of plants is a key agronomic factor determining seed yield, and, thus, it is a major target of crop domestication and improvement. In *Brassica* crops, the evident examples of inflorescence architecture re-shaping associated with domestication are cauliflower (*B. oleracea* ssp. *botrytis*, a hypertrophied mass of inflorescence meristems and floral meristems) and broccoli (*B. oleracea* ssp. *Italica*, a large hypertrophied structure and flower buds) (Carr and Irish 1997; Schilling et al. 2018; Manrique et al. 2019). A previous study found that the phenotype of *ap1-1*/*cal-1* mutant of Arabidopsis has similarity with broccoli and cauliflower, leading to speculation that *APETALA 1* (*AP1*) and *CAULIFLOWER* (*CAL*, the paralog of *AP1*) might be responsible for inflorescence architecture in *B. oleracea* (Smith and King 2000). Surprisingly, the molecular and population genetic studies suggested that the function of *CAL* in *B. oleracea* (*BoCAL*) is compromised both in broccoli and cauliflower (Kempin et al. 1995; Lowman and Purugganan 1999; Purugganan et al. 2000; Smith and King 2000). Although *AP1*-like genes are also associated with the phenotype of inflorescence architecture in *B. oleracea*, while the situation for *AP1*-like genes is less clear due to several copies of *AP1*-like genes in *B. oleracea* (Lowman and Purugganan 1999, Labate et al. 2006; Duclos and Björkman 2008; Schilling et al. 2018). In *B. napus*, a previous study also found candidate gene *BnTFL1* (*Terminal Flower 1*) for control of the inflorescence trait located on chromosome A10 (Li et al. 2018).

Flowering Time

Flowering time is an important agronomic trait, and the phase transition from vegetative to reproductive growth is a pivotal step in plant's life cycle, which directly influences crop yield and quality (Flowers et al. 2009; Purugganan and Fuller 2009; Blackman et al. 2011; Andres and Coupland 2012). Vernalization depicts the ability to promote flowering after prolonged exposure to cold, and it has been the target of artificial selection during crops domestication (Schilling et al. 2018). In the vernalization pathway, *FLOWERING LOCUS C (FLC)*, encoding a MADS-box transcription factor, plays a profound role in floral transition as a potent repressor. *FLC* prefers to express in the vegetative apex, and it controls flowering through repressing the activity of pivotal flowering promoters, such as *SUPPRESSOR OF OVEREXPRESSION OF CONSTANS1* (*SOC1*), *FLOWERING LOCUS T* (*FT*), and *FLOWERING LOCUS D* (*FD*) in Arabidopsis (Amasino and Michaels 2010). The role of *FLC* genes has also been extensively studied in *Brassica* crops. In *B. rapa*, four *FLC* orthologs (*BrFLC1*, *BrFLC2*, *BrFLC3*, and *BrFLC5*) have been identified, and the differences in flowering times in different cultivars were caused by variations in some of them (Schranz et al. 2002; Yuan et al. 2009; Kakizaki et al. 2011; Hou et al. 2012; Wu et al. 2012; Xiao et al. 2013; Dong et al. 2016). For example, a naturally occurring deletion variation in *BrFLC2* gene resulted in great variation in flowering time (Wu et al. 2012). In *B. napus*, studies detected a 2.833-kb insertion in

BnFLC.A2 and its homeologous exchange with *BnFLC.C2* during domestication generated early-flowering *B. napus* ecotype (Chen et al. 2018). Besides, studies also found that the gain-of-function mutation (through inserting a MITE transposable element upstream of *BnFLC10*) exhibited strongly vernalization requirement in winter varieties of rapeseed (Long et al. 2007; Hou et al. 2012). Like *FLC*, *FRIGIDA (FRI)* is a key regulator which inhibits floral transition through activation of *FLC* in *Arabidopsis thaliana*. Wang et al. (2011a) found that six SNPs located within the exons of *BnaA.FRI.a* affect its expression in aerial plant organs, leading to flowering time variation in *B. napus*. Therefore, domestication-related flowering time variation through mutations in orthologs of *FLC* or its upstream regulator *FRIGIDA (FRI)* is typical for members of the *Brassica* lineage. Additionally, in *Arabidopsis*, the histone chaperones SSRP1 in the FACT complex assist the progression of RNA polymerase II and promote *FLC* expression (Van Lijsebettens and Grasser 2010). In contrast, *VERNALIZATION 2 (VRN2)* plays a critical role in maintaining transcriptional repression of *FLC* after vernalization through regulating the deposition of H3K27me3 and H3K9me3 repressive marks at the *FLC* locus (Bastow et al. 2004). The homologous genes for *SSRP1* and *VRN*, respectively, *BnaC02g37430D*, and *BnaC01g21540D*, experienced extensive domestication selection for ecotype improvement, via regulation of *FLC* expression (Lu et al. 2019).

3.5 Population Epigenomics

The epigenomic variation is the set of heritable modifications able to alter gene expression without changing the nucleotide sequence. These variations include DNA methylation, histone modifications, and genomic imprinting, which could cause heritable changes in a trait or phenotype by mechanisms other than DNA mutation (Moler et al. 2019). Epigenomic variation is widespread in wild populations of plants and animals, and it is important evolutionarily but limited transgenerational inheritance information exists (Moler et al. 2019). In the *Brassica* genus, previous studies found that epiallelic variation could extend the range of phenotypes available for selection and demonstrated that chemically-induced hypomethylation has the potential to generate valuable and novel variation for crop improvement, in *B. rapa* (Amoah et al. 2012). Moreover, their work presented that pachytene chromosomes exhibited reduced immuno-staining of 5mC under 5-Azacytidine treatment in a hypomethylated population lines, and phenotypic variation among these lines was coincident, with seed yield, seed protein content, oil content, erucic acid, and linoleic and/or palmitic acid (Amoah et al. 2012).

Like GWAS, population epigenomics studies, such as epigenome-wide association studies (EWAS) and epigenome environment association analysis (epiEAA), have concentrated on associations between epigenomic variation and phenotypic, ecological, disease, and other traits in plants, animal, and humans. And substantial EWAS works have been done to identify the association of DNA methylation with human disease in the past few years. However, there is very little or nothing is known for population epigenomics in the *Brassica* genus. Epigenomics studies in *Brassica* (Hauben et al. 2009; Gupta et al. 2019) have used the recombinant inbred lines (RILs), resynthesized lines and not yet the natural populations. Population epigenomics studies provide novel data that will be useful for identifying natural variation, determining the changes in epigenetic states during abiotic or biotic stress, and understanding the epigenetic states related to agronomically important traits using the segregation of loci. We believe that related population epigenomics studies will continue to be carried out in *Brassica* in the future, as we discuss in Sect. 4.

3.6 Population Transcriptomics

Transcriptomics is the study of all the RNA transcripts (transcriptome) that can be transcribed by a living cell. Like the phenotype traits, transcriptome profiles also can be considered as phenotypes because they are the products of both environmental and genetic variation (Hansen 2010). Population transcriptomics is used in transcriptome-wide profiles to explore the variation in gene expression within and among populations (Ghalambor et al. 2015). Based on population transcriptomics data, pan-transcriptome, molecular markers, and transcriptome-wide association studies (TWAS) have been successfully applied for genetic background selection of varieties and candidate genes selection for target traits (He et al. 2015, 2017; Kim et al. 2016; Harper et al. 2020; Tang et al. 2020).

For a low-quality complex genome assembly, researchers typically integrate relatively high-quality genomes and coding sequence models from ancestral species to generate an associative transcriptome. Associative transcriptomics (AT) is a rapid and cost-effective transcriptome-based technology using RNA-seq, and it has been developed for and applied in the identification of SNPs (Trick et al. 2009), the construction of linkage mapping and genome characterization (Bancroft et al. 2011), and the quantification of transcripts (Higgins et al. 2012). For *Brassica* genus, AT was first used in *B. napus*, resulting in the identification of two QTLs associated with the glucosinolate content of seeds (Harper et al. 2012). These QTLs contained orthologs of an *Arabidopsis* gene encoding the transcription factor HAG1 (At5g61420), which regulates aliphatic glucosinolate biosynthesis (Harper et al. 2012). Subsequently, pan-transcriptome platform has been developed to understand the nascent *B. napus* A and C genomes (He et al. 2015, 2017). Using a new AT platform, Harper et al. (2020) established a pan-transcriptome reference for *B. juncea* based on population transcriptomics. These authors quantified the abundance of 93963 transcripts and identified 355050 single nucleotide polymorphism (SNP) variants and revealed a candidate gene controlling seed weight, *Bja.TTL*, and markers linked to seed color and vitamin E content using an association analysis of function-specific genotypes (Harper et al. 2020).

Molecular markers are still the most common type used in genetic diversity analysis, particularly SNP markers. A high-throughput SNP marker assay could improve genetic background selection in crops. In *Brassica* genus, population transcriptomics has been also applied in molecular marker-assisted breeding. For example, Kim et al. (2016) sequenced the transcriptomes of 20 Chinese cabbage accessions, which presented diversity in many traits, such as maturity, head type, disease resistance, and inner leaf color. In total, they identified 13,976 SSRs and 380,198 SNPs and screened 189 SNPs makers based on criteria, such as the level of polymorphism and segregation ratio (Kim et al. 2016). Their further experiments validated these SNP markers as a credible breeding tool to distribute or select Chinese cabbage accessions (Kim et al. 2016). Like GWAS, TWAS has also been applied to predict candidate genes for complex traits in *Brassica* species. In *B. napus*, Tang et al. (2020) sequenced seed transcriptomes at two developmental stages, including 309 accessions at 20 days after flowering (DAF) and 274 accessions at 40 DAF, and identified 605 genes at 20 DAF and 148 genes at 40 DAF which were significantly associated with seed oil contents (SOC) (Tang et al. 2020). Among these genes, sixty-one genes were found expressed at both stages, which were enriched in the phenylpropanoid metabolic process and the secondary metabolite biosynthetic process, and many flavonoid and secondary metabolism-related genes, such as *TT4*, *TT5*, *TT8,* and *C4H*, were found to be negatively associated with SOC (Tang et al. 2020).

Population transcriptomics data has also reflected the origin and diversification of *Brassica* species. In *B. napus* accessions, An et al. (2019) determined genetic diversity and subgenome variation using population transcriptomics data, and they identified 8,187 differentially expressed genes (DEGs) in leaf tissue between WEAm (Winter rapeseed in Europe and America) and five other genetic clusters of *B. napus* accessions (An et al. 2019). Combined nucleotide diversity and gene expression levels, these authors found that *B. napus* has an asymmetric subgenome evolution and further identified several candidate genes that underwent selective sweeps underlying vernalization (An et al. 2019). Thus, population transcriptomics application has great potential for understanding complex traits, genetic variation, and diversification in the *Brassica* genus.

3.7 Population Genomics for Selection and Breeding in **Brassica**

3.7.1 Genome Wide Association Studies

Although QTL-seq has been widely used in *Brassica* genus to identify genetic basis of key agronomic traits, such as flowering, green curd, seed weight, and others (Tudor et al. 2020; Tan et al. 2020; Fu et al. 2015), GWAS has a higher resolution than the QTL approach in identifying potential candidate genes associated with desired traits. With the advent of the 60K SNP arrays and the publication of reference genomes for many *Brassica* species, GWAS has been widely applied to the fine-mapping of genotype-trait associations on a genome-wide scale.

To date, a search of the Web of Science Core Collection (http://www.webofknowledge.com/WOS) using the keywords "*Brassica*" and "GWAS" yields over 70 published articles, with most publication following the publication of the Darmor-*bzh* genome in 2014 (Chalhoub et al. 2014). A maximum number of GWAS publications in *Brassica* were found in 2020: 1 in 2014, 6 in 2015, <10 yearly for 2016–2019, and 21 in 2020. Here, we listed main GWAS results published in Table 2. It is clear that oil content is the most studied trait, perhaps because oil is the most important product of *Brassica* crops. In addition, the traits of seed quality, seed yield, disease resistance, plant morphology, nutrient absorption, and utilization have also been studied; however, almost all of such studies were performed on *B. napus*, the most valuable *Brassica* crop (Table 2).

In *Brassica*, GWAS is typically performed by genotyping all accessions using the *Brassica* 60K SNP array. For example, Zhao et al. (2019) genotyped 375 low-erucic-acid *B. napus* accessions using the 60K SNP array and performed a GWAS. Their findings resulted in the identification of a novel QTL on chromosome A09, which explained 11.25%, 5.72%, and 6.29% of the phenotypic variation in three consecutive growing seasons. Through fine-mapping and gene expression analysis, they identified three candidate genes and verified their function through the homologous gene mutant of *Arabidopsis* (Zhao et al. 2019). This led to the identification of the gene BnaA09g39570D, which encodes a metal ion-binding protein, as the most likely candidate gene controlling fatty acid metabolism in *B. napus* (Zhao et al. 2019). In addition, 520 *B. napus* accessions with different lodging traits and stem lignin contents were genotyped using the 60K SNP array and a GWAS was performed (Wei et al. 2017). This study led to the identification of 92 SNPs and 50 SSRs associated with the breaking force, breaking strength, lodging coefficient, acid detergent lignin content, and the syringyl/guaiacyl monolignol ratio of *B. napus*. Using RNA-seq, Zhao et al. (2019) obtained four candidate genes regulating lignin.

Next-generation sequencing has enabled the genome-wide discovery of structural variation, InDels, and copy number variants, providing new insights into genotypes and genetic mechanisms in *Brassica*. For example, using whole-genome resequencing for 588 *B. napus* accessions, Lu et al. (2019) identified 206,001 effectively independent SNPs. A total of 60 loci were found to be associated with 10 desired traits, including 5 related to seed yield, 3 to silique length, 4 to oil content, and 48 to seed quality. And the authors identified candidate genes that could be targeted to improve the traits of environment adaption, seed quality, plant morphology, oil content, and seed yield (Lu et al. 2019). Drought stress is a major abiotic stress that can substantially decrease crop yields, and the diminishing availability of water resources means it is important to decipher the mechanisms of drought tolerance in *B. napus* seedlings. A total of 228 accessions *B. napus* were recently phenotyped under normal and water-stressed conditions and genotyped using SLAF-seq (Khanzada et al. 2020). The resulting 201,187 high-quality SNPs were analyzed in a GWAS to identify 314 marker-trait associations strongly linked with drought indices. Finally, 85 genes, with *Arabidopsis* orthologs involved in drought tolerance, were found to underlie these QTLs (Khanzada et al. 2020). With the development of long-read sequencing technology, large SVs among several genomes could be

Table 2 The summary of main GWAS publications in *Brassica* species

Species	Traits	SNP	Chromosome	Position (kb)	Candidate gene	Authors, year
B. napus	Erucic acid content	snp1348001	A08	0	*BnaA08g11130D*	Wang et al. (2018)
		snp1440614	A09	127	*BnaA09g07080D*	
		snp2429000	C03	24	*BnaC03g65980D*	
		snp2503765	C04	22	*BnaC04g16670D*	
	Glucosinolate content	snp285415	A02	21	*BnaA02g33040D*	
		snp774819	A05	91	*BnaA05g15560D*	
				23	*BnaA05g15660D*	
		snp846562	A05	21	*BnaA05g23340D*	
		indel160911	A06	300	*BnaA06g14930D*	
				296	*BnaA06g14950D*	
				291	*BnaA06g14960D*	
		snp971782	A06	109	*BnaA06g16980D*	
		snp1003512	A06	180	*BnaA06g20740D*	
		snp1064558	A06	191	*BnaA06g31890D*	
		snp1431371	A09	24	*BnaA09g01260D*	
				20	*BnaA09g01270D*	
				18	*BnaA09g01280D*	
		snp1438406	A09		*Bra035929 Lost in An genome*	
		indel240 140	A09	27	*BnaA09g08410D*	
				78	*BnaA09g08470D*	
		snp1873764	C01	36	*BnaC01g24060D*	
				51	*BnaC01g24080D*	
				53	*BnaC01g24090D*	
		snp2169599	C02	192	*BnaC02g40090D*	

Species	Trait	Marker	Chr		Gene
B. napus	Seed oil content			195	BnaC02g40120D
				206	BnaC02g40130D
		indel338983	C02		Bo2g161590 Lost in An genome
		indel479620	C09	0	BnaC09g05300D
		indel485167	C09	244	BnaC09g23540D
				243	BnaC09g23550D
		snp152386	A02	211	BnaA02g00830D
				35	BnaA02g01260D
				286	BnaA02g01770D
		snp430455	A03	82	BnaA03g32450D
		snp546567	A04	186	BnaA04g04400D
				190	BnaA04g04410D
				211	BnaA04g04450D
		indel122313	A05	217	BnaA05g06400D
				17	BnaA05g06580D
				12	BnaA05g06620D
				40	BnaA05g06710D
				59	BnaA05g06750D
		snp733528	A05	17	BnaA05g08550D
				190	BnaA05g09070D
		snp842906	A05	184	BnaA05g22260D
		snp878606	A05	182	BnaA05g31780D
		snp949496	A06	268	BnaA06g11630D
				53	BnaA06g12060D
				144	BnaA06g12430D
		indel179669	A06	32	BnaA06g35100D
				105	BnaA06g35290D
		indel239189	A09	101	BnaA09g06090D

(continued)

Table 2 (continued)

Species	Traits	SNP	Chromosome	Position (kb)	Candidate gene	Authors, year
		snp1444416	A09	243	*BnaA09g08660D*	
		snp1448880	A09	196	*BnaA09g10940D*	
		snp1778098	A10	175	*BnaA10g23290D*	
				2	*BnaA10g23670D*	
				58	*BnaA10g23790D*	
				106	*BnaA10g23950D*	
		snp2649172	C04	226	*BnaC04g45430D*	
				145	*BnaC04g45500D*	
				128	*BnaC04g45550D*	
				48	*BnaC04g45690D*	
				136	*BnaC04g45790D*	
				140	*BnaC04g45800D*	
				173	*BnaC04g45860D*	
		snp2988719	C07	138	*BnaC07g17300D*	
		snp3250360	C09	226	*BnaC09g11200D*	
	Flowering time		A01_random	548193	*BnaA01g35360D*	Wu et al. (2019)
			A01_random	548193	*BnaA01g35210D*	
			A02	158168	*BnaA02g00380D*	
			A02	158168	*BnaA02g00370D*	
			A02	158168	*BnaA02g00360D*	
			A02	158168	*BnaA02g00350D*	
			A02	158168	*BnaA02g00310D*	
			A02	3273984	*BnaA02g06840D*	
			A02	3273984	*BnaA02g06850D*	
			A02	3273984	*BnaA02g06600D*	
			A02	3273984	*BnaA02g07010D*	
			A02	6303095	*BnaA02g12130D*	

B. napus						
			A02	20516539	BnaA02g27680D	
			A02	21890868	BnaA02g30040D	
			A02_random	168357	BnaA02g35200D	
			A02_random	280667	BnaA02g35530D	
			A03	6196996	BnaA03g13630D	
			A04	8562234	BnaA04g09490D	
			A04	8787736	BnaA04g09950D	
			A04	8787736	BnaA04g09910D	
			A04	9669582	BnaA04g11120D	
			A07	7372690	BnaA07g07910.1T	
			A07	22791461	BnaA07g33120D	
			A09	25516400	BnaA09g34810D	
			A09	25805610	BnaA09g35310D	
			A10	14920175	BnaA10g21750D	
			A10	14920175	BnaA10g22080D	
			A10	14920175	BnaA10g22090D	
			A10	15803962	BnaA10g23980D	
			A10	15940015	BnaA10g24480D	
			C02	2721235	BnaC02g05300D	
			C02	6760353	BnaC02g11400D	
			C04	28181178	BnaC04g26930D	
			C05	18747617	BnaC05g24380D	
			C06	27055007	BnaC06g25500D	
			C09	47609629	BnaC09g49210D	
			C09	47609629	BnaC09g49060D	
	Silique number	Bn-A01-S16094037	A01	14045897		Li et al. (2020)
		Bn-A02-S7482915	A02	4511692		

(continued)

Table 2 (continued)

Species	Traits	SNP	Chromosome	Position (kb)	Candidate gene	Authors, year
		Bn-A02-S8767145	A02	5739739		
		Bn-A02-S10323998	A02	7170926		
		Bn-A03-S20398340	A03	19267907		
		Bn-A03-S26437883	A03	24829819		
		Bn-A03-S29118954	A03	26262031		
		Bn-A05-S474257	A05	585628		
		Bn-A02-S588093	A07	13217992		
		Bn-A09-S3097420	A09	3031274		
		Bn-A09-S19941634	A09	16914495		
		Bn-A09-S29186255	A09	27063736		
		Bn-A09-S25899875	A09	28211039		
		Bn-A09-S33976464	A09	31210585		
		Bn-A10-S3921433	A10	900559		
		Bn-A10-S10298013	A10	11677067		
B. napus		Bn-A10-S13789579	A10	13825021		
		Bn-scaff_15712_6-S1336179	C02	38045422		
		Bn-scaff_17109_1-S557859	C02	41808474		
		Bn-scaff_16614_1-S1480092	C03	660236		
		Bn-scaff_18936_1-S102755	C03	2745162		
		Bn-scaff_15877_1-S926737	C03	4656329		
		Bn-scaff_22466_1-S754489	C03	14826392		
		Bn-scaff_18602_1-S270185	C03	51667421		
		Bn-scaff_20901_1-S369010	C05	3670200		
		Bn-scaff_15705_3-S436841	C07	36219357		
		Bn-scaff_17487_1-S512535	C09	6772003		

					Tang et al. (2020)
Seed oil content	qOC.A01.1	A01	21240857	BnaA01g31120D	
	qOC.A02.1	A02	3668843	BnaA02g07780D	
	qOC.A02.2	A02	24570634	BnaA02g34550D	
	qOC.A03.1	A03	2368373	BnaA03g05080D	
	qOC.A04.1	A04	13892805	BnaA04g17080D	
	qOC.A05.1	A05	11324697	BnaA05g16720D	
	qOC.A05.2	A05	11324697	BnaA05g28570D	
	qOC.A05.3	A05	20223692	BnaA05g28570D	
	qOC.A07.1	A07	7060737	BnaA07g06820D	
	qOC.A07.2	A07	23923367	BnaA07g35680D	
	qOC.A08.1	A08	10480586	BnaA08g11690D	
	qOC.A09.1	A09	26564814	BnaA09g36810D	
	qOC.A09.2	A09	28172165	BnaA09g40020D	
	qOC.A09.3	A09	28298136	BnaA09g40020D	
	qOC.A09.4	A09	31212153	BnaA09g45790D	
	qOC.A09.5	A09	31316048	BnaA09g45790D	
	qOC.A10.1	A10	728690	BnaA10g01280D	
	qOC.A10.2	A10	6648809	BnaA10g08110D	
	qOC.A10.3	A10	12685470	BnaA10g17050D	
	qOC.C05.1	C05	39950517	BnaC05g42890D	
	qOC.C05.2	C05	40018013	BnaC05g42890D	
	qOC.C05.3	C05	40020140	BnaC05g42890D	
	qOC.C05.4	C05	41002230	BnaC05g44840D	
	qOC.C06.1	C06	7620296	BnaC06g07070D	
	qOC.C06.2	C06	103117589	BnaC06g08720D	
	qOC.C07.1	C07	36650573	BnaC07g33400D	
	qOC.C09.1	C09	37470483	BnaC09g34100D	

(continued)

Table 2 (continued)

Species	Traits	SNP	Chromosome	Position (kb)	Candidate gene	Authors, year
B. napus	Glucosinolate accumulations in leaves and seeds	Bn-A03-p21329715	A03	20095857	BnaA03g40190D	Liu et al. (2020)
		Bn-A03-p21669774	A03	20452811		
		Bn-A01-p9004629	A09	2580835	BnaA09MYB28	
		Bn-A09-p2733282	A09	2677575		
		Bn-scaff_17177_1-p441984	C02	44768013	BnaC02MYB28	
		Bn-scaff_18181_1-p1849246	C07	34322798	BnaC07MYB28	
		Bn-scaff_15705_1-p2274493	C07	35279702		
		Bn-scaff_19783_1-p379086	C09	2850069	BnaC09g05300D	
		Bn-A01-p970103	A01	588064		
		Bn-A09-p10577283	A02	24468610	BnaA02g33530D	
		Bn-A02-p745468	A07	13377023		
		Bn-A10-p6896063	A10	8474298		
		Bn-scaff_22749_1-p67780	C02	26260892	BnaC02g27590D	
		Bn-scaff_16217_1-p181427	C04	22294107		
		Bn-A08-p8426380	C08	11962388		
		Bn-scaff_22835_1-p619832	C09	11113843	BnaC09g14380D	
		Bn-A10-p10454385	A10	11834653		
		Bn-scaff_23432_1-p217818	C04	19102451		
		Bn-scaff_17799_1-p3050608	C09	39518182		
B. oleracea	Apical length		C02	26787029	Bol035969	Thorwarth et al. (2018)
			C03	20986662	Bol035507	
			C06	4914166	Bol032997	
			C06	34494415	Bol040102	
	Cluster width		C02	5063181	Bol007138	
			C02	3528844	Bol021232	
			C02	5063181	Bol007138	

		Length of nearest branch			C09	25012587	Bol012235	
		Number of branches			C06	2323306	Bol035509	
					C07	42	Bol024369	
		Number of days to budding			C01	37688065	Bol023068	
					C02	2708156	Bol024638	
					C02	2708163		
					C02	2708182		
					C06	2949314	Bol026132	
					C07	936738	Bol027177	
					C07	936770		
					C07	41524584	Bol024369	
B. juncea	Oil	A04_6252658		A04	6252658	CER26	Akhatar et al. (2020)
		A06_26456373, 97, 466		A06	26456373-6466	GDSL	
		A09_14737522		A09	14737522	GDSL	
		A09_24082148		A09	24082148	LACS5	
		B05_5272794, 95		B05	5272794-2795	FAD6	
		B06_19315729		B06	19315729	At1g06090	
		B08_34645905, 5977, 5991, 6006		B08	34645905-6006	DIR1	
	Proteins	A04_12420732, 5239		A04	12420732-5239	PASTICCINO 1	
		A06_9257535, 68		A06	9257535-7568	AtP4H3	
		A06_15534622, 4718, 4750		A06	15534622-4750	ASN1	

(continued)

Table 2 (continued)

Species	Traits	SNP	Chromosome	Position (kb)	Candidate gene	Authors, year
		A06_17974981, 5007, 5009, 5010	A06	17974981-5010	GTR2	
		A06_27490668, 71, 93	A06	27490668-0693	At2g41640	
		A09_30208138	A09	30208138	PSP1	
		B03_3166770, 6905, 6907, 6926, 6935, 6980, 7018, 7058	B03	3166770-7058	PSP1	
	Glucosinolates	A04_11732900, 2912, 2989, 3020, 3023, 3024, 3028, 3029	A04	11732900-3029	HB16, SK1	
		A05_5070339	A05	5070339	AT2G35450	
		A06_24763290	A06	24763290	CM1	
		A06_24773388	A06	24773388	CM1	
		B03_19283625, 26	B03	19283625-3626	JMT	
		B04_2285588, 589, 789	B04	2285588-5789	CM1	
		B06_8990634, 55, 70	B06	8990634-0670	LINC4	
		B06_9279577, 727, 731	B06	9279577-9731	CYP81G1, MYB44	

explored in the polyploid species. Many studies have shown that SVs could have a more significant influence on traits than SNPs, and the genome-wide association studies based on SNP genotyping (SNP-GWASs) could miss detecting structural variants in reference genomes. Through PAV-GWAS in *Brassica* species, a 3.9-kb CACTA-like TE inserted upstream of the *BnaA9.CYP78A9* promoter region was found as a key factor to be associated with longer siliques and larger seeds, and a 4,421-bp hAT inserted in the promoter region of *BnaA10.FLC* could be associated with early flowering (Wu et al. 2019; Song et al. 2020). Genotyping can be achieved using many sequencing technologies; however, there are still some deficiencies in the development of high-throughput phenotyping technologies.

3.7.2 Genomic Selection

Genomic selection (GS) models have been applied in animal breeding programs, and the main objective of GS is to reduce phenotyping costs by using markers and accelerate the breeding cycle. To date, GS has been applied in several important crop species, such as rice (*Oryza sativa*), maize (*Zea mays*), wheat (*Triticum aestivum*), and rapeseed (Fikere et al. 2020; Werner et al. 2018; Würschum et al. 2014). Contrary to QTL and association mapping, GS employs genome-wide molecular marker to estimate the breeding values (BVs) of individuals for quantitative traits (Goddard and Hayes 2007). Therefore, GS models can be used to predict the genomic estimated breeding values (GEBVs) of individuals in a testing population based only on their genotypic data, using marker effects estimated in a training population that have been genotyped and phenotyped for the predictive model (Meuwissen et al. 2001).

Genomic selection in *Brassica* crop breeding has been applied using mainly two different approaches. One focuses on predicting additive effects in early generations of a breeding program, like pure breeds based on marker information. The other approach predicts the complete genetic values of individuals considering both additive and nonadditive (dominance and epistasis) effects, such as hybrid performance prediction in hybrid breeding programs. Here, we summarize some applications of genomic selection in *Brassica* species and present that genomic selection is a robust and promising method for assisting breeding of *Brassica* crop species.

Direct Genomic Prediction for Traits

Curd architecture is one of the most important characters determining the curd morphology of cauliflower. Using 192 cauliflower accessions, Thorwarth et al. (2018) demonstrated that genomic prediction worked best for curd-related traits with a high heritability, such as number of days to budding and cluster width, but less efficiently well for the traits of apical length and length of nearest branch with GBLUP (genomic best linear unbiased prediction) and BayesB models. For seed oil yield and quality studies, Zou et al. (2016) found that seed quality traits can be predicted with high accuracy by implementing genome-wide prediction from a bi-parental population comprising 202 doubled haploid lines and a diverse validation set including 117 inbred lines.

Prediction in Segregating Families

In an applied pure breeding program (a special kind of inbreeding), a population may be composed of multiple segregating families with different family size. Consequently, the selection of lines within single segregating families would be an attractive application of GS. Using a population from 391 elite doubled haploid (DH) winter-type oilseed lines of *B. napus* derived from nine families, Würschum et al. (2014) estimated the prediction accuracies for morphological, quality- and yield-related traits using ridge regression best linear unbiased prediction (RR-BLUP) statistical model. Their results illustrated that the prediction in segregating families resulted in somewhat lower accuracies, but for some traits still it remained attractive due to among-family variance (Würschum et al. 2014).

Genomic Hybrid Prediction

Hybrid breeding has been proved to be an important tool for improving yield, and the prediction for the performance of parental combinations is pivotal to improve hybrid breeding programs. However, hybrid performance may not be a reliable indicator if only considering per se performance of parents, due to the phenomenon of heterosis. In addition, selection of well-paired parents based on their phenotype requires the testing of large number of hybrid combinations in extensive field trials. Namely, the number of potential hybrids, in general, dramatically exceeds breeding capacity and budget. Therefore, GS is expected to be even more promising when applied to hybrid breeding, because genotypes of hybrids are predetermined by their parents. In order to predict hybrid performance through incorporating dominance and epistasis effect in *B. napus*, Werner et al. (2018) used a set of 448 hybrids produced from unbalanced crosses between 220 paternal DH lines and five maternal male-sterile inbred lines. The results underlined that genomic prediction could provide a good and robust estimation for hybrid performance based on the genome-wide markers of potential parents (Werner et al. 2018).

4 Future Perspectives

With the development of genome sequencing technology, all *Brassica* genomes in the "triangle of U" have been successfully sequenced. Based on these genomic and resequencing data, the ancestral genomes, evolutionary history, phylogenetic relationships and domestication, genetic diversity and population structure, subgenome differentiation, and transcriptomics in *Brassica* species and populations are well understood. Population genomics could unravel many insights into biology, evolution, population genetics, and domestication of *Brassica* species. We think the future population genomics research could explore the following key topics. Genome sequencing of other *Brassica* species should be conducted to confirm the occurrence and timing of the single WGT event in the evolution of this genus. To date, we have studied the genetic diversity and domestication of some *Brassica* species, but the evolutionary history and population genetic basis of domestication for some varieties and traits, and their differentiation and global geographical distribution need to be investigated in depth using an increased collection of *Brassica* populations, especially those of *B. nigra* and *B. carinata* in the "triangle of U" theories. Although several transcriptomics studies have been conducted, especially to identify genes and their expression patterns, including paralogs and their syntenic genes in *Brassica* genomes at the subgenome and individual gene levels, very little is done on population transcriptomics in *Brassica* species. Population transcriptomics studies should be undertaken to explore population dynamics at the transcriptome levels, as well as use the transcriptome information for TWAS. Recently, single-cell RNA-seq technology has been developed and has been used in the model plant *A. thaliana* and *O. sativa*. Compared with traditional transcriptome profiling, scRNA-seq provides the opportunity to elucidate the developmental landscape of various organs. In the future, single-cell genomics will likely be widely used in *Brassica* species to explore and decipher the dynamics of single-cell gene expression. The genomic resources developed in *Brassica*, including pangenomes have extensively assisted population genomics and other genetics studies and breeding applications in *Brassica*. However, the genomes and pangenomes have been developed only for a few varieties. Therefore, pangenomes for various varieties should be developed to refine the core genomes and genomic variation of the *Brassica* species. Plant phenotypes and agronomic traits are not only influenced at the transcriptomic and translational levels, but are also determined by other features, such as DNA methylation and acetylation. Therefore, we think that population epigenomics studies will accelerate in future for *Brassica* species and varieties, especially to understand their acclimation and adaptation to different climatic and environmental conditions. Indeed, combined population epigenomics and population transcriptomics will greatly help to identify genetic basis of local adaptation to climate. Application of multi-omics approaches will facilitate understanding the genetic basis of complex traits and assisting breeding superior varieties of *Brassica* crops. GS models will be increasingly used for predicting the genomic estimated breeding value of individuals in *Brassica* genetic improvement programs. In addition, we also believe that the high-quality SNP genotyping arrays of *Brassica* species will be developed and continuously updated for *Brassica* population genomics and breeding in future.

5 Conclusions

Population genomics has provided clearer insights into the origin, evolution, domestication, diversification, population structure of the major *Brassica* species, and facilitating their selection and breeding through GWAS and genomic selection. Pangenomes have been developed, which have provided insights into intraspecific genomic variation. Population transcriptomics and epigenomics studies are emerging. Population transcriptomics has not only allowed determining gene expression variation among accessions and varieties, but also it has allowed identification of differentially expressed genes between subgenomes, facilitating understanding of evolutionary history of *Brassica* species. Population epigenomics in *Brassica* has elucidated the genome-wide DNA methylation and histone modification landscapes and the identification of small RNAs. Population epigenomics in combination with population transcriptomics will help understanding phenotypic plasticity and acclimation and adaptation to different and changing climatic and environmental conditions in *Brassica*. Following the establishment of many high-quality reference genomes, breeders can improve their breeding techniques using tools such as GWAS, and high-throughput sequencing technologies, enabling them to identify more accurate genomic regions and the genes associated with desired traits. Moreover, GS provides good models to predict the breeding values and select individuals in *Brassica* breeding programs based on their genotypic data. Population genomics will continue to provide key insights into all of the above aspects at much higher resolution as well as novel insights, while facilitating breeding of *Brassica* species.

Acknowledgement This work was supported by the National Natural Science Foundation of China (31871653), the National Training Program of Innovation and Entrepreneurship for Undergraduates (202110635099), Chongqing Postgraduate Research Innovation Project (CYB21115), the Natural Science Foundation of Chongqing, China, and the 111 project (B12006).

References

Aissiou F, Laperche A, Falentin C, Lodé M, Deniot G, Boutet G, et al. A novel *Brassica rapa* L. genetic diversity found in Algeria. Euphytica. 2018;214:241.
Akhatar J, Singh M, Sharma A, Kaur H, Kaur N, Sharma S, et al. Association mapping of seed quality traits under varying conditions of nitrogen application in *Brassica juncea* L. Czern Coss Front Genet. 2020;11:744.
Al-Shehbaz IA. A generic and tribal synopsis of the Brassicaceae (Cruciferae). Taxon. 2012;61:931–54.
Amasino RM, Michaels SD. The timing of flowering. Plant Physiol. 2010;154:516–20.
Amoah S, Kurup S, Rodriguez-Lopez CM, Welham SJ, Powers SJ, Hopkins CJ, et al. A Hypomethylated population of *Brassica rapa* for forward and reverse Epi-genetics. BMC Plant Biol. 2012;12:193.
An XH, Chen BY, Fu TD, Liu HL. Genetic diversity of Chinese landraces in *Brassica juncea* was analysed by RAPD markers. J Huazhong Agric U. 1999;18:524–7.

An H, Qi XH, Gaynor ML, Hao Y, Gebken SC, Mabry ME, et al. Transcriptome and organellar sequencing highlights the complex origin and diversification of allotetraploid *Brassica napus*. Nat Commun. 2019;10:2878.

Andres F, Coupland G. The genetic basis of flowering responses to seasonal cues. Nat Rev Genet. 2012;13:627–39.

Attia T, Röbbelen G. Cytogenetic relationship within cultivated *Brassica* analyzed in amphihaploids from the 3 diploid ancestors. Can J Genet Cytol. 1986;28:323–9.

Attia T, Busso C, Röbbelen G. Digenomic triploids for an assessment of chromosome relationships in the cultivated diploid *Brassica* species. Genome. 1987;29:326–30.

Ban Y, Khan NA, Yu P. Nutritional and metabolic characteristics of *Brassica carinata* Co-products from biofuel processing in dairy cows. J Agric Food Chem. 2017;65:5994–6001.

Bancroft I, Morgan C, Fraser F, Higgins J, Wells R, Clissold L, et al. Dissecting the genome of the polyploid crop oilseed rape by transcriptome sequencing. Nat Biotechnol. 2011;29:762–6.

Bastow R, Mylne JS, Lister C, Lippman Z, Martiensen RA, Dean C. Vernalization requires epigenetic silencing of *FLC* by histone methylation. Nature. 2004;427:164–7.

Bayer PE, King GJ. Assembly and comparison of two closely related *Brassica napus* genomes. Plant Biotechnol. 2017;15:1602–10.

Belser C, Istace B, Denis E, Dubarry M, Baurens FC, Falentin C, et al. Chromosome-scale assemblies of plant genomes using nanopore long reads and optical maps. Nat Plants. 2018;4:879–87.

Bird KA, An H, Gazave E, Gore MA, Pires JC, Robertson LD, et al. Population structure and phylogenetic relationships in a diverse panel of *Brassica rapa* L. Front Plant Sci. 2017;8:321.

Blackman BK, Rasmussen DA, Strasburg JL, Raduski AR, Burke JM, Knapp SJ, et al. Contributions of flowering time genes to sunflower domestication and improvement. Genetics. 2011;187:271–87.

Bonnema G. In: Edwards D, Batley J, Parkin I, Kole C, editors. Genetics, genomics and breeding of oilseed *Brassicas*. Boca Raton: CRC Press; 2012. p. 47–72.

Bonnema G. Molecular mapping and cloning of genes and QTLs in *Brassica rapa*. The *Brassica rapa* genome. Berlin: Springer; 2015.

Bornet B, Branchard M. Use of ISSR fingerprints to detect microsatellites and genetic diversity in several related *Brassica taxa* and *Arabidopsis thaliana*. Hereditas. 2004;140:245–8.

Branca F, Cartea E. *Brassica*. In: Kole C, editor. Wild crop relatives: genomic and breeding resources, oilseeds. Berlin: Springer; 2011. p. 17–36.

Burton WA, Ripley VL, Potts DA, Salisbury PA. Assessment of genetic diversity in selected breeding lines and cultivars of canola quality *Brassica juncea* and their implications for canola breeding. Euphytica. 2004;136:181–92.

Buzza GC. Plant breeding. In: Kimber DS, McGregor DI, editors. *Brassica* oilseeds: production and utilization. Wallingford: CABI Publishing; 1995. p. 153–75.

Cai CC, Wang XB, Liu B, Wu J, Liang JL, Cui YA, et al. *Brassica rapa* genome 2.0: a reference upgrade through sequence re-assembly and gene re-annotation. Mol Plant. 2017;10:649–651.

Cai X, Wu J, Liang JL, Lin RM, Zhang K, Cheng F, et al. Improved *Brassica oleracea* jzs assembly reveals significant changing of LTR-RT dynamics in different morphotypes. Theor Appl Genet. 2020;133:3187–99.

Carr SM, Irish VF. Floral homeotic gene expression defines developmental arrest stages in *Brassica oleracea* L.vars. *botrytis* and *italica*. Planta. 1997;201:179–88.

Cavell AC, Lydiate DJ, Parkin IAP, Dean C, Trick M. Collinearity between a 30-centimorgan segment of *Arabidopsis thaliana* chromosome 4 and duplicated regions within the *Brassica napus* genome. Genome. 1998;41:62–9.

CFIA (Canadian Food Inspection Agency). The biology of *Brassica rapa* L. A comparison document to the assessment criteria for determining environmental safety of plants with novel traits. Regulatory Directive Dir 1999-02, CFIA; 1999.

Chalhoub B, Denoeud F, Liu SY, Parkin IAP, Tang HB, Wang XY, et al. Early allopolyploid evolution in the post-neolithic *Brassica napus* oilseed genome. Science. 2014;345:950–3.

Chen S, Nelson MN, Ghamkhar K, Fu T, Cowling WA. Divergent patterns of allelic diversity from similar origins: the case of oilseed rape (*Brassica napus* L.) in China and Australia. Genome. 2008;51:1–10.

Chen S, Wan ZJ, Nelson MN, Chauhan JS, Redden R, Burton WA, et al. Evidence from genome-wide simple sequence repeat markers for a polyphyletic origin and secondary centers of genetic diversity of *Brassica juncea* in China and India. J Hered. 2013;104:416–27.

Chen L, Dong FM, Cai J, Xin Q, Fang CC, Liu L, et al. A 2.833-kb insertion in *BnFLC.A2* and its homeologous exchange with *BnFLC.C2* during breeding selection generated early-flowering rapeseed. Mol Plant. 2018;1:222–5.

Chen XQ, Tong CB, Zhang XT, Song AX, Hu M, Dong W, et al. A high-quality *Brassica napus* genome reveals expansion of transposable elements, subgenome evolution and disease resistance. Plant Biotechnol J. 2020a;19:615–30.

Chen RK, Shimono A, Aono M, Nakajima N, Ohsawa R, Yoshioka Y. Genetic diversity and population structure of feral rapeseed (*Brassica napus* L.) in Japan. PLoS One. 2020b;15: e0227990.

Cheng F, Mandakova T, Wu J, Xie Q, Lysak MA, Wang XW. Deciphering the diploid ancestral genome of the mesohexaploid *Brassica rapa*. Plant Cell. 2013;25:1541–54.

Cheng F, Wu J, Wang XW. Genome triplication drove the diversification of *Brassica* plants. Hort Res. 2014;1:14024.

Chiang MS, Chong C, Landry RS, Crete R. Cabbage *Brassica oleracea* sub sp. *capitata* L. In: Kalloo G, Bergh BO, editors. Genetic improvement of vegetable crops. Oxford: Pergamon Press; 1993. p. 113–55.

Choi SR, Teakle GR, Plaha P, Kim JH, Allender CJ, Beynon E, et al. The reference genetic linkage map for the multinational *Brassica rapa* genome sequencing project. Theor Appl Genet. 2007;115:777–92.

Christensen S, von Bothmer R, Poulsen G, Maggioni L, Phillip M, Andersen BA, et al. AFLP analysis of genetic diversity in leafy kale (*Brassica oleracea* L. convar. acephala (DC.) Alef.) landraces, cultivars and wild populations in Europe. Genet Resour Crop Evol. 2011;58:657–66.

Chung H, Jeong YM, Mun JH, Lee SS, Chung WH, Yu HJ, et al. Construction of a genetic map based on high-throughput SNP genotyping and genetic mapping of a TuMV resistance locus in *Brassica rapa*. Mol Genet Genomics. 2014;289:149–60.

Chyi YS, Hoenecke ME, Sernyk JL. A genetic linkage map of restriction fragment length polymorphism loci for *Brassica rapa* (syn campestris). Genome. 1992;35:746–57.

Denford KE, Vaughan JG. A comparative study of certain seed isoenzymes in the ten chromosome complex of *Brassica* campestris and its allies. Ann Bot. 1977;41:411–8.

Doebley JF, Gaut BS, Smith BD. The molecular genetics of crop domestication. Cell. 2006;127:1309–21.

Dong XS, Yi H, Han CT, Nou IS, Swaraz AM, Hur Y. Genomewide analysis of genes associated with bolting in heading type chinese cabbage. Euphytica. 2016;212:65–82.

Duclos DV, Björkman T. Meristem identity gene expression during curd proliferation and flower initiation in *Brassica oleracea*. J Exp Bot. 2008;59:421–33.

Edwards D, Batley J, Snowdon RJ. Accessing complex crop genomes with next-generation sequencing. Theor Appl Genet. 2013;126:1–11.

El-Esawi MA. Taxonomic relationships and biochemical genetic characterization of *Brassica* resources: towards a recent platform for germplasm improvement and utilization. Annu Res Rev Biol. 2015;8:1–11.

El-Esawi MA. Genetic diversity and evolution of *Brassica* genetic resources: from morphology to novel genomic technologies – a review. Plant Genet Resour. 2017;15:388–99.

El-Esawi MA, Germaine K, Bourke P, Malone R. Genetic diversity and population structure of *Brassica oleracea* germplasm in Ireland using SSR markers. C R Biol. 2016;339:133–40.

Falk DA, Knapp E, Guerrant EO. Introduction to restoration genetics. Washington: Society for Ecological Restoration; 2001. http://www.nps.gov/plants/restore/pubs/restgene/restgene.pdf

Fikere M, Barbulescu DM, Malmberg MM, Maharjan P, Salisbury PA, Kant S, et al. Genomic prediction and genetic correlation of agronomic, blackleg disease, and seed quality traits in canola (*Brassica napus* L.). Plants (Basel). 2020;9:719.

Flowers JM, Hanzawa Y, Hall MC, Moore RC, Purugganan MD. Population genomics of the *Arabidopsis thaliana* flowering time gene network. Mol Biol Evol. 2009;26:2475–86.

Franzke A, Lysak MA, Al-Shehbaz IA, Koch MA, Mummenhoff K. Cabbage family affairs: the evolutionary history of Brassicaceae. Trends Plant Sci. 2011;16:108–16.

Fu Y, Wei DY, Dong HL, He YJ, Cui YX, Mei JQ, et al. Comparative quantitative trait loci for silique length and seed weight in *Brassica napus*. Sci Rep. 2015;5:14407.

Gaeta RT, Chris PJ. Homoeologous recombination in allopolyploids: the polyploid ratchet. New Phytol. 2010;186:18–28.

Gazave E, Tassone EE, Ilut DC, Wingerson M, Datema E, Witsenboer HMA. Population genomic analysis reveals differential evolutionary histories and patterns of diversity across subgenomes and subpopulations of *Brassica napus* L. Front Plant Sci. 2016;7:525.

Ghalambor CK, Hoke KL, Ruell EW, Fischer EK, Reznick DN, Hughes KA. Non-adaptive plasticity potentiates rapid adaptive evolution of gene expression in nature. Nature. 2015;525:372–5.

Goddard M, Hayes B. Genomic selection. J Anim Breed Genet. 2007;124:323–30.

Golicz AA, Bayer PE, Barker GC, Edger PP, Kim HR, Martinez PA, et al. The pangenome of an agronomically important crop plant *Brassica oleracea*. Nat Commun. 2016;7:13390.

Gómez-Campo C, editor. Biology of *Brassica* coenospecies. Amsterdam: Elsevier; 1999. p. 59–106.

Gupta S, Sharma N, Akhatar J, Atri C, Kaur J, Kaur G, et al. Analysis of epigenetic landscape in a recombinant inbred line population developed by hybridizing natural and resynthesized *Brassica juncea* (L.) with stable C-genome introgressions. Euphytica. 2019;215:174.

Hansen MM. Expression of interest: transcriptomics and the designation of conservation units. Mol Ecol. 2010;19:1757–9.

Harbered DJ. Cytotaxonomic studies of *Brassica* and related genera. In: Vaughan JG, MacLeod AJ, Jones MG, editors. The biology and chemistry of the cruciferae. London: Academic Press; 1976. p. 47–68.

Harper AL, Trick M, Higgins J, Fraser F, Clissold L, Wells R, et al. Associative transcriptomics of traits in the polyploid crop species *Brassica napus*. Nat Biotechnol. 2012;30:798–802.

Harper AL, He ZS, Langer S, Havlickova L, Wang LH, Fellgett A, et al. Validation of an associative transcriptomics platform in the polyploid crop species *Brassica juncea* by dissection of the genetic architecture of agronomic and quality traits. Plant J. 2020;103:1885–93.

Hauben M, Haesendonck B, Standaert E, Van Der KK, Azmi A, Akpo H, et al. Energy use efficiency is characterized by an epigenetic component that can be directed through artificial selection to increase yield. Proc Natl Acad Sci U S A. 2009;106:20109–14.

He YT, Tu JX, Fu TD, Li DR, Chen BY. Genetic diversity of germplasm resources of *(Brassica campestris* L.) in China by RAPD markers. Acta Agron Sin. 2002;28:693–703.

He ZS, Cheng F, Li Y, Wang XW, Parkin IAP, Chalhoub B, et al. Construction of *Brassica* A and C genome-based ordered pan-transcriptomes for use in rapeseed genomic research. Data Brief. 2015;4:357–62.

He ZS, Wang LH, Harper AL, Havlickova L, Pradhan AK, Parkin IAP, et al. Extensive homoeologous genome exchanges in allopolyploid crops revealed by mRNAseq-based visualization. Plant Biotechnol J. 2017;15:594–604.

Herve Y. Choux. In: Pitrat M, Foury C, editors. History de legumes, des origins a l'oree du XXI siecle. Paris: INRA; 2003.

Heywood V. Flowering plants of the world. London: B.T. Batsford Ltd.; 1993.

Higgins J, Magusin A, Trick M, Fraser F, Bancroft I. Use of mRNA-seq to discriminate contributions to the transcriptome from the constituent genomes of the polyploid crop species *Brassica napus*. BMC Genomics. 2012;13:247.

Hong CP, Kwon SJ, Kim JS, Yang TJ, Park BS, Lim YP. Progress in understanding and sequencing the genome of *Brassica rapa*. Int J Plant Genomics. 2008;2008:582837.

Hosaka K, Kianian SF, McGrath JM, Quiros CF. Development and chromosomal localization of genome specific DNA markers of *Brassica* and evolution of amphidiploids and n = 9 diploid species. Genome. 1990;33:131–42.

Hou JN, Long Y, Raman H, Zou XX, Wang J, Dai ST, et al. A Tourist-like MITE insertion in the upstream region of the *BnFLC.A10* gene is associated with vernalization requirement in rapeseed (*Brassica napus* L.). BMC Plant Biol. 2012;12:238.

Husband BC, Schemske DW. Evolution of the magnitude and timing of inbreeding depression in plants. Evolution. 1996;50:54–70.

Ignatov AN, Artemyeva AM, Hida K. Origin and expansion of cultivated *Brassica rapa* in Eurasia: linguistic facts. Acta Hortic. 2010;867:81–8.

Iniguez-Luy FL, Federico ML. The genetics of *Brassica napus*. In: Schmidt R, Bancroft I, editors. Genetics and genomics of the brassicaceae. New York: Springer; 2011. p. 585–96.

Jain A, Bhatia S, Banga SS, Prakash S, Laxmikumaran M. Potential use of random amplified polymorphic DNA (RAPD) technique to study genetic diversity in Indian mustard (*B. juncea*) and its relatedness to heterosis. Theor Appl Genet. 1994;88:16–122.

Judd WS, Campbell CS, Kellogg EA, Stevens PF. Plant systematics, a phylogenetic approach. Sinauer Associates, Inc: Sunderland; 1999.

Jung SK, Tae YC, King GJ, Jin M, Yang TJ, Jin YM, et al. A sequence-tagged linkage map of *Brassica rapa*. Genetics. 2006;174:29–39.

Kagale S, Robinson SJ, Nixon J, Xiao R, Huebert T, Condie J, et al. Polyploid evolution of the Brassicaceae during the Cenozoic era. Plant Cell. 2014;26:2777–91.

Kakizaki T, Kato T, Fukino N, Ishida M, Hatakeyama K, Matsumoto S. Identification of quantitative trait loci controlling late bolting in Chinese cabbage (*Brassica rapa* L.) parental line Nou 6 gou. Breed Sci. 2011;61:151–9.

Kaur P, Banga S, Kumar N, Gupta S, Akhtar J, Banga SS. Polyphyletic origin of *Brassica juncea* with *B. rapa* and *B. nigra* (Brassicaceae) participating as cytoplasm donor parents in independent hybridization events. Am J Bot. 2014;101:1157–66.

Kebede B, Cheema K, Greenshields DL, Li CX, Selvaraj G, Rahman H. Construction of genetic linkage map and mapping of QTL for seed color in *Brassica rapa*. Genome. 2012;55:813–23.

Kempin SA, Savidge B, Yanofsky MF. Molecular basis of the cauliflower phenotype in arabidopsis. Science. 1995;267:522–5.

Khanzada H, Wassan GM, He HH, Mason AS, Keerio AA, Khanzada S, et al. Differentially evolved drought stress indices determine the genetic variation of *Brassica napus* at seedling traits by genome-wide association mapping. J Adv Res. 2020;24:447–61.

Khedikar Y, Clarke WE, Chen LF, Higgins EE, Kagale S, Koh CS, et al. Narrow genetic base shapes population structure and linkage disequilibrium in an industrial oilseed crop, *Brassica carinata* A. Braun Sci Rep. 2020;10:12629.

Kim HR, Choi SR, Bae J, Hong CP, Lee SY, Hossain MJ, et al. Sequenced BAC anchored reference genetic map that reconciles the ten individual chromosomes of *Brassica rapa*. BMC Genomics. 2009;10:432.

Kim J, Kim DS, Park SY, Lee HE, Ahn YK, Kim JH, et al. Development of a high-throughput SNP marker set by transcriptome sequencing to accelerate genetic background selection in *Brassica rapa*. Hortic Environ Biotechnol. 2016;57:280–90.

Labate JA, Robertson LD, Baldo AM, Björkman T. Inflorescence identity gene alleles are poor predictors of inflorescence type in broccoli and cauliflower. J Am Soc Hort Sci. 2006;131:667–73.

Lagercrantz U, Lydiate DJ. Comparative genome mapping in *Brassica*. Genetics. 1996;144:1903–10.

Lázaro A, Aguinagalde I. Genetic Diversity in *Brassica oleracea* L. (Cruciferae) and Wild Relatives (2n=18) using Isozymes. Ann Bot. 1998a;82:821–8.

Lázaro A, Aguinagalde I. Genetic Diversity in *Brassica oleracea* L. (Cruciferae) and Wild Relatives (2n=18) using RAPD Markers. Ann Bot. 1998b;82:829–33.

Lenser T, Theissen G. Molecular mechanisms involved in convergent crop domestication. Trends Plant Sci. 2013;18:704–14.

Li RG, Zhu L, Wu NF, Fan YL, Wu XM, Qian XZ. Genetic diversity among oilseed cultivars of *Brassica juncea* Czern. Coss in China. J Biotechnol. 1997;5:26–31.

Li W, Zhang JF, Mou YL, Geng JF, McVetty PBE, Hu SW, et al. Integration of Solexa sequences on an ultradense genetic map in *Brassica rapa* L. BMC Genomics. 2011;12:249.

Li KX, Yao YM, Xiao L, Zhao ZG, Guo SM, Fu Z, et al. Fine mapping of the *Brassica napus Bnsdt1* gene associated with determinate growth habit. Theor Appl Genet. 2018;131:193–208.

Li SY, Zhu YY, Varshney RK, Zhan JP, Zheng XX, Shi JQ, et al. A systematic dissection of the mechanisms underlying the natural variation of silique number in rapeseed (*Brassica napus* L.) germplasm. Plant Biotechnol J. 2020;18:568–80.

Liu B, Wang Y, Zhai W, Deng J, Wang H, Cui Y, et al. Development of InDel markers for *Brassica rapa* based on whole-genome re-sequencing. Theor Appl Genet. 2013;126:231–9.

Liu SY, Liu YM, Yang XH, Tong CB, Edwards D, Parkin IAP, et al. The *Brassica oleracea* genome reveals the asymmetrical evolution of polyploid genomes. Nat Commun. 2014;5:3930.

Liu ST, Wang RH, Zhang ZG, Li QY, Wang LH, Wang YQ, et al. High-resolution mapping of quantitative trait loci controlling main floral stalk length in Chinese cabbage (*Brassica rapa* L. ssp. *pekinensis*). BMC Genom. 2019;20:437.

Liu S, Huang HB, Yi XQ, Zhang YY, Yang QY, Zhang CY, et al. Dissection of genetic architecture for glucosinolate accumulations in leaves and seeds of *Brassica napus* by genome-wide association study. Plant Biotechnol J. 2020;18:1472–84.

Liu J, Rana K, McKay J, Xiong ZY, Yu FQ, Mei JQ, et al. Investigating genetic relationship of *Brassica juncea* with *B. nigra* via virtual allopolyploidy and hexaploidy strategy. Mol Breed. 2021;41:5.

Long Y, Shi J, Qiu D, Li R, Zhang C, Wang J, et al. Flowering time quantitative trait loci analysis of oilseed brassica in multiple environments and genomewide alignment with arabidopsis. Genetics. 2007;177:2433–44.

Lowman AC, Purugganan MD. Duplication of the *Brassica oleracea APETALA1* floral homeotic gene and the evolution of domesticated cauliflower. J Hered. 1999;90:514–52.

Lu K, Wei LJ, Li XL, Wang YT, Wu J, Liu M, et al. Whole-genome resequencing reveals *Brassica napus* origin and genetic loci involved in its improvement. Nat Commun. 2019;10:1154.

Luikart G, Kardos M, Hand BK, Rajora OP, Aitkin SN, Hohenlohe PA. Population genomics: advancing understanding of nature. In: Rajora OP, editor. Population genomics: concepts, approaches and applications. Cham: Springer Nature Switzerland AG; 2019. p. 3–79.

Lysak MA, Koch MA, Pecinka A, Schubert I. Chromosome triplication found across the tribe Brassiceae. Genome Res. 2005;15:516–25.

Mabberley DJ. The plant-book, a protable dictionary of the vascular plants. 2nd ed. Cambridge: Cambridge University Press; 1997.

Maggioni L, Bothmer R, Poulsen G, Aloisi KH. Survey and genetic diversity of wild *Brassica oleracea* L. germplasm on the Atlantic coast of France. Genet Resour Crop Evol. 2020;67:1853–66.

Malmberg MM, Shi F, Spangenberg GC, Daetwyler HD, Cogan NOI. Diversity and genome analysis of Australian and global oilseed *Brassica napus* L. germplasm using transcriptomics and whole genome re-sequencing. Front Plant Sci. 2018;9:508.

Mandakova T, Lysak MA. Chromosomal phylogeny and karyotype evolution in x = 7 crucifer species (Brassicaceae). Plant Cell. 2008;20:2559–70.

Manrique S, Friel J, Gramazio P, Hasing T, Ezquer I, Bombarely A. Genetic insights into the modification of the pre-fertilization mechanisms during plant domestication. J Exp Bot. 2019;70:3007–19.

Meffe GK, Carroll CR. Principles of conservation biology. Sunderland: Sinauer Associates; 1994.

Meuwissen THE, Hayes BJ, Goddard ME. Prediction of total genetic value using genome-wide dense marker maps. Genetics. 2001;157:1819–29.

Mittell EA, Cobbold CA, Zeeshan IU, Kilbride EA, Moore KA, Mable BK. Feral populations of *Brassica oleracea* along Atlantic coasts in western Europe. Ecol Evol. 2020;10:11810–25.

Moler ERV, Abakir A, Eleftheriou M, Johnson JS, Krutovsky KV, Lewis LC, et al. Population epigenomics: advancing understanding of phenotypic plasticity, acclimation, adaptation and diseases. In: Rajora OP, editor. Population genomics: concepts, approaches and applications. Cham: Springer Nature Switzerland AG; 2019. p. 179–260.

Nagaharu U. Genome analysis in Brassica with special reference to the experimental formation of *B. napus* and peculiar mode of fertilization. Jpn J Bot. 1935;7:389–452.

Navabi Z. Genetic analysis of the B-genome chromosomes in the *Brassica* species. PhD thesis. Edmonton: Department of Agricultural, Food and Nutritional Science, Faculty of Graduate Studies and Research, University of Alberta; 2009.

Navabi ZK, Huebert T, Sharpe AG, O'Neill CM, Bancroft I, Parkin IAP. Conserved microstructure of the *Brassica* B genome of *Brassica nigra* in relation to homologous regions of *Arabidopsis thaliana*, *B. rapa* and *B. oleracea*. BMC Genomics. 2013;14:250.

Negi MS, Sabharwal V, Bhat SR, Lakshmikumaran M. Utility of AFLP markers for the assessment of genetic diversity within *Brassica nigra* germplasm. Plant Breed. 2004;123:13–6.

Odongo GA, Schlotz N, Herz C, Hanschen FS, Baldermann S, Neugart S, et al. The role of plant processing for the cancer preventive potential of Ethiopian kale (*Brassica carinata*). Food Nutr Res. 2017;61:1271527.

Oduor AMO, Gomez JM, Herrador MB, Perfectti F. Invasion of *Brassica nigra* in North America: distributions and origins of chloroplast DNA haplotypes suggest multiple introductions. Biol Invasions. 2015;17:2447–59.

Ojiewo C, Ebert A, Oluoch MO. *Brassica carinata* (Brassicaceae). A versatile crop for tomorrow. In: Nono-Womdim R, Achigan-Dako E, Pichop GN, Maundu P, Baudoin W, Lutaladio NB, Aphane J, Noorani A, Ghosh K, Hodder A, editors. Indigenous fruit and vegetables of tropical Africa. A guide to a sustainable production of selected underutilized crops, vol. 1. Rome: Food and Agriculture Organization of the United Nations; 2014. p. 123–36.

Pang WX, Li XN, Choi SR, Dhandapani V, Im S, Park MY, et al. Development of a leafy *Brassica rapa* fixed line collection for genetic diversity and population structure analysis. Mol Breeding. 2015;35:54.

Paritosh K, Yadava SK, Gupta V, Panjabi-Massand P, Sodhi YS, Pradhan AK, et al. RNA-seq based SNPs in some agronomically important oleiferous lines of *Brassica rapa* and their use for genome-wide linkage mapping and specific-region fine mapping. BMC Genomics. 2013;14:436.

Parkin IAP, Koh C, Tang HB, Robinson SJ, Kagale S, Clarke WE, et al. Transcriptome and methylome profiling reveals relics of genome dominance in the mesopolyploid *Brassica oleracea*. Genome Biol. 2014;15:R77.

Persson K, Falt AS, von Bothmer R. Genetic diversity of allozymes in turnip (*Brussicu rupa L*. var. *rap*) from the Nordic area. Hereditus. 2001;134:43–52.

Perumal S, Koh CS, Jin L, Buchwaldt M, Parkin IAP. A high-contiguity *Brassica nigra* genome localizes active centromeres and defines the ancestral *Brassica* genome. Nat Plants. 2020;6:929–41.

Pradhan A, Nelson MN, Plummer JA, Cowling WA, Yan GJ. Characterization of *Brassica nigra* collections using simple sequence repeat markers reveals distinct groups associated with geographical location, and frequent mislabelling of species identity. Genome. 2011;54:50–63.

Prakash S, Wu XM, Bhat SR. History, evolution, and domestication of *Brassica* crops. Plant Breed Rev. 2011;35:19–84.

Pu XB, Wang ML, Luan L, Wang XJ, Zhang JF, Li HJ, et al. Genetic diversity analysis of *Brassica juncea* landraces in southwest China. Sci Agric Sin. 2007;40:1610–21.

Purugganan MD, Fuller DQ. The nature of selection during plant domestication. Nature. 2009;457:843–8.

Purugganan MD, Boyles AL, Suddith JI. Variation and selection at the *CAULIFLOWER* floral homeotic gene accompanying the evolution of domesticated *Brassica oleracea*. Genetics. 2000;155:855–62.

Qi XS, An H, Ragsdale AP, Hall TE, Gutenkunst RN, Pires JC, et al. Genomic inferences of domestication events are corroborated by written records in Brassica rapa. Mol Ecol. 2017;26 (13):3373–88.

Qian W, Meng J, Li M, Frauen M, Sass O, Noack J, et al. Introgression of genomic components from Chinese Brassica rapa contributes to widening the genetic diversity in rapeseed (*B. napus* L.), with emphasis on the evolution of Chinese rapeseed. Theor Appl Genet. 2006;113:49–54.

Rakow G. Species origin and economic importance of *Brassica*. In: Pua EC, Douglas CJ, editors. Biotechnology in agriculture and forestry, vol. 54. Berlin: Springer; 2004. p. 3–11.

Raman R, Qiu Y, Coombes N, Song J, Kilian A, Raman H. Molecular diversity analysis and genetic mapping of pod shatter resistance loci in *Brassica carinata* L. Front Plant Sci. 2017;8:1765.

Reiner H, Holzner W, Ebermann R. The development of turnip-type and oilseed-type *Brassica rapa* crops from the wild type in Europe – an overview of botanical, historical and linguistic facts. Proceedings of 9th international rapeseed congress, vol. 4. 1995, p. 1066–1069.

Sauer JD. Historical geography of crop plants: a selective roster. Boca Raton: CRC Press; 1993.

Schilling S, Pan S, Kennedy A, Melzer R. *MADS-box* genes and crop domestication: the jack of all traits. J Exp Bot. 2018;69:1447–1469.

Schranz ME, Quijada P, Sung SB, Lukens L, Amasino R, Osborn TC. Characterization and effects of the replicated flowering time gene *FLC* in *Brassica rapa*. Genetics. 2002;162:1457–68.

Schranz ME, Lysak MA, Mitchell-Olds T. The ABC's of comparative genomics in the Brassicaceae: building blocks of crucifer genomes. Trends Plant Sci. 2006;11:535–42.

Seepaul R, George S, Wright D. Comparative response of *Brassica carinata* and *B. napus* vegetative growth, development and photosynthesis to nitrogen nutrition. Ind Crop Prod. 2016;94:872–83.

Sekhon MS, Gupta VP. Genetic-distance and heterosis in Indian mustard - developmental isozymes as indicators of genetic-relationships. Theor Appl Genet. 1995;91:1148–52.

Shahidi F. Rapessed and canola: global production and distribution. In: Sahidi F, editor. Canola and rapeseed: production, chemistry and processing technology. New York: Van Nostrand Reinhold; 1990. p. 3–13.

Singh KH, Shakya R, Thakur AK, Chauhan DK, Chauhan JS. Genetic diversity in Indian Mustard (*Brassica juncea* (L.) Czernj&Cosson) as revealed by agronomic traits and RAPD markers. Nat Acad Sci Lett. 2013;36:419–27.

Smith LB, King GJ. The distribution of BoCAL-a alleles in *Brassica oleracea* is consistent with a genetic model for curd development and domestication of the cauliflower. Mol Breed. 2000;6:603–13.

Snowdon RJ, Luy FLI. Potential to improve oilseed rape and canola breeding in the genomics era. Plant Breed. 2012;131:351–60.

Soengas P, Cratea ME, Francisco M, Lemaand M, Velasco EP. Genetic structure and diversity of a collection of *Brassica rapa subsp. rapa* L. revealed by simple sequence repeat markers. J Agric Sci. 2011;149:617–24.

Soltis DE, Visger CJ, Marchant DB, Soltis PS. Polyploidy: pitfalls and paths to a paradigm. Am J Bot. 2016;103:1146–66.

Song KM, Osborn TC, Williams PH. *Brassica* taxonomy based on nuclear restriction fragment length polymorphism (RFLPs). 2. Preliminary analysis of sub-species with in *B. rapa* (syn. campestris) and *B. oleracea*. Theor Appl Genet 1988;76:593–600.

Song KM, Osborn TC, Williams PH. *Brassica* taxonomy based on nuclear restriction fragment length polymorphism (RFLPs). 3. Genome relationship in *Brassica* and related genera and the origin of *B. oleracea* and *Brassica rapa* (syn. campestris). Theor Appl Genet. 1990;79:497–506.

Song KM, Suzuki JY, Slocum MK, Williams PM, Osborn TC. A linkage map of *Brassica rapa* (syn. campestris) based on restriction fragment length polymorphism loci. Theor Appl Genet. 1991;82:296–304.

Song K, Slocum MK, Osborn TC. Molecular marker analysis of genes-controlling morphological variation in *Brassica-rapa* (syn campestris). Theor Appl Genet. 1995;90:1–10.

Song JM, Guan ZL, Hu JL, Guo CC, Yang ZQ, Wang S, et al. Eight high-quality genomes reveal pan-genome architecture and ecotype differentiation of *Brassica napus*. Nat Plants. 2020;6:34–45.

Song XM, Wei YP, Xiao D, Gong K, Sun PC, Ren YM, et al. *Brassica carinata* genome characterization clarifies U's triangle model of evolution and polyploidy in *Brassica*. Plant Physiol. 2021;186:388–406.

Srivastava A, Gupta V, Pental D, Pradhan AK. AFLP-based genetic diversity assessment amongst agronomically important natural and some newly synthesized lines of *Brassica juncea*. Theor Appl Genet. 2001;102:193–9.

Stansell Z, Hyma K, Fresnedo-Ramírez J, Sun Q, Mitchell S, Bjorkman T, et al. Genotyping-by-sequencing of *Brassica oleracea* vegetables reveals unique phylogenetic patterns, population structure and domestication footprints. Hortic Res. 2018;5:38.

Stephens JM. Mustard collard – *Brassica carinata* L. University of Florida, Institute of Food and Agricultural Sciences; 2009. Retrieved from https://edis.ifas.ufl.edu/mv096.

Sun FM, Fan GY, Hu Q, Zhou YM, Guan M, Tong CB, et al. The high-quality genome of *Brassica napus* cultivar 'ZS11' reveals the introgression history in semi-winter morphotype. Plant J. 2017;92:452–68.

Sun DL, Wang CG, Zhang XL, Zhang WL, Jiang HM, Yao XW, et al. Draft genome sequence of cauliflower (*Brassica oleracea* L. var. *botrytis*) provides new insights into the C genome in *Brassica* species. Hortic Res. 2019;6:82.

Suwabe K, Suzuki G, Nunome T, Hatakeyama K, Mukai Y, Fukuoka H, et al. Microstructure of a *Brassica rapa* genome segment homoeologous to the resistance gene cluster on arabidopsis chromosome 4. Breed Sci. 2012;62:170–7.

Tan HQ, Wang X, Fei ZJ, Li HX, Tadmor Y, Mazourek M, et al. Genetic mapping of green curd gene Gr in cauliflower. Theor Appl Genet. 2020;133:353–64.

Tang S, Zhao H, Lu SP, Yu LQ, Zhang GF, Zhang YT, et al. Genome- and transcriptome-wide association studies provide insights into the genetic basis of natural variation of seed oil content in *Brassica napus*. Mol Plant. 2020;14:470–87.

Tao YF, Zhao XR, Mace E, Henry R, Jordan D. Exploring and exploiting pan-genomics for crop improvement. Mol Plant. 2019;12:156–69.

Tautz D. Hypervariability of simple sequences as a general source for polymorphic DNA markers. Nucleic Acids Res. 1989;17:6463–71.

Teutonico RA, Osborn TC. Mapping of RFLP and qualitative trait loci in *Brassica rapa* and comparison to the linkage maps of *B. napus*, *B. oleracea*, and *Arabidopsis thaliana*. Theor Appl Genet. 1994;89:885–94.

Thorwarth P, Yousef E, Schmid K. Genomic prediction and association mapping of curd-related traits in gene bank accessions of cauliflower. G3. 2018;8:707–18.

Town CD, Cheung F, Maiti R, Crabtree J, Haas BJ, Wortman JR, et al. Comparative genomics of *Brassica oleracea* and *Arabidopsis thaliana* reveal gene loss, fragmentation, and dispersal after polyploidy. Plant Cell. 2006;18:1348–59.

Trick M, Long Y, Meng J, Bancroft I. Single nucleotide polymorphism (SNP) discovery in the polyploid *Brassica napus* using solexa transcriptome sequencing. Plant Biotechnol J. 2009;7:334–46.

Tsunoda S. Eco-physiology of wild and cultivated forms in *Brassica* and allied genera. In: Tsunoda S, Hinata K, Gomez-Campo C, editors. *Brassica* crops and wild allies. Tokyo: Japan Science Society Press; 1980. p. 109–20.

Tudor EH, Jones DM, He ZS, Bancroft I, Trick M, Wells R, et al. QTL-seq identifies *BnaFT.A02* and *BnaFLC.A02* as candidates for variation in vernalization requirement and response in winter oilseed rape (*Brassica napus*). Plant Biotechnol J. 2020;2466–81.

Van Lijsebettens M, Grasser KD. The role of the transcript elongation factors *FACT* and *HUB1* in leaf growth and the induction of flowering. Plant Signal Behav. 2010;5:715–7.

Vavilov NI. The origin, variation, immunity and breeding of cultivated plants. Chron Bot. 1951;13:1–364.

Vinu V, Singh N, Vasudev S, Yadava DK, Kumar S, Naresh S, et al. Assessment of genetic diversity in *Brassica juncea* (Brassicaceae) genotypes using phenotypic differences and SSR markers. Rev Biol Trop. 2013;61:1919–34.

Wang NA, Qian W, Suppanz I, Wei LJ, Mao BZ, Long Y, et al. Flowering time variation in oilseed rape (*Brassica napus* L.) is associated with allelic variation in the *FRIGIDA* homologue *BnaA. FRI.a.* J Exp Bot. 2011a;62:5641–58.

Wang H, Wu J, Sun SL, Liu B, Cheng F, Sun RF, et al. Glucosinolate biosynthetic genes in *Brassica rapa*. Gene. 2011b;487:135–42.

Wang XW, Wang HZ, Wang J, Sun RF, Wu J, Liu SY, et al. The genome of the mesopolyploid crop species *Brassica rapa*. Nat Genet. 2011c;43:1035–40.

Wang Y, Sun SL, Liu B, Wang H, Deng J, Liao YC, et al. A sequence-based genetic linkage map as a reference for *Brassica rapa* pseudochromosome assembly. BMC Genomics. 2011d;12:239.

Wang N, Li F, Chen BY, Xu K, Yan GX, Qiao JW, et al. Genome-wide investigation of genetic changes during modern breeding of *Brassica napus*. Theor Appl Genet. 2014;127:1817–29.

Wang B, Wu ZK, Li ZH, Zhang QH, Hu JL, Xiao YJ, et al. Dissection of the genetic architecture of three seed-quality traits and consequences for breeding in *Brassica napus*. Plant Biotechnol J. 2018;16:1336–48.

Wang WL, Guan R, Liu X, Zhang HR, Song B, Xu QW, et al. Chromosome level comparative analysis of *Brassica* genomes. Plant Mol Biol. 2019;99:237–49.

Warwick SI, Francis A, Al-Shehbaz IA. Brassicaceae: species checklist and database on CD-Rom. Plant Syst Evol. 2006;259:249–58.

Warwick SI, Mummenhoff K, Sauder CA, Koch MA, Al-Shehbaz IA. Closing the gaps: phylogenetic relationships in the Brassicaceae based on DNA sequence data of nuclear ribosomal ITS region. Plant Syst Evol. 2010;285:209–32.

Wei LJ, Jian HJ, Lu K, Yin NW, Wang J, Duan XJ, et al. Genetic and transcriptomic analyses of lignin- and lodging-related traits in *Brassica napus*. Theor Appl Genet. 2017;130:1961–73.

Wendel JF. Genome evolution in polyploids BT – plant molecular evolution. Plant Mol Biol. 2000;42:225–49.

Werner CR, Qian LW, Voss-Fels KP, Abbadi A, Leckband G, Frisch M, et al. Genome-wide regression models considering general and specific combining ability predict hybrid performance in oilseed rape with similar accuracy regardless of trait architecture. Theor Appl Genet. 2018;131:299–317.

Wu J, Wei KY, Cheng F, Li SK, Wang Q, Zhao JJ, et al. A naturally occurring InDel variation in *BraA.FLC.b* (*BrFLC2*) associated with flowering time variation in *Brassica rapa*. BMC Plant Biol. 2012;12:151.

Wu DZ, Liang Z, Yan T, Xu Y, Xuan LJ, Tang J, et al. Whole-genome resequencing of a worldwide collection of rapeseed accessions reveals the genetic basis of ecotype divergence. Mol Plant. 2019;12:30–43.

Würschum T, Abel S, Zhao Y. Potential of genomic selection in rapeseed (*Brassica napus* L.) breeding. Plant Breed. 2014;133:45–51.

Xiao D, Zhao JJ, Hou XL, Basnet RK, Carpio DPD, Zhang NW, et al. The *Brassica rapa FLC* homologue *FLC2* is a key regulator of flowering time, identified through transcriptional co-expression networks. J Exp Bot. 2013;64:4503–16.

Xu AX, Ma CZ, Xiao ES, Quan JC, Ma CZ, Tian GW, et al. Genetic diversity of *Brassica juncea* from western China. Acta Agron Sin. 2008;34:754–63.

Yang YW, Lai KN, Tai PY, Li WH. Rates of nucleotide substitution in angiosperm mitochondrial DNA sequences and dates of divergence between *Brassica* and other angiosperm lineages. J Mol Evol. 1999;48:597–604.

Yang JH, Liu DY, Wang XW, Ji CM, Cheng F, Liu BN, et al. The genome sequence of allopolyploid *Brassica juncea* and analysis of differential homoeolog gene expression influencing selection. Nat Genet. 2016;48:1225–32.

Yu SC, Zhang FL, Zhao X, Yu YJ, Zhang DS, Zhao XY, et al. An improved *Brassica rapa* genetic linkage map and locus-specific variations in a doubled haploid population. Plant Mol Biol Rep. 2013;31:558–68.

Yu HF, Wang JS, Sheng XG, Zhao ZQ, Shen YS, Branca F, et al. Construction of a high-density genetic map and identification of loci controlling purple sepal trait of flower head in *Brassica oleracea* L. italica. BMC Plant Biol. 2019;19:1–8.

Yuan YX, Wu J, Sun RF, Zhang XW, Xu DH, Bonnema G, et al. A naturally occurring splicing site mutation in the *Brassica rapa FLC1* gene is associated with variation in flowering time. J Exp Bot. 2009;60:1299–308.

Zada M, Zakir N, Rabbani MA, Shinwari ZK. Assessment of genetic variation in Ethiopian mustard (*Brassica carinata* A. Braun) germplasm using multivariate techniques. Pak J Bot. 2013;45:583–93.

Zhang XJ, Chen HY, Channa SA, Zhang YX, Guo Y, Klima M, et al. Genetic diversity in Chinese and exotic *Brassica rapa* L. accessions revealed by SSR and SRAP markers. Braz J Bot. 2017;40:973–82.

Zhang L, Cai X, Wu J, Liu M, Grob S, Cheng F, et al. Improved Brassica rapa reference genome by single-molecule sequencing and chromosome conformation capture technologies. Hortic Res. 2018;5:50.

Zhao JY, Becker HC. Genetic variation in Chinese and European oilseed rapa (*B. napus*) and Turnip rapa (*B. campestris*) analysed with isozymes. Acta Agron Sin. 1998;24:213–20.

Zhao J, Wang X, Deng B, Lou P, Wu J, Sun R, et al. Phylogenetic relationships within *Brassica rapa* inferred from AFLP fingerprints. In: Joint meeting of the 14th crucifer genetics workshop and 4th ISHS symposium on *Brassicas*, Daejeon, Korea: Chungnam National University; 2004. 24-28 Oct 2004, p. 128.

Zhao Q, Wu J, Cai GQ, Yang QY, Shahid M, Fan CC, et al. A novel quantitative trait locus on chromosome A9 controlling oleic acid content in *Brassica napus*. Plant Biotechnol J. 2019;17:2313–24.

Zhou WJ, Zhang, GQ, Tuvesson S, Dayteg C, Gertsson B. Genetic survey of Chinese and Swedish oilseed rape (*Brassica napus* L.) by simple sequence repeats (SSRs). Genet Resour Crop Evol. 2006;53:443–7.

Zou J, Zhao YS, Liu PF, Shi L, Wang XH, Wang M, et al. Seed quality traits can be predicted with high accuracy in *Brassica napus* using genomic data. PLoS One. 2016;11:e0166624.

Population Genomics of Peanut

Ramesh S. Bhat, Kenta Shirasawa, Vinay Sharma, Sachiko N. Isobe,
Hideki Hirakawa, Chikara Kuwata, Manish K. Pandey,
Rajeev K. Varshney, and M. V. Channabyre Gowda

Abstract Population genomics envisages studying numerous loci and genome regions simultaneously to understand the roles of evolutionary processes such as mutation, genetic drift, gene flow, and natural selection that influence variation across genomes and populations thereby unfolding the mechanisms of inbreeding and outbreeding depression, adaptive gene flow, population demographic history and the genomic basis of local adaptation and speciation, ecology, evolution, and conservation biology. Since the population genomics also represents the complete description of the genetic variation that exists at the population levels by collating the information generated through various omics like genomics, epigenomics, transcriptomics, proteomics, and metabolomics, it helps delineate the evolution, domestication, adaptation, genotype–phenotype relationships in the populations. Population genomics of peanut (*Arachis hypogea*), an important oilseed and food crop, is being taken up with the tremendous advancements made in its omics in recent years. The detailed analyses of these omics resources have not only shed light on genome organization and genome features, but also highlighted the peanut origin,

R. S. Bhat (✉)
Department of Biotechnology, University of Agricultural Sciences, Dharwad, India
e-mail: bhatrs@uasd.in

K. Shirasawa · S. N. Isobe
Department of Frontier Research and Development, Kazusa DNA Research Institute, Chiba, Japan

V. Sharma · M. K. Pandey · R. K. Varshney
Center of Excellence in Genomics and Systems Biology, International Crops Research Institute for the Semi-Arid Tropics, Hyderabad, India

H. Hirakawa
Facility for Genome Informatics, Kazusa DNA Research Institute, Chiba, Japan

C. Kuwata
Chiba Prefectural Agriculture and Forestry Research Center, Chiba, Japan

M. V. C. Gowda
Department of Genetics and Plant Breeding, University of Agricultural Sciences, Dharwad, India

Om P. Rajora (ed.), *Population Genomics: Crop Plants*,
Population Genomics [Om P. Rajora (Editor-in-Chief)],
https://doi.org/10.1007/13836_2021_88, © Springer Nature Switzerland AG 2021

spread, domestication, adaptation, and the important genotype–phenotype relationships. Further, the global gene expression atlases for both the subspecies have provided insights into the gene expression landscape. The future developments in peanut population genomics may greatly contribute for understanding the importance of structural polymorphisms in fitness and adaptation, identifying whether the transgenerational epigenetic variations contribute substantially to adaption to changing environments, analyzing gene expression to detect loci under selection and chromosomal islands of adaptive divergence and the alleles associated with various phenotypic traits towards unravelling parallel adaptation. An effort is made in this chapter to review the overall progress towards population genomics in peanut.

Keywords Adaptation · Breeding · Conservation · Diversity · Evolution · Genome-wide associations · Omics · Parallel evolution · Peanut origin · Population differentiation and structure

1 Introduction

Peanut (*Arachis hypogaea* L. 2n=4x=40) is an important oilseed, legume food, and fodder crop. It is grown globally on an area of 28.5 million hectares with a production of 46.0 million tons (http://www.fao.org/faostat/en/#data/QC/visualize) and productivity of 1,611 kg/ha. Globally, over half of the peanut produce goes for oil extraction while the remaining is consumed as raw and processed food. Peanut has earned the name "poor man's almond" because of its rich nutrient contents in terms of oil, proteins, fibers, polyphenols, antioxidants, vitamins, and minerals. In addition, peanut is an excellent source of compounds like resveratrol, phenolic acids, flavonoids and phytosterols, co-enzyme Q10, and amino acids (all 20, with the highest content of arginine) (Arya et al. 2016). With these nutrient profiles, peanut is being considered as a functional food (Arya et al. 2016). Numerous studies have identified the value-added attributes of peanut skins and/or peanut skin extracts (Toomer 2020) as an antioxidant, functional food ingredient, animal feed ingredient, and antimicrobial agent. Enormous efforts are being made to improve the peanut for its productivity, quality, and tolerance to biotic and abiotic stresses (Pandey et al. 2020b), and peanut genomics, epigenomics, transcriptomics, proteomics are contributing for the improvement of peanut.

With the surge in peanut genome sequencing projects and the comparison of multiple related individuals within and between the species of genus *Arachis*, now peanut population genomics is being attempted for understanding the importance of structural polymorphisms in fitness and adaptation, identifying whether the transgenerational epigenetic variations contribute substantially to adaptation to changing environments, analyzing gene expression to detect loci under selection and chromosomal islands of adaptive divergence, and identifying the alleles associated with various phenotypic traits under the influence of evolutionary processes, such as mutation, genetic drift, gene flow, and natural selection. This knowledge might elucidate the population demographic history and the genomic basis of local adaptation and speciation, ecology, evolution, and conservation biology.

High degree of genomic variation exists between the individuals of most of the species (see Hu et al. 2020). Therefore, a true representation of the diversity within a species can be understood with the study of genomic variation in populations. Thus, the approach of genomics gained a new dimension of pan-genomics to represent the genomic diversity of a species in terms of core genes and variable genes. Further, the pan-genome was extended to super-pangenome (Khan et al. 2020) to represent all species with a genus. Currently, pan-genomes are reported in *Oryza sativa* (Zhou et al. 2020), *Hordeum vulgare* (Gao et al. 2020), *Brassica napus* (Song et al. 2020), *Glycine max* (Liu et al. 2020b), *Helianthus annuus* (Hubner et al. 2019), and *Lycopersicum esculentum* (Gao et al. 2019) and which might soon become available in peanut as well.

Genetic variation existing at the population level as revealed by partial or complete genome analysis has now turned the population genetics into population genomics though the concept of which was introduced for the first time with the idea that large-scale polymorphism data could be used to explore the genetic origins of human diseases (Gulcher and Stefansson 1998). The emerging field of population genomics comes along with its specific biological questions and statistical models (Luikart et al. 2019; Dutheil 2020) to provide a key understanding of the microevolution, phylogenetic history, and demography of a population such as genetic diversity, origin, evolution, and domestication. Population genomics aims at studying the evolutionary and population genetics by integrating the advances in "-omics" tools, sequencing technologies, bioinformatics tools, statistical methods, and software (Luikart et al. 2019; Rajora 2019). This helps in understanding the genomic basis of fitness, local adaptation, and phenotypes by unravelling the locus-specific effects from genome-wide effects (Cortinovis et al. 2020). Population genomics also helps in understanding genome differentiation and identifying the domestication-related genes, which might form the basis for future genomics-enabled breeding (Li et al. 2020). Also, the population genomics helps identify the genome-wide effects contributing to various phenotypic traits. For example, population genomics approaches have helped in understanding genome differentiation and identifying a set of domestication genes in lotus (*Nelumbo nucifera*) (Li et al. 2020), association of high-recombination regions with fiber quality traits than those of yield and early maturity traits in cotton (*Gossypium* spp.) (Shen et al. 2019), evolutionary insights with the large-scale characterization of key agronomic traits in peach (*Prunus persica*) (Cao et al. 2019), transition to selfing in Brassicaceae model systems (Mattila et al. 2020), and the genetic basis of adaptation and plasticity in *Zea mays* (Lorant et al. 2020). The phylogenomics and population genomics platforms in plants greatly depend on the reduced cost of high-throughput sequencing and the development of gene sets with wide phylogenetic applicability.

Peanut is taxonomically rich with nine sections, 81 species (Stalker 2017), two subspecies, and six botanical varieties. Currently, the Sequence Read Archive (SRA) of National Center for Biotechnology Information (NCBI) database (https://www.ncbi.nlm.nih.gov/sra) has the whole genome resequencing (WGRS) data of

231 genotypes comprising wild diploids, tetraploids, and botanical varieties; and many more will become available in the coming years. These enormous genome-wide sequence data have contributed to the peanut population genomics for studying its origin, phylogeny, evolution into subspecies and botanical varieties, domestication, genetic diversity, population structure, genome re-shaping, and genotype–phenotype associations. In this regard, the current chapter reviews the latest developments in the area of peanut population genomics encompassing the advancements in genomics and other omics to understand the origin, phylogeny, genome evolution, and genotype–phenotype associations for embarking on the next level understanding on the importance of structural polymorphisms in fitness and adaptation, identifying whether the transgenerational epigenetic variations contribute substantially to adaption to changing environments, analyzing gene expression to detect loci under selection and chromosomal islands of adaptive divergence, and the alleles associated with various phenotypic traits.

2 Resources for Peanut Population Genomics

2.1 Peanut Taxonomy and Genetic Resources

The genus *Arachis* of the family Fabaceae consists of nine sections (*Arachis, Erectoides, Heteranthae, Caulorrhizae, Rhizomatosae, Extranervosae, Triseminatae, Procumbentes,* and *Trierectoides*) based on morphology, cross-compatibility, viability of the hybrids, geographic distribution, and cytogenetics. Across these nine sections, 81 species have been described; of which 69 were described by Krapovickas and Gregory (1994), 11 were described by Valls and Simpson (2005), and one was described by Valls et al. (2013) and Stalker (2017). Of the 81 species, 72 are diploids (2n=2x=20), five are tetraploids (2n=2x=40), and four are aneuploids (2n=2x=18) (Friend et al. 2010). Two tetraploids (*A. hypogaea* and *A. monticola*) are in the section *Arachis*, and the remaining three (*A. glabrata, A. pseudovillosa,* and *A. nitida*) are in the section *Rhizomatosae* (Krapovickas and Gregory 1994; Valls and Simpson 2005). Of the four aneuploids, three (*A. decora, A. palustris,* and *A. praecox*) belong to the section *Arachis* (Peñaloza and Valls 1997; Lavia 1998), and one (*A. porphyrocalyx*) belongs to the section Erectoides (Peñaloza and Valls 2005).

The widely cultivated peanut or groundnut, *Arachis hypogaea*, is among the 31 species belonging to the section *Arachis* along with the forage species *Arachis pintoi, Arachis glabrata,* and *A. sylvestris* (Valls and Simpson 1994) and ornamental species *Arachis repens* (Stalker and Simpson 1995). Occurrence of diploid parental and wild allotetraploid species (*A. monticola*) indicates that the eastern foothills of the Andes in the region of northern Argentina and southern Bolivia could be the area of origin of the cultivated peanut *A. hypogaea* (Krapovickas and Rigoni 1957; Kochert et al. 1996; Ravi et al. 2010). Allotetraploid *Arachis* species was then put under selection for domestication (Kochert et al. 1996), and the domesticated peanut dispersed to South and Central America where both natural and artificial selection produced many landraces of domesticated peanut.

Several efforts were made to identify the probable ancestors of cultivated peanut. Now there is a consensus among the researchers that the cultivated peanut has originated from a hybridization between the diploid species, *A. duranensis* (female parent with A genome under the section *Arachis*) and *A. ipaensis* (male parent with B genome under the section *Arachis*) (Kochert et al. 1996) followed by a single allotetraploidization event involving either chromosome doubling in diploid F_1 (Favero et al. 2006) or unreduced gametes in F_1 (Harlan and de Wet 1975). Recent findings indicate that the cultivated peanut originated about 0.40 million years ago (Zhuang et al. 2019; Zhuang et al. 2020), though there is another view that it originated about 10,000 years back (Bertioli et al. 2019, 2020).

Based on several morphological differences, two subspecies (*A. hypogaea* ssp. *hypogaea* and *A. hypogaea* ssp. *fastigiata*) have been recognized in the cultivated peanut. *A. hypogaea* ssp. *hypogaea* is characterized by alternate branching, absence of flowers on main axis, long duration (120–160 days), and presence of seed dormancy, while *A. hypogaea* ssp. *fastigiata* is recognized by sequential branching, presence of flowers on main axis, short duration (85–130 days), and lack of seed dormancy. Based on the morphological differences, the two subspecies are subdivided into botanical varieties. *A. hypogaea* ssp. *hypogaea* is divided into var. *hypogaea* (Virginia bunch/runner) and var. *hirsuta* (Peruvian runner), whereas *A. hypogaea* ssp. *fastigiata* is classified into var. *fastigiata* (Valencia), var. *vulgaris* (Spanish bunch), var. *aequatoriana,* and var. *peruviana* (Krapovickas 1969, 1973). *A. hypogaea* ssp. *hypogaea* is thought to have originated by a mutation from within *A. hypogaea* ssp. *fastigiata* (Singh 1988).

Genetic diversity is the most important resource for trait mapping and crop improvement. The germplasm collections maintained in India (15,445 accessions), USA (9310 accessions), and China (7,837 accessions) make the germplasm repository. These germplasm collections have been characterized for important agronomic traits and have different diversity panels for use in breeding. ICRISAT, USA, and China developed core collection with 1,704, 831, and 576 accessions, respectively, from the total germplasm collection, which were further reduced to minicore collections of 184, 112, and 298 genotypes, respectively. These accessions with novel alleles, heritability, combining ability, and trait correlations have significantly contributed for the better understanding of the peanut population genomics and overall genetics to develop appropriate breeding strategies for target traits. Population structure, linkage disequilibrium, and marker-trait associations have been studied in these minicore accessions. Several improved lines and sources of variability have been identified or developed for various economically important traits through conventional breeding (see Desmae et al. 2019). These genetic resources have helped the advancement of the peanut genomics including genome sequencing, marker development and genetic and trait mapping, discovery of genes/variants for traits of interest, and integration of marker-assisted breeding for selected traits.

2.2 Genome and Genetic Maps

Population genomics emphasizes the understanding of patterns of genetic variation and evolutionary processes in all genome regions by plotting population genetic statistics across each chromosome using many mapped loci (Hohenlohe et al. 2010). Population genomics requires a sufficient density of DNA markers to detect forces affecting any particular genomic region, e.g., genome regions under selection, genes under selection, regions of reduced recombination. Peanut genomics was not much explored until 1980s due to its large size (~2.7 Gb), high fraction of receptive DNA, and allotetraploidy with two closely related genomes. However, with the sequencing of expressed sequence tags (ESTs) (Wang et al. 2006; Proite et al. 2007; Guo et al. 2008; Bi et al. 2010), peanut genomics research was initiated during the late 1980s to characterize species relationships and investigate more efficient methods to introgress genes from wild species to *A. hypogaea*. Relatively low-density genetic maps were developed initially from inter- and intra-specific crosses to map disease resistance genes. With the development of more markers, construction of high-density maps was reported later. These developments marked the start of peanut genomics (Paterson et al. 2004; Stalker et al. 2009; Pandey et al. 2012; Stalker et al. 2013; Ozias-Akins et al. 2017) and picked the pace in post-genome sequencing era.

As a first step towards characterizing the genome of cultivated peanut, the genomes of the two diploid ancestors (*A. duranensis* V14167 and *A. ipaensis* K30076) of cultivated peanut were sequenced and analyzed (Bertioli et al. 2016) to overcome the challenge in assembling of chromosomal pseudomolecules. Both these accessions were collected from the most likely geographic region of origin for the cultivated peanut. In the same year, Chen et al. (2016a) sequenced *A. duranensis* (accession PI475845 from Bolivia) as well as four synthetic tetraploids and their six diploid parents [two A genome and four B genome, including the suspected B genome progenitor, *Arachis ipaensis*] to gain insight into peanut evolution. Based on the draft genome of *A. duranensis*, the gene models with 50,324 protein-coding genes were proposed. Also, Lu et al. (2018) sequenced *A. ipaensis* and recorded ~1.39 Gb genome with 39,704 predicted protein-encoding genes.

The first reference quality assembly of the *A. monticola* (PI263393) genome was developed with a genome size of ~2.62 Gb (Yin et al. 2018). The efficiency of the current state of the strategy for *de novo* assembly of the highly complex allotetraploid species based on whole-genome shotgun sequencing, single molecule real-time sequencing, high-throughput chromosome conformation capture technology, and BioNano optical genome maps was demonstrated. Subsequently, Yin et al. (2020) re-sequenced 17 wild diploids from AA, BB, EE, KK, and CC groups and 30 tetraploids and compared the previously sequenced genome of *A. monticola* (Yin et al. 2018).

During 2019, two reference genomes; one for the *A. hypogaea* ssp. *fastigiata* and the other for *A. hypogaea* ssp. *hypogaea* of the cultivated tetraploid were reported. The IPGI-led initiative (Bertioli et al. 2019) completed the sequencing of Tifrunner (PI644011 with registration number CV-93) (Holbrook and Culbreath 2007), a runner-type belonging to *Arachis hypogaea* ssp. *hypogaea* by deploying several modern sequencing and assembly technologies such as PacBio and Hi-C data/ technology. A genome of ~2.56 Gb with 20 pseudomolecules and 66,469 predicted genes was reported. Similar advanced technologies were deployed by two independent efforts in China leading to the development of high-quality reference genome assemblies for "Shitouqi" (Zhuang et al. 2019) and "Fuhuasheng" (Chen et al. 2019c) both belonging to *A. hypogaea* ssp. *fastigiata*. The variety "Shitouqi" (zh. h0235) is a well-known Chinese landrace and breeding parent belonging to *A. hypogaea* ssp. *fastigiata* and botanical type *vulgaris* (agronomic type Spanish), while "Fuhuasheng" is a landrace from North China. For Shitouqi, a genome of ~2.54 Gb with 83,709 predicted genes across 20 pseudomolecules was reported, and the heterozygosity was very low (1/6,537 nucleotides); while a genome of ~2.55 Gb with 83,087 predicted genes across 20 pseudomolecules was reported for Fuhuasheng.

2.2.1 Genome Features

It is important to understand the key genome features of peanut for the detailed analysis of population genomics. With the availability of complete genome sequence, it is possible to identify the "footprints" of natural selection in genome-wide patterns of genetic variation. Prediction of genes (both protein-coding and RNA-coding genes), gene structure (promoter, UTRs, exons, and introns), and non-coding regions help in understanding the roles of evolutionary processes such as mutation, genetic drift, gene flow, and natural selection. For this, the salient features of Tifrunner and Shitouqi genomes are presented here. The chromosomes are named from Arahy.01 to Arahy.20 where the first ten (Arahy.1–Arahy.10) represent the A subgenome while the latter ten (Arahy.11–Arahy.20) represent the B subgenome. As it has been reported in other recent polyploids (Chalhoub et al. 2014; Zhang et al. 2015), the number of homeologous gene pairs with expression biased towards the A subgenome was not significantly different from the number biased towards the B subgenome. No significant bias was observed for the expression of the genes between A and B subgenomes. The average length of the transcripts was 1,589.5 bp with 6.8 exons which translate the proteins of 403 amino acids. These sizes were comparable to other legumes but longer than those of A and B genomes. Approximately 76.6% of predicted genes were functionally annotated. In total, 39,127 non-coding RNA (ncRNA) including 4,723 transfer RNAs (tRNAs), 3,107 ribosomal RNAs (rRNAs), 480 microRNAs (miRNAs), and 30,817 small nuclear RNAs have been identified. Most of the small RNA-coding regions corresponded to repeat-rich regions. Among the small RNAs corresponding to gene-rich regions, those originating from the B subgenome was more than that of A subgenome. Studies have indicated their significant role in the formation and functioning of the nodule (Figueredo et al. 2020) and response to *Aspergillus flavus* (Zhao et al. 2020a).

Overall, 77.65% (1.97 Gb) of the genome consisted of repeat elements, of which 64.74% (1.67 Gb) comprised of retrotransposons and 4.49% (114 Mb) comprised of DNA transposons. LTRs are highly abundant in pericentromeric regions, whereas DNA transposons are more frequent in euchromatic arms. Most of the transposable elements [like Gypsy and unclassified Long terminal repeat (LTRs)] expanded after tetraploidization, especially in A subgenome, while those of the B subgenome were derived from the progenitor B genome. However, this observation was in contrast to that of Bertioli et al. (2019), who could not find any major transpositions or insertions after tetraploidization in Tifrunner.

Five chromosomes (Arahy.01, Arahy.05, Arahy.06, Arahy.07, and Arahy.11) showed inversions. Genetic exchange between ancestral genomes could be noticed towards the ends of collinear pairs of homeologous chromosomes. Two patterns of homeologous recombination were evident; the first involved the transfer of chromosome segments between distal collinear regions of chromosomes mostly resulting in tetrasomic genome structures, while the second involved the transfer of alleles dispersed in the chromosomes. However, the latter was strongly biased where the transfer of alleles from B subgenome to A subgenome was more frequent.

The chloroplast genome of *A. hypogaea* and a chloroplastic plasmid were inherited from *A. duranensis*, the female parent of peanut (see Paterson et al. 2004). Sequencing of the chloroplast genome of *A. hypogaea* (Wang et al. 2018) reported a length of 156,354 bp (var. *hypogaea*), 156,878 bp (var. *hirsuta*), 156,718 bp (var. *fastigiata*), and 156,399 bp (var. *vulgaris*). Comparative analysis of the sequences revealed that their gene content, gene order, and GC content were highly conserved since only 46 single nucleotide polymorphisms (SNPs) and 26 insertions/deletions were identified. Most of these variations were found in the non-coding sequences, especially the trnI-GAU intronic region.

2.3 Epigenome

Low DNA sequence polymorphism despite enormous phenotypic variations in peanut indicates the possible role of epigenetic variations. Detection of the epigenetic marks (along with associated expression) provides high power to identify genomic regions associated with traits or evolutionary processes such as fitness, phenotypes, and selection. In peanut, Bertioli et al. (2019) observed lower methylation in the transcribed regions and characteristic decline in methylation at transcription start and end sites like in most plant genomes. Genome-wide methylation per cytosine content was higher in pericentromeric regions than the chromosome arms. Methylation was lower in the A subgenome than the B subgenome; with 76.0% and 80.5% methylation at CG sites, 61.7% and 65.1% methylation at CHG sites (where H is an A, T, or C) and 5.14% and 5.51% methylation at CHH sites, respectively. All of this information provides a good resource for population epigenomics studies.

2.4 Transcriptome

All RNA transcripts (transcriptome) produced by the genome are studied (transcriptomics) for gene annotation, gene discovery, and marker development. However, population transcriptomics uses transcriptome-wide data to study variation in gene expression within and among populations to understand mechanisms underlying evolutionary change, response to environmental change, and plasticity in gene expression. Numerous efforts have been made in peanut to collect and study the transcriptome (see Chen and Liang 2014). Though the initial efforts employed microarrays (Chen et al. 2012; Zhu et al. 2014), the current studies embark on RNA-sequencing. *Arachis hypogaea* gene expression atlas (AhGEA) has been reported for the widely cultivated *A. hypogaea* ssp. *fastigiata* based on RNA-Seq data using 20 diverse tissues across five key developmental stages from an early-maturing, high-yielding, drought-tolerant variety, ICGV 91114 (Sinha et al. 2020). AhGEA sheds light on the complex regulatory networks of gravitropism and photomorphogenesis, seed development, allergens, and oil biosynthesis.

Transcriptomic analysis conducted so far could identify the differentially expressed genes for growth and development, response to biotic and abiotic stresses, and quality (see Pandey et al. 2020b). Transcriptomes have been reported for seed dormancy release and germination (Xu et al. 2020), plant height (Guo et al. 2020), pod development (Yang et al. 2017), nodulation (Peng et al. 2017; Peng et al. 2020), seed coat cracking and pigmentation (Zhang et al. 2016; Huang et al. 2019; Xia et al. 2020), light-mediated embryo and pod development (Zhang et al. 2016), Ca^{2+} regulation in pod development (Yang et al. 2017), regulation of plant triacylglycerol metabolism (Zheng et al. 2017), calcium- and hormone-related responses during different stages of pod development (Hao et al. 2017), seed development (Zhang et al. 2012b), seed abortion (Zhu et al. 2014), geocarpy (Chen et al. 2016b), early pod development in darkness (Xia et al. 2013), and oil accumulation and profile (Yin et al. 2013; Huai et al. 2020).

Similarly, the genes involved in cold tolerance (Jiang et al. 2020), drought tolerance (Guimarães et al. 2012; Bhogireddy et al. 2020; Zhao et al. 2020b), nematode resistance (Guimaraes et al. 2015), response to *Ralstonia solanacearum* (Chen et al. 2014), response of roots to the beneficial and pathogenic fungi (Hao et al. 2017) including late leaf spot (LLS) (Han et al. 2017) and leaf rust (Rathod et al. 2020), response to low temperature during germination among high oleate types (Chen et al. 2019b), and embryo abortion under calcium deficiency (Chen et al. 2019a) have been identified. Bosamia et al. (2020) used RNA-Seq to unravel the mechanisms of resistance to stem rot caused by *Sclerotium rolfsii* using a resistant (NRCG-CS85) and susceptible (TG37A) genotype. Several annotated transcripts are known to have a role in channelizing the downstream of pathogen perception like receptor-like kinases, jasmonic acid pathway enzymes, and transcription factors (TFs), including WRKY, zinc finger protein, and C2-H2 zinc finger showed higher expression in resistant genotype upon infection.

Gene network regulating cadmium uptake and translocation in roots under iron deficiency (Huang et al. 2019) and differential cadmium transport and retention in roots (Yu et al. 2018) have also been identified. Transcriptome analyses have also been used for assessing the genetic variability (Chopra et al. 2014, 2015, 2016), mining tissue-specific contigs for promoter cloning (Geng et al. 2014) and understanding alternative splicing (Ruan et al. 2018). Key genes involved in the regulation of symbiotic pathways induced by *Metarhizium anisopliae* in roots (Wang et al. 2020a) and the members of monosaccharide transporter (MST) gene family have been identified by transcriptional profiling (Wan et al. 2020). A transcriptomic study was conducted to check the variation in the expression of the genome-wide genes due to variation in the DNA methylation across 11 genotypes (Bhat et al. 2019a) (for details, see the Sect. 3.4).

2.5 Proteome

Regulations during transcription, translation, and posttranslational modifications are widespread; therefore, the mRNA content not necessarily corresponds with the protein content (Dhingra et al. 2005). Proteomics is constantly advancing to bridge the gap between DNA sequence-transcriptome-phenotype under the diverse and dynamic stages of growth and development. Since proteins influence important phenotypes and are the products of genes and epigenetic or posttranslational mechanisms, population proteomics has the potential to provide key insights into functional and metapopulation ecology, adaptation, and acclimation processes under various climate and environment conditions. Population proteomics approaches also help identify genetic loci underlying risk of disease and for clinical biomarkers for many human disease conditions. In peanut, the low DNA polymorphism coupled with high morphological variation might involve differences in the proteomics. Katam et al. (2014) reviewed the progress made on the proteomics in peanut especially on peanut allergens and adaptive responses to various stresses. Proteomic analyses have identified the specific proteins produced in response to high-oleic acid content in seed (Liu et al. 2020a), cadmium detoxification and translocation (Yu et al. 2019), allergen production (Mamone et al. 2019), response to water stress (Kottapalli et al. 2009, 2013; Katam et al. 2016), gynophore development (Sun et al. 2013; Zhao et al. 2015), development of aerial and subterranean pods (Zhu et al. 2013), response to salinity (Jain et al. 2006), and polyphenol content (Muralidharan et al. 2020). The advances of stress proteomics, in recent years, have enabled plant biologists to investigate the molecular events associated with stress adaptation in legumes including peanut (Rathi et al. 2016).

2.6 Development of Markers

Availability of genome-wide genetic markers is essential for population genomics and its applications. In peanut, the initial efforts on isozyme and seed protein analyses identified only limited variability among the cultivated peanut accessions (see Lu and Pickersgill 1993; Stalker et al. 1994) though substantial diversity exists within the cultivated peanut genotypes for various morphological, physiological, and agronomic traits (Stalker 1992). Random amplified polymorphic DNA (RAPD) and restriction fragment length polymorphism (RFLP) approaches also failed to detect any DNA variation in the cultivated peanut (Halward et al. 1991, 1992; Kochert et al. 1991; Paik-Ro et al. 1992). However, these approaches could identify genetic variability among the wild types (Halward et al. 1991). Later, He and Prakash (1997) reported polymorphic DAF and AFLP markers in cultivated peanut. SSR markers were developed in the cultivated peanut using DNA library (Hopkins et al. 1999) and the polymorphism was detected. SSRs were also developed from *Arachis pintoi* to identify the variation in *Arachis pintoi* (Palmieri et al. 2002) and the accessions belonging to the section *Caulorrhizae* (*Arachis*, Fabaceae) (Palmieri et al. 2005). Fifty-six SSRs were developed from the cultivated peanut from SSR enriched library (He et al. 2003), of which 19 showed polymorphism. STMS markers were developed in cultivated peanut to detect variation (Ferguson et al. 2004a). Subsequently, several efforts were made to develop genic and non-genic SSRs (Moretzsohn et al. 2004, 2005; Hong et al. 2008; Zhang et al. 2012a; Huang et al. 2016; Peng et al. 2016). A single-locus marker offers an advantage over a multi-locus marker in genetic and breeding studies since its alleles can be assigned to a specific genomic locus. Zhou et al. (2016) developed 1,790 single-locus SSR markers from the *de novo* assembly of peanut sequence reads. Using the reference genome sequences of *A. duranensis* and *A. ipaensis*, Luo et al. (2017) identified 264,135 and 392,107 SSRs from which 84,383 and 120,056 SSR markers were developed. CAPS markers were developed for detecting the mutations at *AhFAD2A* and *AhFAD2B* (Chu et al. 2009). Diversity Array Technology (DArT) marker platform has also been developed for peanut (Kilian 2008).

When compared to aforementioned marker systems with low polymorphic rate (5–6%), the transposable element-based marker system with higher polymorphic rate (up to 22%) was developed (see Bhat et al. 2019b). This was named as *Arachis hypogaea* transposable element (AhTE) marker system, which detects the polymorphism for the insertion of 205 bp long *Arachis hypogaea* miniature inverted-repeat transposable element (*AhMITE1*). AhTE marker system was proposed by developing just one marker (Bhat et al. 2008; Gowda et al. 2010, 2011); and subsequently a large number of such markers were developed in peanut. Shirasawa et al. (2012a) developed 504 AhTE markers using *AhMITE1*-enriched libraries. The representative *AhMITE1* exhibited a mean length of 205.5 bp and a GC content of 30.1%, with AT-rich, 9 bp target site duplications and 25 bp terminal inverted repeats. Later, Shirasawa et al. (2012b) developed additional 535 AhTE markers using transposon-enriched libraries of other cultivars. Since these AhTE markers were highly polymorphic and user-friendly (Kolekar et al. 2016), they were successfully used to construct linkage maps (Shirasawa et al. 2013; Kolekar et al. 2016) and to identify QTL for resistance to LLS and rust (Kolekar et al. 2016). Later, AhTE markers were also used for marker-assisted backcross breeding in peanut (Yeri and Bhat 2016).

With the availability of the genome sequences of the diploid progenitors of peanut (Bertioli et al. 2016), efforts were made to identify the genome-wide distribution of *AhMITE1* (Gayathri et al. 2018). For this, a set of diverse genotypes (33), including the genetically unstable peanut mutants which show hyperactivity of *AhMITE1* (Hake et al. 2018), were used to discover the *AhMITE1* insertion polymorphic sites. The whole genome resequencing (WGRS) reads from these diverse genotypes were analyzed using the computational method polymorphic TEs and their movement detection (PTEMD) (Kang et al. 2016) for the *de novo* discovery polymorphic sites and to develop 2,957 *AhMITE1* markers (Gayathri et al. 2018).

Currently, the advent of next-generation sequencing and genotyping technologies have enabled the detection of SNPs, which have emerged as the marker of choice in mapping (Bertioli et al. 2014) and population genomics studies (Holliday et al. 2018), and several studies (see the section on trait mapping) have reported identifying the SNP markers for mapping and population genomics. Hale et al. (2020) reviewed the methods to reduce per-sample costs in high-throughput targeted sequencing projects, minimal equipment, and consumable requirements for targeted sequencing while comparing several alternatives to reduce bulk costs in DNA extraction, library preparation, target enrichment, and sequencing. A cost calculator was developed for researchers considering targeted sequencing.

We attempted to analyze the WGRS data of 231 genotypes available in the public domain (NCBI SRA BioProject accession numbers: PRJDB4621, PRJDB5785, PRJDB5787, PRJDB0473, PRJNA340877, PRJNA490832, PRJNA490835, PRJNA511348, and PRJNA525866) for the SNPs as an effort towards population genomics and peanut pan-genomics (unpublished data). In comparison with the reference genome of Tifrunner, as high as 4,309,724 SNPs were detected (Table 1) with a range of 113,363 (chromosome 18) to 433,957 (chromosome 03). On an average, a greater number of SNPs were noticed for A subgenome than that of B subgenome.

3 Progress in Population Genomics of Peanut

3.1 *Phylogenomics and Evolution*

Phylogenetic relationships among taxa can be estimated from a wide range of genetic data types, including genomic data. Therefore, the phylogenomics greatly depend on the reduced cost of high-throughput sequencing and the development of gene sets with wide phylogenetic applicability. However, many genetic markers spread across the genome reflect different evolutionary histories because of recombination. This is particularly true in recently diverged species and where incomplete

Table 1 SNP, CNV, and clusters among the genotypes[a] of the genus *Arachis*

Chromosome	WSS		Silhouette		K value	SNP	CNV 1	CNV 2	CNV 3	CNV 4
	Clusters	Index	Clusters	Index						
Arahy.01	2	25.3	2	25.3	14	301,560	2	6,157	20	1,824
Arahy.02	2	23.4	2	23.4	10	252,051	2	5,684	5	787
Arahy.03	2	25.5	2	25.5	11	433,957	33	7,645	19	1,178
Arahy.04	2	19.2	2	19.2	8	342,736	0	7,324	151	175
Arahy.05	2	18.9	2	18.9	15	259,152	12	6,342	11	713
Arahy.06	2	28.8	2	28.8	12	311,306	10	6,496	0	318
Arahy.07	2	28.1	2	28.1	8	252,448	5	4,394	5	183
Arahy.08	2	34.9	2	34.9	8	211,112	0	2,202	3	402
Arahy.09	2	16.6	2	16.6	11	309,496	0	6,892	0	853
Arahy.10	2	11.7	2	11.7	8	261,546	1	6,671	23	2,197
Arahy.11	2	3.2	2	3.2	11	121,154	503	12	5	451
Arahy.12	2	5.2	2	5.2	14	136,277	315	8	5	1,207
Arahy.13	2	7.0	2	7.0	11	172,124	762	927	7	657
Arahy.14	2	3.9	2	3.9	11	130,890	424	8	34	15,636
Arahy.15	3	4.1	3	4.1	11	132,760	611	29	20	252
Arahy.16	2	3.7	2	3.7	11	140,985	950	672	4	6,855
Arahy.17	6	3.6	6	3.6	11	129,843	360	64	5	877
Arahy.18	6	4.6	6	4.6	9	113,363	440	5	0	291
Arahy.19	2	3.9	2	3.9	9	164,841	518	34	8	3,328
Arahy.20	3	6.8	3	6.8	11	132,123	455	1	13	2,009

[a]WGRS data from 179 genotypes were used. CNV 1: CNVs between *Arachis duranensis* (Dur_14167) and Tifrunner, CNV 2: CNVs between *A. ipaensis* (Ipa_K300076) and Tifrunner, CNV 3: CNVs between *A. monticola* (PI_263393) and Tifrunner and CNV 4: CNVs between GPBD 4 and Tifrunner. Only significant CNVs with $\log_2$ of >5 or <−5 are listed

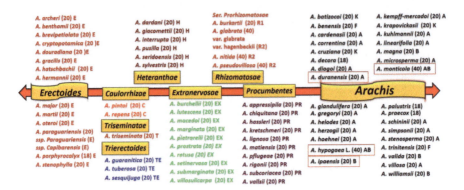

Fig. 1 Species and sections diversity in *Arachis* genus. There are 81 species belonging to 9 sections with two ploidy levels (diploid and tetraploid) and 15 genome types (A, B, AB, D, F, K, EX, T, PR, H, C, T, E, R_1, and R_2) (Stalker 2017). The reference genomes have become available for *A. duranensis* (diploid progenitor A genome), *A. ipaensis* (diploid progenitor B genome), *A. monticola* (wild tetraploid), and both the subspecies (*A. hypogaea* ssp. *hypogaea* and *A. hypogaea* ssp. *fastigiata*) of *A. hypogaea* (cultivated tetraploid). The majority of species in the genus are diploids, but tetraploids are present in the section of *Arachis* and *Rhizomatosae*, with some species being aneuploid in sections of *Arachis* and *Erectoides*. Polyploidy has developed independently in *Arachis* and *Rhizomatosae* sections (Smartt and Stalker 1982). Of the 81 species, *A. hypogaea* species is the only cultivated peanut species that are commercially grown in the field

lineage sorting and admixture are important (Edwards et al. 2016). Methods to estimate phylogeny from large SNP datasets have been developed (Hohenlohe et al. 2018; McKain et al. 2018). Care should be taken to estimate a consensus tree among taxa and reveal patterns of hybridization and admixture during phylogenetic analysis (Pickrell and Pritchard 2012).

In peanut, efforts have been made in the past to investigate the genetic evolutionary relationships between wild types and the botanical types (see Bertioli et al. 2011; Stalker et al. 2016) (Fig. 1). These studies were conducted using the nuclear rDNA internal transcribed spacers (ITS) (Bechara et al. 2010; Wang et al. 2011a), EST-SSRs (He et al. 2014), ITS and plastid trnT–trnF sequences (Friend et al. 2010), tunable genotyping-by-sequencing (tGBS) (Zheng et al. 2018), and SLAF-Seq (Zhang et al. 2017). WGRS-based genomics approach helped to gain a clear understanding of the exact origin of peanut. Sequencing of the diploid ancestors (*A. duranensis* V14167 and *A. ipaensis* K30076), collected from the most likely geographic region for the origin of cultivated peanut, showed that these A and B genomes are similar to the A and B subgenomes of cultivated peanut (Bertioli et al. 2016). Comparisons showed one-to-one correspondences between the chromosomes of the diploids and linkage groups of the cultivated peanut. For the genome-wide comparisons between the sequences of the diploids and the cultivated tetraploid, the long-sequence reads (5.74 Gb with 2× coverage) were produced from *A. hypogaea* cv. Tifrunner. The corrected median identity between the *A. hypogaea* and the pseudomolecules of *A. ipaensis* was higher (99.96%) than that between *A. hypogaea* and *A. duranensis* (98.36%). Thus, the diploid A-genome chromosomes were distinctly less similar to *A. hypogaea* sequences than the B-genome chromosomes. On the basis of remarkably high DNA similarity between the *A. ipaensis* genome and the B subgenome of the cultivated peanut and biogeographic evidences, it was concluded that *A. ipaensis* may be a direct descendant of the same population that contributed the B subgenome to cultivated peanut. To identify the accession of *A. duranensis* closest to the A subgenome donor, 55 accessions representing all the known major populations of *A. duranensis* were sequenced using the DNA enriched for genic regions. The accessions (KGBSPSc 30065, PI 468201 and KGBSPSc 30067, PI 468202) from Rio Seco (Argentina), the likely region of origin of the A subgenome ancestor on the basis of chloroplast and ribosomal DNA haplotypes (Grabiele et al. 2012), were most (99.76%) similar to A subgenome of the Tifrunner. However, in some cases, the ranking of similarity changed by chromosome, indicating the differing levels of variations in chromosomes in different accessions of *A. duranensis*.

To estimate the genome-wide mutation rate, the pseudomolecules of *A. ipaensis* and *A. duranensis* were compared. Over a period of 2.16 million years since the A and B genomes diverged, an average of 1.6×10^{-8} mutations per base per year was observed, while the average for other plants was $1-2 \times 10^{-8}$ (Jiao et al. 2012). With these estimates, the divergence time of *A. duranensis* V14167 from the A subgenome of *A. hypogaea* was ~247,000 years and for the divergence time of *A. ipaensis* from the B subgenome of *A. hypogaea* was just ~9,400 years.

After the hybridization between the A and B genome diploids, peanut underwent polyploidization, evolution, and domestication. The first-evolved allotetraploid species *Arachis monticola* is a unique link between the wild diploid species and the cultivated tetraploid species.

Genomic reconstruction of 17 wild diploids from AA, BB, EE, KK, and CC groups and 30 tetraploids demonstrated a monophyletic origin of A and B subgenomes in the allotetraploid peanut *A. monticola* (Yin et al. 2020) when its ~2.6 Gb genome was sequenced and annotated (Yin et al. 2018). This study revealed that the wild and cultivated tetraploids underwent asymmetric subgenome evolution, including homoeologous exchanges, homoeolog expression bias, and structural variation (SV), leading to subgenome functional divergence during peanut domestication. Significantly, SV-associated homoeologs showed expression bias and correlation with pod size increase from diploids to wild and cultivated tetraploids.

At the whole-genome scale, the effects of homeologous recombination appear similar in diverse peanut accessions (Clevenger et al. 2017; Bertioli et al. 2019). Most of the tetrasomic structures were present in all *A. hypogaea* and *A. monticola* accessions. Fingerprint-like fine-scale patterns of interspersed homeologous alleles within the distal tetrasomic regions were also found to be uniform. In contrast, homeologous recombination patterns in allotetraploid hybrids from the ancestral species were completely distinct. This emphasizes the close relationship of *A. hypogaea* and *A. monticola* and favors a single polyploid origin of both the species. Homeologous recombinations in *A. hypogaea* generated new diversity with some tetrasomic regions differing in different accessions; some accessions showed AAAA genome, while others showed BBBB in some genomic regions.

Allotetraploids derived by colchicine treatment of hybrids between the ancestral diploid species (Favero et al. 2006) of peanut were used to investigate genomic changes during the polyploidization process (Bertioli et al. 2019). In total, 37 different lineages developed from two independently induced polyploidy events were studied. Genetic exchanges occurred between the subgenomes in large blocks and interspersed alleles. These events were partly stochastic, and therefore, differed between the lineages, while being different from those of *A. hypogaea*. Spontaneous changes in flower color in some lineages were related to genetic exchange between the subgenomes; the A genome region that confers the yellow flower color had been replaced by the homeologous B genome region that confers orange flower color, providing a demonstration of phenotypic change as a consequence of genetic exchange between the subgenomes.

Recent phylogenetic analysis based on the resequencing of 52 accessions of 12 species including *A. duranensis* and *A. ipaensis*, as well as the wild tetraploid *A. monticola*, and 30 diverse peanut indicated *A. monticola* as the founder genotype to have evolved after tetraploidization with subsequent divergence into two subspecies and six varieties Zhuang et al. (2019). The origin of *A. monticola* from *A. ipaensis* and *A. duranensis* was supported by high DNA similarity in *A. ipaensis*, *A. monticola* and many peanut varieties than in *A. duranensis*. The study also hypothesized the origin of peanut domestication from *A. monticola*. The phylogenetic tree could classify most of the cultivated peanut accessions into two groups representing the two subspecies. However, some of the accessions of the subspecies *hypogaea* were grouped along with the vulgaris-type cultivars probably due to the inter-subspecies hybridizations. The synthetic amphidiploids were grouped along with their respective diploid. The A genome enrichment resulting probably from non-random retention of the parental chromosomes in the offspring owing to incompatibility hints at an emerging hypothesis that a species other than *A. duranensis*, but more compatible with the B genome, could be the A genome donor. It was also hypothesized that the diploids *A. stenophylla* (with E genome) and *A. pintoi* (with C genome) might have separately evolved into diploid A and B genomes, respectively, forming the progenitor genomes of the peanut. Most of these findings need further investigations to confirm the results.

Polyploidization has been recognized as an important feature of plant evolution, though "wondrous cycles" have played an important part in diversification and adaptation during plant evolution (Blanc and Wolfe 2004; Soltis et al. 2004; Wendel 2015) and polyploids are favored for domestication (Hilu 1993; Salman-Minkov et al. 2016). Similarly, the polyploid *A. hypogaea* was favored for domestication over its diploid relatives according to the available evidence (Bertioli et al. 2016). In spite of diploid species having evolved earlier and possessing greater genetic diversity, they were not domesticated, but the allotetraploid *A. hypogaea* became the crop of worldwide cultivation, probably owing to the following reasons: (1) allotetraploids have larger plant size, leaves, stomata, and epidermal cells than their diploid progenitors (Leal-Bertioli et al. 2017), (2) cultivated peanut has larger pods than the primitive tetraploid (*A. monticola*) (Yin et al. 2020), (3) tetraploids have different transpiration characteristics (Leal-Bertioli et al. 2012) and produce more photosynthetic pigments (Leal-Bertioli et al. 2017), (4) tetraploids probably have wider adaptation, and (5) genome changes after polyploidy that could have generated variation.

It was also hypothesized (see Hake et al. 2018) that the genetic instability followed by genome re-organization contributed significantly to the evolution of peanut. Dharwad Early Runner (DER), a taxonomically unique plant derived from a cross between two *A. hypogaea* ssp. *fastigiata* varieties, showed considerably high genetic instability (Gowda et al. 1996). Some of the DER-derived mutants continued to show genetic instability either spontaneously or by induction with mutagens (Hake et al. 2018). DNA analysis among the several mutants and their derivatives for the transpositional activity of *AhMITE1*, SNP, and copy number variation (CNV) (Xie and Tammi 2009) indicated remarkably high DNA re-organization when compared to genetically stable genotypes. Some of these DNA structural variations also led to phenotypic changes in a few taxonomical, morphological, and productivity traits, including resistance to LLS disease. Hence, it was considered that the genetic instability triggered by genomic shock due to a rare event of hybridization has resulted in enormous genome-wide DNA rearrangements and structural mutations contributing to the evolution of peanut.

3.2 Genetic Diversity, Population Structure, and Sub-specific Genetic/Genomic Divergence

Quantifying genetic diversity and understanding population structure and levels of genetic differentiation among populations distributed spatially is important in order to assess genome-wide genetic variation at the population level and to assess genotype–phenotype associations. Though traditional population genetic tools have been successfully used for studying population structure and genetic differentiation, the genomic techniques provide greater statistical power and precision for estimating parameters. In addition, use of a large number of markers from the genomic data allows estimating population structure and levels of genetic differentiation from fewer individuals.

The cultivated peanut consists of six botanical varieties; var. *hypogaea* (Virginia bunch/runner) and var. *hirsuta* (Peruvian runner) under *A. hypogaea* ssp. *hypogaea*, and var. *fastigiata* (Valencia), var. *vulgaris* (Spanish bunch), var. *aequatoriana* and var. *peruviana* under *A. hypogaea* ssp. *fastigiata*. Some varieties, classified as "irregular type" due to lack of specific description as a result of hybridizations between subspecies or between botanical types, have also been described (Hui-Fang et al. 2008). Probable mechanisms that have contributed for generating variation in peanut have been recognized. Allopolyploidization involving interspecific hybridization followed by whole-genome duplication could be a major source to induce an array of genetic, epigenetic, and gene-expression changes (see Zhang et al. 2020b). These changes can occur as early as in the somatic cells of interspecific

F$_1$ hybrids and might continue in the generations following genome doubling. These early changes contribute to the initial stabilization and establishment of nascent allopolyploids into new species. Generally, these changes are brought through transposable element (TE) activation, sequence elimination, changes in cytosine methylation, and small RNA profiles leading to expression variation. The F$_1$ progeny of peanut with diverse genomes may experience genomic shock and can lead to altered patterns of homoeologous gene expression, including transinteractions, subgenome dominance, and regional partitioning in response to developmental and environmental cues. In the allopolyploids, recombination between homoeologous chromosomes can also generate genomic changes (Zhang et al. 2020b).

Assessing genetic variation is crucial for peanut population genomics studies to estimate population structure and levels of genetic differentiation apart from cultivar improvement and germplasm utilization. In order to enhance the efficiency of the conservation and utilization of diverse genetic resources for peanut improvement, Ferguson et al. (2004b) attempted to understand the distribution of genetic variation among peanut botanical varieties in relation to the geographical origin using 12 simple sequence repeats (SSRs) and 188 peanut landraces representing six botanical varieties from 10 countries in three continents (South America, Asia, and Africa). The study showed more differentiation between the varieties ($F_{ST}=0.33$) than between the continents ($F_{ST}t=0.016$), indicating that the landraces from the three continents were more closely related to each other.

Belamkar et al. (2011) used 32 SSR markers to investigate genetic variation among 96 peanut genotypes consisting of 92 accessions of US peanut minicore, a tetraploid variety (Florunner), diploid progenitors *A. duranensis* (AA) and *A. ipaensis* (BB), and synthetic amphidiploid accession TxAG-6. Wild-species accessions and the synthetic amphidiploid grouped separately from the minicore accessions. Unweighted pair group method with arithmetic average (UPGMA) dendrogram showed four subgroups; one each for the *A. hypogaea* ssp. *hypogaea* and *A. hypogaea* ssp. *fastigiata*, the third group consisting of individuals from both the subspecies (mixed ancestry) and the fourth group consisting of majority of the individuals from var. *fastigiata, peruviana* and *aequatoriana*.

In an attempt to understand the genetic basis of tomato spotted wilt (TSW) virus, Li et al. (2018) genotyped the US peanut minicore collection with 133 SSR markers, and found four subpopulations, generally corresponding to botanical varieties. US minicore was also genotyped with 81 SSR markers and two functional SNP markers from fatty acid desaturase 2 (FAD2) (Wang et al. 2011b). The results showed four major subpopulations that were related to four botanical varieties (var. *hypogaea,* var. *fastigiata,* var. *peruviana,* and var. *vulgaris*). Similarly, Ren et al. (2014) employed 146 SSR markers to study 196 major peanut cultivars extensively planted in different regions in China. A neighbor-joining tree constructed based on pairwise Nei's genetic distances could differentiate the cultivars from the southern region from those of northern region. Wang et al. (2016a) applied 111 SSR markers to examine 79 peanut cultivars and breeding lines from China, India, and the USA. These initial studies included relatively few, primarily SSR markers or a limited number of peanut accessions for assessing the genetic diversity and population structure. Results showed that the alleles per locus in three countries were similar.

Khera et al. (2013) studied the relationship among 280 germplasm lines that originated from a reference set of 300 genotypes representing 48 countries using 73 Kompetitive Allele Specific PCR (KASP) genotyping assays. Germplasm lines belonging to the two subspecies (along with some admixtures) formed two separate groups, while the wild species (diploids) formed a distinct cluster. The same reference set was genotypes with high-density single nucleotide polymorphism (SNP) array "Axiom_Arachis" (58 K SNPs) (Pandey et al. 2017). Diversity analysis with 44,424 polymorphic SNPs identified four clusters. The wild diploids, *A. hypogaea* ssp. *hypogaea* and *A. hypogaea* ssp. *fastigiata* formed separate clusters, while the admixtures of both the subspecies formed a separate cluster.

Zhang et al. (2017) applied next-generation sequencing technology to explore the molecular footprint of agronomic traits related to domestication using 158 Chinese peanut accessions and specific-locus amplified fragment sequencing (SLAF-seq) markers. The markers showed highly differentiated genomic regions between *A. hypogaea* ssp. *hypogaea* and *A. hypogaea* ssp. *fastigiata* accessions. Population structure divided the accessions into two groups: *A. hypogaea* ssp. *hypogaea* and *A. hypogaea* ssp. *fastigiata*.

Another study with 195 peanut accessions collected from 20 peanut-growing provinces of China used 13,435 SNPs generated from GBS (Wang et al. 2019). A neighbor-joining (NJ) tree showed two major groups differentiated at the subspecies level. The first group consisted of the accessions belonging to *A. hypogaea* ssp. *hypogaea* (var. *hypogaea* and var. *hirsuta*), while the second group comprised of *A. hypogaea* ssp. *fastigiata* (var. *vulgaris* and var. *fastigiata*) and a few accessions from both the subspecies. The highest level of nucleotide diversity was observed in var. *hirsuta*, while the lowest in var. *hypogaea*.

But a clear molecular genetic evidence of the differences and relationships among cultivated subspecies and botanical varieties resulted when high-density polymorphic loci were used on diverse peanut germplasms (Zheng et al. 2018). In this study, tunable genotyping-by-sequencing (tGBS) was used to generate 37,128 high-quality SNPs on a total of 320 peanut accessions, including four of the six botanical varieties to assess the genetic and evolutionary relationships at the genome-wide level. The accessions were grouped into three clusters. Almost all of the accessions in the first cluster belonged to var. *fastigiata*, and the other two clusters mainly consisted of accessions from var. *vulgaris* and *A. hypogaea* ssp. *hypogaea*, respectively. Population structure analysis showed that var. *fastigiata* and var. *vulgaris* could be differentiated from each other, while var. *hypogaea* and var. *hirsuta* could not be distinguished. The fixation index (F_{ST}) value showed that the var. *fastigiata* and var. *vulgaris* were closely related to each other ($F_{ST}=0.284$), while both of them were clearly distinct from the var. *hypogaea* ($F_{ST} > 0.4$). This study confirmed the traditional botanical classifications of cultivated peanut using the genome-wide markers.

In a special case, use of additional markers failed to classify the accessions into distinct groups. Genotyping of US minicore collection using Affymetrix version 2.0 SNP array could provide the data on 13,382 SNP markers after filtering (Zhang et al.

2019b). Predicting the genetic structure among the 120 accessions using PCA could not identify any distinct clusters indicating that these minicore collections did not represent a highly structured population. When Otyama et al. (2019) genotyped the US minicore collection with the 58 K SNP array to get 13,527 SNPs for structure analysis, they found long distance LDs with a half decay distance at 3.78 Mb. Structure analysis could separate market type and subspecies with several exceptions, probably supporting the results of Zhang et al. (2019b).

Analysis on genetic distance and clustering with the resequencing of 52 accessions of 12 species including *A. duranensis* and *A. ipaensis*, as well as the wild tetraploid *A. monticola*, and 30 diverse peanut samples (Zhuang et al. 2019) showed the grouping of accessions into three clusters based on the SNP data. The study suggested higher genetic distances and low genome exchanges associated with ploidy difference. The phylogenetic tree could classify most of the cultivated peanut accessions into two groups representing the two subspecies. However, some of the accessions of the *A. hypogaea* ssp. *hypogaea* were grouped along with the vulgaris-type cultivars probably due to the inter-subspecies hybridizations. The synthetic amphidiploids were grouped along with their respective diploid.

Using the WGRS data of 231 genotypes (comprising wild diploids, tetraploids, and botanical varieties) available in the public domain, we tried to group the genotypes. "vcfr" and other packages of R computing environment were used to work out the genetic distances (Euclidean) based on the SNPs of individual chromosomes to group the genotypes, and the number of groups was decided based on the within-cluster sum of square (WSS) and silhouette indices (Charrad et al. 2015). Both WSS and silhouette indices showed clear differences between the A and B subgenomes to form the clusters. Varying number of groups were identified when individual chromosomes were considered separately. Most of the chromosomes could distinguish the genotypes to group them into two clusters, while chromosome Ah15 and Ah20 could classify the genotypes into three groups. But Ah17 (B07) and Ah18 (B08) were extra-ordinary to group the genotypes into six clusters. This could be due to their recent evolution from the crossing over in the B genome after its separation from the A genome (Zhuang et al. 2019). The genotypes belonging to *A. hypogaea* ssp. *hypogaea* could be distinguished from those of *A. hypogaea* ssp. *fastigiata*. The diploid species, accessions of *A. monticola* and the synthetic amphidiploids were also clustered separately.

Population structure was also investigated among the aforementioned 231 genotypes using STRUCTURE analysis with ADMIXTURE model (Alexander et al. 2009). First, the Neighbor-Joining dendrograms were constructed, and then the branch order of the trees was employed to draw ADMIXTURE plots. The best K values ranged from 8 to 15 depending on the chromosomes (Table 1) which was in accordance with ADMIXTURE's cross-validation error suggesting that the number of peanut founder lines varied with the chromosomes. The results of ADMIXTURE and dendrogram analyses were consistent (Figs. 2 and 3). When all 20 chromosomes were considered together for the analysis, nine groups ($K = 9$) were detected. Again, the genotypes belonging to the two subspecies were separated with a few exceptions.

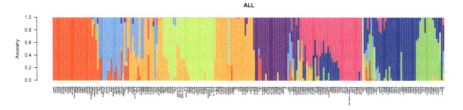

Fig. 2 ADMIXTURE plot for peanut genotypes based on the WGRS data. WGRS sequences of the peanut genotypes (231) were downloaded from https://www.ncbi.nlm.nih.gov/sra, and the reads were mapped on to the reference genome of Tifrunner to identify the SNPs. Branch order of the Neighbor-Joining dendrogram was used to draw the plot

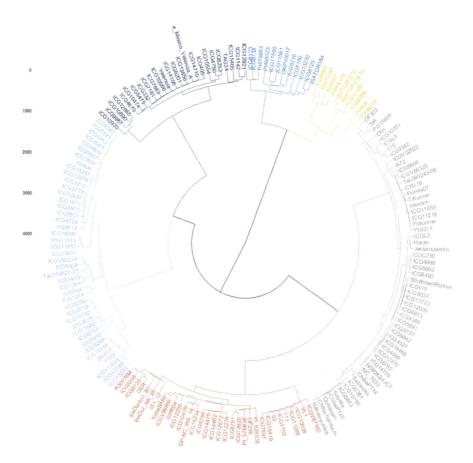

Fig. 3 Dendrogram for peanut genotypes based on the WGRS data. WGRS sequences of the peanut genotypes (231) were downloaded from https://www.ncbi.nlm.nih.gov/sra, and the reads were mapped on to the reference genome of Tifrunner to identify the SNPs. Branch order of the Neighbor-Joining dendrogram was used to draw the plot

The diploid species, accessions of *A. monticola* and the synthetic amphidiploids formed separate subpopulations.

Haplotype diversity was analyzed using 171 peanut accessions in terms of allelic polymorphism at *AhGPAT9A* and *AhGPAT9B* loci associated with triacylglycerol (TAG) synthesis and oil content. At *AhGPAT9A*, 118 polymorphic sites formed 64 haplotypes (a1 to a64), while 94 polymorphic sites in *AhGPAT9B* formed 75 haplotypes (b1 to b75). The haplotype analysis identified elite haplotypes and haplotype combinations related to high oil content (Lv et al. 2020). Since artificial selection contributes to the increase in seed oil during soybean domestication (Zhang et al. 2019a), it is interesting to undertake phylogenetic study in peanut using these polymorphic sites.

3.2.1 Genomic Diversity and Dynamics of Transposable Elements

Transposable elements (TEs), despite the genetic load they may impose on their host genome, play an important role in shaping genomic organization and structure and may cause dramatic changes in phenotypes. TE abundance, diversity, and activity are highly variable among the individuals of a population. Thus, understanding the dynamics of TEs in populations in terms of which factors explain the number of TEs that belong to a particular TE family, order or class, the diversity of TEs present in a given genome, and the frequency distribution of individual TE insertions, is crucial while analyzing the complex organization, function, and evolution of genomes (Barrón et al. 2014). Modelling of TE dynamics was suggested in a population genomics context by incorporating recent advances in TEs into the SNP-based information on the demography, selection, and intrinsic properties of genomes (Bourgeois and Boissinot 2019). A better understanding of the evolutionary dynamics of TEs might be favored by sequencing the individuals on a population scale and comparing the TE abundance between the individuals. Novel bioinformatics tools, such as PoPoolationTE2 (Kofler et al. 2016) and PTEMD (Kang et al. 2016), have been developed for studying the abundance and variation in the transposition of TEs.

Our effort on the population genomics of TEs using the WGRS data for 33 genotypes of peanut indicated a wide range in the copy number of *AhMITE1*, with PI 576638 (*A. hypogaea* ssp. *hypogaea* var. *hirsute*) carrying 133 copies and Chibahandachi (a Japanese *A. hypogaea* ssp. *hypogaea* var. *hypogaea*) recording 637 copies. The A and B subgenomes did not differ much for the copy number of *AhMITE1*. Thus, understanding the dynamics of transposable elements over the evolutionary time scales for their abundance and frequency could reveal the genome evolution and possibly the influence on the phenotype.

3.2.2 Copy Number Variations (CNVs)

We attempted to analyze the WGRS data of 231 genotypes available in the public domain for the CNVs (Xie and Tammi 2009) as an effort towards population genomics and peanut pan-genomics (unpublished data). CNV analysis with a

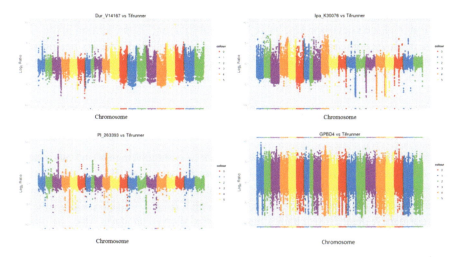

Fig. 4 CNVs in *Arachis duranensis* (Dur_14167), *A. ipaensis* (Ipa_K300076), *A. monticola* (PI_263393) and GPBD 4 as compared to Tifrunner. WGRS sequences of the peanut genotypes were downloaded from https://www.ncbi.nlm.nih.gov/sra, and the reads were mapped on to the reference genome of Tifrunner to identify the CNVs following the standard procedure. Log2 Ratio of >5 or <−5 indicates the significant variations

mean window size of 7,672 bases and a range of 703 (ICG8490) to 56,052 (Dur_V14167) revealed significant CNVs among the 230 genotypes when compared to the reference genome of Tifrunner. Comparison of the ancestral genomes with the reference genome indicated fewer CNVs between the A genome (Dur_14167) and the subgenome of *A. hypogaea* than between the B genome (Ipa_K300076) and the subgenome of *A. hypogaea* (Table 1) (Fig. 4). An elite variety (GPBD 4), with diverse background, showed high copy number variations when compared to the reference genome of Tifrunner as against the primitive amphidiploid *A. monticola* showing variations with Tifrunner. Presence and absence variations were also observed at a few loci. Overall, the results indicated an opportunity to use CNVs to study peanut evolution and domestication apart from studying their influence on presence or absence variations of genes and gene copy number variations.

In another study, Yin et al. (2020) compared the genomes of the *A. monticola* and *A. hypogaea* and identified structural variations (SVs) including deletions, insertions, inversions, duplications and CNVs. Natural selection influenced the SV genes of B subgenome in wild *A. monticola*, while domestication had larger effects on SV genes of A subgenome in the cultivated *A. hypogaea*. Such SVs were also observed among the resistant gene analogs (RGAs). Differential expression between the resistant and the susceptible genotypes was more pronounced among the RGAs with SV than those RGAs without SVs. Domestication for pod size also involved the SV-associated genes. A few putative seed development-related genes with SVs in upstream/exon regions showed higher expression in *A. hypogaea* than in *A. monticola*. A gene on chromosome A08 coding for AUXIN RESPONSE FACTOR

(ARF), playing important roles in auxin-mediated growth and development, including fruit and seed development, has a 275 bp deletion and 7 bp insertion in the 12th exon in *A. hypogaea*. Twelve wild species with small pod size do not possess deletions in ARF2.

3.3 Domestication

Domestication syndrome involving the differences in morphological and physiological traits resulting from evolution has been identified in peanut. Historically, though several species of *Arachis* were under cultivation for their edible pods, they show little signatures of domestication. After allotetraploidization, traits like seed dispersal and dormancy, plant growth habits, and plant size showed changes. Also, the plants became more attractive because of an increase in plant size, different transpiration characteristics, higher photosynthetic capacity, or other characteristics (see Leal-Bertioli et al. 2017). But increase in seed size might not be important in the initial domestication. This was also demonstrated using the artificially synthesized allotetraploids which showed larger leaves, stomata, and epidermal cells than did their diploid parents. Changes in several traits of peanut productivity and adaptation were also mapped using the advanced backcross population developed with a synthetic amphidiploid (Fonceka et al. 2012). Allotetraploids (primitive peanut) produced more photosynthetic pigments, and they were put under cultivation under various selection pressures where they evolved as domesticated plants. Then the domesticated peanut might have been dispersed by trade or migration to various areas of South and Central America. In each new environment, landraces must have evolved due to selection to adapt to local conditions.

Based on the RFLP and cytogenetic evidences, Kochert et al. (1996) proposed that the morphological variation observed in domesticated peanut must have been caused by differences in a few major genes. For example, the characters like upright habit, short growing season, time to maturity, and concentration of fruit production on or near the main stem generally seen in *A. hypogea* ssp. *fastigiata* are governed by a few genes. These traits involved adaptations for cultivation on riverine sandbars since the cultivators selected such types for expanding cultivation to these environments. In domesticated peanut, selection has eliminated the isthmus and pod has only a single chamber, which contains all seeds unlike *A. monticola* where each seed is contained in a single chamber of the pod and adjacent chambers are connected by an elongated isthmus. But *A. monticola* is very closely related and cross-compatible with the domesticated peanut (Grabiele et al. 2012).

Since very little or no introgression from related diploid species occurred after allopolyploidization, the cultivated groundnut shows limited DNA polymorphism. However, recombination is proposed to have played an important role in peanut diversification and domestication (Bertioli et al. 2019). Mapping of recombination hot spots using a RIL population with 200 lines and a natural population with 49 cultivars along with 4,837 SLAF markers showed that the B subgenome had more crossover events than the A subgenome. On an average, 4.34% and 7.86% of the genome contained large numbers of crossover events (>50 cMMb^{-1}) along chromosomes in the RILs and natural population, respectively. The genes located within the recombination hot spot regions were important for peanut response to environmental stimuli like cadmium ion, stress, auxin stimulus, fructose stimulus, wax biosynthetic process, fungus, suggesting that recombination plays a crucial role in peanut adaptation to changing environments (Wang et al. 2020b). Analysis of the homeologous exchange junctions revealed that about 65.3% of the recombination regions were distributed in genic regions in peanut (Zhang et al. 2020b). Genome-wide SNP genotyping of a few US runner germplasm resolved signatures of selection and tetrasomic recombination in peanut (Clevenger et al. 2017). The genomic regions (119–127 Mb on B08 and 75–124 Mb on B09), carrying the genes related to maturity, defense response, and pollen recognition, were found to be positively selected. Based on the structural variant analysis of the *A. monticola* and *A. hypogaea* genomes Yin et al. (2020) reported that domestication had larger effects on SV genes of A subgenome in the cultivated *A. hypogaea*. On the other hand, the natural selection influenced the SV genes of B subgenome in wild *A. monticola*.

Zhang et al. (2017) identified a total of 1,429 genes for domestication. Highly differentiated genomic regions were observed between *A. hypogaea* ssp. *hypogaea* and *A. hypogaea* ssp. *fastigiata* accessions using F_{ST} values and sequence diversity (π) ratios. Among the 1,429 genes, 662 were located on chromosome A03, indicating the selective sweeps on this chromosome. Genes related to height of the main stem, seed weight, disease resistance, the basic transcription factor bHLH linked to seed shape and seed coat color, and *Aradu PBR53* for pod weight were part of these sweeps.

3.4 Population Epigenomics

Understanding the epigenomic variation at the population level and its contributions to phenotypic variations and evolution have been partly possible due to the advances in chromatin state mapping, high-throughput DNA sequencing, and bioinformatics (Moler et al. 2018). Spontaneous epimutations giving rise to epigenomic variation can arise stochastically in plant genomes independently of DNA sequence changes (Johannes and Schmitz 2019). Some forms of epigenomic variation are stable and may show trans-generation inheritance. Understanding the sources of epigenomic variation, influence on the phenotype and the evolution will have profound insights on the phenotypic plasticity and ecological adaptation.

Low DNA sequence polymorphism despite enormous phenotypic variation indicates the possible role of epigenetic variation in peanut. A genome-wide DNA methylation pattern and its influence on gene expression were reported across 11 diverse peanut genotypes (Bhat et al. 2019a). Bisulfite sequencing of 11 genotypes after 21 days of sowing (DAS) produced an average of 127,852,977 reads with a mapping range of 1–1,658 at each DNA methylated site (Bhat et al. 2019a). A total of 75,973,928 sites across the genotypes belonged to the category where all the mapped reads (100%) showed cytosine methylation, while 101,137,805 sites belonged to the category where at least 50% of the mapped reads showed cytosine methylation and 126,487,183 sites showed less than 50% of the mapped reads showing cytosine methylation. The B subgenome exhibited higher DNA methylation sites (46,294,063) than the A subgenome (30,415,166) across the genotypes. A total of 177,574 sites were found in the scaffolds. CHG (where H is A, C, or T) region showed the highest methylation sites (30,537,376) regions, followed by CpG (30,356,066) and CHH (15,993,361) regions across the genotypes. This observation is in line with the previous reports (Feng et al. 2010; Zemach et al. 2010) that the DNA methylation in plants is found both in CpG and non-CpG (CHG and CHH, where H is A, C, or T) contexts in contrast to mammals where DNA methylation occurs predominantly at CpG dinucleotides.

Genotype JL 24 and TMV 2 showed the highest (82,137,767) and the lowest methylation sites (69,044,110), respectively. Such a natural epigenetic variation was also observed among the different ecotypes of *Arabidopsis* (Vaughn et al. 2007). Bhat et al. (2019a) found that a total of 5,379,101 sites were conserved across the 11 genotypes, while the unique sites ranged from 6,575,363 (TMV 2) to 9,190,780 (JL 24). On an average, inter-genic regions (70,464,637 sites) were more prone for DNA methylation than the genic regions including 2 kb upstream and 2 kb downstream regions (6,422,166 sites). Within the genic regions, the introns (1,590,263) showed a greater number of DNA methylation sites than the exonic regions (971,274). The 2 kb upstream and 2 kb downstream regions had 3,860,629 DNA methylation sites, indicating higher proportion of DNA methylation at the upstream and downstream regions than the gene body region. The distribution of DNA methylation within the genome especially in the promoter and gene body regions is very important because it influences the gene expression (Suzuki and Bird 2008).

The number of genes showing at least one methylated site ranged from 51,179 (ICGV 86855) to 55,497 (ICGV 99005) indicating the variation in DNA methylation at the genic regions between the individuals (Bhat et al. 2019a). Of them, *Arahy.0DU9MH*, a 342,359 bp long gene on chromosome 11, showed the highest number of methylated sites, which ranged from 11,488 (ICGV 86855) to 14,026 (JL 24). Within *Arahy.0DU9MH*, the promoter region had 131 methylated sites, while the gene body (142 in exons and 12,573 in introns) had 12,715 sites. The expression of ~53,740 genes varied widely among the 11 genotypes. *Arahy.0DU9MH* with the highest DNA methylations sites did not show any expression at 21 DAS in the leaves of the 11 genotypes. Fifty genes with a wide range of fragments per kilobase of transcript per million mapped reads (FPKM) across the genotypes were selected and checked for DNA methylation to check its influence. Many of these genes showed a negative association between the number of DNA methylation sites and FPKM among the genotypes. *Arahy.FHUH7B* on chromosome 10 showing the highest FPKM of 54,951 had a maximum of 102 DNA methylation sites.

Fourteen C5-MTase coding genes and ten DNA demethylase coding genes identified in the diploid peanut earlier (Wang et al. 2016b) were analyzed for DNA methylation and expression. A considerable variation was observed for methylation across the genes, however, not much variation was observed for methylation across the genotypes.

An attempt was made to identify the differentially DNA methylated sites between the foliar disease-resistant (GPBD 4, VG 9514, ICGV 86855, ICGV 99005, and ICGV 86699) and susceptible (TAG 24, TMV 2, and JL 24) genotypes (Bhat et al. 2019a). Foliar disease-resistant genotypes showed significant differential DNA methylation at 766 sites corresponding to 25 genes. Of them, two genes (*Arahy.1XYC2X* on chromosome 01 and *Arahy.00Z2SH* on chromosome 17) coding for senescence-associated protein showed differential expression where the resistant genotypes recorded higher FPKM at their epialleles.

3.5 Population Transcriptomics

The studies on population transcriptomics are limited in peanut though good quality reference transcriptomes are available for the two diploid progenitors and the tetraploid (Chopra et al. 2014), and a large number of studies have reported the transcriptomics of the contrasting genotypes for a few traits as listed under Sect. 2.4. Recent study showed that the expression patterns of a number of lncRNAs could vary between two closely related inbred lines and they might have significant roles in seed development (Ma et al. 2020). These studies indicate the scope for the population transcriptomics in peanut for describing the expressional variation at the population level. Chopra et al. (2016) reported transcriptome sequencing of diverse wild species and the cultivated species to test whether the cultivated peanut possesses almost no molecular variability. Pools of three tissues from a diverse panel of 22 *Arachis* accessions representing *Arachis hypogaea* botanical varieties, A-, B-, and K- genome diploids, a synthetic amphidiploid and a tetraploid wild species were sequenced using RNASeq and assembled *de novo*. Diversity analysis based on the SNPs discovered from the alignment indicated grouping of diploids according to genome classification and cultivated tetraploids by subspecies. Substantial genetic variability was observed among wild species, but lesser variability was reported for the accessions of the cultivated species. Such transcriptome-derived SNP diversity in peanut may explain some of the phenotypic differences observed in germplasm surveys. Apart from exploring allelic and transcript diversity among the accessions of different species within the genus *Arachis*, the study revealed the possible number of genes in peanut, phylogenic relationships of the Arachis accessions, specificities of SNPs to one accession or to a group of market types or to a group of different genome affiliations and validation of the SNPs obtained from the bioinformatics calls that can be used in a breeding program.

Cluster analysis of variants indicated that sequences of B genome species were the most similar to the tetraploids, and the next closest diploid accession belonged to the A genome species. *A. monticola* accession grouped closer to the *A. hypogaea* ssp. *fastigiata* accessions than to the *A. hypogaea* ssp. *hypogaea* accessions. Overall, the study identified substantial genetic variability among the wild species, while limited variability was noticed among the accessions of the cultivated species at the transcript level.

3.6 Genotype–Phenotype Associations, Selection and Breeding

Genotype–phenotype associations and genetic mapping of desirable traits in peanut using various methods have progressed quite well (Pandey et al. 2020b). Biparental mapping populations have been widely used for mapping the taxonomic, disease resistance, productivity and nutritional traits. A wide array of small-scale molecular markers to genome-wide high-throughput markers has been employed for mapping (Pandey et al. 2020b). In addition, the transcriptomics connecting genome to the gene function and ultimate phenome in biology was also used in peanut for performing genetic mapping of traits in biparental populations (Huang et al. 2020). The study used RNA-Seq on the RILs to identify a major QTL responsible for purple testa color in peanut. The QTL mapping work in peanut has transitioned to population genomics approaches for identifying genotype–phenotype associations and genomic selection. The progress on these aspects is briefly discussed below.

3.6.1 Genome-Wide Association Studies (GWAS)

Knowledge of genetic diversity and population structure along with understanding of the degree of linkage disequilibrium (LD) in a population is of great importance and prerequisite for LD-based mapping. Since peanut offers a great opportunity to constitute diverse association panels required for GWAS, such studies have been conducted in peanut to identify genotype–phenotype associations for foliar disease resistance, productivity traits, nutritional traits, drought tolerance.

The initial study used 96 accessions of the US minicore collection, which were genotyped with 81 SSR markers and two functional SNP markers from fatty acid desaturase 2 (FAD2) (Wang et al. 2011b), and measured for the seed quality traits including oil content, fatty acid composition, flavonoids, and resveratrol. Significant associations of FAD2A gene with the oleic acid and linoleic acid contents were observed.

GWAS in the US peanut minicore collection of 120 accessions was conducted using the Affymetrix version 2.0 SNP array, measuring response to early leaf spot (ELS) and late leaf spot (LLS) diseases and identified 18 QTLs for ELS and 28 QTLs for LLS (Zhang et al. 2020a). Two and four QTLs were major for ELS and LLS with the phenotypic variance explained (PVE) of 16–24%, respectively. Of the six major

QTLs, five were located on the B subgenome and one was located on the A subgenome, suggesting the importance of B subgenome for ELS and LLS resistance. Two genomic regions (SNPs at AX-177643393 and AX-177643343) on chromosome B09 were found to provide significant resistance to both ELS and LLS (PVE of 22–24%). A few candidate genes were also identified from these regions

In order to understand the genetic basis underlying the variations in the mineral composition, GWAS was attempted using the US peanut minicore collection, which was genotyped with SNP array (Affymetrix) (Zhang et al. 2019b). A total of 36 QTLs were identified for five mineral compositions with PVE of 18.35–27.56% in which 24 QTLs were for boron (B), two QTLs for copper (Cu), six QTLs for sodium (Na), three QTLs for sulfur (S), and one QTL for zinc (Zn). A total of 110 nonredundant candidate genes were identified for elemental accumulation. Among them, the elemental/metal transporter gene *arahy.KQD4NT* located on LGB04 was very important

US peanut minicore collection was studied for tomato spotted wilt (TSW), which is a serious virus disease of peanut in the USA (Li et al. 2018). Genotyping of the accessions with 133 SSR markers followed by GWAS indicated that five markers: pPGPseq5D5, GM1135, GM1991, TC23C08, and TC24C06, were consistently associated with TSW resistance. These markers together explained 36.4% of the phenotypic variance.

US peanut minicore collection was also used for GWAS using the 58 K SNP array and 12 seed composition and quality traits (Otyama et al. 2019). The study revealed significant differences for LD estimates between the two subspecies. Estimates were higher in *A. hypogaea* ssp. *hypogaea* than in the *A. hypogaea* ssp. *fastigiata*. Unlike the *A. hypogaea* ssp. *hypogaea*, mean LD values were not significantly different between the *A. hypogaea* ssp. *fastigiata*. The decay distance was significantly longer in *A. hypogaea* ssp. *hypogaea* (average decay distance = 13.52 Mb) than in *A. hypogaea* ssp. *fastigiata* (average decay distance = 3.41 Mb). GWAS revealed significant marker associations for arachidic and behenic fatty acid compositions and blanchability. However, the authors of this study (Otyama et al. 2019) suggested to increase the population size and reduce the population structure.

High-density SNPs were used to dissect the molecular footprint of agronomic traits related to domestication using 158 peanut accessions (Zhang et al. 2017), which included 36 *A. hypogaea* ssp. *hypogaea* type and 122 *A. hypogaea* ssp. *fastigiata* type from different provinces of China. A total of 17,338 high-quality SNPs obtained from the whole peanut genome using specific-locus amplified fragment sequencing (SLAF-seq method) were used to analyze 11 agronomic traits using GWAS. Candidate genes responsible for these traits were identified in the genomic regions surrounding the peak SNPs. *Aradu 52T5J* gene coding for malate dehydrogenase and relating to biomass and plant height was found to be associated with height of the main stem. Similarly, *Aradu J85DC* encoding cytochrome P450 superfamily protein reportedly involved in the strigolactone synthetic pathway in rice was associated with total number of branches. A gene encoding a bHLH transcription factor with pleiotropic effect on seed development showed association

seed length and 10-seed weight. Auxin response factor (ARF) coding gene *Aradu PZ2UH* involved in plant growth and seed development was also associated with 10-seed weight. *Aradu CVC5Q* gene coding for microtubule-associated protein (MAP), which influences seed shape by regulating microtubule growth, was associated with pod width. Another bHLH transcription factor coding gene on B03 was associated with seed coat color.

High-resolution genotyping with SSR (154) and DArT (4,597) markers and multiple season phenotyping data for 50 important agronomic, disease and quality traits were generated on the "reference set" of 304 accessions (Pandey et al. 2014). The SSR markers were found to be superior in differentiating the subgroups of tetraploids, while DArTs were superior in resolving tetraploids from the diploids. Three subgroups with high level of admixture within and between subgroups were detected. A total of 524 highly significant marker trait associations (MTAs) with the PVE of 5.81–90.09% were observed for 36 traits including yield, drought tolerance, and quality parameters

GWAS for the yield-related traits was carried out using 195 peanut accessions which were genotyped by GBS to get 13,435 high-quality SNPs (Wang et al. 2019). The accessions showed two major groups corresponding to the two subspecies (*A. hypogaea* ssp. *hypogaea* and *A. hypogaea* ssp. *fastigiata*). Ninety-three non-overlapping SNP peaks significantly associated with four traits (hundred-seed weight, hundred-pod weight, yield per plant and pod branch number per plant) were identified. In addition, 311 unique candidate genes were detected in the surrounding regions of these peaks.

3.6.2 Nested Association Mapping (NAM) and Multiparent Advanced Generation Intercross (MAGIC)

The initial efforts on identifying genes underlying phenotypes depended on the biparental populations. Such efforts have successfully mapped several traits, including foliar disease resistance, oil content and quality, aflatoxin contamination, yield and seed features (see Vishwakarma et al. 2017). To overcome the drawbacks of biparental populations, multiparent genetic populations, namely NAM (nested-association mapping) and MAGIC (multiparent advanced generation intercross) populations segregating for multiple traits, have been developed in peanut. Two NAM populations (one each for Spanish and Virginia types) and three MAGIC populations (for drought, aflatoxin and multiple agronomic, disease and nutrition traits) have been developed at ICRISAT (see Pandey et al. 2020b). The two NAM populations, NAM_Tifrunner and NAM_Florida-07, were used for dissecting genetic control of 100-pod weight (PW) and 100-seed weight (SW). Two high-density SNP-based genetic maps were constructed with 3,341 loci and 2,668 loci for NAM_Tifrunner and NAM_Florida-07, respectively. QTL analysis in NAM_Tifrunner identified 12 and 8 major effect QTL for PW and SW, respectively. Similarly, 13 and 11 major effect QTL were identified in NAM_Florida-07 for PW and SW, respectively (Gangurde et al. 2020). This study demonstrated the utility of

NAM populations for genetic dissection of complex traits and performing high-resolution genetic mapping of traits in peanut. These genetic-mapping of traits efforts would contribute towards genotype–phenotype association and selection approaches based on population genomics concepts and approaches to describe the fitness consequences of naturally occurring variation, where a large number of molecular markers are scored in individuals from different environments with the goal of identifying markers showing unusual patterns of variation, potentially due to selection at linked sites. This would require a large number of genetic markers and the ability to scan the genome with/without measuring the phenotype. This approach may be attractive because of its simplicity of sampling individuals without knowledge of their breeding history (Stinchcombe and Hoekstra 2008).

3.6.3 Genomic Selection

Genomic selection uses the linkage disequilibrium between genomic markers and quantitative trait phenotypes for prediction and selection of those quantitative trait phenotypes and has been increasingly used in animals and plants (see Jonas et al. 2019). A training population with 340 diverse peanut genotypes, including the elite breeding lines and popular cultivars has been constituted for genomic selection. The population was screened for resistance to LLS and rust and yield traits at multi-locations during multi-seasons (Chaudhari et al. 2019). Genome-based prediction is being optimized for this training population, and the model will be deployed for breeding in ICRISAT and Indian groundnut breeding programs (Pandey et al. 2020a). Also, it will be interesting to integrate multiple new technologies/approaches such as genomic selection, rapid generation advancement, gene editing, and seed-chipping based genotyping for early generation selection along with SNP-based quality control panel for accelerated variety development and making them available to the farmers for cultivation.

3.6.4 Breeding

Outcomes of genomics and genetic mapping of traits have been assisting peanut breeding by increasing the efficiency and precision as it has been demonstrated for improving resistance to root-knot nematode, LLS and rust and oleic acid content (see Nawade et al. 2018; Pandey et al. 2020b). Majority of these genotypes are under multi-location testing or large-scale farm testing for variety development and commercialization.

4 Conclusions and Future Perspectives

Efforts towards peanut population genomics are on the increasing trend over the last 10–15 years with the development of extensive genomic data from various species within *Arachis*. The genomic, transcriptomic, epigenomic, and germplasm resources are fueling the application of peanut population genomics for a variety of basic and applied studies. The structural and population genomics and pangenomics approaches have significantly contributed to our understanding of the origin (conclusively deciding the diploid progenitors), and unravelling genome evolution, genetic/genomic diversity, and population structure in peanut. Detailed genotype–phenotype associations have been worked out for the important traits, and some of these outcomes have been applied in peanut breeding. Genome-wide DNA and RNA landscapes have been used to assess the genetic/genomic diversity in a wide array of peanut germplasm.

However, the population genomics need to unravel new key insights in redefining the concept of minicore population based on the genome-wide sequence data for peanut conservation. Likewise, genetic diversity and population structure analysis need to be worked out integrating extensive genotypes considering the factor of hybridization/outcrossing. Simultaneous population genomics analysis with other omics data might explain the frequent observation of low genetic polymorphism despite greater morphological variation in peanut. The vast knowledge on peanut population genomics may help understanding parallel evolution, which is one of the striking patterns in nature, where the presence of repeated evolution of the same phenotypes, suites of traits, and adaptations suggests a strong role for natural selection in shaping biological diversity (Yuan and Stinchcombe 2020).

Though there is an indication that *A. monticola* is more similar to *A. hypogaea* ssp. *fastigiata* type than to *A. hypogaea* ssp. *hypogaea* type based on transcript-derived SNP data, inclusion of data from a large number of accessions might be necessary before drawing broader conclusions. Transcriptome analysis among cultivated accessions and comparison to wild species will open many avenues for future peanut population genomics: understanding relationships of cultivated tetraploid species to their diploid ancestors, evolutionary development of the cultigen, distinguishing homologous vs. homeologous SNPs for genetic mapping and QTL analysis, study of genome rearrangements after polyploidization and expression bias among genes and genomes.

References

Alexander DH, Novembre J, Lange K. Fast model-based estimation of ancestry in unrelated individuals. Genome Res. 2009;19(9):1655–64.

Arya SS, Salve AR, Chauhan S. Peanuts as functional food: a review. J Food Sci Technol. 2016;53 (1):31–41.

Barrón MG, Fiston-Lavier A-S, Petrov DA, González J. Population genomics of transposable elements in Drosophila. Annu Rev Genet. 2014;48:561–81.

Bechara MD, Moretzsohn MC, Palmieri DA, Monteiro JP, Bacci M, Martins J, et al. Phylogenetic relationships in genus *Arachis* based on ITS and 5. 8 S rDNA sequences. BMC Plant Biol. 2010;10(1):255.

Belamkar V, Selvaraj MG, Ayers JL, Payton PR, Puppala N, Burow MD. A first insight into population structure and linkage disequilibrium in the US peanut minicore collection. Genetica. 2011;139(4):411.

Bertioli DJ, Seijo G, Freitas FO, Valls JFM, Leal-Bertioli S, Moretzsohn MC. An overview of peanut and its wild relatives. Plant Genet Resour. 2011;9(1):134–49.

Bertioli DJ, Ozias-Akins P, Chu Y, Dantas KM, Santos SP, Gouvea E, et al. The use of SNP markers for linkage mapping in diploid and tetraploid peanuts. G3 (Bethesda). 2014;4(1):89–96.

Bertioli DJ, Cannon SB, Froenicke L, Huang G, Farmer AD, Cannon EK, et al. The genome sequences of *Arachis duranensis* and *Arachis ipaensis*, the diploid ancestors of cultivated peanut. Nat Genet. 2016;48(4):438–46.

Bertioli DJ, Jenkins J, Clevenger J, Dudchenko O, Gao D, Seijo G, et al. The genome sequence of segmental allotetraploid peanut *Arachis hypogaea*. Nat Genet. 2019;51(5):877–84.

Bertioli DJ, Abernathy B, Seijo G, Clevenger J, Cannon SB. Evaluating two different models of peanut's origin. Nat Genet. 2020:1–3.

Bhat RS, Patil VU, Chandrashekar TM, Sujay V, Gowda MVC, Kuruvinashetti MS. Recovering flanking sequence tags of miniature inverted-repeat transposable element by thermal asymmetric interlaced-PCR in peanut. Curr Sci. 2008;95(4):452–3.

Bhat RS, Rockey J, Shirasawa K, Tilak IS, Brijesh Patil MP, Reddy VB. DNA methylation and expression analyses reveal epialleles for the foliar disease resistance genes in peanut (*Arachis hypogaea* L.). BMC Res Notes. 2019a;13(1):20.

Bhat RS, Shirasawa K, Monden Y, Yamashita H, Tahara M. Developing transposable element marker system for molecular breeding. In: Jain M, Garg R, editors. Legume genomics: methods and protocols. New York: Humana; 2019b. p. 233–52.

Bhogireddy S, Xavier A, Garg V, Layland N, Arias R, Payton P, et al. Genome-wide transcriptome and physiological analyses provide new insights into peanut drought response mechanisms. Sci Rep. 2020;10(1):4071.

Bi Y-P, Liu W, Xia H, Su L, Zhao C-Z, Wan S-B, et al. EST sequencing and gene expression profiling of cultivated peanut (*Arachis hypogaea* L.). Genome. 2010;53(10):832–9.

Blanc G, Wolfe KH. Widespread paleopolyploidy in model plant species inferred from age distributions of duplicate genes. Plant Cell. 2004;16:1667–78.

Bosamia TC, Dodia SM, Mishra GP, Ahmad S, Joshi B, Thirumalaisamy PP, et al. Unraveling the mechanisms of resistance to Sclerotium rolfsii in peanut (*Arachis hypogaea* L.) using comparative RNA-Seq analysis of resistant and susceptible genotypes. PLoS One. 2020;15(8): e0236823.

Bourgeois Y, Boissinot S. On the population dynamics of junk: a review on the population genomics of transposable elements. Genes. 2019;10(6):419.

Cao K, Li Y, Deng CH, Gardiner SE, Zhu G, Fang W, et al. Comparative population genomics identified genomic regions and candidate genes associated with fruit domestication traits in peach. Plant Biotechnol J. 2019;17(10):1954–70.

Chalhoub B, Denoeud F, Liu S, Parkin IA, Tang H, Wang X, et al. Early allopolyploid evolution in the post-Neolithic Brassica napus oilseed genome. Science. 2014;345(6199):950–3.

Charrad M, Ghazzali N, Boiteau V, Niknafs A. Determining the best number of clusters in a data set. 2015. Recuperado de: https://cran.rproject.org/web/packages/NbClust/NbClust.pdf.

Chaudhari S, Khare D, Patel SC, Subramaniam S, Variath MT, Sudini HK, et al. Genotype × environment studies on resistance to late leaf spot and rust in genomic selection training population of peanut (*Arachis hypogaea* L.). Front Plant Sci. 2019;10:1338.

Chen X, Liang X. Peanut transcriptomics. Genet Genomics Breeding Peanuts. 2014:139.

Chen X, Hong Y, Zhang E, Liu H, Zhou G, Li S, et al. Comparison of gene expression profiles in cultivated peanut (*Arachis hypogaea*) under strong artificial selection. Plant Breed. 2012;131 (5):620–30.

Chen Y, Ren X, Zhou X, Huang L, Yan L, Lei Y, et al. Dynamics in the resistant and susceptible peanut (*Arachis hypogaea* L.) root transcriptome on infection with the *Ralstonia solanacearum*. BMC Genomics. 2014;15(1):1078.

Chen X, Li H, Pandey MK, Yang Q, Wang X, Garg V, et al. Draft genome of the peanut A-genome progenitor (*Arachis duranensis*) provides insights into geocarpy, oil biosynthesis, and allergens. Proc Natl Acad Sci. 2016a;113(24):6785–90.

Chen X, Yang Q, Li H, Li H, Hong Y, Pan L, et al. Transcriptome-wide sequencing provides insights into geocarpy in peanut (*Arachis hypogaea* L.). Plant Biotechnol J. 2016b;14(5):1215–24.

Chen C, Cao Q, Jiang Q, Li J, Yu R, Shi G. Comparative transcriptome analysis reveals gene network regulating cadmium uptake and translocation in peanut roots under iron deficiency. BMC Plant Biol. 2019a;19(1):35.

Chen H, Yang Q, Chen K, Zhao S, Zhang C, Pan R, et al. Integrated microRNA and transcriptome profiling reveals a miRNA-mediated regulatory network of embryo abortion under calcium deficiency in peanut (*Arachis hypogaea* L.). BMC Genomics. 2019b;20(1):1–17.

Chen X, Lu Q, Liu H, Zhang J, Hong Y, Lan H, et al. Sequencing of cultivated peanut, *Arachis hypogaea*, yields insights into genome evolution and oil improvement. Mol Plant. 2019c;

Chopra R, Burow G, Farmer A, Mudge J, Simpson CE, Burow MD. Comparisons of de novo transcriptome assemblers in diploid and polyploid species using peanut (*Arachis* spp.) RNA-Seq data. PLoS One. 2014;9(12):e115055.

Chopra R, Burow G, Farmer A, Mudge J, Simpson CE, Wilkins TA, et al. Next-generation transcriptome sequencing, SNP discovery and validation in four market classes of peanut, *Arachis hypogaea* L. Mol Genet Genomics. 2015;290(3):1169–80.

Chopra R, Burow G, Simpson CE, Chagoya J, Mudge J, Burow MD. Transcriptome sequencing of diverse peanut (*Arachis*) wild species and the cultivated species reveals a wealth of untapped genetic variability. G3. 2016;6(12):3825–36.

Chu Y, Holbrook CC, Ozias-Akins P. Two alleles of *ahFAD2B* control the high oleic acid trait in cultivated peanut. Crop Sci. 2009;49(6):2029–36.

Clevenger J, Chu Y, Chavarro C, Agarwal G, Bertioli DJ, Leal-Bertioli SC, et al. Genome-wide SNP genotyping resolves signatures of selection and tetrasomic recombination in peanut. Mol Plant. 2017;10(2):309–22.

Cortinovis G, Frascarelli G, Di Vittori V, Papa R. Current state and perspectives in population genomics of the common bean. Plants (Basel). 2020;9(3)

Desmae H, Janila P, Okori P, Pandey MK, Motagi BN, Monyo E, et al. Genetics, genomics and breeding of groundnut (*Arachis hypogaea* L.). Plant Breed. 2019;138(4):425–44.

Dhingra V, Gupta M, Andacht T, Fu ZF. New frontiers in proteomics research: a perspective. Int J Pharm. 2005;299(1–2):1–18.

Dutheil JY. Statistical population genomics. New York: Springer; 2020.

Edwards SV, Potter S, Schmitt CJ, Bragg JG, Moritz C. Reticulation, divergence, and the phylogeography–phylogenetics continuum. Proc Natl Acad Sci. 2016;113(29):8025–32.

Favero AP, Simpson CE, Valls JFM, Vello NA. Study of the evolution of cultivated peanut through crossability studies among *Arachis ipaensis*, *A. duranensis*, and *A. hypogaea*. Crop Sci. 2006;46(4):1546–52.

Feng S, Cokus SJ, Zhang X, Chen P-Y, Bostick M, Goll MG, et al. Conservation and divergence of methylation patterning in plants and animals. Proc Natl Acad Sci. 2010;107(19):8689–94.

Ferguson M, Burow M, Schulze S, Bramel P, Paterson A, Kresovich S, et al. Microsatellite identification and characterization in peanut (*A. hypogaea* L.). Theor Appl Genet. 2004a;108(6):1064–70.

Ferguson ME, Bramel PJ, Chandra S. Gene diversity among botanical varieties in peanut (*Arachis hypogaea* L.). Crop Sci. 2004b;44(5):1847–54.

Figueredo MS, Formey D, Rodriguez J, Ibanez F, Hernandez G, Fabra A. Identification of miRNAs linked to peanut nodule functional processes. J Biosci. 2020;45

Fonceka D, Tossim HA, Rivallan R, Vignes H, Faye I, Ndoye O, et al. Fostered and left behind alleles in peanut: interspecific QTL mapping reveals footprints of domestication and useful natural variation for breeding. BMC Plant Biol. 2012;12(1):26.

Friend S, Quandt D, Tallury S, Stalker H, Hilu K. Species, genomes, and section relationships in the genus *Arachis* (Fabaceae): a molecular phylogeny. Plant Syst Evol. 2010;290(1–4):185–99.

Gangurde SS, Wang H, Yaduru S, Pandey MK, Fountain JC, Chu Y, et al. Nested-association mapping (NAM)-based genetic dissection uncovers candidate genes for seed and pod weights in peanut (*Arachis hypogaea*). Plant Biotechnol J. 2020;18(6):1457–71.

Gao L, Gonda I, Sun H, Ma Q, Bao K, Tieman DM, et al. The tomato pan-genome uncovers new genes and a rare allele regulating fruit flavor. Nat Genet. 2019;51(6):1044–51.

Gao S, Wu J, Stiller J, Zheng Z, Zhou M, Wang YG, et al. Identifying barley pan-genome sequence anchors using genetic mapping and machine learning. Theor Appl Genet. 2020;133(9):2535–44.

Gayathri M, Shirasawa K, Varshney RK, Pandey MK, Bhat RS. Development of new *AhMITE1* markers through genome-wide analysis in peanut (*Arachis hypogaea* L.). BMC Res Notes. 2018;11(1):10.

Geng L, Duan X, Liang C, Shu C, Song F, Zhang J. Mining tissue-specific contigs from peanut (*Arachis hypogaea* L.) for promoter cloning by deep transcriptome sequencing. Plant Cell Physiol. 2014;55(10):1793–801.

Gowda MVC, Nadaf HL, Sheshagiri R. The role of mutations in intraspecific differentiation of groundnut (*Arachis hypogaea* L.). Euphytica. 1996;90(1):105–13.

Gowda MVC, Bhat RS, Motagi BN, Sujay V, Varshakumari and Bhat, S. Association of high-frequency origin of late leaf spot resistant mutants with *AhMITE1* transposition in peanut. Plant Breed. 2010;129(5):567–9.

Gowda MVC, Bhat RS, Sujay V, Kusuma P, Varshakumari Bhat S, Varshney RK. Characterization of *AhMITE1* transposition and its association with the mutational and evolutionary origin of botanical types in peanut (*Arachis* spp.). Plant Syst Evol. 2011;291(3–4):153–8.

Grabiele M, Chalup L, Robledo G, Seijo G. Genetic and geographic origin of domesticated peanut as evidenced by 5S rDNA and chloroplast DNA sequences. Plant Syst Evol. 2012;298(6):1151–65.

Guimarães PM, Brasileiro AC, Morgante CV, Martins AC, Pappas G, Silva OB, et al. Global transcriptome analysis of two wild relatives of peanut under drought and fungi infection. BMC Genomics. 2012;13(1):387.

Guimaraes PM, Guimaraes LA, Morgante CV, Silva OB Jr, Araujo ACG, Martins AC, et al. Root transcriptome analysis of wild peanut reveals candidate genes for nematode resistance. PLoS One. 2015;10(10):e0140937.

Gulcher J, Stefansson K. Population genomics: laying the groundwork for genetic disease modeling and targeting. Clin Chem Lab Med. 1998;36(8):523–7.

Guo B, Chen X, Dang P, Scully BT, Liang X, Holbrook CC, et al. Peanut gene expression profiling in developing seeds at different reproduction stages during *Aspergillus parasiticus* infection. BMC Dev Biol. 2008;8(1):12.

Guo F, Ma J, Hou L, Shi S, Sun J, Li G, et al. Transcriptome profiling provides insights into molecular mechanism in peanut semi-dwarf mutant. BMC Genomics. 2020;21(1):211.

Hake AA, Shirasawa K, Yadawad A, Nadaf HL, Gowda MVC, Bhat RS. Genome-wide structural mutations among the lines resulting from genetic instability in peanut (*Arachis hypogaea* L.). Plant Gene. 2018;13:1–7.

Hale H, Gardner EM, Viruel J, Pokorny L, Johnson MG. Strategies for reducing per-sample costs in target capture sequencing for phylogenomics and population genomics in plants. Appl Plant Sci. 2020;8(4):e11337.

Halward TM, Stalker TH, Larue EA, Kochert G. Genetic variation detectable with molecular markers among unadapted germplasm resources of cultivated peanut and related wild species. Genome. 1991;34(6):1013–20.

Halward TM, Stalker TH, Larue EA, Kochert G. Use of single-primer DNA amplifications in genetic studies of peanut (*Arachis hypogaea* L.). Plant Mol Biol. 1992;18(2):315–25.

Han S, Liu H, Yan M, Qi F, Wang Y, Sun Z, et al. Differential gene expression in leaf tissues between mutant and wild-type genotypes response to late leaf spot in peanut (*Arachis hypogaea* L.). PLoS One. 2017;12(8):e0183428.

Hao K, Wang F, Nong X, McNeill MR, Liu S, Wang G, et al. Response of peanut *Arachis hypogaea* roots to the presence of beneficial and pathogenic fungi by transcriptome analysis. Sci Rep. 2017;7(1):1–15.

Harlan JR, de Wet JMJ. On Ö. Winge and a prayer: the origins of polyploidy. Bot Rev. 1975;41(4):361–90.

He G, Prakash C. Identification of polymorphic DNA markers in cultivated peanut (*Arachis hypogaea* L.). Euphytica. 1997;97(2):143–9.

He G, Meng R, Newman M, Gao G, Pittman RN, Prakash CS. Microsatellites as DNA markers in cultivated peanut (*Arachis hypogaea* L.). BMC Plant Biol. 2003;3:3.

He G, Barkley NA, Zhao Y, Yuan M, Prakash C. Phylogenetic relationships of species of genus *Arachis* based on genic sequences. Genome. 2014;57(6):327–34.

Hilu K. Polyploidy and the evolution of domesticated plants. Am J Bot. 1993;80(12):1494–9.

Hohenlohe PA, Bassham S, Etter PD, Stiffler N, Johnson EA, Cresko WA. Population genomics of parallel adaptation in threespine stickleback using sequenced RAD tags. PLoS Genet. 2010;6(2):e1000862.

Hohenlohe PA, Hand BK, Andrews KR, Luikart G. Population genomics provides key insights in ecology and evolution. In: Rajora OP, editor. Population genomics. Cham: Springer; 2018. p. 483–510.

Holbrook CC, Culbreath AK. Registration of 'Tifrunner' peanut. J Plant Regist. 2007;1(2):124.

Holliday J, Hallerman E, Haak D. Genotyping and sequencing technologies in population genetics and genomics. In: Rajora OP, editor. Population genomics. Cham: Springer; 2018. p. 83–125.

Hong Y, Liang X-Q, Chen X-P, Liu H-Y, Zhou G-Y, Li S-X, et al. Construction of genetic linkage map based on SSR markers in peanut (*Arachis hypogaea* L.). Agril Sci China. 2008;7(8):915–21.

Hopkins MS, Casa AM, Wang T, Mitchell SE, Dean RE, Kochert GD, et al. Discovery and characterization of polymorphic simple sequence repeats (SSRs) in peanut. Crop Sci. 1999;39(4):1243–7.

Hu Z, Wei C, Li Z. Computational strategies for eukaryotic pangenome analyses. Pangenome. 2020:293–307.

Huai D, Xue X, Li Y, Wang P, Li J, Yan L, et al. Genome-wide identification of peanut KCS genes reveals that AhKCS1 and AhKCS28 are involved in regulating VLCFA contents in seeds. Front Plant Sci. 2020;11:406.

Huang L, Wu B, Zhao J, Li H, Chen W, Zheng Y, et al. Characterization and transferable utility of microsatellite markers in the wild and cultivated *Arachis* species. PLoS One. 2016;11(5):e0156633.

Huang J, Xing M, Li Y, Cheng F, Gu H, Yue C, et al. Comparative transcriptome analysis of the skin-specific accumulation of anthocyanins in black peanut (*Arachis hypogaea* L.). J Agric Food Chem. 2019;67(4):1312–24.

Huang L, Liu X, Pandey MK, Ren X, Chen H, Xue X, et al. Genome-wide expression quantitative trait locus analysis in a recombinant inbred line population for trait dissection in peanut. Plant Biotechnol J. 2020;18(3):779–90.

Hubner S, Bercovich N, Todesco M, Mandel JR, Odenheimer J, Ziegler E, et al. Sunflower pan-genome analysis shows that hybridization altered gene content and disease resistance. Nat Plants. 2019;5(1):54–62.

Hui-Fang J, Xiao-Ping R, Bo-Shou L, Jia-Quan H, Yong L, Ben-Yin C, et al. Peanut core collection established in China and compared with ICRISAT mini core collection. Acta Agronomica Sinica. 2008;34(1):25–30.

Jain S, Srivastava S, Sarin NB, Kav NN. Proteomics reveals elevated levels of PR 10 proteins in saline-tolerant peanut (*Arachis hypogaea*) calli. Plant Physiol Biochem. 2006;44(4):253–9.

Jiang C, Zhang H, Ren J, Dong J, Zhao X, Wang X, et al. Comparative transcriptome-based mining and expression profiling of transcription factors related to cold tolerance in peanut. Int J Mol Sci. 2020;21(6):1921.

Jiao Y, Zhao H, Ren L, Song W, Zeng B, Guo J, et al. Genome-wide genetic changes during modern breeding of maize. Nat Genet. 2012;44(7):812–5.

Johannes F, Schmitz RJ. Spontaneous epimutations in plants. New Phytol. 2019;221(3):1253–9.

Jonas E, Fikse F, Rönnegård L, Mouresan EF. Genomic selection. In: Rajora OP, editor. Population genomics. Cham: Springer; 2019. p. 427–80.

Kang H, Zhu D, Lin R, Opiyo SO, Jiang N, Shiu S-H, et al. A novel method for identifying polymorphic transposable elements via scanning of high-throughput short reads. DNA Res. 2016;23(3):241–51.

Katam R, Gottschalk V, Survajahala P, Brewster G, Paxton P, Sakata K, et al. Advances in proteomics research for peanut genetics and breeding. In: Genetics, genomics and breeding of peanuts. Boca Raton: CRC Press; 2014. p. 161–77.

Katam R, Sakata K, Suravajhala P, Pechan T, Kambiranda DM, Naik KS, et al. Comparative leaf proteomics of drought-tolerant and-susceptible peanut in response to water stress. J Proteomics. 2016;143:209–26.

Khan AW, Garg V, Roorkiwal M, Golicz AA, Edwards D, Varshney RK. Super-pangenome by integrating the wild side of a species for accelerated crop improvement. Trends Plant Sci. 2020;25(2):148–58.

Khera P, Upadhyaya HD, Pandey MK, Roorkiwal M, Sriswathi M, Janila P, et al. Single nucleotide polymorphism–based genetic diversity in the reference set of peanut (spp.) by developing and applying cost-effective Kompetitive Allele Specific Polymerase Chain Reaction genotyping assays. Plant Genome. 2013;6.

Kilian A. DArT-based whole genome profiling and novel information technologies in support system of modern breeding of groundnut. Proceedings of. 3rd international conference for peanut genomics and biotechnology on "Advances in Arachis through Genomics and Biotechnology" (AAGB), 4–8 November 2008, Patancheru, India: ICRISAT; 2008, p. 2.

Kochert G, Halward T, Branch WD, Simpson CE. RFLP variability in peanut (*Arachis hypogaea* L.) cultivars and wild species. Theor Appl Genet. 1991;81(5):565–70.

Kochert G, Stalker HT, Gimenes M, Galgaro L, Lopes CR, Moore K. RFLP and cytogenetic evidence on the origin and evolution of allotetraploid domesticated peanut, *Arachis hypogaea* (*Leguminosae*). Am J Bot. 1996;83(10):1282–91.

Kofler R, Gómez-Sánchez D, Schlötterer C. PoPoolationTE2: comparative population genomics of transposable elements using Pool-Seq. Mol Biol Evol. 2016;33(10):2759–64.

Kolekar RM, Sujay V, Shirasawa K, Sukruth M, Khedikar YP, Gowda MVC, et al. QTL mapping for late leaf spot and rust resistance using an improved genetic map and extensive phenotypic data on a recombinant inbred line population in peanut (*Arachis hypogaea* L.). Euphytica. 2016;209(1):147–56.

Kottapalli KR, Rakwal R, Shibato J, Burow G, Tissue D, Burke J, et al. Physiology and proteomics of the water-deficit stress response in three contrasting peanut genotypes. Plant Cell Environ. 2009;32(4):380–407.

Kottapalli KR, Zabet-Moghaddam M, Rowland D, Faircloth W, Mirzaei M, Haynes PA, et al. Shotgun label-free quantitative proteomics of water-deficit-stressed midmature peanut (*Arachis hypogaea* L.) seed. J Proteome Res. 2013;12(11):5048–57.

Krapovickas A. The origin, variability and spread of the groundnut (*Arachis hypogaea*). In: Ucko RJ, Dimbledy WC, editors. The domestication and exploitation of plant and animals. London: Greald Duckworth Co. Ltd.; 1969. p. 427–41.

Krapovickas A. Evolution of the genus *Arachis*. In: Moav R, editor. Agricultural genetics-selected topics. New York: Wiley; 1973. p. 135–51.

Krapovickas A, Gregory WC. Taxonomía del género *Arachis* (*Leguminosae*). Bonplandia. 1994;8:1–186.

Krapovickas A, Rigoni VA. Nuevas especies de *Arachis vinculadas* al problema del origen del mani. Darwiniana. 1957;11:431–55.

Lavia GI. Karyotypes of *Arachis palustris* and *A. praecox* (section *Arachis*), two species with basic chromosome number x= 9. Cytologia. 1998;63(2):177–81.

Leal-Bertioli SC, Bertioli DJ, Guimarães PM, Pereira TD, Galhardo I, Silva JP, et al. The effect of tetraploidization of wild *Arachis* on leaf morphology and other drought-related traits. Environ Exp Bot. 2012;84:17–24.

Leal-Bertioli SC, Moretzsohn MC, Santos SP, Brasileiro AC, Guimaraes PM, Bertioli DJ, et al. Phenotypic effects of allotetraploidization of wild *Arachis* and their implications for peanut domestication. Am J Bot. 2017;104(3):379–88.

Li J, Tang Y, Jacobson AL, Dang PM, Li X, Wang ML, et al. Population structure and association mapping to detect QTL controlling tomato spotted wilt virus resistance in cultivated peanuts. Crop J. 2018;6(5):516–26.

Li Y, Zhu FL, Zheng XW, Hu ML, Dong C, Diao Y, et al. Comparative population genomics reveals genetic divergence and selection in lotus, *Nelumbo nucifera*. BMC Genomics. 2020;21(1):146.

Liu H, Hong Y, Lu Q, Li H, Gu J, Ren L, et al. Integrated analysis of comparative lipidomics and proteomics reveals the dynamic changes of lipid molecular species in high-oleic acid peanut seed. J Agric Food Chem. 2020a;68(1):426–38.

Liu Y, Du H, Li P, Shen Y, Peng H, Liu S, et al. Pan-genome of wild and cultivated soybeans. Cell. 2020b;182(1):162–176 e113.

Lorant A, Ross-Ibarra J, Tebnailon M. Genomics of long- and short-term adaptation in maize and teosintes. Methods Mol Biol. 2020;2090:289–312.

Lu J, Pickersgill B. Isozyme variation and species relationships in peanut and its wild relatives (*Arachis* L. – Leguminosae). Theor Appl Genet. 1993;85(5):550–60.

Lu Q, Li H, Hong Y, Zhang G, Wen S, Li X, et al. Genome sequencing and analysis of the peanut B-genome progenitor (*Arachis ipaensis*). Front Plant Sci. 2018;9

Luikart G, Kardos M, Hand BK, Rajora OP, Aitken SN, Hohenlohe PA. Population genomics: advancing understanding of nature. In: Rajora OP, editor. Population genomics. Cham: Springer; 2019. p. 3–79.

Luo H, Xu Z, Li Z, Li X, Lv J, Ren X, et al. Development of SSR markers and identification of major quantitative trait loci controlling shelling percentage in cultivated peanut (*Arachis hypogaea* L.). Theor Appl Genet. 2017:1–14.

Lv Y, Zhang X, Luo L, Yang H, Li P, Zhang K, et al. Characterization of glycerol-3-phosphate acyltransferase 9 (AhGPAT9) genes, their allelic polymorphism and association with oil content in peanut (*Arachis hypogaea* L.). Sci Rep. 2020;10(1):1–15.

Ma X, Zhang X, Traore SM, Xin Z, Ning L, Li K, et al. Genome-wide identification and analysis of long noncoding RNAs (lncRNAs) during seed development in peanut (*Arachis hypogaea* L.). BMC Plant Biol. 2020;20(1):192.

Mamone G, Di Stasio L, De Caro S, Picariello G, Nicolai MA, Ferranti P. Comprehensive analysis of the peanut allergome combining 2-DE gel-based and gel-free proteomics. Food Res Int. 2019;116:1059–65.

Mattila TM, Laenen B, Slotte T. Population genomics of transitions to selfing in Brassicaceae model systems. Methods Mol Biol. 2020;2090:269–87.

McKain MR, Johnson MG, Uribe-Convers S, Eaton D, Yang Y. Practical considerations for plant phylogenomics. Appl Plant Sci. 2018;6(3):e1038.

Moler ER, Abakir A, Eleftheriou M, Johnson JS, Krutovsky KV, Lewis LC, et al. Population epigenomics: advancing understanding of phenotypic plasticity, acclimation, adaptation and diseases. In: Rajora OP, editor. Population genomics. Cham: Springer; 2018. p. 179–260.

Moretzsohn M, Hopkins M, Mitchell S, Kresovich S, Valls J, Ferreira M. Genetic diversity of peanut (*Arachis hypogaea* L.) and its wild relatives based on the analysis of hypervariable regions of the genome. BMC Plant Biol. 2004;4(1):11.

Moretzsohn MC, Leoi L, Proite K, Guimaraes PM, Leal-Bertioli SCM, Gimenes MA, et al. A microsatellite-based, gene-rich linkage map for the AA genome of *Arachis* (*Fabaceae*). Theor Appl Genet. 2005;111(6):1060–71.

Muralidharan S, Poon YY, Wright GC, Haynes PA, Lee NA. Quantitative proteomics analysis of high and low polyphenol expressing recombinant inbred lines (RILs) of peanut (*Arachis hypogaea* L.). Food Chem. 2020;334:127517.

Nawade B, Mishra GP, Radhakrishnan T, Dodia SM, Ahmad S, Kumar A, et al. High oleic peanut breeding: achievements, perspectives, and prospects. Trends Food Sci Technol. 2018;78:107–19.

Otyama PI, Wilkey A, Kulkarni R, Assefa T, Chu Y, Clevenger J, et al. Evaluation of linkage disequilibrium, population structure, and genetic diversity in the U.S. peanut mini core collection. BMC Genomics. 2019;20(1):481.

Ozias-Akins P, Cannon EK, Cannon SB. Genomics resources for peanut improvement. In: Varshney R, Pandey M, Puppala N, editors. The peanut genome, Compendium of plant genomes. Cham: Springer; 2017. p. 69–91.

Paik-Ro OG, Smith RL, Knauft DA. Restriction fragment length polymorphism evaluation of six peanut species within the *Arachis* section. Theor Appl Genet. 1992;84(1):201–8.

Palmieri DA, Hoshino A, Bravo J, Lopes C, Gimenes M. Isolation and characterization of microsatellite loci from the forage species *Arachis pintoi* (Genus *Arachis*). Mol Ecol Notes. 2002;2(4):551–3.

Palmieri D, Bechara M, Curi R, Gimenes M, Lopes C. Novel polymorphic microsatellite markers in section Caulorrhizae (*Arachis*, Fabaceae). Mol Ecol Notes. 2005;5(1):77–9.

Pandey MK, Monyo E, Ozias-Akins P, Liang X, Guimaraes P, Nigam SN, et al. Advances in *Arachis* genomics for peanut improvement. Biotechnol Adv. 2012;30(3):639–51.

Pandey MK, Upadhyaya HD, Rathore A, Vadez V, Sheshshaye MS, Sriswathi M, et al. Genomewide association studies for 50 agronomic traits in peanut using the reference set comprising 300 genotypes from 48 countries of semi-arid tropics of the world. PLoS One. 2014;9(8):e105228.

Pandey MK, Agarwal G, Kale SM, Clevenger J, Nayak SN, Sriswathi M, et al. Development and evaluation of a high density genotyping 'Axiom_Arachis' array with 58 K SNPs for accelerating genetics and breeding in groundnut. Sci Rep. 2017;7:40577.

Pandey MK, Chaudhari S, Jarquin D, Pasupuleti J, Crossa J, Patil SC, et al. Genome-based trait prediction in multi-environment breeding trials in groundnut. Theor Appl Genet. 2020a;133:3101–17. https://doi.org/10.1007/s00122-020-03658-1.

Pandey MK, Pandey AK, Kumar R, Nwosu CV, Guo B, Wright GC, et al. Translational genomics for achieving higher genetic gains in post-genome era in groundnut. Theor Appl Genet. 2020b;133(5):1679–702.

Paterson AH, Stalker HT, Gallo-Meagher M, Burrow M, Dwivedi SL, Crouch JH, et al. Genomics and genetic enhancement of peanut. Legume Crop Genomics. 2004:97–109.

Peñaloza APS, Valls JFM. Contagen do número cromossômico en assos de *Arachis decora* (Legumonsae). In: Vega RFA, Bovi MLA, Betti JA, Voltan RBQ, editors. Simpósio Latino Slericanno de Recursos Genéticos Vegetais. Campanias: IAC/Embrapa-Cenargen; 1997. p. 21.

Peñaloza APS, Valls JFM. Chromosome number and satellited chromosome morphology of eleven species of *Arachis* (Leguminosae). Bonplandia. 2005;14(15):65–72.

Peng Z, Gallo M, Tillman BL, Rowland D, Wang J. Molecular marker development from transcript sequences and germplasm evaluation for cultivated peanut (*Arachis hypogaea* L.). Mol Genet Genomics. 2016;291(1):363–81.

Peng Z, Liu F, Wang L, Zhou H, Paudel D, Tan L, et al. Transcriptome profiles reveal gene regulation of peanut (*Arachis hypogaea* L.) nodulation. Sci Rep. 2017;7(1):1–12.

Peng Z, Zhao Z, Clevenger JP, Chu Y, Paudel D, Ozias-Akins P, et al. Comparison of SNP calling pipelines and NGS platforms to predict the genomic regions harboring candidate genes for nodulation in cultivated peanut. Front Genet. 2020;11:222.

Pickrell J, Pritchard J. Inference of population splits and mixtures from genome-wide allele frequency data. Nat Precedings. 2012:1.

Proite K, Leal-Bertioli SC, Bertioli DJ, Moretzsohn MC, da Silva FR, Martins NF, et al. ESTs from a wild *Arachis* species for gene discovery and marker development. BMC Plant Biol. 2007;7 (1):7.

Rajora OP. Population genomics: concepts, approaches and applications. Cham: Springer Nature Switzerland AG; 2019.

Rathi D, Gayen D, Gayali S, Chakraborty S, Chakraborty N. Legume proteomics: progress, prospects, and challenges. Proteomics. 2016;16(2):310–27.

Rathod V, Hamid R, Tomar RS, Patel R, Padhiyar S, Kheni J, et al. Comparative RNA-Seq profiling of a resistant and susceptible peanut (*Arachis hypogaea*) genotypes in response to leaf rust infection caused by *Puccinia arachidis*. 3 Biotech. 2020;10(6):284.

Ravi K, Hari U, Sangam D, David H, Varshney RK. Genetic relationships among seven sections of genus *Arachis* studied by using SSR markers. BMC Plant Biol. 2010;10(1):15.

Ren X, Jiang H, Yan Z, Chen Y, Zhou X, Huang L, et al. Genetic diversity and population structure of the major peanut (*Arachis hypogaea* L.) cultivars grown in China by SSR markers. PLoS One. 2014;9(2):e88091.

Ruan J, Guo F, Wang Y, Li X, Wan S, Shan L, et al. Transcriptome analysis of alternative splicing in peanut (*Arachis hypogaea* L.). BMC Plant Biol. 2018;18(1):139.

Salman-Minkov A, Sabath N, Mayrose I. Whole-genome duplication as a key factor in crop domestication. Nat Plants. 2016;2(8):1–4.

Shen C, Wang N, Huang C, Wang M, Zhang X, Lin Z. Population genomics reveals a fine-scale recombination landscape for genetic improvement of cotton. Plant J. 2019;99(3):494–505.

Shirasawa K, Hirakawa H, Tabata S, Hasegawa M, Kiyoshima H, Suzuki S, et al. Characterization of active miniature inverted-repeat transposable elements in the peanut genome. Theor Appl Genet. 2012a;124(8):1429–38.

Shirasawa K, Koilkonda P, Aoki K, Hirakawa H, Tabata S, Watanabe M, et al. *In silico* polymorphism analysis for the development of simple sequence repeat and transposon markers and construction of linkage map in cultivated peanut. BMC Plant Biol. 2012b;12(1):80.

Shirasawa K, Bertioli DJ, Varshney RK, Moretzsohn MC, Leal-Bertioli SC, Thudi M, et al. Integrated consensus map of cultivated peanut and wild relatives reveals structures of the A and B genomes of *Arachis* and divergence of the legume genomes. DNA Res. 2013;20 (2):173–84.

Singh AK. Putative genome donors of *Arachis hypogaea* (*Fabaceae*), evidence from crosses with synthetic amphidiploids. Plant Syst Evol. 1988;160(3):143–51.

Sinha P, Bajaj P, Pazhamala LT, Nayak SN, Pandey MK, Chitikineni A, et al. *Arachis hypogea* gene expression atlas (AhGEA) for *fastigiata* subspecies of cultivated groundnut to accelerate functional and translational genomics applications. Plant Biotechnol J. 2020; https://doi.org/10.1111/pbi.13374.

Smartt J, Stalker HT. Speciation and cytogenetics in *Arachis*. In: Pattee HE, Young CE, editors. Peanut science and technology. Yoakum: American Peanut Research and Education Society; 1982. p. 21–45.

Soltis DE, Soltis PS, Tate JA. Advances in the study of polyploidy since plant speciation. New Phytol. 2004;161:173–91.

Song JM, Liu DX, Xie WZ, Yang Z, Guo L, Liu K, et al. BnPIR: Brassica napus Pan-genome information resource for 1,689 accessions. Plant Biotechnol J. 2020; https://doi.org/10.1111/pbi.13491.

Stalker HT. Utilizing *Arachis* germplasm resources. In: Nigam SN, editor. Groundnut, a global perspective. Patancheru: ICRISAT; 1992. p. 281–96.

Stalker HT. Utilizing wild species for peanut improvement. Crop Sci. 2017;57(3):1102–20.

Stalker HT, Simpson CE. Genetic resources in *Arachis*. In: He P, Stalker HT, editors. Advances in peanut science. Stillwatter: APRES Inc.; 1995. p. 14–53.

Stalker H, Phillips T, Murphy J, Jones T. Variation of isozyme patterns among *Arachis* species. Theor Appl Genet. 1994;87(6):746–55.

Stalker HT, Weissinger AK, Milla-Lewis S, Holbrook CC. Genomics: an evolving science in peanut. Peanut Sci. 2009;36(1):2–10.

Stalker H, Tallury S, Ozias-Akins P, Bertioli D, Bertioli SL. The value of diploid peanut relatives for breeding and genomics. Peanut Sci. 2013;40(2):70–88.

Stalker HT, Tallury SP, Seijo GR, Leal-Bertioli SC. Biology, speciation, and utilization of peanut species. In: Stalker HT, Wilson RF, editors. Peanuts: genetics, processing, and utilization. Champaign: AOCS Press; 2016. p. 27–66.

Stinchcombe JR, Hoekstra HE. Combining population genomics and quantitative genetics: finding the genes underlying ecologically important traits. Heredity. 2008;100(2):158–70.

Sun Y, Wang Q, Li Z, Hou L, Dai S, Liu W. Comparative proteomics of peanut gynophore development under dark and mechanical stimulation. J Proteome Res. 2013;12(12):5502–11.

Suzuki MM, Bird A. DNA methylation landscapes: provocative insights from epigenomics. Nat Rev Genet. 2008;9(6):465.

Toomer OT. A comprehensive review of the value-added uses of peanut (*Arachis hypogaea*) skins and by-products. Crit Rev Food Sci Nutr. 2020;60(2):341–50.

Valls J, Simpson C. Taxonomy, natural distribution, and attributes of *Arachis*. Biol Agron Forage Arachis. 1994:1–18.

Valls JF, Simpson CE. New species of *Arachis* (*Leguminosae*) from Brazil, Paraguay and Bolivia. Bonplandia. 2005;14:35–63.

Valls JFM, Da Costa LC, Custodio AR. A novel trifoliolate species of *Arachis* (*Fabaceae*) and further comments on the taxonomic section Trierectoides. Bonplandia. 2013:91–7.

Vaughn MW, Tanurdžić M, Lippman Z, Jiang H, Carrasquillo R, Rabinowicz PD, et al. Epigenetic natural variation in *Arabidopsis thaliana*. PLoS Biol. 2007;5(7):e174.

Vishwakarma MK, Nayak LSN, Guo B, Wan L, Liao B, Varshney RK, et al. Classical and molecular approaches for mapping of genes and quantitative trait loci in peanut (*Arachis hypogaea* L.). In: Varshney RK, Pandey MK, Puppala N, editors. The peanut genome. New York: Springer; 2017. p. 93–116.

Wan L, Ren W, Miao H, Zhang J, Fang J. Genome-wide identification, expression, and association analysis of the monosaccharide transporter (MST) gene family in peanut (*Arachis hypogaea* L.). 3 Biotech. 2020;10(3):130.

Wang X, Su L, Quan X, Shan L, Zhang H, Bi Y. Peanut (*Arachis hypogaea* L.) EST sequencing, gene cloning an Agrobacteria-mediated transformation. Proceedings of the international groundnut conference on groundnut aflatoxin and genomics. 2006. p. 59–60.

Wang CT, Wang XZ, Tang YY, Chen DX, Cui FG, Zhang JC, et al. Phylogeny of *Arachis* based on internal transcribed spacer sequences. Genet Resour Crop Evol. 2011a;58(2):311–9.

Wang ML, Sukumaran S, Barkley NA, Chen Z, Chen CY, Guo B, et al. Population structure and marker–trait association analysis of the US peanut (*Arachis hypogaea* L.) mini-core collection. Theor Appl Genet. 2011b;123(8):1307–17.

Wang H, Khera P, Huang B, Yuan M, Katam R, Zhuang W, et al. Analysis of genetic diversity and population structure of peanut cultivars and breeding lines from China, India and the US using simple sequence repeat markers. J Integr Plant Biol. 2016a;58(5):452–65.

Wang P, Gao C, Bian X, Zhao S, Zhao C, Xia H, et al. Genome-wide identification and comparative analysis of cytosine-5 DNA methyltransferase and demethylase families in wild and cultivated peanut. Front Plant Sci. 2016b;7:7.

Wang J, Li C, Yan C, Zhao X, Shan S. A comparative analysis of the complete chloroplast genome sequences of four peanut botanical varieties. PeerJ. 2018;6:e5349.

Wang J, Yan C, Li Y, Li C, Zhao X, Yuan C, et al. GWAS discovery of candidate genes for yield-related traits in peanut and support from earlier QTL mapping studies. Genes. 2019;10(10):803.

Wang F, Nong X, Hao K, Cai N, Wang G, Liu S, et al. Identification of the key genes involved in the regulation of symbiotic pathways induced by Metarhizium anisopliae in peanut (*Arachis hypogaea*) roots. 3 Biotech. 2020a;10(3):124.

Wang X, Xu P, Ren Y, Yin L, Li S, Wang Y, et al. Genome-wide identification of meiotic recombination hot spots detected by SLAF in peanut (*Arachis hypogaea* L.). Sci Rep. 2020b;10(1):1–11.

Wendel JF. The wondrous cycles of polyploidy in plants. Am J Bot. 2015;102(11):1753–6.

Xia H, Zhao C, Hou L, Li A, Zhao S, Bi Y, et al. Transcriptome profiling of peanut gynophores revealed global reprogramming of gene expression during early pod development in darkness. BMC Genomics. 2013;14(1):517.

Xia H, Zhu L, Zhao C, Li K, Shang C, Hou L, et al. Comparative transcriptome analysis of anthocyanin synthesis in black and pink peanut. Plant Signal Behav. 2020;15(2):1721044.

Xie C, Tammi MT. CNV-seq, a new method to detect copy number variation using high-throughput sequencing. BMC Bioinformatics. 2009;10(1):80.

Xu P, Tang G, Cui W, Chen G, Ma CL, Zhu J, et al. Transcriptional differences in peanut (*Arachis hypogaea* L.) seeds at the freshly harvested, after-ripening and newly germinated seed stages: insights into the regulatory networks of seed dormancy release and germination. PLoS One. 2020;15(1):e0219413.

Yang S, Li L, Zhang J, Geng Y, Guo F, Wang J, et al. Transcriptome and differential expression profiling analysis of the mechanism of Ca^{2+} regulation in peanut (*Arachis hypogaea*) pod development. Front Plant Sci. 2017;8:1609.

Yeri SB, Bhat RS. Development of late leaf spot and rust resistant backcross lines in JL 24 variety of groundnut (*Arachis hypogaea* L.). Electron J Plant Breed. 2016;7(1):37–41.

Yin D, Wang Y, Zhang X, Li H, Lu X, Zhang J, et al. De novo assembly of the peanut (*Arachis hypogaea* L.) seed transcriptome revealed candidate unigenes for oil accumulation pathways. PLoS One. 2013;8(9):e73767.

Yin D, Ji C, Ma X, Li H, Zhang W, Li S, et al. Genome of an allotetraploid wild peanut *Arachis monticola*: a *de novo* assembly. GigaScience. 2018;7(6):giy066.

Yin D, Ji C, Song Q, Zhang W, Zhang X, Zhao K, et al. Comparison of *Arachis monticola* with diploid and cultivated tetraploid genomes reveals asymmetric subgenome evolution and improvement of peanut. Adv Sci. 2020;7(4):1901672.

Yu R, Ma Y, Li Y, Li X, Liu C, Du X, et al. Comparative transcriptome analysis revealed key factors for differential cadmium transport and retention in roots of two contrasting peanut cultivars. BMC Genomics. 2018;19(1):938.

Yu R, Jiang Q, Xv C, Li L, Bu S, Shi G. Comparative proteomics analysis of peanut roots reveals differential mechanisms of cadmium detoxification and translocation between two cultivars differing in cadmium accumulation. BMC Plant Biol. 2019;19(1):137.

Yuan M, Stinchcombe JR. Population genomics of parallel adaptation. Mol Ecol. 2020;29:4033–6.

Zemach A, McDaniel IE, Silva P, Zilberman D. Genome-wide evolutionary analysis of eukaryotic DNA methylation. Science. 2010;328(5980):916–9.

Zhang J, Liang S, Duan J, Wang J, Chen S, Cheng Z, et al. De novo assembly and characterisation of the transcriptome during seed development, and generation of genic-SSR markers in peanut (*Arachis hypogaea* L.). BMC Genomics. 2012a;13:90.

Zhang J, Liang S, Duan J, Wang J, Chen S, Cheng Z, et al. *De novo* assembly and characterisation of the transcriptome during seed development, and generation of genic-SSR markers in peanut (*Arachis hypogaea* L.). BMC Genomics. 2012b;13(1):90.

Zhang T, Hu Y, Jiang W, Fang L, Guan X, Chen J, et al. Sequencing of allotetraploid cotton (*Gossypium hirsutum* L. acc. TM-1) provides a resource for fiber improvement. Nat Biotechnol. 2015;33(5):531–7.

Zhang Y, Wang P, Xia H, Zhao C, Hou L, Li C, et al. Comparative transcriptome analysis of basal and zygote-located tip regions of peanut ovaries provides insight into the mechanism of light regulation in peanut embryo and pod development. BMC Genomics. 2016;17(1):606.

Zhang X, Zhang J, He X, Wang Y, Ma X, Yin D. Genome-wide association study of major agronomic traits related to domestication in peanut. Front Plant Sci. 2017;8:1611.

Zhang D, Zhang H, Hu Z, Chu S, Yu K, Lv L, et al. Artificial selection on GmOLEO1 contributes to the increase in seed oil during soybean domestication. PLoS Genet. 2019a;15(7):e1008267.

Zhang H, Wang ML, Schaefer R, Dang P, Jiang T, Chen C. GWAS and coexpression network reveal ionomic variation in cultivated peanut. J Agric Food Chem. 2019b;67(43):12026–36.

Zhang H, Chu Y, Dang P, Tang Y, Jiang T, Clevenger JP, et al. Identification of QTLs for resistance to leaf spots in cultivated peanut (*Arachis hypogaea* L.) through GWAS analysis. Theor Appl Genet. 2020a;133(7):2051–61.

Zhang Z, Gou X, Xun H, Bian Y, Ma X, Li J, et al. Homoeologous exchanges occur through intragenic recombination generating novel transcripts and proteins in wheat and other polyploids. Proc Natl Acad Sci. 2020b;117(25):14561–71.

Zhao C, Zhao S, Hou L, Xia H, Wang J, Li C, et al. Proteomics analysis reveals differentially activated pathways that operate in peanut gynophores at different developmental stages. BMC Plant Biol. 2015;15(1):188.

Zhao C, Li T, Zhao Y, Zhang B, Li A, Zhao S, et al. Integrated small RNA and mRNA expression profiles reveal miRNAs and their target genes in response to *Aspergillus flavus* growth in peanut seeds. BMC Plant Biol. 2020a;20(1):215.

Zhao N, He M, Li L, Cui S, Hou M, Wang L, et al. Identification and expression analysis of WRKY gene family under drought stress in peanut (*Arachis hypogaea* L.). PLoS One. 2020b;15(4):e0231396.

Zheng L, Shockey J, Guo F, Shi L, Li X, Shan L, et al. Discovery of a new mechanism for regulation of plant triacylglycerol metabolism: the peanut diacylglycerol acyltransferase-1 gene family transcriptome is highly enriched in alternative splicing variants. J Plant Physiol. 2017;219:62–70.

Zheng Z, Sun Z, Fang Y, Qi F, Liu H, Miao L, et al. Genetic diversity, population structure, and botanical variety of 320 global peanut accessions revealed through tunable genotyping-by-sequencing. Sci Rep. 2018;8(1):14500.

Zhou X, Dong Y, Zhao J, Huang L, Ren X, Chen Y, et al. Genomic survey sequencing for development and validation of single-locus SSR markers in peanut (*Arachis hypogaea* L.). BMC Genomics. 2016;17(1):420.

Zhou Y, Chebotarov D, Kudrna D, Llaca V, Lee S, Rajasekar S, et al. A platinum standard pan-genome resource that represents the population structure of Asian rice. Sci Data. 2020;7(1):113.

Zhu W, Zhang E, Li H, Chen X, Zhu F, Hong Y, et al. Comparative proteomics analysis of developing peanut aerial and subterranean pods identifies pod swelling related proteins. J Proteomics. 2013;91:172–87.

Zhu W, Chen X, Li H, Zhu F, Hong Y, Varshney RK, et al. Comparative transcriptome analysis of aerial and subterranean pods development provides insights into seed abortion in peanut. Plant Mol Biol. 2014;85(4–5):395–409.

Zhuang W, Chen H, Yang M, Wang J, Pandey MK, Zhang C, et al. The genome of cultivated peanut provides insight into legume karyotypes, polyploid evolution and crop domestication. Nat Genet. 2019;51(5):865–76.

Zhuang W, Wang X, Paterson AH, Chen H, Yang M, Zhang C, et al. Reply to: evaluating two different models of peanut's origin. Nat Genet. 2020:1–4.

Population Genomics of Yams: Evolution and Domestication of *Dioscorea* Species

Yu Sugihara, Aoi Kudoh, Muluneh Tamiru Oli, Hiroki Takagi, Satoshi Natsume, Motoki Shimizu, Akira Abe, Robert Asiedu, Asrat Asfaw, Patrick Adebola, and Ryohei Terauchi

Abstract Yam is a collective name of tuber crops belonging to the genus *Dioscorea*. Yam is important not only as a staple food crop but also as an integral component of society and culture of the millions of people who depend on it. However, due to its regional importance, yam has long been regarded as an "orphan crop" lacking a due global attention. Although this perception is changing with recent advances in genomics technologies, domestication processes of most yam species are still ambiguous. This is mainly due to the complicated evolutionary history of *Dioscorea* species caused by frequent hybridization and polyploidization, which is possibly caused by dioecy that imposed obligate outcrossing to the species of *Dioscorea*. In this chapter, we provide an overview of the evolution of *Dioscorea* and address the domestication of yam from population genomics perspectives by

---

Yu Sugihara and Aoi Kudoh contributed equally to this work.

Y. Sugihara · A. Kudoh
Laboratory of Crop Evolution, Graduate School of Agriculture, Kyoto University, Kyoto, Japan

M. T. Oli
Iwate Biotechnology Research Center, Kitakami, Iwate, Japan

Department of Animal, Plant, and Soil Sciences, School of Life Sciences, La Trobe University, Melbourne, VIC, Australia

H. Takagi
Iwate Biotechnology Research Center, Kitakami, Iwate, Japan

Ishikawa Prefectural University, Nonoichi, Ishikawa, Japan

S. Natsume · M. Shimizu · A. Abe
Iwate Biotechnology Research Center, Kitakami, Iwate, Japan

R. Asiedu · A. Asfaw · P. Adebola
International Institute of Tropical Agriculture (IITA), Ibadan, Nigeria

R. Terauchi (✉)
Laboratory of Crop Evolution, Graduate School of Agriculture, Kyoto University, Kyoto, Japan

Iwate Biotechnology Research Center, Kitakami, Iwate, Japan
e-mail: terauchi@ibrc.or.jp

focusing on the processes of hybridization and polyploidization. A review is given to the recent population genomics studies on the hybrid origin of *D. rotundata* in West and Central Africa, the global dispersion of *D. alata* through human migrations, and the whole-genome duplication of the South America species of *D. trifida*. In the end, we give a summary of current understanding of sex-determination system in *Dioscorea*.

Keywords Dioecy · *Dioscorea* · Domestication · Evolution · Genetic diversity · Hybridization · Polyploidy · Population genomics · Sex determination · Yam

1 Introduction

Yam is a collective name of tuber crops belonging to the genus *Dioscorea*. In 2018, the global yam production was around 72.6 million tons (FAOSTAT 2018). The major yam species include *Dioscorea rotundata*, *D. alata*, *D. trifida*, *D. polystachya*, and *D. esculenta* (Arnau et al. 2010). White Guinea yam (*D. rotundata*) is the most important yam worldwide, mainly grown in West and Central Africa, especially Côte d'Ivoire, Ghana, Togo, Benin, Nigeria and Cameroon, the region known as the "yam belt", which accounts for ∼92.5% of the total world yam production (FAOSTAT 2018). Yam is a staple crop in many tropical countries, and it also plays important roles in society and culture of the people in the major yam-growing regions (Coursey 1972; Obidiegwu and Akpabio 2017; Obidiegwu et al. 2020). However, due to its localized importance, yam has been regarded as an "orphan crop" and received considerably less research attention compared to the major crop species.

The genetic improvement of yam is urgently needed for the food security of yam-growing regions, but it is constrained by various abiotic and biotic factors (Mignouna et al. 2003). For example, the entire genus *Dioscorea* is characterized by dioecy, with male and female flowers borne on separate individuals, which imposes obligate outcrossing to the species in the genus. Due to its dioecy, farmers clonally propagate yams to maintain its germplasms, and true seeds are rarely used as the starting materials for planting. However, this clonal propagation reduces the genetic diversity, which causes the vulnerability to plant diseases. Also, the clonal propagation causes the difficulty of purging deleterious mutations from the germplasms like in cassava (Ramu et al. 2017). To achieve effective yam improvement by overcoming these constraints, we need to answer key questions in yam genetics and genomics including: (1) what is the genetic relationships between cultivated yams and their wild relatives, and how the domestication of yam happened? (2) how to deploy *Dioscorea* genetic diversity to improve agronomic traits of cultivated yams? (3) how dioecy of *Dioscorea* is genetically controlled and how we can manipulate it to make an efficient cross breeding? Thanks to the recent development of genome sequencing technologies, we can now address these questions using population genomics approaches. In this chapter, we review the latest findings of the domestication of yam from population genomics perspectives.

2 The Genus *Dioscorea*: Its Origin and Botanical Characteristics

The genus *Dioscorea*, which consists of approximately 630 species, is the largest one in the family Dioscoreaceae of monocotyledons (WCSP 2020). It is widely distributed in the tropical and temperate regions and occurs in diverse environments from forests to grasslands (Wilkin et al. 2005; Maurin et al. 2016; Viruel et al. 2016). Several studies have been conducted on the phylogenetic relationships of species in *Dioscorea*. Previously, intrageneric taxa have been proposed based on morphological characters (Uline 1898; Knuth 1924; Prain and Burkill 1936, 1939; Burkill 1960; Huber 1998). However, diagnostic keys and delineation of taxa varied according to the authors. Recently, phylogenetic analyses have been conducted based on chloroplast DNA (cpDNA) sequences and nuclear gene sequences (for review, see Noda et al. 2020). Noda et al. (2020) provided a large-scale phylogenetic tree containing 183 species and proposed dividing *Dioscorea* into two subgenera (*Dioscorea* and *Helmia*), with 11 major clades and 27 sections/species groups.

Dioscorea likely originated in the Laurasian Palaearctic between the Late Cretaceous and the Early Eocene (Fig. 1). In the Eocene and Oligocene, *Dioscorea* expanded to the southern region by long-distance dispersal or migration by land bridges. In the Oligocene and Miocene, main *Dioscorea* lineages experienced divergence events on a worldwide scale. In the Miocene and Pliocene, some lineages dispersed into new areas. The number of biogeographical speciation events seems to have decreased after the Quaternary period began (Maurin et al. 2016; Viruel et al. 2016; Couto et al. 2018).

The majority of *Dioscorea* species are perennial herbaceous climbers with simple or compound leaves and reproduce sexually and/or clonally (Fig. 2). Flowers in *Dioscorea* are mostly dioecious with male and female flowers borne on separate individuals, and multiple sex-determination systems were reported in the genus (see Table 2 and Sect. 6). Most species produce winged seeds and capsular, six-seeded fruits, while some species have wingless seeds, samaroid or berry fruits (Caddick et al. 2002; Noda et al. 2020). In addition to sexual reproduction, *Dioscorea* species propagate clonally by bulbils, rhizomes, or tubers. Bulbils are aerial tubers that are formed in the axils of leaves or bracts of some *Dioscorea* species (Fig. 2f). They are mainly consumed as food, but also used as folk medicine in many cultures (Ikiriza et al. 2019). Bulbils are generally brown-colored and have small tubercles over their surface, but their shape and size vary in the different species (Murty and Purnima 1983). *D. bulbifera* (also known as aerial yam) is the major bulbil-producing species and is characterized by considerable bulbil shape diversity (Terauchi et al. 1991).

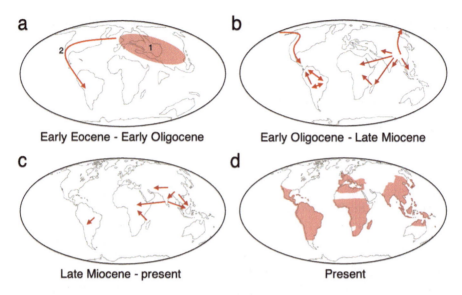

Fig. 1 Biogeographical origin and distribution of *Dioscorea* species (Viruel et al. 2016). (**a**) *Dioscorea* likely originated in the Laurasian Palaearctic in the Late Cretaceous and the Early Eocene (1) and then dispersed from Asia to South America (2). (**b**) In the Oligocene and Miocene, *Dioscorea* mainly expanded to the southern region. (**c**) Some lineages dispersed into new areas in the Miocene and Pliocene, but speciation events decreased in the Quaternary. (**d**) Geographical distribution in the present era. (Maps are based on C. R. Scotese's PALEOMAP project; www.scotese.com)

Fig. 2 Morphological diversity of the above-ground parts of *Dioscorea* species. (**a**) *D. tokoro*, (**b**) *D. quinqueloba*, (**c**) *D. rotundata*, (**d**) a stem of *D. mangenotiana* with thorns, (**e**) flowers of *D. japonica*, (**f**) a bulbil of *D. bulbifera*

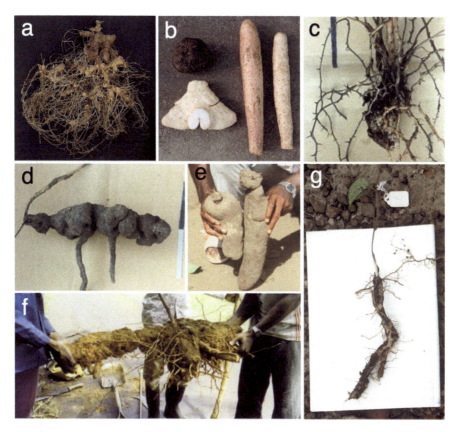

Fig. 3 Rhizomes and tubers of *Dioscorea* species. (**a**) rhizomes of *D. tokoro*, (**b**) Tukuneimo group (left top), Ichoimo group (left bottom), Nagaimo group (right) in *D. polystachya*, (**c**) *D. praehensilis*, (**d**) *D. minutiflora*, (**e**) *D. rotundata* (left), *D. cayenensis* (right), (**f**) *D. mangenotiana*, (**g**) *D. abyssinica*

Rhizomes and tubers represent morphologically diverse structures that serve as underground starch storage organs (Fig. 3). Because these storage organs serve as food sources for various wild animals, they have evolved defense traits. For example, *D. praehensilis* has crown roots with spines to protect tubers from burrowing or digging animals (Fig. 3c). Some species of the African clade have thick corky barks covering the pachycaul structure that may provide protection against fire and herbivores (Scott 1897; Maurin et al. 2016). In addition, *Dioscorea* species produce diverse secondary metabolites such as saponins, alkaloids, and tannins that serve a variety of functions including defense against herbivores (Coursey 1967). Chemical components of some species have medicinal values (Liu et al. 2008; Dutta 2015).

3 Ploidy in *Dioscorea*

Basic chromosome number of *Dioscorea* is $\times = 10$ in the section Stenophora. However, in the section Enantiophyllum that includes the major cultivated species, the basic chromosome number is $\times = 20$ (Scarcelli et al. 2005; Arnau et al. 2009). Our survey of the available literature shows that at least 33% of the species in the genus *Dioscorea* have variable ploidy levels including within the same species (Table 1). As we discuss in the later sections, *Dioscorea* species have undergone several hybridization events that sometimes led to changes in the ploidy levels. Information on ploidy is important to apply the correct population genomics approaches for the study of *Dioscorea*. For example, if the sequence reads obtained from a polyploid individual are aligned to a consensus haploid reference genome, variant calling based on the assumption of diploid genome may lead to wrong genotyping results.

Table 1 The various ploidy levels of *Dioscorea* species

Species[a]	Chromosome number	References
D. abyssinica Hochst.	2n = 40	Miège (1954)
D. aculeata Webstar	2n = 40	Martin and Ortiz (1963)
D. alata L.	2n = 40, 60, 80	Arnau et al. (2009)
D. alata var. *purpurea* Roxb.	2n = 80	Ramachandran (1968)
D. althaeoides Knuth	2n = 20	Kiangsu Institute of Botany (1976)
D. asclepiadea Prain & Burkill	2n = 40	Takeuchi et al. (1970)
D. aspersa Prain & Burkill	2n = 40	Chin et al. (1985)
D. balcanica Košanin	2n = 20	Miège (1954)
D. banzhuana Pei & Ting	2n = 20	Kiangsu Institute of Botany (1976)
D. batatas Decne.	2n = 140, 144	Nakajima (1933), Smith (1937)
D. belophylla Voigt	2n = 50, 80, 100	Raghavan (1958, 1959)
D. benthamii Prain & Burkill	2n = c. 100	Chin et al. (1985)
D. bernoulliana Prain & Burkill	2n = 36	Martin and Ortiz (1966)
D. biformifolia Pei & Ting	2n = 20	Kiangsu Institute of Botany (1976)
D. bulbifera L.	2n = 36, 40, 54, 60, 80, 98–100	Smith (1937), Miège (1954), Raghavan (1958)
D. bulbifera var. *bulbifera* L.	2n = 40	Paul and Debnath (2019)
D. caucasica Lipsky	2n = 20	Meurman (1925)
D. cayenensis Lam.	2n = 36, 54, 80, 140	Smith (1937), Miège (1954), Martin and Ortiz (1963)
D. ceratandra Uline	2n = 36	Martin and Ortiz (1963)
D. chingii Prain & Burkill	2n = 20	Kiangsu Institute of Botany (1976)

(continued)

Table 1 (continued)

Species[a]	Chromosome number	References
D. cirrhosa Lour.	2n = 40	Chin et al. (1985)
D. cirrhosa var. *cylindrica* Ting & Chang	2n = 40	Chin et al. (1985)
D. collettii Hook.f.	2n = 20	Kiangsu Institute of Botany (1976)
D. collettii var. *hypoglauca* (Palibin) Pei & Ting	2n = 40	Kiangsu Institute of Botany (1976)
D. composita Hemsl.	2n = 36, 54	Martin and Ortiz (1963, 1966)
D. convolvulacea Schlect. & Cham.	2n = 36	Martin and Ortiz (1963)
D. deltoidea Wall.	2n = 20, 40	Raghavan (1958), Mehra and Sachdeva (1976)
D. discolor	2n = 40	Smith (1937)
D. dumetorum Pax	2n = 36, 40, 45, 54	Miège (1954), Baquar (1980)
D. escuintlensis Matuda	2n = 36	Martin and Ortiz (1966)
D. esculenta Burkill	2n = 40, 60, 80, 90, 100	Miège (1954), Cox et al. (1958), Raghavan (1958), Chin et al. (1985)
D. esculenta (Cherukizhengu)	2n = 100	Ramachandran (1968)
D. esculenta (Nanakizhengu)	2n = 90	Ramachandran (1968)
D. exalata Ting & Chang	2n = 60, 80	Chin et al. (1985)
D. fargesii Franch.	2n = 64	Smith (1937)
D. floribunda Mart. & Gal.	2n = 36, 54, 72, 144	Martin and Ortiz (1963, 1966)
D. fordii Prain & Burkill	2n = 40	Chin et al. (1985)
D. friedrichsthalii R.Knuth	2n = 36	Martin and Ortiz (1963)
D. futschuensis Uline	2n = 40	Kiangsu Institute of Botany (1976)
D. glabra Roxb.	2n = 40	Mehra and Sachdeva (1976)
D. gracillima Miq.	2n = 20, 40	Nakajima (1933), Takeuchi et al. (1970)
D. hamiltonii	2n = 40	Paul and Debnath (2019)
D. hastata J. Miège	2n > 120	Miège (1954)
D. hemsleyi Prain & Burkill	2n = 60	Pei et al. (1979)
D. henryi (Prain & Burkill) Ting	2n = 40	Chin et al. (1985)
D. hirtiflora Benth.	2n = 40	Miège (1954)
D. hispida Dennst.	2n = 40	Raghavan (1958)
D. hondurensis R. Knuth	2n = 36	Martin and Ortiz (1966)
D. izuensis Akahori	2n = 20	Takeuchi et al. (1970)
D. japonica Thunb.	2n = 40, 80, 100	Nakajima (1933), Araki et al. (1983), Chin et al. (1985)
D. kamoonensis Kunth	2n = 40, 60	Chin et al. (1985)
D. macroura Harms.	2n = 40	Smith (1937)

(continued)

Table 1 (continued)

Species[a]	Chromosome number	References
D. mangenotiana J. Miège	2n = 72, 80	Miège (1954)
D. melanophyma Prain & Burkill	2n = 40	Chin et al. (1985)
D. mexicana Guillemin	2n = 36	Martin and Ortiz (1966)
D. minutiflora Engl.	2n > 120	Miège (1954)
D. nentaphylla L.	2n = 120	Chin et al. (1985)
D. nipponica Makino	2n = 20, 40	Takeuchi et al. (1970), Kiangsu Institute of Botany (1976)
D. nipponica var. *rosthornii* Prain & Burkill	2n = 40	Kiangsu Institute of Botany (1976)
D. nitens Prain & Burkill	2n = 60	Chin et al. (1985)
D. nummularia Lam.	2n = 60, 100, 120	Lebot et al. (2017)
D. opposita Thunb.	2n = 140	Araki et al. (1983)
D. oppositifolia L.	2n = 40, 140, 138–142	Smith (1937), Raghavan (1958), Chin et al. (1985)
D. oppositifolia var *linnaei* Prain & Burkill	2n = 40	Ramachandran (1968)
D. paniculata Michx	2n = 36	Martin and Ortiz (1966)
D. panthaica Prain & Burkill	2n = 40	Kiangsu Institute of Botany (1976)
D. parviflora C. T. Ting	2n = 20	Pei et al. (1979)
D. pentaphylla L.	2n = 40, 50, 60, 80, 120, 144	Smith (1937), Raghavan (1958, 1959), Chin et al. (1985), Paul and Debnath (2019)
D. pentaphylla var. *jacquemontii* Prain & Burkill	2n = 40, 80	Raghavan (1958)
D. pentaphylla var. *linnaei* Prain & Burkill	2n = 40, 80	Raghavan (1958)
D. pentaphylla var. *rheedei* Prain & Burkill	2n = 40	Ramachandran (1968)
D. persimilis Prain & Burkill	2n = 40	Chin et al. (1985)
D. poilanei Prain & Burkill	2n = 20	Kiangsu Institute of Botany (1976)
D. polygonoides Humb. & Bonpl.	2n = 36, 54	Martin and Ortiz (1963, 1966)
D. polystachya Turcz. (Ichoimo group)	2n = 100	Babil et al. (2013)
D. polystachya Turcz. (Nagaimo group)	2n = 140	Babil et al. (2013)
D. polystachya Turcz. (Tsukuneimo group)	2n = 100	Babil et al. (2013)
D. praehensilis Benth.	2n = 40	Miège (1954)
D. preussii Pax	2n = 40	Miège (1954)
D. pubera Blume	2n = 40	Raghavan (1958)
D. pyrenaica Bord.	2n = 24	Miège (1954)

(continued)

Table 1 (continued)

Species[a]	Chromosome number	References
D. quaternata Gmel.	2n = 36, 54	Jensen (1937), Martin and Ortiz (1966)
D. quinqueloba Thunb.	2n = 20	Smith (1937)
D. reticulata C. Gay	2n = 61	Smith (1937)
D. rotundata Poir.	2n = 40, 60	Martin and Ortiz (1963), Baquar (1980)
D. sansibarensis Pax.	2n = 40	Miège (1954)
D. sativa L.	2n = 40	Sharma and De (1956)
D. sativa Thunb.	2n = 80	Ramachandran (1962)
D. septemloba Thunb.	2n = 20, 40	Takeuchi et al. (1970), Kiangsu Institute of Botany (1976)
D. septemloba var. *sititoana*	2n = 40	Takeuchi et al. (1970)
D. shimperana Hochst ex Kunth	2n = 80	Baquar (1980)
D. simulans Prain & Burkill	2n = 20	Kiangsu Institute of Botany (1976)
D. sinuata Vell.	2n = 24, 34 (male), 36 (female)	Meurman (1925), Suessenguth (1921)
D. smilacifolia de Wild.	2n > 120	Miège (1954)
D. spiculiflora	2n = 36	Martin and Ortiz (1963)
D. spinosa Roxb.	2n = 90	Ramachandran (1962)
D. subcalva Prain & Burkill	2n = 60	Pei et al. (1979)
D. subcalva var. *submollis* (Knuth) Ling & Ting	2n = 60	Chin et al. (1985)
D. tentaculigera Prain & Burkill	2n = 40	Pei et al. (1979)
D. tenuipes Franch. & Savat.	2n = 20, 30, 40	Takeuchi et al. (1970)
D. tokoro Makino	2n = 20	Nakajima (1933)
D. tomentosa Koenig	2n = 40, 60	Raghavan (1959)
D. trifida	2n = 40, 60[b], 80	Bousalem et al. (2006, 2010)
D. villosa L.	2n = 60	Smith (1937)
D. wallichii Hook.f.	2n = 40	Raghavan (1959)
D. yunnanensis Prain & Burkill	2n = 40	Chin et al. (1985)
D. zingiberensis Wright	2n = 20, 30, 40	Kiangsu Institute of Botany (1976), Xiaoqin et al. (2003)
Tamus communis L.[c]	2n = 48	Meurman (1925)
Tamus edulis Lowe[c]	2n = 96	Borgen (1974)

[a]Species name used in the references
[b]2C DNA content suggests the existence of the triploid *D. trifida* (Bousalem et al. 2010)
[c]Synonym of *D. communis* (L.) Caddick & Wilkin

4 Reference Genome Sequences for *Dioscorea*

A reliable reference genome sequence is indispensable for genome diversity studies of a species. The first chromosome-level reference genome obtained for *Dioscorea* species was that of *D. rotundata* (Tamiru et al. 2017). The size of *D. rotundata* genome was estimated to be ~570-Mbp using flow cytometry and k-mer analyses of genome sequences. The genome contained 26,198 protein coding genes. White Guinea yam genome sequence is distant from those of other monocotyledon species including Poales (*Oryza* and *Brachypodium*), Arecales (*Elaeis* and *Phoenix*), and Zingiberales (*Musa*), indicating that *Dioscorea* lineage was split from other monocotyledons early in the evolution. Publication of the reference genome, which also reported a DNA marker for sex identification in *D. rotundata*, has served as a catalyst for further studies into genomes of not only *D. rotundata* (Scarcelli et al. 2019; Bhattacharjee et al. 2020; Zhang et al. 2020; Sugihara et al. 2020) but also of *D. alata* (Cormier et al. 2019a, b; Sharif et al. 2020) and the pathogens associated with yams such as yam mosaic virus (Silva et al. 2019). *D. rotundata* scaffolds were generated by combining two types of Illumina short reads: paired-end and mate-pair jump reads. The mate-pair jump reads bridged the contigs assembled using paired-end short reads, with the gaps between the contigs represented with "N" for missing bases. The scaffolds were then ordered to generate a chromosome-level reference genome guided by a linkage map generated by the DNA markers of restriction site associated DNA (RAD)-tags (Baird et al. 2008) using the pseudo-testcross mapping method (Grattapaglia and Sederoff 1994). This reference genome has been shared with the wider scientific community via Ensembl (Howe et al. 2019; ftp://ftp.ensemblgenomes.org/pub/plants/release-48/fasta/dioscorea_rotundata).

Recently, the reference genome of *D. rotundata* has been updated with long reads generated by Oxford Nanopore Technologies (Sugihara et al. 2020). Long read-based de novo genome assembly resulted in longer contigs with a minimum number of missing bases, unlike those constructed by mate-pair jump reads. The newly generated contigs were also ordered using a linkage map generated by massive number of single nucleotide polymorphism (SNP) markers, instead of RAD-markers, obtained from whole-genome re-sequencing of 156 F_1 progeny. This reference genome has been made available via Ensembl (Howe et al. 2019) as *D. rotundata* reference genome ver. 2 (https://plants.ensembl.org/Dioscorea_rotundata/Info/Index).

Reference genomes of two other *Dioscorea* species are also available in the public database. The recent chromosome-level reference genome of *D. alata* is yet to be published, but it is accessible on YamBase (https://yambase.org/). Additionally, the reference genome of *D. dumetorum*, which was generated by sequencing with Oxford Nanopore Technologies, has been reported (Siadjeu et al. 2020; https://pub.uni-bielefeld.de/record/2941469). Both *D. rotundata* and *D. alata* belong to the botanical section Enantiophyllum. The *D. dumetorum* reference genome represents the first genome sequenced from the section Lasiophyton. These multiple reference genomes will facilitate the comparative genomics and pangenome of *Dioscorea* species and be especially useful for the analysis of the transition of the sex-determination locus within *Dioscorea* (Cormier et al. 2019a; see Sect. 6).

5 Origin and Domestication of *Dioscorea* Species

Yams of different *Dioscorea* species are believed to be independently domesticated in different continents: *D. rotundata* and *D. cayenensis* in West and Central Africa, *D. alata* in Southeast Asia, and *D. trifida* in South America. However, our knowledge of their origins has been limited until recently. This is mainly due to the frequent hybridization and polyploidization of many species including *D. rotundata* (Terauchi et al. 1992; Scarcelli et al. 2006; Chaïr et al. 2010; Girma et al. 2014; Scarcelli et al. 2017; Sugihara et al. 2020) and *D. alata* (Chaïr et al. 2016; Sharif et al. 2020). The recent population genomics studies have started unveiling the domestication processes of the major species (Scarcelli et al. 2019; Sugihara et al. 2020; Sharif et al. 2020).

5.1 Origin and Domestication of Guinea Yam

White Guinea yam, *D. rotundata*, is a true cultigen and the wild species *D. abyssinica* and *D. praehensilis* have been proposed as its two candidate progenitors (Coursey 1976a, b; Terauchi et al. 1992; Scarcelli et al. 2006; Chaïr et al. 2010; Girma et al. 2014; Scarcelli et al. 2017; Magwé-Tindo et al. 2018). *D. abyssinica* and *D. praehensilis* are distributed in savannah and rainforest areas, respectively, of West and Central Africa. By comparing whole-genome sequences of 80 *D. rotundata*, 29 *D. abyssinica*, 26 *D. praehensilis* from West Africa (Western *D. praehensilis*), and 18 *D. praehensilis* from Cameroon (Cameroonian *D. praehensilis*), Scarcelli and colleagues recently proposed that *D. rotundata* was domesticated from *D. praehensilis* in the northern part of Benin within the Niger River basin (Scarcelli et al. 2019). This report is a major contribution towards elucidating the origin of Guinea yam. However, after a careful reassessment of the Scarcelli et al. (2019) data and results, we reached at a different conclusion (Sugihara et al. 2020).

Our study included genome sequences of 336 *D. rotundata* accessions in addition to the *D. rotundata* and wild species accessions analyzed by Scarcelli et al. (2019). First, we conducted clustering analysis and flow cytometry analysis of our 336 *D. rotundata* accessions. Based on these results, we classified 308 accessions as diploid and 28 accessions as triploid. Focusing on the diploid *D. rotundata* accessions, we attempted to elucidate their phylogenetic relationships with the wild relative species; *D. abyssinica*, Western *D. praehensilis*, and Cameroonian *D. praehensilis* as reported by Scarcelli et al. (2019). We reconstructed a rooted neighbor-joining tree of four African yam taxa using *D. alata,* an Asian species, as an outgroup (Fig. 4a). In our result, *D. rotundata* was genetically closer to *D. abyssinica* than to *D. praehensilis*, which was not consistent to Scarcelli's hypothesis indicating that *D. rotundata* was directly domesticated from *D. praehensilis*. To test whether *D. rotundata* was derived from *D. abyssinica* or

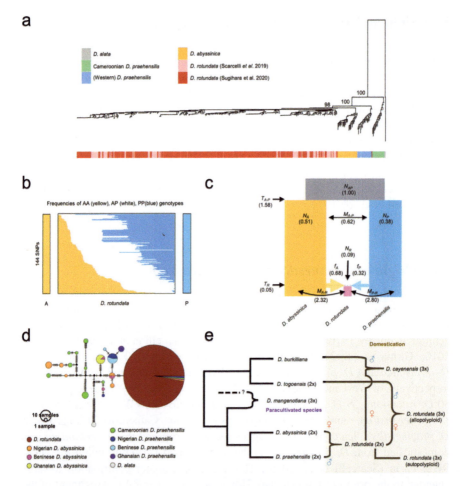

Fig. 4 Domestication history of Guinea yam. (**a**) Neighbor-joining tree of four African yam taxa reconstructed using *D. alata* as an outgroup (adopted from Sugihara et al. 2020, Fig. 1c). (**b**) Frequencies of individuals homozygous for *D. abyssinica* allele (A: indicated by yellow color), homozygous for *D. praehensilis* allele (P: indicated by blue color), and heterozygous for A and P (indicated by white color) among the 388 *D. rotundata* sequences as studied for 144 SNPs (adopted from Sugihara et al. 2020, Fig. 2b). (**c**) Demographic history of *D. rotundata* as inferred by ∂a∂i (Gutenkunst et al. 2009). *N*, *T*, *M*, and *f* represent a relative population size from the ancestral population (N_{AP}), divergence time, migration rate, a fraction of genomic contribution, respectively. The ∂a∂i analysis inferred the hybrid origin of *D. rotundata* with the high migration rates between *D. rotundata* and the two wild relatives (adopted from Sugihara et al. 2020, Fig. 2c). (**d**) Haplotype network of the whole plastid genomes of 416 *D. rotundata* (including the triploid accessions), 68 accessions of wild relative species, and two *D. alata* accessions used as the outgroup. The number of vertical dashes represents the number of mutations (adopted from Sugihara et al. 2020, Fig. 3a). (**e**) Reticulated evolutionary history of Guinea yam (Sugihara et al. 2020; Girma et al. 2014). *D. rotundata* (white Guinea yam) is derived from a homoploid hybridization between *D. abyssinica* and *D. praehensilis*. *D. cayenensis* (yellow Guinea yam) and the majority of triploid *D. rotundata* are derived from the polyploid hybridization between a female diploid *D. rotundata* and a male African wild yam. At least four species contributed to the gene pool of Guinea yam

D. praehensilis, we focused on the allele frequencies on 144 SNPs which were oppositely fixed in *D. abyssinica* and *D. praehensilis* (Fig. 4b). If *D. rotundata* was derived from either of *D. abyssinica* or *D. praehensilis*, allele frequencies of the analyzed SNPs should be highly skewed to either of the candidate progenitors. However, the allele frequencies of *D. rotundata* were intermediate. This observation suggested a hybrid origin of *D. rotundata* between *D. abyssinica* and *D. praehensilis*. To conform this hypothesis, we compared the three evolutionary models by $\partial a \partial i$ which assesses evolutionary models with their likelihood based on the site frequency spectrum (Gutenkunst et al. 2009). The first model was that *D. rotundata* had been derived from *D. abyssinica*. The second model was that *D. rotundata* had been derived from *D. praehensilis* (Scarcelli's hypothesis). The third model was that *D. rotundata* had been hybrid-derived between *D. abyssinica* and *D. praehensilis*. As a result, $\partial a \partial i$ showed the highest likelihood in the third model (the hybrid origin of *D. rotundata*) out of the three models.

Our finding suggested that *D. rotundata* is most likely a homoploid hybrid between *D. abyssinica* and *D. praehensilis* (Fig. 4b, c; Sugihara et al. 2020). Homoploid hybrid speciation is the formation of a new hybrid species without altering the ploidy levels of the parents (Mallet 2007). The origin of *D. rotundata* by hybridization seems to be recent when compared with timing of the speciation of *D. praehensilis* from *D. abyssinica* (Fig. 4c, d). Genomic contributions from *D. abyssinica* and *D. praehensilis* during the hybridization event were estimated to be ~68% and 32%, respectively (Fig. 4c). The relative population size of *D. rotundata* is much smaller than those of its wild relatives, which indicates that *D. rotundata* was affected by domestication bottleneck. Chloroplast DNA is predominantly inherited maternally in angiosperms. We extracted chloroplast DNA sequences from the whole-genome sequence reads and studied its polymorphisms in our samples. A chloroplast DNA haplotype network suggested that *D. abyssinica* and *D. praehensilis* were the maternal and paternal parents of *D. rotundata*, respectively (Fig. 4d, e; Sugihara et al. 2020). The hybrid origin of white Guinea yam was initially proposed by Coursey in 1976 based on morphological comparisons (Coursey 1976a). Our results from genome analyses support his hypothesis that spontaneous hybridization between wild yams could have occurred at the artifactual "dump heaps" created by people living in the savannah between the forest and the Sahara (Coursey 1976b).

The most common cases of origin of crops by hybridization accompanies polyploidization (also known as allopolyploidization) as exemplified by bread wheat (*Triticum aestivum*) (Peng et al. 2011), banana (*Musa acuminata*) (Heslop-Harrison and Schwarzacher 2007), cotton (*Gossypium* spp.) (Zhang et al. 2015), and canola (*Brassica napus*) (Chalhoub et al. 2014). The homoploid hybrid speciation in *D. rotundata* is unique in that its domestication did not involve polyploidization.

We hypothesize that Guinea yam has been established by the process of "ennoblement" (Dumont and Vernier 2000; Mignouna and Dansi 2003; Scarcelli et al. 2006; Chaïr et al. 2010). "Ennoblement" is a traditional farmers' practice that involves collecting tubers of wild yams from the bush and forest and planting them in their fields, and it likely contributes to the genetic diversity of yam through hybridization and introgression (Jarvis and Hodgkin 1999; Scarcelli et al. 2006;

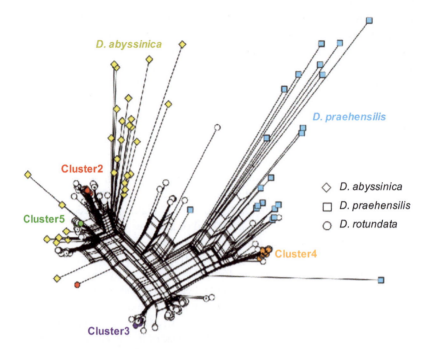

Fig. 5 Genetic relationship among *D. rotundata* and its wild progenitors around the *SWEETIE* gene. A Neighbor-Net (adopted from Sugihara et al. 2020, Fig. 4c) was constructed by SplitsTree (Huson and Bryant 2006). Based on STRUCTURE analysis, five clusters (cluster 1–5) were defined in Sugihara et al. (2020). In this analysis, the accessions in cluster 1 were removed because they were triploid. Based on this result, the gene of cluster 2 and 5 were genetically close to *D. abyssinica*, while that of cluster 4 was genetically close to *D. praehensilis*. This different genetic relationship in the different clusters indicate the introgressions from both wild progenitors

Chaïr et al. 2010). Similar farmers' practices of sympatrically planting crop wild relatives and cultivars often occur in the centers of origin of domesticated plants (Jarvis and Hodgkin 1999).

The locus of the *SWEETIE* gene is an example that the crop wild relatives contributed to the genetic diversity of *D. rotundata*. The *SWEETIE* gene is known to be involved in sugar flux (Veyres et al. 2008a, b). In the diploid *D. rotundata*, this locus showed the signature of extensive introgressions from *D. abyssinica* and *D. praehensilis* (Fig. 5). We identified the introgressions by applying f_4 statistic (Reich et al. 2009) of population genomics to our diploid *D. rotundata* accessions and its wild progenitors as analyzed in Scarcelli et al. (2019). The f_4 statistic can detect genomic regions showing discordant topologies among different genetic groups of *D. rotundata* with respect to *D. abyssinica* and *D. praehensilis*. We hypothesize that this locus was under selection.

In addition to the majority of diploid accessions, triploid accessions of *D. rotundata* have been identified by flow cytometry analysis (Girma et al. 2014; Sugihara et al. 2020). These triploid *D. rotundata* accessions seem to have been derived from a hybridization between the diploid *D. rotundata* and the wild yam species *D. togoensis* (Fig. 4e; Girma et al. 2014). Since the chloroplast sequences of the triploid *D. rotundata* were shared with the diploid *D. rotundata*, their maternal and paternal parents are likely the diploid *D. rotundata* and *D. togoensis*, respectively (Fig. 4d; Sugihara et al. 2020). Successful interspecific crosses between *D. rotundata* and *D. togoensis* were also reported (Girma et al. 2014). The triploid *D. rotundata* formed by autopolyploidization seems to be the minority as compared to those formed by allopolyploidization (Girma et al. 2014). It is not easy to distinguish the triploid *D. rotundata* accessions from the diploid accessions based on morphology. However, some morphological traits, such as the presence of barky patches, absence of waxiness, and dark green leaf color, have been shown to correlate with ploidy level (Girma et al. 2014). Agricultural importance of the triploid *D. rotundata* is yet to be studied.

D. cayenensis (yellow Guinea yam) is another species that is likely a triploid hybrid between the diploid *D. rotundata* and the rainforest-adapted wild species *D. burkilliana* (Fig. 4e; Terauchi et al. 1992; Girma et al. 2014). Based on chloroplast DNA polymorphisms, it was suggested that diploid *D. rotundata* is the maternal parent of *D. cayenensis* (Terauchi et al. 1992). Using nuclear ribosomal DNA polymorphisms (Terauchi et al. 1992) and Genotype-by-Sequencing (GBS) analysis (Girma et al. 2014), it was inferred that *D. burkilliana* is the paternal parent. *D. cayenensis* has a woody corm above the fleshy tuber, and this trait is shared with *D. burkilliana*. Interestingly, *D. burkilliana* is also subjected to "ennoblement" together with *D. abyssinica* and *D. praehensilis* (Mignouna and Dansi 2003), which probably contributed to its hybridization with *D. rotundata*.

D. mangenotiana (syn. *D. baya*) is genetically close to Guinea yam (Girma et al. 2014; Magwé-Tindo et al. 2018). Previously, *D. mangenotiana* and *D. baya* were considered different species. However, a recent study proposed *D. mangenotiana* as the adult form of *D. baya* (Magwé-Tindo et al. 2018). *D. mangenotiana* is a rainforest-adapted wild species characterized by spiny stems and the production of very big tubers (Figs. 2d and 3f; Dounias 2001). The bases of its fleshy edible tubers are attached to very large woody corms, which probably provide protection from herbivores. In Southern Cameroon, *D. mangenotiana* has long been subjected to "paracultivation" by the Baka Pygmies (Dounias 2001). "Paracultivation" is the exploitation and maintenance of wild plants in their original/natural environments (Dounias 2001). Intriguingly, *D. mangenotiana* was reported as a triploid species (Girma et al. 2014) with a large number of heterozygous DNA markers, which probably suggests that it may be an allopolyploid. However, no research has been carried out to date to identify the ancestors of *D. mangenotiana*.

5.2 Origin and Domestication of D. alata

D. alata (greater yam) is widely cultivated in pantropical regions (Asia, the Pacific, Africa, and the Caribbean) in contrast to *D. rotundata* that is restricted to Africa and the Caribbean. This worldwide dispersion of *D. alata* occurred mainly through human migrations (Sharif et al. 2020). As discussed above, the domestication process of *D. rotundata* seems mainly driven by hybridization and introgression. By contrast, the domestication of *D. alata* was mainly driven by vegetative propagation and autopolyploidization, which caused erratic or no flowering in this species (Arnau et al. 2010; Sharif et al. 2020). This predominant vegetative propagation in *D. alata* is in line with its very low nucleotide diversity ($\pi = 0.96 \sim 1.29 \times 10^{-5}$ in Fig. 6: Sharif et al. 2020) as compared with the higher level of nucleotide diversity in *D. rotundata* ($\pi = 1.48 \times 10^{-3}$) (Sugihara et al. 2020).

Sharif et al. (2020) used genome sequence analysis of 643 accessions spanning four continents to study the dispersion route of *D. alata*, which confirmed the region around tropical Eastern Asia as the geographical origin of *D. alata*. This finding supports the previous proposal by Burkill (1960) that *D. alata* was originated in the tropical Eastern Asia. Additionally, Sharif and colleagues hypothesized that *D. alata* was domesticated independently in the Mainland Southeast Asia and the Pacific (① in Fig. 6). Their hypothesis is based on the early divergence between the two regions estimated for *D. alata* (Sharif et al. 2020) as well as the estimated date of early human settlement of Sahul (at least 50,000 years ago) (Bird et al. 2019). After the divergence between the subgroups of Mainland Southeast Asia and the Pacific, *D. alata* reached the Indian subcontinent (② in Fig. 6). According to their demographic inference, there was a continuous migration between the Indian subcontinent

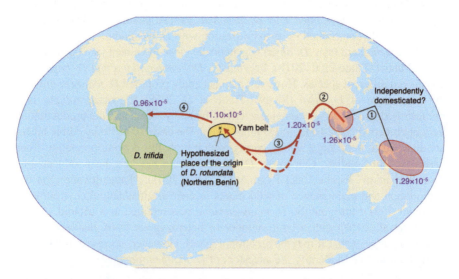

Fig. 6 Global dispersion of *D. alata* (Sharif et al. 2020) and distributions of other cultivated yam species. Figures in purple color represent nucleotide diversities of the diploid *D. alata* accessions in different regions. *D. alata* was originated in Southeast Asia (①). After its domestication in Southeast Asia, it reached Africa via the Indian Peninsula (② and ③), and finally reached the Caribbean (④) from Africa. The introduction to Africa might be through either Madagascar (broken line in ③) or East Africa via the Swahili coast (solid line in ③). The gradual reduction in *D. alata* nucleotide diversities clearly suggests the founder effects

and the Pacific. From the Indian subcontinent, *D. alata* reached Africa (③ in Fig. 6). Since the African and Madagascar *D. alata* are genetically closer to that of the Indian subcontinent than to that from the Mainland Southeast Asia, this dispersion seems to have occurred via the Indian subcontinent (Sharif et al. 2020). The detailed route of how *D. alata* reached Africa is still unclear. However, it might be through either Madagascar or East Africa via the Swahili coast (③ in Fig. 6). The relatively low nucleotide diversity of African accessions suggests a founder effect from the Indian subcontinent (Sharif et al. 2020). From Africa, *D. alata* was introduced into the Caribbean (④ in Fig. 6) most probably during the Colombian exchange, which introduced African crops to the tropical America (Boivin et al. 2012). The lowest nucleotide diversity of the Caribbean accessions suggests a strong founder effect from Africa (Sharif et al. 2020; Fig. 6). Taken together, the decreasing nucleotide diversity of *D. alata* clearly reflects the route for its dispersion.

The distribution of the clonal and polyploid accessions of *D. alata* also reflects the route of dispersion of the species (Sharif et al. 2020). Approximately 68% of the inferred clonal clusters (multi-locus lineages in Sharif et al. 2020) had an intercontinental distribution. The Caribbean accessions had the highest number of the intercontinentally shared clonal clusters, while those of Asia had the lowest. Therefore, the distribution of the clonal clusters reflects dispersion of this species through vegetative propagation. The dispersion route of *D. alata* was also reflected in the distribution of its polyploid accessions. Most of the polyploid accessions are genetically close to either the Asian or the Pacific accessions. This suggests that the polyploidization of *D. alata* occurred several times before migration of the species from Asia and the Pacific to Africa and the Caribbean. Moreover, most triploid accessions were genetically closer to the Asian lineages rather than to the Pacific lineages. This suggests that most triploid accessions were derived from Asia.

The wild progenitor of *D. alata* is still unknown. However, some studies revealed that *D. alata* is phylogenetically close to the wild species *D. nummularia*, *D. transverse*, and *D. hastifolia* (Malapa et al. 2005; Chaïr et al. 2016). Especially, *D. nummularia*, a species native to Melanesia and Island Southeast Asia, is also subjected to a "paracultivation"-like practice (Dounias 2001; Chaïr et al. 2016). Interestingly, cultivars of natural hybrid between *D. alata* and *D. nummularia* have been reported and known as "strong yam" by farmers of Vanuatu (Chaïr et al. 2016). These interspecific hybrids were treated as unidentified taxa or erroneously assigned to *D. transverse* (Malapa et al. 2005; Chaïr et al. 2016). This mis-assignment is probably related to their potential hybrid status (Chaïr et al. 2016). Recently, researchers have started paying attention to the diversity of *D. nummularia* (Lebot et al. 2017) and attempts are made to introduce the resistance trait of *D. nummularia* against the anthracnose disease caused by *Colletotrichum* pathogens into *D. alata* by artificial hybridization (Lebot et al. 2019). Although the cross between *D. alata* and *D. nummularia* resulted in fertile seeds, *D. nummularia* is not regarded as a direct ancestor of *D. alata* (Chaïr et al. 2016). Further population genomics study is needed by including multiple accessions of *D. nummularia*, *D. transverse,* and *D. hastifolia* to clarify the origin of *D. alata*.

5.3 Origin and Domestication of D. trifida

D. trifida of the section Macrogynodium was domesticated in South America and is commonly referred to as the indigenous "Amerindian" yam. Genetic segregation study and cytogenetics suggested that the cultivated *D. trifida* is autotetraploid ($2n = 4\times = 80$) (Bousalem et al. 2006), while a wild *D. trifida* with diploid genome ($2n = 2\times = 40$) was found in French Guyana (Bousalem et al. 2010). A phylogenetic study based on amplified fragment length polymorphism (AFLP) clearly separated the cultivated ($4\times$) and wild ($2\times$) *D. trifida* (Bousalem et al. 2010). A region where the diploid *D. trifida* predominate was also found. *D. trifida* represents a clear case that whole-genome duplication may have played an important role in the domestication of *Dioscorea*. The autopolyploidization is not a preferred subject in population genomics, but this may help to calibrate the time when the polyploidization happened.

5.4 Origin and Domestication of D. dumetorum

D. dumetorum, known as the "trifoliate yam" because of its trifoliate leaves, belongs to the section Lasiophyton. *D. dumetorum* is widely consumed in West and Central Africa and occurs in both cultivated and wild forms. Despite its high yielding nature and nutritional richness, severe postharvest hardening of the tubers makes *D. dumetorum* a minor crop (Sefa-Dedeh and Afoakwa 2002). Genetic diversity study of *D. dumetorum* revealed that the accessions from Togo and Nigeria had the highest genetic diversity (Sonibare et al. 2010). This indicates that the center of genetic diversity and the possible origin of *D. dumetorum* might be around Togo and Nigeria. Another study focusing on the accessions from Cameroon identified gene flow and admixture among the accessions, which is probably caused by farmers' breeding practices (Siadjeu et al. 2018). The same study revealed different ploidy levels in *D. dumetorum* and showed that diploids and triploids have different geographical distributions in Cameroon. Interestingly, population genomics study showed that the distribution of the triploid *D. dumetorum* was positively correlated with the region containing a higher level of gene flow (Siadjeu et al. 2018).

6 Evolution of Sex in *Dioscorea* Species

Most *Dioscorea* species are dioecious, bearing male and female flowers on separate individuals. This sexual system affects genetics and population genomics of the genera and consequently deserves a special attention. Dioecious plants account for ~5–6% of angiosperm species (Renner 2014). Based on its scattered taxonomic distribution, dioecy is suggested to have evolved recently and independently from hermaphroditic co-sexual ancestors (Renner and Ricklefs 1995; Charlesworth 2002). Many studies have focused on understanding the process of this large-scale convergent evolution from co-sexual to dioecy in plants. To elucidate the process, sex-determination systems have been studied in several taxa (Akagi et al. 2016, 2019; Harkess et al. 2020).

In *Dioscorea*, multiple sex-determination systems have been reported by cytological observations and molecular analyses (Table 2). As part of our work on the genetics of *D. tokoro*, a wild species from East Asia, we studied the inheritance and segregation pattern of AFLP markers in an F_1 family derived from a cross between male and female *D. tokoro* plants, which suggested an XY/XX (male/ female) sex-determination system in this species (Terauchi and Kahl 1999). Although most *Dioscorea* species have male heterogametic sex-determination system, female heterogametic sex-determination system (ZZ/ZW) and extra chromosomes in female (XO/XX) have also been reported (Table 2). Sex change of individuals was observed in *D. rotundata* having female heterogametic sex-determination system (ZZ/ZW) (Tamiru et al. 2017). Here, we review the sex-determination systems of ZZ/ZW and XY/XX that occur in *D. rotundata* and *D. alata*, respectively.

Sex in *D. rotundata* is regulated by a female-specific genomic region that we recently identified by QTL-seq analysis of an F_1 progeny segregating for sex (Tamiru et al. 2017). QTL-seq is an NGS-based bulked segregant analysis (BSA) method to identify the genomic regions underlying traits of interest using progeny derived from crosses made between cultivars/lines showing contrasting phenotypes for the traits (Takagi et al. 2013; Itoh et al. 2019). To identify the genomic region associated with sex in *D. rotundata*, QTL-seq was applied to sequences generated for male and female DNA pools prepared from an F_1 progeny derived from a cross between male and female plants. Accordingly, a candidate genomic region was detected on chromosome 11 using SNP markers that were heterozygous in the female parent whereas no candidate regions were detected when SNP markers heterozygous in the male parent were used. The candidate genomic region showed significant structural differences between the male and female sequences. PCR amplification and short read mapping analysis further identified a female-specific region within the candidate genomic region delineated by QTL-seq. These results suggest that the sex-determination system of *D. rotundata* is ZZ/ZW (male/female), not XO/XX (male/female).

A DNA maker, "sp16," was developed within the female-specific (*W*-) genomic region of *D. rotundata* for prediction of sex of the plant at the seedling stage (Tamiru et al. 2017). The usefulness of the DNA marker has been demonstrated in diverse *D. rotundata* accessions (Agre et al. 2020; Denadi et al. 2020). However, the marker type was not perfectly associated with sex, suggesting the manifestation of sex is unstable in *D. rotundata* over a time period (Tamiru et al. 2017). Sex change is widely known in plants and animals (Policansky 1982). Interestingly, this sex change was rarely observed in the male *D. rotundata* plants, suggesting a gene on the *W*-region with "sp16" marker seems to suppress maleness, and its effect is unstable.

Table 2 Reported sex-determination systems in *Dioscorea*

Species	Sex-determination system	Methods	Reference
D. alata L.	XY	QTL detection	Cormier et al. (2019a)
D. bulbifera L.	XY	Cytological observations	Ramachandran (1962)
D. deltoidea Wall.	ZW	Cytological observations	Bhat and Bindroo (1980)
D. discolor	a	Cytological observations	Smith (1937)
D. fargesii Franch.	a	Cytological observations	Smith (1937)
D. gracillima Miq.	XY	Cytological observations	Nakajima (1937)
D. japonica Thunb.	XY	Cytological observations	Nakajima (1942)
D. macroura Harms.	a	Cytological observations	Smith (1937)
D. pentaphylla L.	XY	Cytological observations	Ramachandran (1962)
D. reticulata C. Gay	XO	Cytological observations	Smith (1937)
D. rotundata	ZW	QTL detection	Tamiru et al. (2017)
D. sinuata Vell.	XO	Cytological observations	Meurman (1925)
D. spinosa Roxb.	XY	Cytological observations	Ramachandran (1962)
D. tokoro Makino	XY	AFLP analysis	Terauchi and Kahl (1999)
D. tomentosa Koenig	XY	Cytological observations	Ramachandran (1962)

[a]Heteromorphic chromosomes were reported but no information about sex determination system

In our recent study, we also investigated the genomic contribution of *D. abyssinica* and *D. praehensilis* to *D. rotundata* chromosome-wise (Fig. 7; Sugihara et al. 2020). The chromosome 11 of *D. rotundata* harboring the sex-determination locus was highly skewed towards that of *D. abyssinica*. As described in the previous Sect. 5.1 above, *D. rotundata* is likely a hybrid species derived from *D. abyssinica* and *D. praehensilis*, and the genetic divergences from both wild progenitors are basically similar across the genome (Sugihara et al. 2020). However, of all the chromosomes, chromosome 11 of *D. rotundata* had the shortest genetic distance from that of *D. abyssinica* and the largest genetic distance from that of *D. praehensilis* (Sugihara et al. 2020). Similar interspecies divergence differences between autosomes and sex chromosome have also been reported in the dioecious plant species of the genus *Silene* (Hu and Filatov 2016).

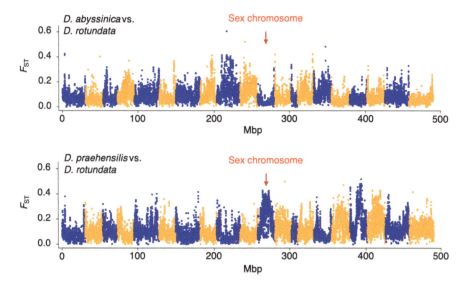

Fig. 7 Genomic scan of F_{ST} values between cultivated yam (*D. rotundata*) and two wild species *D. abyssinica* (top) and *D. praehensilis* (bottom) (adopted from Sugihara et al. 2020, Fig. 2d). Sliding window analysis was conducted with 100-kb window and 20-kb step. Chromosome 11 of *D. rotundata* containing the sex-determining locus shows the shortest distance to that of *D. abyssinica* and the largest distance to that of *D. praehensilis*

D. alata has a male heterogametic sex-determination system (XY/XX) (Cormier et al. 2019a), which is like the majority of *Dioscorea* species (Table 2). *D. alata* belongs to the section Enantiophyllum with *D. rotundata,* a species with a female heterogametic sex-determination system (ZZ/ZW). Interestingly, the genomic region associated with sex of *D. alata* was identified on linkage group 6, which also corresponded to chromosome 6 of *D. rotundata* (Cormier et al. 2019a). The sex-determination locus of *D. rotundata* was identified on chromosome 11 (Tamiru et al. 2017), suggesting the transition of sex-determination system from XY to ZW occurred in the section Enantiophyllum. A similar transition of sex-determination system was reported in the genus *Populus*. Most species *Populus* have the XY sex-determination system, while *P. alba* has the ZW sex-determination system (Müller et al. 2020).

As discussed above, the genus *Dioscorea* contains many dioecious species with divergent sex-determination systems. Transition of sex-determination systems are presumably involved in species divergence (Kumar et al. 2014). Further genomics research in *Dioscorea* species will help identify the causative genes involved in sex determination and to clarify how the sex-determination systems have diversified in the genus. Such information will also allow us to manipulate sex of plants for effective cross breeding of yam crops.

7 Conclusions and Future Perspectives

Population genomics and cytogenetics studies have revealed important domestication processes in *Dioscorea* species, but many questions still remain. For example, we still do not know the key traits and the genes involved in yam domestications, although some studies have identified genes showing signature of selection in *D. rotundata* including *SWEETIE* gene in our study (Akakpo et al. 2017; Scarcelli et al. 2019; Sugihara et al. 2020). *D. abyssinica* and *D. praehensilis*, the wild relatives of *D. rotundata*, are subjected to an ongoing practice of "ennoblement." Additionally, it has been shown that the most cultivars introduced by "ennoblement" are indeed hybrids between the cultivated and wild yams (Scarcelli et al. 2006). These findings probably indicate that the wild species cannot directly be domesticated to become cultivars and that hybridization was necessary to generate white Guinea yam cultivars. Similar interspecific hybridization was also reported in *D. alata* (Chaïr et al. 2016). Consequently, analyzing hybridization is important to understand what attributes characterize *D. rotundata* and other cultivated yams. Probably, *D. rotundata* was established as a cultivar as a result of heterosis derived from the hybridization between *D. abyssinica* and *D. praehensilis*.

Understanding the genomes of crop wild relatives would facilitate efficient breeding programs. Crop wild relatives are expected to have potentially beneficial alleles that are not available in the cultivars. The farmers unconsciously introduce these beneficial alleles to the cultivars presumably by "ennoblement." Since the genomic regions containing the beneficial alleles should be affected by selective sweeps, population genomics analyses may be able to identify these regions (Akakpo et al. 2017; Scarcelli et al. 2019; Sugihara et al. 2020). Currently, there is no evidence that these candidate selective sweeps affected any phenotypes. However future functional studies of the identified genes would reveal their impact on the change of traits in the crops.

Another standing question is how many times the domestication processes occurred in the various cultivated *Dioscorea* species. A recent study hypothesized multiple domestication processes of *D. alata* in separate regions (Sharif et al. 2020). The cultivated yam landraces from Southern Ethiopian are phylogenetically close to the cultivated gene pools of *D. rotundata*, but they were clearly separate from Nigerian *D. rotundata* (Tamiru et al. 2007). Although the model-based population genetics/genomics is needed to infer the detailed demographic history, this result may suggest independent domestication processes of *D. rotundata* in Ethiopia (or East Africa) and Nigeria.

In this chapter, the importance of hybridization and polyploidization for the domestication of *Dioscorea* species has been discussed. Some of these events appear to have played an important role in yam domestication. In recent years, our knowledge of yam domestication has dramatically improved thanks to the advances in sequencing technologies and statistical methods for population genomics analysis. These developments also allowed us to identify, among others, the transition of the sex-determination system in the section Enantiophyllum. Future studies should further unravel the complex evolutionary history of *Dioscorea* species including hybridization, polyploidization, and sexual/asexual propagation.

Acknowledgements This study was supported by AfricaYam Project funded by the Bill and Melinda Gates Foundation (BMGF, grant number OPP1052998). This work is dedicated to the memory of Günter Kahl who has pioneered genetics and molecular biology studies of *Dioscorea*.

References

Agre P, Nwachukwu C, Olasanmi B, Obidiegwu J, Nwachukwu E, Adebola P, et al. Sample preservation and plant sex prediction in white Guinea yam (*Dioscorea rotundata* Poir.). J Appl Biotechnol Rep. 2020;7(3):145–51.

Akagi T, Henry IM, Kawai T, Comai L, Tao R. Epigenetic regulation of the sex determination gene MeGI in polyploid persimmon. Plant Cell. 2016;28(12):2905–15.

Akagi T, Pilkington SM, Varkonyi-Gasic E, Henry IM, Sugano SS, Sonoda M, et al. Two Y-chromosome-encoded genes determine sex in kiwifruit. Nat Plants. 2019;5(8):801–9.

Akakpo R, Scarcelli N, Chaïr H, Dansi A, Djedatin G, Thuillet AC, et al. Molecular basis of African yam domestication: analyses of selection point to root development, starch biosynthesis, and photosynthesis related genes. BMC Genomics. 2017;18(1):782.

Araki H, Harada T, Yakuwa T. Some characteristics of interspecific hybrids between *Dioscorea japonica* Thunb. and *Dioscorea opposita* Thunb. J Japan Soc Hort Sci. 1983;52(2):153–8.

Arnau G, Nemorin A, Maledon E, Abraham K. Revision of ploidy status of *Dioscorea alata* L. (Dioscoreaceae) by cytogenetic and microsatellite segregation analysis. Theor Appl Genet. 2009;118(7):1239–49.

Arnau G, Abraham K, Sheela MN, Chaïr H, Sartie A, Asiedu R. Yams. In: Bradshaw J, editor. Root and tuber crops. Handbook of plant breeding, vol. 7. New York: Springer; 2010. p. 127–48.

Babil P, Kondo S, Iwata H, Kushikawa S, Shiwachi H. Intra-specific ploidy variations in cultivated Chinese yam (*Dioscorea polystachya* Turcz.). Trop Agr Develop. 2013;57(3):101–7.

Baird NA, Etter PD, Atwood TS, Currey MC, Shiver AL, Lewis ZA, et al. Rapid SNP discovery and genetic mapping using sequenced RAD markers. PLoS One. 2008;3(10):e3376.

Baquar SR. Chromosome behaviour in Nigerian yams (Dioscorea). Genetica. 1980;54(1):1–9.

Bhat BK, Bindroo BB. Sex chromosomes in *Dioscorea deltoidea* Wall. Cytologia. 1980;45(4):739–42.

Bhattacharjee R, Agre P, Bauchet G, De Koeyer D, Lopez-Montes A, Kumar PL, et al. Genotyping-by-sequencing to unlock genetic diversity and population structure in White yam (*Dioscorea rotundata* Poir.). Agronomy. 2020;10(9):1437.

Bird MI, Condie SA, O'Connor S, O'Grady D, Reepmeyer C, Ulm S, et al. Early human settlement of Sahul was not an accident. Sci Rep. 2019;9:8220.

Boivin N, Fuller DQ, Crowther A. Old World globalization and the Columbian exchange: comparison and contrast. World Archaeol. 2012;44(3):452–69.

Borgen L. Chromosome numbers of Macaronesian flowering plants II. Norw J Bot. 1974;21:195–210.

Bousalem M, Arnau G, Hochu I, Arnolin R, Viader V, Santoni S, et al. Microsatellite segregation analysis and cytogenetic evidence for tetrasomic inheritance in the American yam *Dioscorea trifida* and a new basic chromosome number in the *Dioscoreae*. Theor Appl Genet. 2006;113(3):439–51.

Bousalem M, Viader V, Mariac C, Gomez RM, Hochu I, Santoni S, et al. Evidence of diploidy in the wild Amerindian yam, a putative progenitor of the endangered species *Dioscorea trifida* (Dioscoreaceae). Genome. 2010;53(5):371–83.

Burkill IH. The organography and the evolution of the Dioscoreaceae, the family of the yams. J Linn Soc. 1960;56:319–412.

Caddick LR, Wilkin P, Rudall PJ, Hedderson TAJ, Chase MW. Yams reclassified: a recircumscription of Dioscoreaceae and Dioscoreales. Taxon. 2002;51(1):103–14.

Chaïr H, Cornet D, Deu M, Baco MN, Agbangla A, Duval MF, et al. Impact of farmer selection on yam genetic diversity. Conserv Genet. 2010;11(6):2255–65.

Chaïr H, Sardos J, Supply A, Mournet P, Malapa R, Lebot V. Plastid phylogenetics of Oceania yams (*Dioscorea* spp., Dioscoreaceae) reveals natural interspecific hybridization of the greater yam (*D. alata*). Bot J Linn Soc. 2016;180(3):319–33.

Chalhoub B, Denoeud F, Liu S, Parkin IAP, Tang H, Wang X, et al. Early allopolyploid evolution in the post-Neolithic *Brassica napus* oilseed genome. Science. 2014;345(6199):950–3.

Charlesworth D. Plant sex determination and sex chromosomes. Heredity. 2002;88(2):94–101.

Chin HC, Chang MC, Ling PP, Ting CT, Dou FP. A cytotaxonomic study on Chinese *Dioscorea* L. the chromosome numbers and their relation to the origin and evolution of the genus. Acta Phytotax Sin. 1985;23(1):11–8.

Cormier F, Lawac F, Maledon E, Gravillon MC, Nudol E, Mournet P, et al. A reference high-density genetic map of greater yam (*Dioscorea alata* L.). Theor Appl Genet. 2019a;132(6):1733–44.

Cormier F, Mournet P, Causse S, Arnau G, Maledon E, Gomez RM, et al. Development of a cost-effective single nucleotide polymorphism genotyping array for management of greater yam germplasm collections. Ecol Evol. 2019b;9(10):5617–36.

Coursey DG. Yams. An account of the nature, origins, cultivation and utilisation of the useful members of the Dioscoreaceae. In: Rhind D, et al., editors. Tropical agricultural series. London: Longmans; 1967. p. 108–29.

Coursey DG. The civilization of the yam: interrelationships of man and yams in Africa and the Indo-Pacific region. Archaeol Phys Anthropol Ocean. 1972;7(3):215–33.

Coursey DG. Yams, *Dioscorea* spp. (Dioscoreaceae). In: Simmonds NW, editor. Evolution of crop plants. London: Longman; 1976a. p. 70–4.

Coursey DG. The origins and domestication of yams in Africa. In: Harlan JR, editor. Origins of African plant domestication. Boston: De Gruyter Mouton; 1976b. p. 383–408.

Couto RS, Martins AC, Bolson M, Lopes RC, Smidt EC, Braga JMA. Time calibrated tree of *Dioscorea* (Dioscoreaceae) indicates four origins of yams in the Neotropics since the Eocene. Bot J Linn Soc. 2018;188(2):144–60.

Cox DK, Corzo AH, Matuda E, Durán JGG. Estudio de las Dioscoreas Mexicanas. I. *Dioscorea spiculiflora* Hemsl. Bol Soc Bot Mex. 1958;22:12–27.

Denadi N, Gandonou C, Missihoun AA, Zoundjihékpon J, Quinet M. Plant sex prediction using genetic markers in cultivated yams (*Dioscorea rotundata* Poir.) in Benin. Agronomy. 2020;10(10):1521.

Dounias E. The management of wild yam tubers by the Baka pygmies in southern Cameroon. Afr Study Monogr Suppl. 2001;26:135–56.

Dumont R, Vernier P. Domestication of yams (*Dioscorea cayenensis-rotundata*) within the Bariba ethnic group in Benin. Outlook Agric. 2000;29(2):137–42.

Dutta B. Food and medicinal values of certain species of *Dioscorea* with special reference to Assam. J Pharmacogn Phytochem. 2015;3(4):15–8.

FAOSTAT (2018) Food and Agriculture Organization. http://www.fao.org/statistics. Accessed 10 May 2020

Girma G, Hyma KE, Asiedu R, Mitchell SE, Gedil M, Spillane C. Next-generation sequencing based genotyping, cytometry and phenotyping for understanding diversity and evolution of guinea yams. Theor Appl Genet. 2014;127(8):1783–94.

Grattapaglia D, Sederoff R. Genetic linkage maps of *Eucalyptus grandis* and *Eucalyptus urophylla* using a pseudo-testcross: mapping strategy and RAPD markers. Genetics. 1994;137(4):1121–37.

Gutenkunst RN, Hernandez RD, Williamson SH, Bustamante CD. Inferring the joint demographic history of multiple populations from multidimensional SNP frequency data. PLoS Genet. 2009;5(10):e1000695.

Harkess A, Huang K, van der Hulst R, Tissen B, Caplan JL, Koppula A, et al. Sex determination by two Y-linked genes in garden asparagus. Plant Cell. 2020;32(6):1790–6.

Heslop-Harrison JS, Schwarzacher T. Domestication, genomics and the future for banana. Ann Bot. 2007;100(5):1073–84.

Howe KL, Contreras-Moreira B, De Silva N, Maslen G, Akanni W, Allen J, et al. Ensembl genomes 2020 – enabling non-vertebrate genomic research. Nucleic Acids Res. 2019;48(D1):D689–95.

Hu XS, Filatov DA. The large-X effect in plants: increased species divergence and reduced gene flow on the *Silene* X-chromosome. Mol Ecol. 2016;25(11):2609–19.

Huber H. Dioscoreaceae. In: Kubitzki K, editor. The families and genera of vascular plants, Flowering plants: monocotyledons, vol. III. Berlin: Springer; 1998. p. 216–35.

Huson DH, Bryant D. Application of phylogenetic networks in evolutionary studies. Mol Biol Evol. 2006;23(2):254–67.

Ikiriza H, Ogwang PE, Peter EL, Hedmon O, Tolo CU, Abubaker M, et al. *Dioscorea bulbifera*, a highly threatened African medicinal plant, a review. Cogent Biol. 2019;5(1):1631561.

Itoh N, Segawa T, Tamiru M, Abe A, Sakamoto S, Uemura A, et al. Next-generation sequencing-based bulked segregant analysis for QTL mapping in the heterozygous species *Brassica rapa*. Theor Appl Genet. 2019;132(10):2913–25.

Jarvis DI, Hodgkin T. Wild relatives and crop cultivars: detecting natural introgression and farmer selection of new genetic combinations in agroecosystems. Mol Ecol. 1999;8:S159–73.

Jensen HW. Meiosis in several species of dioecious Monocotyledoneae I. The possibility of sex-chromosomes. Cytologia. 1937;1:96–103.

Kiangsu Institute of Botany. Studies on Chinese *Dioscorea* sect. Stenophora Pr et Burk and their chromosome numbers. Acta Phytotax Sin. 1976;14(1):65–72.

Knuth R. Dioscoreaceae. In: Engler HGA, editor. Das Pflanzenrich, 87 (IV. 43). Leipzig: H. R. Engelmann (J. Cramer); 1924. p. 1–387.

Kumar S, Kumari R, Sharma V. Genetics of dioecy and causal sex chromosomes in plants. J Genet. 2014;93(1):241–77.

Lebot V, Malapa R, Abraham K. The Pacific yam (*Dioscorea nummularia* Lam.), an under-exploited tuber crop from Melanesia. Genet Resour Crop Evol. 2017;64(1):217–35.

Lebot V, Abraham K, Kaoh J, Rogers C, Molisalé T. Development of anthracnose resistant hybrids of the greater yam (*Dioscorea alata* L.) and interspecific hybrids with *D. nummularia* Lam. Genet Resour Crop Evol. 2019;66(4):871–83.

Liu XT, Wang ZZ, Xiao W, Zhao HW, Hu J, Yu B. Cholestane and spirostane glycosides from the rhizomes of *Dioscorea septemloba*. Phytochemistry. 2008;69(6):1411–8.

Magwé-Tindo J, Wieringa JJ, Sonké B, Zapfack L, Vigouroux Y, Couvreur TLP, et al. Guinea yam (*Dioscorea* spp., Dioscoreaceae) wild relatives identified using whole plastome phylogenetic analyses. Taxon. 2018;67(5):905–15.

Malapa R, Arnau G, Noyer JL, Lebot V. Genetic diversity of the greater yam (*Dioscorea alata* L.) and relatedness to *D. nummularia* Lam. and *D. transversa* Br. as revealed with AFLP markers. Genet Resour Crop Evol. 2005;52(7):919–29.

Mallet J. Hybrid speciation. Nature. 2007;446(7133):279–83.

Martin FW, Ortiz S. Chromosome numbers and behavior in some species of *Dioscorea*. Cytologia. 1963;28(1):96–101.

Martin FW, Ortiz S. New chromosome numbers in some *Dioscorea* species. Cytologia. 1966;31(1):105–7.

Maurin O, Muasya AM, Catalan P, Shongwe EZ, Viruel J, Wilkin P, et al. Diversification into novel habitats in the Africa clade of *Dioscorea* (Dioscoreaceae): erect habit and elephant's foot tubers. BMC Evol Biol. 2016;16(1):238.

Mehra PN, Sachdeva SK. Cytological observations on some W. Himalayan Monocots IV. Several families. Cytologia. 1976;41(1):31–53.

Meurman O. The chromosomal behavior of some dioecious plants and their relatives with special reference to the sex chromosomes. Soc Scient Fenn Comm Biol II. 1925;3:1–105.

Miège J. Nombres chromosomiques et rèpartition gèographique de quelques plantes tropicales et èquatoriales. Rev Cytol Biol Vég Bot. 1954;15(4):312–48.

Mignouna HD, Dansi A. Yam (*Dioscorea* spp.) domestication by the Nago and Fon ethnic groups in Benin. Genet Resour Crop Evol. 2003;50(5):519–28.

Mignouna HD, Abang MM, Asiedu R. Harnessing modern biotechnology for tropical tuber crop improvement: yam (*Dioscorea* spp.) molecular breeding. Afr J Biotechnol. 2003;2(12):478–85.

Müller NA, Kersten B, Leite Montalvão AP, Mähler N, Bernhardsson C, Bräutigam K, et al. A single gene underlies the dynamic evolution of poplar sex determination. Nat Plants. 2020;6 (6):630–7.

Murty YS, Purnima. Morphology, anatomy and development of bulbil in some dioscoreas. Proc Indian Acad Sci (Plant Sci). 1983;92(6):443–9.

Nakajima G. Chromosome numbers in some angiosperms. Jap J Genet. 1933;9(1):1–5.

Nakajima G. Cytological studies in some dioecious plants. Cytologia. 1937;1:282–92.

Nakajima G. Cytological studies in some flowering dioecious plants, with special reference to the sex chromosomes. Cytologia. 1942;12(2–3):262–70.

Noda H, Yamashita J, Fuse S, Pooma R, Poopath M, Tobe H, et al. A large-scale phylogenetic analysis of *Dioscorea* (Dioscoreaceae), with reference to character evolution and subgeneric recognition. Acta Phytotax Geobot. 2020;71(2):103–28.

Obidiegwu JE, Akpabio EM. The geography of yam cultivation in southern Nigeria: exploring its social meanings and cultural functions. J Ethn Foods. 2017;4(1):28–35.

Obidiegwu JE, Lyons JB, Chilaka CA. The *Dioscorea* genus (Yam) – an appraisal of nutritional and therapeutic potentials. Foods. 2020;9(9):1304.

Paul C, Debnath B. A report on new chromosome number of three *Dioscorea* species. Plant Sci Today. 2019;6(2):147–50.

Pei C, Ting CT, Chin HC, Su P, Tang SY, Chang HC. A preliminary systematic study of *Dioscorea* L. sect. Stenophora Uline. Acta Phytotax Sin. 1979;17(3):61–72.

Peng JH, Sun D, Nevo E. Domestication evolution, genetics and genomics in wheat. Mol Breeding. 2011;28(3):281–301.

Policansky D. Sex change in plants and animals. Annu Rev Ecol Syst. 1982;13(1):471–95.

Prain D, Burkill IH. An account of the genus Dioscorea in the east, part 1: the species which twine to the left. Ann Roy Bot Gard. 1936;14:1–210.

Prain D, Burkill IH. An account of the genus Dioscorea in the east, part 2: the species which twine to the right. Ann Roy Bot Gard. 1939;14:211–528.

Raghavan RS. A chromosome survey of indian dioscoreas. Proc Indian Acad Sci B. 1958;48(1):59–63.

Raghavan RS. A note on some south Indian species of the genus *Dioscorea*. Curr Sci. 1959;28 (8):337–8.

Ramachandran K. Studies on the cytology and sex determination of the Dioscoreaceae. J Indian Bot Soc. 1962;41:93–8.

Ramachandran K. Cytological studies in Dioscoreaceae. Cytologia. 1968;33(3–4):401–10.

Ramu P, Esuma W, Kawuki R, Rabbi IY, Egesi C, Bredeson JV, et al. Cassava haplotype map highlights fixation of deleterious mutations during clonal propagation. Nat Genet. 2017;49 (6):959–63.

Reich D, Thangaraj K, Patterson N, Price AL, Singh L. Reconstructing Indian population history. Nature. 2009;461(7263):489–94.

Renner SS. The relative and absolute frequencies of angiosperm sexual systems: dioecy, monoecy, gynodioecy, and an updated online database. Am J Bot. 2014;101(10):1588–96.

Renner SS, Ricklefs RE. Dioecy and its correlates in the flowering plants. Am J Bot. 1995;82 (5):596–606.

Scarcelli N, Daïnou O, Agbangla C, Tostain S, Pham JL. Segregation patterns of isozyme loci and microsatellite markers show the diploidy of African yam *Dioscorea rotundata* (2n=40). Theor Appl Genet. 2005;111(2):226–32.

Scarcelli N, Tostain S, Vigouroux Y, Agbangla C, Daïnou O, Pham JL. Farmers' use of wild relative and sexual reproduction in a vegetatively propagated crop. The case of yam in Benin. Mol Ecol. 2006;15(9):2421–31.

Scarcelli N, Chaïr H, Causse S, Vesta R, Couvreur TLP, Vigouroux Y. Crop wild relative conservation: wild yams are not that wild. Biol Conserv. 2017;210:325–33.

Scarcelli N, Cubry P, Akakpo R, Thuillet AC, Obidiegwu J, Baco MN, et al. Yam genomics supports West Africa as a major cradle of crop domestication. Sci Adv. 2019;5(5):eaaw1947.

Scott DH. On two new instances of spinous roots. Ann Bot. 1897;11(42):327–32.

Sefa-Dedeh S, Afoakwa EO. Biochemical and textural changes in trifoliate yam *Dioscorea dumetorum* tubers after harvest. Food Chem. 2002;79(1):27–40.

Sharif BM, Burgarella C, Cormier F, Mournet P, Causse S, Van KN, et al. Genome-wide genotyping elucidates the geographical diversification and dispersal of the polyploid and clonally propagated yam (*Dioscorea alata*). Ann Bot. 2020;126(6):1029–38.

Sharma AK, De DN. Polyploidy in *Dioscorea*. Genetica. 1956;28(1):112–20.

Siadjeu C, Mayland-Quellhorst E, Albach DC. Genetic diversity and population structure of trifoliate yam (*Dioscorea dumetorum* Kunth) in Cameroon revealed by genotyping-by-sequencing (GBS). BMC Plant Biol. 2018;18(1):359.

Siadjeu C, Pucker B, Viehöver P, Albach DC, Weisshaar B. High contiguity de novo genome sequence assembly of trifoliate yam (*Dioscorea dumetorum*) using long read sequencing. Genes. 2020;11(3):274.

Silva G, Bömer M, Rathnayake AI, Sewe SO, Visendi P, Oyekanmi JO, et al. Molecular characterization of a new virus species identified in yam (*Dioscorea* spp.) by high-throughput sequencing. Plants. 2019;8(6):167.

Smith BW. Notes on the cytology and distribution of the Dioscoreaceae. Bull Torrey Bot Club. 1937;64(4):189–97.

Sonibare MA, Asiedu R, Albach DC. Genetic diversity of *Dioscorea dumetorum* (Kunth) Pax using amplified fragment length polymorphisms (AFLP) and cpDNA. Biochem Syst Ecol. 2010;38(3):320–34.

Suessenguth. Bemerkungen zur meiotischen und somatischen Kernteilung bei einigen Monokotylen. Flora Allg Bot Ztg. 1921;114(3–4):313–28.

Sugihara Y, Darkwa K, Yaegashi H, Natsume S, Shimizu M, Abe A, et al. Genome analyses reveal the hybrid origin of the staple crop white Guinea yam (*Dioscorea rotundata*). Proc Natl Acad Sci U S A. 2020;117(50):31987–92.

Takagi H, Abe A, Yoshida K, Kosugi S, Natsume S, Mitsuoka C, et al. QTL-seq: rapid mapping of quantitative trait loci in rice by whole genome resequencing of DNA from two bulked populations. Plant J. 2013;74(1):174–83.

Takeuchi Y, Iwao T, Akahori A. Chromosome numbers of some Japanese *Dioscorea* species. Acta Phytotax Geobot. 1970;24(4–6):168–73.

Tamiru M, Becker HC, Maass BL. Genetic diversity in yam germplasm from Ethiopia and their relatedness to the main cultivated *Dioscorea* species assessed by AFLP markers. Crop Sci. 2007;47(4):1744–53.

Tamiru M, Natsume S, Takagi H, White B, Yaegashi H, Shimizu M, et al. Genome sequencing of the staple food crop white Guinea yam enables the development of a molecular marker for sex determination. BMC Biol. 2017;15(1):86.

Terauchi R, Kahl G. Sex determination in *Dioscorea tokoro*, a wild yam species. In: Ainsworth CC, editor. Sex determination in plants. Oxford: Bios Scientific Publishers; 1999. p. 163–71.

Terauchi R, Terachi T, Tsunewaki K. Intraspecific variation of chloroplast DNA in *Dioscorea bulbifera* L. Theor Appl Genet. 1991;81(4):461–70.

Terauchi R, Chikaleke VA, Thottappilly G, Hahn SK. Origin and phylogeny of Guinea yams as revealed by RFLP analysis of chloroplast DNA and nuclear ribosomal DNA. Theor Appl Genet. 1992;83(6–7):743–51.

Uline EB. Eine monographie der Dioscoreaceen. Bot Jahrb Syst. 1898;25:126–65.

Veyres N, Danon A, Aono M, Galliot S, Karibasappa YB, Diet A, et al. The Arabidopsis sweetie mutant is affected in carbohydrate metabolism and defective in the control of growth, development and senescence. Plant J. 2008a;55(4):665–86.

Veyres N, Aono M, Sangwan-Norree BS, Sangwan RS. Has Arabidopsis SWEETIE protein a role in sugar flux and utilization? Plant Signal Behav. 2008b;3(9):722–5.

Viruel J, Segarra-Moragues JG, Raz L, Forest F, Wilkin P, Sanmartín I, et al. Late Cretaceous–Early Eocene origin of yams (*Dioscorea*, Dioscoreaceae) in the Laurasian Palaearctic and their subsequent Oligocene–Miocene diversification. J Biogeogr. 2016;43(4):750–62.

WCSP. World checklist of selected plant families. Kew: Facilitated by the Royal Botanic Gardens; 2020. http://wcsp.science.kew.org/. Accessed 10 Dec 2020

Wilkin P, Schols P, Chase MW, Chayamarit K, Furness CA, Huysmans S, et al. A plastid gene phylogeny of the yam genus, *Dioscorea*: roots, fruits and Madagascar. Syst Bot. 2005;30(4):736–49.

Xiaoqin Z, Guolu L, Xiaolin L. A comparision among natural variations of *Dioscorea zingiberensis*. J Trop Subtrop Bot. 2003;11(3):267–70.

Zhang T, Hu Y, Jiang W, Fang L, Guan X, Chen J, et al. Sequencing of allotetraploid cotton (*Gossypium hirsutum* L. acc. TM-1) provides a resource for fiber improvement. Nat Biotechnol. 2015;33(5):531–7.

Zhang YM, Chen M, Sun L, Wang Y, Yin J, Liu J, et al. Genome-wide identification and evolutionary analysis of NBS-LRR genes from *Dioscorea rotundata*. Front Genet. 2020;11:484.

Open Access This chapter is licensed under the terms of the Creative Commons Attribution 4.0 International License (http://creativecommons.org/licenses/by/4.0/), which permits use, sharing, adaptation, distribution and reproduction in any medium or format, as long as you give appropriate credit to the original author(s) and the source, provide a link to the Creative Commons license and indicate if changes were made.

The images or other third party material in this chapter are included in the chapter's Creative Commons license, unless indicated otherwise in a credit line to the material. If material is not included in the chapter's Creative Commons license and your intended use is not permitted by statutory regulation or exceeds the permitted use, you will need to obtain permission directly from the copyright holder.

Population Genomics of Sweet Watermelon

Padma Nimmakayala, Purushothaman Natarajan, Carlos Lopez-Ortiz, Sudip K. Dutta, Amnon Levi, and Umesh K. Reddy

Abstract Sweet watermelon (*Citrullus lanatus* var. *vulgaris*) belongs to the genus *Citrullus* Schrad and has six known species including *Citrullus naudinianus, C. colocynthis, C. ecirrhosus, C. rehmii, C. amarus,* and *C. mucosospermus* which exist in the wild. All these *Citrullus* spp. have wide variety of fruits with various shapes, size, rind thickness, colors and patterns, and flesh texture. Population genomics studies such as whole-genome sequencing of cultivars and related species, resequencing to generate genomewide SNPs and INDELs, characterization of the chromosomal organization by fluorescence in situ hybridization of rDNA probes, transcriptome analyses, QTL analyses for fruit traits, biotic and abiotic stresses have addressed multiple questions related to the biology and evolution of watermelon. Genome-wide genetic diversity, distribution of linkage disequilibrium, model-based approaches for population structure, and admixture allowed to infer shared ancestries involving global collections of sweet watermelon. Similarly, published studies in watermelon genetic diversity have not systematically prioritized role of *C. amarus* genes in diversification of watermelon gene pool. Phylogeny analyses based on whole-genome sequencing and conserved ITS and chloroplast markers of various *Citrullus* species indicated that *C. mucosospermus* and its derivative *C. lanatus* subsp. *cordophanus* might be the ancestors of sweet watermelon. However, knowledge is elusive to figure out after losing bitterness in fruit, if the fruit size, rind thickness, flesh softening, lycopene accumulation, sucrose synthesis and ripening simultaneously were the part of domestication syndrome or there had been a logical order of the evolutionary histories of the mutations in underlying genes that played roles in the domestication. This chapter is a review of evolution, genetic diversity, phylogenies, selected QTL and transcriptome studies, and genomics in perspective of domestication and diversification of cultivated watermelon clade.

P. Nimmakayala · P. Natarajan · C. Lopez-Ortiz · S. K. Dutta · U. K. Reddy (✉)
Gus R. Douglass Institute, Department of Biology, West Virginia State University, Institute, WV, USA
e-mail: ureddy@wvstateu.edu

A. Levi
USDA, ARS, U.S. Vegetable Laboratory, Charleston, SC, USA

Keywords Citrullus · Domestication syndrome · Egusi · Genetic diversity · Linkage disequilibrium · Origin and evolution · Phylogeny · Selective sweeps · Sweet watermelon

1 Introduction

Watermelon (*Citrullus lanatus*) ($2n = 11$) belongs to the genus *Citrullus* Schrad and the center of origin of this genus is in Africa (Whitaker and Bemis 1976). The genus comprises seven known diploid ($n = 11$) species (Dane and Liu 2007; Jeffrey 1975), of which *Citrullus naudinianus, C. colocynthis, C. ecirrhosus, C. rehmii, C. amarus,* and *C. mucosospermus* (egusi) exist in the wild and have wide variety of fruits with various shapes, size, rind thickness, colors and patterns, and flesh texture colors, sugar content, tastes, and blends (Fig. 1). The genus *Citrullus* is placed closest to a clade formed by bottle gourd (*Lagenaria* spp.) (Chomicki and Renner 2015; Jobst et al. 1998; Schaefer and Renner 2011). *C. naudinianus*, of the Kalahari region, is a sister species to the other six species, followed by *C. colocynthis* and *C. rehmii*, an annual Kalahari species (Jarret and Newman 2000). *C. amarus* (Bailey 1930) is a sister species to tendril-less *C. ecirrhosus*; both are native to the Kalahari region (De Winter 1990).

Citrullus amarus was previously named as *C. lanatus* var. *citroides* and is popularly known as tsamma, cow, or citron melon and cultivated for pickle making or animal fodder in southern Africa (Meeuse 1962). *Citrullus lanatus* and *C. amarus* produce fertile offspring and are of great importance in modern breeding for resistance to various diseases (Levi et al. 2000). Fursa (1972) and Fursa (1983) reported that *C. mucosospermus*, the egusi melon, has white, oleaginous seeds with a black margin is the closest relative to *C. lanatus*. The *C. mucosospermus* exists in the regions of Nigeria and Senegal and is cultivated mainly for its tasty white seeds that are consumed as a raw snack in West Africa (Meeuse 1962; Okoli 1984; Robinson and Decker-Walters 1999). *C. mucosospermus* is closest to the annual *C. lanatus* (Thunb.) Matsum & Nakai, the progenitor of sweet watermelon (Chomicki and Renner 2015; Fursa 1983). The Plant Genetic Resources Conservation Unit (Griffin, GA), US Department of Agriculture–Agricultural Research Services, maintains more than 1,650 US plant introductions of *C. lanatus* and several semidomesticated forms that will be of immense importance for future studies on the domestication process (Levi et al. 2013, 2017). A critical question that remains unsolved is whether the domestication process in watermelon was a single event or multiple events involving look-alike subspecies that are primarily egusi derivatives and Kordofan melon types with nonbitter whitish pulp. To date, it is not yet known if the key mechanisms that led to (1) increased fruit size and flesh and rind softness; (2) replacing the bitter taste with sweetness; and (3) coloration of white flesh occurred all at once or involved several generations. It is also not yet known if the domestication is a single event of evolution or many simultaneous events of domestication that led to the currently cultivated watermelon.

Fig. 1 Morphological features of various species in *Citrullus* genus

Use of population genetic and genomic inference is becoming a strategic tool for linking biology, evolution, phylogeny, speciation, phylogeography, domestication, genotype-phenotype association, and conservation to accelerate plant breeding and basic biological research (Birkner and Blath 2009; Pool et al. 2010; Rajora 2019; Luikart et al. 2019). Among various methods, comparison of whole-genome assemblies from various species and assembling multiple genomic reads to growing pan-genome reference are currently choice for trait discovery including plant type, maturity time, fruit traits, flavor, aroma, and additional copies of biotic and abiotic resistance genes that are lost in process of domestication and breeding (Crisci et al. 2012). More recently, there have been rapid developments in graph-based pan-genome assembly, where a graph representing genomic diversity and conservation is constructed for tomato (*Solanum lycopersicum*), soybean (*Glycine max*), and rice (*Oryza sativa*) (Bayer et al. 2020b).

Comparative pan-genome analysis involves examining the similarities and unique differences to shed light on the underlying genome evolution and identify economically important alleles (Golicz et al. 2016a). Defining presence and absence variations (PAVs) by comparative mapping has shown that numerous alterations in diverse genomes contributed to genetic diversity among the related species. Over time, chromosomes are broken, reassembled, partially or wholly duplicated, and even eliminated, ultimately resulting in reproductive isolation and speciation (Bayer et al. 2020a; Golicz et al. 2016b; Montenegro et al. 2017). For example, comparative genome analysis to understand conserved syntenic blocks in *Citrullus* genomes holds promise for clarifying the selection pressures driving genetic changes. PAVs are an important source of genetic and phenotypic diversity. Precisely speaking, structural variations that include insertions, deletions, duplications, inversions, and translocations resolved with optical mapping strategies have been associated with stress tolerance, resistance, increase in yields, reproductive morphology, adaptation, and speciation (Gupta 2021). Such investigations will elucidate alterations at the level of the whole genome, for diversifying cultivars with narrow genetic backgrounds (Tirnaz et al. 2020). Delineating *Citrullus* species based on gene content remains a challenge, particularly considering the significant gene PAV between individuals of a species.

2 Origin, Evolution, and Phylogeny

Paintings of watermelon served on a tray found in Egyptian tombs suggested that these fruits were semi-ripened (Janick et al. 2007). Several 5,000-year-old seeds of wild watermelon, *C. lanatus*, found at the archaeological site Uan Muhuggiag in southwest Libya, are the oldest discovered seeds belonging to Neolithic pastoralists. *Tractatus de herbis* (British Library manuscript, Egerton 747, 2003) from southern Italy around the year 1,300 contains an accurate, informative image of watermelon (Paris 2015). In addition, Paris (2015) explored a dozen more original illustrations from Italy. These images depicted watermelon plant with long internodes, alternate leaves with pinnatifid leaf laminae, and small, round, and striped fruits with both red and white flesh (Carrara Herbal, British Library manuscript Egerton 2020, 1974). The variation in watermelon fruit size, shape, and coloration depicted in the illustrations represents at least six cultivars of watermelon, three of which probably had red, sweet flesh and three that appear to have been citrons. Evidently, citron watermelons were more common in Mediterranean Europe in the past than they are today (Wasylikowa and van der Veen 2004).

Dane and Liu (Dane and Lang 2004; Dane and Liu 2007) studied chloroplast DNA by using PCR-RFLP and sequencing analysis of several noncoding regions of cultivated egusi and citron watermelon types. They identified distinct chlorotype lineages, separating the cultivated and egusi-type watermelon from citron accessions. The authors further suggested that cultivated and egusi appear to have diverged independently from a common ancestor, possibly *C. ecirrhosus* from Namibia. Chomicki and Renner (2015) confirmed the tendril-less South African endemic species *C. ecirrhosus* as an ancestor for *C. mucosospermus* and *C. lanatus*.

Achigan-Dako et al. (2021) analyzed noncoding chloroplast DNA sequences (trnT-trnL and ndhF-rpl32 regions) from a global collection of 135 accessions and identified 38 haplotypes in *C. lanatus, C. mucosospermus, C. amarus, and C. colocynthis*. The least diverse species was *C. mucosospermus*, with 5 haplotypes, and the most diverse was *C. colocynthis*, with 16 haplotypes. *C. lanatus* and

C. mucosospermus shared haplotypes, which support the hypotheses that the center of origin for cultivated watermelon is West Africa, where *C. mucosospermus* thrives. Wild and primitive domesticated forms of watermelon have been observed repeatedly in Sudan and neighboring countries of northeastern Africa. A second hypothesis for watermelon domestication suggests a northeastern African origin, where the white-fleshed Kordophan melon which belongs to the under-explored "cordophanus" from Sudan, resembling cultivated watermelon, is found (Renner et al. 2017, 2019). Chomicki and Renner (2015) resolved the phylogeny using nuclear ITS region (ITS1, 5.8S rDNA, ITS2; the trnL intron; the trnL-trnF, rpl20-rps12, trnR-atpA, trnG-trnS, Ycf9-trnG, and Ycf6-PsbM spacer; and the genes ndhF, rbcL, and matK of various *Citrullus* species along with silica-dried leaves or herbarium material of *C. lanatus*, prepared by Linnaeus's disciple and collector Carl Peter Thunberg in South Africa in 1773. This study confirmed the genetic closeness of cultivated watermelon with *C. mucosospermus* from western Africa. In this study, a 30-bp deletion in the plastid trnS-trnG intergenic spacer was used as a genetic marker, which is ideal for barcoding because it reliably distinguishes the citron melon *C. amarus* from the sweet watermelon.

2.1 Cytogenetic Analysis of Species Divergence in Citrullus *Genus*

The 18S–28S and 5S rDNA sites are useful chromosome landmarks and provide valuable evidence about genome organization and evolution. Aryal (2011), Aryal et al. (2010), Reddy et al. (2013) resolved the dynamics, distribution, and directionality of rDNA gains and losses and aimed to understand the contribution of site number variation in the speciation of the genus *Citrullus*. In this study, fluorescent in situ hybridization was used with the 18S–28S and 5S rDNA gene loci to evaluate the differences among species. FISH analyses revealed a similar rDNA configuration in sweet watermelon and colocynth. The sweet watermelon and colocynth genomes contain two 18S–28S rDNA gene loci, each located on a different chromosome, and one 5S rDNA locus, which co-localizes with one of the 18S–28S rDNA gene loci. However, *C. rehmii* has one 18S–28S rDNA locus and one 5S rDNA locus positioned on different chromosomes, whereas *C. amarus* has one 18S–28S rDNA and two 5S rDNA loci, each located on a different chromosome. A FISH analysis of F_1 (citron × sweet watermelon) chromosome spreads revealed uniparental homologous rDNA gene copies pertaining to sweet watermelon versus citron chromosomes. The sweet watermelon chromosome contains the 18S–28S and 5S rDNA loci, whereas the citron homologue chromosome has the 5S rDNA locus but not the 18S–28S rDNA locus. Genomic in situ hybridization, with the entire citron genome used as a probe to be differentially hybridized on sweet watermelon chromosome spreads, revealed that the citron genomic probes mainly hybridize to subtelomeric and pericentromeric regions of sweet watermelon chromosomes, which suggests extensive divergence between the citron and sweet watermelon genomes.

Li et al. (2016) used FISH for comparative mapping of 5S and 45S rDNA to identify cultivated watermelon. *C. mucosospermus, C. colocynthis,* and *C. naudinianus* (or *Acanthosicyos naudinianus*) had two 45S rDNA loci and one 5S rDNA locus, which was located syntenic to one of the 45S rDNA loci. *C. ecirrhosus* and *C. lanatus* had one 45S rDNA locus and two 5S rDNA loci, each located on a different chromosome. *C. rehmii* had one 5S and one 45S rDNA locus on different chromosomes. In addition, this cytogenetic analysis suggested that *A. naudinianus* was more closely related to *Cucumis* than *Citrullus* or *Acanthosicyos* but with a unique position and may be a bridge between *Citrullus* and *Cucumis*.

3 Genomic and Transcriptomic Resources for Population Genomics Studies

Genomes, genetic maps, and transcriptomes provide highly useful genome-wide genetic markers as well as mapped genetic markers for population genomics studies and applications (see Luikart et al. (2019) for details). Here we summarize substantial genomic and transcriptomic resources developed in watermelon.

3.1 Genomes

The development of whole-genome sequence drafts has provided a foundation for population genomics studies, widening a narrow genetic background, marker-assisted selection and to understand intricate genome rearrangements to study genetics and breed improved varieties in less-important crops like watermelon. Furthermore, by using the reference genome maps of various watermelons, we can identify major genes affecting agronomic characters found in different species. Guo et al. (2013) analyzed the syntenic relationships between watermelon, cucumber, melon, and grape to identify 3,543 orthologous relationships covering 60% of the watermelon genome. This study further resolved complicated syntenic patterns using detailed chromosome-to-chromosome relationships within the Cucurbitaceae family and identified orthologous chromosomes among watermelon, cucumber, and melon. The insights of high degree of complexity of chromosomal evolution and rearrangement by using chromosome-to-chromosome orthologous relationships unveiled genomic relationships of these three important crop species of the Cucurbitaceae family. Integration of independent analyses of duplications within, and syntenies between, the four eudicot genomes (watermelon, cucumber (*Cucumis sativus*), melon, and grape (*Vitis vinifera*)) led to the precise characterization in watermelon of the seven paleotriplications identified recently as the basis for defining seven ancestral chromosomal groups in eudicots (Abrouk et al. 2010). With the ancestral hexaploidization (γ) reported for the eudicots, Guo et al. (2013) proposed an evolutionary scenario that has shaped the 11 watermelon chromosomes from the 7-chromosome eudicot ancestors through the 21 paleohexaploid intermediates. The authors suggested that the transition from the 21-chromosome eudicot intermediate ancestors involved 81 fissions and 91 fusions to reach the modern 11-chromosome structure of watermelon, represented as a mosaic of 102 ancestral blocks in the watermelon genome. Guo et al. (2013) identified 159.8 Mb (45.2%) of the assembled watermelon genome as transposable element repeats; 68.3% could be annotated with known repeat families. Transposable element divergence rates peaked at 32%. The authors further identified 920 (7.8 Mb) full-length LTR retrotransposons in the watermelon genome. Over the past 4.5 million years, LTR retrotransposons accumulated much faster in watermelon than cucumber; hence, the overall difference in their genome sizes may reflect the differential LTR retrotransposon accumulation (Guo et al. 2013). In watermelon, 45 members belonging to the LOX gene family were arranged in two tandem arrays. Among the 197 receptor-like genes in the watermelon genome, 35 encode receptor-like proteins lacking a kinase domain in addition to the extracellular LRR and transmembrane domains (Guo et al. 2013). In watermelon, 44 NBS-LRR-TIR genes (18 TIRs and 26 coiled-coil NBS-LRR–encoding genes) were identified.

Wu et al. (2019) reported a high-quality genome sequence of the watermelon cultivar 'Charleston Gray', a principal American dessert watermelon, to complement the previous reference genome from '97103', an East Asian cultivar. Chromosome-scale assembly of the 'Charleston Gray' genome is of 382.5 Mb and mapped to 11 pseudomolecules (Fig. 2) (Wu et al. 2019). This genome has an assigned orientation to an additional of 9.9 Mb of scaffolds in comparison with previously sequenced genome, according to two other recently developed genetic maps (Branham et al. 2017; Reddy et al. 2014a). Comparative analyses between genomes of 'Charleston Gray' and '97103' revealed genomic variants that may underlie phenotypic differences between the two cultivars.

3.2 Genetic Maps and Genetic Analysis

There have been linkage mapping efforts for watermelon since the period of RAPDs to tag important traits as well as purely to understand the length of genetic map or to define synteny with the other cucurbit species. Carter (2008) mapped genes for bacterial fruit blotch resistance; QTLseq analysis of semi-dwarfism (Cho et al. 2021); Dou et al. (2018a) mapped candidate gene (ClFS1) for fruit shape; genetic analysis of the yellow skin trait (Dou et al. 2017); QTL analysis of seed size (Gao et al. 2019b); mapping for gummy stem blight resistance (Gimode et al. 2021); QTLs for fruit traits (Guo et al. 2019); Hashizume et al. (1996) built a linkage map with RAPDs; QTL analysis of horticultural traits (Hashizume et al. 2003); mapping for Fusarium Wilt resistance (Hawkins et al. 2001); mapping sweet genes between

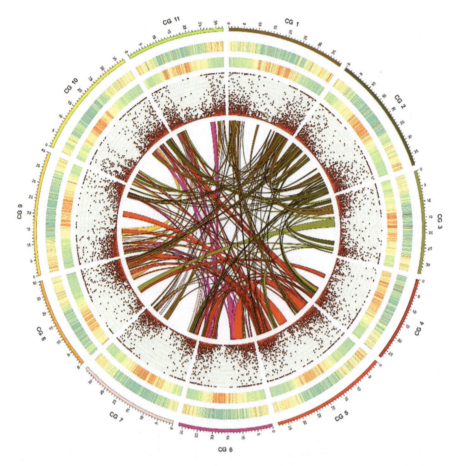

Fig. 2 Genomic landscape of watermelon, 'Charleston Gray'. The outermost circle is the ideogram of 11 chromosomes in Mb scale, followed by circles of gene density and TE density represented by percentage of genomic regions covered by genes and repeat sequences in 200-kb windows, respectively (green to red, low to high), gene expression levels (RPKM; Maximum = 200) and syntenic blocks within the genome depicted by lines. (Reproduced from Wu et al. (2019) https://doi.org/10.1111/pbi.13136)

watermelon and melon (Jayakodi et al. 2019); HRM markers based on SNPs identified from next-generation resequencing of susceptible and resistant parents to gummy stem blight (Lee et al. 2018); Levi et al. (2002) built a reference genetic map; genetic map with diverse set of markers (Levi et al. 2004); an integrated linkage map for watermelon based on SRAP, AFLP, SSR, ISSR, and RAPD markers (Levi et al. 2006); a linkage map of cleaved amplified polymorphic sequence markers (Liu et al. 2016b); Luan et al. (2019) mapped a gene related to seeds in egusi; Meru and McGregor (2013) mapped genes for race 1 of Fusarium Wilt; Nimmakayala et al. (2014a) built a genetic map using veracode technology; Paudel et al. (2019) refined a map with KASP markers; Prothro et al. (2012) performed QTL analysis for seed size; Ren et al. (2015) built a high density map using DArT markers; Rhee et al. (2015a) built a reference map using SSRs; and Zhang et al. (2004) developed RIL population for building a genetic map. Reddy et al. (2014b) mapped 10,480 SNPs into a genetic linkage map using a mapping population that contained 113 progenies generated from a cross of egusi and sweet watermelon. This genetic map defined 3,821 recombination events within the skeletal map. The skeletal map for chromosomes 1 to 11 contained 406, 339, 240, 219, 450, 257, 305, 391, 464, 373, and 388 recombination events, respectively. Reddy et al. (2014b) examined the collinearity of genetic and physical maps for various chromosomes and identified two regions that are not collinear between genetic map and physical map.

3.2.1 Recombination Landscape

Genome-wide recombination rate (GWRR) was estimated using the formula cM/Mb (Fig. 3). We observed wide variation of GWRRs within and among the chromosomes. Mean GWRR for chromosomes 1 to 11 was estimated at 1.25, 1.09, 1.04, 1.25, 1.37, 1.34, 1.06, 1.18, 1.00, 1.15, and 1.49, respectively. GWRR range was 0.32–2.8, 0.03–3.8, 0.09–1.69, 0.28–3.6, 0.02–3.6, 0.21–3, 0.12–2.96, 0.12–3.85, 0.12–1.97, 0.03–3.45, and 0.04–3.80, respectively. Twelve hot spots of recombination containing GWRR in the range of 2 to 4 were distributed on chromosomes 1, 2, 4, 6, 7, 8, and 11. Chromosomes 3, 5, 9, and 10 did not show GWRR > 2; hence, this part of the genome may be less recombinant. However, a trend was noted, whereby the hot spots of recombination (peak of GWRR) correspond to the increase in nucleotide diversity (π) on chromosomes 2, 4, 5, 6, 7, and 11; hence, the recombination landscape was an important factor shaping the cultivar divergence on these chromosomes.

3.3 *Transcriptomes*

There are ample transcriptome resources in watermelon targeting various tissues, fruit development, ripening, flesh color, and biotic and abiotic stress. Genome-wide transcriptome analyses of root stock/scion interactions were studied by Fallik and Ziv (2020); potassium starvation response (Fan et al. 2014); soluble sugar and organic compound accumulation (Gao et al. 2018); and fruit texture (Gao et al. 2020). A study on transcript movement across the reciprocal grafts was carried out by Garcia-Lozano et al. (2020). Garcia-Lozano et al. (2021) performed transcriptome, methylome, and chromatin capture analysis to understand diploid and isogenic tetraploid effects on gene expression and scion-specific long-distance signaling (Gautier et al. 2020). RNAseq studies have been conducted to study flavor accumulation (Gong et al. (2021), carotenoid pathway regulators (Grassi et al. 2013), comparative fruit development (Guo et al. 2015), irradiated pollen analysis (Hu et al. 2019), anthracnose resistance (Jang et al. 2019), Fusarium Wilt (Jiang et al. 2019a),

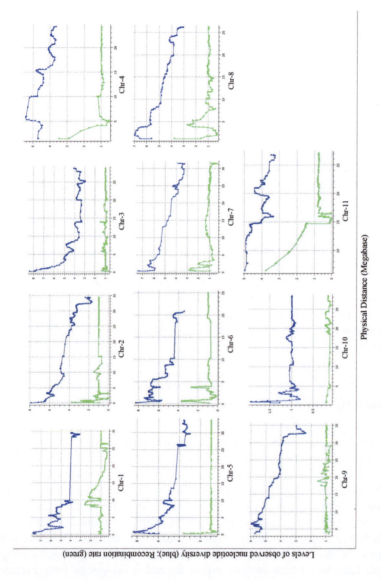

Fig. 3 Distribution of genome-wide recombination rate (GWRR) and observed nucleotide diversity (π) along chromosomes in the watermelon genome. In each plot, the horizontal axis (in Mb) represents the physical distance (PD) along the reference chromosomes and the vertical axis (cM/Mb) the genetic-to-physical distance ratio (green) and log −2 transformed values of nucleotide diversity (π). (Reproduced from Reddy et al. (2015) https://doi.org/10.1534/g3.114.012815)

fruit cracking (Jiang et al. 2019b), citrulline accumulation (Joshi et al. 2019), effects of root knot nematode (Karki et al. 2021), watermelon size (Kim et al. 2019), cucumber green mottle mosaic virus (Li et al. 2017), yellow rind analysis (Liu et al. 2020), graft response (Liu et al. 2016a), microRNA regulation in grated watermelon (Liu et al. 2013, Fusarium Wilt (Lü et al. 2014), response to low nitrogen (Nawaz et al. 2018), SL-type esterases/lipases (Ren et al. 2019), gain of oligosaccharide hydrolysis and sugar transport enhanced carbohydrate partitioning (Ren et al. 2021). Also transcriptome profiling has been conducted for male-sterile and fertile watermelon lines (Rhee et al. 2017; Wang et al. 2020), and to identify differentially expressed genes in floral buds and flowers of male-sterile and fertile lines in watermelon (Rhee et al. 2015b) and impaired biosynthetic networks in defective tapetum that lead to male sterility (Wei et al. 2021). Transcriptomics studies have also revealed short-term salt stress response in watermelon seedlings (Song et al. 2020); candidate genes responsive to cucumber green mottle mosaic virus infection (Sun et al. 2019), key gene networks controlling organic acid and sugar metabolism during watermelon fruit development by integrating metabolic phenotypes (Umer et al. 2020), chilling stress responsiveness in grafted watermelon seedlings (Xu et al. 2016), mechanisms of sugar metabolism and accumulation in sweet and nonsweet watermelon fruits (Xu et al. 2012), and root response to short-term osmotic stress (Yang et al. 2016).

Comparative transcriptome analysis of two contrasting watermelon genotypes during fruit development and ripening (Zhu et al. 2017) and effect of different heat shock periods on the unfertilized ovule in watermelon (ZHU et al. 2020) are the studies involving gene expression and transcriptome. Guo et al. (2015) compared transcriptome profiles of fruit tissues of cultivated watermelon '97103' and wild watermelon 'PI296341-FR' (egusi) and identified 2,452, 826, and 322 differentially expressed genes in cultivated flesh, cultivated mesocarp, and wild flesh, respectively, during fruit development. Comparative transcriptome profiling analysis identified critical genes potentially involved in controlling fruit quality traits, including α-galactosidase, invertase, and UDP-galactose/glucose pyrophosphorylase; sugar transporter genes involved in determining fruit sugar content, including phytoene synthase, β-carotene hydroxylase, 9-*cis*-epoxycarotenoid dioxygenase; carotenoid cleavage dioxygenase genes involved in carotenoid metabolism; and 4-coumarate. In addition, the ethylene biosynthesis and signaling pathway including ACC oxidase, ethylene receptor, and ethylene responsive factor showed highly ripening-associated expression patterns, which indicate a possible role of ethylene in fruit development and ripening of watermelon, a nonclimacteric fruit.

The huge transcriptomic resource has been developed in watermelon, including differentially expressed genes underlying abiotic and biotic factors, biochemical composition, male sterility, fruit size, fruit quality, and other traits and provides an excellent gene repertoire for population-level genomics studies and applications.

4 Molecular Markers, Genetic Diversity, and Structure

Morphological diversity in watermelon pertaining to collections, cultivars, and related wild species has been thoroughly explored across the geographical regions. Mujaju et al. (2010) studied morphological diversity of landraces of Zimbabwe, and Dje et al. (2010) characterized African edible seed type watermelon diversity. Genetic diversity analysis of Indian core collection was performed by multiple researchers (Bhakar et al. 2017; Hajiali et al. 2017; Krishna Prasad et al. 2002; Mahla and Choudhary 2013; Pal et al. 2015; Pandey et al. 2019). Morphological diversity was characterized for watermelon cultivars from Palestine (Alimari et al. 2017); Iran (Kiyani and Jahanbin 2006), Bangladesh (Mohisina et al. 2020); and Sistan (Afghanistan) (Naroueirad et al. 2010). Morphologically, Chinese watermelon genetic resources were thoroughly explored (Ji et al. 2013; Lee et al. 1996; Pan et al. 2015; Shang et al. 2012; Sheng et al. 2012; Ulutürk 2009) while the genetic diversity within the cultivars has also been widely explored. Genetic diversity of Turkey collections revealed that Turkey is a secondary center of origin for watermelon (Sari et al. 2005; Solmaz and Sarı 2009; Solmaz et al. 2010). Morphological characterization of Hungarian watermelons was carried out by Szamosi et al. (2009). Romão (2000) found narrow genetic diversity in watermelon collection from Northeast Brazil (Lima et al. 2017). Intriguingly, watermelon cultivars are morphologically highly diverse and respond well to wide agroclimatic zones across the globe.

First-generation markers are random and do not possess sequence information (RAPD, AFLP, SRAP, and IISR). Many watermelon researchers used these markers to declare watermelon collections have very narrow genetic diversity. ISSR marker analysis of Indian collections (Soghani et al. 2018); SRAP analysis of Chinese collections (Aiping et al. 2008; Li et al. 2013; Yan and Chunqing 2005); AFLP analysis of Chinese collections (Li et al. 2007); and RAPD analysis of US watermelon collections (Levi and Thomas 2007) were some of these, and all of these studies revealed a narrow cultivar diversity.

Simple Sequence Repeats or microsatellites (SSRs) are the second-generation markers that used Sanger method of sequencing for capture, characterize, and design primers for use in PCR-based genotyping. Moreover, SSRs are highly polymorphic and reproducible codominant markers. Northeastern Brazilian collection analysis using microsatellites was performed by Gama et al. (2013). Genetic diversity was resolved in Indian collection by Verma and Arya (2008) and Chinese collection using SSRs were analyzed by Hwang et al. (2011), Wang et al. (2015), Zhang et al. (2010), Kwon et al. (2010), Zhang et al. (2012), Zhao et al. (2010, 2014). US cultivars were studied using SSRs by Levi et al. (2008). Watermelon collections were analyzed using SSRs for diversity in South Africa by Mashilo et al. (2017); citron collection by Mashilo et al. (2016); West African collection by Minsart et al. (2011); Zimbabwe collection by Mujaju et al. (2010); Zimbabwe landraces by Mujaju et al. (2012); Mozambique collection by Munisse et al. (2013); Mali landraces by Nantoume et al. (2013); and Turkish collections by Solmaz et al. (2016). A genome-wide scan of selective sweeps and association mapping of fruit traits using microsatellite markers in watermelon was performed by Reddy et al. (2015). High-frequency oligonucleotides are designed to target functional genes, and HFO-TAG markers revealed wide genetic diversity among *Citrullus* spp. accessions by Levi et al. (2013). All the studies reemphasized the previous findings that there is no widespread genetic diversity in the cultivated germplasm, but there is widespread genetic diversity among the semidomesticated and land races collected from the northeastern Africa to and Southern Africa.

NGS methods involving reduced representation libraries or resequencing and have been employed to understand chromosome-wide genetic diversity among watermelon collections. These are the third-generation markers such as SNPs, structural variants (SVs), and indels (insertions and deletions), which involve high-throughput sequencing. Resequencing of 20 diverse accessions was carried out by Guo et al. (2013) that made available first reference genome sequences for watermelon and related wild species. Many studies involving GBS-generated SNPs for watermelon collections were carried out by Lee et al. (2019), Nimmakayala et al. (2014b), Park et al. (2018), Reddy et al. (2014b), Wu et al. (2019) to resolve chromosome-wide diversity and LD decay of cultivated watermelon. In addition to nuclear markers, chloroplast markers are used to understand cytoplasmic diversity. Chloroplast SNPs were developed by Cui et al. (2020); chloroplast sequence variation of watermelon and related wild species was studied by Dane and Lang (2004), Dane and Liu (2007), and chloroplast-specific SSRs were characterized by Hu et al. (2011) resolving phylogeny and evolution of cultivated watermelon.

Population structure analysis presented in this chapter shows wide admixture of cultivated sweet watermelons with various proportions of *C. amarus* lineages (Figs. 4 and 5) expanding the genetic diversity of cultivated watermelons. Deeper understanding of admixture of *C. amarus* genome in cultivated and landrace watermelon would probably need to resolve if *C. amarus* introgression occurred before or after domestication. Wu et al. (2019) genotyped 1,052 sweet watermelon accessions along with landraces, egusi types, and *C. amarus* collections predominantly collected from Africa, Europe, North America, and Asia and revealed ~62 K SNPs. A subset of these data allowed for estimating genomic diversity across various groups, constructing a neighbor-joining tree, resolving population structure, estimating chromosome-wise LD patterns, and understanding the extent of population differentiation and genetic bottlenecks in sweet watermelon. Using a subset of SNPs that have 0.05 allelic frequency and 70% call rate, the PCA resolved three distinct clusters of sweet watermelons that are admixed with semidomesticated and egusi types (Fig. 6). Three cultivar groups resolved in PCA cannot be explained based on their geographical distribution indicating more complex evolutionary processes such as admixture with *C. amarus* may be underlying. Distance matrix of cultivars, semidomesticated and egusi, were used to construct chromosome-wise neighbor-joining trees (Fig. 7) (Reddy et al. 2015), which revealed differentiation of wild (egusi), semidomesticated (landraces), and fully domesticated cultivars occurred across each chromosome indicating genome-wide divergence of watermelon. Pairwise F_{ST} estimates involving wild versus semidomesticated and

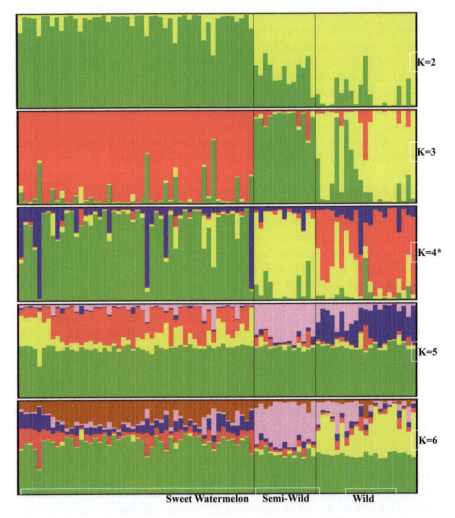

Fig. 4 Population structure for $k = 4$. Clusters are separated by vertical lines with cluster colors indicating various ancestries. Red was predominant in wild, yellow in semi-wild, and green in cultivated watermelon. Cultivated watermelon represented most of the green and to a lesser extent purple, yellow, and red clusters. (Reproduced from Reddy et al. (2015) https://doi.org/10.1534/g3.114.012815)

semidomesticated versus cultivars are presented in Fig. 8. High F_{ST} distribution across the chromosomes can be noted that are under selection across the domestication process. This study (Wu et al. 2019) genotyped 1,365 watermelon plant introduction (PI) lines maintained at the US National Plant Germplasm System and involved 25,000 single nucleotide polymorphisms (SNPs) with genotyping-by-sequencing (GBS). These PI lines belonged to three *Citrullus* species: *C. lanatus*, *C. mucosospermus,* and *C. amarus*. Population genomic analyses with

Fig. 5 Admixture analysis of world collections of watermelon accessions along with *C. amarus* (blue lineage) collections by population structure analysis. (Reproduced from Wu et al. (2019) https://doi.org/10.1111/pbi.13136)

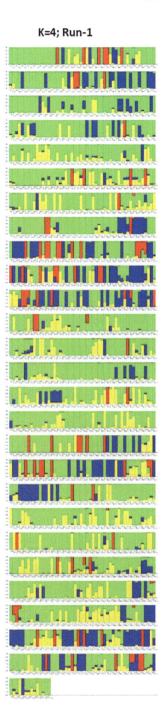

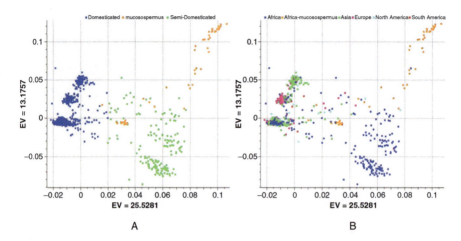

Fig. 6 Principal component analysis (PCA) showing the ordination of the world collections of watermelon accessions (1,209 egusi, landraces, and sweet watermelons along with 17,535 single nucleotide polymorphisms (SNPs) generated by genotyping by sequencing (GBS) on the first two principal components. (Reproduced from Wu et al. (2019) https://doi.org/10.1111/pbi.13136)

these SNPs revealed a close relationship between *C. lanatus* and *C. mucosospermus* and identified three major groups of cultivars and corresponding semidomesticated types.

Consistent with our current understanding of the extent of genetic diversity in sweet watermelons and prior diversity analyses, the most recent analysis revealed 0.037% genetic diversity in worldwide sweet watermelons (Wu et al. 2019). This study revealed high genetic diversity for sweet watermelons in South Africa, and most landraces of South Africa clustered with sweet watermelon, which strengthens the argument that the Kalahari Desert could be the center of diversification if not center of origin for sweet watermelon. In addition, PI22557, PI542122, PI482367, PI542115, PI482318, PI482288, PI270565, PI271774, PI482284, PI270565, PI171392, PI271777, PI271778, and PI271776 are highly diverse landraces belonging to southern Africa that cluster separately from the other sweet watermelon collections (Wu et al. 2019). Alternatively, nonbitter whitish pulp egusi types of western African or Sudanese form, known as the Kordofan melon (*C. lanatus* subsp. *cordophanus*), appears to be the closest relatives of domesticated watermelons and could be the progenitors of sweet watermelon types because they share higher allele frequencies, which indicate a common evolutionary pattern. Similar to this observation, Chomicki and Renner (2015) confirmed observations by Fursa (1972) and Fursa (1983) that egusi and sweet watermelon are closely related. Strong anthropogenic evidence of sweet watermelon is from Egyptian tomb paintings and excavations in Sudan indicating domestication in North and West African germplasm (Paris 2015; Renner et al. 2017).

SNP-based analyses revealed an unused sweet watermelon germplasm in the rest of the world breeding programs (Nimmakayala et al. 2014a, b; Wu et al. 2019). For

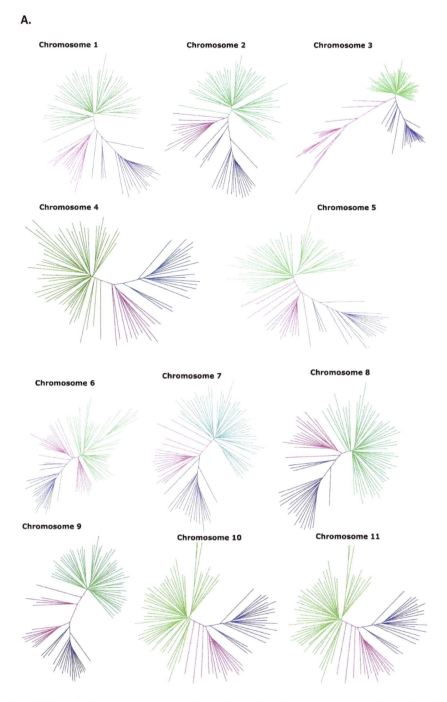

Fig. 7 Phylogenetic trees constructed with neighbor-joining for domesticated, semidomesticated, and cultivars. (Reproduced from Reddy et al. (2015) https://doi.org/10.1534/g3.114.012815)

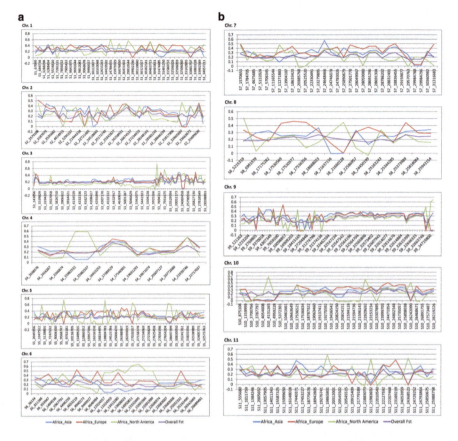

Fig. 8 Genome-wide window-based scans of pairwise F_{ST} for accessions from Asia, Europe, and the Americas compared with those from Africa. Selection signatures can be seen on parts of chromosome 3 and 9, where F_{ST} distribution revealed distinct sweeps (Nimmakayala et al. (2014b) https://doi.org/10.1186/1471-2164-15-767)

example, 20 accessions from the southern African sweet watermelon germplasm were widely distributed and separated from the sweet watermelon cluster. Such accessions can be explored for crossing and diversifying the narrow genetic diversity of sweet watermelon. So far, 105 private SNPs segregating in sweet watermelon are noted that were not found to be polymorphic in related wild species. Such SNPs are valuable because they can be of adaptive importance and would be of immense use for generating passport information for sweet watermelons. Wu et al. (2019) used genome-wide SNPs generated by GBS for genomic-level diversity analysis of 1,135 diverse collections from Africa, Asia, Europe, and the Americas. The analysis identified three distinct clusters of cultivated sweet watermelons. These findings agree with those of previous studies (Levi et al. 2001, 2013; Nimmakayala et al. 2011; Reddy et al. 2014b; Romão 2000; Zhang et al. 2012) concluding a molecular diversity of ~2% among the cultivated watermelon.

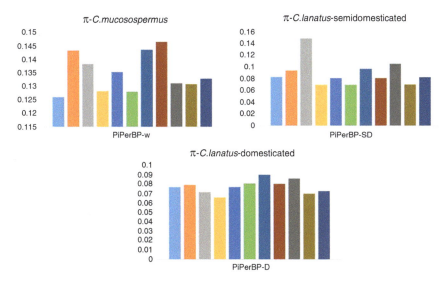

Fig. 9 Frequency spectrum for the chromosomal means of nucleotide diversity (π) across wild, semidomesticated, and cultivars. (Nimmakayala et al. (2014b) https://doi.org/10.1186/1471-2164-15-767)

4.1 Nucleotide Diversity

Nucleotide diversity (π) is a measure of the average number of nucleotide differences per site (polymorphism) in the sample population and differs from common genetic diversity measures (Singh et al. 2017) within a population (Nei and Li 1979). Wu et al. (2019) and Nimmakayala et al. (2014a, b) estimated π by using the entire set of SNPs (unfiltered for minor allele frequency) with a 70% call rate (Fig. 9). The estimated π was 0.0003, 0.003, and 0.003 for sweet watermelon, egusi, and landraces, respectively, and coverage of the genome that harbors various polymorphisms was estimated at 60,454 bp. The π was estimated at 0.0002, 0.0002, and 0.0003 for sweet watermelons of Asia, Europe, and North America, respectively. The π was 0.002 for the entire *C. lanatus* collections from Africa. Within Africa, the π was 0.003, 0.002, and 0.001 for samples from South, West, and North Africa, respectively, indicating that the South or West African types are more diverse than North African types and could be the centers of origin for sweet watermelon. Tajima's D was estimated at -2.94 ($p = 0.003$), -3.58 ($p = 0.0003$), -3.206 ($p = 0.001$), and -3.279 ($p = 0.001$) for sweet watermelons, egusi, and landraces, respectively. (Fig. 10) The less negative Tajima's D for sweet watermelons suggests bottlenecks causing narrow genetic diversity because of the removal of rare alleles due to sampling during the domestication event. Founder populations will have diverse rare alleles, which thus caused the more negative Tajima's D in egusi and landraces.

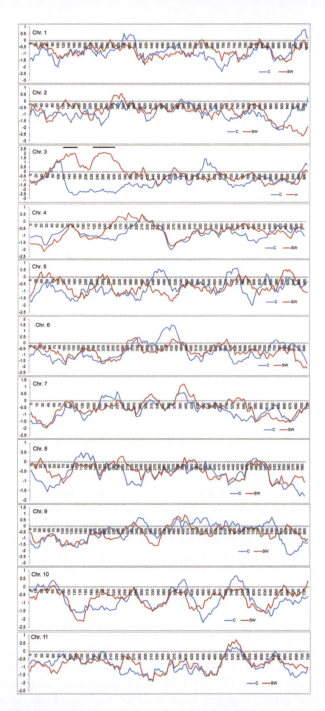

Fig. 10 Genome-wide window-based Tajima's *D* of cultivated (blue) and semi-wild watermelon (red) across various chromosomes. If Tajima's D is negative for blue and positive for red, then that region of the genome is under selective sweep. Note two dark lines on chromosome 3 that showed strong signal for selective sweep. (Nimmakayala et al. (2014b) https://doi.org/10.1186/1471-2164-15-767)

4.2 Linkage Disequilibrium (LD) Patterns for Sweet Watermelons, Egusi, and Landraces

Patterns of LD decay were consistent with the population genetics and population structure analyses. Egusi and landrace groups had the shortest mean physical distances of LD decay (6.32 and 35.7 Kb) as compared with sweet watermelon (97.9 Kb) (Nimmakayala et al. 2014a, b). Various chromosomes of watermelon showed moderate to high LD and as compared with other cucurbits such as melon (Fig. 11) (Nimmakayala et al. 2014a, b) and cucumber (Wang et al. 2018). High and moderate LD decay allows for using high-resolution genome-wide association study to trace trait associations with moderate throughput of SNPs. Haplotype distributions resolving into LD are presented in Figs. 12 and 13 (Nimmakayala et al. 2014a, b). The LD decay (r^2) varied for various chromosomes in different groups (Reddy et al. 2015). For sweet watermelon, the LD decay was 95.3, 94.5, 106.5, 77.99, 93.13, 106.49, 105.46, 98.48, 98.16, 95.74, and 2.89 Kb for chro-1,2, 3, 4, 5, 6, 7, 8, 9, 10, and 11, respectively. The LD was substantially less in all groups versus sweet watermelons, which indicates substantially increased LD for sweet watermelons across the chromosomes. The mean genome-wide LD decay was 97.9, 35.7, 6.32, and 22.64 Kb for sweet watermelons, landraces, egusi types, and *C. amarus*, respectively. Population structure analysis revealed predominant ancestries for wild, semi-wild, and domesticated watermelons as well as admixture of various ancestries that were important for domestication. Sliding window analysis of Tajima's D across various chromosomes was used to resolve selective sweep. LD decay was estimated for various chromosomes. Strong selective sweeps have been reported revealing important genes that may have played a role in sweet watermelon domestication (Nimmakayala et al. 2014b; Reddy et al. 2015). Guo et al. (2019) resolved genome-wide genetic diversity to study LD and selective sweeps, and pairwise fixation index (F_{ST}) distributions were examined among worldwide cultivated watermelons to track signals of domestication. A sliding window analysis using pairwise F_{ST} values used to resolve selective sweeps indicated differential selection on various regions of the genome that might have contributed to the domestication process. Pairwise analysis of adjacent SNPs within a chromosome as well as within a haplotype allowed for estimating genome-wide LD decay. LD was also detected within individual genes on various chromosomes. In this genome-wide SNP analysis, authors inferred population structure, genetic diversity, phylogenetic relationships among the watermelon genotypes and its wild ancestor species. Since phylogenetic approaches would not discriminate the genotyping data based on the ancestral and modern genomes, network approach is more appropriate to understand evolution of various taxa in the population (Fig. 13). Networks depict reticulated processes of evolution such as hybridization, horizontal gene transfer, or admixture within the haplotypes.

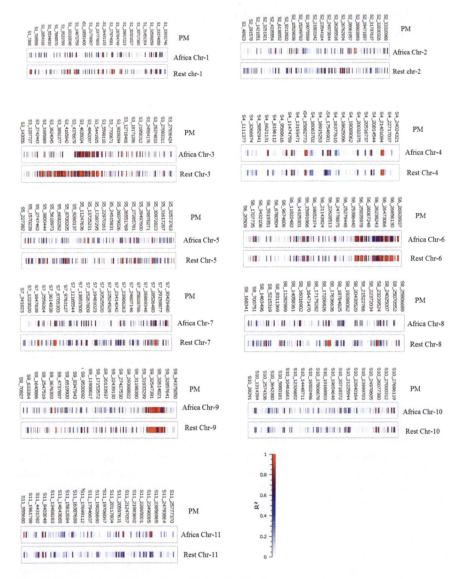

Fig. 11 Comparison of linkage disequilibrium distribution across various chromosomes between watermelon accessions from Africa and the rest of the world. (Nimmakayala et al. (2014b) https://doi.org/10.1186/1471-2164-15-767)

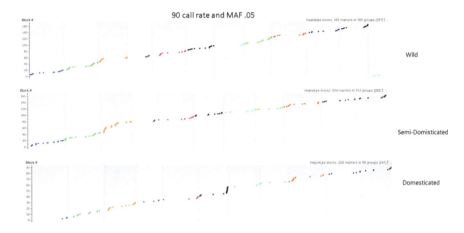

Fig. 12 Haplotype distribution across the chromosomes. (Nimmakayala et al. (2014b) https://doi.org/10.1186/1471-2164-15-767)

5 Domestication Syndrome

Sweet watermelon domestication could have had more than one evolutionary history. Genome-wide shared SNPs resolved the population structure, which has three distinct groups of cultivars (Wu et al. 2019). Three groups could be rooted from South Africa, the Nile valley, or more recently, West Africa, with the oldest archeological evidence from Libya and Egypt (Renner et al. 2019). The geographic origin and domestication of watermelons has been unclear until recently. Renner et al. (2019) showed extensive nuclear and plastid genomic data from a 3,560-year-old *Citrullus* leaf found in a mummy's sarcophagus and skimmed genomes for representatives of the seven extant species of *Citrullus*. Modern cultivars and the ancient plant uniquely shared mutations in a lycopene metabolism gene (LYCB) affecting pulp color and a stop codon in a transcription factor regulating bitter cucurbitacin compounds. Renner et al. (2019) concluded that the white-fleshed, nonbitter melons from southern Sudan are the closest relatives of domesticated watermelon, and the ancient genome shares unique alleles with a red-fleshed, nonbitter domesticated form (but no wild forms), which suggests that 18th Dynasty Egyptians were cultivating domesticated watermelon 3,500 years ago. Ren et al. (2018) resequenced 96 recombinant inbred lines derived from crossing sweet and unsweet accessions to map a putative tonoplast sugar transporter gene (*ClTST2*) for sugar content. Accumulation of soluble sugars in watermelon fruit, a major quality trait, had been selected during domestication. *ClTST2* is regulated by a sugar-induced WRKY transcription factor SUSIWM1. An SNP in the promoter of *ClTST2* and strong selective sweep also showed an association with sugar content in 400 watermelon accessions. Further molecular analyses revealed a binding site for SUSIWM1, a sugar-induced transcription factor, within the *ClTST2* promoter. Wu

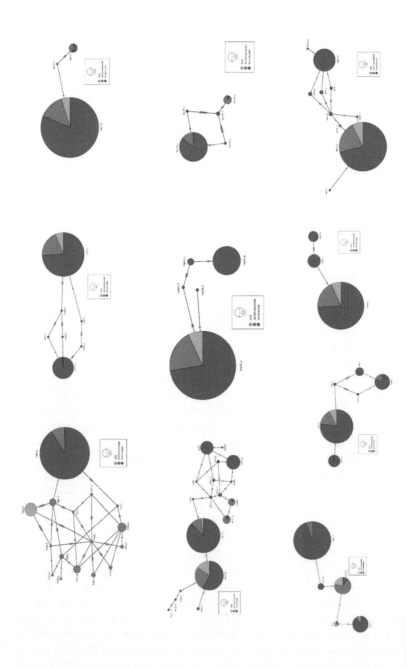

Fig. 13 Haplotype networking within the highly significant LD blocks. Networking manifests the number of segregating sites across the branches resolving differences among wild, semidomesticated, and domesticated watermelons. (Nimmakayala et al. (2014b) https://doi.org/10.1186/1471-2164-15-767)

et al. (2019) performed GWAS for fruit flesh color, fruit shape, and rind stripe pattern in 788, 864, and 695 accessions. A region on chromosome 4 was associated with red coloration of fruit flesh, and this genome region contained lycopene β-cyclase (*LCYB*) (chromosome 4) that converts lycopene to downstream carotenoids (Bang et al. 2010). Rind stripe pattern was identified on chromosome 6 (at ~30.2 Mb) with the "S" locus that controls the foreground stripe pattern (Park et al. 2016). A haplotype of five SNPs on chromosome 3 (at ~31.1 Mb) was associated with elongated fruit shape; the SNPs were located near the *ClFS1* gene (~22.6 Kb distance; chromosome 3: 31 086 958–31 090 579) mapped to control watermelon fruit elongation (Dou et al. 2018b). Two other SNPs associated with fruit shape were on chromosomes 2 and 6. Dou et al. (2018b) identified a locus on chromosome 3 of the watermelon genome controlling fruit shape. Segregation analysis in F_2 and BC_1 populations derived from a cross between two inbred lines 'Duan125' (elongate fruit) and 'Zhengzhouzigua' (spherical fruit) suggested that fruit shape of watermelon is controlled by a single locus, and elongate fruit (OO) is incompletely dominant to spherical fruit (oo), with the heterozygote (Oo) being oval fruit. GWAS profiles of 315 accessions identified a major locus on watermelon chromosome 3, which was confirmed by BSA-seq mapping in the F_2 population. The candidate gene was mapped to a region of 46 kb on chromosome 3. There were only four genes present in the corresponding region in the reference genome. Four candidate genes were sequenced in this region, which revealed that the CDS of Cla011257 had a 159-bp deletion that resulted in the omission of 53 amino acids in elongate watermelon. Cla011257 is annotated as an IQD protein family with a role in Ca_2+-CaM signaling. Dou et al. (2018a) studied yellow skin color by using BSA-seq and GWAS. A segregation analysis in F_2 and BC_1 populations derived from a cross of two inbred lines '94E1' (yellow skin) and 'Qingfeng' (green skin) suggested a 59.8-kb region on chromosome 4.

A three-locus model for rind phenotypes in watermelon (*C. lanatus*) was previously proposed from genetic analysis (Park et al. 2016). These three loci, *S* (foreground stripe pattern), *D* (depth of rind color), and *Dgo* (background rind color), segregate in a Mendelian manner. Oren et al. (2019) fine-mapped APRR2, a transcription factor regulating dark- and light-green rind pigment accumulation in melon and watermelon. Reddy et al. (2015) used single-sequence repeats with association mapping to associate BVWS00358, a GA repeat microsatellite, as the GAGA type of transcription factor located in the 5' untranslated region of the Cys2His2 zinc finger motif for fruit length. Guo et al. (2019) associated flesh sweetness with the sugar transporter gene *ClTST2* by using a robust GWAS involving resequencing 414 accessions. Two additional regions strongly associated with flesh sweetness were found in the sucrose synthase and raffinose synthase genes on chromosome 10. The flesh color gene lycopene β-cyclase was detected on chromosomes 4. In total, 14 quantitative trait loci were found to be associated with fruit shape, including *ClFS1* (*Cla97C03G066390*), which is known to control fruit elongation. Three peaks highly associated with rind color and rind stripe were found on chromosomes 4, 6, and 8, corresponding to the rind trait loci *Dgo*, *S,* and *D*, respectively. The strongest signal associated with rind stripe was found in a

WD40-repeat gene. The strongest associated SNP on chromosome 3 overlapped with seed coat color and was located in *Cla97C03G057100*, which encodes a polyphenol oxidase that polymerizes o-quinones to produce black, brown, or red pigment.

This resequencing study (Guo et al. 2019) involving *C. lanatus* landraces and *C. mucosospermus* resolved 151 domestication sweeps with a cumulative size of 24.8 Mb (containing 771 genes). Similarly, comparison of modern *C. lanatus* cultivars and landraces identified 125 selective sweeps (17.2 Mb and 667 genes) related to sweet watermelon domestication. Some of the important genes located in selective sweeps, thus indicating their role in domestication, included acetolactate synthase (Guo et al.), benzyl alcohol O-benzoyltransferase (*BEBT*), β-galactosidase (*BGAL*), alkaline α-galactosidase (*ClAGA2*), bitter fruit bHLH transcription factor (*ClBt*), tonoplast sugar transporter (*ClTST2*), calmodulin (*CML*), carotenoid isomerase (*CRTISO*), glyoxal oxidase (*GLOX*), heat stress transcription factor (*HSF*), auxin transporter (*LAX*), *LCYB*, purple acid phosphatase (*PAP*), pectinesterase (*PE*), polygalacturonase (*PG*), polyol/monosaccharide transporter (*PMT*), phytoene synthase 1 (*PSY1*), receptor-like protein kinase (*RLK*), slow growth 1 (*SLO1*, encoding an E motif-containing pentatricopeptide repeat protein), sugar transporter (*STP*), and xyloglucan endotransglucosylase/hydrolase (*XTH*). This study involves mapping loci underlying watermelon fruit size, a nonbitter allele combined with positive selection of genes that control raffinose catabolism and sugar transport to lead to enhanced sweetness. This study hypothesized that the fruit flesh coloration and sugar accumulation occurred simultaneously in the watermelon breeding history. Renner et al. (2021) sequenced a Sudanese watermelon form with nonbitter whitish pulp, known as the Kordofan melon (*C. lanatus* subsp. *cordophanus*) to identify that this type to be the closest relative of domesticated watermelons and a possible progenitor based on the gene flow and allele frequencies involving 15,824 genome structural variants (SVs) and other anthropological evidence including Egyptian tomb paintings. This study used previously generated SV data (Guo et al. 2019) to infer shifts in allelic frequencies to conclude that the fruit sweetness has gradually increased over the course of watermelon domestication.

6 Future Perspectives

Watermelon domestication and improvement has been ongoing for almost 6,000 years and has evolved from bitter to bland to sweet. Collectively, findings by the genomics studies reviewed here have helped in understanding the domestication history of sweet watermelon. Beyond basic biology, an understanding of distribution of PAVs among crop wild relatives (CWRs) would be of great use in diversifying cultivars by introgression of broader repertoire of genes that provide a valuable source of genetic diversity. Currently it is possible to construct plant pan-genomes to resolve important structural and gene position information, short variable regions, and presence and absence variation (PAV) across the species.

Domestication alleles and the core resistant mechanisms in cultivated watermelon have undergone long-standing genetic changes that affect how this crop would respond to various pests and diseases. Decoding the genetic basis of natural variation is central to understanding crop evolution and, in turn, improving crop breeding (Liang et al. 2021). Fortunately, it is now feasible to assemble multiple high-quality genomes that can be used to construct high-resolution pan-genomes that capture all the variation. However, assembling, displaying, and interacting with such high-resolution pan-genomes will require the development of new tools (Lei et al. 2021). With the advent of large-scale whole-genome variation data, population geneticists are currently interested in considering increasingly more complex models for inferring population genetic models. Population geneticists are currently turning to deep learning, which is an active area of machine learning research to demonstrate powerful frameworks for population genetic/genomic analysis that can allow accurate inference of previously intractable models (Sheehan and Song 2016). Next-generation sequencing of living wild and primitive watermelons from northeastern Africa, modern sweet dessert watermelons, and other *Citrullus* taxa and state-of-the-art pan-genomic analysis offer opportunities to rigorously mine the loci that have led to domestication and selection in the breeding history of watermelon. Watermelon pan-genome can be a repository of nonredundant collection of genes and/or DNA sequences and their evolution among the species of *Citrullus* genus. Cucurbit genomes being small, future pan-genomes of these crops would be typically much smaller than the pan-genomes of the other cultivated species in general. The construction and interpretation of cucurbit pan-genomes and linking them would be challenging due to large-scale genomic changes that occurred in various genera. Cucurbit pan-genomes can focus not only on protein coding genes but also transposable elements, noncoding and repetitive sequences to understand how noncoding and repetitive DNA play important roles in determining the speciation in various genera of Cucurbitaceae family and underlying mechanisms in genome evolution of various cultivated species in this plant family.

7 Conclusion

Modern genomics tools along with population genetics concepts and approaches have allowed for identifying genes for fruit traits, including bitterless taste, fruit elongation, size, enhanced sweetness, and flesh color that have roles in speciation and domestication in sweet watermelon. The alkaline α-galactosidase ClAGA2, which functions in phloem unloading of stachyose and raffinose, and the sugar transporter ClTST2 have been under selection, facilitating sugar accumulation. Fruit flesh coloration was facilitated by the key biosynthetic enzyme PSY, and further enhancement of carotenoid accumulation was facilitated by ClTST2. Red fruit flesh color was later selected by maintaining an amino acid substitution in the lycopene metabolism enzyme LCYB (Guo et al. 2019). A fruit shape gene, ClFS1 (Cla97C03G066390), an IQD protein family with a role in Ca2+-CaM signaling that is located in the largest LD block is important to control fruit elongation. Watermelon improvement is very crucial to satisfy the world demand in the presence of different challenges like ever increasing pathogens and abiotic stress such as drought and acidity. Knowledge related to the effect of pesticides and heavy metals in soils on fruits of watermelons is still elusive. Though studies on watermelon domestication identified several semidomesticated species, now the challenge is to use these collections for introgression programs to bring important genes that are lost in the process of domestication and improvement. The introduction of cultivated watermelon germplasm across the world during the last 2000 years has a profound impact on the genetic diversity through selecting the desirable fruit traits according to local consumer preferences. Moreover, seedless watermelons have been the choice for all the consumers across the world. This further reduced genetic diversity among the watermelon cultivars that can be converted into tetraploids that can set seed and contribute to desirable traits in triploid breeding. Eventually, genetic erosion is the depletion in cultivar variation because of triploid breeding, which largely causes the endangerment of a majority of cultivars because all the cultivars though they have desirable features cannot be converted into the productive tetraploids. Creating watermelon germplasm with resistance to diseases, drought, soil salinity, and insect damage, which simultaneously have higher nutritional quality, is challenging to conventional breeding due to the complex and diffuse genetic basis of those traits. Recent advances in gene editing technology, such as base editors and prime-editing, coupled with a deeper understanding of the genetic basis of domestication delivered by the analysis of crop "pangenomes," open the exciting prospect of creating novel crops via manipulation of domestication-related genes in wild species (Gao et al. 2019a; Stokstad 2019). A de novo domestication platform may allow rapid and precise conversion of crop wild relatives into crops, while retaining many of the valuable resilience and resistance traits left behind during domestication and breeding. Using the tomato as case in point, we can conclude how a knowledge-driven genomic pipelines could be exploited to contribute to breeding resistant and nutritious watermelons (Gasparini et al. 2021).

Acknowledgments Funding is provided by USDA SCRI contract 2020-51181-32139 and USDA NIFA 2017-38821-26434.

References

Abrouk M, Murat F, Pont C, Messing J, Jackson S, Faraut T, et al. Palaeogenomics of plants: synteny-based modelling of extinct ancestors. Trends Plant Sci. 2010;15:479–87.

Achigan-Dako EG, Degbey H, Hale I, Blattner FR. Georeferenced phylogenetic analysis of a global collection of wild and cultivated Citrullus species. Ecol Evol. 2021;11:1918–36.

Aiping Z, Xiaowu W, Daqing Z. SRAP analysis for the genetic diversity of watermelon varieties resources. Chinese Agric Sci Bull. 2008;24:115–20.

Alimari A, Zaid A, Fadda Z. Genetic diversity in local Palestinian watermelon (Citrullus lanatus) accessions. Int J Agric Policy and Res. 2017;5:157–62.

Aryal N. Cytomolecular characterization of rDNA distribution & copy number variation among various Citrullus species. San Diego, CA: International Plant & Animal Genome XIX; 2011.

Aryal N, Nimmakayala P, Islam-Faridi N, Levi A, Vajja GV, Reddy UK. Molecular and cytogenetic characterization of watermelon using DNA markers and fish. Hortscience. Am Soc Hortic Sci 113 S West St, STE 200, Alexandria, VA 22314, 2010;S303-S303

Bailey L. Three discussions in Cucurbitaceae. Gentes Herbarum. 1930;2:175–86.

Bang H, Davis AR, Kim S, Leskovar DI, King SR. Flesh color inheritance and gene interactions among canary yellow, pale yellow, and red watermelon. J Am Soc Hortic Sci. 2010;135:362–8.

Baumann FA. Das Erbario carrarese und die Bildtradition des Tractatus de herbis: ein Beitrag zur Geschichte der Pflanzendarstellung im Ubergang von Spatmittelalter zur Fruhrenaissance. Benteli; 1974.

Bayer PE, Golicz AA, Scheben A, Batley J, Edwards D. Plant pan-genomes are the new reference. Nat plants. 2020a;6:914–20.

Bayer PE, Golicz AA, Scheben A, Batley J, Edwards D. Plant pan-genomes are the new reference. Nat Plants. 2020b;6:914–20.

Bhakar JK, Sharma R, Singh PK, Singh G. Genetic diversity in natural watermelon (Citrullus lanatus) populations of Rajasthan. Indian J Agric Sci. 2017;87:811–5.

Birkner M, Blath J. Coalescents and population genetic inference. Trends Stoch Anal. 2009;353: 329.

Branham SE, Levi A, Farnham MW, Wechter WP. A GBS-SNP-based linkage map and quantitative trait loci (QTL) associated with resistance to Fusarium oxysporum f. sp. niveum race 2 identified in Citrullus lanatus var. citroides. Theor Appl Genet. 2017;130:319–30.

British Library, Collins M, Raphael S. A medieval herbal: a facsimile of British Library Egerton MS 747. British Library; 2003.

Carter LG. Genetic linkage map for watermelon (Citrullus lanatus var. lanatus) useful in identifying DNA markers associated with bacterial fruit blotch resistance. College of Charleston; 2008.

Cho Y, Lee S, Park J, Kwon S, Park G, Kim H, et al. Identification of a candidate gene controlling semi-dwarfism in watermelon, Citrullus lanatus, using a combination of genetic linkage mapping and QTL-seq. Hortic Environ Biotechnol. 2021;62:447–59.

Chomicki G, Renner SS. Watermelon origin solved with molecular phylogenetics including L innaean material: another example of museomics. New Phytol. 2015;205:526–32.

Crisci JL, Poh Y-P, Bean A, Simkin A, Jensen JD. Recent progress in polymorphism-based population genetic inference. J Hered. 2012;103:287–96.

Cui H, Ding Z, Zhu Q, Wu Y, Gao P. Population structure and genetic diversity of watermelon (Citrullus lanatus) based on SNP of chloroplast genome. 3 Biotech. 2020;10:1–8.

Dane F, Lang P. Sequence variation at cpDNA regions of watermelon and related wild species: implications for the evolution of Citrullus haplotypes. Am J Bot. 2004;91:1922–9.

Dane F, Liu J. Diversity and origin of cultivated and citron type watermelon (Citrullus lanatus). Genet Resour Crop Evol. 2007;54:1255–65.

De Winter B. Notes on African plants. Cucurbitaceae. A new species of Citrullus (Benincaseae) from the Namib Desert, Namibia: Bothalia; 1990.

Dje Y, Tahi C, Bi AZ, Baudoin J-P, Bertin P. Use of ISSR markers to assess genetic diversity of African edible seeded Citrullus lanatus landraces. Sci Hortic. 2010;124:159–64.

Dou J, Zhao S, Lu X, He N, Zhang L, Liu W. Genetic analysis of the yellow skin trait in watermelon. IV international symposium on molecular markers in horticulture 1203, 2017. pp. 117–120

Dou J, Lu X, Ali A, Zhao S, Zhang L, He N, et al. Genetic mapping reveals a marker for yellow skin in watermelon (Citrullus lanatus L.). PLoS One. 2018a;13:e0200617.

Dou J, Zhao S, Lu X, He N, Zhang L, Ali A, et al. Genetic mapping reveals a candidate gene (ClFS1) for fruit shape in watermelon (Citrullus lanatus L.). Theor Appl Genet. 2018b;131:947–58.

Fallik E, Ziv C. How rootstock/scion combinations affect watermelon fruit quality after harvest? J Sci Food Agric. 2020;100:3275–82.

Fan M, Huang Y, Zhong Y, Kong Q, Xie J, Niu M, et al. Comparative transcriptome profiling of potassium starvation responsiveness in two contrasting watermelon genotypes. Planta. 2014;239:397–410.

Fursa TJBZ. K sistematike roda Citrullus Schrad.(On the taxonomy of genus Citrullus Schrad.). Bot. Zhurn. (Moscow & Leningrad). 1972;57:31–41.

Fursa TJGS. Novyi vid arbuza Citrullus mucosospermus (Fursa) Fursa.(A new species of watermelon Citrullus mucosospermus (Fursa) Fursa.) Trudy Prikl. Bot. 1983;81:108–12.

Gama RDS, Santos C, Dias RD. Genetic variability of watermelon accessions based on microsatellite markers. Embrapa Semiárido-Artigo em periódico indexado (ALICE). 2013:12, 747.

Gao L, Zhao S, Lu X, He N, Zhu H, Dou J, et al. Comparative transcriptome analysis reveals key genes potentially related to soluble sugar and organic acid accumulation in watermelon. PLoS One. 2018;13:e0190096.

Gao L, Gonda I, Sun H, Ma Q, Bao K, Tieman DM, et al. The tomato pan genome uncovers new genes and a rare allele regulating fruit flavor. Nat Genet. 2019a;51:1044–51.

Gao M, Liang X, Guo Y, Zhang Y, Gao Y, Liu X, Liu J, Liu Xs, Yu C, Li J. QTL analysis of seed size by a high-density GBS genetic map in watermelon (Citrullus lanatus L.). VI international symposium on cucurbits 1294, 2019b. pp. 121–126

Gao Y, Guo Y, Su Z, Yu Y, Zhu Z, Gao P, et al. Transcriptome analysis of genes related to fruit texture in watermelon. Sci Hortic. 2020;262:109075.

Garcia-Lozano M, Dutta SK, Natarajan P, Tomason YR, Lopez C, Katam R, et al. Transcriptome changes in reciprocal grafts involving watermelon and bottle gourd reveal molecular mechanisms involved in increase of the fruit size, rind toughness and soluble solids. Plant Mol Biol. 2020;102:213–23.

Garcia-Lozano M, Natarajan P, Levi A, Katam R, Lopez-Ortiz C, Nimmakayala P, et al. Altered chromatin conformation and transcriptional regulation in watermelon following genome doubling. Plant J. 2021;106:588–600.

Gasparini K, Moreira JR, Peres LEP, Zsögön A. De novo domestication of wild species to create crops with increased resilience and nutritional value. Curr Opin Plant Biol. 2021;60:102006.

Gautier AT, Cochetel N, Merlin I, Hevin C, Lauvergeat V, Vivin P, et al. Scion genotypes exert long distance control over rootstock transcriptome responses to low phosphate in grafted grapevine. BMC Plant Biol. 2020;20:1–15.

Gimode W, Bao K, Fei Z, McGregor C. QTL associated with gummy stem blight resistance in watermelon. Theor Appl Genet. 2021;134:573–84.

Golicz AA, Batley J, Edwards D. Towards plant pangenomics. Plant Biotechnol J. 2016a;14:1099–105.

Golicz AA, Bayer PE, Barker GC, Edger PP, Kim H, Martinez PA, et al. The pangenome of an agronomically important crop plant Brassica oleracea. Nat Commun. 2016b;7:1–8.

Gong C, Zhu H, Lu X, Yang D, Zhao S, Umer MJ, et al. An integrated transcriptome and metabolome approach reveals the accumulation of taste-related metabolites and gene regulatory networks during watermelon fruit development. Planta. 2021;254:1–12.

Grassi S, Piro G, Lee JM, Zheng Y, Fei Z, Dalessandro G, et al. Comparative genomics reveals candidate carotenoid pathway regulators of ripening watermelon fruit. BMC Genomics. 2013;14:1–20.

Guo S, Zhang J, Sun H, Salse J, Lucas WJ, Zhang H, et al. The draft genome of watermelon (Citrullus lanatus) and resequencing of 20 diverse accessions. Nat Genet. 2013;45:51–8.

Guo S, Sun H, Zhang H, Liu J, Ren Y, Gong G, et al. Comparative transcriptome analysis of cultivated and wild watermelon during fruit development. PLoS One. 2015;10:e0130267.

Guo S, Zhao S, Sun H, Wang X, Wu S, Lin T, et al. Resequencing of 414 cultivated and wild watermelon accessions identifies selection for fruit quality traits. Nat Genet. 2019;51:1616–23.

Gupta PK. Quantitative genetics: pan-genomes, SVs, and k-mers for GWAS. Trends Genet. 2021;37:868.

Hajiali A, Darvishzadeh R, Zahedi B, Abbaskohpayegani J. Exploring genetic diversity of some iranian watermelon (Citrullus vulgaris) accessions in urmia climatic conditions. J Plant Prod (Agron Breed Hortic). 2017;40:29–40.

Hashizume T, Shimamoto I, Harushima Y, Yui M, Sato T, Imai T, et al. Construction of a linkage map for watermelon (Citrullus lanatus (Thunb.) Matsum & Nakai) using random amplified polymorphic DNA (RAPD). Euphytica. 1996;90:265–73.

Hashizume T, Shimamoto I, Hirai M. Construction of a linkage map and QTL analysis of horticultural traits for watermelon [Citrullus lanatus (THUNB.) MATSUM & NAKAI] using RAPD, RFLP and ISSR markers. Theor Appl Genet. 2003;106:779–85.

Hawkins LK, Dane F, Kubisiak TL, Rhodes BB, Jarret RL. Linkage mapping in a watermelon population segregating for fusarium wilt resistance. J Am Soc Hortic Sci. 2001;126:344–50.

Hu J-B, Li J-W, Li Q, Ma S-W, Wang J-M. The use of chloroplast microsatellite markers for assessing cytoplasmic variation in a watermelon germplasm collection. Mol Biol Rep. 2011;38: 4985–90.

Hu Z, Lan S, Zhao N, Su N, Xue Q, Liu J, et al. Soft-X-irradiated pollens induce parthenocarpy in watermelon via rapid changes of hormone-signalings and hormonal regulation. Sci Hortic. 2019;250:317–28.

Hwang J, Jumsoon K, Byeonggu S, Kwanghwan K, Younghoon P. Genetic diversity in watermelon cultivars and related species based on AFLPs and EST-SSRs. Notulae Botanicae Horti Agrobotanici Cluj-Napoca. 2011;39:285–92.

Jang YJ, Seo M, Hersh CP, Rhee S-J, Kim Y, Lee GP. An evolutionarily conserved non-synonymous SNP in a leucine-rich repeat domain determines anthracnose resistance in watermelon. Theor Appl Genet. 2019;132:473–88.

Janick J, Paris HS, Parrish DC. The cucurbits of Mediterranean antiquity: identification of taxa from ancient images and descriptions. Ann Bot. 2007;100:1441–57.

Jarret RL, Newman M. Phylogenetic relationships among species of Citrullus and the placement of C. rehmii De Winter as determined by Internal Transcribed Spacer (ITS) sequence heterogeneity. Genet Resour Crop Evol. 2000;47:215–22.

Jayakodi M, Schreiber M, Mascher M. Sweet genes in melon and watermelon. Nat Genet. 2019;51: 1572–3.

Jeffrey C. Further notes on Cucurbitaceae: III: some southern African taxa. Kew Bull. 1975;30(3): 475–93.

Ji H, Zhang Y, Li Y, Pan C, Chen N. Phenotypic diversity in main characters of watermelon germplasm. J Northwest A & F University-Natural Science Edition. 2013;41:155–60.

Jiang C-H, Yao X-F, Mi D-D, Li Z-J, Yang B-Y, Zheng Y, et al. Comparative transcriptome analysis reveals the biocontrol mechanism of Bacillus velezensis F21 against Fusarium wilt on watermelon. Front Microbiol. 2019a;10:652.

Jiang H, Tian H, Yan C, Jia L, Wang Y, Wang M, et al. RNA-seq analysis of watermelon (Citrullus lanatus) to identify genes involved in fruit cracking. Sci Hortic. 2019b;248:248–55.

Jobst J, King K, Hemleben V. Molecular evolution of the Internal Transcribed Spacers (ITS1 and ITS2) and phylogenetic relationships among species of the family Cucurbitaceae. Mol Phylogenet Evol. 1998;9:204–19.

Joshi V, Joshi M, Silwal D, Noonan K, Rodriguez S, Penalosa A. Systematized biosynthesis and catabolism regulate citrulline accumulation in watermelon. Phytochemistry. 2019;162:129–40.

Karki K, Coolong T, Kousik C, Petkar A, Myers B, Hajihassani A, Mandal M, Dutta B. Micronutrients affect induced resistance in watermelon against Fusarium oxysporum f. sp. niveum and Meloidogyne incognita infection, 2021.

Kim EJ, Noh SJ, Park S-W, Jeon Y-M, Kim YS, Huh YS, Kim TI. Development of vertical cultivation technology for standardized fruits and mass production of small and medium sized watermelon, 2019. 한국육종학회 학술발표회지. pp. 104–104

Kiyani MR, Jahanbin G. Investigation on genetic variation of Iran watermelon accession. Iran J Field Crops Res. 2006;4:334–46.

Krishna Prasad V, Pitchaimuthu M, Dutta O. Adaptive responses and diversity pattern in watermelon. Indian J Hortic. 2002;59:298–306.

Kwon YS, Oh YH, Yi SI, Kim HY, An JM, Yang SG, Ok SH, Shin JS. Informative SSR markers for commercial variety discrimination in watermelon (Citrullus lanatus). Genes & Genomics. 2010;32(2):115–22.

Lee S, Shin JS, Park K, Hong Y. Detection of genetic diversity using RAPD-PCR and sugar analysis in watermelon [Citrullus lanantus (Thunb.) Mansf.] germplasm. Theor Appl Genet. 1996;92:719–25.

Lee ES, Kim J, Hong JP, Kim D-S, Kim M, Huh Y-C, et al. Development of HRM markers based on SNPs identified from next generation resequencing of susceptible and resistant parents to gummy stem blight in watermelon. Korean Society of Breeding Science. 2018;50:424–33.

Lee KJ, Lee J-R, Sebastin R, Shin M-J, Kim S-H, Cho G-T, et al. Genetic diversity assessed by Genotyping by Sequencing (GBS) in watermelon germplasm. Genes. 2019;10:822.

Lei L, Goltsman E, Goodstein D, Wu GA, Rokhsar DS, Vogel JP. Plant Pan-genomics comes of age. Annu Rev Plant Biol. 2021;72:411–35.

Levi A, Thomas CE. DNA markers from different linkage regions of watermelon genome useful in differentiating among closely related watermelon genotypes. HortScience. 2007;42:210–4.

Levi A, Thomas CE, Keinath AP, Wehner TC. Estimation of genetic diversity among Citrullus accessions using RAPD markers. VII Eucarpia meeting on cucurbit genetics and breeding 510, 2000. pp. 385–390

Levi A, Thomas CE, Keinath AP, Wehner TCJGR, Evolution C. Genetic diversity among watermelon (Citrullus lanatus and Citrullus colocynthis) accessions. Genet Resour Crop Evol. 2001;48:559–66.

Levi A, Thies J, Simmons A, Thomas C, Wehner T, Joubeur T, et al. Construction of a reference genetic linkage map for watermelon. Maynard D N Proc Cucurbitaceae. 2002;202:28–35.

Levi A, Thomas CE, Newman M, Reddy O, Zhang X, Xu YJJotASfHS. ISSR and AFLP markers differ among American watermelon cultivars with limited genetic diversity. J Am Soc Hortic Sci. 2004;129:553–8.

Levi A, Thomas C, Trebitsh T, Salman A, King J, Karalius J, et al. An extended linkage map for watermelon based on SRAP, AFLP, SSR, ISSR, and RAPD markers. J Am Soc Hortic Sci. 2006;131:393–402.

Levi A, Ling K, Davis A. EST-SSRs of watermelon (Citrullus sp.) useful in assessing genetic diversity among Lagenaria Siceraria accessions. Plant and animal genome XVI conference proceedings, 2008. p. 643

Levi A, Thies JA, Wechter WP, Harrison HF, Simmons AM, Reddy UK, et al. High frequency oligonucleotides: targeting active gene (HFO-TAG) markers revealed wide genetic diversity among Citrullus spp. accessions useful for enhancing disease or pest resistance in watermelon cultivars. Genet Resour Crop Evol. 2013;60:427–40.

Levi A, Jarret R, Kousik S, Wechter WP, Nimmakayala P, Reddy UK. Genetic resources of watermelon. Genetics and genomics of Cucurbitaceae: Springer; 2017. p. 87–110.

Li Y, Duan H, Ma Z. Study on genetic diversity and relationship in watermelon germplasms by AFLP analysis. Acta Agric Boreali-Sin. 2007;22:177–80.

Li P, Huo X, Cheng Y, Yang B, Duan H. Assessment of genetic diversity in watermelon based on SRAP analysis. J Agric Sci Technol (Beijing). 2013;15:89–96.

Li K-P, Wu Y-X, Zhao H, Wang Y, Lü X-M, Wang J-M, et al. Cytogenetic relationships among Citrullus species in comparison with some genera of the tribe Benincaseae (Cucurbitaceae) as inferred from rDNA distribution patterns. BMC Evol Biol. 2016;16:1–9.

Li X, An M, Xia Z, Bai X, Wu Y. Transcriptome analysis of watermelon (Citrullus lanatus) fruits in response to Cucumber green mottle mosaic virus (CGMMV) infection. Sci Rep. 2017;7:1–12.

Liang Y, Liu H-J, Yan J, Tian F. Natural variation in crops: realized understanding, continuing promise. Annu Rev Plant Biol. 2021;72:357–85.

Lima MNR, de Queiroz MA, da Silva Oliveira AE, da Silva Lima Neto I, de Oliveira RS. Integration of quantitative and qualitative descriptors for genetic diversity studies of watermelon accessions. Aust J Crop Sci. 2017;11:1005.

Liu N, Yang J, Fu X, Zhang L, Tang K, Guy KM, et al. Genome-wide identification and comparative analysis of grafting-responsive mRNA in watermelon grafted onto bottle gourd and squash rootstocks by high-throughput sequencing. Mol Gen Genomics. 2016a;291:621–33.

Liu S, Gao P, Zhu Q, Luan F, Davis AR, Wang X. Development of cleaved amplified polymorphic sequence markers and a CAPS-based genetic linkage map in watermelon (Citrullus lanatus [Thunb.] Matsum. and Nakai) constructed using whole-genome re-sequencing data. Breed Sci. 2016b;66:244–59.

Liu D, Yang H, Yuan Y, Zhu H, Zhang M, Wei X, et al. Comparative transcriptome analysis provides insights into yellow rind formation and preliminary mapping of the Clyr (yellow rind) gene in watermelon. Front Plant Sci. 2020;11:192.

Lü G, Guo S, Zhang H, Geng L, Martyn RD, Xu Y. Colonization of Fusarium Wilt-resistant and susceptible watermelon roots by a green-fluorescent-protein-tagged isolate of Fusarium oxysporum f. sp. niveum. J Phytopathol. 2014;162:228–37.

Luan F, Fan C, Sun L, Cui H, Amanullah S, Tang L, et al. Genetic mapping reveals a candidate gene for egusi seed in watermelon. Euphytica. 2019;215:1–11.

Luikart G, Kardos M, Hand B, Rajora OP, Aitkin S, Hohenlohe PA. Population genomics: advancing understanding of nature. In: Rajora OP, editor. Population genomics: concepts, approaches and applications. Cham, Switzerland: Springer International Publishing AG; 2019. p. 3–79.

Mahla H, Choudhary B. Genetic diversity in seed purpose watermelon (Citrullus lanatus) genotypes under rainfed situations of Thar Desert. Indian J Agric Sci. 2013;83:300.

Mashilo J, Shimelis H, Odindo A, Amelework B. Simple sequence repeat markers reveal genetic diversity within and among landrace collections of citron and dessert watermelon from South Africa. J Am Soc Hortic Sci. 2016;141:598–608.

Mashilo J, Hussein S, Odindo A, Amelework B. Assessment of the genetic diversity of dessert watermelon (Citrullus lanatus var. lanatus) landrace collections of South Africa using SSR markers. Aust J Crop Sci. 2017;11:1392–8.

Meeuse A. The cucurbitaceae of southern Africa. Bothalia. 1962;8:1–111.

Meru G, McGregor C. Genetic mapping of seed traits correlated with seed oil percentage in watermelon. HortScience. 2013;48:955–9.

Minsart L-A, Djè Y, Baudoin J-P, Jacquemart A-L, Bertin P. Set up of simple sequence repeat markers and first investigation of the genetic diversity of West-African watermelon (Citrullus lanatus ssp. vulgaris oleaginous type). Genet Resour Crop Evol. 2011;58:805–14.

Mohosina F, Mehedi M, Mahmud E, Hasan M, Noor M, Rahman M, et al. Genetic diversity of commercially cultivated watermelon (Citrullus lanatus) hybrids in Bangladesh. SABRAO J Breed Genet. 2020;52:418.

Montenegro JD, Golicz AA, Bayer PE, Hurgobin B, Lee H, Chan CKK, et al. The pangenome of hexaploid bread wheat. Plant J. 2017;90:1007–13.

Mujaju C, Sehic J, Werlemark G, Garkava-Gustavsson L, Fatih M, Nybom H. Genetic diversity in watermelon (Citrullus lanatus) landraces from Zimbabwe revealed by RAPD and SSR markers. Hereditas. 2010;147:142–53.

Mujaju C, Werlemark G, Garkava-Gustavsson L, Smulders J, Nybom H. Molecular and farmer-based comparison of a wild-weed and landrace complex of watermelon in Zimbabwe. Aust J Crop Sci. 2012;6:656–61.

Munisse P, Jensen BD, Andersen SB. Genetic differentiation of watermelon landraces in Mozambique using microsatellite markers. Afr J Biotechnol. 2013;12:5513.

Nantoume A, Andersen S, Jensen B. Genetic differentiation of watermelon landrace types in Mali revealed by microsatellite (SSR) markers. Genet Resour Crop Evol. 2013;60:2129–41.

Naroueirad M, Alah DM, Ghasemi A, Fanaei H. Investigation of genetic diversity and broad sense heritability in watermelon accessions of Sistan, 2010.

Nawaz MA, Chen C, Shireen F, Zheng Z, Sohail H, Afzal M, et al. Genome-wide expression profiling of leaves and roots of watermelon in response to low nitrogen. BMC Genomics. 2018;19:1–19.

Nei M, Li W-HJPotNAoS. Mathematical model for studying genetic variation in terms of restriction endonucleases. Proc Natl Acad Sci U S A. 1979;76:5269–73.

Nimmakayala P, Islam-Faridi N, Tomason Y, Lutz F, Levi A, Reddy U. Citrullus. Wild crop relatives: genomic and breeding resources: Springer; 2011. p. 59–66.

Nimmakayala P, Abburi VL, Bhandary A, Abburi L, Vajja VG, Reddy R, et al. Use of VeraCode 384-plex assays for watermelon diversity analysis and integrated genetic map of watermelon with single nucleotide polymorphisms and simple sequence repeats. Mol Breed. 2014a;34:537–548.

Nimmakayala P, Levi A, Abburi L, Abburi VL, Tomason YR, Saminathan T, et al. Single nucleotide polymorphisms generated by genotyping by sequencing to characterize genome-wide diversity, linkage disequilibrium, and selective sweeps in cultivated watermelon. BMC Genome. 2014b;15:767.

Okoli BE. Wild and cultivated cucurbits in Nigeria. Econ Bot. 1984;38:350–7.

Oren E, Tzuri G, Vexler L, Dafna A, Meir A, Faigenboim A, et al. The multi-allelic APRR2 gene is associated with fruit pigment accumulation in melon and watermelon. J Exp Bot. 2019;70: 3781–94.

Pal LR, Yu C-H, Mount SM, Moult JJBg. Insights from GWAS: emerging landscape of mechanisms underlying complex trait disease. BMC Genome. 2015;16:S4.

Pan C-X, Hu Y, Ji H-B, Li Y-M, Chen N-L. Phenotypic diversity and clustering analysis of watermelon germplasm. J Plant Genet Resour. 2015;16:59–63.

Pandey A, Khan MK, Isik R, Turkmen O, Acar R, Seymen M, et al. Genetic diversity and population structure of watermelon (Citrullus sp.) genotypes. 3. Biotech. 2019;9:210.

Paris HSJAob. Origin and emergence of the sweet dessert watermelon. Citrullus lanatus. 2015;116: 133–48.

Park S-w, Kim K-T, Kang S-C, Yang H-B. Rapid and practical molecular marker development for rind traits in watermelon. Hortic Environ Biotechnol. 2016;57:385–91.

Park G, Kim J, Jin B, Yang H-B, Park S-W, Kang S-C, Chung S-M, Park Y. Genome-wide sequence variation in watermelon inbred lines and its implication for marker-assisted breeding, 2018.

Paudel L, Clevenger J, McGregor C. Refining of the egusi locus in watermelon using KASP assays. Sci Hortic. 2019;257:108665.

Pool JE, Hellmann I, Jensen JD, Nielsen R. Population genetic inference from genomic sequence variation. Genome Res. 2010;20:291–300.

Prothro J, Sandlin K, Abdel-Haleem H, Bachlava E, White V, Knapp S, et al. Main and epistatic quantitative trait loci associated with seed size in watermelon. J Am Soc Hortic Sci. 2012;137: 452–7.

Rajora OP, editor. Population genomics: concepts, approaches and application: Springer Nature; 2019. p. 823. ISBN 978-3-030-04587-6; ISBN 978-3-030-04589-0 (eBook)

Reddy U, Aryal N, Islam-Faridi N, Tomason Y, Levi A, Nimmakayala P. Cytomolecular characterization of rDNA distribution in various Citrullus species using fluorescent in situ hybridization. Genet Resour Crop Evol. 2013;60:2091–100.

Reddy UK, Abburi L, Abburi VL, Saminathan T, Cantrell R, Vajja VG, et al. A genome-wide scan of selective sweeps and association mapping of fruit traits using microsatellite markers in watermelon. J Hered. 2014a;106:166–76.

Reddy UK, Nimmakayala P, Levi A, Abburi VL, Saminathan T, Tomason YR, et al. High-resolution genetic map for understanding the effect of genome-wide recombination rate on nucleotide diversity in watermelon. g3. Genomes Genet. 2014b;4:2219. 114.012815

Reddy UK, Abburi L, Abburi VL, Saminathan T, Cantrell R, Vajja VG, et al. A genome-wide scan of selective sweeps and association mapping of fruit traits using microsatellite markers in watermelon. J Hered. 2015;106:166–76.

Ren R, Ray R, Li P, Xu J, Zhang M, Liu G, et al. Construction of a high-density DArTseq SNP-based genetic map and identification of genomic regions with segregation distortion in a genetic population derived from a cross between feral and cultivated-type watermelon. Mol Gen Genomics. 2015;290:1457–70.

Ren Y, Guo S, Zhang J, He H, Sun H, Tian S, et al. A tonoplast sugar transporter underlies a sugar accumulation QTL in watermelon. Plant Physiol. 2018;176:836–50.

Ren R, Yang X, Xu J, Zhang M, Liu G, Yao X. Genome-wide identification and analysis of GDSL-type esterases/lipases in watermelon (Citrullus lanatus). Scientia Horticulturae. 2019;289(6):110461.

Ren Y, Li M, Guo S, Sun H, Zhao J, Zhang J, et al. Evolutionary gain of oligosaccharide hydrolysis and sugar transport enhanced carbohydrate partitioning in sweet watermelon fruits. Plant Cell. 2021;33:1554.

Renner SS, Sousa A, Chomicki GJT. Chromosome numbers, Sudanese wild forms, and classification of the watermelon genus Citrullus, with 50 names allocated to seven biological species. Taxon. 2017;66:1393–405.

Renner SS, Pérez-Escobar OA, Silber MV, Nesbitt M, Preick M, Hofreiter M, et al. A 3500-year-old leaf from a Pharaonic tomb reveals that New Kingdom Egyptians were cultivating domesticated watermelon. bioRxiv. 2019;642785

Renner SS, Wu S, Pérez-Escobar OA, Silber MV, Fei Z, Chomicki G. A chromosome-level genome of a Kordofan melon illuminates the origin of domesticated watermelons. Proc Natl Acad Sci. 2021;118:e2101486118.

Rhee S-J, Han B-K, Jang YJ, Sim TY, Lee GP. Construction of a genetic linkage map using a frame set of simple sequence repeat and high-resolution melting markers for watermelon (Citrullus spp.). Hortic Environ Biotechnol. 2015a;56:669–76.

Rhee S-J, Seo M, Jang Y-J, Cho S, Lee GP. Transcriptome profiling of differentially expressed genes in floral buds and flowers of male sterile and fertile lines in watermelon. BMC Genomics. 2015b;16:1–14.

Rhee S-J, Kwon T, Seo M, Jang YJ, Sim TY, Cho S, et al. De novo-based transcriptome profiling of male-sterile and fertile watermelon lines. PLoS One. 2017;12:e0187147.

Robinson R, Decker-Walters D. Cucurbits. Oxford, UK: CAB International, Wallingford; 1999.

Romão R. Northeast Brazil: a secondary center of diversity for watermelon (Citrullus lanatus). Genet Resour Crop Evol. 2000;47:207–13.

Sari N, Solmaz I, Yetisir H, Unlu H. Watermelon genetic resources in Turkey and their characteristics. III international symposium on cucurbits 731, 2005. pp. 433–438

Schaefer H, Renner SS. Phylogenetic relationships in the order Cucurbitales and a new classification of the gourd family (Cucurbitaceae). Taxon. 2011;60:122–38.

Shang J-L, Wang J-M, Guo L-L, Ma S-W. Genetic diversity and correlation analysis of main botany characters in watermelon genetic resources. J Plant Genet Resour. 2012;13:11–5.

Sheehan S, Song YS. Deep learning for population genetic inference. PLoS Comput Biol. 2016;12:e1004845.

Sheng Y, Luan F, Zhang F, Davis AR. Genetic diversity within Chinese watermelon ecotypes compared with germplasm from other countries. J Am Soc Hortic Sci. 2012;137:144–51.

Singh D, Singh R, Sandhu JS, Chunneja P. Morphological and genetic diversity analysis of Citrullus landraces from India and their genetic inter relationship with continental watermelons. Sci Hortic. 2017;218:240–8.

Soghani ZN, Rahimi M, Nasab MA, Maleki M. Grouping and genetic diversity of different watermelon ecotypes based on agro-morphological traits and ISSR marker. Iheringia, Série Botânica. 2018;73:53–9.

Solmaz I, Sarı N. Characterization of watermelon (Citrullus lanatus) accessions collected from Turkey for morphological traits. Genet Resour Crop Evol. 2009;56:173–88.

Solmaz I, Sari N, Aka-Kacar Y, Yalcin-Mendi NY. The genetic characterization of Turkish watermelon (Citrullus lanatus) accessions using RAPD markers. Genet Resour Crop Evol. 2010;57:763–71.

Solmaz I, Kaçar Y, Sari N, Şimşek Ö. Genetic diversity within Turkish watermelon [Citrullus lanatus (Thunb.) Matsumura & Nakai] accessions revealed by SSR and SRAP markers. Turk J Agric For. 2016;40:407–19.

Song Q, Joshi M, Joshi V. Transcriptomic analysis of short-term salt stress response in watermelon seedlings. Int J Mol Sci. 2020;21:6036.

Stokstad E. The new potato: American Association for the Advancement of Science; 2019.

Sun Y, Fan M, He Y. DNA methylation analysis of the Citrullus lanatus response to Cucumber green mottle mosaic virus infection by whole-genome bisulfite sequencing. Genes. 2019;10: 344.

Szamosi C, Solmaz I, Sari N, Bársony C. Morphological characterization of Hungarian and Turkish watermelon (Citrullus lanatus (Thunb.) Matsum. et Nakai) genetic resources. Genet Resour Crop Evol. 2009;56:1091–105.

Tirnaz S, Edwards D, Batley J. The importance of plant pan-genomes in breeding. Quant Genet Genome Plant Breed. 2020;2:27–32.

Ulutürk Zİ. Determination of genetic diversity in watermelon (Citrullus lanatus (Thunb.) Matsum & Nakai) germplasms. Izmir Institute of Technology (Turkey), 2009

Umer MJ, Safdar LB, Gebremeskel H, Zhao S, Yuan P, Zhu H, et al. Identification of key gene networks controlling organic acid and sugar metabolism during watermelon fruit development by integrating metabolic phenotypes and gene expression profiles. Hortic Res. 2020;7:1–13.

Verma M, Arya L. Development of EST-SSRs in watermelon (Citrullus lanatus var. lanatus) and their transferability to Cucumis spp. J Hortic Sci Biotechnol. 2008;83:732–6.

Wang P, Li Q, Hu J, Su Y. Comparative analysis of genetic diversity among Chinese watermelon germplasmsusing SSR and SRAP markers, and implications for future genetic improvement. Turk J Agric For. 2015;39:322–31.

Wang X, Bao K, Reddy UK, Bai Y, Hammar SA, Jiao C, et al. The USDA cucumber (Cucumis sativus L.) collection: genetic diversity, population structure, genome-wide association studies, and core collection development. Hortic Res. 2018;5:1–13.

Wang Y, Yang X, Yadav V, Mo Y, Yang Y, Zhang R, et al. Analysis of differentially expressed genes and pathways associated with male sterility lines in watermelon via bulked segregant RNA-seq. 3. Biotech. 2020;10:1–15.

Wasylikowa K, van der Veen M. An archaeobotanical contribution to the history of watermelon, Citrullus lanatus (Thunb.) Matsum. & Nakai (syn. C. vulgaris Schrad.). Veg Hist Archaeobotany. 2004;13:213–7.

Wei C, Zhang R, Yue Z, Yan X, Cheng D, Li J, et al. The impaired biosynthetic networks in defective tapetum lead to male sterility in watermelon. J Proteome. 2021;243:104241.

Whitaker TW, Bemis WP. In: Simmonds NW, editor. Cucurbits in evolution of crop plants. London: Longman; 1976. p. 64–69.

Wu S, Wang X, Reddy U, Sun H, Bao K, Gao L, et al. Genome of 'Charleston Gray', the principal American watermelon cultivar, and genetic characterization of 1,365 accessions in the US National Plant Germplasm System watermelon collection. Plant Biotechnol J. 2019;17:2246–58.

Xu Y, Guo S, Liu J, Fei Z, He H, Zhang H, Ren Y, Sun H, Gong G, Zheng Y. Dynamic characteristics of enzymes and transcriptome related to sugar metabolism and accumulation in sweet and non-sweet watermelon fruits. Cucurbitaceae 2012 Proceedings of the Xth EUCARPIA Meeting on Genetics and Breeding of Cucurbitaceae, Antalya, Turkey, 15–18 October, 2012. University of Cukurova, Ziraat Fakultesi, 2012. pp. 103–111

Xu J, Zhang M, Liu G, Yang X, Hou X. Comparative transcriptome profiling of chilling stress responsiveness in grafted watermelon seedlings. Plant Physiol Biochem. 2016;109:561–70.

Yan L, Chunqing Z. Studies on genetic diversity with a molecular marker SRAP of WatermelonHybrids. Acta Horticulturae Sinica. 2005;32:643.

Yang Y, Mo Y, Yang X, Zhang H, Wang Y, Li H, et al. Transcriptome profiling of watermelon root in response to short-term osmotic stress. PLoS One. 2016;11:e0166314.

Zhang R, Xu Y, Yi K, Zhang H, Liu L, Gong G, et al. A genetic linkage map for watermelon derived from recombinant inbred lines. J Am Soc Hortic Sci. 2004;129:237–43.

Zhang F, Luan F, Sheng Y. Analysis of genetic diversity on different ecological watermelon [Citrullus lanatus (Thunb.) Matsum. et Nakai] germplasm using SSR markers. China Vegetables. 2010:36–43.

Zhang H, Wang H, Guo S, Ren Y, Gong G, Weng Y, et al. Identification and validation of a core set of microsatellite markers for genetic diversity analysis in watermelon, Citrullus lanatus Thunb. Matsum Nakai Euphytica. 2012;186:329–42.

Zhao S-J, Liu W-G, Yan Z-H, He N, Bao W-F. Studies of watermelon genetic diversity using SRAP and EST-SSR [J]. Acta Agriculturae Boreali-Sinica. 2010;25(3):76.

Zhao S, Zhu H, Lu X, He N, Liu W. Studies on DNA fingerprinting and genetic diversity of seedless watermelon (Citrullus lanatus) varieties using core simple sequence repeat (SSR) markers. J Agric Biotechnolo. 2014;22:188–94.

Zhu Q, Gao P, Liu S, Zhu Z, Amanullah S, Davis AR, et al. Comparative transcriptome analysis of two contrasting watermelon genotypes during fruit development and ripening. BMC Genomics. 2017;18:1–20.

Zhu Y-C, Sun D-X, Yun D, An G-l, Li W-H, Si W-J, et al. Comparative transcriptome analysis of the effect of different heat shock periods on the unfertilized ovule in watermelon (Citrullus lanatus). J Integr Agric. 2020;19:528–40.

Population Genomics of Perennial Temperate Forage Legumes

Muhammet Şakiroğlu

Abstract Temperate forage legumes are essential components of agricultural systems around the globe with a notable economic value. Being perennials with developed root systems, forage legumes mitigate a number of foremost problems of contemporary agricultural practices including erosion and nutrient leaching. As forage legumes host rhizobium in their roots through symbiotic relationship, they fix a significant amount of atmospheric nitrogen to soil reducing the dependency on synthetic fertilizers, and thus, ensuring the sustainability of the agricultural systems. Despite their widespread cultivation and high ecological and economic value, the genomics studies in forage legumes have lagged behind other crops or domesticated animals. Owing to the recent advances of high-throughput sequencing technologies that accompanied by advanced bioinformatics tools, thousands to millions of single nucleotide polymorphisms (SNP) are easily identified and employed in virtually any crop including low-resource forage legumes. Development of these and other genomics resources, approaches, and tools has allowed to address the key questions in the population genomics of perennial temperate forage that includes examining genetic diversity and population structure of cultivated and wild populations, origin and genetic relationships among species, identification of the center of origin, understanding domestication patterns, unveiling the pattern of acclimation and adaptation to climate change, and investigating the effective use of the genomics tools such as GWAS and genome-wide selection. In the present chapter, key population genomics aspects of alfalfa (*Medicago sativa* L.), red clover (*Trifolium pratense* L.), white clover (*Trifolium repens* L.), sainfoin (*Onobrychis viciifolia* Scop.), and birdsfoot trefoil (*Lotus corniculatus* L.) are synthesized and discussed. Then future perspectives of population genomics research and applications in temperate forage legumes are discussed.

M. Şakiroğlu (✉)
Department of Bioengineering, Adana Alparslan Türkeş Science and Technology University, Adana, Turkey
e-mail: msakiroglu@atu.edu.tr

Keywords Alfalfa · Birdsfoot trefoil · Center of diversity and domestication · Clover · GWAS and genomic selection · Intra and inter-specific genetic diversity and population structure · Perennial forage legumes · Population genomics · Sainfoin

1 Introduction

Nitrogen availability and uptake is a major factor affecting the productivity of agricultural systems. A large portion of the nitrogen demand in any agricultural setting is met by excessive application of synthetic fertilizers (Graham and Vance 2000). While the cost of synthetic nitrogen fertilizers is a major issue in agricultural production, there is a growing concern of nitrogen fertilizers run-offs and leaching to the environment (Wang et al. 2019). The sustainable alternative of the environmentally problematic and high cost synthetic fertilizers is the biological nitrogen fixation through the symbiotic relation between legumes and bacteria (Saha et al. 2017). Only second to cereals, the legumes are widely cultivated around the globe both for direct human consumption and for forage for a wide range of farm animals (Raymond 1969).

A satisfactory amount of dietary protein is required for functional body in humans and a considerable portion of human protein diet is from animals. The animal-based protein is largely supplied from farm animals where the amino acid pattern is very close to human needs (Elmadfa and Meyer 2017). The performance and health of farm animals that are directly used for human consumption depend on high quality nutrition. Therefore, forages play a major role in the livestock farming for their importance in animal nutrition (Corson et al. 1999).

Forages have both direct and indirect economic benefits. Direct economic value of forages includes the seed and hay production and sale whereas indirect benefits include animal by-products, nitrogen fixation, conservation benefits, and secondary metabolites (Bouton 2007; Cornara et al. 2016). Due to direct and indirect economic benefits and their role in animal nutrition, forage legumes are widely planted throughout the world. The legume forages can broadly be classified based on the geographic latitude and climate they are grown as tropical forage legumes or temperate forage legumes (Vendramini 2020). Only a restricted number of predominantly leafy and herbaceous forage legumes that are high yielding in the temperate regions of the globe (<30°C) are classified as the temperate forage legumes, and they play a significant role in animal production systems (Annicchiarico et al. 2015a). Although a number of annual temperate forage legumes (e.g., medics, annual clovers, and vetches) are cultivated in specific regions such as ley farming systems of Australia, a heavy proportion of the temperate forage legumes around the globe are perennials, such as alfalfa (*M. sativa-falcata* complex), clover (*Trifolium* spp.), sainfoin (*Onobrychis* spp.), and birdsfoot trefoil (*Lotus corniculatus*) (Annicchiarico et al. 2015a).

Perennial legumes have higher economic, agronomic, and environmental benefits compared to annual counterparts. They more efficiently increase soil fertility through a higher rate of nitrogen fixation, decrease the consumption of energy and capitalize the production of feed protein per unit area, decrease the need for pesticide and herbicide applications, reduce the risk of soil erosion, elevate the water use efficiency in drought-stressed environments, utilize resources in marginal environments, and reduce the cost for tilling and sowing operations with a flexibility of being utilized as hay, silage, grazing, or high-protein pellet (Annicchiarico et al. 2015a; Bues et al. 2013; Karlen et al. 2006; Obrycki and Karlen 2020; Porqueddu et al. 2005; Rasmussen et al. 2012).

The perennial forage legumes vary for ploidy levels but have similar mating systems and breeding goals and strategies. We have focused on four most widely cultivated perennial temperate forage legumes alfalfa, clover, sainfoin, and birdsfoot trefoil in this chapter. These species have tetraploid/diploid genomes with high similarities in their genomes, outcrossing mating system, and severe inbreeding depression (Ellison et al. 2006; Grant and Small 1996; Small 2010; Williams et al. 2012; Young and Udvardi 2009). Due to inability of obtaining inbreed lines and obligatory outcrossing mating system, all released cultivars are synthetic cultivars or population hybrids (Brummer 1999; Elliott et al. 1972; Şakiroğlu and Brummer 2007). Therefore, here we focus on population genomics of four different species (Fig. 1).

Alfalfa (*Medicago sativa* L.) is a widely cultivated fodder crop across the globe with a considerable economic value. The crop is widely adapted to virtually all the growing conditions. Alfalfa has a wide range of superior traits including high yield,

Fig. 1 The temperate forage legumes considered in this chapter. (**a**) alfalfa (*Medicago sativa* L.), (**b**) red clover (*Trifolium pratense* L.), (**c**) white clover (*Trifolium repens* L.), (**d**) sainfoin (*Onobrychis viciifolia* Scop.), and (**e**) birdsfoot trefoil (*Lotus corniculatus* L.) (source: pxfuel.com)

adaptability and quality, resistance to a number of biotic and abiotic stress conditions, ability to fix a high volume nitrogen, and persistence (the survival of an alfalfa population over time) that allow adoption of alfalfa in nearly every corner of the world (Small 2010).

Clover is the common name for a set of forage and ornamental crops from the genus *Trifolium*. This genus is one of the largest genera in Leguminosae (Fabaceae) and consists of a total of 255 annual or perennial herbaceous species. The members of the genus *Trifolium* have a global distribution. However, the main distribution range of the species is in the temperate northern hemisphere (Ellison et al. 2006; Zohary and Heller 1984). Although a number of clover species are cultivated as fodder crops across the globe, red clover (*Trifolium pratense* L.) and white clover (*Trifolium repens* L.) are the two most widely cultivated forage crops.

Sainfoin (*Onobrychis viciifolia* Scop.) is a very nutritious forage legume that manifests resilience to a number of biotic and abiotic stress factors, such as pests (Morrill et al. 1998), and drought tolerance (Kong et al. 2014), wider adaptability while providing the essential protein source to livestock worldwide (Graham and Vance 2000). Sainfoin has a long history of cultivation across the globe, including Asia, Europe, and North America (Frame et al. 1998). Although sainfoin is cultivated globally as a fodder crop, the main centers of species diversity are in Anatolia (a total of 52 species), Iran (about 53 species) and Caucasia (up to 39 species) (Çelik et al. 2011).

Birdsfoot trefoil (*Lotus corniculatus* L.) is a cultivated forage legume native to grasslands of temperate hemiboreal and boreal zones as well as elevated plains, lowlands, and alpine zones (Steiner 1999). The plant has a number of other common names including eggs and bacon, birdsfoot deervetch. It is a perennial herbaceous plant used as forage largely owing to its nonbloating properties (Seaney and Henson 1970).

Although the demarcations among clover (*Trifolium*) species are clear, the other three forage groups considered here are species complexes. While alfalfa is selected from *M. sativa-falcata* species complex (a number of closely related taxa with ploidy level variations) (İlhan et al. 2016), the birdsfoot trefoil is selected from the *L. corniculatus* complex that contains up to 12 diploid and tetraploid species (Steiner 1999).

1.1 Key Questions That Populations Genomics Can Address in Perennial Temperate Forage Legumes

Although there is an ongoing evolution in the use of the term "population genomics" and a confusion among researchers, Luikart et al. (2019) provide a very eloquent frame for definition and boundaries of the population genomics. The authors have suggested use of broad sense population genomics for the use of new genomics technologies, large-scale and genome-wide information to address questions in population genetics. However, they suggested using narrow sense population genomics which they eloquently defined as *"the use of conceptually novel approaches to address questions intractable by traditional genetic methods by using high-density genome-wide markers (e.g., DNA, RNA, epigenetic marks) to provide high power to detect genomic regions associated with traits or evolutionary processes such as fitness, phenotypes, and selection"* (Luikart et al. 2019). I am following their conceptual definition of population genomics in this chapter.

Throughout this chapter, I include the population genomics analyses of the TFL that are based on whole-genome resequencing data, genotyping-by-sequencing, exome sequencing, and transcriptome sequencing as well as the genome-wide markers to delineate any of the questions below. Relevant studies for genomic analyses of the TFL based on the pre-whole genome sequencing or genome-wide markers were also included when deemed necessary to provide the context. Population genomics could provide key insights into the biology, evolution, domestication, demography, phylogeography, speciation and phylogenomics of temperate forage legumes and assist in their selection and breeding and conservation of their genetic resources as in other taxa (Rajora 2019). Here I am discussing the current key questions that population genomics approaches could address.

1. As the perennial TFLs are main sources of animal nutrition in much of the developed world, plant germplasm for the forage species has been redistributed across the globe extensively. The pattern of genetic diversity among cultivated and wild germplasm and the rate of genetic divergence with respect to geography and breeding efforts have implications for collecting germplasm resources more effectively and for proper utilization of the genetic resources in plant breeding programs (Jin et al. 2003; Vekemans and Hardy 2004). Understanding genetic diversity and population structure is also crucial to understand forces present in the evolution of the wild germplasm including phylogeographic patterns affected from the climatic oscillations, migrations, and recolonization of plant populations after glacial retreats (Avise 2009; Feliner 2011; Sakiroglu and Brummer 2013). Therefore, *examining genetic diversity and population structure, degree of inter/intra species genetic diversity along with the boundaries of the interploidy hybridization in cultivated species and their wild relatives (primary gene pools)* is a key question in population genomics of TFLs.

2. Although the large-scale genomics studies conducted in crop species are essentially dedicated to accelerate the genetic improvement efforts, population genomics in the past decade extended to the ecological and evolutionary studies of wild crop relatives (Hufford et al. 2012). Due to millennia-long anthropogenic effects in forages, their centers of origin and the distribution patterns are not clear (Badr et al. 2008; Sakiroglu and Brummer 2013). There is a common domestication pattern among crop plants that manifested in a number of traits like seed dispersal, winter and summer dormancy, biomass yield, and harvest index. The

modifications in the set of similar human-selected traits accompanied by a genetic bottleneck are collectively denoted as "domestication syndrome" (Hammer 1984). Despite this common pattern, the extent and nature of the genomic alterations differs considerably among different crop species. For instance, increasing the proportion of the non-shattering pods, promotion of rapid germination after sowing, elevation of disease resistance, and stress tolerance were some of the main breeding targets in cereals, such as barley (*Hordeum vulgare* L.) and wheat (*Triticum aestivum* L.), whereas the decrease in defensive toxins was the target in legumes (Badr and El-Shazly 2012). Therefore, the genomic regions and genes under selection might be considerably different in different crops. Thus, identification of the *center of origin for the species of interest, the domestication history, and world distribution along with the effect of selection/domestication on the genome structure* are another set of longstanding key questions in TFLs. Population genomics can effectively address these questions in TFL.

3. With the increase in the human population, elevated consumption patterns, degraded environment, and diminishing resources, genetic improvement efforts are ought to be intensified and accelerated in perennial forage legumes (Woodfield and Brummer 2001). Therefore, improving resource use efficiency for fertilizer and water while maintaining the soil quality, reducing erosion, increasing pest and disease resistance, and enhancing animal nutrition are major goals and challenges in the temperate forage legumes. One of the efficient ways to overcome these challenges is the exploitation of the molecular breeding tools and technologies for the development of new cultivars (Humphreys 2005). Genomic regions and polymorphisms underlying the target phenotype can be used in molecular-assisted selection and breeding of forage species. Furthermore, population genomics approaches, such as genomic selection, have been recognized to be a key methodology to facilitate swift selection of cultivars with desired traits (Andres et al. 2020; Biazzi et al. 2017). Thus, *investigating the effective use of population genomics approaches such as GWAS and genomic selection in TFLs* is yet another key application of population genomics of TFLs.

4. The effect of climate change on distributions of species has already been documented (Bradshaw and Holzapfel 2001; Lobell et al. 2011), and the impact of persistent changes in climate is projected to be increased in many plant species (Hancock et al. 2011) including wild relatives of TFL. The capacity to respond to changing climate would depend on the standing genetic variation and the genetic basis of local adaptation to climate. Therefore, *understanding the pattern of acclimation and adaptation to climate change using landscape genomics, population epigenomics, and population transcriptomics* is also a major key question that population genomics could address in TFLs (Luikart et al. 2019; Rajora 2019; Moler et al. 2019).

2 Polyploidy and Inbreeding Depression in Temperate Forage Legumes and Associated Challenges for Population Genomics Studies

Polyploidy is a common evolutionary mechanism in plant genomes and widely present across the plant kingdom (Soltis et al. 1992). The nature and the role of polyploidy have been extensively investigated owing to availability of the wide range of genomics tools (Chen and Birchler 2013). Perennial temperate forage legumes are largely polyploids. Despite the predictions that autopolyploidy would evolve to a degree that it will behave like disomic inheritance (Soltis et al. 1993), currently alfalfa (Julier et al. 2003; Stanford 1951), sainfoin (Fyfe 1946), and birdsfoot trefoil (Steiner 1999) are behaving as autotetraploids forming quadrivalents. Despite the persistent difficulties of genomics studies and breeding efforts, autotetraploidy is noted to be a positive trait as it promotes a wider environmental adaptation (Annicchiarico et al. 2015a) through the capacity of retaining a higher number of beneficial mutant alleles (Bingham et al. 1994; Gallais 1981).

Alfalfa is improved from a wide gene pool collectively called *M. sativa falcata* species complex in which the chromosome number variation exists and several subspecies are freely hybridizing. Within-species chromosome number variation along with the autotetraploid segregation pattern was reported for alfalfa and allied taxa (Stanford 1951). Diploid subspecies ($2n = 2x = 16$) are listed as *M. sativa* subsp. *glomerata, M. sativa* subsp. *falcata, M. sativa* subsp. *caerulea*, and natural hybrid between *M. sativa* subsp. *falcata* and *M. sativa* subsp. *caerulea* as *M. sativa* subsp. × *hemicycla*. On the other hand, as the direct analogous of the diploid subspecies, a number of tetraploid subspecies ($2n = 4x = 32$) are also (*M. sativa* subsp. *glutinosa, M. sativa* subsp. *sativa, M. sativa* subsp. *falcata*, and *M. sativa* subsp. × *varia*) listed in the complex (Lesins and Lesins 1982; Quiros and Bauchan 1988; Small 2011; Stanford et al. 1972).

Unlike alfalfa, cultivation patterns of sainfoin species, their ploidy levels, and the gene transfer patterns are less clear. About 150 species are currently considered within the genus *Onobrychis*. The cultivated sainfoin populations are largely derived from the species *Onobrychis viciifolia* which has earlier been referred to as *Onobrychis sativa*. The two closely related taxa, *Onobrychis transcaucasica* and *Onobrychis arenaria* are also cultivated in different locations (Akopian 2009; Smirnova et al. 2012). Similarly, a number of different *Onobrychis* species that have basic chromosome numbers of $x = 7$ and $x = 8$ with the $2n = 14, 28, 56, 16, 32$ were reported (Akçelik et al. 2012). Although $2n = 14, 28, 16, 32$ are reported in different studies (Fyfe 1946; Tamas 2006), the octoploid proposition of $2n = 56$ should be viewed cautiously as it was based on the chromosome counts of a single specimen. Nonetheless, the pattern of within-species ploidy variations based on a large number of accessions from each species using more accurate and swift methods, such as flow cytometry, is also absent for sainfoin. The studies targeting the degree of demarcation among *Onobrychis* species are rare, the status of the cultivated sainfoin and allied taxa ought to be established using genomics tools.

Similar to alfalfa and sainfoin, birdsfoot trefoil is also derived from a species complex that contains several diploid ($2n = 2x = 12$) and tetraploid taxa ($2n = 4x = 24$) with unclear degrees of species demarcation. The *L. corniculatus* complex that birdsfoot trefoil belongs to is a complex group that comprises 12 diploid and tetraploid species (Ball and Chrtkova-Zertova 1968) including *L. corniculatus*, *L. alpinus* (DC.) Schleich., *L. borbasii* Ujhelyi, *L. caucasicus* Kupr., *L. delortii* Timb.-Lagr., *L. filicaulis* Desf., *L. glaber* Mill., *L. japonicus* (Regal) Larsen, *L. krylovii* Schisch. and Serg., *L. preslii* Ten., *L. schoelleri* Schweinf., and *L. uliginosus* Schkuhr (Steiner 1999). The cultivated birdsfoot trefoil, *L. corniculatus*, was initially regarded as a sole tetraploid ($2n = 4x = 24$); however, two recent reports of varieties with diploid ($2n = 2x = 12$) cytotypes (Beuselinck and Steiner 1992; Grant and Small 1996) indicated the presence of diploid progenitors of the tetraploid cytotypes. The nature and origin of tetraploidy in *L. corniculatus* complex have been contentious and all earlier proposals were based solely on agro-morphological traits. Although there is discussion in the literature regarding either the autotetraploid nature arisen form putative ancestry of diploid species or an allotetraploid nature by means of an inter-specific hybrid between two putative species (*L. alpinus*, *L. uliginosus*) followed by a whole genome duplication (Fjellstrom et al. 2001; Larsen 1958; Ross and Jones 1985; Steiner 1999), a recent RFLP-based study reported tetrasomic segregation pattern indicating a strong evidence of autotetraploidy of birdsfoot trefoil (Fjellstrom et al. 2001). Since the nature of polyploidization has detectable patterns within species, use of genomics tools could facilitate determining the nature and origin of tetraploidy in the *L. corniculatus* complex.

Among all perennial temperate forage legumes considered in this book chapter, the two clover species have a better understanding of their cytogenetic states. White clover (*Trifolium repens*) has an allotetraploid genome structure ($2n = 4x = 32$) with bivalent formation and manifests disomic inheritance. Based on a number of genomics tools, the two proposed progenitors were identified as *T. occidentale* (Schreb.) and *T. pallescens* (Schreb.) and both species are diploids (Atwood and Hill 1940; Ellison et al. 2006). Red clover (*Trifolium pretense*) on the contrary is a diploid species with a much smaller genome ($2n = 2x = 14$) (Sato et al. 2005) (Table 1).

Table 1 Basic cytogenetic parameters of the temperate forage legumes considered in this book chapter

Forage species	Chromosome number	Ploidy level	2C DNA (pg)	Nature of polyploidy
Alfalfa (*Medicago sativa* species complex)	$2n = 2x = 16$	Diploid	1.8	Autotetraploid
	$2n = 4x = 32$	Tetraploid	3.2	
White clover (*Trifolium repens*)	$2n = 4x = 32$	Tetraploid	2.22	Allotetraploid
Red clover (*Trifolium pratense*)	$2n = 2x = 16$	Diploid	0.91	
Sainfoin (*Onobrychis viciifolia* complex)	$2n = 4x = 28$	Tetraploid	2.6	Autotetraploid
Birdsfoot trefoil (*Lotus corniculatus* complex)	$2n = 2x = 12$	Diploid	?	Autotetraploid
	$2n = 4x = 24$	Tetraploid	2.5	

One of the challenges regarding the polyploids is the complexity in the analyses of the genome that poses a major barrier for effective use of GWAS and genomic selection. The challenges have arisen from the fundamental differences in the patterns of gene segregation and recombination between the two cytogenetic states (diploid vs. tetraploids) along with the high level of heterozygosity promoted by the outbreeding nature of autotetraploids (Chen et al. 2020). The researchers have undertaken two strategies to overcome such problems (Şakiroğlu et al. 2010; Sakiroglu and Brummer 2017). The first strategy was to use diploid counterparts or relatives to avoid polyploid complexities, and it was the primary strategy prior to the widespread availability of the genomics tools. For instance, association mapping of forage yield in alfalfa was conducted using a set of diploid germplasm (Sakiroglu et al. 2012). The second strategy is the use of genomics tools and development of specific statistical methods to address the polyploid genome structure (Doerge and Craig 2000; Wu et al. 2002; Xie and Xu 2000). However, a set of robust statistical methods and software tools have recently been developed and implemented for polyploid organisms that allow proper estimation of the allele dosage of SNPs (Chen et al. 2020; Ferreira et al. 2019). Population genomics studies of TFL can exploit these new strategies to tackle the major questions that have not been addressed yet.

3 Genomic Resources for Population Genomics in Temperate Forage Legumes

Unfortunately, the intensity of the population genomics research on the temperate forage legumes has lagged behind the other cultivated plants and animals despite their high economic value. This could initially be attributed to the fact that the number of researchers working on the TFLs has been much lower compared to other crops and lower interest of private companies in crops that do not require to be planted every year (Barker and Kalton 2015). In addition to the number of researchers, there was a general scarcity of early genomics tools that were available for population genomics studies in TFLs. Many molecular applications had restricted use in population genomics studies, where the discovery and use of a dense and a robust set of genetic markers across many individuals were required (Davey and Blaxter 2010). Ideally, population genomics studies should be executed with molecular tools that allow a large number (hundreds to thousands) of polymorphic markers sufficiently covering the entire genome in a single experimentation (Luikart et al. 2019; Rajora 2019).

Due to polyploidy, obligate cross-pollination, and large genome sizes, marker discovery and assay development was very cumbersome in TFLs (Eujayl et al. 2004). The marker development was therefore costly and required extensive funding resources that were largely nonexistent for TFLs (Barker and Kalton 2015). On the other hand, this limitation in the population genomics research – at least partially –

had been overcome via extensive use of the legume model barrel medic, *Medicago truncatula* (Annicchiarico et al. 2015a; Eujayl et al. 2004; Han et al. 2010).

M. truncatula is an annual temperate forage legume that is congeneric to alfalfa (*M. sativa*) and demonstrates genomic similarity to clovers (*Trifolium* sap.), sainfoin (*O. viciifolia*), and birdsfoot trefoil (*L. corniculatus*). EST sequences of *M. truncatula* genome were widely used for DNA marker discovery in TFLs. In fact, *M. truncatula* genome tools were extensively utilized to develop both pre-genomics SSR markers in many forage species (Avcı et al. 2014; İlhan et al. 2016; Kempf et al. 2016; Şakiroğlu et al. 2010; Sledge et al. 2005; Zhang et al. 2007) and used as a reference for inferring the location of the GBS SNPs in alfalfa (Sakiroglu and Brummer 2017). Therefore, information on genetic variation had been obtained by using markers developed in closely related model legume *M. truncatula* starting from isozymes, then RAPD, RFLP, AFLP, and more recently markers such as SSRs and SNPs along with microarray chips (Annicchiarico et al. 2015a).

As forage yield and forage quality are the paramount traits for breeding along with persistence and resistance to a number of biotic and abiotic stress factors (Li and Brummer 2012), molecular markers were proposed as an aid to accelerate breeding efforts for target traits very early (Brummer 2004). A set of such markers were developed and used in structural genomics approaches such as genome mapping and quantitative trait loci (QTL) mapping in forages (Brummer et al. 1993; Echt et al. 1994; Julier et al. 2003; Kaló et al. 2000; Kiss et al. 1993; Li et al. 2012; Narasimhamoorthy et al. 2007; Robins et al. 2007; Sledge et al. 2005; Tavoletti et al. 1996) which were concomitantly used as tools to study population genetics. The early tedious, non-robust, and low yielding marker systems such as RFLP, AFLP, and RAPD were quickly replaced with the more abundant and robust markers in alfalfa (X. Li and Brummer 2012). SSRs derived from *Medicago truncatula* were also used extensively in alfalfa for construction of genetic maps (Julier et al. 2003; Liu et al. 2013; Sledge et al. 2005).

As the marker technology evolved toward low-cost and highly abundant SNP markers, high resolution melting (HRM) technology based on transcriptome sequencing and finally GBS-based SNPs were used in alfalfa genome mapping (Adhikari et al. 2018; Adhikari and Missaoui 2019; Han et al. 2011; Li et al. 2012).

Besides the tools transferred from the *M. truncatula*, a true breakthrough happened when next generation sequencing technology became widely available. A dramatic reduction in the cost and time combined with the ability to produce gigabases of DNA sequence by the sequencing platforms allowed researchers to use reduced genomes of TFLs be sequenced across many individuals (Biazzi et al. 2019; Davey and Blaxter 2010; Sakiroglu and Brummer 2017). This provided a tremendous opportunity for fast and effective genotyping. These new and efficient population genomics tools have started being used in population genomics studies of alfalfa to understand genetic diversity patterns, population structure and to perform association mapping (Biazzi et al. 2017; Li et al. 2012; Sakiroglu and Brummer

2017). The tools have also been used to understand population structure and genetic diversity in red clover (Jones et al. 2020; Sæther 2018). These new tools ought to be used to address the key questions in population genomics of perennial TFLs that were elaborated above.

4 Progress in Population Genomics of Temperate Forage Legumes

4.1 Population Genomics of Alfalfa

4.1.1 Genetic Diversity, Population Structure, Differentiation, and Relationships Among Subspecies

Since alfalfa is a member of the *Medicago sativa-falcata (M. sativa species)* complex, examining patterns of genetic diversity and population structure among all the members of the complex, degree of intertaxa and interploidy hybridizations and the degree of demarcation between the cultivated species and wild germplasm has been of interest.

All taxonomic units included in the complex are perennials that occur both at the diploid ($2n = 2x = 16$) and tetraploid ($2n = 4x = 32$) levels (Lesins and Lesins 1979). The presence of bidirectional interploidy gene flow through unreduced gametes or (McCoy and Bingham 1988).

To briefly differentiate the taxa considered in the *M. sativa species* complex, we could use the morphological traits combined with the ploidy levels. Three main flower color classes exist among the members of the complex which are yellow, purple, and a spectrum of variegated colors observed in hybrids (Barnes 1972; Gunn et al. 1978; Lesins and Lesins 1982). The other commonly used morphological trait to differentiate taxa in the complex is the legume (pod) shape that is either straight to falcate or coiled. As the pod shape does not have discrete classes, the trait has quantitatively been measured as the number of coils with a ¼ coils units (Lesins and Lesins 1982; Quiros and Bauchan 1988; Şakiroğlu et al. 2010; Small 2011). The presence or absence of the glandular hairs on fruits has also been used as a secondary morphological trait (Small 1986, 2011).

There has been a historical discrepancy regarding the status of the taxonomic units associated or contained within in the *M. sativa* species complex and a broader consensus has been achieved only after population genomics studies complemented the morphological parameters (flower color, pod shape, pollen shape) or the ploidy levels were inadequate to delineate the true clustering pattern among the members of the complex alone (Havananda et al. 2011; İlhan et al. 2016; Şakiroğlu et al. 2010; Small et al. 1990).

The previous studies depending solely on morphology to understand the taxonomic status of alfalfa and allies erroneously resulted in a high number of putative species (Lesins and Gillies 1972). For example, Sinskaya (1950) considered *caerulea, hemicycla, falcata*, and *sativa* as species and proposed a number of new putative species such as *M. praesativa, M. jemenensis, M. tripolitanica, M. syriaco-palestinica, M. eusativa, M. asiatica, M. tetrahemicycla, M. quasifalcata,* and *M. difalcata*. Nonetheless, a large number of putative species suggested earlier by Sinskaya and various other authors were later found to be cultivated ecotypes and largely disregarded (Lesins and Lesins 1979; Sinskaya 1950). As the aforementioned morphological traits (flower color, pod/fruit shape, and pod glandular hairs) that have been employed to understand the relationship among taxa are governed only by a few genes, they are noted to be inadequate to represent the entire genome (Barnes 1972; Şakiroğlu et al. 2010; Small 2011) and their power to discriminate among taxa was further reduced by extensive hybridization within the complex (Small 2011; Small and Brookes 1984). Thus, there has been a need for population genomics tools be employed in alfalfa to resolve the taxonomic relationship and to determine the status of the member taxa. Population genomics tools and methods have recently been used to estimate accurate genome-wide patterns of diversity and structure among different subspecies and populations (Sakiroglu and Brummer 2017).

A wide range of early population genetics studies focused on the genetic diversity of cultivated and wild relatives of alfalfa are reviewed in Hawkins and Yu (2018); Li and Brummer (2012); Şakiroğlu and İlhan (2021). The heavy dominance of genetics studies among the tetraploid cultivated materials (and the nine historical germplasm sources in the development of modern cultivars) was not surprising due to their role in the improvement of alfalfa. Barnes (1977) listed the nine historical germplasm sources as *Medicago sativa* subsp. *falcata, Medicago sativa* subsp. *sativa* Ladak, Flemish, Turkistan, Indian, African, Chilean, Peruvian, and *Medicago sativa* subsp. *varia*. The genetic relationship among these nine germplasm sources along with the extent of the genetic diversity represented the earliest population genetics studies in alfalfa using RFLP (Kidwell et al. 1994; Maureira et al. 2004), AFLP (Segovia-Lerma et al. 2003), and SSR markers (Maureira et al. 2004). Distinctness of *M. sativa* subsp. *falcata* from *M. sativa* subsp. *varia* and *M. sativa* subsp. *sativa* (the other 7 germplasm sources) was confirmed in all three studies.

Recently, the transcriptomes of 27 alfalfa genotypes that include both improved and unimproved wild genotypes were sequenced and used to understand patterns of genetic diversity (Li et al. 2012). Overall, a total of 872,384 SNPs and 31,760 Indels were detected. In order to confirm the results, a subset of SNPs was validated using high resolution melting (HRM) approach and 91% of 192 putative SNPs were validated. The study compared the SNP regions across different germplasm groups and reported that about 95% of SNP sites identified among unimproved germplasm were maintained in cultivated germplasm. This was an indication of the fact that a high amount of genetic diversity was still maintained among cultivated alfalfa and there is no indication of a strong diversity reduction during the domestication or modern plant breeding efforts (Li et al. 2012). One of the objectives was the identification of SNPs potentially involved in the domestication. To achieve that, the authors investigated SNPs that were polymorphic within or among wild populations and fixed in the elite breeding materials. The study identified a set of

2,631 SNPs of which 94 were clustered into 15 contigs. These SNPs were considered as candidate loci for domestication. The paper subsequently analyzed population structure of the cultivated alfalfa and non-cultivated genotypes using a total of 173,947 SNPs that were consistently present across all the germplasm groups. The study revealed that the first PC (explained 9.97% of the total variance) separated the tetraploid *M. sativa* subsp. *sativa*, diploid *M. sativa* subsp. *caerulea*, diploid *M. sativa* subsp. *falcata*, and tetraploid *M. sativa* subsp. *falcata* while the second PC (explained 6.70% of the total variance) separated the European cultivated material from the US originated germplasm (Fig. 2). The results clearly indicated that *M. sativa* subsp. *falcata*, *M. sativa* subsp. *caerulea*, and *M. sativa* subsp. *sativa* (cultivated tetraploid alfalfa) are distinct groups (Li et al. 2012).

The population structure of diploid wild members of the complex was also investigated earlier with 19 cDNA RFLPs and the results indicated that diploid *M. sativa* subsp. *falcata* is distinct from *M. sativa* subsp. *caerulea* (Brummer et al.

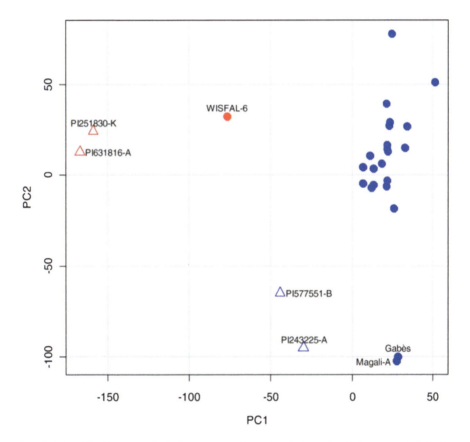

Fig. 2 Plot of the first two principal components from a PCA of SNP variation among 27 alfalfa genotypes reproduced from Li et al. (2012). (Blue solid circles represent tetraploid *M. sativa* subsp. *sativa*; red solid circle represents tetraploid *M. sativa* subsp. *falcata*; blue triangles represent diploid *M. sativa* subsp. *caerulea*; red triangles represent diploid *M. sativa* subsp. *falcata*)

1991). Expanding the bases of plant material has yielded more convincing and conclusive results regarding the subspecies status of the taxa included in the complex along with the hybrid nature of *M. sativa* subsp. *hemicycla* and *M. sativa* subsp. *varia* (Havananda et al. 2010; İlhan et al. 2016; Sakiroglu et al. 2009; Şakiroğlu et al. 2010; Sakiroglu and Brummer 2017).

A broad-based diploid panel of 374 individual genotypes from 104 accessions of the three diploid subspecies was evaluated using an earlier population genetics tools (89 polymorphic SSR markers) (Şakiroğlu et al. 2010). A similar panel was evaluated in accompanying study based on the haplotype network of the cpDNA data at the sequence level. In both studies *M. sativa* subsp. *caerulea* and *M. sativa* subsp. *falcata* formed two separate clusters and *M. sativa* subsp. *hemicycla* indicated a hybrid pattern (Havananda et al. 2010).

The same panel used in Şakiroğlu et al. (2010) was later reevaluated using population genomics tools (Sakiroglu and Brummer 2017). The entire panel of 374 individuals was genotyped using GBS approach. A set of 2,337 randomly genotyped SNPs across 16 chromosomes were obtained and used to understand the patterns of the genetic diversity. This was also the first broad-based population genomics study among diploid subspecies and all the subspecies evaluated (*M. sativa* subsp. *falcata, M. sativa* subsp. *sativa*, and *M. sativa* subsp. × *hemicycla*) were reported to form separate clusters implying discrete units (Sakiroglu and Brummer 2017). The study detected two different types of *M. sativa* subsp. × *hemicycla*. The first cluster of *M. sativa* subsp. × *hemicycla* accessions displayed a separate clustering pattern from the parental subspecies *M. sativa* subsp. *falcata* and *M. sativa* subsp. *caerulea* implying an old and isolated subspecies formation. The second cluster consisted of a number of plant genotypes with a more recent hybrid genome pattern. Therefore, the first cluster has diverged as the core *hemicycla* group and as a true diploid hybrid subspecies (Sakiroglu and Brummer 2017). The study also revealed that each of the two parental subspecies *M. sativa* subsp. *falcata* and *M. sativa* subsp. *caerulea* had further branched into two subclusters (Figs. 3 and 4). This separation of the two *M. sativa* subsp. *caerulea* subclusters was found to be related to the geographical origin of the species, particularly related to the latitude whereas the separation of the *M. sativa* subsp. *falcata* clusters was based on growth habit (lowland falcata and upland falcata ecotypes) (Sakiroglu and Brummer 2017). The results also provided a confirmation for the population structure reported in earlier population genetic studies (Havananda et al. 2010; Şakiroğlu et al. 2010).

To test if a hierarchical clustering pattern similar to diploid taxa would also be present among the tetraploid taxa, a set of SSR markers were employed. A total of 288 wild individual genotypes from 72 accessions of tetraploid *M. sativa* subsp. *falcata, M. sativa* subsp. *sativa,* and *M. sativa* subsp. × *varia* were genotyped using SSR markers. The yellow flowered *M. sativa* subsp. *falcata* accessions were clearly distinguished from purple flowered *M. sativa* subsp. *sativa* and *M. sativa* subsp. × *varia* demonstrated a continuous hybrid genome pattern (İlhan et al. 2016). The lack of a clear separation of *M. sativa* subsp. × *varia* as a unique group from the two progenitor subspecies as opposed to the diploid hybrid *M. sativa* subsp. × *hemicycla* could be regarded as an evidence that diploids have more genetic diversity,

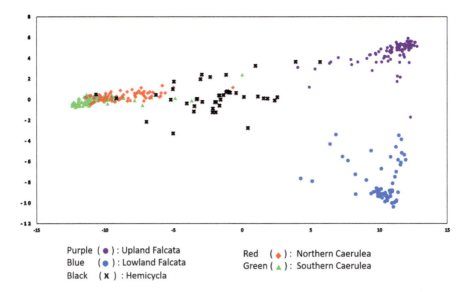

Fig. 3 Population structure among wild diploid alfalfa accessions based on genome-wide SNPs as shown by plotting the first two principal components from a principal components analysis (modified from Sakiroglu and Brummer 2017)

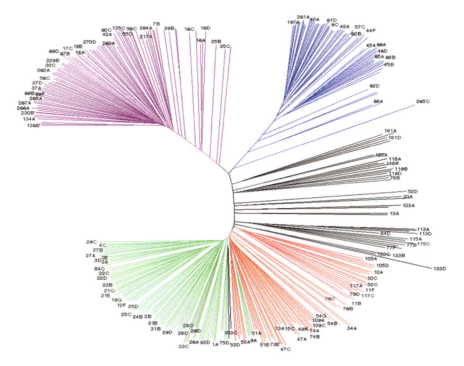

Fig. 4 Neighbor-joining dendrogram of wild diploid alfalfa accessions based on genome-wide SNPs. Purple lines represent Upland Falcata accessions, blue lines represent Lowland Falcata, black color indicates Hemicycla accessions, red color represents Northern Caerulea, and green color indicates Southern Caerulea accessions (modified from Sakiroglu and Brummer 2017)

performed a higher level of population stratification. Therefore, lower level of population stratification and genetic diversity among tetraploid taxa compared to the diploids was interpreted as the evolutionary recentness of the tetraploid taxa (İlhan et al. 2016; Şakiroğlu et al. 2010; Şakiroğlu and İlhan 2021; Sakiroglu and Brummer 2017). A similar broad-based study with the population genomics approaches yet to be conducted.

4.1.2 Origin, Center of Diversity, and Domestication of Alfalfa

One of the fundamental queries in the population genomics of alfalfa was the origin of the alfalfa and allies for which two centers of diversity and origin were proposed as Asia Minor/Caucasia (a region that contains Northwestern Iran, highlands of Armenia, Georgia, and Eastern Turkey) and Central Asia (Kazakhstan, Uzbekistan, and highland of Afghanistan) (Lesins and Lesins 1979; Small 2010; Small and Brookes 1984) (Fig. 5). In order to prioritize one of the two putative centers of diversity for diploid taxon *M. sativa* subsp. *caerulea*, a balanced set of 60 plants (16 accessions) gathered from the two putative diversity centers were genotyped using a set of genome-wide SSR markers. Population genetic analyses indicated that the Central Asian accessions indeed formed a separate cluster from that of the Caucasian/Asia Minor accessions (Sakiroglu and Brummer 2013). The results also indicated that there is a positive correlation between the overall genetic distance and geographical distance of the origin of the accessions. The mean F_{ST} values, allele diversity, and heterozygosity were all higher in Caucasian/Asia Minor accessions than the Central Asian accessions implying that the Caucasian region is the center of diversity. However, despite these early population genetics studies, the large-scale genomic evaluation of the center of diversity for all the member taxa in the *M. sativa* species complex yet to be completed.

Unlike many other crop plants, alfalfa domestication history and the location of domestication is not clear (Prosperi et al. 2014; Small 2010). According to Bolton et al. (1972), the earliest record of alfalfa cultivation is found on the brick tablets of

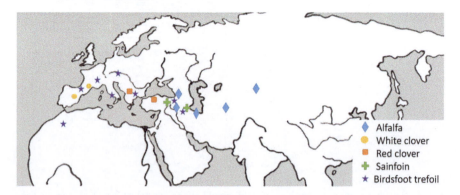

Fig. 5 Proposed centers of origin for the temperate forage legumes

Alacahöyük Ruins Turkey dated as far back as 1,400–1,200 BC. The other earlier record of the cultivation is in the Babylonian texts in 700 BC (Hendry 1923). A number of possible historical routes were provided mostly in the form of speculations (Hendry 1923). The subsequent distribution of alfalfa from the center(s) of origin to Europe, North Africa, Eastern Asia, and finally to North and South America (1600s) and Australia (in the 1800s) is well documented (Barnes 1977). Targeting domestication routes via genomics tools is an important objective for understanding the consequences of domestication on the genetic structure of alfalfa. Unfortunately, population genomics studies targeting domestication history of alfalfa are rare.

The two consecutive studies of the same group targeted domestication history of alfalfa using either neutral nuclear markers or mitochondrial genome. Both studies revealed a similar pattern that there appears to be the presence of at least two independent routes of alfalfa domestication and distribution from its widely accepted center of origin (Asia Minor and Central Asia). A total of 22 mitotypes were reported in 155 individual genotypes indicating a strong geographical structure among cultivated gene pool implying different routes of dissemination. Evaluation of the DNA sequence diversity of putatively neutral nuclear loci (glutamate synthase and pectate lyase) yielded that the domesticated accessions had 31% less diversity than the wild accessions sampled. Based on the coalescent simulations, the authors concluded that their results contradict with the constant population size implying a bottleneck during domestication (Muller et al. 2003, 2006; Prosperi et al. 2014). However, the degree of genomic change during the domestication, location of the genome under the selection during the alfalfa domestication and genome-wide patterns of the bottleneck remain to be investigated as well.

Although a general set of morphophysiological traits, such as seed retention and germination, growth habit, size, coloration, and/or edibility of economically important organs were described for domesticated plants that differ from their wild progenitors (Ladizinsky 2012), major differences in domestication pattern are expected due to the life cycle of plants and domestication traits (Gaut et al. 2015). The TFLs considered here are perennials that are harvested for their biomass as opposed to the annual grain crops which were largely selected for their seed yield and this difference might imply distinction in genomic diversity changes due to domestication and differences in the genes underlying domestication. As whole-genome sequencing and resequencing enables detection and application of a large number of genome-wide markers across multiple populations, they could be invaluable tools in the detection of the genomic regions that played a key role in the adaptations and evolution of TFL.

4.1.3 Population Genomics for Selection and Breeding of Alfalfa

Genome-Wide Association Studies

Application of association mapping as a tool for understanding the degree of linkage disequilibrium (LD) and mapping QTL to a narrow chromosomal region intensified only after availability of a dense set of marker in alfalfa (Sakiroglu and Brummer

2017; Sakiroglu et al. 2012). Within gene LD estimates between SNPs of a broad-based population indicated that LD decayed to <1 kb in diploid and tetraploid populations mainly due to obligatory outcrossing nature of alfalfa (Herrmann et al. 2010; Sakiroglu et al. 2012). When genome-wide estimates of LD were considered, the estimates (2,326 genome-wide SNPs) decreased to (r^2) 0.45 at about 2 kb (Sakiroglu and Brummer 2017). However, a higher LD estimate was also reported (26 kbp) at the same r^2 value (Li et al. 2014). Thus, earlier population genomics studies established both genome-wide and within gene LD estimates in alfalfa.

Association mapping (AM) has been implemented for detecting genomic regions that are associated with traits like forage yield, forage quality, drought and salt stress resistance, and disease and pest resistance in alfalfa. As one of the early works to test the feasibility of AM with SSR markers, Sakiroglu et al. (2012) scored about 89 SSR markers in 374 individual genotypes from 120 unimproved diploid accessions to identify genomic regions controlling forage yield and forage quality. Due to the lower number of genome-wide SSR markers, no significant associations between markers and phenotypes were reported. The study also tested possibility of using candidate gene-based association approach where a dense set of genome-wide markers are absent and found that three SNPs in two different genes (*CCOAOMT/ 111*, *CCOAOMT/1006*, *F5H* exon 2/276) associated with forage yield and quality related traits. With the elevated importance of biofuels, alfalfa stems were suggested to be used for biofuels and genomic regions associated with superior ethanol conversion yield were mapped using a set of SRAP amplification from 619 alfalfa cultivars. A total of 42 markers were identified to have significant associations with cell wall degradability and stem lignification (Dubé et al. 2013). Genome-wide association mapping (GWAS) was also employed using 26,163 SNP markers to understand genomic control of drought tolerance in heterozygous autotetraploid alfalfa. The study identified 19 genomic regions associated with drought tolerance (Zhang et al. 2015). Using different phenotyping approaches, a number of SNPs were found to be associated with the germination rate during the salt stress (Liu and Yu 2017; Yu et al. 2016), with Verticillium wilt (*Verticillium alfalfae*) resistance using different mapping approaches (Yu et al. 2017a, b), and with drought resistance (Lin et al. 2020; Yu 2017). A previously mapped panel of individuals were re-mapped in diploid alfalfa for forage yield and quality using a set of 15,154 GBS-based SNP markers for 23 forage yield and quality traits. A total of 65 SNPs were reported to have significant association with the forage yield and forage quality traits (Sakiroglu and Brummer 2017). GWAS of forage nutritive value was also assessed in a panel of 154 plants using 11,450 SNPs and 83 significant associations with leaf-to-stem ratio, leaf neutral detergent fiber (NDF), leaf acid detergent lignin (ADL), leaf crude protein (CP), leaf in vitro NDF digestibility at 24 h, stem NDF, stem ADL, stem CP, stem in vitro NDF digestibility at 24 h were reported (Biazzi et al. 2017).

Genomic Selection

Genomic selection has gained popularity as a genomics tool for alfalfa genetic improvement recently. As the first genomic selection study in alfalfa, Li et al. (2015) used a set of 175,000 genome-wide GBS SNPs to assist the selection for dry matter yield of a mixture of commercial cultivars. Two cycles of genomic selection were performed after model training with prediction accuracies of 0.43–0.66. The results indicated that the selection efficiency for genomic selection was higher than phenotypic selection in alfalfa and genomic selection has a potential to expedite genetic gain in biomass yield. Genomic selection was also applied to two alfalfa populations of different origins by Annicchiarico et al. (2015b) with 68,972 and 77,610 SNPs, respectively. The accuracies were lower than the first study (e.g., 0.32 and 0.35). In a subsequent study of the same group, about 154 genotypes were genotyped by GBS, and a number of forage quality traits such as leaf-to-stem ratio, protein content, neutral detergent fiber (NDF) and acid detergent lignin (ADL), and NDF digestibility (NFFD) along with forage yield were selected under three growing conditions using genomic selection. The prediction accuracies ranged from 0.3 to 0.4 for all the selected traits (Biazzi et al. 2017). However, genomic prediction accuracies based on 44,757 imputed SNPs in a set of 322 genotypes from 75 alfalfa accessions for a large set of yield and quality traits suggested that moderate to high prediction accuracies could be achieved for a number of quantitative traits such as plant height, flowering date, and plant regrowth while more markers or more genotypes are required for other traits such as fat and crude protein (CP) content (Jia et al. 2018).

4.2 Population Genomics of White Clover (Trifolium repens L.)

4.2.1 Population Structure and Genetic Diversity in White Clover

As white clover (*Trifolium repens*) was found to be an allopolyploid, the overall genome structure of the species was investigated initially to deduce the nature and state of polyploidy. White clover was found to have AABB-type genome with the chromosome number of $2n = 4x = 32$ and a disomic inheritance pattern (Williams et al. 2012). The genome size was reported to be around $1C = \sim 1,1$ Mb (Bennett and Leitch 2011).

One of the basic questions in allopolyploids is the identification of the progenitor genomes and the rise of allopolyploidy (Griffiths et al. 2019; Williams et al. 1987). The two putative progenitor genomes of allopolyploid white clover were proposed as western clover (*Trifolium occidentale*) and pale clover (*Trifolium pallescens*) based on chloroplast trnL intron DNA sequences, nuclear ITS sequences, and reciprocal GISH experiments (Ellison et al. 2006; Williams et al. 2012). Unlike white clover, the two diploid ($2n = 2x = 16$) progenitor species have very narrow and nonoverlapping distribution ranges. The diploid coastal species *T. occidentale* is restricted to ~100 m range of the sea coasts of Europe as opposed to high altitudes (1,800–2,700 m) growth habit of *T. pallescens* restricted to European alpine regions (Griffiths et al. 2019; Williams et al. 2012). The proposed allopolyploidization in white clover is rather unusual in terms of the aerial distribution of the parental species. In general, there is an aerial overlap among extant progenitors in most allopolyploids. However, this is not the case for white clover as *T. occidentale T. pallescens* have nonoverlapping growing areas (Soltis et al. 2016).

The natural distribution areas of white clover range from grasslands of Europe to Western Asia and from North African costs to Northern Russia (Daday 1958; Griffiths et al. 2019) (Fig. 5). Owing largely to its high adaptability and yield in various climates, white clover is cultivated as a forage in moist temperate agriculture globally (Abberton et al. 2007; Daday 1958; Griffiths et al. 2019).

Genome-wide markers are essential tools for genomic analyses and population genomics. Due to unavailability of such marker systems until very recently, a number of pre-genomics marker systems were used in population genetics studies of white clover. Application of the earlier marker systems, such as RAPD, to diverse populations or inbred lines of white clover identified four major groups which corresponded to previously developed inbred lines (Joyce et al. 1999). RAPD markers were also applied to a set of populations collected from the three northeastern states (NY, VT, and PA) of the USA and no clear separation pattern was evident based on the marker profiles (Gustine and Huff 1999). More advanced and reproducible marker systems were also used in white clover and AFLP marker analyses among 52 cultivars and accessions revealed a weak geographic structure (Kölliker et al. 2001b). SSR markers were subsequently developed and used in white clover (Kölliker et al. 2001a). More recently, the universal and easily amplified retrotransposon-based marker system, iPBS (inter primer binding sites), was applied to white clover to understand the population structure at a local scale in urban Riga, Latvia. The study reported that the white clovers formed three clusters where plants from the variety "Daile" formed the first cluster, plants from Rîga urban area formed the second cluster and plants from the different sites of Latvia formed third cluster (Grauda et al. 2015). To provide an online genomics tool for white clover, an ESTs resource containing 42,017 sequences was generated from 16 cDNA libraries (Sawbridge et al. 2003). This EST resource was later used for in silico SNP discovery and a total of 1,409 SNP containing clusters (predicted genes) were reported to have contained 18,517 SNP candidates (Spangenberg et al. 2005). Using the previously generated two-way pseudo-testcross genetic mapping families, a total of 65 SNPs derived from 58 putative genes were detected and declared as suitable markers for genetic mapping in white clover (Cogan et al. 2007).

4.2.2 Origin, Center of Diversity, and Domestication of White Clover

A number of plant species, including white clover, synthesize cyanogenic glycosides and cyanolipids when there is a tissue disruption. These cyanogenic glycosides and cyanolipids are hydrolyzed liberating the respiratory poison, hydrogen cyanide and process known as cyanogenesis (Poulton 1990). The evolution of climate-associated adaptive clines in cyanogenesis in white clover has been investigated recently (Kooyers and Olsen 2013, 2014; Kooyers et al. 2014; Olsen et al. 2007). The global scale large effective population sizes and reduced population structure estimates among populations of the same geography were reported in these studies (Kooyers and Olsen 2013, 2014; Kooyers et al. 2014; Olsen et al. 2007). As the sole genetic study exclusively focused on the genetic diversity of 16 elite white clover cultivars from Europe, North and South America, Australia, and New Zealand, George et al. (2006) used 15 SSR markers and reported no obvious clustering of cultivars.

The origin of the Ladino type of white clover was controversial as it was proposed to be either derived from the German white clover landraces or selected from the indigenous wild germplasm from Lombardy plain of Italy (Annicchiarico and Carelli 2014). The study that aimed to infer possible origin of the Ladino type white clover genotyped a set of diverse germplasm including a landrace, a wild population of Ladino origin, four Italian Non-Ladino wild populations, a Dutch landrace with 32 SSR markers. Polymorphism information content (PIC) of the markers was reported to be high (28 markers out of 32 showed PIC values of 0.5 or above) and hierarchical analyses of the overall genetic variance (AMOVA) indicated that among-populations variation was about 10-fold smaller than that within populations. The SSR-based clustering analyses indicated that Ladino type white clover was closer to Italian germplasm than German one indicating an Italian origin rather than a German one (Annicchiarico and Carelli 2014).

Most of these studies were either proof of concept for application of a specific marker type to white clover or applied to a local or mapping population leaving the pattern of diversity and genetic structure of white clover across the globe to remain largely unknown. The origin and domestication pattern of white clover similarly yet to be investigated using genomics tools. Population genomics studies using genome-wide markers, therefore, could be the key to understand the boundaries of genetic diversity among cultivated germplasm as well as among wild populations. The general effect of domestication on the genomic regions of white clover via identification of candidate adaptive loci is another major goal for white clover population genomics. Identifying loci associated with environmental variation as well as with the species and landscape interactions in the context of the landscape genomics is another research venue for population genomics studies in white clover.

4.2.3 Population Genomics for Selection and Breeding of White Clover

Despite a number of pre-genomics mapping efforts, the number of genomics studies targeting DNA variation associated with the phenotype of interest in white clover is rare.

The only GWAS study in white clover was performed to identify genomic regions associated with the cold tolerance. A total of 192 genotypes from a range of Patagonian populations were selected as the mapping population. EST libraries were generated from all the plants included in the mapping population and subsequently sequenced. The sequence reads were aligned to *M. truncatula* genome (Mt4.0v1) to identify SNPs. The genotyping efforts resulted in 8324 SNPs that passed quality criterion. The genetic structure of the mapping population revealed a weak discrimination of cold sensitive and cold-tolerant clusters while no pattern of differentiation based on geographic origin. Lack of clear separation of white clover population could be attributed to the fact that mapping population originated from a rather narrow geographic region (Chile). The study also concluded that 53 loci were associated with cold-tolerance traits. Out of the 53 SNPs, 17 SNPs were associated more than one trait or were stable across multiple sites (Inostroza et al. 2018). Nonetheless, no genomic selection study has yet been conducted on white clover.

4.3 Population Genomics of Red Clover (Trifolium pratense *L.*)

Unlike all other temperate forage legumes considered here, red clover is a diploid ($2n = 2x = 14$) with 2C DNA content of 0.97 pg (Sato et al. 2005). Nonetheless, a set of artificially induced autotetraploid ($2n = 4X = 28$) cultivars have also been developed and cultivated since 1939 mainly due to their increased biomass, elevated persistency, and enhanced diseases tolerance. However, the artificial tetraploids were noted to have lower seed set compared to diploid counterparts (Kovi et al. 2017).

4.3.1 Population Structure and Genetic Diversity in Red Clover

Compared to white clover, population structure of red clover has been investigated more intensively. There is a wide spectrum of genetics and genomics tools available in red clover providing resources for in-depth population genomics investigations. Early marker systems of RAPD, AFLP, and more intensely SSR markers were widely used in red clover for genetic diversity analyses and deducing population structure (Dias et al. 2008b; Herrmann et al. 2005; Isobe et al. 2012; Sato et al. 2005). In addition, a draft of whole-genome sequence assembly and genome annotation of red clover has recently been published (De Vega et al. 2015; Ištvánek et al. 2014) and de novo assembly of transcriptome is available (Yates et al. 2014) providing a wide range of tools at the disposal of population genomics. More recently, about 72 different genotypes were used to develop an SNP detection and genotyping pipeline in red clover using sequences generated by targeted amplicon sequencing. A total of 69,975 SNPs were identified that could readily be employed in red clover genomics (Li et al. 2019).

The previous studies targeting population structure of red clover used different marker systems. As one of the earliest molecular marker studies, the boundaries of the genetic diversity of 19 landraces (Swiss Mattenklee landraces) and cultivars of red clover were inferred via AFLP markers. Three major groups were identified based on AFLP clustering: Mattenklee landraces, Mattenklee cultivars, and other cultivars (Kölliker et al. 2003). AFLP markers were subsequently employed to understand genetic structure in a broad range of red clover populations, including Swiss wild clover populations, Mattenklee landraces, Mattenklee cultivars, field clover cultivars, Dutch wild clover populations, and Dutch landraces (Herrmann et al. 2005). A total of 12 AFLP primer combinations indicated that the Swiss wild red clover populations were different from the other groups analyzed (Herrmann et al. 2005). Isozyme polymorphism was also used to estimate genetic diversity among 15 unimproved red clover (*Trifolium pretense* L) populations studied from the Caucasus Mountains, and the results indicated a high level of genetic diversity within populations (Mosjidis et al. 2004). A set of 16 Italian natural red clover populations, four landraces, and two varieties were genotyped using two AFLPs primer combinations with 140 polymorphic markers and similarly a high level of genetic diversity within populations was reported (Pagnotta et al. 2011). As an indirect measurement of the genetic diversity, the effects of temporal changes in the population structure of red clover native to three contrasting environments in northern Europe based on AFLP markers were estimated and no consistent changes in average genetic diversity were reported between original and survivor populations (Collins et al. 2012). The red clover core collection maintained in USDA-GRIN system has a total of 85 accessions and the core collection was evaluated for genetic diversity and population structure using SSR markers (Dias et al. 2008a) and RAPD markers (Dias et al. 2008b). In both studies an elevated level of genetic variation was reported, and the largest proportion of variation resided within populations. A set of Ukrainian red clover cultivars were genotyped using 87 microsatellite markers where a much higher proportion of genetic diversity resided within populations (Dugar and Popov 2013).

Recently a more comprehensive global pattern of red clover population structure was performed utilizing the state-of-the-art genomic tools, GBS-based SNPs, in a set of 75 accessions originating largely from Europe along with three Asian accessions and a Middle Eastern one (Jones et al. 2020). A total of 264,927 SNP markers were used in a total of 640 plants. The results provided a set of detailed answers for longstanding population genomics questions in red clover. Different clustering analyses and phylogenetic approach revealed four main groups and nine clusters (Bayesian approach). The four clusters of phylogenetic analyses represented Asian, UK, Iberian, and the European groups. The nine Bayesian clusters further divided the European group into some broad geographic divisions (Fig. 6). The study results also provided a comprehensive picture of the hierarchical sources of the genetic diversity where a large portion was within accessions (56.3%) and 22.9% and 20% were among accessions and among the four main groups, respectively. The results of the structure of the genetic diversity were congruent with studies of earlier marker systems in red clover and other perennial outcrossing forage legumes (Jones et al. 2020; Şakiroğlu et al. 2010).

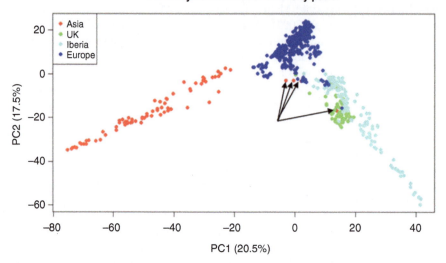

Fig. 6 Population structure of the red clover accessions based on principal component (PC) analyses of 264,927 SNP markers (from Jones et al. 2020). The four red dots highlighted by arrows represent the anomalous Italian accession Aa4445

4.3.2 Origin, Center of Diversity, and Domestication of Red Clover

Red clover has been noted to be a relatively recent addition to the domesticated plant inventory by Moorish Andalusia (Fig. 5) owing to its high nitrogen fixing ability as a response to nutrient depletion in European agricultural system at the beginning of the second millennia (Kjaergaard 2003). Being an integral part of a productive agricultural system, the clover cultivation was dramatically reduced after the groundbreaking discovery of the German chemist Fritz Haber's procedure of extracting nitrogen through ammonia synthesis (Taylor 2008). Most of the current cultivars in the market have an ancestry to European accessions with tall upright stems (Jones et al. 2020). The effect of breeding on genetic structure of red clover has been investigated using a dense set of genome-wide SNPs. Two main methods were employed to conduct a genome-wide scan for detection of the markers under selection. A total of 74 loci (BayeScan) and 1,020 loci (Samβada) were identified in which 60 loci were in common and resided within or near genes. After in-depth analyses, researchers found that these genes were transcription factors (TFs), household genes, and genes involved in growth responses, flowering time, and disease resistance. As the transcription factors are key regulators of gene expression and might have involved in the plants ability to respond to its environment, their role in selection is substantiated (Jones et al. 2020). About 55 of those SNPs were located in a gene and the gene ontology (GO) of about 51 of them were reported. Nonspecific chromosomal regions for selected loci were also reported for persistence in a set of Scandinavian germplasm (Ergon et al. 2019).

4.3.3 Population Genomics for Selection and Breeding of Red Clover

Several genetic linkage and QTL mapping (for persistence, seed yield, and self-incompatibility) studies were conducted in red clover and a consensus linkage map was built (Herrmann et al. 2006; Isobe et al. 2003; Sato et al. 2005) with a map length of 836.6 cM (Isobe et al. 2009).

The GWAS and genomic selection (GS) have not been a regular tool in red clover population genomics. As the short span of LD in red clover is noted to be a barrier to efficient use of GWAS or GS, it has been proposed that Advanced Intercross Line (AIL) populations can be developed and used in red clover to elongate the LD and thus to use these two approaches for improvement of red clover. A total of 1,536 candidate SNPs were used and a number of QTLs were reported for seed yields, traits related to the genetic control of flowering, and morphological traits in red clover (S. Isobe et al. 2013). There has been a paucity of GWAS and GS studies in red clover after this preliminary study.

4.4 *Population Genomics of Sainfoin*

Sainfoin genomics studies have been lagged behind the other forage crops despite its superior set of agronomic properties and increasing cultivation around the globe. Earlier lack of interest in genomics studies has resulted in a paucity of the development of genomics resources solely for this species (Mora-Ortiz et al. 2016). Isozymes, RAPD, AFLP, SSR ISSR, iPBS, and SNP markers have been developed and used only very recently in a number of genomics studies. One of the main objectives of the population genomics in sainfoin is to understand the population structure of the units within *Onobrychis viciifolia* and allies and the organization of the genetic diversity within and among accessions (Bhattarai et al. 2016). A number of studies were conducted in Iran where there is an immense genetic diversity in sainfoin. A set of Iranian accessions from different ecogeographic regions were assessed with RAPD markers (Nosrati et al. 2012; Rasouli et al. 2013) and no clear separation was deduced related to the geography or origin. In a comprehensive global diversity study, ISSR markers were used to determine the genetic diversity patterns and clustering in a global sainfoin collection. Similar to all earlier studies, within-population component of genetic variation was much higher (78–89%) than among accession. There were mainly two clusters representing Iranian accessions and worldwide cultivated material. The genetic diversity in Iranian cultivars alone exceeded the genetic diversity of the entire collection reflecting the primary center of diversity (Zarrabian et al. 2013). Molecular diversity analyses of a set of sainfoin accessions along with other species within genus were inferred from the nuclear internal transcribed spacer (ITS) region and the trnH-psbA and trnT-trnL intergenic spacers of the chloroplast genome. A substantial amount of genetic diversity was reported based on ITS markers, and a genetic distinction of sainfoin populations from western Europe compared to the Eastern Europe and Asia was reported (Hayot Carbonero et al. 2012).

Mainly as an effort to develop SSR markers in sainfoin, about 400 SSR primer combinations were tested in 32 distinct individual sainfoin genotypes. A total of 101 primer pairs revealed polymorphism among 32 sainfoin individuals tested with the amplification of 1,154 allelic variants. The results of the SSR markers-based clustering analyses indicated separation of two groups based on the geographic origin (Kempf et al. 2016). In addition, transcriptome library of RNA sequencing was used to identify SSR and SNP markers in sainfoin and provided a set of tools for population genomics analyses (Mora-Ortiz et al. 2016). Similarly a set of 6,752 EST-SSRs were identified and 2,469 primer pairs were designed using de novo assembly in sainfoin (Shen et al. 2019). However, a comprehensive pattern of genetic diversity based on a balanced representation of worldwide accessions using cutting-edge molecular tools, such as GBS SNPs, is needed to deduce a clear pattern of genetic structure and clustering in sainfoin.

Unfortunately, population genomics studies directly targeting center of diversity, domestication pattern, or effect of domestication on the sainfoin genome are yet to be realized. A wealth of natural sainfoin populations have been gathered from the natural growing areas and subsequently deposited to seed banks. These populations provide an invaluable resource for sainfoin population genomics studies. With the availability of genomics tools such as GBS, targeting a set of broad-based populations will allow understanding organization of genetic diversity as well as relationship among taxa. The genome-wide markers can also be used to understand the landscape genomics of sainfoin.

As a direct reflection of the available tools, the structural genomics studies targeting linkage map and QTL mapping or population genomics studies of association mapping and genomic selection in sainfoin are very rare. The only trait-marker association study in sainfoin was conducted to detect QTL controlling important characters, namely plant height, vigor, and seed yield in a biparental F_1 population. Two independent loci linked to these three vigor-related traits were identified in sainfoin (Kempf et al. 2016).

One of the major limitations of the sainfoin breeding is paucity of genomics studies for identification of genomic regions associated with the yield, quality, and resistance to biotic and abiotic factors. Use of GWAS studies and genomic selection could be a future venue for the population genomics studies in sainfoin.

4.5 Population Genomics of Birdsfoot Trefoil

Birdsfoot trefoil (*Lotus corniculatus* L.) is probably the most poorly understood TFL with very few population genetics studies reported in literature. Indeed, no population genomics study has been reported for this species. Despite economic importance and availability of a very closely related taxa as a model species (*Lotus japonicus*), paucity of genomics studies is rather surprising. As explained above, the birdsfoot trefoil is from the *Lotus corniculatus* complex and is a tetraploid species. However, even the very basic nature of tetraploidy (autotetraploid or allotetraploid) should be confirmed.

Use of molecular markers to delineate the population structure and to understand the extent of genetic diversity started with the use of the isozymes. A set of diverse accessions from diploid Lotus species of *L. alpinus* Schleich., *L. japonicus* (Regel) Larsen, *L. tenuis* Waldst. et Kit, and *L. uliginosus* Schkuhr along with the tetraploid *L. corniculatus* L. were analyzed with seven isozymes. The results indicated that *L. uliginosus* had private alleles that were not detected in *L. corniculatus* in all seven loci implying that *L. uliginosus* cannot be an ancestral taxon. All other taxa shared alleles with the *L. corniculatus* (Raelson and Grant 1988).

Allozymes were also employed alongside with the elevation, morphology, phenology in a number of diploid and tetraploid *Lotus* populations. The results revealed that diploids distributed to a lower altitude while tetraploids inhabited the higher elevations. Tetrasomic inheritance pattern of two allozymes was interpreted as evidence for autotetraploid nature of *Lotus corniculatus* (Gauthier et al. 1998).

A set of high salt-soluble globulin polypeptides (SGPP) were used to understand the genetic pattern of diversity among 128 accessions from diverse geographic regions. The overall 13 protein bands divided accessions into five groups. The two major groups represented highland or lowland ecoregion sites. The third class composed of a number of random accessions and the other two distinct groups included a few unique atypical (possibly misidentified) birdsfoot trefoil accessions (Steiner and Poklemba 1994).

Later RAPD markers were used to identify possible origin of the *L. corniculatus* using 20 polymorphic amplified fragments across diploid and tetraploid members of the *Lotus* complex. Outcome of the RFLP-based clustering indicated that the accessions were primarily grouped along the species lines. It was also reported that *L. uliginosus* is the least close to *L. corniculatus* and should not be considered as a putative progenitor. *Lotus alpinus* was found to be grouped with accessions of *L. corniculatus* (Campos et al. 1994). A total of 130 RAPD fragments were used along with a set of eight ecogeographic characteristics to analyze the genetic structure of a range of birdsfoot trefoil genotypes. Similarity among genotypes was observed based on their morphology, origin site ecology, and RAPD patterns implying that a similar set of phonotypes are acquired by the accessions inhabiting similar ecogeographic regions (Steiner and Garcia de los Santos 2001).

The genetic variation among 11 Italian natural populations was assessed using four AFLP and five SSR markers. The genetic distance estimated with the SSR markers was found to be highly correlated to the geographic distances among Italian accessions (Savo Sardaro et al. 2008). Similarly 4 SSR markers were employed to understand the patterns of genetic diversity among cultivars from Uruguay, and cultivars could be successfully discriminated using even a very few markers (Alem et al. 2011).

More recently a set of ISSR markers were used to deduce the genetic diversity and differentiation among populations of *L. corniculatus* sampled from the same and different habitats. The results indicated that genetic differentiation was present between habitats and among the populations in each habitat (Abraham et al. 2015). Similarly, a set of ISSR markers were used for genetic differentiation among three *L. corniculatus* accessions from a very narrow region in northeastern Greece which revealed that the populations of *L. corniculatus* can be differentiated precisely even in very narrow geographic regions (Merkouropoulos et al. 2017).

The current studies targeting origin of the *L. corniculatus* complex, domestication pattern are inconclusive. The studies conducted to understand the basic population genomics, association mapping, or genomic selection in birdsfoot trefoil (*L. corniculatus*) are practically nonexistent leaving a major gap in the population genomics studies of the birdsfoot trefoil.

5 Future Perspectives and Conclusions

The adverse effects of climate change on the agricultural production are expected to increase with the increasing temperatures and declining precipitation over semiarid regions which projected to reduce agricultural production and global food security. The combined effects of increasing human population and aggregate living standards elevate the demand for animal-derived protein. Moreover, the excessive use of synthetic nitrogen fertilizers has been indicated to be a source of a wide range of environmental complications such as marine eutrophication, global warming, groundwater contamination, and stratospheric ozone destruction (Crews and Peoples 2004). Therefore, there is an urgent need to develop a sustainable agricultural production system for long-term productivity and health of ecosystem (Glover et al. 2007) while providing the required animal nutrition in the context of current challenges of climate change (Boland et al. 2013). One of the most sensible strategies would be mimicking ecological production (Cox et al. 2004). Polycultures of perennial plants (in prairies or agricultural settings) have been producing more yield than agricultural lands while maintaining higher soil quality with a low erosion, higher soil nutrient cycling, and higher biological diversity (Cox et al. 2006; Glover et al. 2007; Sakiroglu et al. 2020). One way to overcome such a circle of challenges is the use of novel perennial sources for animal nutrition in agricultural system that collectively addresses all the challenges above. A more sustainable agriculture would need an elevated proportion of perennial forage legume plants and higher quality animal diet with better nutrient recycling. Perennial temperate forage legumes are plants with a high level of nitrogen fixation ability. All the TFLs described here are very nutritious forage legumes that manifest resilience to a number of biotic and abiotic stress factors, wider adaptability while providing the essential protein source to livestock worldwide. That in turn allows these crops to serve as a backbone of any sustainable animal-based diet for humans while addressing all the contemporary challenges listed above (Giller and Cadisch 1995).

Despite these advantages, the population genomics studies targeting a wide range of wild, semi-improved, and cultivated accessions of TFLs – with the exception of alfalfa – have not been realized yet and a number of persistent key problems are still present in TFLs. Therefore, genetic diversity and population structure, degree of inter/intra species genetic diversity along with the boundaries of the interploidy hybridization in cultivated germplasm and wild populations in all the TFLs need to be investigated. The true patterns of domestication remain largely unknown in virtually all the TFLs and domestication studies lagged behind all the major corps. Comprehensive studies targeting domestication patterns and world distribution along with the effect of selection/domestication on the genome structure in alfalfa, sainfoin, clovers, and birdsfoot trefoil using the state-of-the-art genomics tools would be invaluable. The pattern of acclimation and adaptation to climate change using landscape genomics, population epigenomics, and population transcriptomics has not been realized in TFLs either as done in other plants (Luikart et al. 2019; Moler et al. 2019; Rajora 2019). Given the availability of the current high-throughput genotyping methods and gene expression data, a set of well-designed experiments could address the adaptation to the environmental changes. For instance, the above-mentioned genomic tools could be used to address a set of queries such as the genes and genomic regions involved in evolutionary changes, the epigenetics mechanisms contributing to the evolution in natural populations.

One of the major pitfalls of the population genomics studies in TFLs has been lack of many markers that are assessed cost effectively until very recently. However, high-throughput genotyping methods, such as genotyping-by-sequencing (GBS) (Elshire et al. 2011), have been proved to be an efficient tool for large-scale population genomics studies for the crops with limited genomic resources such as TFLs.

Thus, the sequence-based large-scale genotyping methods such as GBS and transcriptome sequencing could be used to analyze genetic diversity and population structure, degree of inter/intra species genetic diversity along with the boundaries of the interploidy hybridization in both cultivated and wild populations of TFLs.

Compared to biparental genetic mapping, GWAS has been shown to be a promising tool for mapping DNA variations that control agronomic traits owing to the increased access to abundant molecular markers (Elshire et al. 2011). In association mapping, existing allele variation within the mapping panel has a higher representation (Sakiroglu and Brummer 2017). Therefore, GWAS studies could be another future perspective for use of populations genomics tools in the improvement of the TFLs.

The effective use of genomic selection in temperate forage legumes for major agronomic goals such as forage yield, forage quality, and resilience to biotic and abiotic stress factors will require development of models that incorporate autotetraploidy and heterozygosity at least for alfalfa, sainfoin, and birdsfoot trefoil. This is largely due to the requirement of knowledge of allele dosage that could range from nulliplex (aaaa) to simplex (Aaaa), duplex (AAaa), triplex (AAAa), and quadriplex (AAAA). Thus, developing statistical and computational methodology that incorporates autotetraploid pattern of inheritance for TFLs is another future research goal.

Wild populations of temperate forage legumes are important components of the ecosystem diversity and investigating pattern of landscape genomics along with the population epigenomics and population epitranscriptomics is another major goal of population genomics of TFLs.

References

Abberton MT, Fothergill M, Collins RP, Marshall AH. Breeding forage legumes for sustainable and profitable farming systems. Asp Appl Biol. 2007;80:81–8.

Abraham EM, Ganopoulos I, Giagourta P, Osathanunkul M, Bosmali I, Tsaftaris A, et al. Genetic diversity of Lotus corniculatus in relation to habitat type, species composition and species diversity. Biochem Syst Ecol. 2015;63:59–67.

Adhikari L, Missaoui AM. Quantitative trait loci mapping of leaf rust resistance in tetraploid alfalfa. Physiol Mol Plant Pathol. 2019;106:238–45.

Adhikari L, Lindstrom OM, Markham J, Missaoui AM. Dissecting key adaptation traits in the polyploid perennial Medicago sativa using GBS-SNP mapping. Front Plant Sci. 2018;9:934.

Akçelik SE, Avci S, Uzun S, Sancak C. Karyotype analysis of some Onobrychis (Sainfoin) species in Turkey. Arch Biol Sci. 2012;64(2):567–71.

Akopian JA. On some wild relatives of cultivated sainfoin (Onobrychis L.) from the flora of Armenia. Crop Wild Relative. 2009;4:17–8.

Alem D, Narancio R, Dellavalle PD, Rebuffo M, Zarza R, Dalla Rizza M. Molecular characterization of cultivars of Lotus corniculatus using transferable microsatellite markers. Int J Agric Nat Resour. 2011;38(3):453–61.

Andres RJ, Dunne JC, Samayoa LF, Holland JB. Enhancing crop breeding using population genomics approaches. Cham: Springer; 2020.

Annicchiarico P, Carelli M. Origin of Ladino white clover as inferred from patterns of molecular and morphophysiological diversity. Crop Sci. 2014;54(6):2696–706.

Annicchiarico P, Barrett B, Brummer EC, Julier B, Marshall AH. Achievements and challenges in improving temperate perennial forage legumes. Crit Rev Plant Sci. 2015a;34(1–3):327–80.

Annicchiarico P, Nazzicari N, Li X, Wei Y, Pecetti L, Brummer EC. Accuracy of genomic selection for alfalfa biomass yield in different reference populations. BMC Genomics. 2015b;16(1):1020.

Atwood SS, Hill HD. The regularity of meiosis in microsporocytes of Trifolium repens. Am J Bot. 1940:730–5.

Avcı S, Ilhan E, Erayman M, Sancak C. Analysis of Onobrychis genetic diversity using SSR markers from related legume species. J Anim Plant Sci. 2014;24(2):556–66.

Avise JC. Phylogeography: retrospect and prospect. J Biogeogr. 2009;36:3–15.

Badr A, El-Shazly H. Molecular approaches to origin, ancestry and domestication history of crop plants: barley and clover as examples. J Genet Eng Biotechnol. 2012;10(1):1–12.

Badr A, El-Shazly HH, Watson LE. Origin and ancestry of Egyptian clover (*Trifolium alexandrinum* L.) as revealed by AFLP markers. Genet Resour Crop Evol. 2008;55(1):21–31.

Ball PW, Chrtkova-Zertova AA. Lotus L.: 173–176. In: TG Tutin et al. (eds.), Flora Europaeae. Vol 2. Cambridge. 1968.

Barker RE, Kalton RR. Cool-season forage grass breeding: progress, potentials, and benefits. In: Contributions from breeding forage and turf grasses. London: Wiley; 2015. p. 5–20. https://doi.org/10.2135/cssaspecpub15.c2.

Barnes DK. A system for visually classifying alfalfa flower color, Agriculture handbook, 424. Washington: Agricultural Research Service, U.S. Department of Agriculture; 1972.

Barnes DK. Alfalfa germplasm in the United States: genetic vulnerability, use, improvement, and maintenance. Washington: Department of Agriculture, Agricultural Research Service; 1977.

Bennett MD, Leitch IJ. Nuclear DNA amounts in angiosperms: targets, trends and tomorrow. Ann Bot. 2011;107(3):467–590.

Beuselinck PR, Steiner JJ. A proposed framework for identifying, quantifying, and utilizing plant germplasm resources. Field Crop Res. 1992;29(3):261–72.

Bhattarai S, Coulman B, Biligetu B. Sainfoin (Onobrychis viciifolia Scop.): renewed interest as a forage legume for western Canada. Can J Plant Sci. 2016;96(5):748–56.

Biazzi E, Nazzicari N, Pecetti L, Brummer EC, Palmonari A, Tava A, et al. Genome-wide association mapping and genomic selection for alfalfa (Medicago sativa) forage quality traits. PLoS One. 2017;12(1)

Biazzi E, Nazzicari N, Pecetti L, Annicchiarico P. GBS-based genome-wide association and genomic selection for alfalfa (Medicago sativa) forage quality improvement. In: de Bruijn F, editor. The model legume Medicago Truncatula. Hoboken: Wiley; 2019. p. 923–7.

Bingham ET, Groose RW, Woodfield DR, Kidwell KK. Complementary gene interactions in alfalfa are greater in autotetraploids than diploids. Crop Sci. 1994;34(4):823–9.

Boland MJ, Rae AN, Vereijken JM, Meuwissen MP, Fischer AR, van Boekel MA, et al. The future supply of animal-derived protein for human consumption. Trends Food Sci Technol. 2013;29(1):62–73.

Bolton JL, Goplen BP, Baenziger H. World distribution and historical developments. Alfalfa Sci Technol. 1972;15:1–34.

Bouton J. The economic benefits of forage improvement in the United States. Euphytica. 2007;154(3):263–70.

Bradshaw WE, Holzapfel CM. Genetic shift in photoperiodic response correlated with global warming. Proc Natl Acad Sci. 2001;98(25):14509–11.

Brummer EC. Capturing heterosis in forage crop cultivar development. Crop Sci. 1999;39(4):943–54.

Brummer EC. Applying genomics to alfalfa breeding programs. Crop Sci. 2004;44(6):1904.

Brummer EC, Kochert G, Bouton JH. RFLP variation in diploid and tetraploid alfalfa. TAG Theor Appl Genet. 1991;83(1):89–96.

Brummer EC, Bouton JH, Kochert G. Development of an RFLP map in diploid alfalfa. Theor Appl Genet. 1993;86(2–3):329–32.

Bues A, Preissel S, Reckling M, Zander P, Kuhlman T, Topp K, et al. The environmental role of protein crops in the new common agricultural policy. Strasbourg: European Parliament; 2013.

Campos LP, Raelson JV, Grant WF. Genome relationships among Lotus species based on random amplified polymorphic DNA (RAPD). Theor Appl Genet. 1994;88(3–4):417–22.

Çelik A, Karakaya A, Avcı S, Sancak C, Özcan S. Powdery mildews observed on Onobrychis spp. in Turkey. Australas Plant Dis Notes. 2011;6(1):49–53.

Chen ZJ, Birchler JA. Polyploid and hybrid genomics. New York: Wiley; 2013.

Chen J, Leach L, Yang J, Zhang F, Tao Q, Dang Z, et al. A tetrasomic inheritance model and likelihood-based method for mapping quantitative trait loci in autotetraploid species. New Phytol. 2020; https://doi.org/10.1111/nph.16413.

Cogan NOI, Drayton MC, Ponting RC, Vecchies AC, Bannan NR, Sawbridge TI, et al. Validation of in silico-predicted genic SNPs in white clover (*Trifolium repens* L.), an outbreeding allopolyploid species. Mol Genet Genomics. 2007;277(4):413–25.

Collins RP, Helgadóttir Á, Frankow-Lindberg BE, Skøt L, Jones C, Skøt KP. Temporal changes in population genetic diversity and structure in red and white clover grown in three contrasting environments in northern Europe. Ann Bot. 2012;110(6):1341–50.

Cornara L, Xiao J, Burlando B. Therapeutic potential of temperate forage legumes: a review. Crit Rev Food Sci Nutr. 2016;56(Suppl 1):S149–61.

Corson D, Waghorn GC, Ulyatt MJ, Lee J. NIRS: forage analysis and livestock feeding. Proc Conf-New Zeal Grassland Assoc. 1999:127–32.

Cox TS, Picone C, Jackson W. Research priorities in natural systems agriculture. J Crop Improv. 2004;12(1–2):511–31.

Cox TS, Glover JD, Van Tassel DL, Cox CM, DeHaan LR. Prospects for developing perennial grain crops. Herndon: American Institute of Biological Sciences; 2006.

Crews TE, Peoples MB. Legume versus fertilizer sources of nitrogen: ecological tradeoffs and human needs. Agric Ecosyst Environ. 2004;102(3):279–97. https://doi.org/10.1016/j.agee.2003.09.018.

Daday H. Gene frequencies in wild populations of *Trifolium repens* L. Heredity. 1958;12(2):169–84.

Davey JW, Blaxter ML. RADSeq: next-generation population genetics. Brief Funct Genomics. 2010;9(5–6):416–23. https://doi.org/10.1093/bfgp/elq031.

De Vega JJ, Ayling S, Hegarty M, Kudrna D, Goicoechea JL, Ergon Å, et al. Red clover (*Trifolium pratense* L.) draft genome provides a platform for trait improvement. Sci Rep. 2015;5(1):17394. https://doi.org/10.1038/srep17394.

Dias PMB, Julier B, Sampoux J-P, Barre P, Dall'Agnol M. Genetic diversity in red clover (*Trifolium pratense* L.) revealed by morphological and microsatellite (SSR) markers. Euphytica. 2008a;160(2):189–205.

Dias PMB, Pretz VF, Dall'Agnol M, Schifino-Wittmann MT, Zuanazzi JA. Analysis of genetic diversity in the core collection of red clover (Trifolium pratense) with isozyme and RAPD markers. Crop Breed Appl Biotechnol. 2008b;8(3)

Doerge RW, Craig BA. Model selection for quantitative trait locus analysis in polyploids. Proc Natl Acad Sci. 2000;97(14):7951–6.

Dubé M-P, Castonguay Y, Duceppe M-O, Bertrand A, Michaud R. SRAP polymorphisms associated to cell wall degradability in lignified stems of alfalfa. Bioenergy Res. 2013;6(2):644–50.

Dugar YN, Popov VN. Genetic structure and diversity of Ukrainian red clover cultivars revealed by microsatellite markers. Open J Genet. 2013;3(4):235.

Echt CS, Kidwell KK, Knapp SJ, Osborn TC, McCoy TJ. Linkage mapping in diploid alfalfa (Medicago sativa). Genome. 1994;37(1):61–71.

Elliott FC, Johnson IJ, Schonhorst MH. Breeding for forage yield and quality. Alfalfa Sci Technol (alfalfasciencet). 1972:319–33.

Ellison NW, Liston A, Steiner JJ, Williams WM, Taylor NL. Molecular phylogenetics of the clover genus (Trifolium—Leguminosae). Mol Phylogenet Evol. 2006;39(3):688–705.

Elmadfa I, Meyer AL. Animal proteins as important contributors to a healthy human diet. Annu Rev Anim Biosci. 2017;5:111–31.

Elshire RJ, Glaubitz JC, Sun Q, Poland JA, Kawamoto K, Buckler ES, et al. A robust, simple genotyping-by-sequencing (GBS) approach for high diversity species. PLoS One. 2011;6(5): e19379.

Ergon Å, Skøt L, Sæther VE, Rognli OA. Allele frequency changes provide evidence for selection and identification of candidate loci for survival in red clover (*Trifolium pratense* L.). Front Plant Sci. 2019;10:718.

Eujayl I, Sledge MK, Wang L, May GD, Chekhovskiy K, Zwonitzer JC, et al. *Medicago truncatula* EST-SSRs reveal cross-species genetic markers for Medicago spp. Theor Appl Genet. 2004;108 (3):414–22. https://doi.org/10.1007/s00122-003-1450-6.

Feliner GN. Southern European glacial refugia: a tale of tales. Taxon. 2011;60(2):365–72.

Ferreira RCU, Lara LA d C, Chiari L, Barrios SCL, do Valle CB, Valério JR, et al. Genetic mapping with allele dosage information in tetraploid *Urochloa decumbens* (Stapf) RD Webster reveals insights into spittlebug (Notozulia entreriana Berg) resistance. Front Plant Sci. 2019;10:92.

Fjellstrom RG, Beuselinck PR, Steiner JJ. RFLP marker analysis supports tetrasonic inheritance in Lotus corniculatus L. Theor Appl Genet. 2001;102(5):718–25.

Frame J, Charlton JFL, Laidlaw AS. Temperate forage legumes. Wallingford: Cab International; 1998.

Fyfe JL. Polyploidy in sainfoin. Nature. 1946;158(4012):418.

Gallais A. Quantitative genetics and breeding theory of autopolyploid plants. In: Gallais A, editor. Quantitative Genetics and Breeding Methods: Proceedings of the Fourth Meeting of the Section [on] Biometrics in Plant Breeding, Poitiers, France, September 2-4, 1981. Versailles: INRA; 1981.

Gaut BS, Díez CM, Morrell PL. Genomics and the contrasting dynamics of annual and perennial domestication. Trends Genet. 2015;31(12):709–19.

Gauthier P, Lumaret R, Bédécarrats A. Genetic variation and gene flow in Alpine diploid and tetraploid populations of Lotus (L. alpinus (D.C.) Schleicher/L. corniculatus L.). I. Insights from morphological and allozyme markers. Heredity. 1998;80(6):683–93. https://doi.org/10.1046/j.1365-2540.1998.00334.x.

George J, Dobrowolski MP, van Zijll de Jong E, Cogan NO, Smith KF, Forster JW. Assessment of genetic diversity in cultivars of white clover (*Trifolium repens* L.) detected by SSR polymorphisms. Genome. 2006;49(8):919–30.

Giller KE, Cadisch G. Future benefits from biological nitrogen fixation: an ecological approach to agriculture. In: Ladha JK, et al., editors. Management of biological nitrogen fixation for the development of more productive and sustainable agricultural systems. Dordrecht: Kluwer Academic Publishers; 1995. p. 255–77.

Glover JD, Cox CM, Reganold JP. Future farming: a return to roots? Sci Am. 2007;297(2):82–9.

Graham PH, Vance CP. Nitrogen fixation in perspective: an overview of research and extension needs. Field Crop Res. 2000;65(2–3):93–106.

Grant WF, Small E. The origin of the Lotus corniculatus (Fabaceae) complex: a synthesis of diverse evidence. Can J Bot. 1996;74(7):975–89.

Grauda D, Avotiņš K, Fokina O, Kolodinska-Brantestam A, Rashal I. Genetic diversity of white clover (*Trifolium repens* L.) from the urban area of Rīga Baltā Āboliņa (*Trifolium repens* L.) Ģenētiskā Daudzveidība Rīgas Pilsētvidē. Proc Latvian Acad Sci Sect B Nat Exact Appl Sci. 2015;69(3):132–4. Sciendo

Griffiths AG, Moraga R, Tausen M, Gupta V, Bilton TP, Campbell MA, et al. Breaking free: the genomics of allopolyploidy-facilitated niche expansion in white clover. Plant Cell. 2019;31 (7):1466–87.

Gunn CR, Skrdla WH, Spencer HC. Classification of Medicago sativa L. In: Using legume characters and flower colors. Washington: Department of Agriculture, Agricultural Research Service; 1978.

Gustine DL, Huff DR. Genetic variation within among white clover populations from managed permanent patures of the Northeastern USA. Crop Sci. 1999;39(2):524–30.

Hammer K. Das domestikationssyndrom. Die Kulturpflanze. 1984;32(1):11–34.

Han Y, Khu D-M, Torres-Jerez I, Udvardi M, Monteros MJ. Plant transcription factors as novel molecular markers for legumes. In: Huyghe C, editor. Sustainable use of genetic diversity in forage and turf breeding. Dordrecht: Springer; 2010. p. 421–5.

Han Y, Kang Y, Torres-Jerez I, Cheung F, Town CD, Zhao PX, et al. Genome-wide SNP discovery in tetraploid alfalfa using 454 sequencing and high resolution melting analysis. BMC Genomics. 2011;12(1):1–11.

Hancock AM, Brachi B, Faure N, Horton MW, Jarymowycz LB, Sperone FG, et al. Adaptation to climate across the *Arabidopsis thaliana* genome. Science. 2011;334(6052):83–6.

Havananda T, Brummer EC, Maureira-Butler IJ, Doyle JJ. Relationships among diploid members of the Medicago sativa (Fabaceae) species complex based on chloroplast and mitochondrial DNA sequences. Syst Bot. 2010;35(1):140–50.

Havananda T, Brummer EC, Doyle JJ. Complex patterns of autopolyploid evolution in alfalfa and allies (Medicago sativa; Leguminosae). Am J Bot. 2011;98(10):1633–46.

Hawkins C, Yu L-X. Recent progress in alfalfa (Medicago sativa L.) genomics and genomic selection. Crop J. 2018;6(6):565–75.

Hayot Carbonero C, Carbonero F, Smith LMJ, Brown TA. Phylogenetic characterisation of Onobrychis species with special focus on the forage crop Onobrychis viciifolia Scop. Genet Resour Crop Evol. 2012;59(8):1777–88. https://doi.org/10.1007/s10722-012-9800-3.

Hendry GW. Alfalfa in history. Agron J. 1923;15(5):171–6.

Herrmann D, Boller B, Widmer F, Kölliker R. Optimization of bulked AFLP analysis and its application for exploring diversity of natural and cultivated populations of red clover. Genome. 2005;48(3):474–86.

Herrmann D, Boller B, Studer B, Widmer F, Kölliker R. QTL analysis of seed yield components in red clover (*Trifolium pratense* L.). Theor Appl Genet. 2006;112(3):536–45.

Herrmann D, Barre P, Santoni S, Julier B. Association of a CONSTANS-LIKE gene to flowering and height in autotetraploid alfalfa. TAG Theor Appl Genet. 2010;121(5):865–76.

Hufford MB, Bilinski P, Pyhäjärvi T, Ross-Ibarra J. Teosinte as a model system for population and ecological genomics. Trends Genet. 2012;28(12):606–15.

Humphreys MO. Genetic improvement of forage crops–past, present and future. J Agric Sci. 2005;143(6):441–8.

İlhan D, Li X, Brummer EC, Şakiroğlu M. Genetic diversity and population structure of tetraploid accessions of the Medicago sativa–falcata Complex. Crop Sci. 2016;56(3):1146–56.

Inostroza L, Bhakta M, Acuña H, Vásquez C, Ibáñez J, Tapia G, et al. Understanding the complexity of cold tolerance in white clover using temperature gradient locations and a GWAS approach. The Plant Genome. 2018;11(3):1–14.

Isobe S, Klimenko I, Ivashuta S, Gau M, Kozlov NN. First RFLP linkage map of red clover (*Trifolium pratense* L.) based on cDNA probes and its transferability to other red clover germplasm. Theor Appl Genet. 2003;108(1):105–12.

Isobe S, Kölliker R, Hisano H, Sasamoto S, Wada T, Klimenko I, et al. Construction of a consensus linkage map for red clover (*Trifolium pratense* L.). BMC Plant Biol. 2009;9(1):57. https://doi.org/10.1186/1471-2229-9-57.

Isobe SN, Hisano H, Sato S, Hirakawa H, Okumura K, Shirasawa K, et al. Comparative genetic mapping and discovery of linkage disequilibrium across linkage groups in white clover (*Trifolium repens* L.). G3. 2012;2(5):607–17.

Isobe S, Boller B, Klimenko I, Kölliker S, Rana JC, Sharma TR, et al. Genome-wide SNP marker development and QTL identification for genomic selection in red clover. In: Barth S, Milbourne D, editors. Breeding strategies for sustainable forage and turf grass improvement. Dordrecht: Springer Science and Business Media; 2013. p. 29–36.

Ištvánek J, Jaroš M, Křenek A, Řepková J. Genome assembly and annotation for red clover (Trifolium pratense; Fabaceae). Am J Bot. 2014;101(2):327–37. https://doi.org/10.3732/ajb.1300340.

Jia C, Zhao F, Wang X, Han J, Zhao H, Liu G, et al. Genomic prediction for 25 agronomic and quality traits in alfalfa (Medicago sativa). Front Plant Sci. 2018;9:1220.

Jin Y, He T, Lu B-R. Fine scale genetic structure in a wild soybean (*Glycine soja*) population and the implications for conservation. New Phytol. 2003;159(2):513–9.

Jones C, De Vega J, Lloyd D, Hegarty M, Ayling S, Powell W, et al. Population structure and genetic diversity in red clover (*Trifolium pratense* L.) germplasm. Sci Rep. 2020;10(1):1–12.

Joyce TA, Abberton MT, Michaelson-Yeates TPT, Forster JW. Relationships between genetic distance measured by RAPD-PCR and heterosis in inbred lines of white clover (*Trifolium repens* L.). Euphytica. 1999;107(3):159–65.

Julier B, Flajoulot S, Barre P, Cardinet G, Santoni S, Huguet T, et al. Construction of two genetic linkage maps in cultivated tetraploid alfalfa (Medicago sativa) using microsatellite and AFLP markers. BMC Plant Biol. 2003;3(1):9. https://doi.org/10.1186/1471-2229-3-9.

Kaló P, Endre G, Zimanyi L, Csanadi G, Kiss GB. Construction of an improved linkage map of diploid alfalfa (Medicago sativa). TAG Theor Appl Genet. 2000;100(5):641–57.

Karlen DL, Lemunyon JL, Singer JW. Forages for conservation and improved soil quality. In: Barnes RF, Nelson CJ, Moore KF, Collins M, editors. Sixth edition of forages: the science of grassland agriculture, volume II. Ames: Blackwell Publishing, Inc.; 2006. p. 149–66.

Kempf K, Mora-Ortiz M, Smith LM, Kölliker R, Skøt L. Characterization of novel SSR markers in diverse sainfoin (Onobrychis viciifolia) germplasm. BMC Genet. 2016;17(1):124.

Kidwell KK, Austin DF, Osborn TC. RFLP evaluation of nine Medicago accessions representing the original germplasm sources for north American alfalfa cultivars. Crop Sci. 1994;34(1):230.

Kiss GB, Csanádi G, Kálmán K, Kaló P, Ökrész L. Construction of a basic genetic map for alfalfa using RFLP, RAPD, isozyme and morphological markers. Mol Gen Genet MGG. 1993;238(1–2):129–37.

Kjaergaard T. A plant that changed the world: the rise and fall of clover 1000-2000. Landsc Res. 2003;28(1):41–9.

Kölliker R, Jones ES, Drayton MC, Dupal MP, Forster JW. Development and characterisation of simple sequence repeat (SSR) markers for white clover (*Trifolium repens* L.). Theor Appl Genet. 2001a;102(2–3):416–24.

Kölliker R, Jones ES, Jahufer MZZ, Forster JW. Bulked AFLP analysis for the assessment of genetic diversity in white clover (*Trifolium repens* L.). Euphytica. 2001b;121(3):305–15.

Kölliker R, Herrmann D, Boller B, Widmer F. Swiss Mattenklee landraces, a distinct and diverse genetic resource of red clover (*Trifolium pratense* L.). Theor Appl Genet. 2003;107(2):306–15.

Kong J, Pei Z, Du M, Sun G, Zhang X. Effects of arbuscular mycorrhizal fungi on the drought resistance of the mining area repair plant Sainfoin. Int J Min Sci Technol. 2014;24(4):485–9.

Kooyers NJ, Olsen KM. Searching for the bull's eye: agents and targets of selection vary among geographically disparate cyanogenesis clines in white clover (*Trifolium repens* L.). Heredity. 2013;111(6):495–504.

Kooyers NJ, Olsen KM. Adaptive cyanogenesis clines evolve recurrently through geographical sorting of existing gene deletions. J Evol Biol. 2014;27(11):2554–8.

Kooyers NJ, Gage LR, Al-Lozi A, Olsen KM. Aridity shapes cyanogenesis cline evolution in white clover (T rifolium repens L.). Mol Ecol. 2014;23(5):1053–70.

Kovi MR, Amdahl H, Alsheikh M, Rognli OA. De novo and reference transcriptome assembly of transcripts expressed during flowering provide insight into seed setting in tetraploid red clover. Sci Rep. 2017;7(1):44383. https://doi.org/10.1038/srep44383.

Ladizinsky G. Plant evolution under domestication. Berlin: Springer Science and Business Media; 2012.

Larsen K. Cytotaxonomical studies in Lotus IV. Some cases of polyploidy. Bot Tidsskr. 1958;54:44–56.

Lesins K, Gillies CB. Taxonomy and cytogenetics of Medicago. Alfalfa Science and Technology. 1972:53–86.

Lesins KA, Lesins I. Genus Medicago (Leguminosae): A taxogenetic study. Dr. W. Junk bv Publishers; 1979.

Lesins KA, Lesins I. Genus Medicago (Leguminosae), a taxogenetic study. Plant Ecol. 1982;50 (2):92.

Li X, Brummer EC. Applied genetics and genomics in alfalfa breeding. Agronomy. 2012;2 (1):40–61.

Li X, Acharya A, Farmer AD, Crow JA, Bharti AK, Kramer RS, et al. Prevalence of single nucleotide polymorphism among 27 diverse alfalfa genotypes as assessed by transcriptome sequencing. BMC Genomics. 2012;13(1):568.

Li X, Han Y, Wei Y, Acharya A, Farmer AD, Ho J, et al. Development of an alfalfa SNP array and its use to evaluate patterns of population structure and linkage disequilibrium. PLoS One. 2014;9(1):e84329.

Li X, Wei Y, Acharya A, Hansen JL, Crawford JL, Viands DR, et al. Genomic prediction of biomass yield in two selection cycles of a Tetraploid alfalfa breeding population. Plant Genome. 2015;8(2) https://doi.org/10.3835/plantgenome2014.12.0090.

Li W, Riday H, Riehle C, Edwards A, Dinkins R. Identification of single nucleotide polymorphism in red clover (Trifolium pratense L) using targeted genomic amplicon sequencing and RNA-seq. Front Plant Sci. 2019;10:1257. https://doi.org/10.3389/fpls.2019.01257.

Lin S, Medina CA, Boge B, Hu J, Fransen S, Norberg S, et al. Identification of genetic loci associated with forage quality in response to water deficit in autotetraploid alfalfa (Medicago sativa L.). BMC Plant Biol. 2020;20(1):1–18.

Liu X-P, Yu L-X. Genome-wide association mapping of loci associated with plant growth and forage production under salt stress in alfalfa (Medicago sativa L.). Front Plant Sci. 2017;8:853.

Liu Z, Chen T, Ma L, Zhao Z, Zhao PX, Nan Z, et al. Global transcriptome sequencing using the Illumina platform and the development of EST-SSR markers in autotetraploid alfalfa. PLoS One. 2013;8(12):e83549.

Lobell DB, Schlenker W, Costa-Roberts J. Climate trends and global crop production since 1980. Science. 2011;333(6042):616–20.

Luikart G, Kardos M, Hand BK, Rajora OP, Aitken SN, Hohenlohe PA. Population genomics: advancing understanding of nature. In: Rajora OP, editor. Population genomics: concepts, approaches and applications. Cham: Springer Nature Switzerland AG; 2019. p. 3–79.

Maureira IJ, Ortega F, Campos H, Osborn TC. Population structure and combining ability of diverse Medicago sativa germplasms. TAG Theor Appl Genet. 2004;109(4):775–82.

McCoy TJ, Bingham ET. Cytology and cytogenetics of alfalfa. Alfalfa and Alfalfa Improvement. 1988;29:737–76.

Merkouropoulos G, Hilioti Z, Abraham EM, Lazaridou M. Evaluation of Lotus corniculatus L. accessions from different locations at different altitudes reveals phenotypic and genetic diversity. Grass Forage Sci. 2017;72(4):851–6.

Moler ER, Abakir A, Eleftheriou M, Johnson JS, Krutovsky KV, Lewis LC, et al. Population epigenomics: advancing understanding of phenotypic plasticity, acclimation, adaptation and diseases. In: Rajora OP, editor. Population genomics: concepts, approaches and applications. Cham: Springer Nature Switzerland AG; 2019. p. 179–260.

Mora-Ortiz M, Swain MT, Vickers MJ, Hegarty MJ, Kelly R, Smith LMJ, et al. De-novo transcriptome assembly for gene identification, analysis, annotation, and molecular marker discovery in Onobrychis viciifolia. BMC Genomics. 2016;17(1):756. https://doi.org/10.1186/s12864-016-3083-6.

Morrill WL, Ditterline RL, Cash SD. Insect pests and associated root pathogens of sainfoin in western USA. Field Crop Res. 1998;59(2):129–34.

Mosjidis JA, Greene SL, Klingler KA, Afonin A. Isozyme diversity in wild red clover populations from the Caucasus. Crop Sci. 2004;44(2):665–70.

Muller M-H, Prosperi JM, Santoni S, Ronfort J. Inferences from mitochondrial DNA patterns on the domestication history of alfalfa (Medicago sativa). Mol Ecol. 2003;12(8):2187–99.

Muller M-H, Poncet C, Prosperi JM, Santoni S, Ronfort J. Domestication history in the Medicago sativa species complex: inferences from nuclear sequence polymorphism. Mol Ecol. 2006;15(6):1589–602.

Narasimhamoorthy B, Bouton JH, Olsen KM, Sledge MK. Quantitative trait loci and candidate gene mapping of aluminum tolerance in diploid alfalfa. Theor Appl Genet. 2007;114(5):901–13.

Nosrati H, Feizi MAH, Tarrah SS, Haghighi AR. Population genetic variation in sainfoin (Fabaceae) revealed by RAPD markers. Analele Universitatii Din Oradea Fascicula Biologie. 2012;19(1)

Obrycki JF, Karlen DL. Forages for conservation and improved soil quality. In: Moore KJ, Collins M, Nelson CJ, Redfearn DD, editors. Forages: the science of grassland agriculture, vol. 2. 7th ed. Hoboken: Wiley; 2020. p. 227–47.

Olsen KM, Sutherland BL, Small LL. Molecular evolution of the Li/li chemical defence polymorphism in white clover (*Trifolium repens* L.). Mol Ecol. 2007;16(19):4180–93.

Pagnotta MA, Annicchiarico P, Farina A, Proietti S. Characterizing the molecular and morphophysiological diversity of Italian red clover. Euphytica. 2011;179(3):393–404.

Porqueddu C, Maltoni S, McIvor JG, McGilloway DA. Strategies to mitigate seasonality of production in grassland-based systems. Grassland: A Global Resource; 2005. p. 111–22.

Poulton JE. Cyanogenesis in plants. Plant Physiol. 1990;94(2):401–5.

Prosperi J-M, Jenczewski E, Muller M-H, Fourtier S, Sampoux J-P, Ronfort J. Alfalfa domestication history, genetic diversity and genetic resources. Legume Perspect. 2014;4:13–4. https://doi.org/hal-01216251v2f

Quiros CF, Bauchan GR. The genus Medicago and the origin of the Medicago sativa complex. Alfalfa and Alfalfa Improv. 1988;29

Raelson JV, Grant WF. Evaluation of hypotheses concerning the origin of Lotus corniculatus (Fabaceae) using isoenzyme data. Theor Appl Genet. 1988;76(2):267–76.

Rajora OP. Population genomics: concepts, approaches and applications. Cham: Springer Nature Switzerland AG; 2019.

Rasmussen J, Søegaard K, Pirhofer-Walzl K, Eriksen J. N2-fixation and residual N effect of four legume species and four companion grass species. Eur J Agron. 2012;36(1):66–74.

Rasouli M, Jafari AA, Tabaei-Aghdaei SR, Shanjani PS, Darvish F. Assessment of genetic variability of 36 populations of sainfoin (*Onobrychis sativa*) based on RAPD markers. Int J Biosci. 2013;3:15–26.

Raymond WF. The nutritive value of forage crops. In: Advances in agronomy, vol. 21. Amsterdam: Elsevier; 1969. p. 1–108.

Robins JG, Luth D, Campbell TA, Bauchan GR, He C, Viands DR, et al. Genetic mapping of biomass production in tetraploid alfalfa. Crop Sci. 2007;47(1):1.

Ross MD, Jones WT. The origin of Lotus corniculatus. Theor Appl Genet. 1985;71(2):284–8.

Sæther VE. Identification of red clover (*Trifolium pratense* L.) SNPs whose allelic versions appear with different frequency in pure stands and in mixtures with grasses, using GBS and CAPS-markers (Master's thesis). Norwegian University of Life Sciences, AAs; 2018.

Saha B, Saha S, Das A, Bhattacharyya PK, Basak N, Sinha AK, et al. Biological nitrogen fixation for sustainable agriculture. In: Agriculturally important microbes for sustainable agriculture. Berlin: Springer; 2017. p. 81–128.

Şakiroğlu M, Brummer EC. Little heterosis between alfalfa populations derived from the midwestern and southwestern United States. Crop Sci. 2007;47(6):2364–71.

Sakiroglu M, Brummer EC. Presence of phylogeographic structure among wild diploid alfalfa accessions (Medicago sativa L. subsp. Microcarpa Urb.) with evidence of the center of origin. Genet Resour Crop Evol. 2013;60(1):23–31.

Sakiroglu M, Brummer EC. Identification of loci controlling forage yield and nutritive value in diploid alfalfa using GBS-GWAS. Theor Appl Genet. 2017;130(2):261–8. https://doi.org/10.1007/s00122-016-2782-3.

Şakiroğlu M, İlhan D. Medicago sativa species complex: re-visiting the century old problem in the light of molecular tools. Crop Sci. 2021;61:827–838. https://doi.org/10.1002/csc2.20316.

Sakiroglu M, Doyle JJ, Brummer EC. The population genetic structure of diploid Medicago sativa L. germplasm. Dordrecht: Springer; 2009.

Şakiroğlu M, Doyle JJ, Brummer EC. Inferring population structure and genetic diversity of broad range of wild diploid alfalfa (Medicago sativa L.) accessions using SSR markers. Theor Appl Genet. 2010;121(3):403–15.

Sakiroglu M, Sherman-Broyles S, Story A, Moore KJ, Doyle JJ, Brummer EC. Patterns of linkage disequilibrium and association mapping in diploid alfalfa (M. sativa L.). Theor Appl Genet. 2012;125(3):577–90.

Sakiroglu M, Dong C, Hall MB, Jungers J, Picasso V. How does nitrogen and forage harvest affect belowground biomass and non-structural carbohydrates in dual use Kernza intermediate wheatgrass? Crop Sci. 2020; https://doi.org/10.1002/csc2.20239.

Sato S, Isobe S, Asamizu E, Ohmido N, Kataoka R, Nakamura Y, et al. Comprehensive structural analysis of the genome of red clover (*Trifolium pratense* L.). DNA Res. 2005;12(5):301–64.

Savo Sardaro ML, Atallah M, Tavakol E, Russi L, Porceddu E. Diversity for AFLP and SSR in natural populations of Lotus corniculatus L. from Italy. Crop Sci. 2008;48(3):1080–9.

Sawbridge T, Ong E-K, Binnion C, Emmerling M, Meath K, Nunan K, et al. Generation and analysis of expressed sequence tags in white clover (*Trifolium repens* L.). Plant Sci. 2003;165(5):1077–87.

Seaney RR, Henson PR. Birdsfoot trefoil. Adv Agron. 1970;22:119–57.

Segovia-Lerma A, Cantrell RG, Conway JM, Ray IM. AFLP-based assessment of genetic diversity among nine alfalfa germplasms using bulk DNA templates. Genome. 2003;46(1):51–8.

Shen S, Chai X, Zhou Q, Luo D, Wang Y, Liu Z. Development of polymorphic EST-SSR markers and characterization of the autotetraploid genome of sainfoin (Onobrychis viciifolia). PeerJ. 2019;7:e6542. https://doi.org/10.7717/peerj.6542.

Sinskaya EN. Flora of cultivated plants of the USSR: XIII. Perennial leguminous plants. (Part 1, translated 1961). Jerusalem: Israel Program of Scientific Translations; 1950.

Sledge MK, Ray IM, Jiang G. An expressed sequence tag SSR map of tetraploid alfalfa (Medicago sativa L.). Theor Appl Genet. 2005;111(5):980–92.

Small E. Taxonomy of glandular wild alfalfa (Medicago sativa). Can J Bot. 1986;64(9):2125–9.

Small E. Alfalfa and relatives: evolution and classification of Medicago. Ottawa: NRC Research Press; 2010.

Small E. Alfalfa and relatives: evolution and classification of Medicago. Ottawa: CAB International; 2011.

Small E, Brookes BS. Taxonomic circumscription and identification in theMedicago sativa-falcata (alfalfa) continuum. Econ Bot. 1984;38(1):83–96.

Small E, Brookes B, Lassen P. Circumscription of the genus Medicago (Leguminosae) by seed characters. Can J Bot. 1990;68(3):613–29.

Smirnova E, Gusev A, Zaytseva O, Sheina O, Tkachev A, Kuznetsova E, et al. Uptake and accumulation of multiwalled carbon nanotubes change the morphometric and biochemical characteristics of *Onobrychis arenaria* seedlings. Front Chem Sci Eng. 2012;6(2):132–8.

Soltis PS, Doyle JJ, Soltis DE. Molecular data and polyploid evolution in plants. In: Soltis PS, Soltis DE, Doyle JJ, editors. Molecular systematics of plants. Boston: Springer; 1992. p. 177–201.

Soltis DE, Soltis PS, Rieseberg LH. Molecular data and the dynamic nature of polyploidy. Crit Rev Plant Sci. 1993;12(3):243–273.

Soltis DE, Visger CJ, Marchant DB, Soltis PS. Polyploidy: pitfalls and paths to a paradigm. Am J Bot. 2016;103(7):1146–66.

Spangenberg GC, Sawbridge T, Ong EK, Love CG, Erwin TA, Logan EG, et al. Clover ASTRA: a web-based resource for trifolium EST analysis. Molecular breeding for the genetic improvement of forage crops and turf: proceedings of the 4th international symposium on the molecular breeding of forage and turf, a satellite workshop of the XXth international grassland congress, July 2005, Aberystwyth, Wales. Wageningen: Wageningen Academic Publishers; 2005.

Stanford EH. Tetrasomic inheritance in alfalfa. Agron J. 1951;43:222–5.

Stanford EH, Clement WM, Bingham ET. Cytology and evolution of the Medicago sativa-falcata complex. Alfalfa Science and Technology. 1972:87–101.

Steiner JJ. Birdsfoot trefoil origins and germplasm diversity. In: Beuselinck PR, editor. Trefoil: the science and technology of the lotus, CSSA special publications, vol. 28. Hoboken: Wiley; 1999. p. 81–96.

Steiner JJ, Garcia de los Santos G. Adaptive ecology of Lotus corniculatus L. genotypes: I. Plant morphology and RAPD marker characterizations. Crop Sci. 2001;41(2):552–63.

Steiner JJ, Poklemba CJ. Lotus corniculatus classification by seed globulin polypeptides and relationship to accession pedigrees and geographic origin. Crop Sci. 1994;34(1):255–64.

Tamas E. Citological aspects of the Onobrychis genus. Bull Univ Agric Sci Vet Med Cluj-Napoca Agric. 2006;62

Tavoletti S, Veronesi F, Osborn TC. RFLP linkage map of an alfalfa meiotic mutant based on an F1 population. J Hered. 1996;87(2):167–70.

Taylor NL. A century of clover breeding developments in the United States. Crop Sci. 2008;48 (1):1–13.

Vekemans X, Hardy OJ. New insights from fine-scale spatial genetic structure analyses in plant populations. Mol Ecol. 2004;13(4):921–35.

Vendramini JM. Legumes for tropical and subtropical areas. In: Forages, volume 2: the science of grassland agriculture. 7th ed. Hoboken: Wiley; 2020. p. 277.

Wang Y, Ying H, Yin Y, Zheng H, Cui Z. Estimating soil nitrate leaching of nitrogen fertilizer from global meta-analysis. Sci Total Environ. 2019;657:96–102.

Williams WM, Baker MJ, Williams WM. Taxonomy and biosystematics of Trifolium repens. White Clover. 1987;323:342.

Williams WM, Ellison NW, Ansari HA, Verry IM, Hussain SW. Experimental evidence for the ancestry of allotetraploid Trifolium repens and creation of synthetic forms with value for plant breeding. BMC Plant Biol. 2012;12(1):55.

Woodfield DR, Brummer EC. Integrating molecular techniques to maximise the genetic potential of forage legumes. In: Spangenberg G, editor. Molecular breeding of forage crops. Developments in plant breeding, vol. 10. Dordrecht: Springer; 2001. p. 51–65.

Wu R, Ma C-X, Casella G. A bivalent polyploid model for linkage analysis in outcrossing tetraploids. Theor Popul Biol. 2002;62(2):129–51. https://doi.org/10.1006/tpbi.2002.1608.

Xie C, Xu S. Mapping quantitative trait loci in tetraploid populations. Genet Res. 2000;76 (1):105–15.

Yates SA, Swain MT, Hegarty MJ, Chernukin I, Lowe M, Allison GG, et al. De novo assembly of red clover transcriptome based on RNA-Seq data provides insight into drought response, gene discovery and marker identification. BMC Genomics. 2014;15(1):453. https://doi.org/10.1186/1471-2164-15-453.

Young ND, Udvardi M. Translating *Medicago truncatula* genomics to crop legumes. Curr Opin Plant Biol. 2009;12(2):193–201. https://doi.org/10.1016/j.pbi.2008.11.005.

Yu L-X. Identification of single-nucleotide polymorphic loci associated with biomass yield under water deficit in alfalfa (Medicago sativa L.) using genome-wide sequencing and association mapping. Front Plant Sci. 2017;8:1152.

Yu L-X, Liu X, Boge W, Liu X-P. Genome-wide association study identifies loci for salt tolerance during germination in autotetraploid alfalfa (Medicago sativa L.) using genotyping-by-sequencing. Front Plant Sci. 2016;7:956.

Yu L-X, Zheng P, Bhamidimarri S, Liu X-P, Main D. The impact of genotyping-by-sequencing pipelines on SNP discovery and identification of markers associated with verticillium wilt resistance in autotetraploid alfalfa (Medicago sativa L.). Front Plant Sci. 2017a;8:89.

Yu L-X, Zheng P, Zhang T, Rodringuez J, Main D. Genotyping-by-sequencing-based genome-wide association studies on Verticillium wilt resistance in autotetraploid alfalfa (Medicago sativa L.). Mol Plant Pathol. 2017b;18(2):187–94.

Zarrabian M, Majidi MM, Ehtemam MH. Genetic diversity in a worldwide collection of sainfoin using morphological, anatomical, and molecular markers. Crop Sci. 2013;53(6):2483–96.

Zhang Y, Sledge MK, Bouton JH. Genome mapping of white clover (*Trifolium repens* L.) and comparative analysis within the Trifolieae using cross-species SSR markers. Theor Appl Genet. 2007;114(8):1367–78. https://doi.org/10.1007/s00122-007-0523-3.

Zhang T, Yu L-X, Zheng P, Li Y, Rivera M, Main D, et al. Identification of loci associated with drought resistance traits in heterozygous autotetraploid alfalfa (Medicago sativa L.) using genome-wide association studies with genotyping by sequencing. PLoS One. 2015;10(9): e0138931.

Zohary M, Heller D. The genus Trifolium. Jerusalem: Israel Academy of Sciences and Humanities; 1984.

Index

A

Adaptation, 4, 54, 66, 88, 114, 227, 291, 388, 459, 480, 536, 574, 609, 695, 753, 794, 868, 908
Adaptive introgression, 80, 81, 101, 131, 134, 236–239, 389, 465, 497, 500
Admixture, 69–71, 78, 81, 152, 156, 162, 164, 186, 234, 236–238, 354, 365, 432, 460, 494, 496, 497, 499, 592, 806, 811–813, 822, 854, 877, 879, 885
African and Asian rice, 16–18, 23, 49, 50, 52, 66, 68, 72, 74–76, 78–80, 98, 462
Agricultural weeds, 88, 89, 101, 102, 127
Alfalfa, 614, 904–906, 909–921, 930, 931
Approximate Bayesian computation, 71–73, 77, 81
Autotetraploid and polyploid, 478, 479

B

Bioinformatics, 25, 38, 45, 55, 149, 158, 160, 183, 201, 210, 342, 362, 399, 424, 506, 795, 814, 817, 819
Biomass and sweet sorghum, 303, 308
Birdsfoot trefoil, 904–906, 909, 910, 912, 928–931
Bottleneck, 20, 22, 49, 66–70, 72, 75–80, 82, 93, 126, 159, 160, 186, 193, 197, 227, 229–231, 233, 240, 260, 383, 388, 463, 480, 503, 504, 535, 550–552, 560, 585, 590, 622, 624, 632–635, 637, 638, 640, 668, 710, 849, 877, 883, 908, 919

Brassica, 10–12, 18–19, 741–782, 795, 849
Breeding, 14, 54, 88, 124, 144, 180, 226, 291, 343, 378, 458, 480, 534, 574, 610, 692, 744, 795, 838, 866, 908
(Pre-)breeding, 225–268, 291, 303, 321, 666, 669

C

Candidate genes, 92, 94–95, 99, 103, 104, 134, 135, 156, 161, 165, 181, 183, 185, 191–195, 210, 211, 228, 238, 239, 265–267, 295, 297, 302, 314–316, 320, 350, 387, 402, 413, 419, 420, 469, 470, 545, 547, 579, 595, 639, 658–660, 694, 702–704, 713, 720–723, 727, 765, 767–770, 772, 774, 776, 778, 821, 822, 871, 875, 889, 920
Center of diversity and domestication, 918–919, 922–923, 926, 928
Chloroplast, 38–42, 44, 46, 48–55, 462, 485, 492–495, 497, 539, 630, 645, 696, 697, 750, 761, 800, 807, 839, 849, 851, 868, 877, 921, 927
Chromosomal structural evolution, 133
Citrullus, 866–870, 877, 878, 887, 891
Climate adaptation, 54, 546–548
Clover, 614, 904–906, 910, 912, 913, 921–927, 931
Coalescent models, 65–82, 541
Comparative population genomics, 459, 713, 718

Conservation, 4, 6, 14, 54, 55, 146, 147, 152, 164, 262, 292, 305, 343, 365, 389, 393, 429–432, 459, 460, 490, 491, 510–511, 534, 548, 561, 574, 596, 609, 610, 632, 695, 699, 724–727, 755, 795, 810, 824, 867, 904, 907

Convergent evolution, 640, 663–665

Copy number variations (CNVs), 4, 12, 14, 16, 17, 19–21, 24, 238, 355, 397, 420, 490, 541, 583, 705, 713, 714, 805, 809, 814–816

Crop plants, 3–27, 37–55, 144, 146, 147, 150, 151, 158, 180, 185, 188, 225–268, 318, 353, 403, 413, 534, 542, 552, 589, 608, 627, 742–744, 907, 918

Crops, 5, 38, 66, 88, 115, 144, 180, 226, 290, 342, 378, 458, 478, 534, 574, 608, 692, 742, 794, 838, 870, 905

D

Data integration, 158, 268

De-domestication, 92, 93, 96–98, 100, 101

Demographic history, 65–82, 94, 162, 240, 542–543, 548, 795, 848, 858

De novo mutation, 101, 634, 664

Dioecy, 838, 839, 854–857

Dioscorea, 462, 837–859

Diversification, 22, 50, 88, 117, 118, 164, 344, 356, 540, 544, 548, 555, 590, 591, 613–626, 630, 640, 644, 650, 654, 663, 693, 695, 696, 700, 701, 744, 753, 768, 782, 808, 816, 880

Diversity, 4, 38, 66, 92, 115, 144, 186, 227, 290, 347, 378, 458, 479, 534, 574, 609, 692, 744, 795, 838, 867, 906

Domestication, 20, 49, 66, 88, 114, 144, 185, 227, 290, 347, 378, 458, 479, 535, 574, 608, 692, 743, 795, 838, 866, 908

Domestication syndrome, 549, 574, 589, 595, 608, 639, 646, 649–660, 816, 887–890, 908

Domestication triangle, 608, 644

E

Ecological adaptation and speciation, 548

Egusi, 866, 868, 872, 873, 875, 877, 880, 883, 885–887

Epigenomics, 266–267, 361–362, 400, 402, 403, 472, 505–507, 561, 574, 579, 596, 698, 719, 744, 766–767, 781, 782, 794, 800, 817–819, 824, 908, 931

Evolution, 18, 46, 71, 90, 114, 144, 230, 302, 347, 378, 459, 478, 535, 574, 608, 692, 744, 795, 846, 866, 906

Evolution and resequencing, 354

F

Farmer selection, 81

Fine mapping, 184–185, 241, 318, 360, 383, 464, 501, 554, 708, 722, 748, 768, 769

Functional genomics, 16, 352, 432, 576, 578, 579, 583, 710, 726, 728

G

Gene banks, 143–166, 431, 490, 608, 612, 636

Gene flow, 49, 50, 66, 68, 70, 72, 76–79, 81, 118–121, 123, 124, 130–133, 154, 156, 186, 233, 236, 237, 239, 356, 388–390, 395, 431, 432, 458, 460, 462, 465, 500, 544, 589, 618, 626, 634–645, 669, 710, 753, 794, 799, 854, 890, 913

Genetic architecture of agronomic traits, 554, 557–559

Genetic diversity, 4, 45, 66, 93, 126, 144, 186, 227, 290, 347, 383, 458, 479, 534, 574, 609, 692, 744, 796, 838, 867, 907

Genetic diversity and demography, 814

Genetic diversity and population structure, 290, 291, 302–305, 352, 353, 389–397, 458, 461, 465–466, 471–472, 501–505, 585, 586, 694, 708–713, 749, 753–761, 780, 811, 820, 824, 907, 913, 931

Genetic resource conservation, 14, 164

Genetic structure, 154–158, 233, 240, 327, 353, 378, 463, 465, 466, 472, 555–557, 585, 586, 588, 749, 758, 812, 919, 923–926, 928

Genome and transcriptome, 537–539

Genome and transcriptome resources, 483

Genomes, 4, 38, 66, 90, 117, 146, 180, 227, 291, 347, 378, 458, 479, 534, 576, 611, 692, 742, 867, 905, 938

Genome-wide associations, 544, 557–559, 718

Genome-wide association studies (GWAS), 126, 132, 150, 151, 155, 158, 159, 180, 181, 185–191, 197, 203, 211, 212, 259, 265, 294, 297, 310, 312, 314, 315, 318, 320–322, 327, 351, 353, 364, 409–424, 433, 465, 469, 470, 487–489, 507–509, 535, 538, 552, 557–560, 584, 592–596, 651, 664, 665, 692, 694, 708, 710, 713, 714, 719–723, 728, 747, 754, 766, 768–779, 782, 820–822, 885, 889, 908, 911, 919–920, 923, 927, 928, 931

Index 945

Genomic diversity, 6, 7, 14, 19, 22, 24, 291, 292, 378, 405, 407, 459, 465, 574, 609, 634–639, 643–644, 695, 795, 814, 824, 867, 877, 919
Genomic predictions, 165, 180, 202–206, 208, 210, 241, 243–262, 265, 267, 278, 294, 324–326, 427, 428, 489, 779–780, 921
Genomic resources, 104, 346–352, 534, 543, 561, 611–613, 701–705, 726, 744–749, 781, 911–913, 931
Genomics, 38, 66, 89, 114, 146, 180, 228, 291, 344, 378, 458, 478, 534, 574, 608, 692, 743, 794, 838, 867, 905
Genomic selection (GS), 14, 180, 197–211, 228, 243–247, 250, 252, 254, 257–261, 264, 294, 323–327, 353, 365, 407, 424–429, 433, 460, 461, 471, 509–511, 559–560, 595, 596, 665, 668, 723–724, 727, 744, 754, 779–782, 820, 823, 908, 921, 924, 927, 928, 930, 931
Genotype by environment interaction (GEI), 256–260, 305, 326, 345, 425, 470, 558
Genotyping, 96, 99, 146–151, 153, 154, 158–160, 162, 181, 183–185, 191, 197, 200, 201, 211, 228, 244, 245, 294, 309, 316, 320, 323, 326, 327, 351, 353, 364, 395, 409, 425, 431, 432, 460, 461, 464–466, 471, 480, 488–490, 502, 540, 542, 558, 559, 580, 585, 586, 595, 612, 633, 643, 702, 709, 710, 720, 722, 723, 727, 728, 758, 762, 769, 779, 781, 806, 811, 817, 821, 823, 876, 880, 885, 912, 924, 931
Germplasm conservation, 491
Gossypium, 55, 692–697, 699–702, 708, 710, 711, 723–727, 795, 849
Growth habit, 549, 590, 591, 616, 635, 636, 642, 646–648, 650–654, 657, 658, 663, 665, 669, 757, 816, 919, 922
GWAS and genomic selection, 180, 595, 754, 782, 908, 911, 927

H
Haplotypes, 14, 50, 75, 95, 97, 131, 132, 148, 150, 151, 153–155, 159, 162, 166, 186, 190, 197, 230, 232–234, 236, 249, 251, 252, 294, 302, 303, 387, 410, 426, 431, 479, 484, 486, 487, 499, 502, 503, 511, 542, 547, 551, 552, 558, 559, 584, 590–592, 624, 655, 721, 722, 807, 814, 848, 849, 868, 869, 885, 887–889, 916

Helianthus, 114–118, 120, 123, 124, 126, 127, 134, 135, 795
High-throughput sequencing (HTS), 344, 361, 362, 471, 535, 549, 581, 720, 748, 754, 782, 795, 804, 877
Hybridization, 9–11, 14, 15, 17, 21, 45, 54, 75, 78, 81, 92, 93, 97, 99–101, 103, 114, 117, 120, 121, 127, 131, 133, 134, 157, 297, 298, 348, 356, 411, 431, 462, 493, 495, 497, 504–506, 534, 535, 540, 544, 548, 575, 617, 620, 637, 666, 692, 694–696, 717, 719, 743, 762, 797, 806–809, 812, 824, 842, 847–849, 851–853, 858, 859, 869, 885, 907, 914, 931

I
Intra and inter-specific genetic diversity and population structure, 931
Introgression, 49, 71, 91, 117, 151, 195, 229, 294, 388, 459, 479, 536, 590, 612, 713, 746, 816, 849, 877
Invasive species, 54, 126, 127, 129, 130

L
Landscape genomics, 239–240, 305–306, 543–549, 596, 908, 923, 928, 931
Linkage disequilibrium (LD), 20, 69, 70, 131, 150, 153, 155, 156, 158, 180, 185–187, 189, 191, 201, 206, 207, 230, 232–233, 236, 238, 244, 249, 251, 316, 321, 322, 326, 353, 408, 409, 412, 413, 459, 460, 465, 470–472, 479, 507–509, 552, 555, 557–560, 589, 590, 592, 600, 610, 630, 634, 639, 640, 711, 762, 797, 820, 821, 823, 877, 885–887, 919, 920, 927
Linkage mapping (LM), 125, 181–184, 309, 359, 360, 409, 461, 464, 468, 470, 486, 594, 701, 720, 721, 723, 748, 767, 871
Long-distance dispersal, 619, 621, 622, 625, 626, 839
Long-term genetic gain, 227, 228, 251, 252, 254, 255, 261, 268

M
Maize, 8, 66, 189, 227, 290, 378, 458, 549, 589, 610, 779
Mating design, 248
Mitochondria, 38, 42–43, 630

Molecular evolution, 144, 147, 151, 162, 657, 664, 696
Multiple domestications, 50, 78, 609, 628, 640, 858
Multi-trait, 226, 247, 254, 255, 257, 263, 320, 323–325

N

Next-generation sequencing (NGS), 4, 38, 41, 44–51, 53–55, 104, 117, 146, 183, 201, 265, 320, 347, 353, 360, 361, 364, 399, 407, 420, 424, 425, 488, 496, 536, 576, 703, 727, 728, 744, 755, 769, 804, 811, 877, 891, 912

O

Omics, 144, 166, 260, 264–268, 344, 349, 351, 360, 364, 549, 560, 700, 727, 795, 796, 824
Origin, 7, 46, 66, 88, 115, 153, 191, 238, 290, 343, 378, 462, 479, 534, 574, 608, 693, 743, 795, 839, 866, 907
Origin and evolution, 347, 354–356, 480, 574, 695, 762, 764

P

Pan-genomes, 5–19, 21, 22, 24–27, 126, 150, 159, 162, 291–294, 321, 365, 397, 400, 403–407, 431–433, 496, 498, 499, 502, 541–542, 582–584, 595, 695, 701, 705–707, 745–748, 758, 781, 782, 795, 846, 867, 890–892
Pangenomics, 3–27, 824
Parallel evolution, 46, 100, 101, 610, 824
Peanut origin, 797, 798, 806–808
Pearl millet, 66, 76–78, 81, 239, 240, 457–472
Perennial forage legumes, 905, 908, 930
Photoperiod sensitivity, 235, 241, 308, 382, 419, 590, 591, 611, 635, 636, 642, 646–648, 650, 657–659, 661
Phylogeny, 39, 50–53, 115, 117, 129, 132, 294, 301, 347, 365, 490, 491, 494, 535, 537, 539–541, 548, 577, 589, 613, 696, 697, 796, 806, 867–870, 877
Plant genetic resources (PGR), 144–151, 153, 165, 166, 263, 267, 291–298, 459, 724–726
Plants, 4, 38, 66, 88, 114, 144, 180, 226, 290, 343, 380, 458, 486, 534, 574, 608, 697, 742, 795, 838, 866, 906

Pod dehiscence, 646–648
Polyploidy, 11, 23, 54, 55, 356, 489, 491, 692, 693, 695–700, 702, 719, 727, 806, 808, 809, 909–911, 921
Population epigenomics, 266–267, 361–362, 472, 505–507, 561, 596, 719, 744, 766–767, 781, 782, 817–819, 931
Population genetics, 46, 49–51, 67, 70–72, 80, 81, 94, 97, 104, 114, 152, 153, 156, 157, 240, 252, 327, 353, 378, 462, 466, 472, 478, 483, 540, 542, 547, 548, 559, 561, 580, 622, 669, 699, 718, 749, 755, 757, 758, 781, 795, 798, 809, 867, 885, 891, 912, 914, 916, 918, 922, 928
Population genetic variation, 70
Population genomics, 27, 37, 66, 90, 118, 146, 179, 228, 290, 345, 378, 458, 479, 536, 574, 610, 694, 743, 794, 839, 870, 905
Population structure, 70, 75, 79, 151, 156, 180, 185–189, 191, 233, 234, 239, 291, 302–305, 316, 320, 321, 347, 352–354, 357, 360, 389–397, 409, 420, 421, 424, 431, 458, 459, 461, 465, 466, 487–489, 501–505, 538, 542, 547, 555, 558, 574, 585–589, 592, 595, 634, 694, 708–713, 715, 726, 744, 749, 753–763, 780, 782, 796, 797, 809–816, 820, 821, 824, 877–879, 885, 887, 907, 912–918, 921–927, 929, 931
Presence/absence variations (PAV), 4, 12, 14, 15, 17, 19–21, 24, 26, 27, 134, 238, 292, 294, 542, 583, 584, 701, 706, 747, 779, 868, 890

Q

QTL mapping, 101, 103, 119, 146, 189, 197, 199, 200, 202, 294, 308–320, 359–360, 407–411, 464, 487–489, 535, 544, 547, 554, 581, 692, 707–708, 720, 722, 820, 912, 927, 928
Quantitative trait loci (QTLs), 12, 94, 119, 151, 180, 227, 290, 356, 384, 460, 487, 535, 577, 639, 692, 748, 804, 855, 871, 912

S

Sainfoin, 904–906, 909, 910, 912, 927–928, 931
Selection signatures, 230–240, 263, 882
Selective sweeps, 80, 158, 162, 230–232, 549–551, 574, 588–590, 639, 695, 768, 817, 858, 877, 884, 885, 887, 890

Sequential Markovian coalescent, 68–70, 77
Sex determination, 839, 846, 855–858
Solanum lycopersicum, 11, 483, 534–536, 539, 544, 545, 548, 550, 551, 556, 557, 654, 867
Solanum tuberosum, 11, 23, 238, 365, 478, 483–485, 490, 492–495, 497, 499, 500, 502, 503, 507, 561
Sorghum bicolor, 11, 22, 88, 145, 247, 290, 292, 297–302, 379, 399, 406, 459, 462
Speciation, 113–135, 236, 365, 432, 497, 540, 541, 543–549, 560, 640, 693, 696–701, 749, 751, 795, 839, 840, 849, 867–869, 891, 907
Standing genetic variation, 101, 497, 908
Structural variations, 4, 104, 131, 134, 150, 162, 291, 292, 397, 403–406, 432, 484, 490, 502, 541, 583, 584, 693, 701, 706, 713, 722, 747, 809, 815, 868
Sunflower, 113–135
Sweet watermelon, 865–892
Syngameon, 618, 643, 668

T
Tea improvement, 345–347
Tomato, 22, 23, 229, 266, 483, 485, 491, 507, 533–561, 654, 810, 821, 867, 892

Transcriptomes, 9, 15, 121, 126, 160, 201, 267, 268, 295, 348–350, 358, 362, 383, 413, 461, 462, 466, 471, 483, 485–486, 509, 535, 537–540, 548, 550–552, 561, 578, 579, 702–705, 718, 727, 728, 762, 767, 768, 781, 801–802, 819, 824, 870, 873–875, 907, 912, 914, 928, 931
Transcriptomics, 260, 265, 268, 295, 302, 348–352, 363–365, 378, 419, 471, 483–490, 536–539, 545, 552, 560, 574, 576–582, 596, 695, 698, 702, 704, 727, 744, 758, 767–768, 781, 782, 794, 801, 802, 819–820, 824, 870–875, 908, 931

W
Weediness trait, 88, 90, 101, 103, 104
Weedy rice, 87–104

Y
Yam, 50, 66, 76, 228, 235, 462, 615, 837–859

Z
Zea mays, 8, 9, 14, 76, 81, 189, 227, 229, 247, 290, 378, 381, 382, 390, 403, 462, 549, 610, 779, 795

Printed by Libri Plureos GmbH
in Hamburg, Germany